SMALL CETACEANS OF JAPAN

EXPLOITATION AND BIOLOGY

SMALL CETACEANS OF JAPAN

EXPLOITATION AND BIOLOGY

TOSHIO KASUYA

Translation by Toshio Kasuya
Translation edited by William F. Perrin

CRC Press is an imprint of the
Taylor & Francis Group, an informa business

CRC Press
Taylor & Francis Group
6000 Broken Sound Parkway NW, Suite 300
Boca Raton, FL 33487-2742

First issued in paperback 2021

Printed on acid-free paper

ISBN-13: 978-1-4987-7900-5 (hbk)
ISBN-13: 978-0-367-65801-4 (pbk)

Library of Congress Cataloging-in-Publication Data

Names: Kasuya, Toshio, 1937-, author. | Perrin, William F.
Title: Small cetaceans of Japan : exploitation and biology / Toshio Kasuya and William F. Perrin.
Description: Boca Raton : Taylor & Francis, 2017.
Identifiers: LCCN 2016027063 | ISBN 9781498779005 (hardback : alk. paper) | ISBN 9781315395425 (e-book)
Subjects: LCSH: Cetacea--Japan. | Cetacea--Effect of human beings on--Japan. | Whaling--Japan.
Classification: LCC QL737.C4 K264 2017 | DDC 599.50952--dc23
LC record available at https://lccn.loc.gov/2016027063

Visit the Taylor & Francis Web site at
http://www.taylorandfrancis.com

and the CRC Press Web site at
http://www.crcpress.com

Contents

Preface

This book presents the current status of our knowledge on biology, exploitation, and management of several key species of small cetaceans around Japan, together with my views on conservation and management issues. The International Convention for the Regulation of Whaling (ICRW) signed in 1946 forms the basis for the activity of the International Whaling Commission (IWC). The commission, primarily interested in the management of large cetaceans, has been concerned about the increasing catch of smaller cetaceans in concert with the decline of large cetaceans, and in 1974 convened a meeting to achieve a "Review of Biology and Fisheries for Smaller Cetaceans" in Montreal. This meeting covered pygmy right whales and minke whales as well as the currently termed "small cetaceans," such as ocean dolphins, porpoises, beaked whales, pygmy sperm whales, and river dolphins. Currently, there is no agreement among the member countries to place these small cetaceans under the control of the IWC, and the Scientific Committee of the IWC is unable to propose the regulation of catches to the commission.

Since the summer of 1960, I have studied the life history, population dynamics, and management of cetaceans exploited or incidentally killed by Japanese fisheries. Such conservation biology cannot be pure science. It requires understanding of not only the life of the targeted species but also the history of interactions between humans and the species in order to foresee the future of the species and to search for ways to achieve compromise between human activities and the species. A solution that satisfies human interests, in particular the economy, might mean a disaster to the species, and a solution that is desirable for the species may not be accepted by the human community. Because our sense of value changes with time and varies between communities, conservation biology is influenced by the personal history of the concerned people, including scientists. Therefore, it seems to be important for the people involved in these issues to share common scientific information when discussing a conservation issue. This book is aimed at that end.

Out of more than 30 species of small cetaceans known to inhabit Japanese coastal waters, I have selected 7 species most heavily affected by human activities to explain the current status of conservation biology of small cetaceans in Japan and to present the needed direction of management of those species, for conservation specialists as well as students who are interested in the field. The sections of the book are as follows.

The coverage in this book is very limited. Here I will explain the status of the seven selected species in the Cetacea and the conservation issues that will be covered among those faced by cetaceans.

SECTION I: HISTORY OF JAPANESE CETACEAN FISHERIES

Japan may have taken more dolphins and porpoises for human consumption than any other nation. The history and current status of interactions between the Japanese people and small cetaceans are described here. Historical knowledge is necessary for understanding the present status of the fisheries as well as for preparation for the future. The current biological status of small cetaceans off Japan is a product of interaction between human activities and the specific characters of each species, so knowledge of past fisheries is important for understanding the current biology of the small cetaceans.

Operations of Japanese fisheries for small cetaceans were unregulated from the fifteenth century to the 1980s, but even since the nineteenth century they experienced considerable fluctuation as seen in some available quantitative records. It was in the 1980s and early 1990s when hunting of striped dolphins, Dall's porpoises, and Baird's beaked whales first received some sort of regulation in response to international criticisms that started with the Montreal meeting in 1974. I participated in the Scientific Committee of the IWC beginning in 1982 and worked at the Whale Section of the Far Seas Fisheries Research Laboratory (FSFRL) of the Fisheries Agency of Japanese government from 1983 through 1997, and I was able to see actions of the government toward the regulation of small-cetacean fisheries. Therefore, it seems appropriate to record the process for future generations. Japanese fisheries for small cetaceans have remained subordinate to whaling for large cetaceans but have been strongly affected by it. Therefore, this book will address some important incidents that occurred in whaling activities.

While working for FSFRL, I was obliged to learn about fishing history and legal aspects of fishing regulations, but I have never been a specialist in the field, so readers are cautioned there may be some incompleteness or misunderstandings in the description of the fisheries. For the convenience of the readers, I have appended a list of further readings on the history of Japanese whaling.

SECTION II: BIOLOGY

Using seven species of small cetaceans inhabiting the coastal waters of Japan as examples, I explain the methods used in my studies and the biological information thus obtained. Readers will understand that the lives of small cetaceans are extremely variable among species and that a single uniform approach would not be suitable for the management of all the species. Another aim here is to help future biologists in considering the direction of their work.

Finless porpoises and Dall's porpoises nurse their young for less than 1 year, calve every 1–2 years, and apparently possess relatively simple social structure. In contrast, females of the short-finned pilot whale probably calve once every 5–10 years and appear to live with their daughters (and perhaps sons) for almost their entire lives. This is the extreme case of extended maternal care. Several species of toothed cetaceans have developed such an extended matrilineal society, where there are opportunities to accumulate their own experience and behavior patterns within each group, that is, the development of culture. Striped dolphins and bottlenose dolphins fall between these two extremes. The offspring usually start to live apart from their mothers after nursing 2–4 years, but nothing is known about the fission and fusion systems of the group of these species. Baird's beaked whales have a strange life history, where males live 30 years longer than females, and it is suspected that they have a social structure that is quite different from those of any of the species mentioned earlier.

The seventh species dealt with in this book, the Pacific white-sided dolphin, develops a secondary sexual characteristic in the shape of the dorsal fin in adult males. I included a chapter for this species with a belief that it will help in the studies of its behavior. Among the seven species mentioned earlier, six, not including the finless porpoise, are hunted in Japanese small-cetacean fisheries.

CONCLUDING CHAPTER: CETACEAN BIOLOGY AND CONSERVATION

This is a summary of my research activity created based on a lecture given at the Biennial Conference on the Biology of Marine Mammals held in Cape Town in 2007. It includes a list of questions I have not been able to answer and hypotheses proposed to interpret my research results. I also cover here both the biological aspects of the animals and the problems of human communities that must be considered in the conservation of cetaceans, as I have encountered them during my experience as a whale biologist.

Many readers will feel it most convenient just to concentrate on particular chapters of interest. For this reason, I did not expend any particular effort to completely remove all redundancy between chapters.

Toshio Kasuya
April 28, 2009

Introduction: Whales, Dolphins, and Men

The zoological order Cetartiodactyla contains the whales, dolphins, and porpoises as well as the artiodactyls (even-toed ungulates). Humans have had significant influence on the life of cetaceans since about 1000 years ago and have greatly changed the status of some cetacean populations. For example, hunting decreased the abundance of Antarctic blue whales (*Balaenoptera musculus*) from the initial number of about 250,000 to about 1,000 by 1965 when protection started, but since then the population has only increased to about 2,000 (IWC 2009). The Yangtze River dolphin or baiji (*Lipotes vexillifer*) became extinct in the beginning of this century due to the destruction of its environment (Turvey 2008). While some people consider cetaceans as fishery resources and try to manage them as such, others consider them as an important element of our environment and hope to protect them completely. For either purpose, our current knowledge of cetaceans is insufficient.

Using seven species of small cetaceans around Japan as examples, this book presents topics on their life history and problems I experienced while trying to manage them as fishery resources. In other words, this book targets the conservation biology of small cetaceans around Japan. These seven species include the striped dolphins (*Stenella coeruleoalba*), which were hunted in great numbers in the last century, and the Dall's porpoises (*Phocoenoides dalli*), short-finned pilot whales (*Globicephala macrorhynchus*), and Baird's beaked whales (*Berardius bairdii*), which are still hunted in large numbers. Another example, the Japanese population of the narrow-ridged finless porpoises (*Neophocaena asiaeorientalis*), has been damaged by human activities other than direct hunting, and its future seems to be dubious. The broad variety of the life histories of these species will help readers understand the diversity of cetacean biology as well as the diverse conservation problems.

This chapter outlines the content and perspective of the book on the biology and conservation of cetaceans.

I.1 CETACEAN BIOLOGY

I.1.1 Archaeoceti and the Origin of Cetacea

The Cenozoic Era of geology started about 6.5 million years ago and continues to the present. This era is also called the era of mammals because it is during this period that mammals diverged into the various groups that we now see on the planet. About 5.5 million years ago, or during the early Eocene Epoch in the beginning of the Cenozoic Era, a group of animals called *Pakicetus* spp., with their body size about that of a goat, inhabited warm freshwater beaches at the southern foot of the Himalayas, which had just started to rise at the time. They are the oldest known members of the Cetacea. Quite different from recent cetaceans, they probably swam in the water and at the same time were able to run around the shore on four legs. They gradually lost their posterior legs, changed the front legs into fins (flippers), and formed a pair of horizontal fins (tail flukes) at the end of the tail. These animals are grouped into the Archaeoceti. The shape of their teeth differed with the position on the jaw (heterodonty), and their milk teeth were replaced by permanent teeth after birth (diphyodonty). Their dental formula was close to that of primitive mammals, that is, 3 incisors, 1 canine, 4 premolars, and 2–3 molars, namely, 10–11 teeth on each jaw (Uhen 2009).

The Archaeoceti flourished and expanded their habitats to the world oceans, disappearing by the end of the Eocene Epoch, or 3.5 million years before the present. To replace the then declining Archaeoceti, two new groups of cetaceans emerged: the Mysticeti (baleen whales) and Odontoceti (toothed whales), which survive to the present. Although these two taxa are believed to have arisen from the Archaeoceti, details of the process are still unknown. Fossil records indicate that the Archaeoceti arose from the species close to the ancestral ungulates and DNA analyses of recent mammals find great similarity between cetaceans and the even-toed ungulates, particularly the hippopotami (Thewissen 1998). For this reason, the cetaceans now are not recognized as a separate order but included with the even-toed ungulates in the order Cetartiodactyla.

Both Odontoceti and Mysticeti normally produce only one calf at a time. This was probably the case also for the Archaeoceti. This book does not deal with the biology of Archaeoceti.

I.1.2 Mysticeti

The oldest mysticetes, or baleen whales, are known from around the boundary between the Eocene and Oligocene Epochs, or about 3.5 million years before present. They had heterodont dentition and possibly fed on marine organisms by filtering them with serrated posterior teeth (Uhen 2009). Later, they lost their teeth, developing baleen plates (or whale bone) that could function better than the serrated teeth in filtering food organisms in the water. The process of having both teeth and baleen plates might have had a stage similar to the one we currently see on the Dall's porpoise, which has horny protuberance between small rudimental teeth (Miller 1929). Numerous tooth buds, the number being greater than the primitive mammalian figures mentioned earlier, are formed in each jaw of recent baleen whales during their fetal stage but are resorbed to disappear around the time when baleen plates start to grow (Kükenthal 1893; Slijper 1984, in Japanese). We currently identify 14 species of recent baleen whales of 6 genera and 4 families, but there may be additional species recognized when taxonomic questions about the "Bryde's whale group" are resolved. This number is fewer than that of the Odontoceti, which relates to their broader geographical range of distribution. Their body size ranges from 6 to 7 m in the pygmy right whale (*Caperea marginata*) to as much as 30 m in the blue whale in the southern hemisphere. In Japanese coastal waters, 10–11 species of baleen whales of 5 genera and 3 families have been recorded, including a vagrant bowhead whale (*Balaena mysticetus*) encountered in Osaka Bay (see Appendix B; Ohdachi et al. 2009).

By developing baleen plates, baleen whales acquired an efficient way of feeding on relatively small organisms. Such food species are at a low trophic level and are abundant in the ocean. The abundant and stable food supply, even if limited seasonally, could be an environmental factor that allowed augmentation of body size. The whales' seasonal feeding cycle and parturition season located in the starvation period of about 6 months requires large nutritional storage in their body. This is facilitated by large body size, particularly of females, and could have been a result of selective pressure for greater body size (Kasuya 1995). Females of baleen whales are usually larger than males of the same species. Baleen whales are known to be long-lived, with the maximum longevity ranging from about 60 years in the Antarctic minke whale (*Balaenoptera bonaerensis*) to 150–200 years in the bowhead whale. All the recent baleen whales are marine species; it is unlikely that any freshwater species existed during the history of the Mysticeti.

The earlier studies of the biology of baleen whales relied on whaling industries for specimens as well as for research opportunities. Examples will be found among classical studies of the behavior and morphology of the North Atlantic right whale (*Eubalaena glacialis*) and gray whale (*Eschrichtius robustus*) in the nineteenth century using materials from sailing-ship whaling, or studies on the growth and reproduction of balaenopterids using great numbers of carcasses from Norwegian-type whaling in the twentieth century. However, such industry-dependent studies decreased following the decline of commercial whaling in the late twentieth century and its nearly complete closure around 1987 (see Section 1.2) and the emergence of a new type of whale science. For example, individual identification using photographs helps to get long-term information on individual whales, and biopsy samples for DNA and physiological analyses help in individual recognition and determination of sex and reproductive status of living whales. However, as baleen whales are long-lived, it takes many years before we obtain a complete picture of their life history, and it is also highly possible that environmental factors (both natural and anthropogenic) surrounding a species will change before the observations can cover one generation of the species.

Among baleen whales, some species are particularly suited for such nonlethal techniques. They are coastal species and species that decreased in abundance due to past whaling, for example, the humpback whales (*Megaptera novaeangliae*), bowhead whales, right whales, and gray whales. Applying such techniques to balaenopterids, such as the minke whales, will be harder because they often inhabit offshore waters and their numbers are still large.

It might appear to us that the social structure of baleen whales is less developed than that of group-living toothed whales. However, it is well known that baleen whales produce low-frequency sounds that propagate for several hundred kilometers and that they use the sounds for within-species communication, including reproductive activity (Frankel 2009). They can identify the presence of conspecifics at a distance that is greater than the limit of our visual perception. Currently, we do not know how baleen whales locate schools of their food organisms, but I suspect that they may use their voices also as a tool for locating their food organisms or as a tool to exchange feeding information. At present, our methodology for studying the social behavior of baleen whales is inadequate.

This book does not deal further with the biology of the baleen whales.

I.1.3 Odontoceti

The Odontoceti, or toothed whales, have teeth in their jaws and no baleen plates. This group includes more than 70 currently identified species in 35 genera and 9 families and shows broad variation in morphology and life history. In Japanese waters, a total of 30 odontocete species in 22 genera and 5 families have been recorded, excluding an old record of a vagrant narwhal (*Monodon monoceros*) in the Sea of Japan and several recent records of belugas (*Delphinapterus leucas*) off Hokkaido in northern Japan (see Appendix Table; Ohdachi et al. 2009). Most toothed whales are marine species, but finless porpoises (*Neophocaena* spp.), Irrawaddy dolphins (*Orcaella* spp.), and tucuxis (*Sotalia fluviatilis*) have invaded freshwater while leaving their conspecifics or congenerics in the marine environment. The so-called river dolphins, that is, the South Asian river dolphin (*Platanista gangetica*), Yangtze River dolphin (or baiji), and Amazon River dolphin (*Inia geoffrensis*), adapted to freshwater probably much earlier and have already lost their marine counterparts (Kasuya 1997, in Japanese).

Some of the early toothed whales were heterodont in dentition, but recent toothed whales have single-rooted teeth and are homodont (having no morphological differentiation between positions in the jaw), and their first set of teeth is retained for their entire life (monophyodonty). However, toothed whales have achieved broad variation in dental formula and tooth morphology reflecting their feeding habits and social behavior. The narwhal usually has an erupted tooth only on the upper left jaw of an adult male (if we count unerupted teeth, there is a pair of teeth in the upper jaw of both sexes). Many species of Ziphiidae have one pair of teeth at various locations in the lower jaw that often erupt only in adult males. The sperm whale (*Physeter macrocephalus*) has functional teeth only on the lower jaw, which erupt at around the attainment of sexual maturity. These observations suggest that teeth are not indispensable for feeding in some toothed whale species but rather have social functions. The number of teeth varies among species of Delphinidae from Risso's dolphin (*Grampus griseus*) having only 2–3 pairs of mandibular teeth to *Delphinus* spp. and some *Stenella* spp. having over

40 teeth on each jaw. Five species of river dolphins (4 genera in 3 families), including the franciscana (*Pontoporia blainvillei*) that inhabits a coastal marine environment, also have a large number of teeth in each jaw. These numerous teeth are used for grasping or tearing food.

Toothed whales are usually smaller than baleen whales, but the variation in body sizes among the species appears to be greater. The smallest toothed whales measure less than 1.5 m, for example, the finless porpoises in the Indian Ocean, the spinner dolphins (*Stenella longirostris*) in the Gulf of Siam, and the tucuxis in the Amazon and Orinoco Rivers. Sperm whale adult males are the largest among recent toothed whales and may reach 18 m. Twenty-one species of Ziphiidae (5 genera) are between these two extremes and measure 4–11 m.

The 38 species of Delphinidae also have a broad range in body size, from tucuxis mentioned earlier to killer whales (*Orcinus orca*), in which adult males reach 9 m. Some toothed whale species have worldwide distributions, but others are limited to single ocean basins or smaller geographical areas such as particular rivers, inland waters, or polar regions.

Male toothed whales usually grow larger than females. Such sexual dimorphism is particularly developed in some species, for example, sperm whales, killer whales, and two species of pilot whales (*Globicephala* spp.). Sexual selection could have resulted in the dimorphism in these species. However, in some toothed whales, females are larger, for example, in Baird's beaked whales (*Berardius bairdii*, see Chapter 13) and the franciscana. Males of these species do not show an accelerated growth spurt at around puberty, and their growth pattern resembles that of females (Kasuya 2002, in Japanese).

Most of the marine species of toothed whales rely, at least to some degree, on squids for their nutrition. Squids are abundant and a stable nutritional source in the ocean. Among the toothed whales, the larger species tends to be more oceanic and rely more on squids. This tendency is seen among several odontocete taxa. Such large toothed whale species include both sexually dimorphic species (e.g., sperm whales) and other species (e.g., Baird's beaked whales). Utilization of such a rich and stable nutrition source was assisted by acquisition of deep diving ability and at the same time could have enabled augmentation of body size. This suggests again that availability of abundant food resources was a factor that allowed increased body size.

Our knowledge of the longevity of toothed whales is limited to a few species, in which it varies from around 20 years in harbor porpoises (*Phocoena phocoena*) and finless porpoises (see Chapter 8), which are both members of the family Phocoenidae, to 70–80 years in Baird's beaked whales (see Chapter 13) and sperm whales.

Toothed whales capture food organisms one by one, which is a primitive or undeveloped way of feeding compared with that of the baleen whales. However, toothed whales have achieved the ability to use underwater sound for locating food (Au 2009). They further developed the use of sound for communication between individuals of the same species, which was probably followed by the elaboration of social life and the development of their cognitive ability (Würsig 2009). These could have evolved independently in multiple lineages of toothed whales. This book does not deal much with social behavior and with the use of sound by toothed whales.

Studying the social structure of toothed whales started in Japan using carcasses from fisheries. Scientists analyzed the group structure of dolphins taken by drive fisheries or sperm whales killed under special permit for scientific purposes. By combining information thus obtained, scientists estimated the social structure of striped dolphins, spotted dolphins (*Stenella attenuata*), common bottlenose dolphins (*Tursiops truncatus*), short-finned pilot whales, and Baird's beaked whales. Most of the results of such studies are discussed in this book. The same method was also applied on the long-finned pilot whales (*Globicephala melas*) off the Faroe Islands. The greatest disadvantage in such methods is the discontinuity of information. It does not yield much about life before capture and nothing about how the animals would have lived in the following days if they had not been killed.

These problems disappear if we identify individuals in a group and observe their behavior for years. Although this type of study requires many years of hard work, effort continues on several toothed whale species such as the two species of bottlenose dolphins, killer whales, and sperm whales and has produced information on the reproduction and social structure for all of these (Mann et al. 2000). This type of study is hard to apply for some Japanese cetaceans, because hunting continues on them.

The recent Odontoceti contain numerous species of a broad morphological variety, demonstrating a broad radiation. Although our knowledge about the life history and social structure of toothed whales is limited to only about 10 species, it is yet sufficient to reveal that they are at various stages of development in their life history, social structure, and perhaps cognitive ability. This leads me to consider that cetaceans, including both Odontoceti and Mysticeti, are like "primates in the water."

This book, based on the point of view cited earlier and using the seven species of toothed whales around Japan as examples, attempts to present the details of their life history and to clarify the differences among them. The fact that much of the biological information has been collected by myself using constant methodology allows easy comparisons.

I.2 FISHERIES FOR CETACEANS

I.2.1 Old Whaling Activities

It is believed that the oldest large-scale commercial whaling was established in the Basque region by the twelfth century, for the North Atlantic right whales (Francis 1990; Ellis 2009). Following the decline of the species along the European coasts, the industry moved offshore and reached the Newfoundland coast in the middle of the sixteenth century to hunt North Atlantic right whales as well as bowhead whales.

Immediately following the discovery of bowhead whales in Spitsbergen in 1607, the British and Dutch sent fleets of sailing ships equipped with whale boats for chasing and harpooning whales in that region. The Basque whalers were employed in the enterprise. The operation further moved to Davis Strait, to Baffin Bay, and then to Hudson Bay and ended in the early twentieth century due to the collapse of the whale stocks. The fisheries produced whale meat as well as oil and other products when they operated close to the European market but produced only oil and baleen plates after their operation expanded to distant grounds. Whale oil was tried out on the coast near the whaling ground or in the fleets' mother ports from blubber stored in casks. The latter method was applicable only in a cold climate.

In North America, immigrants started whaling in the mid-seventeenth century for baleen whales that were then abundant on the east coast (Starbuck 1964). They took North Atlantic right whales and humpback whales. The story went that one of their whaling vessels was blown offshore in 1712, met a group of sperm whales, and killed one of them. This was the start of the sperm whale fishery called "American whaling" or, later, "south sea whaling." This whaling rapidly expanded, accelerated by the timing when whales in the coastal ground were becoming scarce and the demand increased for spermaceti used in the candle industry, and the fishery was joined by various countries including the United Kingdom and the Netherlands. The blubber was removed from the carcasses alongside the ship, cut into smaller pieces on the deck, and cooked in try-pots placed on try-works made of bricks on the ship's deck. The scrap blubber pieces from which the oil was tried out were used as fuel for the try-works. The oil was stored in barrels, and the cruise continued sometimes for years until the ship had a full load.

These sperm whaling vessels rushed from one newly found ground to another. This type of operation is the most destructive from the management point of view, although such an operation pattern is not limited to whaling. The whalers entered the South Pacific in 1789, arrived in Hawaii in 1819, and first entered the western North Pacific region called the Japan ground in 1820 or 1821 (Starbuck 1964). The Japan ground was the last sperm whale ground reached by *American whaling* that cruised globally. By the 1840s, it expanded operations to the Sea of Japan, Okhotsk Sea, Gulf of Alaska, and Bering Sea, mainly on North Pacific right whales, and entered the Arctic Ocean via the Bering Strait in 1848, hunting the population of bowhead whales there nearly to extinction. *American whaling* ceased in the early twentieth century (Tower 1907; Park 1994, in Japanese), perhaps due, at least to some degree, to the depletion of whale stocks.

American whaling off Japan is believed to have contributed, by depleting baleen whale stocks, to the decline of Japanese coastal net whaling. Near the end of *American whaling*, the profitability of sperm whale hunting declined, which was affected by new industrial products such as coal gas for lighting and oils from fish, cottonseeds, and rape seeds. Petroleum discovered in Pennsylvania had a particularly strong effect on whaling. Thus, the target shifted from sperm whales to North Pacific right whales and bowhead whales (Francis 1990), mainly for their baleen. The damage to sperm whale stocks in the Japan ground caused by *American whaling* has not been well understood.

Kasuya (2009) reviewed the Japanese whaling activities within and outside Japanese territory. The oldest record of Japanese commercial whaling is from around 1570 in Mikawa Bay (34°45′N, 137°00′E) on the Pacific coast of central Japan, but it is generally thought that the operation started earlier. The whalers used hand harpoons as those used for dolphin and swordfish hunting. This technology spread eastward to become Baird's beaked whaling at Katsuyama (35°07′N, 139°50′E), Chiba Prefecture, at the entrance of Tokyo Bay, and continued up to the late nineteenth century. It also spread westward along the Pacific coast to the Shima Peninsula (around 34°20′N, 136°45′N), Kii Peninsula (33°26′N–34°14′N, 135°05′E–136°15′E), and Tosa (southern Shikoku in 32°45′N–33°30′N, 132°45′E–134°30′E) and to the southwestern coast of the Sea of Japan or Tsushima Strait area in longitudes 128°30′E–131°30′E. The fisheries took North Pacific right whales, humpback whales, and gray whales. Then, around 1675–1677, the hand-harpoon whalers started net whaling, in which whales were entangled in nets to limit their movement and then harpooned. This method rapidly spread to central and western Japan (not to Baird's beaked whaling at Katsuyama). The method enabled capture of fast-moving balaenopterids but required large numbers of workers. For example, Tsuro Whaling in Tosa, Shikoku, employed 464 workers during the season, which was in winter, in addition to the temporary workers employed for particular catches (Yamada 1902, in Japanese). Such inefficiency and operations limited to the coast were the weak points of these fisheries, which ended operations in the late nineteenth century.

Section I of this book mentions some of the old Japanese whaling in relation to the dolphin fisheries, and Appendix B lists some useful readings about the history of Japanese whaling (Appendix B is omitted in this English version).

I.2.2 Norwegian-Type Whaling

Most of the old whaling methods, except net whaling, were able to capture only slow-swimming targets of low specific gravity. Such species had become scarce by the middle of the nineteenth century. However, a large number of balaenopterids left in the ocean were hard to capture using rowing boats and hand harpoons assisted by small firearms because these whales swam fast and the carcasses sank.

A new technology to take balaenopterids was invented in 1868 by Sven Foyn, a Norwegian. It is called Norwegian-type whaling or modern whaling, in which tethered whaling harpoons are shot from a canon mounted on a steam or diesel-driven catcher boat. If the harpoon penetrates a whale, two things occur: a grenade on the tip of the harpoon explodes to kill or injure the whale, and the arms on the harpoon head open to prevent the harpoon from drawing out. This combined two technologies of the time, that is, steam sailing and an explosive harpoon fired from a canon. This whaling method rapidly spread to many oceans, for example, to the western North Pacific in 1889 by a Russian and to the Antarctic Ocean in 1904 by a Norwegian.

After several short attempts by pioneers, the Nihon Enyo Gyogyo Co. Ltd. was founded in 1899 in Senzaki (34°23′N, 131°12′E), Yamaguchi Prefecture, and succeeded in establishing Norwegian-type whaling in Japan. The operation began along the coast of the Korean Peninsula and expanded to northern Kyushu and Kii Peninsula and then to the Sanriku Region (Pacific coast of northern Honshu in 38°N–41°N). Japanese whaling before World War II (WWII) also operated off the coasts of Taiwan, Korea, and the Kuril Islands. Japanese factory ship whaling first operated in the Antarctic in the 1934/1935 season and in the North Pacific in the 1940 summer season. Japanese postwar whaling under foreign jurisdiction operated in Taiwan (1957–1959), Okinawa (1958–1965), Brazil (1959–1984), Canada (1962–1972), Chile (1964–1968), South Georgia (1963/1964–1965/1966), Peru (1967–1985), and the Philippines (1983–1984) (Tato 1985, in Japanese; Kasuya 2009).

Let us try to understand the progress of overfishing through looking at the statistics of Antarctic pelagic whaling. The total catch reached a peak in the 1937/1938 season, when 31 fleets caught a total of 28,871 blue whale units (BWU) (14,826 blue whales, 26,457 fin whales, 2,039 humpback whales, 6 sei whales). The BWU is a conversion system based on average oil production; one BWU equals 1 blue whale, 2 fin whales, 2.5 humpback whales, or 6 sei whales. Whalers of the time did not consider Antarctic mike whales as targets, so they did not have a BWU conversion for them. (Taxonomists distinguish this species, *B. bonaerensis*, from the northern minke whale *Balaenoptera acutorostrata*, but this book does not necessarily follow the usage if confusion can be avoided). The year of peak catch for each species tells of the shift from larger to smaller species. The blue whale catch peaked in the 1930/1931 season (28,325 blue whales) and take of the species was prohibited beginning in 1965/1966, fin whale in the 1937/1938 season (26,457 fin whales), which was followed by occasional peaks at a similar level during the 1953/1954 to 1961/1962 seasons, with protection beginning in the 1976/1977 season, and humpback whale in the 1958/1959 season (2,394 humpback whales) with protection since the 1963/1964 season. Humpback whales in the breeding grounds in lower latitudes were overfished earlier, and various catch regulations existed, so the catch trend of the species in the Antarctic pelagic whaling does not reflect the status of stocks. A peak of sei whale catch occurred in the 1964/1965 season (19,874 sei whales) and that species was protected beginning in the 1978/1979 season. Fishing of the smallest species, the Antarctic minke whale, peaked in the 1976/1977 season (7,900 minke whales). Against the decision of the IWC to prohibit the commercial take of minke whales beginning in the 1985/1986 season, Japan continued it until the 1986/1987 season and then started taking the species for scientific purposes (since the 1987/1988 season, see Chapter 7). Similar shifts in targeted species occurred in the North Pacific (Kasuya 2009).

As the number of fleets increased in the Antarctic, overproduction of whale oil became evident beginning in the late 1920s, and British and Norwegian whalers made a 5-year agreement to lay up some whaling fleets to control oil production. Under these circumstances, there were concerns whether the balaenopterid species might also follow the fate previously met by the right whales, the gray whales, and the sperm whales. With the leadership by the League of Nations, an international meeting was held in Geneva, Switzerland, and 26 countries including Switzerland signed the Geneva Convention for the Regulation of Whaling in September 1931 (which came into effect only in January 1936). This was followed by the London Agreement signed in June 1937, which was to apply until June 30, 1938, but was not ratified until May 7, 1938. The agreement of 1937 was amended by another London meeting in June 1938 (which came into effect in December 1938). These agreements included minimum body size by species, prohibition of taking mothers accompanied by calves, full utilization of the carcass, protection of "right whales" (right whales of both hemispheres, bowhead whales, pygmy right whales), and regulated take of gray whales (listed in the 1937 agreement). Japan did not sign these agreements (Omura et al. 1942; Omura 2000, both in Japanese). WWII started in Europe in September 1939, and the number of fleets operating in the Antarctic decreased from 28 in the 1939/1940 season to 11 in 1940/1941 and to 0 in 1941/1942 (Japan sent 6 fleets in both the 1939/1940 and 1940/1941 seasons).

The first postwar Antarctic whaling was carried out with nine fleets in 1945/1946 season (Japan made its first postwar Antarctic operation in the 1946/1947 season with two fleets). Prior to these events, with the expectation that Antarctic whaling must be resumed very soon to supply edible oil, seven allied countries met in London in January 1944 and agreed to set a catch limit of 16,000 BWU for the first postwar pelagic whaling in the Antarctic. They again met in London in November 1945 to agree on the same catch limit and some amendments of the whaling season. Then in November 1946, the whaling nations met again in Washington, DC, and 15 nations signed the current ICRW in December 1946. The convention was structured along the lines of the earlier agreements, including a catch limit of 16,000 BWU (Brandt 1948). The catch limit of 16,000 BWU was maintained until the 1952/1953 season and became one of the major causes of the depletion of whale stocks. This convention came to effect in November 1948, and Japan joined it in April 1951. As of March 2009, the convention had 84 member countries.

The Convention of 1946, which is composed of a preamble and 11 articles, identifies the equal rights of the member countries and agrees on 4 principles that (1) whales are the common property of human communities, (2) whales are fishery resources, (3) whale resources should be scientifically managed, and (4) a harvest is allowed only on whales that can sustain exploitation. The schedule attached to the convention describes the details of whaling regulation such as catch quotas and whaling seasons and can be amended with a 3/4 majority of the voting members of the IWC, which is composed of commissioners nominated by the contracting governments. Any contracting government will not be bound by a decision if it presents an objection to the decision within 90 days (further detailed rule exists). The commission now meets every 2 years to determine the regulations for the next two seasons.

It seems to me that this convention is based on an understanding that whaling is a fishing activity taking whales, but it does not seem to have a definition of whales. For example, it is unclear whether "whales" in the convention means any species of Cetacea or excludes small species such as dolphins and porpoises. This ambiguity is still causing arguments about the competence of the convention. It was likely that many contracting governments did not imagine in 1946 that the small cetaceans would be a problem of management in the future and that the hastily created convention could not later clean up structural deficits. The current IWC works on the tentative understanding that all the toothed whales other than the sperm whales are "small cetaceans" and that the IWC considers their scientific aspects but not their management, such as setting quotas. The Japanese government in its domestic documents defined "small cetaceans" as cetaceans other than sperm whales and two species of the genus *Hyperoodon*. The intention seems to exclude from "small cetaceans" all the species on which Japan has accepted the competence of the IWC.

In 1982, the IWC decided to temporarily set catch the quota at zero for all commercial whaling operations starting with the 1985/1986 Antarctic season and the 1986 coastal season. This action has been called the moratorium. Some whaling operations were excluded from the ban. The first was the whaling by countries that are not members of the ICRW. For example, Canadian Inuit hunt bowhead whales and the people on Lembata Island in Indonesia hunt sperm whales. These operations are of small scale and apparently similar to the aboriginal subsistence whaling of the member countries. The second type of whaling is "aboriginal subsistence whaling" by the ICRW members, which is small-scale whaling for local consumption allowed by the ICRW for Inuit in Alaska and Greenland, Chukchi in northeastern Siberia, and some other communities (Table I.1). The third is whaling under objection, which is allowed for the member countries of the ICRW that opposed the decision made in 1982 to end commercial whaling. Thus, Norway continues hunting several hundred minke whales in the North Atlantic under objection to the moratorium. The fourth is whaling with reservation. Iceland opposed the IWC decision to close commercial whaling and left the ICRW. When it rejoined the ICRW, Iceland made a reservation about the prohibition of commercial whaling. In the 2010 season, Iceland caught 59 minke whales and 142 fin whales. The fifth exclusion from the moratorium is whaling for scientific purposes under a special national permit, which is detailed in the next section. In addition to whaling under scientific permit, Japan allows small-type whaling and dolphin and porpoise fisheries to take nine species of small cetaceans, with the understanding that small cetaceans are outside the IWC competence and thus the fisheries are not affected by the IWC decision to end commercial whaling.

This book deals with Japanese dolphin and porpoise fisheries in Chapters 1 through 3, Japanese small-type whaling in Chapters 4 through 7, and the biology of small cetaceans exploited by these fisheries in Chapters 8 through 15.

I.2.3 Scientific Whaling in Japan

Article VIII of the ICRW signed in 1946 states the following:

> Notwithstanding anything contained in this Convention any Contracting Government may grant to any of its nationals a special permit authorizing that national to kill, take and treat whales for purpose of scientific research… .

It also states that "any whales taken under these special permits shall so far as practicable be processed and the proceeds shall be dealt with in accordance with directions issued by the Government." The Japanese government entered objection to the 1982 decision to stop commercial whaling but informed the IWC in July 1986 of its intention to withdraw the objection at the end of the 1986/1987 whaling season (the opening and closing dates of the season varied with whaling

TABLE I.1
Quotas for Aboriginal Subsistence Whaling for 2008–2012 Seasons Agreed by the International Whaling Commission, Which Are Either Single-Year or Block Quotas for 5 Years with Additional Annual Maximum Limits

Region (Country)	Bowhead Whale	Gray Whale, East Stock[a]	Humpback Whale	Fin Whale	Minke Whale
Alaska (USA)	280/5 years[b]				
Chukotka (Russia)	Part of the Alaska quota	620/5 years[c]			
Washington State (USA)		Part of the Chukotka quota			
Greenland (Denmark)	2/5 years			19/year	212/year
Bequia (St. Vincent and Grenadines)			20/5 years		

Note: From the schedule of International Convention on Regulation of Whaling.
[a] Feeding aggregation on the Asian side is estimated at around 130 and is not hunted.
[b] A few to be allocated to the Chukchi Tribe with the agreement between the two countries.
[c] A few to be allocated to the Makah Tribe with the agreement between the two countries.

types and targeted species) (Chapter 7). About 1 year before this, under a request made in April 1985 by Kazuo Shima, the then IWC commissioner of Japan as well as the vice-director general of the Fisheries Agency, and Ikuo Ikeda, the then director of the Far Seas Fisheries Research Laboratory (FSRL), established and chaired a team to plan Japanese scientific whaling. The team proposed a scientific whaling project that aimed at estimating the age-dependent natural mortality of Antarctic minke whales, and the scientific whaling for the first stage of the project started in the 1987/1988 Antarctic season. The current second-stage program started in the autumn of 2005 and aims to understand the marine ecosystem surrounding whales. It includes take of a maximum of 1415 whales of 7 species in the North Pacific and Antarctic. The project has no stated time limit (see Chapter 7). Iceland also had a short scientific whaling program after the cessation of commercial whaling.

Japanese scientific whaling has received various criticisms from around the world, which can be grouped into the following five categories: (1) interpretation of the convention, that is, Article VIII of the convention does not allow such large-scale and long-term operations; (2) suspicion about real intention, that is, the real and hidden intention of the project might be for economics and survival of the whaling industry; (3) scientific and technical aspect, that is, such studies are unnecessary, because nonlethal methods can be used, or the objectives cannot be achieved with the methods proposed; (4) scientific ethics, that is, it violates the ethics of scientists to kill so many large wild animals for so many years for scientific research; (5) whales are no longer fisheries resources, that is, 60 years have passed since the ICRW and divergent views on whales have emerged. Some details of these criticisms are in Kasuya (2003, 2005, both in Japanese, 2007, 2008, in Japanese) and Ishii (2011, in Japanese). The Scientific Committee of the IWC reviews the scientific whaling program only from the scientific point of view, that is, the third item mentioned earlier.

I.2.4 Dolphin and Porpoise Fisheries in Japan

The history of Japanese fisheries for dolphins and porpoises is dealt with in Section I of this book, including Japanese terminology used in the fisheries and their relationships with whaling. Until the mid-nineteenth century, hunting dolphins and porpoises in Japan was commercially minor and of small scale to satisfy only local demand. Since then, there were several occasions when the dolphin and porpoise fisheries attracted the attention of the country.

The first was during the second half of the nineteenth century or just after the Meiji Revolution, when the new government stimulated the development of the economy and searched for possibilities for new industries. The government investigated the status of dolphin and porpoise fisheries in Japan in an attempt to stimulate them. However, the small-cetacean fisheries could not cope with whaling for large cetaceans in terms of both quantity of products and economic value produced, and they lost the attention of the government and industries and returned to a local concern when Norwegian-type whaling was successfully established in Japan around the turn of the century.

The second rise of Japanese dolphin and porpoise fisheries occurred during the period from the beginning of the full-scale China–Japan war (which started in 1937) to the post-WWII period. The government promoted hunting of sperm whales and various species of small cetaceans to obtain materials for leather, which was important for the military. During WWII, male labor was drafted and fishing vessels were commandeered by the military. Offshore fishing became risky because of air raids and submarine attacks, so women and elderly people on the Izu coast (Pacific coast of central Japan in 34°36′N–35°05′N, 138°45′E–13910′E) shifted their attention more to dolphin hunting in nearshore waters. A shortage of food during and after the war also helped the rise of new small-cetacean hunting in Hyogo Prefecture (coast of southern Sea of Japan in longitude 134°20′E–135°30′E) and on the Sanriku and Izu coasts. Many of these fishery operations shrank during the 1960s.

The third increase in prices occurred from the 1980s to the 1990s, which included a period of gradual decline of Japanese commercial whaling and its final cessation by March 1988. The price of small-cetacean meat increased as it was sold to fill the vacancy left by drying up of the whale meat supply, and the catch of small cetaceans increased. More recently, an oversupply of whale meat from Japanese scientific whaling has caused reduction in the price of small-cetacean meat. These developments are detailed in Section I of this book.

As Japanese fisheries for small cetaceans were economically minor compared to whaling, the government did not pay much attention to their management until the mid-1980s, when it started to manage the fisheries in response to criticisms emanating from the IWC since 1975. The action was too late, because the dolphin fishery off the Izu coasts, which once recorded an annual take of 10,000–20,000 striped dolphins, had already crashed due to overfishing. Now it is rare to find striped dolphins in Japanese coastal waters.

Currently, Japanese fisheries are allowed an annual take of over 15,000 small cetaceans of 8 species. Most of them, with the exception of some live specimens that are sold to aquariums, are used for human consumption. In addition to these, the Japanese government permits a small-type whaling take of 66 Baird's beaked whales, which is considered a species outside of the competence of the ICRW.

This book describes the conservation biology of six of the nine species of small cetaceans currently hunted in Japan.

I.3 OTHER THREATS TO CETACEANS

Threats to cetaceans by Japanese whaling and dolphin and porpoise fisheries were outlined in the previous sections. Cetaceans are also threatened by other human activities and marine pollution (Marsh et al. 2003; Reeves et al. 2003). The threats are briefly summarized as follows and are only occasionally mentioned further in this book.

I.3.1 Intentional Takes

People in the northern polar regions hunt white whales and narwhals for food, and people in Sri Lanka consume dolphins taken with hand harpoon or found in their gill nets. Hand-harpoon hunting of dolphins for human consumption or for use as fish bait is known also from the coasts of Africa and South America (Read 2008; IWC 2011).

Statistics of these catches are often incomplete or absent, and abundance estimates for the affected cetacean populations are often lacking. Because of the usual reticent behavior of fishermen and the technical limitations in measuring these fishing activities and takes, managing the affected cetacean populations is not an easy task. In particular, the capture of freshwater dolphins for fishing bait in the Amazon and Ganges-Brahmaputra Rivers is an important ongoing conservation concern.

I.3.2 Incidental Mortality of Cetaceans in Fisheries

Large-scale pelagic gill-net fisheries for salmon, squids, and tunas operated or continue to operate in various regions. The nets are called drift nets because they are not anchored to the bottom. These fisheries incidentally catch cetaceans, seals, seabirds, and turtles. Even if fishermen discard or lose these fishing nets, they remain on the surface and continue to kill marine animals before they sink to the bottom or become stranded on the beach. Attempts were unsuccessful to effectively reduce such mortality, and at the United Nations General Assembly in December 1989, a resolution was passed to ban high-seas large-scale gill-net fisheries. Japan accepted this in December 1992 and discontinued large drift-net operations in the North Pacific.

Minke whales are often killed in fixed trap nets along the Japanese coasts, and the carcasses have long been utilized for human consumption without the issue being paid much attention by the surrounding community. The Japanese government twice changed the regulation of such takes. In 1990, it became legal to sell minke whales taken in trap nets for local consumption only; then in 2001, the rule changed to permit free trade in such takes (see Section 6.3). This resulted in a sudden increase in the number of minke whales taken in the Japanese trap-net fisheries from 10–20 to 120–130 per year (Table 1.6). The new regulation provided an opportunity for hidden economic activities to surface. A similar magnitude of minke whale mortality is also recorded for the Korean trap-net fisheries. Any coastal cetaceans are vulnerable to capture in trap-net fisheries. A particular problem is catches of species or populations that are currently recovering from past overexploitation, for example, the western stock of gray whales and the North Pacific right whale and regional humpback whale populations. Although it is difficult to regulate the net fisheries, it may be possible to modify fishing gear or to establish a rescue system. If there were an insurance system to cover damage to fishing gear due to whale-rescue activities, it would help promote such rescue. Because the fisheries utilize the oceans for private enterprise, they must have responsibility for conservation.

Fixed bottom gill nets in the Inland Sea of Japan are causing mortality of finless porpoises, but there are no attempts to reduce the mortality. The Japanese government collects mortality data from the fishermen, but they document only a minor fraction of the real mortality. Heavy mortality of finless porpoises is not limited to Japan. Korean scientists reported to the Scientific Committee of the IWC the bycatch of over 200 finless porpoises in coastal gill-net fisheries (IWC 2010).

Drift and fixed gill-net fisheries of small scale are still operated in various coastal seas as well as in large rivers of the world and remain a great threat to small cetaceans, such as the two species of finless porpoises, Irrawaddy dolphins, Indo-Pacific bottlenose dolphins, and several river dolphins. Sound emitters called "pingers" attached to gill nets have successfully reduced the mortality of small cetaceans on the European and North American coasts. Further information on incidental mortality of cetaceans is available in Perrin et al. (1994).

I.3.3 Ship Strikes

Dolphins and porpoises are often attracted to vessels and play in the bow wave, but in other cases, vessels may strike cetaceans and injure or kill them. Such incidents can be unnoticed by the vessels but may be later confirmed through examination of stranded carcasses. Coastal species of whales such as gray whales and right whales often have healed scars evidently caused by vessel strikes. Increasing vessel speed and the development of new hull designs increase the frequency of ship strikes. The Scientific Committee of the IWC has started an effort to estimate the mortality of cetaceans inflicted by ship strikes.

Attempts to reduce ship strikes are made on the east coast of the United States by notifying vessels of the positions of whales to vessels and by restricting transit routes in particular coastal portions of the habitat of North Atlantic right whales. As far as I know, attempts by Japanese ship builders have failed to devise effective ways to detect submerged whales by sonar or to scare whales away from a high-speed passenger ship by sound. It should be noted that these kinds of acoustic devices can also contribute to the deterioration of the environment of whales by possibly displacing them from their preferred habitat.

I.3.4 Chemical Pollution

We have discharged numerous chemicals into the environment through the activities of industry, agriculture, and urban life. They ultimately enter the ocean, and some of them are very stable in the ocean and can be accumulated in cetaceans through the food web. To make the situation worse, cetaceans often have limited ability to process those chemicals.

Mercury has been discharged into the environment in great amounts through burning of coal since the beginning of the industrial revolution. Although we do not have firm evidence on the effect on cetaceans, we know that for laboratory animals and humans its accumulation in their systems has adverse effects on growth and the nervous system. The total mercury contents in the muscle of toothed whales often exceed the allowable limit of 0.4 ppm set by the Japanese government for human consumption; there are concerns for the health of the consumers of cetacean meat.

Chlorinated organic compounds such as pesticides, herbicides, PCBs, and dioxins are known to adversely affect immunity and reproduction of laboratory animals. Some of these chemicals are known to be accumulated in the body of cetaceans.

Urban sewage and agricultural wastes cause increase in nitrogen and phosphorous level in coastal waters and can result in harmful algal blooms. Plankton organisms in such blooms are often poisonous, and the toxin moves to oysters and fishes that eat the plankton. There are cases of death of humpback whales and bottlenose dolphins due to eating fish carrying planktonic toxins.

Chapter 8 of this book discusses the chemical pollution of Japanese finless porpoises.

I.3.5 Whale Watching

Watching whales, dolphins, and porpoises has become popular worldwide and was once thought of solely as a benign use of cetaceans. However, it is now known that cetaceans change behavior responding to the close approach of vessels, and there is no reason to believe that whale watching activity is entirely harmless to the cetaceans being watched.

It is common to have rules for whale watching vessels about how to approach the animals or the minimum distance to be kept between the animals and the vessels, and it is usual to prohibit feeding the animals. In Japan, such rules are only voluntary regulations of local whale watching operations, and departure from them is a problem.

I.3.6 Sound Pollution

While cetaceans use underwater sound for communication between individuals of the same species and detection of food or underwater obstacles (see Section 1.1), there are numerous human activities that emit underwater sound that is strong enough to harm the acoustic environment of cetaceans, for example, cruising vessels, ice breakers, underwater construction, underwater explosions, air guns used for seismic survey, certain scientific equipment, and military sonar (Tyack 2008).

The sound made by baleen whales overlaps with vessel noise in the range of 20–200 Hz. The acoustic environment of baleen whales has been adversely affected since preindustrial days, and their communication range is believed to have diminished. Underwater explosions can damage the hearing organ of cetaceans permanently or temporarily or otherwise operate to dislocate them temporarily.

Our understanding of the background of mass stranding of cetaceans is limited. However, military sonar is now known to be as a cause of some mass stranding of cetaceans. Examination of stranded animals resulted in an interpretation that military sonar forces some cetaceans to divert from their normal diving pattern and contract compression sickness followed by stranding. The beaked whales are believed to be particularly vulnerable to the sounds.

I.3.7 Other Interactions with Fisheries: Depredation and Ecological Competition

Earlier I mentioned cases where cetaceans are killed or wounded in fishing activities. There are also cases where cetaceans steal fish from fishing gear, damage fishing gear, or disturb fishing operations (depredation). Because such behavior often involves learning by cetaceans, the use of sounds to scare them away loses effect rapidly, and the development of effective defensive methods seems to be difficult. A case of dolphin and fishery interactions in the Iki Island area is dealt with in Section 3.4 of this book.

A further basic problem is the possible interaction between fisheries and cetaceans through consumption of marine organisms. Fishermen often consider that fewer cetaceans in the ocean will result in more fish for fishermen and have started a political movement in Japan pushing this view. We have seen such a case in the Iki situation and still see it in propaganda produced by the Japanese whaling industry. However, we should understand that such an idea is still only a hypothesis. The Scientific Committee of the IWC has stated that they do not have an appropriate ecosystem model to examine this hypothesis and that they do not have sufficient data to put in the model if they had one. It is worth noting that marine fish populations were probably in better condition in the nineteenth century or before the depletion of rorqual populations due to heavy exploitation by modern whaling in the twentieth century.

I.3.8 Climate Change

The last ice age ended about 10,000 years ago and was followed by a short period of climate that was warmer than at present. During this warm period, the Arctic Ocean had open water in summer and cetaceans probably found opportunities to move between the North Atlantic and the North Pacific. Some tropical delphinids could have moved between the South Atlantic and the Indian Ocean around Cape Agulhas (Section 12.3.7).

Because the existing cetacean species have experienced such past climate changes, it would be possible for most of them to survive, at least as species, through future climate change due to anthropogenic causes. However, we should note that the speed of expected climate change could be over 10 times faster than in the past. Climate change may proceed within a few generations of the long-lived whales, and these species have not yet recovered from damage done by human activities. Under these conditions, we have to be uncertain about the outcome for the whales.

To make the situation worse, the expected climate change will be accompanied by an exploding human population and destructive human technology. My particular concern is for the populations of riverine and coastal cetaceans. They have already been affected by human activities, and their future survival will be most influenced by habitat modifications as a result of human response to climate change, such as river water extraction for irrigation and coastal construction. Extinction of the baiji around the turn of the century is believed to have been due to the destruction of its habitat by humans, such as by water pollution, incidental mortality in fisheries, decline in prey fish population due to fisheries and coastal construction, and vessel traffic.

Future climate change might change ocean circulation and marine productivity, which influences fisheries as well as the survival of cetaceans, and we do not know how we might be able to respond to them.

REFERENCES

[*In Japanese Language]

Au, W.W.L. 2009. Echolocation. pp. 348–357. In: Perrin, W.F., Würsig, B., and Thewissen, J.G.M. (eds.). *Encyclopedia of Marine Mammals.* Academic Press, Amsterdam, the Netherlands. 1316pp. (2nd edn.).

Brandt, K. 1948. *Whaling and Whale Oil During and after World War II.* Food Research Institute, Stanford University, Stanford, CA. 47pp.

Ellis, F. 2009. Whaling, aboriginal. pp. 1227–1235. In: Perrin, W.F., Würsig, B., and Thewissen, J.G.M. (eds.). *Encyclopedia of Marine Mammals.* Academic Press, Amsterdam, the Netherlands. 1316pp. (2nd edn.).

Francis, D. 1990. *A History of World Whaling.* Viking, Markham, Ontario, Canada. 288pp.

Frankel, A.S. 2009. Sound production. pp. 1056–1071. In: Perrin, W.F., Würsig, B., and Thewissen, J.G.M. (eds.). *Encyclopedia of Marine Mammals.* Academic Press, Amsterdam, the Netherlands. 1316pp. (2nd edn.).

*Ishii, A. (ed.) 2011. *Dissection of the Whaling Issue.* Shin Hyoron, Tokyo, Japan. 322pp.

IWC (International Whaling Commission). 2009. Report of the sub-committee on other southern hemisphere whale stocks. *J. Cetacean Res. Manage.* 11(Suppl.): 220–247.

IWC. 2010. Progress reports. *J. Cetacean Res. Manage.* 11(Suppl. 2): 352–398.

IWC. 2011. Report of the scientific committee. *J. Cetacean Res. Manage.* 12(Suppl.): 1–75.

Kasuya, T. 1995. Overview of cetacean life histories: An essay in their evolution. pp. 481–497. In: Blix, A.S., Walloe, L., and Ultang, O. (eds.). *Whales, Seals, Fish and Man.* Elsevier Science, Amsterdam, the Netherlands. 720pp.

*Kasuya, T. (ed.) 1997. *River Dolphins: Their Past, Present and Future.* Toriumi Shobo, Tokyo, Japan. 214pp.

*Kasuya, T. 2002. Life history of toothed whales. pp. 80–127. In: Miyazaki, N. and Kasuya, T. (eds.). *Mammals of the Sea—Their Past, Present and Future.* Saientisuto Sha, Tokyo, Japan. 311pp. (First published in 1990).

*Kasuya, T. 2003. Sea of whales and sea of man. pp. 61–72. In: *Environmental Yearbook.* Sodo Sha, Tokyo, Japan. 339pp.

*Kasuya, T. 2005. On the whaling issue. *Ecosophia* 16: 56–62.

Kasuya, T. 2007. Japanese whaling and other cetacean fisheries. *Environ. Sci. Pollut. Res.* 14(1): 39–48.

*Kasuya, T. 2008. My thoughts on the whaling issue—From 46 years experience of a whale biologist. *Hito-to Dobutsu-no Kankei Gakkaishi* 20: 38–41.

Kasuya, T. 2009. Japanese whaling. pp. 643–649. In: Perrin, W.F., Würsig, B., and Thewissen, J.G.M. (eds.). *Encyclopedia of Marine Mammals.* Academic Press, Amsterdam, the Netherlands. 1316pp. (2nd edn.).

Kükenthal, W. 1893. Die Bezahnung. pp. 385–448 and pls. XXV. In: Kükenthal, W. (ed). *Vergleichend-Anatomische und Entwickelungsgeschichtliche Untersuchungen an Walthieren.* Verlag von Guster Fisher, Jena, Germany. 448 pp.

Mann, J., Connor, R.C., Tyack, P.L., and Whitehead, H. (eds.) 2000. *Cetacean Societies, Field Studies of Dolphins and Whales.* University of Chicago Press, Chicago, IL. 433pp.

Marsh, H., Arnold, P., Freeman, M., Haynes, D., Laist, D., Read, A., Reynolds, J., and Kasuya, T. 2003. Strategies for conserving marine mammals. pp. 1–19. In: Gale, N., Hindell, M., and Kirkwood, R. (eds.). *Marine Mammals: Fisheries, Tourism and Management Issues.* CSIRO Publishing, Collingwood, Victoria, Australia. 446pp.

Miller, G.S. 1929. The gums of the porpoise *Phocoenoides dalli* (True). *Proc. U.S. Natl. Mus.* 74(26): 1–4 and pls. 1–4.

Ohdachi, S.D., Ishibashi, Y., Iwasa, M.A., and Saitoh, T. (eds.) 2009. *The Wild Mammals of Japan.* Shoukadoh, Tokyo, Japan. 543pp.

*Omura, H. 2000. *Journal of an Antarctic Whaling Cruise, Record of a Whaling Expedition in Its Infancy in 1937/38.* Toriumi Shobo, Tokyo, Japan. 203pp. (with notes by Kasuya, T.).

*Omura, H., Matsuura, Y., and Miyazaki, I. 1942. *Whales—Their Biology and the Practice of Whaling.* Suisansha, Tokyo, Japan. 319pp.

*Park, K. 1994. Arrival of American whalers at the Sea of Japan and discovery of Take-shima Island. *Rekishi-to Minzoku* 11: 101–138.

Perrin, W.F., Donovan, G.P., and Barlow, J. (eds.) 1994. *Gillnet and Cetaceans, Incorporating the Proceedings of the Symposium and Workshop on the Mortality of Cetaceans in Passive Fishing Nets and Traps. Rep. Int. Whal. Commn.* (Special Issue 15): 629pp.

Read, A. 2008. The looming crisis: Interactions between marine mammals and fisheries. *J. Mammal.* 89(3): 541–543.

Reeves, R.R., Smith, B.D., Crespo, E.A., and Notabartolo di Sciara, G. 2003. *Dolphins, Whales and Porpoises: 2002–2010 Conservation Action Plan for the World Cetaceans.* IUCN, Gland, Switzerland. 139pp.

*Slijper, E.J. 1984. *Whales.* Tokyo Daigaku Shuppan Kai, Tokyo, Japan. 403pp. (translated by Hosokawa, H. and Kamiya, T.).

Starbuck, A. 1964. *History of American Whale Fishery from Its Earliest Inception to the Year 1876.* Sentry Press, New York. Vols. 1: 407pp.; 2: 779pp. (First published in 1878).

*Tato, Y. 1985. *History of Whaling and Its Data.* Suisan Sha, Tokyo, Japan. 202pp.

Thewissen, J.G.M. (ed.) 1998. *The Emergence of Whales: Evolutionary Patterns in the Origin of Cetacea.* Plenum Press, New York. 477pp.

Tower, W.S. 1907. *A History of the American Whale Fishery.* The John C. Winston, Philadelphia, PA. 145pp.

Turvey, S. 2008. *Witness to Extinction, How We Failed to Save the Yangtze River Dolphin.* Oxford University Press, New York. 233pp.

Tyack, P. 2008. Implications for marine mammals of large-scale changes in the marine acoustic environment. *J. Mammal.* 89(3): 549–558.

Uhen, M.D. 2009. Evolution of dental morphology. pp. 302–307. In: Perrin, W.F., Würsig, B., and Thewissen, J.G.M. (eds.). *Encyclopedia of Marine Mammals.* Academic Press, Amsterdam, the Netherlands. 1316pp. (2nd edn.).

Würsig, B. 2009. Intelligence and cognition. pp. 616–62 3. In: Perrin, W.F., Würsig, B., and Thewissen, J.G.M. (eds.). *Encyclopedia of Marine Mammals.* Academic Press, Amsterdam, the Netherlands. 1316pp. (2nd edn.).

*Yamada, S. 1902. *Whaling at Tsuro.* Tsuro Hogei Kabushiki Kaisha, Kochi, Japan. 284pp.

Authors

Toshio Kasuya (author and translator) joined the Whales Research Institute (Tokyo) after graduating from a fisheries course at the University of Tokyo in 1961 and started studying life history of great whales being exploited in Japan. His subsequent scientific activities and affiliations are the life history of small cetaceans at the Ocean Research Institute, University of Tokyo (1966–1983); the life history and management of small cetaceans at the Far Seas Fisheries Research Laboratory, Fisheries Agency (1983–1997); and marine mammal biology and teaching at Mie University (1997–2001) and Teikyo University of Science and Technology (2001–2006). He participated in the activities of the Scientific Committee of the International Whaling Commission (1982–2013). He studied sperm whales, eight species of small cetaceans under the threats of human activities in Japan, river dolphins, and dugong, and then he retired from research and education activities in 2013. Among his achievements is finding the extended postreproductive lifetime of some female cetaceans. He received awards from the Society for Conservation Biology (1994), the Society for Marine Mammalogy (2007), and the Japanese Society of Mammalogists (2013).

William F. Perrin (edited translation), after 4 years in the Air Force as a Czech linguist, received his PhD in zoology from the University of California, Los Angeles in 1972. He served for 46 years as fishery biologist at the NOAA's Southwest Fisheries Science Center in La Jolla, California, working on systematics, ecology, and conservation of tropical cetaceans. He received the NOAA Scientific Research and Achievement Award, 1979; Department of Commerce Bronze Medal, 1994; the K.S. Norris Lifetime Achievement Award of the Society for Marine Mammalogy, 2011; and the NOAA Distinguished Career Award, 2013. He served at various times as appointed councilor for aquatic mammals of the Convention on Migratory Species, chair of the Scientific Advisory Committee of the U.S. Marine Mammal Commission, editor of *Marine Mammal Science*, chair of the Cetacean Specialist Group of the International Union for Conservation of Nature and Natural Resources, member of the Scientific Committee of the International Whaling Commission, and research associate in vertebrate zoology of the Smithsonian Institution. He is now retired and working with his wife, Dr. Louella Dolar, on biology and conservation of marine mammals in Southeast Asia.

Section I

History of Japanese Cetacean Fisheries

1 Outline of the History of Small-Cetacean Fisheries

1.1 IMPORTANCE OF HISTORY

Humans have influenced the lives of wild animals, either intentionally or unintentionally. We have often noticed damage to wildlife after causing it and will continue to do so in the future. The tasks of conservation biology include understanding how wildlife has attained the current status, predicting its future, and proposing countermeasures to minimize further damage or to allow recovery from the past damage. Therefore, understanding the history of interactions between wildlife and humans is necessary for people who study wildlife, for scientists who interpret results of their research on wildlife, and for those who utilize existing scientific information for conservation.

It is relatively easy for us to understand interactions between terrestrial or freshwater animals and our activities on land such as logging, reclamation, and irrigation. More difficult is to measure the effects of our activities on the aquatic environment. We have discharged numerous chemicals into the marine environment. Of particular concern are heavy metals such as mercury and persistent poisonous organic compounds such as PCBs and dioxin, which accumulate in the bodies of some cetacean species to levels that threaten their health. This is not only a problem of cetacean conservation but a health problem for humans who rely on marine products for nutrition. Something must have been wrong in our past way of handling nature.

The marine acoustic environment of cetaceans is deteriorating due to various human activities, including military sonar and mineral resources development. Collision of cetaceans with vessels is also a conservation problem for particular cetacean populations that are on the way to recovery from past overexploitation or inhabit waters of heavy vessel traffic. Whale watching may also cause trouble for targeted cetaceans if conducted inappropriately. The Scientific Committee of the International Whaling Commission (IWC) is collecting information on such environmental problems of cetaceans (IWC 2008).

Hunting is the most direct interaction between cetaceans and humans and the most easily identifiable. We measure it in terms of numbers of animals and not by weight as usually is the case in exploiting fish populations. This is a useful point in evaluating the effect of a hunt on a cetacean population. The statistics are available relatively easily from legally controlled fishing activities, but problems exist for other fishing activities. Japanese dolphin and porpoise fisheries have been one of the latter cases. Cetaceans may be killed by accident through various fishing activities not targeting cetaceans, for example, seine net, gill net, trap net, longline, trawl net, and even by nets discarded in the ocean. The incidental takes can cause serious damage to cetacean populations (Anon. 1994), but it depends on social circumstances whether such incidental mortality is ignored by fishermen or recorded in statistics.

It is also important to remember that most fisheries have their hidden side, and there are great difficulties in getting information on fishing activities pursued beyond regulations. Poaching prevailed in Japanese coastal whaling after World War II, but the Fisheries Agency is not willing to accept the fact, and most of the retired whalers do not disclose their past activities, possibly in fear of reaction from their community. Thus, reconstruction of true whale catch statistics has not yet been achieved (see Section 1.4). Under such circumstance, conservation biologists cannot carry out important analyses to estimate the status of whale populations before whaling, or to understand the response of whale populations to the exploitation.

In Japan today, it is not easy to find publications on the history of fisheries for dolphins and porpoises, in contrast with the availability of numerous publications on whaling (list of recommended reading on Japanese whaling is appended to the Japanese version). The reasons are that people did not pay much attention to the small-cetacean fisheries because they were economically minor compared to whaling and that regulation of the fisheries started as recently as the late 1980s. Compared with large cetaceans, small cetaceans tend to inhabit smaller geographical areas and are likely to form numerous local populations, even within one ocean basin. This, together with the fact that their reproductive capacity is often similar to that known for large cetaceans, that is, low (see Section 1.2), means that small-cetacean populations may have a greater risk of depletion due to direct hunting or bycatch. This is the reason why I believe that knowledge of the history of small-cetacean fisheries is important for their conservation.

1.2 *IRUKA* VERSUS *KUJIRA*, CONSERVATION MEASURES

The Japanese often classify cetacean species into either *iruka* (mainly dolphins and porpoises) or *kujira* (mainly whales), but the boundary between them is ambiguous and disagrees with usage in other languages. For example, *iruka* is almost identical with the English term "dolphins and porpoises," but it usually includes *Delphinapterus leucas* and *Monodon monoceros*, in English called beluga whale and narwhal. The usage for large delphinids such as killer and pilot whales varies among communities within Japan. Oceanaria like to use *kujira* for even smaller species such as Risso's dolphins.

It should be noted that this classification of *iruka* and *kujira* has nothing to do with the zoological classification, where we divide all the cetaceans into Odontoceti or toothed whales and Mysticeti or baleen whales. All the baleen whale species, about 14 species in total, are large in body size and called *kujira* in Japan. The toothed whales include about 70 species of various body sizes and are grouped into several families, but there is no family that matches the grouping of *iruka*, that is, "dolphins and porpoises" in the English language.

Classification in zoology is expected to change from time to time and taxonomic opinion can vary among taxonomists. Toothed whales are currently grouped into 9 families: Delphinidae, Phocoenidae, Monodontidae, Ziphiidae, Kogiidae, Physeteridae, Platanistidae, Pontoporiidae, and Iniidae. Delphinidae is the largest group and includes numerous species of various body sizes, for example, striped dolphin, short-finned pilot whale, and killer whale. Dall's porpoise and the finless porpoise are members of Phocoenidae. The finless porpoise in the Inland Sea is called *ze-gondo. Gondo* is apparently used for all species with round foreheads, such as pilot whales. In English, "porpoise" refers to species of Phocoenidae, distinguished from so-called dolphins of Delphinidae and several other families. Seagoing people, for example, fishermen in the United States and historically whalers, often use "porpoise" to refer to any dolphin- or porpoise-sized small cetacean.

Taxonomists combine the three families Delphinidae, Phocoenidae, and Monodontidae into the superfamily Delphinoidea. Three families of fresh water dolphins, that is, Platanistidae, Pontoporiidae, and Iniidae, were once combined into another super family, Platanistoidea, which included the two Amazon river dolphins (= botos), South Asian river dolphin (Ganges and Indus river dolphins), Yangtze river dolphin (= baiji, which became extinct around the turn of the century), and La Plata dolphin (= franciscana). They inhabit coastal waters of the western South Atlantic and large rivers of Asia and South America that were not covered by ice during the last ice age. In this book, I will use "dolphins and porpoises [i.e., *Iruka* in Japanese]" for all the species of Delphinoidea, distinguished from "river dolphins" representing the five species mentioned.

As explained earlier, in Japan the concept of "whales" and "dolphins and porpoises" has little to do with their biology or taxonomy. The distinction must be sought in the memory of our community. Our ancestors might have identified one group of cetaceans as large graceful animals and the other as small agile animals, and this became a common understanding in numerous Japanese communities that lived with the ocean. Such understandings change with time, and there has been no criterion of body size to determine the boundary between the two groups. Nowadays, some who pay more attention to their movements consider the killer whale a whale, but others who put importance on the skeletal structure will deal with it as a dolphin. The two species of Kogiidae cause more difficulty. Their body structure is closer to that of the sperm whale, which grows to 18 m, than to that of many species of Delphinidae. The pygmy sperm whale, the larger species of Kogiidae, measures 3.5 m, and the smaller dwarf sperm whale measures about 1 m less. Although they are small in body size, they dive long and behave like a whale. Because they are not members of Delphinoidea, they are not dealt with as "dolphins" in this book.

There is a general rule among mammals that larger species grow slower, live longer, and have lower reproductive rates (Schmidt-Nielsen 1995, in Japanese translation). Until the 1960s when a technique was introduced to estimate the age of dolphins using growth layers in the teeth, there was an expectation that dolphins and porpoises might mature at a younger age and reproduce at a higher rate than cetaceans of large body size such as whales. However, age determination of delphinids revealed that they matured later and lived longer than anticipated, and scientists were puzzled by the suggestion that their reproductive rate might not be as high as expected. Nowadays, there is a common understanding among whale biologists that there is no single relationship in cetaceans between body size and life history parameters such as longevity or mortality rate. Any such relationship that might exist in a single group such as Balaenopteridae does not necessarily apply to a larger taxonomic group, such as all baleen whales. Several factors are at work in determining cetacean body size.

Now let us compare some life history parameters of dolphins and porpoises with those of whales, using data for toothed whales from Kasuya (1991, in Japanese) and for baleen whales from Kato (1990, in Japanese), limiting the discussion to females. Both Japanese finless porpoises and common porpoises, members of Phocoenidae, grow to around 2 m or less, are sexually mature at a young age of 3–6 years, reproduce every 1–2 years, and live to around 20 years. Two delphinid species, the striped dolphin (2.3 m) and short-finned pilot whale (4–5 m), attain sexual maturity at a similar age of 8–9 years, but their average calving intervals differ considerably, that is, 2–3 years and 7–8 years, respectively. Females (11–12 m) of the sperm whale grow to over twice the size of short-finned pilot whales (up to 4 or 5 m depending on the population), but both species live to about 60 years and calve at an average interval of about 7 years. These calving intervals are an average for all the sexually mature females in the population, including post-reproductive females. About 25% of adult females are post-reproductive in two populations of short-finned pilot whales off Japan, and the presence of such post-reproductive females is suggested also for sperm whales by a fecundity trend declining with increasing age (Best et al. 1984). Humpback whales, baleen whales of about 17 m, live to a similar age as sperm whales, attain body size greater than sperm whales, and calve every 1–4 years (2–3 years on average), which is similar to striped and bottlenose dolphins. Blue and fin whales, balaenopterid species of greater body size, seem to have life histories similar to that of the humpback whale. These examples are sufficient to deny the plausibility of estimating life history from body size alone or assuming high reproductive rates for dolphins and porpoises because they are small.

The typical annual life cycle of a baleen whale comprises seasonal feeding, followed by several months of starvation, when calving takes place. This reproductive pattern is possible due to a great nutritional storage capacity, which is helped by large body size. Thus, selection may have favored larger female size, which may have also resulted in the augmentation of male size because some genes controlling body size may be autosomal. Another factor contributing to evolution of large size was feeding at a low trophic level, such as on zooplankton or small fish, which are available in great quantity and consistently over years (albeit with some seasonal variation). Large body size alone could have also benefited whales in a stable environment of abundant nutritional supply in their survival and reproduction.

Another factor that contributed to evolution of great size in some toothed whales was sexual selection. Male sperm whales visit female schools in the mating season and compete for opportunities to mate, and larger males are likely to father more offspring. This could have worked to increase male body and subsequently affected female size too. Such a male reproductive strategy works better if females live in groups. Therefore, it is likely that females started to live in groups or with evolution toward the present-day matrilineal social structure, and males adapted their behavior to utilize the situation. An environmental factor that supported great body size in sperm whales was probably squid as forage. Squid are available over a broad geographical area and in large quantities. With the possible exception of killer whales and false killer whales, toothed whales of large body size rely mainly on squid for their nutrition, for example, sperm whales, beaked whales, pilot whales, and Risso's dolphins.

Large body size is not always a good thing for toothed whales. In some environments where production was limited or unstable, individuals of smaller body size may have done better (island forms). We find such examples in dolphin populations in rivers, where they have smaller body than their oceanic counterparts.

It seems that life history and reproduction of cetaceans are not single functions of body size but relate more with numerous factors such as migration, food habit, social structure, and reproductive strategy (Kasuya 1991, in Japanese, 1995). Baleen whales live a less social life than some toothed whales, wean their calves at a young age (0.5–1 year), and reproduce at shorter intervals. A short calving cycle is also seen in some small toothed whales in the Phocoenidae (Dall's porpoise and common porpoise) and Delphinidae (Commerson's dolphins in an aquarium). Toothed whales differ from baleen whales in their social structure, showing the development of various stages of group living, underlain by the use of sound for communication and development of cognition for social life. Female calves of such group-living toothed whales have a stronger tendency than male calves to continue to live with their mother. Examples are communities of pilot whales, killer whales, and sperm whales, which have a matrilineal social structure where females of several generations live together and have adapted a reproductive strategy toward extended maternal care and production of a small number of offspring. This is an extreme case of extended maternal care. Such social structure can evolve without affecting body size but can greatly affect calving interval, nursing period, and longevity of females. This explains why there is little correlation between body size and life history parameters in the toothed whales. The behavior of adult males in matrilineal communities varies among species.

The next topic is the rate of increase of population or how much we can take from the population without further reducing it, a question frequently asked when managing small cetaceans as fishery resources. In a simplistic understanding, increase rate is the difference between birth and natural mortality rates, and multiplying it times population size will yield the number of individuals that can be harvested without further reduction of the population (we tentatively leave problems of social structure for later discussion). As the expected rate of increase is only several percent for cetaceans, mortality and birth rates must be estimated with accuracy to one or more decimal places if we attempt to calculate the increase rate as the difference between the two parameters. Such a procedure is almost impossible with our current level of knowledge. Estimation of population size also incorporates broad uncertainty, with 95% confidence intervals often exceeding 50% on both sides of the mean estimate. Small-cetacean species may contain numerous local populations, but the identification of their seasonal geographical ranges is often difficult.

The simplest logic of fishery resource management assumes equal birth and natural mortality rates for an unexploited population, that is, perhaps an unrealistically stable population. If population size is reduced by hunting, then population density decreases and per capita food availability increases. Responding to this, the population increases birth rate and decreases natural mortality rate, attaining a positive population increase rate. The value of the increase rate (R) increases with decreasing population level and reaches a maximum value when the population level is close to zero. The maximum increase rate is called the innate capacity of increase or maximum reproductive rate (R_{max}). Population level (P) is the ratio of current population size relative to initial (or pre-exploitation) population size. The product of population level and increase rate, PR, is the sustainable yield (SY). If this theory is correct, SY will reach a maximum value (maximum sustainable yield, MSY) at a certain population level (called MSY level, or MSYL). The increase rate at which MSY is produced is called the MSY rate. If R can be expressed by a linear function of P, then the recovery curve of a depleted population is sigmoid and MSYL is 0.5. However, the R–P curve for higher animals such as mammals is thought to be convex, and the MSYL is believed to be somewhere between 0.5 and 1.0. The Scientific Committee of IWC assumes MSYL = 0.6 for baleen whales.

The hypothesis assumes that the population level remains stable during a particular length of time that is required for the cetacean population to improve its increase rate. The increase rate is a function of various life history parameters such as mortality rate, pregnancy rate, and age at sexual maturity. Although we do not know exactly how long the population takes to adjust these parameters for the new population density, it will take several years to fully adjust the age at sexual maturity. It is also a possibility that a newly emerged cetacean fishery operates so rigorously that it does not allow time for the cetacean population to respond with a density-dependent change, which results in a rapid decline of the exploited population.

In order to apply this theory, or hypothesis, for any real cetacean population, we must know the values in the R–P equation, the value of R_{max}, initial population size, and current population level. Such information is not available for most cetacean populations. Furthermore, a whale population could have been influenced by human activities other than whaling, for example, exploitation of fish that is also consumed by cetaceans or destruction of cetacean habitat. It is also a possibility that more than one species of cetaceans compete for resources. The single species model is totally unable to handle multi-species or multi-fishery situations.

Therefore, the Scientific Committee of the IWC developed the so-called Revised Management Procedure (RMP), which is a system to determine a safe catch level based on past catches, population estimates to be obtained at several-year intervals, and their trend. Because this method puts more weight on safe management of whale populations than on exploitation (a "precautionary approach"), the resultant catch limits are expected to be so small that they will not satisfy the industries willing to operate commercial whaling. The RMP has been completed for baleen whales, but the results have not been applied for management because many countries are now unwilling to reopen commercial whaling.

Some readers may wonder if the RMP can be used for toothed whales. This task has not begun because of the understanding that social structure and mating systems of toothed whales are more complicated than those of baleen whales, and we do not have enough knowledge about them. Although the Scientific Committee had sperm whales in mind when considering this problem, similar problems must also arise for other toothed whales. Japanese readers who are interested in the management of large whales are recommended to refer to Kitahara (1996, in Japanese) and Sakuramoto et al. (1991, in Japanese). The former describes the revised management procedure and principles of management of whale stocks, whereas the latter describes the biology of whales and management of whale stocks.

There is no doubt that the maximum rate of increase, R_{max}, varies by species or population of small cetaceans, about which we do not have reliable knowledge. However, if we have some rough idea of the value, it will be possible to suggest safe levels of catch or to judge if a current level of human caused mortality is acceptable. For example, if we are sure that a current take rate is less than half of the R_{max}, we can conclude that there will be no risk of further reduction of the stock size. When using such a method it is necessary to understand the accuracy or bias of fishery-related data and to incorporate them into the process. This idea is used in the management method called Potential Biological Removal (PBR) of Wade (1998). It must be cautioned that the method assumes 4% for R_{max} for all small-cetacean species and an arbitrary safety factor parameter. Results of simulations suggest that it allows safe management of small-cetacean stocks, provided that the value of the "safety factor" is adequate. The amount of removal allowed by this method is usually about 1% of the current abundance, although the figure may change with input data. If this method has any problems, one of them would be the flexibility allowed for the value of the safety factor. It would be possible to adopt a particular figure in order to create a desired catch limit. A critical review of the PBR process is required before accepting a particular management proposal. An example of PBR is given in Section 12.6.3.

Most toothed whales live in groups, but the formation of groups and degree of stability of the groups vary among species. Matrilineal communities of killer whales, sperm whales, and perhaps short-finned pilot whales, which are known to have old post-reproductive females in the school, will be placed at the extreme end of such evolutionary tendencies. The function of such old females may not be limited to uniting group members as core individuals or allo-parenting offspring of her kin, but also to function as a carrier of culture, that is, to accumulate information on the school's environment and experience and to transmit the information to younger generations by example or training. The existence of culture is accepted for communities of apes and monkeys, and some toothed whales will be additional examples. The marine environment is less predictable than the terrestrial environment of primates. For example, primates will have less difficulty in daily search for food resources if they remember the location and season of particular fruit trees, but locating fish schools in the ocean will require more complicated information on the environment and the ability to analyze it. Attacking a fish school once located will require more cooperation of group members than picking fruit from a tree. Thus, experience and knowledge accumulated by elderly members of a toothed-whale school may have an important function for daily life in the less predictable marine environment.

If a toothed whale species contains communities of broad cultural diversity, this increases the adaptive ability of the species. Current conservation biology recognizes the need to conserve genetic variability, but the importance of conserving cultural variability should be considered in future management. The effect of hunting on dolphins and porpoises populations must differ with the method of hunting (i.e., drive fishery vs. hand-harpoon fishery) and the social structure of the hunted species.

1.3 SMALL-CETACEANS AND RELATED FISHERIES

Some fragmentary records of early fisheries for small cetaceans were collected by the Japanese government in the late 1800s after completion of the Meiji revolution in 1864–1871. A systematic effort to collect such statistics started in 1957 (Ohsumi 1972) and continues in the Ministry of Agriculture, Forestry and Fisheries, which records the annual number of animals killed in each prefecture but does not record species. Another series of statistics was started in 1972 by the Whaling Section of the Fisheries Agency and continues to the present (see Section 1.4). In addition to these government statistics, some scientists have published statistics based on information collected at fish markets at the landing ports. These are cited in the following chapters.

Around 1887, the Japanese government requested that the prefecture governments report the status of fisheries within their territories, including fisheries for whales, dolphins, and porpoises. The intention was probably to obtain baseline information for the establishment of the Fisheries Acts, which were to control and promote fishing industries in Japan. One of the results was the publication of *Suisan Chosa Yosatu Hokoku* [Report of the Preliminary Survey of Fisheries] of the *Noshomu-sho* [Ministry of Agriculture and Commerce] (1890–1893, in Japanese). This reviewed Japanese fishery operations of 1888–1891 in a standardized way. Hokkaido was not included for some reason, so the whaling operation then existing at Haboro (44°22′N, 141°42′E) on the Sea of Japan coast of Hokkaido was not described. Cetacean-related portions of the publication were reproduced by Mr. K. Takeuchi in Vol. 30 of his magazine *Hogeisen* (Takeuchi 1999, in Japanese). Table 1.1 shows the location of Japanese whaling and dolphin and porpoise fisheries extracted from that summary.

According to an old manuscript *Geiki* [Records of Whales], Japanese commercial whaling started in Mikawa Bay (34°45′N, 137°00′E) on the Pacific coast of central Honshu, around 1570–1572 (Hashiura 1969, in Japanese), which is supported by numerous records of whaling in Morosaki (34°42′N, 136°58′E) on the coast of Mikawa Bay (Minami Chita Choshi Hensan Iinkai [Editorial Committee of History of Minami Chita Town] 1991 a,b, in Japanese). This was hand-harpoon whaling, where many harpooners harpooned a whale, attached lines, and killed it. This method was transmitted east to Katsuyama (35°07′N, 139°50′E) in Chiba Prefecture (34°55′N–36°06′N, 139°45′E–140°53′E) and used in the Baird's beaked whale fishery until the middle of the nineteenth century (Chapter 13). The method was also transmitted west to Ise area (34°15′N–34°30′N, 136°30′E–136°50′E), Kii coast (33°25′N–33°45′N, 135°45′E–136°00′E), Kochi Prefecture, coasts of northern Kyushu and nearby islands in the Tsushima Strait, and Yamaguchi (33°45′N–34°45′N, 130°45′E–132°15′E), where it was soon replaced by net whaling.

The so-called net whaling was a modification of hand-harpoon whaling in which nets were placed in front of whales to entangle them before the process of harpooning and slaughter. This method made it possible to take balaenopterids. There are several views on the origin of net whaling. Although the basis of the statement is unclear, Hashiura (1969) stated that net whaling was first attempted in Taiji (33°36′N, 135°57′E) in Wakayama Prefecture (33°25′N–33°45′N, 135°00′E–136°00′E) in 1675 using nets made of rice straw rope and succeeded in

TABLE 1.1
Location of Whaling and Small-Cetacean Fisheries Operated in 1888–1891 as Published in Noshomu-sho [Ministry of Agriculture and Commerce] (1890–1893, in Japanese)

Prefecture	Traditional and New Whaling	Trap-Net Whaling	Driving Small Cetaceans	Hand-Harpooning Small Cetaceans
Iwate			Akasaki, Yamada	
Miyagi	(Off Kinkazan I.)[a]			
Tokyo	(Off Oshima I.)[d]			Offshore[b]
Kanagawa			Manazuru	
Shizuoka			Inatori, Suruga Bay area (inc. Tago)	
Wakayama	Koza, Taiji, Miwasaki		Southeast coast	East and west coasts
Kochi	Iburi, Kubotsu, Mitsu, Tsuro, (Ukitsu)			
Okinawa			Nago	
Nagasaki	Oshima, Arikawa, Ikitsuki, Iki, (Hirado), (Sakito), (Uematsu)		Arikawa Bay	
Saga	Ogawajima			
Yamaguchi	Setozaki, Ushirobata, Kayoi-ura, Kawajiri, Kiwado		O-hibi in Misumi Bay	
Kyoto	Kameshima in Ine		Hirata in Ine	
Ishikawa		Hizue, Kazanashi, Ushizu, Mikawa	Kazanashi,[c] Mawaki, Ogi, Ushizu	
Yamagata			Nedugaseki	
Akita			Kosagawa	

Note: The underlined localities were noted to be still operating in a later publication by Noshomu-sho Suisan-kyoku [Bureau of Fisheries of Ministry of Agriculture and Commerce] (1900, in French), and those in parentheses were new enterprises only noted in the latter reference, which used modern imported technology. The rise and fall of small-cetacean fisheries are not covered in these references.

[a] Attempts at sailing-ship whaling.
[b] Taken by tuna longline fishermen.
[c] Taken by trap-net fishery.
[d] Likely an attempt using whaling cannon (Section 12.7.1).

the next year using stronger *ramie* nets. However, Mizuno (1885, in Japanese) stated that an owner of the Taiji whaling team, Kakuemon Yoriharu Taiji, introduced the method in 1675 from Tango, Sea of Japan coast of Kyoto Prefecture, and then improved the nets to *ramie* nets in 1677. Another manuscript of the nineteenth-century *Taiji-ura Hogei Enkaku-shi* [History of Whaling in Taiji] reprinted in Hashiura (1969, in Japanese) stated that net whaling started in Taiji in 1667 as a technological improvement in hunting humpback whales, after the decline of North Pacific right whales and gray whales migrating to the coastal whaling ground.

The method of net whaling quickly spread to other whaling locations: soon to Koza (a village next to Taiji), westward to Iki Island in 1677 and Goto Island in 1678 (both in the Tsushima Strait) (Kijima 1944; Hashiura 1969; Nakazono 2009, all in Japanese), to Setozaki in 1677 and Kayoi-ura in the 1670s (both locations in Yamaguchi Prefecture) (Yoshitome 2009, in Japanese), and to Kochi in 1683. Although these records identify Taiji as the pioneer of net whaling, there are some conflicting records; that is, legends in Omura (northern Kyushu), Kayoi-ura, and Mijima (both in Yamaguchi Prefecture) state that net whaling was a local invention (Hashiura 1969; Yoshitome 2009, both in Japanese). As net whaling is an adaption of existing and rather simple fishing methods using nets, it could not have been very difficult to devise adaption to whaling.

Japanese net whaling thus was established around 1677. It met extreme difficulty surviving in the middle of the nineteenth century due to decline of whales (Maeda and Teraoka 1958, in Japanese), and attempts were made beginning around 1887 to restructure the industry by increasing effort for hunting still abundant balaenopterids and introducing some new methods. The attempts of *Dai-ami* (see Table 1.1), also called *Kujira-shiki-ami*, was a modification of net whaling that used a fixed trap having four sides covered with net of rice straw. The bottom was 100 fathoms long and closed with net, and one side was open as an entrance 62 fathoms wide. The net was placed in an area of expected passage of whales. Watch boats at the nets waited for whales entering the net, or attempted to drive whales coming close. After getting whales in the net, it was partially taken in and the whales were harpooned for slaughter. This method was attempted in northern Kyushu and off the Noto Peninsula (36°50′N–37°30′N, 136°40′E–137°20′E) on the Sea of Japan coast of central Honshu (Tachihira 1992; Kitamura 1995 (manuscript dated 1838); Torisu 1999, all in Japanese). Other attempts included use of imported bomb-lance-guns or use of the technique of American-type sail-ship whaling (Torisu 1999, in Japanese). These old-fashioned imported techniques failed to achieve the objectives, and the restoration of the Japanese whaling industry had to await the introduction of Norwegian-type whaling, or modern whaling.

Early drive fisheries for dolphins and porpoises, called *tatekiri-ami*, were operated often as a cooperative activity of a local community when suitable cetacean schools were found close to the coast. They drove the school into harbor or into net enclosures placed near the coast, closed the entrance, and slaughtered them. Among the operations listed in Table 1.1, the operation at Yuri-gun, Akita Prefecture (39°07′N–40°20′N, 139°40′E–140°05′E) on the Sea of Japan coast of northern Honshu placed spotters in lookouts established at vantage points during March and April, but another nearby group at Nezuga-seki, Yamagata Prefecture (38°33′N–39°07′N, 139°35′E–139°53′E) south of Akita Prefecture, was more opportunistic and operated only when dolphin schools were found near the shore. These two operations did not appear in the record of Dainihon Suisankai [Fishery Association of Great Japan] (1980, in Japanese) cited below and were not seen in statistics of catch composition or oil production (Tables 1.7 and 1.8). They were probably small-scale opportunistic operations.

Along the coast of Wakayama Prefecture on the Pacific coast of central Honshu, small cetaceans were hunted by both drive and hand-harpoon fisheries (Table 1.1). The latter was stated to have been a side business of net whalers (Noshomu-sho [Ministry of Agriculture and Commerce] 1890–1893, in Japanese) and seemed to be different from the drive fisheries operated there by local communities.

Table 1.1 shows that a great change occurred in Japanese whaling in the late 1890s to early 1900s, when net whaling and hand-harpoon whaling ceased operation and a new attempt at modern whaling started. Both Miwasaki and Kozaura in Wakayama Prefecture are recorded as operating net whaling by the Noshomu-sho [Ministry of Agriculture and Commerce] (1890–1893, in Japanese), but several years later they were classified as locations of past operations by Noshomu-sho Suisan-kyoku [Bureau of Fisheries of Ministry of Agriculture and Commerce] (1911, in Japanese). Table 1.1 also lists a new whaling enterprise off Kinkazan on the Pacific coast of northern Honshu, which was an unsuccessful attempt at American-type whaling using sailing ships and whale boats.

Enyo-hogei [Offshore Whaling] was the first Japanese company that took whales using the Norwegian-type method, but it failed to survive in the enterprise (it took three fin whales near Thushima Island in April 1898). It was Nihon Enyo Gyogyo [Japan Offshore Fisheries] established in 1899 in Senzaki in Yamaguchi Prefecture that first succeeded in the enterprise of modern whaling (Akashi 1910, in Japanese). Table 1.2 cites the catch composition of net whaling during 1893–1897 from a book prepared for an exposition in Paris (Noshomu-sho Suisan-kyoku [Bureau of Fisheries of Ministry of Agriculture and Commerce] 1900, in French).

The common names of baleen whales listed in Table 1.2 are not always the same as those currently used in Japan. Hattori (1887–1888, in Japanese) stated that both *noso* in Wakayama Prefecture and *shiro-nagasu* in northern Kyushu represented his *nagasu-kujira* (fin whale in English) and that *nagasu* of Wakayama Prefecture and *nitari* of northern Kyushu represented his *shiro-nagasu-kujira* (blue whale in English). I observed during North Pacific pelagic whaling in the 1960s that gunners from Wakayama Prefecture used *noso* for the fin whale, which supports the statement of Hattori. Kasuya and Yamada (1995, in Japanese) stated that local names for

TABLE 1.2
Whales Taken by Traditional Whaling in 1893-1897

Prefecture	*Semi*[a]	*Nagasu*	*Zato*[b]	*Ko-kusira*[c]	*Iwashi-kusira*[d]
Nagasaki	0	180[e]	41	5	10
Saga	0	54[e]	20	5	0
Yamaguchi	0	100[e]	34	8	0
Kochi	2	35[f]	47	27	44
Total	2	369	142	45	54
Annual mean	0.4	73.8	28.4	9.0	10.8

Source: Noshomu-sho Suisan-kyoku, *Histoire de L'industrie de la Peche Maritime et Fluviale*, Noshomu-sho Suisan-kyoku, Tokyo, Japan, 1900, 153pp., in French.
Note: Species names are given in the literature (see text for further details).
[a] North Pacific right whale.
[b] Humpback whale.
[c] Gray whale.
[d] "Bryde's whale complex."
[e] Likely represents fin whale.
[f] Likely represents blue whale.

fin and blue whales could have been different among locations even within northern Kyushu and nearby Yamaguchi Prefecture and that other common names were in use in Kochi and Wakayama Prefectures on the Pacific coast. They stated that the fin whale (*nagasu* or *nagasu-kujira* of present day) was called *noso* at Setozaki and Kayoi-ura, Yamaguchi Prefecture, Shikoku, and Wakayama; *nagaso* at Kiwado and Kawajiri, Yamaguchi Prefecture; *shiro-nagasu* at Ikituki, northern Kyushu. The same authors stated that the blue whale (*shiro-nagasu* or *shiro-nagasu-kujira* of present day) was called *nagasu* or *nagaso* at Setozaki, Yamaguchi Prefecture, Shikoku, and Wakayama; *shiro-nagaso* at Kiwado and Kawajiri, Yamaguchi Prefecture; *nitari-nagasu* at Ikitsuki, northern Kyushu; *hai-iro-nagasu* at Kiwado, Yamaguchi Prefecture. English translations of these Japanese words are "long pleats" for *nagasu* and *nagaso*, "white" for *shiro*, "gray" for *hai-iro*, and "enigmatic" or "similar to" for *nitari*.

The latest known use of old names for blue and fin whales was by Yamada (1902, in Japanese), and the earliest use of the current system was by Akashi (1910, in Japanese). Japanese common names for blue and fin whales seem to have been standardized following usage at Kiwado and Kawajiri, Yamaguchi Prefecture, at the beginning of the last century. This relates with the establishment of the Nihon Enyo Gyogyo in 1899 in Senzaki, which became the leading company of Japanese whaling (Kasuya and Yamada 1995, in Japanese). Catch statistics of Japanese modern whaling have been established since 1911 (Kasahara 1950, in Japanese), with the use of current common names.

Accepting this interpretation of old common names of some balaenopterids, Table 1.3 presents geographical and historical changes in the catch composition of baleen whales. Net whaling on the Sea of Japan coast is represented by Kawajiri, Yamaguchi Prefecture, for the years 1698–1840 and 1845–1901 and that on the Pacific coast by Tsuro, Kochi Prefecture, for the years 1849–1865, 1874–1890, and 1891–1896. The catches by modern whaling in the Sea of Japan (1911–1920) and off the Pacific coast (1911–1919) are also listed for comparison. Table 1.3 shows that catches of North Pacific right whales and humpback whales declined with time on both sides of Japan, which was accompanied by the increase of fin whale catches in the Sea of Japan and of catches of blue whales in the Pacific. This trend of catch composition shift continued into modern whaling. In the early twentieth century, catches of fin

TABLE 1.3
Japanese Vernacular Names of Whales and Historical Change in the Target Species of Japanese Whaling

English Name	Blue Whale	Fin Whale	Bryde's Whales	Gray Whale	Humpback Whale	Right Whale	
Current Japanese Name	*Shironagasu-kujira*	*Nagasu-kujira*	*Nitari-kujira*[a]	*Koku-kujira*	*Zato-kujira*	*Semi-kujira*	
Local names at: Kawajiri (S/West. S. of Japan) Tsuro (Shikoku, Pacific Coast)	*Shironagaso* *Nagasu*	*Nagaso* *Noso*	*Iwashi* *Iwashi*	*Koku* *Koku*	*Zato* *Zato*	*Semi* *Semi*	Total Number
Kawajiri, 1698–1840[b]	0%	2.3%	0%	12.3%	65.0%	20.4%	1070
Kawajiri, 1845–1901[b]	1.8%	36.9%	0%	8.7%	44.8%	7.8%	504
Sea of Japan, 1911–1920[c]	0.3%	65.9%	0.7%	29.7%	3.4%	0.0%	2627
Tsuro, 1849–1865[d]	1.4%	0%	9.5%	27.4%	56.6%	5.1%	369
Tsuro, 1874–1890[d]	8.4%	3.2%	14.4%	28.8%	37.9%	7.3%	285
Tsuro, 1891–1896[d]	18.2%	5.1%	31.3%	17.2%	26.2%	2.0%	99
Pacific, 1911–1919[e]	52.2%	8.1%	32.0%	0%	7.2%	0.5%	634

[a] *Nitari* and *nitari-nagasu* of the nineteenth century were used for blue whale (Kasuya and Yamada 1995, in Japanese).
[b] Traditional whaling (Tada 1978, in Japanese).
[c] Modern whaling for years 1911, 1914–1920 off northern Kyushu and Sea of Japan (Kasahara 1950, in Japanese).
[d] Traditional whaling (Yamada 1902, in Japanese).
[e] Modern whaling for years 1911, 1914–1919 in the Pacific between 32°N and 36°N (Kasahara 1950, in Japanese).

whales were more frequent than of blue whales in the Sea of Japan, and the composition was reversed along the Pacific coast. If we consider Table 1.2 with the same interpretation of the old common names, we can say that *nagasu* recorded for Nagasaki (32°10′N–34°40′N, 128°40′E–130°20′E) and Saga (33°15′N–33°40′N, 129°45′E–130°05′E) (both in northern Kyushu), and Yamaguchi Prefectures represents fin whales and the same name recorded for Kochi (Pacific coast) represents blue whales.

The interpretation of old common names of blue and fin whales was based on drawings of external morphology and descriptive text and did not consider the descriptions of baleen plates exhibited at the second fishery exposition in Japan recorded by Oku and Kajikawa (1899, in Japanese), which are extracted in the following as additional evidence of this interpretation.

1. Baleen plates labeled *nagasu-kujira* and the description match the blue whale. These baleen plates were from Tsuro Hogei Co. and Ukitsu Hogei Co. (both of Kochi Prefecture) and from a person in Saga Prefecture in northern Kyushu. The description stated "large, thick and pure black." North Pacific right whales were well known as *semi-kujira* and cannot be confused with this species.
2. Baleen plates labeled *nagasu-kujira* and the description match the fin whale. These baleen plates were from a person in Saga Prefecture and were described as "white with blue stripes." The second specimen from another person in the same prefecture was described as "similar to the above but smaller" also are identifiable as from a fin whale. The third specimen was from Kawajiri whaling group, Yamaguchi Prefecture, described as "medium size, thin thickness, pure black, and accompanied by smaller white plates" may have been from a fin whale.
3. Baleen plates labeled *noso-kujira* and the description match the fin whale. These baleen plates were from Tsuro Hogei Co. and Ukitsu Hogei Co. (both in Kochi Prefecture, Pacific coast) and were described as "medium size with beautiful black and white stripes."

These three interpretations agree with those made by Kasuya and Yamada (1995, in Japanese) for the old Japanese common names. Oku and Kajikawa (1899, in Japanese) list four additional ambiguous examples. One baleen specimen presented by Ogawajima Whaling Co. in Saga Prefecture, northern Kyushu, was labeled as *nagasu-kujira* and described by the authors as "medium size, slightly thin and black." Another baleen specimen presented by a person in the same prefecture was also labeled in the same way and described as "similar to the already mentioned Ogawajima specimen." The third specimen from Goto Hogei Co. off northern Kyushu was not labeled with species name and was described as "thin and pure black." The fourth specimen labeled *nagasu-kujira* presented by a person in Ishikawa Prefecture (36°18′N–37°33′N, 136°15′E–137°22′E, on the Sea of Japan coast of central Honshu where trap-net whaling was operated) was described "lower end faded into white as if from a gray whale." These four specimens have the possibility of being from balaenopterids, including the minke whale. It should be noted that none of the Japanese references before the twentieth century mentioned the minke whale. The species could have been confused with some other baleen whales (Kasuya and Yamada 1995, in Japanese).

Although the common name *iwashi-kujira* is for the sei whale (*Balaenoptera borealis* Lesson, 1828) in present Japan, the baleen whale species represented by that common name has also changed with time. The word first appeared in a list of whales taken by net whaling in western Japan before the nineteenth century. The early whale books *Nitto-gyofu* [Pictorial Catalogue of Japanese Fish] by Kanda (1731, in Japanese), *Geishi-ko* [On Natural History of Whales] by Otsuki (1808, in Japanese), and *Isanatori Ekotoba* [Pictorial Description of Whaling] by Oyamada (1832, in Japanese) state that *iwashi-kujira* and *Katsuo-kujira* are the same species, but *Geishi* [Natural History of Whales] by Yamase (1760, in Japanese) states that they are different species. It should be noted that the sei whale (*B. borealis*) does not occur off the coast of western Japan, where net whaling operated (Kasuya and Yamada 1995, in Japanese). Thus, in those days *iwashi-kujira* must have represented some of the Bryde's whale group complex, in which I include the recently described Omura's whale (*Balaenoptera omurai* Wada, Oishi, and Yamada, 2003) and an unsettled group of species (cf. *Balanenoptera edeni* and *B. brydei*). The latter are thought to have three parallel longitudinal ridges on the dorsal surface of the rostrum. This use of *iwashi-kujira* also applies to Table 1.2. Later around 1905 modern whaling expanded operations to Sanriku (Pacific coast of northern Honshu in 37°54′N–41°35′N) and the Pacific coast of Hokkaido (41°30′N–43°20′N) and started to take the sei whale (*B. borealis*). Because the sei whale and Bryde's whale were not distinguished in those days, the common name *iwashi-kujira* was used also for the sei whale (*B. borealis*) off northern Japan. After about 50 years, in the 1950s, scientists recognized the separate existence of sei and "Bryde's whales" where *B. brydei* was considered as a junior synonym of *B. edeni* (see Table 1.5) and gave *iwasahi-kujira* to the sei whale and a half-forgotten old common name *nitari-kujira* to the Bryde's whale. Although it was possible to create a new common name for the sei whale, that did not happen. Thus, *iwashi-kujira* was used for either Bryde's whale of the coastal form or Omura's whale until the nineteenth century, for both sei and Bryde's whale (mostly of offshore form) in the early twentieth century, and for the sei whale (*B. borealis*) since the middle of the twentieth century.

In 1946, soon after World War II, Japan started pelagic whaling in Bonin Islands waters and took "sei whales," which were soon noticed to be different from the sei whales (*B. borealis*) that occur in the Antarctic and off the Pacific coast of northern Japan. Therefore, they were recorded in Japanese statistics as "southern form sei whale" (1955–1960) and then as *nitari-kujira* as mentioned earlier (from 1961 to the present). The international whaling statistics distinguished between the two species beginning in 1968 and used *B. edeni* for Bryde's whales from the Bonin Island area. This was based on the assumption that *B. brydei* Olsen (1913) hunted off the coast of South Africa was synonymous with *B. edeni* Andersen (1879) described from an individual stranded near Rangoon (1879 is the correct publication date of this scientific name).

The taxonomy of *B. edeni* and *B. brydei* is still unresolved. It is known from both Japan and South Africa that the complex contains two forms; a small coastal form and a larger offshore form. Yamada and Ishikawa (2009) concluded that the latter (i.e., the larger offshore form known from the Bonin Islands area and currently called *nitari-kujira*) corresponds to *B. brydei* and the former (smaller coastal form) to *B. edeni* and gave the old half-forgotten Japanese common name *katsuo-kujira* to *B. edeni* without presenting evidence of validity of the Japanese name. They also proposed the English name of Bryde's whale for the former species and Eden's whale for the latter. However, we are still waiting for the scientific evidence to be published to determine whether the two forms really represent two species and if this use of scientific and common names is valid. The technical problem in solving this question is the absence of a type specimen for *B. brydei* (the type specimen of *B. edeni* is in Calcutta).

Table 1.4 shows the geographical distribution of fisheries for dolphins and porpoises in Japan. For reference purposes, the production of whaling of the old type in 1891 (no modern whaling had started) and the production of coastal modern whaling in 1924 (no traditional whaling operated) are also included. The figures for 1957 are the number of "dolphins and porpoises" and that of "whales" taken by fisheries other than whaling. Japanese coastal land-based whaling, Antarctic pelagic whaling, and North Pacific pelagic whaling are not included in Table 1.4 but in Table 1.5. The distinction between "whales" and "dolphins and porpoises" is unclear in Table 1.4. However, the facts that there were only 85 "whales" taken by fisheries other than whaling and that most of them were from Ibaraki (40 whales) and Chiba (18 whales) Prefectures on the Pacific coast of central Honshu suggest that they were mostly large delphinids and some baleen whales incidentally taken in coastal trap nets or seine-net fisheries in 1957. The products of whaling can have a large effect on the economy of dolphin and porpoise fisheries. Therefore, I have listed in Table 1.5 the catches of whales by Japanese modern whaling (1910–2008) and in Table 1.6 the incidental catch of large whales by various fisheries other than whaling (1991–2008).

The distribution of dolphin and porpoise fisheries in the middle 1880s differed slightly from that in the later years, with drive fisheries recording large catches in Shizuoka (34°36′N–35°05′N, 137°30′E–139°10′E), Ishikawa on the Noto Peninsula and Nagasaki Prefectures. The fisheries in Iwate (38°59′N–40°27′N) and Miyagi (37°54′N–38°59′N) Prefectures on the Pacific coast of northern Honshu were drive fisheries and not hand-harpoon fisheries as now (Table 1.1). The catches of dolphins and porpoises were small in this region in 1924 (Table 1.4), suggesting that the hand-harpoon fisheries were not started or were still at a small scale in 1924. The hand-harpoon fishery for dolphins and porpoises had to await an introduction of motor-driven fishing vessels.

TABLE 1.4
Geographical and Historical Comparison of Dolphin and Porpoise Fisheries in Years 1891, 1924, and 1957 Compared with Production from Whaling (Only Available for 1891 and 1924)

	Dolphins and Porpoises		Whales	
Prefecture and Year	Weight (Tons) or No. of Individuals	Yen	Weight (Ton) or No. of Individuals	Value (Yen)
Hokkaido				
1891	0	27	63	4,081
1924	2	40	4602	673,132
1957	306 ind.			
Aomori				
1924	—	—	45	13,117
1957	4 ind.			
Iwate				
1891	38	540	—	—
1924	6	26,653	1097	163,643
1957	4,021 ind.			
Miyagi				
1891	3	290	1	85
1924	—	—	967	289,485
1957	3,365 ind.			
Fukushima				
1957	239 ind.			
Ibaraki				
1891	7	190	—	—
1924	0	4,440	—	—
1957	148 ind.			
Chiba				
1891	2	60	86	4,952
1957	1,729 ind.			
Tokyo				
1924			660	105,785
1957	5 ind.			
Kanagawa				
1891	7	442	—	—
1924	0	150	—	—
1957	15 ind.			
Shizuoka				
1891	78	4,402	10	252
1957	5,012 ind.			
Mie				
1891	271 ind.	622	6 ind.	3,000
1924	0	93	0	1,676
Wakayama				
1891	15	1,335	33	1,895
1924	0	458	37	5,806
Kochi				
1891	1	40	408	28,381
1924	0	2,410	1	28,420
1957	25 ind.			
Ehime				
1891	7	169		
Miyazaki				
1924	—	—	0	99,500
1957	11 ind.			

(*Continued*)

TABLE 1.4 (*Continued*)
Geographical and Historical Comparison of Dolphin and Porpoise Fisheries in Years 1891, 1924, and 1957 Compared with Production from Whaling (Only Available for 1891 and 1924)

Prefecture and Year	Dolphins and Porpoises: Weight (Tons) or No. of Individuals	Dolphins and Porpoises: Yen	Whales: Weight (Ton) or No. of Individuals	Whales: Value (Yen)
Kagoshima				
1891	3	78	99	6,001
1924			0	2,000
1957	6 ind.			
Akita				
1957	3 ind.			
Niigata				
1957	87 ind.			
Toyama				
1891	9	350	35	1,687
1924	0	2,150	0	5,700
1957	33 ind.			
Ishikawa				
1891	1,700 ind.	2,949	28 tons + 4 ind.	3,470
1924	0	1,930	6	14,252
1957	100 ind.			
Fukui				
1891	0	7		
1957	68 ind.			
Kyoto				
1891	—	—	6	357
1957	1 ind.			
Hyogo				
1891	2	103	—	—
1957	65 ind.			
Tottori				
1891	0	8	—	—
Shimane				
1891	0	4	—	—
1957	28 ind.			
Yamaguchi				
1891	—	—	792	62,947
1924	—	—	768	226,053
1957	7 ind.			
Oita				
1891	0	1	—	—
Nagasaki				
1891	246	3,525	1728	110,368
1924	4	1,108	53	230,986
1957	13 ind.			
Saga				
1891	8	155	749	32,917
1924	—	—	26	124,424
Osaka				
1891			101	350
Kagawa				
1957	7 ind.			
Okayama				
1957	7 ind.			
Okinawa				
1924	0	1,487		
Total				
1891	434 +1,971 ind.	15,300	4144 +10 ind.	260,385
1924	16	40,919	8268	1,983,882
1957	15,305 ind.			

Note: Only years with production are listed (in tons or as number of individuals). The figures for 1957 include blackfish, dolphins, and porpoises taken by all fisheries other than whaling (i.e., pelagic whaling, large-type whaling, and small-type whaling are excluded). The 1891 whaling was operated by traditional methods (net whaling, hand-harpoon whaling, and trap-net whaling), and the 1924 whaling was operated by Norwegian-type or modern whaling. The 1891 data are from Noshomu-sho [Ministry of Agriculture and Commerce] (1890-1893, in Japanese), the 1924 data are from Norindaijin Kanbo Tokei-ka [Statistics Division of Ministry of Agriculture and Forestry] (1926, in Japanese), and the 1957 statistics are from Ohsumi (1972). Fractional figures have been disregarded. See Table 1.5 for whaling statistics.

The dolphin fishery in Wakayama Prefecture, where porpoises are extremely rare, declined at least temporarily sometime around 1891 and 1924 accompanying a decline in net whaling. This agrees with the statement of the Noshomu-sho [Ministry of Agriculture and Commerce] (1888–1893, in Japanese) that the dolphin fishery there is a side operation of net whaling. There was a collapse of net whaling in Taiji in Wakayama Prefecture around 1879 following a catastrophic shipwreck of whaling crews in December 1878. In those days, the dolphin drive fishery was operated opportunistically by the local community and depended on the chance of dolphin schools closely approaching the coast, which was irregular (Chapter 3).

A distinction between "whales" and "dolphins and porpoises" for the year 1891 in Table 1.4 is not stated, and it is unclear if the distinction followed the same standard among prefectures, but the figures show that the dolphin and porpoise fisheries produced only about 10% of the total products (in weight) by old-fashioned whaling of the time and about 6% of the amount of sales. The same table shows that the traditional types of whaling were operated in Nagasaki Prefecture (42%) in northern Kyushu, Yamaguchi Prefecture (19%) at the western tip of Honshu, Saga Prefecture (18%) in northern Kyushu, Kochi (32°40′N–33°30′N, 132°40′E–134°20′E) Prefecture (9%) on the Pacific coast of Shikoku, and Osaka Prefecture (2%). These five prefectures produced about 91% of the total

Japanese whaling products of the time. Osaka, one of the four prefectures, did not operate whaling, and it is unclear why Osaka recorded the production of whale products. The average price of whaling products was 62 yen/ton for all of Japan including Osaka, but it was only 3.4 yen/ton for Osaka. It is likely that some particular part of the whaling products were sent to Osaka for further processing. Cotton industries in Osaka Prefecture imported vertebral tendons of dolphins for bow strings (Dainihon Suiankai [Fishery Association of Great Japan] 1890, in Japanese). Cotton fiber removed from seeds was beaten before further processing. A mallet and a bow with string of whale tendon were used for the purpose. Thus, it is likely that some factories in Osaka imported vertebral tendons of whales or tendons surrounding the spermaceti organ of sperm whales and processed them for the cotton industry.

Catch of whales in the same year (1891) by Japanese traditional whaling in northern Kyushu was 52 whales in Nagasaki Prefecture and 3 in Fukuoka Prefecture (33°25′N–33°55′N, 130°00′E–131°00′E), with species composition of 35 fin whales, 9 humpback whales, 4 gray whales, and 2 sei whale (presumable Bryde's-whale complex) (Torisu 1999, in Japanese). Torisu (1999, in Japanese) reported additional catch statistics by species in Nagasaki Prefecture for 13 years from 1884 to 1896. The total annual catch ranged between 38 and 79 with a mean of 56 whales. The species composition of the 731 whales taken in the 13 seasons was 422 fin whales (58%), 162 humpback whales (22%), 44 sei whales (6%, presumable Bryde's-whale complex), 29 gray whales (4%), 5 blue whales (1%), 6 right whales (1%), and 63 other whales (9%).

The year 1891 was just prior to the establishment of modern whaling, and there were efforts to continue old-fashioned net whaling with some attempts using technology introduced from American whaling (see preceding text in this section). Under these circumstances, there might have been a temporary increase in the catch of small cetaceans to compensate for declining production from whaling in the 1890s. The total landing of dolphins and porpoises in 1924 declined to only 16 tons, which was only 4% of the figure in 1891 (Table 1.4), suggesting suppression by increased whaling products from coastal whaling. Japanese Antarctic whaling started only in the 1934/1935 season and North Pacific pelagic whaling in 1940. Importation of whale meat from the Antarctic was first permitted in the 1937/1938 season (Kasuya 2000, in Japanese). During the period from 1911 to 1924, Japanese modern whaling annually caught about 1,000 baleen whales and 300 sperm whales (Table 1.5) and recorded 6,052 tons or about 1,020,000 yen worth of products other than whale oil (Norin-daijin Kanbo Tokei-ka [Statistics Division of Ministry of Agriculture and Forestry] 1926, in Japanese). These baleen whales were mainly fin and sei whales (including Bryde's whales), with some blue, North Pacific right, and gray whales.

Around 1957, we saw an increased take of dolphins and porpoises. If body weight of 50–100 kg is assumed, the total weight could have been 75–150 tons, which was several times greater than the annual catch in the nineteenth century. Explosion of the dolphin and porpoise fisheries during World War II (Chapters 2 and 3) probably continued to the 1950s, assisted by the improved transportation system in Japan.

Another temporary increase in small-cetacean catches was recorded in the late 1980s, when coastal whaling companies promoted dolphin and porpoise hunting and purchased the catch to be used for "processed whale meat" (Chapter 2). It seems that dolphin and porpoise fisheries in Japan survived under continuous influence of the whaling industries. Most Japanese consumers preferred whale meat over dolphin and porpoise meat, and the whaling industry also could easily overwhelm the dolphin and porpoise fisheries in quantity of products. Therefore, with the possible exception of local areas where people had a preference for small cetaceans, hunting of dolphins and porpoises was unable to compete with whaling. Tables 1.5 and 1.6 list catch statistics for whales taken by whaling and other fisheries in Japan to assist understanding of the environment surrounding fisheries for dolphins and porpoises.

Table 1.4 lists prefectures where dolphin and porpoise fisheries were operated in 1891. The major operations in monetary yield were in Shizuoka (28%), Nagasaki (23%), Ishikawa (19%), and Wakayama (8%) (statistics are incomplete for weight). These four prefectures are known to have had drive fisheries (Chapter 3), and their production comprised about 80% of the total in value sold in Japan. Although it was not probably technically impossible, it would not have been economic to capture dolphins and porpoises from rowed boats or sailing ships.

According to Yanaginara (1887, in Japanese), villagers on the coast of Noto Peninsula on the Sea of Japan coast of central Honshu drove dolphins found near shore into their harbor and used them for agricultural fertilizer (this statement will in reality apply only to the waste portion; see observation of Takenaka (1890) cited in following part of this paragraph). He stated that an exception to this was for false killer whales, of which meat, blubber, tail flukes, and fins were salted and sold for human consumption. He also wrote that tendons along the vertebrae were sold for use as strings in bows used as in cotton beaters in Japan and for human consumption in China. The situation was different for the operation on the Pacific coast. Fishermen in Kanagawa Prefecture (35°08′N–35°32′N, 139°10′E–139°45′E) sold fresh meat to Tokyo and dried meat either locally or to Nagano (centered in 36°15′N, 138°00′E) and Yamanashi (centered in 35°35′N, 138°30′E) Prefectures, located inland. Villagers at Tago on the west coast of the Izu Peninsula (34°36′N–35°05′N, 138°45′E–139°10′E) on the Pacific coast of central Honshu sold fresh meat locally and also extracted oil from the catch (Yanaginara 1887, in Japanese). According to the same author, the common dolphin was sold at 3 yen/animal and false killer whale at 12 yen/animal. Takenaka (1890, in Japanese) recorded similar average prices of dolphins at Noto in 1887: 1.9 yen for common dolphin and 8 yen for false killer whale. The usage of dolphins recorded by him was mostly for human consumption (fresh or salted), with additional use for extraction of oil, fertilizer (viscera and bones), and tendons for bows used in cotton beaters. He did not mention the use of meat for fertilizer and commented only on the future

TABLE 1.5
Whales Taken by Japanese Whaling

	Antarctic Pelagic Whaling[b]						North Pacific Whalings[c,d]									
Year[a]	Minke	Sei[d]	Fin	Blue	Humpback	Sperm	Minke	Sei	Bryde's	Fin	Blue	Humpback	Sperm	Baird's Beaked	Short-F. Pilot	Killer
1910								156		217	97	29	57	—	—	—
1915								723		817	57	105	252	—	—	—
1920								393		438	35	83	245	—	—	—
1925								492		410	30	158	479	—	—	—
1930							(13)	411		400	56	62	753	—	—	—
1935			174	456	9		(24)	392		273	21	78	1005	35	—	—
1940		6	3661	3225	2399	657	(79)	432		544	49	141	1488	25	—	—
1945							(68)	74		169	10	11	266	—	—	—
1950			2052	271	9	409	259(202)	539		141	7	5	1305	197	715	—
1955		7	4524	383	240	1308	427(374)	509	(91)	1714	100	126	2590	258	61	18
1960		1,773	8912	1144	211	1552	253(257)	991	(406)	1524	71	2	3908	147	168	77
1965		11,310	910			482	334(312)	1864	(8)	1477	49	43	4260	172	288	169
1970	4	4,137	1607			1334	320(307)	3792	(73)	595			6184	118	152	12
1975	3017	1,316	118			592	370(328)	484	804	129			4110	46	53	3
1980	3120						379(364)		307				1192	31	1	2
1985	1941						327(319)		317				400	40	62	
1987	273[300]						304		317				200	40		
1988	241[330]													57	128	
1989	330													54	58	9
1990	327													54	18	3
1991	288													54	59	
1992	330													54	81	
1993	330													54	91	
1994	330						21[100]							54	55	
1995	440[440]						100							54	100	
1996	440						77							54	100	
1997	438						100							54	77	
1998	389						100		1[e]					54	84	
1999	439						100							62	104	
2000	440						40		43[50]				5[10]	62	106	
2001	440						100	1[e]	50				8	62	87	
2002	440						150[150]	39[50]	50				5	62	83	

(Continued)

TABLE 1.5 (*Continued*)
Whales Taken by Japanese Whaling

	Antarctic Pelagic Whaling[b]						North Pacific Whalings[c,d]									
Year[a]	Minke	Sei[d]	Fin	Blue	Humpback	Sperm	Minke	Sei	Bryde's	Fin	Blue	Humpback	Sperm	Baird's Beaked	Short-F. Pilot	Killer
2003	440						150	50	50				10	62	69	
2004	440						159[210]	100[100]	50				3	62	42	
2005	853[935]		10[10]				220[220]	100	50				5	66	46	
2006	505		3				187	100	50				6	63	17	
2007	551		[50]		[50]		157	100	50				3	67	16	
2008	679		1				169	100	50				2	64	20	
2009	506		1				162	100	50				1	67	22	
2010	170		2				119	100	50				3	66	10	
2011	266		1				126	95	50				1	61		
2012	103						182	100	34				3	71	16	
2013[f]	251						95	100	28				1	62	10	

Notes: Catch limits for scientific whaling, which started in autumn 1987, are shown in square brackets only for the first year. From Kasahara (1950, in Japanese), Tato (1985, in Japanese), Nishiwaki and Handa (1958), Maeda and Teraoka (1958, in Japanese), Fisheries Agency statistics (see Section 1.4), and home pages of Fisheries Research Agency (FRA) of Fisheries Agency. FRA statistics for the North Pacific minke whale (Status of International Fisheries Resources 2012.) are given by sex, but the annual totals before 1987 disagree with figures in Fisheries Agency publications (sexes combined). Therefore, former figures are shown in parentheses to indicate the degree of disagreement. This table has been revised with the addition of new data available since the original Japanese language edition appeared in 2011.

[a] Statistics cover period from April of the year indicated to March of the next year, but pelagic whaling off the Bonin Islands (1946–1951) covers period from February or March to May or June of the indicated year.

[b] Includes sperm whales taken in the lower latitudes of the southern hemisphere on the way to the Antarctic and excludes land-based whaling under foreign jurisdiction, such as the operation from South Georgia (1963/1964–1965/1966). The scientific whaling program almost reached the catch limit in the earlier seasons, but after 2006 the catch often did not reach the quota, and the take of humpback whales was suspended for political reasons.

[c] North Pacific whaling includes pelagic whaling off the Bonin Islands (1946–1951), pelagic whaling in the northern North Pacific (1940–1941 and 1952–1979), large-type whaling in coastal Japanese waters (ended in March 1988), operation of catcher/factory ship for minke whales and small cetaceans (1973–1975), and small-type whaling (currently in operation). It also includes operations in the former colonies (until 1945).

[d] Bryde's whales were included in sei whales until the 1975 season (in parentheses is the number of Bryde's whales included in the sei whale catch). The following figures are additional to the table: 6 gray whales and 436 whales of unknown species in 1910; 139 gray whales and 7 North Pacific right whales in 1915; 75 gray whales and 10 North Pacific right whales in 1920; 10 gray whales and 9 North Pacific right whales in 1925; 30 gray whales and 5 North Pacific right whales in 1930; 2 North Pacific right whales in 1935; and 58 gray whales and 1 North Pacific right whale taken in 1940 by a Japanese prewar pelagic whaling expedition to the North Pacific.

[e] Accidental catch in scientific whaling due to misidentification of species.

[f] In addition small-type whaling caught one false killer whale in 2013.

TABLE 1.6
Incidental Mortality of Large Cetaceans in Japanese Coastal Fisheries Other than Whaling

Year	Minke	Humpback	Gray	Bryde's	Right	Fin	Sperm	Unknown	Total
1991	5	1(1)					(1)		6(2)
1992	8(2)						(1)	2	10(3)
1993	14(3)						(1)	17(1)	31(5)
1994	16						(4)	6	22(4)
1995	20(1)	(1)	(1)				1	(1)	21(4)
1996	27(11)	2	1[a]				1(3)	(1)	31(14)
1997	27(9)	1	(1)		(1)	1	(5)		29(7)
1998	24(3)	1(1)				(1)	(1)	(3)	25(9)
1999	19(10)	1(1)							20(11)
2000	29(4)	1		(1)			(5)		30(10)
2001	79(10)	(1)		(1)		1	(3)	1(3)	81(18)
2002	109(7)	3	(1)	(1)	(1)	(2)	(22)		112(35)
2003	125(12)	3		1	1		(12)		130(24)
2004	113(8)	5(1)		2[b](1)		(2)	(4)		120(16)
2005	122(9)	3(2)	3	(1)[b]	(1)	(2)	(9)		128(24)
2006	141(10)	4(3)				(2)	1(4)	1[c]	156(21)
2007	156(5)	1(1)	1(1)	(2)		1	(8)		160(17)
2008	133(7)	3(1)				1(1)	(1)		137(10)

Sources: Fisheries Agency statistics; Kato, H. et al., Status report of conservation and researches on the western gray whales in Japan, May 2007–April 2008, IWC/SC/60/O8, 9pp., Document presented to the 60th *Meeting of the Scientific Committee of IWC* (available from IWC Secretariat, Cambridge, U.K.), 2008 for gray whale.

Notes: Additional numbers of stranded whales in parentheses. A change in government policy in 2001 could have improved the reporting rate for minke whales found in fishing gear and resulted in the sudden increase in the reported catch (Sections 5.8 and 6.3).

[a] This gray whale was reported by the Fishery Agency as stranded, but it should be considered a poached whale, because it was the anterior portion of a carcass carrying numerous hand harpoons that stranded at Suttsu (42°47′N, 140°14′E) on the Sea of Japan coast of Hokkaido. The posterior portion was not found.

[b] These whales (one incidental take in 2004 in Yamaguchi Prefecture and another stranding in 2005 in Miyazaki Prefecture) were identified as *B. omurai*.

[c] Reported as a sei whale.

possibility of exporting tendons to China, which is different from what Yanaginara said (1887, in Japanese). It is known that the thick tendon along the whale vertebrae was exported to China and consumed as dried abalone (Ishida 1917, in Japanese). However, whether dolphin tendons were sold for the same purpose will need further confirmation.

Several years after this investigation, Dai-nihon Suisankai [Fishery Association of Great Japan] (1890, in Japanese), which was an association for the promotion of Japanese fisheries, published results of inquiries sent to major dolphin drive fisheries including opportunistic operations as well as more active operations using dolphin lookouts. This survey seems to be the same as that mentioned by the No-shomusho [Ministry of Agriculture and Commerce] (1890–1893, in Japanese, see Table 1.4). The resultant catch statistics included the period from 1870 to 1889, but most were from the periods 1886–1887 or 1887–1889, and it was unclear whether there was no catch in the remaining years or the statistics were missing. It was possible that opportunistic dolphin drivers did not leave records for years when they did not operate. The recorded operation of dolphin drives are as follows:

Ishikawa Prefecture, Noto Peninsula on Sea of Japan coast
- Ushitsu (37°18′N, 137°09′E): 1887–1889
- Ogi (37°18′N, 137°14′E): 1889
- Mawaki (37°18′N, 137°12′E): 1887–1889

Izu coast in Shizuoka Prefecture, Pacific coast of central Honshu
- Arari (34°50′N, 138°46′E): 1882–1889
- Tago (34°48′N, 138°46′E): 1886–1889
- Ito (34°58′N, 139°06′E): 1886–1887
- Inatori (34°46′N, 139°02′E): 1885–1888
- Kawana (34°57′N, 139°07′E): 1888–1889
- Nishiura (25°53′ E): 1887—1888
- Uchiura (35°14′N, 138°54′E): 1887–1888
- Heda (34°58′N, 138°47′E): 1888

Sanriku region, Pacific coast of northern Honshu
- Akasaki (39°05′N, 141°43′E): 1870–1885
- Kamaishi (39°16′N, 141°53′E): 1888–1889
- Funakoshi (39°26′N, 141°57′E): 1887–1888

TABLE 1.7
Dolphin Drive Fisheries around 1887–1889 as Documented in the Literature

Prefecture	Ishikawa	Iwate	Shizuoka
No. of drive groups	3	2	8
Year of Statistics	1887–1889	1887–1888	1887–1888
Month of operation	March–July	October–April	March–October, September–March, or year-round
Kama-iruka	—	—	90
Ma-iruka	181	374	3153
Nyudo-iruka	202	—	97
Nezumi-iruka	—	53	—

Source: Dainihon Suisankai, *Dainhon Suisankai Hokoku*, 98, 240, 1890, in Japanese.
Note: Species are as noted in the references.

These reports list catch statistics by year and by species. I have extracted those covering the common period to obtain the prefectural total (Table 1.7), using species name as they appeared in the reports. The problem of species name will be handled in a later part of this chapter and in Chapter 3.

The report also list markets for the products. Hunters in Ishikawa Prefecture sold salted meat to Niigata (37°55′N) and Sakata (38°50′E) Cities on the Sea of Japan coast and dried tendons to Tamatsukuri in Osaka City. Oil was tried out from the head and blubber but the market was not mentioned. The viscera, dry or wet, were sold nearby for fertilizer. The tendons could have been used in the cotton industry in Osaka. Nothing is stated about selling tendons to China.

The same report recorded the use of dolphins from the Izu coast. Fresh meat was sold within the prefecture, that is, Shizuoka, or in the nearby prefectures of Kanagawa and, occasionally, Tokyo. Dried meat was prepared only when rough weather did not permit transport to the markets. Tendons were not produced in Izu, and oil was not extracted (possibly blubber was consumed with the meat, the present method of cooking in the region).

Markets of the Sanriku fishery were several. The Akasaki fishermen sold fresh meat to Mogami (inland part of Yamagata Prefecture centered in 38°25′N, 140°20′E), Aizu (inland area centered in 37°35′N, 139°50′E), and Akita (Sea of Japan coast in around 39°40′N, 140°05′E) regions. Kamaishi fishermen sold fresh meat to the nearby coastal city of Shiogama (39°19′N, 141°01′E) and salted meat to the Akita and Yamagata regions. Kamaishi fishermen extracted oil from *nezumi-irukas* [species undetermined; see Chapter 3], and sold bones and viscera to local farmers as fertilizer or discarded them at sea. Whether they sold meat as fresh or after salting probably depended on the season of the catch and distance from the markets. If they caught dolphins in the summer season, they would have been obliged to salt or dry the meat for distant markets. I did not find reliable records of dolphin and porpoise fisheries operated solely for the production of fertilizer.

I earlier stated that some dolphin fisheries extracted oil from their catch. Statistics on dolphin oil are available in Noshomu-sho Nomu-kyoku [Bureau of Agriculture of Ministry of Agriculture and Commerce] (1892, in Japanese) (Table 1.8).

TABLE 1.8
Production of Oil from Whales, Dolphins, and Porpoises by Prefecture, Combined with Market and Type of Consumption in 1884

Prefecture	Whale Oil	Dolphin and Porpoise Oil	Produced At	Market and Use
Kyoto	9		Yosa-Odajuku	Lighting, Insecticide[a]
Kanagawa	38.4		Yokohama	Export
Nagasaki	39		Arikawa, Uonome, Ikitsuki	Northern Kyushu
Chiba	54		Kachiyama	Tokyo and nearby area
Ishikawa[b]		188	Fugeshi, Suzu	Ishikawa and Niigata
	20		Fugeshi	Yokohama Port
Yamaguchi	67.5		O-tsu	Shimonoseki and others
Wakayama	49		Taiji	Osaka
		6	Taiji	Local
		12 (pilot whale oil)	Miwasaki	Osaka and Ise
Saga	57.6		Ogawa-jima	Local
Kagoshima	8		Kawanabe	Northern Kyushu
Total	342.5 koku (61.8 kL)	206 koku (37.1 kL)		

Source: Noshomu-sho Nomu-kyoku, *Fish Oil and Wax*, Yurindo, Tokyo, Japan, 139pp+5 pls., 1892, in Japanese.
Note: Quantity of oil is given in *koku*, which is about 180 L.
[a] Whale oil was used to kill leaf hoppers and other insects that damage rice plants in the paddy.
[b] In addition, Fugeshi County of Ishikawa Prefecture produced seal oil of 14 koku, which could be from sea lions.

In 1884, nine prefectures produced whale oil, two of which, Ishikawa and Wakayama, also produced dolphin oil. Nagasaki, Yamaguchi, Wakayama, and Saga operated net whaling and possibly other traditional whaling. Ine (35°40′N, 135°18′E, with the three villages of Hizu, Kameshima, and Hirata) in Yosa County, Kyoto, had a system of opportunistic hunting of whales and small cetaceans found in their bay (Hattori 1887–1888; Yoshihara 1976, both in Japanese). Fishermen in Chiba Prefecture hunted Baird's beaked whales with hand harpoons (Yoshihara 1982, in Japanese). In Ishikawa Prefecture, there was a kind of trap-net whaling, *dai-ami*, off the Noto Peninsula (Saito 1981, in Japanese) and an attempt at hand-harpoon whaling in the Kaga region in 1878 (Hattori 1887–1888, in Japanese). Only half of the whale oil was exported; the remaining half was for domestic use, for lighting, and for insecticide in rice paddies. The production of dolphin oil was only about one-third of that of whale oil and was for both export and domestic use as in the case of whale oil. Whale oil became an important export item for Japan only after large production by modern whaling.

No-shomu-sho [Ministry of Agriculture and Commerce] (1890–1893, in Japanese) conducted the first nationwide survey on fisheries for dolphins and porpoises. The second survey, on a smaller scale, was done by the Fisheries Agency in the mid-twentieth century (Suisan-cho Chosa-kenkyu-bu [Investigation and Research Department of the Fisheries Agency] 1968, 1969, in Japanese). In order to decide its response to the request of Iki fishermen of northern Kyushu for solution of a conflict between dolphins and a line-and-hook yellow-tail fishery, the Fisheries Agency attempted to get information from the Kyushu area on the magnitude of small-cetacean fisheries and of conflict between fisheries and small cetaceans. The Fisheries Agency sent questionnaires to 263 fishery cooperative unions in northern, western, and southern Kyushu except for the two prefectures of Oita and Miyazaki on the eastern shore of Kyushu. The 1968 report was based on 251 replies received from September 1967 to March 1968. The 1969 report included an additional 10 replies received later, but there was no significant difference between the two.

According to the Research Division of Fisheries Agency (1969, in Japanese), 57 fishery cooperative unions reported past incidents of taking dolphins and porpoises (only finless porpoises are known to occur in this area), of which 22 reported the incidents as having occurred within the previous 12 months. The total number of animals taken in the 12 months was 239. The species composition was unknown because of uncertainties about the common name of small cetaceans used by the fishermen.

Of the 239 individual accounts, 236 were accompanied by description of the hunting method. The methods were hand harpoon (141 animals), fixed trap net (52), longline for sharks (17), drift gill net for marlin (15), bottom gill net (7), drive fishery (2), and line-and-hook fishery (2). This is likely based on sales records kept by the cooperative unions and does not include catches personally consumed by fishermen. The composition reflects fisheries off the Kyushu region, and surveys in other regions would result in different methods and composition.

The fixed trap net, or *set net*, a literal translation, is a fishing gear of little selectivity, and there is no distinction between direct and incidental take. Any animals found in the net will be sold if they have some commercial value, and others will be discarded or consumed by the fishermen.

A surface gill net or drift net has buoys and small leads and remains at the surface. In the Kyushu area, it may be operated during the day for flying fish or horse mackerel or at night for squid. Salmon drift nets were once operated in great numbers in the Sea of Japan, Okhotsk and Bering Seas, and squid drift nets in the North Pacific. The price of Dall's porpoise meat once exceeded that of squid in the 1980s, so the fishermen landed the meat of Dall's porpoises incidentally caught in their squid gill nets. Gill nets for tunas and swordfish were called large-mesh gill nets because of the mesh size and were known to have incidentally caught dolphins, porpoises, sea birds, and sea turtles (Yatsu et al. 1994). Such incidental mortality caused international concern, and a resolution was passed at the General Assembly of the United Nations in December 1989 to ban operation of large-scale high-seas drift-net fisheries. Japan accepted the ban and stopped operations in international waters in December 1992.

A bottom-set gill net is called *tate-ami*, in Kyushu and the western Inland Sea, which is different from a trap net that may be referred to by the same word in some area of Japan (Uda et al. 1962, in Japanese). The bottom-set gill net has small floats and leads and is placed on the sea floor. It usually operates at night to capture bottom fish and is known to kill finless porpoises in Japan (Chapter 8). In the North Sea, bottom-set gill nets of similar principle kill harbor porpoises, but sound emitters called pingers attached to the net are known to decrease the mortality. However, a possible adverse effect of eliminating porpoises from their feeding ground has also been hypothesized (Teilmann et al. 2006).

Longlines and hook-and-line fisheries may kill dolphins and porpoises. Japanese longline fishermen have complained about dolphins stealing tunas from their hooks. The dolphin species was the false killer whale, not the killer whale as stated by the fishermen. The whales stole the tuna's body and left the head on the hook and occasionally became entangled in the line and drowned. I had thought that toothed whales could detect fish hooks and avoid swallowing them. However, direct mortality of cetaceans by swallowing fish hooks need to be considered. A Tasmanian beaked whale (*Tasmacetus shepherdi*) stranded in Argentine was found with fish hooks from bottom longlines in the stomach and intestines, and these hooks are believed to be the direct cause of death (IWC 2009). In the United States, false killer whales are known to become entangled in longlines or swallow hooks (Marine Mammal Commission 2010).

Japanese purse seine fisheries can kill dolphins and porpoises (Ohsumi 1972). In the eastern tropical Pacific, an international seine fishery uses dolphin schools as an indicator of yellow-fin tuna swimming below the dolphins and sets the net on them, resulting in high incidental mortality of dolphins. However, such mortality was not recorded for the purse seine fishery operated for yellowtail in the Iki Island area in the information collected by Suisan-cho Chosa-kenkyu-bu (1969, in Japanese). It is not certain whether the fishery cooperative unions did not operate the purse seine fishery, the net fishermen did not land the dolphins killed in their nets, or dolphins were not killed in the nets.

Both trawl-net and drag-net fisheries have reported take of finless porpoises. The investigation cited earlier by the Fisheries Agency recorded a case of a drive fishery for dolphins with only two dolphins. Dolphin driving in the Kyushu area was operated in those days cooperatively by neighboring villages, and the catch was shared among the participants. The operation could have caught more animals and shared by villagers, leaving only a few individuals to be sold and recorded by the fishery cooperative union.

1.4 RECENT STATISTICS FOR SMALL CETACEAN FISHERIES

Japan started collecting statistics on small-cetacean fisheries in 1972. The Whaling Team of the Offshore Division of the Fisheries Agency requested prefectural governments to supply the statistics of such fisheries and compiled the reports into national statistics. To respond to the request of the central government, the prefectural governments forwarded the request to fishery cooperative unions in the prefectures. Therefore, the statistics were, in principle, based on the sales records of each fishery cooperative union.

Annually from 1972 to 1987, the Whaling Team created mimeographed tables with a title that was translated as "Report of Status of Direct and Incidental Catches of Small Cetaceans." It also listed the number of operating bodies as well as amounts of sales for each year (monthly figures not included). This task was taken up by the Coastal Division of the Fisheries Agency in 1986. Thus, there were two sets of similar statistics for the years 1986 and 1987, which differed in details. Those of the Coastal Division appear to be closer to the truth. Any statistics must have some errors, and this case gives us information about the magnitude of such errors. In recent years, the whaling-related task has returned to the Whaling Team of the Fisheries Agency. In the early stage of collecting the statistics, there were problems in the identification of local names of small cetaceans. A Whale Research Team of the Far Seas Fisheries Research Laboratory, Fisheries Agency, worked to unify common names and edited the statistics since 1979 to form a part of the National Annual Report to the IWC.

The full text of the National Progress Report was published in the Report of the International Whaling Commission for statistics up through 1997. For the statistics since 1998, only abbreviated summaries have been printed in the Supplement volume of the *Journal of Cetacean Research and Management* (first issued in 1999). Japan has stopped presenting information on small-cetacean fisheries to the IWC, but full recent statistics are available through the homepage of the Japanese Fisheries Agency. I refer in this book to the various types of statistics as "Fisheries Agency statistics."

Table 1.9 shows catch statistics for small cetaceans in Japan cited from the original Fisheries Agency statistics. It includes takes of drive fisheries, hand-harpoon fisheries, small-type whaling, culling, and mortality incidental to other fisheries within the 200-nautical-mile zone. Statistics on stranding have been collected since 1988 but are not included in this table except for finless porpoises. Species names in Table 1.9 are as they appeared in the Fisheries Agency records, so catch figures for certain species (particularly before 1979) disagree with those given for each fishery types in subsequent chapters, where species names are corrected based on my interpretation. Table 1.9 shows that among 13 species of small cetaceans taken in Japan, the most significant are fewer than 10 species, including Dall's porpoise, *ma-iruka* [which basically means *Delphinus* spp., but may also include other species], striped dolphin, pantropical spotted dolphin, common bottlenose dolphin, and short-finned pilot whale.

Dall's porpoises were taken in the greatest numbers, including the two-color morphs *dalli*-type and *truei*-type. They are distinguished by the size of a white patch on the flank. The *dalli*-type has a smaller white patch extending from the anal region to the level of the dorsal fin, while the *truei*-type has a larger white patch extending to the base of the flipper. Within the *dalli*-type, those wintering in the Sea of Japan have slightly smaller white patches compared with *dalli*-types that winter in the Pacific (Amano and Hayano 2007). The two types, once dealt with as separate species, are currently considered to comprise a single species (Chapter 9).

Although the difference between the two types of Dall's porpoise is distinct, I have found it rather difficult to record the two types correctly at busy landing ports. Some early statistics of the Fisheries Agency occasionally recorded proportions of the two color types, but they were often quite different from the proportions I recorded at the landing ports. Therefore, I did not list the records of types reported for the early years. Fisheries Agency statistics have fully distinguished between the forms since 1989, as shown in Table 1.9. It should also be noted that deliberate misreporting of color types can occur for the fishing operations, because the current fishing quota are set by color type. The two species of *Delphinus* off Japan have had a short history of being called by the single common name *ma-iruka* by scientists who dealt with the long-beaked type (*hase-iruka*) and short-beaked type (*ma-iruka*) as the same species. *Delphinus capensis* is found in the Sea of Japan and East China Sea and was often recorded by the local name of *hase-iruka*, and *D. delphis* from the Pacific coast was recorded as *ma-iruka* (Kasuya and Yamada 1995, in Japanese), which makes the identification of the species slightly difficult. A greater problem occurred because local fishermen often used *ma-iruka* for a species that was most common in their region, because it meant "the correct dolphin species to be taken" or "the right species of dolphin." Prior to 1978, there were large take of *ma-iruka*, mostly from two fisheries. One was a hand-harpoon fishery in Iwate Prefecture on the Pacific coast of northern Honshu, and the other was a drive fishery in Shizuoka Prefecture, on the Izu Peninsula on the Pacific coast of central Honshu. Our whale-sighting records did not indicate such a

TABLE 1.9
Catch of Small Cetaceans by Japanese Coastal Fisheries, Including Small-Type Whaling, Drive Fisheries, Hand-Harpoon Fisheries, Culling, and Incidental Mortality within the 200-Mile Zone (Excluding Released Individuals)

	Dall's Porpoise			Pantropical			Short-Finned Pilot Whale		False	Baird's		Pacific				Others and
			Striped	Spotted	*Tursiops*	Risso's	South.	North.	Killer	Beaked	Finless	White-Sided	*Delphinus*	Killer	Harbor	Unknown
Species	*Dalli*-Type	*Truei*-Type	Dolphin	Dolphin	spp.	Dolphin	Form	Form	Whale	Whale	Porpoise	Dolphin	spp.	Whale	Porpoise	Species
1972	5,198		55		15	1	287	4	25	86		17	8552	4	5	1257
1973	5,003		51		37	0	424	10	12	32			9032	10		107
1974	5,105		209		121	11	226	6	10	33		22	8799	2	417	1705
1975	4,977		848	1298	36	2	516	1	4	46		2	8423	1	313	135
1976	9,899		382	17	135	55	436		1	13	10	101	6451	1	108	502
1977	7,689		1,192	1287	986	6	478		44	44	1	83	5252	1	60	22
1978	8,052		747	4184	1104	199	420		637	37	2	82	2611		72	9
1979	5,766		912	427	687	508	101		340	28	3	350	1474	5	1151	183
	6,878		2,212	427	666	935	104		339	28	1	424	89	5		190
1980	6,767		16,344	1460	3493	3	691		377	31	153	73	124	2	40	9
1981	9,962		4,803	169	328	17	569	3	8	39	14	245	76	5	46	67
1982	12,994		2,039	3799	834	6	311	87	1	60		172	184	5	123	428
1983	12,952		2,249	2945	751	200	378	125	290	37	8	1627	94	2	20	487
1984	10,217		3,741	743	464		512	160	60	40		2765	6		36	31
1985	10,885		3,230	484	849		668	62	127	40		40	331	7	66	110
1986	10,534		2,981	693	230	2	347	28	3	40		22	2	8	66	15
1987	13,406	—	2,173		1812	3	393		2	40		17	10	3	29	669
1986*	16,515	—	2,770	891	238		347	29	4	40	1		28	10		7
1987*	25,600	—	389	1815	1,810	6	386		33	40	3		33	3		15
1988	40,422	—	2,294	1879	828	130	482	98	72	57	15		168	7	6	15
1989	15,954	13,095	1,226	189	408	20	202	50	32	54	4		117	9	10	204
1990	9,363	12,448	749	11	1364	116	157	10	156	56	9(2)	39	235	3	5	283
1991	11,177	6,457	1,022	153	438	410	312	43	54	54	4	19	30			266
1992	3,400	8,009	1,122	637	173	121	312	48	97	54	1(1)	136	283			88
Quota(1993)	9,000(9,000)	8,700(8,420)	725(700)	950(925)	1100(1075)	1300(570)	450(450)	50(50)	50(50)	54						
1993	5,735	8,588	544	565	215	505	293	44	21	54	7(12)	2	5			9
1994	8,097	7,854	548	449	362	312	170	26		54	5(12)	6	7			2
1995	7,002	5,394	619	105	975	412	190	50	49	54	7(20)	2	4			5
1996	8,040	8,062	303	67	314	377	434	50	40	54	3(24)	21	6			
1997	8,534	10,007	602	23	352	242	297	50	43	54	1(16)			1		
1998	5,305	6,082	451	460	266	445	194	35	48	54	7(54)	1				
1999	6,535	8,441	597	38	749	489	336	60	5	62	1(92)	11				
2000	7,513	8,658	300	39	1426	506	254	50	8	62	20(92)	1				2

(Continued)

TABLE 1.9 (*Continued*)
Catch of Small Cetaceans by Japanese Coastal Fisheries, Including Small-Type Whaling, Drive Fisheries, Hand-Harpoon Fisheries, Culling, and Incidental Mortality within the 200-Mile Zone (Excluding Released Individuals)

	Dall's Porpoise						Short-Finned Pilot Whale									
Species	*Dalli*-Type	*Truei*-Type	Striped Dolphin	Pantropical Spotted Dolphin	*Tursiops* spp.	Risso's Dolphin	South. Form	North. Form	False Killer Whale	Baird's Beaked Whale	Finless Porpoise	Pacific White-Sided Dolphin	*Delphinus* spp.	Killer Whale	Harbor Porpoise	Others and Unknown Species
2001	8,430	8,220	484	10	259	478	344	47	37	62	9(76)	6			1	2
2002	7,614	8,335	642	419	801	388	134	47	7	63	8(86)	9			5	1
2003	8,308	7,412	450	132	181	379	118	42	21	62	9(114)	9				1
2004	4,614	9,175	637	2	653	512	163	13	7	62	7(81)	12	1		1	1
2005	6,880	7,784	457	13	363	394	154	22	1	66	5(109)	14			3	3
2006	4,212	7,802	515	405	375	345	264	7	35	63	10(114)	12	1		1	1
2007	4,070	7,287	470	16	405	519	338		4	67	14(117)	20			7	
2008	2,594	4,632	600	329	391	339	181		5	64	20(157)	26	2		11	5

Notes: Figures from 1972 to the first line for 1979 have not been reported to the IWC, but those from the second line for 1979 to 1999 have been reported. Statistics beginning in 2000 are from the webpage of the Fisheries Agency. The Fisheries Agency statistics are summarized and presented here without my interpretation on species identification, which could cause some discrepancies with figures in other tables. Stranding of finless porpoises is given in parentheses for reference purposes only because it is likely to have originated from incidental mortality (available only since 1988).

1972–1988 seasons: Two color types of Dall's porpoises not separated.

1979 season: In addition to these figures, the mother-ship salmon drift-net fishery reported 685 Dall's porpoises and 3 harbor porpoises and the land-based salmon drift-net fishery reported 127 Dall's porpoises incidentally taken in their nets.

1980 season: The table includes 153 finless porpoises and 30 harbor porpoises reported as taken by the hand-harpoon fishery of Iwate Prefecture, which is likely to be misidentification of species. In addition to this figure, the mother-ship salmon drift-net fishery reported 1000 Dall's porpoises and 4 harbor porpoises and the land-based salmon drift-net fishery reported 139 Dall's porpoises taken in their nets.

1986–1987 seasons: The Offshore Division of the Fisheries Agency collected statistics for 1972–1987; then the Coastal Division took up the task in the 1988 season and later reported revised figures for the 1986 and 1987 seasons. The upper lines represent the statistics of the Offshore Division and the lower line those of the Coastal Division. Thus, if two figures are given for 1 year the latter one is likely closer to the truth.

1987 season: The figures do not include a minimum of 1668 small cetaceans (including 188 Pacific white-sided dolphins, 261 northern right whale dolphins, 819 Dall's porpoises, and others) reported by the Offshore Division as taken incidental to various drift-net fisheries of Japan.

1988 season: The figures do not include a total of 2999 cetaceans (including 1663 Dall's porpoises, 116 striped dolphins, 68 northern right whale dolphins, and others) incidentally killed in the Japanese squid drift-net fishery and large-mesh drift-net fishery and landed at Japanese ports.

1989 season: The figures do not include 331 Dall's porpoises discarded by the Japanese offshore salmon drift-net fisheries and a total of 707 small cetaceans (including 150 Dall's porpoises, 241 striped dolphins, and others) incidentally taken by the large-mesh drift-net fishery and landed at Japanese ports.

1990 season: The figures do not include a total of 16,635 small cetaceans estimated as incidentally killed by the Japanese high-seas squid drift-net fisheries, which includes 3,093 Dall's porpoises, 562 common dolphins, 7,909 northern right whale dolphins, 4,447 Pacific white-sided dolphins, and others.

1991 season: The figure of 11,177 *dalli*-type Dall's porpoises includes 6,637 Dall's porpoises of unknown color type. The figures do not include a total of 18,142 small cetaceans estimated as incidentally killed by Japanese high-seas salmon drift-net and squid drift-net fisheries, which includes 3,338 Dall's porpoises, 9,320 northern right whale dolphins, 3,784 Pacific white-sided dolphins, 1,036 common dolphins, and others.

1993 season: This was the first year when all the small-cetacean fisheries were capped with quotas by species. The figures in parentheses are the quotas actually given to each fishery and were often smaller than the national quotas reported to the IWC. See Table 4.3 and Section 6.5 for further details of the quotas.

large population of *D. delphis* off either Iwate or Shizuoka. It was difficult to encounter *Delphinus* off Iwate in the winter when *ma-iruka* was reported to be hunted, and my field experience on fisheries in Iwate Prefecture confirmed that they mostly hunted Dall's porpoises. Therefore, *ma-iruka* reported by the Iwate fishermen must be Dall's porpoises. It was common knowledge by field scientists that dolphin fishermen on the Izu coast (Shizuoka Prefecture) and at Taiji (Wakayama Prefecture) used *ma-iruka* for the striped dolphin (*S. coeruleoalba*). We observed that hand-harpoon fishermen off Taiji made large takes of striped dolphins in the early 1970s, but this was supported by the reported statistics only when *ma-iruka* in the statistics was recognized to represent striped dolphins. The Whale Research Team of the Far Seas Fisheries Research Laboratory almost completed the task of unifying the common names among the small-cetacean fishermen and reported correct statistics to the IWC since 1979.

The statistics for cetacean strandings are incomplete compared with those for direct fisheries. Records of strandings increase as public interest increases. Most noteworthy would be the recent increase in the strandings of finless porpoises. Carcasses of species such as this in inland waters have a high probability washing ashore and being subsequently recorded by voluntary observers. Therefore, the high incidence of finless porpoise strandings may relate to both higher mortality and greater public interest in report strandings. I have placed strandings of finless porpoises in parentheses in Table 1.9 (this table does not list stranding of other species).

I believe that fishery statistics need to be viewed with caution. They contain both careless errors and incompleteness due to technical difficulties, and they are likely to be manipulated by persons with vested interests. If fishermen feel pressure from criticisms of the slaughter or rapid increase in the catch, they will be tempted to under-report the catch. Once a quota system is introduced, the fishermen may wish to fill it completely, leading to possible over-reporting, or to exceed it, leading to possible under-reporting. Government employees may be tempted to ignore such misreporting behavior in the interest of reducing their workload or of offering benefit to the industry. I have witnessed all of these cases for Japanese cetacean fisheries (Chapters 12 and 16).

In Table 1.10, I have included the geographical distribution of particular fisheries for small cetaceans in Japan in some years as such information is missing in Table 1.9. Both of these tables are from the same sources. Japanese coastal whaling after World War II was classified into two categories of "large-type" and "small-type" whaling. The former hunted sperm whales and baleen whales other than mike whales and processed them at land stations. The latter used catcher boats smaller than 50 gross tons, took minke whales and other small cetaceans, and processed them at land stations. In addition to these two types of whaling, Japan operated pelagic whaling in the Antarctic and North Pacific. Catches of these whaling operations are listed in Table 1.5.

Statistics of the pelagic whaling seem to be more or less reliable, at least in number and species, because they were reported by industries under inspection by three government inspectors. However, coastal whaling was different, and it could have taken two or three times greater number of whales than reported (Watase 1995; Kasuya 1999; Kondo 2001, these three in Japanese; Kasuya 1999, Kasuya and Brownell 1999). Japanese commercial whaling diminished in scale rapidly in the 1970s and ceased operation almost completely in the late 1980s. The last pelagic whaling was carried out in the 1979 summer season in the North Pacific and the Antarctic in the 1986/1987 season. Coastal whaling ceased by March 31, 1988. The current Japanese whaling operation of Norwegian type is limited to Baird's beaked whales and some delphinid species taken by small-type whaling (Chapter 4) and to several species of baleen whales and sperm whales taken for scientific purposes (Chapter 7).

The decline of Japanese whaling in the 1970s and 1980s caused a shortage of whale products, increased the price of small-cetacean products, and caused an increase in catch of dolphins and porpoises not regulated by quotas. Japanese small-type whaling, after prohibition of take of minke whales in 1988, subsisted on Baird's beaked whales and large delphinids off the Pacific coasts of central and northern Honshu.

Hand-harpoon fisheries operate in Sanriku (Pacific coast of Northern Honshu), Wakayama (Pacific coast of central Honshu), and Nago (26°35′N, 127°59′E) in Okinawa in southern Japan. The greatest increase in takes occurred in the first two of these. Most of the community-based dolphin drives dissolved much earlier than the prohibition of commercial whaling, and the currently operating teams are of recent establishment or, in a few cases, relicts of the past operations (Chapter 3). The drive fishery in Shizuoka has a long history but was not able to respond to the increase in demand change with larger catches, because the population of striped dolphins, the main target species of the fishery, had already been badly depleted. Iki in Nagasaki Prefecture, northern Kyushu, at one time recorded a large cull of dolphins, but almost ceased the practice by 1988. In Taiji in Wakayama Prefecture, a modern dolphin-drive team made the first attempt in 1969 and recorded large takes for only for a short period. Drive fisheries may be too destructive to operate at sustainable levels.

It is difficult to obtain correct statistics of mortality of cetaceans in trap nets, gill nets, and purse seines. Carcasses discarded at sea are usually unreported, and legal regulation may also cause biases. The reported mortality of minke whales incidental to trap-net fisheries was less than 20 individuals per year when the government prohibited selling of such by caught minke whales, and the accuracy of the statistics was questioned (Tobayama et al. 1992). However, there was a sudden increase in reported catches in 2001 to over 100 individuals coincident with a regulation change allowing the sale of such takes (Table 1.6). Earlier, most of the minke whales previously found in trap nets must have been utilized without being reported.

Hand-harpoon fisheries, drive fisheries, and small-type whaling take small cetaceans today in Japan, which will be dealt with further in the following chapters.

TABLE 1.10
Historical Change in Locations and Fisheries That Killed Minke Whales and Small Cetaceans (Toothed Whales Other than Sperm Whale), as Represented by the Number of Whales Taken in Some Selected Years

									Included in the Left Columns	
Prefecture	**Small-Type Whaling**	**Hand Harpoon**	**Drive**	**Trap Net**	**Seine Net**	**Gill Net**	**Others**	**Total**	**Baird's Beaked Whale**	**Minke Whale**
1972										
Hokkaido	237							237		231
Iwate		5,122						5,122		
Miyagi	110						30	140		110
Fukushima						143		143		
Ibaraki		29						29		
Chiba	93	1,169[a]				+	5	1,267[a]	86	5
Shizuoka			7,703					7,703		
Mie						20		20		
Wakayama	100	896[b]	+					996[b]		
Okinawa			170					170		
Ishikawa				14				14		1
Kyoto				12				12		
Yamaguchi		17						17		
Nagasaki			8					8		
Total	540	7,233[a,b]	7,881	26	0	163	35	15,878[a,b]	86	347
1980										
Hokkaido	229			15		20		264	8	221
Aomori				1				1		
Iwate		6,966		14				6,980		
Miyagi	158	16				77		251		158
Ibaraki		11			4	99		114		2
Chiba	24							24	23	
Shizuoka			6,660					6,660		
Wakayama	2		11,981					11,983		
Kochi				20				20		
Okinawa		107						107		
Niigata						1		1		
Ishikawa				11				11		
Fukui				5				5		
Kyoto				7				7		
Yamaguchi				1				1		1
Nagasaki		15	3,471[c]	31				3,517[c]		
Total	413	7,115	22,112	105	4	197[d]	0	29,946[c,d]	31	382
1990										
Hokkaido	2	1,747		5				1,754	2	
Aomori		10		2				12		2
Iwate		19,888		24				19,912		12
Miyagi	38	314		1				353	25	1
Chiba	34	67						101	27	
Shizuoka				1				1		1
Mie		13						13		
Wakayama	15	50	2,542					2,607		
Kochi				1				1		
Oita		1						1		
Miyazaki				30				30		
Okinawa		79						79		
Niigata				1				1		1

(*Continued*)

TABLE 1.10 (*Continued*)
Historical Change in Locations and Fisheries That Killed Minke Whales and Small Cetaceans (Toothed Whales Other than Sperm Whale), as Represented by the Number of Whales Taken in Some Selected Years

Prefecture	Small-Type Whaling	Hand Harpoon	Drive	Trap Net	Seine Net	Gill Net	Others	Total	Included in the Left Columns: Baird's Beaked Whale	Included in the Left Columns: Minke Whale
Ishikawa				2				2		1
Kyoto				13				13		
Shimane				1				1		1
Fukuoka				2				2		
Nagasaki	2			4			131[f]	137[f]		1
Total	91	22,169	2,542	87	0	0[e]	131[f]	25,020[e,f]	54	20
2000										
Hokkaido	10	1,272		2				1,284	10	
Aomori							1	1		1
Iwate		14,695						14,695		
Miyagi	76	204						280	26	
Chiba	33			2[g]				35[g]	26	1
Aichi							3	3		
Wakayama	69	275	2,077	4				2,425		4
Kochi				3				3		3
Okinawa		105						105		
Niigata				1				1		1
Toyama				7				7		6
Ishikawa				11				11		11
Yamaguchi							5	5		
Fukuoka							9	9		
Saga							1	1		
Nagasaki				2				2		2
Kumamoto							1	1		
Total	188	16,551	2,077	32[g]	0	0	20	18,868[g]	62	29
2008										
Hokkaido	13	533		14		5		565	13	8
Aomori				11				11		10
Iwate		6,513		12[h]				6,525[h]		11
Miyagi	25	180		7				212	25	6
Chiba	26			1				27	26	1
Shizuoka				2				2		1
Aichi						3	1	4		
Mie				8[h]				8[h]		7
Wakayama	20	280	1,497	5[i]				1,802[i]		4
Kochi				9				9		8
Oita				1			1[h]	2[h]		
Miyazaki				1				1		1
Kagoshima				5				5		5
Okinawa		63		3				71		3
Akita				1				1		1
Niigata				7				7		7
Toyama				9				9		9
Ishikawa				25				25		24
Fukui				3				3		3
Kyoto				7				7		6
Shimane				1				1		1
Yamaguchi				4		7		11		3

(*Continued*)

TABLE 1.10 (*Continued*)
Historical Change in Locations and Fisheries That Killed Minke Whales and Small Cetaceans (Toothed Whales Other than Sperm Whale), as Represented by the Number of Whales Taken in Some Selected Years

									Included in the Left Columns	
Prefecture	Small-Type Whaling	Hand Harpoon	Drive	Trap Net	Seine Net	Gill Net	Others	Total	Baird's Beaked Whale	Minke Whale
Fukuoka				1		2		2		
Saga						1		1		
Nagasaki				20				20		17
Kumamoto				2[h]		2		4[h]		
Total	84	7,574	1,497	158[j]		20	2[h]	9,335[j]	64	134

Source: Fisheries Agency statistics.
Note: Data for incidental mortality in offshore fisheries are incomplete. The bottom totals may include catches not allocated to prefecture. Scientific whaling is not included.
[a] Includes catch in large-mesh drift-net fishery.
[b] This is the period when the fishery shifted from hand harpoon to drive, and the figure includes about 100 short-finned pilot whales taken by the drive fishery.
[c] Includes 2120 killed in culling operation at Katsumoto.
[d] Excludes incidental mortality of 1143 estimated as killed in the high-seas salmon drift-net fishery.
[e] Excludes a total of 16,635 Dall's porpoises, Pacific white-sided dolphins, and northern right whale dolphins, and other species estimated as killed by Japanese squid drift-net fisheries in the North Pacific.
[f] Includes 31 dolphins of 3 species killed in culling.
[g] Includes one humpback whale.
[h] Includes one humpback whale.
[i] Includes one fin whale.
[j] Includes three humpback whales and a fin whale.

REFERENCES

[*In Japanese Language]

*Akashi, K. 1910. *History of Norwegian-Type Whaling in Japan.* Toyo Hogei Kabushiki Kaisha, Osaka, Japan. 280+40pp.

Amano, M. and Hayano, A. 2007. Intermingling of *dalli*-type Dall's porpoises into a wintering *truei*-type population off Japan: Implications from color patterns. *Mar. Mamm. Sci.* 23(1): 1–14.

Anon. 1994. Report of the workshop on mortality of cetaceans in passive fishing nets and traps. pp. 6–71. In: Perrin, W.F., Donovan, G.P., and Barlow, J. (eds.). *Gillnets and Cetaceans, Incorporating the Proceedings of the Symposium and Workshop on the Mortality of Cetaceans in Passive Fishing Nets and Traps. Rep. Int. Whal. Commn.* (Special Issue 15), 629pp.

Best, P.B., Canham, P.A.S., and Macleod, N. 1984. Patterns of reproduction in sperm whales, *Physeter macroephalus. Rep. Int. Whal. Commn.* (Special Issue 6): 51–79.

*Dainihon Suisankai. 1890. Survey of hunting and utilization of dolphins and porpoises. *Dainhon Suisankai Hokoku* 98: 240–251.

*Hashiura, Y. 1969. *History of Whaling at Taiji in Kumano.* Heibonsha, Tokyo, Japan. 662pp.

*Hattori, T. 1887–1888. *Miscellanea of Whaling.* Dainihon Suisankai, Tokyo, Japan. Vols. 1: 109pp.; 2: 210pp.

*Ishida, T. 1917. *Considerations on Fishery Products.* Kokko Insatsu, Tokyo, Japan. 392pp.

IWC. 2008. Report of the standing working group on environmental concerns. *J. Cetacean Res. Manage.* 10(Suppl.): 247–292.

IWC. 2009. Report of the sub-committee on small cetaceans. *J. Cetacean Res. Manage.* 11(Suppl.): 311–333.

*Kanda, G. 1731 (date of preface). *Pictorial Catalogue of Japanese Fish.* Vol. 4 (Fishes without Scales). Unpublished manuscript.

*Kasahara, A. 1950. *Whaling and Whale Stocks around Japan: Nihonsuisan K.K. Kenkyusho Hokoku No. 4.* Nihonsuisan K.K. Kenkyusho, Tokyo, Japan. 103pp.+51pp. figs.

*Kasuya, T. 1991. Life history parameters and reproductive strategies of whales. pp. 67–86. In: Sakuramoto, K., Kato, H., and Tanaka, S. (eds.). *Management and Investigation of Whale Stocks.* Koseisha Koseikaku, Tokyo, Japan. 273pp.

Kasuya, T. 1995. Overview of cetacean life histories: An essay in their evolution. pp. 481–497. In: Blix, A.S., Walloe, L., and Ultang, O. (eds.). *Whales, Seals Fish and Man.* Elsevier, Amsterdam, the Netherlands. 720pp.

Kasuya, T. 1999. Examination of reliability of catch statistics in the Japanese coastal sperm whale fishery. *J. Cetacean Res. Manage.* 1(1): 109–122.

*Kasuya, T. 1999. Manipulation of statistics by the Japanese Coastal Sperm Whale Fishery. *IBI Rep.* (Kamogawa, Japan) 9: 75–92.

*Kasuya, T. 2000. Explanatory Notes. pp. 155–203. In: Omura, H. *Journal of an Antarctic Whaling Cruise—Record of Whaling Operation in Its Infancy in 1937/38.* Toriumi Shobo, Tokyo, Japan. 204pp.

Kasuya, T. and Brownell, R.L. Jr. 1999. Additional information on the reliability of Japanese coastal whaling statistics. IWC/SC/51/O7. 15pp. Document presented to the *51st Meeting of the Scientific Committee of IWC* (available from IWC Secretariat, Cambridge, U.K.).

*Kasuya, T. and Yamada, T. 1995. *List of Japanese Cetaceans.* Gieken Series 7. Nihon Geirui Kenkyu Sho (Institute of Cetacean Research), Tokyo, Japan. 90pp.

*Kato, H. 1990. Life history of baleen whales with special reference to Antarctic minke whale. pp. 128–150. In: Miyazaki, N. and Kasuya, T. (eds.). *Mammals in the Sea—Their Past, Present and Future.* Saientisutosha, Tokyo, Japan. 311pp.

Kato, H., Ishikawa, H., Miyashita, T., and Takaya, S. 2008. Status report of conservation and researches on the western gray whales in Japan, May 2007–April 2008. IWC/SC/60/O8. 9pp. Document presented to the 60th *Meeting of the Scientific Committee of IWC* (available from IWC Secretariat, Cambridge, U.K.).

*Kijima, T. 1944. *Consideration of History of Japanese Fisheries.* Seibi Shokaku, Tokyo, Japan. 314pp.

*Kitahara, T. (ed.) 1996. *Lessons of Whaling: International Atmosphere Surrounding Fishery Resources.* Seizando Shoten, Tokyo, Japan. 233pp.

*Kitamura, K. 1995. Illustrated fisheries in Noto. pp. 117–239. In: *Collection of Japanese Agricultural Literatures.* Vol. 58. Nosan Gyoson Bunka Kyokai, Tokyo, Japan. 406+I–XIIIpp. (Manuscript dated 1838).

*Kondo, I. 2001. *Rise and Fall of Japanese Coastal Whaling.* Sanyosha, Tokyo, Japan. 449pp.

*Maeda, K. and Teraoka, Y. 1958. *Whaling.* Isana Shobo, Tokyo, Japan. 346pp. (First published in 1952).

Marine Mammal Commission. 2010. Annual report to congress 2009. Marine Mammal Commission, Bethesda, MD. 291pp.

*Minami Chita Choshi Hensan Iinkai. 1991a. *Records of Minami Chita Town, Main Text.* Minamichita-cho, Minamichita, Japan. 954+4pp.

*Minami Chita Choshi Hensan Iinkai. 1991b. *Records of Minami Chita Town, Data Section 6.* Minamichita-cho, Minamichita, Japan. 501pp.

*Mizuno, M. 1885. Examination report of items exhibited at the fishery exposition. Section I, Part 1. Noshomu-sho Suisan-kyoku, Tokyo, Japan. 246pp.

*Nakazono, N. 2009. Whaling in Kyushu. pp. 22–36. In: Kojima, T. (ed.). *On Whales and Japanese—Revisit to Coastal Whaling.* Tokyo Shoten, Tokyo, Japan. 255pp.

Nishiwaki, M. and Handa, C. 1958. Killer whales caught in the coastal waters off Japan for recent 10 years. *Sci. Rep. Whales Research Inst.* (Tokyo) 13:85–96.

*Norin-daijin Kanbo Tokei-ka. 1926. *First Statistics of Ministry of Agriculture* (Text reprinted in *Hogeisen*, Vol. 30, published in 1999 by Takeuchi, K.).

*Noshomu-sho. 1890–1893. Report of preliminary survey on fisheries (Text reprinted in *Hogeisen*, Vol. 30, published in 1999 by Takeuchi, K.).

*Noshomu-sho. 1894. *Special Investigation of Fisheries* (Text reprinted in *Hogeieen*, Vol. 30, published in 1999 by Takeuchi, K.).

*Noshomu-sho Nomu-kyoku. 1892. *Fish Oil and Wax.* Yurindo, Tokyo, Japan. 139pp+5 pls.

*Noshomu-sho Suisan-kyoku. 1911. *Fishery Activities of Japan.* Vol. 1. Suisan Shoin, Tokyo, Japan. 424pp.

Noshomu-sho Suisan-kyoku. 1900. *Histoire de L'industrie de la Peche Maritime et Fluviale.* Noshomu-sho Suisan-kyoku, Tokyo, Japan. 153pp. (in French).

Ohsumi, S. 1972. Catch of marine mammals, mainly of small cetaceans, by local fisheries along the coast of Japan. *Bull. Far Seas Fish. Res. Lab.* 7: 137–166.

*Oku, K. and Kajikawa, A. 1899. Examination report of items exhibited at the second fishery exposition. Vol. 2, Part 3. Noshomu-sho Suisan-kyoku, Tokyo, Japan. 325pp.

*Otsuki, K. 1808 (manuscript). Natural history of whales. pp. 7–152. In: *Geishi Ko [Natural History of Whales].* Printed in 1983 by Kowa Shuppan, Tokyo, Japan. 538+31pp.

*Oyamada, O. 1832. *Pictorial Description of Whaling.* Tatamiya, Hirado, Japan. Vols. 1: 40pp., 2: 40+6pp.

*Saito, K. 1981. *History of Noto Town*, Vol. 2: Fisheries. Noto-cho Yakuba, Noto-cho, Japan. 1040pp.

*Sakuramoto, K., Kato, H., and Tanaka, S. 1991. *Whale Stocks, Their Management and Research.* Koseisha Koseikaku, Tokyo, Japan. 273pp.

*Schmidt-Nielsen, K. 1995. *Scaling: Why is Animal Size So Important*? Koronasha, Tokyo, Japan. 302pp. (Translated by Shimozawa, T., Ohara, M., and Urano, T.).

Status of International Fisheries Resources. 2012. Suisan Sogo Kenkyu Center, Fisheries Agency. http://kokushi.job.affrc.go.jp/index-2.html. Accessed January 30, 2013.

*Suisan-cho Chosa-kenkyu-bu. 1968. Report of basic investigations in search of counteraction for fishery damages caused by small cetaceans in western Japan (42nd year of Showa). Suisan-cho Chosa-kenkyu-bu, Tokyo, Japan. 96pp.

*Suisan-cho Chosa-kenkyu-bu. 1969. Comprehensive report of basic investigations in search of counteraction for fishery damages caused by small cetaceans in western Japan (42nd and 43rd year of Showa). Suisan-cho Chosa-kenkyu-bu, Tokyo, Japan. 108pp.

*Tachihira, S. (ed.) 1992. *Pictorial Description of Fisheries in the Goto Islands.* Nagasaki-ken Gyogyoshi Kenkyukai, Nagasaki, Japan. 96pp. (Publication of a manuscript by an anonymous author dated 1882, with notes by Tachihira, S.).

*Takenaka, K. 1890. Methods of hunting dolphins and porpoises. *Dainhon Suisankai Hokoku* 95: 87–91.

*Takeuchi, K. 1999. *Hogeisen.* Vol. 30. Takeuchi Kenji, Yokohama, Japan. No pagination.

*Tada, H. 1978. *Study of the Whaling History of Yamaguchi Prefecture in the Meiji Era.* Matsuno Shoten, Tokuyama, Japan. 256pp.

Teilmann, J., Miller, L., Kirketerp, T., Hansen, K., and Brando, S. 2006. Reactions of captive harbor porpoises (*Phocoena phocoena*) to pinger-like sounds. *Mar. Mammal Sci.* 22(2): 240–260.

Tobayama, T., Yanagisawa, F., and Kasuya, T. 1992. Incidental take of minke whales in Japanese trap nets. *Rep. Int. Whal. Commn.* 42: 433–436.

*Torisu, K. 1999. *Historical Study of Whaling in Western Japan.* Kyushu Daigaku Shuppankai, Fukuoka, Japan. 414+XXVIIIpp.

*Uda, M., Oshima, Y., Kumagori, T., Suehiro, Y., Takahashi, Y., Higashi, H., Hiyama, Y., and Watari, S. (eds.) 1962. *Fishery Handbook.* Toyo Keizai Shinposha, Tokyo, Japan. 725pp.

Wade, P.R. 1998. Calculating limits to the allowable human-caused mortality of cetaceans and pinnipeds. *Mar. Mammal Sci.* 14(1): 1–37.

*Watase, S. 1995. Hidden stories of whaling. pp. 235–240. In: Anon. (ed.). *Now We Shall Speak Up—Time is Over.* Vol. 1. Sei Sei Shuppan, Tokyo, Japan. 328pp.

*Yamada, S. 1902. *Whaling at Tsuro.* Tsuro Hogei Kabushiki Kaisha, Kochi, Japan. 284pp.

Yamada, T.K. and Ishikawa, H. 2009. *Balaenoptera brydei* Olsen, 1913 and *Balaenoptera edeni* Andersen, 1878. pp. 325–327. In: Ohdachi, S.D., Ishibashi, Y., Iwasa, M.A., and Saitoh, T. (eds.). *The Wild Mammals of Japan.* Shoukadoh, Kyoto, Japan. 544pp.+4 maps.

*Yamase, H. 1760. *Natural History of Whales.* Osaka Shorin, Osaka, Japan. 16+54pp.

*Yanaginara, E. 1887. Miscellanea on dolphins and porpoises. *Dainihon Suisankai Hokoku* 63: 18–23.

Yatsu, A., Shimada, H., and Murata, M. 1994. Distribution of epipelagic fishes, squids, marine mammals, seabirds and sea turtles in the central North Pacific. *Int. North Pacific Fish. Commn. Bull.* 53(II): 111–146.

*Yoshihara, T. 1976. Whaling at Ine Harbor of Tango. *Tokyo Suisan Daigaku Ronshu* 11: 145–184.

*Yoshihara, T. 1982. *Whaling in Southern Boshu [Chiba Prefecture], with Notes on Whale Tombs.* Aizawa Bunko, Ichikawa, Japan. 227pp.

*Yoshitome, T. 2009. Whaling in the Chugoku-chiho. pp. 53–69. In: Kojima, T. (ed.). *On Whales and Japanese—Revisit to Coastal Whaling.* Tokyo Shoten, Tokyo, Japan. 255pp.

2 Hand-Harpoon Fisheries for Dolphins and Porpoises

2.1 PREHISTORIC OPERATIONS

Hand-harpoon fishing is defined as fishing for fish and marine mammals by throwing a harpoon with a detachable head from a boat. It is called a fishery if operated commercially. The Japanese term *tsukin-bo* currently means both the fishing method and the hand harpoon. The instrument is called *kite* or *hanare* by the Ainu (Natori 1945, in Japanese) and *choki* or *tsukin-bo* by Japanese fishermen. The last portion of the latter name, *bo*, is possibly a word originating from Awa in the southern part of Chiba Prefecture (34°55′N–35°44′N, 139°45′E–140°53′E) on the Pacific coast of central Honshu (Japanese main island), meaning "a person who carries out the activity" (Tamura 1996, in Japanese). Thus, *tsukin-bo* itself would initially have meant "harpooner." Although archaeologists classify detachable harpoon heads into several types (Watanabe 1984; Yamaura 1996, both in Japanese), they fall into two major types based on the way the head and the shaft are connected, that is, either the tail of the harpoon head is inserted into a socket at the tip of the harpoon shaft or the tip of the harpoon shaft is inserted into a socket at the base of the harpoon head. Current Japanese hand harpoons are of the latter type (Figure 2.1). The harpoon head is tied to a line, which is then tied to the shaft and to the ship or to a buoy. If an animal is hit by the harpoon, the head turns 90° in the body and prevents detachment. The shaft and buoy are recovered together with the prey.

The hand harpoon appeared in Japan at the beginning of the Jomon Era (12,000–1,000 years BC) and was widely used in two separate regions. One region was from Hokkaido (northern island, which is one of the four major islands of Japan) to Mikawa Bay (*c.* 34°45′N, 137°E) on the Pacific coast of central Japan, and the other was from the Korean Peninsula to northern Kyushu (western island, which is one of the four major islands of Japan) and nearby islands on the coasts of the eastern East China Sea and southern Sea of Japan. The shape of the harpoon head was different between the two locations, suggesting separate origin (Anraku 1985; Yamaura 1996, both in Japanese). Hand-harpoon fishing is also known from the Okhotsk Culture that flourished in the eighth to twelfth centuries along the coasts of the Okhotsk Sea including Hokkaido and forms the subsequent Ainu Culture in Hokkaido. The harpoon has been used in recent Japanese fisheries for various marine animals such as small baleen whales, dolphins, porpoises, seals, swordfish, and sunfish.

In Volcano Bay (42°20′N, 140°30′E) on the Pacific coast of southern Hokkaido, the Ainu hunted minke whales by throwing several aconite-poisoned hand harpoons at a single whale (Natori 1945, in Japanese). Whaling with hand harpoons of this type is known from aboriginal communities in the northern North Pacific from the northwestern coast of North America, Alaska, Siberia, the Kuril Islands, and Hokkaido (Kasuya 1981, in Japanese). Heizer (1943) stated that people in the Alaskan Peninsula, Aleutian Islands, Kuril Islands, and Hokkaido used aconite-poisoned lances, but not harpoons with detachable heads. However, it appears from a description by Natori (1945, in Japanese) that the Ainu in Hokkaido used hand harpoons with detachable heads together with aconite poison.

Japanese archaeological sites of the Jomon Era that contained numerous cetacean remains are as follows (from north to south, with name of nearby town and prefecture): (1) Higashi-kushiro (Kushiro, Hokkaido), (2) Irie (Abuta, Hokkaido), (3) Miyashita (Satohama, Miyagi at 37°54′N–38°59′N), (4) Natagiri (Tateyama, Kanagawa at 35°08′N–35°32′N, 139°10′E–139°45′E), (5) Yoshii (Yokosuka, Kanagawa), (6) Shomyoji and Aogadai (Yokohama, Kanagawa), (7) Idokawa (Ito, Shizuoka at 34°36′N–35°05′N, 137°30′E–139°10′E), (8) Asahi (Himi, Toyama at 36°45′N–36°55′N, 137°00′E–137°43′E), and (9) Mawaki (Noto, Ishikawa at 36°18′N–37°33′N, 136°15′E–137°22′E). The last two are on the Sea of Japan coast of central Honshu and in a range of *c.* 36°50′N–37°20′N, 136°50′N–137°15′E, and the others are on the Pacific coasts of central and northern Japan (north of 35°N). These sites often contained detachable harpoon heads as well as remains of Dall's porpoises, which cannot be caught with the drive-in method. Thus, it would be reasonable to assume that at least some of the porpoises were taken with hand harpoons.

It is worth noting that the (9) Mawaki Site (37°20′N, 137°10′E) on Noto Peninsula (36°50′N–37°30′N, 136°40′E–137°20′E), well known for evidence of dolphin hunting, has not produced detachable harpoon heads but numerous slender lance heads made of stone, some of which were found inserted in dolphin bone and deduced to have been used in dolphin fishing (Hiraguchi 1986, in Japanese). The lance was certainly used in the dolphin fishery, but it must have been used in killing of dolphins secured in the bay by some other method. It could be technically possible to kill dolphins with such lances, but it might be difficult to secure the carcasses, which usually sink. Mawaki and nearby villages, where a dolphin drive fishery was operated until the twentieth century (Chapter 3), possibly have oceanographic or geographical conditions suitable for such a fishery. The people of the Mawaki Site in the Jomon Era probably drove dolphins into the bay and slaughtered them using the stone lances.

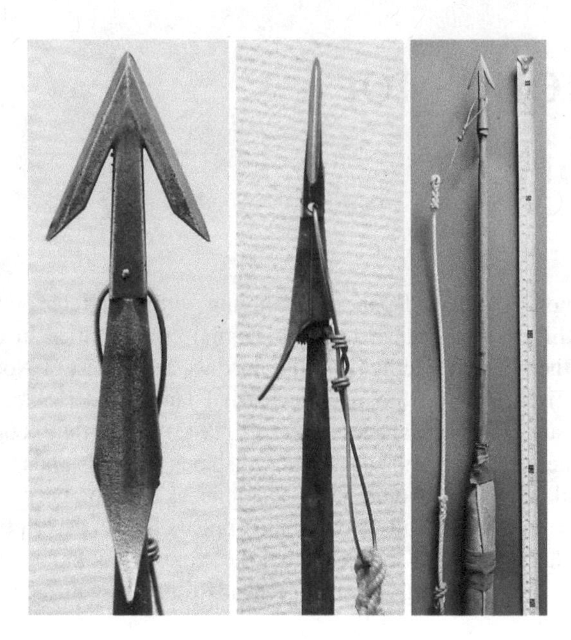

FIGURE 2.1 Hand harpoon used by Japanese dolphin and porpoise hunters. This specimen was purchased in the Sanriku Region on the Pacific coast of northern Japan, latitude 37°54′N–41°35′N, and used during the research cruise of the *Hoyo-maru No. 12* to the western North Pacific in 1982 (Section 9.3.3). *Left*: Molded stainless steel harpoon head. The distance between barb tips is 32 mm. *Middle*: Side view of the harpoon head showing direction of the third barb. Linear distance from the tip of the harpoon head to the tip of the third barb is 102 mm. Attachment of the harpoon head to the foreshaft (the iron of the harpoon shaft) and its connection to harpoon rope via stainless steel wire are shown. The foreshaft is made of soft iron and is about 58 cm long excluding the socket portion. *Right*: Connections among the harpoon head, harpoon rope, foreshaft, and wooden harpoon shaft that is inserted into the socket of the foreshaft. The harpoon rope is tied once or twice to the harpoon shaft and then to a float. The foreshaft is also tied to the wooden shaft with a rope independent of the harpoon rope. If electric shocker is to be used, an electric cable from the battery is connected to the stainless steel wire on the harpoon head and another cable placed in the sea. The distance between black marks on the scale is 5 *sun* or 15.15 cm. The height of the ship's bow determines the length of the harpoon shaft.

The identification of the species of dolphins hunted in the Jomon Era is still incomplete. Bone fragments of cetaceans were reported by Kasuya et al. (1985, in Japanese): numerous harbor porpoises from (1) Higashi-kushiro Site, numerous Pacific white-sided dolphins and a false killer whale from (2) Irie Site, some finless porpoises and a harbor porpoise from (3) Miyashita Site, numerous common dolphins and some bottlenose dolphins and pilot whales from (4) Natagiri Site, and Pacific white-sided dolphins, common dolphins, and bottlenose dolphins from (5) Shomyoji Site. Skulls of six common dolphins together with two wild boars and a Japanese deer were found in the (6) Idokawa Site (Kurino and Nagahama 1985, in Japanese). From many of these archaeological sites, detachable harpoon heads were recovered as well as remains of harbor porpoises and Dall's porpoises (Kabukai-A Site), for which the driving method is not suitable. So it is reasonable to assume that the inhabitants of many of these locations took marine mammals with hand harpoons. It is also clear that hunting of dolphins and porpoises with hand harpoons was more common in northeastern Japan (Kasuya et al. 1985, in Japanese).

The Okhotsk Culture flourished in the eighth to twelfth centuries in a broad area of the Okhotsk Sea coast where ice floes approached in winter (Oba and Oi 1973, in Japanese). It left numerous archaeological sites in northern Hokkaido: the Sea of Japan coast of northwestern Hokkaido, the Okhotsk Sea coasts of Hokkaido, and the Nemuro Peninsula (43°20′N, 145°40′E) on the Pacific coast of eastern Hokkaido. Cetacean remains and harpoon heads found in these sites suggest that the inhabitants relied greatly on hunting marine mammals. The Onkoromanai Site near Wakkanai (45°25′N, 141°40′E), on the northwestern tip of Hokkaido at the southwest opening of the Okhotsk Sea, covered a long historical period from the late Jomon Culture, Okhotsk Culture, and Ainu Culture (Kaneko 1973, in Japanese) and produced relicts of acorn barnacles, sea urchins, sea shells, fish (16 species), birds (21 species), and mammals (24 species). The mammal specimens include one individual each of the killer whale, bottlenose dolphin, Pacific white-sided dolphin, and Dall's porpoise. Although these alone may not be enough evidence of a marine mammal fishery, the conclusion is supported by artifacts found together with the remains.

Among the Okhotsk Culture sites, two sites, Kabukai-A on Rebun Island (45°20′N, 141°00′E) and the Matawakka shell mound on Rishiri Island (45°10′N, 141°14′E) in the northeastern Sea of Japan near the southwest opening of the Okhotsk Sea, have been well examined for marine mammal remains. The Kabukai-A Site has produced detachable harpoon heads and remains of the 10 species of cetaceans (with minimum numbers of individuals in parentheses): Pacific white-sided dolphin (11), harbor porpoise (4), Dall's porpoise (5), pilot whale (25), false killer whale (5), sperm whale (2), ziphiids (1), humpback whale (6), minke whale (1), and North Pacific right whale (2) (Kasuya 1975, 1981, in Japanese). Many of these species were found in multiple strata; pilot whales and Pacific white-sided dolphins were found in almost every stratum. No gray whale remains were found. While some of the cetacean remains may represent stranded animals, there can be no doubt that people of the site lived by hunting marine mammals with hand harpoons.

It would be difficult to drive large strong-swimming cetaceans into a harbor with rowed boats. Driving Dall's and harbor porpoises, which live in small schools and swim in a way difficult to predict, would be lost even if driven with numerous boats (except at locations geographically and behaviorally suited for driving). However, if one is patient, it is possible to harpoon these porpoises in nearshore waters. The Inuit hunt seals and dolphins with hand harpoons, and my Canadian friend has used a shotgun to collect harbor porpoises for study. Other more gregarious coastal species are easier targets of driving, for example, bottlenose and common dolphins.

Yahata (1943, in Japanese) reported 24 needle cases, believed to be made of bird bones, from the Bentenjima Site of the Okhotsk Culture on the Nemuro Peninsula in eastern Hokkaido. Two of these had on them drawings depicting scenes of cetacean hunting. The top panel of Figure 2.2 shows a whale and a small boat with a standing harpooner and six oarsmen. The whale and boat are connected with two lines. This scene reminds me of the minke whale hunt by Ainu in Volcano Bay described by Natori (1945, in Japanese). The identity of the cetacean drawn on this needle case is uncertain, but it could be a minke or gray whale. Another needle case from the same site also depicts a whaling scene (second panel of Figure 2.2).

The two whaling scenes do not show use of a buoy, but such a scene is on one of the 21 needle cases of the Okhotsk Culture reported by Tsuboi (1908, 1909, both in Japanese) from Aniwa Bay, southern Sakhalin (third panel of Figure 2.2). Seven or eight persons are in the boat, which is connected to a submerged baleen whale with a line. Another line connects the ship and a buoy. The shape of the lower lip suggests that the whale on the left is a gray whale and the whale on the right could be the same (Tsuboi assumed the right-side one was a humpback whale). It was usual for such needle cases to have two similar drawings arranged symmetrically. Gray whales regularly migrated to the coasts of Hokkaido in the nineteenth century. They were particularly common in the Teshio (*c.* 44°40′N) and Oshima (41°30′N–42°40′N) regions on the Sea of Japan coasts of western Hokkaido and the Kitami Region (*c.* 144°E)

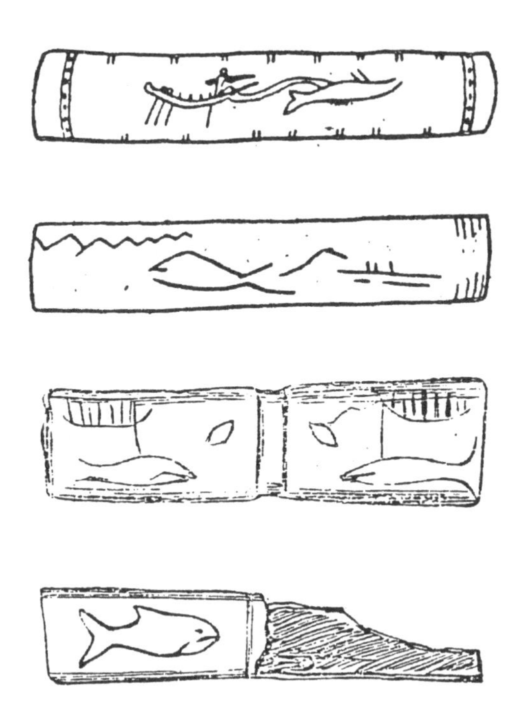

FIGURE 2.2 Cetaceans and cetacean fishery depicted on needle cases of the Okhotsk Culture from the eighth to twelfth century. Two specimens on the top are from the Bentenjima Shell Mound near Nemuro (43°20′N, 145°35′E) (partial reproduction of a figure from Yahata 1943, in Japanese), and the bottom two are from the Aniwa Bay coast in southern Sakhalin (partial reproduction of Figure 2 of Tsuboi 1908, in Japanese). The second specimen from the bottom was later reproduced in Tsuboi (1909, in Japanese) and Yahata (1943, in Japanese) with an additional line depicting a right flipper.

on the Okhotsk coast of Hokkaido (Hokusui Kyokai 1977, in Japanese). Sato (1900–1902, in Japanese) reported an old-style whaling operation from Haboro (44°20′N, 141°40′E) on the Sea of Japan coast of northwestern Hokkaido that took 26 gray whales in 1889 and 27 in 1890. The fishing season was from April to June.

Tsuboi (1908, in Japanese) reported another needle case from southern Sakhalin with an engraving depicting a Dall's porpoise (bottom panel of Figure 2.2). The right half was damaged. He reported that other six needle cases depict something that looks like a buoy, but certain interpretation was not possible. The needle cases discussed earlier suggest that people of the Okhotsk Culture hunted cetaceans with hand harpoons.

2.2 HAND-HARPOON FISHERY FOR SMALL CETACEANS (1): BEFORE WORLD WAR II

Records of hand-harpoon fisheries for dolphins and porpoises before the Meiji Revolution in the 1860s are scanty. Such records increased after the revolution, possibly reflecting the government's attempt to promote fishing industries. Mizuno (1883, in Japanese) stated that Japanese fisheries for dolphins and porpoises used nets in coastal waters and lances in offshore waters and that in the waters near the Izu Peninsula (34°36′N–35°05′N, 138°45′E–139°10′E, which is a part of Shizuoka Prefecture) people used lances for fishing for cetaceans. This statement does not match well with my understanding that the dolphin fishery on the Izu coasts was an important large-scale fishing activity of the region since the seventeenth century and that the method was driving (Chapter 3).

Hattori (1887, in Japanese) listed the following four major regions of hunting dolphins and porpoises:

1. Hokkaido, northern island of Japan at 41°20′N–45°30′N, 139°20′E–145°50′E
2. Rikuchu (i.e., Iwate Prefecture, 38°59′N–40°27′N), Rikuzen (i.e., Miyagi Prefecture, 37°54′N–38°59′N), and Iwaki (i.e., Fukushima Prefecture, 36°52′N–37°54′N) on the Pacific coast of northern Honshu
3. The Awa Region (i.e., southern part of Chiba Prefecture), the Izu Peninsula, and Kii (i.e., Wakayama Prefecture, 33°25′N–34°15′N, 135°00′E–136°00′E) on the Pacific coast of central Honshu
4. Noto Peninsula (Sea of Japan coast of central Honshu) and Hizen (32°30′N–34°45′N, 128°30′E–130°00′E), which includes the Saga and Nagasaki Prefectures in northwestern Kyushu

He then stated that large catches were obtained with nets such as the *haya-uchi-ami*, *sanbyaku-ami*, and *tome-ami* but harpoons were also used. Although he did not describe the nature of these nets or of the harpoons, such names appeared in the operation of the drive fishery on the Izu coast (Chapter 3). It seems to me that commercial fisheries for dolphins or porpoises in the nineteenth century mostly used nets in drive fisheries.

As implements for a hand-harpoon fishery were inexpensive and could be used with limited experience for a broad variety of targets, fishing for marine mammals with hand harpoons could have occurred widely. However, it would not have been an important method for commercial hunting of dolphins and porpoises in the nineteenth century, because of the limited efficiency carried out on small rowed boats or sailboats. Thus, the large-scale hand-harpoon fisheries for small cetaceans probably had to await the introduction of motor-driven fishing vessels.

Japanese traditional whaling could have had its origin in a hand-harpoon fishery, and we see similar equipment in both operations. Early Japanese whaling in Mikawa Bay (at 34°45′N, 137°00′E), which was mentioned in the history of Japanese old-type whaling (Chapter 1), used hand harpoons with detachable harpoon heads. *Choshu-zasshi* (Miscellaneous Records of Choshu) by Naito (1770, in Japanese) stated that they first threw a *haya-mori* (or quick harpoon) to connect the whale and boat. The *haya-mori* was lightweight and equipped with a detachable head. This action was followed by the use of the heavier *denchu-mori*, which had a thick asymmetrical harpoon head fixed to a heavy wooden shaft with a rod of soft iron that could be bent easily. The heavy *denchu-mori* would have had greater power but a shorter throwing range. *Gei-ki* (*Records of Whales*) written in the late eighteenth century and reprinted by Hashiura (1969, in Japanese) showed drawings of whaling tools used by whalers in Taiji (33°36′N, 135°57′E), Wakayama Prefecture. One of the drawings depicts a hand harpoon named *chokiri*. It had a detachable head, and the name is similar to *choki* used by recent Japanese fishermen for their hand harpoon. However, I could not confirm the use of harpoons with detachable heads in whaling operations of the subsequent period. Nakazono (2009, in Japanese) made a similar observation. Net whaling in Ikitsuki (33°20′N, 129°25′E), northeastern Kyushu, and in Taiji in the early nineteenth century (Oyamada 1832, Hashiura 1969, both in Japanese) did not use harpoons with detachable heads. Drawings of harpoons called *haya-mori* showed a fixed head, that is, the harpoon was more like the *denchu-mori*. Net whaling started around 1775–1777 (Chapter 1). The whales were entangled in nets before being harpooned. This probably made it unnecessary to connect the boat to the whale at a distance with a lightweight harpoon with detachable head.

Matsuura (1942, in Japanese) described the history of the Japanese hand-harpoon fishery in the nineteenth to early twentieth century. In the middle of the nineteenth century, the fishery was operated in the Awa Region in the southern part of Chiba Prefecture, for swordfish. Later, with the introduction of motor-driven vessels, Wakayama and Oita Prefectures (32°45′N–33°37′N, 131°11′E–132°06′E, northeastern Kyushu facing the Inland Sea) joined the fishery. Sometime later, H. Iwasa of Mie Prefecture (33°45′N–35°00′N, 136°00′E–136°55′E, Pacific coast of central Honshu) invited a technical advisor from Chiba (presumably the Awa region: TK) and started a hand-harpoon fishery for dolphins and porpoises. He also stated that around 1923 a group of fishermen led by K. Koyama of Miyagi Prefecture started hunting dolphins and porpoises using both hand harpoons and shotguns, which was copied by many followers. However, most of these fishermen took many species, including small cetaceans, fur seals, seals, tunas, and swordfish (Matsuura 1942, in Japanese).

Otsuchi-cho Gyogyo-shi (*History of Fisheries of Otsuchi Town*) by Otsuchi-cho Gyogyo-shi Hensan Iinkai (Editorial Committee for History of Fisheries of Otsuchi Town) (1983, in Japanese) noted that fishermen in Otsuchi (39°21′N, 141°54′E), Iwate Prefecture, made a great contribution to the establishment of hand-harpoon fisheries for dolphins and porpoises. They learned the method from hand-harpoon fishermen from Chiba Prefecture (presumably the Awa region: TK) who started to visit the Otsuchi Area around 1921 for hunting of dolphins, porpoises, and swordfish. However, the book states on another page that fishermen in Kirikiri, a village next to Otsuchi, invited a technical advisor from Chiba Prefecture and started dolphin and porpoise hunting around 1917 and that a similar attempt was made by fishermen in Hakozaki, another village next to Otsuchi. These records suggest that fishermen of Iwate Prefecture on the Pacific coast of northern Honshu established the hand-harpoon fishery for dolphins and porpoises around 1920.

In the 1920s, they used a detachable harpoon head on a wooden shaft 2.4–3.0 cm in diameter and 3.6–4.5 m long. The *History of Fisheries of Otsuchi Town* states that E. Shozushima introduced the use of a shotgun to the hand-harpoon fishery around 1932–1933 and greatly increased the efficiency of the fishery and that it was also around this time that a harpooner's platform was installed on the bow of the harpoon vessel, which had an increased speed of over 10 knots. Hirasima and Ono (1944, in Japanese) also stated that the use of a shotgun began around 1924 (*sic*) by E. Shozushima of Iwate Prefecture. These improvements were followed by expansion of the fishing range from Chiba in the south to the Kuril Islands and Sakhalin in the north. In 1933, 90 fishing families out of 128 in the Akahama Area of Otsuchi Town operated the hand-harpoon fishery. It seems to be true that the use of a shotgun followed by harpooning improved the efficiency of hunting, but it is also true that such operations have had large hit-but-lost rates (see Section 2.3).

Compared with rowed boats, motor-driven vessels increased the efficiency of hunting in two ways. One benefit of the latter was the increase in the operation range associated with faster speed. The other was in the response of dolphins and porpoises coming to the bow wave where they could be harpooned. The *History of Fisheries of Otsuchi Town* recorded that the governor of Iwate Prefecture promoted the use of engines in the fishery by inviting fishermen to an education class about the machines in 1912 and that 14 fishing vessels in Otsuchi Town were equipped with engines in the next year. The number of motor-driven fishing vessels in Iwate Prefecture increased rapidly in the early 1910s. This date was, at least in Iwate Prefecture, earlier than that in the conclusion by Kasuya (1978) that establishment of the hand-harpoon fishery for Dall's porpoises could have been in the 1920s when the number of motor-driven vessels increased rapidly in Japanese costal fisheries. When World War II ended in 1945, there were 26 hand-harpoon vessels in the Akahama Area, where most of the hand-harpoon fishermen of Otushi Town lived (Committee for History of Fisheries of Otsuchi Town 1983, in Japanese).

Dainihon Suisan Kai (Fishery Association of Great Japan) promoted Japanese fisheries and has published a journal since 1882 for the development of fisheries and introduction of new technology, including new technology for marine mammal fisheries and utilization of the catch. All the cetacean-related articles in the journal are available in "Hogeisen," a private publication by K. Takeuchi. There were numerous articles published in the journal both on whaling and fisheries for dolphins and porpoises before 1906, but there were only whaling articles after that date. This indicates that the Japanese fishery community lost interest in dolphin and porpoise fisheries around that time. Modern whaling was first attempted in Japan by *Enyo Hogei* (Offshore Whaling) Co. in 1898, and *Nihon Enyo-gyogyo* (Japan Offshore Fishery) Co. established in the next year succeeded in the modern whaling enterprise. This was followed by the establishment of too many whaling companies, and the industry and government made an effort to merge 12 of them into 1, *Toyo Hogei* (Oriental Whaling) Co. in 1909, which was the first step in the government action to reduce whaling vessels that continued after World War II (Chapter 7). The establishment of the modern whaling industry was possibly the major reason why Japanese fishing communities lost interest in small-cetacean fisheries, which suffered by comparison to whaling in producing meat or oil. Most of the small-cetacean fisheries probably lost public attention and developed into local fisheries for the subsequent period of nearly 20 years.

Dolphin and porpoise fisheries again received public attention when Japan started a move toward World War II (also called the Pacific War). Japan established a puppet government in Manchuria in northeastern China in 1932, withdrew from the League of Nations in 1933, and started war with China in July 1937. Japan called the conflict an "incident" rather than war, because the government was afraid of a ban on trade of military materials from the United States. However, Japan finally started war with the United States in December 1941, which almost completely stopped importation of military materials greatly needed for the war.

To prepare for this situation, the Japanese government established the "Tosei Kyoku" (Control Bureau [of Materials]) in 1937 and decided in July 1938 to allow use of leather only for military purposes. The prohibition on leather for public use included shark and whale leathers (Kamiyama 1943, in Japanese). Responding to this situation, the whaling companies in 1938 established a company aimed at production of whale leather. Then in January 1939, the government organized one guild by uniting the then existing 6 companies producing marine leathers (i.e., shark, dolphin, and whale leathers), and an additional 13 companies were added to this guild in June of the same year. Japan carried out the last prewar North Pacific pelagic whaling in the summer of 1941 and the last Antarctic whaling in the 1940/1941 season. Thus, the sources for whale skin became limited to coastal whaling, which was becoming increasingly difficult with U.S. submarines and air raids. Under these circumstances, the demand for dolphin and porpoise skin probably increased.

The Japanese Antarctic whaling fleets produced oil for export to Europe, and importation of whale meat was prohibited until the 1936/1937 season to protect domestic industries (Kasuya 2000, in Japanese). Thus, it seems that the Japanese market for whaling products was almost fully met with coastal whaling until around 1937. However, importation of Antarctic whale meat was permitted with a maximum limit in the 1937/1938 season, and it was unregulated after the 1938/1939 season. The military was the main consumer of whale meat. Yoshida (1939, in Japanese) stated that the governor of Mie Prefecture canceled the then existing prohibition of dolphin hunting in offshore waters. Thus, it seems to be reasonable to consider that dolphin and porpoise fishing received increased attention not only as a source of leather but also as a source of food and was promoted by the government.

Yoshida (1939, in Japanese) wrote an article with a title that translates as *Recent Rise of Offshore Dolphin and Porpoise Fishery with Hand Harpoons* and reported a case where Taiheiyo Gyogyo (Pacific Fishery) Co. Ltd. based at Kesen-numa (38°54′N, 141°34′E) in Miyagi Prefecture on the Pacific coast of northern Honshu started a small-cetacean fishery in October 1938 using the mother ship *Tenjin-maru* (200 gross tons), two survey ships (each about 70 gross tons), and two chartered local hand-harpoon vessels previously used for swordfish (each about 15–19 gross tons). The fleet used hand harpoons purchased from Awa in the southern part of Chiba Prefecture. The fleet took 436 dolphins and porpoises by November 15, 1938. Operation of this fleet was also described in Matsuura (1942, in Japanese), which will be mentioned later. The objective of this company was to obtain materials for dolphin leather, selling the salted skins to a leather factory. The meat was salted or dried for human consumption but was not welcomed by the market.

Yoshida (1939, in Japanese) recorded the landing of dolphins and porpoises at some selected ports in 1935–1937, which was before the operation of the earlier mentioned Taiheiyo Gyogyo Co. Ltd. The landings at Kesen-numa were 69,623 kg in 1935, 98,775 kg in 1936, and 60,735 kg in 1937. The average of the 3 years was 76,376 kg. If we use his average figure of 56 kg/animal, the annual average landings were of 1360 individuals. These were taken by tuna longline fishermen on the way to the fishing grounds and sold for human consumption as well as bait for a longline shark fishery. In addition to these, the author recorded a total of 29,685 kg (i.e., estimated 530 individuals) landed at Miyako (39°38′N, 141°57′E), Kamaishi (39°16′N, 141°53′E), and Ofunato (39°05′N, 141°43′E) on the Pacific coast of northern Honshu during January to March 1937, which could have been taken by local fishermen operating in nearshore waters in winter. Combining the two yields an annual take of about 1900 individuals. This figure covers landings only at four selected ports in Iwate Prefecture. We do not know the number of ports where small cetaceans were landed in those days in the Sanriku Region (Pacific coast of northern Honshu from 37°54′N to 41°35′N), including the three prefectures of Miyagi, Iwate, and Aomori (40°27′N–41°30′N, 139°42′E–141°40′E). Kasuya (1982) recorded that the landing ports in the region varied with time, and the number was between 6 and 7 ports in 1963–1968, 7 ports in 1969–1971, and 8 ports in 1972–1975. During this 13-year period, the total winter landings varied annually between 4600 and 8000 individuals. I would suspect that the total landings of dolphins and porpoises in the Sanriku Region prior to the expansion around the wartime period could have been 1.5–2 times that figure (around 3000 or at most 4000 individuals).

In response to the rapid expansion of small-cetacean fisheries in Japan, Matsuura (1942, in Japanese) wrote a long article titled *Iruka-no Hanashi* (*Story about Dolphins and Porpoises*) in a Japanese journal. He was a cetacean biologist and in a position to be up on new information on Japanese fisheries as a staff person of the Whaling Section of the Fisheries Agency. In the article, he described the small-cetacean fauna off Japan, the current status of small-cetacean fisheries in Japan, and the future of such fisheries. He explained that hunting of small cetaceans using a small catcher boat and small-caliber harpoon cannon (small-type whaling of today) was regulated only in Chiba Prefecture (Chapter 4), that drive fisheries for dolphins and porpoises were licensed by the Fishery Act for opportunistic operations of particular local communities in waters facing them (Chapter 3), and that other types of hunting of dolphins and porpoises were unregulated (e.g., hand-harpoon fishery). Then, he reported that in response to the recent increase of the price of meat and needs for leather, catches of small cetaceans saw a great increase, that is, a total catch of 1,000 large delphinids such as pilot whales and 45,000 smaller dolphins and porpoises sold for a total of 2,500,000 yen in 1941 in all of Japan, and that the fishery in Shizuoka Prefecture (which is almost equal in size to the fishery on the Izu Peninsula) recorded an extremely rare catch of several tens of thousands in the spring hunt in 1942 (Chapter 3). The total small-cetacean catch of this period was comparable to the level we later experienced in the late 1980s, although the species composition was much different.

Matsuura (1942, in Japanese) further stated that the Taiheiyo Gyogyo Co. Ltd. using the *Tenjin-maru* fleet made the first attempt at a dolphin and porpoise fishery covering nearly 12 months from November 2, 1938, to December 7, 1939, but discontinued it due to economic failure. It was before the wartime boom recorded in 1941. The *Tenjin-maru* fleet used a total of nine vessels, including four hand-harpoon boats (10–19 gross tons) that also used shotguns, two small-type whalers of the present-day definition (70–71 gross tons), and three transporters (40–200 gross tons). All these vessels did not work at the same time, but usually one transporter and 1–5 catcher boats worked together. The operation ranged from Cape Muroto (33°15′N, 134°10′E) on the Pacific coast to Iturup Island (45°N, southern Kuril Islands), with the main operation areas off Taiji (Wakayama Prefecture), Kesen-numa (Miyagi Prefecture), and Abashiri (44°01′N, 144°16′E) on the Okhotsk coast of Hokkaido, with a short segment off the Sea of Japan coast of Hokkaido.

According to *Iruka Gyogyo Jisseki* (*Results of the Small Cetacean Fishing*) by Taiheiyo Gyogyo Co. Ltd. (unpublished, dated 1940, in Japanese), the hunting equipment used by the fleet included hand harpoons, 1 rifle, 11 shotguns, 3 Greener guns, 1 shoulder gun for a harpoon with line, 1 Norwegian-type whaling cannon (55 mm), 1 triple-barreled whaling cannon, and 1 five-barreled whaling cannon. The company chartered the vessels with whaling cannon (small-type whaling vessels of the present day). A hand harpoon was used after a shotgun, as was usual at the time. The catch composition is in Table 2.1. The record does not distinguish between the two color types of Dall's porpoise, most of which were taken in March to September in waters within 20–30 nautical miles from the Sanriku coast and at sea surface temperature below 19°C. They were stated to be infrequent in offshore waters. The large number of takes of *ma-iruka*, believed to be the short-beaked common dolphin, *Delphinus delphis*, was stated to have been made mostly from September to December at surface water temperature of 12°C–19°C found 50–100 nautical miles off the Sanriku coast. It was stated that the *ma-iruka* was found in warmer and more offshore waters than the Dall's porpoise. The offshore operation may explain the higher proportion of common dolphins in the catch. I did not encounter such distant operations when I worked on the hand-harpoon fishery for Dall's porpoises in the 1970s.

The skin (with blubber) was salted and sold to a leather factory. The meat was sold fresh, salted, or dried. Oil was tried out from the head and mandible but was not sold at the time of the report. The company had total sales of 44,618 yen (Matsuura 1942 reported a total sale of 55,855 yen, which possibly includes later sales) against total expenditures of

TABLE 2.1
Takes in Small-Cetacean Fishery by Taiheiyo Gyogyo Kaisha (Pacific Fisheries) from November 2, 1938 to December 7, 1939

Species		No. Taken
Cetacea	Common dolphin	2503
	Striped dolphin	290
	Dall's porpoise	1163
	Pacific white-sided dolphin	32
	Northern right whale dolphin	98
	Bottlenose dolphin	29
	Finless porpoise	2
	Blackfish	67
	Total	4184
Others	Marlins	280
	Tunas	30
	Sharks	127

Source: Prepared from text and a table on pages 100–101 in Matsuura, Y., *Kaiyo Gyogyo*, 71, 53, 1942, in Japanese.

187,096 yen. Although the operation was an economic failure, the information obtained helped the Iwate hunters operate the fishery off the Hokkaido coasts in the summer of later years.

Matsuura (1942, in Japanese) recorded the magnitude of the hand-harpoon fishery around 1941. There were 267 hand-harpoon vessels that used only hand harpoons (no shotguns). The boats were 5–6 gross tons, used for family business, and did not specialize in dolphin and porpoise hunting; they sought other species depending on season or opportunity of encounter. They were grouped into two regions:

1. 250 vessels along the coast of the southern Sea of Japan: 246 in Hyogo Prefecture (134°20′E–135°50′E) and 4 in Tottori Prefecture (133°20′E–134°20′E).
2. 17 vessels on the Pacific coast of central Honshu: 7 in Mie Prefecture, 6 in Wakayama Prefecture, 3 in Shizuoka Prefecture, and 1 in Kanagawa Prefecture.

These figures for Hyogo Prefecture agree with Matsui and Uchihashi (1943, in Japanese) and Noguchi (1943, in Japanese), who reported a new hand-harpoon fishery for dolphins and porpoises that operated from March to June off the coast of their prefecture.

In addition to what was given earlier, there were 87 vessels that were equipped with hand harpoons and shotguns and considered by Matsuura (1942, in Japanese) to be more specialized for hunting small cetaceans (Table 2.2). There were 48 such vessels in Iwate Prefecture, and operational details were recorded for 24 of them by the same author. I will try using the information to estimate the total catch by the fishery of the time. Out of the 24 Iwate vessels, 21 belonged to the Kirikiri Area of Otsuchi Town and operated from December 1941 to May 1942 in waters from Choshi (35°42′N, 140°51′E) to Miyako on the Pacific coast of northern Honshu (i.e., in the area and season when Dall's porpoises were expected to predominate) and got an average of 338 animals/vessel for the season (the maximum was 700) with an average of 56 animals/vessel/month. These figures allow one to calculate the total catch of the 21 vessels at 7098 animals. In addition to these 21 vessels, another 3 boats operated longer (about 9 months from December 1941 to August 1942) in waters extending from Choshi to the Okhotsk Sea, including all the waters along the Pacific coast of northern Japan, and recorded an average take of 1160 animals/vessel (the maximum was 1500) and monthly average of 128 animals/vessel. Thus, the total take of the three vessels could have been 3480 animals. The 24 Iwate vessels with statistics took 7,098 + 3,480 = 10,578 animals that were sold at 479,700 yen. The three vessels that expanded operation to the Pacific and Okhotsk Sea coasts of Hokkaido got an over three times greater monthly average compared with operations off the Pacific coast of northern Honshu. One reason was that the northern operation was in summer when weather and sea conditions were better than during the winter off the Pacific coast of Honshu, and another reason would have been that there were present *dalli*-type Dall's porpoises that wintered in the Sea of Japan and were possibly less depleted.

One way of estimating total catch by the 87 vessels in Table 2.2 is to extrapolate catch of the 21 Iwate vessels operating in the winter (338 animals/vessel/season) to 63 vessels of similar equipment but not accompanied by operation data. This calculation of 10,578 + 338 × 63 yields 31,872 animals for the 1941/1942 season. Another possible calculation is to assume that the proportion of vessels of longer operation was the same for the whole fleet (i.e., 3:21). This assumption leads to an estimate of the total catch by the 87 vessels of (1,160 × 3 + 338 × 21) × 87/24 = 38,345 animals for the 1941/1942 season. The latter figure will be an overestimate of the catch, because the Iwate vessels were known to have pursued the fishery most vigorously. I suspect that the former figure, about 32,000 animals, would be closer to the number of small cetaceans taken by the 87 vessels using hand harpoons and shotguns in waters from Chiba to Hokkaido in the 1941/1942 season. This figure does not include those taken by over 260 hand-harpoon vessels that did not specialize in marine mammals.

Most of the earlier estimated 32,000 individuals would have been Dall's porpoises, with some warmwater species such as common dolphins, if they were taken in the fall season in offshore waters, which is less likely. In addition to this, the drive fishery along the coast of the Izu Peninsula recorded a catch close to 10,000 (Table 3.14). The sum of these figures results in a figure that is close to the 45,000 stated by Matsuura (1942, in Japanese) as the total catch of small cetaceans in Japan in 1941. The Izu fishery recorded a peak catch of about 20,100 in 1942 and declined thereafter, presumably due to the difficulty of operating in the war environment (Chapter 3).

TABLE 2.2
Magnitude of Small-Cetacean Fisheries Using Hand Harpoon and Shotgun in 1941

	In Operation			In Preparation
	No. of Operating Bodies and the Targets			
Prefecture[a]	Small Cetaceans	Fish and Cetaceans	Total No. of Vessels	No. of Operating Bodies
Hokkaido	0	8[b]	8	2
Iwate	40	6	48	2
Miyagi	14	Unknown	16	Unknown
Chiba[c]	0	10	12	Unknown
Ibaraki	0	3	3	0
Kochi	0	0	0	3
Total	54	27+	87	7+

Source: Reproduced from a table on page 95 in Matsuura, Y., *Kaiyo Gyogyo*, 71, 53, 1942, in Japanese.

[a] Although these prefectures are on the Pacific coast, operation of hand-harpoon fishery also known for that time off the southern coast of the Sea of Japan (see Sections 2.2 and 9.3.1).

[b] Main targets were tunas and marlins.

[c] Includes only those based in Choshi Port.

Hirashima and Ono (1944, in Japanese) stated about the earlier mentioned Taiheiyo Gyogyo Co. Ltd. that it carried out a small-cetacean fishery in 1937 (the correct date was 1938/1939: TK) using the *Kinyo-maru* and three other Iwate vessels and found a good fishing ground off the Pacific coast of Hokkaido, off the Shiretoko Peninsula (northeastern part of Hokkaido between the southeastern Okhotsk Sea and Nemuro Strait), and in the Okhotsk Sea and that this triggered annual operation of the fishery by Iwate vessels using Abashiri on the Okhotsk Sea coast of Hokkaido as a base port. In 1943, they operated off Abashiri from late June to late September using vessels of 20 tons equipped with 80 hp engines and landed and processed the catch at Abashiri. The meat was sold at 57 yen/animal or 0.97 yen/kg, and the blubber at 32 yen/animal or 1.31 yen/kg. The total was about 90 yen/animal. During the 1943 season mentioned earlier, they fished on 62 days, resulting in 351 vessel days, and caught a total of xxxx (*sic*) animals (the number was expressed in this way presumably to hide economic information during the war). The maximum catch of 26.4 animals/vessel/day was recorded on August 22. This daily catch was only about 10% of the figure reported for the post–World War II operation off Iwate Prefecture reported by Wilke et al. (1953).

Using monthly figures of operation days and vessel days recorded by Hirashima and Ono (1944, in Japanese), the monthly averages of fishing vessels are calculated at 4.0 in June, 4.4 in July, 7.6 in August, and 5.9 in September. This assumes that all the vessels have operated on days of suitable weather and underestimates the real number of vessels engaged in the fishery. However, the calculation seems to suggest that the number of Japanese hand-harpoon vessels for small cetaceans that operated in the Okhotsk Sea in 1943 was less than 10 even in the peak month. This is greater than the three vessels recorded for the 1941/1942 season (Matsuura 1942, in Japanese) but smaller than the recent magnitude of the fishery. The upper range of the annual catch in the Okhotsk Sea operation is estimated by multiplying the maximum figure of 26.4 animals/vessel/day by 351 vessel days of the season, which results in 9266 animals. I suspect that the real catch of the season was probably about half that figure, that is, around 5000 animals. During the operation, fishermen first shot the surfacing animal and then retrieved it with a hand harpoon. If we accept the statement of Wilke et al. (1953) that one-third of the carcasses were lost before retrieval, the real number of dolphins and porpoises killed by the fishery could have been 1.5 times the given figure.

A new hand-harpoon fishery for small cetaceans was recorded in Hyogo Prefecture in the southern Sea of Japan (Matsui and Uchihashi 1943, in Japanese). It started in 1940 following an effort by the Fishery Experimental Laboratory of Hyogo Prefecture to improve the utilization of small cetaceans. The blubber was used for leather after extracting oil by pressing or with a solvent, meat for human consumption, and bone for fertilizer. The extracted oil was reportedly to be made into a lubricant and a substitute for gasoline (Noguchi 1943, in Japanese). The third volume of *Hogei-benran* (*Whaling Handbook*) by Nihon Hogei-gyo Suisan Kumiai (Japanese Association of Whalers) (1943, in Japanese) lists edible oil, fuel oil, lubricant oil, and glycerin for dynamite as the whale products of the time. Although I am unable to confirm a gasoline substitute created from cetacean oil, it could have been mixed with diesel oil as practiced by Japanese Antarctic whaling fleets in the 1980s. Although Noguchi (1943, in Japanese) wrote that the fishery took Dall's porpoises and common dolphins, the accompanying photograph shows a Pacific white-sided dolphin, which was supported by the statistics of Noguchi (1946, in Japanese).

The dockside price of small cetaceans increased in Hyogo Prefecture from 8–9 yen/animal in 1940 to 35 yen in 1941 and 70–80 yen in 1942, which invited a rapid increase of the catch (Matsui and Uchihashi 1943, in Japanese). Noguchi (1946, in Japanese) reported monthly statistics of small cetaceans landed at Kinosaki Port (35°39′N, 134°50′E), Hyogo Prefecture. The annual total was 211 animals in 1941, 785 in 1942, 1622 in 1943, and 636 in 1944. The decline in 1944 might reflect increased difficulty in operations due to increased enemy activities. The total of 2998 animals was composed of 2731 Dall's porpoises (91%) and 267 Pacific white-sided dolphins (9%). During the early part of the season (middle February to early May), the catch of Dall's porpoises exceeded that of Pacific white-sided dolphins (1841:20), the two species were taken in almost equal numbers in mid-May (18:15), and only Pacific white-sided dolphins were taken in late May to late June (0:14).

Statistics on small cetaceans are scarce for the period near the end of the war. Censorship might have prevented publication of statistics, but there are cases where authors voluntarily omitted such information from their articles. Matsuura (1943, in Japanese) stated that due to the increased demand for meat and leather materials, catches of small cetaceans by hand harpoon and so-called small-type whaling increased greatly in 1941, to 1,000 large delphinids and 45,000 small delphinids, and that large portions of the small-delphinid catch were obtained from drive fisheries on the Izu coast in Shizuoka Prefecture. Noguchi (1946, in Japanese) reported a similar figure for 1941 and further stated that the fishery in Shizuoka Prefecture (which is almost identical with the Izu fishery) recorded landing of about 28,000 animals, of which about 20,000 were from Arari (34°50′N, 138°46′E), 4,000 from Tago (34°48′N, 138°46′E), and 4,000 from Heda (34°58′N, 138°47′E). Both Matsuura (1943, in Japanese) and Noguchi (1946, in Japanese) expressed their concern about depletion of dolphin stocks by the Izu fishery.

Leather was made from the external dense layers of cetacean skin left after removing the inner loose layer of high fat content. Fat also has to be removed from the external layers, a laborious process. Further details of the process are in Doi (1902, in Japanese) and Ishida (1917, in Japanese). Kurakami (1925, in Japanese) stated an optimistic view of the early time that dolphin and porpoise skin could be used for shoes, but there was another view that among cetacean leathers those made from the head skin of sperm whales was the best (Omura et al. 1942; Matsuura 1943, both in Japanese). Maeda and Teraoka (1958, in Japanese) stated that leather made from the head skin of sperm whales was of high quality and used since 1938 for shoe soles, personal, and industrial belts and that other cetacean leathers were less common due to the high cost of production. Wilke et al. (1953) recorded the use of dolphin and porpoise skin for low-quality leather in the post–World War II period in Japan. Leather made from the skin of small cetaceans was of inferior quality.

This should not be confused with beluga leather, which was highly valued for durability and water resistance and once used for shoes and bags (Goode 1884). Production of cetacean leather probably ended in Japan sometime after recovery from the postwar economy. In the 1980s, I could find products of cetacean leather sold only in a gift shop in Ayukawa (38°18′N, 141°31′E), a whaling town in Miyagi Prefecture. They were of novelty value only, such as purses made of the penis skin of sperm whales or ventral grooves obtained from fetuses of *Balaenoptera* sp.

2.3 HAND-HARPOON FISHERY FOR SMALL CETACEANS (2): AFTER WORLD WAR II

After the end of World War II in August 1945, Japan suffered from severe shortage of food and high inflation, and the government encouraged various fishery activities, including whaling and small-cetacean fisheries. In those days, drive fisheries were licensed by the Fishery Act of the time to particular communities having the tradition of monopolizing particular nearshore waters for the purpose (Section 2.2 and Chapter 3), but hand-harpoon fisheries for marine animals including small cetaceans were left unregulated and operated everywhere off the Japanese coasts (Matsuura 1943, in Japanese; Wilke et al. 1953).

The postwar operation of hand-harpoon fisheries for small cetaceans was recorded by Wilke et al. (1953) and Noguchi (1946, in Japanese), but the latter has some ambiguity in dates dealt with. The former was authored by the United States and Japanese scientists who jointly worked on a fur seal research vessel and will be cited in this section. The hand-harpoon fishing of the time operated in the Pacific waters off northern Honshu and Hokkaido and in the southern Okhotsk Sea coast of Hokkaido. Hand-harpoon fishermen started the season in March off Choshi, Chiba Prefecture, reached Hosoura (38°40′N, 141°30′E), Miyagi Prefecture, by the middle of March, and operated off Iwate Prefecture (38°59′N–40°27′N) until June. From June to summer, the fishing ground moved to the coast of Hokkaido (Pacific and Okhotsk Sea coasts). In autumn, the fishermen resumed operation off the Pacific coast of Ibaraki (35°45′N–36°50′N) and Fukushima (36°50′N–37°52′N) Prefectures. They operated within 30 nautical miles of the coast but mostly within 10–15 nautical miles. This pattern differs slightly from the current pattern of operation for the Dall's porpoise fishery, which operates in early summer off the west coast of Hokkaido (northeastern Sea of Japan), presumably to avoid foggy seas off eastern Hokkaido. Wilke et al. (1953) did not mention the winter operation off Iwate Prefecture for Dall's porpoises, but Uchida (1954, in Japanese) reported that a good season for Dall's porpoises started in February off Miyako in Iwate Prefecture and moved south with the progress of the seasons to off Choshi.

Uchida (1954, in Japanese) stated that the fishing grounds for Pacific white-sided dolphins, right whale dolphins, and common dolphins differed from those for Dall's porpoises. Because it was known from offshore surveys of the North Pacific that the first two species were distributed in latitudes between the striped dolphin and the Dall's porpoise (Miyashita 1993), the fishing ground for these two species was to the south of the ground for Dall's porpoises. The common dolphins were reported to be found further offshore (Section 2.2).

Wilke et al. (1953) cited statements of fishermen that the demand for small-cetacean meat was far greater than the supply and that the number of operating vessels increased annually and in 1949 reached the level of prewar operations. Because the per-vessel catch exceeded the prewar level, the total Japanese take of small cetaceans in those days could have exceeded 46,000 animals, that is, the total of all species taken by various fisheries in 1941 (Matsuura 1943, in Japanese). The Japanese government placed all fishery products under a rationing system until 1949 and set the price of dolphins and porpoises at 2000–3000 yen/animal, but Wilke et al. (1953) stated that most of the catch entered the black market, where it got 2.5 times the government price. Thus, it is impossible to expect correct fishery statistics for the time. The statistics for 1941 given by Matsuura (1943, in Japanese) were also obtained while the government controlled the circulation and would not have included the fraction that entered the underground market. However, the government would have controlled the circulation better during the war because of higher social cooperation, so I expect the statistics would be less biased from that period. Wilke et al. (1953) stated that all fishery products including small cetaceans were removed from the rationing system in 1949 and that the prices dropped. Decline in price and free circulation could have followed improvement of the balance between demand and supply.

Wilke et al. (1953) reported the species composition of small cetaceans landed at Onahama (36°57′N, 140°54′E) on the Pacific coast of Fukushima Prefecture from May to June 1949 as 465 right whale dolphins, 697 Pacific white-sided dolphins, and 1 *truei*-type Dall's porpoise, for a total of 1163. In 1966, I purchased a right whale dolphin in Ayukawa on the Pacific coast of Miyagi Prefecture. It had been in a freezer for a long period unsold because flavor of the species was believed to be inferior to that of the Dall's porpoise. However, even such nonpreferred species were hunted shortly after the war. The use of small cetaceans taken by the hand-harpoon fishery in around 1949 was as follows. The skin and blubber were used mostly for extracting oil and some for low-quality leather. Melon and mandibles were for extraction of high-quality machine oil. Meat, heart, liver, and kidney were for human consumption, and the remaining portion of the viscera and the bones were used as fertilizer. However, the viscera were discarded in the sea before returning to port in northern Japan (Wilke et al. 1953). When I investigated the Dall's porpoise fisheries in Iwate Prefecture in the 1970s, the blubber was not used for leather but sold together with meat for human consumption, which was also the way the animal was consumed in Shizuoka Prefecture and surrounding regions.

Wilke et al. (1953) recorded the use of shotguns in the hand-harpoon fishery. Some of the vessels caught as many as 200 animals in a day, but one-third of the animals shot with a gun sank and were lost before retrieval with a hand harpoon. This means that their operation was wasteful; 1.5 times as many as landed were killed. Hand-harpoon boats of the time had a platform on the bow for multiple harpooners, were about 20–30 gross tons, were equipped with hot-bulb engine (also called semidiesel engine, which had a heated hollow bulb on the cylinder head for ignition), and had a speed of 7–10 knots. Each vessel had 10–12 crewmen, 2–3 shotguns, and 20–30 hand harpoons. However, Matsuura (1943, in Japanese) and Noguchi (1946, in Japanese) recorded cases of smaller vessels of 7–10 gross tons with 2–3 crewmen.

Some Dall's porpoises will come to the bow of a harpoon boat and play in the bow wave, but others will not approach the vessel (Kasuya 1978). It is not difficult, even without a shotgun, to harpoon individuals playing in the bow wave. There are more young animals among those that approach the bow, while adult individuals tend to be at a distance from the vessel. Mothers accompanied by their calves avoid the vessel and are hard to approach by chasing (Kasuya and Jones 1984). Porpoise hunters of the time could have used shotguns on such individuals that stayed away from the boat and put a line on them before they sank (Wilke et al. 1953).

Shotguns were also used by fur seal hunters in the Sanriku Region on the Pacific coast of northern Honshu. The 1911 agreement by four countries (Japan, the United Kingdom for Canada, the United States, and USSR) prohibited pelagic hunting of fur seals because it killed both sexes and was thought to be harmful for the recovery of the population. However, in 1940 Japan canceled the agreement and allowed pelagic hunting by Japanese fishermen, presumably as one of the preparation measures for the coming war. After the war, the Japanese government prohibited the operation, accepting an order by the General Headquarter of the Allied Nations, and this policy was retained after the peace treaty that came into effect in April 1952. However, many fishermen of the Sanriku Region still owned shotguns with the putative purpose of hunting small cetaceans but were believed to actually have hunted fur seals.

Following the new treaty for the protection of the northern fur seal, which was signed in February 1957 and came to effect in October of the same year (Chapter 5), the Japanese government purchased the guns from these fishermen and prohibited the use of guns for hunting of dolphins and porpoises (Ohsumi 1972). The prohibition of guns for small cetaceans, which came to effect in September 1959, was limited to waters north of 36°N and for the period from February 20 to June 20 (Chapter 5).

Uchida (1954, in Japanese) described details of the gear of the hand-harpoon fishery of the time used off the Sanriku Region. Each vessel had 2 or 3 shotguns of 10 or 12 caliber, but 8 caliber guns and larger balls were preferred. Length of the barrel was 32–50 in.; the longer barrel was preferred. The harpoon consisted of wooden shaft, metal foreshaft, detachable harpoon head, line, and buoy. Each vessel had over 30 harpoon shafts made of poles 30–36 mm in diameter and 3.6–4.2 m in length and equipped with a metal foreshaft. The shaft was made of evergreen oak. A foreshaft with one or two prongs was used for hunting dolphins and porpoises. A foreshaft with three prongs was used for swordfish. For the multiple prongs, short lines tied to each harpoon head were united into a single line. Each prong of the foreshaft was inserted into the base of the harpoon head. Length of the line depended on target species and water depth, for example, 30 fathoms (off Abashiri) or 50 fathoms (off Sanriku) for Dall's porpoises, or 100 fathoms for common dolphins and Pacific white-sided dolphins off Sanriku. The gunners stood on the platform projecting from the bow of the vessel and shot the animal just before it dived. The harpooners stood behind the gunners on the platform and threw the harpoon to secure the animal before it sank.

In the 1970s and 1980s when I studied the fishery, oak was no longer in use for the shaft; cheaper imported luan replaced it. A gun was not openly used, although some fishermen might have used a shotgun in secret. I could identify bullet holes on the bodies of some porpoises landed. Length of the wooden shaft depended on vessel size. The shaft should be of sufficient length for the harpoon head to reach the sea surface when the harpooner reached down. One end of the wooden shaft was inserted into a socket at the base of the foreshaft and secured with a nail, and the tip of the foreshaft was inserted into the socket of the harpoon head together with a piece of soft wood to help firm attachment. A harpoon head was made of a shaft, a pair of barbs extending posteriorly from the tip of the shaft, and another barb extending from the base of the shaft. The two sets of barbs were placed perpendicular to each other (Figure 2.1). A thin wire was attached to the harpoon head through a hole at the center of the shaft and connected to a line. The line was connected to the wooden shaft and then to a buoy.

Once a harpoon hit a porpoise or dolphin, the fishermen threw line and buoy overboard and chased another animal in the school. If conditions permitted, they attempted to chase windward while harpooning and then move downwind for retrieving the catches. In the 1980s, an electric shocker of 50 V was commonly used between the harpoon strike and throwing the buoy to immobilize the target. The electric shocker often caused tetanus and damaged the quality of the meat on part of the body (Chapter 9).

2.4 EXPANSION OF DALL'S PORPOISE FISHERY TOWARD THE 1980s

From the late 1960s to the 1980s, I visited fishing ports to examine small cetaceans taken by various fisheries, to collect catch statistics, and to gain opportunities to go out on the vessels. In those days, almost every small fishing boat had on board several hand harpoons as a contingency. If they found sunfish, dolphins, or porpoises, they would harpoon them to add some variety to their menu on the ship. If the crew were too small to consume a whole dolphin or porpoise, only internal organs would be consumed and the rest discarded, which I observed on a small-type whaling boat off Abashiri, Okhotsk Sea coast of Hokkaido, that took Dall's porpoises for food.

Occasional use of hand harpoons for hunting marine animals was also seen on research vessels, for example, the *Tansei-maru*, a research vessel of the Ocean Research Institute, University of Tokyo, and the *Hoyo-maru No. 12* a fishing vessel chartered by the Fisheries Agency. I was on board the latter vessel chartered in 1982 for the purpose of observation and hand-harpoon sampling of Dall's porpoises in the offshore North Pacific south of the western Aleutian Islands (Chapter 9). Our breakfast on the vessel was only steamed rice, salted vegetables, and miso soup in a kettle, which never left the cooking stove during the 1-month cruise (water and materials were added daily), and the other meals of the day were also of minimum quality and constant in content. In order to get some fresh food, we brought to the table a dish of raw meat and blubber from Dall's porpoises hunted for our research purpose. Such noncommercial marine mammal consumption was common in Japanese coastal fisheries but not recorded by fish markets at the landing ports and escaped the fishery statistics. Wholesale fish markets in the fishing ports were usually operated by the fishery cooperative union (FCU) of the region, and in other cases, landing records of the fish markets and also payment by agreement were forwarded to the FCU to which each fisherman belonged. FCUs compiled the record into base data for the fishery statistics, put the payment into the account of each fisherman, and collected a levy.

When social disorder after World War II ended, commercial operation of Japanese hand-harpoon fisheries for dolphins and porpoises became limited to particular regions. The largest was operated for Dall's porpoises by fishermen in Iwate, Miyagi, and Hokkaido, which was followed by hand-harpoon fisheries for other species in Choshi in Chiba Prefecture and Taiji in Wakayama Prefecture, both on the Pacific coast of central Honshu, and in Nago in Okinawa Prefecture in southwestern Japan. Table 1.10 shows numbers of dolphins and porpoises for each prefecture and fishery type for selected years, and Tables 2.3 and 2.4 show numbers of operating bodies in hand-harpoon fisheries for dolphins and porpoises.

The large-scale and multispecies hand-harpoon fisheries in the postwar period described by Wilke et al. (1953, see Section 2.3) in the 1949 season probably ended before 1957, when government statistics as analyzed by Ohsumi (1972, see Section 2.5) originated. We do not have reliable information on the operations during the period from 1949 to 1957.

Miyazaki (1983) analyzed the geography of small-cetacean fisheries in Japan for 6 years from 1976 to 1981 and indicated that the drive fisheries and hand-harpoon fisheries took about the same number of small cetaceans but that the species composition greatly differed between them. The hand-harpoon fisheries obtained 98% of their catch from Dall's porpoises, but the drive fisheries got 97% of their catch from striped dolphins and took no Dall's porpoises. Three prefectures in the Sanriku Region took 95% of the Dall's porpoises, followed by Ibaraki Prefecture on the Pacific coast south of the Sanriku Region and Hokkaido. Although the three prefectures of Aomori, Iwate, and Miyagi are often combined for statistical purposes into the Sanriku Region, the participation of Aomori in the hand-harpoon fishery for small cetaceans was almost negligible during that period (see Tables 2.3 and 2.6). I observed the hand-harpoon fishery of the Iwate and Miyagi fishermen in the 1970s, when the Iwate fishermen took most of the catch and operated the fishery for Dall's porpoises in the winter (December to April) when there were no other suitable fish species to take (Kasuya 1982). The vessels left port before sunrise with a crew of three or four, searched for porpoises in the rough and cold sea, and returned to port with the catch around sunset.

The Dall's porpoise carcasses, with viscera removed at sea, were landed at the fish market to be weighed and auctioned by dealers. The dockside price in 1973 was below 60 yen/kg in the warm season (April to November), higher in December (160 yen), reached a maximum in January (200 yen), and again went down in March to 70 yen. In April, their daily take increased, but it could not compensate for the price decline, so there were fewer boats operating. In June, they shifted their target to swordfish that migrated there in the warm season and were sold at a higher price than Dall's porpoises. Thus, the landing of Dall's porpoises in the Sanriku Region was highest in February and March and lowest in September (Figure 2.3).

Figure 2.4 shows annual changes in the magnitude of hand-harpoon fisheries in Iwate and Miyagi Prefectures in the 10 years beginning in 1962 using statistics of the peak months (January to April). Statistics are not available for the number of vessels that operated the hand-harpoon fisheries for porpoises. Instead, I used the number of occasions of landings as an indicator of number of operating vessels. There is no known explanation on the very low figure in 1962. In the following years, the number of landings fluctuated annually but showed a slight overall increase. When landing occasions reached small peaks in 1965, 1970, and 1973, the total catch also showed a small peak. One would expect a correlation between effort and the catch. However, catch per fishing effort may have been lower in years of high fishing effort because the amount of fishing effort may change independent of porpoise density on the fishing ground. In order to look at this, I calculated average weight per landing (total weight landed/no. of occasions of landing) (bottom of Figure 2.4). This parameter, a substitute for catch per effort, showed an apparent decline from 1960s to the 1970s, while effort showed an apparent increase.

This measurement of fishing effort has some problems. If a hand-harpoon vessel returns without catch, its fishing effort is totally ignored. Also, if a vessel owner considers his catch of the day too small to stop to land it, he will keep the catch of the day on board to combine it with the catch the following day. These factors will cause underestimation of fishing effort and overestimation of catch per effort. This could result in underestimating a declining trend in catch per effort. If we take into account this bias in the data, the real

TABLE 2.3
Numbers of Operating Bodies and Vessels That Engaged in Hand-Harpoon Hunting of Dolphins and Porpoises, by Prefecture

Year	Hokkaido	Aomori	Iwate	Miyagi	Fukushima	Ibaraki	Chiba	Shizuoka	Mie	Wakayama	Oita	Ehime	Nagasaki	Okinawa
1972	0	0	145	0	0	6	11 (26)	0	0	0	0	0	0	0
1973	0	0	145	0	1	6	11 (26)	0	0	0	0	0	0	0
1974	0	0	145	0	1	6	11 (26)	0	0	0	0	0	0	0
1975	0	0	145	0	1	6	11 (26)	0	0	0	0	0	0	0
1976	0	0	156	0	1	6	2	0	0	0	0	0	0	0
1977	0	0	123	+	0	3	2	0	0	0	0	0	4	0
1978	0	0	143	18 (25)	0	9	0	0	0	0	0	0	1	7
1979	0	0	110	101	0	0	0	0	1	0	0	0	0	87
1980	0	0	92 (93)	9	0	1	0	0	0	0	0	0	+	40
1981	0	0	88	13	0	1 (5)	0	0	0	0	0	0	0	10
1982	0	0	124	17	0	11	9	0	0	0	0	0	4	6
1983	0	0	130	15	0	11	12	0	0	0	0	0	0	6
1984	2 (1)	0	146	15	0	3	0	0	0	0	0	0	0	4
1985	0	0	159	26	0	6	0	0	0	0	0	0	0	4
1986	15	0	168	20	0	0	0	0	0	0	0	0	0	5
	20	3	207	43	0	0	0	0	0	0	28	3	0	0
1987	1	0	208	21+	0	0	0	0	0	0	0	0	0	4
	24	6	287	39	0	0	0	0	0	0	25	3	0	0
1988	61	8	333	39	0	0	2	0	0	37	23	3	1	6
1989	60	12	360	40	0	0	29	0	0	37	19	3	0	6
1990	29	12	263	31	0	0	16	0	6	15	1	0	0	4
1994														6
1996	42	11	219	8			14			116				6
1999/2000	17	10	222	7			14			100				
2000/2001	17	8	223	7			16			100				6
2007/2008	16	8	196	7			11			100				6
2008/2009	16	8	196	7			11			100				6
2009/2010	16	8	196	7			11			100				6

The number of operating bodies usually agreed with that of vessels; the latter is indicated in parentheses only when they disagreed. These statistics were collected by the Whaling Section of the Offshore Division of the Fisheries Agency (1972–1987) and the Coastal Division (since 1986); there are two sets of statistics for 1986 and 1987. Figures by the Offshore Division are in the upper line and those by the Coastal Division in the lower line.

In *1971*, several hand-harpoon fishermen at Taiji, Wakayama Prefecture, established a dolphin-drive team, and the hand-harpoon fishery almost ceased operation until 1987. In *1979*, sudden increase of vessels at Okinawa was probably due to counting of vessels that participated in dolphin drive at Nago. In *1981*, the 10 vessels in Okinawa included 8 from Nago and 2 from On-na-son; the former were likely crossbow hunters. In *1984*, excluding two vessels of Takahama-cho, Fukui, that were reported to have driven a Baird's beaked whale.

TABLE 2.4
Hand-Harpoon Vessels Usually Obtained Operation Licenses from Their Prefecture of Registration, but Some of Them Obtained Additional Licenses from Other Prefectures to Broaden the Scope of Their Operation

Prefecture of Vessel Registration	Prefecture from Which License Was Issued to the Vessel					
	Hokkaido	Iwate	Aomori (Pacific)	Aomori (Sea of Japan)	Miyagi	Total
Hokkaido	49	9	4	0	2	64
Iwate	58	303	63	11	36	471
Aomori	0	0	12	0	0	12
Miyagi	5	21	9	0	39	74
Chiba	0	0	1	1	0	2
Ehime	0	2	0	0	0	2
Oita	0	19	0	0	0	19
Total no. of license	112	354	89	12	77	644

Such a license was issued based on previous activity. This is shown here using the 1990 season as an example. The reason for disagreement of the figures with those in Table 2.3 is unexplained. From records of the Coastal Division of the Fisheries Agency.

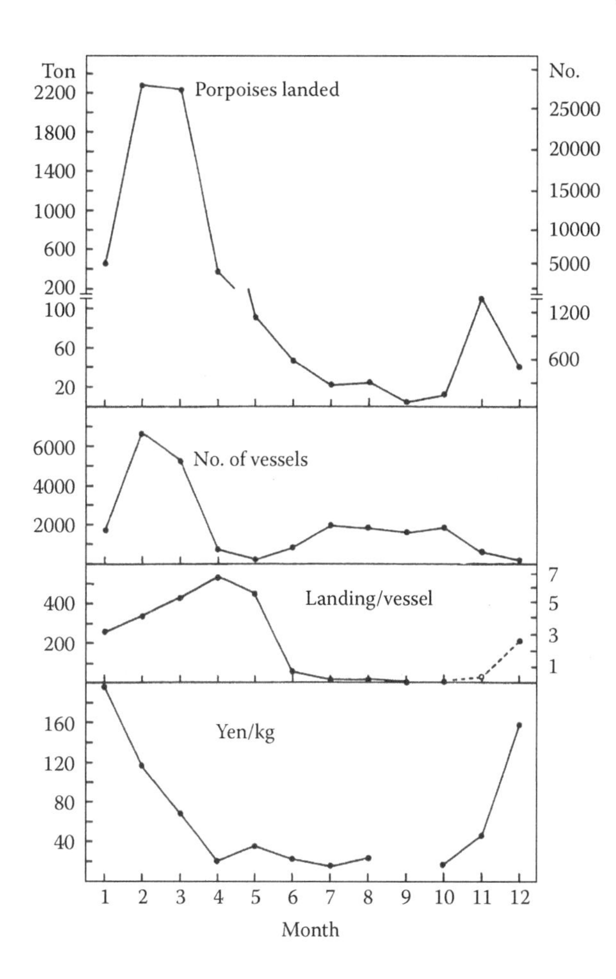

FIGURE 2.3 Seasonality of the Dall's porpoise hand-harpoon fishery in 1962–1973 appeared in the records of wholesale fish markets in Miyagi (37°54′N–38°59′N) and Iwate (38°59′N–40°27′N) Prefectures, in the Sanriku Region. Fishermen of Aomori Prefecture, one of the three prefectures in the Sanriku Region, are not active in the porpoise fishery. *Top panel*: Total monthly landing of gutted porpoises (left scale, in tons) and number of individuals (right scale) calculated using 80.8 kg/gutted porpoise. *Second panel*: Cumulative number of vessels that landed the catch. *Third panel*: Weight of the catch per landing (left scale, in kg) and number of individuals calculated (right scale). *Fourth panel*: Mean dockside price (yen/kg of gutted porpoise) of Dall's porpoises in 1973. (Reproduced from Figure 7 of Kasuya, T., Preliminary report of the biology, catch and population of *Phocoenoides* in the western North Pacific, pp. 3–19, in: Clark, J.G. et al., eds., *Mammals of the Sea*, Vol. IV: Small Cetaceans, Seals, Sirenians and Otters, FAO Fisheries Series, No. 5, FAO, Rome, Italy, 1982, 531pp. With permission of FAO.)

change in the catch per effort in the Dall's porpoise fishery could have been greater than suggested earlier. I interpret the result in Figure 2.4 to indicate that the Dall's porpoise fishery off the Sanriku Region maintained the catch nearly stable between 4500 and 7500 individuals by increasing the amount of fishing effort to compensate for a decline in catch per effort during the 1960s to the early 1970s. There may have been a decrease in the abundance of Dall's porpoises migrating into the fishing grounds.

The available dockside price of Dall's porpoises is expressed as yen/kg for carcasses from which only the viscera were removed; it was approximately 100 yen/kg in February 1973 (Figure 2.3). The average weight of Dall's porpoises without viscera in the Sanriku Area was 80.8 kg (Kasuya 1982). Thus, an average daily catch of 4–5 porpoises (Figure 2.4) was sold at 30,000–40,000 yen, which would leave several thousand yen per crewman after subtracting the cost of fuel and fishing gear. If fishing boats that returned without catch and are not recorded in the statistics are included, the average daily earnings per crew member must have been less than this figure. As the sea was often rough in the peak winter season of the fishery and nearly half the days were unsuitable for hunting, the income of the porpoise hunter was unstable and at a lower level. An owner of a hand-harpoon boat who lived in the fishing town of Otsuchi near the large city of Kamaishi on the Pacific coast told me in the early 1970s that he would rather be a temporary winter worker at any factory in Kamaishi rather than operate the porpoise fishery in winter. Thus, a slight change in the dockside price of porpoises could have affected the motivation of the hunters.

Over a period of years, I visited fish markets, which were operated by FCUs or in tight cooperation with them, to collect landing records of dolphins and porpoises in Iwate and Miyagi Prefectures (Kasuya 1978, 1982). The catch was believed to be mostly Dall's porpoises, and the total annual landings were in a range between 416 tons (1967) and 763 tons (1964), which I estimated to be equivalent to 5150 animals and 9440 animals, respectively, by assuming 80.8 kg/carcass

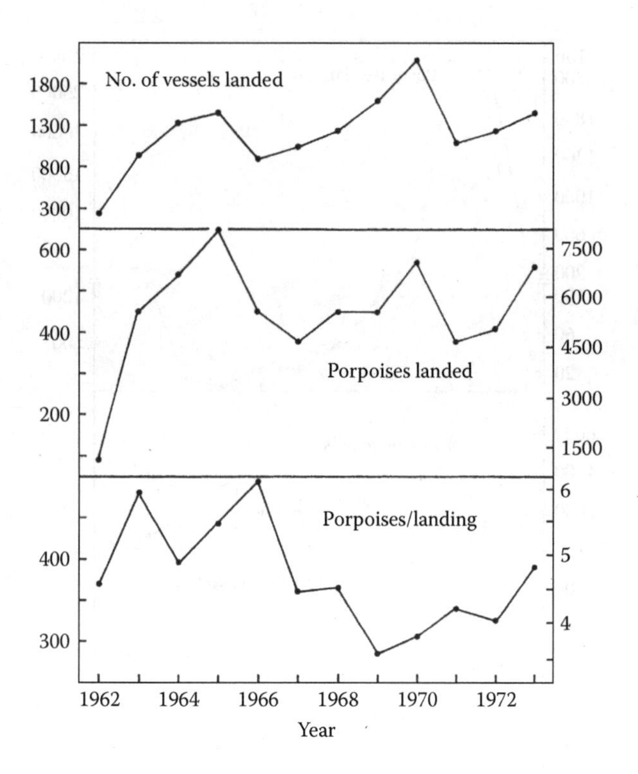

FIGURE 2.4 Annual fluctuation in the Dall's porpoise hand-harpoon fishery in Miyagi and Iwate Prefectures seen through statistics of the peak months from January to April 1962–1973. *Top panel*: Cumulative number of vessels that landed the catch. *Second panel*: Weight of gutted porpoises landed (left scale, in kg) and number of individuals calculated from the weight (right scale). *Third panel*: Weight (left scale, in kg) and number of porpoises/landing (right scale) calculated from the weight. See Figure 2.3 for additional explanation. (Reproduced from Figure 9 of Kasuya, T., Preliminary report of the biology, catch and population of *Phocoenoides* in the western North Pacific, pp. 3–19, in: Clark, J.G. et al., eds., *Mammals of the Sea*, Vol. IV: Small Cetaceans, Seals, Sirenians and Otters, FAO Fisheries Series, No. 5, FAO, Rome, Italy, 1982, 531pp. With permission of FAO.)

without viscera. The annual mean for 13 years (1963–1975) was 7180 animals.

A local porpoise dealer told me that Dall's porpoises were once sold to the mountainous inland areas of Akita (38°50′N–40°20′N, 139°40′E–141°00′E) and Yamagata (37°45′N–38°50′N, 139°35′E–140°40′E) Prefectures (see Section 1.3 for similar records in the late nineteenth century). People in these snowy areas used to store a whole porpoise under the snow for food in the winter. Such markets had already disappeared when I visited the Sanriku Region in the early 1970s, and one major dealer stationed in Otsuchi bought most of the porpoises landed on the Iwate and Miyagi coasts and sent them daily by truck to Numazu (35°05′N, 138°51′E) and other fish markets in Shizuoka Prefecture. Dealers in Shizuoka who bought the Dall's porpoises separated bones from meat and blubber and sold the edible portion to retailers in the prefecture. There was a large demand in Shizuoka and nearby regions for Dall's porpoises from the Sanriku Region because the supply of striped dolphins from the drive fishery on the Izu coast in Shizuoka Prefecture had declined.

On a winter day, a friend of mine in Shizuoka Prefecture visited the Numazu Fish Market to get skulls of Dall's porpoises for specimens, but he found that all the skulls he examined had a portion of the mandible missing and were unsuitable for his purpose. He unfortunately encountered porpoises that I had examined in landing ports in the Sanriku Region from which I extracted teeth for age determination.

I found (Kasuya 1982) that landings of Dall's porpoises in the Sanriku Region were lower in years when landings of striped dolphins on the Izu coast were high and suggested that there might be some common oceanographic factor behind the negative correlation. However, there would be another social factor to be considered, that is, the price of porpoises. In years when catches of striped dolphins were low, the prices of both striped dolphins and Dall's porpoises could have been high in Shizuoka markets. This could have been reflected in the dockside price of Dall's porpoises in the Sanriku Region and increased the motivation of the hand-harpoon fishermen to go Dall's porpoise hunting. Thus, the catches of Dall's porpoises could have been negatively correlated with those of striped dolphins because of market conditions.

The hand-harpoon fisheries for Dall's porpoises showed signs of expansion of catches as well as geographical range in the 1980s (Kasuya and Miyashita 1989, in Japanese). The first change was around 1980, when some Iwate fishermen expanded the winter operation to close to the southern limit of the wintering grounds of the *truei*-type Dall's porpoise, that is, off the Pacific coast of Ibaraki Prefecture. This made it impossible for them to return to their home port daily but forced them to make a cruise of about one week's duration, staying at night at the sea (or perhaps in a nearby port). The catch was stored on board and brought back to ports in Iwate Prefecture. The fishermen told me that longer trips started in order to save fuel and subsequently the operational range broadened. Even if this explanation were accepted, there could have been another reason, for example, to respond to the increased demand of porpoise meat while avoiding competition among vessels. This interpretation is supported by the fact that at almost the same time, some of the Iwate fishermen expanded the season to the summer and the fishing grounds to further north, that is, the Pacific coast of Hokkaido. Those who operated the fishery in summer off the Pacific coast of Hokkaido returned to Iwate ports with their catch stored on board with ice. Some of them were observed operating the hand-harpoon fishery off the Sea of Japan coast of Akita Prefecture (39°07′N–40°20′N) in summer (A. Kawamura, pers. comm.). Then, around 1985, Iwate fishermen further expanded the summer operation to the Okhotsk Sea. In the summer of 1988, a total of about 60 vessels, including about 8 Hokkaido vessels, hunted Dall's porpoises in the Okhotsk Sea (see Section 2.5). From around 1980 to 1985, both season and grounds for the Dall's porpoise fishery showed an expansion, and year-round operations resumed as seen in the last peri-war period.

There are two major populations of the Dall's porpoise being fished around Japan: *dalli*-types that winter in the Sea of Japan and summer in the southern Okhotsk Sea and off the Pacific coast of Hokkaido and *truei*-types that winter off the Pacific coast of northern Honshu and summer in the central Okhotsk Sea (Chapter 9). The earlier mentioned historical changes in the operation pattern of the Dall's porpoise fisheries accompanied change in the targeted populations of the species. From the 1920s to the middle 1930s (first stage), Japanese hand-harpoon fisheries exploited the *truei*-types in their wintering ground off the Pacific coast of northern Honshu. From the late 1930s to perhaps the early 1950s (second stage), they operated on both types, *dalli*-types in their summering ground off the coasts of Hokkaido and in the wintering ground off Hyogo in the southern Sea of Japan and *truei*-types in their wintering ground. From sometime in the 1950s to the early 1980s (third stage), the operations returned to a single population of *truei*-types in their wintering ground. However, since the early 1980s (fourth stage), the hand-harpoon fishery resumed operation on both stocks as in the peri–World War II period.

While some fishermen hunted Dall's porpoises only in the winter season, others pursued the fishery year-round. The general operation pattern of the latter in the 1980s was as follows:

From February to April: Sanriku and nearby waters (*truei*-types)
April: Beginning of move to north
From May to June: Northeastern Sea of Japan off Aomori and Hokkaido (*dalli*-types in the Sea of Japan)
From July to September: Southern Okhotsk Sea coast of Hokkaido (*dalli*-types from the Sea of Japan)
From October to December: Pacific coast of Hokkaido (mainly *dalli*-types from the Sea of Japan) and then to Iwate coast (*truei*-types)

They apparently avoided foggy Pacific waters off eastern Hokkaido in summer. Some of the vessels had freezers on board and operated in cooperation with other vessels. Other vessels flensed the catch on board and stored the products in a freezer on land in Hokkaido (e.g., Shari, 43°55′N, 144°40′E). The frozen products were either transported on land to Iwate markets (e.g., Otsuchi Fish Market) for auction or sold directly to coastal whaling companies such as Nitto Hogei or Nihon Hogei. A director of the land station of the latter company in Ayukawa once told me that his station had been purchasing several hundred Dall's porpoises annually in the 1980s to be processed at the station.

Even if the switch by some Iwate fishermen from daily trips to longer fishing trips is explained by economizing on fuel, the expansion of the season to summer and to the Hokkaido coast in the 1980s has another explanation, that is, increased demand and summer price still remaining attractive to the fishermen. I believe this is due to a new demand for porpoise meat from a food industry, which had previously sold whale meat. Production of whale meat was diminishing in those days, and some whaling companies contacted porpoise hunters or purchased Dall's porpoises in an attempt to substitute small-cetacean meat for whale meat.

Dall's porpoise was sold as *iruka* (dolphin or porpoise) in places where people had customarily consumed it, for example, Shizuoka and Wakayama Prefectures, but for other regions the meat, fins, and blubber of small cetaceans were processed into various "whale product" before sending them to the retail market. This was confirmed through market sampling (Cipriano and Palumbi 1999). Such a case was also reported in a newspaper in 2002. The Dall's porpoises sold directly from fishermen to food processors were unlikely to be included in the Fisheries Agency statistics, because the fishery statistics are compiled by prefectural governments from records of FCUs or related fish markets.

Japanese coastal whaling on large cetaceans gradually shrank and finally ended by the end of March 1988, leaving only scientific whaling in the Antarctic as a source of whale meat for the Japanese market (Chapter 7). Two years later, in 1990, the Environmental Investigation Agency (EIA) investigated the market for Dall's porpoises taken by the hand-harpoon fishery off Sanriku (EIA 1990). The results seem reasonable judging from my earlier experience with the fishery. EIA (1990) identified four types of porpoise dealers who purchased porpoises at landing ports. The first were local retailers who bought porpoises at wholesale fish markets at the landing ports and sold the meat to local consumers as "porpoise meat." Local supermarkets also belonged to this category. The second group consisted of three local porpoise flensers, who flensed carcasses into meat and sold it to food processors. Hosaka Shoten in Kesen-numa City purchased porpoises at Otsuchi, Rikuchu Suisan in Yamada Town (39°27′N, 141°57′E) purchased porpoises landed at a nearby fish market, and Sanriku Kokusai Suisan in Kamaishi City purchased porpoises landed in Kamaishi. The third group consisted of local food processors; EIA identified 10 of them in the Sanriku Region. They processed porpoise meat into *nanban-zuke* (a kind of marinated meat). The fourth group contained companies that previously operated coastal whaling. Del Mar (previous Nitto Whaling Co. Ltd.) and New Nippo (previous Nihon Whaling Co. Ltd.) purchased porpoises directly from the harpoon vessels, processed them, and sold the edible portion to food processors. This agrees with my observations on the latter company. One of the food processors, Kyoshoku, was a member company of Taiyo and operated several factories. It produced *nanban-zuke* at two factories in Chiba City (35°38′N, 140°06′E) and Nagoya City (35°11′N, 136°54′E), and *soft-kujira* (sausage made of porpoise meat and fish meat) and *kujira-bacon* from porpoise blubber at a factory in Yamagata. The price of Dall's porpoises at Kamaishi in April 1990 according to EIA (1990) was 200 yen/kg for carcasses weighed after removing the viscera, 45% of which was edible.

EIA repeated another survey in 1999 (EIA 1999). Only a small percentage of porpoises landed at wholesale fish markets went to local fish retailers. Some of them would purchase one animal to divide among them. Most of the remaining carcasses were purchased by local large-scale porpoise flensers, who separated edible portions such as meat, blubber, and fin to sell to local supermarkets and food processors in Tokyo and Shizuoka. The porpoises were usually flensed near the landing ports, but some of them were sent out whole to an unknown destination. Four flensers, Kokusai Suisan, Otuchi Ichirei, Kamaki Shoten and Kobayashi Shoji, were identified as having purchased porpoises at wholesale markets in Otsuchi and Kamaishi. Food processors purchase the products from the flensers and processed them into canned meat, salted meat, or marinated meat, which was sold as "whale products." Kokusai Suisan, one of the flensers, processed about 30 carcasses daily.

I find some differences between the two reports by EIA. The list of final products was more detailed in the 1990 report. Kokusai Suisan was noted in both reports, but some of the dealers disappeared in the second report. The member companies of the whaling group, New Nippo, Del Mar, and Kyoshoku, are not seen in the second report. It is unclear if these differences reflect true changes or are artifacts. However, it seems to be true that there were some changes in the dealers in the business. For example, I observed that Kikudai Suisan at Otuchi Town, which once purchased almost all the porpoises landed on the Sanriku coast to send them to Shizuoka Prefecture in the 1970s, ceased business in the 1980s.

In March 2011 in the small town of Susuma (34°07′N, 131°52′E) in Yamaguchi Prefecture, western Honshu, I saw a supermarket that was selling packed salted dry meat of Dall's porpoises. It was labeled as salted red meat of *kujira* (whale), the material as Dall's porpoise from Sanriku, and the name of manufacturer as Maruko Shoji in Shimonoseki City (at 33°57′N, 130°56′E). This example clearly indicates that products from dolphins and porpoises still have a wide circulation in Japan. The same company also processes and sells products of scientific whaling.

Behind the expansion of the hand-harpoon fishery in the 1980s was a decrease in the catch of the whaling industry (Table 1.5) and a subsequent increase in the price of dolphins and porpoises (Table 3.21). It is known that under such circumstances some whaling companies contacted hand-harpoon fishermen to purchase their catch. The Japanese government had accepted the International Whaling Commission (IWC) competence on all the baleen whales, sperm whales, and bottlenose whales (*Hyperoodon* spp.) and stopped commercial hunting of these species as of March 1987 (Antarctic whaling) and March 1988 (North Pacific). The scientific whaling program started in the late 1987 in the Antarctic Ocean, expanded to the North Pacific in 1994, and has continued to the present (2014). The accuracy of official whaling statistics in Table 1.5 is variable among operations. It was a common understanding among whale scientists that the true catches in postwar whaling off the Japanese coast (large-type and small-type whaling) were often greater than the reported figures (Watase 1995; Kasuya 1999; Kondo 2001, three in Japanese; Kasuya 1999; Kasuya and Brownell 1999, 2001; Kondo and Kasuya 2002). The true figures were usually twice or three times the official statistics for Bryde's and sperm whales, but for some particular months, the reported figures could have exceeded the true catches. This happened when the company did too much underreporting in early months of the season and found that the quota would not be reached by the end of the season. Small-type whalers, which were not allowed to take sperm whales, were well known to have poached them in large numbers. They could have underreported takes of minke whales when the fishery was operated with a quota. For these reasons, the decline of the supply of whale meat in the 1980s could have been greater than estimated from the official catch statistics in Table 1.5.

Under the shortage of whale meat, coastal whalers contacted small-cetacean fishermen to obtain cetacean meat from them (see the preceding text in this section). Furuki (1989, in Japanese) stated in a monthly magazine about the increase in dockside price of porpoises off Sanriku: "In the recent few years porpoise meat has caught the attention of the people as a replacement for meat from the whaling industries that have been prohibited, and the dockside price of porpoise is over 200 yen/kg, which is nearly twice the price of two years ago. The price sometimes reaches 300–400 yen/kg." The price of 300–400 yen/kg equates to an average price of about 30,000 yen/animal. He also stated that "as a response to this price increase, some hand-harpoon fishermen moved from opportunistic winter operations to year-round operations" and that "the governor of Iwate Prefecture placed the fishery under a license system in December 1989, and 350 vessels obtained licenses" (Furuki 1989). His statement agrees with the decision of the Fisheries Agency to place all the hand-harpoon fisheries for dolphins and porpoises under control of prefecture governments or Regional Fishery Coordination Committees by the end of 1989 (Kishiro and Kasuya 1993). This was later modified, placing all the hand-harpoon fishermen for small cetaceans under a licensing system of the prefecture governor by April 2002 (Chapter 6).

The Institute of Cetacean Research (ICR 1990, 1991, 1992, 1993, all in Japanese) with financial support from the Fisheries Agency studied the hand-harpoon fishery for Dall's porpoises in 1989–1992. The study included a market survey at the Otsuchi Fish Market and examination of the catch on board the fishing vessels. Landings of Dall's porpoises at the Otsuchi Fish Market represented 52%–93% of the total landings of the species in Japan (Table 2.5). Thus, information from the Otsuchi Fish Market roughly represents the features of the fishery of the time in Japan.

Seasonal patterns of the fishery identified by ICR scientists in 1989–1992 were as follows:

From December to April: Off the Pacific coasts of Ibaraki to Aomori on the Pacific coast in latitudes of 35°45′N–41°30′N
From May to June: Off the Pacific coast (41°30′N–43°25′N) and the Sea of Japan coast (41°30′N–45°30′N) of Hokkaido
From August to October: Southern Okhotsk Sea

This pattern represents that of vessels observed on board by the biologists or of vessels that landed catches at the Otsuchi Fish Market while the observers were there. Thus, more data could result in detection of other more minor details. However, the observed pattern of the late 1980s to early 1990s does not differ from the pattern in the 1980s.

TABLE 2.5
Hunting Grounds of Hand-Harpoon Fishermen Who Landed Dall's Porpoises at Otsuchi Fish Market, Iwate Prefecture, and the Composition of the Catch by Color Type

Sea of Operation	Color Type	1989	1990	1991[a]	1992[a]
Okhotsk Sea	*Dalli*-type			6,249–7,382	1158–1311
	Truei-type			12–14	3
Sea of Japan	*Dalli*-type			2,577–2,612	1491–1498
	Truei-type[b]			64	32
Pacific off Sanriku	*Dalli*-type			255	266
	Truei-type			2,672–2,673	3260
Other area	*Dalli*-type			76–85	51–52
	Truei-type			343	462
Total	*Dalli*-type			9,157–10,334	2966–3127
	Truei-type			3,091–3,094	3757–3757
	Otsuchi, total	15,211[c]	13,394[c]	12,248–13,428	6723–6884
	% of entire Japan	52.3%	76.0%	85.5%–93.8%	50.9%–60.4%

Sources: From Nihon Geirui Kenkyusho (Institute of Cetacean Research) (1990, 1991, 1992, 1993, in Japanese).

Otsuchi Fish Market was the largest wholesale market for Dall's porpoises in Japan, as indicated by the weight data.

[a] The values of 53.4 and 44.9 kg/individual used to convert some catches landed as meat caused a range in the number of porpoises landed. The area of operation was estimated from the general operation pattern of the fishery and fishery regulations of the time, and there remain some uncertainties about true location of the operations.

[b] These figures do not prove that the *truei*-type occurs in the Sea of Japan.

[c] Meat was converted into a number of porpoises using 53.4 kg/individual. These figures represent total landing at Otsuchi during April to March of the next year, while the catch for Japan represents the annual figures in Tables 2.7 and 2.8.

ICR scientists examined the prefecture composition of vessels in the landing statistics at the Otsuchi Fish Market. Out of 136 vessels identified during the 1989/1990 season (April to March), 124 were licensed by Iwate Prefecture, 9 by Hokkaido, and 3 by Miyagi. It is evident that the Dall's porpoise fishery of the time was dominated by the Iwate vessels. The total number of landings was 1428 in 1989/1990 (vs. 870 from January to December 1992), with a maximum of 247 in June, followed by 202 in February. No landings were recorded in August and September. In 1992, the peak of landings was in February (231 landings), followed by 201 in March. There were only 0–6 monthly landings in July to October, 1992.

The low summer landing at Otsuchi was due to the creation of a closed season in summer off Hokkaido. The fishing season was set beginning in 1991, and the fishery for small cetaceans was allowed from November 1 to April 30 for waters off Iwate, Miyagi, Aomori, and Chiba and from May 1 to June 15 and from August 1 to October 30 for waters off Hokkaido. Closing the summer operation off Hokkaido was to protect animals in the calving season, an idea originating in the Fisheries Agency.

The landings of Dall's porpoises at the Otsuchi Fish Market were in two forms. One was as a whole carcass with head and fins but without viscera. The other was as flensed products, that is, meat and other parts of commercial value. Visual identification of species and color type was possible only in the first case. The ICR reported that the landings of carcasses without viscera in 1989/1990 season consisted of 12,013 individuals (5,949 in 1992), with 11,895 Dall's porpoises (99%) and 118 small cetaceans of other species (1%). In 1992, the Dall's porpoises were 99.6% of the total and other species 0.4% or 81 individuals. The peak months in 1989/1990 were June (3023) and May (2440). There were no landings in August and September. The May and June landings seem to be from operations off the Hokkaido coasts. Landings as whole carcasses during December to March, which likely represented traditional winter operations off the Sanriku coast, were only 2992 individuals or 24.9% of the annual landings.

Species landed other than Dall's porpoises in the 1989/1990 season were 99 Pacific white-sided dolphins (62 in 1992), 13 common bottlenose dolphins (1 in 1992), fewer than 10 Pacific right whale dolphins (18 in 1992), and common dolphins (*Delphinus* sp.) (none in 1992). Kawamura et al. (1983, in Japanese) surveyed the seasonal small-cetacean faunal change in the Tsugaru Strait Area between Aomori (northern Honshu) and Hakodate (southern Hokkaido) using ferry services between the ports and recorded Dall's porpoises, Pacific white-sided dolphins, common dolphins, and striped dolphins. They recorded high rates of encounter with these species in May and June but were unable to clarify seasonal variation.

Flensed porpoise products auctioned at the Otsuchi Fish Market were brought in either on harpoon vessels or by truck and weighed 182.2 tons in the 1989/1990 season (45.7 tons in 1992). They were composed of meat, dorsal fins, tail flukes, and hearts. Most of them (99.6%) were landed in October (151.8 tons or 83.3%) and in June (29.7 tons or 16.3%). Although this seasonal variation was slightly different from that in 1992, that is, 9.7 tons (21.2%) in October and 32.6 tons (71.3%) in September, in general most of the flensed products were sold in autumn.

If we use a value of 53.4 kg/Dall's porpoise, the average meat production for the species in the Okhotsk Sea, the total weight of meat 177.06 tons (excluding other edible products) auctioned at the Otsuchi Fish Market in 1989/1990 season represents 3,316 Dall's porpoises, which together with the number of whole carcasses mentioned earlier gives 15,211 individuals as an estimate of Dall's porpoises brought to the Otsuchi market in 1989/1990 (April to March). A similar procedure for the 1992 season (January to December) gives 856 Dall's porpoises as estimated from meat. This together with 6724 Dall's porpoises sold as whole carcass yields an estimate of 7580 Dall's porpoises auctioned in 1992 at the Otsuchi Fish Market. Although this series of studies presented a different figure of 44.9 kg/Dall's porpoise as average meat production for the Okhotsk Sea, it is unclear which figure would be more suitable for the earlier calculations (see Table 2.5).

The monthly average price for carcasses of Dall's porpoises auctioned in the 1989/1990 season (April to March) at the Otsuchi Fish Markets reached a peak of 316 yen/kg in January, while the peak figure for 1992 (January to December) was 550 yen/kg, recorded in December. The lowest record for the 1989/1990 season was 129 yen/kg in June and that for 1992 was 200 yen/kg in August. The monthly average price for meat exhibited a similar trend, that is, in 1989/1990 there was a peak of 341 yen/kg in August and a low of 205 yen/kg in June, while in 1992 the peak price was 642 yen/kg in October and the low of 520 yen/kg in August. These figures indicate a rapid increase from 1989/1990 to 1992 in the price of Dall's porpoise products as well as a large seasonal fluctuation in price with a peak in winter and lower prices in summer.

Using the data outlined earlier, scientists of the ICR (ICR 1990, 1991, 1992, 1993, all in Japanese) estimated the number of Dall's porpoises auctioned at the Otsuchi Fish Market in 1989–1992 (Table 2.5). The identification of fishing ground and porpoise stocks relies mostly on the interpretation of landing records of all vessels based on the information on fishing behavior of sample vessels observed by scientists. The results show that the difference in landing at Otsuchi Fish Market between 1989/1990 and 1992 was due to fewer landings of *dalli*-type Dall's porpoises from the Hokkaido coast in 1992. The Japanese hand-harpoon fishery for Dall's porpoises operated without quota until the 1992 season. However, there were various types of pressure or requests from the Fisheries Agency to decrease the catch, and I have identified cases where the fishery responded with manipulation of the statistics (see Section 2.5).

Table 2.5 presents the proportion of Dall's porpoise auctioned at the Otsuchi Market in total take of the species in Japan, as well as the estimated proportions of the two color types of the species auctioned at the Otsuchi Market. It includes a method for converting the weight of meat into number of porpoises, using two conversion factors, 53.4 or 44.9 kg/animal. The former figure came from 109 *dalli*-type Dall's porpoises processed on board two vessels that operated in the Okhotsk Sea in August 1989 and were observed by ICR scientists. The cruises processed their catch on board and stored the meat and fins in a freezer at a port before sending them to Otsuchi. The latter figure was calculated from particular cases of landing where meat was received by the market together with information on number of porpoises that produced the meat. The accuracy has not been examined by scientists.

2.5 STATISTICS OF THE DALL'S PORPOISE HAND-HARPOON FISHERY

Since at least 1970, only the hand-harpoon boats hunted Dall's porpoises, and the small cetaceans taken by the boats were nearly limited to Dall's porpoises. Therefore, the Dall's porpoise fishery and hand-harpoon fishery for small cetaceans in northeastern Japan can be dealt with synonymously for at least the time being. In the future, some fishermen may wish to take other species such as Pacific white-sided dolphins, a small quota for which was allowed by the Fisheries Agency since the 2007/2008 season for existing small-cetacean fisheries. This appears to have been a response of the Fisheries Agency to the desire of aquariums or animal dealers. No existing small-cetacean fisheries in Japan have actively responded to the new quota of Pacific white-sided dolphins.

Now, we will review the history of Japanese hand-harpoon fisheries using the existing fishery statistics. It was in 1957 that the Division of Statistics of the Ministry of Agriculture started collecting statistics on marine mammals taken by coastal fisheries other than whaling. These statistics, which continue to the present with some modifications, originally gave the total number of animals taken by type of fishery. Marine mammal species are grouped into three categories of "seals," "whales," and "dolphins and porpoises," and the fishery types were grouped into "drive fishery," "hand-harpoon fishery for dolphins and porpoises," "seals taken by gun," and "incidental mortality." The species classification limits the biological value of the statistics. The category of incidental mortality includes only those animals taken by seine, trap net (also called "set net"), trawl, longline, and gill net. Thus, these statistics do not include whaling, which is dealt with in other statistics. The Fisheries Agency started collecting cetacean statistics by species in 1972 (Chapter 1).

The interpretation of incidental mortality of marine mammals requires some caution. Most incidental kills that were discarded in the sea were likely to be unrecorded and missing from the statistics. Incidental kill could also have been hidden. For example, in the 1980s the reported incidental mortality of minke whales in the trap-net fishery was at most 10–20 individuals/year, and underreporting was suspected

(Tobayama et al. 1992). In July 2001, the regulations changed to allow commercial use of minke whales incidentally taken in trap nets, and this resulted in a sudden increase of the bycatch to over 100 whales (Table 1.6). When the commercial utilization of the incidental catch was earlier prohibited, the bycatch of minke whales was not recorded or was recorded under dubious categories such as "other miscellaneous species" or even as "sharks."

Ohsumi (1972) summarized these mentioned statistics collected by the Division of Statistics of the Ministry of Agriculture (Figure 2.5). Ten prefectures exceeded 100 animals in the average annual take of dolphins and porpoises in the 14 years (1957–1970): 9251 in Shizuoka, 6040 in Iwate, 2411 in Miyagi, 1322 in Chiba, 489 in Wakayama, 241 in Hokkaido, 334 in Fukushima, 244 in Nagasaki (32°10′N–34°40′N, 128°40′E–130°20′E), 241 in Ibaraki, and 103 in Aomori. Out of these 10 prefectures, 7 had not operated drive fisheries, so the catch must have been mainly by hand harpoon (some incidental takes could be included); these prefectures were Iwate, Miyagi, Chiba, Hokkaido, Fukushima, Ibaraki, and Aomori. The remaining three prefectures of Shizuoka, Wakayama, and Nagasaki had dolphin drives (see Chapter 3), so the catch could have been by driving, hand harpoon, or incidental takes. Several villages of Nagasaki Prefecture had opportunistic dolphin drives. Taiji Town in Wakayama Prefecture had opportunistic drives. This was replaced by a more active system of driving, with a first attempt in 1969, constant operation for short-finned pilot whales beginning in 1971, and expansion to striped dolphins in 1973. Several groups of fishermen on the Izu coast of Shizuoka Prefecture drove striped dolphins using scouting vessels.

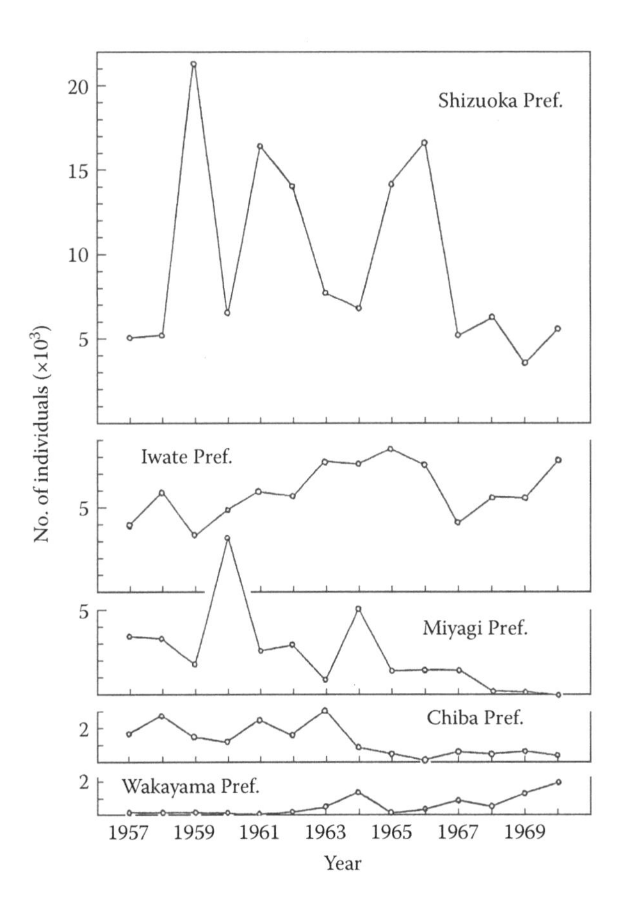

FIGURE 2.5 Landings of all the species of dolphins and porpoises by prefecture in the statistics of the Ministry of Agriculture and Forestry, which includes takes by drive fisheries, hand-harpoon fisheries, and incidental takes in other fisheries. This shows prefectures that were active in dolphin and porpoise fisheries in Japan. (Reproduced from Figure 4 of Ohsumi, S., *Bull. Far Seas Fish. Res. Lab.*, 7, 137, 1972. With permission of the author.)

Ohsumi (1972) further compared catches among fishery types. The catch of marine mammals (whales, dolphins and porpoises, and seals) made by fisheries other than whaling in the 6 years from 1957 to 1962 amounted to 12,335 tons, of which 7,126 tons (57.7%) were by dolphin-drive fisheries together with shooting of seals, 3,361 tons (27.2%) by hand-harpoon fisheries, and 742 tons (6.0%) by seine net. The total of the catch by these three categories of fisheries was 90%. The statistics analyzed by Ohsumi (1972) do not distinguish between dolphin drives and seal shooting. However, I consider that the proportion of seal hunting can be ignored, because only 8,901 seals (about 3% in number of animals) were taken during the period 1957–1970 compared with 299,575 dolphins, although the period covered by the statistics for this figure is slightly different from what was given earlier (1957–1962).

Ohsumi (1972) indicated that about 15% of the total for marine mammals was taken by methods other than hand harpoon or drive (whaling was excluded from the analysis). However, there could have been uncertainties about the actual method of fishing for porpoises, and the license system before the hand-harpoon fisheries for small cetaceans came under the new license system in 1989. I once noticed at a wholesale fish market in the Sanriku Region in the early 1970s a system of statistics where a porpoise taken using a hand harpoon by a licensee in the longline fishery was recorded as a catch of the longline fishery. Such cases would not have been rare, because hand-harpoon fishing was a simple technique and not regulated. The statistics analyzed by Ohsumi (1972) compiled the catch by location of registration of the fishing vessel. This would have caused only minor disagreement between the actual place of catch and the location of the registering prefecture. Ohsumi (1972) reported that there were 3973 vessels operating hand-harpoon fisheries (for any marine animals) and that most of them sought swordfish as the main target.

Based on Ohsumi (1972), we can look at the geographical distribution of two major types of fisheries for small cetaceans, hand harpoon, and drive. The period covered by statistics is from 1957 to 1962 as given previously. During the 6-year period, various fisheries other than whaling landed marine mammals of 12,335 tons (see preceding text in this section), of which 7,126 tons were dolphins and porpoises taken by drive fisheries (with a minor component from seal shooting). Drive fisheries (and seal shooting) recorded the highest catch of 5714 tons (80.1%) in group-D waters, which covered waters from Chiba to Mie Prefectures on the southern coast of middle Honshu (33°45′N–35°45′N, 136°00′E–140°55′E), followed by 1230 tons (17.3%) in group-G waters of western to northern Kyushu on the East China Sea coast (32°45′N–34°45′N, 128°30′E–130°50′E). Although the statistics nominally include catches of seals, these two regions are outside the current habitat of any pinnipeds. The group-D waters include the Izu coast of Shizuoka Prefecture, and group-G waters include Nagasaki Prefecture. Of the total take of 3361 tons by hand-harpoon fisheries, the majority of 2471 ton (91.9%) was taken in group-B waters on the east coast of northern Honshu (35°45′N–41°30′N, 140°35′E–142°05′E), followed by 777 tons (23.1%) in region-D, 95 tons (2.8%) of region-A or the Hokkaido coast (41°20′N–45°30′N, 139°20′E–145°59′E), and 43 tons (1.3%) in region-F on the coast of the southern Sea of Japan (33°55′N–41°15′N, 130°50′E–136°15′E).

Analyses by Ohsumi (1972) of the Statistics of the Ministry of Agriculture revealed that hand-harpoon fisheries for small cetaceans were mainly operated in Miyagi and Iwate Prefectures on the Pacific coast of northern Honshu (37°55′N–40°25′N). The share of the two prefectures was almost equal in the 1950s, and then Miyagi Prefecture's share decreased and became almost negligible after 1968 with an annual take of fewer than 1000 (Table 2.6 and Figure 2.5). On the other hand, Iwate Prefecture started with an annual take of around 5000 in the 1950s and increased the catch annually to about 8500 animals/year by around 1965. Interpretation of the subsequent decline of the catch around 1967 will be discussed later.

In a 1982 paper, I analyzed the statistics for the hand-harpoon fisheries for small cetaceans in Iwate and Miyagi Prefectures in 1963–1975. This study was first presented to a FAO meeting held in Bergen in 1976 and covered several years subsequent to those studied by Ohsumi (1972). I visited

TABLE 2.6
Catch of Dolphins and Porpoises by Prefecture in Northeastern Japan

	Ohsumi (1972)[a]					Kasuya (1982)[b]	Coastal Division, Fish. Agency[c]	Offshore Division, Fish. Agency[d]
Year	Hokkaido	Aomori	Iwate	Miyagi	Total	Iwate + Miyagi	Iwate	Iwate
1957	303	4	4020	3365	7,692			
1958	440	0	5891	3720	10,051			
1959	375	0	3398	1905	5,678			
1960	505	2	4918	9292	14,717			
1961	1841	4	5064	2491	9,400			
1962	184	1	5695	3022	8,902			
1963	194	0	7766	862	8,822	9040 (5540)		
1964	114	441	7600	5094	13,249	9440 (6630)		
1965	416	2	8465	1328	10,211	9180 (8040)	4204 (49)	
1966	258	644	7576	1440	9,918	7980 (5590)	4315 (90)	
1967	61	0	4105	1400	5,566	5150 (4710)	4276 (51)	
1968	56	259	5619	121	6,055	6020 (5540)	4094 (77)	
1969	22	0	5632	112	5,766	7020 (5550)	4384 (90)	
1970	15	84	7813	51	7,963	8060 (7150)	4269 (89)	
1971						5210 (4640)	4039 (93)	
1972						5190 (5020)		5119 (145)
1973						7230 (6930)		4866 (145)
1974						6470 (5630)		4967 (145)
1975						7350 (6920)		4859 (145)

Species and method of hunting were not specified, but hand harpoon was the only method of hunting and Dall's porpoise was the principal target of the fishery at the time in the region.

[a] Based on official statistics of the Ministry of Agriculture and Forestry.

[b] In parentheses are landings in the peak season (from January to April), which are included in the annual total. I collected landing records of "dolphins and porpoises" at wholesale fish markets in these prefectures and converted the weight into number of porpoises using an average of 80.8 kg/gutted carcass obtained from my observation of Dall's porpoises at the landing market. The fishermen gutted the carcasses before landing them at the market (other parts remained intact).

[c] These statistics (dated December 15, 1981) were part of reports from three prefectures (Iwate, Shizuoka, and Nagasaki) responding to a request from the Coastal Division. The sampled operating bodies (in parentheses, almost equal to fleet size) are small and the statistics must be downwardly biased.

[d] These statistics were extracted from statistics of the Whaling Section of the Offshore Division collected via prefecture governments, where only Iwate Prefecture was known to hunt Dall's porpoises (see Tables 1.9, 1.10, and 2.3). Iwate recorded a take of only four small cetaceans other than Dall's porpoises. In parentheses are the numbers of operating bodies, which are almost identical with the number of operating vessels.

wholesale fish markets in Miyagi and Iwate Prefectures to copy the landing records for dolphins and porpoises (Table 2.6) and spent six winters at these landing ports to examine the catch composition. The catches examined revealed that 98.8% of the small cetaceans landed at these ports in the high season of autumn to spring were Dall's porpoises. Therefore, all the small cetaceans recorded in the previous statistics are safely assumed to have been Dall's porpoises, although the fish market statistics did not always record species. For the period when the statistics reported by Ohsumi (1972) and Kasuya (1982) are comparable, the two bodies of data show reasonable agreement for some years (1965, 1967, 1968, 1970) but a differential greater than 1000 individuals (–18% to +28%) in other years (1964, 1966, 1969). The reason for the disagreement is unknown (Table 2.6).

Kasuya's (1982) statistics showed that the average number of porpoises/landing/vessel was in the range of 4.5–6.5 during the 1962–1968 seasons but decreased to 3.5–5.0 during the 1969–1973 seasons (Figure 2.4). It is reasonable to postulate that this reflected a decline in abundance. The annual catch ranged from 5100 to 9400 porpoises for the years 1963–1975, which was nearly the same as reported by Ohsumi (1972). The low catch around 1967 can be seen in both sets of statistics. In general, an annual catch of over 8000 was recorded until the 1966 season, but in later years, it remained at a lower level of 5200–7400 (1970 was the exception). The annual fluctuation in the catch showed a negative correlation with the catch of striped dolphins on the Izu coast, and the presence of some economic factors was suggested (see Section 2.4).

The Whaling Section of the Offshore Division, a Fisheries Agency section responsible for the management of whaling, in 1972, started to collect statistics on small cetaceans taken by local fisheries. The statistics were collected through the prefectural governments and were based on landing records of member fishermen kept by their FCUs, which all Japanese fishermen belonged to. These Fisheries Agency statistics agree quite well with those in Kasuya (1982) for the first year, 1972, but figures for later years disagree with Kasuya's (i.e. smaller catches in the Fisheries Agency statistics). The disagreement cannot be explained by inclusion of species other than the Dall's porpoise or inclusion of catches in Miyagi Prefecture (Tables 2.6 and 2.7). The Fisheries Agency statistics included the number of vessels operating in hand-harpoon fisheries (Tables 2.3 and 2.6). The Fisheries Agency statistics for Iwate Prefecture recorded a constant fleet size of 145 vessels and almost constant catches for the years 1972–1975, while Kasuya (1982) reported an increasing catch by Iwate and Miyagi fishermen (the catch in Miyagi were almost negligible). Then, in 1976, the hand-harpoon vessels in Iwate showed a sudden increase to 156, accompanied by an increase in catch to 9700 (Tables 2.3 and 2.7). The number of operating vessels in Iwate exhibited strange fluctuation in 1977–1985 (Table 2.3). From these observations, I have an impression that the disagreement of catch figures for Iwate between Kasuya (1982) and the Fisheries Agency statistics was due to some technical problem in collecting the latter statistics, for example, selecting sample vessels.

Miyazaki (1983) reviewed the status of Japanese small-cetacean fisheries using the Whaling Section statistics for 6 years (1976–1981) and his own data on the hand-harpoon fishery at Yamada Fish Market in Iwate Prefecture and on drive fisheries on the Izu coast in Shizuoka Prefecture and Taiji in Wakayama Prefecture. The hand-harpoon fishery took a total of 51,700 animals during the 6 years, which was comparable to 58,200 taken by drive fisheries. The hand-harpoon fishery got 98% of its catch from Dall's porpoises, followed by 200–299 each of striped dolphins, bottlenose dolphins, and Pacific white-sided dolphins and 100–199 short-finned pilot whales. These figures include dolphins and porpoises taken by small-type whaling vessels and by the crossbow fishery in Okinawa. The small-type whaling vessels occasionally took such dolphins and porpoises, particularly off Taiji where demand existed for such species, and some small-type whalers hunted Dall's porpoises with hand harpoons in the 1987 and 1988 seasons (footnote to Table 2.7; Chapter 4). The crossbow fishery operated for short-finned pilot whales and some other species (Table 2.13).

According to Miyazaki (1983), vessels from the Sanriku coast on the Pacific coast of northern Honshu contributed 95.5% of the 51,200 Dall's porpoises taken by Japanese fisheries in the 6 years mentioned earlier, followed by 3% each for Ibaraki and Fukushima and 1.5% for Hokkaido. The number of vessels that operated drive fisheries is shown in Tables 2.3 and 2.4.

After a chaotic postwar period, the Japanese hand-harpoon fisheries for dolphins and porpoises operated almost solely for Dall's porpoises and mainly off the Pacific coast of northern Honshu, including Sanriku, and to lesser degree off Hokkaido and took annual catches between 5,000 and 10,000 during the 20 years of the 1960s and 1970s. Kasuya and Miyashita (1989, in Japanese) analyzed trends of the Dall's porpoise catch from this period to the 1980s (Figure 2.6) and concluded that the catch probably declined slowly from 1963 to the mid-1970s and then increased until 1987. They identified two troughs in the catch during the latter increasing phase, in 1979–1980 and 1984–1986. These troughs in the catch agree with periods of change in the operation pattern. The decline in catches could have been due to a decline in abundance of the *truei*-type Dall's porpoises wintering off the Sanriku coasts. Responding to this, some Sanriku fishermen changed their pattern of winter operation from daily trips to longer trips of several days. The fishing ground also changed from near the port area to further south off the coast of Ibaraki Prefecture. This change started in the early 1980s and improved operational economy of the fishery and decreased competition among fishing vessels by decreasing vessel densities. The second action by the Sanriku fishermen was to extend operation to the Hokkaido coast and the fishing season to summer. This change occurred in the mid-1980s and led to exploitation of a new population of Dall's porpoises, the *dalli*-type Dall's porpoise population that wintered in the Sea of Japan and summered off the Pacific coast of Hokkaido and in the southern Okhotsk Sea. Thus, in order to continue high catches of Dall's porpoises, the Japanese hand-harpoon fishery increased fishing effort

TABLE 2.7
Catch Statistics for Dall's Porpoises by Prefecture of Vessel Registration and by Calendar Year[a]

							Total Hand-Harpoon Catch	
Year	**Hokkaido**	**Aomori**	**Iwate**	**Miyagi**	**Fukushima**	**Ibaraki**	**Dall's Porpoises**	**Other Species**
1972			5,119 (3)	*0 (30)*	*81 (62)*	0 (29)	5,119	32
1973			4,866 (4)	*47 (31)*	0 (2)	0 (10)	4,866	16
					90 (37)	*0 (3)*		
1974			4,967 (2)		80 (25)	0 (153)	5,047	180
				54 (36)	*0 (11)*	*0 (36)*		
1975			4,859 (0)		0 (1)	0 (38)	4,859	39
				25 (17)	*90 (27)*	*0 (53)*		
1976			9,780 (500)		83 (7)	0 (1)	9,863	508
				36 (24)	*0 (127)*	*0 (37)*		
1977			8,625 (337)	100 (30)		0 (14)	8,725	381
			0 (21)	*361 (161)*	*72 (111)*	*0 (22)*		
1978			8,098 (111)	128 (127)		0 (38)	8,226	276
		15 (0)	*0*	*200 (54)*	*0 (54)*			
1979			6,795 (0)	77 (50)			6,872	50
	2 (0)		*4 (7)*	*0 (41)*		*0 (13)*		
1980			6,855 (111)	16 (0)		0 (11)	6,871	122
	0 (35)	*0 (1)*	*0 (14)*	*49 (28)*		*0 (103)*		
1981			9,764 (129)	39 (0)		0 (45)	9,803	174
	73 (111)	*3 (0)*	*24 (0)*	*62 (0)*		*0 (56)*		
1982			12,791 (137)	27 (0)		0 (114)	12,818	251
	86 (146)			*9 (43)*		*15 (29)*		
1983	0		12,775 (20)	3 (31)		0 (46)	12,778	97
	93 (4)			*8 (16)*	*0 (13)*	*0 (44)*		
1984	0 (45)		10,131 (46)	15 (231)		0 (41)	10,146	363
	194 (61)			*0 (28)*		*5 (10)*		
1985	0		10,375 (58)	0 (387)		0 (35)	10,375	482
	238 (67)	*0 (3)*	*0 (55)*	*0 (224)*		*3 (19)*		
1986[b]	0 (30)		10,082 (11)	452 (0)			10,534	41
	189 (177)			*0 (2000)*		*1 (16)*		
1987[c]	0 (8)		13,198 (135)	208 (667)			13,406	810
	363 (216)	316 (221)	*0 (8)*	*0 (305)*		*6 (0)*		
Year[d]	**Hokkaido**	**Aomori**	**Iwate**	**Miyagi**	**Ehime**	**Oita**	**Dall's porpoise**	**Others**
1986	330 (27)	14 (0)	15,068 (0)	593 (0)	15 (0)	487 (0)	16,498[e]	28
	507 (694)		*30 (0)*	*0 (839)*				
1987	408 (33)	12 (0)	24,168 (0)	586 (0)	10 (0)	553 (0)	25,733[e]	33
	545 (322)		*50 (118)*	*0 (517)*				
1988[f]	3906 (153)	12 (0)	35,011 (0)	857 (0)	38 (0)	533 (0)	40,357	153
	409 (445)	*0 (9)*	*25 (440)*					
1989[g]	D3068	D12	D12,485	D388			D15,953	386
	T466(33)[h]		T12,482(120)[h]	T23(233)	T26	T98	T13,095	
							D/T153[h]	
1990[i]	D1463	D10	D7,629	D258			D9,360	158
	T284		T12,154 (105)	T4(52)		(1)	T12,442	
	D195 (207)		*D88 (121)*	*(D39)*				
1991[j]	D1220	D8	D3,259	D184		D0	D4,671	59
	T39	*0 (1)*	T6,406	T9(5)		T3	T6,457	
	0 (1)		6,506(54)				D/T6,506	
			0 (2)					
1992	D1024	D8	D2,360	D2			D3,394	160
	T65(105)		T7,942(55)	T2			T8,009	

(*Continued*)

TABLE 2.7 (*Continued*)
Catch Statistics for Dall's Porpoises by Prefecture of Vessel Registration and by Calendar Year[a]

Figures in roman represent catches by hand-harpoon fishery and those in italics catches by other fisheries. The distinction between fishing method and fishery type could be sometimes ambiguous. For example, I witnessed at a fish market in Iwate that Dall's porpoises taken using hand harpoon during the return of vessels from the large-mesh drift-net fishery were recorded as taken by the drift-net fishery. In the 1987 and 1988 seasons, some small-type whalers hunted Dall's porpoises using their catcher boats and hand harpoons in an attempt to survive in the situation where minke whaling would terminate. The operation was in the Okhotsk Sea off Abashiri, Hokkaido (10 *dalli*-types in 1987, 266 *dalli*-types in 1988) and in the Pacific off Ayukawa in Miyagi Prefecture (140 small cetaceans of unidentified species in 1987 and 227 *dalli*-type Dall's porpoises in 1988). I have included these catches made by small-type whalers in the catch of the hand-harpoon fishery because they have not been listed in the whaling statistics. Figures in parentheses are for unknown species or species other than Dall's porpoise (additional). I have estimated that some species identifications in the Fisheries Agency statistics are erroneous and actually should be referred to the Dall's porpoise. These are, for examples, all the 1340 *ma-iruka*, which means *Delphinus* spp., landed in 1977 at Taro, Iwate Prefecture, and all the 1077 *nezumi-iruka*, which means common porpoise, landed from January to March 1979 at Otsuchi, Iwate Prefecture. There were similar additional cases in the early stages of the statistics. These markets discontinued such suspicious use of species names in subsequent years. Some of the earlier statistics apparently distinguished between the two color types of the species, but I ignored the designations because the distinction was made for a minor proportion of the entire take of the species and the accuracy was questionable (also see footnote g to this table). D represents *dalli*-type and T *truei*-type as recorded in the statistics. This table covers the period 1972–1992 when the fishery operated without a quota and statistics were collected by the Whaling Section of the Offshore Division of the Fisheries Agency (1972–1987) or the Coastal Division (1986–present).

[a] Commercial take of dolphins and porpoises off the Sea of Japan coast was limited to Hokkaido and Aomori during the period covered by the statistics. Dall's porpoise does not inhabit the Nagasaki waters where culling occurred.

[b] Incidental takes by drift-net fisheries could have occurred in waters outside the prefecture of registration. Incidental take of small cetaceans has also been reported by drift-net fisheries registered at prefectures not listed in this table. For further details, see Japanese Progress Report to IWC.

[c] Statistics of the Whaling Section of the Offshore Division, Fisheries Agency, end here. See Table 1.9 for further details on the disagreement of Dall's porpoise statistics with those of the Coastal Division.

[d] Statistics of the Coastal Division of the Fisheries Agency start here. Incidental mortality in trap nets of some years is not given by species.

[e] Due to an unknown reason, the Japanese total for Dall's porpoises taken by the hand-harpoon fishery shows minor disagreement with the totals for each prefecture.

[f] These figures do not include a total of 1134 small cetaceans landed by salmon drift-net fisheries in 1988 and a total of 1864 small cetaceans landed by squid and large-mesh drift-net fisheries in the same year. The species composition was 1663 Dall's porpoises, 268 northern right whale dolphins, 116 striped dolphins, and other species.

[g] Responding to the recommendation of the IWC, the Japanese Fisheries Agency in 1989 started an attempt to collect Dall's porpoise statistics by color type. I have listed their classification as reported. Color type was not available when the catch was landed as meat; then, the number had to be estimated from weight.

[h] The original statistics did not clarify whether these figures represent Dall's porpoises of unknown color type or small cetaceans of unknown species. The 90 common dolphins reported from Miyagi are listed here in other species.

[i] These figures do not include incidental mortality of small cetaceans in 1990 that occurred outside the Japanese EEZ. Those takes were estimated from data from sampled vessels; 1 individual in the salmon drift-net fishery and 16,635 in the squid drift-net fishery. The species composition was 3094 Dall's porpoises, 7909 northern right whale dolphins, 4447 Pacific white-sided dolphins, 562 common dolphins, and other species.

[j] These figures do not include incidental mortality of small cetaceans in 1991 outside the Japanese EEZ: 135 in the salmon drift-net fishery and 18,007 in the squid drift-net fishery. The species composition was 3342 Dall's porpoises, 9320 northern right whale dolphins, 3784 Pacific white-sided dolphins, and 1035 common dolphins and other species.

and expanded the targeted populations. It is likely that the *truei*-type population was unable to support the annual catch of 5,000–10,000 recorded in the 1960s and 1970s and that the current catch quota for the *truei*-type population, 8,700 individuals/season, cannot be sustained by the population.

Kasuya and Miyashita (1989, in Japanese) importantly suggested that the catch statistics on the Japanese Dall's porpoises were likely to be biased downward and apparently stimulated the Fisheries Agency to take action to improve them. Their paper was published in a popular scientific journal in April 1989 and referred to in the IWC Scientific Committee meeting that year. They did not submit the paper to a scientific journal because publication in such a journal could take up to a year. Neither did they attempt to submit the paper to the IWC Scientific Committee formally as a document from the Japanese delegation because of fearing refusal by the Fisheries Agency, which must approve any such document in advance.

Kasuya and Miyashita (1989, in Japanese) questioned the Fisheries Agency statistics for the Dall's porpoise fisheries based on expected annual catch per vessel and the number of harpoon vessels observed on the fishing grounds. As one of the authors, I learned from an Iwate fisherman that they took 80–110 porpoises/month in the winter operation off Sanriku and 240–360 porpoise/month in the summer operation off

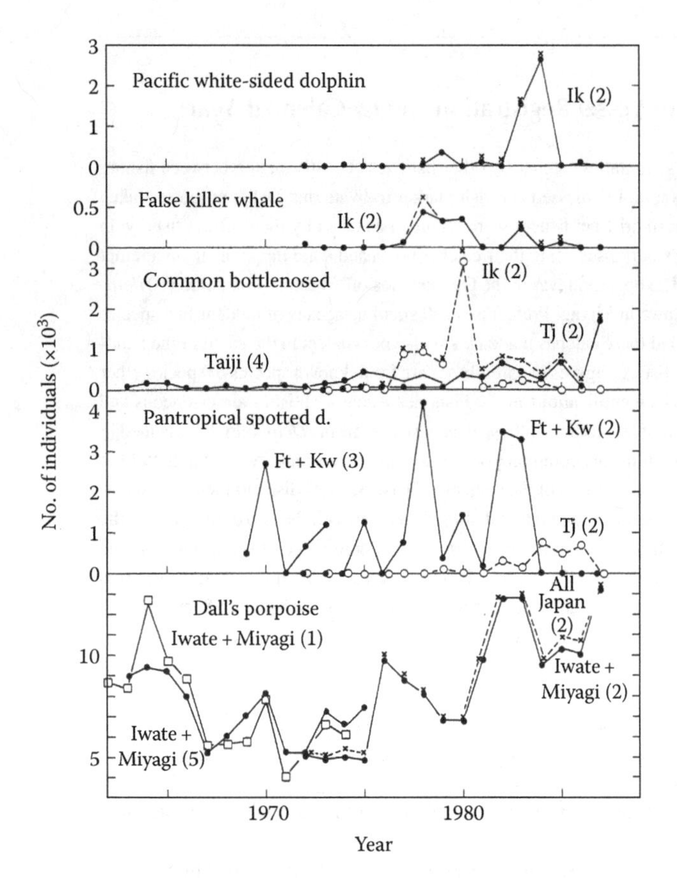

FIGURE 2.6 Operation of dolphin fisheries and their catches compared between different types of statistics. *Location*: Ft, Futo; Ik, Iki; Kw, Kawana; and Tj, Taiji. *Statistics*: 1, Ministry of Agriculture and Forestry; 2, Offshore Division of the Fisheries Agency; 3, Miyazaki et al. (1974); 4, Miyazaki (1980); and 5, Kasuya (1982). (Reproduced from Figure 2 of Kasuya, T. and Miyashita, T., *Saishu-to Shiiku*, 51(4), 154, 1989, in Japanese.)

Hokkaido. He actually took a total of 1158 porpoises from February to September (excluding a total of 60 days in April, June, and August, when he did not operate the fishery for some reason) and in areas off Sanriku and Hokkaido. So if he in addition operated during the remaining season from October to December, it would have been easy for him to catch 1200 porpoises in 1 year. These catch figures of 80–110 porpoise/month for the Sanriku operation might be hard to achieve but would not be impossible if the fishermen operated 20 days per month and took 4–5 porpoises per day. Other fishermen informed me that high-powered vessels with a speed of over 20 knots could catch 2000 porpoises per year. The number of operating vessels was obtained during my trip to Hokkaido in August 1988, when I counted at least 21 hand-harpoon vessels from Iwate in ports along the Okhotsk Sea coast and 9 vessels in Kiritappu (43°05′N, 145°10′E, Pacific coast of Hokkaido). Of the nine boats in Kiritappu, eight were Hokkaido boats. Other informants on the Okhotsk Sea coast of Hokkaido said that there were 40 vessels in the summer of 1988 (obtained at Abashiri in July–August 1988) or 64 vessels (obtained at Monbetsu, 44°20′N, 143°20′E, in January 1989).

The porpoise hunters encountered in the Hokkaido ports must have operated year-round. Therefore, only the 30 vessels confirmed by myself could have captured over 30,000 porpoises in the 1988 season. In addition to these, there must have been numerous fishermen who operated off Sanriku only during the winter season. It was hard to assume that the operating vessels suddenly increased in the 1988 season, so the total catch of 13,000 Dall's porpoises reported by the Fisheries Agency for 1987 season seemed to be too small. We reported these questions in a popular article (Kasuya and Miyashita 1989, in Japanese) and at the same time reported them to the Fisheries Agency, without expectation of a particular response.

The Coastal Division of the Fisheries Agency separately from the Whaling Section of the Offshore Division started collecting statistics of dolphin and porpoise fisheries in Japan and reported the 1988 statistics to the Scientific Committee of the IWC in May 1989. The Coastal Division reported a take of about 40,000 Dall's porpoises in 1988, which agreed with "over 30,000" estimated by Kasuya and Miyashita (1989, in Japanese). Responding to a question by the Scientific Committee about the reason for such a rapid increase and the reliability of the previously reported statistics, the Coastal Division reported to the following year's Scientific Committee that the previously reported statistics had involved underreporting and presented revised figures for the 1986 and 1987 seasons (Tables 2.7 and 2.8). The statistics for 1989 and afterward classified the catch of Dall's porpoises into the two color morphs. This was an improvement, although the accuracy needed to be confirmed.

The Scientific Committee of 1989 concluded that (1) the 1988 catch was clearly unsustainable by the stock and (2) it was urgent to decrease the catch to the previous level (which itself might be unsustainable) and requested Japan to clarify which of the two sets of statistics should be accepted and if the catch really exhibited such a rapid increase.

TABLE 2.8
Disagreement between Catch Statistics for Dall's Porpoises Issued by Two Government Agencies, Presumably Due to Difference in Coverage and Manipulation by the Industry

Sources	1985	1986	1987	1988	1989	1990	1991
Offshore section[a]	10,378	10,534	13,406				
Coastal section[b]		16,515	25,600	40,367	29,048	17,634	14,318
Kasuya (1992)[c]			37,200	45,600			

[a] Compiled by the Whaling Section of the Offshore Division of the Fisheries Agency based on reports from prefecture governments that relied on the records of fisheries cooperative unions.

[b] Compiled by the Coastal Division of the Fisheries Agency in principle in the same way as used by the Offshore Division.

[c] Correction made by Kasuya (1992) to the Coastal Division statistics by examining records of the fishery cooperative unions in Iwate Prefecture.

Responding to the last two requests by the Scientific Committee, I made a visit accompanied by a staff employee of Iwate Prefecture to FCUs of the prefectures, to which most of the Dall's porpoise hunters belonged, examined their landing records, and reported the result to the Scientific Committee in Kasuya (1992). The report concluded that the older statistics contained intentional underreporting but that there were some cases of overreporting, as summarized in the following:

1. Both the statistics of the Coastal and Offshore Divisions did not include porpoises sold directly to dealers by the fishermen, that is, not auctioned at wholesale markets in the landing ports. This biased the catches downward.
2. Some fishermen who fished in remote areas might sell their catch through fish markets operated by local FCUs of which the fishermen were not member. Most such landings were not included in the statistics. This also biased the catches downward.
3. The statistics of the Coastal Division were based on original landing records kept at the FCU and not from statistics compiled by the FCU as in the case of the Offshore Division.
4. Meat received from Hokkaido was converted into number of animals assuming 80 kg/individual, which was an average weight of a whole carcass without viscera taken off Sanriku. The use of 53.4 kg/individual would have been more appropriate. This biased the catch statistics downward.
5. Apart from biases caused by the previous factors, Iwate Prefecture, with cooperation of the FCUs, intentionally underreported catches in 1987 and overreported in 1988. The former would have been to avoid external criticism or to delay expected regulations, and the latter would have been to obtain a higher quota under the quota system expected to start in 1989.
6. The Otsuchi FCU, which handled about half of the total Japanese catch of porpoises (Table 2.5), rejected my request to examine the records of large catches (either whole carcasses or meat) received from Hokkaido for some unexplained reason. This leaves a large uncertainty in evaluating the existing statistics.
7. Correction of the previous fourth and fifth factors added 37,200 porpoises in 1987 or a 49% increase over the statistics reported by the Coastal Division and 45,600 individuals in 1988 or a 13% increase (Table 2.8). The real figures would be higher if we took into account the first, second, and sixth factors.

The catch statistics currently available to us do not take account of the first and second factors. Any fisheries operated under government control are likely to have some misreporting in the statistics, but we experienced an unacceptable level of deception assisted by the prefecture government for the short-term benefit of the fishermen. Such activity must be eliminated from the management of common properties such as fishery resources.

The history of government regulation of the Dall's porpoise fisheries is outlined briefly here (for details see Chapter 6). In January to March 1989, the Fisheries Agency decided that all the hand-harpoon fisheries for small cetaceans had to be licensed by the prefectural government (Iwate) or approved by the Regional Fishery Coordination Committee (Hokkaido, Aomori, and Miyagi). The licensing would apply only to vessels that had actually operated in the previous season and required that they operate with a quota and within a fishing season and that they land their catches only at particular fishing ports designated by the government. About 300 vessels gained a license in 1989 in Iwate Prefecture, where the majority of the Dall's porpoise hunters were registered. Later, the rule was changed to place all the operations under a prefectural license system by April 2002 (website of the Fisheries Agency). At the time of introduction of the license system in early 1989, when a quota system was not yet introduced, the Fisheries Agency requested Iwate Prefecture to decrease the Dall's porpoise catch to 70% (24,500) of that of the previous year (about 35,000) in the 1989 season and further to 85% (20,800) of the catch expected in 1989 in the 1990 season. This was further strengthened on July 13, 1989, with a request to Iwate Prefecture for a plan to decrease the catch to 10,000 by 1991. Similar requests were also sent to Miyagi and Hokkaido (Fisheries Agency document).

These actions of the Fisheries Agency agree with the following statement of M. Morimoto at the IWC Scientific Committee meeting in 1990 (IWC 1991): "The catch of Dall's porpoise increased from 25,600 in 1987 to 40,367 in 1988. Responding [to] this situation, we introduced regulations to the fishery in early 1989 and decreased the catch to 29,048. This was 28% below the catch in 1988. We consider this catch still too high, and will make effort to decrease the catch in 1991 to a level of 85% of the 1989 season."

The Fisheries Agency set a quota of 17,600 Dall's porpoises (no distinction of the two color types that represent two stocks off Japan) for the 1991 and 1992 seasons, and then a similar quota was divided equally into two parts for the two color types for the 1993 season, that is, 9,000 *dalli*-types and 8,700 *truei*-types. The total of the two types was 17,700. However, the sums of the figures given to each prefecture for the years 1996–2006 were 9,000 *dalli*-types and 8,420 *truei*-types (total 17,420) (Kasuya unpublished). The difference, 280, could have been retained by the Agency to hide unexpected catches.

The Dall's porpoise quota decreased slightly in the 2007/2008 season to 8707 *dalli*-types and 8168 *truei*-types (Table 2.9) and was divided among Hokkaido (16 vessels), Aomori (8 vessels), Iwate (196 vessels), and Miyagi (7 vessels). Total of the two color types was 16,875, shared among 227 fishing vessels. Minor modifications of the quotas for Japanese dolphin and porpoise fisheries apparently continued in later years.

In 2007, the Japanese Fisheries Agency allowed drive and hand-harpoon fisheries off the Pacific coast of Honshu to

TABLE 2.9
Catch of Dall's Porpoises by the Hand-Harpoon Fishery Compiled by Prefecture of Vessel Registration and by Calendar Year

Prefecture	Hokkaido		Aomori	Iwate		Miyagi		All of Japan		
Color Types	D	T	D	D	T	D	T	D	T	Total
Quota	1500	100	20	7200	8300	280	20	9,000	8,420	17,420[a]
										17,700[b]
1993	1083	49	15	4624	8536	9	2	5,731	8,587	14,318
1994	1423	53	14	6627	7801	29	0	8,093	7,854	15,947
1995	1234	54	0	5713	5340	55	0	7,002	5,394	12,396
1996	1222	50	0	6705	8010	111	2	8,038	8,062	16,100
1997	999	31	2	7433	9976	99	0	8,533	10,007	18,540
1998	994	69	0	4116	6013	193	0	5,303	6,082	11,385
1999	670	57	0	5632	8371	77	0	6,379	8,428	14,807
2000	1203	69	0	6106	8589	204	0	7,513	8,658	16,171
2001	1413	100	0	6960	8120	57	0	8,430	8,220	16,650
2002	1328	89	0	6057	8243	229	3	7,614	8,335	15,949
2003	1655	84	0	6427	7325	226	3	8,308	7,412	15,720
2004	647	66	0	3796	9106	171	0	4,614	9,175	13,789
2005	1240	51	0	5394	7733	246	0	6,880	7,784	14,664
2006	719	44	0	3312	7758	181	0	4,212	7,802	12,014
Quota for 2007/2008[c]	1451	98	18	6969	8054	269	16	8,707	8,168	16,875
2007	841	44	0	2975	7243	254	0	4,070	7,287	11,357
2008	467	66	0	1947	4566	180	0	2,594	4,632	7,226
2009	308	0	0	1362	7769	103	0	1,773	7,767	9,540
2010[d]	116	2	0	1140	3532	0	129	1,256	3,663	4,919
2011[d]	0	0	0	89	1855	0	8	89	1,863	1,952
2012	0	0	0	29	376	0	0	29	376	405
2013	0	0	0	77	1198	18	0	95	1,198	1,293

D, *dalli*-type; T, *truei*-type. Quota was given by calendar year (1993–1995) or season of operation (1996/1997–present), and minor modifications were made annually since 2007/2008 season (Table 6.4). Based on Fisheries Agency statistics. This table has been revised with the addition of new data available since the original Japanese language edition appeared in 2011.

[a] From Iwasaki in website of Offshore Division of the Fisheries Agency.

[b] Japanese Progress Report to IWC (National Research Institute of Far Seas Fisheries of Japan 1995) gives the Japanese quota as 8,700 *truei*-types and 9,000 *dalli*-types for a total of 17,700 Dall's porpoises, which is greater than the sum of the quotas given to the prefectures, 17,420. The difference could have been a reserve kept by the Fisheries Agency in case of any excess catch. Licenses for the Dall's porpoise fishery were limited to these four prefectures.

[c] See Table 6.4 for quotas for subsequent seasons.

[d] A large number of fishing vessels and their operating records in the Iwate and Miyagi Regions were lost in the earthquake and tsunami on March 11, 2011, which could have caused downward bias in the statistics for 2010 and 2011.

catch Pacific white-sided dolphins. Out of the total quota of 360 individuals, 154 were given to the hand-harpoon fishery off Iwate, 36 to the drive fishery off Shizuoka, and 170 to the hand-harpoon and drive fisheries off Wakayama (Tables 6.3 and 6.4). Pacific white-sided dolphins in Japanese coastal waters and those in the offshore western North Pacific are thought to belong to separate populations (Hayano et al. 2004). However, there is no positive evidence that the white-sided dolphins in the East China Sea, Sea of Japan, and Okhotsk Sea and those along the Pacific coast of Japan belong to a single population. Under such circumstances, it is necessary to confirm whether the abundance of the species off the Pacific coasts of Japan alone is large enough to support the catch.

The government procedure for deciding catch quotas is further considered in Chapter 6. Here, I will comment on how quotas are allocated to fishermen and the compliance with quotas. The Fisheries Agency decides a quota for each species and divides it among the prefectures. Then, the prefectural governments divide them among the fishery types and in some cases the operating entities. The drive fisheries have no difficulties with this procedure, because there is only one operating entity in each prefecture (Shizuoka and Wakayama). The allocation can be a problem in the case of hand-harpoon fisheries.

For example, 227 hand-harpoon vessels hunted Dall's porpoises in the 2007/2008 season (Table 2.3). If a total quota for the season, that is, 17,029 animals representing three populations of two species (including Pacific white-sided dolphins), were divided among them, the average share would be only 74 animals/vessel. Such a small quota can be fully subscribed within less than 1 month. A full month's operation by a single vessel could catch 80–110 individuals in winter off Sanriku and 240–360 in summer in the Okhotsk Sea where vessels of higher efficiency operated (see the preceding text in this section). No fishermen will be satisfied with such a small quota.

In order to avoid this situation, a system introduced was similar to that in earlier Antarctic whaling, that is, every vessel had an equal right to hunt until all vessels received a simultaneous order to stop operations. The Fishery Division of the Iwate Prefecture Government in November 1993 presented a document to the Japanese Fisheries Agency reporting that in the 1991 season it monitored the progress of fishing and successfully stopped the fishery operation when the catch reached the quota and stated that it would repeat the system in the next season. The Fishery Division of Iwate Prefecture did not have its own information system and must have relied on catch data offered by the fisheries cooperative unions, so the key to stop hunting was in the hands of the fishermen. This situation was similar to what was described by EIA (1999), which reported that in the Iwate case it was a voluntary group established by the hand-harpoon fishermen that decided when hunting was to stop. A similar method was used in Hokkaido, where the Regional Fishery Coordination Committee gave the order to stop hunting in an early stage of the quota system after which the task moved to the prefecture government. In both Hokkaido and Iwate, the participation of the prefecture government seems to have been only nominal; it determined the closure date for the season based on information presented by the fishermen and in consultation with them and sent the order through the FCUs. In the 1960s each local fish market or FCU prepared tables from their landing records, and this formed the basis of the prefecture statistics. We know from informants that there was misreporting of various kinds, intentional or unintentional and with or without consent of the local government.

The situation in Miyagi Prefecture was apparently more difficult. The number of hand-harpoon vessels decreased from a maximum of 40 to the recent 7 vessels (Table 2.3) and had a quota of about 300 porpoises. They had a license to hunt in the nearby waters off Miyagi or Iwate but not in the more distant Hokkaido waters (Table 2.4). It would be unprofitable for them to send the fleet to the Hokkaido coast with such a small quota. Their near-water operation was apparent also because their fishing season was from November to April (Table 6.4). In the 2007/2008 season, the Miyagi fishermen had a Dall's porpoise quota of 269 *dalli*-types and 30 *truei*-types and operated in the nearby grounds where about 95% of Dall's porpoises were represented by *dalli*-types (Kasuya 1978). They reported a take of 254 *dalli*-types (Table 2.9). If the reports were accurate, they must have had an extremely difficult operation in selecting *dalli*-types, which were rare in the fishing grounds. (The Fishery Agency recently agreed that the fishermen misidentified the color types: N. Kurasawa, pers. comm. 2011.). However, this is not reflected in the quota of subsequent seasons (see Table 6.4).

These are questions on the officially reported statistics. There have been reports of hidden catches. A newspaper reported that five or six minke whales that were putatively taken in trap nets in Iwate and Aomori Prefectures were sent to Ishinomaki City, Miyagi Prefecture, and that two to three of them had hand-harpoon heads in the body (reported by a newspaper Nihon Keizai Shinbun [Japan Economy News] dated July 10, 1990). In the 1990s, a small-type whaler found a hand-harpoon head in the body of a northern-form short-finned pilot whale killed by his ship. On May 16, 1996, the anterior half of a gray whale with numerous hand-harpoon heads in the body was stranded on the coast of Suttsu (42°35′N, 140°15′E) on the western coast of Hokkaido (Brownell and Kasuya 1999). Its posterior part was not found. The Fisheries Agency explained that the harpoon heads were made of stainless steel and differed from those used by Japanese fishermen. As far as I could tell from the photographs, they were not different from those used by Japanese fishermen (Figure 2.1). Japanese fishermen prefer stainless steel harpoon heads for use with an electric shocker because of rust resistance. Gray whales could be killed if several hand-harpoon vessels with electric shocker cooperated. It is almost impossible to obtain statistics on these illegal operations.

Table 2.10 compares catches of small cetaceans by various fisheries between 1988 and 2004. The year 1988 was marked by the end of commercial whaling in March and a peak catch of dolphins and porpoises in Japan. In 2004, this situation almost ended and there was some change in the species composition of the catch. The alarming high level of Dall's porpoise catch returned to a lower level, although this did not necessarily mean sustainability and some questions on the reliability of statistics still remained. Catches of striped dolphins continued, suggesting further depletion of the stock. The southern-form short-finned pilot whale population, which has been hunted by the crossbow fishery in Okinawa (classified as a hand-harpoon fishery by the Fisheries Agency), drive fishery and small-type whaling, showing a decline in landings. Recent increases in the catches of Risso's dolphins and common bottlenose dolphins need to be watched with caution.

2.6 HAND-HARPOON FISHERY OFF CHOSHI

A hand-harpoon fishery for dolphins and porpoises has been operated at a low level by members of the Sotokawa and Nishizaki Fishery Cooperative Unions in Choshi, Chiba Prefecture, and has landed the catch at the Choshi Fish Market. Dolphins taken by other fisheries such as in large-mesh drift nets were also landed at the Choshi Fish Market. Table 2.11 distinguishes the catch of the hand-harpoon fishery from that of other fisheries only since 1977. Kasuya (1976, in Japanese) reported information obtained from the Choshi Fish Market that indicated that the hand-harpoon fishery mostly took striped dolphins and estimated the total annual landing of the species by various fisheries at around 1500, although the basis of the estimation was unclear.

TABLE 2.10
Take of Small Cetaceans in Japanese Fisheries, a Comparison between 1988 and 2004

Species	Hand Harpoon	Drive	Small-Type Whaling	Live Capture[a]	Salmon Drift Net	Trap Net	Other Fisheries	Stranded	Total
Common porpoise									
1988	6	0	0	0	0	0	71	0	77
2004	0	0	0	0	0	1	0	1	2
Dalli-type Dall's									
1988[b]	40,367	0	0	0	1134	54	530	0	42,085
2004	4,614	0	0	0	0	0	0	1	4,615
Truei-type Dall's									
1988[b]	+	0	0	0	0	0	0	0	+
2004	9,175	0	0	0	0	0	0	0	9,175
Finless porpoise									
1988	0	0	0	0	0	14	1	0	15
2004	0	0	0	0	0	1[c]	6	81	88
Common dolphins									
1988	153	0	0	0	0	7	12	0	172
2004	0	0	0	0	0	1	0	1	2
Pac. white-sided d.									
1988	0	0	0	0	0	0	0	0	0
2004	0	0	0	(12)	0	12[c]	0	2	14
Striped dolphin									
1988	104	2,123	0	0	0	17	166	0	2,410
2004	83	554	0	0	0	0	0	4	641
Pantropic. spot. d.									
1988	38	1,837	0	4	0	0	0	0	1,879
2004	2	0	0	0	0	0	0	18	20
Bottlenose dolphins									
1988	32	729	0	51	0	2	14	0	828
2004	53	594[d]	0	(115)	0	6[e]	0	3	656
N. right whale. d.									
1988	0	0	0	0	0	0	268	0	268
2004	0	0	0	0	0	0	0	0	0
Risso's dolphin									
1988	0	109	0	15	0	6	0	0	130
2004	60	444[f]	7	(7)	0	1	0	4	516
Northern form[g]									
1988	0	0	98	0	0	0	1	0	99
2004	0	0	13	0	0	0	0	0	13
Southern form[g]									
1988	116	327	30	0	0	9	13	0	495
2004	72	62	29	0	0	0	0	4	167
False killer whale									
1988	6	42	0	22	0	2	0	0	72
2004	3	0	0	(4)	0	4[c]	0	1	8
Killer whale									
1988	0	0	7	0	0	0	0	0	7
2004	0	0	0	0	0	0	0	0	0
Baird's b. whale									
1988	0	0	57	0	0	0	0	0	57
2004	0	0	62	0	0	2	0	4	68

(Continued)

TABLE 2.10 (*Continued*)
Take of Small Cetaceans in Japanese Fisheries, a Comparison between 1988 and 2004

Species	Hand Harpoon	Drive	Small-Type Whaling	Live Capture[a]	Salmon Drift Net	Trap Net	Other Fisheries	Stranded	Total
Others and unknown									
1988	1	0	0	2	0	8	988	0	999
2004	0	0	0	0	0	1	0	27	28

Japanese pelagic whaling and large-type whaling ceased operation by March 31, 1988, and the catch of small cetaceans underwent a peak in 1988. Based on the Fisheries Agency statistics.

[a] Live-captured animals were selected for aquariums from the take of drive fisheries or other coastal fisheries. Live captures in 1988 could not be traced back to the source fisheries and were not included in the statistics for them, but those in 2004 (total in parentheses) were able to be traced back to the source fisheries and included in their catches.

[b] The two color types of Dall's porpoise segregate geographically. The 1988 statistics did not distinguish between the two types, and both were combined in the *dalli*-type.

[c] All the individuals were sold as live specimens.

[d] Of which 110 were sold as live specimens.

[e] Of which 5 were sold as live specimens.

[f] Of which 7 were sold as live specimens.

[g] Short-finned pilot whales off the Pacific coast of Japan contain two geographical forms with an approximate distribution boundary at 35°N latitude.

Table 2.11 shows as an indication of magnitude of the hand-harpoon fishery the numbers of operating entities and vessels that landed small cetaceans. After 1977, there were declines both in the number of operating entities and in their catches. This means that small-cetacean hunting has changed into a side business of other fisheries.

Since 1993, the fishery has obtained an annual quota of 80 striped dolphins but reported only a few catches. In the 2008/2009 season, 11 hand-harpoon vessels in Chiba Prefecture obtained a quota of 64 striped dolphins. This quota was eight animals fewer than in the previous season while the fleet size was the same.

2.7 HAND-HARPOON FISHERY AROUND TAIJI

The Taiji Area has a long tradition of hunting and consuming cetaceans. In the early 1960s, a small-type whaler *Katsu-maru* based at Taiji operated daily for small cetaceans. Other small fishing vessels that worked in various fisheries also occasionally landed hand-harpooned dolphins at the Taiji Fish Market, which was operated by the Taiji Fishery Cooperative Union. The small-type whalers took larger species such as the short-finned pilot whale, which was particularly preferred by local people, the killer whale, and ziphiids, and the smaller fishing vessels caught various smaller species including striped dolphins.

The small-cetacean fishery of Taiji recorded three successive changes in a short period around 1970:

1. First drive of short-finned pilot whales located by scouting vessel(s) in offshore waters in 1969
2. Expansion of the hand-harpoon fishery for striped dolphins in 1970 (the hand-harpoon hunters formed a cooperative group and became the founding group of the drive fishery, which began in 1971 on short-finned pilot whales and later expanded to striped dolphins and other species)
3. Start of driving striped dolphins in 1973 by the earlier mentioned cooperative group

Miyazaki (1980) analyzed records of cetaceans landed at the Taiji Fish Market during 17 years (1963–1979). Statistics for subsequent years are available in Kishiro and Kasuya (1993). The following are the average annual landings of dolphins compared between before and after the previously mentioned changes in the cetacean fisheries. Ranges are in parentheses (statistics are based on Myazaki 1980).

Year	1964–1969	1970–1979
Striped dolphin	578 (331–819)	1152 (562–2397)
Short-finned pilot whale	96 (52–134)	215 (91–479)
Risso's dolphin	54 (33–83)	15 (0–62)
Bottlenose dolphin	34 (14–66)	39 (3–103)

Other small-cetacean species landed at Taiji during this period were 53 killer whales and 40 Cuvier's beaked whales, which are most likely to have been taken by the small-type whalers. The striped dolphin was the most numerous in the catch through the whole period, followed by the short-finned pilot whale. The catches of these two species doubled after the changes in the fishery around 1970. As a result of the establishment of such efficient ways of hunting, other fishermen almost ceased harpooning dolphins (Kishirao and Kasuya 1993). In the 1980s, the drive fishery further increased catches of the striped dolphin, short-finned pilot whale, and bottlenose dolphin (Chapter 3).

Small-type whaling is a type of modern whaling using vessels smaller than 50 gross tons equipped with a whaling cannon of caliber 50 mm or smaller. It is allowed to take minke

TABLE 2.11
Dolphins Landed at Choshi in Chiba Prefecture[a]

Year[b]	Striped Dolphin	Common Dolphin	Pantropical Spotted Dolphin	Pacific White-Sided Dolphin	Black Fish	Others and Unknown	Total	No. of Operating Bodies and Vessels: Hand Harpoon	No. of Operating Bodies and Vessels: Large Mesh Drift Net
1972	1					1168	1169	11 (26)	6 (12)
1973					217	89	306	11 (26)	9 (12)
1974	117				*18*	1544	1679	11 (26)	9 (12)
1975	756				*5*	90	851	11 (26)	9 (12)
1976	295	13	17		1	16	342[c]	9 (—)	6 (—)
1977		31				1	32[d]	2 (2)	—
1978				*15*			15[e]	—	—
1981					*8*	*63*	71	—	7 (7)
1982						29; *66*	93	(9)	(6)
1983						63; *64*	127	(12)	(5)
1984						*22*	22	0	6
1986	*82*				*3*	1	86	0	6 (6)
1987[f]	*73*				*2*		75	0	(5)
1986[g]	10						10	—	—
1987	20					*1*	21	—	—
1988	38					*1*	39	—	—
1989	48					*1*	49	—	—
1990	67						67	—	—
1991	14						14	—	—
1992						6	6	—	—
Quota since 1993	80	0	0	0	0		80		
1993	6						6	—	—
1994	7						7	—	—
1995	6				*1*		7	—	—
1996						*2*	2	—	—
2002				*1*		*1*	2	—	—
2003						*1*	1	—	—
2004						*1*	1	—	—
2005–2007									
Quota for 2007/2008	72	0	0	0	0		72		

Figures in roman are catches by the hand-harpoon fishery and in italics catches identified as from other fisheries. In parentheses are number of vessels that operated. From Fisheries Agency statistics. See Table 6.3 for details of the quotas.

[a] No distinction was given until 1976 between catches by hand-harpoon and large-mesh drift-net fisheries. A quota of 80 striped dolphins was given in 1993 to the drive fishery.

[b] No landing of cetaceans in the 1979, 1980, 1985, 1997–1999, 2000, 2001, and 2005–2013.

[c] Includes 11 common dolphins and 5 finless porpoises taken in trap nets in the prefecture.

[d] Includes one common dolphin and one dolphin of unknown species taken in trap nets in the prefecture.

[e] Catch of seine net landed at Kamogawa Port, Chiba Prefecture.

[f] Statistics of the Offshore Division (1972–1987) end here. Striped dolphins in 1986 and 1987 were reported as taken by other fisheries, but they were likely taken by the hand-harpoon fishery.

[g] Statistics of Coastal Division (1986–) start here. Figures disagree between the two statistics.

whales and toothed whales other than sperm whales. The operation was placed under a license system in December 1947 by the Ministry of Agriculture and Forestry (Ohsumi 1975). The fishery received a catch quota of minke whales first in 1977 (Tato 1985, in Japanese), Baird's beaked whales in 1984 (Kasuya 1995b, in Japanese), and northern-form short-finned pilot whales in 1986 (Kasuya and Tai 1993; Kasuya 1995a, in Japanese). Hunting of other species was unregulated for a longer period. A small catcher boat of the category, *Katsu-maru*, operated in Taiji until 1979, but it moved operations elsewhere during the period 1980–1987 because it could not compete with the drive fishery in the hunting of pilot whales (Kasuya 1976, in Japanese; Kishiro and Kasuya 1993).

Rough-toothed and spotted dolphins have been identified in sighting surveys off Taiji, and skeletons of these species were located in the bone yards of small-type whalers in Taiji in the 1970s. Thus, there is no doubt that hand-harpoon fishermen or small-type whalers have taken these species. However, we do not find records of these species in the earlier catch statistics, suggesting that these species have not been distinguished in the statistics from striped or bottlenose dolphins.

Dolphin hunters in Taiji once identified a type of dolphin called *haukasu*. According to K. Shimizu, gunner captain of the small-type whaler *Katsu-maru*, the name meant "hybrid species with blotches (or spots)." Kasuya and Yamada (1995) took into account a statement by a local fish dealer S. Mizutani that it "had no spots on the body," but "had dark color on the belly, white lips, and a rostrum longer than a bottlenose dolphin" and concluded that it was a rough-toothed dolphin (Chapter 11). A firm conclusion would require more information.

As noted earlier, the hand-harpoon fishery for small cetaceans required a license beginning in 1989, and the license was issued only for vessels that had a record of recent operation. This was an early step toward control of the fishery with quotas and was a response of the Coastal Division of the Fisheries Agency to the criticisms against leaving small-cetacean fisheries uncontrolled. Another factor behind the government effort was a fear of uncontrollable expansion of small-cetacean fisheries following the cessation of commercial whaling for large cetaceans. In 1989, 15 Taiji vessels obtained a license for the hand-harpoon fishery for small cetaceans with a season from February 1 to August 31. I do not have a list of these vessels, but the number agrees with the number of vessels that belonged to the cooperative group of drive fisheries in the 1980s. Shortly later, fishermen in other nearby villages such as Katsuura and Miwasaki found the prospect attractive and started hand-harpoon fisheries. Such unlicensed operations increased to 147 vessels by the year 1991 and became a political issue. Finally, all the vessels obtained licenses (Kishiro and Kasuya 1993).

The hand-harpoon fishermen in Wakayama Prefecture obtained a catch quota of 300 dolphins of unspecified species (including short-finned pilot whales, but excluding killer whales) in 1991 and reported a total take of 456 dolphins, which included 253 Risso's dolphins, 57 bottlenose dolphins, 45 spotted dolphins, 10 striped dolphins, 4 rough-toothed dolphins, 3 short-finned pilot whales, and 84 unidentified dolphins (Kishiro and Kasuya 1993). In 1992, the quota increased to 400 dolphins (short-finned pilot whales not included). In 1993, a total quota of 520 dolphins was given by species: 100 striped dolphins, 70 spotted dolphins, 100 bottlenose dolphins, and 250 Risso's dolphins. The Taiji driving team probably objected to sharing their quota for the valuable short-finned pilot whales with the newcomers (dolphin hunters in nearby locations) and instead agreed to give them Risso's dolphins. The Japanese government replaced the given catch quotas with new ones in 2007/2008 season for the 100 licensees. The total number was 547 dolphins, which was an increase of 27 dolphins as a result of an addition of 36 Pacific white-sided dolphins and a subtraction of 5 bottlenose dolphins and 4 Risso's dolphins (Table 2.12).

The license for hand-harpoon hunting for dolphins is not renewed if the licensee does not operate within a certain period.

TABLE 2.12
Dolphins Taken by Hand-Harpoon Fishery in Wakayama Prefecture (Which Includes Taiji and Other Locations) since 1993, When the Fishery Operated with Species Quotas Given by the Government

Dolphin Species	Striped Dolphin	Pantropical Spotted Dolphin	Bottlenose Dolphin[a]	Risso's Dolphin	Pacific White-Sided Dolphin
Quota since 1993	100	70	100	250	0
1993	88	50	40	232	
1994	98	49	40	141	
1995	83	32	64	185	
1996	92	67	98	279	
1997	57	23	57	148	
1998	73	63	95	265	
1999	76	38	68	227	
2000	65	12	79	119	
2001	66	10	44	107	
2002	77	18	38	154	
2003	68	30	52	168	
2004	83	2	43	60	
2005	60	13	66	46	
2006	36	5	75	105	
Quota for 2007/2008	100	70	95	246	36
2007	86	16	97	185	0
2008	65	0	93	122	0
2009	98	3	77	94	7
2010	100	7	38	126	0
2011	96	2	40	104	0
2012	94	12	73	52	2
2013	67	14	68	38	0

From Fisheries Agency statistics. See Tables 3.17 and 3.18 for statistics before 1993 and Table 6.3 for the quotas. This table has been revised with the addition of new data available since the original Japanese language edition appeared in 2011.

[a] Presence of Indo-Pacific bottlenose dolphin has not been confirmed.

Thus, the number of licensees decreased with time (Table 2.3). The total number of licensees in Wakayama Prefecture was 116 in 1994, 116 in 1996, and 100 in the 2000/2001 to 2007/2008 seasons. According to another record of the Coastal Division of the Fisheries Agency, a total of 104 licenses in 1994 were distributed to 10 FCUs, from the largest of 32 at Katsuura and 28 at Taiji to the smallest of 2 at Shingu and 1 at Shimo-tawara and Tanabe. It would seem to be difficult to control operations of so many hand-harpoon vessels working with rather small catch quotas.

In the past, Taiji people did not favor meat of the Risso's dolphin, and fishermen even discarded the meat overboard after removing the internal organs for personal consumption. However, since 1988 when Japanese commercial whaling for large cetaceans ended, the price of Risso's dolphins increased in Taiji to over 200,000 yen/animal. The increase rate (1991/1987) was 3.9 times for the species, which was much higher than the value of 2.6 times for short-finned pilot whales, the most favored species by Taiji people, or 2.1 times for other smaller delphinids (Table 3.21). Reportedly, the color of the meat, which resembles minke whale meat, attracted the eyes of consumers. This situation might be changing now because of an oversupply of baleen whale meat from Japanese scientific whaling programs (Sakuma 2006, in Japanese).

2.8 CROSSBOW FISHERY IN NAGO

Nago (26°35′N, 128°00′E) on Okinawa Island in the southwestern islands of Japan was known to have traditional opportunistic dolphin driving cooperatively operated by the community. This tradition almost disappeared in the 1970s (Chapter 3), but the tradition of dolphin meat consumption remained. The crossbow fishery started in Nago to respond to the demand in 1975. The Japanese Fisheries Agency classified this fishery as a hand-harpoon fishery for regulatory purposes, although the technique used was quite different from that in the hand-harpoon fishery.

It is my understanding that this technique was invented to take advantage of a loophole in Japanese whaling regulation. In Japan, whaling was defined as an activity of fishing for whales with *harpoon discharged from cannon mounted on motor-powered vessel*, and the activity was strictly regulated and finally prohibited for commercial purpose as of April 1988 (Chapter 7). One of the loopholes was to shoot the harpoon from a crossbow, and another was the use of a sailing vessel. A person with the latter plan or idea of using a sailing vessel was reported in a newspaper around 1988 but did not carry through with it.

The equipment of the crossbow fishery is described by Okinawa-ken Suisan Shikenjo (Fishery Experimental Station of Okinawa) (1986, in Japanese). The fishermen use a fishing vessel of 3–5 gross tons with a crew of 2–3. A swivel is placed on the bow for the crossbow. The crossbow is T-shaped; the vertical arm works as a slide for the harpoon and the bottom end as a handle. A rubber belt attached to both ends of the transverse arm propels the harpoon when the trigger is released. The harpoon is made of a steel pipe 2.7 m in length and 16 mm in diameter, equipped with a detachable harpoon head on the front end, and is fitted to the middle of the rubber belt. The rubber belt is stretched by hand with the help of a gear to the trigger cock. The shooting range is about 20 m.

The crossbow fishery at Nago was first started by 6–7 vessels (Kishiro and Kasuya 1993) in 1975 when no license was required. Six vessels obtained a license from the Regional Fishery Coordination Committee in February 1, 1989, following the policy of the Fisheries Agency (see Chapter 6). In 1994, the fishery had six vessels in the range from 2.5 gross tons (with engine of 35 hp) to 8.5 gross tons (with 120 hp engine).

The group of vessel owners set a voluntary quota of a total of 100 dolphins/year in 1991 (Kishiro and Kasuya 1993). In 1993, the Fisheries Agency set a quota by species: 100 short-finned pilot whales, 10 false killer whales, and 10 bottlenose dolphins (Kasuya 1997, in Japanese). This was replaced in the 2007/2008 season (i.e., the fishing season that started in 2007) by a new quota of 92 short-finned pilot whales, 20 false killer whales, and 9 bottlenose dolphins for a total of six vessels (Tables 2.13 and 3.22). The statistics of number and species have to rely on the reports of fishermen or the weight landed, because most of the catch is processed offshore. Compliance with the quotas appears to me as a problem in this fishery as for other hand-harpoon fisheries.

The Nago fishery is allowed to operate from February 1 to October 31, with an annual quota determined, for management purpose, for the 12-month period that starts on October 1.

TABLE 2.13
Dolphins Taken by Crossbow Fishery in Nago in Okinawa Prefecture since 1993, When the Fishery Operated with Species Quotas Given by the Government

Dolphin Species	Southern Form[a]	False Killer Whale	Bottlenose Dolphins[b]	Risso's Dolphin	Others and Unknown	Total
Quota since 1993	100	10	10			120
1993	89	2	9			100
1994	81		10			91
1995	90	9	10	14		123
1996	84	10	4			98
1997	66	3	8		1[c]	78
1998	61	8	7			76
1999	79	5	8			92
2000	89	8	8			105
2001	92	8	8			108
2002	38		3			41
2003	36	4	7			47
2004	72	3	10			85
2005	90	1	10			101
2006	56	5	12			73
Quota for 2007/2008	92	20	9			121
2007	79	4	4			87
2008	62	5	1			68
2009	54	1	4			59
2010	34	0	1			35
2011	46	3	3			52
2012	25	0	3			28
2013	47	0	3			50

From Fisheries Agency statistics. See Table 3.22 for earlier catches and Table 6.3 for details of the quotas. This table has been revised with the addition of new data available since the original Japanese language edition appeared in 2011.

[a] One of the two geographical forms of short-finned pilot whale distributed approximately south of 35°N latitude off the Pacific coast of Japan.

[b] May include both species of the genus.

[c] Killer whale.

However, the main operation lasts from February to June. During the fishing season, crossbow vessels repeat trips, extending for a maximum of 5-6 days each. The catches are flensed on board and the meat stored on ice until return to port. The catch used to be boiled at dockside in Nago Port and sent to Nago City (Kishiro and Kasuya 1993). When I visited the Nago Port in February 1991, there were several half oil drums for cooking whale meat but no trace of recent use, perhaps because of the beginning of the season or change in the distribution system. The price of dolphin meat was reportedly 1000 yen/kg in 1991. Endo (2008, in Japanese) reported a system that started around 1993, in which meat of short-finned pilot whales and false killer whales was sent through the Nago Fishery Cooperative Union to Fukuoka Chuo Wholesale Market in Fukuoka (33°35′N, 130°25′E), northern Kyushu, and meat of other dolphin species was sold locally. Kasuya (1997, in Japanese) also recorded that when the supply of whale meat declined around 1988, some of the meat of short-finned pilot whales was transshipped by sea to Osaka (34°40′N, 135°30′E) and to Shimonoseki (34°55′N, 130°55′E) at the western end of Honshu.

Table 2.13 shows the statistics of operation by species quota that started in 1993. The earlier statistics are in Table 3.22, where drive and crossbow fisheries are not separated and perhaps some real hand-harpoon catch is also included before 1980 when there are too many vessels recorded. The record of 140 melon-headed whales in 1979 could be from a driving operation conducted with the cooperation of a number of vessels.

In the mid-1980s, with news that the crossbow fishery in Okinawa was going well, some fishermen in Ayukawa on the Pacific coast of northern Honshu planned to use this method for hunting northern-form short-finned pilot whales. In 1982, small-type whalers resumed hunting the species off Ayukawa and were eager to increase the annual catch, but we scientists refused the request with a belief that the population was not large enough to sustain added take (Chapters 7 and 12). We feared repetition of the mismanagement of the stock that happened after World War II (Kasuya and Tai 1993). The idea of using the then unregulated crossbow method for the fishery appeared to be a way to escape government regulation. This idea did not proceed further, probably because of adverse pressure by the Fisheries Agency.

REFERENCES

[*In Japanese Language]

*Anraku, T. 1985. Fishing activities around Saikai and Goto Islands. *Kikan Kokogaku* 11: 39–42.

Brownell, R.L. Jr. and Kasuya, T. 1999. Western gray whale captured off western Hokkaido, Japan. Paper SC/51/AS25. 7pp. Document presented to the *51st Meeting of the Scientific Committee of IWC* (available from IWC Secretariat, Cambridge, U.K.).

Cipriano, F. and Palumbi, S.R. 1999. Rapid genotyping techniques for identification of species and stock identity in fresh, frozen, cooked and canned whale products. IWC/SC/51/9. 25pp. Document presented to the *51st Meeting of the Scientific Committee of IWC* (available from IWC Secretariat, Cambridge, U.K.).

*Doi, S. 1902. *Ordinary Methods of Fishery Production*. Kozanbo, Tokyo, Japan. 373pp.

EIA (Environmental Investigation Agency). 1990. The global war against small cetaceans. EIA, London, U.K. 57pp.

EIA. 1999. Japan's senseless slaughter: An investigation into the Dall's porpoise hunt—The largest cetacean kill in the world. EIA, London, U.K. 22pp.

*Endo, A. 2008. *Small Scale Coastal Whaling and Changing Utilization of Whale Resources—Ways of Whale Meat Consumption and the Function*. Doctoral thesis, Hiroshima University, Higashihiroshima, Japan. 149pp.

*Furuki, M. 1989. International atmosphere and the problem of porpoise hand-harpoon fishermen. *Gekkan Weeks* 8: 154–159.

Goode, G.B. 1884. The whales and porpoises. pp. 7–32. In: Goode, G.B. (ed.). *The Fisheries and Fishery Industries of the United States, Section I. Natural History of Useful Aquatic Animals*. U.S. Commission of Fish and Fisheries, Washington, DC. 895pp.

*Hashiura, Y. 1969. *History of Whaling at Taiji in Kumano*. Heibonsha, Tokyo, Japan. 662pp.

*Hattori, T. 1887. Methods of hunting dolphins and porpoises. *Dainihon Suisankaiho* 66: 39–41.

Hayano, A., Yoshioka, M., Tanaka, M., and Amano, M. 2004. Population differentiation in the Pacific white-sided dolphin *Lagenorhynchus obliquidens* inferred from mitochondrial DNA and microsatellite analyses. *Zool. Sci.* 21: 989–999.

Heizer, R.F. 1943. Aconite poison whaling in Asia and America, an Aleutian transfer to the New World. *Smithsonian Inst. Bull.* 133: 419–468, pls. 19–23.

*Hiraguchi, T. 1986. Dolphins, from capture to disposal. pp. 373–382. In: Yamada, Y. (ed.). *Mawaki Archaeological Site in Noto-cho, Ishikawa Prefecture*. Noto Kyoiku Iinkai Mawaki Iseki Hakkutsu Chosadan, Noto-cho, Japan. 482pp.

*Hirashima, Y. and Ono, S. 1944. Dolphin and porpoise fisheries in the Abashiri area. *Hokusuishi Geppo* 1(2): 82–90.

*Hokusui Kyokai. 1977. *Draft History of Fisheries in Hokkaido*. Kokusho Kankokai, Tokyo, Japan. 874pp. (First published in 1935).

*ICR (Institute of Cetacean Research). 1990. Report of investigation of porpoises in Japanese waters, 1st year of Heisei. Institute of Cetacean Research, Tokyo, Japan. 78pp.

*ICR. 1991. Report of investigation of porpoises in Japanese waters, 2nd year of Heisei. Institute of Cetacean Research, Tokyo, Japan. 100pp.

*ICR. 1992. Report of investigation of porpoises in Japanese waters, 3rd year of Heisei. Institute of Cetacean Research, Tokyo, Japan. 101pp.

*ICR. 1993. Report of investigation of porpoises in Japanese waters, 4th year of Heisei. Institute of Cetacean Research, Tokyo, Japan. 44pp.

*Ishida, T. 1917. *Considerations on Fishery Products*. Kokko Insatsu, Tokyo, Japan. 392pp.

IWC. 1991. Report of the small cetacean subcommittee. *Rep. Int. Whal. Commn.* 41: 172–190.

*Kamiyama, S. 1943. *Marine Leather*. Suisan Keizai Kenkyusho, Tokyo, Japan. 356pp.

*Kaneko, M. 1973. Animal remains from the Onkoromanai shell mound. pp. 187–246. In: Oba, T. and Oi, H. (eds.). *Study of Okhotsk Culture 1. Onkoromanai Shell Mound*. Tokyo Daigaku Shuppankai, Tokyo, Japan. 291pp.+40 pls.

Kasuya, T. 1975. Past occurrence of *Globicephala melaena* in the western North Pacific. *Sci. Rep. Whales Res. Inst.* 27: 95–110.

*Kasuya, T. 1976. Stock of striped dolphins. *Geiken Tsushin* 295: 19–23, 296: 29–35.

Kasuya, T. 1978. The life history of Dall's porpoise with special reference to the stock off the Pacific coast of Japan. *Sci. Rep. Whales Res. Inst.* 30: 1–63.

*Kasuya, T. 1981. Whale remains from the Kabukai A-site. pp. 675–783. In: Oba, T. and Oi, H. (eds.). *Study of Okhotsk Culture-3, Kabukai Site.* Vol. 2. Tokyo Daigaku Shuppan-kai, Tokyo, Japan. 727pp.+pls. 135–249.

Kasuya, T. 1982. Preliminary report of the biology, catch and population of *Phocoenoides* in the western North Pacific. pp. 3–19. In: Clark, J.G., Goodman, J., and Soave, G.A. (eds.). *Mammals of the Sea*, Vol. IV: Small Cetaceans, Seals, Sirenians and Otters. FAO Fisheries Series, No. 5. FAO, Rome, Italy. 531pp.

Kasuya, T. 1992. Examination of Japanese statistics for the Dall's porpoise hand harpoon fishery. *Rep. Int. Whal. Commn.* 42: 521–528.

*Kasuya, T. 1995a. Short-finned pilot whale. pp. 542–551. In: Odate, S. (ed.). *Basic Data on Rare Wild Aquatic Organisms of Japan (II).* Nihon Suisan Shigen Hogo Kyokai, Tokyo, Japan. 751pp.

*Kasuya, T. 1995b. Baird's beaked whale. pp. 521–529. In: Odate, S. (ed.). *Basic Data on Rare Wild Aquatic Organisms of Japan (II).* Nihon Suisan Shigen Hogo Kyokai, Tokyo, Japan. 751pp.

*Kasuya, T. 1997. Current status of Japanese whale fisheries and management of stocks. *Nihonkai Cetol.* 7: 37–49.

Kasuya, T. 1999. Examination of reliability of catch statistics in the Japanese coastal sperm whale fishery. *J. Cetacean Res. Manage.* 1(1): 109–122.

*Kasuya, T. 1999. Manipulation of statistics by the Japanese Coastal Sperm Whale Fishery. *IBI Rep.* (Kamogawa, Japan) 9: 75–92.

*Kasuya, T. 2000. Explanatory notes. pp. 155–203. In: Omura, H. *Journal of an Antarctic Whaling Cruise—Record of Whaling Operation in Its Infancy in 1937/38.* Toriumi Shobo, Tokyo, Japan. 204pp.

Kasuya, T. and Brownell, R.L. Jr. 1999. Additional information on the reliability of Japanese coastal whaling statistics. IWC/SC/51/O7. 15pp. Document presented to the *51st Meeting of the Scientific Committee of IWC* (available from IWC Secretariat, Cambridge, U.K.).

Kasuya, T. and Brownell, R.L. Jr. 2001. Illegal Japanese coastal whaling and other manipulations on catch records. IWC/SC/53/RMP24. 4pp. Document presented to the *53rd Meeting of the Scientific Committee of IWC* (available from IWC Secretariat, Cambridge, U.K.).

Kasuya, T. and Jones, L.L. 1984. Behavior and segregation of the Dall's porpoise in the northwestern North Pacific Ocean. *Sci. Rep. Whales Res. Inst.* 35: 107–128.

*Kasuya, T., Kaneko, M., and Nishimoto, T. 1985. Archaeology and related science 7: Zoology. *Kikan Kokogaku* 11: 91–95.

*Kasuya, T. and Miyashita, T. 1989. Japanese small cetacean fisheries and problems in management. *Saishū to shiiku* 51(4): 154–160.

Kasuya, T. and Tai, S. 1993. Life history of short-finned pilot whale stocks off Japan and a description of the fishery. *Rep. Int. Whal. Commn.* (Special Issue 14): 339–473.

*Kasuya, T. and Yamada, T. 1995. *List of Japanese Cetaceans.* Gieken Series 7. Nihon Geirui Kenkyu Sho (Institute of Cetacean Research), Tokyo, Japan. 90pp.

*Kawamura, A., Nakano, H., Tanaka, H., Sato, O., Fujise, Y., and Nishida, K. 1983. Results of small cetacean surveys from railway ferry boat between Aomori and Hakodate. *Geiken Tshushin* 351+352: 29–52.

Kishiro, T. and Kasuya, T. 1993. Review of Japanese dolphin drive fisheries and their status. *Rep. Int. Whal. Commn.* 43: 439–452.

*Kondo, I. 2001. *Rise and Fall of Japanese Coastal Whaling.* Sanyosha, Tokyo, Japan. 449pp.

Kondo, I. and Kasuya, T. 2002. True catch statistics for a Japanese coastal whaling company in 1965–1978. IWC/SC/54/O13. 23pp. Document presented to the *54th Meeting of the Scientific Committee of IWC* (available from IWC Secretariat, Cambridge, U.K.).

*Kurakami, M. 1925. *Fishery Items of Animals and Plants.* Sugiyama Shoten, Tokyo, Japan. 472+23pp.

*Kurino, K. and Nagahama, M. 1985. Dolphin hunting in Sagami Bay. *Kikan Kokogaku* 11: 31–34.

*Maeda, K. and Teraoka, Y. 1958. *Whaling.* Isana Shobo, Tokyo, Japan. 346pp. (First published in 1952).

*Matsui, K. and Uchihashi, K. 1943. On the Dall's porpoise off Tajima (preliminary report). *Hyogoken Chuto Kyoiku Hakubutsugaku Zasshi* 8+9: 53–109.

*Matsuura, Y. 1942. Story about dolphins and porpoises. *Kaiyo Gyogyo* 71: 53–103.

*Matsuura, Y. 1943. *Marine Mammals.* Ten-nensha, Tokyo, Japan. 298pp.

Miyashita, T. 1993. Distribution and abundance of some dolphins taken in the North Pacific driftnet fisheries. *Int. North Pacific Fish. Commn. Bull.* 53(III): 435–450.

Miyazaki, N. 1980. Catch records of Cetaceans off the coast of the Kii Peninsula. *Mem. Natl. Sci. Mus. (Tokyo)* 13: 69–82.

Miyazaki, N. 1983. Catch statistics of small cetaceans taken in Japanese waters. *Rep. Int. Whal. Commn.* 33: 621–631.

Miyazaki, N., Kasuya, T., and Nishiwaki, M. 1974. Distribution and migration of two species of *Stenella* in the Pacific coast of Japan. *Sci. Rep. Whales Res. Inst.* 26: 227–243.

*Mizuno, M. 1883. Questions and answers on catch, products and markets of dolphins and porpoises. *Dainihon Suisankai Hokoku* 21: 26–29.

*Naito, T. 1770. On whaling equipment. pp. 308–312. In: *Miscellanea of Choshu.* Vol. 13. Reprinted in 1975 by Aichiken Kyodo Shiryo Kankokai, Nagoya, Japan (total page not available).

*Nakazono, N. 2009. Whaling in Kyushu. pp. 22–36. In: Kojima, T. (ed.). *On Whales and Japanese—Revisit to Coastal Whaling.* Tokyo Shoten, Tokyo, Japan. 255pp.

National Research Institute of Far Seas Fisheries of Japan 1995. Japan Progress Report on Cetacean Research, April 1993 to March 1994. Report of the International Whaling Commission, Cambridge, U.K., vol. 45, pp. 239–244.

*Natori, T. 1945. *Whaling of Ainu in Volcano Bay.* Hoppo Bunka Shuppansha, Sapporo, Japan. 306pp.

*Nihon Hogei-gyo Suisan Kumiai. 1943. *Whaling Handbook.* Vol. 3. Nihon Hogeigyo Suisan Kumiai, Tokyo, Japan. 205pp.

*Noguchi, E. 1943. On the common dolphin and Dall's porpoise and their utilization. *Hyogo-ken Chuto Kyoiku Hakubutsugaku Zasshi* 8+9: 17–22.

*Noguchi, E. 1946. Dolphins and porpoises and their utilization. pp. 3–36. In: Noguchi, E. and Nakamura, A. (eds.). *Scomber Fishery and Utilization of Dolphins and Porpoises.* Kasumigaseki Shobo, Tokyo, Japan. 70pp.

*Oba, T. and Oi, H. (eds.) 1973. *Study of Okhotsk Culture*, Vol. 1: Onkoromanai Shell Mound. Tokyo Daigaku Shuppan Kai, Tokyo, Japan. 291pp.+40 plates.

Ohsumi, S. 1972. Catch of marine mammals, mainly of small cetaceans, by local fisheries along the coast of Japan. *Bull. Far Seas Fish. Res. Lab.* 7: 137–166.

Ohsumi, S. 1975. Review of Japanese small-type whaling. *J. Fish. Res. Board Canada* 32(7): 1111–1121.

*Okinawa-ken Suisan Shikenjo. 1986. *Methods and Instruments of Fisheries in Okinawa*. Okinawa-ken Gyogyo Shinko Kikin, Naha, Japan. 241pp.

*Omura, H., Matsuura, Y., and Miyazaki, I. 1942. *Whales: Their Biology and the Practice of Whaling*. Suisan Sha, Tokyo, Japan. 319pp.

*Otsuchi-cho Gyogyo-shi Hensan Iinkai. 1983. *History of Fisheries of Otsuchi Town*. Otsuchi-cho Gyogyo Kyodo Kumiai, Otsuchi, Japan. 1360pp.

*Oyamada, O. 1832. *Pictorial Description of Whaling*. Tatamiya, Hirado, Japan. Vols. 1: 40pp., 2: 40+6pp.

*Sakuma, J. 2006. *Survey of Consumption of Scientific Whaling Products (Whale Meat)—Supply, Price and Stock Piles*. Iruka Ando Kujira Akushon Nettowaku, Tokyo, Japan. 30pp.

*Sato, T. 1900–1902. Records of whaling in Hokkaido. *Danihon Suisankaiho* 216: 292–296; 221: 587–592; 233: 1318–1325.

*Tamura, I. 1996. *History of Fishery Culture*. Yuzankaku, Tokyo, Japan. 193pp.

*Tato, Y. 1985. *History of Whaling with Statistics*. Suisansha, Tokyo, Japan. 202pp.

Tobayama, T., Yanagisawa, F., and Kasuya, T. 1992. Incidental take of minke whales in Japanese trap nets. *Rep. Int. Whal. Commn.* 42: 433–436.

*Tsuboi, S. 1908. Tubes of bird bone found at a stone age site in Sakhalin. *Tokyo Jinruigaku Zasshi* 263: 157–164.

*Tsuboi, S. 1909. Stone age relics carved with whaling scene. *Toyogaku Zasshi* 25(330): 101–106.

*Uchida, I. 1954. *Marine Mammal Fisheries (Suisan Koza Gyogyo Hen)*. Vol. 9. Dainihon Suisan Kai, Tokyo, Japan. 26pp.

*Watanabe, M. 1984. *Fisheries in the Jomon Era*. Yuzan Kaku, Tokyo, Japan. 248pp.

*Watase, S. 1995. Hidden stories of whaling. pp. 235–240. In: Anon. (ed.). *Now We Shall Speak Up—Time is Over*. Vol. 1. Sei Sei Shuppan, Tokyo, Japan. 328pp.

Wilke, F., Taniwaki, T., and Kuroda, N. 1953. *Phocoenoides* and *Lagenorhynchus* in Japan, with notes on hunting. *J. Mammal.* 34(4): 488–497.

*Yahata, I. 1943. Needle case of bone. *Kodai Bunka* 14(8): 251–259.

*Yamaura, K. 1996. Fishery activities of the Jomon Era. pp. 103–148. In: Ohayashi, T. (ed.). *The Archaic Time in Japan 8, Tradition of Fishing People*. Chuo Koronsha, Tokyo, Japan. 478pp.

*Yoshida, S. 1939. Recent rise of the offshore dolphin and porpoise fishery with hand harpoon. *Suisan Kenkyushi* 34(1): 39–43.

3 Drive Fisheries for Dolphins

In a drive fishery, a school of fish or small cetaceans is captured by driving them into a bay or an area surrounded by a net or fence. This kind of fishery has been carried out in Japan at numerous sites and for various groups of living marine animals but is used now only to drive dolphins at limited locations. Historically, various levels of commercialism determined the method of operation. It changed from a primitive opportunistic operation with the cooperation of the entire local community and pursued only when a school of dolphins was found in a suitable location to a more active community-based operation with spotters at vantage points, and further to the most aggressive recent operation by a limited group of fishermen that send high-speed scouting vessels daily up to a maximum of 100 km offshore (third colored photo).

The method of killing also changed with time. In earlier operations at Kawana (34°56′N, 139°07′E) on the coast of the Izu Peninsula (34°37′N–35°05′N, 138°45′E–139°10′E; Figure 3.1) first, a group of dolphins were separated from the main school and herded to shore. Then young men entered the water and brought the live dolphins up on shore for slaughter and processing. In the 1960s, the Kawana fishermen used hooks on long poles to take live animals to the shore. In the 1970s, they cut the dolphin's throat at the boat side, tied a rope to the tail peduncle, and brought the dolphins up on the shore with a crane (fourth–eighth colored photo) (Kasuya and Miyazaki 1982; Kasuya and Miyashita 1989, in Japanese). In the 1980s, the drive fishermen at Taiji (33°37′N, 135°55′E) hand harpooned dolphins in the net, drew them up, and then killed them by lancing. More recently, they insert a knife into the medulla oblongata for a quicker death. This method might be seen more humane by outsiders, but pulling the live dolphin to the boat remains a difficult task.

Drive fisheries for small-cetaceans target schooling species. However, the school has to be of sufficient size to compensate for the labor spent in the operation. This explains why there have been no records of driving porpoises in Japan (they do not occur in large schools). The method could have originated from an opportunistic operation at a local community when a suitable animal school was found near the shore. Examples of such a dolphin fishery that are not limited to Japan are known from Newfoundland, the Orkney Islands, and the Shetland Islands (e.g., Tudor et al. 1883; Madsen 1992), and the method is still used in the Faroe Islands (Joensen 1976; Bloch 2007), where catch statistics have been recorded since the sixteenth century (Zachariassen 1993).

A cove or inlet with a deep and narrow entrance and broad interior is believed to be suitable for a dolphin drive. Inlets at Arari (34°50′N, 138°46′E) and Heda (34°58′N, 138°47′E) on the Izu coasts and Yamada Cove (39°25′N, 142°00′E) in Iwate Prefecture (38°55′N–40°27′N) on the Pacific coast of northern Honshu (which is a Japanese main island extending *c.* 33°25′N–45°30′N and 130°57′E–142°05′E) are such examples. Some particular shores are more likely to see strandings of cetaceans. A hypothesis of local geomagnetic anomalies has been proposed to explain this (Kirschvink 1990; Klinowska 1990). It assumes that dolphins use geomagnetic anomalies as a supplemental clue in navigation, and they are likely to be stranded in places where lines of equal geomagnetic anomaly cross the coast. Among several explanations proposed for causes of cetacean stranding, the hypothesis of geomagnetic anomalies is the one most supported by scientists. Villagers near such locations would have more opportunities for driving and for developing a drive fishery. If some members of a dolphin school are stranded, other members tend to remain in the vicinity. Such behavior helped the emergence of dolphin drive fisheries. It would be worthwhile to examine the geomagnetic environment of places with a long history of dolphin drive fisheries in Japan.

3.1 COASTS OF NOTO PENINSULA, SEA OF JAPAN

The Mawaki (37°18′N, 137°13′E) site on the east coast of Noto Peninsula (*c.* 36°50′N–37°30′N, 136°40′E–137°20′E) is a well-known archaeological site ranging from 6,000 to 2,000 years BC, a later period of the Jomon Era (14,000–1,000 years BC; various opinions exist about the dates) that produced a large number of dolphin skeletons. Hiraguchi (1986, in Japanese) identified them using the morphology of the cervical vertebrae and reported that the majority were Pacific white-sided dolphin and *Delphinus* sp. (cf. *D. capensis*) representing 143 individuals, followed by bottlenose dolphins (12), a pilot whale, and a Risso's dolphin. In addition, he identified a mandible attributable to a false killer whale. Identification of skulls and mandibles led to similar tallies, with the highest number being Pacific white-sided dolphins and common dolphins followed by a smaller number of bottlenose dolphins (Hiraguchi 1993, in Japanese). Because the excavation covered only a limited portion of the expected range of the site, caution was required before reaching firm conclusions about the species composition in the fishing activity at the time. However, the presence of such gregarious dolphin species suggests that there were dolphin drives, and the multiple species suggest that the bones represent more than one operation. However, it is unclear whether the people assisted dolphins to be stranded, drove dolphin schools found in the inlet, or even drove them from outside the inlet.

Numerous lance heads made of stone were recovered from the Mawaki site. Hiraguchi (1990, in Japanese) correctly interpreted them as used to stab the dolphins. They were likely to have been used purely for killing purposes, not for connecting

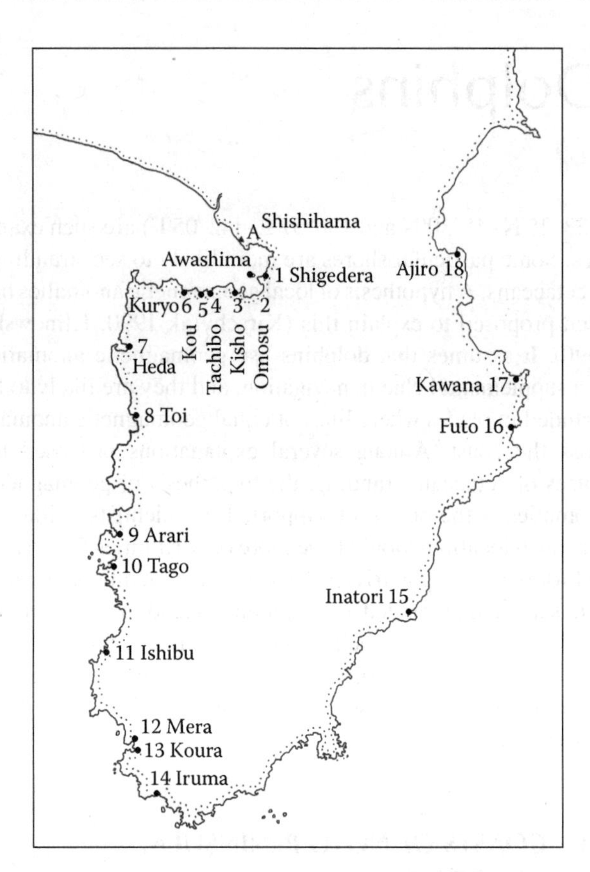

FIGURE 3.1 Locations of dolphin drive fisheries in the late nineteenth century on the coast of the Izu Peninsula (34°36′N–35°05′N, 138°45′E–139°10′E) recorded by Kawashima (1894, in Japanese). Shigedera (1) also operated at Awashima on the coast of an off-lying island. Omosu (2) and Nagahama (not illustrated) and Kuryo (6) and Ashibo (not illustrated) operated drive fisheries cooperatively. Shishihama (A) left records of sale of salted dolphin meat but no records of driving. Its participation in the fishery could have been limited to processing of dolphins taken by other villages. Only Arari (9), Futo (16), and Kawana (17) operated drive fisheries in the 1960s.

animals to the hunter. Since a dolphin carcass usually sinks, it is necessary to attach a line to the animal if it is to be killed at a distance from the shore, and a detachable harpoon head or barbed lance is necessary for the purpose; these have not been recovered from the Mawaki site. Thus, I conclude that the Mawaki people of the Jomon Era used stone lances to kill dolphins in shallow waters when the animals were unable to escape. Such a situation could occur either with naturally stranded animals or those driven to the shore.

This scenario is similar to the dolphin drive fishery at Mawaki village in the nineteenth century as described in *Suisan Hosai Shi* (Records of Fishery Activities) by Noshomu-sho Suisan-kyoku (Bureau of Fisheries of Ministry of Agriculture and Commerce) (1911, in Japanese). It describes a series of steps: (1) drive the dolphin school close to shore, (2) encircle the school with a large-mesh net, (3) hold a dolphin in the water and direct it toward the beach (as a dolphin cannot swim backward, it proceeds to the shallows), and (4) kill the dolphin with a lance. This method avoids the problem of retrieving dolphin carcasses from the bottom in deeper water and rather easily brings the animal alive to the shore for killing. A similar procedure is also found in an illustration depicting the dolphin drive fishery in Nagasaki Prefecture, northern Kyushu, published in *Gyogyo Shi* (Records of Fisheries [in Nagasaki Prefecture]) (Nagasaki-ken 1896, in Japanese), which states that fishermen in the water hold dolphins, direct them toward the shore, and make them strand (to be killed).

Kasuya and Yamada (1995, in Japanese) interpreted old common names of dolphins along the coasts of the Sea of Japan and East China Sea (northern Kyushu) based on descriptions and illustrations of Kizaki (1773, in Japanese): *niudou* is oki-gondo (current name of false killer whale), *handou* is hando-iruka (bottlenose dolphin), and *hase* is hase-iruka (*Delphinus capensis*). *Niudou* and *nyudo* mean an enlightened Buddhist with shaved head. There is uncertainty about the meaning of *handou* and *hando*, which could have meant a state before an enlightenment but could also have meant the comical actor in kabuki drama, a crowded noisy situation, or even a water tub (also see Section 3.8.2). They were unable to identify dolphins called *nezumi* and *shiratago* (also see Section 3.6). In the 1970s, fishermen in northern Kyushu and nearby Iki Island (33°47′N, 129°43′E) called every dolphin species with a long beak *nezumi* or *nezumi-iruka*, because they say they produce sounds like *nezumi* (mouse or rat). *Phocoena phocoena* is called currently in Japan *nezumi-iruka*, but the species has not been reported off northern Kyushu. Similar old names used in the drive fishery on the Izu coasts might have meant other species (see Section 3.8.2). We need to be cautious in identifying the species names of the 1700s.

The Record of Fisheries mentioned earlier (Nagasaki-ken 1896, in Japanese) gave a list of dolphins taken in the prefecture *Nyu-dou-iruka* (popularly known as *bouzu-iruka*,) and *hase-iruka* (popularly known as *ma-wiruka* [sic]). The former was stated to resemble the *goto-kujira* (pilot whale, cf. *Globicephala*) but measures 6–9 m and has small taro-tuber-like teeth. This is the *oki-gondo* (false killer whale) of the present day. The latter is stated to be found with the former species and measures from 5–6 shaku (a shaku is almost equal to a foot) to 7–8 shaku. It seems to represent *Delphinus* sp. Bottlenose dolphins often form mixed schools with false killer whales but grow to over 3 m (10 ft).

As cited earlier, the Records of Fishery Activities by Noshomu-sho Suisan-kyoku (Bureau of Fisheries of Ministry of Agriculture and Commerce) (1911, in Japanese) stated that people along Uonome Nishimura Bay in Arikawa (32°58′N, 129°07′E) in Goto Islands (see Section 3.5) operated a drive fishery since around 1688–1703 with known targets of two dolphin species, that is, *ma-iruka* or *hase-iruka* and *nyudo-iruka* or *bauzu-iruka*, and that villagers of Mawaki and other nearby villages on Noto Peninsula drove three species of dolphins: *ma-iruka*, *nyudo-iruka*, and *kamairuka* or *shisumi-iruka*.

The previously mentioned old names of dolphins and current interpretations are in Table 3.1. Pilot whales (*Globicephala* spp.) are very rare in the Sea of Japan and off northern Kyushu. Instead, false killer whales (Kasuya 1975; Tamura et al. 1986, in Japanese; Kasuya et al. 1988) are often found there. Therefore, aquarium staff in northern Kyushu

TABLE 3.1
Vernacular Names of Small Cetaceans in the Sea of Japan and East China Sea Areas That Appeared in Old Literature

Old Vernacular Name	Location[a]	Reference[b]	Current Common Name
Hase	Hizen	A	*Hase-iruka*, i.e., long-beaked common dolphin
Ma-iruka, i.e., *Hase-iruka*	Arikawa and Uonome	B	*Hase-iruka*
Mairuka	Noto	B	*Hase-iruka*
Nyudo	Hizen	A	*Oki-gondo*, i.e., false killer whale
Nyudo-iruka	Noto	B	*Oki-gondo*
Nyudo-iruka, i.e., *Bozu-iruka*	Uonome and Arikawa	B	*Oki-gondo*
Hando	Hizen	A	*Hando-iruka*, i.e., bottlenose dolphins
Kama-iruka, i.e., *Shisumi-iruka*	Noto	B	*Kama-iruka*, i.e., Pacific white-sided dolphin
Nezumi[c]	Hizen	A	Unknown
Shiratago	Hizen	A	Unknown

Source: Kasuya, T. and Yamada, T., *List of Japanese Cetaceans*, Gieken Series 7, Nihon Geirui Kenkyu Sho (Institute of Cetacean Research), Tokyo, Japan, 1995, 90pp., in Japanese.

[a] *Arikawa* and *Uonome* are in the Goto Islands, which are located in the East China Sea off northwestern Kyushu, *Hizen* covers Saga and Nagasaki Prefectures in northern Kyushu, and *Noto* or Noto Peninsula is on the middle Sea of Japan coast of Honshu.

[b] A, Kisaki (1773, in Japanese); B, Noshomu-sho Suisann-kyoku (Bureau of Fisheries of Ministry of Agriculture and Commerce) (1911, in Japanese).

[c] Kasuya and Yamada (1995, in Japanese) were unable to identify current common names for these two old vernacular names. In the 1960s, fishermen in some locations in Nagasaki Prefecture called dolphins they caught *nezumi* or *nezumi-iruka* because they produced sounds like *nezumi* (mice or rats). This suggests that the vernacular name is likely to have represented more than one dolphin species.

in the 1970s meant false killer whales by the word *gondo* (pilot whales or blackfishes). Similar cases are seen in recent records of fishermen and statistics.

The genus *Delphinus* was once considered to contain a single species with two geographical forms, but now the geographical variation is interpreted to represent two species of worldwide distribution, *D. delphs* and *D. capensis* with subspecies *D. c. tropicalis* (Jefferson and Van Waerebeek 2002), or three species of *D. delphis*, *D. capensis*, and *D. tropicalis*, with distribution of the third species limited to coastal waters extending from the Arabian Sea to the South China Sea (Rice 1998).

It was known in Japan that the *hase-iruka* from the Tsushima Islands (see Section 3.3) area, off northern Kyushu, has a longer beak and more teeth than the *ma-iruka* off the Sanriku (see Section 3.7) region on the Pacific coasts of northern Honshu. The former is currently attributed to *D. capensis*, inhabiting the Pacific coast of southwestern Japan, East China Sea and perhaps Sea of Japan, and the latter to *D. delphis* off the Pacific coast of northern Japan.

The Records of Fishery Activities by Noshomu-sho Suisan-kyoku (Bureau of Fisheries of Ministry of Agriculture and Commerce) (1911, in Japanese) stated that the dolphin drive method was best known at Mawaki in Takakura village, which was followed by Ogi (37°18′N, 137°14′N), Ushitsu (37°18′N, 137°09′E), Nakai (37°14′N, 137°14′E), and others. Nakai village is situated at the northwestern corner of Nanao Cove, which is broad inside and opens to the west coast of Toyama Bay (*c.* 37°N, 137°15′E). Another two villages known to have dolphin drive fisheries are on the northwestern shore of Toyama Bay. These locations are on the east side of Noto Peninsula in the southern Sea of Japan.

The fishing season was from late March to late July, with a peak in May and June. The dolphin species caught were *ma-iruka* (present *hase-iruka* or *D. capensis*), *nyudo-iruka* (present *oki-gondo* or false killer whale), *kama-iruka*, or *shisumi-iruka* (present *kama-iruka* or Pacific white-sided dolphin). The Tsushima Current, which is a branch of the warm Kuroshio Current, flows northward along the Sea of Japan coast of Honshu and increases in strength during March to July. The fishing season coincides with the time when these species expand their range northward from spring to early summer.

The Mawaki site of dolphin hunting in the Jomon era and Mawaki Village that carried out modern dolphin drives in the nineteenth and early twentieth centuries are identical in location, and all the other villages with modern dolphin drives on the Noto coasts are located close to each other. It is still to be confirmed whether the fishery continued for thousands of years at the same location. The species composition in the two periods has the two common elements of *Delphinus* sp. and Pacific white-sided dolphin, but they disagree in the limited abundance of false killer whales in the catch of the Jomon Era. The reason for the disagreement cannot be concluded at present, because the Mawaki site has been only partially excavated and we are uncertain how the currently available species composition represents the catch composition of the time.

Descriptions of the dolphin drives on the coasts of Noto Peninsula in modern time are available in three references: (1) *Noto-koku Saigyo Zue* (Illustrated Fisheries in Noto) manuscript dated 1838 and reprinted in Kitamura (1995, in Japanese), (2) *Dai-2-kai Suisan Hakurankai Shinsa Hokoku* (Examination Report of Items Exhibited at the Second Fishery Exposition) by Kaneda and Niwa (1899, in Japanese), and (3) *Suisan Hosai Shi* (Records of Fishery Activities) by Noshomu-sho Suisan-kyoku (Bureau of Fisheries of Ministry of Agriculture and Commerce) (1911, in Japanese). These descriptions are summarized as follows. Dolphins were searched for by telescope from one or two lookouts at vantage points on the coast. Scouting vessels were also sent out because false killer whales were in small schools and hard to spot from land. There were 3–4 scouting vessels carrying 6–7 persons (source 1 above) or 20 vessels with 3 persons (3).

If they found dolphin schools within 16–20 km from the shore (2), the message was sent to the lookouts and then to the village. Then the villagers rowed out *haya-fune* (quick driving boats) crewed by 2–3 with *haya-uchi-ami* (quick setting nets). Six to seven scouting vessels were used in the early nineteenth century (1), 40–50 vessels in the late nineteenth century (2), or 70 to over 100 (3). The difference between the three sources suggests an increase in the fishing effort during the nineteenth to early twentieth century. The driving vessels and scouting vessels in cooperation drove the dolphin school to the port with the help of nets placed behind the school and sounds produced by pounding the water and the vessel sides. Dolphin schools found while moving toward the harbor were easily driven, but other schools were difficult to make change their direction of swimming.

After the dolphin school entered into the port, a *tome-ami* (securing net) was placed at the entrance to prevent escape. Then, using another net *yose-ami* (herding net), dolphins were herded to the shallows at the beach where the fishermen could work. The fishermen entered the shallow water, took a dolphin in their arms, and brought it to ashore to slaughter. The dolphins were violent if hooked, but calm in the arms of naked men; the fishermen compared them to prostitutes. The same comparison was made in the 1960s by drive fishermen at Kawana on the Izu coasts on the Pacific coast of central Honshu. One of the sources mentioned earlier (2) stated that false killer whales were not herded to the shallows with the herding net but brought to the beach one by one.

The drive fishery at Mawaki had good catches until the 1890s and then declined and ceased operation in 1896, according to source (3) mentioned earlier. However, Yamada (1995, in Japanese) stated that the fishery lasted until the early 1920s, and in 1969, I interviewed a local old villager who told me that the drive fishery was carried out even after World War II and ended when a breakwater was constructed on the beach and damaged the beach for driving (Kishiro and Kasuya 1993). Admitting some uncertainty regarding the real reason for the end of the operation, it seems to be true that Mawaki village operated the dolphin drive, perhaps at a low level, for some time after World War II.

Statistics are almost absent for the drive fishery along the coast of Noto Peninsula. Only Takenaka (1890, in Japanese) copied catch statistics and sales records for some limited years in the late nineteenth century from records kept by the prefecture (Table 3.2). The Mawaki people did not process the catch but sold the whole carcasses to Ogi and Ushitsu at a price of 1.4–1.8 yen/individual for *Delphinus* sp., which weighed around 60 kg, and 5.5–7.7 yen/individual for false killer whales, which weighed around 225 kg. The Ushitsu people flensed their catch, salted the meat, and sold it to the cities of Niigata (37°55′N, 139°02′E) and Sakata (38°55′N, 139°50′E). The head and blubber were rendered for oil. Tendons were sliced, rinsed, dried, and sold to the Tamatsukuri area of Oska to be used as strings of cotton beating bows. Internal organs and skeletons were sold locally or to Toyama Prefecture on the southern coast of Toyama Bay raw or dried on the beach. The Ushitsu people probably processed the carcasses purchased from Mawaki in a similar manner. This is supported by the higher price of dolphin sold at Ushitsu, that is, 2.0–2.6 yen for *Delphinus* sp. and 5.3–9.3 yen for false killer whales.

TABLE 3.2
Dolphins Taken by Three Drive Groups at Ushitsu, Ogi, and Takakura in the Noto Region

Year	Location	*Ma-iruka*	*Nyudo-iruka*	Total
1887	Ushitsu	26	71	97
	Mawaki	0	75	75
	Total	26	146	172
1888	Ushitsu	0	53	53
	Mawaki	286	110	396
	Total	286	163	449
1889	Ushitsu	0	79	79
	Ogi[a]	0	121	121
	Mawaki	231	97	328
	Total	231	297	528
1887–1889	Total	543	606	1149

Source: Takenaka, K., *Dainihon Suisankai Hokoku*, 98, 241, 1890, in Japanese.

[a] Species was not given. I estimated the species from accompanying per capita price (two other villages sold larger *nyudo-iruka* [false killer whales] at higher prices than the smaller *ma-iruka* [common dolphins]).

3.2 INE, SEA OF JAPAN

Ine (35°40′N, 135°18′E) is located on the western corner of Wakasa Bay (*c.* 35°35′E, 135°35′N) on the southern Sea of Japan coast and comprised of the three villages Hidu, Kame-shima, and Hirata. Kame-shima traditionally authorized taking of whales found in Ine Cove by closing the entrance and then harpooning them. The catch records from 1655 to 1913 have been reprinted in a slightly edited form in the second volume of *Ine Choshi* (*History of Ine Town*) by Ine Choshi Hensan Iinkai (Editorial Committee of the History of Ine Town) (1985, in Japanese), but the value of the source is diminished by the uncertainty in the process of converting old species names into modern names. The original catch records for whales were analyzed by Yoshihara (1976, in Japanese).

The village of Hirata had the rights to take dolphins in Ine Cove. *Kyoto-fu Gyogyo Shi* (Fisheries of Kyoto) (Kyoto-fu Suisan Koshujo [Kyoto Fishery Training Center] 1909, in Japanese) described the fishery and the gear. The fishing season was from March to April. As soon as a dolphin school was sighted in the Cove, the people of Hirata surrounded it with rowed boats, drove it in an inlet in front of Hirata, closed the entrance with a net, and slaughtered the dolphins. Records of catches exist for 1732–1901 (Wakuta 1989, in Japanese). This valuable resource for studying the cetacean fauna and historical change has not been analyzed by scientists.

3.3 TSUSHIMA ISLAND, TSUSHIMA STRAIT

Tsushima Island (*c.* 34°05′N–34°40′N, 129°13′E–129°30′E) stretches about 70 × 15 km and appears to consist of two islands because of an artificially created canal. It is located in the middle of Tsushima Strait, which connects the East China Sea and the Sea of Japan. Many villages there have a long history of dolphin drives. Oyama and Shibahara (1987, in Japanese) recorded the history of the dolphin drive fishery, which they believed was going to be lost from the memory of the islanders. The oldest record on the dolphin fishery dates from 1404. This was a letter from a local governor to the head of O-ura on Asaji Cove (*c.* 34°20′N, 129°15′E) encouraging dolphin hunting as a taxable activity. The feudal lord of the islands in 1641 recorded the appointment of a magistrate in charge of the dolphin fishery and taxation of the dolphin fishery at Ina (34°34′N, 129°19′E) on the northwestern coast of Tsushima Island. There is also a seventeenth century record of a move of a village to the beach at O-ura that was considered to be better for the dolphin fishery. These records suggest two facts: (1) the villagers of Tsushima cooperated in hunting dolphins found in the inlet since the fifteenth century, and (2) the dolphin fishery was important for the economy of the country, which was mountainous and had little farm land. Although firm evidence of net use in the dolphin fishery was available only from 1846, in view of the earlier records of the nearby Goto Islands (see Section 3.5), a net must have been used earlier.

In Tsushima Island, the dolphin drive was operated cooperatively by several closely associated villages and had rules on obligations of the villagers and the shares to be received from the catch. Five such written agreements have survived to the present. They are dated between 1895 and 1923 and provide evidence of a minimum of five dolphin drive groups in Tsushima. Three of the five groups were at Asaji Cove, which opens on the East China Sea, and the remaining two were on the Oroshika (34°24′N, 129°23′E) and Miura (34°30′N, 129°25′E) inlets, which open on the Sea of Japan. If we add to these five locations another three groups at Ina, Nita (34°33′N, 129°21′E), and Shushi (34°36′N, 129°28′E) that were orally recorded as having dolphin drives, there were a total of eight drive groups in the late nineteenth to early twentieth century in Tsushima Island. Each of these groups was formed by 3–7 villages; a total of about 20 villages participated in the dolphin drives (Oyama and Shibahara 1987, in Japanese).

Suisan-cho Chosa-kenkyu-bu (Investigation and Research Department of the Fisheries Agency) (1968, 1969, in Japanese) made a survey on dolphin hunting along the northern, western, and southern coasts of Kyushu using a questionnaire. Its main objective was to collect information on recent operations to be used in finding a way to resolve the dolphin-fishery conflict at Iki Island (*c.* 33°47′N, 129°43′E) in Tsushima Strait. The objective was different from that of the previously cited study by Oyama and Shibahara (1987, in Japanese). The survey also identified eight groups or fishery cooperative unions (FCUs) that had been involved in dolphin drives in recent years. Three of the eight were located in Asaji Cove, which opens on the East China Sea (Ozaki FCU and Mitu-shima-saikai FCU in Mitu-shima Town and Kara-saki FCU in Toyotama Village), two of them (Sasuna FCU at 34°38′N, 129°23′E, and Ina FCU) were facing to the East China Sea, and the remaining three (Toyotama-mura-higashi and Hinode FCUs of Toyotama Village at 34°24′N, 129°23′E, and Kamoise FCU of Mitsu-shima Town at 34°16′N, 129°20′E) were on the Sea of Japan side of the islands. Of the eight FCUs, only the Ina FCU recorded a drive of 23–30 dolphins in 1950. Since the 1920s, the system of FCUs has changed, and some small villages could have merged into larger ones, so strict comparison of the two sets of driving teams is difficult. However, I could identify some common names between the two lists and some identical geographical locations, and I suspect that the two sets of eight drive groups in the two lists are probably identical.

Oyama and Shibahara (1987, in Japanese) described the dolphin drive. The driving started with news of sighting of dolphin school by someone in the cove, who got a reward for the sighting. Receiving the news, all the villagers rushed to their boats and steered them to the school with a *kazura* on board. The *kazura* was a traditional tool used to scare fish and dolphins in the drive fishery, a rope with about 1 m long wood pieces hanging from it at about 2 m intervals and a weight on one end. It was hung vertically in the water from the boat while driving the target animals. They also hit the side of the boat to scare the dolphins with the sound. When the dolphin school entered a suitable area, they closed the inlet with an *ohiki-ami* (large stretched net), whose procedure was called *harikiri* (shut down).

After closing the inlet, there was a ceremony of the "first lance" by one or two women dressed as *hazashi* (harpooners) who lanced the first animal. The killing tool "lance" was not a hand harpoon with a detachable head as used by the Japanese hand-harpoon fishery (Chapter 2) but had an asymmetrical iron head with barbs. The iron head was fixed to a wooden shaft. After this ceremony, the participants either killed the animals in the water with lances and then pulled them up onto the beach or pulled up several dolphins at a time with a net and killed them on the beach. Miyamoto (2001, in Japanese) reported that the ceremony of "first harpoon" was also performed on the occasion of driving tunas. As far as is known to me, the ceremony was a unique feature of the drive fishery in Tsushima. Another interesting point is the use of a lance to kill the dolphins, which was quite different from the killing method used in the Noto fishery in modern times (see Section 3.1) and from the hand-harpoon fishery in northeastern Japan that used a hand harpoon with detachable head. The Tsushima method might represent an old method of hunting in the islands but more probably was introduced from old-type Japanese whaling that once operated in the islands. The latter idea comes from two facts, that is, the shape of the lance that is similar to that of the *haya-mori* (quick lance) used in Japanese net whaling (see Section 2.2) and the use of the old whaling term *hazashi*.

An example of the economic contribution of this fishery to the local community of Tsushima is related by Oyama and

Sibahara (1987, in Japanese). In 1901, one team in Oroshika drove a school of over 918 dolphins and sold them at an average price of 2.95 yen/individual (2.5 yen per carcass found dead in the net). After subtracting expenditures of 1063 yen, they had a total profit of 1701 yen to be shared among participants from four villages. Another example was at Asaji in March 1948, when they drove a school of over 1000 dolphins and killed several hundred with the cooperation of three villages. The profit was sufficient to bring electricity lines to every family in the villages.

The last recorded dolphin driving in Tsushima is found in the statistics for dolphins in Nagasaki Prefecture published by Suisan-cho Chosa-kenkyu-bu (Investigation and Research Department of the Fisheries Agency) (1969, in Japanese). In the statistics covering the period from 1944 to 1966, there is only one record of a drive of 20–30 dolphins at Ina in 1950. The previously mentioned large catch in March 1948 is not in the records; the statistics of the early period may be incomplete. It is safe to conclude that dolphin driving almost ceased around 1950 in Tsushima Island.

The dolphin drive fishery in Tsushima resumed in 1970 (Table 3.3), stimulated by a subsidy by Nagasaki Prefecture (since 1966) and later by the Fisheries Agency (beginning in 1976) for hunting of dolphins as a way of solving the dolphin-fishery conflict in the Iki Island area. Tamura et al. (1986, in Japanese) collected statistics on dolphin hunting in Nagasaki Prefecture (cited in Tables 3.3 and 3.4) in relation to the dolphin-fishery conflict at Iki Island. The catch was 762 *Delphinus* sp. in 1970, 26 bottlenose dolphins in 1975, 167 dolphins in 1978, and 56 dolphins in 1976. The species was not identified for the last 2 years because the statistics were inseparable from those for takes in other locations in Nagasaki Prefecture (Table 3.4). No Phocoenidae species are known from the Tsushima Island area.

Although the resumption of dolphin hunting in Tsushima had nothing to do with local demand for human consumption, it was also true that there was still some demand for dolphin meat. For example, some of the dolphins taken at Iki Island were sent to Tsushima around 1977 to ease the burden on the Iki people of disposing of the dolphins culled.

TABLE 3.3
Annual Take of Small Cetaceans by Fishery Type and Region in Nagasaki Prefecture, Northern Kyushu

	Drive				Hand Harpoon				
Year	Tsushima	Iki	Goto	Others	Tsushima	Iki	Goto	Others	Total of Nagasaki
1965						2	7		29
1966			380			1	12		500
1967							62		90
1968						9	14	1	96
1969						2			5
1970	762					2			764
1971			22			5			27
1972		8							8
1973			84						84
1974			455						455
1975	26		315						341
1976		55	108						163
1977		943	530		5				1,481
1978	167	1,398	27		2				1,634
1979	56	1,646	29						1,815
1980		2,120	1351				15		3,517
1981		142							170
1982		163					8		177
1983		1,880	251	83					2,224
1984		2,924							2,931
1985		198	72						274
Total	1011	11,477	3624	83	7	21	118	1	16,785

Source: Compiled from records in Tamura, T. et al., eds., Report of contract studies in search of countermeasures for fishery hazard (culling of harmful organisms) (56th to 60th year of Showa), Suisan-cho Gyogyo Kogai (Yugaiseibutsu Kujo) Taisaku Chosa Kento Iinkai (Fisheries Agency), Tokyo, Japan, 1986, 285pp., in Japanese.

Species information not available in this source. The total figures include takes by other fisheries such as in trap nets.

TABLE 3.4
Annual Take of Small Cetaceans by Species in Nagasaki Prefecture, Northern Kyushu

Year	Bottlenose Dolphins[a]	False Killer Whale	Risso's Dolphin	Pacific White-Sided Dolphin	Pantropical Spotted Dolphin	Striped Dolphin	Common Dolphins[b]	Nezumi-Iruka[c]	Other Species[d]	Total
1965		2						7	20	29
1966	460							15	25	500
1967	13							62	15	90
1968	17	5	1	8	12		52	1		96
1969	2			1		1		1		5
1970	1	1					762			764
1971		5						22		27
1972		2	1					5		8
1973	84									84
1974	38							417		455
1975	28							313		341
1976			55					108		163
1977	908	35	6		530			2		1481
1978	25	424	199		954			3	29	1634
1979	333	340	499		596			44	3	1815
1980	3035	371	3					103	5	3517
1981	18		23	108				10	11	170
1982	163		2		8			3	1	177
1983	273	241	151	1559						2224
1984	176			2754				1		2931
1985	114	84			72			4		274

Fishery type and location are not available in these data. Based on the same source as Table 3.3.

[a] Common bottlenose dolphin is common and dominant in this area (see Chapter 11), but the possibility of occurrence of Indo-Pacific bottlenose dolphins in low abundance cannot be excluded.

[b] Geography suggests that they are most likely long-beaked common dolphins.

[c] Local name *nezumi-iruka* of the time is believed to have represented several dolphin species, and current Japanese names remain to be investigated (see footnote to Table 3.1).

[d] All the "other species" of 1965–1967 (60 individuals in total) were recorded as *sunameri* (finless porpoise). Some of the listed fishing methods (hand harpoon, gill net, trap net, seine net, and drive) may be able to take the species, but further information is needed to confirm the species identification. All the "other species" of 1978–1982 (60 in total) were recorded as *ma-gondo* (short-finned pilot whale), but the species is uncommon in the region, and I experienced cases where false killer whales were recorded with the name.

Catches of dolphins by drive, hand harpoon, and trap nets in Nagasaki Prefecture in the post–World War II period were published in Tamura et al. (1986, in Japanese) and Kasuya (1985b). These are cited in Tables 3.3 and 3.5. Table 3.5 lists catches of dolphins by all these types of fisheries in the Fisheries Agency statistics (partly unpublished).

When dolphin drive fisheries came under a license system in 1982 (Chapter 6), none of the FCUs in Tsushima Island obtained licenses, which meant the complete end of the dolphin drive fisheries there.

3.4 IKI ISLAND, TSUSHIMA STRAIT

Iki Island (*c.* 33°47′N, 129°43′E), about 15 × 18 km in size, is located in Tsushima Strait, about 50 km south of Tsushima Island and about 20 km north of the northern coast of Kyushu. Suisan-cho Chosa-kenkyu-bu (Investigation and Research Department of the Fisheries Agency) (1968, 1969, both in Japanese) reported from a questionnaire survey that the people of Iki Island, Goto Islands, and Tsushima Island used to drive dolphin schools with the cooperation of nearby villages. The driving system involved two or three FCUs and nearby agricultural villages and controlled everything from driving to allocation of share to the participants. This is apparently similar to the structure described by Oyama and Shibahara (1987, in Japanese) in the Tsushima fishery. The earlier reports by the Fisheries Agency in 1968 and 1969 listed three such drive groups at Iki Island located in Gono-ura (33°45′N, 129°41′E), Ishida (33°45′N, 129°45′E), and Yahata-ura (33°48′N, 129°48′E) and recorded several dolphin drives made by them. The Gono-ura FCU led three dolphin drives into the inlet of Watara-osaki in 1944, 1945, and 1956, and the Iki-tobu FCU led two drives into the inlet of Yahata-ura in 1962 and 1963. The total taken in these five drives was 1229 dolphins. The species was known only for the last three drives, which contained Risso's dolphins, false killer whales, and bottlenose dolphins. These were

TABLE 3.5
Small Cetaceans Taken by Drive Fisheries in Nagasaki Prefecture, Northern Kyushu, since 1972

Year	False Killer Whale	Risso's Dolphin	Bottlenose Dolphins[a]	Pantropical Spotted Dolphin	Pacific White-Sided Dolphin	Others and Unknown[b]	Location or Operating Bodies
1972	2	1				5 *nezumi*	Katsumoto
1973			84				Uonome
1974			38			417 *nezumi*	Arikawa, Uonome
1975			28			313 *nezumi*	Arikawa, Mitsushima, Sasuhara
1976		55				108 *nezumi*	Katsumoto, Uonome
1977	35	*6*	908	530		*2 nezumi*	Katsumoto, Miiraku
1978	424 + 27[c]	167	949		25		Katsumoto, Miiraku, Mitsushima-
	2[c]	*32*	*5*			*3 nezumi*	tokai, Toyotama-hinode
1979	339 + 3[c]	465	565		333	26 *nezumi*	Aso, Fukue, Katsumoto, Miiraku,
	1	*34*	*31*			*18 nezumi*	Minecho-tobu, Toyosaki,
1980	351		3035			85 *nezumi*	Arikawa, Katsumoto, Kita-uonome-
	20 + 5[c]	*3*				*18 nezumi*	daiichi, Miiraku, Uonome
1981		16	18		108		Katsumoto
	11[c]	*7*				*10 nezumi*	
1982	*1*[c]	*2*	163	*8*		*3 nezumi*	Katsumoto
1983	241	151	263		1559		Arikawa, Ikitsuki, Katsumoto
			10				Shin-uonome, Tateura, Urakuwa
1984			170		2668		Katsumoto
1985	84		114			72 Delphinidae	Katsumoto, Miiraku
						4 Dall's porpoises	
1986		*2*	55				Location unknown
1987[d]						*4 nezumi*	
1986[e]						*6*	
1987						*25*	
1988						*19* with *6 nezumi*	
1989						*6*	
1990	45	22					Katsumoto
	2[f]						
1991			2			*6* common dolphins	
						2 Risso's dolphins	
1993	18		32			*4* finless porpoises	Katsumoto
1995			49			*2* finless porpoises	Katsumoto
1996						*3* finless porpoises	
1998						*3* finless porpoises	
1999						*1* finless porpoises	
2002						*1* finless porpoises	
2004						2 finless porpoises	
2005			*1*			*1* finless porpoises	
2006						2 finless porpoises	
2007						2 finless porpoises	
2008						*3* finless porpoises	

Figures in italics are for those taken by other fisheries such as hand harpoon and trap net and are additional to the catch by drive fisheries. From Fisheries Agency statistics. Years without catch are not listed.

[a] All the confirmed takes were common bottlenose dolphins, but the possibility of some Indo-Pacific bottlenose dolphins being included cannot be fully excluded.

[b] See footnote to Table 3.1 for *nezumi-iruka* in Nagasaki Prefecture.

[c] Recorded as *ma-gondo*. Along the Pacific coast of Japan, this vernacular name has a long history of being used for the short-finned pilot whale (particularly for the southern form), but the short-finned pilot whale is uncommon off northern Kyushu, and the vernacular name could have been used for the false killer whale, which was common there. My telephone contact with the Miiraku Fishery Cooperative Union in 1979 on the identity of "27 ma-gondo" taken there in 1978 failed to obtain information to resolve the question.

[d] Statistics of the Offshore Division, Fisheries Agency, end here.

[e] Statistics of the Coastal Division, Fisheries Agency, start here. The two sets of statistics may disagree.

[f] Taken by a small-type whaler (Gaibo Hogei) who sent an unsuccessful expedition to the East China Sea in search of short-finned pilot whales (Sections 3.4 and 12.2).

the last operations by the traditional dolphin drive system at the Iki Island, after which the groups dissolved and the custom of consuming dolphin meat almost disappeared. This is supported by the following incidents.

After several attempts at culling dolphins as a means of resolving the dolphin-fishery conflict in the Iki Island area (see the following text in this section), Katsumoto (33°51′N, 129°42′E), the FCU in Iki Island first succeeded in driving dolphins in 1976, which was followed by a large number of drives in 1977 and later. One of the problems the Katsumoto FCU met during the procedure was how to dispose of the carcasses. The Katsumoto FCU did not find consumers of dolphin meat at Iki Island and sent some of the catches to consumers at Tsushima Island. Another problem faced by the Katsumoto FCU was lack of knowledge on how to drive the dolphins, which was not available at Iki Island, and they requested advice from the driving team in Taiji (33°36′N, 135°57′E) on the Pacific coast of central Honshu. This suggests that the technique of driving was lost at Iki Island or that it was not suitable for driving dolphins from a distance of 20 km from the port or from midway between Iki Island and Tsushima Island.

Iki Island fishermen using line and hook for yellowtail identified trouble with dolphins in the 1910s (Kasuya 1985b). I had an opportunity to examine the daily records of the yellowtail fishery kept by an Iki fisherman in the 1960s and 1970s; they showed that he had to abandon his favorite Shichiriga-sone ground and move to another ground or return to Katsumoto port if he encountered a dolphin school on the fishing ground. Even under such circumstances, they could fish successfully if there were abundant yellowtail. However, they would not be able to tolerate the presence of dolphins if the yellowtail stock became depleted for some reasons. This could have been a key factor in the response of the Katsumoto FCU in the dolphin-fishery conflict, which is detailed in Suisan-cho Chosa-kenkyu-bu (Investigation and Research Department of the Fisheries Agency 1968, 1969, in Japanese), Kasuya (1985b), and Tamura et al. (1986, in Japanese) and summarized as follows.

1. Sometime after World War II, the Katsumoto FCU attempted to cull dolphins using hand harpoons and a shoulder gun that shoots a small harpoon attached to a line, with no success.
2. In 1964, the Katsumoto FCU requested Nagasaki Prefecture to help resolve the problem, and in 1965, all the FCUs on the island joined in the request.
3. In 1965, the Katsumoto FCU introduced sound emitters that were used by the dolphin drive fishery along the Izu coasts, but they lost effect after continued use. Beginning in 1966, Nagasaki Prefecture paid a subsidy for the culling of dolphins.
4. In 1967, Nagasaki Prefecture requested that the Fisheries Agency help with a solution. The Fisheries Agency established a research group, which produced two reports Suisan-cho Chosa-kenkyu-bu (Investigation and Research Department of the Fisheries Agency, 1968, 1969, in Japanese).
5. In 1968, an attempt to scare dolphins with shotguns showed no effects, and a small-type whaling vessel was introduced in 1970. It captured 20 dolphins in the subsequent 6 years but did not have the expected effect. I find it unusual that there were so few catches.
6. In 1976, an attempt to capture dolphins with a large-mesh drift net failed. Then attempts at driving dolphins, assisted by Taiji fishermen, caught two Risso's dolphin schools. This was followed by large-scale culling and international criticism of the culling. The Fisheries Agency started to pay a subsidy in addition to that of Nagasaki Prefecture.
7. In 1978, the Science and Technology Agency of the Japanese government established a research team aimed at detecting dolphin schools and expelling them from the fishing grounds. This team was transferred to the Fisheries Agency in 1979 and produced no significant results.
8. In 1979–1981, using funding from the conservation community, my group started studying dolphins culled by the operation, which was reported by Kasuya and Miyazaki (1981, in Japanese), Kasuya (1985b), and Chapters 11 and 14 of this book.
9. In 1981–1982, the previously mentioned Fisheries Agency research team switched to biological studies aimed at determining an allowable removal level, which produced the report by Tamura et al. (1986, in Japanese). My group discontinued its activities.

In addition to small-scale beach fishing, the Iki fishermen carried out various other types of fisheries over time (Yoshida 1979, in Japanese). Net whaling started around 1710 and continued to be an important fishery during the eighteenth and nineteenth centuries. It ended operation in the late 1880s due to decline in the catch. Then net fisheries for sardines started, reached a peak around 1920, and ceased operation around 1942 due to a crash of the sardine stock. After the sardine fishery, their major fishing targets moved to squid and yellowtail, which had been fished since the nineteenth century. In 1956, when the Katsumoto fishermen fished for yellowtail and squid, the total sales of yellowtail were about seven times those for squid (sum of raw and dried squid). The proportion of yellowtail declined with time, and the balance was reversed in 1971, that is, yellowtail sold for a total of 168 million yen and squid for 454 million yen. I ascribe this to a decline in the yellowtail fishery.

The yellowtail ground of Shichiriga-sone, a bank situated about 28 km north of Iki Island or between Iki Island and Tsushima Island, was found around 1910 by the Katsumoto fishermen of Iki Island and exclusively used by them for the kai-tsuke fishery (feeding fishery), in which the fishermen scattered sardines to attract yellowtail before line and hook fishing season started. This required a great investment for feeding over several months, but it had the effect of collecting the fish from the surrounding area and allowed the Katsumoto fishermen to monopolize the ground (Yoshida 1979, in Japanese). This fishery peaked in the 1930s and ended in 1940 for the stated reasons that sardines became unavailable for the feeding operation and the yellowtail dispersed. With this event, the Shichiriga-sone ground was opened to every fisherman, but it was still almost exclusively fished by Iki fishermen.

The Katsumoto fishermen limited their method of yellowtail fishing to hook and line until the 1970s, when the dolphin-fishery conflict developed. The gear was a line with several tens of hooks with lures. One or two anglers worked in a small boat, each with a line, waiting for a fish school to surface. Such fishing opportunities occurred only once or twice a day and each lasted 2 hours at the longest. If the surfacing school was disturbed by something, for example, dolphins, it was a great loss to the fishermen.

The Katsumoto fishermen did not accept the use of seine nets or trap nets, with the belief that such efficient fishing gears were likely to overfish the yellowtail, and stuck to the hook-and-line fishery. They thought that the presence of dolphins on the fishing ground adversely affected the fishery in several ways:

1. Dolphins stole fish on their hooks.
2. Dolphins scared the yellowtail school causing them to sink (loss of fishing opportunity).
3. Dolphins consumed various fish taken by the fishermen.

At the beginning, fishermen put more stress on the first two factors, which could be easily identified (Tamura et al. 1986, in Japanese). But with the progress of time they placed more stress on damage of the third type and asked the government for a solution. Both the identification of competition between fisheries and dolphins for food and the solution of the competition would be extremely difficult.

Kasuya (1985b) examined the proportion of days when fishermen encountered dolphins on the yellowtail ground and found that it increased from 25% in 1973 to 79% in 1979, at a rate of 21%/year. Then it declined to 40% in 1981. He concluded that this change was not due to change in the dolphin population, but that it could be due to variation in the aggregation of dolphins on the fishing ground. If the abundance of food of dolphins decreased in the broader Tsushima Strait area, then the dolphins would concentrate on some small area such as Shichiriga-sone that had a relatively high food density and would increase opportunities to encounter the fishermen there. In addition to the previously mentioned possibility of environmental change, a possibility of decline in the yellowtail population must be considered. The yellowtail stock was also exploited by trap nets, seine nets, and fry fisheries to supply seed for fish culture. If the yellowtail stock declined, the effect will appear first in the least efficient fishery, that is, the hook-and-line fishery, and perceived trouble with dolphins would arise (Kasuya 1985b).

The hook-and-line fishery of Katsumoto went through some change. While the number of fishing vessels increased 17% from 413 to 485 in the 13 years from 1964 to 1977 and the average vessel size increased from 2.68 (excluding vessels without motors) to 3.73 gross ton (all had motors), the annual landing of yellowtail declined from about 2500 (1965–1970) to 500 tons (since 1971). Although the fish price also affected the economy of the fishermen, the decline of the catch accompanied with an increase in the fishing fleet could be an indication of a decline in economy of the Katsumoto fishermen (Kasuya 1985b).

While the Katsumoto fishery experienced this change, the annual landings of yellowtail at Iki Island by fisheries other than hook and line more than doubled from 200–300 (in 1965–1967) to 600–800 tons (in 1973–1977). The decline of total yellowtail landings at the Iki Island was also identified for the period. The fishermen in Katsumoto in fear of overfishing refused to adopt a more effective fishing method and used only hook and line, but the fishermen in the other four FCUs on the island introduced more effective fishing methods and increased their share of the Iki Island fishery (Kasuya 1985b). It is necessary to understand such features of the fishery to better understand the dolphin-fishery conflict in the Iki Island area.

The research team of the Fishery Agency did not pay attention to these fundamental problems of the fishery but attempted to search for a quick solution, such as finding a way to expel dolphins from the fishing ground, culling of dolphin populations, and estimation of an allowable culling level based on the study of life history and abundance of the dolphins as attempted by Tamura et al. (1986, in Japanese). Tamura et al. (1986, in Japanese) recorded the annual catch of dolphins by species in Nagasaki Prefecture during 1965 to 1985 (all fishing methods combined) and total annual catch by region and by fishing gear of the same period (no distinction among dolphin species), reproduced in Tables 3.3 and 3.4.

While Japanese scientists were studying the problem, the fishermen of Iki Island continued culling dolphins under the leadership of the Katsumoto FCU and with subsidy from the prefecture and the Fisheries Agency. They removed a total of over 10,000 dolphins during the 10-year period from 1976 to 1985. This was not a traditional drive fishery for human consumption but an activity to reduce the abundance of dolphins for expected, but unconfirmed, benefit of the fishery. Species composition of the culled dolphins is in Table 3.6 from Kasuya (1985b) based on figures recorded by scientists on site. The same figures are also in Kasuya and Miyazaki (1981, in Japanese). Also listed are the official statistics of the Katsumoto FCU (Harada and Tsukamoto 1980, in Japanese) and those collected by the Fisheries Agency. There is some disagreement between the scientists' figures and those of the Katsumoto FCU. For example, 27 carcasses of Risso's dolphins were counted by the scientists from a drive on February 8, 1979, but 17 was the corresponding figure recorded by the Katsumoto FCU. The Pacific white-sided dolphins that were driven on February 9, 1979, were 247 in number as recorded by the scientists against 283 by the FCU. A mixed school of Risso's dolphins and bottlenose dolphins was driven on February 12, 1979, and killed subsequently. The scientists recorded the composition as 10 Risso's and 73 bottlenose dolphins versus the almost reversed composition of 83 Risso's and 10 bottlenose dolphins that was recorded

TABLE 3.6
Dolphin Culling by the Katsumoto Fishery Cooperative Union, Iki Island, Nagasaki Prefecture

		Kasuya (1985b)			
Year[a]	Dolphin Species	No. of Individuals	School Composition[e]	Harada and Tsukamoto (1980)	Fisheries Agency
1972	False killer whale				2
	Risso's dolphin				1
	Nezumi-iruka[b]				5
1976	Risso's dolphin	55(2)		55(2)	55
1977	Bottlenose dolphin	899(4)	Including 2 drives on BND plus	899(4)	908
	False killer whale	35(2)	FKW	35(2)	35
		Total 4 drives			
1978	Bottlenose dolphins	958(5)	Including 3 drives on BND plus	949(4)	949
	False killer whale	349(4)	FKW	358(3)	424
	Pacific white-sided d.	25(1)		25(1)	25
	Ma-gondo[c]	0		66(2)	0
		Total 7 drives			
1979	Risso's dolphin[d]	388(6)	Includes 2 drives on BND plus FKW	454(7)	454
	Bottlenose dolphins	604(4)	and a drive on RSD plus BND	541(4)	541
	False killer whale	318(3)		318(3)	318
	Pacific white-sided d.	297(2)		333(2)	333
		Total 12 drives			
1980	Bottlenose dolphins	1574(5)	Includes 3 drives on BND plus FKW	1855(5)	1855
	False killer whale	245(3)		265(3)	265
		Total 5 drives			
1981	Risso's dolphin	16(1)			16
	Bottlenose dolphins	18(1)			18
	Pacific white-sided d.	108(2)			108
		Total 4 drives			
1982	Bottlenose dolphins	157(4)	Includes a drive on BND plus FKW		163
	False killer whale	6(1)			0
		Total 4 drives			
1983	Bottlenose dolphins				263
	False killer whale				241
	Pacific white-sided d.				1376
1984	Bottlenose dolphins				170
	Pacific white-sided d.				2668
1985	Bottlenose dolphins				114
	False killer whale				84
1986[b]	Bottlenose dolphins				55
	Pacific white-sided d.				4
1990	Bottlenose dolphin				64
	False killer whale				45
	Risso's dolphin				22
1993	Bottlenose dolphins				32
	False killer whale				18
1995	Bottlenose dolphins				49

Sources: Kasuya, T., Fishery-dolphin conflict in the Iki Island area of Japan, pp. 253–273, in: Beddington, J.R. et al., eds., *Marine Mammals and Fisheries*, George Allen and Unwinn, London, U.K., 1985b, 354pp.; Harada, Y. and Tsukamoto, Y., Dolphin and porpoise fishery, pp. 411–435, in: Kumamoto, H., ed., *History of Fisheries in Katsumoto*, Katsumoto Gyogyo Kyodo Kumiai, Katsumoto, Japan, 1980, 576pp., in Japanese; Fisheries Agency statistics (see Section 1.4).

These figures are included in other statistics of the prefecture (Tables 3.3 through 3.5). The figures do not include individuals that escaped from the net before slaughter. In parentheses is the number of drive operations (one drive on mixture of two species was counted for both species).

[a] Years of no operation are not tabulated.

[b] See footnote b to Table 3.1 for the identification of *nezumi* or *nezumi-iruka*.

[c] Recorded as *ma-gondo*, but this incident was not recorded in other statistics. For interpretation of this vernacular name, see footnotes to Table 3.5.

[d] Kasuya (1985b) recorded take of 100 Risso's dolphins in one drive on March 7, 1979, while Harada and Tsukamoto (1980, in Japanese) recorded it as two drives of 50 individuals each.

[e] Species key. BND, bottlenose dolphins (all the individuals examined by scientists were common bottlenose dolphins); FKW, false killer whale; and RSD, Risso's dolphin.

by the FCU. There were some cases where the scientists recorded greater numbers. There are no explanations for the disagreements. In those days, the subsidy was paid based on the number of dolphins (irrespective of species), and we had the impression that the Nagasaki Prefecture staff as well as the FCU staff recorded the numbers carefully but that they were reluctant to correct a number once reported. Any statistics should be considered a kind of relative indicator, and arguing about minor disagreement does not have much value.

The species composition of dolphins killed during 7 years from 1976 to 1982 (Kasuya 1985b) and the same number for years from 1983 to 1995 (Fisheries Agency statistics) was as follows:

Year	1976–1982	1983–1995
Pacific white-sided dolphin	466 (7.7%)	4048 (77.7%)
Bottlenose dolphin	4147 (68.1%)	747 (14.4%)
False killer whale	953 (15.6%)	388 (7.5%)
Risso's dolphin	525 (8.6%)	22 (0.4%)
Total	6091 (100%)	5205 (100%)

Participation in the culling was an obligation of union members of all the FCUs at the Iki Island; therefore, the dolphin drives usually involved 300–500 vessels. Drives of Pacific white-sided dolphins often failed in the early stages of the culling activity. Some drives started on a large school of the species but ended with far fewer individuals in the net enclosure, because most of the school members escaped before closure of the net. However, in the 1980s, the driving technique improved and often drove large groups of Pacific white-sided dolphins. This increased the proportion of Pacific white-sided dolphins in the culls. It is unclear whether the decline in the bottlenose dolphins in the later part of the culling operation was due to local depletion of the stock or due to shift of the fishermen's effort to Pacific white-sided dolphins.

Uncertainty still remains about the identification of animals recorded as short-finned pilot whales. The *History of Fisheries in Katsumoto Town* by Kumamoto (1980, in Japanese) recorded two drives, of 50 and 16 individuals, of the species on April 21 and 23, 1978, respectively. These drives were not reported by scientists who worked there but by prefecture staff who identified the species of captured dolphins using a manual prepared by a group of biologists including myself, or by the biologists who received photographs of dolphins taken in the vicinity. Although short-finned pilot whales had been taken off northern Kyushu and sighted in the Sea of Japan (Kasuya 1975), the species was believed to be rare in the area, and most of the identifications as *gondo* by local people represent the *oki-gondo* (false killer whale), not the *kobire-gondo* (short-finned pilot whale). Thus, I suspect the 66 individuals could have been false killer whales, as they were listed in the Fisheries Agency statistics.

Food habits and behavior can differ among dolphin species. However, Nagasaki Prefecture and the Fisheries Agency did not consider whether damage to the yellowtail fishery was different among the dolphin species involved and paid equal subsidies for culling of any small cetacean. Examination of stomach contents of dolphins taken in the region revealed that Risso's dolphins ate only squid and out of the three other species that ate fish and squid, only false killer whales and Pacific white-sided dolphins were found with remains of yellowtail in their stomachs. If the government was targeting the difficult task of solving problems relating to ecological interaction between the fishery and dolphins, it should have collected more information on the food habits of the dolphin species.

How to dispose of the dolphin carcasses was the most difficult problem the Katsumoto FCU encountered during the dolphin culling. The Katsumoto FCU kept the live dolphins until slaughter in a net enclosure in an inlet of a nearby island called Tatsuno-shima, which was used in summer by sea-bathing visitors. Although Tsushima and Goto Islands had customarily consumed dolphins and received some of them, the demand was soon oversaturated. All the available area on Tatsuno-shima became full of pits for burying the carcasses. Disposal in the ocean was prohibited in 1979 by the Coast Guard as a violation of an act against marine pollution. The demand for live animals for aquariums was also limited. The last solution was to send the whole carcasses to a fertilizer factory in 1979, as shredded carcasses beginning in 1980. My research team of volunteer students made an enormous effort to collect biological materials from carcasses being shredded. In 1981, the Fisheries Agency team took up the task.

The dolphin drive fishery was placed under a license system in 1982, but none of the Iki fishermen obtained licenses. The Katsumoto FCU continued the culling until 1995 using a prefecture permit for culling of dolphins for the purpose of eliminating animals that are harmful for fisheries. The total culling permitted for fisheries in Nagasaki Prefecture was 104 dolphins in 1990. Statistics of the cull are available in the Japanese Progress Report to the International Whaling Commission. The culled animals were sold to aquariums or kept alive in an enclosure near the Katsumoto FCU for tourists.

I visited Taira on Ukushima Island (33°15′N, 129°08′E), a small island 90 km to the southwest of Katsumoto in 1990. I arrived on February 13 for a purpose to be mentioned later and heard a story at the Ukushima FCU the next day. In late January or early February of the same year, the Katsumoto FCU drove a group of false killer whales and bottlenose dolphins as part of the culling operation. Four live false killer whales were sold to an aquarium (230,000 yen/individual), and carcasses of the remaining 27 false killer whales and 16 bottlenose dolphins were sold to the Gaibo Hogei Co. Ltd. (200,000 yen/individual for the false killer whales and 10,000 yen/individual for the bottlenose dolphins). The Gaibo Hogei transported them to Taira to be processed and sold the meat to the Shimonoseki Fish Market. Average meat production was 130 kg/individual false killer whale. The quality of some of the meat was said to have been poor because the animals were kept in the net for about a week, but the higher quality portion of the false killer whale meat was sold at 3200 yen/kg and the rest at 1600 yen/kg. I did not get information on the proportions of the two kinds of meat, but if it is assumed to be 1:2, then the proceeds would be 277,000 yen/individual, resulting in a profit of 70,000–80,000 yen/animal. This occurred when the supply of whale meat diminished after the end of commercial whaling. I did not have any information on the meat from the bottlenose dolphins. I had barbecued false killer whale at Iki Island, and it tasted better than sperm whale.

Earlier, Gaibo Hogei constructed a whale processing platform on a small uninhabited island near Ukushima Island with an expectation of taking some short-finned pilot whales in the East China Sea and placed a team of flensers at Taira on Ukushima Island. Therefore, I went there to collect biological specimens and to observe the operation. The enterprise discontinued within two to three seasons on account of failure. The company purchased dolphin carcasses from Katsumoto, probably to create work for the flensing team. When I was told about the plan for Ukushima whaling by Gaibo Hogei, I advised the owner not to pursue the plan because short-finned pilot whales were scarce there, but I gave up offering further advice with his rebuttal that nobody could be sure that none would be found there. The operation ended with almost zero catch (see footnote "f" to Table 3.5). It was well known through various surveys conducted in relation to the dolphin-fishery conflict in the Iki Island area that short-finned pilot whales were scarce in the vicinity, and I still do not understand why Gaibo Hogei pursued the project.

The dolphin-fishery conflict in the Iki Island area stimulated studies on distribution, abundance, and life history of small cetaceans in the East China Sea and Sea of Japan. Results of these studies were summarized in Tamura et al. (1986, in Japanese). The record of conflict was published in Kasuya (1985b), life history of the bottlenose dolphin in Kasuya et al. (1997, in Japanese) and Chapter 11, and growth of the Pacific white-sided dolphin is covered in Chapter 14. Life history of the false killer whale will be treated in a future publication.

In my understanding, in 1978 the Japanese community realized through the experience of the dolphin-fishery conflict in the Iki Island area to them the astonishing fact that it had a view of nature that was quite different from that of many other modern communities, particularly in Europe and North America. In 1978, the third year of attempts at dolphin drives, the Iki fishermen were delighted with their first successful drive of over 1000 false killer whales and bottlenose dolphins. However, the "success" drew criticism from all over the world, and Japanese overseas establishments in over 11 countries received an avalanche of telegrams of protest (Harada and Tsukamoto 1980, Tamura et al. 1986, both in Japanese). The Japanese government responded to this by changing its policy from culling the dolphins to driving them away from the fishing ground and established a research team for the purpose (1978–1980). Facing with failure of this objective, this research team moved toward estimating allowable culling level (1981–1982). However, the government continued the subsidy for killing dolphins, together with advising the fishermen to hide the killing scenes from the eyes of the public. This was a behavior to put up a good front without any real improvement. One night during this period, a U.S. citizen, D. Kate, went to Tatsuno-shima Island and released dolphins held in a net there. Legal action would not have any effect on this kind of radical activity, and scientific explanation of the need for culling, even if it were possible, would not satisfy the public. Hiding the killing scene could only increase distrust by the critics and create difficulty in achieving mutual understanding.

The Iki fishermen believed that they could cull dolphins because dolphins disturbed their fishing activities or because the presence of dolphins in the ocean was thought to be harmful for their fisheries. This came from misunderstandings by the Iki fishermen and perhaps the Japanese community. How to view dolphins will vary among current world communities, but an increasing number of people feel friendship toward dolphins and wish to handle them with care. Another misunderstanding was about the ownership of dolphins as an element of our environment. It is becoming widely accepted that nature on this planet is a common property of all human beings. Although the Iki fishermen are temporarily allowed to utilize their surrounding seas for their fishing activities, it does not necessarily mean that they have a free right of use. Marine fishery resources and cetaceans living on them are not an exception. In recent years, some people have identified the right of survival for wildlife. The Japanese people should have understood such new concepts and the importance of explaining why the culling was necessary and how it could be appropriate before the culling was carried out. And the effect of the culling should have been evaluated after its execution.

The Iki fishermen have erected a cenotaph on a beach of Tatsuno-shima Island where over 10,000 dolphins were slaughtered. To me it is only an eyesore. Although it might express the fishermen's repentance for the slaughter, it does not answer the concerns of the world's people about the destruction of common property. In any case, similar behavior is still common among Japanese whalers.

3.5 GOTO ISLANDS, EAST CHINA SEA

Goto Islands (*c.* 32°35′N–33°20′N, 128°35′E–129°15′E) are a group of islands extending about 90 km from southwest to northeast and located about 50 km to the west of Kyushu. The previously mentioned Ukushima Island is the northernmost element of the group. Uonome (33°00′N, 129°09′E) and Arikawa (32°58′N, 129°07′E), both known for a history of whaling and dolphin fisheries, are in Nakadori Island in the northern part of the Goto Islands group and are located on both sides of Arikawa Bay, which opens northward. Phocoenidae species have not been recorded around this islands group, as in the case of Tsushima and Iki Islands.

Shibusawa (1982, in Japanese) reviewed historical documents on fisheries at Arikawa and stated that the oldest document with the word *yuruka-ami* (dolphin net) appeared in 1377 and that *iruka-ryo* (dolphin fishery) and *iruka-oikomi* (dolphin drive) appeared in another document dated 1691. These indicate that the people in Arikawa operated a dolphin drive fishery in the fourteenth century.

Suisan Hosai-shi (Records of Fishing Activities) by Noshomu-sho Suisan-kyoku (Bureau of Fisheries of Ministry of Agriculture and Commerce) (1911, in Japanese) stated that "netting dolphins" at Arikawa and Uonome was known since before the late seventeenth century (this does not disagree with the earlier statement) and that the target included *ma-iruka* (or *hase-iruka*) and *nyudo-iruka* (or *bauzu-iruka*). The former name corresponds to the current *hase-iruka* (long-beaked common dolphin, *Delphinus capensis*) and the latter to the false killer whale (also see Section 3.8.2). The operation did not have lookouts or scouting vessels, which was similar to the fishery in Tsushima Islands, but differed from those along the coast of Noto Peninsula, which probably had a more commercial element. Anybody who found a school of dolphins while working in any kind of fishery raised a garment or straw mat at the top of their ship's mast as a signal of finding dolphins. This action was relayed until it reached the port. The first spotter and subsequent reporters received a share for their contribution. The share could cover more than the economic loss due to discontinued fishing activity. Everybody who sighted the signal hurried to where the school was spotted, and tens to a hundred fishing boats surrounded three sides of the school and drove it to their inlet. After driving the dolphins into the inlet, the entrance was closed with straw nets of two mesh sizes (the smaller-mesh net was on the inside). Then they brought dolphins close to the beach with a hemp net and slaughtered them. The catch was divided among all the families. This fishery lasted until after World War II. K. Mizue of Nagasaki University published numerous scientific studies based on bottlenose dolphins, false killer whales, Risso's dolphins, and spotted dolphins taken by this fishery, as seen in Ohsumi (1972).

Suisan-cho Chosa-kenkyu-bu (Investigation and Research Department of the Fisheries Agency) (1968, 1969, both in Japanese) was a response by the Fisheries Agency to the dolphin-fishery conflict in the Iki Island area that occurred in the 1960s (see Section 3.4). The reports listed, in addition to Arikawa and Uonome, Miiraku (32°45′N, 128°42′E) and Tomie (32°37′N, 128°46′E) in the southern part of Goto Islands as places having a history of dolphin drive fisheries. The dolphin species reported by the Fisheries Agency reports were similar to those examined by Mizue's group, except for Pacific white-sided dolphins, which were not examined by Mizue's group, and pantropical spotted dolphins, which were confirmed by Mizue's group but absent in the Fisheries Agency list (Table 3.7). The spotted dolphin could have been included under the unconfirmed local name *nezumi-iruka*. The drive fishery in Goto Islands mainly targeted three species, namely, bottlenose dolphin, Risso's dolphin, and false killer whales, with some additional takes of spotted dolphins and Pacific white-sided dolphins. Driving Pacific white-sided dolphins was considered difficult by Izu fishermen and Iki Islanders, but it was also true that the species was driven by several driving groups in Iki (Section 3.4) and on the coasts of Sanriku (Section 3.7) and Izu (Section 3.8).

The history of dolphin drive fisheries of Goto Islands ended in 1982, when the fisheries were placed under a license system and nobody in the islands obtained a license. Subsequent dolphin

TABLE 3.7
Some Earlier Additional Records of Dolphin Drives in Nagasaki Prefecture, Northern Kyushu

Species	Arikawa & Uonome[c]	Miiraku[c]	Tomie[c]	Iki Island	Tsushima I.	Ao & Tobishima[c]
Bottlenose dolphins	398 (July 1960)	658 (January 1955)		20 (December 1962)[b]		300 (February 1960)
	300 (December 1960)	530 (October 1958)		70 (December 1963)[b]		
		400 (December 1958)				
		200 (July 1959)				
		97 (August 1960)				
		108 (July 1966)				
		272 (August 1966)				
False killer whale	200 (December 1960)	400 (December 1958)				300 (February 1960)
Risso's dolphin	20 (September 1959)	30 (July 1959)		39 (November 1956)		
	20 (September 1960)	380 (October 1960)[a]				
		142 (January 1961)				
		23 (October 1963)				
Pacific white-sided d.	1200 (1964/1965)					
Unknown species	39* (October 1954)	27 (October 1959)	10 (August 1964)	300 (1944)	25 (June 1950)	2000–3000 (1916)
	83* (November 1962)	76 (November 1959)		800 (1945)	Number unknown (October 1963)	10 (1950)

Source: Compiled from Suisan-cho Chosa-kenkyu-bu, Comprehensive report of basic investigations in search of counteraction for fishery damages caused by small cetaceans in western Japan (42nd and 43rd year of Showa), Suisan-cho Chosa-kenkyu-bu, Tokyo, Japan, 1969, 108pp., in Japanese.

Species recorded as *gondo* was dealt with as unknown species (with asterisk), which was most likely the false killer whale.

[a] Includes bottlenose dolphins driven together (proportion of the two species is unknown).

[b] Includes false killer whales driven together (proportion of the two species is unknown).

[c] Arikawa, Miiraku, Tomie, and Uonome are in the Goto Islands in the East China Sea, and Arikawa and Uonome are located close to each other. Ao and Tobishima are located on islands that are close to the northern coast of Kyushu.

drives were likely to have been done with a permit for culling marine animals that were harming fishery operations.

3.6 NORTHERN COASTS OF KYUSHU AND YAMAGUCHI PREFECTURE

Iruka-ryo-no Koto (About Dolphin Fisheries) by Kizaki (1773, in Japanese, printed in Hawley 1958–1960) described dolphin drives in Karatsu (33°28′N, 129°58′E) in Saga Prefecture. Local people drove dolphin schools from nearby coasts to their inlets, shut the entrance with several layers of nets, and then slaughtered them. Such dolphin drive fisheries were operated at several villages along the coast of northern Kyushu. Kizaki (1773, in Japanese) listed five species of dolphins as their targets, that is, *niudou*, *handou*, *hase*, *nezumi*, and *shiratago*, and stated that the last two species swam fast and were harder to drive than the first three species. An accompanying drawing depicts simultaneous driving of two dolphin species identifiable as false killer whales and bottlenose dolphins, *niudou* and *handou*, respectively. These two species were often driven together in Iki Island (Table 3.6). *Hase* corresponds to the current Japanese name *hase-iruka* or long-beaked common dolphin, but *Nezumi* and *shiratago* have not been identified (Kasuya and Yamada 1995, in Japanese). However, if we take note that (1) *nezumi* and *shiratago* were fast and hard to drive and (2) among historical records of dolphin fisheries along the coasts of Sea of Japan and northern Kyushu, local names that indicated the Pacific white-sided dolphin appeared only on the Noto coast, that is, *kama-iruka* (*shisumi-iruka*) (Noshomu-sho Suisan-kyoku [Bureau of Fisheries of Ministry of Agriculture and Commerce] 1911, in Japanese, see Section 3.1), it is likely that both or one of the two names (*nezumi* and *shiratago*) indicated Pacific white-sided dolphins.

Suisan-cho Chosa-kenkyu-bu (Survey and Research Department of the Fisheries Agency) (1969) sent questionnaires to 261 FCUs in Yamaguchi, Fukuoka, Saga, Nagasaki, Kumamoto, and Kagoshima Prefectures in northern, western, and southern Kyushu (excluding Miyazaki and Oita Prefectures on the east side of Kyushu) asking for information on the known records of past dolphin hunting. The replies included information on dolphin drives from 1916 to 1966. If we exclude records for Goto, Iki, and Tsushima Islands that have been covered earlier, there remain three FCUs having a history of dolphin drives on the coast of northern Kyushu, that is, 10 dolphins of unknown species driven at Ikitsuki (33°20′N, 129°25′E) in 1950, 2000–3000 *Delphinus* sp. or Pacific white-sided dolphins driven at Ao-ura (33°27′N, 129°46′E) in 1916, and 300 each of false killer whales and bottlenose dolphins (presumably in a mixed school) at Tobishima (33°24′N, 129°46′E) in February 1960 (Table 3.7).

Tsutsumi et al. (1961, in Japanese) analyzed stomach contents of false killer whales and bottlenose dolphins taken at Tobishima in February 1960, which confirms corresponding species records in Table 3.7. Their materials also included five individuals each of false killer whales and bottlenose dolphins obtained from a drive in February 1960 at Omi-shima in Senzaki (34°23′N, 131°12′E) in Yamaguchi Prefecture, which were transported to an aquarium and subsequently died. This drive was not listed in the Fisheries Agency statistics and seemed to be a case where a group of fishermen at Omi-shima who had experience in driving dolphins drove a school at the request of an aquarium that needed live specimens (Table 3.8).

TABLE 3.8
Dolphin Drives Recorded along the Coast of Yamaguchi Prefecture and Northern Kyushu (excluding Nagasaki Prefecture) for Years 1972–2004

Year	Species	Number Taken	Location[a]	Comments
1972	Pacific white-sided dolphin	17	Senzaki	Sent alive to Shimonoseki Aquarium
1974	Pacific white-sided dolphin	22	Senzaki	Sent alive to Shimonoseki Aquarium

Source: Fisheries Agency statistics (see Section 1.4).
[a] Senzaki is located on the coast of the southern Sea of Japan, Yamaguchi Prefecture.

Although there were records of old-type whaling on the coast of the Inland Sea, for example, in Onoda (34°00′N, 131°11′E) in Yamaguchi Prefecture and Kan-nonji (34°08′N, 133°39′E) in Ehime Prefecture in the nineteenth century (Shindo 1968, in Japanese; Omura 1974), whaling flourished to a greater degree in the islands area off northern Kyushu and in Yamaguchi Prefecture along the coasts of the southern Sea of Japan and eastern East China Sea. Dolphin drive groups are also known to have existed in the whaling area on the coasts of northern Kyushu and Yamaguchi Prefecture (no porpoise inhabits this water except for the finless porpoise *Neophocaena* sp.), but these locations that were close to the mainland apparently ceased operations earlier than the previously mentioned remote islands.

The Fisheries Agency statistics list a case of a drive of two Baird's beaked whales at Takahama-cho (35°29′N, 135°37′E) in Fukui Prefecture on the Sea of Japan coast, which I interpret as a case of assisted stranding (see last line of footnote to Table 2.3).

No fishermen of this region acquired licenses for dolphin drives in 1982, when the fishery was placed under a license system by the Fisheries Agency.

3.7 SANRIKU COASTS, NORTH PACIFIC

The Sanriku Region is a Japanese geographical term for the eastern slope area (area facing the North Pacific) of northern Honshu (Japanese main island), which includes the three prefectures of Miyagi (37°44′N–38°59′N), Iwate (38°59′N–40°27′N), and Aomori (40°27′N–41°30′N). The Sanriku Coasts are the Pacific coasts of this region. Of the three prefectures that constitute the Sanriku Region, the areas of active dolphin and porpoise fisheries are limited to the two southern prefectures, which apparently agrees with the coastline having numerous *rias* (narrow inlets formed by submerged river valleys). The warm Kuroshio Current and cold Oyashio Current alternate off the coast seasonally and cause sea surface temperature fluctuation between over 25°C and below 5°C.

The Sanriku Coasts are currently known for hand-harpoon fisheries for porpoises, but there was a time when dolphin drive fisheries were operated instead. A survey of fishery activities in the late nineteenth century conducted by the Japanese government recorded four places in Iwate Prefecture as operating drive fisheries: Aka-saki (39°05′N, 141°43′E), Kamaishi (39°16′N, 141°53′E), Funakoshi (39°26′N, 141°57′E), and O-ura (or Yamada Cove) (39°28′N, 142°00′E) (Noshomu-sho [Ministry of Agriculture and Commerce] 1890–1893; Takenaka 1890, both in Japanese). Kawabata (1986, in Japanese) in *Yamada Choshi* (History of Yamada Town) summarized the history of dolphin drive fisheries in the region and also listed the four places (Table 3.9). These three references contain elements similar to each other, which suggests that Takenaka (1890, in Japanese) and Kawabata (1986, in Japanese) were based on old materials, at least in part, as was the government report of 1890–1893.

O-ura is located on the southwestern corner of the broad Yamada Cove, which is connected to the Pacific with a narrow opening. In the context of the dolphin drive, *Yamada* meant a village of O-ura. Akasaki is located at the end of a long narrow inlet (a *ria*) of Ofunato. Kamaishi and Ofunato are also located at the head of narrow inlets. Table 3.9 gives the years with actual dolphin drives around the time when the government survey was conducted. The histories of the dolphin drives must have started earlier and lasted longer, probably until around the time when the hand-harpoon fisheries for dolphins and porpoises started in around 1920 (see Section 2.2).

According to Kawabata (1986, in Japanese), the dolphin drive fishery at O-ura started in 1727 using technology introduced from To-ni (39°12′N, 141°52′E), south of Kamaishi, and capital from Otsuchi (39°21′N, 141°54′E) and O-ura. This suggests that To-ni, at the head of Toni inlet, was another location of a dolphin drive fishery. The system used by the drive group of O-ura changed with time, and the drives are believed to have ended sometime between 1912 and 1926. The driving started with sighting of dolphin schools in Yamada Cove. Everybody in O-ura Village rowed out with their boats loaded with nets or stones, drove the school into a small inlet in front of O-ura, closed the entrance with nets, and landed the dolphins. The operation was carried out cooperatively by all the villagers. In the late nineteenth century, classes of the junior school were closed to allow participation in the drive. The operation was an important activity of the local community as in the Okinawa (Section 3.10) and Nagasaki regions (Sections 3.3 through 3.5). After paying taxes, the remaining proceeds were distributed to investors, driving participants, families, the Buddhism temple, the Shinto Shrine, and the junior school in the village. Nearby villagers called this accounting custom "accounting of O-ura style." The O-ura people once had to make additional nets after sighting a large school of dolphins. This allowed description of the operation (fishery of O-ura style) by outsiders. I heard these tales in the 1970s, when I often stayed at Tanohama in Funakoshi (south of O-ura across a small peninsula) to study the Dall's porpoise fishery there. I detected envious feelings expressed by surrounding villagers; the dolphin drive fishery was once very prosperous in O-ura.

Kawabata (1986, in Japanese) reprinted historical documents on the dolphin drive fishery in the region. In that report, I identified several local names for dolphins: *iruka* (which is a general name for dolphins and porpoises and has appeared in documents of O-ura), *goto-iruka* (Funakoshi), *kama-iruka* (Akasaki, Kamaishi), *ma-iruka* (Akasaki, Funakoshi, Kamaishi, O-ura), *nezumi-iruka* (Funakoshi, O-ura), and *nyudo-iruka* (Akasaki, Kamaishi). The number of individuals driven is available only for *nezumi-iruka* and *mairuka* (Table 3.10). One of the documents on Akasaki was accompanied by the following descriptions of *kama-iruka* and *ma-iruka*: "They are taken together and the former has a tail fin which is bent, and the latter has a straighter tail fin and is female. Thus, they are not separated in either driving operations or records of statistics." If we assume that this description is true and the "tail fin" indicates "dorsal fin," then we can conclude that the Pacific white-sided dolphin was called *ma-iruka* (right kind of dolphin) possibly because it was most common in their operation and also called *kama-iruka* (sickle dolphin) because the adult male had a sickle-shaped dorsal fin as a secondary sexual character (Chapter 14). Although we have no firm evidence on which to consider that the earlier usage of *ma-iruka* and *kama-iruka* was common among the driving groups on the Sanriku coast, I think it likely because of broad use of the names in the region. I have seen several photographs at the O-ura FCU of a big drive made there on July 1, 1913. The estimated number of dolphins was 2000–3000, and I identified the school as composed of only Pacific white-sided dolphins (the photographs were not accompanied by the local species name). One of these photographs is reproduced in Kawabata (1986, in Japanese).

TABLE 3.9
Past Operations of Dolphin Drive Fisheries in Iwate and Miyagi Prefectures, Pacific Coasts of Northern Honshu

Prefecture	Location	Date of Operation	Reference
Miyagi	Akasaki Village	1870–1885	K, N, T
Iwate	Kamaishi Town	1888–1889	T
	Kamaishi Town	1888	K
	Funakoshi Village	1887–1888	T
	Funakoshi Village	1887–1889	K
	Yamada Cove		N
	Oura, Yamada Cove	1728–1913	K

Location names are as they appeared in the references.
Reference key: K, Kawabata (1986, in Japanese); N, Noshomu-sho (Ministry of Agriculture and Commerce) (1890–1893, in Japanese); and T, Takenaka (1890, in Japanese).

TABLE 3.10
Dolphins Taken by Drive Fisheries in Iwate Prefecture

Date	Location	*Iruka*	*Nezumi-iruka*	*Ma-iruka*	Price
March 1857	Oura		2200	3590	*Nezumi*: 1 ryo (=yen)/8 individuals *Ma-iruka*: 1 ryo (=yen)/7 individuals
March 1870	Oura			111	About 3.5 ryo (=yen)/individual
1870	Akasaki			54	2.31 yen/individual[a]
1871	Akasaki			198	0.67 yen/individual[a]
1874	Akasaki			200	0.70 yen/individual[a]
February 1882	Oura	2385			Over 1.72 yen/individual
1885	Akasaki			50	0.15 yen/individual[a]
1888	Kamaishi			300	4.18 yen/individual (meat, oil, and others)
1887[b]	Funakoshi		70	200	Production *Ma-iruka*: 63.75 kg of meat/ind. valued at 3.50 yen/ind. *Nezumi-iruka*: 52.5 kg of meat/ind. valued at 2.20 yen/ind. and 18 L of oil/ind. valued at 1 yen/ind.
1888[b]	Funakoshi		60	180	
1889[b]	Funakoshi		30	68	

Source: Assembled from Kawabata, H., Dolphin and porpoise fisheries, pp. 636–664, in: Yamada Choshi Hensan Iinkai, ed., *History of Yamada Town*, Vol. 1, Yamada-cho Kyoiku Iinkai, Yamada, Japan, 1986, 10+1095pp., in Japanese.

Species names are as they appeared in the literature. Species names are as recorded in the literature and may disagree with current Japanese names. The author cited a document of the time stating that "females of *ma-iruka* (right species of dolphin) have a straight tail fin (likely dorsal fin), while males of the species have a curved tail fin and are called *kama-iruka* (sickle dolphin)," and that "they are caught together and are not commercially distinguished." This suggests that both *kama-iruka* and *ma-iruka* of the time in the region indicated current *kama-iruka* (Pacific white-sided dolphin) and that the species was the most popular species for the fishermen (see Chapter 14 for sexual dimorphism of the dorsal fin).

[a] Meat production was stated as 37.5 kg/individual, which was different from records in Funakoshi.

[b] Price and production per individual are uniform for the 3 years.

According to documents reprinted in Kawabata (1986, in Japanese), average meat productions for *ma-iruka*, 37.5–67.5 kg, was not very different from 52.5 kg for *nezumi-iruka* (see the following text in this section). It did not mention meat production of *goto-irula* but stated that the species was the most tasty, similar to "whales." Taste does not help species identification. I would say that both short-finned pilot whale and harbor porpoise taste best among small cetaceans. The common name *nezumi-iruka* has been recorded from the eighteenth to nineteenth century in various places along the Pacific coast of central Japan and northern Kyushu apparently for various small-cetacean species. Nagasawa (1916 in Kasuya and Yamada 1995, both in Japanese) was the first to use the name for *Phocoena phocoena* (*sic*), but with an insufficient basis for doing so. Currently, there is no evidence on which to conclude that over 2000 *nezumi-iruka* driven at villages in Iwate Prefecture in the nineteenth century (Table 3.10) were common porpoises (*P. phocoena*), although evidence against the supposition is also lacking. The name *goto-iruka* and *nyudo-iruka* are judged based on similarity to correspond to the present short-finned pilot whale (which is called *ma-gondo*, *gondo* or *gondo-kujira*) or false killer whale (which is called *bo-zu* or *oki-gondo*) (see Sections 3.1 and 3.8.2 for the interpretation of the words). It is worth noting that these old records do not list *kamiyo* or *kameyo*, which are common names for the Dall's porpoise currently used in the Sanriku Region. It is unlikely that such an offshore species would enter into inlets or coves in large enough numbers to attract drive fisheries.

Kawabata (1986, in Japanese) in *Yamada Choshi* (History of Yamada Town) stated that "common dolphins, striped dolphins, Pacific white-sided dolphins, Dall's porpoises and common porpoises could have been the main target of the drive fishery at O-ura, and catches of Pacific right whale dolphins seemed rare." This was not supported by the original historical documents but was possibly based on his interpretation of the cetacean fauna off Iwate and recent small-cetacean fisheries in Japan and should be considered with caution.

Kawabata (1986, in Japanese) listed the fishing season, products, and markets of the drive fisheries in Iwate Prefecture based on old documents. The seasons for Akasaki were from May to June, from September to October, and from December to January, and there was no clear peak season. Kamaishi had its season from May to December with a peak season from July to August. Funakoshi had a season from October to February. Determining the peak season would not be easy, because the drives operated opportunistically when group of dolphins chanced to come near the inlets.

The catch was sold as raw meat in autumn and winter or as salted meat in spring and summer. The price also varied by season, that is, 2–3 yen/individual in winter and 0.15–0.5 yen in summer, possibly due to the difficulty of preservation. In the late nineteenth century, 1 yen was equivalent to one silver dollar.

The meat was sold in the Mogami area (*c.* 38°45′N, 140°15′E) in Yamagata Prefecture and Akita Prefecture (*c.* 38°50′N–40°25′N, 140°45′E–141°00′E) and Aizu (*c.* 37°35′N, 139°50′E) in Fukushima Prefecture. These are either hilly inland areas or basins surrounded by mountains. Raw meat was also shipped to Shiogama (38°19′N, 141°01′E) in Miyagi Prefecture. These demands also supported the subsequent hand-harpoon fishery for Dall' porpoises (see Sections 1.3 and 2.4).

Approximate amounts of products were also found in the old records reprinted in Kawabata (1986, in Japanese). *Ma-iruka* produced 37.5 kg of meat per dolphin on average (at Akasaki), 63.8 kg (Funakoshi), and 67.5 kg (Kamaishi). *Nezumi-iruka* produced 52.5 kg of meat (at Funakoshi). These figures are close to the 50 kg per Dall's porpoise (Section 2.4). Dolphin oil production varied depending on how the carcass was used. At Funakoshi, the fishermen tried out oil from the head and blubber and got 18 L/*nezumi-iruka*, while Kamaishi fishermen took oil from the head, bone, and viscera and obtained 1.2 L/*ma-iruka* (they sold blubber attached to meat). The production of meat could have changed if blubber was weighed with meat or not. In summary, per head production of *nezumi-iruka* was 52.5 kg of meat and 18 L of oil at Funakoshi and that of *ma-iruka* was 67.5 kg of meat plus blubber and 1.2 L of oil at Kamaishi.

The dolphin drive fisheries in Iwate Prefecture have been operated since the Meiji revolution with licenses that allowed particular fishery groups to monopolize certain sea area as in the case of net whaling. Kawabata (1986, in Japanese) recorded anecdotes at O-ura that the last dolphin drive with net(s) was in July 1920, when 70–80 dolphins were taken. This was around the time when the hand-harpoon fishery for small cetaceans emerged in this region. The dates of the last drives in other Iwate villages are unknown.

3.8 IZU PENINSULA, NORTH PACIFIC

3.8.1 Methods of the Drive Fisheries

The Izu Peninsula (*c.* 34°36′N–35°05′N, 138°45′E–139°10′E) is a part of Shizuoka Prefecture. It is a hilly land about 55 km long and 35 km wide that projects to the south from the Pacific coast of central Honshu. The east coast faces Sagami Bay and the west coast faces Suruga Bay; both are under the influence of the warm Kuroshio Current for the whole year. The names and locations of dolphin driving teams on the Izu coast are in Figure 3.1.

Archaeological sites known to have cetacean remains are distributed from Hokkaido in the north to the Southwestern Islands of Japan. An area from Tokyo Bay to Sagami Bay on the Pacific coast of central Japan is known to have a high density of such sites. The cetacean species found in these sites include pilot whales and common dolphins (Section 2.1; Anraku 1985; Kasuya et al. 1985; Kawaguchi and Nishinaka 1985; Kurino and Nagahama 1985, all in Japanese).

The oldest record of a dolphin drive fishery on the coasts of the Izu Peninsula is from 1619 (Shibusawa 1982, in Japanese). It describes the shares of catches by *iruka-ami* (dolphin netting) between the two neighbor cooperating villages of Nagahama and Omosu (35°01′N, 138°53′E) (Figure 3.1). There must have been other driving groups in those days on the Izu coast. *Shizuoka-ken Suisan Shi* (Fisheries of Shizuoka Prefecture) by Kawashima (1894, in Japanese) was a comprehensive description of fishing activities in Shizuoka Prefecture in the 1880s and reported that there were a total of 18 dolphin drive teams on the Izu coast (Table 3.11 and Figure 3.1); joint operations of two villages were counted as one. Two such cases of cooperation were known between Ashibo and Kuryo (35°01′N, 138°50′E) and between Nagahama and Omosu. Although *Suisan Hosai Shi* (Records of Fishery Activities) by Noshomu-sho Suisan-kyoku (Bureau of Fisheries of Ministry of Agriculture and Commerce) (1911, in Japanese) listed only eight driving groups, it was possibly not a comprehensive list. It listed Shin-shuku (35°18′N, 139°34′E) in Hiratsuka in Kanagawa Prefecture as operating a dolphin fishery using a drag net, because the area was located at the middle of a shallow beach of Sagami Bay and had no inlet suitable for driving. The *Kebuki-kusa* by Matsue (1638, not seen, cited in Hawley 1958–1960), a guide book of *haiku* (kind of poetry) with information on tourist spots and local products, wrote that Eno-shima (35°18′N, 139°30′E) on the northeastern Sagami Bay was known for dolphin meat. This suggests a possibility that dried dolphin meat produced in Hiratsuka was brought to Eno-shima, a tourist spot about 13 km east of Hiratsuka.

Records of dolphin drive fisheries on the Izu coast are almost absent for the 1910s to the 1930s, but changes in driving groups probably occurred. Although there were some new enterprises such as at Kawana (34°57′N, 139°07′E) that started operation in 1888 and Futo (34°55′N, 139°08′E) that started in 1897 (see this section below), the number of driving teams on the Izu coast probably declined since the late nineteenth century. This was deduced from the following records of Shizuoka-ken Kyoiku Iinkai [Board of Education of Shizuoka Prefecture (1986, 1987, both in Japanese)]. When Inatori Village (34°46′N, 139°03′E) reopened the fishery during World War II, it learned the method from Arari (34°50′N, 138°46′E) because the technique was lost after Inatori ceased driving around the turn of the century. Heda Village (34°58′N, 138°47′E) also reopened driving during the war after a period of pause. It is uncertain whether Tago Village (34°48′N, 138°46′E) completely stopped the operation before it reportedly reopened the fishery during the war. Nakamura (1988, in Japanese) reported that Toi (34°55′N, 138°48′E) reopened its drive fishery in 1941 or 1942 (Figure 3.2).

The driving team of Arari had a long history of opportunistic driving, but it was only during the war that it started sending out scouting vessels for dolphin schools. During the war, young men and large vessels were drafted by the military, and there were left in the village only elderly men, women, and small vessels. In addition, night fishing risked attack by U.S. submarines, and daytime offshore operation risked U.S. air raids. So the people available in the village worked on dolphin fishing using small vessels and in nearshore waters.

Due to the food crisis after the war, the drive fisheries on the Izu coast experienced resurgence. In December

TABLE 3.11
Historical Change in Dolphin Drive Groups on the Coast of the Izu Peninsula

Year	East Coast (North of Kawazu)	South Coast	West Coast (North of Matsuzaki)	No. of Groups	References[a]
1619			Nagahama + Omosu		Sb
Early nineteenth century	Inatori (in 1827), Yukawa + Matsubara				It, Sz
1890s	Inatori, Kawana	Iruma, Koura Mera	Arari, Heda, Tago, Toi, Nishiura (Ashibo + Kuryo, Enashi, Hirasawa, Kisho, Kou, Kuzure, Tachibo),[b] Omosu + Nagahama, Shigedera	18	Ka
Unspecified[c]	Ajiro, Futo, Inatori, Kawana, Shimo-kawazu	Ishibu	Arari, Heda, Tago, Toi		Sz
1900s	Inatori, Ito, Komuro (i.e., Kawana)		Heda, Nishiura, Tago, Uchiura, Ugusu	8	No
1940s–1950s[d]	Futo, Inatori, Kawana		Arari, Heda, Tago, Toi	7	Na, Sz
1960s	Futo, Kawana		Arari	3	Z
1970s[e]	Futo, Kawana		Arari	2–3	Z
1984~	Futo			1	Z

Two locations combined with "+" represent a joint operation.

[a] *Reference keys (all in Japanese)*: It, Ito-shi (Records of Ito; not seen); Ka, Kawashima (1894); Na, Nakamura (1988); No, Noshomu-sho Suisan-kyoku (Bureau of Fisheries of Ministry of Agriculture and Commerce, 1911); Sb, Shibusawa (1982); Sz, Shizuoka-ken Kyoiku Iinkai (Board of Education of Shizuoka Prefecture, 1986, 1987); and Z, (Kasuya unpublished).

[b] Nishiura included these eight villages that operated dolphin drive fisheries.

[c] Inatori started operations before 1827, when a memorial monument was erected. Kawana started operations in December 1888 and Futo in around 1897 (Kasuya 1985a; Shizuoka-ken Kyoiku Iinkai 1987, in Japanese).

[d] During World War II, Heda, Inatori, and Toi (probably Tago too) resumed dolphin drives, and Arari switched from opportunistic operations to more active operations that accompanied daily searching activities. Inatori ceased operations in the nineteenth century and resumed them during the war. An operation license by the governor of Shizuoka Prefecture, which covered October 27, 1942, to December 31, 1947, is preserved at Heda.

[e] Arari made the last regular drives in 1961 but made opportunistic drives until 1973. Kawana and Futo operated jointly for 1967–1983 seasons.

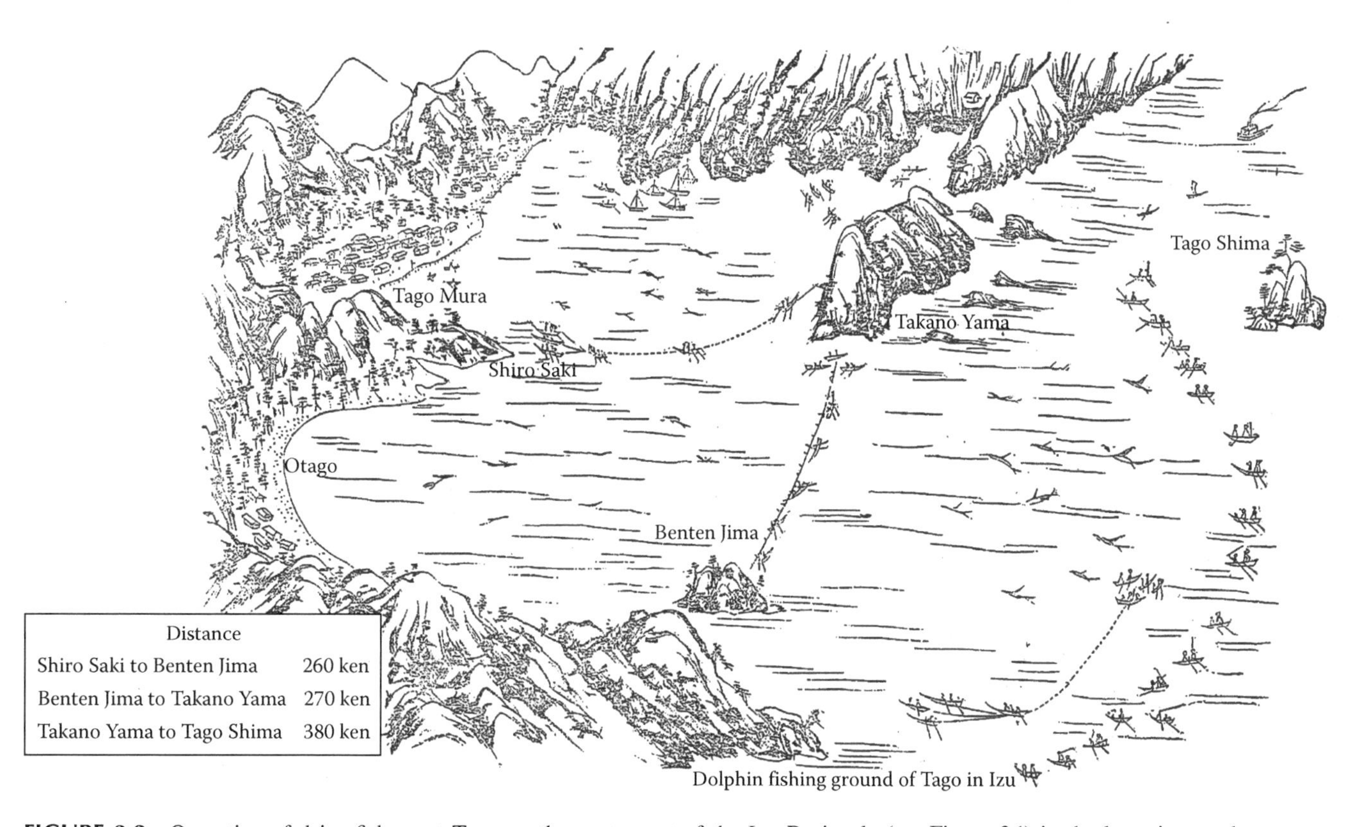

FIGURE 3.2 Operation of drive fishery at Tago on the west coast of the Izu Peninsula (see Figure 3.1) in the late nineteenth century. Distance between locations is measured in unit 'ken' (1 ken = 1.8 m). (Reproduced from p. 48, Vol. 2 of Kawashima, T., *Fisheries of Shizuoka Prefecture*, Shizuoka-ken Gyogyo-kumiai Torishimarijo, Shizuoka, Japan, 1894, Vols. 1: 288pp.; 2: 182pp.; 3: 406pp.; 4: 362pp., in Japanese.)

1949, fishermen of the Arari FCU caught over 3500 dolphins and spent over 2 weeks processing the catch. They sold the catch for a total of 25 million yen and received the amount in 100 yen bills, which was so bulky that it covered the floor of the account room of the FCU where the amount was confirmed (Shizuoka-ken Kyoiku Iinkai [Board of Education of Shizuoka Prefecture] 1986, in Japanese). This statement allows us to estimate the dockside price of dolphin at 7000 yen/individual, which was a black market price in view of the information available in Wilke et al. (1953), which stated that the officially decided price was 2000–3000 yen/individual in 1948 but that the black market price on the Sanriku coast was 2.5 times the official price. Dolphins sold on the black market were not included in the official statistics.

The favorable period for the Izu fishermen soon ended. When I visited Kawana the first time in 1960 to assist M. Nishiwaki and S. Ohsumi working on striped dolphins, only the three groups of Arari, Futo, and Kawana were operating dolphin drives, and the operation of Inatori was only a memory. The Arari team last used scouting vessels in 1961, returned to the prewar operation of opportunistic driving with low level of takes, and made the last drive of 37 bottlenose dolphins in 1973. In this year, all the members of the drive group *O-ami* (large net) of the Arari FCU sold their shares to the FCU and stopped operations. The Arari fishermen gave the reasons of (1) decline in the number of dolphins migrating to the fishing ground, (2) decline in the market price of dolphins, and (3) shift of their fishing activities to offshore waters using larger vessels (Shizuoka-ken Kyoiku Iinkai 1986, in Japanese).

The fishermen of Futo and Kawana lived in close proximity (Figure 3.1) and fished dolphins in the same area with hostility toward each other. They had a big conflict at sea in 1903 and also after World War II (Shizuoka-ken Kyoiku Iinkai 1987, in Japanese). It was around 1966 when fishermen from both villages raced to drive with a belief that "they will take this dolphin school if we do not take it now." As a result, the market was saturated, the price went down to 300 yen/individual, and carcasses were left to drift outside the Futo port. This incident led them to agree in February 1967 on cooperative driving of dolphins through the mediation of the *Shizuoka-ken Suisan Shiken-jo* (Shizuoka Prefecture Fisheries Experimental Laboratory), and they started joint driving in autumn of 1967 (Shizuoka-ken Kyoiku Iinkai [Board of Education of Shizuoka Prefecture] 1987, in Japanese). This date agrees with a footnote on a statistical table prepared by the Coastal Division of the Fisheries Agency (dated February 15, 1981, unpublished; also see footnote to Table 3.14). In this respect, Kasuya (1985a) and Kishiro and Kasuya (1993), who gave autumn of 1968 as the date of the first cooperative drive, seem to be wrong. Each of the two FCUs of Futo and Kawana offered two scouting vessels (total four) to work jointly in driving dolphins, which were driven into one of the two harbors alternatively. This joint operation continued through the season that started in autumn 1983. The Kawana FCU retired from the fishery in autumn 1984. The Kawana FCU took a loss in the dolphin fishery since around 1978 (Shizuoka-ken Kyoiku Iinkai [Board of Education of Shizuoka Prefecture] 1987, in Japanese) and finally ended its operations.

The decline of dolphin drive fisheries on the Izu coast cannot be attributed to a decline in demand for dolphin meat, because Shizuoka Prefecture imported large numbers of Dall's porpoises from the Sanriku coast in the late 1970s to make up for the lost supply of dolphin meat from the Izu coast (Section 2.4; Kasuya 1985a; Kishiro and Kasuya 1993). However, if Dall's porpoises had not been imported, perhaps the Kawana fishermen could have survived longer financially. As of 2008, the Futo FCU was the only group operating a dolphin drive fishery on the Izu coast, but no dolphins have been caught since 2005 (Table 3.16). The history of this fishery is also dealt with in Chapter 10.

Kawashima (1894, in Japanese) reported in his *Shizuoka-ken Suisan Shi* (Fisheries of Shizuoka Prefecture) that dolphin drive fisheries along the Izu coast used various types of nets in different places and different times. The use of the *nekosai-ami* (fixed bottom net), which was a kind of fixed trap net, was limited to villages along the coastal stretch of about 23 km from Futo northward to Izusan (35°42′N, 139°06′E). In other parts of the Izu coast, they used the *tatekiri-ami* (Shut out net), which was also called *O-ami* (large net) for the dolphin fishery. In this fishery, a school of dolphins or tuna found in nearshore waters was driven into an inlet or net set close to shore, and then the entrance was closed with a net before capture.

Kawashima (1894, in Japanese) stated that the dolphin fishery at Tago was the largest on the Izu coast and described the operation. A similar but shorter description was also given in Noshomu-sho Suisan-kyoku (Bureau of Fisheries of Ministry of Agriculture and Commerce) (1911, in Japanese). The village of Tago is located at the head of a deep inlet with a small entrance. The net used for the dolphin fishery, *O-ami* (large net), was jointly owned by the villagers, and they all participated. They did not send out scouting vessels. Upon receiving news of a dolphin school sighted by vessels working outside the inlet, the villagers went out of the harbor and drove the dolphins from a distance of several *ri* (which is equal to 4 km) into the inlet of Tago by various methods, that is, banging the side of their vessel, dipping the sail in the water, or throwing the anchor into the water. They closed the entrance of the harbor after the dolphins entered. If the dolphin school was small, they pulled all the animals up at one time, but a large school was divided into several smaller groups and landed. Special care was needed for the Pacific white-sided dolphin to prevent them from escaping. A net was placed behind the school of Pacific white-sided dolphins while they were still being driven. After they entered the harbor, the entrance was closed with two sets of nets, and a boat was placed to scare dolphins away from the nets by hanging ropes from the boat.

The method of driving Pacific white-sided dolphins used at Tago was to use an extra net and place a guard boat to keep them far from the closure nets. I do not know if the Pacific white-sided dolphins would actually jump over a net to escape. The Tago fishermen did not seem to have placed dolphin spotters at vantage points. It was probably because numerous fishing vessels operated outside the port.

Along the coast of Uchiura Bay (*c.* 35°03′N, 138°50′E) in the northeastern corner of Suruga Bay (or off the northwestern corner of the Izu Peninsula), the fishermen placed watchmen at vantage points to spot schools of dolphins and fish and to provide advice during the driving operation. These operations are illustrated in Kawashima (1894, in Japanese) and Yamamoto (1884, in Japanese). Fishermen on the southern coast of Uchiura Bay started in the late nineteenth century to place a long net stretching offshore to guide fish schools into their harbor or net placed near the shore. This caused conflicts between neighboring fishing groups. Dolphin schools also occurred off the northern coast of Uchiura Bay, but there was no dolphin drive north of Shishi-hama Village, presumably due to the absence of suitable inlets on the coast (Figure 3.1).

The use of scouting vessels differed among driving groups. Arari started it during the last war (see the preceding text in this section), but Kawana sent two or three vessels for dolphins off the port before the use of motor-driven vessels that may have started in the early 1920s. A similar system continued after the introduction of motor-driven vessels, but the number of scouting vessels is unknown. In 1962, the Kawana FCU introduced a high-speed scouting vessel and later moved to three vessels. The speed increased stepwise from 13 knots for the first vessel in 1962 to 40 knots for the third stage in 1983 (Shizuoka-ken Kyoiku Iinkai [Board of Education of Shizuoka Prefecture] 1987, in Japanese). Kasuya and Miyazaki (1982) recorded the following information from Mr. Suda, President of the Kawana FCU of the time. When rowed or sailing boats were used in the fishery, dolphins were driven from within several miles from the port. With the introduction of motors in the 1920s, the range increased to the entrance of Sagami Bay (37 km from Kawana Harbor), and the use of high-speed scouting vessels in 1962 further expanded the range to close to Miyake Island (34°05′N, 139°30′E), which was about 100 km south of Kawana. N. Miyazaki rode on some of these high speedboats to observe the behavior and distribution of striped dolphins in the early 1970s. It is certain that the searching range expanded with time and resulted in the increase in the fishing effort. The Futo FCU followed Kawana in the use of high-speed scouting vessels, which led to races between the two FCUs. This ended in February 1967 when the two FCUs agreed on joint operation of dolphin drives and in ending the hostile contacts of many years (see above).

Kawana started its dolphin drives in 1888 using techniques introduced from Inatori (Shizuoka-ken Kyoiku Iinkai [Board of Education of Shizuoka Prefecture] 1987, in Japanese); this is noted on a stone monument constructed at Kawana in 1889 (Kasuya 1985a). Discontinuation of the fishery was proposed at the general meeting of the Kawana FCU in February 1984 and was passed after amendment for a "pause" (Shizuoka-ken Kyoiku Iinkai [Board of Education of Shizuoka Prefecture] 1987, in Japanese). This was the end of the Kawana dolphin drive fishery that lasted almost 100 years.

Futo started to drive dolphins around 1897, or about 10 years later than Kawana, and had a short pause in operations after a big battle with the Kawana fleet in 1903. It reopened the fishery around 1930–1931, still operating competitively with Kawana, carried out cooperative operations with Kawana in 1967–1983, and returned to independent operation in autumn 1984. The number of scouting vessels used changed during the last phase of the operations from two (1984–1986) to one (1987–present) (Kishiro and Kasuya 1993).

A summary of dolphin fisheries along the coasts of the Izu Peninsula is also given in Kojima (2009, in Japanese).

3.8.2 Dolphin Species Taken by the Izu Fishery

In the past, Japanese fishermen used the name *ma-iruka* (right species of dolphin) for any species that was the most popular to them. This naming caused trouble for T. Ogawa when he started collecting materials for his study of brain anatomy in the 1930s and made study of the taxonomy of small cetaceans around Japan difficult (Ogawa 1950, in Japanese).

Kawashima (1894, in Japanese) listed 10 cetacean species either known around the Izu Peninsula or hunted by the Izu fishery: sperm whale, Baird's beaked whale, *iwashi-kujira* (sei or Bryde's whale, see Section 1.3), and seven delphinoids (*tsubame-iruka*, *ma-iruka*, *kama-iruka*, *nyudo-iruka*, *nezumi-iruka*, and *shachi*). In addition to these, he recorded species names collected from each driving group. The only phocoenid likely inhabiting waters along the Izu coast is *Neophocaena* sp., which is not known to be hunted in the area (Chapter 8).

Takenaka (1890, in Japanese) listed species, number of animals, weight, and price of dolphins taken at Arari during the 8 years from 1882 to 1889. Another source of information on the local names of dolphins is a handwritten table in my possession prepared by the Arari FCU and titled *Iruka Shurui-betsu Hokaku Shirabe, Showa-25-nen 1-gatsu Yori 32-nen 4-gatsu Made* (Catch of Dolphins by Species, from January 1950 to April 1957), which is a photocopy of the original copy possessed by S. Ohsumi. This table was possibly prepared at the request of the Fisheries Agency and lists species, number of animals, total weight, and date of drive by month for each drive. These figures have been published in Kasuya (1976, in Japanese) with the permission of S. Ohsumi.

Local names of dolphins that appeared in the previously mentioned sources are in Table 3.12 together with my interpretations. Among these, old local names, *shachi* and *kama-iruka*, must be the same as today, meaning killer whale and Pacific white-sided dolphin, respectively. The former is commonly used throughout Japan, and the latter seems to have derived from the sickle-shaped dorsal fin and is widely recorded at Taiji and Sanriku on the Pacific coast of Japan. The statement by Kawashima (1894, in Japanese) that *kama-iruka* weighs 28–129 kg and inhabits coastal waters also supports this conclusion.

Next to be considered are *o-iruka* (large dolphin) in the above mentioned statistics table at the Arari FCU and *nyudo-iruka* (Takenaka 1890; Kawashima 1894, both in Japanese). The latter name is translated roughly as "Buddhist monk dolphin." Nyudo is similar in meaning with *bozu* (master of a Buddhist temple or Buddhist monk). Because these people used to have shaved heads, both *nyudo* and *bozu* have also meant a bald person. Therefore, *bozu* or *nyudo* could have been used for dolphins that have a round head such as false killer or short-finned pilot whales. The false killer whale was called *nyudo-iruka* or *bozu-iruka* in the Noto and Goto regions (Table 3.1).

The Arari FCU made 18 drives of 1891 individual *o-iruka* in 6 years and 4 months from 1950 to 1957. The group size in one

TABLE 3.12
Vernacular Names of Dolphins Used along the Coast of the Izu Peninsula as Recorded in Early Literature and Records of the Futo Fishery Cooperative Union Cited in Kasuya (1976, in Japanese)

Vernacular Name	Location[a]	References (in Japanese)	Current Common Name and English Name
Shyachi	Izu	Kawashima (1894)	*Shyachi* Killer whale
Kama-iruka	Ajiro, Izu	Kawashima (1894), Takenaka (1890)	*Kama-iruka* Pacific white-sided dolphin
Ma-iruka	Izu	Kawashima (1894), Takenaka (1890)	*Suji-iruka* Striped dolphin
Ma-yuruka	Inatori	Kawashima (1894)	*Suji-iruka* Striped dolphin
Ma-iruka	Arari	Arari Fish. Coop. Union (unpublished)	*Suji-iruka* Striped dolphin
Hasunaga	Arari	Arari Fish. Coop. Union (unpublished)	*Hando-iruka* Common bottlenose dolphin
O-iruka	Arari	Arari Fish. Coop. Union (unpublished)	*Kobire-gondo* Short-finned pilot whale
Nyudo-iruka (also called *gonzo-iruka*, or *Bozu-iruka*)	Izu	Kawashima (1894), Takenaka (1890)	*Kobire-gondo* Short-finned pilot whale
Matsuba-iruka	Toi	Kawashima (1894)	*Hana-gondo* Risso's dolphin
Tsubame-iruka	Ajiro	Kawashima (1894)	Unknown

[a] "Izu" indicates the whole Izu Peninsula, and other geographical names indicate a portion of the Izu coast.

drive ranged from 16 to 226 animals; the average was 105. The average weight (of meat?) was 421 kg /individual, which was the largest among three species of dolphins recorded as driven in the period by the Arari FCU, *ma-iruka*, *hasunaga-iruka*, and *o-iruka*. Takenaka (1890, in Japanese) recorded the take of 760 *nyudo-iruka* in 1882 at Arari (the number of drives yielding this catch is unknown) and the meat production of 480 kg/individual. The two weight figures are similar if both are assumed to be meat weight and are almost twice the figure of meat production of false killer whales recorded in the Noto region. Thus, it is likely that the common name *nyudo-iruka* was used for different species between the Noto and Izu regions.

Kawashima (1894, in Japanese) stated that body weight of the *nyudo-iruka* off Izu coasts was between 170 and 1330 kg. The relationship between body weight (W, kg) and body length (L, cm) is given by $\log W = 2.8873 \log L + \log(2.377 \times 10^{-5})$ for southern-form short-finned pilot whales off Japan (Kasuya and Matsui 1984). This gives a body weight of 1270 kg for a male at the modal body length of 4.75 m and 1470 kg for the largest recorded males of 5 m. Little is known about the body weight of false killer whales, but a 475 cm female weighed 773 kg (Odell and McClune 1999). Based on meat production, the *nyudo-iruka* off the Izu coasts are more likely to have been short-finned pilot whales.

Miyashita (1993a,b) presented abundance estimates for six species of warmwater dolphins hunted by Japanese drive fisheries using sighting data collected from June to September in the 9 years from 1983 to 1991 (Table 3.13). This study covered latitudes 20°N–50°N and longitudes 127°E–180°E of the East China Sea and northern North Pacific. The six species were sighted only in latitudes 23°N–42°N. At the time of the survey, the stock of striped dolphins had already collapsed due to overfishing, and the species was scarce in the near-shore waters of the Pacific coast. His analyses showed that current population of false killer whales off the Pacific coasts of Japan was smaller than that of the more heavily exploited short-finned pilot whales in the same area. The ratio of the former species to the latter was 1:7 in the coastal Pacific, 2:5 in the offshore Pacific, and 1:3 in the southern waters in the survey range. This was quite the reverse of the situation for the East China Sea and Sea of Japan (see Tables 3.4 and 3.5).

Discussed earlier are the major reasons why I consider that the *nyudo-iruka* and *o-iruka* of the Izu coast were short-finned pilot whales, in particular the southern-form short-finned pilot whales. The northern form of the species is usually found north of Choshi (*c.* 35°45′N) (Chapter 12). An anecdote that supports this conclusion was related to me in Arari in June 1967 when I was there to examine a school of short-finned pilot whales driven into the harbor. A villager told me that Arari fishermen drove a large group of the same species sometime soon after the last war, that is, World War II, and spent several days processing them. He also told that some (or one) of the animals were found ingesting a straw sandal, supposedly due to hunger. Aside from the ingestion of sandals, his description of the fishery agrees with the available catch records.

The next problem is the *hasu-naga* (of unknown literal meaning) found in catch statistics of Arari in 1950–1957 (Table 3.12). The Arari FCU recorded six drives of 300 *hasu-naga*.

TABLE 3.13
Abundance of Tropical and Temperate Dolphins in the Western North Pacific Estimated by Sighting Surveys, Together with a Number of Sighted Dolphins on Which the Estimates Are Based

	Estimated Abundance[a] with 95% Confidence Interval			No. of Individuals Sighted[a]
Species	**Coastal Area**	**Offshore Area**	**Southern Area**	**Whole Areas**
Striped dolphin	19,631	497,725	52,682	18,610
	5,727–67,288	351,416–704,949	10,940–253,700	
Pantropical spotted dolphin	15,900	294,321	127,843	13,519
	7,459–33,892	202,705–427,344	64,959–251,604	
Common bottlenose dolphin	36,791	100,281	31,720	6,562
	22,699–59,630	52,537–191,412	7,665–13,261	
Risso's dolphin	31,012	45,233	7,044	4,043
	20,600–46,686	26,894–76,077	1,802–27,542	
Short-finned pilot whale (southern form)	14,012	20,884	18,712	2,172
	8,996–21,824	11,081–39,361	8,448–41,445	
Short-finned pilot whale (northern form)	4,239 (CV = 0.61)[b]			
	5,344[c]			
False killer whale	2,029	8,569	6,070	511
	907–4,541	4,497–16,327	2,308–15,965	

[a] Estimates for species other than northern-form short-finned pilot whale are from Tables 9 through 15 of Miyashita (1993a). Area divisions are coastal area (30°N–42°N and west of 145°E), offshore area (30°N–42°N and 145°E–180°E), and southern area (23°N–30°N and 127°E–180°E).

[b] This was estimated by Japanese scientists using sighting data of 1982–1988 (IWC 1992). This population inhabits the northern part of the coastal waters defined earlier.

[c] This was estimated by Japanese scientists using the data of 1984–1985 (IWC 1992). See Section 12.6 and Appendix to Chapter 12 for recent abundance estimates and trends in abundance.

Average group size was small at 50 individuals. The average weight was 184.9 kg/individual, which is much greater than the weight of the striped dolphin. Responding to my inquiry in 1976, a member of the staff of the Arari FCU reported that the *hasu-naga* was a dolphin species performing tricks in Japanese aquaria and that the Arari FCU used to drive them for aquaria. Thus, it is safe to conclude that *hasu-naga* was used for bottlenose dolphins.

The *nezumi-iruka* (rat [colored] dolphin) or (mouse [colored] dolphin) was recorded by Kawashima (1894, in Japanese). He stated that weight was 94–188 kg/individual, which was greater than for striped dolphins and Pacific white-sided dolphins, and that they were found with *nyudo-iruka* (short-finned pilot whale). Off the Pacific coast of Japan, the bottlenose dolphin is the most frequent species with which short-finned pilot whales are found together, and the next is the Pacific white-sided dolphin (Kasuya and Marsh 1984). Thus, *nezumi-iruka* off the Izu coasts was also likely a name for the bottlenose dolphin, which is called *kuro* (black) at Taiji. This interpretation does not apply to the cetacean fauna in the Tsushima Strait area, including northern Kyushu.

It is generally believed that the Izu fishermen used *ma-iruka* and *ma-yuruka* (right species of dolphin) for striped dolphins, at least after the World War II period (Kasuya and Yamada 1995, in Japanese). However, there is no firm evidence that the usage was maintained throughout the history of drive fisheries on the Izu coast, which lasted for over four centuries. Kawashima (1894, in Japanese) stated that this species is found in large schools and that the body weighs 45–94 kg, which was smaller than the 28–129 kg weight of *kama-iruka* (Pacific white-sided dolphin) also hunted at Tago on the west coast of the Izu Peninsula. I will reserve conclusions on whether *ma-iruka* of the Izu coasts in the nineteenth century represented striped dolphins or some other smaller species such as common dolphins until additional information becomes available.

It was autumn of 1960 when I first worked on studying the catches of dolphin drive fisheries on the Izu coast. In those days, the fishermen used *ma-iruka* for the striped dolphin, the main target of the operation, and none of them used *suji-iruka* for the species. In the 1980s, it became more common for them to use the Japanese standard name *suji-iruka* for the striped dolphin, presumably affected by the work of the Fisheries Agency to collect statistics.

Izu fishermen after the World War II period believed that it was technically impossible to drive Pacific white-sided dolphins. However, in his description of dolphin drive fisheries on the Izu coasts, Kawashima (1894, in Japanese) gave an example of driving of 30 Pacific white-sided dolphins at Toi in 1890 and mentioned the use of a special type of net for Pacific white-sided dolphins, which was different from the net used for driving other species. *Suisan Hosai Shi* (Records of Fishery Activities) by Noshomu-sho Suisan-kyoku (Bureau of Fisheries of Ministry of Agriculture and Commerce) (1911, in Japanese) stated that fishermen at Tago replaced rice straw nets partially with stronger hemp nets to drive Pacific white-sided dolphins, which were believed to be intelligent and agile. In O-ura village in Yamada Cove, there are photos of a large group of Pacific white-sided dolphins driven there (see Section 3.7). We have seen similar examples in the recent operations at Iki Island. These records show that driving Pacific white-sided dolphins was practiced by the Izu fishermen in the nineteenth century with some modification of gear and methods and that the technique was lost in the drive fisheries on the Izu coast.

3.8.3 Catch Statistics and Magnitude of Operation

Kawashima (1894, in Japanese) reported catches of dolphins in 1877–1891 at Tago Village, which he thought to be the top dolphin drive locality on the Izu coast. The same figures are also in Nakamura (1988, in Japanese). During the 15-year period, Tago Village made drives only in 13 seasons (no catch in the remaining 2 seasons). The unstable catch level was due to the opportunistic nature of the operation; they did not send out scouting vessels. The total catch in the 13 seasons was 8707 with an annual average of 580. *Gonzo*, that is, short-finned pilot whales, represented only 460 individuals or 5.3% of the total. The remaining 94.7% were recorded as "dolphins," which could have been made up of striped dolphins as well as other species. Extrapolating the previously mentioned catch to all the dolphin drive teams (18) known to have operated on the Izu coast (Kawashima 1894, in Japanese), I estimate an annual average take of dolphins on the entire Izu coast at about 10,000 for the 1880s. This figure is an overestimate, because the catch of the most prosperous team, Tago Village, is extrapolated to other teams, including the Mera and Koura teams, which are recorded to have almost ceased driving dolphins (Kawashima 1894, in Japanese). The average price in 1882 on the Izu coast was 9.21 yen for a short-finned pilot whale and 2.86 yen for a "dolphin."

Matsuura (1942, 1943, both in Japanese) stated that increased demand for meat and skin of dolphins caused a sharp increase in the catch of dolphins and porpoises in the 1941 season, to a record 1,000 large delphinoids such as pilot whales and 45,000 smaller dolphins and porpoises (including in all types of cetacean fisheries such as driving, hand harpooning, and small-type whaling by the definition of 1947). He also stated that the Izu fishery contributed greatly to the figure, particularly to the latter category. Noguchi (1946, in Japanese), in addition to citing the same figures for 1941, further stated that a total of 28,000 dolphins were caught in Shizuoka Prefecture in the 1942 season, including 20,000 at Arari, 4,000 at Tago, and 4,000 at Heda on the Izu coasts. The catch figure for Arari agrees with that listed in Table 3.14, but those for Tago and Heda are not in the table, which means that Table 3.14 is incomplete for the 1940s. Both Matsuura (1943, in Japanese) and Noguchi (1946, in Japanese) feared depletion of the dolphin stocks by overfishing.

Some additional statistics are available for the operation at Arari Village. Nakamura (1988, in Japanese) copied annual catches of dolphins (species not specified) from an epitaph on a monument in Arari (Table 3.14). It covered the years 1934–1949 but made no mention of six seasons in 1935 and 1937–1941, presumably meaning no catches in those years. The same is also found in a manuscript containing statistics for Arari collected by S. Ohsumi and provided for my study (Kasuya 1976, in Japanese). It covers the years 1942–1957, where figures for 1942–1949 are monthly statistics without species identification and those for 1950–1957 are given for each drive and by species (see Section 3.8.2). The two sets of statistics agree for 1942–1949. The total catch in the 1950–1957 period by Arari Village was 1,891 short-finned pilot whales (17 drives), 300 bottlenose dolphins (6 drives), and 40,093 striped dolphins (33 drives). These figures are tabulated in Table 3.14 in annually aggregated form. Arari Village made rather small catches in 1934–1936, left no statistics for 1937–1940, and then recorded a peak of 20,131 dolphins in 1942 that was followed by a decline to about 5,000/year in 1945 and 1946 (Table 3.14). The peak agrees with the statement that the village moved from an opportunistic drive to a more active one using dolphin scouting vessels during the war (Section 3.8.1). The statistics suggest that the change in the operation probably occurred in 1941 or a few years before.

Japan started a full-scale invasion of China in July 1937 and a war with the United States in December 1941 and was defeated in August 1945. It was to prepare for the war that the Japanese Antarctic whaling fleets were allowed to bring back whale meat for the first time in the 1937/1938 season (Kasuya 2000; Omura 2000, both in Japanese). The Japanese small-cetacean fisheries also expanded rapidly during the peri-war period (Section 2.2). Threats by the enemy and shortage of materials could have impacted operations of the drive fisheries during the war. Shizuoka-ken Kyoiku Iinkai (Board of Education of Shizuoka Prefecture) (1986, in Japanese) described such an effect on the drive fisheries along the Izu coast. As seamen and fishing vessels were drafted by the military beginning in 1941 and only one skipjack vessel became available to them, the Arari fishermen on the Izu coast directed their fishing effort to small cetaceans in nearshore waters. In Tago, there remained only several small fishing vessels of less than 10 tons, and women, elderly men, and children after other workers and better vessels were drafted for the war effort, and the villagers worked on dolphin drives and hook-and-line fisheries in nearshore waters safe from the enemy.

This may explain, at least in part, the rapid increase in small-cetacean catches during the peri-war period along the Izu coast. However, there remain some unexplained inconsistencies in the statistics. After the Arari fishermen reported an annual catch of 4,500–7,700 dolphins during the peri-war period of 1943–1946, they reported a catch of only several hundred in each of the next 2 years (1947 and 1948), followed by a large annual catch of over 13,000 dolphins since 1949. Even if such a change (decline followed by an increase) might be explained by natural fluctuation, that cannot be the sole reason to be considered. In the case of the hand-harpoon fishery in the Sanriku region of the Pacific coasts of northern Honshu (37°44′N–41°35′N), most of the catch of small cetaceans entered the black market when the government controlled the price and distribution of the fishery products, but the black market ended when government control ended in 1949 (Section 2.3). I suspect that a similar situation may

TABLE 3.14
Dolphins Taken by Drive Fisheries on the Coast of the Izu Peninsula

	Futo[a]	Kawana[b]	Futo + Kawana[c]	Arari[d]			
Year	All Species	All Species	Striped Dolphin	Short-Finned P. W. (Southern Form)	Bottlenose Dolphins	Striped Dolphin	Total
1934							196
1936							1,027
1942		1460					20,131[e]
1943		0					7,761
1944		1081					6,579
1945		2846					4,473
1946		2710					5,470
1947		0					395
1948		5311					581
1949		1511					11,930
1950		1516		224	1	13,171	13,396
				224	*1*	*13,671*	*13,896*
1951		2235		425	0	11,207	11,632
				425	*0*	*10,962*	*11,387*
1952	—	2405		650	25	5,700	6,375
	120			*650*	*25*	*5,627*	*6,302*
1953	—	2947		349	71	1,143	1,563
	31			*232*	*188*	*1,081*	*1,501*
1954	—			31	0	298	329
	0			*31*	*0*	*298*	*329*
1955	—			86	15	2,459	2,560
	117			*86*	*15*	*2,552*	*2,653*
1956	—		2,755	126	188	5,752	6,066
	484			*126*	*188*	*5,748*	*6,062*
1957	—			—	—	2,751(257)	—
	421			*866*	*293*	*2,726*	*3,897*[f]
1958	—		2,114	—	—	1,567	—
	1060			*365*	*48*	*1,504*	*1,940*[g]
1959	—		15,649	—	—	6,304(83)	—
	2848			*138*	*60*	*4,214*	*4,412*
1960	—		14,351	—	—	67	—
	3172			*0*	*0*	*248*	*248*
1961	—		9,794	—	—	775	—
	8606			*0*	*0*	*775*	*907*[h]
1962	—		8,554				
	4132						*0*
1963	—		8,509				
	4599						*0*
1964	—		6,428				
	3795						*0*
1965	—		9,696				
	8757			*20*	*0*	*0*	*47*[i]
1966	—		8,371				
	7154						
1967	—		3,664				
	1250						
1968	—		9,160			90	—
	3382						
1969	—		3,130(435)				
	1775						

(*Continued*)

TABLE 3.14 (*Continued*)
Dolphins Taken by Drive Fisheries on the Coast of the Izu Peninsula

	Futo[a]	Kawana[b]	Futo + Kawana[c]	Arari[d]			
Year	**All Species**	**All Species**	**Striped Dolphin**	**Short-Finned P. W. (Southern Form)**	**Bottlenose Dolphins**	**Striped Dolphin**	**Total**
1970	—		5,307(2,697)			41	—
	2867						
1971	—		3,315(0)				
	3131						
1972	3824[j]	3824[j]	7,235(662)				55[k]
1973	3997	3997	6,799(1,162)		37		37
1974	8000(120)		11,715				0
1975	5310(1298)		5,996				0
1976	5204(90)		5,175				0
1977	1675(562)	1823(520)	4,020				0
1978	1668(1691)	425(2643)	2,028				0
1979	1300(25)	0(332)	1,300				0
1980	3821(310)	1399(1130)	5,278				0
1981	73(189)	0(169)	73				0

This table lists statistics privately collected by scientists and those collected through early activities of the Fisheries Agency. Most of the unspecified catches are believed to represent striped dolphins. In parentheses are additional figures for pantropical spotted dolphins (1957–1973) or additional figures for species other than striped dolphins taken at Kawana and Futo (since 1974). See Table 3.16 for full set of Fisheries Agency statistics.

[a] *Futo*. 1952–1971 records in italics are from a document of the Coastal Division of the Fisheries Agency (dated December 15, 1981), which states that Futo and Kawana started joint operation in 1967. The 1972–1981 records in roman are the statistics of the Offshore Division of the Fisheries Agency.

[b] *Kawana*. The 1942–1953 records are from Kasuya (1976, in Japanese, which lists monthly figures), and those of 1972–1981 are the statistics of the Offshore Division of the Fisheries Agency.

[c] *Kawana + Futo*. Catch of striped dolphins in Miyazaki et al. (1974, which erroneously assumed *hasunaga* [common bottlenose dolphin] as pantropical spotted dolphin) and Miyazaki (1983, which gives monthly figures). Although records of Futo are missing here for the 1942–1956 seasons, it operated the fishery. Shizuoka-ken Kyoiku Iinkai (Board of Education of Shizuoka Prefecture) (1987, pp. 3, 92–93, in Japanese) stated that "Futo started the fishery in around 1897, about 10 years after Kawana, it paused operation till around 1930 after a great at-sea conflict with Kawana in 1903, and then it resumed competing with Kawana until February 1967 when the two group agreed on a joint operation."

[d] *Arari in roman*. Figures not in brackets represent catch of striped dolphins and those in parentheses pantropical spotted dolphins, from Nakamura (1988, in Japanese) for 1934–1936, Kasuya (1976, in Japanese, which gives monthly figures) for 1942–1956, Miyazaki (1983, only striped dolphins) for 1957–1970, and Offshore Division of the Fisheries Agency (as Table 3.16) for 1972–1981. *Arari in italics*. Figures for 1950–1965 are from a document of the Coastal Division of the Fisheries Agency (dated December 15, 1981). Although it states "no catch since 1966," some sporadic catches occurred after that year.

[e] Matsuura (1943, in Japanese) and Noguchi (1946, in Japanese) wrote about dolphin hunting on the Izu coast and stated that in 1942 they caught "about 20,000 at Arari, 4,000 at Tago and 4,000 at Heda, with a total of about 28,000."

[f] Includes 12 Risso's dolphins.

[g] Includes 23 Risso's dolphins.

[h] Includes 132 Risso's dolphins.

[i] Includes 27 rough-toothed dolphins.

[j] Kawana and Futo started joint operations in 1967, and each recorded half of the total catch in the year. Figures in parentheses are catches of other species, which are additional to striped dolphins (not in brackets). See Table 3.16 for details.

[k] These are rough-toothed dolphins.

have occurred in the various fishing villages on the Izu coast and that several thousand dolphins caught at Arari could have entered the black market annually until 1948 and not be recorded in the statistics (see Section 3.8.1).

Scientists started to examine the catches of drive fisheries on the Izu coast in the 1960s. While he was working for an aquarium in Ito City (34°58′N, 139°07′E) near Futo and Kawana villages, Tobayama (1969, in Japanese) studied stomach contents and species composition of dolphins taken at the two villages on the east coast of the Izu Peninsula (on the west coast of Sagami Bay), where two of the three dolphin drive teams operated on the Izu coast during his study period of 1963–1968. Another drive team was at Arari on the west coast of the peninsula (on the east coast of Suruga Bay). During the 6-year period (coverage of 1963 and

TABLE 3.15
Species Composition of Dolphins Taken at Foto and Kawana on the Izu Coast in 1963–1968 and Identified by Scientists

Species	Striped Dolphin	Pantropical Spotted Dolphin	Common Dolphin[a]	Common Bottlenose Dolphin[b]	Short-Finned Pilot Whale[c]	Pygmy Killer Whale	Total
No. of drives	143	8	4	8	1	1	165
%	86.7	4.9	2.4	4.8	0.6	0.6	100%
No. of dolphins driven	62,655	1202	697	201	28	14	64,797
%	96.7	1.8	1.0	0.3	0.0	0.2	100%

Source: Table 1 of Tobayama, T., *Geiken Tsushin*, 217, 109, 1969, in Japanese.
[a] The long-beaked common dolphin is not known in this region.
[b] The Indo-Pacific bottlenose dolphin has not been confirmed in the catch.
[c] Northern form is not known in this region.

1964 was incomplete), he observed 165 drives on six species and confirmed that 96.7% of the catch was composed of striped dolphins, followed by pantropical spotted dolphins and other minor species (Table 3.15). He also noted that common dolphins and spotted dolphins were first driven on the east coast of the Izu Peninsula in 1963. On the west coast, a drive of spotted dolphins was first recorded in 1959 at Arari (Nishiwaki et al. 1965). Although we are uncertain that these two species, common dolphin and spotted dolphin, were not driven in earlier years, it seems to be true that the fishermen had no memories of it.

Tobayama (1969, in Japanese) indicated that bottlenose dolphins were hunted to supply live specimens for aquaria and that short-finned pilot whales were not preferred for driving. He believed that the proportions of these species in the drive catches did not represent the relative abundances of the species in the fishing ground, Sagami Bay. I remember a statement by several Izu fishermen that the meat of these two species did not taste good. The fishermen of Izu and Taiji agree in disliking the tough bottlenose dolphin meat, but their opinion differed on pilot whale meat. The Izu fishermen disliked pilot whale meat, saying that the *tare* (salted-dry meat like jerky) was too oily, but the Taiji fishermen liked the oily pilot whale meat either raw or dry salted. I agree with the preference of the Taiji fishermen on the taste, although the dark color of the meat may not be attractive.

Kasuya (1976, in Japanese) and Tobayama (1969, in Japanese) present the monthly distribution of the catches on the Izu coast. The operation in 1942–1953 on the west coast of the Izu Peninsula conducted drives in every month of the year, with fewer in February and from August to November and the highest peak from March to July, followed by another peak from December to January. Thus, the drive fishery on the west coast of the Izu Peninsula, or in Suruga Bay, had two indistinct peaks in winter and from spring to early summer. Contrary to this, the drive fishery on the east coast of the peninsula, in Sagami Bay, in 1942–1953 reported catch only in September and from November to April, with most of the catch occurring in December (Kasuya 1976, in Japanese). In later years, 1963–1968, the operations of Kawana and Futo villages on the east coast were limited from mid-October to late January and had a peak in November (Tobayama 1969, in Japanese). Such differences in fishing season between the west and east coasts of the Izu Peninsula have been reported in several studies (Ohsumi 1972; Miyazaki and Nishiwaki 1978; also see Table 10.3).

The data given earlier also suggest that the driving season on the east coast of the Izu Peninsula narrowed with time. Such a change could be due to various reasons: operational reasons, alternation of target species or stocks, and decline in the abundance of a targeted population. The fishermen prefer to operate in a season of higher price for better sales or in the migration peak for lower operation costs. The effect was more pronounced in later years when driving teams stopped competing. If the abundance of a target population declines, driving effort is likely to concentrate in the peak migration season. The fishing season is analyzed in relation to the migration of striped dolphins in Section 10.4 (Table 10.3).

Table 3.16 presents recent statistics for the dolphin drive fisheries, based on earlier statistics in Kishiro and Kasuya (1993) that summarized previously published records and later statistics available in the annual progress reports by the Japanese government to the International Whaling Commission and home pages of the Fisheries Agency. It is considered to cover all operations on the Izu coast. The basic data for these statistics are the landings and sale records of the FCUs, the only available source of such statistics. For the purpose of comparison, I have listed the catches of dolphins and porpoises in the entire Shizuoka Prefecture, a part of which is the Izu coast, reported by the Statistic Division of the Ministry of Agriculture and Forestry for earlier seasons (Ohsumi 1972). Some of the disagreement between the two set of statistics might come from takes at some locations outside the Izu Peninsula (which must be minor) or timing of the statistics, that is, calendar year (from January to December) versus fiscal year (from April to March), but some disagreement cannot be thus explained, that is, 1965–1967.

Apart from the technical problems presented earlier, we find in the statistics a change of biological interest, that is, an increase of spotted dolphins accompanied by a decline of striped dolphins. The change started in the 1970s, and relative catches almost reversed in the 1980s. This was before the start of quotas by species in 1993 (70 striped dolphins, 455 spotted dolphins, and 75 bottlenose dolphins), which reflected the catch composition of the time (see Section 6.5). Spotted dolphins were

TABLE 3.16
Dolphins Taken by Drive Fisheries on the Izu Coast, Shizuoka Prefecture

Year	Short-Finned P. W. (Southern Form)	False Killer W.	Bottlenose Dolphins[a]	Striped Dolphin	Pantropical Spotted D.	Others	Total	Total, Shizuoka Prefecture[b]
1958	365	—	48	3,681	—	23	4,051	5,321
1959	138	—	60	19,863	—	—	20,061	21,342
1960	—	—	—	14,599	—	—	14,599	6,497
1961	—	—	—	10,569	—	132	10,701	16,407
1962	—	—	—	8,554	—	—	8,554	8,968
1963	—	—	—	8,509	—	—	8,509	7,690
1964	—	—	—	6,428	—	—	6,428	6,838
1965	33*	—	—	9,696	—	27	9,756	14,105
1966	—	—	—	8,371	—	—	8,371	16,668
1967	30*	—	—	3,664	—	—	3,694	5,250
1968	—	—	—	9,250	—	—	9,250	6,303
1969	—	—	—	3,130	435	—	3,565	3,601
1970	—	—	—	5,348	2697	—	8,045	5,601
1971	—	—	—	3,315	—	—	3,315	
1972	_0_	_0_	_0_	_7,235_	_662_	_55[c]_	_7,952_	
	0	0	0	7,648	0	55[c]	7,703	
1973	_0_	_0_	_37_	_6,799_	_1162_	_0_	_7,998_	
	0	0	37	7,994	0	0	8,031	
1974	_0_	_0_	_120_	_11,715_	_0_	_0_	_11,835_	
	0	0	120	8,000	0	0	8,120	
1975	_0_	_0_	_0_	_5,996_	_1298_	_0_	_7,294_	
	0	0	0	15,302	1298	0	16,600	
1976	_0_	_0_	_90_	_5,175_	_0_	_0_	_5,265_	
	0	0	90	5,204	0	0	5,294	
1977	_73*_	_0_	_0_	_4,020_	_757_	_0_	_4,777_	
	0	0	0	3,823	757	0	4,580	
1978	_80*_	_123_	_27_	_2,028_	_4184_	_0_	_6,442_	
	0	123	27	2,093	4184	0	6,427	
1979	0	0	0	1,300	357	0	1,657	
1980	0	0	0	5,220	1440	0	6,660	
1981	20	0	0	73	169	0	262	
1982	0	0	0	246	3498	0	3,744	
1983	0	0	0	40	2789	0	2,829	
1984	0	0	0	925	0	2[d]	927	
1985	0	43	101	578	0	46[e]	768	
1986	0	0	0	198[f]	0	0	198	
1987	0	0	0	1,815[f]	0	0	1,815	
1988	0	0	143	356	191	0	690	
1989	0	0	66	102	0	0	168	
1991	0	0	0	32	0	0	32	
1992	0	0	0	0	0	280[f]	0	
Quota for 1993[g]	[0]	[0]	[75]	[70]	[455]	[0]	[0]	
1993					95		95	
1994	0	0	35	0	0	0	35	
1996	0	5	43	0	0	0	48	
1999	0	0	71	0	0	0	71	
2004[h]	0	0	24	0	0	0	24	

(Continued)

TABLE 3.16 (*Continued*)
Dolphins Taken by Drive Fisheries on the Izu Coast, Shizuoka Prefecture

Figures for 1958–1971 are the sum of catches at Kawana and Futo (Miyazaki et al. 1974) and those at Arari (in Coastal Division document dated December 5, 1981), where the former includes only striped and pantropical spotted dolphins (other species are not included) and latter includes all the species taken. Italics for 1972–1978 are from Kishiro and Kasuya (1993) and roman figures beginning in 1972 are Fisheries Agency statistics that may include small numbers of catches by other fisheries. Short-finned pilot whales with asterisks were examined by me and analyzed by Kasuya and Marsh (1984) but are missing in the official statistics. Years of no catch after 1993 are omitted from the table.

[a] Indo-Pacific bottlenose dolphin has not been identified in the catch.
[b] Statistics of the Ministry of Agriculture and Forestry cited from Ohsumi (1972). These statistics do not distinguish dolphin species.
[c] Unknown species (Kishiro and Kasuya 1993) or rough-toothed dolphins (Offshore Division of Fisheries Agency).
[d] Pacific white-sided dolphins.
[e] Common dolphins.
[f] While stated as striped dolphin on a document received by the Offshore Division from Shizuoka Prefecture, Coastal Division lists them as pantropical spotted dolphins.
[g] Minor annual modifications were made on quotas since 1987 (Table 6.3).
[h] No catch was recorded from 2005 to 2013.

less welcomed by the fishermen as well as consumers because they were smaller and the blubber was tougher than that of the striped dolphin. Because the accuracy of catch statistics had already improved in the 1970s and 1980s, the alternation of the species composition in the drive fisheries off the Izu coast must have been real and can be taken to indicate decline in the relative abundance of striped dolphins on the fishing grounds.

The historical trend in the dolphin drive fisheries on the Izu coasts is shown in Figure 3.3. There is some ambiguity of species identification, but it is safe to assume that most of the dolphins taken on the Izu coast since the 1960s were striped dolphins (Table 3.15). Figure 3.3 presents the catches of striped dolphins on the Izu coast (bottom) and at Taiji (top). The bottom panel contains catches of (1) all species of delphinoids taken in Shizuoka Prefecture (1972–1987, Ministry of Agriculture and Forestry), (2) striped dolphins taken in all of Japan (1972–1987, Fisheries Agency statistics), (3) striped dolphins taken on the Izu coast (Futo and Kawana operated during the period of the statistics) (1972–1987, Fisheries Agency statistics), (4) striped dolphins taken at Arari in earlier years (1949–1961, Fisheries Agency statistics), and (5) striped dolphins taken at Futo (1955–1961) and Futo and Kawana (1961–1973) (Miyazaki et al. 1974). The top panel of Figure 3.3 shows the catches of striped dolphins at Taiji (drive and hand harpoon) (1) in 1963–1979 (Miyazaki 1980) and (2) in 1972–1987 (Fisheries Agency statistics).

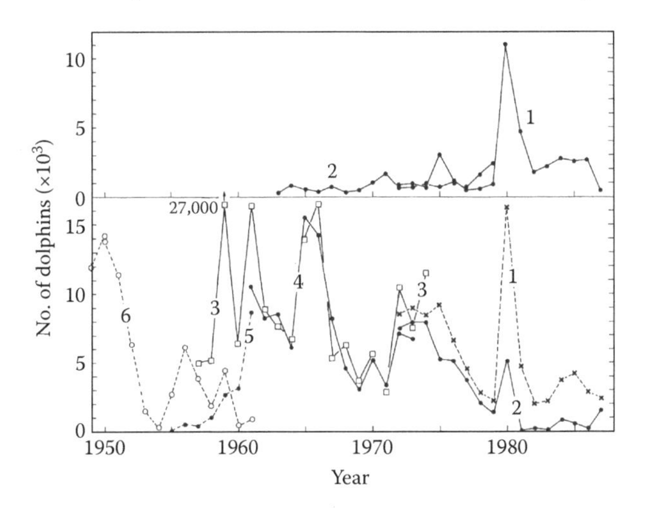

FIGURE 3.3 Outline of the annual trend of striped dolphin fisheries in the early period of collection of statistics. *Top panel*: Take of striped dolphins by Taiji fisheries (hand-harpoon and subsequent drive fisheries) in the statistics of (1) Fisheries Agency and (2) Miyazaki (1980). *Bottom panel*: (1) National total of striped dolphins recorded in the Fisheries Agency statistics, (2) striped dolphins taken by Kawana and Futo groups in the Fisheries Agency statistics, (3) prefecture total of all dolphins in the statistics of Ministry of Agriculture and Forestry, (4) striped dolphins taken at Futo and Kawana in Miyazaki et al. (1974), (5) striped dolphins taken at Futo (Miyazaki et al. 1974), and (6) striped dolphins taken at Arari (Kasuya 1976, in Japanese). (Reproduced from Figure 6 of Kasuya, T. and Miyashita, T., *Saishu-to Shiiku*, 51(4), 154, 1989, in Japanese.)

Some disagreements in catch figures between sets of statistics are unavoidable, but it is worth noting the general agreement of the statistics concerning the annual trend in the 1960s. From this, I conclude that there were large catches of striped dolphins during the early 1960s and before, and then with considerable annual fluctuation, the catch of the species declined since the 1970s. These figures also show that most of the Japanese catch of striped dolphins was obtained from drive fisheries along the Izu coast, with Taiji contributing to lesser degree.

Catch quotas by species were first set for Japanese dolphin drive fisheries in 1993, remained almost unchanged until 2006, and underwent minor annual modifications since 2007. The quotas given to the drive fishery on the Izu coast in Shizuoka Prefecture for the season that started in the autumn of 2008 were as follows (figures in parentheses are quotas for the 2007 season): 36 Pacific white-sided dolphins (36), 56 striped dolphins (63), 67 bottlenose dolphins (71), 365 pantropical spotted dolphins (409), and 10 false killer whales (10) (see Tables 6.3 and 6.4). The total was 534 dolphins/small whales (589). The Pacific white-sided dolphin and false killer whale became fishery targets in 2007, presumably because of expected demand from aquaria. In the 2008 season, only Futo was expected to operate, with 50 vessels, but no catch resulted (Table 3.16).

3.9 TAIJI, NORTH PACIFIC

3.9.1 History of the Dolphin Fisheries

Taiji (33°36′N, 135°57′E) belongs to Wakayama Prefecture and is located on the Pacific coast of the Kii Peninsula on central Honshu, the main island of Japan. It is about 300 km to the southwest of the Izu Peninsula and is under the influence of the warm Kuroshio Current throughout the year.

Hashiura (1969, in Japanese), a useful book with numerous old references on the history of Japanese whaling reproduced in it, stated that the oldest record of systematic Japanese whaling activity was of hand-harpoon whaling in Mikawa Bay (34°45′N, 137°00′E) on the Pacific coast of central Japan, in 1570–1573. However, he also noted that such operations could have started earlier because there were records of circulation of whale meat in the market at earlier dates. This hand-harpoon whaling spread eastward to the Miura (35°15′N, 139°40′E) and Boso (35°15N, 140°00′E) Peninsulas, westward to the Shima (34°10′N–34°30′N, 136°20′E–136°50′E) and Kii (33°26′N–34°10N, 135°05′E–136°20′E) Peninsulas, Tosa (present Kochi Prefecture, 32°40′N–33°30′N, 132°25′E–134°20′E) and Northern Kyushu. Taiji started a whale fishery in 1606 together with several villages on the Kii coasts. Then Taiji introduced the use of nets around 1675–1677, the start of "net whaling," and continued such whaling until 1880. The technique of net whaling is also believed to have spread to western Japan (Section 1.3). The Taiji whalers took short-finned pilot whales by harpooning on days when whaling was not going on and took pleasure in eating the meat and drinking in the front yard of their whaling chief (Hashiura 1969, in Japanese). Hamanaka (1979, in Japanese) stated in the *Taiji Choshi* (History of Taiji Town) that Taiji whalers started the pilot whale hunt as a side business when net whaling declined due to depletion of whale stocks in the nineteenth century. Noshomu-sho (Ministry of Agriculture and Commerce) (1890–1893, in Japanese) also stated that dolphins were hunted in the late nineteenth century as a side business of net whalers in Taiji.

Hashiura (1969, in Japanese) in 1958 learned of past "night hunting for pilot whales" in Taiji that operated on a dark night from a rowed boat called *tento-sen*. In this fishing method, the fishermen harpooned pilot whales using as a marker the animal's fins luminescent with *Noctiluca*. If one animal was wounded other members of the school would approach, so the fishermen sometimes had the opportunity to capture three to four whales in one night. Hashiura (1969, in Japanese) stated that this fishery operated while net whaling was conducted but also after the end of net whaling in Taiji in 1880. Any fishermen could adopt such a technique as a personal fishing activity. The name of *tento-sen* was thought by Hamanaka (1979, in Japanese) to indicate a type of small fishing vessel that became common throughout Japan in the nineteenth century, although the reason for the name was unknown to him. Kawashima (1894, in Japanese) illustrated the construction of a *tento-sen* widely used along the west coast of the Izu Peninsula, a wooden open boat about 10.5 m long and 1.9 m wide.

An attempt to hunt short-finned pilot whales using a small whaling cannon mounted on the *Tento-sen* was made at Taiji in 1900. Success was attained in 1903 with the three-barreled whaling cannon invented by Kenzo Maeda (Hamanaka 1979, in Japanese). He further improved it the next year by inventing the five-barreled whaling cannon. These whaling canons fire three or five whaling harpoons at the same time. Harpoon lines were attached to one main line, which was attached to a whaling boat. T. Isone, gunner captain of a small-type whaling vessels, told me that the barrels of this type of whaling cannon were so aligned to hit one point at a distance of 50 *ken* (about 90 m). Such multibarrel structure could function to (1) increase the probability of a hit, like a shotgun, (2) increase damage to the game, and (3) increase total harpoon weight, thus increasing power to carry the harpoon line. However, I do not know why it was preferred in those days over the single barreled whaling cannon of greater caliber that is used by present-day small-type whalers of Japan.

The small whaling boats at Taiji started to be equipped with engines in 1913, and this rapidly increased, which was the start of evolution to the current small-type whaling vessel (still called *tento-sen*) (Hamanaka 1979, in Japanese). None of the 11 *tento-sen* were equipped with engines in 1912 and took a total of 120 "pilot whales." In 1913, at least some of them had engines and took 381 "pilot whales." Catch records were available for 11 seasons out of the 19 seasons from 1913 to 1931 and recorded take of a total of 5320 "pilot whales" with the annual range of 144–708 (Hamanaka 1979, in Japanese). In the early 1900s, this fishing method was introduced to Inatori on the Izu coasts and took "pilot whales" until around 1955 (Nakamura 1988, in Japanese). "Pilot whale" mentioned in the previous text represents the vernacular name *gondo*, which is presumed to indicate short-finned pilot whales.

In the Taiji area, there was a strong demand for short-finned pilot whales. Therefore, there was an attempt to take "pilot whales" using a seine net with *tento-sen* used for scouting in 1933 (Hamanaka 1979, in Japanese). It proved technically possible but failed economically due to unstable catch and resultant price fluctuation and was discontinued in May 1935. This seine-net whaling took, in addition to "pilot whales," several sperm whales and a "sei" whale (presumably a species of the "Bryde's whale group"), which is interesting as an example of live-capture methodology for large cetaceans.

Driving was another way that responded to the local demand for pilot whales in Taiji. It was operated cooperatively by the local community since early on as part of opportunistic driving of various marine animals such as pilot whales, skipjack, tunas, and other fish species that approached the harbor. The equipment was common property, and the proceeds were used for common purposes by the village. A document was prepared in 1906 by the village to explain the system for the central government (Hamanaka 1979, in Japanese). This traditional fishery was identified as a fishing right of the local community, and the license was issued by the prefecture governor. According to S. Matsui of the Office of Taiji Town, the last occurrence of this type for a pilot whale hunt was from the late 1950s to the early 1960s, when the office staff also joined the operation. This is supported by the notation of proceeds from pilot whales of about 290,000 yen recorded in the balance sheet of the town for fiscal year 1951. The last license identified was issued for the period of 10 years from September 1, 1983, to August 31, 1993. Licenses appeared to have been discontinued subsequently, presumably because there had been no operations during the preceding 10-year period. This change reflects the fact that a group of Taiji fishermen in 1971 started a new fishery to drive dolphin schools from offshore waters, which was different from the traditional community-based driving that targeted dolphin schools coming close to a particular harbor. Driving of short-finned pilot whales by the Taiji community was reported to have also been frequent in the early twentieth century (Taiji 1937, in Japanese) and confirmed by photographs taken in the late 1920s, around 1935, and around 1955 published by Hori (1994, in Japanese).

In the summer of 1969, K. Shimizu, then gunner captain of the small-type whaling vessel *Katsu-maru* (first generation), and several other Taiji fishermen succeeded in driving a school of short-finned pilot whales from offshore waters. This was to obtain live dolphins for the "Taiji Whale Museum" that opened in April of that year. Following this incident, T. Hashimoto and his group of Taiji fishermen started driving short-finned pilot whales in 1971 and expanded the operation to striped dolphins in 1973. The precursor of this driving group was a group of Taiji fishermen that had hand-harpooned striped dolphins for several years. They learned the driving method from the Izu coast and used similar sound emitters, a sealed steel pipe about 2 m long filled with oil with a steel cone of about 30 cm in diameter attached at one end. The cone is placed in the water and the other end in the air was hammered to produce underwater sound during the drive (see third colored photo). In present operations, high-speed dolphin scouting vessels of 13–14 tons are sent out in the early morning to search within a radius of 15–20 nautical miles from Taiji. If a suitable dolphin school is located, it is driven with the cooperation of other member boats called from the port into a small inlet of Hatakejiri near Taiji harbor for subsequent slaughter (Kishiro and Kasuya 1993). The dolphin drive fishery in Taiji presently operates regularly on short-finned pilot whales, false killer whales, Risso's dolphins, bottlenose dolphins, striped dolphins, and pantropical spotted dolphins found in offshore waters and is different from the earlier traditional opportunistic driving.

Success by this driving team led to the establishment of a second team in Taiji in 1979. The two groups raced against each other, expanded the target species to Risso's dolphins and bottlenose dolphins, and caused concerns about conservation. In 1982, all the Japanese dolphin drive fisheries were placed under a licensing system of the prefecture, and the Taiji driving fishermen obtained the license with a condition that the two teams merge into one. The two Taiji driving teams merged into one, but the number of boats participating remained the same. It seems to have been difficult to change the system whereby each boat got an even share. The number of boats was 15 until 1983 and decreased to 14 in 1984 due to the retirement of one person. Half of the vessels participated in the daily scouting activity for the 1982/1983 to 1989/1990 seasons, but all 14 boats have joined in since the 1990/1991 season. This has meant almost doubling of the scouting effort in recent years (Kishiro and Kasuya 1993).

The Taiji people like the meat of the short-finned pilot whale, *tare* (salted-dried meat) or *sashimi* (sliced raw meat), as well as boiled internal organs such as liver, kidney, lung, and intestine. The tail meat of adult males is particularly preferred as *sashimi* because of its high fat content. The catch of short-finned pilot whales increased with the recent establishment of the drive fishery at Taiji, and the fishermen began an effort to expand the market by increasing the number of consumers, which included sending whole carcasses to supermarkets in Shikoku and western Honshu and butchering them in front of customers. However, in 1988 Japanese commercial whaling for large cetaceans ended and the supply of whale meat decreased. This situation opened an opportunity for dolphin meat to be distributed as "like whale" meat. The meat of short-finned pilot whales was suited for this because of its taste despite the dark color, and a shortage of pilot whale meat developed in Taiji. The situation has now changed because of the recent expansion of scientific whaling, importation of whale meat from Iceland, and decreased consumer demand.

The Taiji fishermen hunted dolphins using various methods including small-type whaling, hand harpoon, and driving, as summarized as follows in a chronology. The traditional opportunistic dolphin drive in nearshore waters once allowed for the Taiji people is not included in the following summary, because the details are unknown and the operation ended in the 1950s.

1969: Success in driving short-finned pilot whales from offshore waters.

1970: Rapid increase in the catch of striped dolphins by the hand-harpoon fishery.

1971: A dolphin drive team was established by eight hand-harpoon fishermen, each member with a fishing boat (eight in total).

1973: Expansion of the drive fishery to striped dolphins and then to other species.

1979: A second dolphin drive team was established in Taiji by seven members, each member with one vessel (seven in total). The last small-type whaling vessel surviving in Taiji shifted operation elsewhere about this time to avoid competition with the drive fisheries (it caught three short-finned pilot whales in 1979, two killer whales in 1980, and nothing afterward off Taiji).

1983: All dolphin drive fisheries in Japan were placed under a licensing system of the prefectures. With the advice of Wakayama Prefecture, the two Taiji driving teams merged and obtained a license, but the number of operating vessels remained at 15. The fishing season was determined to be from October 1 to April 30. The fishermen set an autonomous catch limit of 500 short-finned pilot whales and 5000 other dolphins (no species limitations).

1984/1985 season: One driving vessel retired, and 14 vessels participated subsequently (this change was reflected in the license in 1986; see Kishiro and Kasuya 1993).

1986: The autonomous catch limits became a condition of the license of the drive fishery (catch quotas).

1989: Fifteen hand-harpoon vessels obtained approval for dolphin hunting from the Regional Fishery Coordination Committee (later moved to a license by the prefecture). The fishing season was from February 1 to August 31, and the autonomous catch limit was 100 dolphins (species not specified). Operation of a small-type whaling vessel resumed in Taiji with a catch quota of 50 short-finned pilot whales.

1991: The quota for the dolphin drive team decreased to a total of 2900 delphinoids (including not more than 500 short-finned pilot whales) and not more than 5 killer whales for aquaria. The hand-harpoon dolphin fishery increased to a prefecture total of 147 vessels including those from ports other than Taiji. They received a quota of 300 delphinoids (excluding killer whales).

1992: The quota for the dolphin drive team decreased to a total of 2500 delphinoids (including no more than 1000 striped dolphins and no more than 300 short-finned pilot whales). The quota for the dolphin hand-harpoon fishery in the prefecture increased to 400 delphinoids (take of short-finned pilot whales was prohibited). Two small-type whaling vessels operated off Taiji (increase of one vessel) with a total quota of 50 short-finned pilot whales and 30 Risso's dolphins.

1993: The Fisheries Agency for the first time set quotas by species for each type of small-cetacean fishery. Responding to this action, Wakayama Prefecture sent a letter to the director of Taiji FCU. It was titled "On introduction of catch quotas to dolphin fisheries" and stated: "While we have permitted the activity of dolphin drive fishery by your FCU with certain limitations of catch and season, now we have received the following figures from the Fisheries Agency on appropriate catch limits. So we request you to pay adequate attention to operate the fishery within the limits; 450 striped dolphins, 940 bottlenose dolphins, 420 pantropical spotted dolphins, 350 Risso's dolphins, 300 short-finned pilot whales, and 40 false killer whales."

The Taiji drive fishery received a quota of 2500 of 6 delphinoid species for the calendar year 1993 and 2380 for 1994 (Table 6.3). The system changed in 1996/1997 to determine quota by fishing season, but the totals remained the same until the 2007/2008 season. Beginning then, catch of the Pacific white-sided dolphin was allowed with a new set of quotas, and minor quota modifications continued annually (Tables 6.3 and 6.4). Quotas received by the Taiji drive fishery for the 2008/2009 season (with those for the previous season in parentheses) were 134 (134) Pacific white-sided dolphins, 450 (450) striped dolphins, 795 (842) bottlenose dolphins, 400 (400) pantropical spotted dolphins, 290 (295) Risso's dolphins, 254 (277) short-finned pilot whales, and 70 (70) false killer whales, for a total of 2393 (2468) of 7 delphinoid species.

In the following text are translations of documents licensing the dolphin drive fishery at Taiji (issued to the FCU). These documents, issued by the governor of Wakayama Prefecture in 1995, retained the same number of animals to be taken and the same season to be operated as in the first license issued on September 14, 1983, with amendments of the regulations made in a letter from the Director of the Division of Agriculture, Forestry, and Fishery, Wakayama Prefecture, following advice from the Fisheries Agency of Japan.

[Page 1]

Permit No. 1

PERMIT FOR OPERATION OF CETACEAN DRIVE NET FISHERY

Address: Taiji 3165-7, Taiji Town, Higashi-muro County, Wakayama Prefecture
Name: Taiji Fisheries Cooperative Union

1. Type of Fishery: Cetacean drive net fishery
2. Area of Operation: Waters off Wakayama Prefecture east of longitude of Cape Kashino, with a condition that the driving is done into the area reserved for common use of the Taiji FCU, which is exempted if the owner of another common fishing area agrees
3. Period of Operation: From October 1 to the end of February for dolphins, from October 1 to April 30 for pilot whales, and from January 1 to December 31 for killer whales
4. Vessels
   (1) Name of vessels: On the 2nd page
   (2) Registration nos. of vessels: On the 2nd page
   (3) Gross tons of vessels: On the 2nd page
   (4) Type and horsepower rating of the engines: On the 2nd page
5. Period of validity: From October 1, 1995, to September 30, 1999
6. Other conditions: On the 2nd page

September 29, 1995
Shiro Kariya, governor of Wakayama Prefecture [Official Stamp]

[2nd page]

4. Vessels

Name of Vessel	Registration No.	Gross Ton	Horsepower
Fusamaru no. 5	WK3-12680	4.99	Diesel 90
Motomaru	WK3-13468	4.98	Diesel 90
Shokomaru	WK3-12027	4.07	Diesel 90
Daiyumaru no. 3	WK3-14285	4.99	Diesel 90
Kishumaru	WK#-13407	4.98	Diesel 90
Kanemaru no. 3	WK3-13441	4.99	Diesel 90
Sachimaru	WK2-3155	14.02	Diesel 120
Eikomaru	WK2-2731	8.95	Diesel 110

(Continued)

Name of Vessel	Registration No.	Gross Ton	Horsepower
Koshinmaru	WK3-17847	4.0	Diesel 90
Takamaru	WK2-3421	5.80	Diesel 70
Yoshimaru	WK2-3260	9.96	Diesel 120
Isomaru	WK3-16370	4.94	Diesel 80
Ryoyumaru	WK2-3807	7.9	Diesel 130
Kiyomaru	WK3-15750	4.79	Diesel 90

6. Other conditions
   (1) The catches in one fishing season shall be within the following numbers, which will be superseded by other figures if such are indicated: a total of 5,000 dolphins and killer whales and an additional 500 pilot whales.
   (2) Individuals below 2 m in body length shall be released.
   (3) The fishery shall not interfere with operations of other licensed fisheries.
   (4) After completion of a fishing season, by May 30 at the latest, catch results of the operation shall be presented to the governor of the prefecture using a form determined elsewhere.
   (3) Other conditions may be added if considered necessary for fishery management.

In accordance with item 6(1) on catch limits, there was a following notification by Wakayama Prefecture.

Fishery No. 77
April 30, 1995

To: President of the Taiji Fishery Cooperative Union

From: Director of the Division of Agriculture, Forestry and Fishery, Wakayama Prefecture

ON SETTING CATCH LIMITS FOR DOLPHIN FISHERIES

We have previously given your fishery cooperative union a license for a dolphin drive fishery with regulation of catch limits and operation periods. Now we have received from the Fisheries Agency of Japan the following figures as appropriate catch limits and request you not to exceed in your operation the number indicated for each species.

	Figures given
Striped dolphin	450
Bottlenose dolphin	940
Pantropical spotted dolphin	420
Risso's dolphin	350
Short-finned pilot whale, southern form	300
False killer whale	40

In accordance with item 6(4) on catch reports, the following document was prepared by the Taiji Fishery Cooperative Union, where these catch limits are under the heading "autonomous catch limit."

CETACEANS CAUGHT BY DRIVE FISHERY IN THE 1994 SEASON (JANUARY TO DECEMBER)

	Autonomous Catch Limits	Numbers Taken
Striped dolphin	450	450
Bottlenose dolphin	940	420
Pantropical spotted dolphin	420	420
Risso's dolphin	350	243
False killer whale	40	0
Short-finned pilot whale	300	157
Total	2,500	1,404

(Autonomous catch limits were set by species since 1994.)

3.9.2 Catch Statistics

Since 1972, the Fisheries Agency has been collecting annual catch statistics of dolphins by species through the prefectures (Sections 1.4 and 2.5). Some scientists collected statistics for the earlier period at local fish market and FCUs, as seen in Miyazaki et al. (1974) and Miyazaki (1980, 1983) for the Taiji fishery. Kishiro and Kasuya (1993) combined these statistics and constructed longer catch series.

Miyazaki estimated the landings of dolphins from the number of sets of viscera sold by species at the wholesale market operated by the Taiji FCU. This was possible because the internal organs were sold by numbers for local consumption, but it suffered from the fact that similar species such as striped dolphins and pantropical spotted dolphins would not have been separated. This sales record usually does not distinguish the type of fishery such as small-type whaling, hand-harpoon fishery, and drive fishery that were operated from Taiji. The annual trend of landings of dolphins thus obtained is shown in Figure 3.3 (striped dolphins), Figure 3.4 (short-finned pilot whales), and Figure 2.6 (other species).

After the previous work, I was able to receive the landing records of cetaceans for the years 1963–1994 from the Taiji FCU and analyze them with the cooperation of T. Kishiro in order to understand the catch trends in the fisheries. This project revised earlier statistics as presented in Table 3.17, but further analysis of the catch trends was discontinued due to my transfer to a university. Most of the original records of the statistics given in Table 3.17 were recorded as number of individuals by species, but some of the sales in 1980–1983 were recorded by the weight of meat by species. In such cases, the weight of meat was converted into a number of individuals using the average meat production of landings on other dates. The number of individuals thus calculated sometimes disagreed with the number of sets of viscera of the species sold on the same day. In such cases, I used the greater figure as the landings of the species for the day, accepting some possibility of overestimation. The meat production per individual used for the previous estimates and the sample sizes are 36 kg for pantropical spotted dolphin (n = 27), 39 kg for striped dolphin (n = 41), 85 kg for bottlenose dolphin (n = 153), 65 kg for Risso's

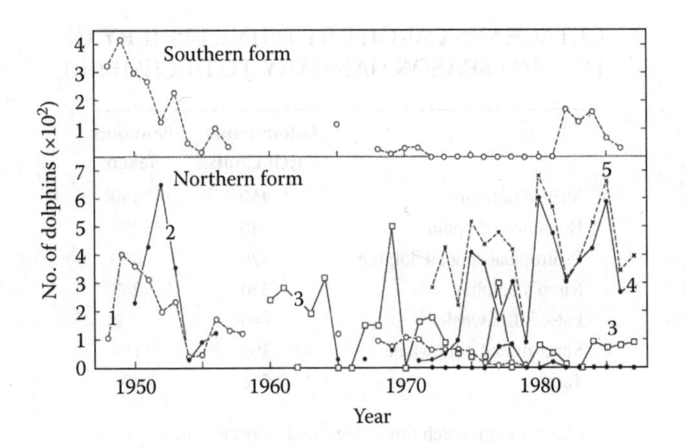

FIGURE 3.4 Outline of the annual catch of short-finned pilot whales in Japan before 1993, when the fisheries operated without regulation by quota. *Top panel*: Take of the northern form by small-type whaling (national total). *Bottom panel*: Take of the southern form by (1) small-type whaling (national total), (2) drive fishery on the Izu coast, (3) drive and crossbow fishery in Okinawa, (4) small-type whaling and drive fishery at Taiji, and (5) national total as appeared in the Fisheries Agency statistics. (Based on statistics in Kasuya and Marsh (1984); Fisheries Agency statistics (see Section 1.4). Reproduced from Figure 4 of Kasuya, T. and Miyashita, T., *Saishu-to Shiiku*, 51(4), 154, 1989, in Japanese.)

dolphin (n = 11), 175 kg for false killer whale (n = 78), and 106 kg for southern-form short-finned pilot whale (n = 205). These figures possibly include only harvested meat and may appear small compared with the muscle weight expected for the species. I used the figure for the striped dolphins to estimate the number of rough-toothed dolphins, for which data were not available. Pacific white-sided dolphins and spotted dolphins appeared in the statistics beginning in 1980, false killer whales in 1982, and rough-toothed dolphins in 1984. The catches of these species were considered small earlier on, and it is unclear if they were recorded in the statistics. The identity of the vernacular name *haukasu* used by elderly Taiji fishermen is not completely settled yet (Section 2.7). The statistics thus constructed and used as the basis of Table 3.17 did not include catches of small-type whaling in 1988 and subsequent years, so I have added these figures from other Fisheries Agency statistics.

Comparison of the statistical figures in Table 3.17 with corresponding figures for 1963–1979 in Miyazaki (1980) shows good agreement. The figures for the short-finned pilot whale, Risso's dolphin, bottlenose dolphin, and killer whale are in perfect agreement. The catch of striped dolphins in Table 3.17 was only four animals greater than in Miyazaki (1980). The figures in Table 3.17 show reasonable agreement with those in Table 3.18, which were collected by the Fisheries Agency and reported to the Scientific Committee of the International Whaling Commission. However, some significant disagreement between the two bodies of statistics exists in the annual catches of striped dolphins and short-finned pilot whales during the 1980s. Figures in my statistics (Table 3.17) are often 200–2000 higher than in the Fisheries Agency statistics for the same year (Table 3.18). I feel that the disagreement is too great to be attributable to simple error of the sort that is expected for my process of preparation of statistics and that there have been attempts at manipulating the statistics at some stage between the landings and reporting to the international body. The purpose of the manipulation could have been either political or economic (Table 3.19).

3.9.3 Fishing Season

Using the original figures in Table 3.17, I analyzed the fishing season of the Taiji dolphin fisheries (Table 3.20). The seasonality of landing was compared between the early years when the fishing season was not regulated (1963–1981) and the later years when operation was allowed for particular months (1982–1994). In the early period of no regulation, the peak catch was in the 4 months from October to January, when 81.0% of the annual catch was landed. The peak months remained the same in the later years with seasonal regulation of the operation, and the only minor change was the slight increase in the proportion of landings in the 4 months to 85.3%. This indicates that the regulation of fishing season just reinforced the seasonal pattern of the fishery and had little function in regulating the fishery. In 1970–1980, when there was no regulation of the fishing season, I observed that the fishermen ended the drive fishery for the year to switch the operation to trolling for skipjacks as soon as the fish species arrived off Taiji.

3.9.4 Price of Dolphins

The price of dolphins is influenced by the balance of supply and demand, and the price has a strong influence on motivation of the fishermen and thus on the annual trend of the catch. Table 3.21 shows the annual change in the average dockside price of dolphins in Taiji. The price of dolphins increased rapidly in the 1980s while the catches did not decline. The price in the early 1990s was several times of that of the late 1970s. This is not explained by change in the consumer price, that is, inflation, during the period. It can be interpreted as a reflection of the elevated price of whale meat due to the decreased supply of whaling products during the period. The price increase for short-finned pilot whales was particularly rapid. The meat of short-finned pilot whales was preferred by consumers and used as a substitute for baleen whale meat. At one time in

TABLE 3.17
Dolphins Landed at the Taiji Fish Market, Calculated Using Records of the Taiji Fishery Cooperative Union That Operated the Wholesale Market (See Text for Further Details)

Year	Striped Dolphin	Bottlenose Dolphins[a]	Risso's Dolphin	Short-Finned Pilot Whale[b]	Killer Whale	False Killer W.	Rough-Toothed Dolphin[c]	Pantropical Spotted D.[d]	Pacific White-Sided D.
1963	331	29	59	98	6	—	—	—	—
1964	934	50	73	146	15	—	—	—	—
1965	642	23	58	134	6	—	—	—	—
1966	422	14	36	52	1	—	—	—	—
1967	819	28	83	68	1	—	—	—	—
1968	400	66	37	96	4	—	—	—	—
1969	503	31	33	77	4	—	—	—	—
1970	992	34	21	116	2	—	—	—	—
1971	1,717	63	26	110	3	—	—	—	—
1972	700	65	62	91	2	—	—	—	—
1973	727	55	20	155	1	—	—	—	—
1974	967	24	15	193		—	—	—	—
1975	759	103	7	479	1	—	—	—	—
1976	1,053	18		369	1	—	—	—	—
1977	562	18		192		—	—	—	—
1978	1,644	3		322		—	—	—	—
1979	2,397	6		121	6	—	—	—	—
1980	12,835	412		841	2		100	10	1
1981	5,742	926		820			6		6
1982	1,757	740	4	342	2	1	48	229	
1983	2,179	539	36	110	2	37		156	
1984	2,812	229		424		55	63	677	3
1985	2,859	819		589	8		60	484	2
1986	2,720	98		264	10			693	
1987	358	1670	2	294	3				11
1988	1,767	586	109	347(20)	2(2)	42		1646	
1989	1,000	233	10(5)	76(5)	1(1)		80	120	
1990	796	1274	88(6)	83(8)		70(1)	80(2)	6(3)	
1991	985	404	395(92)	254(10)		45(3)	32	180	1
1992	973	167	114(30)	228(32)		91		636	
1993	507	147	281(30)	200(43)				432	
1994	496	281	176(20)	78(18)				408	

Catches of small-type whaling, hand-harpoon fishery, drive fishery, and trap nets were not separated and are included in the catch before 1988, but those of small-type whaling were available separately (shown in parentheses) and included in the total figure as for the earlier period. Taiji was the only major dolphin landing port in Wakayama Prefecture.

[a] Indo-Pacific bottlenose dolphin has not been confirmed in the catch and we do not have a resident population of the species near the coast, but the possibility of take of some individuals of the species cannot be excluded.

[b] Southern form.

[c] It is unknown how this species was classified before 1980. Mixing with bottlenose dolphins was a possibility.

[d] It is unknown how this species was classified before 1980. Mixing with striped dolphins was a possibility.

TABLE 3.18
Landings of Dolphins at Taiji and Other Locations in Wakayama Prefecture

Species	Striped Dolphin	Pantropical Spotted Dolphin	Bottlenose Dolphins	Risso's Dolphin	Short-Finned Pilot Whale	False Killer Whale	Killer Whale	Rough-Toothed Dolphin	Others and Unknown	Fishery Types
1972	850		15		30					D, H
	(12)				(85)		(1)		(2 + 1[a])	S
1973	981				52					D, H
	(4)				(68)				(6 + 2[a])	S
1974	763		11	11	94					D, H
					(56)		(2)		(6 + 1[a])	S
1975	3,097		7		410					D, H
	(2)		(1)	(1)	(52)	(1)	(1)		(5)	S
1976	1,183		1		370					D, H
	(3)		(1)		(10)	(1)	(1)			S
1977	490		18		170					D
					(7)		(1)			S
1978	613		48		309					D
					(11)					S
1979	893	70	50	9	87		5			D
					(3)					S
1980	11,017		345		605				14[b]	D
							(2)			S
1981	4,710		297	4	476		5		7[b]	D
1982	1,758	293	668	4	305	1	4			D
1983	2,179	156	462	49	378	32	2			D
1984	2,812	740	229		424	55			3	D
1985	2,639	484	715		589		7	60	2	D
1986	2,720	693	98		264		8			D
1987	358		1670	2	294		3		11[b]	D
1988	1,767	1646	586	109	327	42				D
					(22)		(2)			S
1989	1,000	120	264	11	71				80	D
					(5)		(1)		(5)	S
1990	682	9	1286	82	75	69		161	228[c]	D, H
				(6)	(8)	(8)				S
1991	971	153	388	301	244	42		4	244[d]	D, H
			(1)	(92)	(10)	(3)				S
1992	1,045	636	154	88	200	81				D, H
				(30)	(29)					S

This table covers the period 1972–1992 when fisheries other than small-type whaling operated without a quota from the government. The fisheries involved were drive (D), hand-harpoon (H), and small-type whaling (S). The catch of small-type whaling is in parentheses and is additional to the catch of the two other fisheries (not bracketed). From the statistics of the Offshore Division of the Fisheries Agency (1972–1987), those of Coastal Division (since 1986) and records of the Small-Type Whaling Association. In addition to the figures listed, a small-type whaling vessel recorded a total of 24 Baird's beaked whales taken off Taiji, 8 in 1975, 1 in 1976, 8 in 1977, and 7 in 1978. These are catches made by a Taiji-based vessel operated off Wadaura, Chiba Prefecture (report of the Small-Type Whaling Association).

[a] Recorded as Cuvier's beaked whale, but the possibility of it being *Mesoplodon* species is not excluded.
[b] Pacific white-sided dolphin.
[c] 148 common dolphins and 80 melon-headed whales.
[d] 100 Fraser's dolphins, 1 Pacific white-sided dolphin, 1 *Kogia* sp., 60 melon-headed whales, and 82 individuals of unknown species.

TABLE 3.19
Dolphins Taken by the Drive Fishery at Taiji since 1993, When It Operated with By-Species Quotas of the Fisheries Agency

Species	Striped Dolphin	Pantropical Spotted Dolphin	Common Bottlenose Dolphin	Risso's Dolphin	Short-finned pilot whale (south. form)	False killer whale	Pacific White-Sided Dolphin
Quota since 1993	450	420	940	350	300	40	
1993	450	420	134	243(30)	157(43)	0	
1994	440	400	261	151(20)	60(18)	0	
1995	450	73	852	186(20)	49(46)	40	
1996	211	0	143	73(20)	300(46)	25	
1997	545	0	287	60(20)	204(22)	40	
1998	376	397	164	160(20)	84(46)	40	
1999	520	0	596	250(12)	211(31)	0	
2000	235	27	1339	367	109(49)	0	
2001	418	0	195	350(17)	212(36)	29	
2002	565	400	760	221(12)	55(35)	7	
2003	382	102	121	191(19)	55(27)	17	
2004	554	0	570	444(7)	62(29)	0	
2005	397	0	285	340(8)	40(24)	0	
2006	479	400	285	232(7)	198(10)	30	
Quota for 2007/2008	450	400	842	295	277	70	134
2007	384	0	300	312(20)	243(16)	0	0
2008	575	329	297	216	99(20)	0	21
2009	321	0	352	336	219(22)	0	14
2010	458	125	395	271	0(10)	0	27
2011	406	106	76	273	74	17	24
2012	508	98	186	188	172(16)	0	2
2013	498	126	190	298	88(10)	0(1)	39

Taiji was the only license holder for a drive fishery in Wakayama Prefecture. Additional cetaceans landed at Taiji by small-type whaling are given in parentheses. From Fisheries Agency statistics. See Table 2.12 for the catch of hand-harpoon fishery in Wakayama Prefecture during the period and Table 6.3 for quota, which had minor annual modifications since 2007/2008.

the 1980s, pilot whale meat almost disappeared from the local market in Taiji. A full-grown male short-finned pilot whale (4.5 in body length) was sold for about 5 million yen and an adult female (3.5 m) sold for 2–3 million yen in the late 1980s. The price difference was due to the fact that males are about 36% heavier than females (Table 12.2) and that adult males of the southern form had particularly preferred tail meat with high fat contents (see Sections 3.8.3 and 3.9.1). Therefore, the drive fishermen where possible tended to select a school containing more adult males for driving (Kasuya and Marsh 1984).

3.10 NAGO, OKINAWA ISLAND

Nago (26°35′N, 127°59′E) is on the East China Sea coast of Okinawa Island, which is located in the middle of the island chain of Nansei Shoto (Southwestern Islands) of Japan between Kyushu and Taiwan. It is about 670 km from Goto Islands, the nearest location of a past dolphin fishery, and about 1100 km from Taiji, the location of the currently existing dolphin fishery. The island chain is under strong influence of the warm Kuroshio Current.

In the Okinawa Islands, *pitu* means dolphins as well as the activity of driving dolphins. This dolphin drive fishery, which starts with a dolphin school coming into a suitable inlet and is operated with cooperation of the villagers, was believed to have been operated in earlier times at various locations in the islands (Nago Hakubutsu-kan [Nago Museum] 1994, in Japanese), but the details are currently known only for Nago. At Nago, the dolphin drive was an important activity of the local community, and there were recorded rituals for gaining good fishing opportunities Nago Hakubutsu-kan (Nago Museum) (1994, in Japanese).

When a school of dolphins was found at the entrance of Nago harbor, school classes were interrupted and everyone stopped work. Boat owners cooperated in driving the school

TABLE 3.20
Seasonality of Dolphin Fishery at Taiji Shown by Monthly Number of Dolphins Landed

						Total	
Species	Striped Dolphin	Pantropical Spotted D.	Common Bottlenose Dolphin	Risso's Dolphin	Short-Finned Pilot Whale[a]	No. of Dolphins Landed	%
1963–1981: Operated without regulation of fishing season							
January	4,116		51	10	630	4,807	11.7
February	1,609		409	24	710	2,752	6.7
March	758		307	141	159	1,365	3.3
April	151		183	128	304	766	1.9
May	333		220	60	479	1,092	2.7
June	508		121	36	455	1,120	2.7
July	146		21	28	240	435	1.1
August	7		6	8	127	148	0.4
September	4		10		83	97	0.2
October	5,047		127	14	376	5,564	13.5
November	7,434		253	56	369	8,112	19.7
December	14,033		260	25	545	14,863	36.1
Total	34,146		1,968	530	4,477	41,121	100
1982–1994: Operated with closed season							
January	4,066	487	745	90	351	5,739	15.8
February	1,157	619	892	131	325	3,124	8.6
March	202		97	65	198	562	1.6
April	20	8	222	92	98	440	1.2
May	3	23	74	72	19	191	0.5
June	696	52	30	163		941	2.6
July	7	1	3	3	3	17	0.0
August	1	11	8	20		40	0.4
September							0
October	6,059	2,907	1,249	48	791	11,054	30.5
November	1,811	617	2,987	171	854	6,440	17.8
December	5,187	939	879	177	517	7,699	21.2
Total	19,209	5,664	7,186	1,032	3,156	36,247	100

[a] Southern form.

into the harbor by throwing stones, hitting a steel rod with a hammer, or hanging *shirushika* into the water (Miyazaki 1982; Nago Hakubutsu-kan [Nago Museum] 1994, both in Japanese). The *shirushika* is a rope with numerous palm leaves attached to it, which is similar to the *kazura* used in the dolphin drive in Tsushima Island (Section 3.3). A boat from the mayor's office signaled the completion of driving with a flag. With the signal, everybody went to the beach with a knife, lance or harpoon in hand and joined in the slaughter. Every participant received an equal share. Those who were unable to consume all of their share sold the excess, raw or boiled.

The drive fishery in Nago apparently declined during the 1970s. Uchida (1985, in Japanese) listed possible reasons for the decline: (1) the beach suitable for driving disappeared from Nago harbor as a result of expansion of the fishing port, (2) cooperation within the local community weakened due to a change in the social atmosphere, and (3) there was a decline in dolphin schools approaching the harbor. Kishiro and Kasuya (1993) thought that Nago did not obtain a license in 1982 when all drive fisheries were placed under a licensing system of the prefecture governments and ended the fishery in that year. There are other views (see Section 6.5). Actually, several incidents of driving after 1982 were reported in newspapers (Kishiro and Kasuya 1993), for example, 9 short-finned pilot whales in 1982 and 20 bottlenose dolphins in 1989. The drive of bottlenose dolphins in 1989 could have marked the real end of the dolphin drive fishery in Nago.

Statistics for the Nago fishery are available in Miyazaki (1982, in Japanese) and Uchida (1985, in Japanese) and are based on records kept at the Nago FCU. They are cited in

TABLE 3.21
Dockside Price of Dolphins at the Taiji Fish Market, in 1000 Yen

Year	Short-Finned Pilot Whale	False Killer Whale	Risso's Dolphin	Common Bottlenose Dolphin	Striped Dolphin	Pantropical Spotted Dolphin	Sum of Small Species[a]	Price Index
1972	89						12(732)	100
1973	31						17(460)	112
1974	56						15(895)	138
1975	64						12(960)	154
1976	80						13(1,230)	168
1977	82						18(368)	182
1978	158						25(464)	190
1979	30						13(1,206)	197
1980	83						8(10,359)	212
1981	99						11(5,018)	222
1982	101						18(2,682)	228
1983	90						21(2,878)	233
1984	161						17(3,859)	238
1985	146			26	14	15	16(3,838)	243
1986	191			31	18	16	19(3,511)	244
1987	318		59	47	28		43(2,030)	244
1988	522	286	131	70	35	12	36(4,150)	246
1989	477		136	70	25	23	33(1,358)	252
1990	979	486	153	89	56	29	91(2,227)	260
1991	826	1,197	233	104	39	26	91(1,490)	—

Source: Kishiro, T. and Kasuya, T., *Rep. Int. Whal. Commn.*, 43, 439, 1993.

[a] Average per capita price in 1000 yen for all dolphin species other than short-finned pilot and killer whales. In parentheses are number of individuals involved in the calculation.

Miyazaki (1983) and Kishiro and Kasuya (1993). Another series of statistics is that collected by the Whaling Section, Offshore Division of the Japanese Fisheries Agency, for 1972–1987, and by the Coastal Division since 1986. These statistics are summarized in Table 3.22.

The statistics of Uchida (1985, in Japanese) disagree with those of the Fisheries Agency for 1976 and later years. The reason seems to be that the former listed only catches by the drive fishery while the latter included other fishing methods such as the hand-harpoon and crossbow fisheries. Miyashiro (1987, in Japanese), who recorded the history of folk customs and fishing activities in Nago, stated that the crossbow fishery started in Nago in 1957 responding to the shortage of dolphin meat due to a decline in the drive fishery. The Fisheries Agency statistics did not distinguish the catch of the crossbow fishery from that of the drive fishery until 1977. The crossbow fishery was classified with the "hand-harpoon fishery" beginning in 1978. In this book, I have separated the crossbow fishery from the hand-harpoon fishery (see Section 2.8).

3.11 MIURANISHI, WESTERN SHIKOKU

Miuranishi (33°10′N, 132°29′E), Ehime Prefecture, is located on the west coast of Shikoku and faces the Bungo Channel that is located between Kyushu and Shikoku and connects the Inland Sea and the North Pacific. Miyawaki and Hosokawa (2008, in Japanese) recorded the history of the dolphin drive fishery in Miuranishi. This group learned the method of dolphin drive from Saiki (32°59′N, 131°54′E) in 1775. There are records remaining of 13 drives that occurred in the 28-year period from 1844 to 1871. The species and number of dolphins driven are unknown. If a dolphin school was in their harbor, the fishermen placed a closing net at the entrance and captured them, which was apparently similar to the opportunistic dolphin drives carried out in numerous locations in Japan.

Saiki, which taught the dolphin drive method to Miuranishi, belongs to Oita Prefecture and is located opposite Miuranishi across the Bungo Channel. So it is reasonable to assume that there were yet unidentified dolphin drive fisheries in Oita Prefecture.

TABLE 3.22
Dolphins Taken at Nago, Okinawa Prefecture

Year, Methods, and Vessels	Short-Finned Pilot Whale[a]	False Killer Whale	Bottlenose Dolphins[b]	*Delphinus* sp.[c]	Others	Total
1960	243[d]		<77			243
1961	281					281
1963	189					189
1964	318					318
1967	150					150
1968	150					150
1969	500					500
1971	165[e]	<19	<45			165
1972	170					170
(D: 59 vessels)	*170*					*170*
1973	87[d]		<87			87
(D: 59 vessels)	*87*					*87*
1974	53					53
(D:59 vessels)	*53*					*53*
1975	49					49
(D: 59 vessels)	*49*					*49*
1976	0			0	23[f]	23
(D: 59 vessels)	*36*			*23*		*59*
1977	298			0		298
(D: 79 vessels)	*301*			*37*		*338*
1978[g]		0		0		0
(C: 7 vessels)		*90*		*25*		*115*
1979				0	140[h]	140
(D + C: 87 vessels)				*49*	*140*[h]	*189*
1980	91	0		0		91
(D + C: 40 vessels)	*80*	*5*		*22*		*107*
1981[i]	0	0		0		0
(C: 8 vessels)	*50*	*7*	*2*	*25*		*82*
(H: 2 vessels)						*2*
1982	9			0		9
(C: 6 vessels)	*5*			*82*		*87*
1983 *(C: 6 vessels)*	*131*	*17*	*16*			*164*

(Continued)

TABLE 3.22 (*Continued*)
Dolphins Taken at Nago, Okinawa Prefecture

Year, Methods, and Vessels	Short-Finned Pilot Whale[a]	False Killer Whale	Bottlenose Dolphins[b]	*Delphinus* sp.[c]	Others	Total
1984 (*C: 4 vessels*)	*88*	*5*	*63*		*3*[j]	*159*
1985 (*C: 4 vessels*)	*70*		*31*			*101*
1986 (*C: 5 vessels*)	*82*	*2*	*77*			*161*
1987 (*C: 4 vessels*)[k]	*92*	*2*	*18*		*2*[l]	*114*
1986[m]	*82*		*77*		*2*	*161*
1987	*92*				*22*	*114*
1988	*116*				*17*	*133*
1989	*93*	*25*			*26*	*144*
1990	*74*	*3*			*2*	*79*
1991	*52*	*9*	*16*		*3*	*80*
1992	*79*	*0*	*1*			*80*

Figures in roman represent catch by drive fishery in 1960–1982 reported by Uchida (1985, in Japanese), while those in italics represent statistics collected by the Fisheries Agency (since 1972). The latter may be greater than the former for the same year, because of inclusion of catches by the crossbow and hand-harpoon fisheries. In parentheses are fishing method and number of vessels that operated as given in the Fisheries Agency statistics. The large number of vessels operated in 1979 and 1980 is probably due to dolphin drives by the local community. Years without catch are not listed. See Table 2.13 for statistics in 1993 and later years, when only the crossbow fishery was allowed to operate under a quota system. Fishing types are C, crossbow; D, drive; and H, hand-harpoon.

[a] Southern form.
[b] May include two species.
[c] Presumed to be *D. capensis*, long-beaked common dolphin.
[d] Includes bottlenose dolphins.
[e] Includes bottlenose dolphins and false killer whales.
[f] Rough-toothed dolphin. This catch is recorded as common dolphins in the Fisheries Agency statistics.
[g] Crossbow fishery started at Nago in 1975 (Section 2.8) and first recorded in 1978 statistics of Fisheries Agency under hand-harpoon fishery.
[h] Melon-headed whale.
[i] Fisheries Agency statistics list number of hand-harpoon vessels as 8 at Nago and 2 at On-nason. The former will be crossbow vessels and the latter the ordinary hand-harpoon fishery operated opportunistically.
[j] Pantropical spotted dolphin.
[k] Statistics of the "Offshore Section" end here.
[l] One each Risso's dolphin and rough-toothed dolphin.
[m] Statistics of the "Coastal Section" start here.

TABLE 3.23
Summary of Recent Dolphin Drive Fisheries in Japan

Year	Shizuoka	Wakayama	Nagasaki	Yamaguchi	Okinawa	Notes (for Years with Recorded Changes)
1972	3(41)	1(7)	1(400)	1(12)	1(59)	*Shizuoka*, Futo (3) and Kawana (3) jointly operated since 1967, Arari (35); *Wakayama*, Taiji; *Nagasaki*, Katsumoto; *Yamaguchi*, Ohibi; *Okinawa*, Nago
1973	3(41)	1(7)	1(?)	0	1(59)	*Nagasaki*, Uonome
1974	2(6)	1(7)	2(?)	1	1(59)	*Shizuoka*, Futo and Kawana; *Nagasaki*, Arikawa and Uonome
1975	2(6)	1(7)	3(?)	0	1(59)	*Nagasaki*, Arikawa, Mitsushima, Sasuhara
1976	2(6)	1(7)	2(400+)	0	1(59)	*Nagasaki*, Katsumoto (400), Uonome
1977	2(6)	1(7)	2(684)	0	1(79)	*Nagasaki*, Katsumoto (660 for culling), Miiraku (24)
1978	2(5)	1(8)	4(1293)	0	0	*Shizuoka*, Futo (2) and Kawana (3); *Nagasaki*, Katsumoto (660), Miiraku (13), Mitsushima (571), Yokoura (49); *Okinawa*, Nago operated crossbow fishery with 7 vessels
1979	2(6)	2(13)	6(728)	0	0	*Nagasaki*, Katsumoto (660), Miiraku (20), Mitsushima (26), Kamitsushima (18), Minecho (3), Fukue (1); *Okinawa*, Nago was recorded with the use of 87 vessels for hand harpooning, but it was likely to have been a drive
1980	2(5)	2(14)	5(660+)	0	0	*Nagasaki*, Katsumoto (660), Miiraku, Uonome, Arikawa; *Okinawa*, Nago used 40 vessels for hand harpooning, which was likely to have been a drive
1981	2(5)	2(14)	1(350)	0	0	*Nagasaki*, Katsumoto (350) *Okinawa*, Nago (8 crossbow vessels) and On-na-son (2 hand-harpoon vessels) operated dolphin fishery
1982	2(32)	1(15)	1(300)	0	0	*Shizuoka*,[b] Futo (15) and Kawana (17); *Nagasaki*, Katsumoto (300)
1983	2(17)	1(15)	6(594)	0	0	*Shizuoka*, Futo (15) and Kawana (2); *Nagasaki*, Katsumoto, Arikawa, Ikitsuki, Uraguwa, Shin-uonome, Tachiura
1984	1(2)	1(14)	1(700)	0	0	*Shizuoka*, Futo (2); *Nagasaki*, Katsumoto (700)
1985	1(1)	1(14)	2(1355)	0	0	*Nagasaki*, Katsumoto (1350), Miiraku (5)
1986	1(1)	1(14)	1(350)	0	0	*Nagasaki*, Katsumoto (350)
1987	1(1)	1(14)	0	0	0	
1988[a]	1(1)	1(14)	1	0	0	*Nagasaki*, Katsumoto
1989[a]	1	1(14)	0	0	0	
1990[a]	1	1(14)	0	0	0	*Shizuoka*, Futo did not operate the drive fishery in this year
1996[a]	(13)	1(14)				*Shizuoka*, likely to represent Futo
1999/2000	1(7)	1(14)	0	0	0	
2000/2001	1(7)	1(13)	0	0	0	
2007/2008	1(50)	1(17)	0	0	0	*Shizuoka*, principle of counting vessels could have changed[b]
2008/2009	1(50)	1(17)	0	0	0	
2009/2010	1(50)	1(17)				

Source: Fisheries Agency Statistics collected by Offshore and Coastal Divisions of Fisheries Agency (see Section 1.4).
Number of operating bodies is given for prefectures. The number of vessels employed in the operation is given in parentheses.
[a] Collected by the Coastal Division of the Fisheries Agency. Other figures were collected by the Offshore Division.
[b] The sudden increase of vessels employed was likely due to a change in counting method. In earlier times, they counted only vessels that were used in daily searching and subsequent driving toward the harbor, but here they probably counted all the vessels used in the operation for searching, driving, netting, and landing procedures.

REFERENCES

[*In Japanese Language]

*Anraku, T. 1985. Fishing activities around Saikai and Goto Islands. *Kikan Kokogaku* 11: 39–42.

Bloch, D. 2007. *Pilot Whales and the Whale Drive in the Faroes.* H.N. Jacobsens Bakahandil, Torshavn, Faroe Islands. 64pp.

*Hamanaka, E. 1979. *History of Taiji Town*. Taiji-cho, Taiji, Japan. 952pp.

*Harada, Y. and Tsukamoto, Y. 1980. Dolphin and porpoise fishery. pp. 411–435. In: Kumamoto, H. (ed.). *History of Fisheries in Katsumoto*. Katsumoto Gyogyo Kyodo Kumiai, Katsumoto, Japan. 576pp.

*Hashiura, Y. 1969. *History of Whaling at Taiji in Kumano.* Heibonsha, Tokyo, Japan. 662pp.

Hawley, F. 1958–1960. *Miscellanea Japonica, II. Whales and Whaling in Japan.* Published by the author, Kyoto, Japan. 354+Xpp.

*Hiraguchi, T. 1986. Outline of animal relics. pp. 346–365. In: Yamada, Y. (ed.). *Mawaki Archaeological Site in Noto-cho, Ishikawa Prefecture.* Noto Kyoiku Iinkai Mawaki Iseki Hakkutsu Chosadan, Noto-cho, Japan. 482pp.

*Hiraguchi, T. 1990. Paleolithic "spear shaped pointed stone tool" compared with "stone spear" of Jomon Era—With special reference to relics from Hokuriku region. pp. 51–66. In: Ito Nobuo Sensei Tsuito Ronbunshu Kankokai. (ed.). *Ito Nobuo Sensei Tsuito Kokogaku Kodaishi Ronko.* Ito Nobuo Sensei Tsuito Ronbunshu Kankokai, Sendai, Japan. 727pp.

*Hiraguchi, Y. 1993. Study on dolphin fisheries in Jomon Era based on identification of individual animals. pp. 1–52. In: Anon. (ed.). Report of studies conducted with Ministry of Education, Culture and Science (Ippan Kenkyu C) in 4th year of Heisei.

*Hori, T. 1994. *My Homeland Taiji—Photographs in the Meiji, Taisho and Showa Eras.* Shito Kai, Taiji, Japan. 62pp.

*Ine Choshi Hensan Iinkai. 1985. *History of Ine Town.* Vol. 2. Ine Cho, Ine, Japan. 988+137pp.

IWC. 1992. Report of the sub-committee on small cetaceans. *Rep. Int. Whal. Commn.* 42: 178–228.

Jefferson, T.A. and Van Waerebeek, K. 2002. The taxonomic status of the nominal dolphin species *Delphinus tropicalis* van Bree, 1971. *Mar. Mammal Sci.* 18(4): 787–818.

Joensen, J.P. 1976. Pilot whaling in the Faroe Islands. *Ethnol. Scand.* (Lund) 1976: 1–42.

*Kaneda, K. and Niwa, H. 1899. Examination report of items exhibited at the second fishery exposition. Vol. 1, Part 2. Noshomusho Suisan-kyoku, Tokyo, Japan. 219pp.

Kasuya, T. 1975. Past occurrence of *Globicephala melaena* in the western North Pacific. *Sci. Rep. Whales Res. Inst.* 27: 95–110.

*Kasuya, T. 1976. Stock of striped dolphins. *Geiken Tsushin* 295: 19–23, 296: 29–35.

Kasuya, T. 1985a. Effect of exploitation on reproductive parameters of the spotted and striped dolphins off the Pacific coast of Japan. *Sci. Rep. Whales Res. Inst.* 36: 107–138.

Kasuya, T. 1985b. Fishery-dolphin conflict in the Iki Island area of Japan. pp. 253–273. In: Beddington, J.R., Beverton, R.J.H., and Lavigne, D.M. (eds.). *Marine Mammals and Fisheries.* George Allen and Unwinn, London, U.K. 354pp.

*Kasuya, T. 2000. Explanatory Notes. pp. 155–203. In: Omura, H. (ed.). *Journal of an Antarctic Whaling Cruise—Record of Whaling Operation in Its Infancy in 1937/38.* Toriumi Shobo, Tokyo, Japan. 204pp.

*Kasuya, T., Izumisawa, Y., Komyo, Y., Ishino, Y., and Maeda, Y. 1997. Life history parameters of the common bottlenose dolphin off Japan. *IBI Rep.* (Kamogawa, Japan) 7: 71–107.

*Kasuya, T., Kaneko, M., and Nishimkoto, T. 1985. Archaeology and related science 7: Zoology. *Kikan Kokogaku* 11: 91–95.

Kasuya, T. and Marsh, H. 1984. Life history and reproductive biology of short-finned-pilot whale, *Globicephala macrorhynchus*, off the Pacific coast of Japan. *Rep. Int. Whal. Commn.* (Special Issue 6): 259–310.

Kasuya, T. and Matsui, S. 1984. Age determination growth of the short-finned pilot whales off the Pacific coast of Japan. *Sci. Rep. Whales Res. Inst.* 35: 57–91.

*Kasuya, T. and Miyashita, T. 1989. Japanese small cetacean fisheries and problems in management. *Saishu-to Shiiku* 51(4): 154–160.

Kasuya, T., Miyashita, T., and Kasamatsu, F. 1988. Segregation of two forms of short-finned pilot whales off the Pacific coast of Japan. *Sci. Rep. Whales Res. Inst.* 39: 77–90.

*Kasuya, T. and Miyazaki, N. 1981. Dolphins in the Iki Island area and conflict with fisheries—Preliminary report of three year study. *Geiken Tsushin* 340: 25–36.

Kasuya, T. and Miyazaki, N. 1982. The stock of *Stenella coeruleoalba* off the Pacific coast of Japan. pp. 21–37. In: Clark, J.G., Goodman, J., and Soave, G.A. (eds.). *Mammals of the Sea,* Vol. IV: Small Cetaceans, Seals, Sirenians and Otters. FAO Fisheries Series, No. 5. FAO, Rome, Italy. 531pp.

*Kasuya, T. and Yamada, T. 1995. *List of Japanese Cetaceans.* Gieken Series 7. Nihon Geirui Kenkyu Sho (Institute of Cetacean Research), Tokyo, Japan. 90pp.

*Kawabata, H. 1986. Dolphin and porpoise fisheries. pp. 636–664. In: Yamada Choshi Hensan Iinkai. (ed.). *History of Yamada Town.* Vol. 1. Yamada-cho Kyoiku Iinkai, Yamada, Japan. 10+1095pp.

*Kawaguchi, S. and Nishinaka, K. 1985. Mammal bones from archaeological shell mounds in Kagoshima Prefecture. *Kikan Kokogaku* 11: 43–47.

*Kawashima, T. 1894. *Fisheries of Shizuoka Prefecture.* Shizuoka-ken Gyogyo-kumiai Torishimarijo, Shizuoka, Japan. Vols. 1: 288pp.; 2: 182pp.; 3: 406pp.; 4: 362pp.

Kirschvink, J.L. 1990. Geomagnetic sensitivity in cetaceans: An update with live stranding records in the United States. pp. 639–649. In: Thomas, J.A. and Kastelein, R.A. (eds.). *Sensory Abilities of Cetaceans: Laboratory and Field Evidence.* Plenum Press, New York. 710pp.

Kishiro, T. and Kasuya, T. 1993. Review of Japanese dolphin drive fisheries and their status. *Rep. Int. Whal. Commn.* 43: 439–452.

*Kitamura, K. 1995. Illustrated fisheries in Noto. pp. 117–239. In: Sato, T., Tokunaga, M., and Eto, A. (eds.). *Collection of Japanese Agricultural Literatures.* Vol. 58. Nosan Gyoson Bunka Kyokai, Tokyo, Japan. 406+I–XIIIpp. (Manuscript dated 1838).

*Kizaki, M. 1773 (unpublished). "Iruka-ryo-no Koto" (About Dolphin Fisheries). In: Kisaki, M. (ed.). "Hizen Koku Karatu-ryo Bussan Zu-ko" (Illustrated Consideration of Products of Karatsu, Hizen Country) (hand written unpaginated scroll).

Klinowska, M. 1990. Geomagnetic orientation in cetaceans: behavioral evidence. pp. 651–681. In: Thomas, J.A. and Kastelein, R.A. (eds.). *Sensory Abilities of Cetaceans: Laboratory and Field Evidence.* Plenum Press, New York. 710pp.

*Kojima, T. 2009. The dolphin fishery in the Izu Region. pp. 86–98. In: Kojima, T. (ed.). *On Whales and Japanese—Revisit to Coastal Whaling.* Tokyo Shoten, Tokyo, Japan. 255pp.

*Kumamoto, H. (ed.) 1980. *History of Fisheries of Katsumoto Town.* Katumoto-cho Gyogyo Kyodo Kumiai, Katumoto, Japan. 576pp.

*Kurino, K. and Nagahama, M. 1985. Dolphin hunting in Sagami Bay. *Kikan Kokogaku* 11: 31–34.

*Kyoto-fu Suisan Koshujo. 1909. *Fisheries of Kyoto,* Vol. 2: Ine Village of Yosa County. Kyoto-fu Suisan Koshujo, Kyoto, Japan. 114pp.

Madsen, H. 1992. *Grind, Faeroernes Hvalfangst.* Skuvanes, Torshavn, Faroe Islands. 188pp.

*Matsuura, Y. 1942. Story about dolphins and porpoises. *Kaiyo Gyogyo* 71: 53–103.

*Matsuura, Y. 1943. *Marine Mammals.* Ten-nensha, Tokyo, Japan. 298pp.

*Miyamoto, J. 2001. *Ethnology of Women.* Iwanami Shoten, Tokyo, Japan. 324pp.

*Miyashiro, M. 1987. *Miscellanea of Nago People.* Published by the author, Nago, Japan. 292pp.

Miyashita, T. 1993a. Abundance of dolphin stocks in the western North Pacific taken by the Japanese drive fishery. *Rep. Int. Whal. Commn.* 43: 417–437.

Miyashita, T. 1993b. Distribution and abundance of some dolphins taken in the North Pacific driftnet fisheries. *Int. North Pacific Fish. Commn. Bull.* 53(III): 435–450.

*Miyawaki, K. and Hosokawa, T. 2008. *The Mind of the Japanese Viewed from Whale Monuments—Memory of Whales in the Bungo Pass Area.* Norin Tokei Shuppan, Tokyo, Japan. 281pp.

Miyazaki, N. 1980. Catch records of Cetaceans off the coast of the Kii Peninsula. *Mem. Natl. Sci. Mus.* (Tokyo) 13: 69–82.

*Miyazaki, N. 1982. Dugongs and whales. pp. 157–166. In: Kisaki, K. (ed.). *Natural History of the Ryukyu Islands.* Tsukiji Shokan, Tokyo, Japan. 282p.

Miyazaki, N. 1983. Catch statistics of small cetaceans taken in Japanese waters. *Rep. Int. Whal. Commn.* 33: 621–631.

Miyazaki, N., Kasuya, T., and Nishiwaki, M. 1974. Distribution and migration of two species of *Stenella* in the Pacific coast of Japan. *Sci. Rep. Whales Res. Inst.* 26: 227–243.

Miyazaki, N. and Nishiwaki, M. 1978. School structure of the striped dolphin off the Pacific coast of Japan. *Sci. Rep. Whales Res. Inst.* 30: 65–116.

*Nagasaki-ken. 1896. *Records of Fisheries* [of Nagasaki Prefecture]. Tsuruno Goro, Nagasaki. 169pp.+pls.

*Nagasawa, R. 1916. Scientific names of 11 Japanese of dolphins and porpoises. *Dobutsugaku Zasshi* (Tokyo Dobutsu Gakkai) 28(327): 45–47.

*Nago Hakubutsu-kan. 1994. *Pitu and Nago People—Dolphin Fisheries of Nago, Okinawa.* Nago Hakubutsu-kan, Nago, Japan. 154pp.

*Nakamura, Y. 1988. On dolphin and porpoise fisheries. pp. 92–136. In: Shizuoka-ken Minzoku Geino Kenkyukai. (ed.). *Marine Folklore of Shizuoka—Culture in Kuroshio Current.* Shizuoka Shinbunsha, Shizuoka, Japan. 379pp.

Nishiwaki, M., Nakajima, M., and Kamiya, T. 1965. A rare species of dolphin (*Stenella attenuata*) from Arari, Japan. *Sci. Rep. Whales Res. Inst.* 19: 53–64, pls. I–VI.

*Noguchi, E. 1946. Dolphins and porpoises and their utilization. pp. 3–36. In: Noguchi, E. and Nakamura, A. (eds.). *Scomber Fishery and Utilization of Dolphins and Porpoises.* Kasumigaseki Shobo, Tokyo, Japan. 70pp.

*Noshomu-sho. 1890–1893. Report of preliminary survey on fisheries (Text reprinted in *Hogeisen*, Vol. 30, published in 1999 by Takeuchi, K.).

*Noshomu-sho Suisan-kyoku. 1911. *Fishery Activities of Japan.* Vol. 1. Suisan Shoin, Tokyo, Japan. 424pp.

Odell, D.K. and McClune, K.M. 1999. False killer whale *Pseudorca crassidens* (Owen, 1846). pp. 213–243. In: Ridgway, S.H. and Harrison, R. Sir. (eds.). *Handbook of Marine Mammals*, Vol. 6: The Second Book of Dolphins and the Porpoises. Academic Press, San Diego, CA. 486pp.

*Ogawa, T. 1950. *Tale of Whales.* Chuo Koron Sha, Tokyo, Japan. 260pp.

Ohsumi, S. 1972. Catch of marine mammals, mainly of small cetaceans, by local fisheries along the coast of Japan. *Bull. Far Seas Fish. Res. Lab.* 7: 137–166.

Omura, H. 1974. Possible migration route of the gray whale on the coast of Japan. *Sci. Rep. Whales Res. Inst.* 26: 1–14.

*Omura, H. 2000. *Journal of an Antarctic Whaling Cruise—Record of Whaling Operation in Its Infancy in 1937/38.* Toriumi Shobo, Tokyo, Japan. 204pp.

*Oyama, H. and Shibahara, C. 1987. *Dolphin Fisheries by Villagers in Tsushima Island.* Mitsushima-no Shizen-to Bunka-wo Mamoru Kai, Mitsushima, Japan. 47pp.

Rice, D.W. 1998. *Marine Mammals of the World*, Spec. Publ. No. 4. Society for Marine Mammals, Lawrence, Kansas. 231pp.

*Shibusawa, K. 1982. *Japanese Fishery Technology before the Meiji Revolution.* Noma Kagaku Igaku Kenkyu Shiryokan, Tokyo, Japan. 701pp.

*Shindo, N. 1968. *Study of Whales in the Inland Sea.* Kobe-shi Ishi Kyodo Kumiai, Kobe, Japan. 135pp.

*Shizuoka-ken Kyoiku Iinkai. 1986. Report on the cultural heritage of Shizuoka Prefecture 33: Manners and customs of fisheries in Izu, part 1. Shizuoka-ken Bunkazai Hozon Kyokai, Shizuoka, Japan. 211pp.

*Shizuoka-ken Kyoiku Iinkai. 1987. Report on the cultural heritage of Shizuoka Prefecture 39: Manners and customs of fisheries in Izu, part 2. Shizuoka-ken Bunkazai Hozon Kyokai, Shizuoka, Japan. 193pp.

*Suisan-cho Chosa-kenkyu-bu. 1968. Report of basic investigations in search of counteraction for fishery damages caused by small cetaceans in western Japan (42nd year of Showa). Suisan-cho Chosa-kenkyu-bu, Tokyo, Japan. 96pp.

*Suisan-cho Chosa-kenkyu-bu. 1969. Comprehensive report of basic investigations in search of counteraction for fishery damages caused by small cetaceans in western Japan (42nd and 43rd year of Showa). Suisan-cho Chosa-kenkyu-bu, Tokyo, Japan. 108pp.

*Taiji, G. 1937. *On Whaling at Taiji.* Kyushujinsha, Wakayama, Japan. 139pp.

*Takenaka, K. 1890. Utilization and catch statistics of dolphins and porpoises. *Dainihon Suisankai Hokoku* 98: 241–249.

*Tamura, T., Ohsumi, S., and Arai, S. (eds.) 1986. Report of contract studies in search of countermeasures for fishery hazard (culling of harmful organisms) (56th to 60th year of Showa). Suisan-cho Gyogyo Kogai (Yugaiseibutsu Kujo) Taisaku Chosa Kento Iinkai (Fisheries Agency), Tokyo, Japan. 285pp.

*Tobayama, T. 1969. School size of striped dolphins and its variation viewed from fishing data in Sagami Bay. *Geiken Tsushin* 217: 109–119.

*Tsutsumi, T., Uemura, J., and Mizue, K. 1961. Study on small toothed whales off the West Coast of Kyushu—V. On food habits of small toothed whales. *Bull. Fac. Fish. Nagasaki Univ.* 11: 19–28.

Tudor, J., Peach, B.J., Horne, J., Fortescue, W.I., and White, P. 1883. *The Orkneys and Shetland; Their Past and Present State.* Edward Stanford, London, U.K. 703pp.

*Uchida, S. 1985. Report of surveys on small cetaceans for aquarium display, part 2. Kaiyohaku Kinen Koen Kanri Zaidan, Motobu, Japan. 36pp.

*Wakuta, M. 1989. *Past and Present of Funaya [Fishing Boat Hanger].* Amano Hashidate Shuppan, Kyoto, Japan. 88pp.

Wilke, F., Taniwaki, T., and Kuroda, N. 1953. *Phocoenoides* and *Lagenorhynchus* in Japan, with notes on hunting. *J. Mammal.* 34(4): 488–497.

*Yamada, Y. 1995. *Mawaki Archaeological Site with Illustration.* Noto-cho Kyoiku Iinkai, Noto-cho, Japan. 41pp.

*Yamamoto, Y. 1884. Report of examination of items exhibited at the fishery exposition, section I, part 2: Equipment of marine fisheries. Noshomu-sho Nomu-kyoku, Tokyo, Japan. 134pp.

*Yoshida, T. (ed.) 1979. *The Social Anthropology of a Fishing Village—Transformation of Katumoto in Iki Island.* Tokyo Daigaku Shuppankai, Tokyo, Japan. 251+4pp.

*Yoshihara, T. 1976. Whaling at Ine Harbor of Tango. *Tokyo Suisan Daigaku Ronshu* 11: 145–184.

Zachariassen, P. 1993. Pilot whale catch in the Faroe Islands, 1709–1992. *Rep. Int. Whal. Commn.* (Special Issue 14, Biology of Northern Hemisphere Pilot Whales): 69–88.

4 Small-Type Whaling

4.1 ORIGIN

Hunting minke whales and toothed whales other than the sperm whale using harpoons discharged from small-caliber cannons mounted on small vessels with screw propulsion was defined on December 5, 1947 (Ministry Order No. 91, see Section 5.4) as *kogata-hogei-gyo* [small-type whaling] and placed under a license system of the Japanese government (Ministry Order No. 91). Recent regulations restrict the targets to Baird's beaked whales, short-finned pilot whales, Risso's dolphins, and false killer whales for commercial purpose (a special permit to take minke whales for scientific purpose is additional). Regulations on the caliber of the cannon and vessel size have also changed several times. For simplicity, I use the term "small-type whaling" for similar operations before the year of official recognition of the method.

This small-type whale fishery and the vessels used for it have been called *minku-sen* [minke whaling ship], *gondo-sen* [pilot whaling ship], or *tento-sen* (see Section 3.9.1) in Wakayama Prefecture (33°26′N–34°19′N, 135°00′E–136°00′E) and on the Pacific coast of central Japan (Maeda and Teraoka 1958, in Japanese). The governor of Chiba Prefecture (34°55′N–35°44′N, 139°45′E–140°53′E) on the Pacific coast of central Japan introduced regulations on this fishery before the Ministry Order mentioned earlier.

This fishery has its origin in Japan in the late nineteenth century, when the Greener's harpoon gun invented in American whaling was introduced to Japan (Matsubara 1896, in Japanese) and attempts for its improvement began (it was also termed a mid-sized whaling cannon) (Kaneda and Niwa 1899, in Japanese). A. Sekizawa tried hunting Baird's beaked whales off Chiba Prefecture in 1891 with a Greener's gun mounted on a rowed boat (Sekizawa 1892a, in Japanese) and caught the first whale the next year (Sekizawa 1892b, in Japanese). In 1908, Tokai-Gyogyo Co. Ltd. succeeded in the Baird's beaked whale fishery off Chiba using a Greener's gun of 37 mm caliber imported from Norway (see Section 13.7.1). This success triggered the establishment of numerous operations for hunting Baird's beaked whales in Chiba.

The governor of Chiba Prefecture recognized the importance of this whaling for the local economy and the risk of overcompetition among whalers and placed the fishery under a license system of the prefecture in 1920. At the time, the fishery was defined as Whaling with Rowed Boats or Motor Driven Vessels and 26 vessels got the license. In the fishery of the time, the whales were taken from rowed catcher boats, and motor-driven vessels were used for towing whale carcasses and the catcher boats. Chiba Prefecture further continued its effort to decrease the number of licensees (Komaki 1996, in Japanese). By 1942, the fleet declined to 15 vessels (Matsuura 1942, in Japanese).

In Taiji (33°36′N, 135°57′E) in Wakayama Prefecture, a particular kind of whaling cannon was invented and was in use for many years. K. Maeda of Taiji started the idea when he was shown a whaling cannon in Korea by A. Sekizawa. Later, in 1903, Maeda made a three-barreled whaling cannon and used it in hunting short-finned pilot whales off Taiji. In 1904, he modified it into a five-barreled cannon, which was called "Maeda's multi-barreled pilot whaling gun" and used it for hunting pilot whales in Taiji until around 1967, when the last pilot whale vessel in Taiji switched from Maeda's multi-barreled cannon to a single-barrel, small-caliber cannon. The pilot whale vessels of Taiji (also called *tento-sen*) first introduced engines in 1913 (Hamanaka 1979, in Japanese).

Thus, a type of Japanese whaling that was later defined as small-type whaling was established around 1910. This fishery was operated by 22 vessels in 1941 and expanded to the Sanriku coast on the Pacific coast of northern Honshu in latitudes 37°54′N–41°35′N (Matsuura 1942, in Japanese). The number further increased to 31 or 33 vessels (including those in the planning stage) in 1942 and to 46 in 1943, including expansion to more prefectures, that is, 2 vessels in Saga (northern Kyushu), 15 in Wakayama, 4 in Mie (Pacific coast in 33°45′N–35°00′N, 136°00′E–136°55′E), 2 in Shizuoka (34°36′N–35°08′N, 137°29′E–39°10′E), 9 in Tokyo (Pacific coast in 26°45′N–35°40′N, 139°07′E–142°06′E with southern islands), 10 in Miyagi (southern Sanriku Region in latitudes 37°54′N–38°59′N), and 4 in Iwate (middle Sanriku Region in latitudes 38°59′N–40°27′N), for a total of 46 (Nihon Hogei-gyo Suisan Kumiai [Japanese Association of Whale Fisheries] 1943, in Japanese; Table 4.1). As in the case of other cetacean fisheries in Japan, this reflected the government policy during the war (Matsuura 1942, 1943, both in Japanese; Chapter 2). The trend of increase continued after World War II, and the government identified a need to place the operation under a license system in 1947.

The small-type whalers usually made daily trips within the prefecture of registration, but it was technically and legally possible to operate in any other waters. In the postwar period, minke whales became the main target of the fishery, and vessels registered in Chiba or the Sanriku Region visited the Sea of Japan and the Okhotsk Sea for minke whales (which were scarce off the Pacific coast west of Chiba Prefecture). In some cases, their catch records may have been compiled in the prefecture of registration, not by the position of the actual catch, which must be identified before analyzing the statistics.

In the 1941 season, 22 small-type whaling vessels took a total of 734 whales (Table 4.1). The average was 26 per vessel. Six vessels from the Sanriku coast caught a total of

TABLE 4.1
Peri–World War II Expansion of Japanese Small-Type Whaling Indicated by Number of Vessels and Whales Taken

Fishing Ground[a]	Year	No. of Vessels[b]	Minke Whale	Baird's Beaked Whale	Short-Finned Pilot Whale	Killer Whale	Other Species[c]	Total
Hokkaido	1941	0	0	0	0	0	0	0
	1948		81	30	33	23	0	167
	1951	3	102	30	10	0	0	142
Sanriku	1941	6	22	0	397	15	C. 100	C. 535
	1948		81	0	208	3	0	392
	1951	17	128	101	274	0	0	503
Chiba	1941	11	0	24	2	0	5[e]	31
	1948		1	46	2	0	0	49
	1951	15	0	108	0	0	0	108
Wakayama	1941	5	no data	0	118	no data	C. 50	C. 168
	1948		23	0	418	22	19	482
	1951	18	1	0	220	0	0	221
Kochi	1941	0	0	0	0	0	0	0
	1948		1	0	64	0	21	86
	1951	6	0	0	84	0	0	84
Kyushu	1941	0	0	0	0	0	0	0
	1948		98	0	0	0	0	98
	1951	10	22	0	13	0	0	35
Sea of Japan	1941	0	0	0	0	0	0	0
	1948		0	0	0	0	0	0
	1951	0	81	3	17	0	0	101
Total Japan	1941	22	22	24	517	15	C. 156	C. 734
	1948		285	76	725	48	40	1174
	1951	75	334	242	618	0	0	1194

Sources: Matsuura, Y., *Kaiyo Gyogyo*, 71, 53, 1942, in Japanese; Matsuura, Y., *Marine Mammals*, Ten-nensha, Tokyo, Japan, 1943, 298pp., in Japanese for 1941; Maeda and Teraoka (1958, in Japanese) for 1948 and 1951.

Note: There is some disagreement with Table 4.2.

[a] Maeda and Teraoka (1958, in Japanese) stated that there were 32 land stations conducting small-type whaling in 1948–1951 located at the following 26 locations (by prefecture): **Hokkaido**: 6 locations (Akkeshi, Abashiri, Esashi, Kushiro, Monbetsu, and Urakawa); **Iwate**: 2 (Kamaishi and Osawa); **Miyagi**: 3 (Ayukawa, Hosoura, and Onagawa); **Chiba**: 2 (Chikura and Shirahama); **Wakayama**: 1 (Taiji); **Kochi**: 2 (Shi-ina and Shimizu); **Nagasaki**: 1 (Koyagi); **Saga**: 3 (Yobuko, Nagoya, and Uchiage); **Fukuoka**: 1 (Maebaru); **Miyazaki**: 3 (Kadokawa, Nango, and Nobeoka); **Ishikawa**: 2 (Ogi and Ushitsu).

[b] Fleet size of April 20, 1952 was assumed for 1951. There were 6 additional vessels that received operation licenses in Tokyo (included in the total for Japan).

[c] Take of "dolphins and porpoises" is likely underrepresented.

534 cetaceans: 397 pilot whales, 15 killer whales, 100 dolphins, and 22 minke whales. This means that only 27% of the vessels took 72% of the whales and dolphins reported by the entire small-type whaling fleet. Their catch was mostly short-finned pilot whales of the northern form; minke whales were a minor element. The catch of 11 Chiba boats was 24 Baird's beaked whales (the official statistics recorded 23: see Table 13.14), 7 pilot whales, and a small number of presumably dolphins. I question whether they made a profit by taking only 2–3 Baird's beaked whales per season, or whether these Chiba whalers worked in other fisheries in the rest of the year. Five Wakayama boats took 118 pilot whales and 50 dolphins (Table 4.1).

Small-type whaling expanded operations during the peri-war period, and it also underwent a considerable change in the target species as seen in the fragmentary catch statistics. Table 4.1 was constructed to describe the changes, from Matsuura (1942, 1943, both in Japanese) that recorded catch and operating vessels by regions (for the 1941 season) and from Maeda and Teraoka (1958, in Japanese) that gave catch statistics of 4 seasons from 1948 to 1951. Table 4.2 shows operation of the fishery in a later period based on various official statistics. Ohsumi (1975) is useful for understanding the operations of small-type whaling in Japan for the years 1948–1972.

4.2 OPERATION PATTERN

The catch of minke whales by Japanese small-type whaling was small until 1941, when 22 minke whales were taken by 22 vessels (Table 4.1; Matsuura 1942, 1943, both in Japanese), but the catch increased to 285 in 1948 and about 500 per year in the late 1950s (Table 4.2). The minke whale remained the major target species of the fishery until 1987, the last year of commercial hunt of minke whales (Table 4.3). There could have been two reasons for the lower catch of minke whales in

TABLE 4.2
Whales Taken by Commercial Small-Type Whaling during 1941–1987 (Official Statistics)

Year	Minke Whale[a]		Baird's Beaked Whale	Short-Finned Pilot Whale[b]	Killer Whale	Others[c]	Total
1941	22		24	530[d]	9	3[e]	588
1948	285		76	725	48	41	1175
1949	184		95	890	44	49	1262
1950	259		197	715	24	18	1213
1951	334		242	618	66	34	1294
1952	485		322	335	58	39	1239
1953	406		270	460	66	37	1239
1954	365		230	75	100	33	803
1955	427		258	61	85	40	871
1956	532		297	297	38	46	1210
1957	423		186	174	78	37	898
1958	513		229	197	73	178	1189
1959	280		186	144	36	391	1037
1960	253		147	168	48	178	794
1961	332		133	133	54	305	957
1962	238		145	80	47	377	887
1963	220		160	228	43	179	830
1964	301		189	217	99	434	1240
1965	334		172	288	169	71	1034
1966	365		171	199	137	220	1092
1967	285		107	237	101	294	1024
1968	239		117	166	22	274	818
1969	234		138	130	16	56	574
1970	320		113	152	12	29	626
1971	285		118	181	10	25	619
1972	341		86	91	3	17	538
1973	423(541)[f]		31(32)[f]	77	0	2	533
1974	291(372)[f]		32	62	2	1	388
1975	290(370)[f]		41(46)[f]	53	3	5	392
1976	360		13	11	1	3	388
1977[g]	248(541)		44	6	1	1	300
1978	400(400)		36	11	0	0	447
1979	P392; J15 (400)		28	3	0	0	438
1980	P364(421)	J15(15)	31	1	2	0	413
1981	P359(421)	15(15)	39	0	0	0	413
1982	P309(421)	J15(15)	60	85[h]	0	0	469
1983	P279(421)	J11(15)	37(40)	125(175)	0	1	453
1984	P367(367)	J0(0)	38(40)	160(175)	0	0	565
1985	P320(320)	J7(10)	40(40)	62	0	0	429
1986	P311(320)		40(40)	29(50)	2	0	382
1987	P304(320)		40(40)	0	0	0	344

Sources: Nihon Hogei-gyo Suisan Kumiai, *Whaling Handbook*, Vol. 3, Nihon Hogeigyo Suisan Kumiai, Tokyo, Japan, 1943, 205pp., in Japanese for 1941; Ohsumi, S., *J. Fish. Res. Board Canada*, 32(7), 1111, 1975 for 1948–1972; Fisheries Agency publications for 1951–1987 (Enyo-gyogyo-ka (Suisan-cho) [Offshore Fishery Division of the Fisheries Agency], *Outline of Whaling*, Suisan-cho, Tokyo, Japan, 1975, 46pp.; Enyo-gyogyo-ka (Suisan-cho), *Outline of Whaling*, Suisan-cho, Tokyo, Japan, 50pp., 1976; Enyo-gyogyo-ka (Suisan-cho), *Information on Whaling*, Suisan-cho, Tokyo, Japan, 1977a, 359pp., Enyo-gyogyo-ka (Suisan-cho), *Outline of Whaling*, Suisan-cho, Tokyo, Japan, 1977b, 58pp.; Enyo-ka (Suisan-cho) [Offshore Division of the Fisheries Agency], *Outline of Whaling*, Suisan-cho, Tokyo, Japan, 1979, 64pp.; Enyo-ka (Suisan-cho), *Outline of Whaling*, Suisan-cho, Tokyo, Japan, 1982, 56pp.; Enyo-ka (Suisan-cho), *Outline of Whaling*, Suisan-cho, Tokyo, Japan, 1984, 62pp.; Enyo-ka (Suisan-cho), *Outline of Whaling*, Suisan-cho, Tokyo, Japan, 1987, 62pp.; Enyo-ka (Suisan-cho), *Outline of Whaling*, Suisan-cho, Tokyo, Japan, 1988, 62pp.).

Note: Figures in these sources reasonably agree for the years 1951–1972, when cross-check is possible (see footnote a for some disagreements).

(*Continued*)

TABLE 4.2 (*Continued*)
Whales Taken by Commercial Small-Type Whaling during 1941–1987 (Official Statistics)

[a] Ohsumi (1975) recorded 512. Home pages of Fisheries Agency give figures by sex, but those before 1986 are significantly smaller than those given here (see Table 1.5). Reason for the disagreement not explained.
[b] Likely to include some other blackfish species.
[c] Includes smaller dolphins and porpoises.
[d] Out of the 530, 407 were recorded as *shio-goto* taken off Sanriku and are believed to be of the northern form, and 41 recorded as *naisa-goto* taken off Wakayama Prefecture are believed to be of the southern form. Nihon Hogei-gyo Suisan Kumiai (1943, in Japanese) gives monthly landings of short-finned pilot whales (710 off Sanriku and 293 off Wakayama) in 1939–1941 and those of minke whales in 1939–1942. The statistics show that the main minke whale ground was off Sanriku and the season was from January to July (a peak in March–May), that *shio-goto* was taken off Sanriku in June–December with a peak in August–October, and that *naisa-goto* was taken off Wakayama year-around with a peak in May–August.
[e] These 3 whales are reported as Cuvier's beaked whales.
[f] Figures in parentheses are totals including whales taken in "test operation" of the catcher factory ship *Miwa-maru*.
[g] Figures in parentheses since 1977 represent the minke whale quota given by IWC or a Japanese autonomous quota for other species. A 5-year block quota was given for minke whaling for 1980–1984. The minke whale quota since 1980 was given separately for the East China Sea–Sea of Japan stock (J) and the western North Pacific–Okhotsk Sea stock (P).
[h] For this figure, see footnote f to Table 4.3.

the early period. The more plausible reason was that large-type whaling (for the definition see Section 5.4) operated off the Japanese coasts and supplied a large amount of meat of "sei whales" (which included both sei and Bryde's whales) and fin whales. The second possible reason would be the difficulty of distribution of products using the still undeveloped transportation system of the time from remote land stations to whale meat consumers, who were mostly located in western Japan. Importation of whale meat from Japanese Antarctic whaling fleets was permitted experimentally for the first time in the 1937/1938 season, or at the beginning of the Sino-Japan war.

The Japanese market for whale meat of the time was probably not large enough to satisfy the entire whaling industry. A whaler's wife recorded the hard life of small-type whalers who had to survive competing with large-type whaling and Antarctic factory ship whaling (Shoji 1988, in Japanese and English).

The major minke whaling grounds were off the coasts of Sanriku and Hokkaido during the period 1948–1972. Each of these regions recorded 40%–45% of the total Japanese catch of the species; only about 15% was taken in other waters (Ohsumi 1975). The minke whaling grounds were located on both sides of the Japanese islands and shifted northward with the progress of the season (Omura and Sakiura 1956). The Sea of Japan ground on the west side of Japan extended from northern Kyushu to western Hokkaido and to the Okhotsk Sea. The fishing season ranged from February/March to October, with a peak in March–April in the southwestern area and in June/July off Hokkaido in northern Japan. The grounds on the east side of Japan on the Pacific coast extended from Sanriku to eastern Hokkaido and to the Okhotsk Sea and were whaled from January to October with a peak in May–June off Sanriku and July/August off eastern Hokkaido (Omura and Sakiura 1956).

Only a few minke whales have been taken off Chiba; instead, Baird's beaked whales have been the major target of small-type whalers there. The species was hunted only off the Chiba coast and consumed locally until the beginning of the 1940s (Kasuya 1995, in Japanese). Statistics on Baird's beaked whales taken by small-type whalers are available by region only for 1948–1952, 1965–1969, and from 1982 (Table 13.14). Whalers off Abashiri (44°01′N, 144°16′E) in the southern Okhotsk Sea recorded a catch of 24 Baird's beaked whales in 1948 and continued the operation thereafter, suggesting that they started the hunt perhaps in 1942–1947, that is, during World War II or shortly after it. In other regions, catches of more than 10 Baird's beaked whales have appeared in the records for the Sea of Japan since around 1949, for off Sanriku since around 1950, and off Kushiro (42°59′N, 144°23′E) on the Pacific coast of eastern Hokkaido since around 1951. Thus, I conclude that in areas other than Chiba and Abashiri, hunting of Baird's beaked whales started around 1950, or several years after the end of World War II. Fishing grounds for the species in the postwar period were off Chiba, off Sanriku, Toyama Bay (*c.* 37°N, 137°15′E), off the Oshima Peninsula (41°20′N–43°20′N, 139°40E′–140°00′E, off southwestern Hokkaido), and off Abashiri (Omura et al. 1955). During the years from 1948 to 1972, the operations off Sanriku and Chiba each recorded about 40%–45% of the Baird's beaked whales taken in the whole of Japan; the remaining regions (Hokkaido coasts and the Sea of Japan) took only about 15%.

The season of the Baird's beaked whale fishery was from early summer to autumn off Chiba and Sanriku with a peak in July–August. The small catch of the species off Kushiro had a peak in October–November, the Okhotsk Sea operation had two peaks in May–June and September–October, and the Sea of Japan operation had a peak around July. These peaks in Baird's beaked whale takes reflected two factors other than migration of the species through the fishing grounds. One was the presence of minke whales. Minke whales were preferred over Baird's beaked whales because they sold at higher price and were easier to capture. Another plausible factor was an unevaluated effect of poaching of sperm whales. It has been reported that some small-type whalers off the Pacific coast of Japan poached sperm whales and reported some of them as

TABLE 4.3
Operation of Japanese Small-Type Whaling before and after Cessation of Commercial Whaling on Minke Whales with the 1987 Season

Year	Licensee (No. of Vessels)	No. of Vessels Operated[a]	Minke Whale	Baird's Beaked Whale[b]	Short-Finned Pilot Whale, N. Form[c]	Short-Finned Pilot Whale, S. Form[d]	Risso's Dolphin[e]	Others	Amount (10^6 yen)	
									Proceed	Balance
1982	8	8	(436) 324	60	85[f]	7			843	62
1983	9	9	290	(40) 37	(175) 125			1	928	57
1984	9	9	(367) 367	38	(175) 160				1033	109
1985	9	9	(330) 327	40	(Open quota) 62				1039	130
1986	9	9	(320) 311	40	(50) 28	1		2	1245	204
1987	9	9	304	40	(0) 0				1317	238
1988	9	6		(60) 57	(100) 98	30		7	447	−39
1989	9	4		54	(50) 50	(50) 8		14	464	71
1990	9	4		(54) 54	10	8		19	489	37
1991	9	5		54	43	16	92	5	565	97
1992	9	5		54	48	33	(30) 30		610	96
1993	9	5		54	44	47	30		660	109
1994	9	5		54	26	29	(20) 20		620	79
1995	9	5		54	50	50	20		701	101
1996	9	5		54	50	50	20		725	115
1997	9	5		54	50	27	20		691	86
1998	9	5		54	35	49	20		711	56
1999	9	5		(62) 62	60	44	12		790	62
2000	9	5		62	50	56	20		732	45
2001	9	5		62	47	40	17		628	−35
2002	9	5		62	47	36	12		503	−114
2003	9	5		62	42	27	19		448	−130
2004	9	5		62	13	29	7		417	−123
2005	9	5		(66) 66	(36) 22	24	8		402	−126
2006	9	5		63	7	(36) 10	7		324	−171
2007	9	5		67	0	16	20		322	−154
2008	9	5		64	0	20	0	(20) 0[g]	289	−151
2009	7	5		67	0	22	0			
2010	7	5		66	0	10	0			
2011	7	5		61	0	0	0			
2012	7	5		71	0	16	0			
2013	7	5		62	0	10	0	1[h]		

Note: Quota (the quota could be carried over to the next year; thus, the catch in a particular year could apparently exceed an annual quota) is shown in parentheses only for years with a new quota. The minke whale quota was decided by the IWC, but quotas for other species were based on agreement between the whalers and the government. From Fisheries Agency statistics and report of the Small-Type Whaling Association. This table has been revised with the addition of new data available since the original Japanese language edition appeared in 2011.

[a] A small-type whaling vessel *Katsu-maru* operated from Taiji in Wakayama Prefecture until 1980, left Taiji for other grounds including Wadaura in Chiba Prefecture during 1981–1987, and revived the Taiji operation in 1988 while operating in other waters (see also Table 3.18) together with the *Seishin-maru*, which was a Taiji boat that entered into small-type whaling in 1983 with a minke catch in northern Japan.

[b] The total national quota for Baird's beaked whales was set at 40 in 1983, 60 in 1988, 54 in 1989 (adjustment for retirement of one vessel), 62 in 1999 (creation of a quota of 8 for the Sea of Japan), and at 66 in 2005. The area allocation, which had some flexibility to meet the operational requirements of the industry, started with 5 for Abashiri and 35 for Wadaura in 1983, was changed in 1988 and 1989, and as of 2005–2013 was 4 for Abashiri, 26 for Wadaura, 26 for Ayukawa, and 10 for the Sea of Japan to be landed at Hakodate (see Table 13.15).

[c] Operated from Ayukawa.

[d] Mainly operated off Taiji but some operated off Wadaura. The 7 in 1982 were taken off Ayukawa (Kasuya and Tai 1993).

[e] Operated off Taiji.

[f] The Small-Type Whaling Association amended this figure to 172 responding to suspicion by scientists, but this was not followed up with formal amendment by the government (Kasuya and Tai 1993).

[g] A quota of 20 false killer whales was established.

[h] False killer whale.

Baird's beaked whales (Komaki 1996, Kasuya 1999, both in Japanese; see Section 13.7.2). Such distortion will be small in the statistics of the Sea of Japan and Okhotsk Sea operation because sperm whales are uncommon there. Japanese small-type whaling off the Pacific coast of Japan reported takes of about 4400 Baird's beaked whales during the 25 years from 1948 to 1972. Recent abundance of the species in the region was estimated at about 5000 (CV = 0.56) (IWC 2001). It should be investigated whether the abundance of Baird's beaked whales in the early 1940s could have been large enough to explain the catch history and current abundance estimate (Kasuya 1995, in Japanese).

Although the term *gondo* [pilot whales] is mostly likely to represent two forms of short-finned pilot whales off Japan, it was also possibly applied in many of the early statistics to small numbers of false killer whales and Risso's dolphins (Chapter 3). The small-type whalers of Taiji hunted short-finned pilot whales of the southern form, supported by the food preferences of the local community, which identified Risso's dolphins and false killer whales as inferior to short-finned pilot whales. This situation was comparable to the Baird's beaked whale fishery off Chiba that was supported by local demand for meat of that particular species. Ohsumi (1975) using published statistics of *gondo* [pilot whales] by prefecture in 1948–1972 showed that about 50% of the Japanese pilot whale catch was obtained in Wakayama Prefecture (Taiji included) and another 35% in the Sanriku Region. The current understanding of distribution of stocks of pilot whales suggests that the former was of the southern-form short-finned pilot whales and the latter of the northern form of the species. The analysis of Kasuya (1975) of monthly operational reports of small-type whalers presented to the Fisheries Agency in 1949–1952 reached a similar conclusion. Matsuura (1942, 1943, both in Japanese) reported a catch of 397 pilot whales off Sanriku and 118 off Wakayama in 1941. He also reported a rapid increase in the number of small-type whalers off Sanriku, that is, 7 vessels in operation and 4 vessels in preparation (as of July 1942). From these bits of information, I have the impression that large-scale hunting of pilot whales started off Sanriku possibly during the war as in the case of other small-cetacean fisheries. Catch statistics and quotas for small-type whaling are given in Tables 4.2 and 4.3.

REFERENCES

[*In Japanese Language]

*Enyo-gyogyo-ka (Suisan-cho). 1975. *Outline of Whaling*. Suisan-cho, Tokyo, Japan. 46pp.

*Enyo-gyogyo-ka (Suisan-cho). 1976. *Outline of Whaling*. Suisan-cho, Tokyo, Japan. 50pp.

*Enyo-gyogyo-ka (Suisan-cho). 1977a. *Information on Whaling*. Suisan-cho, Tokyo, Japan. 359pp.

*Enyo-gyogyo-ka (Suisan-cho). 1977b. *Outline of Whaling*. Suisan-cho, Tokyo, Japan. 58pp.

*Enyo-ka (Suisan-cho). 1979. *Outline of Whaling*. Suisan-cho, Tokyo, Japan. 64pp.

*Enyo-ka (Suisan-cho). 1982. *Outline of Whaling*. Suisan-cho, Tokyo, Japan. 56pp.

*Enyo-ka (Suisan-cho). 1984. *Outline of Whaling*. Suisan-cho, Tokyo, Japan. 62pp.

*Enyo-ka (Suisan-cho). 1987. *Outline of Whaling*. Suisan-cho, Tokyo, Japan. 62pp.

*Enyo-ka (Suisan-cho). 1988. *Outline of Whaling*. Suisan-cho, Tokyo, Japan. 62pp.

*Hamanaka, E. 1979. *History of Taiji Town*. Taiji-cho, Taiji, Japan. 952pp.

IWC. 2001. Report of the standing subcommittee on small cetaceans. *J. Cetacean Res. Manage.* 3(Suppl.): 263–282.

*Kaneda, K. and Niwa, H. 1899. Examination report of items exhibited at the second fishery exposition. Vol. 1, Part 2. Noshomusho Suisan-kyoku, Tokyo, Japan. 219pp.

Kasuya, T. 1975. Past occurrence of *Globicephala melaena* in the western North Pacific. *Sci. Rep. Whales Res. Inst.* 27: 95–110.

*Kasuya, T. 1995. Baird's beaked whale. pp. 521–529. In: Odate, S. (ed.). *Basic Data on Rare Wild Aquatic Organisms of Japan (II)*. Nihon Suisan Shigen Hogo Kyokai, Tokyo, Japan. 751pp.

*Kasuya, T. 1999. Manipulation of statistics by the Japanese Coastal Sperm Whale Fishery. *IBI Rep.* (Kamogawa, Japan) 9: 75–92.

Kasuya, T. and Tai, S. 1993. Life history of short-finned pilot whale stocks off Japan and a description of the fishery. *Rep. Int. Whal. Commn.* (Special Issue 14):339–473.

*Komaki, K. (ed.) 1996. *Master of Yellowtail and Wada Fishery*. Shoji Hirotsugu, Wada, Japan. 308pp.

*Maeda, K. and Teraoka, Y. 1958. *Whaling*. Isana Shobo, Tokyo, Japan. 346pp. (First published in 1952).

*Matsubara, S. 1896. *Whaling*. Dainihon Suisankai, Tokyo, Japan. 298+10pp.

*Matsuura, Y. 1942. Story about dolphins and porpoises. *Kaiyo Gyogyo* 71: 53–103.

*Matsuura, Y. 1943. *Marine Mammals*. Ten-nensha, Tokyo, Japan. 298pp.

*Nihon Hogei-gyo Suisan Kumiai. 1943. *Whaling Handbook*. Vol. 3. Nihon Hogeigyo Suisan Kumiai, Tokyo, Japan. 205pp.

Ohsumi, S. 1975. Review of Japanese small-type whaling. *J. Fish. Res. Board Canada* 32(7): 1111–1121.

Omura, H. and Sakiura, H. 1956. Studies on the little piked whale from the coast of Japan. *Sci. Rep. Whales Res. Inst.* 11: 1–37.

Omura, H., Fujino, K., and Kimura, S. 1955. Beaked whale *Berardius bairdii* of Japan, with notes on *Ziphius cavirostris*. *Sci. Rep. Whales Res. Inst.* 10: 89–132.

*Sekizawa, A. 1892a. Experiment with a whaling gun. *Dainihon Suisankai Hokoku* 117: 4–25.

*Sekizawa, A. 1892b. Whaling in waters off Chiba and Oshima Island. *Dainihon Suisankai Hokoku* 123: 606–607.

*Shoji, M. 1988. For people waiting for summer. pp. 66–75. In: Takahashi, J. (ed.). *Women's Tales of Whaling: Life Stories of 11 Japanese Women Who Live with Whaling*. Nihon Hogei Kyokai, Tokyo, Japan. 131pp.

5 Management of Cetacean Fisheries in Japan

5.1 FISHERY ACTS

The Fishery Act stipulates principles for the national management of fisheries and functions like a constitution for fishery management. *Shorei* [ministerial orders], *kokuji* [notifications], *menkyo* [licenses], and *kyoka* [permits] issued by the Minister in charge of fisheries or actions by the prefectural governors on fishery management must be consistent with this act.

Prior to the Meiji Revolution of 1864–1868, fishery management by feudal governors was mostly for the purpose of taxation and less for conservation. In 1875, the new government declared that Japan's coastal seas belong to the government and that those who fish must rent an area from the government (Dajokan Declaration No. 195) [the *Dajokan* of the time functioned like a prime minister or cabinet member]. However, this caused trouble with fishermen who practiced traditional fishing customs, and in 1876, the government replaced Declaration No. 195 with Dajokan Notification No. 74, which approved traditional fishing rights and made it an obligation of the fishermen to pay taxes to their prefectural government (Katayama 1937; Matsumoto 1977, both in Japanese). This action also was aimed at taxation.

The first bill that resulted in a Fishery Act was presented in 1893 to the Imperial Diet by Diet members with an intention of protecting fishermen's rights and fishery resources. After numerous amendments, the bill passed the Diet in 1901, was promulgated in the same year, and came into effect on July 1, 1902 (Katayama 1937, in Japanese). This first Fishery Act, now called the "old Fishery Act," comprised 36 articles. Article 13 stated that governors of the prefectures could issue, with approval of the Minister, orders necessary for the protection of fishing resources and for the regulation of fisheries and that the Minister could send orders to local governors to carry out such actions. Thus, the Act had the new goal of conservation of fishery resources. With amendments in 1910 and 1933, this Fishery Act functioned as the basis of Japanese fishery management until 1950.

The Fishery Act of 1901 did not have language relating to whaling or dolphin and porpoise fisheries. However, it stated that those who operate a fishery using fixed fishing gear, those who operate in a particular limited sea area (Article 3), or those who monopolize a particular sea area (Article 4) must obtain a license from the local governor. This offered a legal basis for dolphin-drive fisheries and old-type whaling operations using hand harpoon and net or a trap-net set at an expected whale passage. This was after unsuccessful attempts at importing old whaling technology, that is, American-type whaling with sailing ships attempted by Manjiro in 1859–1863 and bomb-lance-gun whaling by several coastal whaling groups in 1872–1888 (Yamashita 2004, in Japanese), and before the establishment of modern Norwegian-type whaling. Although Japanese Norwegian-type whaling took the first whale in 1898, success in the business had to wait the efforts of Nihon Enyo Gyo-gyo [Japan Far Seas Fishery] Co. Ltd., established in 1899 in Senzaki (34°23′N, 131°12′E), Yamaguchi Prefecture (Akashi 1910, in Japanese).

The Fishery Act was amended in 1910, and the changes came into effect in April 1911. The new Article 35 stated that "steamship trawling and steamship whaling require a ministerial permit." Another amendment of March 1933 had the additional words "mother-ship fisheries" in Article 35, which meant that mother-ship whaling required a permit from the Minister. The complete text of the Act is printed in Akashi (1910, in Japanese) and Katayama (1937, in Japanese).

The new and current Fishery Act was promulgated on December 15, 1949, and came into effect in March 1950 (Law No. 267). As of 2006, the amendment of June 23, 2006 (Law No. 93) was the last. The current Act contains the following elements that relate to small-cetacean and whale fisheries.

1. *Fishery Right*: This is the right to operate the following three particular types of fishery: (a) a fishery using a trap net that is fixed at a particular location, (b) various kinds of aqua-culture operated in a particular sea area, and (c) various fishing activities by local communities such as taking seaweed and animals on the beach, driving animals that enter into a particular harbor, or into a particular fenced area (dolphin-drive fisheries were once managed by this), and (d) fishing in a particular fishing ground created by continued feeding of fish for subsequent fishing activities. These fisheries require the license of the prefecture governor. This fishing right is usually owned by Fishery Cooperative Unions (FCU) for use by the member fishermen.
2. *Specified Fishery*: Particular types of fisheries determined by Government Ordinance, which include mother-ship whaling, large-type whaling, and small-type whaling.
3. *Fisheries Adjustment Commission (FAC)*: There are three types of FAC: (a) A Regional FAC is established for each prefecture; the members are elected by fishermen. The prefectural governor, who can give necessary instructions to the Regional

FAC, must consult the Regional FAC before issuing a fishery license. The Regional FAC can issue orders prohibiting or restricting particular fisheries, restricting the number of fishermen, restricting fishing grounds, and other items considered necessary. (b) The Multiple Regional FAC is to handle problems common to multiple prefectures and is composed of members representing the Regional FACs. (c) A Broad Area FAC is established for each for three marine areas: the Pacific, the Sea of Japan and western Kyushu, and the Seto Inland Sea. It is composed of the Regional FACs, fishery specialists, and representatives of fisheries, and presents its opinion to the Minister when requested.

To have representatives of fisheries in the FAC is democratic, but the structure has an important drawback because it is insufficiently structured to reflect opinion of citizens other than fishermen on management of fishing resources or conservation of the marine environment.

4. *Fishery Policy Council*: The Council responds to requests for consultation by the Minister and sends opinions to the Minister on issuing permits for specific fisheries or changing the conditions of permits.
5. *Fishery Adjustment*: After consulting the Fishery Policy Council, the Minister can issue ministerial orders on the following four items and decide penalties for offences. The governor of a prefecture can issue regulations on the same matters with the advice of the Regional Fishery Adjustment Committee and with consent of the Minister.
   a. Restriction or prohibition of take or processing of aquatic animals and plants
   b. Restriction or prohibition of selling or possession of aquatic animals and plants or their products
   c. Restriction or prohibition on fishing gear or fishing vessels
   d. Restriction on the number or qualifications of fishermen

Based on these provisions of the Fishery Act, the Minister in charge of fisheries has issued ministerial orders and notifications and announced details of fishery management. In addition, various documents, called *tsutatsu* or *tsuchi* hereinafter translated as "circular," have been dispatched from the Director General of the Fisheries Agency or from the Director or General Manager of various Departments or Divisions of the Fisheries Agency. They functioned to supplement the rules determined by the minister in issuing additional regulations for the fisheries and had the same function as regular orders from the minister when they reached the prefecture or FCUs. If any prefecture or FCU ignored them, it could receive an adverse response from the issuing section. This custom caused a loss of transparency in fishery management.

The following are brief summaries of various regulation measures for cetacean fisheries.

5.2 RULE FOR REGULATION OF WHALING

The government promulgated the Rule for Regulation of Whaling (Ministerial Order No. 41) on October 21, 1909, and placed it into effect on November 1 of the same year, which was just before promulgation of the previously mentioned amendment of the Fishery Act in 1910. Article 1 of the Rule reads that those who hunt whales using steam or sailing vessel must obtain a ministerial permit for each vessel. Article 3 states that the same rule applies to land stations. The permit was to be invalidated if the fishery was not operated for 2 years without justifiable reason (Article 8). The Minister could regulate or prohibit whaling for particular whale species, seasons, and areas (Article 9).

Based on Article 9, the government limited the number of "steamship whalers" to 30 vessels on October 21, 1909 (Ministerial Notification No. 418). This was altered to 25 vessels on June 27, 1934, and maintained at that number until 1963 (see Section 7.2.1). This coincided with actions of the Japanese whaling community in 1909 and 1934 that reorganized the whaling companies and reduced the number. There could have been government pressure behind this activity. However, the government did not take direct action to protect depleted whale species.

The Rule for Regulation of Whaling had a minor insignificant amendment in 1911. Subsequently, it was amended twice in July 1934 and June 1936, as stated by Omura et al. (1942, in Japanese). The amended rule stated that it applied to fisheries that take sperm whales and baleen whales other than the minke whale (Article 1.1). It was made clear that the type of fishery which was later called small-type whaling was not regulated (see Section 5.4). The amendments defined "steamship whaling" as "whaling that uses a harpoon cannon from vessels equipped with screw propeller and excludes mother-ship whaling" (Article 1.2) and stated that the operation required a ministerial permit. Although whaling from sailing vessels was also regulated by this rule (Article 1.3), there were none in operation at the time. Mother-ship whaling was outside of the scope of this rule because it was regulated by the Rule of Regulation of Mother-ship Fisheries (promulgated in July 1934).

To harmonize the Rule with the international regulation of whaling of the time, Japan modified the Rule for Regulation of Whaling on June 8, 1938 (Ministerial Notification No. 200), prohibiting steamship whaling from taking certain whales: mothers accompanied by suckling calves (all species), and blue, fin, humpback, sei and sperm whales of a size below certain limits. In those days, the Bryde's whale was not distinguished from the sei whale. The minimum size limits for blue and fin whales were 2 or 1.5 m smaller, respectively, than in the international regulations of the time that came into effect on May 7, 1938. Although the gray whales and right whales were internationally protected, they were allowed to be taken by Japanese whalers in the North Pacific. These species became protected in Japanese North Pacific whaling after World War II (Table 5.1).

There was a period of about 10 years from the time a Japanese whaler first whaled with the Norwegian-type

TABLE 5.1
Protected Species and Minimum Size Limit of Whales: Comparison between International and Domestic Regulations[a]

	International Agreement of 1938		Japanese Regulation in 1938[d]		Japanese Regulation after World War II[f]	
Species	**Human Consumption or Cattle Feed[b]**	**Other Uses or Factory Ship Whaling**	**Land-Based Whaling**	**Factory Ship Whaling**	**Human Consumption or Cattle Feed**	**Other Uses or Factory Ship Whaling**
Gray whale	Prohibited	Prohibited	Permitted[e]	Permitted[e]	Prohibited	Prohibited
Right whales	Prohibited	Prohibited	Permitted	By regions[e]	Prohibited	Prohibited
Blue whale	65 ft (19.8 m)	70 ft (21.3 m)	18.18 m	19.81 m	19.8 m	21.3 m
Fin whale	50 ft (15.2 m)	55 ft (16.8 m)	15.15 m	16.76 m	15.2 m	16.8 m
Humpback whale	30 ft (9.1 m)	35 ft (10.7 m)[c]	10.60 m	10.66 m	10.7 m	10.7 m
Sei whale			10.60 m	10.66 m	10.7 m	12.2 m
Sperm whale	30 ft (9.1 m)	35 ft (10.7 m)	9.09 m	10.66 m	10.7 m	10.7 m

Note: Take of whales below the size indicated was prohibited.
[a] International regulations were given in feet. Metric conversion in parentheses is only for reader's reference.
[b] Limited to land-based whaling where meat used for human consumption or cattle feed.
[c] One-year prohibition was agreed for the Antarctic.
[d] Notification No. 200 of Ministry of Agriculture and Forestry dated June 8, 1938.
[e] Take was permitted in the North Pacific north of 20°N, that is, prohibited in the Antarctic.
[f] Ministry of Agriculture and Forestry notification no. 263 (December 14, 1948) on coastal whaling and Ministerial Order No. 112 (December 14, 1948) on factory ship whaling. These figures agreed with international size limits of the time.

method in 1898 to the time the government started to control the fishery in 1910. The number of whaling vessels rapidly increased during the period, and the government had to continue its efforts to reduce the expanded fishery even into the postwar period.

The Rule for Regulation of Whaling functioned to control one type of coastal whaling that was later called "large-type whaling" by the Rule for Regulation of Steam Vessel Whaling of 1947 (Section 5.4).

5.3 RULE FOR REGULATION OF MOTHER-SHIP FISHERIES

In conjunction with the amendment of the Fishery Act in 1933, the Ministry of Agriculture and Forestry enacted the Rule for Regulation of Mother-ship Fisheries (Ministerial Order No. 19) in July 1934. This new rule placed mother-ship whaling being planned for the 1934/1935 Antarctic season under its control together with then-existing mother-ship fisheries for salmon and crab, which had been controlled by another rule and required ministerial permits for their operation.

The Ministerial Notification No. 200 of June 8, 1938, decided minimum size limits and prohibition of certain whales for mother-ship whaling as well as land-based whaling in Japanese coastal waters, which was later called "large-type whaling" (Table 5.1). This was to match the international agreement that came into effect on May 7, 1938 (Omura et al. 1942, in Japanese; Section 7.2.1). Although the minimum size limits for mother-ship whaling were slightly greater than those for coastal whaling, that for the most important blue whale was still 1.3 m smaller than the international size limit of the time, presumably for the protection of the Japanese fleet (Kasuya 2000, in Japanese). This Notification prohibited take of gray whales and right whales with the exception of the North Pacific north of 20°N. Thus, these species could be hunted by Japanese coastal whaling and mother-ship whaling operated in the North Pacific.

After World War II, the Rule for Regulation of Mother-ship Fisheries was amended by Ministerial Orders No. 52 of September 30, 1946 (not seen by the author) and No. 112 of December 14, 1948. Chapter 4 of the latter amendment regulated mother-ship whaling and determined operation areas (Articles 40 and 41), species allowed, and body-length limits (Article 41) as in the international agreement of the time. Before this amendment, the General Head Quarters of the Allied Forces (GHQ) ordered the Japanese government to respect the international rules of whaling (Section 5.7).

The Nihon Hogei [Japan Whaling] Co. Ltd. purchased the whaling factory ship *Antarctic* from Norway in 1934 and on the way to Japan operated whaling in the Antarctic from December to February of the 1934/1935 Antarctic season. This was the start of Japanese Antarctic whaling. Amendment of the Fishery Act and promulgation of the Rule for Regulation of Mother-ship Fisheries were to prepare for this movement in the whaling industry.

5.4 RULE FOR REGULATION OF STEAMSHIP WHALING

On December 5, 1947, the Japanese government abolished the Rule for Regulation of Whaling and promulgated and brought into effect a new Rule for Regulation of Steamship Whaling on the same day (Ministerial Order No. 91). This new rule probably anticipated the fundamental renewal of the Fishery Act that happened 2 years later (December 15, 1949) and contained quite important provisions for the management of whaling.

One important change was deletion of sailing-ship whaling, presumably to recognize the status quo. However, it created an environment where free operation was allowed for such whaling. Soon after an announcement by the Japanese government in 1986 closing all commercial whaling by March 1988 after accepting the international agreement of a moratorium of commercial whaling (Section 7), there appeared a newspaper article stating that somebody had announced a plan for sailing-ship whaling. Such an operation could have been legally possible under Japanese law, but economic and technical problem might have interfered. Based on the same idea of circumventing the regulations, some Okinawa fishermen in the 1970s started a dolphin fishery using harpoons discharged from a crossbow or catapult powered with rubber cords (see Section 2.8). Actions taken to resolve this loophole are described in Section 5.6.

A second important action was to classify all Norwegian-type whaling in coastal waters into two types: large-type whaling and small-type whaling. The so-called small-type whaling had been unregulated with exception of the Chiba area (Sections 4.1 and 13.7.1), and this new rule brought it for the first time under government control. It defined "steamship whaling" as "fishing activity to catch whales using a harpoon cannon mounted on a vessel equipped with a screw propeller." Large-type whaling, which is one of the types of steamship whaling, was allowed to take sperm whales and baleen whales other than minke whales. The other type of steamship whaling—small-type whaling—was allowed to take cetacean species other than those allowed for large-type whaling by using a whaling cannon of less than 40 mm caliber. The regulations on cannon size and vessel size have been amended several times.

The third point in the Rule was assignment to the Minister the right to issue permits for whaling vessels and whaling stations and to determine whale species to be taken, whaling season, whaling ground, and number of whaling vessels to be operated (Article 15). Based on this article, ministerial notification determined details of whaling operations, including quota and size limit. For information on the regulation of factory ship whaling, see Section 5.3.

5.5 RULE FOR REGULATION OF SMALL-TYPE WHALING

The Rule for Regulation of Steamship Whaling of December 5, 1947, covered both large-type and small-type whaling, but the new Fishery Act of 1949, which came into effect in 1950, classified the former fishery as one of the Specified Offshore Fisheries (see Section 5.6) and only small-type whaling was left under the control of the Rule for Regulation of Steamship Whaling. This prompted the government to move regulation of small-type whaling to a newly created Rule for Regulation of Small-type Whaling (Ministerial Order No. 19 of March 14, 1950). Caliber of the cannon was not to exceed 43 mm.

5.6 RULE FOR REGULATION OF SPECIFIED OFFSHORE FISHERIES

Article 52 of the new Fishery Act (Law No. 267), which was promulgated on December 15, 1949 and came into effect on March 14, 1950, stated that fisheries using vessels and specified by the order would require ministerial permits (Specified Offshore Fishery). Based on Article 52, the government promulgated the Rule for Regulation of Specified Offshore Fisheries (Ministerial Order No. 17) on March 14, 1950. Chapter 2 of the rule regulated large-type whaling in a way that was almost identical to that in the earlier Rule for Regulation of Steamship Whaling of 1947. Regulation on size limits, protected species, and number of catcher boats could be decided by ministerial notification. The amendment of this rule in September 1962 seemed to have been made on matters not relating to whaling (Matsumoto 1980, in Japanese).

In 1983, there arouse a suspicion that some of the Japanese whalers were taking Bryde's whales in Philippine waters and processing them on board, that is, a possibility that they were violating the decision of the International Whaling Commission (IWC) to prohibit factory ship whaling in the North Pacific (any vessel used for on-board flensing was considered a factory ship). In order to stop such illegal operations, in 1984 the Japanese government amended Ministerial Order No. 5 of 1963 on Permission and Regulation of Specified Fisheries and prohibited Japanese nationals from participating, without ministerial permit, in taking and processing of baleen whales and sperm whales by non-Japanese vessels operating in the western North Pacific.

On February 22, 1990, the Ministry of Agriculture, Forestry and Fisheries further amended Ministerial Order No. 5 and placed it into effect on April 1, 1990 (Ministerial Order No. 2). The amendment was an addition of an article that stated "no one can take sperm whales and baleen whales with harpoon (with the exemption of harpoons discharged from cannon), gun (with exemption of harpoon cannon), or by the method of drive fishery." This was a new prohibition on hunting large cetaceans by dolphin fishermen. Such incidents were known previously among dolphin fishermen.

5.7 ALTERATIONS OF WHALING REGULATIONS AND INTERNATIONAL RELATIONSHIPS AFTER WORLD WAR II

Japanese whaling during and before World War II operated with regulations that were less strict than in the international agreement of the time, but the surrender of Japan to the Allied Forces on August 15, 1945, triggered actions to accept the international standards on minimum size limits, protected whales, and the season of Antarctic whaling.

On November 3, 1945, the GHQ sent memorandum SCAPIN 233 to the Japanese government ordering it to respect the international agreements on the regulation of whaling of the time. The international agreements included (1) the Geneva Accord of September 24, 1931, (2) the International Agreement for the Regulation of Whaling signed in London on June 8, 1937, and (3) the International Agreement for the Regulation of Whaling signed in London on June 24, 1938. In those days, any directions or orders by the GHQ superseded every Japanese legal regulation, so SCAPIN 233 prohibited Japanese whaling from taking gray and right whales. Following this memorandum, the Minister of Agriculture and Forestry issued Ministerial Notification No. 126 on September 30, 1946, mentioning in Article 9 (which authorized the Minister to regulate whaling operations) of the Rule for Regulation of Whaling and Ministerial Notification No. 200 of 1938 (describing prohibition and size limits) (see Section 5.2). This notification (not seen by the author) could have incorporated the order of the GHQ.

Ministerial Order No. 91 dated December 5, 1947, repealed Ministerial Order No. 200 of 1938; prohibited steam-ship whalers from taking gray, right, and bowhead whales; and declared new minimum size limits for other whale species. Then Ministerial Notification No. 263 of December 14, 1948, altered the minimum size limit from two decimal places to a single decimal place, for example, from 21.34 to 21.3 m for the Antarctic blue whale (Table 5.1).

On November 23, 1948 [*sic*], the GHQ issued Memorandum No. 1942, repealing its earlier memorandums on Japanese whaling and ordering the Japanese government to respect the international agreement on regulation of whaling signed in Washington, DC on 27 [*sic*] December 1946, which was the current International Convention for the Regulation of Whaling (ICRW). This was the date when Japan was bound to the current ICRW and was earlier than April 4, 1951, when Japan joined the convention or September 8, 1951, when the peace treaty was signed (it came into effect on April 28, 1952).

On February 12, 1959, the Ministry of Agriculture and Forestry issued Ministerial Order No. 4 titled Rules for Regulation of Dolphin and Porpoise Fisheries. This was the first legal action of the Japanese government on such fisheries and prohibited the take of dolphins and porpoises with firearms other than harpoon cannon in waters north of 36°N during February 20 to June 20. The prohibition period of 4 months was the migration season of the fur seal. Small-type whaling was not bound by this Ministerial Order.

The Japanese government promoted the pelagic fur seal fishery with some restrictions, as seen in the Rule of Issuing Licenses for the Fur Seal Hunt (Dajokan Declaration No. 16 of 1884) and Rule for Licensing of Fur Seal and Sea Otter Hunts (1895). However, on July 7, 1911, the North Pacific Fur Seal Convention for protection of fur seals was signed (and ratified in the same year) by four countries: Japan, Russia, the United Kingdom (Canada), and the United States. It contained a prohibition on pelagic hunting of fur seals (Section 2.4). In 1912, the Japanese government placed in force a law prohibiting the pelagic fur seal hunt and paid compensation to the hunters. Then, in 1940, probably as one of various actions taken in preparation for war, the Japanese government announced its cancelation of participation in the fur seal convention (and the convention automatically lost validity) and issued a ministerial permit for the pelagic fur seal fishery. Although the GHQ ordered prohibition of the fur seal hunt after World War II, some fishermen continued the hunt fur seals with guns, by disguising it as dolphin and porpoise hunting.

A new convention, the Interim Convention on the Conservation of North Pacific Fur Seals, was signed in February 1957 by Canada, Japan, the United States, and the USSR (ratified in October of the same year), which prohibited commercial hunting of fur seals in pelagic waters. Under these circumstances, the Japanese government paid compensation and put a stop to the pelagic hunt by September 1959. The earlier mentioned Ministerial Order No. 4 of 1959 was for this purpose. The Japanese government was afraid of a situation where firearms used in dolphin and porpoise fisheries would be used also in the pelagic fur seal fishery (Ohsumi 1972).

5.8 CURRENT REGULATION OF CETACEAN FISHERIES

Ministerial Order No. 92 of April 20, 2001 on amendment of Ministerial Order No. 5 of 1963 (on Permit and Regulation of Specified Fisheries) came into effect on July 1, 2001. This was the first action of the Japanese government to place hunting of all cetacean species under its control. Article 90.1 of the earlier Order No. 5 stated that "Whale species designated separately by the Minister shall not be taken in Antarctic waters south of 60°S," but it was amended by Order No. 92 that stipulated that "Whale species other than *baleen whales and such* and designated separately by the Minister shall not be taken [italics by the author]." The Minister defined *baleen whales and such* as baleen whales, sperm whale, and two species of the genus *Hyperoodon*, on which Japanese government accepted the competence of the ICRW and accepted prohibition of commercial hunts by the IWC. I interpret this to mean that this amendment had in mind the regulation of hunting of small cetaceans. Ministerial Order No. 92 of 2001 stated in Article 90.8 that "Any persons other than large-type whalers, small-type whalers, and mother-ship whalers are not allowed to take *baleen whales and such*," which intended to close the above mentioned loopholes in the whaling regulations (Section 5.4). Further, Article 90.9 of Order 92 stated that "Take of toothed whales other than sperm whales and the genus *Hyperoodon* in waters north of 60°S is allowed for mother-ship whalers and small-type whalers, with the exception of toothed whale species designated by the Minister and the exception of operations of fishermen permitted by a prefectural governor." This provided a legal cover for dolphin and porpoise fisheries currently in operation with prefectural license and quota by species determined by the Fisheries Agency of the Ministry of Agriculture, Forestry and Fisheries. Penalty for violation of the rule given earlier is imprisonment for 2 years or less, a fine of 500,000 yen or less, or both (Article 106.1.1). Those who wish to take small cetaceans for scientific purpose must acquire a permit from the Minister of Agriculture, Forestry and Fisheries (Article 1 of the Rule for Enforcement of the

Fishery Act, and Circular 13 Suikan No. 1004 of Director General of Fisheries Agency dated July 1, 2001).

Article 90.8 of Ministerial Order No. 5, with amendment by Ministerial Order No. 92 (2001), stated that "Any persons other than large-type whalers, small-type whalers and mother-ship whalers are not allowed to take *baleen whales and such*, with the exception of incidental take made during the operation of fisheries determined by the Minister [*italics* by the author]." Ministerial Notification No. 563 of the same date (April 20, 2001) stated that fisheries of the exception were fixed trap net of large type and small type and that violation of this would be subject to penalty. These two types of net fishery must analyze the DNA of the whales taken incidentally and report it to the Minister (Article 97.2 of the amended Ministerial Order No. 5).

The amendment apparently opened a way for utilization and free circulation of baleen whales and sperm whales incidentally taken in the two types of fixed trap nets, which were not allowed previously (see also Sections 6.3 and 6.4). However, the situation was not so clear. On the same day when the amendment of the old Ministerial Order No. 5 by a new Ministerial Order No. 92 came into effect, a Circular (13 Suikan No. 1004, dated July 1, 2001) was sent out by the Director General of the Fisheries Agency to the prefecture governors. It had the lengthy title "On a Principle to Control Hunting and Incidental Take of Whales (including Small Cetaceans) after Enforcement of the Ministerial Order that Amends Part of the Ministerial Order on Permit and Regulation of Specified Fisheries." This circular gave a new guideline on the handling of incidentally taken cetaceans and stated that previous guidelines on the same subject would be abolished between June 30, 2001, and March 31, 2002. The earlier guidelines were following circulars No. 2-1039 on Whales Found in Fixed Trap-nets (dated June 26, 1990), No. 10-2638 On DNA Sampling from Stranded Minke Whales (dated September 29, 1998), and No. 3-1022 on Small Cetaceans Captured, Incidentally Taken, or Stranded (dated March 28, 191), some of which will be discussed further in Chapter 6.

The circular from the Director General of Fisheries Agency dated July 1, 2001, detailed the procedures required before selling cetaceans incidentally taken in the trap-net fishery, which I interpret as creating additional hurdles to Ministerial Order No. 92. It is summarized as follows:

1. Ministerial Order No. 92 of 2001, acknowledging that the fixed trap-net fishery does not operate with the intention of taking *baleen whales and such*, and taking into account such facts as that incidental capture of whales can cause damage to fishing gears and other targeted species in the net and incur costs in disposing of the carcasses of captured whales, intends rational use of fisheries resources. (The "baleen whales and such" is defined in the same manner as by the Ministerial Order [see above].)
2. This amendment has no intention to promote the utilization of *baleen whales and such* found in fixed trap nets. This amendment does not alter the hitherto practiced effort of releasing whales found in the fixed trap net, when the situation of the entrapment and condition of trapped whales suggest it is reasonable to release them.
3. The whale species listed in Table 5.1 are of low abundance and should be dealt with adequate consideration as noted earlier. [The table lists five species of whales: right whale, gray whale, humpback whale, East China Sea stock of fin whale, and East China Sea stock of Bryde's whale.]
4. Any incidents of incidental take of *baleen whales and such* shall be reported to the Minister of Agriculture, Forestry and Fisheries, irrespective of presence or absence of utilization (Article 90.8.2).
5. If an incidentally taken whale is utilized, a sample for DNA analysis must be presented to a scientific institute (Article 97.2.2), through which the incident of entrapment can be reported to the Minister. The Institute of Cetacean Research (ICR) in Tokyo is recommended for the analysis. [WWW of the Whaling Section of Fisheries Agency stated that payment of 100,000 yen must accompany the DNA sample for analysis.]
6. Those who possess, sell, or process such whales in violation of the guideline given earlier shall receive imprisonment not exceeding 6 months or a penalty of 300,000 yen (Article 107.1).

I had difficulty in creating this summary and expect that readers will have problems in understanding it. In particular, the original statements of (1) and (2) presented earlier are extremely ambiguous, and that of (3) is also unclear to me as to whether it is an obligation, recommendation, or request. It seems to me that Circular No. 1004 of the Director General of the Fisheries Agency intends to twist the interpretation of Ministerial Order No. 92 toward the opposite direction, or at the best to reflect honest interpretation of the Ministerial Order. The ambiguity of the Circular was probably intended to avoid criticism on inconsistency between the two documents (Ministerial Order No. 92 and Circular No. 1004).

It is difficult to punish violation of Circular No. 1004 of the Director General of Fisheries Agency if it is not supported by Ministerial Order No. 92. Even if a fisherman kills a baleen whale found alive in his fixed trap net, there will be difficulty punishing him. Japanese official statistics before 2001 reported annual incidental mortality of minke whales in fixed trap nets to be at most around 10–20, but the number increased to over 100 since 2002. The earlier figures must have been due to underreporting (Kasuya 2007).

Article 90.9 of Ministerial Order No. 5 (with amendments by Ministerial Order No. 92 of April 20, 2001) allowed the take of certain species of small cetaceans in certain fisheries (Chapter 6). Ministerial Notification No. 564 of 2001 listed seven species of dolphins and porpoises to be allowed in such fisheries: Dall's porpoise, striped dolphin, common bottlenose dolphin, pantropical spotted dolphin, Risso's dolphin, short-finned pilot whale, and false killer whale. This list is identical to the species list accompanied by catch quotas of the Fisheries Agency in 1993. Later, in December 2006, the Fisheries Agency added the Pacific white-sided dolphin to the species list to be hunted beginning in the 2007 season.

The previously mentioned Circular No. 1004 of the Director General of the Fisheries Agency (July 1, 2001) also mentioned small cetaceans stranded or incidentally taken in fishing gear:

1. In principle, live animals stranded or found in fishing gear should be released.
2. It is acceptable to consume carcasses found in fishing gear, with hygienic caution. Consumption of stranded carcasses requires special hygienic caution, and it is desirable that they be buried or cremated.
3. Every incident of (1) and (2) mentioned earlier shall be reported to the Fisheries Agency via the prefectural governor.
4. Those who use small cetaceans stranded or found in fishing gear must report each case to the Fisheries Agency. However, the use of the following nine species for such purposes shall be limited to animals found dead: killer whale, white whale [beluga], Cuvier's beaked whale, Blainville's beaked whale, ginkgo-toothed whale, Hubbs' beaked whale, Stejneger's beaked whale, harbor porpoise, and long-beaked common dolphin.

These guidelines do not deal with stranded baleen whales. Therefore, Ministerial Order No. 77 was promulgated on October 12, 2004, and brought into effect on the same day, to further amend Ministerial Order No. 5 of 1963 on "Permit and Regulation of Specified Fisheries" (Ministerial Order No. 5 had been already amended by Ministerial Order No. 92 of April 20, 2001). The amendment was on the handling of stranded *baleen whales and such* (see above for the definition), and Article 81 (which corresponds to Article 90.8 of amendment of 2001) stated that "Persons other than large-type whalers, small-type whalers or mother-ship whalers are not allowed to take *baleen whales and such*. This rule is exempted for incidental take in fisheries determined by the Minister and take of stranded or drifted *baleen whales and such* whale species in a condition defined by the Minister (underlining by the author)." Then Ministerial Notification No. 1834 of the same year defined such conditions of whales as follows:

1. Dead animals
2. Animals that have the possibility to harm humans
3. Animals that are unlikely to recover from wounds
4. Animals that do not change position on their own for over 48 hours

To accommodate this change, the previously mentioned Circular No. 1004 (dated 2001) of the Director General of the Fisheries Agency was also amended on the same day to allow selling of stranded *baleen whales and such* whales with similar procedures as for those taken incidentally in particular fishing activities. A request was made to release stranded or incidentally captured gray whales in a Circular of the Director General of Fisheries Agency in April 2006 (Kato et al. 2008).

5.9 OTHER LAWS ON CONSERVATION OF CETACEANS

5.9.1 Act on Protection of Fisheries Resources

This act, promulgated on December 17, 1951, and last modified on June 2, 2010, differs from the legal actions mentioned earlier that intend to manage cetacean stocks while allowing exploitation. This act authorizes the Minister and prefectural governors to designate prohibited species, protected areas, and prohibited fishing gears and to quarantine and regulate marine constructions and makes it their obligation to conduct the release of artificially incubated juvenile fish and maintenance of fish ladders. The Minister and prefectural governors can prohibit take of some aquatic animals and plants (Article 4).

Ministerial Order No. 15 of April 1, 1993, modified the existing Rule for Enforcement of the Act on Protection of Fisheries Resources (Ministerial Order No. 44 in 1952) and prohibited the hunting of blue whales, bowhead whales, finless porpoises, and dugongs. A Circular of the Director General of Fisheries Agency of the same day (not seen) requested the prefectures to release (if found alive) and bury or incinerate (if found dead) finless porpoises incidentally taken in fishing gear and to report the incidents to the Ministry of Agriculture, Forestry and Fisheries. Scientists who wish to use stranded carcasses of the species mentioned earlier must request a permit.

The action given earlier does not prohibit incidental takes of such animals or request effort to decrease the number of such incidents; it only prohibits intended takes. It could have functioned to stop the live capture of finless porpoises by Japanese aquaria, but there were no fisheries in Japan targeting any of the four marine mammal species mentioned earlier. It is a mystery why the bowhead whale was listed; it does not inhabit Japanese waters.

Gray whales in the western North Pacific have been hunted to an extremely low level; the number of 1-year-old or older individuals was estimated at about 130 in 2007. Western gray whales have occasionally been killed in Japanese fixed trap nets on their migration between feeding grounds in the Okhotsk Sea and yet unconfirmed wintering ground in more southern waters, and the adverse effect on the population is of great concern (IWC 2008). Under these circumstances, the Japanese government on December 3, 2007, amended the Rule for Enforcement of Act on Protection of Fisheries Resources and listed the gray whale as a protected species (came into effect on January 1, 2008). Four cetacean species are now protected by this law; their capture, possession, or sale is prohibited.

5.9.2 Act for Conservation of Endangered Species of Wild Fauna and Flora

This act established on June 5, 1992, was aimed at the domestic implementation of the Convention on Trade in Endangered Species (CITES) and bilateral agreements on protection of migratory birds. It also develops regulations on capture and transportation of wild animals and plants and rules for conservation of their habitats and artificial breeding activities.

CITES was signed in Washington, DC on March 3, 1973, and came into effect on July 1, 1975. Japan ratified it on November 4, 1980. The convention lists species in three Appendices. Listed on Appendix I are the most endangered species, whose survival is affected or potentially affected by international trade. These species cannot be traded internationally for commercial purpose, and import and export permits are required for scientific exchange. Appendix II lists species that are likely to become listed on Appendix I if international trade is not regulated. International trade in such species must be accompanied by export permit of the state of origin. Appendix III lists species that member countries request be listed for the protection of the species within their territory, soliciting the cooperation of other member countries in regulating international trade. International trade in such species must be accompanied by export permit or certificate of origin prepared by the exporting country. The CITES regulations also apply to transfer of a species into national territory from international waters. Each member country is allowed to have a reservation on the classifications presented earlier to avoid trade regulation.

Appendix I lists all the baleen whales and toothed whales of 10 genera in 6 families, of which 9 species of baleen whales and 3 species of toothed whales inhabit Japanese waters. These are the North Pacific right whale, gray whale, blue whale, fin whale, humpback whale, sei whale, Bryde's whale, Omura's whale, minke whale, sperm whale, Baird's beaked whale, and finless porpoise. Appendix II lists the rest of the cetacean species. There are no cetacean species listed in Appendix III.

The Government of Japan took reservations on the fin whale, sei whale (with exception of North Pacific and southern hemisphere waters between longitudes 0°E and 70°E), Bryde's whale, common minke whale, Antarctic minke whale, sperm whale, Baird's beaked whale, and Irrawaddy dolphin. Thus, Japan is allowed to manage these reserved species as listed in Appendix II. The Japanese reservation on the Irrawaddy dolphin could function to camouflage its intention to reserve all the commercially important cetacean species.

The Japanese "Act for Conservation of Endangered Species of Wild Fauna and Flora" classifies endangered species into four categories: (1) internationally endangered fauna and flora, which should be conserved with international cooperation; (2) domestic endangered fauna and flora, which inhabit Japanese territory; (3) specific domestic endangered fauna and flora, which are not included in item (1) and are domesticated and commercially reproduced; and (4) species newly identified or found to inhabit Japan and require protection. In general, capture of listed wildlife is controlled by the Minister of Environment, but take of small cetaceans for scientific purposes requires permits from the Ministry of Agriculture, Forestry and Fisheries (Section 5.8).

Government Order No. 134 dated April 20, 2012, included most of the CITES Appendix I cetacean species in Category 1 (Internationally Endangered Species) and listed no cetacean species in Categories 2–4. The exceptions are cetacean species commercially exploited by Japanese fisheries. Among the cetacean species that are listed in CITES Appendix I and found in Japanese waters, only five species (North Pacific right whale, gray whale, blue whale, humpback whale, and finless porpoise) are listed in Category 1. The seven remaining species are not listed in Category 1, although six of them are known from Japanese waters: fin whale, sei whale, Bryde's whale, common minke whale, Antarctic minke whale, sperm whale, and Baird's beaked whale. These seven species are reserved by Japan under CITES and are hunted in Japanese whaling.

In Japan, the Fisheries Agency functions as the scientific authority as well as the management authority of CITES for marine fauna and flora, and it issues export and import permits for them. The government reservation to CITES Appendix I enables Japan to carry out commercial trade in whale products. Further, this reservation together with the government definition that "cetaceans taken by Japan-flagged vessels outside Japanese territory can be considered as obtained within Japan" eliminates problem for Japanese cetacean fisheries in landing their catches made outside territorial waters. However, it can hinder the importation of cetacean samples collected by biopsy by Japan-flagged vessel in the exclusive economic zones of other countries, because Japan does not issue an import permit and therefore the export country does not issue an export permit. Such a case occurred between Russia and Japan in August 2009. In the case of exporting biopsy samples taken by Japan-flagged vessels in international waters, there is no such problem because Japanese export permits copy the CITES classification as it appears on the import permits (as in the case of export of blue whale, fin whale, and sei whale sample to the United States in June 2012). CITES has a rule allowing for an institutional permit, where international transfer of scientific specimens between the permit-holding institutes is exempt from other limitations. This system is meant to promote scientific research by reducing the documentation task for scientists but has not been practiced in Japan.

5.9.3 Cultural Heritage Protection Act

This act can classify a particular animal or plant or its habitat as a "Natural Monument" and take measures necessary for its protection. The finless porpoise in the Inland Sea is the only example of cetaceans dealt with in this way. In November 1930, the Ministry of Education designated the area within a 1.5 km radius of the southern tip of Abashima Island in the Inland Sea (34°19′N, 132°57′E) a natural monument as an "Area of Finless Porpoise Migration" and prohibited capture and disturbance to the species there. Three reasons were given for the designation: (1) Japan is located at the northern limit of the species, (2) aggregation of numerous finless porpoises is a spectacle, and (3) there is an angling fishery that was believed to be assisted by the presence of finless porpoises (Kaburagi 1932, in Japanese).

The line-and-hook fishery uses aggregating finless porpoises as a marker for fishing sea bream and other fishes. It was based on understandings that both finless porpoise and sea bream fed on sand lance which were abundant in the Inland Sea and that sand lance attacked by finless porpoises from the sea surface would descend to escape and would attract sea bream to ascend for feeding, which would make them available for the fishermen. According to the local fishermen, this fishing ended in the late 1960s following the disappearance of sand lance in the vicinity (Kasuya and Kureha 1979). I was able to see numerous finless porpoises near the area of the natural monument in the 1970s, but they had almost disappeared from the area by the late 1990s (Kasuya et al. 2002; Section 8.4.5). For the conservation of finless porpoises, it is necessary to protect the broader habitat of the species as well as that of their prey species.

REFERENCES

[*In Japanese Language]

*Akashi, K. 1910. *History of Norwegian-Type Whaling in Japan.* Toyo Hogei Kabushiki Kaisha, Osaka, Japan. 280+40pp.

IWC. 2008. Report of the sub-committee on bowhead, right and gray whales. Unpublished manuscript. 32pp. (Published in *J. Cetacean Res. Manage.* 11(Suppl.): 169–192, 2009.)

*Kaburagi, T. 1932. A location of finless porpoise aggregation and fishing operation. pp. 72–75. In: Monbu-sho (ed.). Report of investigation of natural monument, animals, part 2. Monbu-sho, Tokyo, Japan.

*Kasuya, T. 2000. Explanatory Notes. pp. 155–203. In: Omura, H. *Journal of an Antarctic Whaling Cruise—Record of Whaling Operation in Its Infancy in 1937/38.* Toriumi Shobo, Tokyo, Japan. 204pp.

Kasuya, T. 2007. Japanese whaling and other cetacean fisheries. *Environ. Sci. Pollut. Res.* 14(1): 39–48.

Kasuya, T. and Kureha, K. 1979. The population of finless porpoise in the Inland Sea of Japan. *Sci. Rep. Whales Res. Inst.* 31: 1–44.

Kasuya, T., Yamamoto, Y., and Iwatsuki, T. 2002. Abundance decline in the finless porpoise population in the Inland Sea of Japan. *Raffles Bull. Zool.* 10(Suppl.): 57–65.

*Katayama, F. 1937. *History of Fisheries of Great Japan.* Nogyo-to Suisansha, Tokyo, Japan. 1102+9pp.

Kato, H., Ishikawa, H., Miyashita, T., and Takaya, S. 2008. Status report of conservation and researches on the western gray whales in Japan, May 2007–April 2008. IWC/SC/60/O8. 9pp. Document presented to the 60th *Meeting of the Scientific Committee of IWC* (available from IWC Secretariat, Cambridge, U.K.).

*Matsumoto, I. 1977. *Annotated Chronology of Modern Japanese Fisheries before World War II.* Suisansha, Tokyo, Japan. 97pp.

*Matsumoto, I. 1980. *Noted Chronology of Modern Japanese Fisheries, after World War II.* Suisansha, Tokyo, Japan. 119pp.

Ohsumi, S. 1972. Catch of marine mammals, mainly of small cetaceans, by local fisheries along the coast of Japan. *Bull. Far Seas Fish. Res. Lab.* 7: 137–166.

*Omura, H., Matsuura, Y., and Miyazaki, I. 1942. *Whales: Their Biology and the Practice of Whaling.* Suisan Sha, Tokyo, Japan. 319pp.

*Yamashita, S. 2004. *Whaling.* Vol. 2. Hosei Daigaku Shuppan Kyoku, Tokyo, Japan. 295+5pp.

6 Regulation of Fisheries for Dolphins and Porpoises

6.1 HISTORICAL BACKGROUND AND EARLY GOVERNMENT ACTIONS

It was July 1, 2001, when Ministerial Order No. 92 placed various Japanese cetacean fisheries under a single legal system (Section 5.8), by incorporating dolphin and porpoise fisheries into the then-existing regulation system for whaling. Responding to international and domestic concerns and criticisms of the management of Japanese dolphin and porpoise fisheries since around 1975, the Fisheries Agency and prefecture governments attempted to control these fisheries before the issuance of Ministerial Order No. 92 using various administrative techniques. I worked for the Fisheries Agency of the Ministry of Agriculture, Forestry and Fisheries as a cetacean scientist from 1983 to 1997, had opportunities to watch the attempts in the Agency, and felt that the reason Japan took almost 25 years to establish a control system for small-cetacean fisheries was because of lack of consensus within the Agency on the regulation needs and perhaps because of technical difficulties of the regulation. In other words, the earlier less efficient actions taken by some divisions of the Fisheries Agency could have been the products of their efforts under a limited mandate. The following is a brief history of small-cetacean management before 2001.

The old Fishery Act (Section 5.1) came into effect in 1901 and offered a way to manage Japanese fisheries in an orderly way. This act allowed a particular group of fishermen to monopolize a limited area in coastal waters for particular fishing activities, the "common fishing right." This gave a basis for the opportunistic driving of dolphins found in or near the harbor by local communities that had continued since before the Meiji Revolution around 1864–1868.

Kishiro and Kasuya (1993) stated that in 1951 Shizuoka Prefecture placed all the drive fisheries in the prefecture under its license system. Before this action, the dolphin-drive fisheries on the coast of the Izu Peninsula (34°36′N–35°05′N, 138°45′E–139°10′E) in Shizuoka Prefecture operated as one of the drive fisheries for various marine animals that were approved as the common fishing right mentioned earlier (see Section 3.8 and footnote to Table 3.11). In 1959, Shizuoka Prefecture changed the license from "drive fishery" to "dolphin-drive fishery," which meant narrowing the fishing target, and regulated the operational area, fishing season, and number of vessels allowed (Kishiro and Kasuya 1993; unpublished document of Suisan-cho Kaiyo-gyogyo-bu Sanjikan [Councilor for Marine Fisheries Department of Fisheries Agency] 1986, in Japanese). In the autumn of 1960, I started investigating dolphins taken by the drive fishery on the Izu coast and confirmed that drive fisheries surviving there were for dolphins only. The number of vessels allowed by the prefectural license was 2 for Kawana (34°57′N, 139°07′E), 2 for Futo (34°55′N, 139°08′E) and 12 for Arari (34°50′N, 138°46′E), and the fishing season was from September 1 to March 31. The operational area was not limited to inshore waters but was limited to waters in front of Shizuoka Prefecture on the west side of the Izu Peninsula for the Arari team and on the east side for the Kawana and Futo teams. The boundary between the two grounds was at the longitude of Tsumeki Cape (138°59′E). The permit was issued to each FCU with a statement that the catch should not exceed the past level, but the statement seemed to be nothing more than a request. Although the Fisheries Agency made some earlier requests to prefectures to decrease their catches of dolphins, the first functional catch limit was established in 1991 when Shizuoka Prefecture had a catch limit of a total of 730 of all species (Kishiro and Kasuya 1993).

At Taiji (33°36′N, 135°57′E) in Wakayama Prefecture there was also a traditional dolphin-drive fishery in the local community that operated when dolphins were found near the harbor. This fishery was named *gondo-tatekiri-ami* [stopping pilot whales in the net] and licensed as a common fishing right. The last recorded operation was in the 1950s (see Section 3.9), but the license lasted until August 1993, when renewal was discontinued, possibly due to the absence of operations for more than 20 years.

The dolphin-drive fishery currently in operation at Taiji was first attempted in 1969, using methods learned from the Izu coast, and fully established with a drive team in 1971 (Section 3.9). The operation was quite different from the traditional opportunistic dolphin drive of the past, because the team sent out vessels daily to scout for dolphins 15–20 nautical miles offshore. In 1982, Wakayama Prefecture placed this fishery under its license system (Kishiro and Kasuya 1993). At this time, the prefecture advised the then-existing two driving teams to merge, which allowed issuing a single operation permit but allowed all vessels of the two teams (15 vessels in total) to participate in the fishing. This license served to stop new entry into the fishery. The merging of the two teams contributed to ending competition between them and was meant to decrease chances of overhunting, but the retention of the large fleet did lead to overhunting. The license was renewed every 3 years. The fishing season and catch limit were self-determined. The season was from October 1 to the end of April for short-finned pilot whales, from October 1 to the end of February for other dolphins, and lasted the whole year for killer whales. The catch limit was 500 short-finned pilot whales and a total of 5000 dolphins of other species (not specified). These catch limits became a part of the conditions of the operation permit in 1986. Although the catch limits in the permit remained the same, the Fisheries Agency pressed the prefecture for further reduction of the catch (see Section 6.5).

Nago (26°35′N, 127°59′E) in Okinawa Prefecture also had a history of opportunistic dolphin drives (Section 3.10). The Nago group seemed to have acquired a prefecture license for the dolphin drive around the 1980s (see Section 6.2) but ceased operations in the 1980s due to the change in the atmosphere in the local community or destruction of suitable beach by construction. Kishiro and Kasuya (1993) recorded a drive of short-finned pilot whales in 1982 and of bottlenose dolphins in 1989, which could have been the last operation.

Dolphin and porpoise fisheries were operated at numerous locations in earlier Japan, but many of them ceased operation before the Fisheries Agency started regulation around 1990 without being described in detail (Chapters 2 and 3).

The hand-harpoon fishery for dolphins and porpoise operated without regulation until 1988. It was placed under a license system in 1989, and a quota was given in 1990.

6.2 ACTIONS OF THE FISHERIES AGENCY

Dolphin drives have been among the important fisheries along the coast of the Izu Peninsula since the early seventeenth century (Section 3.8). Operations expanded around the time of World War II and reached an annual take of 10,000–15,000 striped dolphins. However, the high catch was followed after the 1960s by a gradual decline to record low catches of around 1000 in the late 1980s, and the decline of the stock became evident (Kasuya 1985). Since the 1970s, the hand-harpoon fisheries off the Sanriku Coast on the Pacific (37°54′N–41°35′N) exported their catches of Dall's porpoises to Shizuoka Prefecture to supplement the decreased supply of striped dolphins, and the hand-harpoon fishery for Dall's porpoises expanded to the Hokkaido coast in the 1980s (Section 2.4).

The IWC is formed by commissioners nominated by each contracting government of the ICRW, has been working for the management of whaling and whale resources, and has established various committees for that objective. The Scientific Committee (SC), one of the committees under the IWC, established the smallcetacean sub-committee in 1973. The subcommittee first met in Montreal in 1974 and became a standing subcommittee of the SC in 1975. Beginning with the first meeting, the subcommittee paid attention to the Japanese dolphin and porpoise fisheries and expressed its concern about the two kinds of Japanese small-cetacean fisheries described earlier (hand-harpoon fisheries in the Sanriku Region and drive fisheries on the Izu coast) and later on the drive fisheries at Taiji and hunting of short-finned pilot whales off Sanriku by small-type whalers (Kasuya 2007). Although there is no agreement in the IWC on its competence to manage small-cetacean fisheries, there is a general agreement that the SC can consider the scientific aspects of small cetaceans and the fisheries.

A moratorium on commercial whaling was first proposed in the IWC in 1972, one year before the establishment of the small-cetacean subcommittee, followed by annual repetition of the proposal and its adoption in 1982 to be implemented beginning with the 1985–1986 Antarctic whaling season and 1986 coastal season. This meant that Japan would end whaling activities with the exception of drive fisheries, hand-harpoon fisheries, and small-type whaling for small cetaceans, that is, Delphinoidea species and most Ziphiidae species (Chapter 7). However, the Japanese whale fishery did not move in this direction but delayed acceptance for 2 years and then started so-called scientific whaling in the Antarctic in the 1987/1988 season and in the North Pacific in 1994 (Sections 6.2 and 7.4). These Japanese scientific whaling programs are debated with suspicion about the scientific value, on plausibility of the objectives, and about commercial elements (Kasuya 2003, 2005, both in Japanese, 2007).

Under these circumstances, the Fisheries Agency probably felt it inappropriate to view Japanese small-cetacean fisheries as minor local issues and to leave them unregulated. In 1982, when the drive fisheries on the Izu coast were already under the license system of Shizuoka Prefecture, the Fisheries Agency pressured Wakayama Prefecture to put the drive fishery at Taiji under its license system. The same action may have taken place for the drive fishery at Nago in Okinawa Prefecture (a memorandum of the Fisheries Agency dated July 7, 1988, which is cited in the next paragraph). These Fisheries Agency actions and subsequent actions of the prefectural governments functioned to stop additional entries into the fisheries. We had to wait several years before the establishment of catch controls with quotas by species.

In February 1988, the Fisheries Agency established a discussion group on the dolphin and porpoise fisheries under the leadership of Kazuo Shima, Councilor of the Fisheries Agency, and with membership of heads of related departments and divisions. The discussion group issued an agreed document titled "On a Desirable Policy on Small-Cetacean Fisheries around Japan, 7 July 1988," which was based on a background document by Shiro Ebisawa of the Department of Fishery Promotion titled "On the Dolphin and Porpoise Problems around Japan, 3 March 1988." The former document is useful in indicating the understanding within the Fisheries Agency on problems of Japanese small-cetacean fisheries and is summarized in the following translation (headings are as in the original text and my notes are in square brackets):

[Background of the Problems]

1. International concern about management and conservation of large and small cetaceans has increased since the United Nations Conference on the Human Environment in 1972. This will continue in the future.
2. Drive and hand-harpoon fisheries for dolphins and porpoises are economically important for some prefectures, and the value of their products has been

increasing along with increases in the price of whaling products.

3. Management actions on the exploited stocks cannot be limited to the Japanese EEZ but must expand to problems of incidental mortality of cetaceans in fisheries in the North Pacific.

[Current status of the fisheries]

4. The drive fisheries operate with permits issued by the prefecture governors and are called *iruka oikomi gyogyo* [dolphin-drive fishery] in Shizuoka Prefecture and *geirui oikomi ami gyogyo* [cetacean drive net fishery] in Wakayama Prefecture. The drive fishery in Okinawa Prefecture is also regulated in the same way. [Kishiro and Kasuya (1993) stated that the Nago fishermen had not obtained a license.]
5. The drive fisheries have targeted mainly striped dolphins, and the catch, which reached 30,000 per year, has recently declined and is now surpassed by other species taken in hand-harpoon fisheries. Drive fisheries in 1987 caught about 2300 cetaceans (including 300 small whales) in Wakayama, 1800 in Shizuoka, and none in Okinawa.
6. The hand-harpoon fishery, which is currently unregulated, recorded a take of 15,000 in 1987. Most (98%) were taken by fishermen in Iwate Prefecture. The Iwate fishermen taking Dall's porpoises are increasing (208 vessels in 1987) and expanding the fishing ground to waters of other prefectures. Their catch also underwent a 50% increase from 1985 to 1987. A decline in the population of *truei*-type Dall's porpoises has been suggested.
7. Nine small-type whaling vessels currently operate with a permit from the Minister of Agriculture, Forestry and Fisheries. The moratorium on commercial whaling was applied to them in 1988 [see Section 7.1.2], so the small-type whalers are now surviving by taking Baird's beaked whales (quota of 40), pilot whales, and other small cetaceans.
8. About 200 small cetaceans are culled annually for the purpose of reducing damage to fisheries. The culling operation has received international criticism, and the Fisheries Agency has requested the fishermen to minimize the cull. [The background paper given earlier by Ebisawa of March 3, 1988 stated that the purpose of culling at Iki Island (33°47′N, 129°43′E) had changed into a commercial fishery and that out of 197 false killer whales supposedly culled, 30 were sold to aquaria and the remaining 10 to meat dealers in Tokyo.]
9. Mortality of dolphins and porpoises incidental to Japanese fisheries was reported at 1500 in drift nets and 50 in fixed trap-net fisheries in 1987. [The background paper by Ebisawa listed the salmon drift-net fishery, large mesh drift-net fishery, large- and small-type purse seine fisheries (302 licensee in total), and fixed trap-net fishery (1204 large-type and 6535 small-type trap nets) as fisheries that are likely to kill dolphins and porpoises incidentally. He stated that of these fisheries, the salmon drift-net fishery in the Sea of Japan (67 vessels), and purse seine and fixed trap-net fisheries were of concern for the management of small cetaceans in Japanese waters. The Sea of Japan was covered by the International Convention for the High Seas Fisheries of the North Pacific Ocean signed by Canada, Japan, and the United States in 1952 and modified in 1978 and was a problem area inhabited by various cetacean populations as well as fur seals and Japanese sea lions (possibly now extinct), but it did not attract attention of the member countries presumably because it was outside the range of waters inhabited by marine mammal populations within the U.S. exclusive economic zone (EEZ) and because Japan wanted to hide the problem from international attention.]

[Current Measures by Japan]

10. Managing small-cetacean populations that are affected by Japanese fisheries exceeds the capacity of the prefecture governments, so it must be done by the Japanese Government, but the Government has not established principles of a management policy.
11. Before deciding catch limits, there is a need to improve abundance estimates and biological information on the cetacean species affected.

[Medium Term Goals]

12. Establish a government system for conservation and management of small-cetacean stocks, which shall be followed by:
13. Establish a system of research and collection of biological data for each cetacean species affected.
14. Conduct technical consideration of catch and effort control for the management of each cetacean species.
15. Obtain consensus within the Fisheries Agency and adjust legal and administrative systems accordingly.
16. Collect international information on the problem.
17. Exchange opinions between the Fisheries Agency and governors of the related prefectures on the direction of the international atmosphere.

[Tentative actions]

18. Place the hand-harpoon fishery in Iwate Prefecture under control of the Regional Fisheries Adjustment Committee (Regional FAC) in 1988, and suppress further expansion of the fishery and increase of the catch.
19. Consider whether government advice is needed for other small-cetacean fisheries based on information to be collected through Item 16.

[Incidental mortality and culling]

20. Keeping in mind the results of ongoing attempts to reduce the incidental mortality of small cetaceans in the North Pacific fishery, apply the solution to the coastal fisheries. [Under the convention on North Pacific high-seas fisheries signed by the three countries in 1952 and with consent of the United States, Japan operated a salmon gillnet fishery in the U.S. EEZ. However, the coverage of the U.S. Marine Mammal Protection Act of 1972 was expanded in 1977 to the U.S. EEZ, and the U.S. Government was required to prove that incidental mortality of Dall's porpoises and fur seals in the gill-net fishery was within allowable level before issuing the fishing permit for Japanese gillnetters. If this was not determined, the Japanese fishery was unable to operate in the U.S. EEZ. Therefore, the two countries in 1978 started biological studies of these marine mammals as well as development of technology to reduce the incidental mortality. Later, this activity expanded in geographical range and countries involved to include the operation of large-mesh and squid gill-net fisheries. However, the technical prospects for reducing incidental mortality or for assessing the status of the affected small-cetacean stocks were dim (more small-cetacean stocks became identified with progress of the research). In 1987, lawsuits were filed by conservationists against the Japanese fishery and the U.S. Government, and the gill-net fisheries did not operate as of the 1988 season. In January 1989, the Japanese fishery was defeated in the lawsuits (Iida et al. 1994, in Japanese), and in December 1989, the General Assembly of the United Nations adopted a resolution to close high-seas large-scale gill-net fisheries. Japan accepted the resolution in December 1992. Thus, the operations of the Japanese high-seas squid gill-net fishery, large-mesh gill-net fishery, and salmon gill-net fishery came to an end. Many of the results of scientific studies conducted in relation to the activities mentioned earlier were published in International North Pacific Fisheries Commission Bulletin No. 53 (I, II and III), which is the report of an international symposium held in Tokyo in November 1991.]
21. Culling is necessary for the time being but should be reconsidered in the future.
22. For further progress in the direction agreed upon, a discussion group should be established in the Fisheries Agency.

Following this agreement and with leadership of the Coastal Division of the Department of Fishery Promotion, there seems to have been continued discussion within the Fisheries Agency. One of the probable products was a document titled "Proposal, Rule for Regulation of Dolphin and Porpoise Fisheries." This was circulated for comment in October 1988 to the Offshore Division (Whaling Section and North Pacific Section) and Coastal Divisions of the Department of Marine Fisheries, the Far Seas Fisheries Research Laboratory (Whale Research Group), and related prefectures. It was amended after comments by the reviewers in January 1989 and finalized as a proposal on January 25, 1989. Although it appeared to me that those who worked for this task expected the rule to be issued as a ministerial order in 1989, this did not happen, presumably because consensus was not reached within the Fisheries Agency.

In December 1988, while the Coastal Division was working on the "Rule for Regulation of Dolphin and Porpoise Fisheries" mentioned earlier, a newspaper in Iwate Prefecture reported that the hand-harpoon fishery for porpoises in the prefecture would be placed under a prefectural license system. This suggests that another alternative was considered in the Fisheries Agency. With the leadership of the Coastal Division, on January 25, 1989, it was decided to place the hand-harpoon fisheries for dolphins and porpoises under a prefectural licensing system (Iwate Prefecture), under approval of the Regional FAC (in most of the other prefectures) or the Multi-regional FAC (Hokkaido and Nagasaki) and to decide fishing seasons and particular ports of landing for most of the prefectures. These fisheries were later moved to be under control of the prefecture governments by April 2002. The date of permit issue, number of licensees, and ports of landing were listed as follows (as of July 1, 1989). The number of vessels was that expected for January 1989 based on the results of operations in 1988 (Hokkaido) or results in past years and may have differed slightly in the actual permits issued.

January	Aomori (January 31, no vessels)
February	Iwate (February 1, 9 ports, 360 vessels)
March	Miyagi (March 1, 8 ports, 60 vessels)
	Okinawa (March 1, 5 vessels)
	Hokkaido (March 20, 9 ports, 50 vessels)
	Tokyo (Bonin Islands: March 25, 1 port; Other islands: March 14, 2 ports; no vessels)
	Shizuoka (March 18, 3 ports, no vessels)
	Chiba (March 28, 1–2 vessels)
	Kanagawa (March 31, 2 ports, no vessels)
	Oita (March 31, 21 vessels)
June	Wakayama (June 1, 7 ports, 50 vessels)
October	Nagasaki (Expected for October 1)
Not decided	Ehime (to be decided in Autumn)

Vessel sizes for these permit holders in Iwate Prefecture were in the range of 4–19 gross tons. The permit holders had an obligation to operate within the given fishing season and within licensed waters and to report their catch. Because each prefecture started the system on a different date, the full-scale regulations came into effect in all of Japan only in 1990, when a total of 644 permits were issued (Table 2.4). The number of permits is greater than the number of vessels holding the permits because some vessels obtained permits for multiple regions.

Expansion of the Japanese hand-harpoon fisheries for small cetaceans preceded the action of the government described earlier, which became clear in statistics presented to the annual meeting of the IWC in June 1989 in San Diego. Japan reported a total of 13,406 Dall's porpoises taken by various fisheries in 1987, but the number presented to the San Diego meeting was 41,555 for the year 1988. The Scientific Committee believed that such a large catch was not sustainable and recommended that the catch be reduced as soon as possible to the previous level and that catch statistics by stocks (*dalli*-type and *truei*-type) be obtained (Section 2.5). A similar recommendation was repeated in 1990.

The response of the Fisheries Agency to this recommendation is found in a document of the Coastal Division titled "Countermeasures to the Hand-Harpoon Fisheries for Dolphins and Porpoises," which was probably prepared soon after the IWC meeting of 1990 for the formation of consensus within the Fisheries Agency. It referred to the statement of the IWC SC and noted that the Japanese hand-harpoon fishery in 1988 took over 40,000 Dall's porpoises, that 34,000 of them were taken by Iwate fishermen, and that 313 out of a total of 520 hand-harpoon vessels in 1989 were from Iwate Prefecture.

The document of the Coastal Division noted that Japan placed the hand-harpoon fishery for small cetaceans under a license system in 1989, controlled the number of operating vessels, decided fishing seasons, and made it obligatory to report the catch. The Japanese Government had requested the prefectures to reduce catches by 30% in 1989 and further by 15% in 1990 through reduction of the fleet size and fishing season, and it already reached agreement with Iwate Prefecture on a final goal of reducing the catch of the prefecture to 10,000 by 1994. However, in view of the discussion in the IWC of this year (1990), it became necessary to achieve the final goal in 1991, and the Coastal Division therefore would request on July 13 that Iwate Prefecture make a plan to reduce their annual catch to 10,000 as soon as possible and would send similar requests to Hokkaido and Miyagi Prefectures after receiving a response from Iwate Prefecture. Thus, Japan would become ready for the 1992 IWC Meeting and the Second Conference on the Human Environment. The actions proposed earlier agreed with the statement of Morimoto of the Fisheries Agency at the Scientific Committee meeting of the IWC in 1990 (Section 2.5).

Culling of dolphins at Iki Island was allowed without limitation of the number until 1989, when an upper limit was set at 104 for 1990 (131 taken) and 96 for 1991 (Coastal Division dated March 1991).

6.3 MEASURES TAKEN WITH REGARD TO INCIDENTAL TAKE AND STRANDINGS

Cetacean fisheries in Japan have been regulated as described earlier, but bycatch of cetaceans in various fisheries such as those using fixed trap nets has long been left unregulated. These fisheries killed cetaceans from the same populations as exploited in dolphin and porpoise fisheries, small-type whaling, large-type whaling, and so-called scientific whaling. They also killed whales of species that had been depleted by past whaling and protected by the IWC (gray whale, North Pacific right whale, and humpback whale). The statistics for such mortality appear to be quite incomplete; only a fraction of the catch is believed to have been reported (Tobayama et al. 1992; Kasuya 2007). In view of our past experience with *dai ami hogei* [trap-net whaling] that once operated in the nineteenth century on the coasts of northern Kyusyu and Noto Peninsula (36°50′N–37°30′N, 136°40′E–137°20′E), it was plausible that some new type of net whaling might be established if incidental takes were left uncontrolled. In order to avoid such a situation, the Fisheries Agency sent out several circulars before a full solution was achieved by Ministerial Order No. 92 of 2001 on hunting and incidental takes and No. 77 of 2004 on strandings (see Section 5.8).

The first action of the Fisheries Agency was Circular No. 2-1039 of June 28, 1990, co-signed by the Director of the Coastal Division of the Department of Fishery Promotion and the Director of the Offshore Division of the Department of Marine Fisheries, which was titled "On Handling of Whales Incidentally Taken in the Fixed Trap-Net Fishery" and sent to fishery sections of the prefecture governments. It requested that the fishermen be guided as follows: (1) this is on handling of all species of Mysticeti and Physeteridae found in fixed trap nets, (2) if the animal is found alive it should be released to the sea, and if found dead it should be buried, to eliminate potential poaching, (3) in an area with a tradition of consuming whales, the carcass found dead can be consumed locally with hygienic precautions, (4) incidents as described in the 2nd and 3rd items given earlier should be reported by the corresponding FCU to the prefecture government to be forwarded to the Coastal Division of the Fisheries Agency. These guidelines did not mention dolphins and porpoises. At that point in time, Physeteridae included two species of the genus *Kogia*, which were excluded in a later circular of the same title (Circular No. 3-1022 of March 28, 1991; see below). In the same year, the 2nd and 3rd circulars were sent out by the Director of the Coastal Division to the prefecture governments: Circular No. 2-1050 of September 20, 1990, and No. 2-1066 of November 30, 1990, intended to cover more species.

The second circular (No. 2-1050) was titled "On Handling of Incidentally Taken Small Cetaceans (Dolphins and Porpoises)" and targeted at incidental take of three particular dolphin species (narwhal, finless porpoise, and killer whale) in fixed trap nets, drag nets, and other fisheries. It stated that (1) the animal should be released to the sea if it is found alive, (2) it should be buried if it is found dead, and (3) incidents of the 2nd case should be reported by the FCU to the prefecture to be forwarded to the Coastal Division of the Fisheries Agency. It did not mention local consumption of the carcass; thus, the consumption was prohibited.

The third circular (No. 2-1066) was titled "On Handling of Small Cetaceans" and targeted cases of incidental take in fishing gear and stranding of species that were not covered by the two circulars (Nos. 2-1039 and 2-1050) mentioned earlier, that is, all the toothed whales other than the three species of

Physeteridae and three particular toothed whales (narwhal, finless porpoise, and killer whale). It stated that (1) if found alive the animal should be released to the sea, (2) if found dead the carcass should be in principle buried or incinerated, (3) it is acceptable to consume carcasses found dead in regions with such customs, but selling of the carcass is not acceptable, (4) the 2nd and 3rd cases should be reported by the FCU to the prefecture to be forwarded to the Coastal Division of the Fisheries Agency.

Through these three circulars handling of incidental take of all cetacean species became regulated. However, the first circular allowed selling of carcasses in the local market, the 3rd circular prohibited selling of carcasses, and the 2nd did not allow even non-commercial consumption of carcasses found in nets. This resulted in a strange situation, where minke whale carcasses found in nets could be circulated in the local market for human consumption, and beaked whale carcasses could be consumed but could not enter local markets. Consumption of beaked whales was prohibited by a later circular (5th circular of March 28, 1991; see below).

The 4th and 5th circulars were issued presumably to streamline the complicated guidelines. There were two similar but different circulars on related subjects, which had the same document number (No. 3-1022) and the same date (March 28, 1991) but were signed differently. One of the circulars, which I tentatively call "4th circular," was signed by both the Directors of the Coastal Division and Offshore Division and titled "On the Handling of Whales Incidentally Taken in Fixed Trap-Nets" and made a minor change in the guideline given in the earlier Circular No. 2-1039 by excluding the two species of the genus *Kogia* from the regulation. The other circular, "5th circular," was signed by the Director General of the Department of Fishery Promotion and titled "On Handling of Small Cetaceans (Dolphins and Porpoises)." The preamble of this circular stated that the earlier circulars Nos. 2-1050 and 2-1066 would be repealed in the near future, which means that the two earlier circulars were still valid at the time of issuance of this circular (March 28, 1991). This circular defined *iruka* [dolphins and porpoises] to include all the toothed whales other than the sperm whale and requested the fishery management of the prefecture governments to advise fishermen as outlined here.

[Three Categories of *Iruka*]

Group 1: Cuvier's beaked whale, killer whale, northern-form short-finned pilot whale, Fraser's dolphin, harbor porpoise, finless porpoise, narwhal, and other toothed whales not included in the following two categories.

Group 2: Baird's beaked whale, false killer whale, southern-form short-finned pilot whale, bottlenose dolphin, Risso's dolphin, pantropical spotted dolphin, striped dolphin, Dall's porpoise (including two color morphs).

Group 3: Melon-headed whale, common dolphins, northern right whale dolphin, Pacific white-sided dolphin, rough-toothed dolphin.

[Take and Cull]

1. It is prohibited to take or cull species in Group 1. They should be released to the sea if found alive while operating a fishery, or buried or incinerated if found dead. The incident should be reported to the Coastal Division. Those who intend to take these species for scientific purpose should consult the Coastal Division.
2. Take or cull of species in Group 2 are permitted within the catch limits given elsewhere and within regulations set by other related rules and government orders.
3. Species in Group 3 should be managed with caution after consulting the Coastal Division.

[Incidental Take and Stranding]

1. Any species of *iruka* should be released if found alive.
2. If found dead, species in Group 1 should be buried or incinerated. Consumption of species of Groups 2 and 3 is acceptable in regions with such customs. Commercial circulation of them is prohibited. Any incidents should be reported by the FCU to the prefecture government to be forwarded to the Coastal Division of the Fisheries Agency.

The circular summarized earlier drive and hand-harpoon fisheries prohibited from taking finless porpoises and harbor porpoises, which are coastal species and require special care for conservation. Although it may be hard to understand why the northern-form short-finned pilot whale was placed in Group 1 and Baird's beaked whales in Group 2, when both species were hunted by small-type whaling with ministerial permit, it probably functioned to protect the northern-form short-finned pilot whale from local fisheries that were under the control of the prefecture government or the FAC. With the exception of Baird's beaked whale, other Group-2 species were hunted by drive and hand-harpoon fisheries under the control of the prefecture government or the FCU and had quotas set in 1993. The Group-3 species were not exploited by Japanese fisheries of the time, and their abundance was unknown.

Circular No. 2-1066 of November 30, 1990, allowed local consumption of Ziphiidae found dead in nets, but the 5th circular of March 28, 1991, prohibited this with the exception of Baird's beaked whales. At this stage, incidentally taken minke whale carcasses were allowed to be sold in local markets by Circular No. 2-1030 (June 28, 1990), but in reality they were often transported to remote markets for higher prices. Such scandals were occasionally reported in newspapers around 1993.

6.4 PROBLEMS IN MANAGEMENT OF INCIDENTAL MORTALITY

The Fisheries Agency made attempts to control incidental take of cetaceans through a series of circular letters before 2001, when Ministerial Order No. 92 was issued (Section 5.8) to modify the earlier Ministerial Order No. 5 of 1963 on

"Permits and Regulation of Specified Fisheries." Requests or advice in circular letters sent out by the Director of the Coastal Division or by the Director General of the Department of Fishery Promotion had, in my view, a weak legal basis, and thus, penalizing violations was probably difficult. It was likely to involve arbitrary fines or penalty by administrative sections of the government and was likely to cause an unfair relationship between the Fisheries Agency and fishing industries. In spite of such deficits and attempts made toward modification of the old Ministerial Order in 1988–1989, this ambiguous method continued for 11 years. Behind this delay, there were difficulties in reaching consensus within the Fisheries Agency on the understanding of the small-cetacean situation in Japan and on the regulation measures to be taken.

Ministerial Order No. 92 of 2001 placed fishery and incidental take of small cetaceans under legal control, and No. 77 of 2004 placed stranded cetaceans under legal control. These actions streamlined the previous patchy management measures for cetacean species affected by whaling, dolphin and porpoise fisheries, incidental mortality, and stranding. However, disagreement or ambiguity of interpretation remains between the Ministerial Orders and circulars of the Director General of the Fisheries Agency issued simultaneously with the orders (Section 5.8).

The fixed trap-net fishery is, in principle, a passive fishery of limited selectivity, where every marine organism found in the net is processed or handled in accordance with its commercial value. The Coastal Division of the Department of Fishery Promotion apparently refused for many years to apply this basic principle of the fixed trap-net fishery to cetaceans. This could have been because the difficulty of controlling what is caught in a trap-net fishery was recognized. Whales are managed in principle to allow takes according to their current population levels and expected increase rates of the stocks. However, fixed trap nets take a particular proportion of marine organisms migrating through the vicinity, and the take rate depends on the behavior of the species as well as characteristics and location of the gear. If a fixed trap-net fishery were forced to modify operations to minimize the catch of most depleted whale species, it could be lethal to the fishery. Japanese fixed trap nets would have to share quotas with small-type whaling (which were actually zero) if whales were identified as a legitimate target. The status of the western gray whale stock was much worse than that of the minke whale, and no incidental takes would be accepted for the trap-net fishery. The solution for this situation was probably to consider that cetaceans were not the target of a fixed trap-net fishery and to prohibit commercial use of the carcasses found in the net.

However, there was another, socially related problem, where a minke whale taken in a fixed trap net could be sold for several million yen (100–150 yen/US$ in the 1990s), and many such whales went into the underground market. This was determined by the observation of a limited number of trap-net operations (Tobayama et al. 1992) and DNA analysis of market samples (Baker et al. 2000; Delebout et al. 2002), and the Scientific Committee of the IWC questioned the reliability of the statistics. Newspapers also reported such cases. As the Fisheries Agency was unable to deny such facts or to further enforce the prohibition rule, it was obliged to accept the status quo and issued Ministerial Order 92 of 2001, which allowed "take and sale" of whales found in fixed trap nets. A circular of the Director General of the Fisheries Agency was issued on the date same as Ministerial Order No. 92, and it had the apparent intention to restrict "take" of whales found in trap nets in a way that would be accepted internationally.

Recent Japanese fixed trap-net fisheries report annual takes of over 120 minke whales and occasional takes of humpback whale, North Pacific right whale, and western gray whales. The population of the last is estimated to consist of about 130 individuals in total, making incidental takes a great concern for its management. Japan has taken almost no action to reduce the incidental take of cetaceans in fixed trap nets, drift gill nets, and bottom gill nets. Since 2008, the Fishery Resource Protection Act has prohibited take, sale, and possession of gray whales, but it does little more than to say "gray whales must be released if found alive in the net," which is insufficient as a measure to reduce incidental mortality of the species (Section 5.9). A considerable number of finless porpoises are incidentally killed by the bottom gill-net fishery in the Inland Sea, and such mortality is believed to be one of the causes of the population decline (Kasuya et al. 2002; Section 8.4.5.1). Not even estimation of the incidental mortality has been attempted by the government. Japanese fishery management policy should give more attention to reducing incidental mortality of cetaceans in fishing gear. This, in my view, is an obligation of fisheries that exploit the marine ecosystem.

6.5 CATCH QUOTAS FOR DOLPHINS AND PORPOISES

In 1989, when the Fisheries Agency placed the hand-harpoon fishery under a license system, the Coastal Division made a request to the prefecture governments to advise hand-harpoon fisheries and drive fisheries under their control not to take finless porpoises and narwhals (original request not seen but recorded in a document of the Coastal Division; Kasuya 1994, in Japanese). Because there were no fishing activities for these species in Japan, this request or advice functioned only as a precautionary action or a first step toward future intervention in the fishery management. This was followed by Circular No. 2-1050 of September 20, 1990, that required that narwhals, finless porpoises, and killer whales be released or buried depending on the condition at the time of capture. The list of prohibited species was further expanded in 1991 (see Section 6.3).

The current dolphin-drive fishery at Taiji started in 1969 with 8 vessels, joined by a second group (7 vessels) in 1979, and then formed one team of 15 vessels in 1982 when the fishery was placed under the license system of Wakayama Prefecture. The fleet decreased to 14 in the 1984/1985 season; this was reflected in the license in 1986. The fishing season has been decided by species since 1982 (for details see Section 3.9). The first autonomous catch limit was decided for the season that started in autumn 1982 as 500 short-finned pilot whales and 5000 other dolphins (species not specified); this became a condition attached to the license issued in 1992 and was retained for later years. However, a statement attached to the license said, "If necessary there will be additional regulations," allowing for occasional modification of catch quotas. The Coastal Division of the Fisheries Agency imposed smaller quotas for dolphins (other than short-finned pilot whales) on the Taiji fishery: 3500 dolphins for 1989, 3324 for 1990, and 3100 for 1991 (Kasuya unpublished record dated January 1991). The last figure was proposed and could have been negotiated at a later stage as noted in Kishiro and Kasuya (1993), which listed the quota for the 1991 season as 2900 dolphins, including up to 500 short-finned pilot whales. In addition, a live take of up to five killer whales was allowed for aquaria. For the next season that started in 1992 the quota was further reduced to 2500 total dolphins, including up to 300 short-finned pilot whales and up to 1000 striped dolphins (Kishiro and Kasuya 1993).

In my view, the fishermen of Taiji totally ignored an additional condition attached to the license stating that "any animals smaller than 2 m must be released." I was once asked by the Director General of the Taiji FCU if such a regulation had any meaning for the dolphin population, but I could not identify any value in it and so replied. I have seen that many individuals, particularly small ones, of some species (e.g., striped dolphin and southern-form short-finned pilot whale) did not live long in the harbor, presumably due to the stress received during the drive, and survival of juveniles seems to be dubious if released alone separated from the mother.

The dolphin-drive fisheries on the Izu coast were placed under the license system of Shizuoka Prefecture in 1951 (Section 6.1). The first catch quota for the fisheries was 730 dolphins in 1990 (no species specification), which was reduced to 657 in 1991. These quotas were given to Futo, the only dolphin-drive group remaining on the Izu coast. It was plausible that the Coastal Division of the Fisheries Agency proposed some catch limits for this drive fishery before 1990, but I was unable to confirm that.

Nago (26°35′N, 127°59′E) in Okinawa Prefecture had a long history of traditional dolphin-drive fishing and seemed to have received a prefecture license in 1982 at the same time as other dolphin-drive teams in Wakayama and Shizuoka Prefectures (see Section 6.2).

It was for the 1990 season that all the dolphin and porpoise fisheries first received quotas based on advice from the Fisheries Agency (Table 6.1). The quota for 1992 could have been similar to that for 1991, but the details are unknown. These quotas were not by species, but Wakayama put in place some loose regulations by species. Cetacean species listed in Group 1 of Circular No. 3-1022 (see Section 6.3) were either the targets of small-type whaling or protected species and were excluded as targets for the driving or hand-harpoon fisheries.

Some prefectures set additional conditions on the licenses for dolphin and porpoise fisheries. For example, Iwate Prefecture prohibited the take of baleen whales, sperm whales, narwhals, killer whales, finless porpoises, suckling calves and cows accompanied by calves in the hand-harpoon fishery (beginning in 1989). Chiba Prefecture in 1991 permitted only *iruka* [dolphins and porpoises] other than Dall's porpoise, finless porpoise, killer whale, and narwhal in the hand-harpoon fishery. (The narwhal was known off Japan by only one case in the eighteenth century [Otsuki 1795, in Japanese].) Wakayama Prefecture required release of animals smaller than 2 m, which was continued up to 1992. However, the real value of such supplemental conditions is dubious other than improving appearances to the outside. There were cases of driving of baleen whales off Taiji and hand harpooning of northern-form short-finned pilot whales off the Pacific coast of northern Japan.

Out of the eight species listed in Group 2 of Circular No. 2-1022, seven species other than Baird's beaked whale were accepted as targets of dolphin and porpoise fisheries, and the catch quotas were decided by stock and allocated to prefectures and type of fisheries beginning in 1993. The principles used in calculating the quotas and the products are listed in Table 6.2. The calculation was based on the mean values of abundance, which were estimated by geographical area by scientists of the Far Seas Fisheries Research Laboratory of the Fisheries Agency. These abundance figures were multiplied by hypothetical annual increase rates, that is, 4% (Dall's porpoise),

TABLE 6.1
Early Stage of Japanese Quotas on Dolphin and Porpoise Fisheries That Started in 1990 without Distinction by Species (Prefectures Are Arranged from North to South)

Fishery Type	Hand Harpoon		Drive		Small-Type Whaling	
Year	1990	1991	1990	1991	1990	1991
Hokkaido	1,287	1,000				
Aomori	10	10				
Iwate	19,000	9,600				
Miyagi	500	300				
Chiba	68	100				
Shizuoka			730	657		
Wakayama	100	100	3224	3100		
Nagasaki[a]			104	96		
Ehime	24	0				
Oita	99	0				
Okinawa[b]	100	100				
Total	21,188	11,210	4058	3953	154[c]	154[d]

Note: For species quotas for small-type whaling that started earlier, see Table 4.3. This table has been revised with the addition of new data available since the original Japanese language edition appeared in 2011.

[a] Quota for culling.
[b] Crossbow fishery.
[c] Quota for 4 vessels.
[d] Quota for 5 vessels.

TABLE 6.2
Procedure of Calculating Species Quotas for 1993 Season for Dolphin and Porpoise Fisheries, from Records of Coastal Division of Fisheries Agency and Kasuya (1997, in Japanese)

Species/Stocks	Stock Size	Rate of Increase (%)	Safety Factor	Allowance	Quota for 1993	Quota (Bold) or Actual Catch (in Parentheses) before 1993			
						1992	1991	1990	1989
Dalli-type[a]	226,000	4			9,000	(4,099)	(4,671)	(9,360)	(15,953)
Truei-type[a]	217,000	4			8,700	(8,166)	(6,457)	(12,442)	(13,095)
Unknown type							(6,506)		
Total, Dall's porpoise					17,600	**17,600**	**17,600**		
						(12,265)	(17,634)	(21,802)	(29,048)
Striped dolphin[b]	22,500	3		+50	725	**1,000**			
						(1,049)	(1,017)	(749)	(1,225)
Common bottlenose dolphin	35,100	3		+50	1,100	(171)	(409)	(1,298)	(377)
Pantropical spotted dolphin	30,100	3		+50	950	(636)	(153)	(6)	(129)
Risso's dolphin	42,000	3		+50	1,300	(91)	(298)	(82)	(13)
Southern-form pilot whale[c]	20,300	2		+50	450	**400**			
						(200)	(296)	(149)	(194)
False killer whale	5,000	2	0.5		50	(91)	(51)	(72)	(30)
Killer whale								(3)	(9)
Northern-form pilot whale[c]	*5,000*	*2*	*0.5*		*50*	*(46)*	*(43)*	*(10)*	*(50)*
Baird's beaked whale[d]	*3,900*	*2*	*0.75*		*60*	*(54)*	*(54)*	*(54)*	*(54)*

Note: Numbers in italics indicate species and stocks taken by small-type whaling and managed by the Offshore Division of the Fisheries Agency. Other species and stocks were managed by the Coastal Division.

[a] Main targets of hand-harpoon fishery in northern Japan were a *dalli*-type stock in the Sea of Japan–southern Okhotsk Sea and a *truei*-type stock in the western North Pacific–central Okhotsk Sea. The total quota for the 1993 season does not match the sum of quotas for the two types.

[b] Quota for striped dolphin started in 1992 at 1000.

[c] The northern and southern forms are geographical forms of the short-finned pilot whale hunted off Japan.

[d] Effort is made here to justify the quota of 60 Baird's beaked whales.

3% (striped dolphin, bottlenose dolphin, pantropical spotted dolphin, Risso's dolphin), and 2% (southern-form short-finned pilot whale, false killer whale, Baird's beaked whale). Then the product was either multiplied by safety factors of 0.5 or 0.75 or supplemented with an adjustment figure of 50, to obtain final quota figures.

The abundance estimates used were from Miyashita (1991) for Dall's porpoise and Miyashita (1993) for other species with modifications to include Okinawa area (see Tables 3.13 and 9.9). One of the basic problems with the abundance estimates is their great confidence intervals, which must be taken into consideration for safe management of stocks (Section 1.2). These figures, particularly for Dall's porpoises, may have been biased upward or downward due to the porpoises being attracted to survey vessels or avoiding them (Section 9.7). The latter study was obliged to combine sighting data for a 10-year period (1982–1991); thus, if abundance declined during the survey period, the estimates could have an upward bias. Another problem in using figures in the latter study for management comes from non-correspondence of the survey area and fishing area. The abundance estimate included waters nearly 200 nautical miles from shore, while driving was usually operated within 15–20 nautical miles from port. Therefore, there was a risk of overestimation of exploited stocks if mixing of coastal and offshore individuals is limited or if there are more than one stock in the area used for the abundance estimate.

Although it is generally accepted that population growth rate will increase with decreasing population level (Section 1.2), this has not been reliably applied to management. When asked by the Fisheries Agency, our scientists without firm scientific evidence suggested three values for the potential rate of increase rate; 3%, 2%, and 1% instead of the figures actually used by the Coastal Division. The Coastal Division calculations resulted in 33%, 50%, and 100% inflation, respectively, of quotas that could be derived from the population growth rate suggested by scientists. For some unstated purpose, 50 individuals were added to the results of calculations for several species: striped dolphin, bottlenose dolphin, pantropical spotted dolphin, Risso's dolphin, and southern-form short-finned pilot whale. This rider further inflated the figures calculated using the optimistic abundance estimates and population growth rates.

I believe that these dubious procedures for calculating quotas for 1993 were introduced to satisfy two needs: (1) to reach some predetermined figures and (2) to pretend that scientific procedures were being used. Safety factors of 0.5 and

TABLE 6.3
Allocation of Quotas to Each Dolphin-Drive Group and Hand-Harpoon Fishery of Prefectures in Central and Southern Japan (See Table 6.4 for Quotas for Hand-Harpoon Fisheries in Northern Japan)

	Chiba	Shizuoka	Wakayama		Okinawa		
Prefecture/Fishery Type	Hand Harpoon	Drive	Hand Harpoon	Drive	Crossbow	Total	Quota in Table 6.2
Period of operation for 2007/2008 season[a]	November 1–April 30	September 1–March 31	January 1–August 31	September 1–April 30	February 1–October 31		
Striped dolphin							
1993-2006	80	70	100	450		700	725
2007/2008	72	63	100	450		685	
2008/2009	64	56	100	450		670	
2009/2010	56	49	100	450		655	
2010/2011	48	42	100	450		640	
2011/2012	40	35	100	450		625	
2012/2013	32	28	100	450		610	
2013/2014	24	21	100	450		595	
Bottlenose d.[b]							
1993		75	50	940	10	1075	1100
1994-2006		75	100	890	10	1075	
2007/2008		71	95	842	9	1017	
2008/2009		67	89	795	9	960	
2009/2010		63	84	748	8	903	
2010/2011		59	79	700	8	846	
2011/2012		55	73	652	7	787	
2012/2013		51	68	604	7	730	
2013/2014		47	63	557	6	673	
Pantropical spotted d.							
1993		455	50	420		925	950
1994-2006		455	70	400		925	
2007/2008		409	70	400		879	
2008/2009		365	70	400		835	
2009/2010		318	70	400		788	
2010/2011		272	70	400		742	
2011/2012		227	70	400		697	
2012/2013		181	70	400		651	
2013/2014		136	70	400		606	
Risso's d.[c]							
1993			200	350		550	1300
1994-2006			250	300		550	
2007/2008			246	295		541	
2008/2009			242	290		532	
2009/2010			238	286		524	
2010/2011			234	280		514	
2011/2012			230	275		505	
2012/2013			226	270		496	
2013/2014			222	265		487	
Southern form[c]							
1993-2006				300	100	400	450
2007/2008				277	92	369	
2008/2009				254	85	339	
2009/2010				230	77	307	
2010/2011				207	69	276	
2011/2012				184	61	245	
2012/2013				161	53	214	
2013/2014				150	46	196	

(*Continued*)

TABLE 6.3 (*Continued*)
Allocation of Quotas to Each Dolphin-Drive Group and Hand-Harpoon Fishery of Prefectures in Central and Southern Japan (See Table 6.4 for Quotas for Hand-Harpoon Fisheries in Northern Japan)

	Chiba	Shizuoka	Wakayama		Okinawa		
Prefecture/Fishery Type	Hand Harpoon	Drive	Hand Harpoon	Drive	Crossbow	Total	Quota in Table 6.2
False killer w.[c]							
1993-2006				40	10	50	50
2007/2008 to 2013/2014		10		70	20	100	
Pacific white-s. dolphin							
2007/2008 to 2013/2014		36	36	134		206[d]	
Total[c]							
1993	80	600	400	2500	120	3700	4575
1994-2006	80	600	520	2380	120	3700	
2007/2008	72	589	547	2468	121	3797	
2008/2009	64	534	537	2393	114	3642	
2009/2010	56	476	528	2318	105	3483	
2010/2011	48	419	519	2241	97	3324	
2011/2012	40	363	509	2165	88	3165	
2012/2013	32	306	500	2089	80	3007	
2013/2014	24	250	491	2026	72	2863	

Note: Wakayama Prefecture quota modified in 1994 in allocation between drive fishery and hand-harpoon fishery. See Table 4.3 for quotas given to small-type whaling. This table has been revised with the addition of new data available since the original Japanese language edition appeared in 2011.

[a] Until the 1995, season quotas were allocated for the calendar year; since 1997 (*truei*-type Dall's porpoise) and 1996 (other stocks), they were given for single continuous operation periods, August 1 to July 31 of the next year (Dall's porpoise) or October 1 to September 30 of the next year (other species). An interim adjustment measure was applied for the earlier half of the 1996 season. In 2006, Wakayama Prefecture altered the season to September 1 to August 31. Prefectures were allowed to modify the actual operation period within these limits. The table shows the actual operation period for the 2007/2008 season. The catch statistics continued to cover a calendar year.

[b] Thought to be common bottlenose dolphin.

[c] Quotas received by small-type whalers are additional to the figures in Table 6.3 (see Table 4.3).

[d] Hand-harpoon fishery in Iwate Prefecture received a quota of 154 Pacific white-sided dolphins (see Table 6.4).

0.75 were applied to false killer whales and Baird's beaked whales, respectively. These safety factors effectively reduced the quota close to the actual catch level of the time (false killer whale) or the prior catch quota (Baird's beaked whale). The Coastal Division seems to have had a hidden objective to create 1993 quotas which were similar to the actual catches of the previous years.

The catch quotas thus created (Table 6.2), which were reported to the Scientific Committee of the IWC, were divided by prefectures and fishery types (Tables 6.3 and 6.4), leaving small reserves for the Fisheries Agency. Thus, figures in Table 6.2 often exceed the totals of Tables 6.3 and 6.4. The reserve for the Fisheries Agency functioned as a kind of safeguard against violation of the catch limit by some prefectures. These quotas, with some minor modification, were retained until 2008. Since the 2007/2008 season, a total of 360 Pacific white-sided dolphins has been added to the quota and allocated to various fisheries (Tables 6.3 and 6.4).

It should be noted that since 1991 the Japanese Dall's porpoise hunters have received a total quota of 17,600 porpoises of both color types, which was identical to the quota "scientifically calculated" for 1993. Here we see the reason why the increase rate of the species had to be 4%. This led to the assumption of a 3% increase rate for the striped dolphin, which has a longer calving interval than the Dall's porpoise. The adjustment of 50 striped dolphins had to be added to reach a figure close to 1000, which was a quota accepted since 1992 for the species. The safety factor of 0.75 was applied in calculating the quota of Baird's beaked whales. It was necessary to get a figure of 60 individuals, which was a quota accepted by the small-type whalers since 1988 (Kasuya 2007).

The mathematics used by the Coastal Division in creating catch quotas for Japanese small-cetacean fisheries in 1993 might appear to be based on biology, but in my view they were a tool invented to maintain earlier catch regulations accepted by the fisheries and, at the same time, to defend against

TABLE 6.4
Allocation of Quotas to Hand-Harpoon Fisheries of Prefectures in Northern Japan (See Table 6.3 for Quota Allocation in Central and Southern Japan)

Prefecture	Hokkaido	Aomori	Iwate	Miyagi	Total	Quota in Table 6.2
Period of operation for 2007/2008 season[a]	May 1–June 15, August 1–October 31	November 1–April 30	November 1–April 30	November 1–April 30		
Dalli-type						
1993-2006	1500	20	7,200	280	9,000	9,000
2007/2008	1451	18	6,969	269	8,707	
2008/2009	1399	16	6,721	260	8,396	
2009/2010	1348	14	6,472	250	8,084	
2010/2011	1296	12	6,225	241	7,774	
2011/2012	1244	10	5,975	231	7,460	
2012/2013	1192	8	5,726	221	7,147	
2013/2014	1141	6	5,478	212	6,837	
Truei-type						
1993-2006	100		8,300	20	8,420	8,700
2007/2008	98		8,054	16	8,168	
2008/2009	95		7,805	16	7,916	
2009/2010	92		7,557	15	7,664	
2010/2011	89		7,108	215	7,412	
2011/2012	86		6,860	214	7,160	
2012/2013	83		6,611	214	6,908	
2013/2014	80		6,363	213	6,656	
Pacific white-s. dolphin						
2007/2008 to 2013/2014			154		154[b]	
Total						
1993-2006	1600	20	15,500	300	17,420	17,600
2007/2008	1549	18	15,177	285	17,029	
2008/2009	1494	16	14,680	276	16,466	
2009/2010	1440	14	14,183	265	15,902	
2010/2011	1385	12	13,487	456	15,340	
2011/2012	1330	10	12,989	445	14,774	
2012/2013	1275	8	12,491	435	14,209	
2013/2014	1221	6	11,995	425	13,647	

Note: See Table 4.3 for quotas given to small-type whaling.
[a] See footnote a to Table 6.3.
[b] A quota of 206 Pacific white-sided dolphins given to fisheries in central and southern Japan.

domestic criticism. Such mathematics would have never been understood by the IWC if presented to the Scientific Committee (see Section 12.6.3.2 for additional information).

The catch quotas have been given to the fishermen on an annual calendar basis since at least 1993. The system was felt inconvenient by some fisheries that operated through the northern winter and was modified to be given for the fishing season since autumn of 1997 (hand-harpoon fishery for *truei*-type Dall's porpoises) or 1996 (most other fisheries), with some transition measures for preceding years (see Tables 6.3 and 6.4 for the fishing season). Although quotas are given for a fishing season that usually starts after summer and ends before the next summer, the catch statistics are compiled by calendar year. Thus, some apparent disagreement between annual quotas and catches does not necessarily indicate violation of the regulations.

The Far Seas Fisheries Research Laboratory of the Fisheries Agency reported the regulations on small-cetacean fisheries and their catch statistics as a part of the Japanese annual progress report on cetacean research to the Scientific Committee of the IWC until 1999 but discontinued this beginning in 2000. They are now available on the website of the Fisheries Agency (e.g., Status of International Fishery Resources. Accessed 2013.).

REFERENCES

[*In Japanese Language]

Baker, C.S., Lento, G.M., Cipriano, F., Dalebout, M.L., and Palumbi, S.R. 2000. Scientific whaling: Source of illegal products for market? *Science* 290: 1695.

Delebout, M.L., Lento, G.M., Cipriano, F., Funahashi, N., and Baker, C.S. 2002. How many protected minke whales are sold in Japan and Korea? A census by microsatellite DNA profiling. *Animal Conserv.* 5: 143–152.

*Iida, K., Yoshida, Y., Yamamoto, K., and Misawa, D. 1994. *History of Catcher Boat Operations [in the North Pacific Salmon Factory Ship Fishery]*. Vol. 3. Nihon Sakemasu Gyogyo Kyodo Kumiai, Tokyo, Japan. 316pp.

Kasuya, T. 1985. Effect of exploitation on reproductive parameters of the spotted and striped dolphins off the Pacific coast of Japan. *Sci. Rep. Whales Res. Inst.* 36: 107–138.

*Kasuya, T. 1994. Finless porpoise. pp. 626–634. In: *Basic Data on Rare Wild Aquatic Organisms of Japan (I)*. Fisheries Agency, Tokyo, Japan. 696pp.

*Kasuya, T. 1997. Current status of Japanese whale fisheries and management of stocks. *Nihonkai Cetol.* 7: 37–49.

*Kasuya, T. 2003. Sea of whales and sea of humans. pp. 61–72. In: *Environment Yearbook*. Sodo Sha, Tokyo, Japan. 340pp.

*Kasuya, T. 2005. On the whaling issue. *Ecosophia* 16: 56–62.

Kasuya, T. 2007. Japanese whaling and other cetacean fisheries. *Environ. Sci. Pollut. Res.* 14(1): 39–48.

Kasuya, T., Yamamoto, Y., and Iwatsuki, T. 2002. Abundance decline in the finless porpoise population in the Inland Sea of Japan. *Raffles Bull. Zool.* 10(Suppl.): 57–65.

Kishiro, T. and Kasuya, T. 1993. Review of Japanese dolphin drive fisheries and their status. *Rep. Int. Whal. Commn.* 43: 439–452.

Miyashita, T. 1991. Stocks and abundance of Dall's porpoises in the Okhotsk Sea and adjacent waters. IWC/SC/43/SM7. 24pp. Document presented to the *43rd Meeting of the Scientific Committee of IWC* (available from IWC Secretariat, Cambridge, U.K.).

Miyashita, T. 1993. Abundance of dolphin stocks in the western North Pacific taken by the Japanese drive fishery. *Rep. Int. Whal. Commn.* 43: 417–437.

*Otsuki, G. 1795. New information on six items. pp. 7–152. In: *Rokubutsu Shinshi and Ikkaku Iko [New Information on Six Items and Miscellanea of the Narwhal]* (Reprinted in 1980 by Kowa Shuppan, Tokyo, Japan. 432+17pp.).

Status of International Fishery Resources. 2012. Comprehensive Fisheries Research Center, Fisheries Agency. http://www.jfa.maff.go.jp/kokushi_hp/toppage.htm. Accessed August 30, 2013.

*Suisan-cho Kaiyo-gyogyo-bu Sanjikan. 1986. *Information on Small Cetaceans (Provisional)*. Suisan-cho Kaiyo-gyogyo-bu, Tokyo, Japan. 46pp.

Tobayama, T., Yanagisawa, F., and Kasuya, T. 1992. Incidental take of minke whales in Japanese trap nets. *Rep. Int. Whal. Commn.* 42: 433–436.

7 Moving toward the End of Commercial Whaling

Pelagic, or mother-ship, whaling was prohibited in the North Atlantic in 1937 and in the North Pacific in 1980, leaving such operations allowed only for minke whales in the Antarctic. There were three types of whaling in Japan when the moratorium on commercial whaling was adopted at the annual meeting of the International Whaling Commission (IWC) in 1982: Antarctic pelagic whaling for minke whales, large-type whaling for sperm whales and Bryde's whales, and small-type whaling for minke whales, Baird's beaked whales, and various small cetaceans. The latter two types of whaling operated using land stations and were called "coastal whaling."

Japanese coastal whaling for sperm whales realized increasing difficulty in operations, which together with increasing distance to the fishing ground gave an impression that the status of the sperm whale stock could be deteriorating. However, the Scientific Committee of the IWC could not reach agreement on the status of the stock, and in July 1981, it decided that member nations could not take sperm whales in the North Pacific unless the Commission reached at an agreed quota for the stock. Japan raised an objection against this decision on November 9, 1981, and declared that it would operate the fishery under its own judgment. Later, in 1984, the IWC was unable to agree on a catch quota for the stock for the 1985 whaling season, and Japan was expected to operate sperm whaling in the season.

The moratorium on commercial whaling was proposed annually to the IWC beginning in 1972 and passed with a three-fourths majority vote in 1982. The moratorium was to come into effect in the 1985/1986 Antarctic season and 1986 summer season in the northern hemisphere. The Japanese government raised an objection to this decision. In June 1984, prior to the moratorium coming into effect, the IWC decided on a quota of 4224 minke whales for the 1984/1985 Antarctic season, which was 63% of the quota for the previous season. Japan also filed an objection to this decision.

However, there was a possibility that the United States would stop importation of Japanese fishery products (based on the "Pelly Amendment") or stop issuing permits for Japanese fisheries in the U.S. EEZ (based on the "Packwood-Magnusson Amendment") if Japan continued whaling despite the decision of the IWC to use the right of objection allowed by the International Convention for the Regulation of Whaling (ICRW).

The Japanese government negotiated with the United States to avoid such a response and agreed with the United States to withdraw the objections. The subsequent actions of Japan are listed as follows:

1. Japan took in the 1984/1985 Antarctic season 1941 minke whales, or a reduction to 64% of the catch of the previous season as determined by IWC, while USSR and Brazil pursued the catch of the previous season. This functioned for Japan as if it had withdrawn the objection to the IWC quota.
2. With regard to the objection of 1981 on the coastal sperm whale fishery, Japan informed the IWC on December 11, 1984, of withdrawal of its objection effective April 1, 1988. Operation of three seasons before the withdrawal, that is, the 1985–1987 summer seasons, had been agreed with the United States (IWC 1986).
3. On the objection to the moratorium on commercial whaling, Japan notified the IWC of its withdrawal on June 1, 1986, effective on May 1, 1987, for Antarctic minke whaling, on October 1, 1987, for North Pacific whaling for minke and Bryde's whales, and on April 1, 1988, for North Pacific sperm whaling. These dates were the next to last days of the 1987/1988 Antarctic season and the 1987 North Pacific summer season.

In summary, Japan decided to end all whaling for large cetaceans by the end of March 1988 (Kasuya 2005, in Japanese, 2007).

7.1 SMALL-TYPE WHALING

7.1.1 Brief History

Japan had a type of whale fishery to take minke whales and toothed whales other than sperm whales using small-caliber harpoon cannons mounted on small vessels. This fishery was called *gondo-sen* or *tento-sen* in Wakayama Prefecture or *minku-sen* in broader Japan and later called *kogata-hogei* [small-type whaling] (Section 4.1). The word "small-type whaling" appeared first in 1947 in the Rule for Regulation of Steam Ship Whaling (Sections 5.4 and 5.5). This fishery was regulated since 1920 in Chiba Prefecture by prefecture license (Section 4.1) but was unregulated in other prefectures until 1947 (Omura et al. 1942; Maeda and Teraoka 1958, both in Japanese; Ohsumi 1975).

Japan joined World War II in December 1941, and all six Japanese whaling factory ships were sunk by the enemy by May 1944 after being commandeered by the navy as oil tankers. Many large-type catcher boats had similar fates; 67 vessels were lost, leaving only 28 for whaling activity at the end of the war, including two vessels salvaged after the war. While this destruction was in progress, beginning in January 1944 the government allocated five small-type whale catcher boats to each of the three fishing companies (15 catchers in total) that were established especially for wartime whaling and allowed these small-type whalers to take large whales that had previously been allowed to be taken only by large-type whalers. The exact date of this allocation was either January 1 (Kushiro-shi Somu-bu Chiiki Shiryo Shitsu [Regional History Section of Kushiro City] 2006, in Japanese) or January 15, 1944 (Maeda and Teraoka 1958, in Japanese). Although the duration of this special allocation was scheduled to last 1 year, record of its termination was not available to me. It could have continued until the end of the war in August 1945. Kajino (1958, in Japanese) wrote that small-type whalers off Taiji and Sanriku (Pacific coast in 37°54′N–41°35′N) started illegal hunting of large whales during the war, which caused a dispute with large-type whalers involving government sections. The wartime special allocation could have been in a sustained status (Section 13.5.3).

The number of small-type whaling vessels was about 20 before World War II but was increasing after the war. In order to control the fishery, the Japanese government issued the Rule for Regulation of Steam Ship Whaling (Ministerial Order No. 91), which defined two types of whaling (small-type whaling and large-type whaling; Sections 5.4 and 5.5) and required small-type whaling vessels and their land stations to obtain permits from the Minister of Agriculture and Forestry.

The maximum size of the cannons allowed was modified several times. It was first 40 mm from March 1, 1948, and increased to 50 mm in 1952 (Ohsumi 1975). However, there were cases of pilot whale hunters that used 3- or 5-barrel cannons of 20–25 mm caliber or double-barreled cannons of 36 mm caliber. These multi-barrel whaling cannons simultaneously discharged multiple harpoons connected to a single main whaling line (see Section 3.9.1).

The maximum vessel size allowed by the regulation of 1947 for small-type whaling was 30 gross tons, with transitional exceptions for larger vessels until December 1951 (Maeda and Teraoka 1958, in Japanese). On condition of combining existing licensed tonnage, in 1963 it was permitted to build larger vessels of up to 40 gross tons, and a later act for the improvement of cabin capacity further increased the size limit to 50 gross tons (Ohsumi 1975). The number of licensed vessels increased slightly from 73 in 1948 to 80 in 1950 and then started to decrease (Ohsumi 1975; also see Table 4.1). The methods used by the government to reduce the fleet of small-type whalers were in principle to shift their hunting to large whales, that is, to (1) construct a large-type whaling catcher boat of 350 gross tons by assembling the licensed tonnage of small-type whalers and letting it participate in mother-ship whaling for sperm whales in the North Pacific (in 1960), (2) increase the tonnage of large-type whaling catcher boats in the same way and let the owners shift to large-type coastal whaling (in 1961), and (3) construct a large-type whaling catcher boat in the same way and let it join the large-type whaling of Japan (in 1968) (Ohsumi 1975). Here, Japanese whaling policy repeated the approach to reducing fleet size that it used for large-type whaling. But the effect on the small-type whaling fleet was worse because it counteracted the continuing effort since the early twentieth century to reduce the fleet of large-type whalers (Section 7.2).

The ICRW regulates the season for baleen whales to be within 6 contiguous months. Thus, small-type whalers were allowed in 1952 to hunt minke whales from February 1 to July 31 (Maeda and Teraoka 1958, in Japanese), but this was changed to 6 months from April 1 to September 30 at some stage before 1975 (Enyo-gyogyo-ka [Division of Offshore Fisheries] 1977, in Japanese). This fishery had no minimum size limit. Based on Article 15 of the Rule for Regulation of Steam Ship Whaling which stated that "the Minister of Agriculture and Forestry can, if required, regulate whale species, whaling season, area of whaling operation, and number of vessels," Ministerial Notifications No. 200 of 1938 and No. 263 of 1948 were issued for the purposes of (1) prohibiting take of gray whales and right whales (right whales and bowhead whale), (2) setting minimum size limits that varied by species and by the use of the product (smaller size was allowed if it was for local consumption as human or animal food), and (3) prohibiting take of suckling calves and females accompanied by calves. Provisions (1) and (2) did not include whale species allowed for small-type whaling, but (3) did not state whale species and so could have been applied to all whale species, including those taken by small-type whaling. However, I have never seen the small-type whalers paying attention to it.

The IWC set the annual quota for minke whales off Japan beginning in the 1977 season. It became a block quota for 5 years in 1985, with limits to the total of the 5 years as well as annual catch limits, where the annual quota was slightly greater than the 5-year average. The Japanese government set autonomous catch limits of 40 Baird's beaked whales in 1983, which was increased several times (see Table 13.15). After a pause of 6 years, in 1982 the small-type whalers recommenced hunting of northern-form short-finned pilot whales and operated without a catch limit. The whalers reported 85 pilot whales taken in 1982, but responding to my distrust in the figure they corrected it to 172, which has not been verified (footnote f to Table 4.3). The catch limit for the 1983 and 1984 seasons was 175 and decreased to 50 for 1986. At the request of the Agency scientists, no catch limit was set for 1985, as they distrusted the catch reports and wished to control the catch by regulating effort. Behind this move was my discovery in Ayukawa (38°18′N, 141°31′E, Pacific coast) in 1984 that pilot whales were processed secretly at night at the land station of Gaibo Hogei Co. Ltd. (Kasuya and Tai 1993).

7.1.2 Actions Related to the Moratorium on Commercial Whaling

In 1986, Japan announced that it would accept the IWC decision for a moratorium on commercial whaling from the 1987/1988 Antarctic season and 1988 summer season, which meant the end of minke whale hunting by small-type whalers. In order to transition to the new situation and with an optimistic expectation of allocation of temporary emergency quotas of minke whales for the fishery in the near future, the small-type whalers decided to refrain from hunting northern-form short-finned pilot whales in 1987 and to carry over the quota of 50 animals to the 1988 season (total of 100 to be taken in 1988). In addition to this decision, they presented the following requests to the Fisheries Agency (based on a document of the Division of Far Seas Fisheries, Fisheries Agency dated August 1988).

1. The Baird's beaked whale quota of 40 should be increased to 90.
2. The catch of 100 northern-form short-finned pilot whales agreed upon for the 1988 season should be increased to 270 and be maintained at that level for subsequent years. The fishing season for the stock (October 5 to November 18) should be altered to 90 days from September 1, and onboard processing of the catch should be permitted.
3. A quota of 300 southern-form short-finned pilot whales should be allocated (no regulation existed then), and onboard processing of the catch should be permitted.

Responding to concerns of the IWC about the management of the Baird's beaked whale fishery, the Japanese government decided on an autonomous catch quota of 40 animals beginning in the 1983 season, which was the average of the previous 10 years (Table 13.14). Responding to the request by the small-type whalers, the Fisheries Agency increased the catch quota of the species to 60 for the 1988 season (57 were taken in the season). At the IWC, the government explained that this was a special case for a single year to remedy the difficulty of the whalers who had to adjust themselves to operation without a minke whale quota, but this inflated figure was retained in subsequent years. The decrease of the Baird's beaked whale quota to 54, which occurred in 1989, was to adjust for a decrease in the number of licensees from 15 to 14 in 1989. A new quota of 8 whales for the operation in the Sea of Japan resulted in a total quota of 62 Baird's beaked whales in 1999. In 2005, the quota increased to a total of 66 whales, of which 10 were allocated to the Hakodate station (41°46′N, 140°44′E, used for operation in the northern Sea of Japan), 4 to the operation off Abashiri (44°01′N, 144°16′E, on the southern coast of the Okhotsk Sea), and 26 to Ayukawa and 26 to Wadaura (35°03′N, 140°01′E on the Pacific coast) (Table 13.15).

Considering the request of the small-type whalers for the two stocks of short-finned pilot whales, the quota increase for the northern-form stock was rejected. Scientists of the Far Seas Fisheries Research Laboratory expressed an objection to the requested increase with the belief that, based on the post–World War II operation, even if 200–300 northern-form short-finned pilot whales might be taken in the first few years, such a large catch would not be sustainable over a longer period. Therefore, the whalers were allowed, as agreed upon earlier, to hunt 2 years' quotas in 1 year, that is, 100 whales in 1988. In order to allocate southern-form short-finned pilot whales to small-type whalers, it was necessary to determine the capacity of the stock and then divide the catch between the small-type whalers and the other existing fishery for the stock, the drive fishery at Taiji (33°36′N, 135°57′E). So, as an experiment, the Fisheries Agency allowed the small-type whalers free operation in the 1988 season for southern-form short-finned pilot whales. Their operation ended with a take of 98 northern-form whales and 30 southern-form, which was indicative of the upper limits available to them with their hunting ability and the population size of the whales. In 1989, the small-type whalers received a quota of 50 southern-form finned pilot whales.

The small-type whalers requested permission to process their catches on board, which appeared reasonable from the point of view of their operational economy, and there was an opinion in the Fisheries Agency to allow it after consulting with the Central Fishery Adjustment Committee of the year, but the result went in the other direction for reasons unknown to the scientists. Before the decision was made, we scientists of the Fisheries Agency expressed our strong opinion that onboard processing of the carcasses was definitely unacceptable because (1) it was almost impossible to find a sufficient number of reliable observers to be placed on all the vessels, (2) it would create difficulty in biological examination of the whales taken, and (3) it would likely create opportunities for poaching and misreporting of the catch. We had seen cases where Fisheries Agency inspectors often did not sufficiently pursue their duty at whaling land stations (Section 15.3.2; Kasuya 1999; Kondo 2001, in Japanese).

In the 1987 season, 9 small-type whaling vessels of 8 companies operated with a quota of 320 minke whales and 40 Baird's beaked whales (a quota of 50 northern-form pilot whales was left for the next season). In 1988, 3 vessels left the fleet, and 8 companies operated 6 vessels. A quota of 60 Baird's beaked whales was allocated to 5 vessels: *Sumitomo-maru No. 31* (Gaibo Hogei), *Yasu-maru No. 1* (Y. Shimomichi and Taiji FCU), *Koei-maru No. 75* (Y. Toba and Seiyo Gyogyo), *Katsu-maru* (T. Isone), and *Taisho-maru No. 2* (K. Okuta and Miyoshi Hogei), and a quota of 100 northern-form pilot whales was allocated to the 4 vessels other than *Sumitomo-maru No. 31* and *Yasu-maru No. 1*. Their operation of 1988 also took 30 southern-form pilot whales under the open quota for the year. Compared with the operation of 9 vessels in the 1987 season that ended with 1318 million yen of proceeds and 238 million yen of profit, the operation of 1988 gained proceeds of 477.6 million yen with expenditures of 516.1 million yen and resulted in a 38-million-yen deficit (Table 4.3). The post-moratorium operation of small-type whaling started to make a profit in 1989, recorded a maximum profit of 115 million yen in 1996, and again started to record a deficit in 2001. Their annual deficit continued to increase through 2006, the last year for which the data were available to me. All the small-type whalers recorded a profit in 2000, some recorded a deficit in 2001, and all the companies reported deficits beginning in 2002 (Table 4.3). Present small-type whaling is operated by 8 whalers using 5 catcher boats (Tables 4.3 and 7.1), but their operation is probably not sustainable with only the traditional commercial hunting of small toothed whales.

TABLE 7.1
Japanese Small-Type Whaling Fleet and Operational Pattern in the 2007 Season

Vessel	Gross Tons	Owner	Jointly Operating with	Area, Season,[a] and Species[b] of Operation	Quota[a]	Scientific Whaling in Coastal Japan for Minke Whales
Koei-maru No. 75	46.24	Toba Hogei	Seiyo Gyogyo	Sanriku: June–August, Bbw, Pwn Sanriku: October–December, Pwn	Bbw: 14 Pwn: 18	Sanriku: April–May Kushiro: September–October
Taisho-maru No. 28	47.31	Nihon Kinkai, Miyoshi Hogei		Sanriku: June–August, Bbw, Pwn Abashiri: August–September, Bbw	Bbw: 14 Pwn: 18	Sanriku: April–May Kushiro: September–October
Katsu-maru No. 7	32.00	I. Isone	Y. Shimomichi	Taiji: May–September, Pws, Rsd Chiba: June–August, Bbw, Pws Abashiri: August, Bbw	Bbw: 14 Pws: 14 Rsd: 7	Sanriku: April–May Kushiro: September–October
Sumitomo-maru No. 31	32.00	Gaibo Hogei	*Sumitomo-maru No. 21*	Taiji: May–September, Pws, Rsd Chiba: June–August, Bbw, Pws	Bbw: 14 Pws:14 Rsd: 6	Sanriku: April–May Kushiro: September–October
Seiwa-maru	15.20	Taiji Fish. Co-op. Union		Taiji: May–September, Pws, Rsd S. of Japan: May–June, Bbw	Bbw: 10 Pws: 8 Rsd: 7	
Taisho-maru	42.35	Seiyo Gyogyo		No operation		
Undecided		Miyoshi Hogei, Nihon Kinkai		No operation		
Yasu-maru No. 1	44.55	Y. Shimomichi		No operation		
Sumitomo-maru No. 21	22.00	Gaibo Hogei		No operation		

[a] The fishery took northern-form short-finned pilot whales in autumn to early winter until 2006, but in 2007 it changed the season as indicated earlier in order to adjust to operations of scientific whaling.

[b] *Species Key*: Bbw, Baird's beaked whale; Pwn, Short-finned pilot whale of northern form; Pws, Short-finned pilot whale of southern form; Rsd, Risso's dolphin.

One cause of the deficits in the operation of small-type whaling vessels is the recent decline of price of their products. In order to demonstrate the price change, I will compare proceeds from operations at Hakodate and Abashiri for the years 1998–2005. These land stations processed only Baird's beaked whales during the period. I have combined the two locations to increase sample size, by ignoring (1) the proportions of animals landed at the two locations that fluctuated between 2 (Abashiri):8 (Hakodate) and 4:10, and (2) the fact that Abashiri always recorded a 10%–20% higher per-animal price. The average proceeds from one Baird's beaked whale came to 10.6 million yen in 1998, the highest in the period. Then, it declined to 8.48 million yen in 2000, the last year when the total small-type whaling fishery recorded a profit, to 8.03 million yen in 2001, and to 4.57 million yen in 2005. The value of a Baird's beaked whale declined by more than half during the 7 years.

Although a decline in demand for whale meat in Japan could have been behind the change in the price for Baird's beaked whale meat, it must not be ignored that there was an increased supply of baleen whale meat from Japanese scientific whaling (see Table 1.5 and Section 7.4.3). Japanese scientific whaling started in the 1987/1988 Antarctic season with a proposed take of 300 minke whales and expanded to 500 ± 40 minke whales, 50 Bryde's whales, and 10 sperm whales in 2000 (North Pacific and Antarctic combined). Assuming that the yield of three minke whales is equivalent to one Bryde's whale, the supply of whale meat from scientific whaling doubled in 12 years. Since 2007, the Japanese scientific whaling has expanded to the proposed take of 1070 ± 85 minke whales, 100 sei whales, 50 fin whales, 50 humpback whales, 50 Bryde's whales, and 10 sperm whales, but the actual takes have often been much lower than proposed (Table 1.5).

Four small-type whaling catcher boats were first hired in 2002 for scientific whaling in the coastal waters to take 50 minke whales off Ayukawa on the Pacific coast. The next year the whaling expanded to Pacific waters off Kushiro (42°59′N, 144°23′E) to take an additional 60 minke whales (120 in total). The Japanese small-type whalers are probably surviving by being involved in the scientific whaling. Although the amount of money they receive from the participation is unknown to me, it will lessen the reported deficit in ordinary fishing activity for small cetaceans. The small-type whalers once enjoyed a profitable business at the beginning of the whaling moratorium but suffered because of the scientific whaling. Now they are unable to survive without being involved in scientific whaling. However, scientific whaling itself is suffering from a price decline and increase in back stock of whale meat, which started in 2000 and became evident in 2004 (Sakuma 2006, in Japanese).

In 2010, the coastal element of scientific whaling for minke whales was transferred from the Institute of Cetacean Research (ICR) to a group of small-type whalers. Seven small-type whalers formed a corporation called Chi-iki Hogei Suishin Kyokai [Association for Promotion of Local Whaling], which gets a catch permit for minke whales for scientific purpose and a financial subsidy of 265 million yen from the government. It entrusts biological studies of the catch to Tokyo Ocean University, the National Institute of Far Seas Fisheries (Fisheries Agency), and the ICR.

When Japan accepted the moratorium on commercial whaling, it could have been thought by some IWC members that Japan would end hunting of minke whales. However, the Japanese government raised the issue that Japanese small-type whaling had a noncommercial element that was similar to the aboriginal subsistent whaling defined by IWC and in 1988 requested a special minke whale quota for small-type whaling with the condition that the products would be consumed within the local communities of Abashiri, Ayukawa, and Taiji. The elements similar to aboriginal subsistent whaling were investigated by several cultural anthropologists with support from the Japanese government and the industry, and several reports were submitted to the IWC. They were printed in two volumes (Institute of Cetacean Research 1996; Government of Japan 1997) and also presented in Komatsu (2001, in Japanese).

The noncommercial elements identified included sharing of the first catch with nearby residents or presenting it to the local shrine. These were generally accepted as noncommercial elements, although they also occur in commercial whaling. However, it was not clear why such activities need a take of 50 minke whales. In response to the criticism by the IWC that the nutritional needs of the local people could be met by the products of the Antarctic scientific whaling, the Japanese government argued that the frozen meat from the Antarctic was not palatable to the local people.

Similar requests for a special minke whale quota were presented in various versions annually to the IWC, but they were not accepted as of 2013. Some criticisms were that the difficulty of the local communities was not due to the moratorium on commercial whaling but was a problem common to the modern community, or that a remedy to the difficulty of the local communities was a responsibility of the Japanese government that accepted the moratorium. However, there seemed to be a more difficult problem in connection with the Japanese government proposal for a special minke whale quota, which was how to restrict circulation of the whaling products within the local communities. It is evident to me that under the current Japanese legal and transportation systems, the movement of whale products across the borders of local communities cannot be blocked. A far greater obstacle may have been suspicion by IWC member countries about the intention of the Japanese government, that is, was Japan really going to accept the moratorium on commercial whaling, or was the special request an attempt to establish another loophole?

7.2 LARGE-TYPE WHALING

7.2.1 Before World War II

The Japanese attempt to take large whales in coastal waters using the Norwegian-type whaling method, large-type whaling as later defined, started in 1896. Enyo Hogei K.K. [Far Seas Whaling Co. Ltd.] of Nagasaki (32°44′N, 129°53′E) in northwestern Kyushu took the first whale in Tsushima Strait in April 1898, but success of the business had to wait until Nihon Enyo Gyogyo K.K. [Japan Far Seas Whaling Co. Ltd.] was established in 1899. Because of the initial success, 14 whaling companies were established by 1907, and the business became unstable due to overproduction and overcompetition. Under these circumstances there was movement toward amalgamation of the companies in 1908, when 12 whaling companies operated using 28 vessels (excluding two sailing-ship whalers). The number of whaling companies was halved in 1909 (Akashi 1910, in Japanese). After consulting the whaling industry and based on Article 9 of the Rule for Regulation of Whaling issued in 1909 (Section 5.2), in October 1909 the Ministry of Agriculture and Commerce issued Ministerial Notification No. 418 setting the upper limit of whaling vessels at 30.

The regulation only capped the whaling fleet at its existing size and did not reduce it. From this time forward, Japanese whaling policy pursued a reduction of the whaling fleet in tandem with the decline of whale populations, until 1988 when commercial hunting of large cetaceans ended. The excess of whaling vessels was exacerbated further by the loss of some whaling grounds, such as Taiwan, the Korean Peninsula, and the Kuril Islands after the war. There were 25 large-type whaling catcher boats licensed in 1934, 22 in 1963, 17 in 1965, 15 in 1967, 14 in 1968, 12 in 1970 (2 licenses were used to increase the tonnage of other vessels), 11 in 1974, 8 in 1976, 7 in 1977, and 5 in 1984 (retained to March 1988). The licensed boats did not all operate in all years (Kondo 2001, in Japanese).

Japanese whaling was classified into the three categories of pelagic whaling, large-type whaling, and small-type whaling (Chapter 5). Pelagic whaling was regulated by the Rule for Regulation of Mother-Ship Fisheries in 1934, and large-type whaling was regulated by the Rule for Regulation of Whaling in 1909, Rule for Regulation of Steam Ship Whaling in 1947, and Rule for Regulation of Specified Offshore Fisheries in 1950. The term *large-type whaling* appeared first in 1947 in the Rule for Regulation of Steam Ship Whaling. Large-type whaling was allowed in principle to take sperm whales and baleen whales other than minke whales, but additional prohibitions could be added depending on stock levels and for adjustment with other whale fisheries. These two whaling types did not have regulations on vessel size or caliber of whaling cannon and usually used catcher boats of several hundred gross tons and 75–90 mm whaling cannons. As large-type whaling used land stations for processing the catch, their fishing grounds were limited to within about 300 nautical miles (500–600 km) from the coast. Therefore, both large-type and small-type whaling were called coastal whaling.

Direct and indirect methods can be used in the management of whale stocks. The former regulate catch of whales by number, area, and species and are the most effective way to manage stocks, but the latter methods, regulating fishing area and season or number of vessels, were used by Japan. Japan first limited fleet size for large-type whaling before World War II and then set minimum body lengths but did not protect any species. Although pelagic whaling in the Antarctic of the time was not allowed to take right whales, the prohibition was not applied to large-type whaling in Japanese coastal waters (Table 5.1).

International regulation of whaling before World War II required the land stations used for baleen whales to be closed for a minimum of 6 contiguous months (4 months were required for sperm whaling) and any whaling within a distance of 1000 miles (1690 km) to operate in the same season. Japan was not a member of the then-existing convention, becoming bound by its rules only in November 1945 by order of the General Head Quarters (GHQ) of the Allied Forces that occupied Japan (Section 5.7). The current ICRW signed in December 1946 has a similar regulation of the whaling season, and contracting government can decide on the season within the rule. The Antarctic whaling season was decided by the IWC.

It was known since before World War II that the history of whaling was a typical example of overexploitation and that international cooperation was necessary to manage the whale stocks. This problem was discussed at the League of Nations in 1924 and 1927 without any agreement reached. With leadership of the League of Nations, the first international convention for the regulation of whaling was signed in Geneva in 1931 and came into effect in January 1936 with ratifications. This convention was called "the Geneva Convention" in Japan. This convention prohibited the take of right whales (bowhead whale, right whales of both hemispheres, and pigmy right whale), suckling calves and cows accompanied by a calf as direct conservation measures. Japan did not join this convention for two reasons: (1) the USSR did not join it and (2) Japan needed to take right whales in the North Pacific (Omura et al. 1942; Kasuya 2000; Omura 2000, three in Japanese). The next convention for the same purpose was signed in London in June 1937 and came into effect in May 1938; it added the gray whale to the list of prohibited species and set minimum size limits. Japan did not send delegates to the conference or join the convention but announced that Japan would send its delegate to the meeting the next year when the Japanese whaling industry, which was still in an infantile state, would be expected to be ready for it. Japan sent delegates to the subsequent conferences in June 1938 and July 1939 but did not join the convention. Japan made an announcement at the 1938 meeting that it would join the convention by the 1939/1940 Antarctic season after adjusting domestic whaling regulations to the convention, and at the 1939 meeting, Japan announced that it had completed domestic legal procedures to join the convention by the 1939/1940 Antarctic season. According to Omura et al. (1942, in Japanese), World War II that started in 1939 was the reason why Japan lost a chance to join the convention.

Japan did not join the two prewar conventions for the regulation of whaling but made some amendments in June 1938 to the Rule for Regulation of Whaling, which included minimum size limits and prohibition on take of calves and cows accompanied by calves. However, there was no ban on the take of right and gray whales in the North Pacific (Table 5.1; Omura et al. 1942, in Japanese). These two species were stated to be still important for the coastal whalers, even though they were considered by the international community to have been heavily depleted. The two species became partly protected by the amendment of the Rule for Regulation of Mother Ship Whaling in June 1938, where take of these species was prohibited in waters other than the northern part of the North Pacific (including a part of the Arctic Ocean) (Section 5.3). This means that these species were protected from Japanese whaling only in the southern hemisphere, tropical Pacific, North Atlantic, and most of the Arctic Ocean (Omura et al. 1942, in Japanese) and that Japanese North Pacific whaling, both large-type and mother-ship whaling, was allowed to take right whales and gray whales until the end of World War II (Section 5.7).

7.2.2 After World War II

The two conventions discussed earlier and an agreement among the Allied Forces shortly before the end of the war formed the skeleton of the current ICRW signed in Washington, DC in December 1946, which came into effect in October 1948 (Section I.2.2). The ICRW has 11 Articles, a Protocol, and an accompanying Schedule. The Protocol consists of later amendments to the main text, which come into effect with ratification by the member countries. Each member country nominated a commissioner to form the International Whaling Commission (IWC), under which there are several committees including the Scientific Committee (SC). With the advice of the SC, the IWC makes decisions necessary for the management of the whale stocks. The Schedule contains details of whaling regulations, including fishing season, protected species, size limits, and catch quotas. Amendment of the Schedule requires a three-fourths majority, but a simple majority applies to other decisions of the IWC. Japan joined the convention on April 21, 1951, but Japanese post–World War II whaling followed the regulations of the prewar conventions and of the ICRW by order of GHQ (Section 5.7). Protected species were first decided by the Geneva Convention in 1937, and the number of such species increased following depletion of the stocks (Table 7.2).

To assist in resolving the postwar food crisis in Japan, the GHQ issued a Notification SCAPIN 233 on November 3, 1945, permitting the reopening of Japanese coastal whaling, with two conditions: (1) respect the international convention for the regulation of whaling of the time and (2) make maximum utilization of whales for human consumption (Maeda and Teraoka 1958, in Japanese). The GHQ orders superseded the Japanese legal system of the time, so SCAPIN 233 functioned to apply the international standard of regulation of whaling for the first time to the Japanese whaling fishery. After this date, the

TABLE 7.2
Whaling Season When Total Prohibition of Hunting Came into Effect with International Agreement for Protection of Depleted Stocks

Species	Southern Hemisphere	North Atlantic	North Pacific
Gray whale	—	—	1937 (Japan accepted in 1945)
Right whales[a]	1936	1936	1936 (Japan accepted in 1945)
Humpback whale	1964	1954	1966
Blue whale	1965	1955	1966
Fin whale	1976/1977 Antarctic season	1976[b]	1976
Sei whale	1978/1979 Antarctic season	1976	1976
Sperm whale	1981/1982 Antarctic season	1986	1980 (eastern NP), 1988 (western NP)

[a] The Geneva Convention was signed in 1931 and came into effect in 1936 with ratification.
[b] Applied only to Western Norway–Faroe Islands–Nova Scotia population.

Japanese government issued Ministerial Notification No. 126 on September 30, 1946, presumably for the purpose of modifying then-existing whaling rules to match SCAPIN 233, but I have been unable to confirm the text. Then Ministerial Order No. 91 issued on December 5, 1947, put forward regulations on steam ship whaling, which included (1) prohibition on take of gray whales and right whales (which was new to Japanese whaling in the North Pacific), (2) new minimum size limits (Table 5.1), (3) prohibition on take of calves or suckling calves and cows accompanied by them (same as before), and (4) cancelation of Ministerial Notification No. 200 of 1938.

The first minimum size limits in Japanese whaling were in Ministerial Notification No. 200 of June 8, 1938, which prohibited the take of calves or suckling calves and cows accompanied by them and the take of whales below a certain body length (Table 5.1). Similar regulations were issued for mothership whaling (see below). These regulations were explained by Omura et al. (1942, in Japanese) as meant to meet the objectives of the international agreement for the regulation of whaling of the time, but the Japanese minimum size limits were slightly smaller than in the international the regulations, which allowed greater operational freedom for Japanese whaling.

After World War II, the Ministry of Agriculture and Forestry issued Ministerial Order No. 91 on December 5, 1947 (see Section 5.4) and decided new minimum size limits, which were given to the second decimal place in meters. Subsequently, Ministerial Notification No. 263 of December 1948 amended these minimum size limits by rounding them off to the first decimal place (Table 5.1). While the Schedule of the ICRW expresses body length of whales to the nearest foot, Japanese regulations expect the industry to measure whales by the metric system to the nearest 10 cm. Therefore, a hypothetical whale of 8.995 m or 29.51 ft can be characterized as 9.0 m (29.53 ft) by the Japanese method or 30 ft (9.14 m) by the ICRW method, leaving a possible disagreement of about 15 cm or 0.5 ft at the most.

The minimum size limit has been altered several times by fishing season or fishery type. For example, the size limit for fin whales was 57 ft for Antarctic whaling, 55 ft for North Pacific pelagic whaling, and 50 ft for Japanese coastal whaling, which reflected the size difference of the species between the two hemispheres and local use of carcasses for human consumption or animal food. Size limits for sei and Bryde's whales were 40 ft for pelagic operation and 35 ft for land-based whaling. These minimum size limits were set well above the body length at weaning but below the length at attainment of sexual maturity and did not function to protect breeding females but probably to increase per capita production.

The minimum size limit for the North Pacific sperm whale was maintained at 38 ft for pelagic whaling and 35 ft for coastal whaling for many years. However, the 1972 meeting of IWC reduced the size limit, except for the North Atlantic, to 30 ft to be applied beginning in the 1972 southern pelagic season and 1973 coastal season, and decided catch quotas by sex. The IWC meeting in 1976 created a maximum size limit of 45 ft for the southern hemisphere north of 40°S in October to January, and the meeting in 1980 decided to apply the same maximum limit to North Pacific whaling beginning in 1981. The scientific justification for the reduction in minimum size limit was to allow reduction of the less exploited female population and that for the upper size limit was to protect overexploited adult males to provide more mating opportunities for females, with the expectation of a greater reproductive rate. The idea behind setting quotas by sex was probably similar. I feel that such management measures were nothing more than speculative moves by population scientists attempted without sufficient knowledge on the reality of whaling operation or the social structure of sperm whales. There is no evidence indicating that such regulation measures resulted in the expected effects in the sperm whale population or that they were respected by the whalers.

The whaling gunners estimated whale size by eye before harpooning them to select large whales for greater production if the opportunity was available. It was well understood that estimating the body length was not very accurate. Therefore, if the whalers really wanted to avoid whales smaller than the size limit, there would be few whales near the minimum size limit in the whaling statistics. On the other hand, if their main objective was to catch a certain number of whales in a limited time period, they would pay less attention to size, and there would be numerous whales in the statistics around the minimum size limit. The reality was that we almost always find in the catch composition a high knife edge peak at the minimum size limit and a negligible number of whales below the limit. This suggests that the main objective was to catch a certain number of whales and the whalers falsified the lengths of undersized whales to meet the regulation.

The Japanese coastal sperm whale fishery had a catch quota since the 1959 season under voluntary regulation by the government or by the industry, and the pelagic operation also had a voluntary quota for sperm whales beginning in 1957, but these catch limits were not based on population dynamics of the species and they were often altered for the benefit of the industry. The fishery operated in 1971 under regulations agreed upon by the four North Pacific whaling countries of Canada, Japan, the United States, and the USSR, covering both land-based and pelagic whaling. This meant that a quota for the species was divided among the four countries and the Japanese share was then divided between coastal and pelagic whaling. The IWC decided the quota for the North Pacific sperm whales beginning in the 1972 season.

The Japanese baleen whale fishery in the North Pacific was regulated voluntarily by the government or industry using the Blue Whale Unit (BWU) (see Section I.2.2 for definition), setting maximum allowable takes for blue whales (beginning with the 1955 pelagic operation) and fin whales (beginning in 1966 for pelagic and 1968 for coastal operations). Beginning in the 1969 season the four North Pacific whaling countries set quotas by species for fin and sei whales (combined with Bryde's whale). Baleen whales were placed under the control of the IWC in the 1971 season (one year before the sperm whale). The blue whale and humpback whale became protected in 1966 (Table 7.2).

After the Japanese government expressed its decision in June 1986 to accept the moratorium on commercial whaling (discussed earlier), the Japanese coastal whaling for large whales that started in 1898 ended 90 years of operation in March 1988. Manipulation of catch statistics (sex and number of animals) by Japanese and USSR whalers has been documented (Kasuya 1999; Kasuya and Brownell 1999, 2001; Yablokov and Zemsky 2000; Kondo and Kasuya 2002).

7.3 PELAGIC WHALING

7.3.1 Establishment, Expansion, and Reduction

The International Agreement for the Regulation of Whaling signed in London in 1937 limited baleen whaling with mothership to the southern hemisphere south of 40°S and to the North Pacific north of 20°N (but north of 35°N in waters east of 150°W) and attached seas, including a small portion of the Arctic Oceans connected to the North Pacific. The season for Antarctic baleen whaling was limited to December 8 to March 7, but no season was set for sperm whales. It also decided minimum size limits and rules for the utilization of carcasses, and it required processing of carcasses within 36 hours of capture. Japan did not sign this earlier convention for various reasons (see Sections 5.2, 5.3, and 7.2.1). The current ICRW signed in 1946 followed the principles mentioned earlier, although it amended the whaling season occasionally, created shorter seasons for blue and humpback whales, and established whale sanctuaries for particular species or for all whale species.

Japan first operated pelagic whaling in the Antarctic in December 1934 with 1 fleet, increasing its fleets to 2 in 1936/1937, to 3 in 1937/1938, and to 6 during the 1938/1939 to 1940/1941 seasons. After a pause during the 1941/1942 to 1945/1946 seasons, Japan reopened Antarctic whaling in the 1946/1947 season with 2 fleets. The Japanese fleet size reached a peak of 7 during the 1960/1961 to 1964/1965 seasons, attaining an annual maximum of 18,259 whales taken in 1964/1965. This was followed by a gradual decline in the number of fleets accompanying the reduction of catch quotas to 5 fleets in 1965/1966 season, 4 in 1966/1967, and 1967/1968 seasons; 3 fleets sent by each whaling company during the 1968/1969 to 1970/1971 seasons, 4 fleets with an additional minke whaling fleet during the 1971/1972 to 1974/1975 seasons, 3 fleets in the 1975/1976 season, 2 fleets in the 1976/1977 season, and 1 fleet during the 1977/1978 to 1986/1987 seasons (Tato 1985; Itabashi 1987, both in Japanese). Since the 1979/1980 season, the Japanese fleets have operated only for minke whales (Section I.2).

Japan started North Pacific pelagic whaling in the 1940 and 1941 seasons with 1 fleet (Table 1.5), discontinuing operations during World War II. Takes during the 2 years totaled 1252 whales, including sperm whales and 5 species of baleen whales (Maeda and Teraoka 1958, in Japanese). After the war, Japan operated pelagic whaling for Bryde's whales (which were misidentified as sei whales) in the Bonin Island area during 6 seasons from 1946 to 1951. With the peace treaty of April 1952, it shifted the grounds to the northern North Pacific. The Japanese North Pacific pelagic whaling reached a maximum of 3 fleets during the 1962–1975 seasons, with an annual maximum of 7548 whales taken in 1968. The operation then declined to 1 fleet from the 1976 season and ended in the 1979 season (Kasuya 2009).

In order to avoid competition with Japanese coastal whaling and based on the Rule for Regulation of Mother Ship Fishery of 1934 (Section 5.3), the government prohibited Japanese pelagic whaling in waters surrounded by 20°N–52°30′N and 118°E–159°E (Omura et al. 1942, in Japanese).

7.3.2 Background of Overfishing

On November 3, 1945, the GHQ ordered Japan to respect the then-existing international agreements on the regulation of whaling, and the government modified domestic rules accordingly (Section 5.7). Attempts of the IWC to improve the management of Antarctic whale resources often met with objections by the whaling countries, or the efficiency of agreement was decreased when some countries raised objections to a decision (which was allowed by the convention), because such amendments of the regulations were usually against the short-term benefit of the whaling industries. Such attempts related to whaling season, reduction of quotas, quotas by species (departure from BWU), protected species, and whale sanctuaries. Although the scientific value of the individual proposals might merit further evaluation, such an attitude by some whaling countries caused delays in conservation measures for the depleted whale stocks. The history of whale management is a reflection of shortcomings of the ICRW and weakness of the international conservation community of the time.

Much worse was the complete disregard of the agreed rules by some countries. It has become well known that the government of the USSR carried out illegal operations in the Antarctic and North Pacific, even taking large numbers of protected species (Yablokov 1995; Yablokov and Zemsky 2000). In the early stage of Japanese whaling, some catcher boats discarded the carcasses of small whales at the end of the day's operation (Watase 1995, in Japanese). Such wasteful operations were probably possible only when whales were abundant in the early period of Antarctic whaling. It was my observation that during the early 1960s, Japanese pelagic whaling may have misreported body length and lactation status of some females but not the number of whales taken. Violation of whaling rules also occurred in Japanese coastal whaling (Sections 12.6.3, 13.7.2, and 15.3.2).

The ICRW of 1946 considered member countries equal and did not in principle allocate quota to countries. Every country was allowed, if it wished, to engage in whaling equally. This led to a situation where the countries started whaling simultaneously and stopped when their total catch reached the annual quota, or the whaling countries decided their share outside the convention. The first method continued until the 1958/1959 Antarctic season, when the International Bureau of Whaling Statistics estimated the date when total catch would reach the year's catch limit based on weekly catch reports received from the whaling fleets and announced closing of the season 2 weeks in advance. The quota varied from 16000 BWU (1946/1947 to 1952/1953) to 15,500/14,500 BWU (1953/1954 to 1959/1960). Norway and the Netherlands left the ICRW in the 1959/1960 season, and the IWC agreed to operate the Antarctic whaling without quotas for the 1960/1961 and 1961/1962 seasons. After a period of disorder of 3 seasons, 5 Antarctic whaling countries (Japan, the Netherlands, Norway, the United Kingdom, USSR) agreed in 1962 to divide the IWC quota among them; details are in Itabashi (1987, in Japanese).

The BWU system was used as a tool to decide the catch quota until the 1971/1972 Antarctic season. Although the science of the time might have been insufficient to set quota by species, it was also a common understanding that the BWU system harmed the rational management of whale stocks. The system lasted so long only because it was convenient for the whaling industries. Statistics on the post–World War II whaling operations are available in Tato (1985, in Japanese).

7.4 APPROACHING THE END OF COMMERCIAL WHALING

7.4.1 Establishment of Nihon Kyodo Hogei

Japanese North Pacific pelagic whaling, which started in 1940, operated with 3 fleets beginning in the 1962 season. Japanese Antarctic pelagic whaling, which started in the 1934/1935 season and once operated with 7 fleets, diminished to 3 fleets with the 1968/1969 season (see above). The Japanese quotas for the 1975/1976 Antarctic season were 132 fin whales, 1331 sei whales, and 3017 mike whales, and the quotas for the 1975 North Pacific whaling were 134 fin whales, 1345 sei and Bryde's whales, and 4275 sperm whales (quotas for pelagic and coastal whaling combined), which was felt to be insufficient to maintain 3 fleets in each hemisphere.

Under these circumstances, the Japanese government recommended merging of the whaling companies, and Nihon Kyodo Hogei K.K. [Japan Union Whaling Co. Ltd.] was established by the merging of six whaling companies—Taiyo Gyogyo, Nihon Suisan, Kyokuyo, Nitto Hogei, Nihon Hogei, and Hokuyo Hogei—with a capital of 3000 million yen and president I. Fujita, former IWC commissioner, invited to serve as the president. The company owned three factory ships and 20 whale catcher boats. It operated Antarctic whaling with 2 fleets in the 1976–1977 season, then with 1 fleet in the 1977–1978 to 1986–1987 seasons, and North Pacific whaling with 1 fleet in the 1976–1979 seasons. In the North Pacific, pelagic whaling was prohibited in 1980, and only coastal whaling remained there (Nihon Kyodo Hogei K.K. Shashi-ko Hensan Iinkai [Editorial Committee of the Draft History of Japan Union Whaling] 2002, in Japanese).

In July 1982, the IWC decided to close commercial whaling after three seasons. Japan raised an objection to the decision but with pressure from other countries withdrew its objection as of May 1, 1987 (Antarctic whaling), October 1, 1987 (coastal baleen whaling), and April 1, 1988 (coastal sperm whaling), which meant that Japan operated its last commercial whaling in the 1986/1987 Antarctic season and 1987/1988 coastal season. The Japanese government considered that the IWC decision did not extend to the fishery for small cetaceans such as Baird's beaked whale and dolphins and porpoises and allowed the hunting of them to continue.

7.4.2 Restructuring of the Whales Research Institute

The Nakabe Kagaku Kenkyu-sho [Nakabe Science Institute] was established in 1941 at Tsukishima, Tokyo, using donations from I. Nakabe, Hayashikane Shoten [Hayashikane Trading], and Taiyo Hogei [Taiyo Whaling]. This became a nonprofit corporation of the same name in 1942 (Tokuyama 1992; Ikeda 1997, both in Japanese). Using this institute as a base and with additional funds from Taiyo Gyogyo [Taiyo Fisheries], a nonprofit corporation Geirui Kenkyu-sho [the Whales Research Institute (WRI)] was established in 1947 and started publishing a journal in 1948, *The Scientific Reports of the Whales Research Institute*. In 1959, the WRI became a section of the Japan Whaling Association (established in 1948). The official funding for the institute became meaningless with postwar inflation, and the institute was operated through donations by Japanese whaling companies. T. Maruyama was the first director, and H. Omura followed in the position in 1952.

In the early 1960s, the whaling industries found it difficult to pay the costs of the WRI, and the government decided to take in some whale scientists from the Institute. Thus, three scientists from the WRI moved to the Tokai Regional Fisheries Research Laboratory in 1966 and in the next year to the Far Seas Fisheries Research Laboratory that was established in 1967. Three other scientists moved to the Ocean Research Institute of the University of Tokyo, and another moved to a fishery research institute in Hawaii. This restructuring left only H. Omura as a scientist working for the WRI and brought the institute to the verge of extinction. However, at the suggestion of the industry, the WRI decided to continue studies on whale behavior and oceanography of the whaling grounds with a few new scientists. The WRI thus survived and made an unexpected contribution to the industry near the end of commercial whaling.

7.4.3 The Institute of Cetacean Research and Scientific Whaling

There were discussions, probably in various parts of the whaling community of Japan, on how Japan should handle the whaling industry after the government accepted the moratorium on commercial whaling. A discussion group privately established by the Director General of the Fisheries Agency concluded on July 31, 1984, that Japan should start whaling for scientific purpose if commercial whaling were discontinued. It was an impression of mine in those days, when I was an employee of the Far Seas Fisheries Research Laboratory of the Fisheries Agency, that the whaling industry expected that the moratorium on commercial whaling would not last too long and that commercial whaling would be reopened within 10 years at the latest. The whaling industry and the Fisheries Agency probably reached the conclusion that after commercial whaling ended Japan should preserve the technology and system of whaling by operating whaling for scientific purposes, which is allowed by Article 8 of the ICRW. K. Shima, the IWC commissioner for Japan, presented the idea at a meeting held on April 12, 1985, at the Far Seas Fisheries Research Laboratory in Shimizu (the date of 1984 stated in Kasuya (2007) is incorrect) and requested the attendees to prepare a proposal for the so-called scientific whaling. The attendees at the meeting were whale scientists of the Shimizu laboratory, staff of Nihon Kyodo Hogei [Japan Union Whaling], and staff of the Whaling Section of the Fisheries Agency. Following the request by Shima, I. Ikeda, director of the Shimizu laboratory, convened and chaired a working group for the purpose. The working group was formed of Fisheries Agency staff, whale scientists of the Shimizu laboratory (including myself), university scientists, WRI scientists, and staff of Nihon Kyodo Hogei [Japan Union Whaling].

While scientists were discussing their plan for scientific whaling, there were activities ongoing to restructure the whaling industry and whale research system. Half of the staff of Nihon Kyodo Hogei [Japan Union Whaling] moved to Kyodo Senpaku K.K. [Union Shipping Co. Ltd.], newly established by the whaling company. The remaining staff of Nihon Kyodo Hogei [Japan Union Whaling] moved to Nihon Geirui Kenkyu-sho [The Institute of Cetacean Research (ICR)], which was established by remodeling the existing WRI. This happened in November 1987 (Ikeda 1997, in Japanese). I. Ikeda was invited to be the Director General of ICR. ICR got a government permit to take whales for scientific purpose and entrusted to Kyodo Senpaku [Union Shipping] associated activities, including taking whales, processing them, and selling the products, except for biological investigation of the captured whales, which was left for scientists from the ICR. The operation costs were met mostly from the proceeds of sale of the products and partially by subsidy by the government. The budget for the 2003/2004 fiscal year, when they planned to take a maximum of 600 whales (550 minke whales, 50 sei whales, 50 Bryde's whales, and 10 sperm whales), listed expected proceeds of 5900 million yen and subsidy of 940 million yen (Kasuya 2007). I identified 790 million yen of subsidy to the ICR in the government budget of the 2010 fiscal year, but it was uncertain if it covered all subsidies from the government. The settlement of the accounts has not been available to me.

It has been the general understanding of the Japanese whaling community that the scientific whaling program is allowed by Article 8 of the ICRW and that it helps preserve whaling technology and the whaling system until the time when commercial whaling will be reopened. The scientists were requested to prepare a scientific whaling program that would last many years and would require enough whales to pay the cost of operation, and they created the research goal to "estimate the age-specific natural mortality rate of Antarctic minke whales." Mathematicians first thought it necessary to take 1500 minke whales annually, but this was rejected by the Fisheries Agency, who thought the number was too high, and they compromised by halving the figure by modifying estimation and sampling methods. However, a final decision was made at higher levels in the Fisheries Agency to start the project with 300 minke whales in the 1987/1988 season (Kasuya 2007). The project was to last for 16 years. Although there was discussion on how the project should be continued if commercial whaling resumed during the period because commercial whaling and the scientific whaling could not coexist on the same minke whale population, such consideration was discontinued at some stage of the discussion, and such a situation has not as yet happened.

While pursuing the scientific whaling program it became evident that the achievement of the original goal to estimate age-specific natural mortality would be difficult, and the goal was modified for the estimation of average natural mortality rate, with addition of an element on ecosystem studies. This first-stage program of Japan ended with the 2004/2005 Antarctic season. In the 2005/2006 Antarctic season, Japan started a new, second-stage program targeting understanding of the Antarctic ecosystem surrounding cetaceans. The ICR started a scientific whaling program with the same purpose in the North Pacific in 1994. Both of the programs are formulated to last indefinitely.

There have been various criticisms of the programs of scientific whaling of Japan (Kasuya 2007). They can be summarized as follows: (1) the scientific value or necessity of the project is doubtful, (2) feasibility to achieve the goal by the method proposed is dubious, (3) killing so many wild animals is against the ethics of scientists, and (4) social or economic objectives are hidden behind the project. The Scientific Committee of the IWC reviewed only the scientific aspects of the project, that is, items 1 and 2. Until around 1993, I participated in planning the scientific whaling program as one of the Fisheries Agency scientists, and I now feel guilty to have overlooked some important points concerning how to make the project truly scientific. First, we should have determined how and when to bring the project to an end. Ending such large projects is harder than starting them, because the community that depends on the project grows bigger and stronger. Second, the financial system of funding the science and the scientists by the proceeds of sale of the whale carcasses should have been avoided. The current system makes unclear the separation of "scientific needs" from "economic needs" for dead whales, which makes the scientists face difficult judgments and damages their credibility (see Sections I.2.3 and 7.1.2).

I had the opinion that it was a good time to discontinue the scientific whaling program when we realized the difficulty of achieving the initial objective of the project, which was to estimate the age-specific natural mortality rates of Antarctic minke whales. However, that did not happen due to political and economic reasons. It was around the time when some scientists who helped the project from outside the Fisheries Agency or the ICR left the discussion group. In December 2006, the Scientific Committee of the IWC had a meeting in Tokyo to review the results of the first-stage scientific whaling program of Japan. The text of the report agreed upon at the end of the review meeting stated that the project almost failed to achieve most of the primary objectives of the project.* To me this seemed another good opportunity to bring the scientific whaling project to an end. However, before the meeting and perhaps expecting such a conclusion, the Fisheries Agency started a very challenging second-stage scientific whaling program in the 2005/2006 Antarctic season, which was directed toward understanding of the marine ecosystem surrounding cetaceans in the Antarctic. This program continued to the 2013–2014 season. Then, after a one-year pause of hunting in the 2014/2015 Antarctic season responding to the ruling of the International Court of Justice in March 2014, Japan in the 2015/2016 season started a revised scientific whaling program with diminished sample size. Japanese whaling policy again lost a chance to end the scientific whaling and direct the ICR to investigate another way to survive.

* This statement does not exist in either of the final version of the report presented to the annual meeting of the Scientific Committee in 2007 (SC/59REP 1) or in the published version (IWC 2008). The statement appears to have been omitted editorially after the workshop meeting.

REFERENCES

[*In Japanese Language]

*Akashi, K. 1910. *History of Norwegian-Type Whaling in Japan.* Toyo Hogei Kabushiki Kaisha, Osaka, Japan. 280+40pp.

*Enyo-gyogyo-ka (Suisan-cho). 1976. *Outline of Whaling.* Suisan-cho, Tokyo, Japan. 50pp.

*Enyo-gyogyo-ka (Suisan-cho). 1977. *Information on Whaling.* Suisan-cho, Tokyo, Japan. 359pp.

Government of Japan. 1997. Papers on Japanese small-type coastal whaling submitted by the Government of Japan to the International Whaling Commission, 1986–1996. Government of Japan, Tokyo, Japan. i–iv+299pp.+4 unnumbered plates.

*Ikeda, I. 1997. Preface. pp. 15–28. In: Nishi, N. (ed.). *Ten Years of the Institute of Cetacean Research.* Nihon Geirui Kenkyusho, Tokyo, Japan. 503pp.

Institute of Cetacean Research. 1996. Report on Japanese small-type coastal whaling submitted by the Government of Japan to the International Whaling Commission, 1986–1995. Institute of Cetacean Research, Tokyo, Japan. 295pp.

*Itabashi, M. 1987. *History of Antarctic Whaling.* Chuo Koron Sha, Tokyo, Japan. 233pp.

IWC. 1986. Chairman's report of the Thirty-Seven Annual Meeting. *Rep. Int. Whal. Commn.* 36: 10–29.

IWC. 2008. Report of the intercessional workshop to review data and results from special permit on minke whales in the Antarctic, Tokyo, December 4–8, 2006. *J. Cetacean Res. Manage.* 10(Suppl.): 411–445.

*Kajino, T. 1958. Poachers. pp. 113–120. In: Kajino, T. 1958. *Rowdies in Ports.* Dojinsha, Tokyo, Japan. 270pp.

Kasuya, T. 1999. Examination of reliability of catch statistics in the Japanese coastal sperm whale fishery. *J. Cetacean Res. Manage.* 1(1): 109–122.

*Kasuya, T. 1999. Manipulation of statistics by the Japanese Coastal Sperm Whale Fishery. *IBI Rep.* (Kamogawa, Japan) 9: 75–92.

*Kasuya, T. 2000. Explanatory Notes. pp. 155–203. In: Omura, H. *Journal of an Antarctic Whaling Cruise—Record of Whaling Operation in Its Infancy in 1937/38.* Toriumi Shobo, Tokyo, Japan. 204pp.

*Kasuya, T. 2005. On the whaling issue. *Ecosophia* 16: 56–62.

Kasuya, T. 2007. Japanese whaling and other cetacean fisheries. *Environ. Sci. Pollut. Res.* 14(1): 39–48.

Kasuya, T. 2009. Japanese whaling. pp. 643–649. In: Perrin, W.F., Würsig, B., and Thewissen, J.G.M. (eds.). *Encyclopedia of Marine Mammals.* Academic Press, San Diego, CA. 1316pp.

Kasuya, T. and Brownell, R.L. Jr. 1999. Additional information on the reliability of Japanese coastal whaling statistics. IWC/SC/51/O7. 15pp. Document presented to the *51st Meeting of the Scientific Committee of IWC* (available from IWC Secretariat).

Kasuya, T. and Brownell, R.L. Jr. 2001. Illegal Japanese coastal whaling and other manipulations on catch records. IWC/SC/53/RMP24. 4pp. Document presented to the *53rd Meeting of the Scientific Committee of IWC* (available from IWC Secretariat).

Kasuya, T. and Tai, S. 1993. Life history of short-finned pilot whale stocks off Japan and a description of the fishery. *Rep. Int. Whal. Commn.* (Special Issue 14): 339–473.

*Komatsu, M. 2001. *Truth of the Whaling Dispute.* Chikyusha, Tokyo, Japan. 326pp.

*Kondo, I. 2001. *Rise and Fall of Japanese Coastal Whaling.* Sanyosha, Tokyo, Japan. 449pp.

Kondo, I. and Kasuya, T. 2002. True catch statistics for a Japanese coastal whaling company in 1965–1978. IWC/SC/54/O13. 23pp. Document presented to the *54th Meeting of the Scientific Committee of IWC* (available from IWC Secretariat).

*Kushiro-shi Somu-bu Chiiki Shiryo Shitsu. 2006. *History of Whaling in Kushiro.* Kushiro City Office, Kushiro, Japan. 379pp.

*Maeda, K. and Teraoka, Y. 1958. *Whaling.* Isana Shobo, Tokyo, Japan. 346pp. (First published in 1952).

*Nihon Kyodo Hogei K.K. Shashi-ko Hensan Iinkai. 2002. *Draft History of Nihon Kyodo Hogei Co. Ltd.* Nihon Kyodo Hogei Kabushiki Kaisha Shashi Henshu Iinkai, Tokyo, Japan. 340pp.

Ohsumi, S. 1975. Review of Japanese small-type whaling. *J. Fish. Res. Board Canada* 32(7): 1111–1121.
*Omura, H. 2000. *Journal of an Antarctic Whaling Cruise—Record of Whaling Operation in Its Infancy in 1937/38*. Toriumi Shobo, Tokyo, Japan. 204pp.
*Omura, H., Matsuura, Y., and Miyazaki, I. 1942. *Whales: Their Biology and the Practice of Whaling*. Suisan Sha, Tokyo, Japan. 319pp.
*Sakuma, J. 2006. *Survey of Consumption of Scientific Whaling Products (Whale Meat)—Supply, Price and Stock Piles*. Iruka Ando Kujira Akushon Nettowaku, Tokyo, Japan. 30pp.
*Tato, Y. 1985. *History of Whaling with Statistics*. Suisan-sha, Tokyo, Japan. 202pp.
*Tokuyama, N. 1992. *History of Whaling Activities of Taiyo Gyogyo*. Published by the author, Dazaifu, Japan. 821+4pp.
*Watase, S. 1995. Hidden stories of whaling. pp. 235–240. In: Anon. (ed.). *Now We Shall Speak Up—Time is Over*. Vol. 1. Sei Sei Shuppan, Tokyo, Japan. 328pp.
Yablokov, A. 1995. *Soviet Antarctic Whaling Data (1947–1972)*. Center for Russian Environmental Policy, Moscow, Russia. 320pp.
Yablokov, A.V. and Zemsky, V.A. 2000. *Soviet Whaling Data (1949–1979)*. Center for Russian Environmental Policy, Marine Mammal Council, Moscow, Russia. 408pp.

FIGURE 1 Small-type whaler *Katsumaru No. 7* capturing a Baird's beaked whale in the Nemuro Strait. August 2007. (Photo by M. Amano.)

FIGURE 2 A hand-harpoon boat searching for Dall's porpoises off Kuji, Iwate Prefecture, Pacific coast of northern Japan. September 1988. (Photo by T. Miyashita.)

FIGURE 3 Taiji fishermen driving a school of Risso's dolphins into an inlet near Taiji. At this point, sound emitters are being struck on the vessel. January 1991. (Photo by T. Kasuya.)

FIGURE 4 A school of about 300 striped dolphins, in Kawana Harbor, Shizuoka Prefecture which was driven on the previous day, being divided into smaller groups by net. Figures 5—8 show operations on the school.

FIGURE 5 The net is partially taken in.

FIGURE 6 Using hand hooks, the fishermen pull dolphins close, cut their throats, and put ropes on their tail peduncles.

FIGURE 7 Dolphins landed with a crane.

FIGURE 8 The fishermen cut open the dolphins to extract internal organs, measure weight, and send them to the fish market. Dolphins may not be killed and gutted until after being landed with a crane. A biological investigation was carried out here between scenes in Figures 7 and 8 in November 7 and 8, 1977. (Photos in figures 4—8 by T. Kasuya.)

FIGURE 9 Two color types of Dall's porpoise: the *dalli*-type (in the rear) and *truei*-type (in the front). The two types may swim together in the overlapping portions of their ranges. Photographed in the central Okhotsk Sea in August 1990. (Photo by T. Miyashita.)

FIGURE 10 A group of *truei*-type Dall's porpoises, off Otsuchi, Iwate Prefecture, Pacific coast of northern Japan. April 2005. (Photo by M. Amano.)

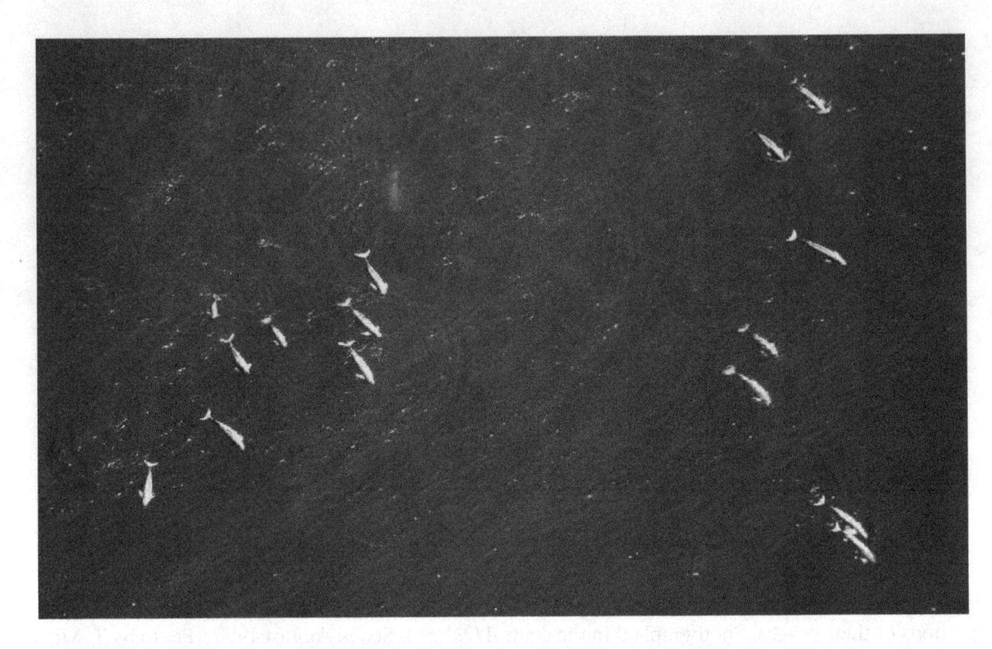

FIGURE 11 An aggregation of finless porpoises off Ibaraki Prefecture, Pacific coast of central Japan. Such large aggregations are uncommon, and neither the function nor the process of the formation is known. Curiously, there were no calves in the group, although the photo was taken after the calving season. August 2005. (Photo by S. Minamikawa.)

FIGURE 12 Small part of a school of about 300 striped dolphins. Western North Pacific, September 2005. (Photo by T. Miyashita.)

FIGURE 13 Part of a pod of short-finned pilot whales that contained 18 individuals that always swam together. The pigmentation pattern is similar to that of the Japanese southern form. Off New Jersey, east coast of the United States, September 2006. (Photo by B. K. Rone, NOAA Fisheries/Northeast Fisheries Science Center, NMFS Scientific Permit 779-1633.)

FIGURE 14 A group of Indo-Pacific bottlenose dolphins. Off Chichijima, Bonin Islands, July 1996. (Photo by M. Shinohara.)

FIGURE 15 A group of Pacific white-sided dolphins. The degree of convexity of the front edge of the dorsal fin increases in adult males. In Volcano Bay, Pacific coast of southern Hokkaido, July 2008. (Photo by T. Yoshida.)

FIGURE 16 A group of Baird's beaked whales off Rausu (eastern Hokkaido), Nemuro Strait, September 1994. (Photo by Asahi Shinbun.)

Section II

Biology

8 Finless Porpoise

8.1 DESCRIPTION OF THE SPECIES

Some languages distinguish between dolphins and porpoises, but the Japanese language does not; the word *iruka* applies to both the Delphinidae and Phocoenidae as well as some species of other families, such as the beluga and the river dolphins. However, in this book, I use *iruka* only for the three families of the Delphinoidea (Delphinidae, Phocoenidae, and Monodontidae) and use *kawa-iruka* (river dolphins) for the taxonomically distinct groups of river dolphins (Chapter I).

The Taxonomy Committee of the Society for Marine Mammalogy proposed recognition of two species in the genus *Neophocaena*, that is, *N. phocoenoides* inhabiting waters from the South China Sea to the Indian Ocean and *Neophocaena asiaeorientalis* in waters of the Yangtze River, Yellow Sea, Korean coast, and Japan. The former has a broad tubercular area on the back and the latter a narrower one. Although I have no intention of rejecting the proposal, this chapter is based on the earlier usage of only one species in the genus. The main reasons for this are that this chapter was drafted before the new taxonomic opinion become available; that classification is less important, in my opinion, for conservation biology than correct identification of units to conserve; and that we will be better able to revisit the systematics of the genus *Neophocaena* when we have accumulated more information on geographical variation within it (also see the "Note" at the end of this chapter).

The finless porpoise, *Neophocaena phocaenoides* (G. Cuvier, 1829), is one of the three species of the family Phocoenidae inhabiting Japanese waters and is the only warmwater species. It has been recorded in coastal and inshore waters, south of Hime near Mawaki (37°18′N, 137°12′E), on the west coast of Toyama Bay (37°N, 137°15′E) on the Sea of Japan coast, and south of Sendai Bay (38°20′N, 141°15′E) on the Pacific coast. It usually inhabits waters shallower than 50 m. The harbor porpoise also inhabits the continental shelf area within the 200 m isobath from northeastern Japan to the Okhotsk and Bering Seas (Taguchi et al. 2010), which partially overlaps with the range of the finless porpoise (Kasuya 1994, in Japanese). Dall's porpoise, another member of the family, also inhabits cold waters like the harbor porpoise, but it occurs in offshore waters, usually outside the continental shelf area.

Among the three species of Phocoenidae off Japan, the finless porpoise is identified by the smallest body size, the absence of a dorsal fin, and a broad rostrum on the skull. The other two species have a more pointed rostrum when the skull is seen in dorsal view (Table 8.1). The relationship between rostrum width and skull length of the finless porpoises is shown in Figure 8.3.

The finless porpoise is one of the smallest species of Cetacea together with the tucuxi, *Sotalia fluviatilis*, in the Amazon River (Caballero et al. 2007) and a subspecies of the spinner dolphin, *Stenella longirostris roseiventris*, in the Gulf of Siam (Perrin et al. 1989, 1999). Although some individuals of the species in Japanese waters exceed 190 cm in body length, those in the Indian Ocean grow to only 150–160 cm. The finless porpoise has a round head and lacks the slender rostrum often seen among dolphins and river dolphins; with its blunt head it resembles the much larger short-finned pilot whale. There are 16–20 teeth in each jaw, equipped with a crown of a flat spade shape, with exception of the anteriormost 1–2 pairs, which are pointed (Kasuya 1999). The tooth crown is often worn off, and the typical morphology is not seen in old individuals. The height and shape of the dorsal ridge are geographically variable. The entire body of the adult Japanese finless porpoise is grayish white with a paler ventral side.

In Japan, the finless porpoise has been known by numerous names, presumably reflecting frequent contact with humans in the coastal habitat. Yamashita (1991, in Japanese) collected over 20 local names of the species that appeared in the literature, which could be classified into five groups (Table 8.2). He attributed one of them, *bozu* (shaven head, or Buddhist monk) group, to the shape of the head, *name* or *nami* (smooth, or wave) group to their swimming style, *suna* (sand) group to the tubercles on the dorsal ridge, and *same* (shark) group to sharks, but he reserved conclusion on the origin of the *gondo* group. I would attribute the origin of the *gondo* group to the shape of the head, which is proportionally large and blunt as also seen in the short-finned pilot whale (*ma-gondo*), false killer whale (*oki-gondo*), and Risso's dolphin (*hana-gondo*).

Yamashita (1991, in Japanese) stated that names of the *suna* group are still in use in Tokyo and the nearby area, *same* group in Ise Bay (34°45′N, 136°45′E), *bozu* group in Okayama Prefecture in an eastern part of the Inland Sea, *gondo* group in Hiroshima Prefecture in a western part of the Inland Sea, and *name* group in the broad Inland Sea area. I had the impression in the late 1970s, when I studied finless porpoises in the Inland Sea, that *name* or *name-no-uo* (*name* fish) were used in the eastern Inland Sea and *de-gon* or *ze-gondo* in the western Inland Sea (Kasuya and Yamada 1995, in Japanese).

8.2 SCIENTIFIC NAME AND TYPE LOCALITY

The finless porpoise was first described, as *Delphinus phocaenoides*, by G. Cuvier in 1829 (*Le Règne Animal* 1: 291, not seen). This was followed by several alternations of the scientific name. Gray (1846) moved this species to

TABLE 8.1
Distinctive Skull Characters of Three Phocoenidae Species around Japan

Character	Finless Porpoise	Harbor Porpoise	Dall's Porpoise
Skull length (cm)	22–25	26–29	31–33
Rostrum length (%)	35–40	39–50	C. 42
Shape of rostrum	Broad	Slender and downwardly directed	Slender and tilted tip
No. of teeth (each jaw)	16–21	22–30	17–35
Tooth diameter at cervix (mm)	>2	>2	<1

TABLE 8.2
Vernacular Names of Finless Porpoise in Japan

Group	Vernacular Name	Location
Bozu group[a]	*Osho-uo*	Nagatao, Su-wo, Fukuyama
	Bozu-uo	Su-wo
	Bozu	Taiji
	Bon-san	Okayama
Gondo group[b]	*Se-kondo, ze-gondo, de-gondo, De-gon*	Aki, Su-wo, Hiroshima
	Den-gui	Kamagari
Suna group[c]	*Sunameri*	Hitachi, Edo
	Sunameri-kujira	Oshu (N. Japan)
	Sunameri-iruka	Nishiwaki (1965)
Mami group[d]	*Namino-uo*	Chikuzen
	Name, name-uo	Bizen, Bicchu, Fukuyama, Wakayama
	Name-kujira	Arakawa (1974)
	Nameso	Okayama
	Nameno-uo, nameri	Kumano, Okayama
	Noso	Okayama
Same group[e]	*Su-same, su-zame*	Ise, Kii, Taiji

Source: Prepared from Yamashita, K., *Kaiyo-to Seibutsu*, 13, 130, 1991.
[a] *Bozu* means bald head or bonze (Buddhist monk).
[b] *Gondo* [pilot whales] have round heads.
[c] Presumably originated from dorsal tubercles.
[d] *Nami* means ripple or wave in the sea.
[e] *Same* means shark.

a newly erected genus *Neomeris*. However, Palmer (1899) found that *Neomeris* had been preoccupied by a species that was believed to belong to the Coelenterata and created another new genus *Neophocaena* for the finless porpoise. Thus, the scientific name of the species is expressed as *N. phocaenoides* (G. Cuvier, 1892), with the author's name in parenthesis.

However, the history of the scientific name of this species was more complicated, with incidents of misprinted use of *Meomeris* (Gray 1847) and *Neomeris* (Coues 1890) or temporary resurrection of *Neomeris*. Thomas (1922) and Allen (1923) agreed with the earlier taxonomists in the view that *Neomeris* is invalid, but believed that *Meomeris* created by Gray (1847) by misprint was valid, which led to a short period of using that genus name. Soon after this, Thomas (1925) found that the *Neomeris*, which was once thought to belong to the Coelenterata, was in reality a calcareous alga and considered that *Neomeris* was valid for the finless porpoise. This was accepted for almost 36 years, until Hershkovitz (1961) found that *Neomeris* was preoccupied by a polychaete in 1844 and once again advanced the once abandoned name *Meomeris*. However, in the same year, due to modification of the International Code of Zoological Nomenclature to reject misspelled names, the current genus name *Neophocaena* came into use. Further details are in Hershkovitz (1966) and Rice (1998).

The type specimen of the finless porpoise that was the basis of the taxonomic arguments was stated to have been from the Cape of Good Hope, but this was questioned and the specimen was thought to have been presumably brought from India (Allen 1923). No evidence exists on the presence of this species along the coast of South Africa (Rice 1977).

The Japanese finless porpoise was first described as *Delphinus melas* by Temminck and Schlegel (1844) on pages 14–16 of *Fauna Japonica Les Mammifères Marins* based on a specimen presumably from Kyushu on the East China Sea. The authors and date of publication of this reference are from Kuroda (1934, in Japanese), but Hershkovitz (1966) considered the author and the date to be Temminck (1841) based on the same publication. I am unable to determine the source of the difference.

8.3 DISTRIBUTION

8.3.1 Worldwide Distribution

The distribution of the finless porpoise has been summarized by Kasuya (1999) and Amano (2003a). Reeves et al. (1997) recorded the occurrence of the species by range country, and Parsons and Wang (1998) offered information on the species in Southeast Asia. The following description of distribution of the species is based on these works. Finless porpoises are marine cetaceans inhabiting the coastal waters of tropical Asia. They enter into mangrove swamps and may enter into some riverine habitats, for example, 60 km above the mouth of the Indus River, 40 km into the Brahmaputra River, and 20 km into the Yalu River. An endemic population of the species occurs in a 1670 km stretch of the middle and lower reaches of the Yangtze River.

Finless porpoises have often been stated to inhabit coastal waters up to a depth of 200 m or up to the outer edge of the continental shelf. This does not correctly describe their habitat in Japanese waters. Shirakihara et al. (1994) recorded the relationship between density of finless porpoises and water depth in Tachibana Bay (32°30′N–32°45′N, 130°45′E–130°10′E) and the connected Ariake Sound (32°30′N–33°20′N, 130°10′E–130°35′E) in western Kyushu. The Ariake Sound is a shallow inland body of water approximately 90 × 20 km (maximum depth of about 50 m, but mostly less than 20 m) and has two narrow openings to the outer sea. One of the openings is through Tachibana Bay, which is connected to the East China Sea through a broader opening. Another narrow passage connects the Ariake Sound to the Yatsushiro Sea, which opens to the East China Sea through several narrow passes and is not inhabited by finless porpoises. In Tachibana Bay, which has an area of about 20 × 30 km, the survey covered waters up to the depth of 70 m and found that finless porpoise density declined at depths greater than 50 m, but some animals occurred at even greater depth. Shirakihara et al. (1994) surveyed the entire Ariake Sound and recorded finless porpoises in the whole depth range but with higher density in shallower waters. This observation suggests that the finless porpoise prefers waters shallower than 50 m.

The Inland Sea, which is the largest habitat of finless porpoises in Japan, has an average water depth of 31 m and a maximum depth of about 98 m near Hayasui Pass (33°20′N, 132°00′E) at the southwestern entrance of the sea. Most of the Inland Sea is shallower than 40 m. From surveys conducted in the Inland Sea in 1976–1978, Kasuya and Kureha (1979) noted that sighting of finless porpoises was limited approximately within a depth of 40 m and that even within this depth range density rapidly declined with an increasing distance from the shore. The sighting rate within 1 nautical mile (1.851 km) from the shore was 2.57/survey of 10 nautical miles. The rate declined to 1.19 in the 1–3 nautical mile range and to 0.24 in waters beyond 3 nautical miles from the nearest shore. The 10-fold density decline within 3 nautical miles suggests that density is not uniform within 1 nautical mile of shore (Kasuya and Kureha 1979).

Based on sightings of finless porpoises during the cruise of Miyashita et al. (1995) in the Yellow Sea in 1994, Reeves et al. (1997) correctly stated that some finless porpoises occurred in waters 240 km from the shore. However, it should be noted that there is shallow water extending from the Chinese coast eastward and the maximum depth in the Yellow Sea is located near the Korean coast or to the east of the middle point of the sea. The midpoint between the two shores is over 200 km from the coasts and is only about 50 m deep. The presence of finless porpoises in the offshore Yellow Sea should be considered to be due to the shallow water depth.

Korean scientists repeated cetacean sighting cruises off the west coast of the Korean Peninsula (eastern part of the Yellow Sea) and reported high density of finless porpoises in the archipelago (e.g., An et al. 2009). While the sighting track lines were evenly distributed from the Korean coast to the longitude of 123°25′E, finless porpoises occurred only to the east of *c.* 125°30′E or about 90 km from the west coast of the Korean Peninsula and within the coastal archipelago. These porpoises were sighted at depths less than 50 m (personal communication from Y. An). This indicates that the habitats of finless porpoises in China and Korea are separated by deeper water in the eastern Yellow Sea. However, this does not preclude a possibility that the habitats meet on the northern Bohai coast.

The distribution of finless porpoise is apparently affected by both distance offshore and water depth; thus, if water is shallow, there is a chance of seeing finless porpoises in waters distant from the shore. We do not know what kind of oceanographic factor might be functioning in the apparent correlation among finless porpoise, water depth, and distance offshore. There have been statements that finless porpoises aggregate in a water channel between islands or near the tip of a cape (Kasuya and Kureha 1979) or that they prefer sand bottom or soft muddy bottom (Jefferson and Hung 2004), which most likely reflects distribution of their prey.

The finless porpoise is known from the Persian Gulf to the coast of northern Japan. Between these two limits, it is known from Pakistan, India, Sri Lanka, Bangladesh, the Malay Peninsula, Indochina, China, and the western and southern coasts of the Korean Peninsula. It is also known from the northern coast of Sumatra, the islands of Bangka and Belitung, the northern coasts of Java, Kalimantan (except for the east coast), and the Tambelan Islands. A record of the species from Palawan Island is now known to not be valid. More survey effort will likely find the species on the Myanmar coast. The incidental mortality of finless porpoises has been reported from fisheries along the west coast of Taiwan and in the Penghu Islands. Although some of the carcasses stranded on the west coast of Taiwan might have represented bycatch by Taiwanese fishermen taken along the continental coast and subsequently discarded, it is also true that finless porpoises inhabit the Taiwanese coast (Wang and Yang 2007, in Chinese and English; Wang et al. 2008). Finless porpoises are known from the western and southern coasts of the Korean Peninsula, but they are absent from the east coast. The eastern boundary of the finless porpoise in Korea seems to be located around Ulsan, which is seen in cruise reports by Korean scientists (e.g., An et al. 2006; An and Kim 2007).

In Japan, finless porpoises are known from western Kyushu to northeastern Japan. Shirakihara et al. (1992a) collected published records of finless porpoises in Japan and at the same time made a questionnaire survey of fishery cooperative unions (FCUs). The survey reported that on the Pacific coast of Japan, the northern limit of records of the species is on the northern coast (38°24′N) of Sendai Bay and along the coast of the Oshika Peninsula (38°16′–38°24′N, 141°25′E–141°33′E), which separates the bay from the Pacific Ocean. On the Sea of Japan coast, there are records from the coast of northern Kyushu to Hime near Mawaki on the west coast of Toyama Bay. The questionnaire survey could obtain information from a broad area in a short period and help create a basis for future surveys, but caution is needed because of possible inaccuracy in species identification, date, and location (location of the FCU was used as location of the finless porpoise). The survey suggested that the northern limit of the species is at Onagawa (38°26′N, 141°27′E) on the Pacific coast and at the northern coast of Sado Island (37°50′N–38°20′N, 138°12′E–138°30′E) off the Sea of Japan coast. These limits show good agreement with the known range, although they slightly expand the range north of the confirmed records.

The following three points should be noted about the results of the questionnaire survey: (1) except for the area around Kanmon Pass (33°56′N, 130°56′E), there were only a limited number of positive replies for the Sea of Japan coast, (2) positive replies were also scarce for the Pacific coast facing the outer seas (i.e., Kyushu, Shikoku, Kii Peninsula, and Shizuoka Prefecture), and (3) there were no positive replies for the islands south of Kyushu and along the islands on Tsushima Strait (i.e., Iki Island and Tsushima Island). These suggest discontinuity in the habitats of Japanese and Korean finless porpoises and discontinuity of distribution even within the Japanese coastal habitat.

8.3.2 Taxonomy and Geographical Variation

Finless porpoises are classified into three subspecies (Rice 1998; also see the "Note" at the end of this chapter): *N. phocaenoides phocaenoides* inhabiting waters from the Indian Ocean to the South China Sea, *N. phocaenoides asiaeorientalis* (Pilleri and Gihr 1972) in the Yangtze River, and *N. p. sunameri* (Pilleri and Gihr 1975) in waters of the East China Sea, Yellow Sea, Bohai Sea, and Japan.

Finless porpoises inhabit shallow coastal waters, and the distribution pattern is almost linear. Additionally, their habitats are often separated by steep coastal topography, which is likely to limit genetic exchange between the habitats. Marine species having such a distribution pattern would be less likely to have genetic exchange between remote habitats compared with other species that have 2D distributions. The finless porpoise, which apparently has its origin in tropical waters, now has expanded its range to cold temperate waters such as the Bohai Sea, the Inland Sea, and Sendai Bay, where winter temperature can be around 5°C or less, and adapted to broadly variable environments. Genetic isolation and selection pressure under different environments could have resulted in distinct local populations.

The known geographical variation in the external morphology of the finless porpoise includes variation in body size, pigmentation, and shape of the dorsal ridge. Individuals in the Indian Ocean have a pale area extending from the chin to the belly (Pilleri and Chen 1980), juveniles in the Yangtze River have a pale lip at the angle of the gape (Reeves et al. 1997), and young individuals off Hong Kong have paler lips and throat region with gray in the remaining area, which darkens with growth (Parsons and Wang 1998). Contrary to this, newborns in Japanese water have an almost black pigmentation that becomes paler at the age of 4–6 months and changes to light gray in adults (Kasuya and Kureha 1979). Some of the other toothed cetaceans, for example, Risso's dolphin, beluga, narwhal, and sperm whale, become paler with age, and the Amazon River dolphin is known to become darker if placed in less turbid water. Before using body pigmentation for taxonomy or identification of local populations of finless porpoises, we need more information on the range of individual variation and on ontogenetic change.

Finless porpoises tend to be smaller in tropical waters than in boreal waters. The largest males known from the Indian Ocean were 150 cm long and the largest female 155 cm. They were 201 and 200 cm long, respectively, in the Yellow Sea and Bohai regions where the sea surface may freeze in winter (Kasuya 1999). Individuals in the Yangtze River (male, 168 cm; female, 151 cm), Inland Sea of Japan (male, 192 cm; female, 180 cm), and western Kyushu of Japan (male, 175 cm; female, 165 cm) are slightly smaller. This suggests that Bergman's rule might apply to the finless porpoise. Although in some of the examples given earlier, females are larger than males, this is due to small sample size. On average, males of this species grow larger than females (Kasuya 1999).

Instead of a dorsal fin, the finless porpoise has a dorsal ridge extending from the level of the shoulder blades to the midlength of the tail peduncle. There is an area of numerous tubercles on the dorsal ridge. Various hypotheses have been proposed for functions of the tubercle area, including "protection of the skin," "for a mother to carry her calf," "sensory organ," or "for contact between individuals" (Kasuya 1999). Geographical variation of the shape of the ridge has attracted the attention of taxonomists (Figure 8.1). Finless porpoises in waters from Japan to the west coast of the Taiwan Strait (continental shore) throughout the Yellow Sea and East China Sea have a narrow ridge with a maximum width of 1–2 cm, height of 2–3 cm, and tubercles arranged in 4–10 rows. The corresponding structure on freshwater individuals in the Yangtze River is much narrower, the 2–3 mm wide ridge has only 2–3 tubercle rows (Howell 1927; Wang 1999, in Chinese), but these porpoises may be grouped with the northern oceanic individuals in a narrow-ridged type. See Section 8.3.4 for additional observations on the morphology of the dorsal ridge area of the narrow-ridged types.

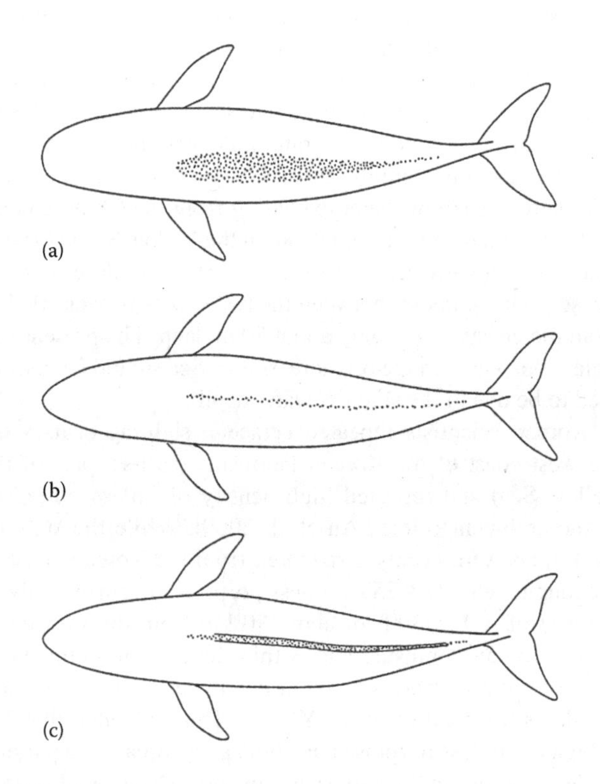

FIGURE 8.1 Geographical variation in the shape of the dorsal ridge and distribution of horny tubercles of finless porpoises: (a) the Indian Ocean and South China Sea, (b) the Yangtze River, (c) the East China Sea and the waters of Japan. (Reproduced from Figure 2 of Amano, M., *Gekkan Kaiyo*, 35(8), 531, 2003b. With permission.)

In the southern habitats from the Taiwan Strait to the Indian Ocean along the coasts of the South China Sea and South East Asia, finless porpoises have a broader tubercle area on the dorsal ridge, with a maximum width of 12 cm and greater number of tubercle rows, often exceeding 20 (Pilleri and Chen 1980; Wang 1999, in Chinese). Both the narrow-ridged and the broad-ridged types occur in waters around the Jinmen and Matsu Islands on the west cost of the Taiwan Strait (continental coast) (Parsons and Wang 1998; Jefferson 2002; Wang and Yang 2007, in Chinese and English). Jefferson (2002) has reported 4 individuals of the narrow-ridged type (0.4–0.6 cm in width) and 17 of the broad-ridged type (ridge 3.0–9.5 cm in width) collected in a fishing port on the Fujian coast, continental China, but it is unclear whether they were obtained in the same season or they might have been seasonally separated. On the east coast of the Taiwan Strait (west coast of Taiwan), most sightings and strandings have been of the narrow-ridged type, with only one stranding of the broad-ridged type; the suspicion is that it may have drifted from the continental side of the strait or have been discarded by fishermen who had been fishing on the continental side (J. Wang personal communication on January 18, 2008). A detailed description of the dorsal ridge and its geographical variation is available in Jefferson (2002).

Jefferson (2002) compared skull morphology between the two ridge types using specimens covering a broad geographical range from the Persian Gulf to Japan. He did not identify a clear cline in skull length across the latitudinal range. However, an analysis using the data in Jefferson (2002) and separating the two ridge types reveals a latitudinal cline in skull size. The narrow-ridged-type porpoises from the waters of Japan and the Yellow Sea and Bohai Sea have greater skull length (about 21.5–25 cm) than the Yangtze individuals (21 cm). Among the broad-ridged types, those from the South China Sea (Vietnam, Hong Kong, Taiwan, Fujian Province) tend to have larger skulls (approximately 21–24.5 cm) than those in the southern region of Malaysia, Thailand, Mekong, India, and Pakistan (19–21.5 cm). Even within the broad-ridged types to the north of the Malay Peninsula (i.e., excluding the Indian Ocean individuals), there seems to be a latitudinal cline from the smallest skulls (18–21 cm) on the Malay coasts to the largest ones (21.5–25 cm) on the coast of Fujian Province in China. Analyses combining specimens of broader geographical ranges or combining the two ridge types could have masked the presence of the clines.

Pilleri and Gihr (1972) were the first to recognize multiple species in this genus. They found statistically significant difference in some of the mean skull measurements between specimens from the Indian Ocean and those from the Yellow Sea and the Yangtze River and proposed to deal with them as two separate species, that is, *Neomeris phocaenoides* and *N. asiaeorientalis*, respectively. The type locality of the latter was the Yangtze River, but their analysis included the Yellow Sea specimens. Subsequently, the same authors studied Japanese specimens and named them *N. sunameri* (Pilleri and Gihr 1975). This taxonomy depended on arbitrarily selected morphological measurements and on statistical difference in mean values. Sample size affects the statistical power to detect difference between two mean values. Some minor morphological difference is likely to exist between geographically isolated populations, and they can be identified as different on average if sample sizes are sufficient, but taxonomic evaluation is another problem; mean differences do not necessarily indicate separate species. Pilleri's group also reported the earlier mentioned difference in the dorsal ridge between *N. phocaenoides* and *N. asiaeorientalis* (Plleri and Chen 1980).

Amano et al. (1992) carried out multivariate analysis of skull measurements among specimens from three geographical areas, namely, the Indian Ocean plus the South China Sea, Yangtze River, and Japan, and found that individuals of the first group were separated from members of the latter two groups and that separation of individuals between the latter two groups was incomplete. They concluded that the three groups are separate subspecies. Although their analysis revealed greater differentiation of the Indian Ocean–South China Sea series (*N. p. phocaenoides*) than that between those from the Yangtze River (*N. p. asiaeorientalis*) and Japanese waters (*N. p. sunameri*), this was not reflected in their taxonomic conclusions. Amano later (2003) reviewed the distribution and taxonomy of the species.

Jefferson (2002) attempted to improve the taxonomy by combining osteology and external morphology. He divided specimens that covered the whole distribution range into two groups by morphology of the dorsal ridge: broad-ridged type (*N. p. phocaenoides*) and narrow-ridged type (*N. p. asiaeorientalis* and *N. p. sunameri*) and carried out multivariate analysis of the skull morphology. His result with the broad-ridged types showed that the Indian Ocean individuals and southern South China Sea individuals were almost inseparable and that individuals in the southern South China Sea and the northern South China Sea were nearly completely separated (with partial overlap). For the narrow-ridged type, he found that the series were only partially separated among three geographical areas (Japan, the Yangtze River, and the Yellow Sea plus the Bohai Sea).

The scientists who place weight on the morphology of the dorsal ridge and on geographical overlap of the range in the boundary area are of the opinion that these two types should be handled as separate species. This idea comes from interpretation of the sympatric distribution as an indication of the presence of some mechanism of reproductive isolation. Such a view was expressed by Wang et al. (2008), who analyzed width of the dorsal ridge, microsatellite DNA, and mitochondrial DNA (mtDNA), based on 18 specimens from Hong Kong waters where only the broad-ridged type was known and 38 from area where both broad-ridged and narrow-ridged types were known. Among the latter were 15 specimens from Fujian Province, 17 from the Matsu Island area, and 6 from the west coast of Taiwan. They found that all the specimens were classified into one of the two ridge-width types with no intermediates. The microsatellite analysis was able to classify all the specimens into the correct ridge types. Out of the 69 alleles, 40 (58%) occurred only in one of the ridge types, which they interpreted as an indication of low genetic interchange.

Chen et al. (2010) reached the same conclusion through microsatellite analysis. However, it is also true that half of the alleles were common between the two ridge types. Wang et al. (2008) interpreted their results to indicate that separation of the two lines of finless porpoises occurred within less than 18,000 years ago or after the last glacial period, but apparently reserved judgment on the taxonomic position of the two lines (broad-ridged and narrow-ridged types) (see the "Note" at the end of this chapter).

In my view, determination of the mechanism of reproductive isolation between partially sympatric populations of cetaceans is an important research objective, and interpretation requires careful consideration. The causes of reproductive isolation between such populations are not limited to physiological mechanisms but can be obtained if the breeding members of two populations are segregated, for example, by two populations overlapping only in the nonbreeding season or by segregation of breeding components (range overlap limited to nonbreeding components). Behavioral or cultural differences in some highly social species can also limit genetic exchange; this is thought to happen in killer and sperm whales.

It appears that finless porpoises in the Yangtze River constitute a single population, but currently there is almost no reliable information on the boundary between the riverine and marine populations, whether their range overlaps, or how the boundary moves seasonally reflecting change in riverine discharge. Many of the past studies classified their specimens simply into river and ocean series without giving geographical location or time of sampling (e.g., Gao and Zhou 1995a,b). Such basic information is important for the improvement of our knowledge on local populations and better understanding of differences between the populations.

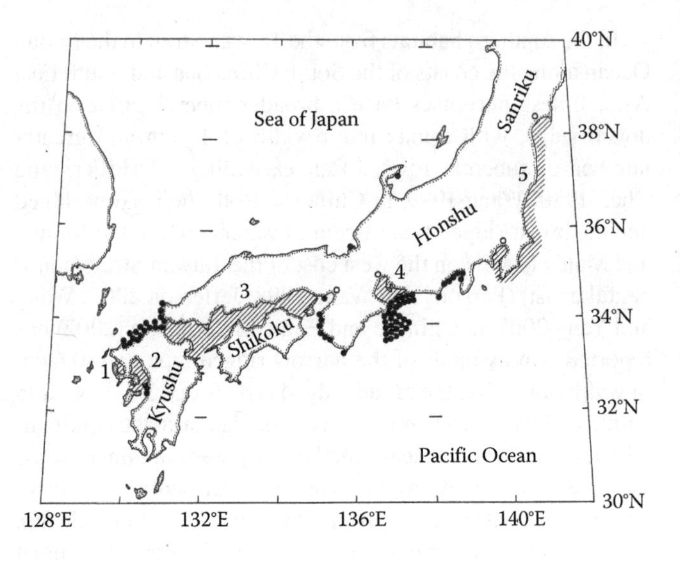

FIGURE 8.2 Distribution of finless porpoise in Japan. Shaded areas surveyed for abundance estimate, with exception of Tokyo Bay, which has been aerially surveyed but has not been included in abundance estimation. Closed circles represent strandings of the species outside the surveyed area. Each closed circle represents two individuals for the Ise Bay–Mikawa Bay area (4) and one individual for the rest of the areas (1–3 and 5). The numerals indicate currently recognized populations and are common among Figures 8.2, 8.3, 8.11, and 8.14 and Table 8.3, 8.5, 8.10, 8.12, and 8.15. (Reproduced from Figure 1 of Kasuya, T. et al., *Raffles Bull. Zool.*, (Supplement 10), 57, 2002. With permission of National University of Singapore.)

8.3.3 Distribution of Finless Porpoises in Japan

Our knowledge on the general distribution of Japanese finless porpoises depends mostly on Shirakihara et al. (1992a), which is based on published records of the species and questionnaires sent to 2053 FCUs covering the entirety of Japan except for Hokkaido, where finless porpoises were not expected. The questionnaire asked if the member fishermen had ever seen the finless porpoise (1382 FCUs replied). The northernmost positive responses were obtained from an FCU on the northern coast of Sado Island in the Sea of Japan and an FCU near Onagawa on the Pacific coast (Section 8.3.1). These records almost agreed with published records from Hime on the west coast of Toyama Bay in the Sea of Japan and the northern shore of Sendai Bay on the Pacific coast but slightly expanded the range to the north. Although such a distribution is highly probable, it seems to be premature to reach any firm conclusions before scientists confirm the distribution.

Shirakihara et al. (1992a) noted that FCUs with positive replies tended to be located on a shallow seashore without rocky bottom and listed five such sea areas: (1) Omura Bay; (2) Ariake Sound–Tachibana Bay; (3) the Inland Sea, Osaka Bay, and the Kii Channel; (4) Ise Bay–Mikawa Bay, and (5) the area extending from Tokyo Bay to Sendai Bay (see Figure 8.2 for the locations). The species does not seem to inhabit the coasts of the Japanese Southwestern Islands (24°N–31°N, 123°E–131°E), which separate the western North Pacific and the East China Sea. Although a white shark landed at Okinawa Island was found with two finless porpoises in different states of digestion in the stomach (Kasuya 1999) and one finless porpoise was stranded at the same island (see Section 8.3.4), these incidents should not be taken as evidence of occurrence of the species in the Southwestern Islands.

Now, we will consider further details of the distribution of finless porpoises in Japan. The Shimabara Peninsula is located between the earlier mentioned two habitats of the finless porpoise, Omura Bay and Ariake Sound–Tachibana Bay, and records of the species are scarce along the coast of the peninsula, which suggests that the two habitats on the west coast of Kyushu are discontinuous. Distribution of finless porpoises in the western Inland Sea is continuous through Kanmon Pass (which connects the northwestern Inland Sea and the southern Sea of Japan) to the coasts of northern Kyushu and western Yamaguchi Prefecture in longitudes from 130°15′E to 130°55′E, where year-round occurrence of the species has been confirmed. Therefore, the habitat of the species in the Inland Sea–Osaka Bay–Kii Channel extends to this area in the southern Sea of Japan (Shirakihara et al. 1992b; Nakamura et al. 2003; Nakamura and Hiruda 2003, last two in Japanese). Akamatsu et al. (2008) placed underwater microphones in Kanmon Pass to detect the movement of finless porpoises for a total of 57 days during February, March, and November from 2005 to 2006. Kanmon Pass has a water depth of about 10–30 m and a minimum width of 670 m. They recorded sounds produced by 37 finless porpoises, including 33 recorded during the nighttime hours of 2000 to 0400. This time coincided with minimum vessel traffic. Finless porpoises seemed to travel through the pass during the night, but it was not certain if this was related with traffic density or if the negative correlation was only an apparent one. They did not find correlation between the current direction and finless porpoise travel, but they noted that out of 16 individuals, 14 swam with the current (the tidal current reverses periodically). This observation agreed with that of Kasuya and Kureha (1979) for Naruto Pass (34°15′N, 134°39′E), which connects the southeastern Inland Sea and Kii Channel and is known to have rapid tidal currents (see Section 8.4.1).

Finless porpoises in Omura Bay are separated by an almost empty area along the coast of northwestern Kyushu from individuals in the Kanmon area–Inland Sea–Osaka Bay–Kii Channel. Thus, the porpoises in Omura Bay and Ariake Sound–Tachibana Bay may constitute two distinct populations. Regarding distribution of finless porpoise along the Sea of Japan coasts, Shirakihara et al. (1992a) recorded in their questionnaire survey only a small number of positive replies and identified only a few confirmed records of the species along the Sea of Japan coast east of the Kanmon area. This means that finless porpoises are uncommon along the coast east of Yamaguchi Prefecture. The limited number of records of the species from the Sea of Japan coast between Yamaguchi Prefecture and Hime in Toyama Bay area is likely to represent stragglers from the Kanmon area, that is, from those inhabiting the Inland Sea–Osaka Bay–Kii Channel area.

Finless porpoises are continuously distributed from the Kanmon area to the Kii Channel via the Inland Sea and Osaka Bay (Figure 8.2), but they are rare on the Pacific coast of southern Kyushu, Shikoku Island (32°45′N–33°30′N, 132°45′E–134°30′E), and the Kii Peninsula (33°26′N–34°10′N, 135°05′E–136°20′E). The only exceptions were one record from Cape Shionomisaki (33°26′N, 135°45′E) and another from Taiji (33°36′N, 135°57′E) on the Pacific coast of the Kii Peninsula (Shirakihara et al. 1992a; see Section 8.3.4). Thus, individuals in the Kanmon area–Inland Sea–Osaka Bay–Kii Channel and those in Ise Bay–Mikawa Bay are discontinuous in distribution. The latter two bays are connected to each other and then open to the Pacific through a common opening, and sighting of finless porpoises are continuous between the two bays. Numerous stranding records and some sightings along the Pacific coast near the exit to the Pacific suggest that individuals in Ise Bay–Mikawa Bay move outside of the bays at least seasonally (Figure 8.2).

A question remains on the identity of finless porpoises in Suruga Bay (35°N, 138°40′E) about 170 km to the east of the Ise Bay–Mikawa Bay opening. The presence of the species there has been known from early times (Kuroda 1940, in Japanese). Although this habitat appears to be separated by an empty area from Ise Bay–Mikawa Bay, there is insufficient information to confirm that.

Another remaining question on the distribution of Japanese finless porpoises relates to the porpoises in the Tokyo Bay to Sendai Bay area. We know that some finless porpoises occur year-round in Tokyo Bay north of Futtsu (35°19′N, 139°47′E) near the entrance, but there are no confirmed records of the species outside (or south) of Tokyo Bay: west coast of the Awa area (34°54′N–35°09′N, 139°45′E–140°13′E) or southern part of the Boso Peninsula (34°54′N–35°30′N, 139°45′E–140°25′E). Aerial sighting surveys by Amano et al. (2003) did not find finless porpoises in this area, and the results of the interview survey by Suzuno et al. (2010, in Japanese) suggest a possibility of past decline of the species in the area as well as its absence along the coast in recent time. This information suggests a hypothesis that finless porpoises in Tokyo Bay are separated from those found along the east coast of the Boso Peninsula and further north to Sendai Bay.

The population structure of finless porpoises inhabiting the broad area extending from the southern tip (34°55′N, 139°53′E) of the Boso Peninsula to the bottom (38°24′N) of Sendai Bay also needs further consideration. Amano et al. (2003) suggested a possible discontinuity of distribution around Cape Shioya (37°00′N, 141°59′E). If this is accepted, there is a possibility of three local populations in the area from Tokyo Bay to Sendai Bay along the Pacific coast. Yoshida et al. (2001) suggested from the analysis of mtDNA a possibility of different populations in Sendai Bay and along the Pacific coast (35°15′N–37°N) south of Sendai Bay. The population structure of finless porpoises in Tokyo Bay to Sendai Bay area remains to be investigated.

8.3.4 Local Populations of Japanese Finless Porpoises

The preceding section discussed possible local populations of Japanese finless porpoises based on the information on their distribution. For the hypotheses thus created to be confirmed, limitation in interchange of individuals between the habitats or the presence of only limited genetic exchange between them must be confirmed. If genetic exchange or mixing of individuals is sufficiently small to allow independence in population dynamics, individuals in each habitat should be dealt with as different populations for the purpose of conservation in management.

Both genetic and osteological studies have been used for the analysis of population structure of Japanese finless porpoises. As morphology can be influenced by both environment and the genome, two sets of samples from a single population in different historical periods might show significant differences if there were some large environmental change in the intervening period, such as in forage. So, caution is needed in analyzing specimens collected over a broad historical time period.

In order to identify local populations of Japanese finless porpoises, Yoshida et al. (1995) analyzed skull morphology and Yoshida et al. (2001) analyzed haplotype frequency of mtDNA; both were reviewed by Yoshida (2002). Yoshida et al. (1995) used a total of 146 skulls from five geographical regions: 8 skulls from Omura Bay, 74 from Ariake

Sound–Tachibana Bay, 48 from the Kanmon area and the Inland Sea, 11 from Ise Bay–Mikawa Bay, and 5 from Tokyo Bay to Sendai Bay area. Samples of both sexes were combined after removing characters found with sexual dimorphism. Season and year of collection of the specimens were not given in their study. They first analyzed relative growth, using all the samples, of 15 measurements against the condylobasal length of the skull (analysis of covariance) and found 6 measurements with geographical difference in pair-wise comparison of the five sample groups. One measurement compared between Omura Bay and Kanmon area–Inland Sea was an exception. The most distinct difference was found in two measurements relating to the width of the skull: width of the rostrum at midlength and width of the skull across the zygomatic processes of the squamosal. Specimens from Ise Bay–Mikawa Bay had these values significantly smaller than those for specimens from the other four geographical regions (Figure 8.3). Canonical discriminant analysis using skulls aged 4 years or older and using the earlier mentioned six characters (Figure 8.3) yielded the following results. The Ise Bay–Mikawa Bay group was separated from all other groups. The Ariake Sound–Tachibana Bay group was separated from the geographically close Omura Bay group and the Inland Sea–Kanmon group of considerable geographical distance away but partially overlapped with the geographically distant Tokyo Bay–Sendai Bay group. The Omura Bay group and Inland Sea–Kanmon group were only partially separated from each other in the discriminant analysis; they are located near each other geographically. The Tokyo Bay–Sendai Bay group, which covers an extended northern habitat of this species in Japan, overlapped partially with the Omura Bay group, Ariake Sound–Tachibana Bay group, and Inland Sea–Kanmon group but separated completely from the Ise Bay–Mikawa Bay group, which was the closest geographically.

The earlier discussion can be summarized thusly: if we exclude animals from Omura Bay in western Kyushu and the Pacific coast north of Tokyo Bay, finless porpoises in each of the three areas of Ariake Sound–Tachibana Bay, Inland Sea–Kanmon, and Ise Bay–Mikawa Bay are individually identifiable to the location of origin based on their skull morphology. Porpoises in Omura Bay show some morphological differences from those in the geographically close Inland Sea–Kanmon area and those in the more distant Tokyo Bay and northern area. This result suggests the presence of a minimum of five local populations of finless porpoises in Japan. Unfortunately, only five specimens from the broad Tokyo Bay–Sendai Bay area were available for this study; they had to be combined into a single group for the analysis.

Yoshida et al. (2001) analyzed 345 base pairs of the control region of mtDNA of 174 finless porpoises from Japan, identified 10 haplotypes, and analyzed haplotype frequency in the earlier mentioned five geographical areas (Table 8.3). As mtDNA is maternally inherited, the analysis can detect males from another habitat visiting temporarily to breed only when they happen to be sampled directly; their offspring will bear the mtDNA of the resident mothers. Detection of sex-biased genetic dispersal must also involve the analysis of nuclear DNA. However, Yoshida et al. (2001) concluded that sex-biased movement did not exist because no between-sex differences in haplotype frequency were detected and therefore combined the sexes for the analysis of haplotype frequency. Difference in haplotype frequency was statistically significant

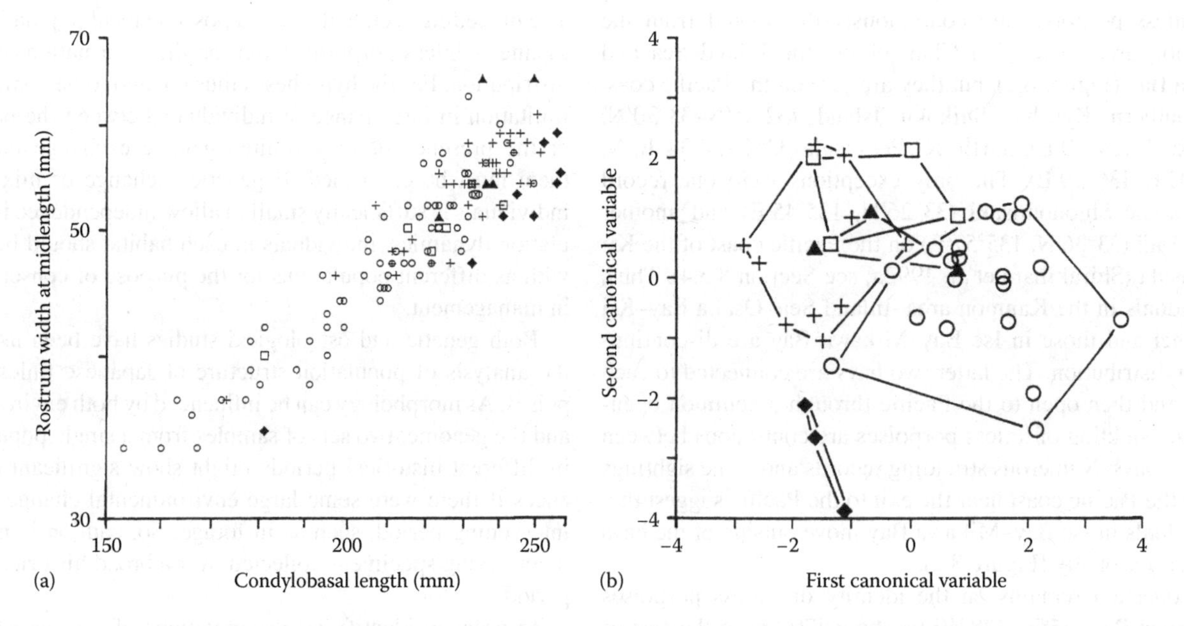

FIGURE 8.3 Comparison of the skull morphology of five currently identified finless porpoise populations in Japan: (a) relationship between condylobasal length and width of rostrum at midlength and (b) scatterplots of scores on the first and second canonical variables derived from six skull characters. (1) Open square, Omura Bay population; (2) open circle, Ariake Sound–Tachibana Bay population; (3) cross, population in the Inland Sea and nearby waters; (4) closed diamond, Ise and Mikawa Bay population; (5) closed triangle, Tokyo Bay to Sendai Bay population. (Reproduced from Figure 3 of Yoshida, H., *Raffles Bull. Zool.*, (Supplement 10), 35, 2002. With permission of National University of Singapore.)

TABLE 8.3
Geographical Differences in Mitochondrial Haplotype Frequencies of Japanese Finless Porpoises

Haplotypes	a	b	c	d	e	f	g	h, i, j	No. of Haplotypes	Sample Size
(1) Omura Bay	0	0	0	8	0	0	0	0	1	8
(2) Ariake Sound–Tachibana Bay	0	0	0	9	0	46	6	4	6	65
(3) Inland Sea and Hibiki-nada[a]	0	0	0	27	3	0	0	0	2	30
(4) Ise and Mikawa Bays	0	4	52	0	0	0	0	0	2	56
(5) Tokyo Bay–Sendai Bay	7	7	0	0	0	0	0	0	2	14

Source: Reproduced from Table 1 of Yoshida, H. et al., *J. Mammalogy*, 82(1), 123, 2001. With permission of Oxford University Press.
[a] Hibiki-nada is a portion of the southern Sea of Japan connected to Kanmon Pass or the northwestern opening of the Inland Sea.

in all the pair-wise combinations of samples except for one case, the comparison between 8 samples from Omura Bay and 30 from the Inland Sea–Kanmon area, which could have been due to a small sample size for the Omura Bay population.

Yoshida et al. (2001) noted the presence of genetic isolation between Omura Bay and Ariake Sound–Tachibana Bay, which are only 60 km apart, as an indication of limited intermingling between the habitats. They also stated that a female finless porpoise obtained at Taiji had a haplotype known from the Ise Bay–Mikawa Bay area, which was about 30–40 km west of the known range of the population. The authors identified only two haplotypes for the 14 porpoises from the Tokyo Bay–Sendai Bay area, noted that all the 7 individuals from Tokyo Bay had a single haplotype (type a) and that all the specimens from Sendai Bay (presumably 6 individuals) had another haplotype (type b), and suggested that there was a possibility of more than one local population in this extended area. Their 14 samples from the Tokyo Bay–Sendai Bay area contained 1 sample from the Ibaraki Museum in Ibaraki Prefecture (35°45′N–36°50′N on the Pacific coast). This museum specimen was likely to have been from the nearby coast, so the total Sendai Bay sample could be six animals.

In 2004, an emaciated finless porpoise (129.6 cm and 19.2 kg) was stranded on the coast of the Motobu Peninsula (26°40′N, 127°50′E) on the East China Sea coast of Okinawa Island. Yoshida et al. (2010) analyzed mtDNA of this animal and found that its haplotype did not match any of the 10 haplotypes found among Japanese specimens but matched that of one of the 17 haplotypes reported from finless porpoises inhabiting the Chinese coast from the Taiwan Strait to the Yellow Sea. They concluded that the porpoise must have been a stray from the Chinese coast. They mentioned only that the stranded individual had a narrow dorsal ridge, but it was clear in an accompanying photograph that there was a longitudinal groove on each side of the dorsal ridge. Thus, the cross section of the ridge area shows a W shape, which is different from the feature in Japanese individuals but known from finless porpoises along the coasts of Korea and the Yellow Sea.

The analyses given earlier lead to the conclusion that there are several local populations of finless porpoises in Japan that are currently almost isolated from each other. However, this does not necessarily mean that the isolation dates from long ago geologically. The planet has had a series of climatic cycles. The last glacial period that peaked in around 18,000–20,000 years before the present was followed by gradual warming and recorded peak warmth 5,000–6,000 years before the present. This is called in Japan the period of *jomon* marine transgression because it occurred in the prehistoric *Jomon* cultural era known from cord-patterned pottery. Then the climate became cooler with retreat of the ocean to the current status in marine geography around Japan. Current global warming may reverse the climate again with a speed much faster than the natural cyclic changes of the past.

While the climate was becoming warmer in the recent geographic past, some finless porpoises in the southern habitat could have immigrated to the northern habitat and established a new local population. Some in the new population could have survived to the present through the subsequent cooling period. There presumably was a founder effect, and there could have been new haplotypes arising only through rare mutations. The currently surviving parent population would be expected to have greater genetic variation than the daughter population.

The data presented by Yoshida et al. (2001) are particularly interesting in this regard. The number of haplotypes tended to decline moving north. If we exclude the Omura Bay sample of small size and also for the time being the Tokyo Bay–Sendai Bay sample, which is likely to be from multiple populations, the Ariake Sound–Tachibana Bay sample from western Kyushu showed the highest variation with six haplotypes (with a ratio of 46:9:6:2:1:1), followed by a decreasing number of haplotypes in the Inland Sea–Kanmon sample (two haplotypes with a ratio of 27:3) and the Ise Bay–Mikawa Bay sample (two haplotypes with a ratio of 52:4) (Figure 8.2 and Table 8.3). This decreasing cline is followed by a single haplotype in the Tokyo Bay sample (n = 7) and another single haplotype in the Sendai Bay sample (presumed n = 6, see the preceding text in this section). This cline in haplotype frequency from southwest to northeast may reflect the history of expansion of habitat from south to north during the period of warming. Another interesting point in the results presented by Yoshida et al. (2001) is the absence of common haplotypes among the three southwestern samples (three populations) and the two northeastern samples (presumed to be from more than two populations). This also reflects the history of formation of local populations and history of low genetic exchange between nearby populations.

8.4 ABUNDANCE AND SEASONALITY OF DISTRIBUTION

8.4.1 Seasonal Movements

In Japan, the term *migration* has been often used without a firm definition and even confounded with "dispersal" or "shift." In a strict sense, migration represents an activity where animals move toward a destination with other activities such as feeding or mating repressed during the travel. Shift means seasonal expansion or retraction of the geographical range to meet seasonal change in the suitable environment. Among cetaceans, gray whales and humpback whales are species known to migrate in the strict sense. However, some gray whales are known to feed on the way to their destination in the Bering Sea. Strict definitions are likely to have exceptions. Seasonal movement of striped dolphins along the Pacific coast of Japan is apparently different from migration. They inhabit the Kuroshio front area and areas to the south of it. In early summer as sea temperature rises, their range moves north to reach latitude of 40°N–41°N in the summer, and in the autumn, it retreats to a southern boundary at 33°30′N–34°30′N. Their reproductive and feeding activities continue throughout the movement. This can be called a seasonal movement or shift rather than migration.

Pilleri and Gihr (1972) were probably the first to mention seasonal movements of finless porpoises. They stated that the density of the species in the Indus delta was highest in the winter season of October–April and then declined; they concluded that the porpoises moved offshore in the spring to feed on shrimps. Wang (1984, in Chinese) reported an opposite case where finless porpoises in the Bohai Sea, Yellow Sea, and East China Sea approached the coast in March–April or May, resulting in density increase in coastal waters. The freshwater population in the Yangtze River is thought to make seasonal shifts as water level changes.

Shirakihara et al. (1994), based on 2.5 years of riding ferries, observed seasonal change in finless porpoise density in the Ariake Sound–Tachibana Bay area of western Kyushu. The geography of this area is described in brief in Section 8.3.1. They recorded a clear seasonal change in sighting density inside the sound from a high of over 4 animals/100 km of searching from January through June to a low of less than 2 from July through November. This seasonal fluctuation has some similarity with that observed in the Inland Sea by Kasuya and Kureha (1979). Shirakihara et al. (1994), however, observed an opposite density trend in Tachibana Bay and in the area connecting to the Ariake Sound, where density was lower from January through May and higher from July through November. The sighting density in the peak month was only 1 individual/100 km of effort, which was much lower than the figure for inside the sound. If we accept this observation, then we can infer that finless porpoises tend to spend winter in the Ariake Sound and move to Tachibana Bay in summer. The low sighting density in Tachibana Bay could be explained by one or more of the following factors: (1) most of the population remained in the sound in summer, (2) the porpoises were dispersed in Tachibana Bay area in summer, or (3) sea conditions in Tachibana Bay were unfavorable for sighting.

The results of aerial surveys by Yoshida et al. (1997) did not apparently agree with the interpretation mentioned earlier (Table 8.4). They repeated the surveys in the Ariake Sound, Tachibana Bay, and the surrounding area for finless porpoises during 5 months from May 1993 to May 1994, and they were able to confirm an offshore limit of distribution of the species at the 50 m isobath and its absence in the coastal waters between Omura Bay and Tachibana Bay and also in the Yatsushiro Sea, which is another enclosed bay connected to the Ariake Sound. Their results confirmed that distribution of finless porpoises was continuous from the Ariake Sound to Tachibana Bay but that this Ariake Sound–Tachibana Bay habitat was unconnected with other habitats of finless porpoises in Japan. Their abundance estimates for the Ariake Sound recorded a peak in May and a low in June, while those for Tachibana Bay reached a peak in February and a low in November. Broad confidence intervals accompanying the estimates preclude firm conclusions on seasonal abundance change in these waters. Thus, Yoshida et al. (1997) could not offer positive support for the inference of Shirakihara et al. (1994) on seasonal movement in the species. Although support for seasonal change in the abundance of finless porpoises is inconclusive for the Ariake Sound or Tachibana Bay, there remains the possibility that winter abundance in Tachibana Bay (2416 individuals in February) is greater than the abundances in the rest of the seasons (458 in August and 398 in November; Table 8.4), which is the reverse of the trend suggested by Shirakihara et al. (1994).

TABLE 8.4
Abundance of the Finless Porpoise in the Ariake Sound–Tachibana Bay Populations and Their Seasonal Fluctuations, Estimated by Sighting Surveys and Assuming g(0) = 1

Area	Month	No. Sighted	Abundance	CV	95% Confidence Interval
Ariake Sound	January	42	1762	0.34	914–3396
	May	69	3111	0.31	1711–5661
	June	27	1214	0.51	474–3110
	August	47	1930	0.31	1074–3470
	All months	185	1983	0.19	1382–2847
Tachibana Bay	February	91	2416	0.49	972–6006
	August	17	458	0.46	193–1087
	November	14	398	0.53	150–1051
	All months	122	1110	0.29	642–1920
Total	Total	307	3093	0.16	2278–4201

Source: Reproduced from Table 2 of Yoshida, H. et al., *Res. Popul. Ecol.*, 39(2), 239, 1997. With permission of the Springer Japan.

Omura Bay in western Kyushu is located to the northwest of Ariake Sound–Tachibana Bay, has an area of about 35 × 15 km, and is connected to the East China Sea by two northern passages with widths of 200 and 10 m. This bay was surveyed by Shirakihara et al. (1994) using vessels and by Yoshida et al. (1998) from the air. Shirakihara et al. (1994) apparently analyzed sightings recorded by the crews of high-speed ferryboats operating in Omura Bay; they concluded that finless porpoises were dispersed throughout the bay in *spring* but aggregated in coastal parts of the bay in other seasons. They also interviewed fishermen along the coast of the East China Sea near the entrance to Omura Bay and found records of rare incidental mortality in fisheries but were unable to find anyone who had seen the species in the wild. Yoshida et al. (1998) flew aerial surveys in the same bay in February, May, August, and November; they found that finless porpoises were dispersed widely in the bay in May but that the density was lower in the central part of the bay in other seasons, which apparently agreed with the results of Shirakihara et al. (1994). However, the lack of a definition of "spring" in Shirakihara et al. (1994) caused me some difficulty in precise comparisons. Abundance estimated by Yoshida et al. (1998) showed an apparent seasonal trend from 343 individuals in May to 203 in August, 104 in November, and 92 in February, but these estimates were not statistically different. Using this result and the earlier mentioned interviews made by Shirakihara et al. (1994), Yoshida et al. (1998) concluded that finless porpoises in Omura Bay usually do not leave the bay. They most likely have small-scale seasonal movements within the bay following prey species and remain within the bay year-round.

The Inland Sea is the largest inland water of Japan, with a size of about 370 × 50 km or surface area of 14,300 km^2, and contains about 3,000 islands. The sea is connected to outer seas by four passes: Kanmon Pass (600 m wide and 47 m deep) at the northwest corner opens to the southwestern Sea of Japan; Hayasui Pass (14 km wide and 195 m deep) at the southwest corner opens to the Bungo Channel, which opens to the Pacific Ocean; Akashi Pass (3.6 km wide and 100 m deep) at the northeast corner opens to Osaka Bay, which then opens to the Kii Channel; and Naruto Pass (1.4 km wide and 90 m deep) at the southeast corner opens to the Kii Channel, which opens to the Pacific Ocean (Figure 8.2).

Over a period of about 2.5 years from April 1976 to October 1978, Kasuya and Kureha (1979) surveyed the distribution of finless porpoises in the Inland Sea. They used low-speed passenger ferryboats or ferries for both passengers and cars but avoided high-speed passenger ferries. The total number of ferry tracks used was 31 including some in Osaka Bay, but this figure depends partly on the method of counting tracks. If a track is defined as from embarkation to disembarkation, the number is 34 tracks served by 28 different vessels. One survey took about 2 weeks, and every track was surveyed as many times as possible within the survey. Thus, the total surveyed distance was several times greater than the sum of the surveyed tracks.

The density of finless porpoises in the Inland Sea varied depending on distance from the shore; the factor behind this is unknown. Combining all the survey data gave an average sighting rate of about 25 individuals per 100 nautical miles for waters within 1 nautical mile from the shore, which decreased to slightly over 10 in the 1–3 nautical mile range and to less than 5 in waters over 3 nautical miles from the shore (Figure 8.4). The sighting density also appeared to vary with distance from the eastern and western openings, with lower density in waters near the two extremities of the sea than in central waters. This was constant among the three strata of distance from the shore and among seasons (Figure 8.5) in the late 1970s (Kasuya and Kureha 1979), but the distribution pattern changed in the late 1990s (Kasuya et al. 2002).

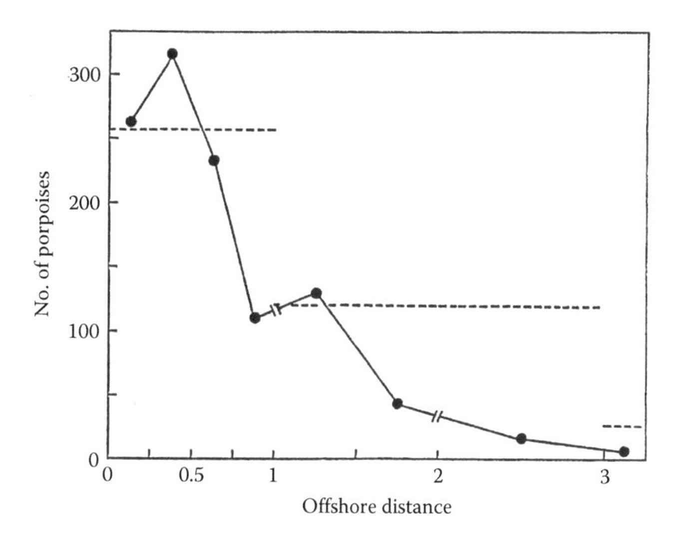

FIGURE 8.4 Relationship between density of finless porpoises and offshore distance (in nautical miles) in the Inland Sea. Closed circles and solid lines indicate number of finless porpoises sighted during the survey in 1976–1978. Dotted lines indicate number of finless porpoises sighted per 1000 nautical miles of the survey cruise. (Reproduced from Figure 17 of Kasuya, T. and Kureha, K., *Sci. Rep. Whales Res. Inst.*, 31, 1, 1979.)

Kasuya and Kureha (1979) observed a seasonal density decline in the nearshore stratum (<1 nautical mile) of the middle part of the sea (areas 2, 3, and 4 in Figure 8.5) from about 1.0 in November–February or 1.2 in March–April to about 0.5 in May–September and felt that the density was likely to be high in the spring. They also observed the estimated abundance of the species in the Inland Sea to be highest in April and lowest in October (Figure 8.6). This seasonal change is mostly a reflection of change in the nearshore stratum (<1 nautical mile). The abundance estimated in this study is now believed to be unreliable due to technical problems (see Section 8.4.2), but the estimated seasonal and geographical trends still reflect reality.

The results of Kasuya and Kureha (1979) indicated that most of the finless porpoise population in the Inland Sea inhabits the nearshore area and the possibility that some porpoises may disperse either to the offshore area or to waters near the outer seas. The latter was similar to the case suggested by Shirakihara et al. (1994), which was not confirmable in data from a subsequent study by Yoshida et al. (2001).

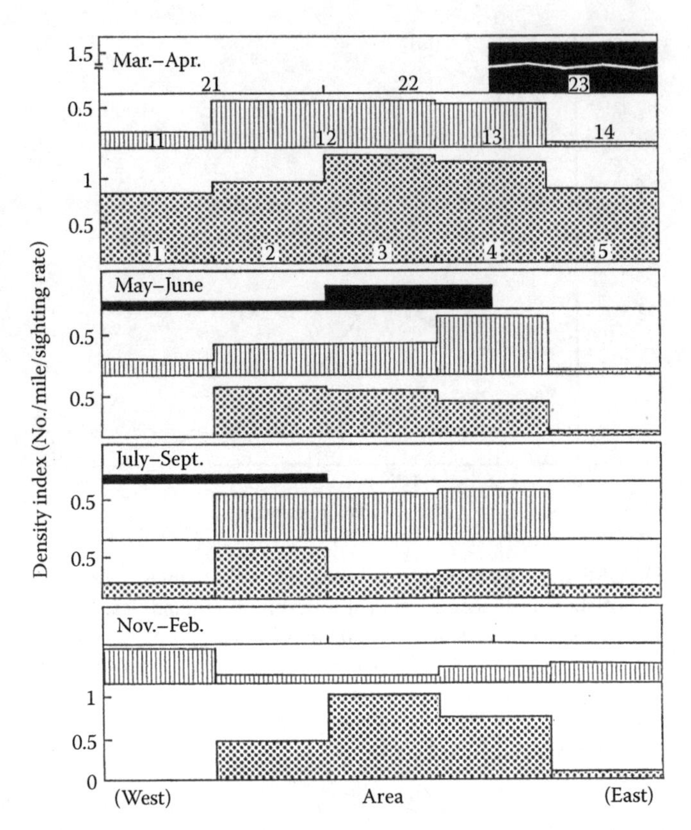

FIGURE 8.5 Density index of finless porpoises (number of individuals/cruised distance corrected for sighting rate) in the Inland Sea in 1976–1978 shown for areas divided by the distance from the shore and the west–east divisions of the sea. Area divisions are small in inshore strata but larger in the offshore strata and are numbered from the western part of the Inland Sea to the eastern part. Dotted squares (area 1–5) represent strata of ≤1 nautical mile from the shore, squares with vertical line (area 11–14) represent strata of 1–3 nautical miles from the shore, and black squares (area 21–23) represent strata ≥3 nautical miles from the shore. (Reproduced from Figure 19 of Kasuya, T. and Kureha, K., *Sci. Rep. Whales Res. Inst.*, 31, 1, 1979.)

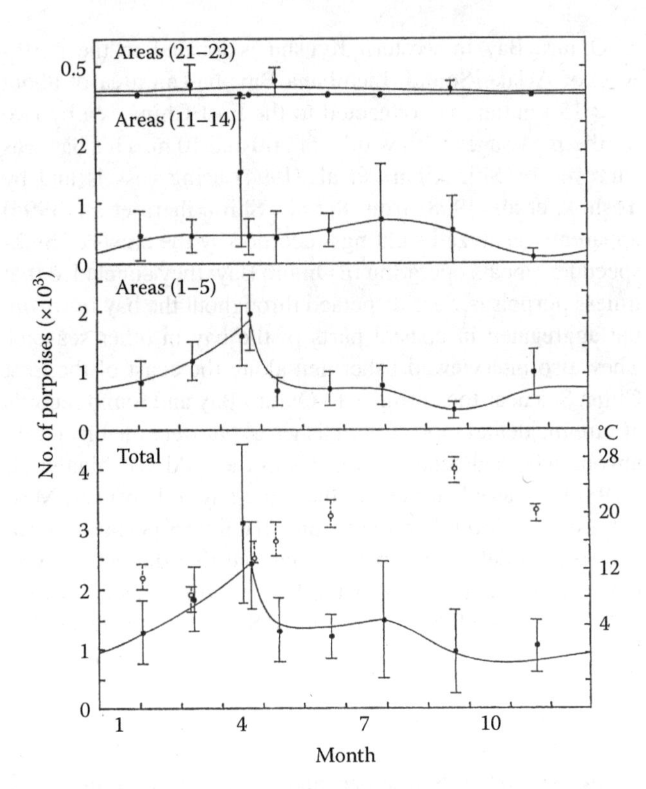

FIGURE 8.6 Apparent seasonal change in the abundance of finless porpoises in the Inland Sea for each stratum, estimated from survey data in 1976–1978. Areas 1–5 represent nearshore strata (≤1 nautical miles from the shore), areas 11–14 intermediate strata (1–3 nautical miles from the shore), and areas 21–23 offshore strata (≥3 nautical miles from the shore). Closed circle represents abundance and accompanying vertical bar the 90% confidence interval of the abundance estimate. The mean (open circle) of the sea surface temperature and the range (vertical bar) observed during each session of the survey are also indicated. Due to technical problems, the estimate of absolute abundance is less reliable than the relative difference in abundance between strata. (Reproduced from Figure 21 of Kasuya, T. and Kureha, K., *Sci. Rep. Whales Res. Inst.*, 31, 1, 1979.)

Kasuya and Kureha (1979) unsuccessfully attempted to confirm whether the finless porpoises in the Inland Sea move out in the early summer through any of the four passes. Land-based observation of 18 hours extending over a 2-day period at Kanmon Pass (northwestern exit) in May and 7 ferry cruises across Hayasui Pass (southwestern exit) in March and April ended with sightings of no finless porpoises. Forty-eight ferry cruises across Akashi Pass (northeastern exit) in March–April and 14 cruises in October across the same pass ended with sightings of only 6 finless porpoises in two groups in March. Naruto Pass, the southeast exit, was surveyed in March and April–May, using ferryboats across the western side (Inland Sea side) of the pass and the eastern side (Kii Channel side). A total of 139.2 nautical miles of survey on both sides of Naruto Pass recorded 233 finless porpoises. Their density was higher in downstream flow in the pass, so the location of high density switched according to the direction of the tidal current, perhaps in relation to the distribution of prey organisms. Our attempt failed to relate swimming direction to seasonal movement of finless porpoises between the Inland Sea and nearby waters. In the late 1970s, I had the impression that the species was extremely abundant in the region of Naruto Pass at least in the March–May season, which was in contrast with near absence observed by Kasuya et al. (2002) in the late 1990s (see Section 8.4.4.1).

8.4.2 Problems in Estimating Abundance

Various methods have been used to estimate the abundance of cetaceans, but they all have some biases due to the technique used or behavior of the animals. A census counts all the members of a population and might be considered the most reliable, but the applicable situation is limited to the cases where the range of the population is small or all the members aggregate in a particular area at a time. If all the members of a population cannot be encountered at a particular time, it becomes necessary to distinguish already identified individuals from new encounters using external morphology, which is applicable to particular species such as humpback or gray whales. Although the eastern stock of gray whales might be thought as easy to census because almost all the members pass by a particular coastal location, the method also requires estimation of the number of individuals that travel during the night or on days of poor visibility and those that pass outside of the observers' visual range. Counting error by observers is also unavoidable.

The mark–recapture method consists of marking or individually identifying animals in the population and then, after a period for mixing, determining the ratio of marked animals to unmarked animals in the population and thereby estimating the population size. Individual identification can be based on external pigmentation, natural scars, artificial marks, or DNA fingerprinting. Full mixing of marked and unmarked individuals is assumed in this method.

Distance-sampling methods are another way to estimate whale abundance. The method first determines the study area, such as the Inland Sea, samples a portion of area within the study area, estimates the density of whales in the sampled area, and extrapolates the density into the total study area. The sampled area is usually a strip along the cruise track of a vessel or an airplane. This method is widely used in cetacean management, and elaborated versions and theories based on them have been developed. The study by Miyashita (2002, in Japanese) is useful for beginners to understand the principle of the method and that by Garner et al. (2002, in Japanese) is useful for more advanced users.

When designing track lines for a sighting survey, seasonal movement of the targeted whales must be considered. For example, if a vessel survey proceeds from south to north within the targeted area during a season when whales are expected to shift their range northward, then there is a risk of the observer traveling with northbound whales, and the survey will result in an overestimate. If the survey proceeds in the opposite direction, the bias will be reversed. Such risks are particularly great for shipboard surveys because they often last for several weeks, and correction of the bias is difficult. It is a good idea to plan a survey in a season when the targeted population reaches the climax of seasonal movement.

Another important point to be considered in designing sighting surveys is any geographical density gradient of the targeted species. The area must be sampled randomly or systematically to avoid sample bias. However, random sampling is usually impractical in track-line design, and unbiased systematic sampling has been applied by arranging the track lines evenly and parallel to the expected density gradient of the whales. The study area can be divided into several subareas to allocate greater effort in a subarea with the expected high whale density. Then the abundance is estimated for each subarea before generating an abundance estimate for the entire study area. This helps increase the precision of the total abundance estimate. This method was followed by aerial surveys for Japanese finless porpoises conducted in 2000 using funds from the Environment Agency (Seibutsu Tayosei Senta [Biodiversity Center] 2002, in Japanese). Track lines for an aerial survey can be designed almost independent of coastlines or distribution of islands. The survey design of Kasuya and Kureha (1979), which used ferryboats connecting the islands in the Inland Sea, did not do this. The track lines tended to be parallel with the coastline, that is, perpendicular to an observed density gradient of the finless porpoise, and were likely to cause bias in the abundance estimation. Their attempt to divide the survey data into strata by offshore distance is believed to be insufficient to remove the bias.

The third point to be remembered for cetacean sighting surveys is the fact that only a certain proportion of individuals present in the sampled strip are identified and recorded. The proportion of undetected whales increases with increasing perpendicular distance from the track line and will be almost zero at a distance of 500 m or more in the case of the finless porpoise. This decreasing trend is used to estimate sighting rate as a proportion of sighting rate at zero distance, which is expressed as g(0) and often assumed as 1. However, in reality g(0) can never be 1, and such an assumption biases the abundance estimate downward. A value of g(0) = 0.3 was estimated for a shipboard survey of harbor porpoises, which are similar to finless porpoises in behavior and in body size aside from the presence of a dorsal fin (Hammond et al. 1995). This g(0) estimate was for a survey employing three observers; the value must be smaller for surveys with only one observer, such as that by Kasuya and Kureha (1979). The value of g(0) was estimated from data for duplicate sightings collected by placing two independent observer teams on a vessel.

The estimate of g(0) is also affected by additional factors such as vessel speed, eye height of observers, number of observers and their sighting ability, sea condition, diving profile of cetaceans, and density and group size. The values for aerial survey and shipboard survey will be totally different. Kasuya and Kureha (1979) surveyed finless porpoises at Beaufort sea states below 3, but they still found that the decline of sighting rate with increasing perpendicular distance was more rapid in November–February (average Beaufort state of 1.2–1.7) than in March–June (average Beaufort state of 0.6–1.2). This was interpreted as an effect of sea-state difference.

The response of cetaceans to survey vessels causes bias to the sighting rate and subsequently the abundance estimate. Some dolphins and porpoises (e.g., weaned young Dall's porpoises) tend to approach sighting vessels and are likely to cause underestimation of perpendicular distance, compared with the case where the porpoises are indifferent to vessels, and to overestimate abundance. On the other hand, a tendency to move away from the vessel (e.g., by adult Dall's porpoises) or to dive underwater before being identified by observers will bias the abundance estimate downward. Usually, finless porpoises are not attracted to vessels but rather avoid them (Pilleri and Gihr 1973–1974) or dive beneath the bow (Zhou et al. 1979). They have not been observed riding the bow, but they have been observed to (1) dive suddenly in front of the bow when approached by a vessel to a distance of 10–20 m or (2) ride the stern wave for a few seconds when the waves pass over the animal (Kasuya and Kureha 1979). Kureha (1976, in Japanese) noted that finless porpoises usually do not ride a ship's bow wave, but some juveniles may occasionally ride the stern wave.

8.4.3 Abundance of Japanese Finless Porpoises

The abundance of Japanese finless porpoises has been estimated by aerial or shipboard surveys, which were not necessarily free from the various types of bias mentioned earlier. In particular, the results of Kasuya and Kureha (1979) using ferryboats in the Inland Sea now seem to have problems. The comparison of two sighting densities obtained by similar methods, one by Kasuya and Kureha (1979) and the other by Kasuya et al. (2002), reveals a drastic decline in density during the 22-year period, which can be interpreted as an evidence of decline in abundance. However, abundance estimated in recent years using aerial surveys (Shirakihara et al. 2007) is greater than that estimated by the early study of Kasuya and Kureha (1979). Accepting the decline in sighting density is reasonable because it derived from surveys using similar and simple methods less affected by methodology than the abundance estimates. This led us to reject one of the two abundance estimates, that is, the smaller figure from the shipboard survey in the late 1970s (Kasuya and Kureha 1979) or the larger figure from the aerial survey in 2000 (Shirakihara et al. 2007). I consider the earlier abundance estimate unreliable, because it was based on undeveloped sighting theory and survey methodology. In particular, the g(0) value assumed at 0.5 by us for the shipboard survey could have been too great, causing downward bias in the estimate. During this survey, I was alone on board almost all the time and collected all the data, including sightings of finless porpoises, vessel position, weather, and sea surface temperature measured using a bucket water sampler. The more recent aerial survey used two observers and one data recorder and probably had greater sighting ability than my shipboard survey. The greater eye height in the aerial survey probably compensated for the bias caused of the greater speed. Additional technical deficits in the abundance estimation of Kasuya and Kureha (1979) are described in the following. Caution is necessary in comparing abundance estimates from aerial and shipboard surveys.

All the recent aerial surveys for Japanese finless porpoises described here have used the same methods and aircraft: a high-wing Cessna with four seats, a pilot in the right front seat, a recorder in the left front, and a pair of observers in the rear seats. The sighting altitude was 500 ft (150 m), and the speed was 80–90 knots (148–167 km/h or 2470–2780 m/min). The most recent aerial survey was carried out by K. Shirakihara's group with the support by the Ministry of Environment and covered all the major habitats of the finless porpoise in Japan. The Seibutsu Tayosei Senta (Biodiversity Center 2002, in Japanese) published the entire results of the project, which was followed by the publication by particular sections in scientific journals by individual participants. The results of this survey are useful for comparing relative values of several finless porpoise habitats in Japan, but any problem in the methodology must affect all of them.

Published estimates of finless porpoise abundance in Japan are summarized in the following. Unless otherwise stated, g(0) is assumed to be 1.

8.4.3.1 Omura Bay

A geographical description of Omura Bay is given in Section 8.4.1. This bay was surveyed twice, the first time in 1993–1994 by Yoshida et al. (1998) and the second in 2000 by Shirakihara and Shirakihara (2002, in Japanese) as described earlier.

Yoshida et al. (1998) carried out four surveys when the Beaufort sea state was 2 or below. They obtained the largest estimate of 343 in May and the smallest of 92 in February, but the four estimates were not statistically different from each other. Therefore, they made a pooled estimate of 187 individuals using the 33 sightings of 53 porpoises in the four surveys. Only sightings by the north-side observer were used, because the south-side observer was subject to glare from the sun, on either the left or the right side of the plane on its east/west tracks at an interval of 1 nautical mile (1852 m).

Shirakihara and Shirakihara (2002) created two sets of track lines across Omura Bay; each set had 13 parallel track lines with an interval of 1 nautical mile, and the track lines of the two sets were 0.5 nautical miles apart. Their plan was to survey Omura Bay twice by randomly choosing one of the two sets for each survey and ended with choosing each of them. They made the two surveys in April, which was close in time to May when the maximum figure was obtained by Yoshida et al. (1998), and recorded a total of 38 sightings of 54 individuals and obtained an abundance estimate of 298 finless porpoises in Omura Bay.

The difference between the two estimates, 187 by Yoshida et al. (1998) and 298 by Shirakihara and Shirakihara (2002), was not statistically significant, and no change in abundance was estimated for the 7-year period. However, such small populations must be managed with great caution for conservation purposes. The surface area of Omura Bay of 320 km^2 gives an average density of 0.58–0.90 porpoises/km^2.

8.4.3.2 Ariake Sound–Tachibana Bay

A geographical description of the Ariake Sound–Tachibana Bay area is available in Section 8.3.1. The two connected sea areas are believed to be inhabited by a single population of finless porpoises, and the proportions of animals in the two areas may vary seasonally, although the details are still unknown. Shirakihara et al. (1994) were the first to estimate the abundance of finless porpoises in the Ariake Sound portion. They collected sighting data from three ferry tracks and from platform-of-opportunity fishing vessels and a university training ship. The number of observers, 1 or 2, varied among surveys and vessels. By combining these data that covered a whole year, they estimated an abundance of 2700 finless porpoises in the Ariake Sound. This figure does not include Tachibana Bay.

Yoshida et al. (1997) made repeated aerial surveys in the Ariake Sound–Tachibana Bay area covering 8 months of the year. The parallel track lines were arranged west/east with an interval of 2 nautical miles. As no seasonal difference in density was identified in either part of the area, they combined all the sighting data to obtain a total abundance of 3093 (Tables 8.4 and 8.5). With a total surface area of 2465 km^2 (Shirakihara and Shirakihara 2002, in Japanese), the density of finless porpoises in the Ariake Sound–Tachibana Bay area is estimated at 1.25/km^2. We should note that the total survey effort and the total surveyed area can be used in estimating abundance; however, for calculating density it is more appropriate to use the area finless porpoises are confirmed to inhabit. Inclusion of offshore water not inhabited by the species biases downward the true density of the species in its habitat.

Using the same method as Yoshida et al. (1997), (Shirakihara and Shirakihara 2002, in Japanese) surveyed this area in 2000, sighted 157 finless porpoises, and estimated the abundance at 3807 (Table 8.5). The density was 1.54/km^2.

The three abundance estimates extending over the past 12 years were not statistically different from each other, but that does not necessarily mean that the population size remained stable during the period. Rather, the interpretation should be that the statistical power was insufficient to detect change in the abundance if any did occur.

8.4.3.3 Inland Sea

A geographical description of the Inland Sea is available in Section 8.4.1. Kasuya and Kureha (1979) made the first attempt to study the distribution and abundance of Japanese finless porpoises there. They made nine series of surveys from ferryboats during the 2 years and 7 months from April 1976 to October 1978. There was a single observer for most of the cruises. The vessel speed of the 28 ferryboats that covered 34 segments (from embarkation to disembarkation) ranged from 8.5 to 18.5 knots with an average of 11.3 knots (simple average) and 12.3 knots weighted by cruised distance. The abundance estimates from the nine surveys were not statistically different from each other (Figure 8.6). Two independent surveys in April resulted in the highest figures of about 3000 porpoises (with approximately 90% confidence interval of 2800–4400) and about 2450 (1700–2800). The estimate of g(0) was 0.5, yielding an estimate of abundance of 4900 finless porpoises in the Inland Sea. It is unclear why they used the smaller figure of the two. However, if the larger figure (3000) was used, the estimated abundance would be about 6000 individuals (Table 8.5).

TABLE 8.5
Abundance of Japanese Finless Porpoises Estimated by Sighting Surveys

Area/Population[a]	Dates of Data	Abundance	95% Confidence Interval	Survey Method	Reference
(1) Omura Bay	August 1993–May 1994	187	127–277	Aerial survey	Yoshida et al. (1998)
(1) Omura Bay	April 2000	298	199–419	Aerial survey	Shirakihara and Shirakihara (2002)
(2) Ariake Sound	August 1988–May 1992	2700	CV = 0.2	Various vessels	Shirakihara et al. (1994)
(2) Ariake Sound–Tachibana Bay	May 1993–May 1994	3093	2,278–4,201	Aerial survey	Yoshida et al. (1997)
Ariake Sound		1983	1,382–2,847		
Tachibana Bay		1110	642–1,920		
(2) Ariake Sound–Tachibana Bay	March 2000	3807	2,767–5,237	Aerial survey	Shirakihara and Shirakihara (2002)
(3) Inland Sea[b]	April 1976–October 1978	4900–6000[c]		Ferryboats	Kasuya and Kureha (1979)
(3) Inland Sea[b]	April–May 2000	7572	5,411–10,596	Aerial survey	Shirakihara et al. (2007)
Middle and east area		1895	1,326–2,708		
Su-wo Nada area		5569	3,692–8,398		
Beppu Bay		108	32–362		
(4) Ise and Mikawa Bays[d]	Apr–Jun 1991–1995	1046	619–1,792	Vessel survey	Miyashita et al. (2003)
Ise Bay		536	241–1,192		
Mikawa Bay		510	261–998		
(4) Ise and Mikawa Bays	May 2000	3743	2,355–5,949	Aerial survey	Yoshioka (2002)
Ise Bay		3038	1,766–5,225		
Mikawa Bay		705	344–1,445		
(5) Awa-Sendai Bay[e]	May–July 2000	3387	1,778–6,452	Aerial survey	Amano et al. (2003)

Note: g(0) = 1 is assumed if not stated otherwise.

[a] Numerals in parentheses indicate currently identified stocks in Table 8.3 and Figure 8.2. See Section 8.3.4 for the possibility of additional stocks.

[b] Excludes Osaka Bay, Hibiki-nada, and most of the area of the Kii Channel, which are inhabited by the same population. *Su-wo Nada* is the major portion of the western Inland Sea west of Yanai Port (33°57′N, 132°08′E).

[c] g(0) = 0.5 is assumed, which is believed to be insufficient for a survey using a single onboard observer compared with aerial surveys with three observers.

[d] This is an estimate made assuming g(0) = 0.899 for data for April–June when density was recorded at the maximum. This estimate does not include porpoises possibly present in the Pacific area adjacent to the two bays (i.e., off the coasts of Shima and Atsumi Peninsulas).

[e] This estimate does not include porpoises in Tokyo Bay. *Awa* indicates southern part of Chiba Prefecture.

The method of Kasuya and Kureha (1979) was certainly inferior to recent methods of survey and analysis, and the resultant abundance estimates will have little value except for historical importance as a record of sighting density and distribution pattern of the species in the Inland Sea. The problems in Kasuya and Kureha (1979) in addition to those stated earlier are as follows: (1) Many of the survey tracks were in parallel with the coastline or perpendicular to the expected density gradient, and more sighting effort was in high-density areas (although the data were stratified by offshore distance to decrease the bias). (2) A view ahead was not available on most of the vessels (left and right sides were observed), and the sighting angle was assumed to be 90° and the radial distance was used as perpendicular distance, which could have caused an underestimate (the true perpendicular distance has to be calculated by r $\sin\alpha$, where r is radial distance and α sighting angle). (3) The problematic estimation of g(0) at 0.5 and surveys by single observers could have overestimated the true g(0) value (underestimating the abundance).

The surveys by Shirakihara et al. (2007) were part of a series of systematic surveys made in 2000 to estimate the abundance of finless porpoises in Japanese waters (Seibutsu Tayosei Senta [Biodiversity Center] 2002, in Japanese). They surveyed the Inland Sea and part of the Kii Channel near Naruto Pass at the southeastern corner of the sea but excluded its southwestern corner, which was deeper than 50 m (finless porpoises were not expected to occur there). The flight track lines were arranged in a north/south direction with an interval of 4′ (about 3.3 nautical miles or 6.1 km). The eastern part of Naruto Pass was added because it was a high-density area recorded by Kasuya and Kureha (1979), but the survey ended with sighting of only one group in that area. They excluded Osaka Bay, most of the Kii Channel, and a small area north of Kanmon Pass (the southern Sea of Japan), where some finless porpoises of the Inland Sea population were known to occur.

It is worthwhile noting that Shirakihara et al. (2007) also found a higher density of finless porpoises in the Suwo Nada area (approximately west of 132°10′E and north of 33°35′N) in the western Inland Sea than in the middle and eastern Inland Sea. This was quite different from the results reported by Kasuya and Kureha (1979) but was the same as reported by Kasuya et al. (2002), which was clear evidence of historical change in the distribution pattern or local depletion of the species between late 1970s and late 1990s in the Inland Sea. In observing a total of 2218.1 km in 60 track lines, Shirakihara et al. (2007) sighted 148 schools of finless porpoises (no other cetacean species were present). The mean school size was 1.56, which allows calculation of the total number of porpoises in the 148 schools at 231. The estimated abundance was 7,572 individuals for the surveyed area of 13,949 km^2 (Table 8.5). The abundance in the western Inland Sea west of 132°10′E, with inclusion of Suwo Nada and Beppu Bay (33°15′N–33°30′N, 131°30′E–131°45′E) southwest of Suwo Nada, was 5677 or 75% of the total abundance in the Inland Sea.

The density of the finless porpoises was 1.31/km^2 in the Suwo Nada area in the western Inland Sea, 0.506 in Beppu Bay, and 0.208 in the central and eastern Inland Sea east of 132°10′E. A relatively high density in the Suwo Nada area seems to reflect a less damaged environment with broad shallow seas. It should be noted that the density was much lower in the islands area further east. The distribution pattern was quite different from that recorded by Kasuya and Kuraha (1979) in the late 1970s.

8.4.3.4 Ise Bay and Mikawa Bay

Ise Bay is about 70 km long and 30 km wide, has a maximum depth of 38 m, and is connected to the Pacific with a 12-km wide opening at the southern end. The shallower Mikawa Bay is 40 km long and 12 km wide, located to the east of Ise Bay and connected to Ise Bay with a 11-km wide entrance near the opening to the Pacific. This Ise Bay–Mikawa Bay area was surveyed twice in the past. The first was a collaborative study between the Toba Aquarium and the whale scientists of the Far Seas Fisheries Research Laboratory. Twelve cruises on small vessels were conducted in a 3.5-year period from 1991 to 1995 (Miyashita et al. 2003, in Japanese). The track lines were arranged in a sawtooth pattern along the coast. Two observers were placed on a sighting platform with eye height of 3.5–6 m, depending on the vessel. The survey speed was 10–13 knots (18.5–24.1 km/h). Sighting was conducted in Beaufort sea state 4 or less. However, later analysis revealed that the sighting rate of 8.0 animals/100 km at sea state 0–2 (252 porpoises) declined to 2.3 at sea states 3 and 4 (23 porpoises), and the authors excluded the data obtained at sea states 3 and 4 (23 porpoises and 1003 km of survey effort) from the subsequent abundance estimation.

Miyashita et al. (2003, in Japanese), using g(0) = 0.899, estimated the finless porpoise abundance in Ise Bay–Mikawa Bay at 1046 (CV = 0.28) in April–June when the highest density was observed (Table 8.5). The value of g(0) was calculated as the probability of a finless porpoise surfacing on a track line within visual distance using surfacing intervals of a porpoise kept in an aquarium tank of 4.5 m depth. It seems to me that the g(0) value was an overestimate (underestimating abundance) due to the following reasons: (1) the diving time of aquarium animals that live in a shallow tank and are fed from the surface must be shorter than that of wild animals that live in deeper water and feed in the water column or on the bottom and (2) it is impossible for observers to identify all the individuals that surface on the track line within their visual distance (some must be overlooked).

Miyashita et al. (2003, in Japanese) found that the densities of finless porpoises in Mikawa Bay were more than twice higher than those in Ise Bay in any season and that the densities tended to be lower from October to March and higher from April to June. They postulated that finless porpoises move out of the bays in winter (Table 8.6). However, the large coefficients of variation that accompany the density estimates throw some doubt on the conclusions. The density differences between the two connected bays disagree with the results of a later study by Yoshioka (2002, in Japanese). Some anthropogenic factors are suspected to be behind the disagreement (see the following text in this section). However, it is premature to exclude the possibility of seasonal density changes in Ise and Mikawa Bays. I observed finless porpoises on the Pacific

TABLE 8.6
Estimated Density of Finless Porpoises in Ise and Mikawa Bays Shown by Population Size/km² and Accompanying CV in %

Area	January–March	April–June	July–September	October–December
Ise Bay	0.14 (54%)	0.34 (41%)	0.19 (35%)	0.02 (88%)
Mikawa Bay	0.24 (67%)	0.99 (34%)	0.61 (53%)	0.60 (25%)

Source: Prepared from text of Miyashita, T. et al., *Gekkan Kaiyo*, 35, 581, 2003.

coast of the Shima Peninsula just west of the entrance of Ise Bay in April–June 2000 (Kasuya unpublished), and there are numerous records of stranding of the species along the coast (Figure 8.2).

After the shipboard survey in 1991–1995 (Miyashita et al. 2003, in Japanese), an aerial survey was conducted in May 2000 by Yoshioka (2002, in Japanese) as a part of a project of the Environmental Agency (Seibutsu Tayosei Senta [Biodiversity Center] 2002, in Japanese). This survey had east/west flight tracks in Ise Bay and north/south tracks in Mikawa Bay. The interval of the track lines was 2 or 3 nautical miles (3.7 or 5.6 km). The survey covered the Pacific coast near the entrance of Ise Bay within the 100 m isobath because earlier information indicated the presence of the species there, but the survey ended with no sighting of finless porpoises there and this area was excluded from subsequent analyses. From the survey, Yoshioka (2002, in Japanese), assuming $g(0) = 1$, estimated the abundance of finless porpoise at 3038 in Ise Bay and 705 in Mikawa Bay (Table 8.5) and the density at 1.95/km² in Ise Bay and 1.38 in Mikawa Bay.

The aerial survey resulted in much higher abundance estimates than those estimated by Miyashita et al. (2003, in Japanese) and the disagreement was greater for Ise Bay than for Mikawa Bay. As stated by Yoshioka (2002, in Japanese), this could be due to differences in methodology, that is, aerial versus vessel surveys. I once participated in a survey of Ise Bay with Miyashita in the winter, when the survey vessel was unable to approach close to the coast because of (1) shallow water depth and (2) the presence of nets for alga culture in the winter season. These factors could have caused downward bias in the abundance estimate, but there is some uncertainty about how the obstacles differed between the two bays.

8.4.3.5 From Tokyo Bay to Sendai Bay

The current Tokyo Bay (35°30′N, 139°50′E) is about 30 km wide and 54 km long from the northern end to the narrowest southern entrance, which is about 9 km wide (at 35°15′N). Tokyo Bay opens through this entrance to eastern Sagami Bay that opens to the Pacific. It is full of vessel traffic, and most of the natural shorelines have disappeared through reclamation for human use. Finless porpoises were known in Tokyo Bay since early times (Otsuki 1808, in Japanese). Strandings and sightings were reported from Yokosuka (35°17′N, 139°41′E) near the entrance of the bay by Nakajima (1963, in Japanese) and Ishikawa (1994, in Japanese). More recent sightings were made in 2000 (Amano et al. 2003) and in 2003 (Kasuya unpublished) at the northern end of the bay (off Kasai), near the central bank (Nakanose), and eastern shores of the bay (off Kisarazu and Futtsu). Amano et al. (2003) made secondary sightings of five porpoises in two groups when returning from their aerial sighting survey transit (Figure 8.7); these

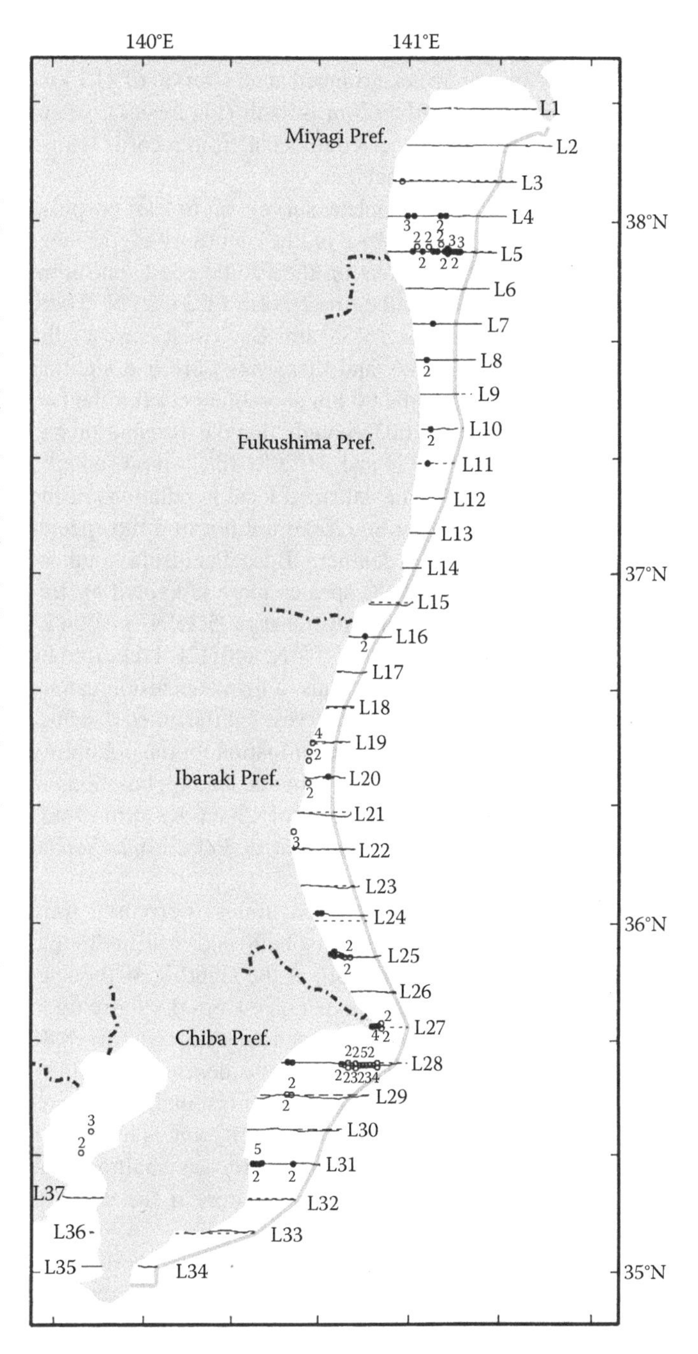

FIGURE 8.7 Flight tracks and positions of sightings of finless porpoises during May–July 2000. Closed circles represent primary sighting, open circles secondary sighting, and numerals school size. A 50 m isobath is indicated. (Reproduced from Figure 1 of Amano, M. et al., *Mammal Study*, 28, 103, 2003. With permission of the Marine Mammal Society of Japan.)

were not used for abundance estimation. Currently, there is no estimate of abundance of finless porpoises in Tokyo Bay.

A survey reported by Amano et al. (2003) was also a part of the project of the Ministry of Environment in 2000 (Seibutsu Tayosei Senta [Biodiversity Center] 2002, in Japanese). Their survey covered an area from western Sagami Bay, or the west coast of the southern part of the Boso Peninsula south of Minato (35°15′N, 139°52′E), and the Pacific coast from the tip (34°54′N, 139°53′E) of the Boso Peninsula to northern Sendai Bay (38°20′N), which is a semicircular bay with north/south length of 18 km and a 40 km wide opening to the Pacific. The total 34 survey tracks, arranged at an interval of 11.1 km, covered waters within the 60 m isobath (Figure 8.7), where finless porpoises were expected to occur from records within the 50 m isobath in other habitats.

This was the first complete survey of finless porpoise habitat north of Tokyo. Most of the sightings of the species occurred in two clusters on the Pacific coast, one from 35°18′N to 36°30′N and the other north of 37°20′N. There were no sightings in eastern Sagami Bay (west coast of the southern Boso Peninsula), and there was only one sighting of two porpoises along the 90 km shoreline between the two clusters. Amano et al. (2003) thought that the distribution gap between latitudes 36°30′N and 37°20′N likely represented a boundary between still unconfirmed local populations of the species. While Amano et al. (2003) did not find the species on the east coast of the southern Boso Peninsula south of 35°15′N, the presence of the species there is known by frequent opportunistic sightings off Ohara (35°15′N, 140°24′E) and a stranding at Kominato (35°07′N, 140°12E′) reported by Ishikawa (1994, in Japanese). Thus, a firm conclusion cannot be reached from a single aerial survey, but the noted distribution gaps are worthy of further examination for the possibility of local populations in the areas, that is, local populations in Tokyo Bay, along the Pacific coast of Chiba–Kashima Nada (35°N–36°30′N) and along the coast of Fukushima–Sendai Bay (37°20′N–38°24′N).

Amano et al. (2003) found that finless porpoises were present only inside the 40 m isobath and within 20 km from the shore. The water depth at the sighting of the species increased with increasing distance from the shore up to 5 km offshore; then, the correlation was lost (Figure 8.8). With the exception of one sighting at a depth of 35–40 m, all the sightings occurred inside the 35 m isobath. I interpret this to mean that finless porpoises were in waters over 5 km from the shore because the depth there was shallow. The range of finless porpoises extends offshore if the depth is shallow (see Section 8.3.1).

Amano et al. (2003), using sighting effort within the 50 m isobath and 51 primary sightings, estimated the abundance of finless porpoises at 3387 (Table 8.5) and the density at 0.502/km². The exclusion of sighting effort between the 40 and 50 m isobaths, where no sightings occurred, could have narrowed the confidence interval of the estimate; however, average abundance would not change, because both effort and size of the study area decreased. However, the use of the area of 6750 km² that includes area between 40 and 50 m isobaths could have biased downward the density of finless porpoises in their habitat.

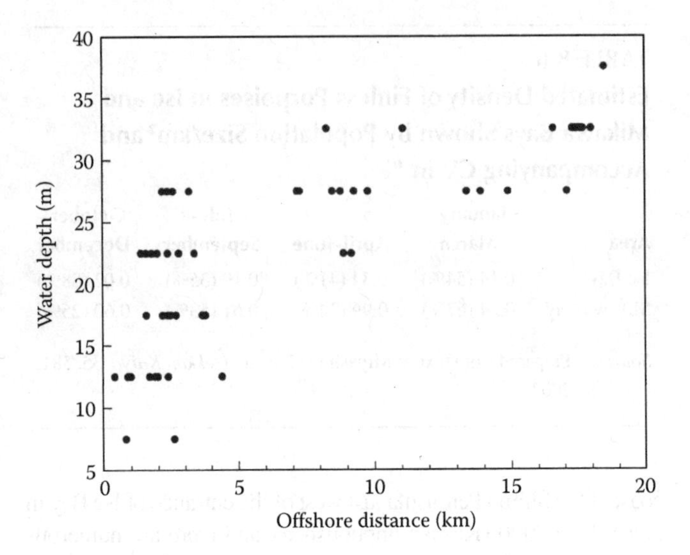

FIGURE 8.8 Relationship between water depth (m) and offshore distance (km) at the position of sighting of finless porpoises (closed circles) along the Pacific coast of Chiba Prefecture to Sendai Bay in Miyagi Prefecture (see Figure 8.7 for the surveyed area). (Reproduced from Figure 5 of Amano, M. et al., *Mammal Study*, 28, 103, 2003. With permission of the Marine Mammal Society of Japan.)

8.4.4 Abundance Decline in the Inland Sea Population

8.4.4.1 Background

It was on September 25, 1997, that I noticed that something strange could have happened to finless porpoises in the Inland Sea. As an adviser to the Ajia Kosoku Co. Ltd (Asian Aero Survey Co. Ltd), which wanted to investigate the feasibility of using an airplane for finless porpoise studies, I joined a Cessna flight. I suspected that the company intended to prepare for a possible future project of the government. We planned to carry out a basic technological test for the aerial survey of finless porpoises. Our plane departed from a small airport near Osaka, crossed the Kii Channel, and flew in the direction of the eastern Inland Sea area that includes Naruto Pass, Shodo-shima Island (34°30′N, 134°15′E), and the Ieshima Islands (34°40′N, 134°30′E), where the species was abundant in the late 1970s (Kasuya and Kureha 1979). The flight of 3 hours and 50 minutes ended with unexpectedly low sightings of only three porpoises in a group near Shodo-shima. I found it hard to believe that I failed to see even one porpoise in the Naruto Pass area, where in the late 1970s I had experienced such a high density that it was difficult to record. I was also astonished by the destruction on the Ieshima Islands, where all the hills and trees behind the villages had been scraped off to sea level, which was a result of mining to produce reclamation materials for the construction of the new Kansai Airport. The flight company made an additional independent flight the next October to Kawanoe (34°01′N, 133°34′E), further west of Shodo-shima Island, but it also ended with no sighting of finless porpoises.

This incident made me revisit the status of finless porpoises in the entire Inland Sea.

The Inland Sea finless porpoises were surveyed twice over an interval of about 22 years using ferryboats and similar sighting methods. Although these surveys were not suitable for abundance estimation, they were able to indicate historical changes in the density and distribution pattern of the species. The first survey covered about 2.5 years from April 1976 to October 1979 (Kasuya and Kureha 1979) and the second about 1 year from March 1999 to April 2000. Sighting data from the second survey were analyzed in comparison with the data from the first survey for changes in density and distribution pattern between the two surveys, and the results were published in Kasuya et al. (2002), with some additional analyses in Kasuya (2008, 2010, both in Japanese).

The Inland Sea was faced with numerous environmental problems in the 1970s, for example, frequent oil spills, increased red tides due to eutrophication by city sewage, damage by the red tides of cultured yellowtail, and progress of chemical pollution by discharges from industry and agriculture. Newspapers concluded that chemical pollution was responsible for reported malformed fishes. To respond to this situation, the government promulgated Provisional Measures for Protection of the Environment of the Inland Sea in October 1973 (came into effect in November of the same year) and then modified it in June 1978 to Special Measures for Protection of the Environment of the Inland Sea (came into effect in June 1979). The major difference between the two regulations was a change from controlling the concentration of pollutants in discharges to controlling total pollutant amount allowed to be discharged. Efforts toward the objective included decreasing phosphorus in sewage by improving detergents for family use and constructing sewage processing facilities to decrease the amount of nitrogen discharged. According to the Setonaikai Kankyo Hozen Kyokai (Association for Conservation of Environment of Inland Sea, 1996, in Japanese), the total number of oil spills reported for the Inland Sea, Osaka Bay, and the Kii Channel increased from 155 in 1970 to a peak of 874 in 1973 and declined to around 100 in 1994. The incidence of red tides increased from 79 in 1970 to a peak of 326 in 1976 and then decreased to 87 or fewer in 1995 and thereafter. Although these improvements reflected effort toward the conservation of the Inland Sea, further improvement remains dubious.

Under such circumstances, it was feared that many of the chemical pollutants would be transferred to the finless porpoises through the food web and be accumulated in the body. Even if individual porpoises did not apparently suffer from acute poisoning by red tides or artificial chemical pollutants, chronic effects might affect the population trend through changes in mortality or reproductive rates. We thought it necessary to record the current status of the finless porpoise population in the Inland Sea in order to identify possible future changes, and myself and K. Kureha, both at the Ocean Research Institute of the University of Tokyo, started the first stage of a survey with the financial support of the World Wildlife Fund, Japan.

8.4.4.2 Survey Plan

The first survey by Kasuya and Kureha (1979) placed high priority on determining the seasonal change in the density and geographical density pattern in relation to the environment, and the plan was to repeat numerous ferry tracks within the limits of time and budget. The recorded information included ferry track, sightings of finless porpoises (position, school size, growth stage, substructure within the school), and sea surface temperature. The ferryboats used were mainly for both passengers and cars, but those for only passengers were also used provided that they were not high-speed ferries. Sighting effort was carried out only when Beaufort sea state was below 3, which is when white caps start to be seen.

Kasuya et al. (2002) chose April–June for the second set of surveys in 1999 and 2000, because more finless porpoises were sighted during those months in the first surveys. However, due to survey scheduling, there was a small amount of effort in late March and early July included. They extracted data from the first set of surveys in the late 1970s only for those months and same ferry tracks for comparison between the first and second sets of surveys (Figure 8.9). During the approximately 22 years between the two surveys, bridges were constructed to connect some islands and some ferry operations were discontinued, but we placed the highest priority on selecting the same tracks and making multiple trips on them. In addition to the second ferry-based survey, we used an additional vessel *Seisui-maru*, a training vessel from Mie University, for a total of 9 days on three cruises.

The *Seisui-maru* started at Naruto Pass, the southeastern entrance, cruised through the Inland Sea toward the west, and completed the survey in Hayasi Pass, the southwestern exit. In the Inland Sea, the track lines of this vessel included both duplication of ferry tracks and entirely new tracks to obtain information in areas of no ferry operation. Sighting effort on the *Seisi-maru* was carried out with the participation of A. Kawamura and K. Shirakihara of Mie University, M. Furuta and staff of the Toba Aquarium, and M. Shirakihara and crew of the vessel; therefore, the sighting efficiency should have been much superior to that in earlier surveys, which were staffed by a single observer. These data were combined for analysis.

If we had found an apparent increase in the density of finless porpoises through comparison of the two sets of data, we could not have reached any conclusion, because the second survey contained cruises with higher sighting efficiency. However, we obtained a lower density in the more recent survey, so a declining trend was judged valid. It was mentioned earlier how the ferry tracks were selected. However, one weak point of the study was the absence of exact data to confirm the assumption of Kasuya et al. (2002) that vessel speeds were similar between the two surveys.

The original sighting data from the first and second surveys are deposited at the National Science Museum in Tokyo for open use by any scientists who may wish to confirm the analysis or carry out new analyses.

8.4.4.3 Diminished High-Density Area

Now, let us compare the positions of sighting of finless porpoises between the first and second surveys, which were

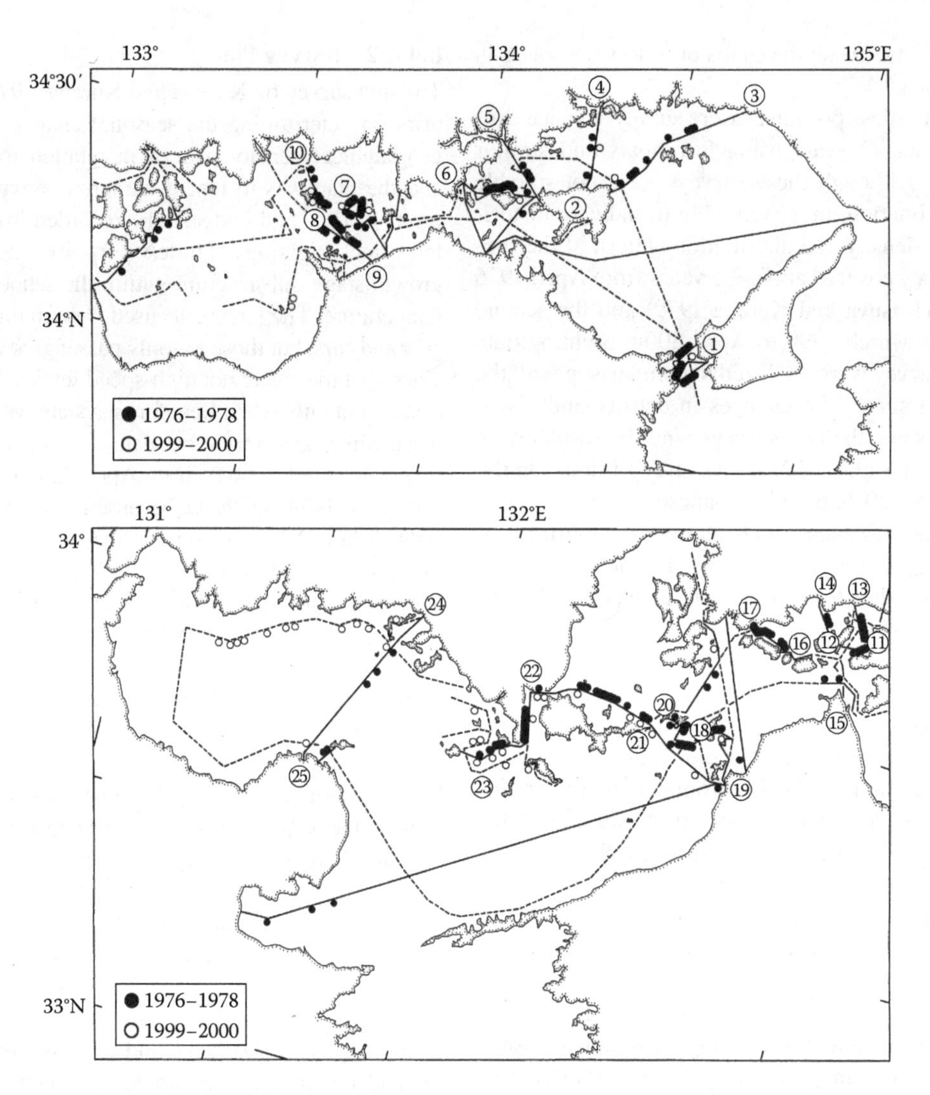

FIGURE 8.9 Distribution of finless porpoise sightings in the Inland Sea, compared between the first shipboard survey in 1976–1978 (closed circles) and the second in 1999–2000 (open circles). Solid lines represent ferry routes observed in both surveys and dotted lines those observed only in the second survey. Tracks observed only in the first survey are not included. Only records of the same season (from March 2 to July 1) have been extracted for this comparison. (Reproduced from Figures 2 and 3 of Kasuya, T. et al., *Raffles Bull. Zool.*, (Supplement 10), 57, 2002. With permission of National University of Singapore.)

about 22 years apart (Figure 8.9). Most of the track lines are expressed by two numbers connected by a hyphen, which indicate ports of embarkation and disembarkation, but short track lines such as that in Naruto Pass are expressed by a single number. The solid lines in Figure 8.9 indicate track lines surveyed in both surveys, black circles indicate sighting positions in the first survey, and the white circles indicate those of the second survey. However, in high-density areas they represent only the geographical range of sightings rather than number of sightings. Some of the track lines are accompanied by almost continuous black circles indicating extended area of high sighting density in the first survey, but white circles (second-survey sightings) are much sparser in those areas. This visualizes high-density areas that disappeared during the 22 years.

Among the cruise tracks in the central and eastern Inland Sea (upper panel of Figure 8.9), Naruto Pass (1), around Shodo-shima Island (tracks 2–3, 2–4, 2–5, and 2–6), from Omi-shima to Kure via Osaki and Kamagari (track 11–17), and Mitsu-hama to Yanai via Nakajima and Hashira-jima (track 18–22) are such areas of drastic decline in density of sightings. The second cruise of the *Seisui-maru* made along the track from Kure (17), to Nasake-jima (20, 21), to Yanai (22) was of particular interest, because numerous spotters including ship's crews searched for finless porpoises under ideal weather condition but ended with so few sightings that I could not believe that we were cruising the same sea that I surveyed 22 years before (bottom frame of Figure 8.9).

Sighting density in the Suwo Nada area was apparently higher than in the central and eastern Inland Sea, and the density decline was less pronounced. This observation agrees with the recent distribution pattern of the species observed through aerial survey by Shirakihara et al. (2007). In addition to the finless porpoise, I had the impression that seabirds had also declined considerably from the late 1970s to 1990–2000, but I did not collect seabird data for analysis.

The above discussion summarizes my impression about historical change in the finless porpoise density in the Inland Sea. Now, I will confirm it using mean number of finless porpoises sighted on each track line (Table 8.7). The survey track from Shodo-shima to Uno (track 2–6), for example, was traveled five times in the survey in the late 1970s with an average of 8.20 finless porpoises sighted (range 0–27), but eight trips on the same track in 1990–2000 resulted in an average sighting of 0.13 (range 0–1). The probability of such a difference arising by chance is 1.2%; thus, it could be interpreted to be due to a real change in the density of the species. If the probability is less than 5%, the difference is accepted as real, stated as significant in Table 8.7. All of the 18 tracks available for this analysis (Table 8.7) showed an apparent density decline, and 11 of them showed a statistically significant difference. The decline was not significant for seven tracks, but this result could be due to small sample size.

Two tracks in the Suwo Nada area showed a relatively minor density decline. The track between Yanai and Iwaishima (track 22–23) showed decline to 88.5% of the density in the late 1970s, and the difference was statistically significant. Another track between Tokuyama and Takedazu (track 24–25) showed decline to 50% from the earlier survey, but the difference was not significant.

8.4.4.4 Density Decline and Distance from the Shore

Finless porpoises usually inhabit waters within the 50 m isobath, so they can be found in such offshore waters as the middle of the East China Sea, where depth is only about 50 m (Section 8.3.1). I have shown that in the Inland Sea of Japan their distribution is apparently influenced also by distance from the shore, although the degrees of contribution of water depth and offshore distance are unknown. Table 8.8 shows the sighting density of finless porpoises in relation to offshore distance for both the first and the second surveys. Data used here are from the same months and include those in Table 8.7 plus those from other cruise tracks available in Kasuya and Kureha (1979) and Kasuya et al. (2002).

In the case of the first survey in the late 1970s, the density in the nearshore stratum (<1 nautical mile) was 18.2/100 km in the central and eastern Inland Sea, 9.1 in the middle stratum (1–3 nautical miles), and 2–3/100 km in the offshore stratum (>3 nautical miles) with limited data. It is apparent from these data that the density of the finless porpoise rapidly declined with increasing distance from the shore. In the western Inland Sea (west of 132°10′E), the density in the nearshore stratum was 19.6, which was close to the corresponding figure from the central and eastern parts of the sea. However, the middle stratum of the same area had a density of 16.5, which was considerably higher than the corresponding figure for the central and eastern Inland Sea. This was likely to be a reflection of a shallow area extending far from the shore. Extremely low density in the offshore stratum was the same in both regions. From the earlier discussion, it is inferred that the density pattern of the finless porpoise in the late 1970s was in principle similar between regions of the Inland Sea.

TABLE 8.7
Density Decline of Finless Porpoises in the Inland Sea Detected through Repetition of Similar Surveys about 22 Years Apart, the First of Which Was Conducted March 2–June 25 in 1976–1978 and the Second March 30–July 1 in 1999–2000

Ferry Track	Survey	No. of Repetitions	No. of Porpoises/Trip	Second/First (%)	p[a] (%)	Significance
1	First	33	7.27	4.5	1.1	Significant
	Second	6	0.33			
2–3	First	11	1.27	0	6.0	
	Second	6	0			
2–4	First	4	0.75	18.7	18.5	
	Second	7	0.14			
2–5	First	4	4.00	0	2.7	Significant
	Second	9	0			
2–6	First	5	8.20	1.6	1.2	Significant
	Second	8	0.13			
7	First	9	7.00	10.0	<1	Significant
	Second	10	0.70			
8–9	First	7	0.86	0	8.2	
	Second	6	0			
8–10	First	6	13.0	6.4	<1	Significant
	Second	6	0.83			
11–13	First	6	7.67	0	<1	Significant
	Second	8	0			
11–12	First	7	4.57	0	7.3	
	Second	2	0			
12–14	First	6	2.83	0	8.8	
	Second	2	0			
12–15	First	8	2.25	0	4.0	Significant
	Second	10	0			
16–17	First	5	7.20	0	<1	Significant
	Second	7	0			
18–19	First	5	2.60	14.6	37.1	
	Second	8	0.38			
18–20	First	5	10.8	0	<1	Significant
	Second	7	0			
21–22	First	7	15.0	4.5	<1	Significant
	Second	9	0.67			
22–23	First	7	7.29	88.5	2.6	Significant
	Second	11	6.45			
24–25	First	4	1.50	50.0	85.1	
	Second	4	0.75			

Source: Reproduced from Table 8 of Kasuya, T. et al., *Raffles Bull. Zool.*, (Supplement 10), 57, 2002. With permission of National University of Singapore.

Note: The relative density is calculated for each ferry track as the average number of individuals sighted per trip. The ferry tracks are indicated by the start and end positions in Figure 8.9. The density change is statistically significant on 11 of the 18 tracks analyzed ($p < 0.05$).

[a] Probability that the observed density difference was due to chance.

TABLE 8.8
Regional and Topographic Effects on the Density Decline of Finless Porpoises in the Inland Sea

Offshore Distance (Nautical Miles)	Surveys	Sea Area (km²)	Survey Tracks (km)	No. of Porpoises Sighted	Density (No. of Finless Porpoises per km × 100) and Second/First Ration (%)	Abundance Index (Density × Area/1000)
Middle and Eastern Inland Sea (Approximately East of Yanai Port[a])						
<1	First	2480	3272	595	18.2	45.10
	Second		2504	15	0.6 (3%)	1.49
1–3	First	3920	1065	97	9.1	35.70
	Second		887	16	1.8 (20%)	7.07
>3	First	1327	117	3	(2.6)	(3.40)
	Second		274	0	(0)	(0)
Western Inland Sea (Approximately West of Yanai Port)						
<1	First	2329	670	131	19.6	45.54
	Second		767	86	11.2 (57%)	26.11
1–3	First	1142	400	66	16.5	18.84
	Second		346	28	8.1 (49%)	9.25
1–3	First	3070	393	2	(0.5)	(1.56)
	Second		159	1	(0.6)	(1.93)

Source: Based on date used by Kasuya, T. et al., *Raffles Bull. Zool.*, (Supplement 10), 57, 2002.
Note: The data sources and the seasons are the same as in Table 8.7, but this analysis includes some data not used in Table 8.7. The density is expressed as the number of finless porpoises sighted per 100 km of survey track. In parentheses are figures thought to be of little confidence due to small sample size.
[a] Yanai Port is located at 33°57′N, 132°08′E.

Now, we compare these results with those of the second survey in the central and eastern Inland Sea. The densities from the second survey were only 3% of those in the first survey in the nearshore stratum, 20% in the middle stratum, and apparently zero in the offshore stratum, which means that the density declined in all the three strata during the 22-year period. The density from the second survey was 0.6/100 km in the nearshore stratum and 1.8 in the middle stratum, and the difference was statistically significant (Kai square test, $p < 1\%$), which indicates that the density decline was drastic in the nearshore stratum and that the density gradient had reversed in recent years. This probably reflects severe destruction of the coastal habitat.

In the western part of the Inland Sea, the density in the nearshore stratum declined to 57% of that in the first survey and in the middle stratum to 49%. These declines of moderate degree were also statistically significant ($0 < 1\%$). For the offshore stratum, we do not have sufficient data to see changes in the 22 years.

8.4.4.5 Decline in Population Size

It is evident that the density of finless porpoises declined in the whole Inland Sea over 22 years. Because the sea constitutes a major habitat of the population, the density decline must have reflected a decline in population size. Kasuya (2008, in Japanese) attempted to estimate the degree of abundance decline using data in Kasuya et al. (2002). The method was to multiply sighting density by the area size to get indices of abundance for each stratum, that is, (no. of individuals sighted)/(distance surveyed, 100 km) × (area size, 1000 km²), and then the indices of abundance for each stratum were added to obtain a total abundance index. Table 8.8 shows the process of the calculation, and Table 8.9 compares the results.

The results suggest that the abundance of finless porpoises in the central and eastern Inland Sea in 1990–2000 was about 10% of the level of the late 1970s, while the corresponding level was 62% in the western Inland Sea. The total abundance index in the late 1970s was 151.2, while that in 1990–2000 was 45.7, which suggested that the Inland Sea population of finless porpoises declined to a level of about 30% of original size in the 22-year period.

The proportion of finless porpoises in the western Inland Sea was 44.3% at the time of the first survey in the late 1970s, but it increased to 81.3% in the second survey in 1990–2000. This latter figure is in reasonable agreement with 75% obtained from the recent aerial survey of Shirakihara et al. (2007) (Section 8.4.3.3). I conclude that the abundance decline of finless porpoises during the 22-year period from the late 1970s to 1990–2000 was greater in the central and eastern Inland Sea and that the major habitat has shrunk to the western part of the Inland Sea and perhaps to the Kanmon area.

8.4.5 Background of the Abundance Decline

The population of Inland Sea finless porpoises in 1990–2000 appears to have been only about 30% of that in the late 1970s, although evaluation of this figure requires some caution because (1) it is not accompanied by a confidence interval, (2) it ignores the possibility of still containing upward bias due to greater number of observers in some recent cruises, and (3) it ignores unknown effects of vessel speed, which might have changed during the period. This change has occurred

TABLE 8.9
Regional Difference in the Abundance Decline of the Inland Sea Finless Porpoises during an Approximate 22-Year Period from 1976–1978 to 1999–2000

		Strata in Offshore Distance[a]			
Geographical Region	**Surveys**	**<1**	**1–3**	**>3**	**Total**
Middle and eastern regions	First	45.10	35.70	3.40	84.20
	Second	1.49 (3.3%)	7.07 (19.8%)	0 (0%)	8.55 (10.2%)
Western region	First	45.54	18.84	1.56	65.94
	Second	26.11 (55.9%)	9.25 (49.1)	1.93 (126%)	37.29 (56.45%)

Note: Based on the abundance indices in Table 8.8.
[a] In nautical miles.

with a particularly intense density decline in the central and eastern part of the Inland Sea, where the decline was greater in the nearshore stratum than outside and the inshore/offshore density gradient reversed during the period.

Finless porpoises live to the age of 20–25 years in both sexes (see Section 8.5.2). Although we do not know their natural mortality rate, it is probably higher than expected for species of longer lifespan such as short-finned pilot whales (females live to 62 years) and sperm whales (to around 70 years). If longevity is defined as the age when a cohort decreases to 1% of the 0-year level and assuming a constant mortality rate for finless porpoises by ignoring the common concept that natural mortality rates of mammals are age dependent with higher values in earlier and later stages of life, then the maximum longevity of 25 years gives an average annual natural mortality rate of 16.8% for the finless porpoise. So, it is probable that the average natural mortality rate in the finless porpoise population is somewhere between 10% and 20%. If a gross annual reproductive rate of the population (no. of births in a year/population size) is equal to the mortality rate, the population will remain stable. However, if the mortality rate increases by 2.5% points and the gross reproductive rate decreases by about 2.5 points, then the population would decline at an annual rate of 5%, which extrapolates to abundance reduction to about 30% of original in 23 years. I suspect that such a situation could have taken place for finless porpoises in the Inland Sea.

Kasuya (1997, 2008, 2010, all in Japanese) and Kasuya et al. (2002) listed several plausible causes of such demographic changes in the finless porpoise population: (1) incidental mortality in fisheries, (2) physiological problems due to chemical pollutants, (3) habitat loss due to reclamation and sand extraction, and (4) ship strikes. They reserved conclusions on the possible effects of (5) interaction with fisheries through food resources, (6) red tides, (7) sound pollution, and (8) epidemics. Shirakihara et al. (2007) noted habitat destruction due to sand extraction. However, it seems to be more reasonable to consider that several factors jointly worked to reduce survival and reproductive rates of the finless porpoise population, as further discussed in the following.

8.4.5.1 Incidental Mortality in Fisheries

Shirakihara et al. (1993) collected finless porpoises killed incidental to fishery operations through contact with FCUs and used them for the study of life history. Among the 114 specimens they collected from the west coast of Kyushu and the western Inland Sea (off the east coast of Kyushu), 84 were known to have been killed in particular fishery operations (58 in bottom-set gillnets, 17 in surface drift gill nets, 7 in trap nets, 1 in a trawl net, and 1 in a drifting discarded net). Because these specimens were obtained through requests to the FCUs in advance of such incidents, there could have been some bias toward fisheries known to have such takes.

In Ise Bay, the Toba Aquarium used drift gill nets for live capture of finless porpoises in the 1970s and 1980s but learned the method was likely to result in high mortality, switched to the use of a sardine seine net, and caught nine porpoises in November 2004. The seine net was 500 m long and operated by two vessels (Furuta et al. 2007, in Japanese). Kasuya et al. (1984) reported the incidental mortality of finless porpoises in seine nets for Spanish mackerel in Ise Bay and reported that the Toba Aquarium hired mackerel fishermen to catch live finless porpoises.

The operation of gill nets, seine nets, and trawl nets is common in the habitats of Japanese finless porpoises. Taking of a large number of finless porpoises, about 300 per year, was reported in fishing operations off the southern coast of the Korean Peninsula, most believed to have been killed in trawl fisheries (An et al. 2010; IWC 2010). Although Shirakihara et al. (1993) listed limited incidents of finless porpoises taken in trawl fisheries, more investigation of this is needed.

We do not know how finless porpoises move and mix within the Inland Sea. However, if finless porpoises are killed by local fisheries, the local density change will be propagated in time to the entire Inland Sea population through the movement of individuals and distribution shift to lower density areas. Thus, local damage to the species can be masked and its identification difficult.

Table 8.10 lists the records of incidental mortality and stranding of finless porpoises in the last 5 years as collected by the Fisheries Agency. Extremely high mortality has been

TABLE 8.10
Recent Estimates of the Abundance of Finless Porpoise in Japan (from Table 8.5) and the Total Reported Figures of Strandings and Incidental Mortality of the Species during a 5-Year Period in 2000–2004

Area and Stocks[a]	Abundance	Incidental Takes	Stranding	Annual Mean of the Total
Chiba-Sendai Bay (5)	3387	4 (0–2)	53 (2–20)[b]	11.4
Tokyo Bay (5)	—	0	0[c]	0
Ise and Mikawa Bays (4)	3743	17 (1–4)	251 (26–64)[d]	53.6
Inland Sea and the nearby area (3)	7572	25 (1–15)	112 (13–28)	27.4
Omura Bay (1)	298	Included below	Included below	Included below
Ariake Sound–Tachibana Bay (2)	3807	6 (0–3)	18 (0–5)	4.8
Others	—	1 (0–1)	4 (0–2)[e]	1.0

Source: From Fisheries Agency statistics (see Section 1.4 for details).
Note: The range of the annual fluctuation is indicated in parentheses.
[a] Numerals in parentheses indicate currently identified stocks in Table 8.3 and Figure 8.2. See Section 8.3.4 for the possibility of additional stocks.
[b] Tokyo Bay, coast of Chiba Prefecture included.
[c] No incidents in the Tokyo Bay area of Tokyo and Kanagawa Prefectures.
[d] Includes two incidental deaths in Shizuoka Prefecture reported in 2000 and 2001.
[e] Includes one incidental death in Kagoshima Prefecture (2002), one stranding in Ishikawa Prefecture (2001), and one stranding in Okinawa Prefecture (2004).

reported from Mie and Aichi Prefectures on the coast of Ise Bay and Mikawa Bay and from Yamaguchi Prefecture in the eastern Inland Sea and Kanmon area. This reflects a high reporting rate due to conservation groups working on those coasts in association with local aquariums and universities. Apparently low incidence of such reports in other regions should be considered to be due to a limited level of observation effort. Yoshida (1994, in Japanese, cited in Yoshida et al. 1997) estimated in his doctoral thesis that 37 finless porpoises were killed annually in fishery operations in the Ariake Sound–Tachibana Bay area in western Kyushu, while Fishery Agency statistics reported the incidental mortality of only around 5 per year in this area plus Omura Bay. I once attempted in the early 2000s to collect the records of incidental mortality from bottom-set gill-net fishermen on Iwaishima Island (33°46′N, 131°58′E), a small island in Suwo Nada in the western Inland Sea, and had the impression that the incidental mortality of the finless porpoise caused by them alone could exceed the annual figure for the prefecture at the time reported by the Fisheries Agency.

Reliable information on the mortality of finless porpoises in various types of fisheries is critically important for establishing conservation plans for the species. It should be considered as an obligation of users of marine resources to collect the mortality data and take necessary mitigation actions. However, it is hard to expect that all the fishermen will respect the Fishery Resources Protection Act (see Section 5.9.1) and report finless porpoises caught in their fishing gear to the government. Rather, most fishermen will avoid the trouble of formalities and discard the carcasses overboard. Some of those discarded carcasses will be stranded and so reported. Government officials working in fishery management are not ignorant of this fact, but it seems to me that they pretend otherwise to escape the burden of mitigation. One example of mitigation is use of sound emitters, called pingers, which have been successfully attached to fishing nets to decrease the incidental mortality of harbor porpoises (Northridge and Hofman 1999).

Caution is required in evaluating figures on incidental mortality or stranding of porpoises because they can be influenced by factors in addition to population size and fishing activities, for example, by the concern of citizens, which is affected by the social atmosphere on conservation. Kasuya et al. (2002) analyzed the trend of incidental mortality in fisheries and stranding of finless porpoises in the Inland Sea in the 29 years from 1970 to 1999, mostly based on data collected by the Institute of Cetacean Research. They observed a low frequency of such incidents during the recent 14 years from 1986 to 1999; 2 cases of incidental mortality and 12 stranded carcasses were recorded with an annual fluctuation of 0–6 and an annual average of only one (6 records in 1998 and 0 in 1999 included). However, they indicated higher figures for the preceding 16-year period from 1970 to 1985 with a total of 62 (37 incidental kill, 16 stranded carcasses, and 9 deaths of unknown cause) with an annual mean of 3.9 and an annual fluctuation of 1–6. Kasuya et al. (2002) felt that the recent decline in such records was apparently inconsistent with recently increasing concern of local communities about cetaceans and suggested the possibility that it could reflect decreased abundance of the species in the Inland Sea that occurred in the mid-1980s.

8.4.5.2 Ecological Interaction with Fisheries

Only limited information is available on the prey items of finless porpoises in Japan. Mizue et al. (1965, in Japanese) reported horse mackerel, sardines, and squid in the stomachs of finless porpoises in western Kyushu, while Kataoka et al. (1976, in Japanese) reported sand lance, squid, and crustaceans in the stomachs of finless porpoises in Ise Bay. Shirakihara et al. (2008) analyzed the stomach contents of finless porpoises incidentally killed in fisheries in two populations off western Kyushu: (1) Omura Bay and (2) Ariake Sound–Tachibana Bay. Their sample contained 9 individuals including 1 juvenile from Omura Bay and 78 individuals including 20 juveniles from Ariake Sound–Tachibana Bay. The prey in the Ariake Sound–Tachibana Bay sample was more diverse than in the Omura Bay sample. Food items in the former sample identified to species included three species of Cephalopoda (mainly Octopodidae, Sepiidae, and Loliginidae) and 26 species of Teleostei (mainly Clupeidae, Engraulidae, and Sciaenidae), in addition to a small number of crustaceans. Prey in the Omura Bay sample contained six species of Teleostei, two species of Cephalopoda, and some crustaceans, with Gobiidae and Atherinidae in most stomachs. The authors thought that the difference in the diversity of prey reflected the marine fauna of the two areas rather than the difference in their sample sizes. They noted for the Omura Bay samples that the food of juvenile finless porpoises below 1 year of age was dominated by small bottom fishes of the families Gobiidae and Apogonidae and Cephalopoda, while adults did not consume Gobiidae, and they concluded that the juveniles of weaning age had food preferences different from those of adults. Although this conclusion could be correct, it could also be true that an interaction between seasonal change of prey fauna and a seasonally limited weaning period could have affected their analysis. Therefore, it is necessary to compare stomach contents between different growth stages using the samples obtained in the same season before firm conclusions can be reached.

Barros et al. (2002) reported the stomach contents of 31 finless porpoises stranded in the Hong Kong area to be composed of fish, squid, shrimp, and octopus (in a decreasing order). Among the fish families, Apogonidae was the most numerous, followed by Sciaenidae, Engraulidae, and Leiognathidae, and Loliginidae was the most common among the cephalopods. These comprised 77% of the stomach contents examined. From these observations, Barros et al. (2002) concluded that finless porpoises in the Hong Kong area (1) feed in coastal waters (food items were all coastal species), (2) feed also on benthic organisms (Sepioidea, Octopoda, and Sciaenidae), and (3) also feed in the water column (Apogonidae, Engraulidae, and Trichiuridae). They also indicated that some of the prey items are species targeted by trawl fisheries in the region.

The coastal marine environment in temperate latitudes inhabited by finless porpoise populations fluctuates greatly with season both in water temperature and prey fauna available to the porpoises. Kasuya and Kureha (1979) recorded the range of surface temperature in the Inland Sea from 6°C to close to 29°C, and a slightly higher temperature was also recorded (Table 10.1). Sendai Bay, the northern limit of the species in Japan, experiences lower winter temperatures than the Inland Sea. Japanese finless porpoises are probably enjoying the high productivity of shallow inland waters through the acquisition of adaptations to a broad temperature range and ability to utilize a broad spectrum of seasonally variable food resources.

Marine animal products from the Inland Sea, excluding products of aquaculture, recorded both a rise and then a decline in the 1965–1994 period (Setonaikai Kankyo Hozen Kyokai [Association for Conservation of Environment of Inland Sea] (1996, in Japanese)). The total increased from slightly under 200,000 tons in 1965 to 350,000 tons in 1982 and then declined to the previous level of 200,000 tons in 1994. Change in the harvested fish fauna accompanied the change in the total amount of fishery production; during the high-production phase, Japanese anchovy *Engraulis japonicus* and sand lance *Ammodytes personatus* increased and jack mackerel *Trachurus* spp. and an octopus *Octopus vulgaris* decreased, but the latter two increased during the later low-production phase. This probably reflected the abundance of phytoplankton feeders such as anchovy that increased responding to eutrophication of the Inland Sea in the 1970s and 1980s and decreased subsequently following decline in phytoplankton in the 1990s due to the improvement of the water quality.

In view of the flexibility of prey preference of finless porpoises, it is unlikely that there was competition at the level of the entire Inland Sea between finless porpoises and fisheries for food resources and that the competition caused the observed decline in the abundance of the porpoises in the 1980s and 1990s. However, this does not preclude the possibility of competition in a particular local habitat.

8.4.5.3 Red Tides

Red tide is so named because the seawater changes to red or brown due to an outbreak of unicellular organisms such as dinoflagellates and diatoms. Fish may suffer mechanical damage to the gills or die from suffocation due to consumption of dissolved oxygen by carcasses of the red tide organisms. However, marine mammals appear to undergo more damages from poisonous chemicals produced by some red tide organisms. There are examples of mortality of humans from consuming mussels that accumulated saxitoxins from plankton and of humpback whales that consumed similarly poisonous mackerel. Gray whales, bottlenose dolphins, manatees, sea lions, and sea otters have been reported to have died from ingesting brevitoxin and domoic acid from red tide. Damage by ciguatoxin to Hawaiian monk seals is suspected (Van Dolah et al. 2003). These acute cases are more easily identified than chronic effects.

Four red tide incidents were recorded in the Inland Sea in 1950, followed by a slow increase to 18 in 1960 and then a linear rapid increase to a maximum of 299 in 1976. After the maximum in 1976, the incidence of red tides in the Inland Sea slowly declined to 188 in 1980, 170 in 1985, and 108 in 1990. The number of red tides subsequently remained almost stable at around 90–100 until 1994 (Setonaikai Kankyo Hozen Kyokai [Association for Conservation of Environment of Inland Sea] (1996, in Japanese). Although I am uncertain how accurate the counting of red tides is, the numbers mentioned earlier give some idea of the annual trend. The red tides were reported from almost the entire range of the Inland Sea during

1970–1975 when the incidents peaked, then the range started to shrink, and as of 1995, they were almost limited to the eastern half of the Inland Sea ranging from Kagawa and Okayama Prefectures to Harima Nada and to Osaka Bay in the east. This improvement could have been a result of regulation of discharge of nutrients into the sea.

Setonaikai Kankyo Hozen Kyokai (Association for Conservation of the Environment of the Inland Sea; 1996, in Japanese) also reported damage by red tide to aquaculture such as that of yellowtail and red sea bream. The amount of the damage was 19.9 billion yen (3.3 billion yellowtail plus 5600 tons of bream) during the 24 years from 1972 to 1995. The red tide organisms that caused the damage were reported to have been *Chattonella* (8 incidents) in 1972–1988 and *Gymnodinium* (15), *Heterosigma* (8), *Noctiluca* (3), and *Gonyaulax* (2) in 1991–1995. Several species of the genus *Chattonella* are known to produce brevitoxin and those of *Gymnodinium* to produce saxitoxins (Van Dolah et al. 2003). However, no incidents have been reported of mortality of marine mammals or seabirds due to ingestion of marine algal toxins in the Inland Sea. It will be valuable to investigate the cause of mortality of aquaculture fishes.

I could not find any cases where marine mammals were impacted by algal toxins in Japan, although there were reported cases where scallops cultured off the Pacific coast of northern Japan became unsuitable for human consumption due to the accumulation of algal toxins. It needs to be confirmed whether Japanese finless porpoises have or have not received acute or chronic damage from algal toxins. If acute poisoning really happened to finless porpoises in the Inland Sea, it is likely to have been noticed. The effect of chronic poisoning, if it has happened, could still be affecting the population dynamics of the current population. Red tides that occurred everywhere in the Inland Sea during the 1970s and 1980s are currently limited to smaller areas, but if they are producing algal toxins, it could affect the density of finless porpoises in the broader range through the movement of prey animals as well as finless porpoises.

8.4.5.4 Chemical Pollution

The Inland Sea of Japan has a geographical structure likely to accumulate pollutants, because it opens to the outer sea through only four small openings and is surrounded by industrial, agricultural, and urbanized areas. Nutrients and chemical pollutants are also discharged by aquaculture. Some of the pollutants discharged through these human activities, such as chlorinated organic compounds (e.g., PCB, DDT, and their derivatives) are poisonous, persist in the sea for long period, and are taken into phytoplankton to be transmitted to zooplankton, fish, and marine mammals through the food chain. The levels of concentration of such molecules increase with a trophic step and become particularly high in cetaceans because they are top predators, live a long life, and have limited physical ability to break down the chemicals.

These pollutants in female cetaceans are transferred to the young through the placenta and milk, which means that females can lower their burden of the pollutants through reproduction and that pollutant level of newborns depends on the reproductive history of their mother. On the other hand, the accumulation process of such pollutants in males tends to be simpler, with levels usually increasing with age, because they do not have such opportunities to shed the pollutants through reproduction. This is the reason why males are often selected as an indicator of pollutant level in a population and why maximum figures are compared between populations.

Kasuya et al. (2002) identified high concentrations of PCB, DDT, and organic tins in finless porpoises in the Inland Sea (Table 8.11) and suggested the possibility to reduce the survival rate and reproductive rate of the population. The pollutant levels exceed levels in toothed whales in the outer seas off Japan or those of belugas in the St. Lawrence estuary, which recorded the level of PCBs in blubber at 100 ppm in females and 250 ppm in males. This beluga population was once hunted down to 600–700 (12%–14% of the initial level) by 1979. It did not show signs of recovery under protection that started in 1980, and the pollutants are suspected to be causing a high incidence of cancer identified (Martineau et al. 1994).

DDT is an insecticide that has been used worldwide and is still in use in some limited countries. The density in marine waters is believed to be declining. However, different from the situation for DDT, the level of PCBs in seawater does not show a declining trend (Reijinders 1996). One to 2 million tons of PCBs have been produced in the world since 1929, and 31% has already been discharged into the environment; 65% is believed to be controlled in some way, but the risk of

TABLE 8.11
Highest Recorded Levels of Persistent Pollutants in the Finless Porpoise in the Inland Sea

Pollutants	Years of Data	Organ	Level	Note	References
PCBs	1968–1975, 1985	Blubber	320 ppm	Production of PCB ended in 1972. The level was 16 times of that in offshore striped dolphin.	O'Shea et al. (1980); Kannan et al. (1989)
DDTs	1970s	Blubber	132 ppm	Marketing ceased in 1971. The level was 4.4 times that in the offshore striped dolphin.	O'Shea et al. (1980)
Organic tin compounds	1985	Liver	10.2 ppm	Several times the level in finless porpoises on outer coasts. Use is regulated.	Tanabe et al. (1998)
Dioxins	1998	Blubber	240 pg/g	Discharge is regulated.	J.M.E. (1999)[a]

[a] From the website of Japanese Ministry of Environment.

leakage is not eliminated. In Japan, the production of PCBs stopped in 1972, and their discharge into the environment is prohibited. Dioxin is a by-product of chemical industries and incineration of city garbage.

These chlorinated organic compounds are known to disrupt the normal function of hormones and the immune system in laboratory animals and to prevent fetuses and suckling calves from the normal development of reproductive functions. Dioxins are the most powerful artificially produced chemicals, and their daily allowance level is 200 pg (10^{-12} g) for a person of 50 kg in Japan. The pollutant levels of some Inland Sea fishes are so high that the consumption of one piece of *sushi* with 30 g of spotted shad is sufficient to exceed the daily limit, because shad in the Inland Sea contains 8–9 pg/g of dioxins. Organic tin compounds have been discharged into the Inland Sea since the 1960s through maritime use for preventing fouling of vessels and nets of aquaculture by organisms and from terrestrial use for plastic products, floor wax, and cleaning industries. One of the well-known environmental effects of organic tin compounds is the formation of a penis in a female snail. It is also known to disrupt the functions of hormones and immune systems of mammals, with strong effects on animals in the early stages of development, which is similar to the effect of chlorinated organic compounds (Colborn and Smolen 2003).

We humans have discharged numerous chemicals not listed earlier into the environment, and some of them are suspected to have adverse effect on marine organisms. Polycyclic aromatic hydrocarbons, one such example, are suspected to be another carcinogen of belugas in the St. Lawrence estuary (Martineau et al. 2003). The effects of persistent marine pollutants last many years after release and cover a greater area of the Inland Sea than red tides. However, it is almost impossible to have a full understanding of the chronic effects of such chemicals on the health of finless porpoises or other cetacean species. If a chemical is identified as harmful to laboratory animals, it is safe and scientific to assume the same effects on cetaceans and to take precautionary actions. Conservation scientists are likely to be requested to provide firm evidence of malignant effects on wildlife, but such an attitude should be strictly rejected because it ends with the sole result of delaying effective action.

8.4.5.5 Reclamation and Sand Extraction

Finless porpoises usually inhabit waters inside the 50 m isobath and are rare in deeper waters, which are limited in the Inland Sea to a small area in the southwestern part of the sea. An analysis of finless porpoise density in relation to offshore distance based on surveys in the late 1970s, when the population size was greater, revealed a sharp density decline from 2.6 individuals/cruise of 10 nautical miles for waters <1 nautical mile from the shore to 0.3 in waters >3 nautical miles from the shore (see Section 8.3.1). This indicates that finless porpoises prefer shallower or nearshore waters even within the 50 m isobath. Such shallow seafloor is likely to have colonies of sea grass or algae and offer habitats for various marine animals consumed by the porpoises. Subsequent surveys in 1999–2000, made about 22 years after the first one, revealed that (1) the density of finless porpoises declined in all parts of the Inland Sea, (2) the decline was greater in the central and eastern portion of the sea and in the nearshore strata (<1 nautical mile from the shore), (3) as a result, the density gradient in the central and eastern region of the sea reversed between nearshore and intermediate (1–3 nautical miles offshore) strata, but (4) the degree of the decline in abundance was smaller in the western Inland Sea, and the earlier gradient pattern was still retained there in the 1999–2000 surveys (see Section 8.4.4).

One would expect the events of environmental destruction that occurred between the late 1970s and the late 1990s in nearshore waters of the Inland Sea and in greater magnitude in the central and eastern part of the sea than in the western part to be behind the finless porpoise decline. Reclamation and sand extraction are such candidates. It is efficient to take sand for construction from shallow coastal waters because of the higher quality and lower cost of the sand. The activity destroys the tidal flat and coastal marine community preferred by finless porpoises, and recovery of the coastal marine community cannot be expected because of the increased water depth. Such destruction further expands with time to nearby areas through sand migration to the deep hollows thus created. As reclamation usually occurred in shallow waters near the coast, its effect was similar to or worse than that of sand extraction, because in some cases fill materials was supplied by sand extracted from nearby areas. I suspect that such activities functioned to decrease the carrying capacity of the central and eastern Inland Sea for finless porpoises.

Arita (1999, in Japanese) stated that sand from weathered granite was preferred as construction material available in the Inland Sea and that it was found in large quantities in narrow channels. The top four prefectures that produced marine sand in the past 30 years were Kagawa, Hiroshima, Okayama, and Ehime in decreasing order, each of which produced 100–130 million m^3 of sand. Yamaguchi Prefecture followed with 25 million m^3. Matsuda (1999, in Japanese) stated that the extraction of marine sand started around 1960 and continued until 1999, when it was completely prohibited. This sand extraction of nearly 40 years resulted in water depth increases ranging from 10–20 to 30–40 m and with destruction of habitat of the sand lance. Fisheries in these prefectures recorded declines in landings of fish species living near the shore or on sandy bottom, such as sand lance, as well as landings of fishes that prey on sand lance (Matsuda 1999, in Japanese).

Shirakihara et al. (2007) identified two areas as the places where most of the sand extraction occurred. These two areas agree with the top four prefectures listed by Arita (1999, in Japanese). One of them (A in Figure 8.10) is an island area between Hiroshima and Ehime Prefectures, and B is another island area between Okayama and Kagawa Prefectures. Shirakihara et al. (2007) believed that the absence of sightings of finless porpoises in these two areas during their aerial surveys in 2000 was due to the destruction of their habitat through sand extraction. When I surveyed from ferryboats connecting these islands in the late 1970s, I was impressed by a high density of finless porpoises between the islands (Figure 8.9; Kasuya and Kureha 1979), and the absence of the

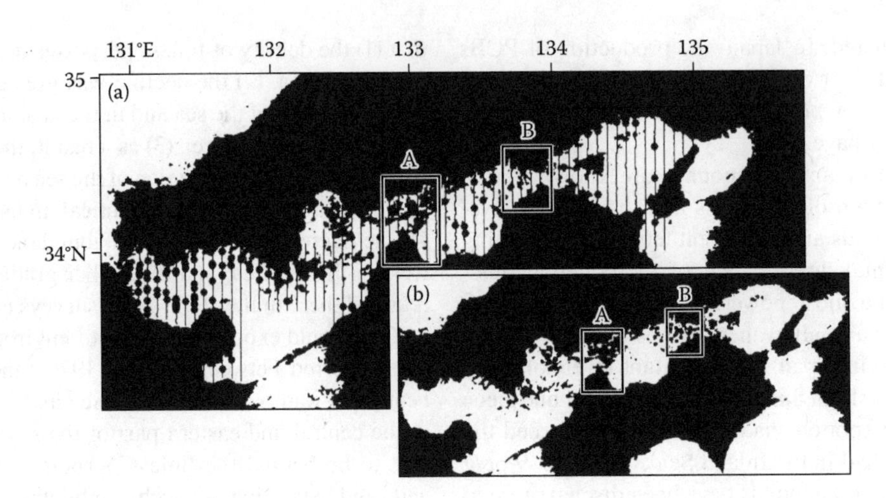

FIGURE 8.10 (a) Track line and position of finless porpoises sighted during the aerial survey (closed circle) in May 2000, (b) Position of heavy sand extraction of the past years (closed circles in frames A and B). (Reproduced from Figure 3 of Shirakihara, K. et al., *Mar. Biol.*, 150, 1025, 2007. With permission.)

species in the outside open waters was interpreted as a reflection of their preference for shallow habitat.

During my survey activity in the late 1970s, I was able to identify sand extraction activities in the northern part of area A in Figure 8.10 (south of Takehara in Hiroshima Prefecture), but there were still numerous finless porpoises present. After about 22 years, when I revisited the area in 1999, finless porpoises had totally disappeared (see track lines connecting points 11, 12, 13, 14, 15, 16, and 17 in Figure 8.9). The water depth south of Takehara was 5–18 m in 1961, but it increased to 35–48 m in 1995 (Shirakihara et al. 2007). Sand extraction for construction materials has increased the water depth of the channels between the islands and has created easy passages for vessels but seems to have destroyed habitats of the finless porpoise in the Inland Sea.

Reclamation started in the Inland Sea in 1898 and created a total of 264 km^2 of land by 1969. The reclamation was particularly intense after 1955 with demand for land for industrial use. A conservation law in 1973 (see Section 8.4.4.1) intensified restrictions, but there was additional reclamation of 97.4 km^2 by 1995 (Setonaikai Kankyo Hozen Kyokai [Association for Conservation of Environment of Inland Sea] 1996, in Japanese). Over 350 km^2 of shallow coastal habitat of the finless porpoise was lost since the late nineteenth century in the Inland Sea. Natural seashores (sandy beaches and rocky shores) remaining in the Inland Sea make up only about 38% of the total coastal length, which is lower than the national average of 57%.

I could not find any difference between the western Inland Sea and the rest of the sea in the proportion of natural seashore or the amount of reclamation. A simple comparison of total amount of destruction would probably not have much meaning, but more important would be the proportion still remaining as a suitable habitat for finless porpoises of the habitat that originally existed in the Inland Sea. The quality of the currently remaining habitat will differ between that of a group of numerous fragmented habitats and that of smaller number of larger habitat even if the total areas are the same. A seaweed and sea grass bed in the Suwo Nada area (40 km^2) is the largest single remaining habitat patch in the Inland Sea. This area is known for a relatively high density of finless porpoises. The next largest habitat areas are found in Hiroshima Bay (23 km^2) and in Iyo Nada (20 km^2) where finless porpoises are rather scarce.

Finless porpoises that have lived in habitat destroyed by human activities will not instantly die but will move to nearby habitats to create an apparent density increase in the new habitat. Because this does not accompany an increase in carrying capacity, the increased density must return to the initial level. Through this process, in the long term, the abundance of the Inland Sea population will decrease to match the new total carrying capacity.

The direct influence of habitat destruction as described earlier might be limited to a particular geographical area. However, we saw a drastic density decline of finless porpoise in the Naruto Pass area of the Inland Sea. The only large-scale construction known in the area during the past 20–30 years was that of a large suspension bridge completed in 1985. It is 41 m above the sea surface and 876 m between piers and is unlikely to have had a significant effect on the distribution of finless porpoises (other than effects of activities during construction). The density decline in the Naruto Pass area must, at least in part, be attributed to physiological disruption due to chemical pollution and mortality in fishing gear, both of which have a final effect on the density in a broader range of the population. An additional cause of the decline will be a still unconfirmed negative effect of local habitat destruction on the carrying capacity of finless porpoises in the broader area. For understanding of these elements, we need better understanding of geographical movement of individual finless porpoises by season and by life cycle within the Inland Sea. Currently, we do not have such information.

8.4.5.6 Other Causes of the Decline

Finless porpoises rarely move away from an approaching vessel but tend to dive in front of it (Kasuya and Kureha 1979).

This behavior is likely to result in ship strikes. In the late 1970s, some fishermen in Te-shima and Manabe-shima in the Shiaku Islands group (34°22′N, 133°40′E) in the Inland Sea at my request, retrieved carcasses of finless porpoises found during their fishing activities and buried them on the beach for my later retrieval during sighting survey trips. Most of these carcasses had apparent wounds from ship's screws, but it was not possible to determine if the injuries were caused before or after death.

Installation of hot-bulb engines (also called "semidiesel engine") on small Japanese fishing vessels started around 1920, and they were replaced by diesel engines in the 1960s. This could have been the start of ship strikes on finless porpoises, but the early history is ignored here to concentrate on vessel traffic after the 1970s. There was an increase of 26% during the period from 1972 to 1994 in the total tonnage of cargo vessels that entered ports in the Inland Sea. In addition to this, there were increased operations of high-speed ferries including jetfoils and hydrofoils. Thus, ship strikes of finless porpoises could have been increasing with time (Kasuya et al. 2002). Underwater noise emitted by vessels, which depends on vessel size and propulsion type, also degraded the underwater acoustic environment of finless porpoises, although the effect cannot be evaluated at present.

Many of the new environmental elements, such as monofilament gill nets, fast-moving ships, and vessel noise, were introduced into marine mammal habitat during the recent two or three generations of the species, and it seems premature to judge how the affected species respond and adjust their life to them (Tyack 2008).

Since the late twentieth century, there have been reports of mass mortalities of cetaceans around the world. The causes include poisoning from red tide (Section 8.4.5.3) and infection by influenza virus and morbillivirus (Geraci 1999). Hundreds of bottlenose dolphins were found dead on the east coast of North America and in the central North Atlantic in 1987–1988, and several thousand striped dolphins that died in the Mediterranean in 1990–1992 were believed to be victims of morbillivirus infection, but disruption of the immune system by accumulated chemical pollutants is suspected to have been a contributing cause.

If mass mortalities of finless porpoises had occurred in the Inland Sea, they would have been noticed because the sea is almost a closed water system, has high fishing activity, and is surrounded by populated coasts. We have no answer at present on why there have been no reports of mass mortalities of any cetaceans along the coast of Japan including the Inland Sea. It is important to investigate the background of such situations and to be prepared for any future mass mortality.

8.4.6 History of the Protection of Finless Porpoises in Japan

8.4.6.1 Industrial Use of Finless Porpoises

The opinion on the suitability of finless porpoises for human consumption varies geographically. Finless porpoises killed incidental to fishing activities in Korea have been sent to market. In the 1960s, fishermen along the coast of Tachibana Bay in western Kyushu discarded most of the finless porpoises found in their small fixed trap nets, but they also sent some of the carcasses to the fish market. Mizue et al. (1965, in Japanese) purchased such carcasses for their osteological study. This confirms that the species has been consumed on the west coast of Kyushu. Later, in the 1970s, Prof. Mizue of Nagasaki University told me that the trap-net fishery was discontinued due to change in fishery regulation and that there was no more incidental mortality of porpoises (Kasuya and Kureha 1979). This explains why Shirakihara et al. (1993) did not include samples from this source.

The people of Ayukawa on the Oshika Peninsula, or on the east coast of Sendai Bay (the northern limit of this species in Japan), hold the species in quite a different light. Nobody there eats finless porpoises because of a belief that it has a strong purgative effect. N. Kimura, the late director of the Ayukawa Whale Museum, once tried meat and blubber cooked together and experienced violent diarrhea within 10 minutes. Then he repeated twice boiling the porpoise meat and poured off the oil before eating; this eliminated the purgative effect. An early nineteenth century scholar in Sendai (west coast of Sendai Bay) wrote in his book *Geishi-ko* (*On Natural History of Whales*) that finless porpoises are unsuitable for human consumption because of high oil content (Otsuki 1808). Such a diarrheic effect of finless porpoise is also known from the Chinese coast (J.Y. Yang 2007, personal information). Kasuya et al. (1985, in Japanese) noted that only the Satohama-Miyashita shell mound (see Section 2.1) produced remains of finless porpoises out of the 22 archaeological sites known with cetacean remains, including sites along the coasts of Tokyo Bay, Mikawa Bay, and the Inland Sea that were inhabited by finless porpoises, and hypothesized that it could be due to the unsuitable nature of the species for human consumption. Out of five volunteer students who ate meat and blubber of finless porpoise off western Kyushu, only one experienced mild diarrhea (M. Shirakihara personal communication, Kasuya 1999). Thus, the possible variation in effect needs to be investigated.

Kaburagi (1932, in Japanese) stated that finless porpoises were hunted for oil in the Inland Sea and that the fishermen of Moriguchi (current Kamiura in 34°16′N, 133°03′E) made expeditions to Aba-shima Island (34°19′N, 132°57′E) area for finless porpoises and had several conflicts with fishermen of Tadanoumi north of Aba-shima. The purpose of the hunt was to obtain oil for use as an insecticide in rice fields, but the Tadanoumi fishermen who fished in the Aba-shima area wanted to protect the porpoises. Otsuki (1808, in Japanese) recorded finless porpoises as a source of whale oil, and Okura (1826, in Japanese) stated that oil from the finless porpoise, as well as other whale oils, can be used to kill locusts (most likely leafhoppers) in rice paddies. It seems likely that finless porpoises were once hunted locally for oil.

After World War II, there was a plan for small-type whaling for finless porpoises, which were considered by the people of the time as a nuisance in the Inland Sea, to make fish paste from the meat (*The Asahi Shinbun* of January 20, 1947). A person at Tonosho in Shodo-shima ordered a double-barreled whaling cannon of 19 mm caliber for *Nihon Seiki Seisakusho* in Kochi, Shikoku. In those days, this type of whaling was not regulated (Sections 4.1, 5.4 and 5.5). I do not know the results of the project, but it would seem to make more sense to operate small-type whaling for pilot whales and other dolphins in the Pacific rather than hunting such a small unprofitable species as the finless porpoise in the Inland Sea.

While I was traveling the Inland Sea in the late 1970s, I often heard from local people that shortly after World War II, they used to extract oil from finless porpoises found in their fishing nets. The oil was used for light during electricity failures. In the early 2000s, I was shown a small glass jar containing finless porpoise oil by a friend at Iwai-shima Island (33°47′N, 131°59′E) in the Suwo Nada region in the western Inland Sea. The foul-smelling oil was extracted from a finless porpoise in early times to be used for burns, scalds, and cuts.

The custom of using finless porpoises in the Inland Sea seems to have ceased in the 1960s or earlier, which is much earlier than the earlier described decline of the finless porpoise population in the sea.

8.4.6.2 Listing of a Natural Monument and Fishery Regulations

Aba-shima Island in Hiroshima Prefecture is a small island of about 2.5 × 0.5 km situated off the northern coast of the central Inland Sea. A sea area of 1.5 km radius centered on the southern tip of the island was listed in November 1930 among natural monument as a "sea where finless porpoises aggregated" (Section 5.9.3). Kaburagi (1932, in Japanese) explained the reason for the listing. Tens or over a hundred finless porpoises aggregated in February to May in a small area surrounded by the islands of Kokuno-shima, Matsushima, and Kara-shima and the southern tip of Aba-shima. The total size of this area is about 2 × 4 km. The scenery was particularly splendid between the tips of Aba-shima and Kara-shima, a rock about 1.5 km to the south of Abashima. About 20 fishing vessels from Tadano-umi Town used to operate a hook-and-line fishery for sea bream and sea bass in this area using finless porpoises as indicators of fish schools. This kind of fishery was said to have lasted for about 200 years. The role of the finless porpoises in the fishery was interpreted as follows: (1) finless porpoises feed on small fish such as sand lance that gather near the sea surface, (2) schools of the small fish descend to escape from predators (i.e., finless porpoises), and (3) sea bream and sea bass attack the small fish from below, giving fishermen the opportunity to place their fishlines. The fishermen erected a small shrine at the southern tip of Aba-shima and had a ritual ceremony for the finless porpoises yearly on January 26 (Kaburagi 1932, in Japanese). Kasuya and Kureha (1979) recorded later information from Tadano-umi FCU that this particular fishery was discontinued in the late 1960s due to disappearance of sand lance in the region. This was shortly after the start of sand extraction (see Section 8.4.5.5).

The Coastal Division of the Fisheries Agency stated that in 1989 it made a request to hand-harpoon and drive fishermen for dolphins and porpoises to refrain from hunting finless porpoises (unpublished information sheet prepared by Coastal Division, dated March 1, 1991). No cetacean fisheries of the time were located near a place that would allow hunting of finless porpoises, and no new fishing projects were planned for the species, so the real reason for the Fisheries Agency request is unclear to me. It could have functioned to prevent local aquariums from hunting the species for display. Then, in September 1990, the Director of the Coastal Division sent out notification No. 2-1050 requesting that small cetaceans found in fishing gears such as fixed trap nets be (1) released if found alive or (2) buried if found dead and that (3) the incident be reported to the Coastal Division (Section 6.3).

The Act for Conservation of Endangered Species of Wild Animals and Plants was enacted in 1992, and its application rule of March 29, 1993, regulated the capture and international and domestic transportation of the finless porpoise (Section 5.9.2). Ministerial Order No. 15 of April 1, 1993, added the finless porpoise, together with bowhead whale, blue whale, dugong, Pacific Ridley turtle, and leatherback turtle, to a list of species protected by the Fishery Resources Protection Act (Section 5.9.1). This prohibited hunting of the species and made it an obligation to report incidental mortality, which did not differ practically from the earlier regulations.

All the Japanese protection measures for the finless porpoise prohibited only intentional capture of the species but did not have any provisions to decrease mortality incidental to fishing operations. Such measures were preferred by the government because they required doing nothing. They amounted to only a piece of paper, with no budget or personnel to implement them. It is my view that such an action is the most passive conservation measure and does not deserve to be called protection. The most urgent actions needed for finless porpoises in the Inland Sea, and perhaps for other populations in Japan, are to conserve their habitats and decrease mortality in fisheries.

8.4.6.3 The Problem Is Not Limited to the Inland Sea

It is believed that there are a minimum of five, and probably more, local populations of finless porpoises in Japanese waters. The areas identified with the populations include (1) Omura Bay, (2) Ariake Sound–Tachibana Bay, (3) the Inland Sea and nearby waters connected to it, and (4) Ise Bay–Mikawa Bay and waters near the entrance of Ise Bay. In addition to these, Tokyo Bay, the Pacific coast of Chiba Prefecture to Kashima Nada (34°54′N to 36°40′N), and the Pacific coast of Fukushima Prefecture to Sendai Bay (36°50′N to 38°24′N) are likely to support populations. The Tokyo Bay–Sendai Bay area has not been well studied because of a paucity of available materials for DNA analysis. The possibility of a small population in Suruga Bay still remains to be confirmed. When there is the possibility of a local population, it is safer to establish conservation actions for it.

Some of the local populations identified or suspected inhabit waters supporting high human activities, for example, Omura Bay, the Ariake Sound, Ise Bay, Mikawa Bay, and Tokyo Bay. At least some of the hazardous human activities discussed earlier take place in all of these waters, including net fisheries, reclamation, chemical pollution, and vessel traffic. It is reasonable to assume that some of the remaining local populations are also experiencing decline in abundance so far unnoticed by scientists. There are still some sporadic sightings of finless porpoises in Tokyo Bay, but it is unreasonable to consider that the abundance was at such a low level in the nineteenth century, when the coast was close to pristine, the waters were less polluted, and the vessel traffic was less. Rather, it should be concluded that the population is one of the most endangered populations of finless porpoises in Japan. The Inland Sea population of finless porpoise just happened to be surveyed by scientists and a decline confirmed there.

I believe it is urgent to improve our understanding of the current status of Japanese finless porpoises and to establish a conservation plan for them. Conserving the coastal marine environment suitable for finless porpoises will also contribute to the health of fisheries as well as the safety of fishery products for humans.

8.5 LIFE HISTORY

8.5.1 Parturition Season

Kasuya and Kureha (1979) concluded that parturitions occur in the Inland Sea from April to August with a peak in April–May. This was based in part on 13 records of juveniles of 100 cm or less during late April to late June in a broad Pacific area from the Inland Sea in the southwest to Sendai Bay in the northeast, including Ise Bay, Mikawa Bay, Shimizu in Suruga Bay, and Tokyo Bay. The additional evidence was the proportion of cow–calf pairs that increased from April to August/September during sighting surveys in the Inland Sea. They felt it would also be of some significance that the proportion of single adults declined during the period; however, this was not independent from the increase in cow–calf pairs.

Iwatsuki (2000, in Japanese) supplemented Kasuya and Kureha (1979) with additional samples and analyzed the seasonality of newborn finless porpoises for a broader geographical area. She confirmed that there was no evidence to suggest different parturition seasons among finless porpoise populations inhabiting the Pacific coast of Japan and the Inland Sea. She defined a newborn as a porpoise larger than the minimum known length of a calf (62 cm) and below the maximum known length of a fetus (91.3 cm). This seems to be a little strict and might have functioned to reduce the estimated range of the parturition season. According to her analysis, all the newborns (11 individuals) occurred only in 3 months from April to July (a peak was identified in May and June) in the Inland Sea, and the 11 newborns were 37% of the total of 27 strandings recorded for the 3 months. In Ise Bay–Mikawa Bay and the nearby Pacific area, all the 56 newborns were recorded in 5 months from April to August (a peak in May); they were 38% of the total of 145 stranded porpoises during the period. I have tabulated in Table 8.12 all the records of strandings and incidental deaths accompanied by date and body length, where newborns were defined as 60–95 cm in body length (see Section 8.5.2).

The estimated parturition peak in April–June for the Inland Sea population shows reasonable agreement with a peak in copulations (April–May) observed in the Toba Aquarium in

TABLE 8.12
Stranding and Incidental Mortality of Japanese Finless Porpoises together with Records of Porpoises of Neonate Size (60–95 cm in Body Length) as an Indication of Parturition Season, Based on Records from November 16, 1929, to May 31, 1999

	Stranding and Incidental Mortality		Neonate-Like Porpoises		
Area and Stock[a]	Number of Porpoises	Month	Number of Porpoises	Month	Peak Month and Number of Porpoises
Ariake Sound–Tachibana Bay (2)	7	Jun, Sep–Jan	2	Oct, Dec	
Hibiki-nada (S. Sea of Japan) (3)	20	Aug–Sep, Nov–Jun	2	Apr–May	
Inland Sea (3)	75	Jan–Dec	11	Apr–Jul	4 in May (3 in June)
Osaka Bay (3)	4	Feb, May, Jul	0		
Kii Channel (3)	8	Jan, Mar–Apr, Jun, Oct, Dec	0		
Ise and Mikawa Bays (4)	195	Jan–Dec	42	Apr–Aug	22 in May (13 in June)
Pacific coast connected to Ise Bay (4)	51	Feb–Dec	14	Apr–Jun, Aug	5 in April (4 in May)
Omaezaki-Numazu[b]	5	Jan, May–Jun	1	May	
Tokyo Bay (N. of Cape Kan-non) (5)	4	May–Jun, Nov	2	May–Jun	
Chiba coast (excluding Tokyo Bay) (5)	16	Mar–Jun, Sep–Nov	4	Mar, May	3 in March
Ibaraki Prefecture (5)	8	Jan, May, Nov	2	May	
Miyagi Prefecture (Sendai Bay) (5)	3	Mar, May, Sep	1	May	

Source: Compiled from Iwatsuki, T., Analyses of stranding records of finless porpoises in Japan, Bachelor thesis presented to Faculty of Bioresources of Mie University, Tsu, Japan, 11pp., 1 Table, 13 Figs., 1 Appendix Table, 2000.

[a] Numerals in parentheses denote stocks currently identified in Table 8.3 and Figure 8.2. See Section 8.3.4 for the possibility of additional stocks.

[b] Suruga Bay is located on the Pacific coast of Shizuoka Prefecture or between Ise and Mikawa Bays and Tokyo Bay.

porpoises captured in Ise Bay (a gestation period of about 11 months is assumed). However, this parturition peak was inconsistent with a trend in cow–calf pairs, which increased in frequency until August/September in the Inland Sea. In order to explain this apparent disagreement, Kasuya and Kureha (1979) assumed that finless porpoises not accompanied by calves tend to move outside the Inland Sea. However, in view of the data collected by Iwatsuki (2000, in Japanese), tabulated in Table 8.12, such an assumption seems unnecessary. We have a total of 68 records of newborns from the Inland Sea plus the Kanmon area connected to the sea and from the Ise Bay–Mikawa Bay area. The monthly distribution of the newborns was 10 individuals (14.7%) in April, 31 (45.6%) in May, 21 (30.9%) in June, 2 (2.9%) in July, and 4 (5.9%) in August. This indicates that the parturition season starts in April, peaks in May, and lasts until August. Sixty percent of total births in a year occur in April and May and the remaining 40% in the 3 months from June to August, which explains the seasonal increase of cow/calf pairs toward August/September observed by Kasuya and Kureha (1979) in the Inland Sea. Now, it is unnecessary to assume a different pattern of seasonal movement between growth and reproductive phases.

Furuta et al. (1989) reported 21 records of large fetuses and neonates of 51.5–90 cm from Ise Bay collected during 1966–1983 that were also included in Iwatsuki (2000, in Japanese). Seventeen of them were over 75 cm long (including two fetuses at 75.0 and 79.0 cm); 2 occurred in March, 11 in April, 3 in May, and 1 in June. This was similar to the seasonality pattern in the Inland Sea. A geographical difference of parturition season, if it existed, would be better analyzed using the larger data set in Table 8.12. All of the neonates reported in a broad area ranging from the Inland Sea area to Sendai Bay along the Pacific coast of Japan occurred during March to August, with a peak in May in two populations with sufficient sample size, a population in the Inland Sea and its adjacent area and another in Ise Bay–Mikawa Bay and the adjacent area. Thus, with an exception of one population off western Kyushu, there is currently no evidence to indicate geographical variation in the parturition season of finless porpoises in Japan.

Mizue et al. (1965) thought that finless porpoises off western Kyushu had a parturition season in late August to early September, which was based on their own observation of one lactating female in late October and the information from fishermen that "newborn caves with umbilical fragments occurred in early September and later, which was preceded by occurrence of large fetuses." The specimen was presumably killed in the Ariake Sound–Tachibana Bay area, but the presence of two populations, one in Omura Bay and another in Ariake Sound–Tachibana Bay, was not known in those days. I once erroneously rejected their conclusion in the analysis of breeding season of the species in the Inland Sea (Kasuya and Kureha 1979), which was due to my careless and inadequate reasoning that all the Japanese finless porpoise populations should have a similar breeding season. I wish to apologize for my impudence and mistake.

Shirakihara et al. (1993) analyzed parturition season by two methods (Figure 8.11), which were (1) to assume neonates 84 cm or shorter as age zero and (2) to estimate parturition

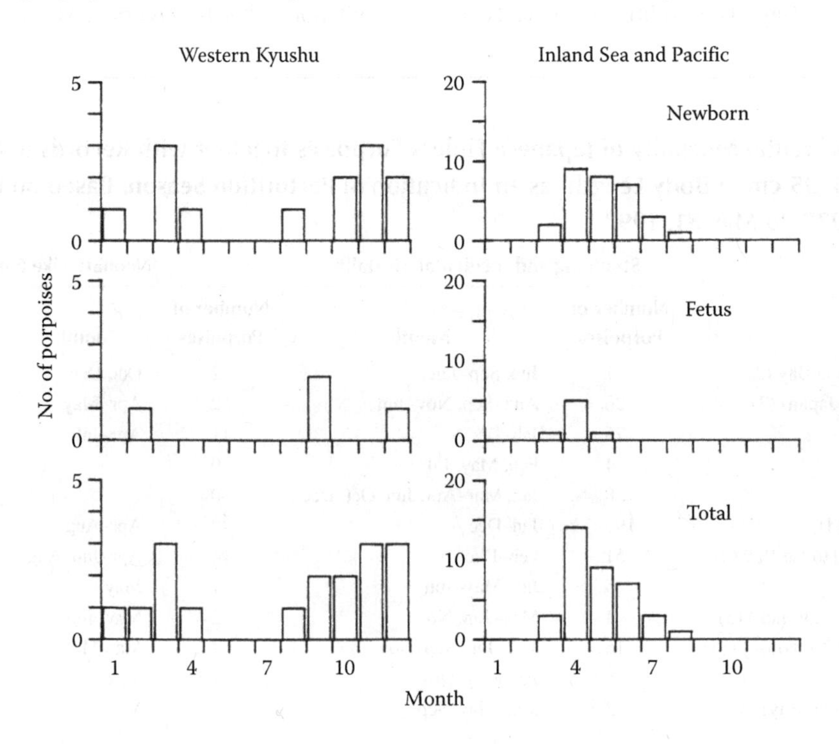

FIGURE 8.11 Parturition season of finless porpoises evidenced by the occurrence of neonates ≤84 cm (top) and that calculated from fetal body length assuming growth of 0.277 cm/day in the linear part of fetal growth and mean neonatal length of 78.6 cm (middle). The bottom panel represents the total of the two sources of parturition season. The left panel is for western Kyushu, which includes the two populations of Omura Bay (1) and Ariake Sound–Tachibana Bay (2), and the right panel is for populations in the Inland Sea and other Pacific areas. (Reproduced from Figure 6 of Shirakihara, M. et al., *Mar. Mammal Sci.*, 9, 392, 1993. With permission of Wiley Press.)

date from fetal length and the date of death by assuming fetal growth rate of 0.277 cm/day and average neonatal length of 78.6 cm (see Section 8.5.2). They applied these methods to two sample groups: western Kyushu (samples from two currently identified populations were combined) and the Inland Sea area plus other Pacific coasts. Results of the two estimation methods agreed reasonably for the two sample groups, and the results for the Inland Sea plus Pacific coasts samples agreed with the conclusion given earlier. However, the parturition season obtained from the western Kyushu sample apparently differed; parturition there extended from August to April with a main peak in November and December, or about 6 months apart from the rest of the populations, and agreed with the conclusions of Mizue et al. (1965, in Japanese). The analysis of Shirakihara et al. (1993) showed a minor peak in March for the western Kyushu sample (Figure 8.11).

The analyses of Shirakihara et al. (1993) suffer by having combined two populations later identified by Yoshida et al. (1995, 2001), those in Omura Bay and those in Ariake Sound–Tachibana Bay. Their sample from the Ariake Sound–Tachibana Bay population was later reanalyzed by the same authors, and they concluded that parturition lasted from August to March with a major peak in November–December and an additional minor peak in March (Shirakihara et al. 2008). Amano (2010, in Japanese) analyzed stranding records in 1901–2008 from the Omura Bay population and found that neonates below 90 cm occurred only in April and May; he concluded that the Omura Bay population had a parturition season that was similar to those in the Inland Sea and to the east. Now, it is known that only one of the two finless porpoise populations in western Kyushu, the Ariake Sound–Tachibana Bay population, has a main parturition peak in autumn/winter and a minor peak in March, only the latter of which matches with the single peak identified for rest of the finless populations in Japan.

Shirakihara et al. (1993) reported a brief observation on seasonal difference in male reproductive system: all 11 adult males sampled in September–January had copious sperm in the epididymides, but 3 adults in June–July had less. All the 14 adults had spermatozoa in the testes (no mention was made of abundance of spermatozoa in the seminiferous tubules). It is a reasonable strategy for males to start production of spermatozoa before the mating season and store a sufficient amount in the epididymides to allow the chance of finding estrous females during any part of the mating season. The observation of Shirakihara et al. (1993) agrees with a mating peak in Autumn known for finless porpoises in the Ariake Sound–Tachibana Bay.

Selection is believed to have worked to place parturitions in a season that would maximize the survival of offspring (Kiltie 1984). Survival of offspring will be affected by various factors, including climate, predators, nutrition of nursing females, and availability of suitable food for the calf during the period of switching from milk to solid food. The two populations off the west coast of Kyushu described earlier inhabit almost similar latitudes, and the climate and oceanography do not differ much. The only difference would be a smaller habitat for the Omura Bay population, which shares the parturition season with the majority of Japanese finless porpoises. It is also strange that individuals in Sendai Bay, the northernmost habitat of the species in Japan, breed in the same season as most of the Japanese populations further south.

8.5.2 Growth

8.5.2.1 Body Length at Birth

The mean body length of neonates was reported as between 78.6 and 81.6 cm (Table 8.13). The difference between these figures, in addition to being due to the limited sample size, is due at least partly to different definitions of neonate. Geographical variation in neonatal length has not been identified in Japan.

The estimation of mean neonatal length by Furuta et al. (1989) was important because it used data from a single population (Ise Bay–Mikawa Bay population) and was based on neonates clearly identified by the presence of a remaining umbilical fragment that was usually 1–1.5 cm long. The body length of neonates thus defined ranged from 76.5 to 85.0 cm (Table 8.13), and their 6 fetal samples ranged from 51.5 to 79.0 cm, which led Furuta et al. (1989) to conclude that finless porpoises in the Ise Bay–Mikawa Bay region were born at 75–80 cm. Shirakihata et al. (1993) combined data from two populations in western Kyushu (Omura Bay and Ariake Sound–Tachibana Bay populations) (Table 8.13). They defined neonates by the presence of at least one of the following characters: (1) vestigial sensory hair on upper jaw, (2) fetal folds, and (3) umbilical fragment. Duration of these characters has not been investigated, but Furuta et al. (1977, in Japanese) stated that the fetal folds last a minimum of 17 days after birth.

The ratio of neonates to fetuses by body-length group increases with body length, and a length where they became equal in proportion has often been taken to be the average body length at birth (e.g., Sections 9.4.3 and 10.5.2).

TABLE 8.13
Records of Neonates of Japanese Finless Porpoises and Mean Neonatal Lengths Calculated from Them

Area of Data	Neonatal Length (Range) (cm)	Sample Size	Mean (cm)	Reference
Inland Sea–Pacific	68.8–97.6	14	78.6	Kasuya and Kureha (1979)
Inland Sea–Ise Bay	77.5–83.0[a]	4	80.0	Kasuya et al. (1986)
Ise Bay	76.5–85.0	6 males	80.1	Furuta et al. (1989)
		4 females	81.6	
Western Kyushu	71.5–84.0[b]	12	78.2	Shirakihara et al. (1993)
Whole Japan	62.0–91.3			Iwatsuki (2000)

[a] Represented by four individuals measured on the day of birth (77.5 cm), 4th (79.0 cm), 16th (81.5 cm), and 28th (83.0 cm).
[b] See Section 8.5.2 for the characters of neonates.

Although this method is less influenced by postnatal growth of newborns, it is not applicable to the finless porpoise due to limitations of sample size. Data used in the previous analyses were obtained mostly from stranded carcasses. The incidence of perinatal mortality is not independent of birth size; for example, smaller newborns might be more liable to die. Therefore, the mean body length at birth as estimated earlier can be biased, but we have no data to confirm this.

8.5.2.2 Fetal Growth and Gestation Time

If body length or the cube root of body weight is plotted on an axis of time measured after conception, mammalian fetal growth can be divided into two phases of a slower curvilinear growth phase in early development and a subsequent linear growth phase. The date t_0, which is the date when the back-extended linear growth cuts the axis of time, is a function of total gestation time (X), that is, $t_0 = 0.3X$ for species with $X = 50–100$ days, $t_0 = 0.2X$ for species with $X = 100–400$ days, and $t_0 = 0.1X$ for species with $X > 400$ days (Hugget and Widdas 1951). This general relationship is applicable to cetaceans (Lockyer 1984; Perrin and Reilly 1984), with the minor modifications of Laws (1959), who proposed 90% of t_0 estimated using the relationship given earlier for cetaceans and using body length instead of the cube root of fetal weight. As the gestation period of finless porpoises is likely to be slightly shorter than 1 year, t_0 is estimated by $t = 0.9 \times 0.15X$. Seasonal fetal size distribution gives an estimate of growth rate during the linear phase, which together with average body length at birth enables the calculation of the value of $(X - t_0)$; then the total gestation time (X) is calculated by dividing the value of $(X - t_0)$ by $(1 - t_0)$. However, we do not have data to use for direct estimation of fetal growth rate and a value of $(X - t_0)$ for the finless porpoise.

An alternative way is to estimate gestation time and fetal growth rate using interspecies relationships of growth parameters. The following relationship between gestation period (Y, month) and average body length at birth (X, cm) is known for the Delphinidae (Perrin et al. 1977).

$$\log Y = 0.4586 \log X + 0.1659$$

Using this equation and assuming average body length at birth of 80 ± 5 cm, Kasuya et al. (1986) estimated the gestation period of the Japanese finless porpoise at 10.9 months, with a range of 10.6–11.2 months.

Another interspecies relationship that has been used for this purpose is that between fetal growth rate during the linear fetal growth phase (Y, cm/day) and average body length at birth (X, cm) for Delphinidae (Kasuya 1977).

$$Y = 0.001462X + 0.1622$$

Assuming 80 cm as the average body length at birth for this equation, Kasuya et al. (1986) obtained 0.279 cm/day or 8.49 cm/month as an average growth rate of fetuses in the linear phase of growth. This is used to calculate mean gestation time:

$$(80/8.49)/(1 - 0.15 \times 0.9) = 10.9 \text{ (month)}$$

In principle, the methods applied earlier cannot be considered very reliable because the average body length at birth is only an estimate for the finless porpoise, but fair agreement of the results by both methods, about 11 months, suggests that they are reasonable.

Izawa and Kataoka (1965, in Japanese) reported the observations of a pair of Ise Bay finless porpoises kept in the Toba Aquarium during September 1963 to March 1965; annual copulation activity started in February and lasted to June. Other observations of 13 individuals of both sexes in a tank at the same aquarium recorded a copulation peak in April and May, and one of the females that had been kept there since 1973 delivered a calf on April 17, 1976 (Furuta et al. 1977, in Japanese). These observations support the analysis presented earlier that Japanese finless porpoises (with exception of the Ariake Sound–Tachibana Bay population in western Kyushu) mate in spring and early summer, with gestation of about 1 year.

Small-cetacean species exhibit common features of gestation time close to one year and neonatal size of slightly less than 1 m. The neonatal length is probably close to a minimum body size required for a homoeothermic species to live safely in the aquatic environment, and the gestation length is an adaptation to the annual cycle of the marine environment. Some cetacean species have achieved greater body size through either of two different ways. Baleen whales have achieved it by increasing the fetal growth rate while retaining gestation time of about one year, which has been required to meet an annual cycle of feeding. Some of the toothed whales such as the sperm whale and killer whale have increased size while retaining a relatively slower fetal growth rate and accepting a gestation period of over 1 year, with a possible cost of a reproductive cycle that does not synchronize with an annual cycle of the marine environment (Kasuya 1995).

8.5.2.3 Nursing Period and Weaning Season

Females of some cetacean species such as the Dall's porpoise (Section 9.4.6) and the harbor porpoise usually enter into estrus and conceive while nursing a calf, but the reproductive cycle of many other cetaceans usually follows the order of separate pregnancy, lactation, and resting. The last two stages have greater variation in duration. If the suckling calf dies, the female usually ceases lactation (adoption of another calf is known to occur in some cases) and enters into estrus in the next mating season for subsequent gestation that may starts earlier than in the normal case. Some healthy well-fed females may conceive during lactation (thus the resting phase is skipped), but such lactation probably ceases before the next parturition, as in short-finned pilot whales (Section 12.4.4). If a calf is allowed to continue suckling for a longer period, the next conception by the mother may be delayed. This can happen to females of great age or to females kept together with their calves in an aquarium.

Furuta et al. (2007, in Japanese) reported the process of weaning for four calves born in an aquarium that lived to weaning. They started taking solid food at the ages of 120, 132, 166, and 223 days, with an average of 160 days (5.3 months). Their daily food consumption increased for 5–6 months from the start of taking solid food. One of the calves, which started solid food on the 120th day (3.9 months), was observed suckling frequency during sampled hours. The number of suckling bouts per hour of observation was 25–30 just after birth, decreased to 10–15 at the age of one month, and further continued to decrease until the 252th day (8.3 months) when observation was discontinued. After an interruption of 104 days, or at the age of 357 days (11.7 months), this individual was observed for 8 hours but showed no sign of suckling. This observation leads to a conclusion that finless porpoises from Ise Bay start taking solid food at the age of 4–7 months, followed by decreasing proportion of milk in the total nutrition with time, and end suckling, in an aquarium condition, at an age between 8.3 and 11.7 months. The aquarium environment would not influence the start of weaning (5.3 months on average). However, the time at completion of weaning will require information on individuals in the wild.

Shirakihara et al. (2008) examined the stomach contents of 17 juveniles from the Ariake Sound–Tachibana Bay population. The body length of animals with only milk was below 99.5 cm, that of those with both solid food and milk was 99.5–107 cm (three individuals), and that of those with only solid food was 93.5–104 cm. These data indicate that the porpoises start ingestion of solid food in the wild at body length 93.5–99.5 cm and that some individuals continue suckling until a body length of 107 cm is reached, although individuals may end suckling earlier. The judgment of the absence of milk in the stomach requires caution because identification of a small amount of milk mixed with solid food is often difficult. Shirakihara et al. (2008) estimated ages at the stages mentioned earlier, using the body length and growth equations of Shirakihara et al. (1993, see Section 8.5.2.4), which suggested that the porpoises start taking solid food at ages between 3–4 and 5–6 months and that some individuals continue suckling up to an age of 9–10 months (from the oldest individual with a trace of milk). The case of extended suckling is close to the observations of Furuta et al. (2007) of an animal in captivity.

Kasuya and Kureha (1979) reported seasonal change in the proportion of cow–calf pairs in the Inland Sea. The proportion of such groups started to increase in spring and reached at a peak of about 30% in late summer and then suddenly decreased to about 10% accompanied by an increase in frequency of single individuals of calf size (Figure 8.12). Although the accuracy of the size estimation was not verified, they also recorded an increase in frequency of juveniles larger than the calves in winter. They inferred from this observation that calves are born in spring and early summer and most of them become independent around October–November, but some accompany their mothers until the next spring or summer.

These three observations on early growth of Japanese finless porpoises do not contradict each other, and it seems to be reasonable to conclude that most females breed synchronously and wean their calves before the mating season of the next spring. This suggests, although inconclusively, that their breeding cycle is likely to be 1 or 2 years long.

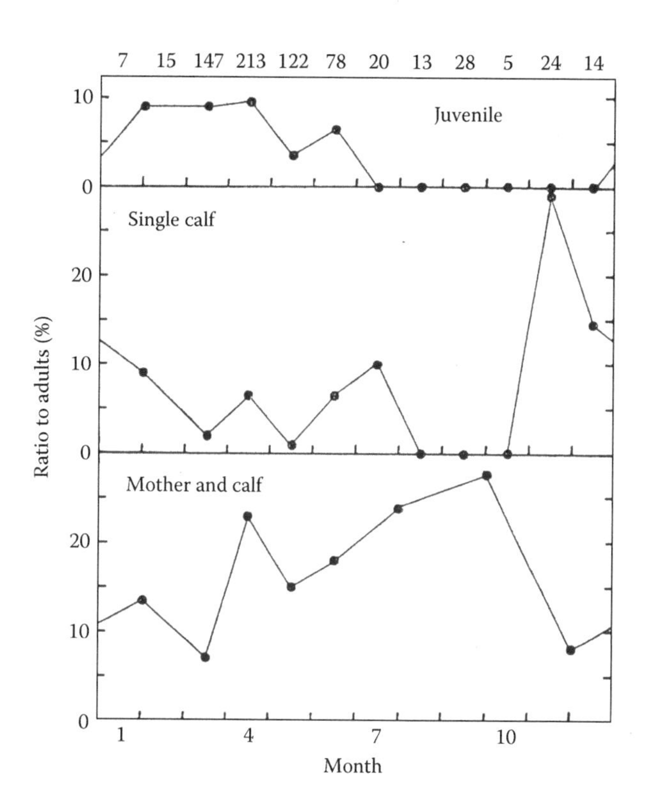

FIGURE 8.12 Seasonal change in the proportion of cow–calf pairs (bottom), small singletons (middle), and larger juveniles (top) visually identified in the Inland Sea during 1976–1978. Their proportions are expressed as percentage of the number of adult individuals sighted, which is shown on the top of the panel. (Reproduced from Figure 6 of Shirakihara, M. et al., *Mar. Mammal Sci.*, 9, 392, 1993. With permission of Wiley Press.)

8.5.2.4 Growth Curve

Furuta et al. (2007) assumed linear growth and got the following regression of body length (Y, cm) on age (X, days) for juvenile finless porpoises born in an aquarium:

$$Y = 98.206 + 0.0976X, \quad 0 \le X \le 600, \quad r = 0.9026$$

This regression was based on 19 points representing an unstated number of calves. If we accept this equation, it suggests a daily increment of 0.0976 cm and an average neonatal length of 98.206 cm, which is unreasonably greater than the 80 cm estimated earlier. The average body length at the age of 1 year of 133.8 cm calculated from the equation presented earlier is only 36% greater than the neonatal length derived from the equation, which is unreasonably small. Thus, it is necessary to investigate whether linear growth can be reasonably fitted to the growth during the period. Six of the 11 points at the age of 0–150 days are below the line and 3 are above, all the 6 points at the age of 150–300 days are above the line, and the 2 points over 450 days are below the line. This is evidence that linear growth is inappropriate for expressing the early growth of finless porpoises; a fit to an upwardly convex growth curve is more likely (Figure 8.13).

Kasuya and Matsui (1984) noted for five species of Delphinidae that body length increments in the first year

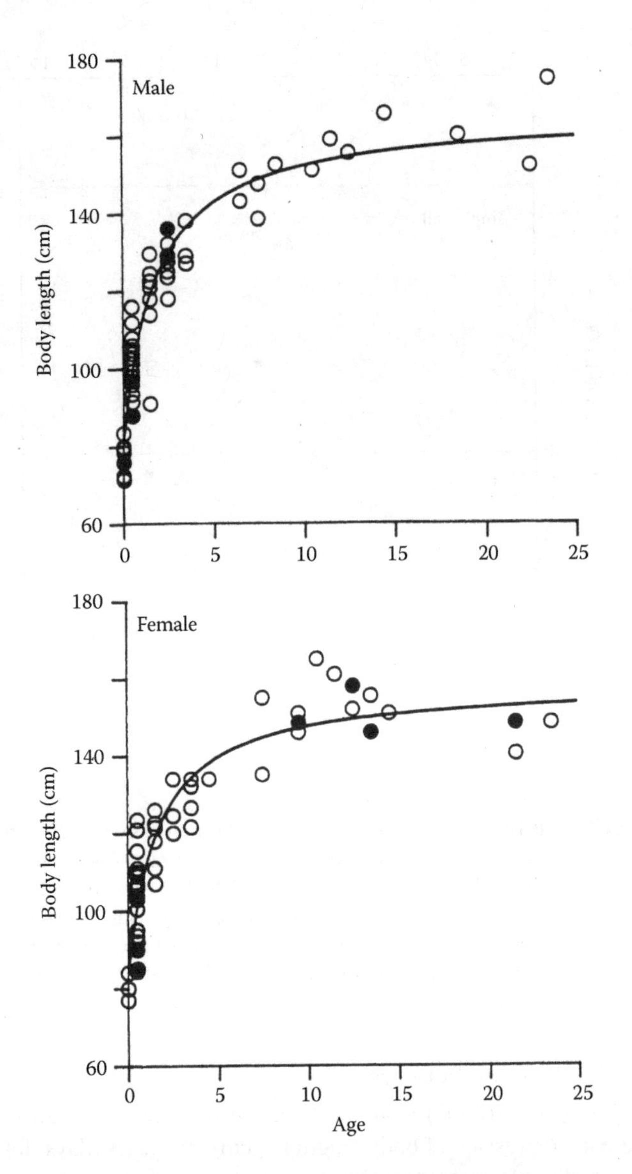

FIGURE 8.13 Mean growth curve of finless porpoises constructed using materials from two populations in western Kyushu (open circles) and the Inland Sea population (closed circle). Age was determined assuming annual deposition of dentinal and cemental layers. (Reproduced from Figure 5 of Shirakihara, M. et al., *Mar. Mammal Sci.*, 9, 392, 1993. With permission of Wiley Press.)

after birth were in the range of 55%–64% of neonatal length: 55% for the bottlenose dolphin, 60% for the spotted dolphin, 61.2% for the short-finned pilot whale, 61.3% for the common dolphin, and 64% for the striped dolphin. Assuming neonatal length of 80 cm and applying this rule to the finless porpoise gives 124–131 cm as the body length at the age of 1 year.

Perrin et al. (1976) obtained the following relationship for five species of toothed whales among average fetal growth rate (X, cm/month), average growth rate of neonate during the period equal to gestation time (Y, cm/month), and average neonatal body length (Z, cm).

$$\text{Log}(X - Y) = -1.33 + 0.997 \log Z$$

Applying $X = 80/11$, $Z = 80$ for this equation gives $Y = 3.58$ for finless porpoise, or 119 cm as the body length at the age of 11 months. Extrapolation of this rate to 12 months gives 123 cm, which is logically expected to be an overestimate.

Kasuya et al. (1986) made a rudimentary attempt to calibrate the accumulation rate of growth layers in the teeth of finless porpoises using six animals of known age: four neonates (age 0–28 days), a male of 153.5 cm that died at the age of 864 days (about 2 years 4 months), and another male of 159.5 cm that died at the age of 1719 days (about 4 years 8 months). The last two individuals were not accompanied by tooth samples for aging. Of the four neonates of known age, the longer-lived individuals always had greater body lengths at death. One possible explanation for this would be growth during their short life. Another was that smaller neonates died earlier. A daily growth rate of 0.2 cm/day was calculated by ignoring the latter possibility. They aged another 8 individuals caught in fishing nets. They were 100–132 cm in body length and aged at 0.5–1.5 years from dentinal growth layers, which almost agreed with the body length at the age of 1 year calculated from the interspecies relationship given earlier. This scanty information is the currently available basis of the estimated annual deposition of dental growth layers for the species.

Later growth of the species was reported by Shirakihara et al. (1993) and Kasuya (1999) using the ages determined by counting growth layers in the teeth and assuming annual deposition of the layers. Growth layers are deposited in both dentine and cementum of young individuals, but deposition of dentine ceases after the age of about 10 layers in this species and aging must rely on reading of cementum layers, which is harder than reading the dentinal layers. Shirakihara et al. (1993) combined specimens mostly from two populations off western Kyushu and a few specimens from the Inland Sea population. They found no growth difference between the combined western Kyushu populations and the Inland Sea population, but they did not test for growth difference between the two populations in western Kyushu (those in Omura Bay and Ariake Sound–Tachibana Bay). The oldest individual in their series was aged 23 years for both sexes. Body lengths of sexually mature male were in the 127–174.5 cm range and those of females in the 135–161 cm range; the modal length was 150–159 cm for both sexes. Male body lengths were slightly greater than those of females. The early fast growth phase seemed to end at around 4–5 years, followed by slower growth that ended at around 10 years in females and at age slightly over that in males.

Shirakihara et al. (1993) tried several growth equations for the finless porpoise and concluded that the following equations fit best to their data on body length (Y, cm) and age (X, years):

$$\text{Male: } Y = 165.5 - 157.2/(X + 1.863)$$

$$\text{Female: } Y = 157.7 - 118.1/(X + 1.624)$$

Kasuya (1999) listed maximum body size in Inland Sea/Pacific coast populations as 192 cm (males) and 180 cm (females).

These are apparently greater than the maximum size of 175 cm (males) and 158 cm (females) known from western Kyushu (two populations combined), but more data are needed before reaching a firm conclusion.

The mean growth curve obtained earlier should be considered as an "apparent" mean growth curve. Imagine a particular case where all human babies born in a town in a particular year are selected as a sample and their height is measured annually till the age of about 30 years when all the individuals of the sample are sure to have ceased growth. If annual means of height were plotted on age, that would yield a "real" mean growth curve for the sampled population. The same procedure can be also done using body weight as an indicator of body size, but the shape of the curve as well as age at cessation of growth will be different from that of body height.

Due to technical reasons, such a real mean growth curve is not available for many cetacean species, including the finless porpoise, and instead we plot body length on age using data obtained in a limited length of time to construct a mean growth curve, which should be called as an "apparent" growth curve and is identical with the "real" mean growth curve only when the growth pattern of the particular population remains the same for a period of about 30 years or more, depending on the species. If such an assumption is violated, the apparent mean growth curve is different from the real one. For example, if sons and daughters grow larger than their parents, perhaps through the improvement of nutrition, the apparent mean growth curve will yield the erroneous impression that the mean body length diminishes at higher age, as observed for North Pacific sperm whales (Kasuya 1991). The opposite will occur if body size in a more recent generation diminishes.

It should also be noted that the real mean growth curve may not be the same as the growth curve of any single individual. In the human example presented earlier, some individuals will have an adolescent growth spurt at the age of 12–13 years and exceed the average size in the sample at that age, which is followed by a stage of slower growth. Other individuals may have the growth spurt at a later age. The presence of an adolescent growth spurt will be obscured on the mean growth curve by broad individual variation in its occurrence at age. The ages at which an individual ceases growth is also variable among individuals, but the mean body length continues to increase until all individuals cease growing. Thus, unless we have supplemental anatomical information such as state of fusion of the vertebral epiphyses (see Section 13.4.5), analysis of a mean growth curve alone is likely to overestimate average age at cessation of growth. Thus, interpretation of the "apparent" mean growth curve requires caution.

I agree with the value of a growth equation as a way of expressing the mean growth curve, the biological value of which was evaluated earlier. But overapplication of the equation is risky. Each growth equation has its own peculiarities. Even when it is judged by a scientist as acceptable to express the mean growth pattern of the whole age range of a cetacean species, the real fit may not be the same for a particular age segment of the sample. Fitted growth equations also depend on the age composition of the sample; therefore, those fitted to samples biased to younger individuals may not adequately show the growth of older individuals, and an equation fitted to samples of higher age may not properly represent the mean growth of younger individuals. Thus, extrapolation of a growth curve to higher ages, where sample size is likely to be limited, for the purpose of estimating asymptotic body length should be done with caution.

8.5.2.5 Body Weight

Several equations have been proposed for the relationship between body weight (W, kg) and body length (L, cm) of finless porpoises. Kasuya (1999) found no significant difference in the relationship among 6 individuals from Pakistan, 15 from China, and 21 from Japan and obtained for the total of 42 specimens a single equation:

$$W = 1.816 \times 10^{-4} L^{2.477} \quad 65.3 < L < 187$$

The equation can be expressed as $\log W = 2.477 \log L - 3.7409$. This equation predicts an average body weight of 37.6 kg for animals of 140 cm and 52.3 kg for animals at 160 cm, or 141% difference, but Kasuya (1999) noted that 10 individuals at 140–160 cm had body weight variation of 26.3–48.2 kg, or 183% difference. This large individual variation in body weight is possibly due to seasonal fluctuation in physiology.

Kataoka et al. (1976, in Japanese) recorded large seasonal variation in the food consumption of finless porpoises (of both sexes) in their aquarium, from a minimum level of 2–3 kg/day/individual in May–September to a high level of 5 kg in December–March. Food consumption was negatively correlated with environmental water temperature. These individuals had been in the aquarium for 3 years and shown active mating behavior (Izawa and Kataoka 1965, in Japanese), which suggests that they were sexually mature. Kataoka et al. (1976, in Japanese) did not give precise biological data for these porpoises but estimated that they were around 160 cm in body length and about 56 kg in body weight.

Furuta et al. (1989) analyzed the body weight of 29 finless porpoises from Ise Bay and nearby waters and found all to lie on a single regression line aside from one fetus of 51.5 cm. They obtained the following equation for the 28 porpoises of both sexes:

$$\log W = 2.8405 \log L - 4.5439 \quad 54.5 < L < 192$$

This equation can also be expressed as $W = 2.858 \times 10^{-5} L^{2.8405}$ and gives a body weight of 52.1 kg at body length 160 cm, which is almost identical with the figure calculated from the equation of Kasuya (1999).

Shirakihara et al. (1993) calculated the following two equations for finless porpoise samples from two populations in western Kyushu, excluding pregnant females:

$$\text{Male:} \quad \log W = 2.4582 \log L - 3.6545 \quad 71.5 < L < 174.5$$

$$\text{Female:} \quad \log W = 2.4174 \log L - 3.5426 \quad 77.0 < L < 151.0$$

The authors stated that there was no between-sex difference in the slopes (2.4582 and 2.4174) but that the Y-intercepts (3.6545 and 3.5426) were significantly different. However, in my opinion, evaluation of the biological value of the difference requires some caution, because it relates to body weight at body length of 1 cm, that is, extrapolation far outside the sample range. The equations can also be expressed as $W = 2.2116 \times 10^{-4}L^{2.4582}$ (males) and $W = 2.8668 \times 10^{-4}L^{2.4174}$. These equations give body weights of 57.0 kg (male) and 61.0 kg (females) at body length of 160 cm, which are greater than the corresponding figures for the Ise Bay population. Interpretation of the difference is left for future investigation.

8.5.2.6 Female Sexual Maturity

Attainment of sexual maturity of female cetaceans has been defined by the first ovulation, which is usually followed by conception. This tradition originated from early cetacean biology that relied on the examination of carcasses stranded or killed in fisheries and received support because the corpus luteum formed after ovulation was believed to remain in the ovary as a corpus albicans for life. Such anatomical information has not always been available for more recent cetacean scientists.

A finless porpoise born in the Miyajima Aquarium from a female taken in the Inland Sea had her first parturition at the age of 3 years (Shirakihara et al. 1993). As pregnancy lasts for about 11 months, the female could have experienced her first ovulation when she was 2 years old. This is the youngest age currently available for female finless porpoises as an age at the attainment of sexual maturity, but it should probably be considered as a special case only achieved under favorable health and nutritional conditions.

Shirakihara et al. (1993) examined the ovaries of 39 female finless porpoises from western Kyushu (see Figure 8.13 for the ages). All the individuals below 135 cm were sexually immature, and those above 145 cm were sexually mature. There was only one individual between the two limits, which was 135 cm in body length and aged 7 years; it had two corpora in the ovaries. Their sample lacked individuals aged 5 and 6 years; all the individuals aged 4 years or younger were immature and those aged 7 years or older were sexually mature. From these data, the authors concluded that sexual maturity was attained at body length 135–145 cm and at the age of 5–6 years. Average age or body length at sexual maturity is somewhere within these ranges. An additional six females from the western Inland Sea and Kanmon area did not change the conclusions obtained earlier (Shirakihara et al. 1993), nor did samples in Kasuya (1999) obtained from the Inland Sea and Ise Bay. The 39 specimens of Shirakihara et al. (1993) lacked females of 5–6 years of age or those at around the attainment of sexual maturity, which will be dealt with in the following.

8.5.2.7 Male Sexual Maturity

A reasonable definition of the attainment of male sexual maturity is the onset of the ability to copulate with and inseminate females, but it is almost impossible or extremely difficult to detect the time when a male attains such a stage. Further, the age at the attainment of such a stage will vary depending on the structure of the community in which the male lives. Testicular histology has been often used for the purpose of identifying sexual maturity, but male maturation is a slow process and it is hard to define a certain histological point at maturity. In order to overcome the difficulty, we need further effort to combine information on testicular histology, level of reproductive hormones, sexual behavior, and degree of reproductive success.

Shirakihara et al. (1993) classified testicular tissue into 4 stages of maturation using the proportion of seminiferous tubules that were spermatogenic: (1) immature, no spermatogenesis; (2) early maturing, over 0% and less than 50%; (3) late maturing, over 50% and less than 100%; and (4) mature, 100% spermatogenic tubules. This classification was the same as that used for short-finned pilot whales (Kasuya and Marsh 1984). Figure 8.4 shows the relationships between testicular weight, age, and the histological criteria. Shirakihara et al. (1993) identified one male (body length, 127.5 cm; mean testis weight, 58.0 g) as sexually mature at the age of 2 years and considered it as an exceptionally precocious case. They concluded that male finless porpoises usually attain sexual maturity at body length 135–140 cm and the age of 4–6 years in both western Kyushu and the Inland Sea.

I feel it particularly interesting to see in their sample (Figure 8.14) a lack of males at early and late maturing stages or males with testes of 50–150 g (mean of both testes). Such a case can happen if the maturation process proceeds very fast, but analysis of the age composition of the sample does not support that interpretation. Sexually immature males were aged 1–3 years in the sample and mature males at over 6 years, and there were no porpoises aged 4 or 5 years that would be likely to represent early and late maturing individuals. The absence of 4–5-year-old males in the sample cannot be ignored for such a fast growing species as the finless porpoise. Noting a similar gap in the age structure of females, I conclude that individuals around puberty are missing in the sample of both sexes and that males enter into the pubertal stage at the age of 2 years at the earliest or at the age of 3–4 years at the latest, attaining sexual maturity by the age of 6 years. During puberty, both males and females must change their behavior in such a way to decrease mortality incidental to net fisheries. This hypothesis awaits further confirmation, but similar behavioral changes around puberty have been reported for striped dolphins (Chapter 10), bottlenose dolphins (Chapter 11), and some other delphinids.

8.5.3 Reproductive Cycle

With the exception of particular species, males tend to spend less energy per offspring than females. Male finless porpoises are the same and participate in reproductive activity in every breeding season to maximize the opportunity of copulation. Gestation lasts slightly less than 1 year in many cetacean species including the finless porpoise. Although this gives them the possibility of annual reproduction, not all the species

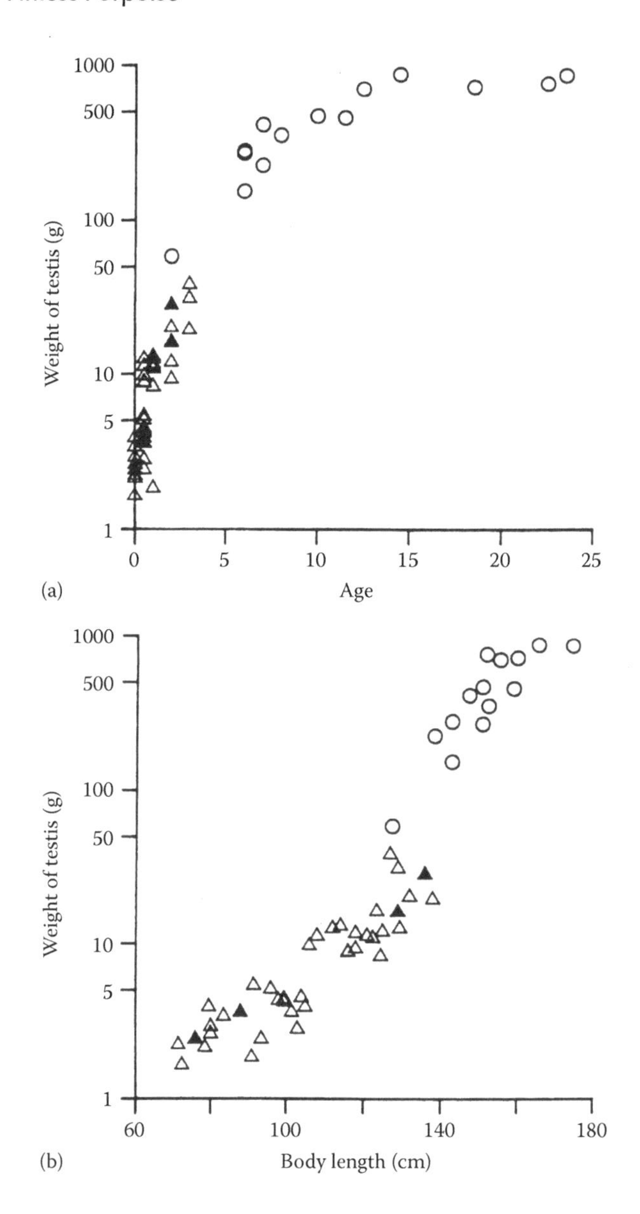

FIGURE 8.14 Weight of single testis with epididymis removed and maturity plotted on age (a) or on body length (b). Triangles represent immature individuals and circles mature individuals. The closed triangles represent the Inland Sea population (3), and open symbols (triangles and circles) represent two populations in western Kyushu (1 and 2 in Figure 8.2). (Reproduced from Figure 7 of Shirakihara, M. et al., *Mar. Mammal Sci.*, 9, 392, 1993. With permission of Wiley Press.)

reproduce annually. For example, 49 calving intervals were observed in female humpback whales (which nurse their calves for 10.5–11 months after gestation of 11.5 months) in the Hawaiian breeding ground, ranging from 1 to 9 years, with annual conception in only nine cases or 18%. The most frequent cycle was 2 years (17 cases or 34.7%), and the mean was 3.2 years (Glockner-Ferrari and Ferrari 1990). Estrus was suppressed during lactation and the time for physiological recovery from lactation.

Currently, almost nothing is known about the reproductive cycle of female finless porpoises. One of the reasons is the difficulty of identifying sex and reproductive status of wild individuals, and another is the difficulty in obtaining unbiased samples of female carcasses. Seasonally unbiased information on reproductive status is ideal for the purpose of analyzing reproductive cycle, but incidental mortality comes from fishing activities that are often seasonal. Stranding of finless porpoise is frequent in the parturition season and biased toward perinatal mortality.

Shirakihara et al. (1993) presented the reproductive composition of 97 finless porpoises from two populations off western Kyushu and the Inland Sea population obtained around Kanmon Pass, where two western Kyushu populations were not distinguished. Fourteen of the 97 specimens were sexually mature males and 10 were sexually mature females (Table 8.14). The reproductive status of the 10 sexually mature females was as follows. One female from western Kyushu was not pregnant in June but had a corpus luteum of ovulation. We do not know if this female had become pregnant during the mating season or would have had to wait until the next mating season. Two western Kyushu specimens in January and February were lactating and also pregnant with small fetuses of 9.7 and 16 cm, respectively. These two females were most likely to have become pregnant in the latest mating season that followed the most recent parturition season, that is, suggesting a 1-year breeding cycle, but the possibility of conception during lactation that lasted for about 1 year could not be fully excluded (2-year breeding cycle). The autumn conceptions suggest that the porpoises were from the Ariake Sound–Tachibana Bay population (see Section 8.5.1). Two females that died in July and September were pregnant with fetuses of 50–60 cm conceived in autumn of the previous year, which suggests that they also were from the Ariake Sound–Tachibana Bay population. As these two pregnant females were not lactating at death, their lactation status at conception or their reproductive cycles were unknown. Two other lactating females from western Kyushu, in March and September, do not allow speculation on their reproductive cycle or population of origin. One adult female from the Inland Sea population had five corpora albicantia in the ovaries and was resting, that is, neither pregnant nor lactating, in July. This female could have lost her calf born in the previous spring (2-year breeding cycle) or was going to have a reproductive cycle of 3 or more years.

The scanty data presented earlier suggest only that Japanese finless porpoises can have reproductive cycles of 1 year as well as 2 or more years.

8.6 SOCIAL STRUCTURE

8.6.1 School

Toothed whales often live with other individuals of the same species. This has been called "group," "aggregation," "school," or "pod" by various scientists and for various cetacean species. Connor (2000) used "group" or "pod" if members of the same species benefit from living together and used "aggregation" for other cases. Acevedo-Gutierrez (2002) followed this usage. Benefits of group living can be cooperation in feeding, nursing, and increased chance of detecting

TABLE 8.14
Reproductive Status of Female Finless Porpoises in Japan

Month of Sample	Ovulated[a]	Pregnant	Pregnant and Lactating[b]	Lactating	Resting[c]	Total
January			1			1
February			1			1
March				1		1
June	1					1
July		1 + (1)			(1)	1 + (2)
September		1		1		2
October		(1)				(1)
Total	1	2 + (2)	2	2	(1)	7 + (3)

Source: Prepared from Table 2 of Shirakihara, M. et al., *Mar. Mammal Sci.*, 9, 392, 1993, which gave data for individual porpoises.

Note: Those noted in parentheses are from the Inland Sea and others from western Kyushu.

[a] Females having a corpus luteum and neither pregnant nor lactating.

[b] Females pregnant and simultaneously lactating.

[c] Sexually mature females neither lactating, pregnant, nor ovulating.

predators or escaping from them, and the group members are likely in close association. Such a group can be identified at sea based on close distance between individuals and swimming speed and direction. However, longer observation is required to identify groups within an aggregation that is formed temporarily by several groups in a suitable feeding ground. Various types of aggregation have been reported for the sperm whale, where the usual group size is around 10 (Connor 2000), including cases of (1) 2 groups forming an aggregation that lasts for several days (feeding cooperation is suspected here), (2) several groups gathering to form a temporary aggregation in a area of several square kilometers, and (3) a large aggregation of almost 1000 individuals assembled in an area of several hundred square kilometers. "School" has been used for cases where the distinction between "pod" or "group" and "aggregation" is difficult or where such distinction is not intended (Acevedo-Gutierrez 2002).

Japanese cetacean scientists used to use school (e.g., Kasuya and Kureha 1979) or group (e.g., Shirakihara et al. 2007) without strictly defining these terms. This was probably either because Japanese cetacean biology started from examination of carcasses taken by various types of fisheries or because onboard observation of live animals was mostly for distribution studies and abundance estimation that did not require distinction between group and aggregations or did not allow time for such distinction. Behavioral studies of cetaceans around Japan are still at an early stage of development, and they will require distinction between groups and temporary aggregations.

8.6.2 School Size

Table 8.15 lists data on school size of Japanese finless porpoises obtained through aerial sighting surveys for abundance estimation, where school is identified only as a unit of sighting (not a unit of social activity) and the distinction between group and aggregation is not required. Although Kasuya and Kureha (1979) made some attempts to distinguish between groups and aggregations, this is not reflected in school size

TABLE 8.15
School Size of Japanese Finless Porpoises

Area	Month	Mean	Mode	Maximum	Sample Size	Data Source	Reference
Omura Bay (1)	Feb, May, Aug, Nov	1.64	1		33	Aerial	Yoshida et al. (1998)
Ariake Sound–Tachibana Bay (2)	Apr–Sep		1	10	256	Vessel	Shirakihara et al. (1994)
	Oct–Mar		1	5	294	Vessel	Shirakihara et al. (1994)
Ariake Sound (2)	May–Aug	1.64	1	≤10	152	Aerial	Yoshida et al. (1997)
	Nov–Jan	1.68	1	≤10	69	Aerial	Yoshida et al. (1997)
Tachibana Bay (2)	May, Jul–Aug	1.80	1	≤10	25	Aerial	Yoshida et al. (1997)
	Nov, Feb	14.24[b]	1	117	27	Aerial	Yoshida et al. (1997)
Inland Sea (3)	Mar–Sep	2.01[c]	1	13	641	Vessel	Kasuya and Kureha (1979)
	Oct–Feb	1.60[c]	1	8	68	Vessel	Kasuya and Kureha (1979)
	Apr–May	1.56	1		148	Aerial	Shirakihara et al. (2007)
Ise and Mikawa Bays (4)	Oct–Mar	1.61	1		54	Vessel	Miyashita et al. (2003)
	Apr–Sep	1.84	1		186	Vessel	Miyashita et al. (2003)
S. Chiba-Sendai Bay (5)	May–Jun	1.81	1	5	70	Aerial	Amano et al. (2003a)

[a] Numerals in parentheses denote stocks currently identified in Table 8.3. See Section 8.3.4 for the possibility of additional stocks.

[b] Included two exceptionally large schools (of 82 and 117 in February). If these are judged as outliers and excluded, the school size was 1–11 individuals with a mean of 1.7.

[c] The 95% confidence interval and the standard error were 1.89–2.13 (SE = 0.06) in March–September and 1.36–1.84 (SE = 0.12) in October–February, the mean school sizes are significantly different. However, the effect of different weather conditions has to be considered before reaching a firm conclusion (wind tended to be stronger in winter).

parameters cited in this table. Yoshida et al. (1997) reported two encounters with large aggregations in Tachibana Bay off western Kyushu in February that were composed of 82 and 117 individuals. The size of other schools ranged from 1 to 11. The mean school size was 1.7, excluding the two large aggregations as outliers and 14.24 if they were included.

The studies cited in Table 8.15 show a feature common to finless porpoises, that is, the upper limit of school size was around 10, the mean school size was less than 2, and the modal school size was 1. Singletons were 48.1% of the total encounters (Kasuya and Kureha 1979) or 44.3% (Amano et al. 2003). Corresponding figures are not available for the other studies cited in Table 8.15. Such a large proportion of singletons suggest that, excluding cow–calf pairs, most individuals of this species live a solitary life and real groups might be limited to cow–calf pairs and mating groups of both sexes. The latter is certainly effective for reproduction but seems to last only for a short period.

Figure 8.15 shows an example of possible interaction among groups photographed in the Inland Sea in July 1976. The top panel shows images of a total of 14 finless porpoises, which I interpret to be composed of 1 group of 3 individuals, 4 groups of 2 individuals, and 2 singletons. The bottom panel, which is an enlargement of the bottom portion of the upper frame, shows seven individuals. Swimming direction of these individuals suggests that at least some of them are probably interacting and can be considered to comprise an aggregation of several groups. The two singletons in the bottom panel are moving toward a cow–calf pair in the middle. The two singletons can perhaps be adult males attempting to approach females in postpartum estrus. One of the individuals in a group of three at the right corner is also directing toward the cow–calf pair.

Kasuya and Kureha (1979) described the internal structure of finless porpoise schools encountered in the Inland Sea using location and distance between individuals in a school (Figure 8.16). They concluded that (1) the majority of singletons were of adult size and juvenile or young individuals were rare as singletons; (2) schools of two individuals were usually formed of two adults or an adult and a juvenile (including cow–calf pairs) and groups of two juveniles or two young individuals were rare; (3) schools of three or more individuals could be a temporary combination of smaller schools or singletons. Thus, finless porpoises live a solitary life in principle, and most groups that are maintained for a period of 6 months or a year will be limited to cow–calf pairs. This is similar to features seen in Dall's porpoises (Section 9.6) but differs from the features of groups of sperm whales or short-finned pilot whales (Chapter 12) where at least females and immature males live in long-lasting matrilineal groups or that of striped dolphins that live in large groups with occasional recombination by growth or reproductive stage (Chapter 10). Male finless porpoises approach adult females or cow–calf pairs for the opportunity to mate, but the association is short-lived because a longer association is unlikely to benefit the male's opportunity for reproduction.

School size in finless porpoises varies seasonally. In the Inland Sea, the proportion of schools of two individuals tended to be high from May to September, when it was equal to or exceeded the frequency of singletons (Table 8.16).

FIGURE 8.15 Aerial photograph of finless porpoises in the Inland Sea taken on July 12, 1976. Numerals in the top panel represent sizes of four schools near the periphery of the photo. The bottom panel is a magnification of the lower portion of the upper panel. (Reproduced from Plate 1 of Kasuya, T. and Kureha, K., *Sci. Rep. Whales Res. Inst.*, 31, 1, 1979.)

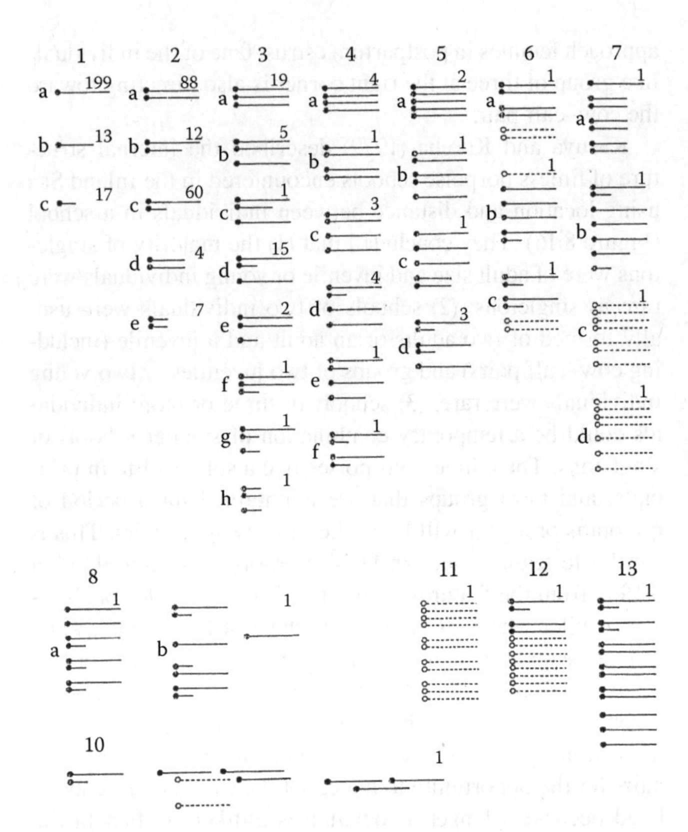

FIGURE 8.16 Schematic depiction of school composition of finless porpoises in the Inland Sea observed during the surveys in 1976–1978. Large numerals at the top are school sizes, and small numerals are numbers of such cases encountered. The length of the bar represents three stages of relative body size, that is, juvenile, intermediate, and adult. Dotted bars represent the individuals without body size estimates. Two-dimensional arrangement of individuals is shown only for 8-b and 10. (Reproduced from Figure 23 of Kasuya, T. and Kureha, K., *Sci. Rep. Whales Res. Inst.*, 31, 1, 1979.)

The ratio of singletons to pairs was 407:263 in October–April, while it was 170:200 in May–September. This is explained by an increased number of cow–calf pairs as well as male and female pairs during the season when parturition proceeds and mating occurs. Further investigation is required to determine if increase in schools of four or more individuals observed during March–September can be explained in the same way.

Seasonal variation in school size has also been observed in finless porpoises in western Kyushu (Shirakihara et al. 1994), where schools of more than six individuals occurred only in the months from April to September, accompanied by a slight increase in mean school size. This seasonal tendency stays the same even when we concentrate on schools of four or more individuals, that is, such schools were limited to April to September with the exception of one school in November. Shirakihara et al. (1994) suggested influence of factors other than reproduction, such as feeding or predators, as a cause of seasonal difference in school size. However, combining the data for two populations with different breeding season made interpretation difficult. The mechanism for the formation of and the function of large aggregations reported from western Kyushu (Table 8.15) and medium-sized aggregations in the Inland Sea (8–10 in Figure 8.16) remain to be investigated.

Note: Jefferson and Wang (2011) proposed a new classification of the genus *Neophocaena* using published biological information also reviewed in this book. They noted that finless porpoises were divided into two distinct groups by the shape of the dorsal keel and that the broad-keel type occurred in the Taiwan Strait and south of it and the narrow type in the strait and north of it. The two types occurring in the Taiwan Strait area have been reproductively isolated since the last glacial period that ended about 18,000 years ago (DNA analysis). The broad-keel type had skull length of 181–245 cm, while the narrow-keel type had skulls of 209–295 cm (reflection of body size). Based on these findings, Jefferson and Wang (2011) proposed classifying finless porpoises into two species containing three subspecies.

One of the two species is *N. phocaenoides* (G. Cuvier 1829) in the Taiwan Strait and south of it. The other is *N. asiaeorientalis* (Pilleri and Gihr 1972) in the Taiwan Strait and north of it including the Yangtze River. They identified the possibility of numerous geographical variations among the latter but recognized only two subspecies: *N. a. asiaeorientalis* (Pilleri and Gihr 1972) in the Yangtze River and *N. asiaeorientalis*

TABLE 8.16
Effect of Season on School Size of Finless Porpoises in the Inland Sea

School Size	Jan	Feb	Mar	Apr	May	Jun	Jul	Aug	Sep	Oct	Nov	Dec
1	83	52	62	47	39	44	18	40	29	60	48	55
2	17	43	25	35	39	33	55	30	43	40	40	36
3	0	5	6	7	14	14	4	10	9	0	8	9
≥4	0	0	7	11	8	9	23	20	19	0	4	0
Max.	2	3	11	11	13	6	7	12	10	2	8	3
Mean	1.17	1.52	1.74	2.03	2.04	1.95	2.73	3.10	2.71	1.40	1.84	1.55
Sample size	6	21	163	271	97	57	22	10	21	5	25	11

Source: Reproduced from Table 3 of Kasuya, T. and Kureha, K., *Sci. Rep. Whales Res. Inst.*, 31, 1, 1979.
Note: School size composition is given in percent.

sunameri (Pilleri and Gihr 1975) in the Taiwan Strait and to the north, including the coasts of Korea and Japan.

As a life history biologist, I am more interested in the interpretation of the distribution. The two types appear to be sympatric in the small Taiwan Strait area, but their general distribution is allopatric. It is worthwhile investigating further details of mechanism of the current reproductive isolation. Reproductive isolation of about 20,000 years, which is rather brief, has been reported between several populations of Dall's porpoises that show partially sympatric distributions. Reproductive isolation between the populations of the Dall's porpoise is thought to be achieved or assisted by the difference in breeding seasons and segregation by age and reproductive status (Sections 9.3 and 9.4). The boundary between two populations of short-finned pilot whales off Japan shifts seasonally, so combining year-round distribution data on a map does not depict the segregation that is retained seasonally. The mechanism inhibiting interbreeding between these two pilot whale populations is likely to be social, and cross-breeding will occur in an environment such as an aquarium.

A classification is a kind of expression of the aesthetic sense of the taxonomists and is destined to be replaced by subsequent new proposals. Until such time, biologists will use the classification proposed by Jefferson and Wang (2011), which places all the Japanese populations into the subspecies *N. a. sunameri* together with individuals off the Korean and Bohai coasts. Japanese members of this subspecies are probably distinguishable morphologically from the continental members (see Section 8.3.4). This taxonomy does not affect the conservation of cetaceans in terms of correctly identifying the unit to conserve, which cannot be greater than the subspecies.

REFERENCES

[*In Japanese Language]

Acevedo-Gutierrez, A. 2002. Group behavior. pp. 537–544. In: Perrin, W.F., Würsig, B., and Thewissen, L.G.M. (eds.). *Encyclopedia of Marine Mammals*. Academic Press, San Diego, CA. 1414pp.

Akamatsu, T., Nakazawa, T., Tsuchiyama, T., and Kimura, T. 2008. Evidence of nighttime movement of finless porpoise through Kanmon Strait monitored using a stationary acoustic device. *Fish. Sci.* 74: 970–976.

Allen, G.M. 1923. The black finless porpoise, *Meomeris*. *Bull. Mus. Comparative Zool.* 65: 233–256.

Amano, M. 2003a. Finless porpoise, *Neophocaena phocaenoides*. pp. 432–435. In: Perrin, W.F., Würsig, B., and Thewissen, L.G.M. (eds.). *Encyclopedia of Marine Mammals*. Academic Press, San Diego, CA. 1414pp.

*Amano, M. 2003b. Geographical variation and taxonomy of the finless porpoise. *Gekkan Kaiyo* 35(8): 531–538.

*Amano, M. June 2010. Strandings of the finless porpoise (*Neophocaena phocaenoides*) in Omura Bay. Abstract of Oral Presentation at a Meeting of Japan Cetology, Towada City, Japan. p. 20.

Amano, M., Miyazaki, N., and Kureha, K. 1992. A morphological comparison of skulls of the finless porpoise *Neophocaena phocaenoides* from the Indian Ocean, Yangtze River and Japanese waters. *J. Mammal. Soc. Jpn.* 17(2): 59–69.

Amano, M., Nakahara, F., Hayano, A., and Shirakihara, K. 2003. Abundance estimate of finless porpoise off the Pacific coast of eastern Japan based on aerial surveys. *Mammal Study* 28: 103–110.

An, Y., Chi, S., and Moon, D. 2010. Korea (Republic of), Progress Report on Cetacean Research, January 2009 to December 2009, with Statistical Data for the Calendar Year 2009. Paper IWC/SC/62/Progress Report of Korea, presented to the Scientific Committee Meeting of IWC. 16pp. (Available from IWC Secretariat, Cambridge, U.K.).

An, Y., Choi, S., Kim, Z., and Kim, H. 2009. Cruise report of the Korean cetacean sighting survey in the Yellow Sea, April–May 2008. Paper IWC/SC/61/NP1, presented to the Scientific Committee Meeting of the International Whaling Commission. 4pp. (Available from IWC Secretariat, Cambridge, U.K.).

An, Y.R. and Kim, Z.G. 2007. Korean (Republic of) Progress Report on Cetacean Research with Statistical Data for the Calendar Year 2005. Paper IWC/SC/59Progress Report of Korea, presented to the Scientific Committee Meeting of the International Whaling Commission. 4pp. (Available from IWC Secretariat, Cambridge, U.K.).

An, Y.R., Kim, Z.G., and Choi, S.G. 2006. Cruise Report of the Korean Cetacean Sighting Survey in the East Sea, April–May 2006. Paper IWC/SC/58/NPM7, presented to the Scientific Committee Meeting of the International Whaling Commission. 4pp. (Available from IWC Secretariat, Cambridge, U.K.).

*Arita, M. 1999. Trend of sand supply from the Inland Sea area for construction. *Setonaikai* 19: 25–39.

Barros, N.B., Jefferson, T.A., and Parsons, E.C.M. 2002. Food habits of finless porpoise (*Neophocaena phocaenoides*) in Hong Kong waters. *Raffles Bull. Zool.* 10: 115–123.

Caballero, S., Trujillo, F., Vianna, J.A., Barios-Garrido, H., Montiel, M.G., Bertran-Pedreros, S., and Marmontel, M. 2007. Taxonomic status of genus *Sotalia*: Species level ranking for "tucuxi" (*Sotalia fluviatilis*) and "costero" (*Sotalia guianensis*) dolphins. *Mar. Mammal Sci.* 23: 358–386.

Chen, L., Bruford, N.W., Xu, S., Zhou, K., and Yang, G. 2010. Microsatellite variation and significant population genetic structure of endangered finless porpoise (*Neophocaena phocaenoides*) in Chinese coastal waters and the Yangtze River. *Mar. Biol.* 157: 1453–1462.

Colborn, T. and Smolen, M.J. 2003. Cetacean and contaminants. pp. 291–332. In: Vos, J.G., Bossart, G.D., Fournier, M., and O'Shea, T.J. (eds.). *Toxicology of Marine Mammals*. Taylor & Francis, London, U.K. 643pp.

Connor, R.C. 2000. Group living in whales and dolphins. pp. 199–218. In: Mann, J., Connor, R.C., Tyack, P.L., and Whitehead, H. (eds.). *Cetacean Societies: Field Studies of Dolphins and Whales*. University of Chicago Press, Chicago, IL. 433pp.

Coues, E. 1890. Phocaena. p. 4449. In: W.D. Whiteney (ed.). *The Century Dictionary: An Encyclopedia Lexicon of the English Language*. The Century Company, New York. (Total page not available).

*Furuta, M., Akagi, D., Komatsu, Y., and Yoshida, H. 2007. *Report on Special Take of Finless Porpoises and Progress of Biological Studies*. Toba Aquarium, Toba, Japan. 127pp.

Furuta, M., Kataoka, T., Sekido, M., Yamamoto, K., Tsukada, O., and Yamashita, T. 1989. Growth of the finless porpoise *Neophocaena phocaenoides* (G. Cuvier, 1829) from the Ise Bay, Central Japan. *Annu. Rep. Toba Aquarium.* 1: 89–102.

Furuta, M., Tsukada, O., and Kataoka, T. 1977. Record of the birth of a finless porpoise called Mary. pp. 1–12. In: Toba Suizokukan (ed.). *Ecology and Breeding of Finless Porpoises*, Toba Aquarium, Toba, Japan. 58pp.

Gao, A. and Zhou, K. 1995a. Geographical variation of external measurements and three subspecies of *Neophocaena phocaenoides* in Chinese waters. *Acta Theriol. Sin.* 15(2): 81–92. (In Chinese with English summary).

Gao, A. and Zhou, K. 1995b. Geographical variation of skull among the populations of *Neophocaena* in Chinese waters. *Acta Theriol. Sin.* 15(3): 161–169. (In Chinese with English summary).

*Garner, G.W., Amstrup, S.C., Laake, J.L., Manly, B.F.J., McDonald, L.L., and Robertson, R.G. 2002. *Marine Mammal Survey and Assessment Methods.* Geiken Series No. 9. Nihon Geirui Kenkyusho (ICR), Tokyo, Japan. 169pp. (Initially published in 1999 by Balkema Publishers, the Netherlands, and translated by Shirakihara, K., Okamura, H., and Kasamatsu, F.)

Geraci, J.R., Harwood, J., and Lounsbury, V.J. 1999. Marine mammal die-offs: Causes, investigation and issues. pp. 367–395. In: Twiss, J.R. Jr. and Reeves, R.R. (eds.). *Conservation and Management of Marine Mammals.* Smithsonian Institution Press, Washington, DC. 469pp.

Glockner-Ferrari, D.A. and Ferrari, M.J. 1990. Reproduction in the humpback whale (*Megaptera novaeangliae*) in Hawaiian waters, 1975–1988: The life history, reproductive rates and behavior of known individuals identified through surface and underwater photography. *Rep. Int. Whal. Commn.* (Special Issue 12): 161–169.

Gray, J.E. 1846. *The Zoological Voyage of H.M.S. Erebus and Terror,* Vol. 1: Mammalia. E.W. Janson, London, U.K. 53pp.+37pls.

Gray, J.E. 1847. *List of the Osteological Specimens of Mammalia in the Collection of the British Museum.* British Museum, London, U.K. 147pp.

Hammond, P.S., Benke, H., Berggren, P., Borchers, D.L., Buckland, S.T., Collet, A., Heide-Jorgensen, M.P. et al. 1995. Distribution and abundance of the harbour porpoise and other small cetaceans in the North Sea and adjacent waters, LIFE92-2/UK/027, Final Report. 240pp.

Hershkovitz, P. 1961. On the nomenclature of certain whales. *Fieldiana Zool.* 39(49): 547–565.

Hershkovitz, P. 1966. *Catalog of Living Whales.* Smithsonian Institute, Washington, DC. 259pp.

Howell, A.B. 1927. Contribution to the anatomy of the Chinese finless porpoise, *Neomeris phocaenoides. Proc. US Natl. Mus.* 70(13): 1–43 and pl.1.

Hugget, A.S.G. and Widdas, W.F. 1951. The relationship between mammalian foetal weight and conception age. *J. Physiol.* 114: 306–317.

*Ishikawa, H. 1994. *Stranding Records of Cetaceans in Japan.* Geiken Series No. 6. Nihon Geirui Kenkyusho (ICR), Tokyo, Japan. 94pp.

*Iwatsuki, T. 2000. Analyses of stranding records of finless porpoises in Japan. Bachelor thesis Presented to Faculty of Bioresources of Mie University, Tsu, Japan. 11pp., 1 Table, 13 Figs., 1 Appendix Table.

IWC (International Whaling Commission). 2010. Report of the Scientific Committee, Annex L: Report of the Sub-Committee on Small Cetaceans. *J. Cetacean Res. Manage.* 12(Supplement 2): 306–330.

*Izawa, K. and Kataoka, T. 1965. Ecology and breeding of finless porpoises. *Nichi Do Sui Shi* 7(3): 57–59.

*Japanese Ministry of Environment. November 1999. Survey of dioxin levels in wildlife in 1999 and 1998. http://www.env.go.jp/press.php3?serial=1837. Accessed December 10, 1999.

Jefferson, T.A. 2002. Preliminary analysis of geographic variation in cranial morphometrics of the finless porpoise (*Neophocaena phocaenoides*). *Raffles Bull. Zool.* (Supplement 10): 3–14.

Jefferson, T.A. and Hung, S.K. 2004. *Neophocaena phocaenoides. Mamm. Species* 746: 12.

Jefferson, T.A. and Wang, J.Y. 2011. Revision of the taxonomy of finless porpoise (genus *Neophocaena*): The existence of two species. *J. Mar. Anim. Ecol.* 4(1): 3–16.

*Kaburagi, T. 1932. Fishing operation and area of finless porpoise aggregation. pp. 72–75. In: Monbu-Sho (ed.). *Report of Investigation of Natural Monuments, Animals, Part 2.* Monbusho, Tokyo, Japan. (Total page not available).

Kannan, S.N., Tanabe, N., Ono, M., and Tatsukawa, R. 1989. Critical evaluation of polychlorinated biphenyl toxicity in terrestrial and marine mammals: Increasing impact on *non-ortho* and *mono-ortho* coplanar polychlorinated biphenyl's from land and ocean. *Arch. Contam. Toxicol.* 18: 850–857.

Kasuya, T. 1977. Age determination and growth of the Baird's beaked whale with a comment on the fetal growth rate. *Sci. Rep. Whales Res. Inst.* 29: 1–20.

Kasuya, T. 1991. Density dependent growth in North Pacific sperm whales. *Mar. Mammal Sci.* 7(3): 230–257.

*Kasuya, T. 1994. Finless porpoise. pp. 626–634. In: Odate, S. (ed.). *Base Data of Rare Wild Aquatic Organisms of Japan* (I). Nihon Suisan Shigen Hogo Kyokai, Tokyo, Japan. 696pp.

Kasuya, T. 1995. Overview of cetacean life histories: An essay in their evolution. pp. 481–497. In: Blix, A.S., Walloe, L., and Ultang, O. (eds.). *Whales, Seals, Fish and Man.* Elsevier, Amsterdam, the Netherlands. 720pp.

*Kasuya, T. 1997. Ecology and conservation of the finless porpoise. *Setonaikai* 10: 27–34.

Kasuya, T. 1999. Finless porpoise *Neophocaena phocaenoides* (G. Cuvier, 1829). pp. 411–442. In: Ridgway, S.H. and Harrison, R. (eds). *Handbook of Marine Mammals,* Vol. 6: Second Book of Dolphins and the Porpoises. Academic Press, San Diego, CA. 486pp.

*Kasuya, T. 2008. Current status of Japanese finless porpoises with particular reference to a population in the Inland Sea. *Isana* 48: 52–70.

*Kasuya, T. 2010. Finless porpoises in the Inland Sea: Their current status and conservation needs. pp. 76–108. In: Nihon Seitai Gakkai Kaminoseki Yobosho Afutakea Iinkai (ed.). *Miraculous Sea: Biodiversity of Kaminoseki Area of the Inland Sea.* Nanpo Shinsha, Kagoshima, Japan. 237pp.

*Kasuya, T., Kaneko, M., and Nishimoto, Y. 1985. Archaeology and related science 7: Zoology. *Kikan Kokogaku* 11: 91–95.

Kasuya, T. and Kureha, K. 1979. The population of finless porpoise in the Inland Sea of Japan. *Sci. Rep. Whales Res. Inst.* 31: 1–44.

Kasuya, T. and Marsh, H. 1984. Life history and reproductive biology of short-finned-pilot whale, *Globicephala macrorhynchus,* off the Pacific coast of Japan. *Rep. Int. Whal. Commn.* (Special Issue 6: Reproduction in Whales, Dolphins and Porpoises): 259–310.

Kasuya, T. and Matsui, S. 1984. Age determination and growth of the short-finned pilot whale off the Pacific coast of Japan. *Sci. Rep. Whales Res. Inst.* 35: 57–91.

Kasuya, T., Tobayama, T., and Matsui, S. 1984. Review of the live capture of small cetaceans in Japan. *Rep. Int. Whal. Commn.* 34: 597–602.

Kasuya, T., Tobayama, T., Saiga, T., and Kataoka, T. 1986. Perinatal growth of delphinoids: Information from aquarium reared bottlenose dolphins and finless porpoises. *Sci. Rep. Whales Res. Inst.* 37: 85–97.

*Kasuya, T. and Yamada, T. 1995. *Catalogue of Japanese Cetaceans.* Geiken Series No. 7. Nihon Geirui Kenkyusho (Institute of Cetacean Research), Tokyo, Japan. 90pp.

Kasuya, T., Yamamoto, Y., and Iwatsuki, T. 2002. Abundance decline in the finless porpoise population in the Inland Sea of Japan. *Raffles Bull. Zool.* (Supplement 10): 57–73.
*Kataoka, T., Kitamura, S., Sekido, M., and Yamamoto, K. 1976. On food habits of the finless porpoise. *Mie Seibutsu* 24/25: 29–36.
Kiltie, R.A. 1984. Seasonality, gestation time, and large mammal extinction. pp. 299–314. In: Martin, P.S. and Klein, R.G. (eds.). *Quaternary Extinctions.* University of Arizona Press, Tucson, AZ. 720pp.
*Kureha, K. 1976. Surf riding of finless porpoises. *Geiken Tsushin* 297: 37–38.
*Kuroda, N. 1934. On mammals in Fauna Japonica by Siebold. pp. 1–12. In: Shokubutu Bunken Kankokai (ed.). *Mammals of Japan*, Vol. 3. Shokubutu Bunken Kankokai, Tokyo, Japan. (Total page not available).
*Kuroda, N. 1940. On cetaceans in Suruga Bay. *Shokubutsu Dobutsu* 8(5): 825–834.
Laws, R.M. 1959. The foetal growth rate of whales with special reference to the fin whale, *Balaenoptera physalus* Linn. *Discov. Rep.* 29: 281–308.
Lockyer, C. 1984. Review of baleen whale (Mysticeti) reproduction and implication for management. *Rep. Int. Whal. Commn.* (Special Issue 6: Reproduction in Whales, Dolphins and Porpoises): 27–50.
Martineau, D., De Guise, S., Fournier, M., Shugart, L., Girard, C., Lagace, A., and Beland, P. 1994. Pathology and toxicology of beluga whales from the St. Lawrence Estuary, Quebec, Canada. Past, present and future. *Sci. Total Environ.* 154 (Special issue on Marine Pollution: Mammals and Toxic Contaminants): 201–215.
Martineau, D., Mikaelian, I., Jean-Martin, L., Labelle, P., and Higgins, R. 2003. Pathology of cetaceans, a case study: Beluga from the St. Lawrence estuary. pp. 333–380. In: Vos, J.G., Bossart, G.D., Fournier, M., and O'Shea, T.J. (eds.). *Toxicology of Marine Mammals.* Taylor & Francis, London, U.K. 643pp.
*Matsuda, O. 1999. Extraction of marine sand and its influence on the marine ecosystem. *Setonaikai* 19: 29–34.
*Miyashita, T. 2002. Estimating abundance of cetaceans: Current status and problems. pp. 167–185. In: Miyazaki, N. and Kasuya, T. (eds.). *Mammals of the Sea: Their Past, Present and Future.* Saientisuto-Sha, Tokyo, Japan. 311pp.
*Miyashita, T., Furuta, M., Hasegawa, S., and Okamura, H. 2003. Sighting survey of finless porpoises in Mikawa Bay and Ise Bay. *Gekkan Kaiyo* 35(8): 581–585.
Miyashita, T., Wang, P., Cheng, J.H., and Yang, G. 1995. Report of the Japan/China joint whale sighting cruise in Yellow Sea and the East China Sea in 1994 summer. Paper SC/47/NP17 presented to Scientific Committee Meeting of the International Whaling Commission. 12pp. (Available from IWC secretariat, Cambridge, U.K.).
*Mizue, K., Yoshida, K., and Masaki, Y. 1965. Studies on the little toothed whales in the west sea area of Kyushu—XII, *Neomeris phocaenoides*, so called Japanese SUNAMERI, caught in the coast of Tachibana Bay, Nagasaki Pref. *Bull. Fac. Fish. Nagasaki Univ.* 18: 7–29.
*Nakajima, M. 1963. A stray finless porpoise. *Geiken Tsushin* 147: 193–196.
*Nakamura, K., Sakakibara, S., Abel, G., Tachkawa, T., Mizushima, K., Wada, M., Doi, T., and Kikuchi, T. 2003. Notes on strandings and incidental catch of finless porpoises *Neophocaena phocaenoides* in coastal waters of Yamaguchi and Fukuoka Prefecture. *Nihonkai Cetol.* 13: 13–18.
*Nakamura, M. and Hiruda, H. 2003. On the drifting, straying and incidental catch of cetaceans on the coast of Fukuoka Prefecture (1997–2001). *Nihonkai Cetol.* 13: 106.
*Northridge, S.P. and Hoffman, R.J. 1999. Marine mammal interaction with fisheries. pp. 99–119. In: Twiss, J.R. Jr. and Reeves, R.R. (eds.). *Conservation and Management of Marine Mammals.* Smithsonian Institution Press, Washington, DC. 469pp.
*Okura, T. 1826. *Locust Control.* Kawachiya, Osaka, Japan. 56pp.
O'Shea, T.J., Brownell, R.L., Clark, D.R. Jr., Walker, W.A., Gay, M.L., and Lamont, T.G. 1980. Organochlorine pollutants in small cetaceans from the Pacific and South Atlantic Oceans, November 1968–June 1976. *Pestic. Monit. J.* 14: 35–46.
*Otsuki, K. 1808 (manuscript). *On Natural History of Whales.* Printed on pp. 7–152 of *Geishi Ko* [*On Natural History of Whales*] (1983). Kowa Shuppan, Tokyo, Japan. 538+31pp.
Palmer, T.S. 1899. Notes on three genera of dolphins. *Proc. Biol. Soc. Wash.* 13: 23–24.
Parsons, E.C.M. and Wang, J.Y. 1998. A review of finless porpoise (*Neophocaena phocaenoides*) from the South China Sea. pp. 287–306. In: Morton, B. (ed). *The Marine Biology of the South China Sea: Proceedings of the Third International Conference on the Marine Biology of the South China Sea.* Hong Kong University Press, Pok Fu Lam, Hong Kong. 612pp.
Perrin, W.F., Coe, J.M., and Zweifel, J.R. 1976. Growth and reproduction of spotted porpoise, *Stenella attenuata*, in the offshore eastern tropical Pacific. *Fish. Bull.* 74(2): 229–269.
Perrin, W.F., Dolar, M.L.L., and Robineau, D. 1999. Spinner dolphins (*Stenella longirostris*) of the western Pacific and Southeast Asia: Pelagic and shallow-water forms. *Mar. Mammal Sci.* 15: 1029–1053.
Perrin, W.F., Holts, D.B., and Miller, R.B. 1977. Growth and reproduction of the eastern spinner dolphin, a geographical form of *Stenella longirostris* in the eastern tropical Pacific. *Fish. Bull.* 75(4): 725–750.
Perrin, W.F., Miyazaki, N., and Kasuya, T. 1989. A dwarf form of spinner dolphin, *Stenella longirostris*, from Thailand. *Mar. Mammal Sci.* 5: 213–227.
Perrin, W.F. and Reilly, S.B. 1984. Reproductive parameters of dolphins and small whales of the family Delphinidae. *Rep. Int. Whal. Commn.* (Special Issue 6: Reproduction in Whales, Dolphins and Porpoises): 97–133.
Pilleri, G. and Chen, P. 1980. *Neophocaena phocaenoides* and *Neophocaena asiaeorientalis*: Taxonomical differences. pp. 25–32. In: G. Pilleri (ed.). *Investigations on Cetacea* XI. Privately published by G. Pilleri, Bern, Switzerland. 220pp.
Pilleri, G. and Gihr, M. 1972. Contribution to the knowledge of the cetaceans of Pakistan with particular reference to the genera *Neomeris*, *Sousa*, *Delphinus*, and *Tursiops* and description of a new Chinese porpoise (*Neomeris asiaeorientalis*). pp. 107–162. In: G. Pilleri (ed.). *Investigations on Cetacea.* IV. Privately published by G. Pilleri, Bern, Switzerland. 297pp.
Pilleri, G. and Gihr, M. 1973–1974. Contribution to the knowledge of the cetaceans of the southwest and monsoon Asia (Persian Gulf, Indus Delta, Malabar, Andaman Sea and Gulf of Siam). pp. 59–149. In: G. Pilleri (ed.). *Investigations on Cetacea.* V. Privately published by G. Pilleri, Bern, Switzerland. 365pp.
Pilleri, G. and Gihr, M. 1975. On the taxonomy and ecology of the finless black porpoise, *Neophocaena* (Cetacea, Delphinidae). *Mammalia* 39: 657–673.
Reeves, R.R., Wang, J.Y., and Leatherwood, S. 1997. The finless porpoise, *Neophocaena phocaenoides* (G. Cuvier, 1829): A summary of current knowledge and recommendations for conservation action. *Asian Mar. Biol.* 14: 111–143.
Reijinders, P.J.H. 1996. Organochlorine and heavy metal contamination in cetaceans: Observed effects, potential impact and future prospects. pp. 203–217. In: Simmonds, M.P.

and Hutchinson, J.D. (eds.). *The Conservation of Whales and Dolphins, Science and Practice.* John Wiley & Sons, Chichester, U.K. 476pp.

Rice, D.W. 1977. A list of the marine mammals of the world. NOAA Technical Report NMFS SSRF-711. National Marine Fish. Service, Seattle, WA. 15pp.

Rice, D.W. 1998. *Marine Mammals of the World, Systematics and Distribution.* Special Publication No. 4. Society for Marine Mammalogy. Lawrence, KS, 231pp.

*Seibutsu Tayosei Senta (ed.) 2002. *Report of Basic Survey for Conservation of the Natural Marine Environment: Survey of Marine Animals (Finless Porpoise Survey).* Kaichu Koen Senta, Tokyo, Japan. 136pp.

*Setonakai Kankyo Hozen Kyokai. 1996. *Environmental Archives of the Inland Sea.* Setonaikai Kankyo Hozen Kyokai, Tokyo, Japan. 161pp.

*Shirakihara, K. and Shirakihara, M. 2002. Abundance [of finless porpoises] in Ariake Sound—Tachibana Bay, Omura Bay, and the Inland Sea. pp. 27–52. In: Seibutsu Tayosei Senta (ed.). *Report of Basic Survey for Conservation of the Natural Marine Environment: Survey of Marine Animals (Finless Porpoise Survey).* Kaichu Koen Senta, Tokyo, Japan. 136pp.

Shirakihara, K., Shirakihara, M., and Yamamoto, Y. 2007. Distribution and abundance of finless porpoise in the Inland Sea of Japan. *Mar. Biol.* 150: 1025–1032.

Shirakihara, K., Yoshida, H., Shirakihara, M., and Takemura, A. 1992a. A questionnaire survey on the distribution of the finless porpoise, *Neophocaena phocaenoides*, in Japanese waters. *Mar. Mammal Sci.* 8(2): 160–164.

Shirakihara, M., Seki, K., Takemura, A., Shirakihara, K., Yoshida, H., and Yamazaki, T. 2008. Food habits of finless porpoises *Neophocaena phocaenoides* in western Kyushu, Japan. *J. Mammal.* 89(5): 1248–1256.

Shirakihara, M., Shirakihara, K., and Takemura, A. 1992b. Records of the finless porpoise (*Neophocaena phocaenoides*) in the waters adjacent to Kanmon Pass, Japan. *Mar. Mammal Sci.* 8(1): 82–85.

Shirakihara, M., Shirakihara, K., and Takemura, A. 1994. Distribution and seasonal density of the finless porpoise *Neophocaena phocaenoides* in the coastal waters of western Kyushu, Japan. *Fish. Sci.* 60(1): 41–46.

Shirakihara, M., Takemura, A., and Shirakihara, K. 1993. Age, growth and reproduction of the finless porpoise, *Neophocaena phocaenoides*, in the coastal waters of western Kyushu, Japan. *Mar. Mammal Sci.* 9: 392–406.

*Suzuno, M., Shirakihara, M., and Furoda, T. 2010. Occurrence of the small toothed whale finless porpoise *Neophocaena phocaenoides* in Chiba area. *Chiba Seibutsu-Shi* 60(1): 11–16.

Taguchi, M., Ishikawa, M., and Matsuishi, T. 2010. Seasonal distribution of harbour porpoise (*Phocoena phocoena*) in Japanese waters inferred from stranding and bycatch records. *Mammal Study* 35: 133–138.

Tanabe, M., Prudente, M., Mizuno, T., Hasegawa, J., Iwata, H., and Miyazaki, M. 1998. Butyltin contamination in marine mammals from North Pacific and Asian coastal waters. *Environ. Sci. Technol.* 32(2): 193–198.

Tenminck, C.F. and Schlegel, H. 1844. *Fauna Japonica, Les Mamminiferes Marins.* Lugdini Batavorum, Arnz. Pt. 3: 1–26 and Plates 21–29.

Thomas, O. 1922. The generic name of the finless-backed porpoise. *Ann. Mag. Nat. Hist.* (Series 9) 9(54): 676–677.

Thomas, O. 1925. The generic name of the finless-backed porpoise. *Ann. Mag. Nat. Hist.* (Series 9) 16(96): 655.

Tyack, P.L. 2008. Implications for marine mammals of large-scale changes in the marine acoustic environment. *J. Mammal.* 89(3): 549–558.

Van Dolah, F.M., Doucette, G.J., Gulland, F.M.D., Roweles, T.L., and Bossart, G.D. 2003. Impacts of algal toxins on marine mammals. pp. 245–269. In: Vos, J.G., Bossart, G.D., Fournier, M., and O'Shea, T.J. (eds.). *Toxicology of Marine Mammals.* Taylor & Francis, London, U.K. 643pp.

Wang, J.Y., Frasier, T.R., Yang, S.C., and White, B.N. 2008. Detecting recent speciation events: The case of the finless porpoise (genus *Neophocaena*). *Heredity* 101: 145–155.

Wang, J.Y. and Yang, S. 2007. *An Identification Guide to the Dolphins and Other Small Cetaceans of Taiwan.* Jen Jen Publishing and National Museum of Marine Biology, Taiwan, Japan. 207pp. (In Chinese and English).

Wang, P. 1984. Distribution, ecology and resources conservation of the finless porpoise in Chinese waters. *Trans. Liaoning Zool. Soc.* 5(1): 106–110. (In Chinese with English summary).

Wang, P. 1999. *Chinese Cetaceans.* Ocean Enterprises, Hong Kong 325pp.+Pl.1-VI. (In Chinese).

*Yamashita, K. 1991. Marine mammals in literature of the 17–19th centuries—5: Finless porpoise. *Kaiyo-to Seibutsu* 13(2): 130–133.

*Yoshida, H. 1994. Ecological study of finless porpoises in coastal waters of Western Kyushu. Doctoral thesis Presented to Nagasaki University, Nagasaki, Japan. 127pp.

Yoshida, H. 2002. Population structure of finless porpoise (*Neophocaena phocaenoides*) in coastal waters of Japan. *Raffles Bull. Zool.* (Supplement 10): 35–42.

Yoshida, H., Higashi, K., Ono, H., and Uchida, S. 2010. Finless porpoise (*Neophocaena phocaenoides*) discovered at Okinawa Island, Japan, with the source population inferred from mitochondrial DNA. *Aquat. Mammals* 36(3): 278–283.

Yoshida, H., Shirakihara, K., Kishiro, H., and Shirakihara, M. 1997. A population size of the finless porpoise, *Neophocaena phocaenoides*, from aerial sighting surveys in Ariake Sound and Tachibana Bay. *Res. Popul. Ecol.* 39(2): 239–247.

Yoshida, H., Shirakihara, K., Shirakihara, M., and Takemura, A. 1995. Geographic variation in the skull morphology of the finless porpoise *Neophocaena phocaenoides* in Japanese waters. *Fish. Sci.* 61(4): 555–558.

Yoshida, H., Shirakihara, K., Shirakihara, M., and Takemura, A. 1998. Finless porpoise abundance in Omura Bay, Japan: Estimation from aerial sighting surveys. *J. Wildl. Manag.* 62(1): 286–291.

Yoshida, H., Yoshioka, M., Chow, S., and Shirakihara, M. 2001. Population structure of finless porpoise (*Neophocaena phocaenoides*) in coastal waters of Japan based on mitochondrial DNA sequences. *J. Mammal.* 82: 123–130.

*Yoshioka, M. 2002. Abundance [of finless porpoises] in Ise Bay and Mikawa Bay. pp. 53–89. In: Seibutsu Tayosei Senta (ed.). *Report of Basic Survey for Conservation of Natural Marine Environment: Survey of Marine Animals (Finless Porpoise Survey).* Kaichu Koen Senta, Tokyo, Japan. 136pp.

Zhou, K., Pilleri, G., and Li, Y. 1979. Observation on the baiji (*Lipotes vexillifer*) and the finless porpoise (*Neophocaena asiaeorientalis*) in the Changjiang (Yangtze) River between Nanjing and Taiyangzhou, with remarks on some physiological adaptations of the baiji to its environment. pp. 109–120. In: G. Pilleri (ed.). *Investigations on Cetacea.* X. Privately published by G. Pilleri, Bern, Switzerland. 366pp.

9 Dall's Porpoise

9.1 DESCRIPTION

The Dall's porpoise, *Phocoenoides dalli* (True 1885), the largest species in the family Phocoenidae, inhabits the northern North Pacific and adjacent seas. It consists of two major pigmentation types, the *dalli*-type and the *truei*-type, and a rare black variation. The *dalli*-type has a white flank patch extending from the anal region to the level of the dorsal fin and is called in Japanese *ishi-iruka* or *ishi-iruka*-type. The *truei*-type has a larger flank patch that reaches to the base of the flipper and is called *rikuzen-iruka* or *rikuzen-iruka*-type. Caution is required because *ishi-iruka* is also used as a species name for the Dall's porpoise. Japanese fishermen along the Pacific coast of northern Japan call a Dall's porpoise of any pigmentation *kamiyo* or *kamiyo-iruka* and use *hankuro* (half black) for *dalli*-type individuals, which are less common in the region. *Suzume-iruka* or *kita-iruka* has also been used for the species.

Among the three species of Phocoenidae found in Japanese waters, the Dall's porpoise is the only offshore inhabitant and swims at high speed, raising rooster-tail splashes. The finless porpoise, another member of the family, inhabits temperate and tropical waters, usually within the 50 m isobath, and is known to the south of 38°24′N on the Pacific coast and 37°30′N on the Sea of Japan coast. The third species of the family, the harbor porpoise, usually inhabits the continental shelf area north of 35°N on the Pacific coast and north of 37°N on the Sea of Japan coast. Dall's porpoises are easily identified by a distinct large white patch on the flank and a profile of small head and elevated thoracic portion. The dorsal fin is almost triangular as in the case of the harbor porpoise, which is in contrast with the falcate dorsal fin of many delphinid species, and the trailing edges of the dorsal fin and flukes are marked with a gray or white margin.

The average body length of full-grown adults is 198.1 cm in males and 189.7 cm in females from the central North Pacific (Ferrero and Walker 1999), but there seems to be large individual variation and some geographical variation. The skull of the Dall's porpoise is the largest among those of the three Phocoenidae species off Japan, and the rostrum is more pointed than in the other two species (Table 8.1). Among the Phocoenidae, the Dall's porpoise has the largest number of vertebrae, 92–98 in total, which are compressed anteroposteriorly and have high neural processes.

9.2 TAXONOMY

The Dall's porpoise has a history of being thought to comprise two species, but it is currently dealt with by the Society for Marine Mammalogy as a single species containing two subspecies represented by the *truei*- and *dalli*-types. This classification assumes greater distances between the *truei*-type population and several *dalli*-type populations than those between the *dalli*-type populations. The species was first described by True (1885) as *Phocaena dalli* based on a specimen collected by D.H. Dall in waters west of Adak Island in the Aleutian Islands. He placed this new species in the genus together with the harbor porpoise. Instead of the genus name *Phocoena*, which was created by G. Cuvier for the harbor porpoise in 1817, Gray used *Phocena* in 1821 or *Phocaena* in 1828, but the original genus name *Phocoena* is currently in use (Hershkovitz 1966). The specimen described by True in 1885 had the pigmentation of the so-called *dalli*-type. Later, in 1911, Andrews described a new species based on a specimen collected off the Sanriku Region (37°54′N–41°35′N) on the Pacific coast of northern Japan and placed it in a newly created genus *Phocoenoides* together with *Ph. dalli* (True 1885) to yield the two species *P. dalli* (True 1885) and *P. truei* Andrews 1911. The specimen described by Andrews was of the *truei*-type. A slightly different spelling, *Phocaenoides*, was once used for the genus, but it has been rejected.

The English names of these two species (or color morphs) were True's porpoise and Dall's porpoise. Japanese cetacean science in its infancy was confused on this. Nagasawa (1916, in Japanese) proposed the Japanese common name *toru-iruka* (True's porpoise) for the former type, but Kishida (1925, in Japanese) used *rikuzen-iruka* for it after personal communication with N. Kuroda. Then, Kuroda (1938, in Japanese) used *toru-iruka* for *P. dallii truei* (sic) and Wogawa-iruka (Ogawa's porpoise) for *P. dallii dallii* (sic), as two subspecies within the single species *P. dalli*. However, Matsu-ura (1943, in Japanese) dealt with them as different species and used the Japanese name *rikuzen-iruka* (or *toru-iruka*) for *P. truei* and *ishi-iruka* (or *Wogawa-iruka* or *kita-iruka*) for *P. dalli*. These scientists were aware of various vernacular names used for the genus by fishermen along the Sanriku coast (*kamiyo-iruka*, *suzume-iruka*, and *hankuro*) but were apparently reluctant to use them as species name (Kasuya and Yamada 1995, in Japanese), presumably because these names did not usually distinguish between the two nominal species.

No objection was raised against moving Dall's porpoise from *Phocoena* to a new genus *Phocoenoides*, but there were various opinions expressed on the taxonomy of the species. This was probably due to the absence of identifiable taxonomic difference between the nominal two species except for the geographically segregating pigmentation pattern.

Kuroda (1954, in Japanese) cruised along the Kuril Islands from off southern Hokkaido at around 42°N to the western Aleutian Islands area and observed that (1) *dalli*-types were present in the whole range, but *truei*-types were found only along the Japanese coast and (2) the two types did not form mixed schools from April to May, which he thought to be their mating season. However, later in July off southern Hokkaido, he observed that the two types rode bow waves together and obtained a *truei*-type fetus from a *dalli*-type female. Kuroda (1954, in Japanese) interpreted this to mean that reproductive isolation of the two types was incomplete and the pigmentation pattern was controlled by two alleles at one locus; he concluded that they should be dealt with as separate subspecies.

Kuroda (1954, in Japanese) did not clarify whether the two pigmentation types joined on the bow of their ship or were in a group before arriving at the bow. Another scientist F. Wilke was on the same vessel with Kuroda and stated that mixed schools of the two types were not common even off Hokkaido (Wilke et al. 1953). The interpretation of the genetics of the pigmentation pattern by Kuroda (1954, in Japanese) seems to be too simplistic (Kasuya 1978), and the taxonomic conclusion is questionable even if his interpretation is accepted. Our current understanding of the species is that the parturition season is followed by the mating season, which is from midsummer to early autumn and differs between populations (Sections 9.4.3 through 9.4.5), and that weaned juveniles are segregated from breeding individuals at least during the season of parturition/mating but possibly also in other seasons (Sections 9.3.3 and 9.4.2).

Miyazaki et al. (1984) reported as results of their sighting survey that (1) only *dalli*-types occurred in July and August in the Bering Sea and (2) that both types were seen in the Pacific area along the Kuril Islands, where 15 schools were of *dalli*-types, 24 schools were of *truei*-types, and 5 schools contained both types. The proportion of mixed schools was only about 10% even in waters where both types occurred. It seems to me that these mixed schools were formed in the area of coexistence after the porpoises arrived there. Such combinations could be considered aggregations as discussed in Section 8.6.1.

Kasuya (1978, 1982) observed in a 100.5 cm full-term fetus obtained from a pregnant *dalli*-type in the Bering Sea that the anterior lateral portion, which was expected to be black on a typical *dalli*-type or white on a typical *truei*-type, had a uniform intermediate gray pigmentation and suggested that this anterior gray area darkens during a period from the late fetal stage to early postnatal stage. I stressed the need for further examination of development of the pigmentation pattern during the perinatal period before accepting the hybrid hypothesis of Kuroda (1954, in Japanese) and Wilke et al. (1953). After this observation, Szczepaniak and Webber (1992) published the photographs of two newborn individuals, a 97 cm *truei*-type and 103 cm *dalli*-type, and stated that distinction of the two types was not difficult at that stage. Distinction of the two types was clearer on their two photographs than for the full-term fetus presented by Kasuya (1978), which confirmed that the two pigmentation patterns become distinct in the early stage of postnatal life.

Houck (1976) questioned the taxonomic significance of the morphological characters used by Andrews (1911) to distinguish *P. truei* (Andrews 1911) from *P. dalli* (True 1885). He pointed out that (1) the greater height of the tail peduncle was a character of the adult male common to both species, (2) the pigmentation patterns in three body areas (anal area, tail flukes, and tail peduncle) varied greatly individually and had little taxonomic value, and (3) only the 5th character, size of the white area on the flank, should be considered as important in evaluating the validity of the two species. He further noted on the variation of the flank patch that entirely black individuals had been reported from both sides of the North Pacific and that *truei*-types occurred only in the western North Pacific and had various degree of individual variation intermediate between *truei*- and *dalli*-types (note: a rare example of the *truei*-type in the eastern North Pacific was later reported by Szczepaniak and Webber (1992)). He concluded that the two types are not reproductively isolated (note: he referred here to the observation of Wilke et al. (1953) on the *truei*-type fetus found in a *dalli*-type female) but the two types could not be considered as two subspecies because they were not geographically isolated. Therefore, he concluded that the two types should be dealt as only two color morphs in one species. It seems to me that he probably overemphasized the absence of reproductive isolation. If reproductive isolation were proven between the two pigmentation types in sympatric distribution, the taxonomic argument could proceed in the opposite direction, that is, to classify them as two distinct species.

Shimura and Numachi (1987) evaluated the genetic distance among toothed whale species based on the variation in 19 loci identified from isozyme analysis of skeletal muscle and liver. Their Dall's porpoise sample included *truei*-types from the Sanriku Region and *dalli*-types from the North Pacific south of the Aleutian Islands at 165°–175′E. They found that the two types were genetically different by only 0.4% and concluded that they should be dealt with as a single species. They also found that the genus *Phocoenoides* was genetically closer to *Neophocaena* than *Phocoena* and that these three genera constitute a single group of Phocoenidae.

Escorza-Treviño et al. (2004) reached a similar conclusion by analyzing mitochondrial DNA (mtDNA) and *microsatellite* DNA. Their specimen series consisted of 23 *truei*-types and 113 *dalli*-types from nearly the entire range of the species. They identified 66 mtDNA haplotypes, many represented by minor differences. Forty-six of the haplotypes were limited to the *dalli*-type and 8 to the *truei*-type and 12 were common to both types. The 66 haplotypes were grouped into two major lineages, each represented by both *dalli*- and *truei*-types. Thus, the evolution of mtDNA did not accord with pigmentation type. They identified several populations of the *dalli*-type but only one of the *truei*-type (Section 9.3.6). The genetic distance between the two pigmentation types was similar to those between populations of the *dalli*-types. Based on these findings, Escorza-Treviño et al. (2004) concluded that the two pigmentation types had a long history of genetic exchange and that the absence of genetic distinction between them was due to genetic exchange, which is still continuing or terminated recently. The high genetic variability of the species, 66 haplotypes represented in 136 individuals, identified by them supported the earlier result of Shimura and Numachi (1987).

Dalli-types are found in the entire range of the species, but *truei*-types are found only in waters extending from the central Okhotsk Sea through the southern Kuril Islands to the Pacific coast of northern Japan (a rare exception from the eastern North Pacific is ignored as an outlier; see the preceding text in this section). The density and proportion of *truei*-types within the range vary geographically and seasonally, reflecting different migration patterns (Section 9.3). There is about a 1-month difference between the mating seasons of nearby populations of the two pigmentation types, and adults from the populations aggregate in different breeding grounds. This functions to suppress free interbreeding between the pigmentation types, but it is questionable if it is sufficient to explain the observed geographical distribution of the two types and the genetic similarity between them. There might be some additional behavioral or physiological elements that restrict successful reproduction between the two types and survival of genetic traits thus introduced. About half of the *truei*-type porpoises landed on the Sanriku coast had various degree of mottling on the anterior part of the white flank patch, but incidents of mottling on the flank patch of *dalli*-type individuals and cases of black forms appeared to be less frequent (Kasuya 1978). Information on the genetic basis of these variations in pigmentation will contribute to understanding the mechanism apparently limiting interbreeding between the *dalli*- and *truei*-types.

9.3 POPULATION STRUCTURE

The Dall's porpoise has several populations, which are also called stocks. Free genetic mixing is expected within a typical population but is suppressed between populations. For management purposes, the identification of populations should be evaluated with more weight on independence of population dynamics rather than on genetic independence. The case of a metapopulation seems to be intermediate, where a group of individuals maintain their own population dynamics but maintain an occasional limited level of exchange with nearby groups of genetic materials or individuals. Such a system enables a species in several patchy habitats to maintain stability as a whole and is expected for cetacean species inhabiting coastal habitats such as *Tursiops aduncus* (see Section 11.3).

Population structure has been rigorously studied in the Dall's porpoise. One of the reasons was international concern about the management of populations of this species incidentally killed in the high-seas salmon drift-net fishery of Japan. Other reasons include the large-scale hunting of the species off Japan and its easily accessible distribution in the northern part of the North Pacific and adjacent seas.

The Japanese mother-ship salmon drift-net fishery started operations in 1952 in the Bering Sea and south of the Aleutian Islands. A group of scientists led by K. Mizue first reported that 10,000–20,000 Dall's porpoises were killed annually in the fishery (Mizue and Yoshida 1965; Mizue et al. 1966, both in Japanese). Later, the U.S. government became concerned about the mortality of the species occurring in its EEZ and started investigating the effect of the fishery on the Dall's porpoise population. A scientific meeting was convened in Seattle in January 1978, and some Japanese university scientists attended by invitation. The Fisheries Agency of Japan was reluctant to investigate the problem, but with increasing pressure by interested countries, it started cooperation with U.S. scientists in the summer of 1978. The activity was transferred to the International North Pacific Fisheries Commission (INPFC), which was established by the International Convention for the High Seas Fisheries of the North Pacific Ocean between Canada, Japan, and the United States. The research activity expanded to the large-mesh drift-net fishery, which targeted tunas and marlins and caused incidental mortality of the Pacific white-sided dolphins, right whale dolphins, and seabirds (INPFC 1993; Sano 1998, in Japanese).

The international research activity aimed to determine whether the incidental mortality was within an allowable level and included studies of population structure, abundance, reproduction, and mortality. At the beginning of the program, the Fisheries Agency of Japan offered the opportunity for U.S. scientists to work on Japanese mother ships to examine porpoise carcasses brought by catcher boats that operated the drift-net fishery, but it did not offer its own scientists for the program, pleading unavailability. This policy was altered in 1982, when Japanese scientists were placed on research vessels (not on mother ships) to collect biological specimens and sighting data. Following a resolution by the United Nations in 1989, Japan in 1992 closed the large-scale drift-net fisheries in the high seas (see Section I). This was the end of all the research activities under international cooperation.

The identification of Dall's porpoise populations was attempted using pigmentation type, level of pollutants, breeding ground, parasite load, and DNA, as outlined in the following.

9.3.1 Pigmentation Pattern

9.3.1.1 Identification of *Truei*-Type Population

The Sea of Japan is inhabited only by *dalli*-types; the southern limit in the winter season is at around latitude 35°N–36°N (see this section). The southern limit along the Japanese coast moves northward from spring to summer (to around 40°N off Akita Prefecture in June and 43°N off Otaru in Hokkaido in July), and the species almost disappears from the Sea of Japan in midsummer. No Dall's porpoises were recorded in October during sighting surveys that covered the eastern half of the Sea of Japan (Miyashita and Kasuya 1988). Dall's porpoises that wintered in the Sea of Japan are believed to move to the Okhotsk Sea through Soya Strait (La Perouse Strait) or to the Pacific through Tsugaru Strait (see the following text in this section).

Kasuya (1978, 1982) reported on pigmentation patterns of Dall's porpoises taken by fisheries off the Sanriku coast on the Pacific coast of northern Japan and found (1) that the incidence of the typical *truei*-type, which has no mottling or an extremely low level in the white flank patch, was only about 50%; (2) that many of the *truei*-types had mottling of varying density in the white patch (most of the mottling occurred in an area that should be black on the typical *dalli*-type but white on the typical *truei*-type); (3) that some of the *dalli*-types, which are scarce off the Sanriku coast, had black spots in the white flank patch; and (4) that rare completely black individuals also exhibited individual variation, particularly in the pigmentation in the inguinal region. An albino individual was reported off the Kuril Islands (Joyce et al. 1982).

Kasuya (1982) detailed the variation of pigmentation in 537 Dall's porpoises taken in the winter fishery off Sanriku and recorded the frequency of types: (1) typical *truei*-types, 51.4%; (2) intermediate types, 44.3%; (3) typical *dalli*-types, 3.9%; and (4) black types, 0.4%. It should be noted that all the individuals of the intermediate types had black spots in the white background, and none of them had white spots in the black background. They could be classified as *truei*-type interpreted broadly. This agrees with visual impressions and allows the conclusion that *truei*-types made up 95.7% of the Dall's porpoises taken by the hand-harpoon fishery off Sanriku in winter.

Classifying the intermediate types as *truei*-types is also reasonable because such intermediate types have been reported only within the known range of the typical *truei*-types and not in waters where *dalli*-types predominantly occur. Dall's porpoises of the *truei*-type in the broad meaning are now believed to constitute one local population of the species that winters off the Pacific coast of northern Japan, and those of the *dalli*-type are thought to be temporal migrants from other populations. Thus, the proportions of the two pigmentation types vary by season and location.

Kasuya (1978) interpreted the then available limited information on the distribution of the species to suggest that *dalli*-types in the southern Okhotsk Sea and those in the offshore North Pacific are separated by the Kuril Islands and by the habitat of *truei*-types along the Pacific coast of northern Japan and the Kuril Islands. I constructed a hypothesis on the population structure of Dall's porpoises around Japan that assumed three populations: two *dalli*-type populations, one in the Okhotsk Sea and Sea of Japan and another to the east of the Kuril Islands and northern Japan, and a *truei*-type population off the east coast of northern Japan and the Kuril Islands. *Dalli*-types that winter in the Sea of Japan are now known to migrate to the southern Okhotsk Sea through two routes: (1) Soya Strait and (2) Tsugaru Strait and southern Kuril Islands. We have no information on the distribution of the species in Tatar Strait, which has a water depth of about 8 m.

Discussion among the INPFC scientists on population structure did not move toward improving the hypotheses but at least temporarily moved in quite a different direction. The scientists from the Japanese Fisheries Agency and the fishing industry accepted the presence of a *truei*-type population off Japan but were against the idea of multiple populations of *dalli*-types. They believed that there was insufficient information supporting multiple populations and proposed to deal with all *dalli*-types in the North Pacific as a single population. The hypothesis of Kasuya (1978) was a plausible argument against the Japanese hypothesis. If multiple populations are managed under an erroneous single-population hypothesis and fishery mortality is geographically biased, then there is a greater risk of depleting a particular population. However, it allows less restricted operations for the fishery industry. This type of controversy still continues in the management of other cetacean populations.

Kasuya (1978) observed Dall's porpoises from the *Ginsei-maru No.2* that operated for minke whales in the southern Okhotsk Sea and confirmed the presence of only *dalli*-types in the area, but my data were limited only to the southern part of the Okhotsk Sea within about 100 km from the coast. This deficit was later covered by T. Miyashita of the Far Seas Fisheries Research Laboratory. He surveyed the whole Okhotsk Sea area during August and September in 1989 and 1990, which agreed with a presumed mating season after parturition and was considered suitable for studying distribution to be used for the study of population structure. Results of his survey were presented to the Scientific Committee of the International Whaling Commission (IWC) (Miyashita and Doroshenko 1990; Miyashita and Berzin 1991), used for abundance estimation (Miyashita 1991), and cited in Yoshioka and Kasuya (1991, in Japanese). These surveys confirmed two breeding areas of *dalli*-types in the northern and southern Okhotsk Sea, separated by a breeding area of *truei*-types in the central Okhotsk Sea. This result leads to a hypothesis of three Dall's porpoise populations in summer in the Okhotsk Sea (Figures 9.1 and 9.2).

The range of the *truei*-type population, which is one of the three Dall's porpoise populations summering in the Okhotsk Sea, extends from the coasts of central and northern Sakhalin Island (latitudes 50°N–56°N) toward the southeast and reaches to the northern and central part of the Kuril Islands (latitudes 47°N–50°N). A total of 24 (89%) mother–calf pairs of *truei*-types out of 27 such pairs found in the Okhotsk Sea were in this central Okhotsk Sea region, and the remaining 3 pairs were further north of the range or in the area occupied by *dalli*-types (Figure 9.2). This was used as evidence of a breeding ground of the *truei*-types in the area. The range of *truei*-types not accompanied by a calf was broader than the range of mother–calf pairs and extended through the Kuril Islands to waters off the coasts of eastern and southern Hokkaido (Figure 9.1). Mother–calf pairs were not found in these Pacific waters. From the similarity in segregation pattern to that between juveniles and breeding population in the Aleutian Islands area of the species (see Section 9.3.3), I believe it is likely that the *truei*-types found in summer in the Pacific waters off Hokkaido are likely to be mostly immature individuals. *Truei*-types do not normally occur in the Pacific waters off the middle and northern Kuril Islands, and the eastern limit of the regular range of *truei*-types is around 153°E (Miyazaki et al. 1984; Miysashita and Doroshenko 1990; Miyashita and Berzin 1991).

Since most of the Okhotsk Sea is covered with ice in the winter, leaving only a small southeastern area ice-free, *truei*-types that summer in the sea are believed to move to the Pacific and winter off the Pacific coast of northern Japan where the hand-harpoon fishery for the species operates. The southern limit of their winter range is off Choshi (35°42′N, 140°51′E) (Miyashita and Kasuya 1988). *Truei*-types do not occur in the Sea of Japan in any season, and the Pacific coast of Hokkaido

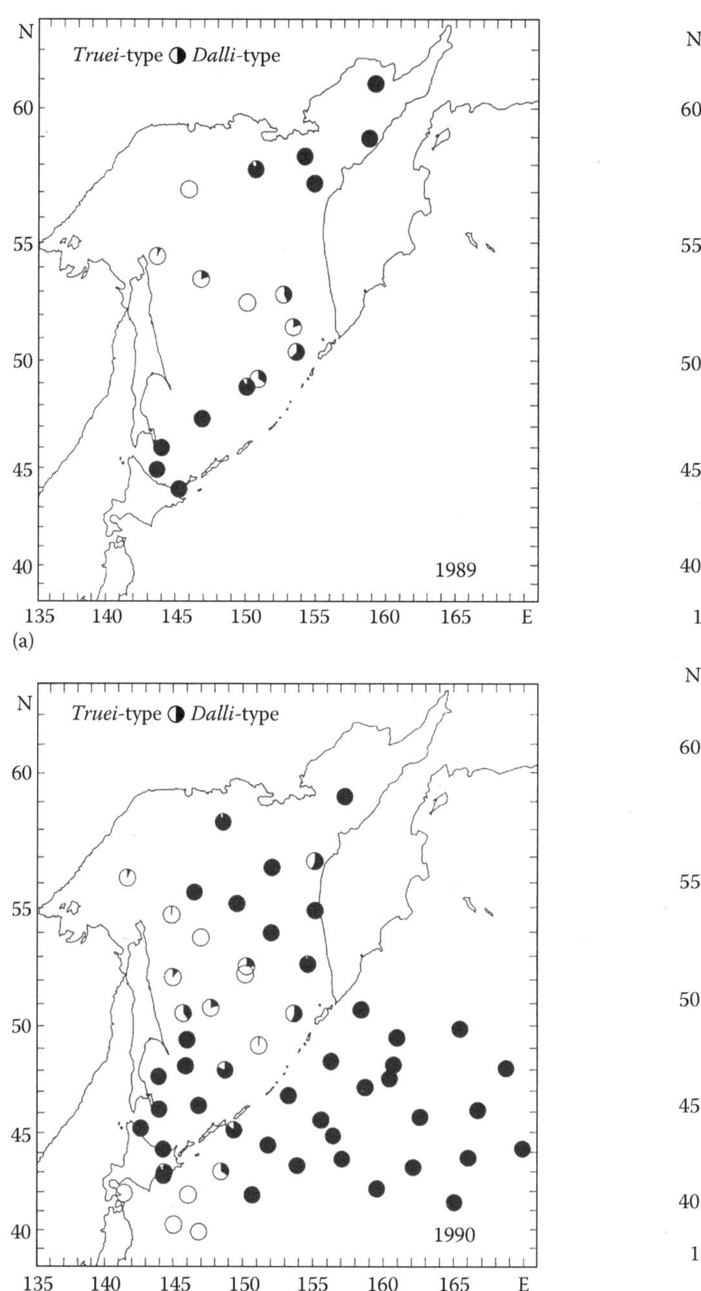

FIGURE 9.1 Proportion of *dalli*- (black) and *truei-types* (white) in daily sightings of Dall's porpoises plotted on the noon position of the survey vessel, based on the data obtained from August to September survey in 1989 (a) and 1990 (b). See Figure 9.2 for the cruise track line. (From Figure 4 of Miyashita, T., Stocks and abundance of Dall's porpoises in the Okhotsk Sea and adjacent waters, Paper IWC/SC/43/SM7 presented to the Scientific Committee Meeting of IWC (unpublished), 24pp, Available from IWC Secretariat, Cambridge, U.K. The abundance estimate is cited in IWC 1992, 1991.)

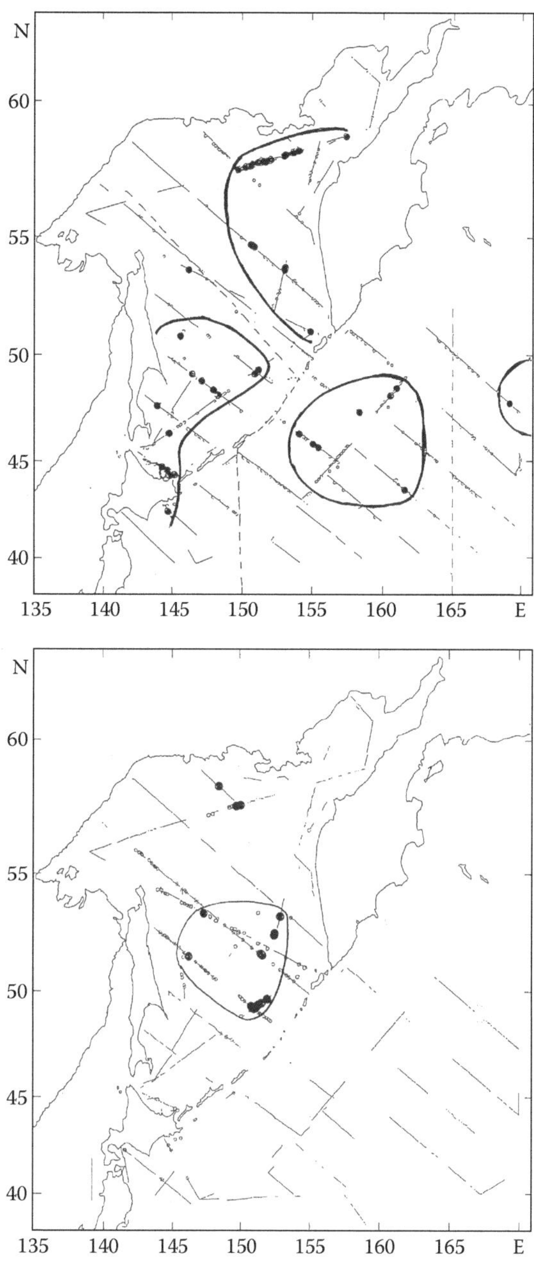

FIGURE 9.2 Cruise track line (solid line) and positions of sightings of Dall's porpoises (small circles) from August to September, 1989 and 1990. Position of cow–calf pairs is indicated by large black circles. Top panel represents *dalli*-type and bottom *truei*-type. Dotted lines in top panel indicate boundary used for abundance estimate (see Table 9.9). (From Figures 5 and 8 of Miyashita, T., Stocks and abundance of Dall's porpoises in the Okhotsk Sea and adjacent waters, Paper IWC/SC/43/SM7 presented to the Scientific Committee Meeting of IWC (unpublished), 24pp, Available from IWC Secretariat, Cambridge, U.K. The abundance estimate is cited in IWC 1992, 1991.)

is predominantly occupied by *dalli*-types in October and presumably also in winter (see Figures 9.4 and 9.5).

Some *dalli*-types occur in the central part of the Okhotsk Sea, which is used by *truei*-types as a breeding ground, but the density is lower than that of the *truei*-types as well as those of *dalli*-types in the northern and southern Okhotsk Sea, that is, a density hiatus of *dalli*-type matches with the breeding ground of *truei*-types (Miyashita 1991). This observation suggests the presence of two separate *dalli*-type populations summering in the northern and southern Okhotsk Sea. Sighting surveys by Miyashita identified 48 mother–calf pairs of *dalli*-types in 1989 and 43 in 1990 in 3 clusters in the northern Okhotsk Sea, southern Okhotsk Sea, and Pacific area south of the Kamchatka Peninsula or east of the central and northern Kuril Islands in 154°E–163°E. These *dalli*-type breeding grounds are further discussed in the following.

Miyashita (1991) observed an absence of mother–calf pairs in waters in the Kuril Islands area and stated that the breeding grounds of *dalli*-types in the Okhotsk Sea and the one in the Pacific south of the Kamchatka Peninsula are discontinuous (Figure 9.2). The two breeding grounds of *dalli*-types in the Okhotsk Sea are separated by that of *truei*-types, and they seem to represent different populations of *dalli*-types. My interpretation of the population structure of Dall's porpoises in the Okhotsk Sea and adjacent waters is the existence of (1) a *truei*-type population, which breeds in the central Okhotsk Sea and migrates to the wintering ground off the Pacific coast of northern Japan through the central and southern Kuril Islands (southern limit is in around 35°N–36°N, and presence in winter off east coast of Hokkaido north of 41°30′N is left to be confirmed); (2) a Sea of Japan *dalli*-type population, which breeds in the southern Okhotsk Sea and migrates to the wintering ground in the Sea of Japan through Soya Strait and via the Pacific coast of eastern Hokkaido and then Tsugaru Strait (presence in winter along the Pacific coast of Hokkaido is still to be determined); (3) a *dalli*-type population, which breeds in the northern Okhotsk Sea and presumably winters in the western North Pacific (summer range in the Pacific is yet to be confirmed); and (4) a *dalli*-type population that breeds in waters south of the Kamchatka Peninsula (summer range and wintering ground are yet to be confirmed). The basis of the reasoning is summarized in the following.

The distribution of *dalli*-types in the northern Okhotsk Sea is apparently continuous into the western North Pacific through the Kuril Islands. However, this cannot be interpreted to necessarily mean there is only a single population. It is more reasonable to pay attention to the presence of two isolated breeding grounds in the area and to assume that two *dalli*-type populations exist (nos. 3 and 4). The distribution of *dalli*-types in the southern Okhotsk Sea continues into the northern Sea of Japan via Soya Strait in the southwestern part of the Okhotsk Sea (Figure 9.4) and also to an area dominated by *dalli*-types off the Pacific coast of eastern Hokkaido (Figures 9.1 and 9.5) through the southern Kuril Islands. Analysis of the white flank patch of *dalli*-types by Amano and Hayano (2007) allocated some of the *dalli*-types off the Sanriku coast to the Sea of Japan population, and a pollutant study by Subramanian et al. (1986) also suggested the strong possibility that *dalli*-types off the Pacific coast of southern Hokkaido were migrants from the Sea of Japan (see Section 9.3.4).

Numerous observations on distribution and pigmentation types of Dall's porpoises in the Sea of Japan have accumulated (Miyashita and Kasuya 1988). Dall's porpoises in the sea are represented only by the *dalli*-type. As shown in Table 9.1, the sea surface temperature at the positions of sightings of these individuals was below 13°C in winter and spring, rising to below 15°C in June and 16°C in July (insufficient data for October). This seasonal change in habitat temperature is smaller than general oceanographic change in the Sea of Japan because Dall's porpoises move north following a suitable temperature environment. This feature is different from that of the finless porpoise in the Inland Sea, which remains in the almost landlocked sea enduring the stress of large seasonal temperature change between 6°C and 29°C (Section 8.4.5.2, Figure 8.6 and Table 10.1). The climax of the southbound movement of Dall's porpoises in the Sea of Japan is in March, when they reach a line that connects Hamada (35°N) in Shimane Prefecture, in Japan, and Pohang (36°N) at the southeastern corner of the Korean Peninsula. Sea surface temperature at this southern limit was 11°C in winter (Miyashita and Kasuya 1988).

Dall's porpoises in the Sea of Japan begin a northward movement in spring. This is evidenced by fishery data. There was a hand-harpoon fishery for small cetaceans off the Tajima coast (*c.* 35°40′N, 134°20′E–135°30′E) (Noguchi and Nakamura 1946, in Japanese). The catch of Dall's porpoises started in February, peaked in April, and ended in May, while

TABLE 9.1
Sea Surface Temperature (in Celsius) at the Position of Sighting of Dall's Porpoise Schools in the Eastern Sea of Japan

	March		May		June		July		October	
SST (n)	NT	D	NT	D	NT	D	NT	D	NT	D
2			1							
3										
4										
5										
6				1						
7						1				
8		2	2	15	3	6				
9		7	1	7		12				
10	3	14			1	10				
11	1	1			1	18			1	
12	3	1			1	7	1			
13	2				1	2		2		1
14					1		1	3	1	
15					1		2	3	5	
16					4		1		2	
17					2					
18					4				1	
19					6					1
20					4				3	
21									1	
22									3	
23									4	
24									1	
Total	9	25	4	23	29	56	5	8	22	2

Source: Reproduced from Table 3 of Miyashita, T. and Kasuya, T., *Sci. Rep. Whales Res. Inst.* (Tokyo), 39, 121, 1988.

Note: SST range is ≥ n and < n + 1 (n being an integer), D denotes the number of *dalli*-type schools (*truei*-type was not encountered), and NT SST at noon position is an indicator of oceanography of the surveyed waters.

the catch of Pacific white-sided dolphins started in February, had a peak in May, and continued to June. The starting date of the fishery could have been affected by weather and the presence of other fishing targets, but the shift from one small-cetacean species to another must reflect a seasonal faunal change. Thus, the last northbound group of Dall's porpoises could have left the Tajima coast in May. Sighting vessels operated by the Far Seas Fisheries Research Laboratory recorded the southern limit of the species at Yamagata (40°N) in early June and at Otaru (43°N) in July (Miyashita and Kasuya 1988).

One of the cruises to which Miyashita and Kasuya (1988) referred started its survey off Yamaguchi Prefecture in the southwestern Sea of Japan in early October, cruised northward in zigzag track lines, reached Soya Strait (45°30′N) after 23 days, and cruised the southern Okhotsk Sea toward Abashiri (44°01′N, 144°16′E). Sightings of Dall's porpoises were recorded only twice in the Sea of Japan, at *c.* 41°N off Aomori Prefecture and at *c.* 42°N off the Oshima Peninsula in southern Hokkaido. The remaining sightings occurred in the Okhotsk Sea. This observation suggests that Dall's porpoises are almost absent during summer and early autumn in the Sea of Japan. The sea surface temperatures at the two sighting positions were 19°C–20°C and 13°C–14°C, respectively, and to the west of the surveyed area off Hokkaido and Aomori there was an area of cold water below 14°C. Therefore, these observations alone cannot exclude the possibility of distribution of Dall's porpoises in the western part of the northern Sea of Japan in this season. However, another sighting cruise that covered the northern half of the Sea of Japan from the Japanese coast to the EEZ of the USSR from May to June 1969 recorded only one school of Dall's porpoises, in the central part of the sea (*c.* 39°N, 135°E) (Kasuya and Kureha 1971). This also supports the conclusion that most Dall's porpoises that winter in the Sea of Japan leave the sea in the summer.

TABLE 9.2
Sea Surface Temperature (in Celsius) at the Position of Sighting of Dall's Porpoise Schools in the Western North Pacific

	May			July			August			September			October		
SST (n)	NT	D	T	NT	D	T	NT	D	T	NT	D	T	NT	D	T
3	1	2	3												
4	1		1												
5	2		10												
6	2	1	8												
7	5	1	12												
8	2	1	9	1											
9	1		7		1										
10			7	2	1									3	
11	1		2			10								2	
12			5			2		2		1	1		3		1
13			5					1			9		4	11	1
14	1					7	1	3		1	2	4	8	13	1
15	1		1	2		1		5		1	1	1	7	2	2
16	1					1		2		2	3	2	7		
17						1		1		2	2	1	10	2	3
18	3		1			1		1		7	3	1	10	2	1
19				3	1	1		1		7		1	4		
20	1			3		1	1	2		11	1				
21						1	5	1		6					
22				3			5		1	6			1		
23				3		1	3	1	1	1					
24				4	1	4	3			3					
25				3			3			4					
26				9			2						1		
27				3			6			2					
Total	22	5	71	36	4	31	29	20	2	54	22	10	55	35	9

Source: Reproduced from Table 4 of Miyashita, T. and Kasuya, T., *Sci. Rep. Whales Res. Inst.* (Tokyo), 39, 121, 1988.

Note: SST range is ≥ n and < n + 1 (n being an integer), D denotes the number of *dalli*-type schools and T *truei*-type schools, and NT SST at noon position is an indicator of the oceanography of the surveyed waters.

Miyashita and Kasuya (1988) analyzed the distribution of Dall's porpoises off the Pacific coast of Japan. Sea surface temperatures there fluctuate seasonally between 3°C and 28°C, but most sightings of Dall's porpoises have occurred in waters below 15°C (Table 9.2). In summer, when surface temperatures are 12°C–28°C, sightings of Dall's porpoises were limited to temperatures below 24°C and mostly below 18°C. There was no difference in temperature preference between *dalli*- and *truei*-types in these waters of the Pacific. Pacific Dall's porpoises also tended to avoid high sea surface temperatures as in the Sea of Japan, but the upper bound tended to be higher in the Pacific than in the Sea of Japan. In summer, the warm Kuroshio Current expands northward over the cold Oyashio Current, raising the surface water temperature, but the feeding activity of Dall's porpoises is probably affected also by subsurface temperatures under the influence of the Oyashio Current.

According to Miyashita and Kasuya (1988), out of 299 Dall's porpoise schools sighted off the Pacific coast of northern Japan during May to October, 96 (32.1%) were schools of *dalli*-types, 195 (65.2%) were schools of *truei*-types, and only 8 schools (2.8%) contained both types. Mixed schools were uncommon and limited to waters where pure schools of both types were present (Figures 9.3 through 9.5). Mixed schools of two distinct species of cetaceans have been recorded with even slightly higher frequencies. For example, Kasuya and Jones (1984) sighted 27 schools of five delphinid species during a sighting cruise in the western North Pacific from August to September 1982, and 10 of them (37%) contained more than one species. This suggests that the mixed Dall's porpoise schools were formed temporarily in waters where the two types overlap as aggregations of two or more schools of pure pigmentation types. Such schools are likely to dissolve into pure schools.

I estimated the proportion of the two pigmentation types using school composition and mean school size in Miyashita and Kasuya (1988), which covered the months from May to October in the Pacific off Sanriku and Hokkaido. By dividing the 8 mixed schools equally into those of each

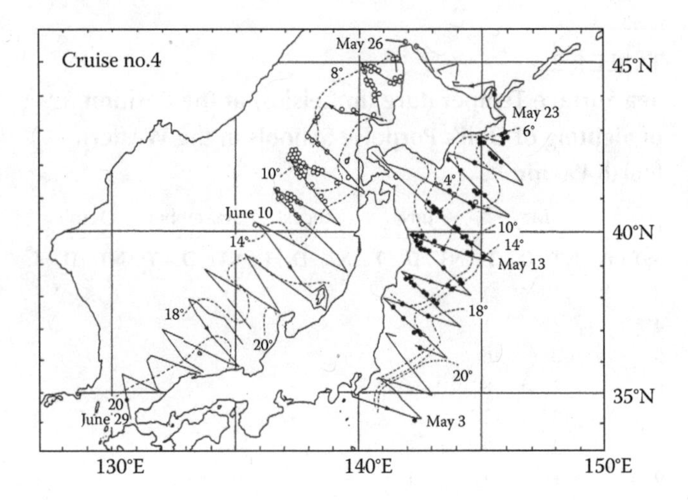

FIGURE 9.3 Positions of sightings of *dalli*-type (open circles) and *truei*-type (closed circles) Dall's porpoises in May (Pacific) and June (Sea of Japan), 1986. Open triangles represent schools of two color types. (From Miyashita, T. and Kasuya, T., *Sci. Rep. Whales Res. Inst.* (Tokyo), 39, 121, 1988.)

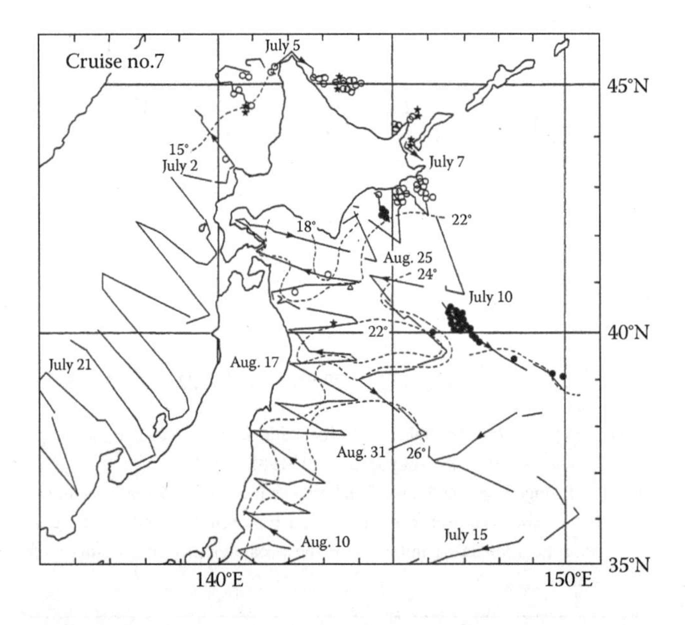

FIGURE 9.4 Positions of sightings of *dalli*-type (open circles) and *truei*-type (closed circles) Dall's porpoises in June (northern Sea of Japan and Soya Strait), July (offshore Pacific), and August (coastal Pacific), 1985. Open triangles represent schools of two color types. (From Miyashita, T. and Kasuya, T., *Sci. Rep. Whales Res. Inst.* (Tokyo), 39, 121, 1988.)

type (4 to each type), I got 100 *dalli*-type schools and 199 *truei*-type schools. I multiplied them by the respective mean school sizes (9.5 individuals/*dalli*-type school and 6.4 individuals/*truei*-type school) to obtain the proportions of the two pigmentation types in the surveyed waters, which were 43% *dalli*-types and 57% *truei*-types. The proportion of *dalli*-type is greater than the corresponding value of less than 10% obtained from the catch of the hand-harpoon fishery off the Sanriku coast during December to April (Kasuya 1978; Amano et al. 1998a,b, both in Japanese). The proportions

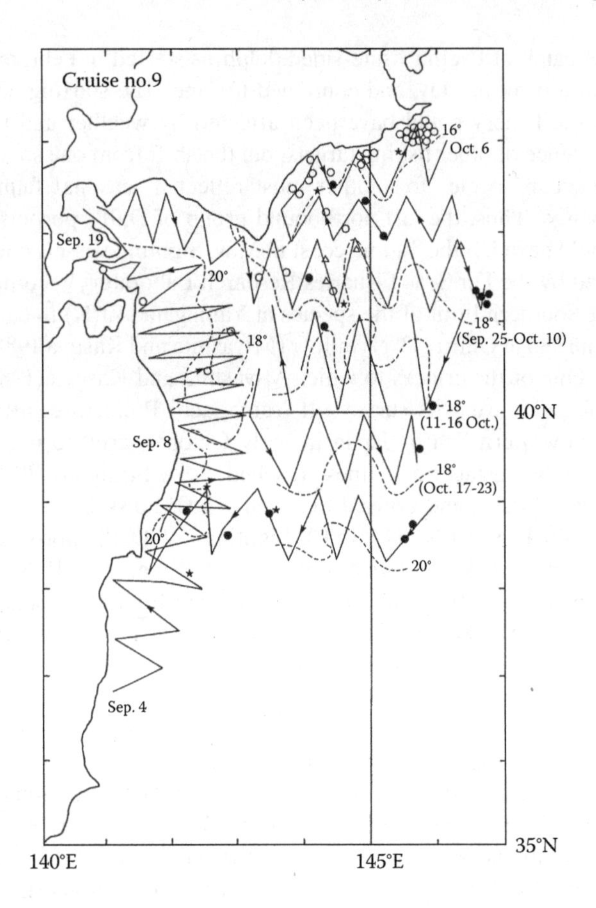

FIGURE 9.5 Positions of sightings of *dalli*-type (open circles) and *truei*-type (closed circles) Dall's porpoises off the Pacific coast of northern Japan in September (western part of the survey area) and October (eastern part of the survey area), 1986. Open triangles represent schools of two color types. (From Miyashita, T. and Kasuya, T., *Sci. Rep. Whales Res. Inst.* (Tokyo), 39, 121, 1988.)

of the two pigmentation types vary by area and season as discussed in the following.

Miyashita and Kasuya (1988) recorded Dall's porpoise schools of unknown type in March about 50 km east of Choshi Point on the Pacific coast of Japan. The sea surface temperature at this position was about 13°C, which was reasonable for the distribution of the species. A *truei*-type school was recorded in May southeast of the earlier sighting at about 34°N, 142°E. This sighting occurred in a water mass of 18°C–19°C surrounded by warmer water of 20°C–21°C (Figure 9.3). Thus, the porpoises could have been isolated from the ordinary range of the species by a seasonal change in oceanography. Other sightings of the species occurred to the north of 36°N at surface water temperatures of less than 16°C. Therefore, Miyashita and Kasuya (1988) considered the southern limit of the species off the Pacific coast of Japan to be around 35°N–36°N. In May, *truei*-types occurred almost continuously, within a range of about 300 km from the coast, from this southern limit to near the Nemuro Peninsula (43°20′N, 145°50′E) (Figure 9.3). However, this distribution pattern was apparently different from that observed in July, August, and October, when *truei*-types occurred in farther offshore waters (Figures 9.4 and 9.5). The pattern of seasonal movement seems to be different between the southbound and northbound movements.

The distribution of *dalli*-types off the Pacific coast is quite different from that of *truei*-types. We know that *dalli*-types were only several percent of Dall's porpoises in the catch of the hand-harpoon fishery off Sanriku operating in the winter. Miyashita and Kasuya (1988) noted a high density of *dalli*-types along the Pacific coast of Hokkaido from Tsugaru Strait (which connects the Sea of Japan and the Pacific) to the Nemuro Peninsula at the southern end of the Kuril Islands. *Dalli*-types were present in this area in early May and apparently increased in density toward a peak in September and October, when *dalli*-types represented almost all the Dall's porpoises within 100 km of the Pacific coast of Hokkaido (Figure 9.5). In the 3 months of May, September, and October for which we have sighting data, their distribution extended from Tsugaru Strait to the Nemuro Peninsula along the Pacific coast of Hokkaido and then to the southern Okhotsk Sea coast. This suggests that they are the *dalli*-type Dall's porpoises that winter in the Sea of Japan (Kawamura et al. 1983, in Japanese; Miyashita and Kasuya 1988), which is supported by the analysis of the white flank patch (see Section 9.3.1.2). In September and October, the density of *truei*-types off Sanriku coast is still low. They have not arrived at the Sanriku ground but appear to be dispersed in more offshore waters (Figure 9.5).

9.3.1.2 Geographical Variation of the White Flank Patch of the *Dalli*-Type

The distribution pattern suggests that *dalli*-types found along the Pacific coast of Hokkaido are of the Sea of Japan origin: this has gained some support from a pollutant study (see Section 9.3.4). The question remained of the origin of *dalli*-types hunted off the Sanriku coast; the answer was obtained from the analysis of the white flank patch. This study was started by Amano and Miyazaki (1996) who compared the white flank patch of *dalli*-types among porpoises from (1) the Sea of Japan, (3) Bering Sea, (5) offshore western North Pacific west of 165°E, and (6) offshore western North Pacific between 165°E and 180°E (numerals in parentheses correspond to those in Figure 9.6) and found that the anterior margin of the white patch was located slightly more posteriorly on the Sea of Japan individuals than on those in other locations. This meant that Sea of Japan Dall's porpoises could be differentiated from those in other regions by the size of the white patch. This finding was reinforced by Amano et al. (2000) with additional data from the eastern North Pacific (4) and by Marui et al. (1996, in Japanese) for the southern Okhotsk Sea (2).

This information was used for the identification of the origin of *dalli*-types hunted off the Sanriku coast and the results published in Marui et al. (1996, in Japanese) and Amano et al. (1998a,b, both in Japanese). These three studies were combined, with no significant changes in the analyses and conclusions, into one scientific publication by Amano and Hayano (2007). Marui et al. (1996, in Japanese) first plotted the distance from the tip of the rostrum to the anterior margin of the flank patch and confirmed that most of the samples were separated into two groups: porpoises from the Sea of Japan and

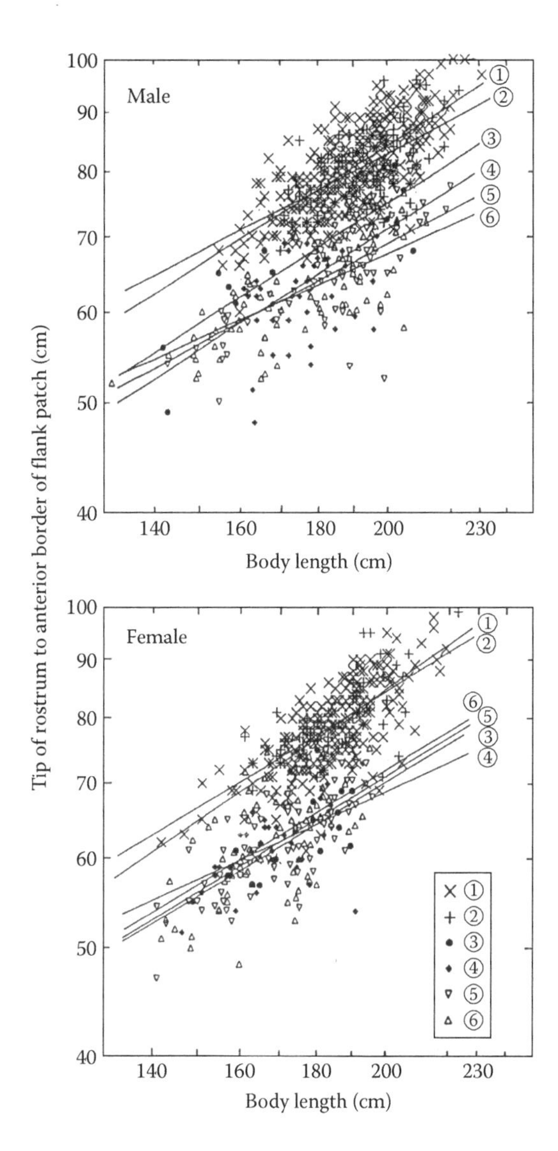

FIGURE 9.6 Relationship between body length and distance from tip of the upper jaw to anterior border of the white flank patch of *dalli*-type males (top panel) and females (bottom). (1) Sea of Japan, (2) southern Okhotsk Sea, (3) Bering Sea, (4) eastern North Pacific east of 180° longitude, (5) western North Pacific west of 165°E, and (6) western North Pacific in 165°E–180°E. The Sea of Japan individuals have the anterior border of the flank patch located more posteriorly (thus, the flank patch is smaller) than *dalli*-type Dall's porpoises of the offshore Pacific. This analysis does not include *dalli*-types known from the northern Okhotsk Sea and those hunted off the Pacific coast of northern Japan (see Figure 9.7 for the latter individuals). (From Marui, M. et al., Between stocks difference of flank patch of *dalli*-type Dall's porpoises, pp. 51–60, in: Miyazaki, N. (ed.), *Report of Contract Studies on Management of Dolphins and Porpoises*, Ocean Research Institute of University of Tokyo, Tokyo, Japan, 1996, 242pp.)

southern Okhotsk Sea and those from other areas including the Bering Sea, eastern North Pacific, offshore western North Pacific west of 165°E, and offshore western North Pacific between 165°E and 180°E. Using this method, Amano et al. (1998a,b, both in Japanese) and Amano and Hayano (2007) confirmed that about 90% of *dalli*-types were correctly classified into those in the geographical locations of the Sea of

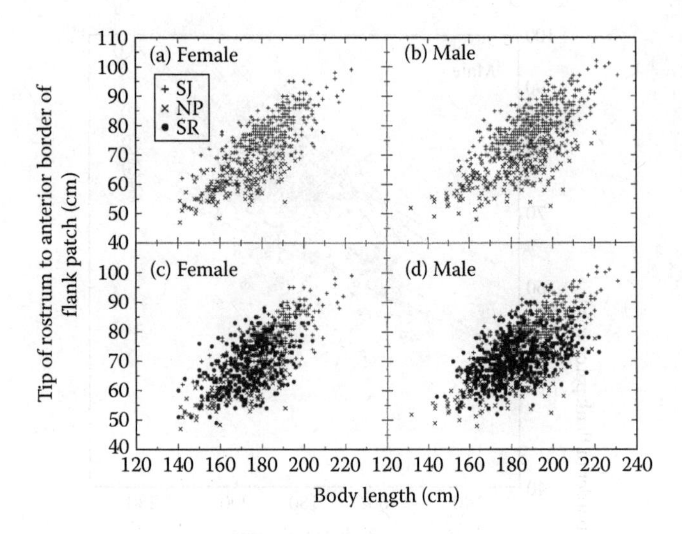

FIGURE 9.7 Attempt to identify small number of *dalli*-type Dall's porpoises found with majority of *truei*-types off the Sanriku Region (37°54′N–41°35′N), Pacific coast of northern Japan, and hunted by Japanese hand-harpoon fishery. The upper two panels (a, females; b, males) compare white flank patch between the Sea of Japan (SJ) and North Pacific (NP) outside of the Sanriku ground (thus, they are almost identical with Figure 9.6). The lower two panels (c, females; d, males) are from the upper panel plus additional *dalli*-type Dall's porpoises taken by Japanese hand-harpoon fishery off the Sanriku Region (SR), to show that the *dalli*-types in the Sanriku Region contain those from the Sea of Japan and offshore Pacific. This analysis like that of Figure 9.6 does not include *dalli*-types identifiable as those from the northern Okhotsk Sea population. (From Figure 2 of Amano, M. and Hayano, A., *Marine Mammal Sci*., 23(1), 1, 2007.)

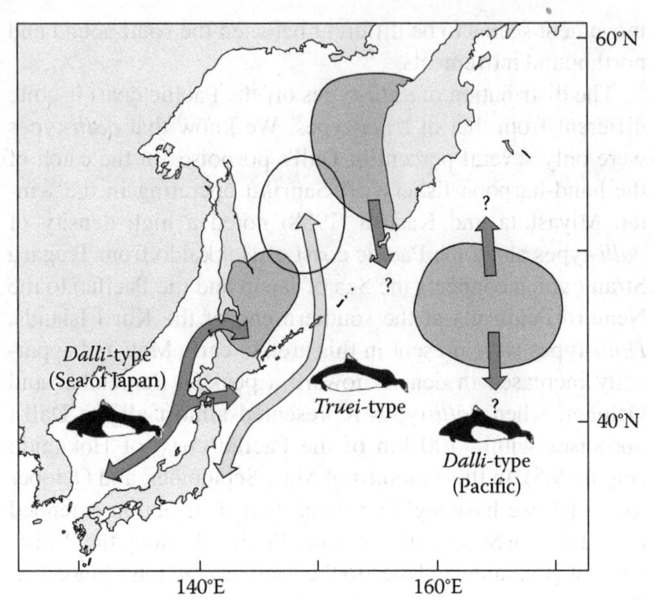

FIGURE 9.8 Main migration routes of Dall's porpoises around Japan deduced from sighting surveys and morphology of flank patch. This diagram ignores a small number of *dalli*-types that winter off the Sanriku coast mixed with a majority of *truei*-types. Summer breeding grounds for the three populations are in the circled areas. (From Figure 1 of Amano, M. and Hayano, A., *Marine Mammal Sci*., 23(1), 1, 2007.)

Japan–southern Okhotsk Sea and offshore North Pacific. They then tested *dalli*-types taken by the hand-harpoon fishery off Sanriku and found that about half of them were classified into the Sea of Japan–southern Okhotsk Sea group and the remaining into the offshore western North Pacific group (Figure 9.7). Their analysis did not cover *dalli*-types from the northern Okhotsk Sea.

The analyses of the flank patch of *dalli*-types show that the current Japanese hand-harpoon fishery in the Sea of Japan and southern Okhotsk Sea targets *dalli*-types of the Sea of Japan–southern Okhotsk Sea population during spring to autumn. The catch of the similar fishery off Sanriku, which operates during the winter, includes about 95% *truei*-types and about 5% *dalli*-types (see Section 9.3.1.1), and the studies by Amano and Hayano (2007) and others have shown that the *dalli*-types (about 5% of the total catch of Dall's porpoise off Sanriku) are divided equally between porpoises from the Sea of Japan–southern Okhotsk Sea population and some other population(s), that is, the northern Okhotsk Sea, North Pacific, and Bering Sea populations. Our current understanding of the distribution pattern of Dall's porpoise populations around Japan is summarized in simplified form in Figure 9.8.

Amano et al. (1998b, in Japanese) made comparisons of sex ratio and maturity between populations of *dalli*-type Dall's porpoises taken in the hand-harpoon fishery off Sanriku. The male proportion was 56% among *dalli*-types from the Sea of Japan–southern Okhotsk Sea population, which was similar to 66% observed for *dalli*-types from other population(s) on the same ground. Male maturity identified from testicular weight varied annually between 5% and 27% for *dalli*-type males from the Sea of Japan–southern Okhotsk Sea population but from 32% to 47% for other *dalli*-types. Averages for the 3 years' data were 14.9% and 40.7%, respectively. Female maturity had to be estimated from body length because the internal organs were often discarded by the fishermen at sea. Female *dalli*-types from the Sea of Japan–southern Okhotsk Sea population showed annual maturity of 0%–10% (3.3% average), and those from other *dalli*-type populations had 29%–67% mature (42.9% average). Thus on the Sanriku ground, *dalli*-type porpoises of both sexes migrating from the Sea of Japan–southern Okhotsk Sea population had apparently lower sexual maturity rates than those from other population(s). However, there remained some uncertainty in the absolute figures, particularly for females.

Before accepting these results, the maturity criterion for females needs to be reexamined. In the analysis, female porpoises of body length of 187 cm or more were judged sexually mature in the Sea of Japan–southern Okhotsk Sea population (Amano and Kuramochi 1992), but 170 cm was used for other population(s). The former figure was based on observations on hand-harpoon vessels, and the latter on incidental mortality in the salmon drift-net fishery in the Bering Sea (Newby 1982) (it was close to the figure of 172.0 cm later published by Ferrero and Walker (1999)). These figures represent body lengths where the probabilities of being immature and mature are expected to be equal, but they are not necessarily the same for samples obtained by different hunting methods or from different habitats. Dall's porpoises respond differentially to hand-harpoon vessels by growth stage and also segregate geographically by growth stage. A sample obtained from the hand-harpoon fishery, which underrepresents sexually mature individuals due to differential response to the vessels, is likely to yield a greater average body length at the attainment of sexual maturity compared with the corresponding estimate for samples from less selective fishing methods such as drift-net fisheries. Segregation by growth stage also has a similar effect. Ferrero and Walker (1999) reported the body-length range of over 5000 individuals as 84–222 cm for males and 86–211 cm for females. Amano and Miyazaki (1992) reported the body-length range of *dalli*-types captured in the hand-harpoon fishery from the Sea of Japan–southern Okhotsk Sea population as 215–220 cm for males and 205–210 cm for females (greater body size to 230 cm was reported by Walker (2001) and Yoshioka et al. (1990) for the southern Okhotsk Sea Dall's porpoises; see Sections 9.3.2 and 9.3.5). Even if the maximum body sizes are similar between the two sample sets, there is a difference of 17 cm between the average body lengths of females at attainment of sexual maturity (170 cm for the drift-net sample vs. 187 cm for the hand-harpoon sample). This can be attributed to gear selection or geographical differences in growth. The mean body length at sexual maturity calculated for samples from the drift-net fishery in the Bering Sea is likely to overestimate to some unknown degree the proportion of sexually mature females in the catch of the hand-harpoon fishery off Japan. See Section 9.3.2 for a possible bias in a similar direction due to a longitudinal cline in body size.

9.3.2 Osteology and Body Length

9.3.2.1 Skull Morphology

Amano and Miyazaki (1992) carried out multivariate analysis of 27 measurements of 289 Dall's porpoise skulls. At least 25 of the specimens were *truei*-types, which were caught off the Pacific coast of Japan, and most of the remaining were *dalli*-types covering a broad area including the Sea of Japan, southern Okhotsk Sea, Bering Sea, and almost the entire longitudinal range in the North Pacific. Specimens from around Japan were obtained from the catch of the hand-harpoon fishery, which included *truei*-types and a small number of *dalli*-types taken off Sanriku in winter and *dalli*-types taken in other areas during spring through autumn. Samples from the Bering Sea and offshore North Pacific were obtained during cruises for scientific purposes using hand harpoons. The sample also included some stranded individuals.

Amano and Miyazaki (1992) found sexual dimorphism in some measurements. They did not detect sexual dimorphism in skull length, which was probably due to limited sample size combined with large individual variation. Their analysis rejected a hypothesis of "no geographical differentiation," which means that the skull morphology exhibited some sort of geographical variation. However, they failed to diagnose the geographical origin of specimens using particular measurements or combinations of them. Such a result can be obtained when individual variation is great compared with difference between populations or when a single sample represents multiple populations, both of which are very likely in this case.

9.3.2.2 Body Length

Kasuya (1978) compared body-length compositions of Dall's porpoises taken in winter by hand-harpoon fisheries and suggested that *dalli*-types in the Sea of Japan were likely larger than *truei*-types off the Pacific coast of northern Japan. A finding of Amano and Miyazaki (1992) was geographical variation in body size of the species in a broader Pacific area. Dall's porpoises of both types taken in Japanese coastal waters west of 155°E had an average skull length of about 33–34 cm (range approximately 31–36 cm), which was similar to the corresponding figure from the California coast. However, those in the central North Pacific between 165°E and 145°W had an average skull length of about 32–33 cm (range approximately 30–35 cm); they were about 1 cm smaller. As the authors did not mention possible body-length difference expected from the different skull size, I attempted to use the equations of Amano and Miyazaki (1993) for this, that is, $Y = 10^{0.056}(X^{0.584})$ for males and $Y = 10^{0.004}(X^{0.616})$ for females, where X is the body length in cm and Y is the length from tip of the rostrum to the ear in cm. These equations suggest that a body-length difference of 15 cm between 205 and 225 cm is equivalent to about 1 cm difference in length from tip of the rostrum to the ear.

Amano and Miyazaki (1992) interpreted their findings on geographical variation in skull size to suggest that Dall's porpoises in coastal waters were larger in body size and that this reflected high productivity of the habitats. The results can also be examined assuming a linear east/west cline. The Bering Sea sample, 8 males and 18 females, offers key information for evaluating the two hypotheses. The sample can be classified as a coastal sample, but it belongs to the central part of the longitudinal range in the North Pacific. Skull length ranged from 30.5 to 34.5 cm, which was similar to the range obtained for the central North Pacific. Thus in my view, the finding of Amano and Miyazaki (1992) could be better interpreted by saying that Dall's porpoise populations on the eastern and western sides of the North Pacific are likely to have larger body size than those in the central portion of the longitudinal range.

The earlier calculation suggests that Dall's porpoises along the Japanese coasts are about 15 cm larger than those in the central Pacific. However, a greater geographical difference in body length was suggested by Yoshioka et al. (1990), who analyzed *dalli*-types (one black type included) taken by the single method of hand harpoon, in the commercial hand-harpoon fishery in the southern Okhotsk Sea and in the harpoon operation for scientific purposes in the offshore North Pacific and Bering Sea outside of the U.S. EEZ during 7 summers from 1982 to 1988. They grouped the sample into four longitudinal strata: (1) the southern Okhotsk Sea, (2) western North Pacific west of 160°E, (3) western–central North Pacific between

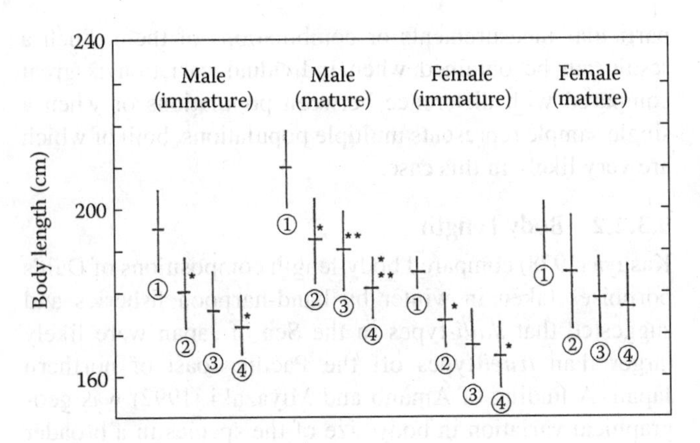

FIGURE 9.9 Geographical variation of body length of *dalli*-type Dall's porpoises that appeared in hand-harpoon samples. The range of one standard deviation is shown on each side of the sample mean for sex and maturity groups. The difference of the mean is significant between pairs of the same dotted symbols; the southern Okhotsk Sea sample was significantly different from every other sample. (1) The southern Okhotsk Sea, (2) western North Pacific west of 160°E, (3) middle-western North Pacific and Bering Sea in 160°E–180°E, (4) and eastern North Pacific in 180°W–137°W. (From Yoshioka, M. et al., Identity of Dalli-type Dall's porpoise stocks in the northern North pacific and adjacent seas, Paper IWC/SC/SM31 presented to the Scientific Committee of IWC (unpublished), 20pp, (Available from IWC Secretariat.), 1990.)

160°E and 180°E and the Bering Sea, and (4) eastern North Pacific between 137°W and 180°W. They compared the mean body lengths between these strata. They found that body length was greater in the western samples, at least within the longitudinal range covered by the strata. Mean body lengths of sexually mature individuals in the westernmost stratum were greater than in the easternmost stratum by 30 cm in males and 25 cm in females, and the values for the 2 strata between them were intermediate (Figure 9.9). Comparison of immature individuals suggested a similar trend. This study did not have materials from the Californian coast, which were included by Amano and Miyazaki (1992), but showed a similar geographical cline as that detected by Amano and Miyazaki (1992) within the latitudinal range analyzed.

9.3.3 Breeding Areas of *Dalli*-Type Dall's Porpoises

Humpback whales provide a typical example of a migration pattern between breeding and feeding grounds (Clapham 2002), where most individuals regularly travel between a particular summer feeding ground and a winter breeding ground (also called wintering ground) where they almost do not feed. Individuals that summer in a feeding ground, for example, Southeast Alaska, depart in autumn for their breeding grounds, which differ among individuals, for example, Hawaii or the Bonin Islands, and those that winter in a breeding ground return to their own feeding grounds. A suckling calf accompanies its mother to her feeding ground in the first summer and is usually weaned there. Calves thus learn their feeding ground from their mother and maintain the round trip for almost their whole life, but they are believed to have more flexibility in breeding ground than feeding ground. Thus, a breeding population contains individuals from several feeding grounds, and a feeding population, whales from several breeding populations. Currently, we do not have such an example in the toothed cetaceans, but Dall's porpoises are known to have some seasonal shift of habitats (see above) and to form breeding aggregation in the summer as described in the following.

We know that the distribution pattern of pigmentation types has revealed the presence of three Dall's porpoise populations around Japan, each of which has its own breeding ground (Figure 9.8) and slightly different breeding season (Section 9.4.5). Several additional breeding grounds of the species are known to the east of the three populations where only *dalli*-types occur and are interpreted as representing local populations.

International cooperation started in the 1970s for studies on the management of Dall's porpoises under the framework of the INPFC (Sections 6.2 and 9.3.3). One of the research activities included collecting biological data from porpoises that were killed in the Japanese salmon drift-net fishery and brought to the mother ships in the summer, but scientists suspected that age structure and reproductive status obtained from such material could be biased. To address this question, scientists conducted several annual cruises beginning in the summer of 1982 for sighting and hand-harpoon sampling in waters outside the drift-net fishery. Scientists on the first cruise were myself, a student Y. Fujise from Japan, a Canadian student and two technical assistants, a hand-harpoon fisherman hired from the Sanriku coast, and a whale spotter hired from a whaling company. Participation of a U.S. scientist was canceled before the cruise due apparently to domestic criticism of killing animals for scientific purposes. The first cruise surveyed waters between the western Aleutian Islands (west of 174°E) and the southern limit of the Dall's porpoise range excluding the EEZs of the United States and USSR. The cruise was on the *Hoyo-maru No. 12* of the Hoyo Suisan Co. Ltd. chartered by the Fisheries Agency of Japan. It was an extremely enjoyable and memorable cruise for me, a university scientist without any previous knowledge on the customs or system of such chartered cruises, but later I realized that it was an extremely low-cost cruise at least partly supported by the crew. For example, the bathtub was always full of warm seawater from the engine, which was nice, but the supply of freshwater for rinsing the body was ice cold, which helped save freshwater. We Japanese scientists were happy if we were able to catch one Dall's porpoise before breakfast, because it offered the only fresh dish available on the vessel, of *sashimi* (uncooked sliced pieces) of skeletal muscle and blubber. The staple food items on the vessel were cooked rice, salted vegetables, salted uncooked saury or chunks of tuna, dry snacks, and *miso* soup with seaweed that never left the cooking oven during the cruise. We scientists thought that the supply of stored food was short, but on returning to Kesen-numa (38°55′N, 141°35′E) after a month, we witnessed unloading of large amounts of frozen food including pork and chickens. After returning to my office, I received a bill for meals on the vessel, which was so big that I suspected it covered the salary of the cook, but the bill was soon canceled by telephone. However, I heard that some people paid the bill.

In this environment our research team worked hard, along predetermined track lines. We approached every cetacean sighting to confirm species and school composition. We sighted 710 Dall's porpoises and took 80 of them by hand harpoon equipped with an electric shocker (50 V) (Figure 2.1). Our strategy in the operation was to chase the porpoises windward while harpooning them one by one and then retrieve the carcasses while running leeward. The electric shocker was sometimes ineffective; the porpoises resumed swimming with the harpoon line and buoy. The electric shocker often caused a fracture of the vertebrae through muscle convulsions, which also destroyed the muscle tissue and rendered it unsuitable for *sashimi*. The porpoises could have drowned rather than been killed directly by electric shock.

This cruise was in a season of peak sea surface temperatures in August and September and was able to confirm the southern limit of Dall's porpoises in summer at around 42°N in the offshore waters in longitudes 155°E–175°E. Dall's porpoises occurred in water temperatures below 20°C, while warmwater species such as the short-beaked common dolphin, striped dolphin, common bottlenose dolphin, and short-finned pilot whale occurred in water temperatures above 17°C. The Pacific white-sided and right whale dolphins occurred in the intermediate temperatures of 12°C–19°C. Combining these data with those obtained from subsequent cruises for the similar purpose, Miyashita (1993) determined that the last two species had the same temperature preference at 11°C–25°C and are distributed in latitudes of 40°N–50°N of the offshore North Pacific and estimated their abundance.

During the cruise, we noticed a distinct geographical difference in the behavior of *dalli*-type Dall's porpoises as detailed in Section 9.6. The difference was felt to be apparently related with surface water temperature. Most individuals in waters above 11°C approached the research vessel and rode the bow wave, but most below that temperature were not attracted to the vessel but avoided it. The area where we encountered the latter type of behavior agreed with the area in which we identified mother–calf pairs, which also avoided the vessel. This means that Dall's porpoises segregate in summer by reproductive or growth stages (Kasuya and Jones 1984). As their mating season comes soon after the parturition season (Sections 9.4.3 and 9.4.4), the area of low water temperature inhabited by mother–calf pairs as well as other presumably adult individuals could also be an area of mating and can be called the breeding ground. The breeding ground identified during this cruise ranged from 45°N to 50°N and from 167°E to 175°N, but its northern and eastern limits were not confirmed.

Kasuya and Ogi (1987), combining data from cruises of similar purpose in 1982, 1983, and 1985, identified a total of three areas inhabited by *dalli*-type mother–calf pairs: the central Bering Sea, south of the Kamchatka Peninsula, and south of the western Aleutian Islands. The cruise of 1984 operated in May and June, or before the parturition season, and did not collect information on the distribution of mother–calf pairs. Sea surface temperature profiles of these three breeding

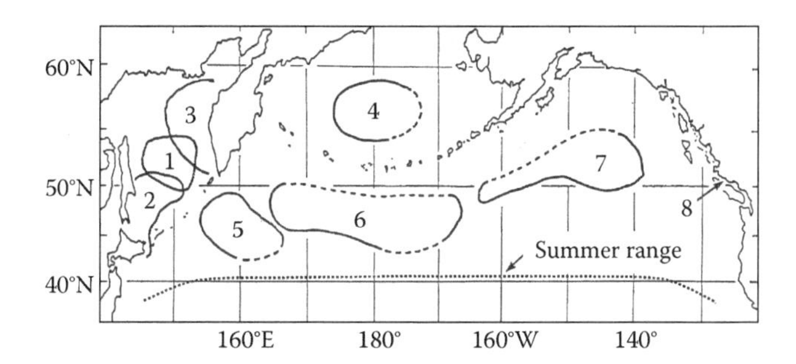

FIGURE 9.10 Breeding grounds of the Dall's porpoise deduced from the location of cow–calf pairs, each of which is thought to represent a population. Breeding ground 1 represents *truei*-types and those numbered 2–7 *dalli*-types. Dotted line represents approximate boundary of summer range of the species. (From Figure Fig. II-4.2 of Yoshioka, M. and Kasuya, T., Population structure of whales viewed from distribution and ecology, pp. 53–63, in Sakuramoto, K., Tanaka, S., and Kato, H. (eds.), *Research and Management of Whale Stocks*, Koseisha Koseikaku, Tokyo, Japan, 1991, 273pp.)

grounds were different from each other, which made it clear that it is incorrect to conclude that a particular surface water temperature determines the location of breeding grounds of the Dall's porpoise.

Yoshioka and Kasuya (1991, in Japanese) made a further effort to collect locations of mother–calf pairs and identified eight breeding grounds in the North Pacific and adjacent seas: (1) the central Okhotsk Sea, (2) the southern Okhotsk Sea, (3) the northern Okhotsk Sea, (4) the central Bering Sea, (5) south of the Kamchatka Peninsula, (6) south of the Aleutian Islands, (7) the central Gulf of Alaska, and (8) coastal waters of North America near Vancouver Island (Figure 9.10). These areas of mother–calf pairs may shift annually, but they are apparently isolated from each other. Breeding ground (1) is used by *truei*-types, (2) by the Sea of Japan–southern Okhotsk Sea population of *dalli*-types, and (3) by the northern Okhotsk Sea population of *dalli*-types. The remaining five breeding grounds (nos. 4–8) are used by *dalli*-types and are believed to represent local populations. We still do not know where the wintering grounds are for the six populations that breed in grounds (3) to (8).

Now, we consider further details on why an area inhabited by mother–calf pairs can be considered a place of mating as well as parturition. Ambiguity remains about the breeding cycle of Dall's porpoises because of the difficulty of obtaining year-round information. Ferrero and Walker (1999) reported that for Dall's porpoises around the Aleutian Islands females mate for conception soon after parturition that occurs in early June to early August, which means that most females conceive annually. This leads to an assumption of 10–11 months gestation. Nursing lasts over 2 months and less than 5 months on average (see Section 9.4.6). It is difficult to assume that parturition and mating occur in different locations; rather it would be reasonable to assume that pregnant females and adult males of a population gather in a particular area for parturition and subsequent mating. This is the reason why the area of mother–calf pairs is considered a breeding ground.

Before confirming the hypothesis that each breeding ground represents a particular population, it is necessary to at least show that each *dalli*-type Dall's porpoise breeds in a particular breeding ground. There is no direct evidence for this, but I feel it likely because there are geographical differences in body size (see Section 9.3.2), parasite load (Section 9.3.5), pollutant level (Section 9.3.4), and genetics (Section 9.3.6). Additional support is available from *truei*-type Dall's porpoises, which adhere to their own breeding ground in the central Okhotsk Sea located between two breeding grounds of *dalli*-types; mother–calf pairs of the two *dalli*-type groups tend to avoid the *truei*-type area (Figures 9.2 and 9.3). Similar behavior can be assumed for *dalli*-types in the Pacific and in the Bering Sea.

If a low level of mixing does occur between the breeding grounds, it will be difficult to detect genetic differences between two neighboring breeding groups even if most of the members adhere to their original breeding ground. However, they should be managed as independent populations, as long as the process of population dynamics remains almost independent and the effect of immigration is smaller than anthropogenic effect on the population.

9.3.4 Accumulation of Pollutants

Subramanian et al. (1986) analyzed geographical variation in PCB and DDE in Dall's porpoises. DDE is DDT and its derivatives. Evaluation of their concentration levels is complicated, because it is a function of total intake, dilution through growth, decomposition in the body, and excretion. The total intake depends on pollutant level in the food and amount of food consumption, dilution occurs in growing young animals, and decomposition and excretion depend on concentration level in the body. In addition, adult females can off-load these pollutants to their offspring through pregnancy and subsequent lactation. In order to avoid the complications of such effects, Subramanian et al. (1986) selected for their analysis only adult males taken during 1979–1985. They obtained the following results (concentration is expressed as ppm/wet weight of blubber).

They showed that the Bering Sea samples (n = 5) had PCB levels of 2.91–6.00 and DDE levels of 4.13–9.97, which were lower than for other samples. Offshore specimens (n = 15) from south of the western Aleutian Islands and east of the Kamchatka Peninsula had similar pollutant levels (PCB 7.05–16.0, DDE 6.64–15.2), but these values were slightly higher than those for the Bering Sea samples. The most interesting was the pollutant load of 8 *truei*-types from off the Sanriku coast and 3 *dalli*-types taken off the Pacific coast of southern Hokkaido between Tsugaru Strait (41°30′N, 141°E) and the Erimo Peninsula (41°50′N, 143°E). Their pollutant levels, PCB 11.2–22.6 and DDE 12.4–36.8, were close to figures for one sample from the Sea of Japan (PCB12.6, DDE 32.4).

Subramanian et al. (1986) also looked at the ratio of PCB/DDE, which was calculated for each individual and then averaged for geographical areas, and obtained the following figures: Bering Sea, 0.60; offshore western North Pacific, 0.96; *truei*-types off Sanriku, 0.91 (individual variation 0.71–1.15); *dalli*-types in the Pacific off Hokkaido, 0.39 (individual variation 0.37–0.41); and *dalli*-types in the Sea of Japan, 0.39. They found that the PCB/DDE ratio was low for the latter two groups, southern Hokkaido and the Sea of Japan. The low figure for the Sea of Japan specimen could be a reflection of high DDE contamination of the sea due to continuing discharge of DDT in China (Tatsukawa et al. 1979).

Subramanian et al. (1986) proposed two possible explanations for the low PCB/DDE ratios of three specimens from the Pacific coast of Hokkaido. The Tsugaru Current runs through Tsugaru Strait to the Pacific, after being diverted from the Tsushima Current in the eastern Sea of Japan. They thought that this current might reach the southern Hokkaido coast and could have polluted Dall's porpoises living there or that *dalli*-types in the Sea of Japan might have migrated to the southern coast of Hokkaido. We do not know what the pollutant levels are of seawater or of prey species consumed by the porpoises in this area. The Tsugaru Current turns southward soon after passing Tsugaru Strait and runs along the Sanriku coast. Therefore, if local seawater pollution is affecting the PCB/DDE level of seasonally migrating Dall's porpoises, the effect could also appear for the *truei*-type individuals that winter off the Sanriku coast, but this was not observed in the data (Subramanian et al. 1986), so I am of the opinion that the latter interpretation is correct.

O'shea et al. (1980) reported extremely high pollutant levels in a *dalli*-type Dall's porpoise off California: PCB 94 ppm and DDE 256 ppm. These figures represent an extreme case of animals in locally polluted waters.

9.3.5 Parasite Load

Species of the genera *Phyllobothrium* and *Monorygma* are tapeworms that use cetaceans as an intermediary host. Their larvae are found in the blubber as a cyst of about soybean size. They are particularly numerous in the blubber in the inguinal region (Dailey and Brownell 1972). Their life history is incompletely known, but sharks are presumably one of their final hosts, receiving larvae through consuming carcasses of dolphins and porpoises (Walker 2001). Nothing is known about their route from the final host to cetaceans. Experiments at sea have revealed that sharks prefer to eat the inguinal region of a dead porpoise. Thus, concentration of the larvae in the region is considered as an adaptation by the parasite. The larvae can survive in the blubber for at least 13 years waiting for a chance to be eaten by sharks. Therefore, if other conditions are the same, the density of the parasite larvae in a porpoise correlates to some degree with the age of the intermediary host.

Walker (2001) investigated the geographical difference in loads of *Phyllobothrium* cysts in *dalli*-type Dall's porpoises, most of which were taken by the Japanese salmon drift-net fishery from June to July in 1983–1986. The sample size was 432 porpoises from the southwestern Bering Sea and 1957 from the western North Pacific adjacent to the Aleutian Islands in 46°N–59°N and 168°E–180°E (Figure 9.11). Additional small series (56 individuals) were obtained from the catch of hand-harpoon fishing by a small-type whaler

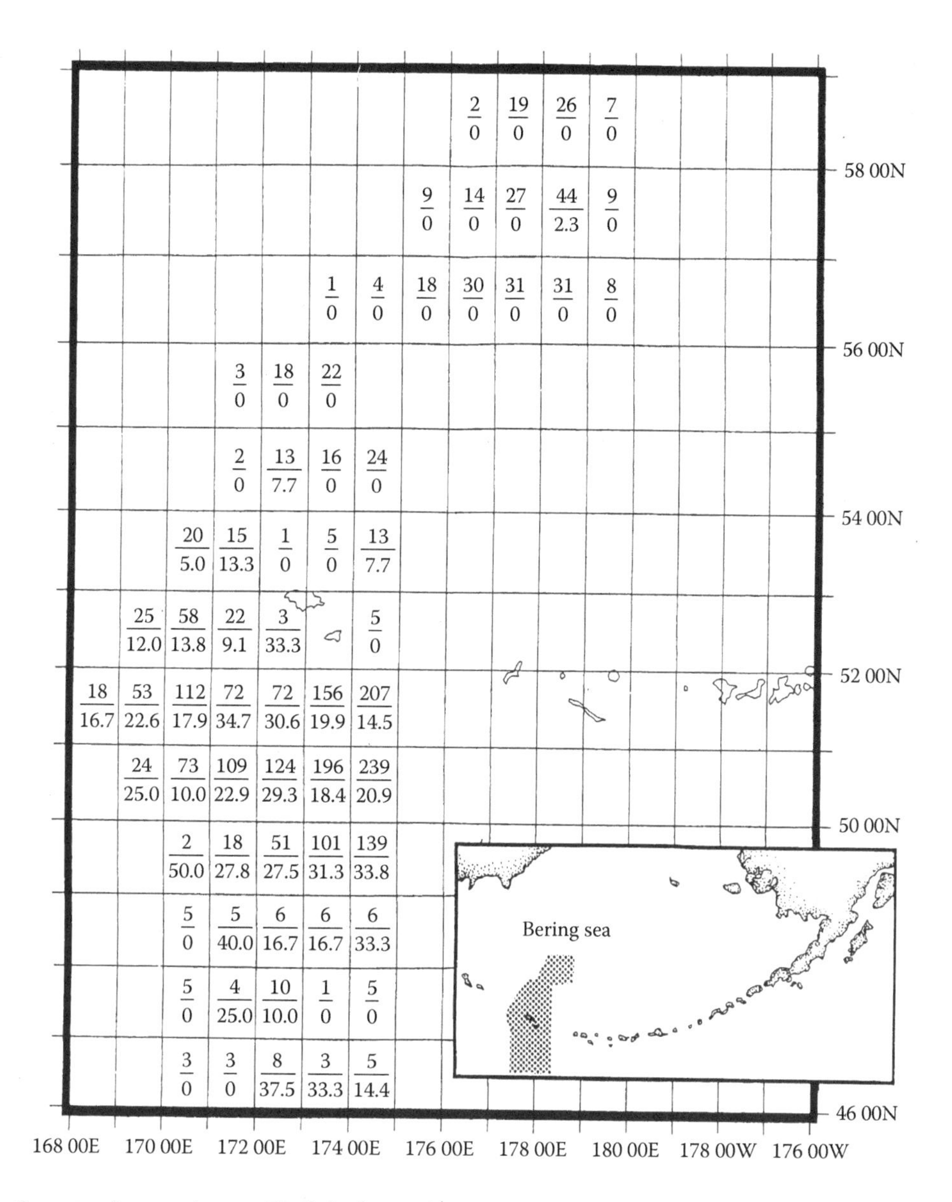

FIGURE 9.11 Infection rate of a cestode cyst, *Phyllobothrium delphini*, in the blubber of the Dall's porpoise around the eastern Aleutian Islands. Upper figure in each square of latitude and longitude represents sample size and lower figure the infection rate (%). (From Figure 1 of Walker, W.A., *Marine Mammal Sci.*, 17(2), 264, 2001.)

in July and August, 1988, in the southern Okhotsk Sea. He removed a piece of blubber from the lateral side centered at the genital area, 35 cm in anterior–posterior length and 17.5 cm in midventral to lateral width, sliced it, and counted the number of cysts.

Most of the drift-net specimens of Dall's porpoise ranged from 160 to 200 cm in body length, but most of the Okhotsk Sea hand-harpoon samples were in the range of 190–230 cm with the modal length range of 180–220 cm. The drift-net sample lacked individuals of 220–229 cm, and the hand-harpoon sample from the Okhotsk Sea had no small individuals below 180 cm, which can be explained mostly by gear selection and to some degree by geographical difference in growth (see Sections 9.3.2 and 9.6.2). The body-length compositions of drift-net samples were the same between north and south of the Aleutian Islands. Approximate body lengths by age in the species are the following: 85–120 cm in the first year, 135–165 cm in the second year, and 140–175 cm in the third year. Dall's porpoises over 160 cm are mostly 2 years old or older, and those below 160 cm are mostly 4 years old or younger (Ferrero and Walker 1999). Walker excluded porpoises below 160 cm from the subsequent analysis because cysts rarely occurred in them.

The proportion of Dall's porpoises with cysts was zero in the Okhotsk Sea, 1.4% in the Bering Sea and Aleutian Islands area, and 22.7% in the Pacific Ocean. The infection rate increased with body length, suggesting correlation with age. The parasite load also increased from north to south: 0.5% (2/378) in the Bering Sea (north of 54°N), 10.8% (18/167) in the Aleutian Islands area in 52°N–54°N, and the highest figure 22.7% in the Pacific (south of 52°N) (Figure 9.11). This latitudinal cline suggests overlap of two populations in the Aleutian Islands area, which agreed with results of isozyme analysis by Winans and Jones (1988) mentioned in the following. It should be noted that combining samples of multiple years or inaccuracy of catch positions, which we are not given, can exaggerate the real level of geographical overlap of two populations in a single year. Walker (2001) speculated that while not denying additional contribution of other intermediary hosts, the distribution pattern of sharks would be behind the geographical difference in parasite load. He noted that only one species of shark *Somniosus pacificus* is known from the Bering Sea, Sea of Japan, and Okhotsk Sea as a species that attacks or scavenges dolphins and porpoises and that the shark species inhabits more coastal waters than the Dall's porpoise and in low density. However, there are more predatory sharks in the offshore waters south of the Aleutian Islands, including *Isurus oxyrinchus* and *Prionace glauca*, which are abundant in latitudes 20°N–50°N and surface water temperature of 7°C–16°C. Other species of sharks abundant in warmer waters also certainly host the parasite, and dolphins in such waters are known to have high load of the cysts.

The findings indicate that Dall's porpoises that summer in the southern Okhotsk Sea or in the Bering Sea spend every summer there and do not move, at least during the summer, to the Pacific.

9.3.6 DNA and Isozymes

DNA exists mostly within the nucleus, but a smaller amount resides in the mitochondria of animal cells, maintains genetic information, and functions to synthesize various proteins including enzymes. It contains numerous minor variations that have no or limited effects on function, which are used for various purposes including individual identification, population structure, and taxonomy. Variations in DNA may appear as minor structural variations of the enzymes produced, called isozymes. Before the technology of studying DNA become available, isozymes were used as a mean of detecting genetic variation.

To illustrate the usefulness and limitations of mtDNA for population study of Dall's porpoises, I will start with a study by Hayano et al. (2003) on the species around Japan, which compares population structure viewed through mtDNA and from pigmentation pattern. Two populations of Dall's porpoises have been identified in the waters around Japan, a *truei*-type population ranging from the Pacific coast of northern Japan to the central Okhotsk Sea and a *dalli*-type population in the Sea of Japan–southern Okhotsk Sea area. The latter is distinguished from the rest of the *dalli*-type populations by a smaller flank patch. The Pacific waters of northern Japan are a major wintering ground of the *truei*-type population and also used by a few members of the Sea of Japan–Okhotsk Sea *dalli*-type population as well as *dalli*-types from other population(s) (Figure 9.8). Hayano et al. (2003) analyzed the mtDNA of a total of 103 specimens including 35 *truei*-types taken by the hand-harpoon fishery of the Sanriku Region (believed to have been taken in winter) on the Pacific coast of northern Japan, 35 *dalli*-types with small flank patch taken by the same fishery (presumably during late spring to early summer) off the west coast of Hokkaido in the northern Sea of Japan, and 33 *dalli*-types of ordinary flank patch taken for scientific purpose in summer (see Section 9.3.3) in the western North Pacific south of the Kamchatka Peninsula (42°N–47°N, 155°E–162°E). Compared with the first two samples, evidence seems to be weaker for the last sample to be represented by a single population.

Hayano et al. (2003) identified a total of 49 mtDNA haplotypes or 20–24 for each sample. Such large genetic variation found in a small sample is evidence of large genetic variation within the population and suggests the population was large in the past. The number of haplotypes unique to each sample (represented by a total of 40 individuals) was also similar among the three samples: 14 for the *truei*-type sample, 12 for the *dalli*-type sample from south of the Kamchatka Peninsula, and 11 for the *dalli*-type sample from the Sea of Japan. A neighbor-joining dendrogram and minimum spanning network of the 49 haplotypes revealed two major clusters and some additional minor ones. Members of the three samples were scattered in these clusters, and the haplotypes unique to each sample were also found intermingled in various clusters and did not constitute clusters that represented particular samples. A similar result was reached by Escorza-Treviño et al. (2004). The authors concluded that separation of these populations occurred rather recently or after the origin of the observed genetic polymorphisms.

Hayano et al. (2003) carried out an analysis of molecular variance to measure genetic differentiation among the three samples, which took account of genetic distance between haplotypes and frequency of haplotypes in each sample. The results indicated some difference among the three samples. They found significant difference between the Sea of Japan *dalli*-type sample and the remaining two samples but no significant difference between the *truei*-type sample and the south-of-Kamchatka *dalli*-type sample. This does not necessarily imply that the latter two samples were identical in haplotype composition. Such a result can be reached due to insufficient sample size, insufficient time after separation of two populations, or continuing low level of genetic flow. Hayano et al. (2003) thought from observation of the geographical distribution of pigmentation patterns that current genetic flow among the populations in Japanese waters is restricted, which is not counter to the conclusion of Escorza-Treviño and Dizon (2000). It seems to me that the observed mtDNA variation has nothing to do with hereditary or behavioral mechanisms that maintain particular pigmentation patterns within the populations; this remains as an important question to be answered for understanding the population structure of the species.

One of the interesting points in the results is the abundance of haplotypes common among samples: 3 haplotypes (represented by 18 individuals) were common among the 3 samples and another 11 haplotypes (represented by 45 individuals) were common in 2 samples. The number of individuals representing the common haplotypes, 4–6 individuals/haplotype, is greater than the corresponding figure 1.1 for 42 individuals representing the 37 unique haplotypes. Common haplotypes are in the majority in the three samples (populations) and unique haplotypes are in the minority. At least some, not necessarily all, of the latter are likely to have evolved after the populations separated.

Hayano et al. (2003) suggested the possibility of some ecological mechanisms that may limit gene flow between neighboring populations. Dall's porpoises segregate by growth and reproductive stage within each population, and the breeding seasons also differ between neighboring populations (Section 9.4.5). These factors will function to diminish opportunities for interbreeding between populations from what might be expected from their overlapping ranges of distribution. However, such behavioral barriers do not seem to be strong enough to explain the apparent genetic isolation between the two major pigmentation types of the Dall's porpoise. If the white flank patch of the Dall's porpoise has a social function, such as identifying mating partners, about which we know nothing, the pigmentation pattern will function to suppress interbreeding between populations, and genes imported through limited interbreeding will be less likely to survive in the population.

Hayano et al. (2003) estimated separation time for the Sea of Japan–southern Okhotsk Sea *dalli*-types and south-of-Kamchatka *dalli*-types at 30,000–40,000 years before present and that between the former and *truei*-types at 10,000–15,000. Their genetic analyses could not distinguish between the south-of-Kamchatka *dalli*-type sample and the *truei*-type sample from off Sanriku. These estimates depend on an assumption of mutation rate, and a low level of gene flow after separation of the populations will bias downward the estimate of time after separation. The latter figure, 10,000–15,000 years before present, agrees with the end of the last glacial age and start of the current warm-climate era. During the glacial period, the Sea of Japan lost direct connection with the Okhotsk Sea, but it is uncertain whether the Sea of Japan was connected to the Pacific through Tsugaru Strait or to the East China Sea through Tsushima Strait. Hayano et al. (2003) did not reach conclusions on the process of formation of the two populations wintering off the east and west coasts of Japan.

The technique employed by Hayanao et al. (2003) is used widely in studies of population structure of cetaceans and is perhaps the most relied upon in the field. However, their results provide a valuable example of the limitation of the methodology. If a founder effect is ignored, we may have to wait at least several thousand years before sufficient genetic differentiation is accumulated between newly formed populations. Our damage to the marine environment can proceed rather fast; populations may decline at an annual rate of several percent. Such rapid decline in abundance can result in reduction of the population range and subsequent split of the distribution into more than one geographical area. Conservation will need to handle these newly formed geographical populations of almost identical genetic structure as independent populations. Thus, the absence of genetic evidence for population structure should not be considered as an indication of no population structure for purposes of management.

Winans and Jones (1988) and Escorza-Treviño and Dizon (2000) studied the genetics of *dalli*-type Dall's porpoises in the vast area of the offshore North Pacific and adjacent seas not covered by Hayano et al. (2003). Information on breeding grounds and parasite load had already suggested the presence of several populations of Dall's porpoises in this region. Winans and Jones (1988) analyzed isozymes of 360 *dalli*-type Dall's porpoises killed in the Japanese salmon drift-net fishery from June to September 1962. The sample came from the area indicated in Figure 9.11 plus the southern area surrounded by 42°N–47°N, 158°E–170°E. They analyzed 26 loci that controlled 14 enzymes and found polymorphisms at 11 loci. The fit to Hardy–Weinberg equilibrium was better if the sample was split into two geographical groups south and north of the Aleutian Islands. This suggested that the sample was not from a uniform population. They determined that this result was influenced by specimens from the Aleutian Islands area at 50°N–52°N and concluded that the ranges of two populations, to the north and south of the Aleutian Islands, overlapped in those latitudes. Their results coincided exactly with those obtained from parasite load analysis by Walker (2001).

Escorza-Treviño and Dizon (2000) used mtDNA to examine the population structure of *dalli*-type Dall's porpoises in the North Pacific and adjacent seas from the Okhotsk Sea to the coast of North America. Their results are briefly presented in the following using the geographical terminology of this chapter. They identified 58 haplotypes and compared their frequencies between the three geographical regions of (1) the Okhotsk Sea, (2) the Bering Sea and central and western North Pacific, and (3) off the coast of North America. The commonest two haplotypes occurred in all three areas, but minor haplotypes, which were thought by them to be derived from the common haplotypes, occurred only in one or two of the three geographical areas. Later, Hayano et al. (2003) made a similar interpretation that common haplotypes were closer to the ancestral type. The common haplotypes were more frequent in the western area than in the east, which was considered by Escorza-Treviño and Dizon to suggest possible expansion of distribution from west to east. As expected, haplotype composition was not uniform in the three areas.

They further divided the sample into smaller geographical areas and found that haplotype structure was different among seven areas, suggesting the presence of different populations: (1) western Bering Sea, (2) eastern Bering Sea, (3) around the central Aleutian Islands, (4) south of the Kamchatka Peninsula, (5) south of the Aleutian Islands, (6) off the coast of North America, and (7) Okhotsk Sea. The genetic difference between the western Bering Sea and the eastern Bering Sea was inconclusive ($p = 0.050$). Evaluation of such difference should be done in reference to other biological information. Out of the these seven putative populations, the two populations south of the Kamchatka Peninsula and south of the Aleutian Islands apparently match the south-of-Kamchatka population and south-of-western Aleutian Islands populations of Kasuya and Ogi (1987) and Yosjhioka and Kasuya (1991), respectively.

Escorza-Treviño and Dizon (2000) did not find significant genetic difference between *dalli*-types in the northern Okhotsk Sea and those in the southern Okhotsk Sea. Amano and Hayano (2007) did not examine the flank patch of *dalli*-types inhabiting the northern Okhotsk Sea in summer, and we do not know if the *dalli*-types in the two distinct areas are morphologically separable or not. However, distribution of pigmentation types and of mother and calf pairs (Figures 9.1 and 9.2) is sufficient to justify an assumption of three populations in the Okhotsk Sea in summer.

Some questions still remain on the population structure of Dall's porpoises in the Bering Sea. Escorza-Treviño and Dizon (2000) thought that there were three populations: (1) western Bering Sea, (2) eastern Bering Sea, and (3) around the central Aleutian Islands. The relationship between the third population and individuals that breed in the central Bering Sea (Kasuya and Ogi 1987) is unclear, because Kasuya and Ogi (1987) did not survey the Russian EEZ (western Bering Sea) and U.S. EEZ (Aleutian Islands area and eastern Bering Sea), and Escorza-Treviño and Dizon (2000) did not have material from the central Bering Sea. If we assume they are all different, there can be four populations in the Bering Sea and central Aleutian Islands area, but it is also likely that the 3rd population breeds in the central Bering Sea (three populations in this case). The question needs to be resolved. Escorza-Treviño and Dizon (2000) did not include in their analysis porpoises from the central Gulf of Alaska where Yoshioka and Kasuya (1991) recorded a breeding ground of the species, thus it seems to be reasonable to expect a population in the central Gulf of Alaska.

Escorza-Treviño and Dizon (2000) presented another interesting result on sexual difference in dispersal pattern of the Dall's porpoise. They found that geographical difference in haplotypes was greater among females than males, which suggests that males move greater distances, perhaps into the range of other nearby populations. However, as inheritance of mtDNA is matrilineal, the analyses do not tell whether such traveling males breed with local females. The answer will be available through analysis of nuclear DNA.

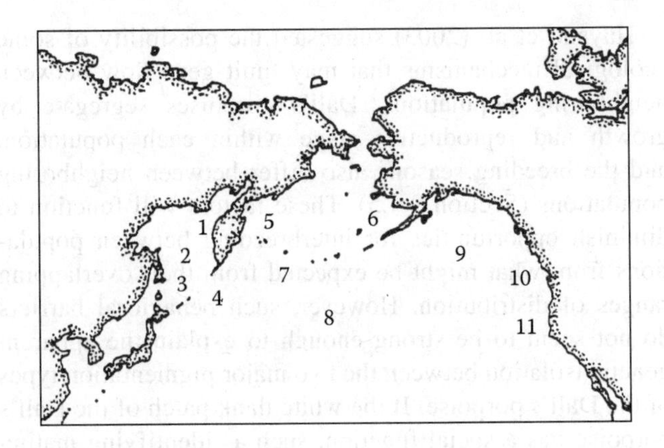

FIGURE 9.12 Dall's porpoise populations identified by the Scientific Committee of the International Whaling Commission. Numerals 1–11 indicate approximate locations of the populations. (From Figure 1 of IWC, *J. Cetacean Res. Manage.*, 4(Suppl.), 325, 2002.)

9.3.7 Population Structure: Summary and Future Studies

The small-cetacean subcommittee of the Scientific Committee of the IWC concluded in 2001 that there are at least 11 populations of the Dall's porpoise in the North Pacific (IWC 2002) (Figure 9.12): (1) a population that breeds in the northern Okhotsk Sea, (2) a *truei*-type population that breeds in the central Okhotsk Sea and winters off the Pacific coast of northern Japan, (3) a population that breeds in the southern Okhotsk Sea and winters in the Sea of Japan, (4) a population in the northwestern North Pacific that breeds south of the Kamchatka Peninsula, (5) a population in the western Bering Sea, (6) a population in the eastern Bering Sea, (7) a population in the western Aleutian Islands area, (8) a population in the central North Pacific that breeds south of the Aleutian Islands, (9) a population that breeds in the central Gulf of Alaska, (10) a population that breeds along the Oregon coast, and (11) a population along the California coast.

Among the 11 Dall's porpoise populations, only the second is represented by *truei*-types; other populations are represented by *dalli*-types. Both wintering ground and breeding ground are known only for the 2nd and 3rd populations. The relationship between the 7th population and individuals that breed in the Central Bering Sea needs further study. It should be noted that the western Aleutian Islands area was shown as an area of mixing of populations.

This conclusion mostly depends on information from pigmentation pattern, mtDNA, and breeding grounds, and its weak point is the absence of linkage of the latter two methodologies. It will be useful for further improvement of our understanding of population structure of the Dall's porpoise to survey distribution of mother–calf pairs in the whole range of the species, including the EEZs of the range countries, and to collect genetic samples from individuals on the breeding grounds, although the tasks will experience some difficulties due to ship-avoiding behavior in the breeding ground. Analysis of nuclear DNA will be useful for the purpose of understanding possible gene flow and the mechanism of genetic isolation between populations.

9.4 LIFE HISTORY

9.4.1 Age Determination

Age is key information for understanding the life history of animals. There have been attempts to age Dall's porpoises taken by the hand-harpoon fishery off Japan and those killed incidentally in the offshore salmon drift-net fishery by reading growth layers in their teeth. The teeth of Dall's porpoise are small, with length of about 1 cm and diameter of 1–2 mm. In adults the crown is often lost through abrasion, and the remaining portion of the tooth is hidden between horny ridges of the gum. The teeth are evolutionarily in the process of degeneration; the function of grasping prey seems to be achieved with the horny surface of the gum. This was interpreted by Miller (1929) to be similar to the process in which ancestral baleen whales evolved baleen plates.

Growth layers are deposited in both the dentine and cementum of dolphins and porpoises, but the layers in dentine are thicker and more regular in spacing and are more suitable for age determination, at least for younger animals. As dentinal layers are accumulated on the pulp wall, the volume of the cavity decreases and dentine deposition finally ceases in most toothed cetaceans, leaving a small space for blood vessels. This stage usually occurs at around the attainment of sexual maturity or soon after when the body growth slows in dolphins and porpoises. After this stage, age has to be determined by reading layers in the cementum. The teeth of Dall's porpoise are probably the most unsuitable for age determination I have ever experienced. Dentine deposition stops at an early age, presumably reflecting the small tooth size or early attainment of physical maturity, and the cemental layers to be relied upon thereafter are irregular. These factors raise a question on the reliability of ages thus determined. Some attempts in the 1970s to use growth layers in the mandible and tympanic bones were unsuccessful, because older layers in the bone tissue are resorbed while new layers are accumulated (Shirakihara and Kasuya unpublished).

In 1984, several scientists attempted to compare their readings on a set of decalcified and hematoxylin-stained tooth sections of Dall's porpoises. The materials were prepared by some of the participants using materials obtained either by U.S. scientists on board Japanese vessels in the salmon drift-net fishery or by Japanese scientists on board the scientific hunting vessels operated in the offshore North Pacific (see Section 9.3.3). The season was summer, or around the parturition season of the species. Readers could not refer to the biological data but were able to identify the origin of the specimen from the specimen number.

Results of the cross reading were reported to the science section of the INPFC meeting in March of the next year (Jones et al. 1985a) but have remained otherwise unpublished. Believing that the experiment is useful for understanding the difficulty of age determination in the species, I have extracted the readings by myself and two anonymous scientists and compared them in Tables 9.3 through 9.5. Table 9.3 compares average ages by body length (sexes combined). Interreader difference in age estimated was unclear for porpoises over 180 cm in body length, which did not mean that interreader difference was negligible. There were greater interreader differences for body lengths of juveniles of 140–159 cm.

TABLE 9.3
Among Readers Disagreement in Age Determination of Dall's Porpoises Experienced in the 1980s

Body-Length Range (cm)	No. of Individuals	Mean Age		
		Reader A	Reader B	Reader C
140–159	29	1.2	2.9	2.8
160–179	43	2.5	3.4	4.0
180–199	34	4.4	4.8	5.9
200–235	9	7.4	8.2	9.1

Note: Materials used were the teeth of *dalli*-types of both sexes. Reader A, Kasuya; B and C, two of several anonymous readers.

Reader A (myself) judged them to be about 1-year old, but the other two readers thought them to be about 3 years old. Porpoises at 160–169 cm were aged at 2–3 years by one reader and 3–4 years or 4 years by the other readers.

In order to examine further details of this disagreement, estimated ages are compared between readers in Table 9.4, where only the results for female porpoises are given. If two readers agreed in their reading, the point should fall on the diagonal line. This did not often happen; the points are scattered over a rather broad range. The only features common among readers were that (1) larger individuals tended to be aged older and (2) all readings were within the range of 0–13 years. Anonymous readers B and C showed better agreement in the readings than comparison between them and reader A. Reader A aged 21 specimens at 1 year, but only 1–2 specimens were aged at 1 year by readers B and C, which suggests that readers B and C counted finer layers compared with reader A, who picked up rather coarse layers.

The next analysis I made was to calculate average body length for each age group in the female sample (Table 9.5). No correlation was found between age and average body length for ages 1–4 years as determined by reader B, which casts some doubt on the age reading. Reader C detected only about 5 cm growth between ages 2 and 3 years; however, it was not accompanied by a reasonable increase in body length below and above this interval. Compared with these readings, the results by reader A gave a better correlation between age and body length, but this alone does not prove that reader A's ages were correct, only that they were correlated with length.

We know from other sources that Dall's porpoises are born at around 100 cm body length, but such newborns were not included in the cross-reading samples. Information on growth during the first year after birth would be a key to evaluate the age readings discussed here. Newborns of several dolphins and porpoises are known to reach 155%–163% of neonatal length in the first year after birth (Section 8.5.2.4; Kasuya and Matsui 1984; Kasuya et al. 1986). Thus, the average body length of 159.5 cm obtained by reader A for animals aged at 1 year seems to be reasonable. If we accept this age reading, then it leads to the conclusion that 1-year-old porpoises were taken in the salmon drift-net fishery in the Bering Sea and Aleutian Islands area and 2-year-olds in the waters south of the Aleutian Islands area of 40°N–45°N where Japanese scientific cruises sampled porpoises with hand harpoons. This further led to the hypothesis of Kasuya and Jones (1984) and Kasuya and Shiraga (1985) that Dall's porpoises segregate by reproductive status as well as growth stage. More importantly, there was general agreement among the INPFC scientists that even if aging by particular readers were accepted, age determination of Dall's porpoises at ages over 6 years would be extremely unreliable.

One of the reasons why the results of the cross-reading experiment were only reviewed at the NPFC meeting and not followed by action would be the absence of firm evidence for choosing one of the several readers who participated. Another reason was the disappointment of the scientists who wanted to use age readings for the analyses of population dynamics.

TABLE 9.4
Cross Reading of Growth Layers in Dall's Porpoise Teeth, Illustrating Trouble Experienced by Scientists in the 1980s

Reader A (Kasuya)																
Reader B	Age	0	1	2	3	4	5	6	7	8	9	10	11	12	13	Total
	0	1														1
	1		1													1
	2		7	4	2											13
	3		7	7	3	1										18
	4		6	2	2	1	1									12
	5				1	3	3									7
	6			1		1	1	1	1	1						6
	7				1							1				2
	8									2						2
	9															
	10			1						1						2
	13													1		1
	Total	1	21	15	9	6	5	1	1	4		1		1		65

Reader A (Kasuya)																
Reader C	Age	0	1	2	3	4	5	6	7	8	9	10	11	12	13	Total
	0	1														1
	1		2													2
	2		8	3	1											12
	3		4	4	3											11
	4		6	3	3	1										13
	5		1	2	1	1	2									7
	6			1		3	2	1								7
	7			2	1		1			2						6
	8															
	9					1			1	1						3
	10									1						1
	11											1				1
	12													1		1
	Total	1	21	15	9	6	5	1	1	4		1		1		65

Reader C																
Reader B	Age	0	1	2	3	4	5	6	7	8	9	10	11	12	13	Total
	0	1														1
	1		1													1
	2		1	6	5	1										13
	3			5	5	5	2	1								18
	4			1	1	6	3	1								12
	5					1	1	4	1							7
	6						1	1	2		2					6
	7								1				1			2
	8										1	1				2
	9															
	10								2							2
	11															
	12													1		1
	13													1		
	Total	1	2	12	11	13	7	7	6		3	1	1	1		65

If we were unable to reliably age animals over 6 years, the age data would be almost useless for such purposes.

Ferrero and Walker (1999) recently published a biological study of Dall's porpoises using large samples from the salmon drift-net fishery collected through the activities of the INPFC. This was very welcome because the only study previously available on the subject was a thesis by Newby (1982). The method of tooth preparation seems to have been identical to that in previous studies, and the analysis is a useful interpretation of the species' life history.

TABLE 9.5
Among Readers Comparison of Early Growth of Dall's Porpoises Estimated by Reading Growth Layers in the Teeth

	Age	No. of Individuals	Mean Body Length (cm)	Body-Length Range	SD
Reader A	1	21	159.5	148–183	9.2
	2	15	171.3	140–206	15.8
	3	9	173.0	160–183	7.3
	4	6	184.7	172–193	8.1
	5	5	185.2	173–200	10.0
	6	1	165.0		
	≥7	7	202.1	172–235	19.6
Reader B	1	1	168		
	2	13	165.5	152–183	12.0
	3	18	167.2	148–183	9.9
	4	12	161.8	140–187	14.2
	5	7	183.1	173–193	7.0
	6	6	187.1	165–210	16.9
	≥7	7	201.9	174–235	18.9
Reader C	1	2	160.0	152–168	11.3
	2	12	162.0	148–183	10.8
	3	11	167.7	154–183	10.1
	4	13	167.8	150–183	11.4
	5	7	170.1	140–200	20.2
	6	7	180.6	161–193	12.5
	≥7	12	195.3	172–235	18.6

9.4.2 Segregation Viewed from Body-Length Composition

The body-length composition of a sample of Dall's porpoises is influenced by sampling season and gear selectivity, but it still offers useful information on behavior and geographical segregation within a population. This source of information is free from the problems attendant on age determinations as described earlier.

Ferrero and Walker (1999) presented the body-length composition of Dall's porpoises taken in the Japanese salmon drift-net fishery in the Pacific and Aleutian Islands area of 46°N–53°N, 170°E–175°E during the months of June and July in 1981–1987 (Figure 9.13). It had two major modes. One was at body length 84–120 cm with a peak around 105 cm for both sexes. The other mode was at 132–220 cm, where length frequencies differed between sexes. There was a female peak at 180–187 cm and two less distinct male peaks at around 158 and 195 cm. The right-hand peaks, 185 cm for females and 195 cm for males, are close to average body length at which growth ceases (see Section 9.4.10). The difference between the sexes of 10 cm is close to that between asymptotic body lengths, which is about 12 cm. These features agree with those identified by Newby (1982). Because the sample was obtained during the parturition season, the smaller individuals constituting the 105 cm peak represent newborns of the year. The hiatus around 115–140 cm represents 1 year's growth.

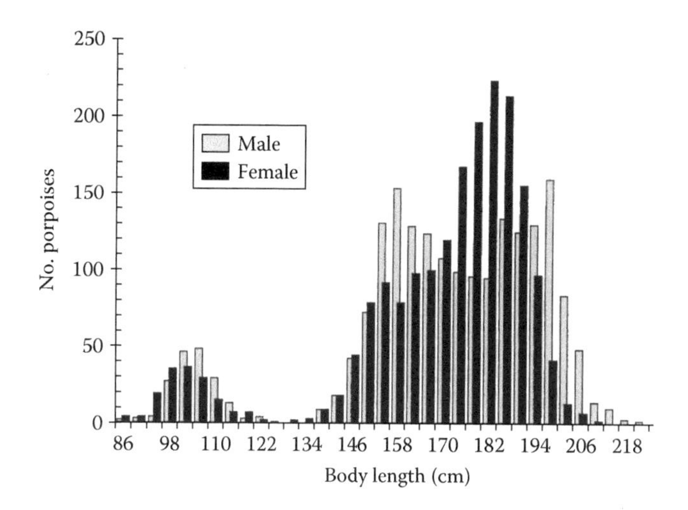

FIGURE 9.13 Body-length composition of *dalli*-type Dall's porpoises incidentally taken in the Japanese salmon drift-net fishery in the western Aleutian Islands area from June to July in 1981–1987. Gray column indicates males and black column females. (From Figure 6 of Ferrero, R.C. and Walker, W.A., *Marine Mammal Sci.*, 15(2), 273, 1999.)

Kasuya and Shiraga (1985) reported body-length composition of Dall's porpoises taken during sighting and hand-harpooning cruises that covered latitudes 40°N–50°N (most of the samples were obtained in latitudes 42°N–47°N) (Figure 9.21), which was south of the area where Ferrero and Walker (1999) and Newby (1982) obtained their samples. The collection period for the hand-harpoon sample, August and September, was about 2 months behind that of the salmon drift-net sample, but the difference was not considered significant in the analyses. The body-length composition in Figure 9.14 lacks calves of the year. Mother–calf pairs were sighted in the northern part of the area covered by the cruise, which was north of 47°N and close to the salmon drift-net ground. It was not possible to harpoon the mother–calf pairs, as well as other individuals in the area due to strong ship-avoiding behavior, which contrasted with the reverse behavior of individuals in the south (Kasuya and Jones 1984). Figure 9.14 compares body-length composition of Dall's porpoises in the breeding ground in the western Aleutian Islands area (i.e., salmon drift-net ground) with that south of the breeding ground (where harpooning was carried out). Similar ship-avoiding behavior by adult individuals has been noted in striped dolphins through the comparison of catches by drive and hand-harpoon fisheries off Taiji (33°36′N, 135°57′E), that is, the latter fishery took mostly weaned immature individuals of both sexes (Kasuya 1978).

Another important point to be noted about body-length composition of the hand-harpoon sample from the southern part of the species' range is a body-length peak at around 165–190 cm in females and 170–195 cm in males, which corresponds to a hiatus in the body-length composition of the drift-net samples of Newby (1982) and Ferrero and Walker (1999). Females over 190 cm and males over 195 cm were less frequent in the hand-harpoon sample than in the drift-net

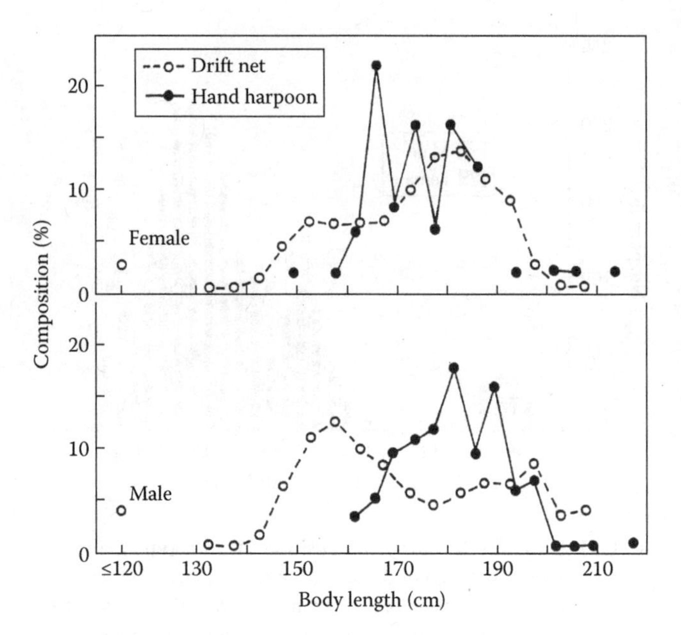

FIGURE 9.14 Body-length composition of *dalli*-type Dall's porpoises compared between individuals killed in the Japanese salmon drift-net fishery in June–July, 1978 and 1980 (Newby 1982), and those taken by hand harpoon in waters south of the drift-net fishery from August to September (1982 and 1983). (From Kasuya, T. and Shiraga, S., *Sci. Rep. Whales Res. Inst.*, 36, 139, 1985.)

sample (Figures 9.13 and 9.14). Combining the two sets of samples better documents the body-length composition of the population.

These observations indicate that Dall's porpoises tend to segregate by sex and growth stage. Inhabitants of the summer breeding ground are mostly calves of the year accompanying their lactating mothers, yearlings, and sexually mature males, while inhabitants of the area south of the breeding ground do not include mother–calf pairs but include numerous juveniles aged 2 years and above (not yet sexually mature) and some adult males and nonlactating adult females (see Section 9.4.6). This was supported by the fact that 88% of the 156 specimens obtained from the latter area (south of the breeding ground) by the dedicated sampling vessel were aged 2–5 years and there were only 2 yearlings and no young of the year (Kasuya and Shiraga 1985). The sex ratio in the southern sample was 2.4 males/female (Kasuya and Shiraga 1985), but the corresponding figure from the northern sample (in the breeding ground) was the reverse at 0.63 males/female (Ferrero and Walker 1999) (Figure 9.18). However, the higher female proportion in the breeding ground could be a reflection of females attaining sexual maturity about 1 year earlier than males.

The earlier discussion on *dalli*-type samples is worth comparing with the body-length composition of *truei*-types off the Sanriku coast, which have been hunted in winter or about half a year after the *dalli*-type samples. The body-length composition of 884 *truei*-types examined by Kasuya (1978) had a mode with a peak at 170–190 cm and was composed of individuals aged 1.5 years and above. The sex ratio was 1.2 males/female, which was similar to that in the earlier mentioned *dalli*-type sample obtained from waters south of the breeding ground. The body-length composition of *truei*-types (n = 1144) landed at the Otsuchi Fish Market (39°21′N, 141°54′E) on the Sanriku coast during November to April in 1995–1996 and examined by Amano et al. (1998c) also showed a similar structure (Figure 9.15). The difference in the peak lengths, 6–7 cm, is a reflection of difference in growth between the sexes. The sex ratio was 2.1 males/female (n = 1979). *Dalli*-types landed at the same market during the same period and examined by Amano et al. (1998c, in Japanese) had the same body-length composition and a similar sex ratio of 1.5 males/female (n = 175). It seems to be true that within-population segregation by growth stage and sex observed in the summer is likely retained in the winter. Amano et al. (1998c, in Japanese) examined Dall's porpoises at the Otsuchi Fish Market nearly throughout the year during 1994–1996 and confirmed that landings from May to October were most likely to have been taken in the northern Sea of Japan and Okhotsk Sea but that landings in winter were from catches off the Sanriku coast as in the past operation examined by Kasuya (1978).

9.4.3 Neonatal Body Length and Parturition Season

9.4.3.1 Neonatal Body Length

Measuring the body length of apparent newborns is the most direct way of obtaining neonatal length, but such an opportunity is limited for cetacean biology and the method is likely to result in some bias due to the time between birth and measurement. Another way is to identify a body length where the proportions of fetuses and neonates are equal in the sample. This method is subject to a bias due to availability differences between fetuses and neonates.

Ferrero and Walker (1999) used materials collected in June and July from the salmon drift-net fishery south of the western Aleutian Islands. They classified the healing of the umbilicus into four stages and calculated the average body length for each stage. The healing stages were (1) umbilicus open with fragment of umbilical cord attached, (2) umbilicus open but without fragment of umbilical cord, (3) progressed stage of healing, and (4) fully healed umbilicus. They presented mean body length for each of these stages (no length distribution available): 99.0 cm at stage 1 (n = 102), 102.7 cm at stage 2 (n = 80), 110.6 cm at stage 3 (n = 31), and 114.1 cm at stage 4 (n = 88). The body-length increment was about 15 cm over all the umbilical stages, but time between the stages was not available. It seems to me that the first figure, 99.0 cm, is closest to the mean neonatal length in the population.

Ferrero and Walker (1999) stated that no individuals in the four categories had deposited postnatal dentine in their teeth as observed in decalcified and hematoxylin-stained preparations. A similar observation has been made on short-finned pilot whales and interpreted as follows (Kasuya and Matsui 1984). The boundary between fetal dentine and postnatal dentine is identified as the neonatal line or neonatal layer. It is a dentinal layer of poor mineralization stained lightly with hematoxylin. Identification of postnatal dentine in a tooth preparation needs to await formation of dark-stained dentine next to the lightly stained neonatal layer. It takes some time for the deposition of

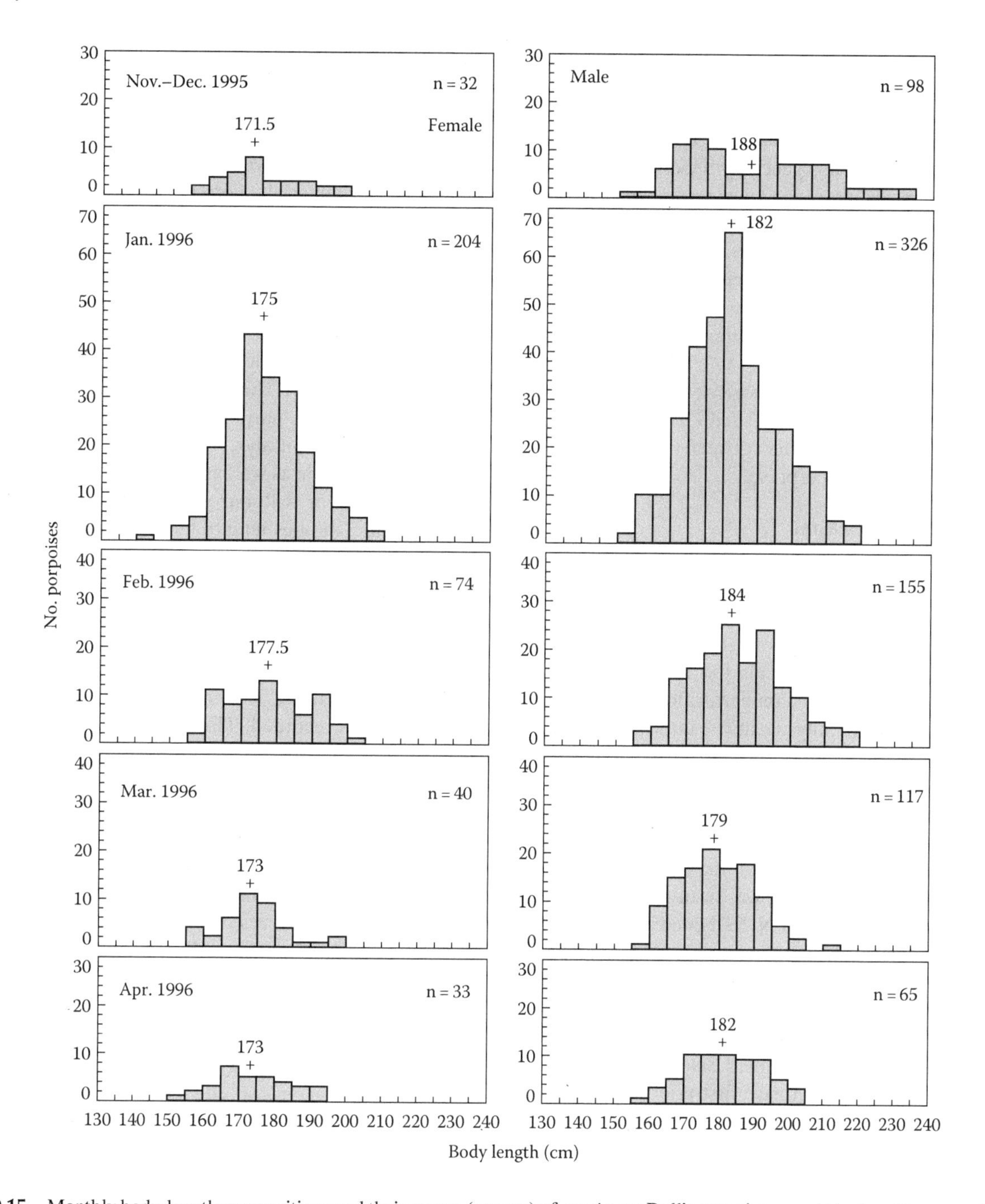

FIGURE 9.15 Monthly body-length compositions and their means (crosses) of *truei*-type Dall's porpoises caught by hand-harpoon fishery off the Sanriku Region and landed at the Otsuchi Fish Market in the period from November 1995 to April 1996 (383 females and 761 males). (From Figure 5 of Amano, M. et al., Biological studies on Dall's porpoises landed at the Otsuchi Fish Market, pp. 33–49, in Miyazaki, N. (ed.), *Report of Contract Studies on Management of Dolphins and Porpoises*, Ocean Research Institute of University of Tokyo, Tokyo, Japan, 1998c, 242pp.)

postnatal predentine to be followed by mineralization. The lag time between birth and full mineralization of postnatal dentine would be sufficiently long for newborn Dall's porpoises to attain a body length of over 114 cm.

The second method used by Ferrero and Waker (1999) was to calculate a point where the proportions of fetuses and neonates were equal in their sample. Fitting two different mathematical models to their data, they calculated mean neonatal length at 101.1 cm with a 95% confidence interval of 100.6–101.6 cm and 103.0 cm (two figures derived from the two models). They also attempted to calculate mean body length of full-term fetuses, which underestimates true neonatal length. It seems to be true that Dall's porpoises in the western Aleutian Islands area are born at an average body length of about 100 cm.

Before the study of Ferrero and Walker (1999), Mizue et al. (1966, in Japanese) examined fetuses and neonates of Dall's porpoises taken in the salmon drift-net fishery around the western Aleutian Islands and reported that body length at birth was between 92 and 109 cm, with an average of about 100 cm. Kasuya (1978) added some supplementary data to those of Mizue et al. (1966, in Japanese) and obtained 100 cm as a body length where the proportions of fetuses and neonate were equal. Thus, there is good agreement among the published estimates of neonatal length of Dall's porpoises.

Adult body size of Dall's porpoises differs between regions (Section 9.3.2). However, it is generally understood that correlation between mean adult size and mean neonatal size is weak among small cetaceans, which can be explained by an interspecies relationship between mean body length at attainment of sexual maturity (X, m) and mean neonatal length (Y, m):

$$Y = 0.532X^{0.916} \quad \text{Ohsumi (1966)}$$

This suggests a 5 cm difference between two species with body lengths at sexual maturity of 2 and 2.1 m. The difference in neonatal length will be smaller between two populations of one species such as Dall's porpoise.

9.4.3.2 Parturition Season

As cetaceans usually deliver their young in waters remote from human habitats, recording seasonal frequency is often impractical. Examined Dall's porpoises were killed in hunting activities or incidental to other fishing activities that were seasonally limited. Thus, materials from such activities did not offer sufficient information on parturition season.

To get around this difficulty and utilize information from the salmon drift-net fishery that operated during the breeding season, Newby (1982) observed seasonal alternation of full-term fetuses, newborns, and lactating females in the incidentally taken samples in order to estimate the seasonality of reproduction. His materials were collected in the U.S. EEZ where the Japanese salmon drift-net fishery was licensed, that is, around the western Aleutian Islands, 46°N–59°N, 168°E–175°E, which was almost the same as the sampling area of Winans and Jones (1988) and Ferrero and Walker (1999) and overlapped with that of Mizue et al. (1966, in Japanese). Newby (1982) obtained his material in the period from June 5 to July 25 of three seasons from 1978 to 1980. Since the United Sates–Japan agreement on the Dall's porpoise project gave first priority to collection of carcasses within the U.S. EEZ, sampling activity was less when the salmon fleets moved outside of the EEZ, including during the previously mentioned period.

Newby (1982) noted that the proportion of pregnant females with full-term fetus decreased in July but that they still occurred through the sampling period, which indicated that the parturition season lasted until late July. Postpartum females started to occur around June 10, the number became equal with that of near-term females around July 10, and then postpartum females exceeded pregnant females in number. From this observation, he concluded that the average parturition date was around July 10. He further stated that the parturition peak was in the last 10 days period of July when he observed a rapid increase in number of newborns (data were grouped for each 10-day period). The difference between average parturition date and parturition peak is unclear to me. His study could not identify the end of the parturition season.

Ferrero and Walker (1999) analyzed a greater sample obtained in 1981–1987, or after Newby (1982) collected his data, covering a slightly longer season from June 2 to July 29. They were apparently able to retrieve most of the incidental takes without distinguishing between inside and outside of the U.S. EEZ as experienced during earlier activities. This sampling scheme improved the quality of the data as well as the quantity of materials. Ferrero and Walker confirmed that almost all the sexually mature females were pregnant or early postpartum. Postpartum females that were lactating started to appear during June 5–9, and pregnant females occurred until July 20–24 and were absent on July 25 and later. A sigmoid curve fitted to frequency of postpartum individuals gave the proportion of such females at 5% on June 11, 50% on July 3, and 95% on July 24. This result indicated that the parturition season of Dall's porpoises around the western Aleutian Islands area lasted for about 50 days from June 5 to July 24, with an average date as well as the peak date of parturition on July 3. This agrees with our general understanding that species in higher latitudes tend to have shorter breeding seasons than those in lower latitudes.

The further analyses of lactating females by Ferrero and Walker (1999) are also worth mentioning. While parturition started in the population on around June 5, females secreting colostrum were already present on June 2 when sampling started and continued to appear on July 24 when the last parturition was recorded. The peak was during June 10–24, which was before the parturition peak on July 3 calculated from a model. Secretion of colostrum starts before parturition and continues until shortly after parturition. Colostrum can be identified through histological examination of mammary glands for the presence of colostrum bodies in the milk (Kasuya and Tai 1993), but it can reasonably be identified as colostrum if milk is secreted by a pregnant female with a near-term fetus. At least at the early stage of the United States–Japan cooperative activities on the Dall's porpoise, there was no instruction for the microscopic identification of colostrum.

9.4.4 Mating Season and Gestation Period

9.4.4.1 Mating Season

The Japanese high-seas salmon drift-net fisheries operated from 1952 to 1992 with two modes: mother-ship (factory ship) fishery and land-based fishery. The operation area changed several times by international agreement, and in the 1980s, mother ships were allowed in the area indicated in Figure 9.11 and the land-based fishery in an area further to the southwest (Sano 1998, in Japanese). The U.S. government placed observers on Japanese mother ships operating in its EEZ to examine Dall's porpoise carcasses brought in by the catcher boats. The results revealed that most of the adult females were pregnant with a full-term fetus or just postpartum, indicating that the parturition season was during the fishing and sampling season of June and July (Composition of 1061 sexually mature females analyzed by Ferrero and Walker (1999) was 50% postpartum, 47% pregnant, and 3% resting).

Ferrero and Walker (1999) observed a seasonal change in Graafian follicles in the samples. The diameter of the largest follicle in most porpoises was below 8 mm during the period from June to the second 10-day period of July, but they suddenly started to see females having follicles of 10–18 mm around July 20 and this continued until July 25, the last day of sampling. This suggested that follicle size at ovulation was not below 10 mm and was around 10–18 mm and that estrus or the mating season started around July 20. The starting day of the mating season was about 45 days after the start of the parturition season (June 5). As the length of the mating season will be almost the same as that of the parturition season, estimated earlier at about 50 days, the mating season lasts approximately from July 20 to September 10.

To further confirm the seasonality of mating, Ferrero and Walker (1999) examined seasonality in spermatogenesis in the testicular tissue. The proportion of inactive males was 20%–30% during early June and declined to less than 5% in the second 10-day period of July, while the proportion of males with testis having very active spermatogenesis increased from 10% in early June to around 40% in the second and third 10-day periods of July. The rest of the males were classified as being at a moderately active stage and constituted 50%–60% of the observed adult males. This observation does not disagree with the mating season deduced earlier, because (1) it takes some time for spermatozoa in the seminiferous tubule to be transported to the epididymis and (2) it is biologically reasonable for some males to be ready for mating before the start of female estrus. A biological explanation needs to be investigated for the 50%–60% of males that were classified as at intermediate activity during the mating season.

9.4.4.2 Gestation Period

Gestation in most cetacean species lasts close to 1 year, with a possible maximum of 15–17 months observed or estimated for killer and sperm whales (Kasuya 1995). The estrous season of Dall's porpoises in the Aleutian Islands area started about 45 days after the start of the parturition season, which suggests that gestation lasts for 10.5 months.

Kasuya (1978) examined the fetal growth of *truei*-types using 39 fetuses measuring 20–50 cm in January–March and obtained 3.3 mm as an estimate of average daily growth rate in the linear phase of fetal growth. By extending this regression line to the right, I obtained a mean parturition date of August 28 when the extended fetal growth reached 100 cm, the estimated mean neonatal length. The time between the mean parturition date and the date when left-side extension of the regression line crossed the axis of time was 300 days. This estimate is downwardly biased because fetal growth in the early stage of gestation is slower. Assuming the bias at 13.5% of the total gestation time for this species (Section 8.5.2.2), I estimated gestation time of 11.4 months and mean conception date at September 17 for *truei*-types off Japan. This figure is based on some assumptions and on a small and seasonally limited sample and is of low reliability. If the gestation time of 10.5 months estimated by Ferrero and Waker (1999) is assumed for the *truei*-type population, peaks of parturition and conception would be expected in middle August and early October, respectively, by shifting both of the dates (August 28 and September 17) by about 0.5 month.

The following relationship between average neonatal length (X, cm) and gestation time (Y, month) was proposed by Perrin et al. (1977):

$$\mathrm{Log}Y = 0.1659 + 0.4586\mathrm{Log}X$$

This equation and neonatal length of 100 cm suggest a gestation time of 12.1 months, which does not agree with the seasonality of reproduction observed earlier and appears to be an overestimate.

9.4.5 Geographical Variation in Breeding Season

Amano and Kuramochi (1992) thought that the Sea of Japan–southern Okhotsk Sea *dalli*-type population had a parturition season in May and June, different from that reported for the western Aleutian Islands area. This conclusion was based on observations on board hand-harpoon vessels from June 15 to July 1, 1988, when the vessels encountered numerous mother–calf pairs in an area from Soya Strait to the southern Okhotsk Sea (44°N–46°N) but were unable to capture them because of their ship-avoiding behavior. They also noted a high proportion of immature individuals, particularly among females, in the catch made in late May 1988 and 1989 by the same vessel in Tsugaru Strait and the nearby eastern Sea of Japan (41°N–43°′N), which suggests a north–south segregation by growth stage within the population. The reproductive data they obtained in these cruises are in Table 9.6.

TABLE 9.6
Reproductive Status of Females in the Sea of Japan–Southern Okhotsk Sea *Dalli*-Type Population

Reproductive Status	May[a]	June[b]	July[c]	August[d]	Total
Immature	27	11	2	2	42
Pregnant		1	5	6	12
Pregnant and lactating[e]			2	0	2
Lactating		1			1
Ovulated[f]		33	9		42
Resting[g]			1		1
Total of mature females		35	17	6	58

Note: The sample was obtained from the hand-harpoon fishery.
[a] Hunted in latitudes 41°20′N–43°30′N, or Tsugaru Strait and the Sea of Japan coast off western Hokkaido in May 1988 and 1989 (Amano and Kuramochi 1992).
[b] Hunted in latitudes 43°50′–45°40′N), or Soya Strait and the southern Okhotsk Sea in the second half of June 1988 (Amano and Kuramochi 1992).
[c] Hunted off Abashiri (44°01′N, 144°16′E), southern Okhotsk Sea in July 13–31 (Yoshioka et al. 1990).
[d] Hunted off Abashiri, southern Okhotsk Sea in August 1–26 (Yoshioka et al. 1990).
[e] Females pregnant and simultaneously lactating.
[f] Nonpregnant females with corpus luteum.
[g] Sexually mature females not pregnant, lactating, nor ovulating.

Conclusions of Amano and Kuramochi (1992) were supported, as detailed in the following, by biological data for 31 *dalli*-types collected by myself and my colleague Aoki during July 13 to August 26, 1988, from the catch of the *Yasu-maru No. 1*, a small-type whaling vessel then operating in the hand-harpoon fishery for Dall's porpoises in the southern Okhotsk Sea. Reproductive data for the 31 carcasses were used in Yoshioka et al. (1990) and are included in Table 9.6. I stayed on the vessel for 4 days in August and recorded numerous Dall's porpoises that could not be approached by the vessel and thus were unable to be harpooned. They were of adult size and did not include mother–calf pairs.

Biological information on *dalli*-types caught in the southern Okhotsk Sea in June through August provides evidence of the breeding season, which is different from that of some other populations of the species (Table 9.6). The only pregnant female in June reported by Amano and Kuramochi (1992) had a full-term fetus of 96 cm, while 7 fetuses in July and 6 fetuses in August (Yoshioka et al. 1990; Kasuya unpublished) were 0.7–3.0 cm (mean: 2.0 cm) and 3.8–9.0 cm (mean: 6.6 cm) long, respectively. The average fetal size increased 4.6 cm in the 1-month period. This suggests only that the parturition season ends by June and the mating season has already started in July. However, an earlier start of the mating season is suggested by the presence of females with a corpus luteum of ovulation (and without identifiable embryo) in the latter half of June. The proportion of such sexually mature females was 94% in the latter half of June and 53% in July (Table 9.6). All of these females must have experienced recent estrus, and many of them could have been in an early stage of pregnancy. Histological examination of endometrial tissue (see Section 12.4.4) was not conducted for these females.

This information allows the conclusion that the Sea of Japan–southern Okhotsk Sea *dalli*-type porpoises mate from mid-June to late July (a 50-day mating season in the Aleutian population is assumed without evidence). The mating season is about 1.5 months earlier than that for a population in the Aleutian Islands area (from July 20 to September 10). Although there are insufficient data to determine the parturition season of the Sea of Japan–Okhotsk Sea population, this mating season, the presence of numerous mother–calf pairs in middle and late June, and an assumption of the same gestation time as in the western Aleutian Islands area (10.5 months) suggest that the parturition season there starts in the middle of April and continues to June, although Amano and Kuramochi (1992) concluded that the population had parturitions in May and June.

Kasuya (1978) fitted a fetal growth curve calculated for the *truei*-type off the Pacific coast of northern Japan to fetal records of *dalli*-types in the western Aleutian Islands area and found about a 1-month difference in the parturition peaks between the two populations, that is, July–August for the Aleutian sample and August–September for the *truei*-types off Japan. Although these absolute dates of parturition may be subject to some uncertainty due to the fetal growth curve having been fitted to small and seasonally limited samples, the resultant difference in mating seasons can be significant. An improved estimation of breeding season also showed a difference of about 1.5 months in the same direction (Sections 9.4.3 and 9.4.4), that is, *dalli*-types in the Aleutian Islands area had a parturition peak in early July and a conception peak in middle August, while *truei*-types off Japan had a parturition peak in mid-August and a conception peak in early October. The two populations have breeding seasons that are about 1 month apart.

Dalli-type porpoises of the Sea of Japan–southern Okhotsk Sea population probably give birth in the northern Sea of Japan off Hokkaido and southern Okhotsk Sea from mid-April to mid-June and mate during the period from mid-June to late July in the southern Okhotsk Sea (off northern Hokkaido), while *truei*-type porpoises mate presumably in September and October (with an estimated conception peak in early October). Although the mating grounds are located close to each other (southern Okhotsk Sea and central Okhotsk Sea), the mating seasons have a difference of about 1 month. This gap between the mating seasons likely functions to limit the chance of interbreeding between the populations. The mating season of a *dalli*-type population that breeds in the northern Okhotsk Sea is still to be investigated. I expect it is likely to have a mating season that is different from that of nearby *truei*-types.

Sea surface temperature in temperate and subarctic waters of the northern hemisphere usually peaks in late August and September. Thus, it can be generalized that Dall's porpoises give birth in early summer and that most of the postpartum females experience estrus in midsummer while nursing the calf. The mating season and gestation time are believed to be so arranged as to maximize the survival of the offspring, but the mating season will also be influenced by maternal physiology that reflects the seasonality of food availability. Marine productivity is high during the short summer in high latitudes where Dall's porpoises summer and offers a suitable feeding environment to the offspring, which are believed to start taking solid food at 2–3 months of age, as well as to their mothers starting the next gestation while nursing. It is reasonable to have parturition just prior to the peak of productivity, because the energetic cost to the mother is believed to be greater for nursing than for maintaining gestation (Lockyer 1981a,b, 1984).

The breeding seasonality and migration pattern of the Sea of Japan–southern Okhotsk Sea population are likely to have a relationship with oceanography in the Okhotsk Sea. The Okhotsk Sea ice floes are formed in the northern part of the sea in November, cover about 75% of the sea in the maximum period, and gradually retreat from the south and entirely disappear in July. They remain off the coast of the southern Okhotsk Sea from January to April. Dall's porpoises that winter in the Sea of Japan arrive at the southwestern and southeastern entrances to the Okhotsk Sea (Soya Strait and Nemuro Pass) in late May, but they have yet to pass into the major part of the southern Okhotsk Seas (Figure 9.3). They enter the southern Okhotsk Sea after retreat of the ice floes that occurs from April to May. Sea surface temperature at the time is around 5°C–6°C, which is close to the minimum sea surface temperature known for the species. Giving birth while they are in the Sea of Japan, that is, before entering the Okhotsk Sea, is likely to benefit the survival of the offspring (Amano and Kuramochi 1992). The winter sea surface temperature off the coasts of Yamaguchi and Shimane (35°N–36°N, 131°E–134°E) where the population is known to winter is around 11°C–13°C, and the temperature off the west coast of Hokkaido where they give birth in May and June is 8°C–13°C; in both cases warmer than the temperature, Dall's porpoises meet when they enter the southern Okhotsk Sea in late May and June (Miyashita and Kasuya 1988).

9.4.6 Reproductive Cycle

One way to work out the reproductive cycle of female cetaceans is to observe individually identified animals for years. The method documents individual variation in the cycle but is feasible for only to a limited number of species. Another method is to estimate the average cycle from the proportion of pregnant females in a sample of sexually mature individuals using gestation length estimated separately as a time calibration. Caution is required because the proportion may change within a year. If there is an ideal situation where we are able to obtain a certain number of unbiased monthly samples from a population for 1 year, we can calculate mean reproductive parameters of the population. "Apparent pregnancy rate" is the proportion of pregnant females in the sample of sexually mature females and may contain biases of various sources. Dividing the "unbiased" apparent pregnancy rate by the gestation period in years gives the annual pregnancy rate, which is a probability of a mature female becoming pregnant in a year, and the reciprocal of the annual pregnancy rate is the mean calving interval. In this method, fetal mortality is incorrectly ignored. The mean lactation period and mean resting period are also calculated assuming that the number of females at each reproductive stage is proportional to the average time in the stage. Such an ideal situation will never happen in our cetacean studies, particularly when samples come from fisheries, because they can be biased due to fishing season, geographical limitation of fishing operation, and gear selection.

Dall's porpoise materials from the western Aleutian Islands area were interpreted to indicate that most of the females give birth during June 5 to July 24 and undergo estrus during July 20 to September 10, or about 45 days after parturition (Section 9.4.3). This suggests that most of the females reproduce annually. Although this will be close to the truth, it has to be remembered that the conclusion was drawn from samples covering only a 2-month period and only part of the geographical range of the population and that it may require correction if some adult nonestrous females are segregating outside the sampled area. Investigation of such a possibility was one of the objectives of the Japanese special cruises for sighting and hunting Dall's porpoises (Section 9.3.3).

To the south of the breeding area around the western Aleutian Islands, these cruises found an area that was inhabited mostly by male Dall's porpoises but contained about 30% females. About 60% of the females, or 17% of the total sample, were sexually mature and were neither pregnant nor lactating, that is, resting. This finding means that some proportion of females in the population do not breed annually and segregate during the summer outside the breeding ground. However, it was not possible to estimate the proportion of such females and estimate annual pregnancy rate for the population.

Kasuya (1978) reported the reproductive status of 71 adult *truei*-type females taken off the Sanriku coast from January to March, which was after the mating season and before the parturition season. There were only 6 in lactation (not pregnant), 5 lactating and simultaneously pregnant, and 58 in pregnancy (not lactating). Ignoring gear selection and segregation by reproductive stage, the proportion of females that became pregnant in the last estrous season was calculated as [5 in lactation and simultaneous pregnancy plus 58 in pregnancy]/71 = 0.89, which is an estimate of the annual pregnancy rate of the *truei*-type population. The remaining 11% represent females that have not become pregnant in the last mating season.

In addition to the 71 females, Kasuya (1978) used other biological information from gutted carcasses to estimate that an additional 34 females were sexually mature and not lactating. These females have a very high probability of having been pregnant, because the fishermen are likely to discard pregnant uteri at sea together with other internal organs. If we assume that these 34 females were pregnant, then the upper limit of their annual pregnancy rate would be given by (63 + 34)/(71 + 34) = 0.923. It is likely that porpoises in the *truei*-type population off Japan reproduce with an annual pregnancy rate of around 90%.

These data allow speculation about the lactation period for the *truei*-type population. The parturition season of this population is assumed to extend for about 45 days centered in mid-August (Section 9.4.4). It was assumed for simplicity that the earlier mentioned 105 likely pregnant females were lactating at the time of conception. Out of the 105 females, 94 females had already ceased lactation by the fishing season from January to March, but 11 females were still lactating. Since parturition of these 11 females likely occurred in the later part of the season (probably in September), they could have been lactating for 4 months (if killed in January) or 6 months (if killed in March). The majority ceased lactation before January. Thus, lactation in *truei*-type population is likely to last 4–6 months at the longest. Amano et al. (1998c) recorded 8 lactating females in 50 sexually mature females killed in the same fishery, which does not alter the earlier conclusion on lactation time.

Ferrero and Walker (1999) found that all the 69 calves of the year collected in the western Aleutian Islands area had only milk in their stomach and concluded that they lived only on milk for a minimum of 2 months after birth.

Harbor porpoises, another species of Phocoenidae, inhabit cold waters of the northern hemisphere and are known to have a life history that is similar to that of Dall's porpoises, that is, early sexual maturity (3–4 years), short longevity (less than 23 years), high annual pregnancy rate (74%–99%), parturition season in summer, and gestation period of 10–11 months (Lockyer 2003). Their lactation length was once thought to last over 8 months based on the presence of lactating females in March, but the possibility of shorter lactation is suggested by a low proportion of lactating females in winter off Iceland (Olafsdottir et al. 2003). Both Dall's porpoise and the harbor porpoise exhibit precocity, short longevity, and high reproductive rate.

9.4.7 Female Sexual Maturity

The age of first parturition is important for the population dynamics of wild animals, but cetacean biologists have used age at the first ovulation as a sign of sexual maturation. The possible reasons could have been because a trace of ovulation, the corpus albicans, remains in the ovary for a considerable length of time, perhaps for life (see also Section 10.5.6) and offers a useful key for identifying maturation and because first ovulations are usually followed by conception in most cetaceans. The age at first parturition can be about 1 year (the gestation time of Dall's porpoise) greater than the age at first ovulation.

Ferrero and Walker (1999) examined materials from 1911 females taken by the salmon drift-net fishery in the western Aleutian Islands area and identified 1061 as sexually mature. This figure may be an overestimate of the maturity rate of the population because it does not take into account individuals segregating south of the salmon drift-net ground (see the following text in this section). The smallest sexually mature female was 147 cm in body length and was pregnant with a near-term fetus; it was considered to have first ovulated in the previous summer. The proportion of sexually mature individuals in the sample increased with increasing body length, and the largest immature female was 193 cm long. The body length at which 50% were sexually mature was 179.7 cm (with 95% confidence interval of 179.6–179.8 cm). This is often stated as the average body length at sexual maturity (Table 9.7). The relationship between sexual maturity and age was analyzed in the same manner. The youngest sexually mature female was aged at 3 years and the oldest immature female at 8 years. Ferrero and Walker (1999) obtained 3.8 years as the unbiased estimate of mean age at the time of sexual maturation using the method of DeMaster (1978) and the age at 50% mature at 4.4 years by fitting a sigmoid curve to the data. The age at 50% mature and the mean age at maturation are different in a strict sense. Considering the example of a Japanese finless porpoise that was born in an aquarium and underwent first estrus when she was 2 years old and gave birth at 3 years of age (Section 8.5.2.6), we can accept such early maturation for the Dall's porpoise.

Kasuya and Shiraga (1985) analyzed hand-harpoon samples obtained in 1982 and 1983 mostly in waters 42°N–48°N, 155°E–180°E, located to the south of the locality of the sample analyzed by Ferrero and Walker (1999). They were unable to harpoon Dall's porpoises that avoided the vessel in waters between the two sampling areas, which included mother–calf pairs and was considered as an extension of the breeding ground in the western Aleutian Islands area studied by Ferrero and Walker (1999). Their sample (185) contained 20 immature and 29 mature females (including two animals taken in latitudes 48°N–50°N, which were females), for a maturity rate of 59%, which was similar to the figure for the western Aleutian Islands area. However, their sample had fewer young and old animals. Female body lengths ranged between 160 and 190 cm, and most were 2–4 years old; only one female was aged at 1 year and the oldest female was 17 years old (Figure 9.19). These females had completed weaning but were most under or around the age at attainment of sexual maturity.

Kasuya and Shiraga (1985) reported that ages of the youngest mature female and oldest immature female were 2 and 4 years (38 females in the age range), respectively, and that the age at 50% maturity was between 2 and 3 years. These figures were slightly lower than the corresponding figures from the western Aleutian Islands area. However, in view of the small sample size of our study and my observation that growth layers in their teeth start to be irregular at the age of 3–4 years, making determination by means of the tooth layers less reliable, a firm conclusion about the significance of the difference between the two estimates is not possible.

Kasuya and Shiraga (1985) analyzed the relationship between female maturity and body length using the hand-harpoon sample from the southern area. The range from the smallest mature female to the largest immature female was 164–187 cm (n = 40), which was similar to the range of

TABLE 9.7
Age (Years) and Body Length (cm) at Attainment of Sexual Maturity of Female Dall's Porpoises

Geography/Population	Around the Western Aleutian Is.	South of the Western Aleutian Is.	Off Sanriku Coast	Sea of Japan–South. Okhotsk Sea Stock
Color Type	*Dalli*-Types	*Dalli*-Types	*Truei*-Types	*Dalli*-Types
Sampling Method	Drift Net	Hand Harpoon	Hand Harpoon	Hand Harpoon
Body length at maturity, range	147–193	164–187	172–203	180–199
Body length at maturity, mean	179.7		186.7	187.0
Age at maturity, range	3–8	2–4	2.5–11.5(?)	
Age at maturity, mean	3.8, 4.4	2–3	6	
Reference and note	Ferrero and Walker (1999)	Kasuya and Shiraga (1985). Segregation by maturity observed.	Kasuya (1978). Ages unreliable after around sexual maturity.	Amano and Kuramochi (1992)

147–193 cm reported by Ferrero and Walker (1999) in a larger sample from the northern area. Kasuya and Shiraga concluded that 50% were sexually mature at around 168–171 cm, which was smaller than the corresponding figure, 179.7 cm, reported by Ferrero and Walker (1999). It was inconclusive whether the disagreement was due to the small sample size of Kasuya and Shiraga (1985) or to a biological factor.

Kasuya (1978) analyzed body length and female sexual maturity using *truei*-type sample obtained from the hand-harpoon fishery off the Pacific coast of northern Japan. I reported that the porpoises attained sexual maturity at body length 172–203 cm and that the length at 50% mature was 186.7 cm. These figures are about 6–7 cm greater than the corresponding figures from the western Aleutian Islands area. It is already known that Dall's porpoises in the western and eastern Pacific are larger than those in the central Pacific (Section 9.3.2). I also analyzed the relationship between sexual maturity and age and found that sexual maturity was reached at 2.5–11.5 years. Such broad individual variation in maturation age throws doubt on the age determinations. It could be attributed to erroneous interpretation of tooth layers that overestimated ages of some animals, particularly those at ages of over 3 or 4 years.

Of the 399 *truei*-type females examined by Kasuya (1978), only 35% (139) were sexually mature, which was lower than the figure from the drift-net samples from the western Aleutian Islands area. The sample contained extremely few individuals below age 2 years and had a peak age at 3–5 years. As there were only two individuals below 160 cm, which were aged at 0.5 year, this skewed age composition, or scarcity of juveniles, is attributed to the real absence of juveniles and not to erroneous age determination. I interpreted this almost complete lack of juveniles born in the previous summer in the *truei*-types sample to indicate that the hand-harpoon fishery hunted mostly weaned individuals before they matured sexually, but it is more correct to say that the fishery targeted mostly individuals between their second winter and sexual maturity. Such a skewed catch can happen due either to geographical segregation by age or to different responses of animals to the harpoon vessel, but the former seems to be more likely because of the similarity to samples in waters south of the western Aleutian Islands area analyzed by Kasuya and Shiraga (1985).

Amano and Kuramochi (1992) reported the sexual maturity of *dalli*-types taken by the hand-harpoon fishery in the Sea of Japan in May and June (38 immature and 35 mature). The proportion of sexually mature was 48%, which was higher than the 35% of the hand-harpoon sample from off Sanriku. The range between the smallest mature and largest immature females was 180–199 cm, and 50% of individuals were sexually mature at 187.0 cm. These figures are close to those for off Sanriku and greater than those for the western Aleutian Islands area.

9.4.8 Male Sexual Maturity

Uncertainty remains about the determination of male sexual maturity. It is more a question of behavioral investigation to judge whether a male has the ability to mate and produce offspring, but cetacean scientists have often attempted to estimate it through histological examination of carcasses taken in fisheries. Maturation of testicular tissue is a gradual process spanning perhaps years, and dividing animals into two categories of mature and immature is a problem on which there is probably no common understanding among scientists. Seasonal breeders such as the Dall's porpoise change in anatomical feature of the gonads with season, which makes it difficult to apply a single criterion to materials collected in different seasons.

Kasuya (1978) analyzed sexual maturity in 486 male *truei*-types taken in winter, or the nonmating season, by the hand-harpoon fishery off Sanriku on the Pacific coast of northern Japan. I sampled tissue from midlength of a testis centrally and peripherally from each male and examined the samples for the presence of spermatids and spermatocytes. A testis was classified as "mature" if either of the cell types was found in both central and peripheral samples, as "maturing" if they were found only in the central sample, and "immature" if they were not found. There were no cases where they were found only in the peripheral sample. I also examined smears from the testis at midlength and from the epididymis for the presence of spermatozoa.

Spermatozoa were present in only 30% of the epididymal smears of mature testes and 10% of maturing testes. Spermatids or spermatocytes in such testes that have no spermatozoa in the winter will be transformed into spermatozoa by the beginning of the next mating season in the autumn. About 3% of testes classified as "immature" by the earlier histological criteria had spermatozoa in the testicular smear, which shows the limitation of detecting small amount of such cells in ordinary histological preparations. These results were in contrast with the summer sample of *dalli*-types obtained by the same hand-harpoon method in waters south of the western Aleutian Islands area (Kasuya and Jones 1984). Twenty-five individuals (80.6%) of 31 histologically mature males had copious spermatozoa in the epididymis, 3 had fewer spermatozoa, and 2 (6%) had no spermatozoa in the epididymis. This difference, that is, 94% of mature testes had sperm in the epididymis in summer compared with 30% in winter, reflects seasonal change in gonadal activity. The maturity criteria of Kasuya (1978) differed slightly from those used by Kasuya and Jones (1984), which classified testicular maturity into four stages of mature, late maturing, early maturing, and immature following a method used for the bottlenose dolphin (Section 11.4.7), short-finned pilot whale (Section 12.4.3.3), and Baird's beaked whale (Section 13.4.2). However, this does not change the earlier conclusion that many mature males do not have spermatozoa in the epididymis during winter.

Mature males almost double their testicular weights in the mating season, while average testicular weight at attainment of sexual maturity does not show such great seasonal change (Table 9.8). The weight of a single mature testis was around 30–100 g in winter off Sanriku (*truei*-types), but it was 40–340 g in the summer sample of *dalli*-types in the western Aleutian Islands area (Figure 9.16), which is the reverse of the geographical cline in body size. Ferrero and Walker (1999)

TABLE 9.8
Weight of Single Testis (g), Body Length (cm), and Age (Years) at Attainment of Sexual Maturity in Male Dall's Porpoises

Geography/Population	Around the Western Aleutian Is.	South of the Western Aleutian Is.	Off Sanriku Coast	Sea of Japan–South Okhotsk Sea Stock
Color Type	*Dalli*-Type	*Dalli*-Type	*Truei*-Types	*dalli*-type
Sampling Method	Drift Net	Hand Harpoon	Hand Harpoon	Hand Harpoon
Testis weight at maturity, range	20–120	c.35–45	13–48	
Testis weight at maturity, mean		40	29.3	40
Body length at maturity, range	166–194	168–199	180–215	180–209
Body length at maturity, mean	179.7	c.184	195.9	192
Age at maturity, range	3–6	2–7	3.5–15.5(?)	
Age at maturity, mean	5.0, 4.5	c.4.5	7.9(?)	
Reference and note	Ferrero and Walker (1999). Summer sample. Testis was weighed with attached epididymis.	Kasuya and Shiraga (1985); Kasuya and Jones (1984). Summer sample. Segregation by maturity observed.	Kasuya (1978). Winter sample. Ages unreliable after around sexual maturity.	Amano and Kuramochi (1992). Sampled in early summer.

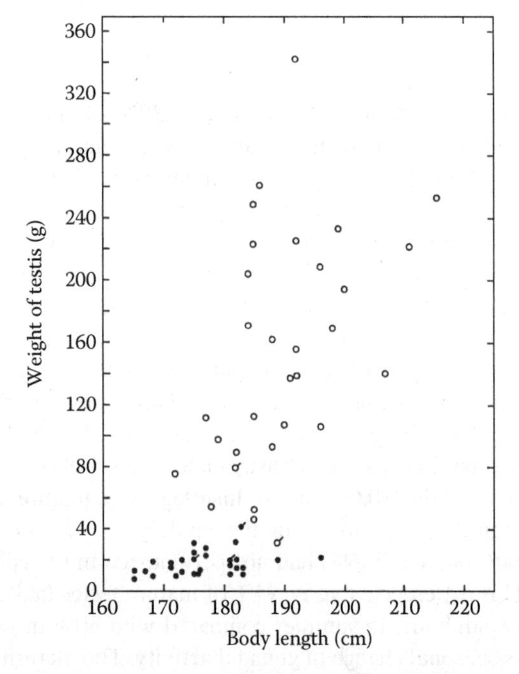

FIGURE 9.16 Weight of single testis (g) with epididymis removed, histological maturity, and body length (cm) of *dalli*-type Dall's porpoises caught by hand harpoon in waters south of the western Aleutian Islands area from August to September 1982. Open circle, mature; open circle with bar, late maturing; closed circle with bar, early maturing; closed circle, immature. (From Kasuya, T. and Jones, L.L., *Sci. Rep. Whales Res. Inst.* (Tokyo) 35, 107, 1984.)

weighed the testis with epididymis attached, so their analyses were not directly comparable with those of other studies.

The diameter of seminiferous tubules of mature males was approximately in a range of 68–120 μm. The relationship between the diameter (Y, μm) and testicular weight (X, g) was shown by a single equation for the entire range of the data for *truei*-types off Sanriku in winter (Kasuya 1978):

$$Y = 33.79\ X^{0.284},\ 5 < X < 100$$

However, the corresponding figure for *dalli*-types seemed to be almost doubled, 125–210 μm, in summer off the western Aleutian Islands area (Ferrero and Walker 1999) (Figure 9.17). The testis-weight data of these two studies are not directly comparable because Kasuya (1978) excluded the epididymis while Ferrero and Walker (1999) weighed the testis together with the epididymis.

Kasuya (1978) obtained 25.77 g and 32.85 g as the weight of a single testis at the attainment of the maturing stage and the mature stage, respectively. However, although the maturing males were in general located intermediate between immature and mature individuals in terms of testicular weight and tubule diameter, the overlap was extremely broad with the two other maturity categories of immature and mature. This casts doubt on his classification of the

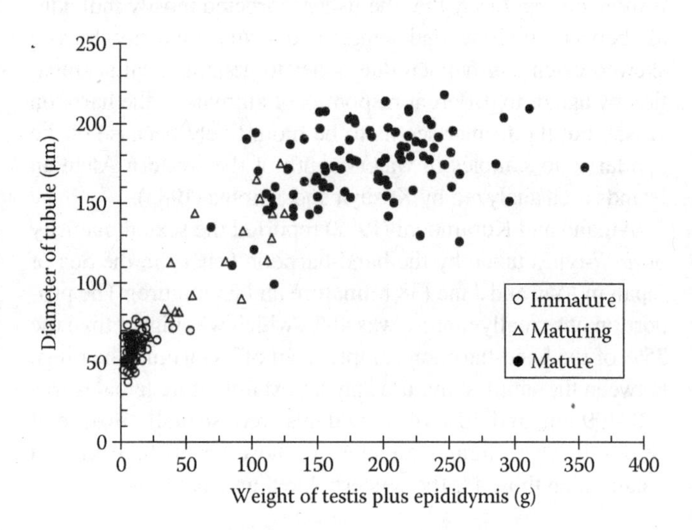

FIGURE 9.17 Histological maturity of testis, diameter of seminiferous tubules, and weight of single testis weighed together with the epididymis. Closed circle, mature; triangle, maturing; open circle, immature. (From Figure 16 of Ferrero, R.C. and Walker, W.A., *Marine Mammal Sci.*, 15(2), 273, 1999.)

stage "maturing," probably due to difficulty in classifying the maturity of reproductively inactive males in winter. The average of the two figures, 29.3 g, did actually separate mature and immature individuals with almost no exception, which suggests that the weight of 29.3 g can be an adequate criterion for classifying winter testes into the two categories of mature and immature. The range in body length of the smallest mature male and the largest immature male was 180–215 cm, and 50% of males were sexually mature at 195.9 cm. The minimum age of a mature male was 3.5 years, and the maximum age of an immature male was 15.5 years. Although the former figure is not unreasonable, the latter should be questioned because of the reason mentioned earlier (problems in aging).

Kasuya and Jones (1984) applied the maturity criteria of testicular histology established for short-finned pilot whales (see Section 12.4.3.3) to the *dalli*-type porpoises hand-harpooned in waters south of the area sampled by Ferrero and Walker (1999). They observed that "mature" and "immature" individuals were separated at around 40 g testis weight and that "early maturing" and "late maturing" males were found near that weight. They concluded that a threshold of 40 g was adequate for separating males into the two large categories of mature and immature (Figure 9.16). This threshold weight was about 10 g greater than the corresponding figure of 29.3 g, for a winter sample of *truei*-types off Sanriku. Kasuya and Shiraga (1985), combining the *dalli*-type samples of 1982 and 1983 from the same area and applying the 40 g criterion to the total of 118 males, found 168–199 cm as the range in length of the smallest mature and largest immature males. They estimated that 50% of males were sexually mature at around 184 cm. The same sample gave an age range of sexual maturation of 2–7 years and age at 50% mature between 4 and 5 years (Table 9.8).

Ferrero and Walker (1999) analyzed the *dalli*-type sample obtained in summer in the western Aleutian Islands area, 46°N–53°N and 170°E–175°E, which was located north of the area sampled by Kasuya and Shiraga (1985). By examining testicular tissue taken from the center at the midlength of the testis, they classified the degree of maturity into three categories: "mature" having spermatozoa, "maturing" having spermatids or spermatocytes but not spermatozoa, and "immature" having none of the three cell types, where the proportion of tubules undergoing spermatogenesis was not taken into account. It should be noted that they weighed the testis together with the epididymis, which I call gonad weight. Tubule diameter correlated positively with gonad weight before maturation, but the correlation was lost in many mature individuals. Tubule diameter ranged from 30 to 80 μm in immature testes (most of the gonads weighed less than 50 g) and increased with gonad weight until about 150 g (Figure 9.17). Gonads of mature males weighed between 70 and 350 g, and tubule diameter was in the range of 120–220 μm in mature males with gonads over 150 g. This diameter is almost twice that (65–120 μm) of mature *truei*-type males in the winter off Sanriku. Mention has been made previously of a similar seasonal change in testis weight.

Ferrero and Walker (1999) analyzed the relationship between body length and sexual maturity based on the large sample obtained from the drift-net fishery. Their figures indicate that the body-length range for smallest mature male and largest immature male was approximately 166–194 cm, which was similar to the corresponding figure of 168–199 cm for the sample from further south, but body length at 50% mature, 179.7 cm, was about 4 cm smaller than for the southern sample (Table 9.8). This disagreement probably could be attributed to selection by hand-harpoon sampling, where mature individuals are less likely to be attracted to the bow wave and are underrepresented compared to immature individuals of the same body length. The age range of the youngest mature male and oldest immature male was 3–6 years, age of 50% mature was calculated at 4.5 years by fitting a sigmoid curve, and average age at sexual maturation was 5.0 years by the method of DeMaster (1978) (Table 9.8).

Amano and Kuramochi (1992) analyzed the sexual maturity of male *dalli*-types taken by the hand-harpoon fishery in the Sea of Japan in May and June, or shortly before the mating season that was believed to be in mid-June to late July (Sections 9.4.4 and 9.4.5), using a method almost identical with that of Kasuya and Jones (1984) and Kasuya and Shiraga (1985). They obtained 40 g as the average weight of the testis at attainment of sexual maturity. This figure was about 10 g greater than the corresponding figure for off the Sanriku coast from January to March but similar to the figure for the offshore population during the period of parturition and mating. They found that immature and mature individuals coexisted in a body-length range of 180–209 cm and half of the individuals were sexually mature at 192 cm, which was about 10 cm greater than in the western Aleutian Islands area and to the south.

9.4.9 Longevity and Age Composition

Kasuya (1978) believed that sexually mature individuals and juveniles below 2 years of age were underrepresented in the hand-harpoon fishery off Sanriku in the winter. This could have been due to differential response to fishing vessels by growth stages or to segregation by growth and reproductive stage. Adult individuals and mother–calf pairs tended to avoid vessels and were found segregated in the western Aleutian Islands area, including in the U.S. EEZ; chase of a maximum of 48 min by a vessel at 10 knots failed to close with any for harpooning. However, behavior was quite different in waters further south, where many approached the bow and offered a chance for sampling with hand harpoon (Kasuya and Jones 1984; Kasuya and Ogi 1987).

The Japanese hand-harpoon fishermen usually give up chasing such shy porpoises and look for other targets, but some of the fishermen recently started using a different strategy, in which they equipped their vessel with powerful engines and continued to chase mother–calf pairs until they were exhausted and then harpooned them (Amano et al. 1998c, in Japanese). This method was started by some fishermen who operated in the Okhotsk Sea in the summer season. In 1995, Amano et al. (1998c, in Japanese) examined the catch of 24 hand-harpoon vessels and recorded 57 lactating females (17.5%) out of 325 sexually mature females, but the rates for 3 vessels included in the 57 were in a range of 30%–60% or 25 out of 54 adult females lactating. Such an operation will inflict greater damage on the population.

The Japanese salmon drift-net fishery operated in the western Aleutian Islands area, which was a breeding ground of the population and was inhabited by *dalli*-type porpoises that were not attracted to the bow. It is unclear if the porpoises did not notice the presence of the net and got entangled or were attracted to prey animals in the net and therefore got entangled. Newby (1982) was the first to analyze Dall's porpoises caught in the salmon drift nets. Ferrero and Walker (1999) carried out similar analyses using larger samples. Their body-length composition data are shown in Figures 9.13 and 9.14 and age composition in Figures 9.18 and 9.19. The abundance of zero-age individuals is quite different between these two studies as shown in Figures 9.18 and 9.19, presumably reflecting a difference in their sampling procedure or a difference in overlap of the sampling period with the parturition season. However, the compositions of ages 1 year and above are similar between them. The frequency decreased with increasing age until the highest age of 15 years. Even allowing some possible uncertainty in age determination of old individuals, the longevity of this species does not exceed 20 years.

Porpoises in the western Aleutian Islands area are believed to reach sexual maturity at the mean age of 3.8 years in females and 5.0 years in males (see Sections 9.4.7 and 9.4.8). The sex ratio was almost 1:1 for ages below 3 years, but the male proportion declined to about 1:2 or less at ages 4 and above, while longevity seemed to be similar for the two sexes at around 14–15 years (Figure 9.18). This general trend is also observed in the age composition reported by Newby (dotted line in Figure 9.19). This was different from the pattern of longevity observed in short-finned pilot whales off Japan, which exhibited a rapid decline of the male proportion after the age

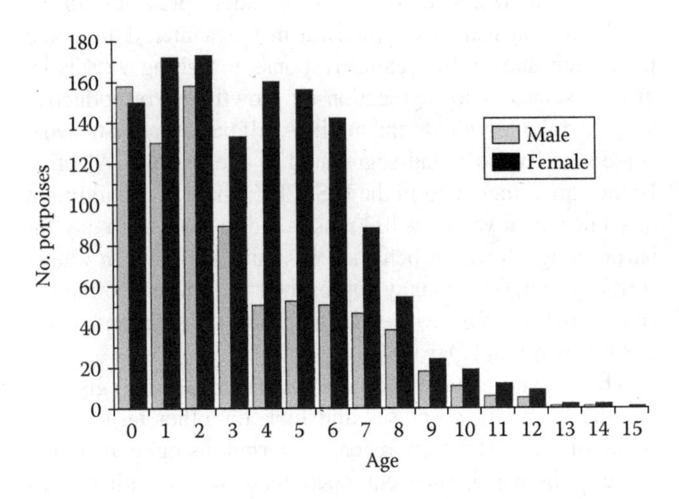

FIGURE 9.18 Age composition of *dalli*-type Dall's porpoises incidentally killed by Japanese salmon drift-net fishery in the western Aleutian Islands area from June to July, 1981–1987. Black column represents females and gray column males. (From Figure 5 of Ferrero, R.C. and Walker, W.A., *Marine Mammal Sci.*, 15(2), 273, 1999.)

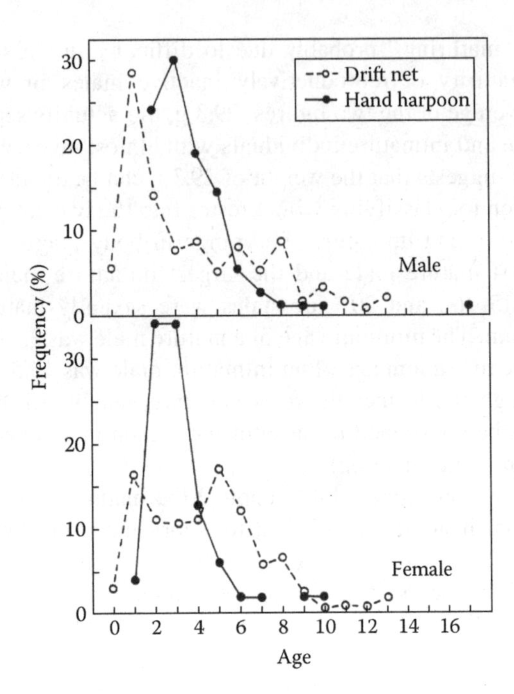

FIGURE 9.19 Age composition of *dalli*-type Dall's porpoises compared between drift-net sample (open circles and dotted line) incidentally killed in the Japanese salmon drift-net fishery in western Aleutian Islands area and aged by Newby (1982) and hand-harpoon sample (closed circles and solid line) taken in waters south of the salmon fishery from August to September in 1982–1983 and aged by Kasuya. These two sample sets are the same as those used in Figure 9.14. (From Kasuya, T. and Shiraga, S., *Sci. Rep. Whales Res. Inst.*, 36, 139, 1985.)

of sexual maturation, accompanied by a great difference in observed longevity between sexes, that is, females lived longer. It follows that the earlier mentioned age structure cannot be attributed to higher male mortality after sexual maturity as occurs in the short-finned pilot whales. It is reasonable to assume that adult males were less vulnerable to the salmon drift-net fishery than adult females and juveniles below the age of 3 years. One of the explanations for this is segregation of males to outside the drift-net ground.

Kasuya and Shiraga (1985) reported the age composition of a hand-harpoon sample obtained in latitudes 42°N–47°N, which is to the south of the salmon drift-net ground off the western Aleutian Islands. Their sample contained far more males, both adult and immature; the adult sex ratio was 55 males to 29 females and immature sex ratio 63 males to 20 females. Figure 9.19 compares age compositions of this hand-harpoon sample with those obtained from the salmon drift-net fishery (Newby 1982). Newby and myself cross-checked our age readings, so the two age readers in Figure 9.19 are expected to have followed similar principles in interpreting tooth layers and are considered comparable. The two drift-net samples are similar in age composition except for the frequency of zero-age animals as mentioned earlier (see Figures 9.18 and 9.19), but the hand-harpoon and drift-net samples show quite different age compositions (see Figure 9.19). Age classes with lower frequency in the northern drift-net sample are abundant in the southern hand-harpoon sample. As mentioned earlier, this allows us to deduce that young individuals of both sexes around the age of sexual maturity tend to segregate in southern waters (note that males mature at a greater age than females).

Similar sex-biased distribution or migration patterns have been reported for humpback whales. Recent nonlethal studies (Brown et al. 1995; Smith et al. 1999) in the breeding ground and on the migration route along the Australian coast identified a high male ratio (2.4 males/female). This was similar to 2.1:1 obtained from the past whaling data in lower latitudes but differed from 0.74:1 obtained from commercial catches in the Antarctic. Brown et al. (1995) deduced from these data that about half the female humpback whales remained in the feeding ground in the Antarctic during the season of parturition and mating. Females of the species breed every 2–3 years, so those females who are not expecting estrus or parturition remain in the higher latitudes for more feeding and recovery for subsequent conception. Such an explanation cannot be applied to the segregation in Dall's porpoise population, because they feed both in the summer breeding ground and wintering grounds.

9.4.10 Body Length and Growth Curve

Kasuya (1978), Newby (1982), Kasuya and Shiraga (1985), and Ferrero and Walker (1999) presented mean growth curves of Dall's porpoises, using materials collected in a relatively short period and assuming that the growth pattern did not change for the years covered by the animals' ages, that is, at least for the previous 15 years in the case of Dall's porpoises. Violation of this assumption of a stable growth pattern would make the obtained mean growth curve unreliable (Section 15.3.2.2). Age determination is another key element in the growth analysis, but it has been my experience that the teeth of Dall's porpoise are the most unsuitable for age determination among those of the odontocetes I have studied. Among the earlier four growth analyses, that of Kasuya (1978) should be considered of less value due to greater uncertainty in age determination, particularly for individuals above the average age at sexual maturation.

Ferrero and Walker (1999) determined physical maturity for 692 Dall's porpoises and made a great contribution to understanding their growth. Physical maturity is a sign of cessation of growth in body length and is characterized by fusion of all the epiphyses to the vertebral centra. The state of maturity can be identified by examining the posterior thoracic and anterior lumbar vertebrae where fusion occurs last. On the vertebrae of physically immature individuals, there is a layer of cartilage between the centrum and the epiphysis; this is replaced by bone tissue in vertebrae that have ceased increasing in length. Ferrero and Walker (1999) showed that the minimum body length of physically mature males was 182 cm and that of the largest immature male 220 cm. The corresponding range for females was 180–205 cm. Mean body length of physically mature males was 198.1 cm (n = 83, SE = 0.8566) and that of females 189.7 cm (n = 164, SE = 0.4002), which showed that males grew about 8–9 cm longer than females. Ferrero and Walker (1999) did not find sexual difference in the age at attainment of physical maturity. The youngest physically mature individual was aged at 5 years and the oldest immature individual at 8 years. The age when 50% of individuals were physically mature was 7.16 years in males with 95% confidence interval of 5.7–8.6 years and 7.24 in females with a confidence interval of 6.3–8.1 years. The interval between average age at sexual maturity and that at physical maturity was about 2–3 years in both sexes.

Ferrero and Walker (1999) fitted Laird–Gompertz models to age-at-body-length data separately for two maturity stages: (1) sexually immature individuals (males <6 years, females <8), forced to pass through average birth length of 101 cm, and (2) sexually mature individuals (males >3 years, females >2 years) with the earlier mentioned body length at attainment of physical maturity as the asymptotic length. The mean growth for each sex is shown by two equations in Figure 9.20. I am unable to understand the value of fitting growth curve separately for two maturity stages of overlapped age ranges. The equations have little meaning for the overlapped age range (males, 3–6 years; females, 2–8 years). However, the equations reasonably suggest that neonates, which are born at a mean length of 101 cm (with a range of 86–118 cm), reach 151 cm in 1 year and 161 cm in 2 years. The average

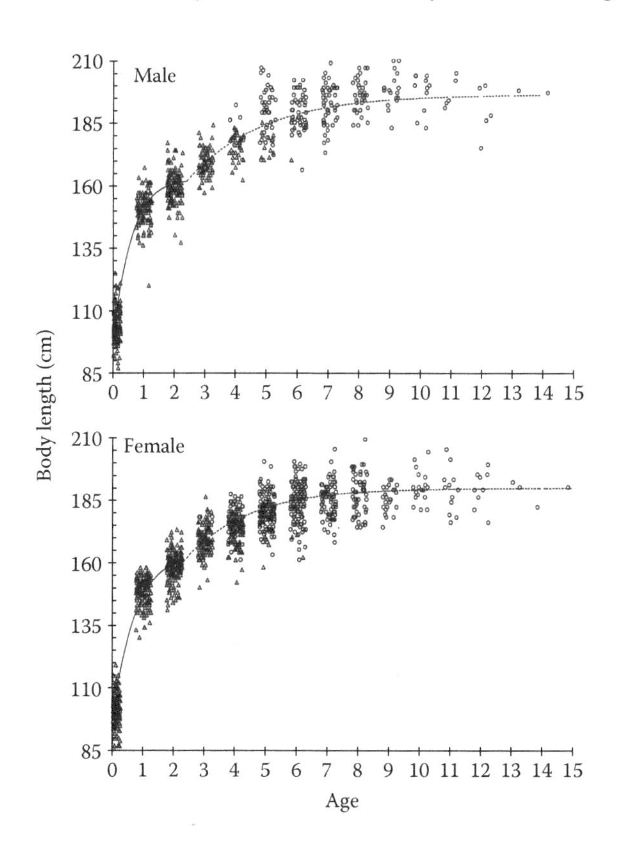

FIGURE 9.20 Mean growth curve of *dalli*-type Dall's porpoises incidentally killed in the Japanese salmon drift-net fishery in the western Aleutian Islands area. Solid curve fitted to immature individuals (triangles) and dotted curve to mature individuals (circles). Top panel males, bottom females. (From Figure 8 of Ferrero, R.C. and Walker, W.A., *Marine Mammal Sci.*, 15(2), 273, 1999.)

growth increment in the first year after birth is about 50% (see Sections 9.4.1 and 9.4.2). Ferrero and Walker (1999) also calculated the relationships between body weight and age, which suggest that a calf born with an average weight of 17 kg attains 158 kg (males) or 123 kg (females) at the time of physical maturation. Plots of body weight on body length suggest that the correlation is almost lost at around the age of 8 years. This should agree with the age when all the individuals have attained physical maturity.

Dall's porpoise males exceed females by 4.4% in average body length and 28% in average weight. Behind the greater weight of males relative to length is sexual dimorphism in body shape (Jefferson 1989). First, volume of dorsal muscle is greater in the thoracic region, which looks vertically swollen when mature males are seen from the side. A second difference is the high tail peduncle of males. These differences contribute to weight difference between the sexes. Other sexual dimorphisms that will not affect weight are shape of the dorsal fin and the tail flukes. The tip of the dorsal fin is located farther forward on adult males than on females, giving an impression that the fin is tilted anteriorly. The posterior margins of the tail flukes expand posteriorly on adult males. The posterior margins of the tail flukes are located anterior to a line connecting the tips of the flukes on juveniles, are nearly on a line with the tips of the flukes on adult females, and are located posterior to the line on adult males. Jefferson (1988) concluded that such sexual dimorphism could function in fighting or deciding rank between males. The large body muscle might have such a function, but further information on behavior is needed to evaluate such a hypothesis.

Koga (1969, in Japanese) obtained the following equation between body weight (W, kg) and body length (L, cm) for *dalli*-types of both sexes taken by the salmon drift-net fishery off the western Aleutian Islands area:

$$\text{Log W} = 2.441\ \text{Log}L - 3.435$$

This equation can also be expressed as $W = 0.000367L^{2.441}$.

9.5 FEEDING HABITS

9.5.1 Food Consumption

Dall's porpoises can make a dash at 55 km/h (Jefferson 1988) and dive to a maximum of 180 m (Berta and Sumich 1999). Such a high activity level and the offshore habitat could have been behind the difficulty of maintaining the species in captivity. In the 1960s, a facility of the U.S. Navy succeeded in keeping Dall's porpoises in captivity for over 2 years, and physiological information thus obtained from 5 porpoises was compared with that for bottlenose dolphins (n = 5), a coastal species, and Pacific white-sided dolphins (5), which inhabit intermediate waters (Ridgway and Johnston 1966). The authors found that the blood volume of Dall's porpoises (143 mL/kg of body weight) was higher than for Pacific white-sided dolphins (108 mL/kg) and bottlenose dolphins (71 mL/kg) and that other parameters followed a similar order, that is, hematocrit values of 57%, 53%, and 45%; hemoglobin density (g/100 mL) 20.3, 17.0, and 14.4; and heart weight as % of body weight 1.31, 0.85, and 0.54. They concluded that these physiological parameters suggest that the Dall's porpoise has the highest physical ability among the three species compared, as evidenced by observations in the ocean. Ridgway and Johnston (1966) further reinforced this conclusion from the daily food requirement to maintain body weight, 14, 8.5, and 6 kg, with the highest for the Dall's porpoise and the lowest for the bottlenose dolphin, and from blubber thickness measured at the same position on the body that had a reverse correlation with activity rank, that is, 1 cm on the Dall's porpoise, 2 cm on the Pacific white-sided dolphin, and 3 cm on the bottlenose dolphin. Generally speaking, Dall's porpoises are the smallest and bottlenose dolphins the largest among the three species. Dall's porpoises seem to have the highest heat production and depend least on the insulation ability of blubber in their thermoregulation. The authors noted high swimming activity of Dall's porpoises during capture operations, as evidenced by "rooster-tail" swimming.

Sergeant (1969) addressed the relationship between daily food consumption and heart weight. He noted that a Dall's porpoise weighing 120 kg (2 m in body length) consumed as much as 15 kg of mackerel daily, or 11.35% of body weight (12.5% when calculated from data in the original text of Sergeant 1969), which was much greater than the corresponding figures for Pacific white-sided dolphins (7.8%) and bottlenose dolphins (4.2%). He found a correlation between daily food intake (kg/body weight) and heart weight/body weight ratio and estimated daily food intakes for other species with known heart weight and body weight. Later, Ridgway and Kohn (1995) with additional data obtained the following relationship between body weight (B, kg) and heart weight (H, kg) of adult Dall's porpoises:

$$\log H = -1.614 + 0.808 \log B$$

They calculated similar equations for adult individuals of the genus *Lagenorhynchus* and bottlenose dolphins and obtained −1.729 and −1.927, respectively, as the intercept. The slope of 0.808 remained the same for the three equations. Their equations gave the heart weight of individuals with an assumed body weight of 190 kg at 1.69 kg for the Dall's porpoise, 1.30 kg for the Pacific white-sided dolphin, and 0.82 kg for the bottlenose dolphin.

Lockyer (1981a) obtained the following single relationship between body weight (B, kg) and heart weight (H, kg) for all species of cetaceans from the harbor porpoise weighing tens of kg to the blue whale weighting over 100 tons:

$$H = 0.00588B^{0.984}$$

This relationship can also be expressed as $\log H = -2.230 + 0.981 \log B$. This expresses the general trend among cetacean species but can be quite different from equations that represent the relationship within smaller taxonomic groups or for single species.

Ohizumi and Miyazaki (1998) questioned the validity of the food consumption rate estimated by Sergeant (1969). They observed for *dalli*-type Dall's porpoises taken in May in the northern Sea of Japan off the west coast of Hokkaido that the maximum contents of the forestomach was 1.68 kg and that the contents of the forestomach tended to decline with time from early morning. They also referred to information for other species suggesting that it took 8 h for a full forestomach to be emptied and calculated that if the stomach was filled three times a day the daily food intake could be only 5.04% of body weight. They concluded that the estimated daily food consumption of 15 kg or 11.35% of body weight reported by Sergeant (1969) represented a case of overfeeding in the captive environment. I would question whether a Dall's porpoise could continue to eat twice as much in captivity as in the wild for over 2 years as reported by Ridgway and Johnston (1966). It should be noted that some Dall's porpoises taken by the salmon drift-net fishery had an amount of undigested soft remains of prey in their stomach that could have represented 4 kg in fresh condition (see Kuramochi et al. 1991).

It is true that there are difficulties in estimating the food requirements of small cetaceans using individuals in captivity, because their activity is different from that of wild animals and change in body weight must be monitored and adjusted carefully. However, if we accept a daily food consumption of 5.04% of body weight proposed for Dall's porpoises by Ohizumi and Miyazaki (1998) and reject the observed daily food consumption of 12.5%, then we also have to reject the corresponding figures 7.8% for Pacific white-sided dolphins and 4.2% for the bottlenose dolphin. However, the last figure was supported by another independent experiment on 11 bottlenose dolphins in a Japanese aquarium, which concluded that the daily food intake of 3.5%–6.1% of body weight was necessary to maintain body weight for 21 months (Section 11.5.2.1). Further, if we accept that Dall's porpoises examined by Ohizumi and Miyazaki (1998) fed less in daytime, which is likely in certain circumstances, then it would be difficult to fill their stomach three times a day at an interval of 8 h. A key problem in their calculation is the assumption that Dall's porpoises fill the forestomach when it is emptied by digestion.

The stomach of dolphins and porpoises has four compartments: (1) forestomach, (2) main stomach, (3) connecting channel, and (4) pyloric stomach. The connecting channel may be divided into two compartments in some species (Harrison et al. 1970). The main function of the forestomach, the largest of the four compartments and lacking in digestive glands, is temporary storage of ingested food, although there is some progress in digestion by powerful mechanical activity and digestive juices migrating up from the main stomach. Food stored in the forestomach gradually moves to the main stomach, the second largest compartment and equipped with digestive glands, then to the connecting channel and pyloric stomach, and finally to the duodenum for subsequent absorption. The feeding environment for dolphins and porpoises is quite different from that of cows feeding in a meadow. For Dall's porpoises, the chance to meet prey animals is less predictable, so they will make an effort to ingest as much food as possible anytime there is an opportunity and there is space to store ingested food in the forestomach. The main stomach and subsequent compartments continue digestion as materials are received from the forestomach. With such feeding habits, the amount of food found in the forestomach at any particular time of the day does not have value for estimating daily food consumption in cetaceans.

9.5.2 Prey

Dall's porpoises consume various epipelagic and mesopelagic squids, fishes, and crustaceans. The species composition apparently varies seasonally and geographically, which is considered a reflection of both availability and food preference. Morejohn (1979) analyzed the stomach contents of carcasses stranded year-round in Monterey Bay. Numerous species of squids and fishes were present in the stomachs throughout the year, but seasonal trends were not clear due to limited sample size. He concluded that the fish species preferred were hake, herring, juvenile rockfish, and anchovy, and the preferred squid species were *Doryteuthis* (then *Loligo*) *opalescens* and *Gonatus* sp. A larger sample would perhaps have identified crustaceans year-round. He reported that the Dall's porpoise occurred year-round in waters outside the 200 m isobath, but in winter the range expanded to waters inside the 100 m isobath, which he suggested reflected seasonal distribution of prey.

Mizue et al. (1966, in Japanese) and Koga (1969, in Japanese) examined the stomach contents of Dall's porpoises killed in the salmon drift-net fishery in the Bering Sea and noted that over half of the individuals had only squid in the stomach and the remaining individuals had squid and fish, squid and crustaceans, or squid, crustaceans, and fish. From these results, they concluded that squid are the main nutritional source for Dall's porpoises (and of salmon) in the region. Mizue et al. (1966) stressed that there was only one Dall's porpoise found with salmon in the stomach even though their sample was obtained in the salmon drift-net fishery.

Mizue et al. (1966) examined food preference by reproductive status of Dall's porpoises killed in the salmon drift-net fishery and found that 53% of 17 pregnant females (with fetuses over 70 cm) had fish in their stomach but only 27% of other individuals had fish in the stomach. From this observation, they concluded that pregnant females prefer fish because they require more nutrition. They did not clarify the sex or reproductive status of the "other individuals," which makes evaluation of their conclusion difficult. Ohizumi et al. (2003) offered more information relating to this question; they found that sexually mature Dall's porpoises of either sex depended significantly more on myctophids than immature individuals of the same sex caught in the salmon drift-net fishery in the Bering Sea. A difference in food preference between sexes was also significant but of less magnitude. The possible factors behind such differences in food preference are still to be investigated. Weaning short-finned pilot whales were found with smaller squid beaks in their stomach than those found in adults in the same school (Section 12.4.4.3).

Lockyer (1981a,b, 1984) indicated that nutritional requirements are greater during lactation than in pregnancy.

Kuramochi et al. (1991) compared the contents of the forestomach between the Bering Sea and offshore western North Pacific based on 32 Dall's porpoises hand-harpooned during research cruises in May through September, 1984 and 1985. Thirty-one had both fish and squid in the forestomach, but the proportions were different between the two geographical areas, north and south of the U.S. EEZ around the Aleutian Islands. The Bering Sea sample had almost equal numbers of squid and fish (45:55), while fish predominated in the Pacific with a ratio of 3 squid to 97 fish. Ohizumi et al. (2003) also found a similar result, that is, 80%–94% of the prey was fish in the offshore western North Pacific west of 135°W while 69% was squid in the Bering Sea. These results are useful in judging geographical difference in prey taken, but the relative degree of nutritional contribution between fish and squid is difficult to evaluate because we do not know how squid beaks and fish otoliths differ in time of passage through the stomach.

The average number of squid (represented by beaks and undigested bodies) found in a single forestomach was 63 in the Bering Sea and 27 in the offshore North Pacific, and squid beaks represented 77% of the Bering Sea total and 91% of the Pacific total (Kuramochi et al. 1991). The squid belonged to 6 families, the majority (89%–97% by the regions) to the Gonatidae. The remaining families were Enoploteuthidae, Cranchiidae, Chiroteuthidae, Onychoteuthidae, and Histiotheuthidae, which comprised only 3%–11% in the total number of individuals identified and were considered almost negligible in the overall diet. The most important family was the Gonatidae; seven species in three genera were identified in the stomachs, in decreasing order of abundance *Gonatopsis borealis*, *Gonatus onyx*, *G. pyros*, *G. berryi*, *G. middendorffi*, *Berryteuthis anonychus*, and *B. magister*. Thus, *G. borealis* was the most important species, comprising 23% (Pacific) or 63% (Bering Sea) of squid identified to species. Fiscus and Jones later (1999) reported cephalopod composition in the stomachs of Dall's porpoises in the Bering Sea and western North Pacific. The sample was obtained from the Japanese salmon drift-net fishery in the area indicated in Figure 9.11 and the Japanese hand-harpoon cruise for scientific purpose in 1982 (Figure 9.21). The results were similar to those reported by Kuramochi et al. (1991) in both the most dominant family being Gonatidae and species composition within the family. Fiscus and Jones (1999) did not find significant difference in the composition of major squid species between the Pacific area and the Bering Sea plus Aleutian Islands area.

Kuramochi et al. (1991) estimated the weight of squid consumed by Dall's porpoises. Using the length of lower beaks, they calculated the mantle length of *G. borealis* consumed by the porpoises to be in a range of 44–405 mm, with 89% in 50–150 mm. The total weight of these squid was estimated at 21–6,460 g/individual in the Pacific and 21–11,413 g/individual in the Bering Sea. They also studied the number and

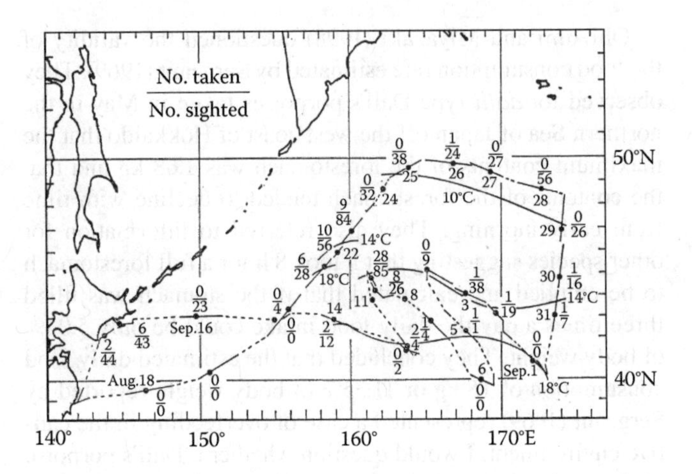

FIGURE 9.21 Results of a research cruise of *Hoyo-maru No. 12* from August to September in 1982. Solid line indicates cruise track with sighting effort, circles noon positions, top numerals number of Dall's porpoises sampled with hand harpoon, and bottom numerals those sighted. Sea surface temperature observed during the cruise was also indicated. (From Kasuya, T. and Jones, L.L., *Sci. Rep. Whales Res. Inst.* (Tokyo) 35, 107, 1984.)

weight of undigested soft parts of squid present in the stomach of 17 Dall's porpoises. The number of squid represented by the soft tissue ranged between 1 and 111 squid per porpoise, and their estimated weight at the time of ingestion ranged between 5 and 4387 g per porpoise. Some of these porpoises could have had over 4 kg of squid, or about 2.9% of body weight in their stomachs at some time before capture (150 kg is assumed for body weight), but we do not know how long it remained in the stomach or how many feeding incidents were represented. It should also be noted that Dall's porpoises often vomit their stomach contents when killed by hand harpoon.

U.S. scientists collected the stomach contents of Dall's porpoises killed in the Japanese salmon drift-net fishery beginning in 1978 and analyzed them subsequently. In 1982, Japan and U.S. scientists planned a cruise for sighting and lethal sampling of the species outside the drift-net operation area, but at a late stage, the United States withdrew from the lethal sampling program (see Section 9.3.3). The stomach contents collected during the cruise from August to September 1982 were transferred to U.S. scientists following an earlier agreement, and it was reported to the INPFC meeting that the cephalopod beaks were analyzed by Cliff Fiscus and fish otoliths by Thomas Crawford (Jones et al. 1985b). The analysis of cephalopod beaks was published by Fiscus and Jones (1999), which is cited previously. However, the results of the study of fish otoliths remain unpublished. The only results of the study available to me are in a table that was handed to me by Crawford's superior at an IWC Scientific Committee meeting and is now deposited in the Zoological Section of the National Science Museum in Japan. It contains the only data on fish species consumed by Dall's porpoises in the offshore North Pacific south of the western Aleutian Islands area.

As Crawford had already retired from his position and I was unable to contact him, I will summarize his results in the following to prevent the valuable record from being lost. His materials were collected by myself and my colleagues on board the *Hoyo-maru No. 12* in August and September 1982, mostly in waters south of the western Aleutian Islands, 40°N–50°N, 158°E–174°E (Figure 9.21), but could have included samples from three individuals killed in the nearshore waters of Hokkaido during the cruise (Fig. 1 in Kasuya and Shiraga 1985). Dall's porpoises in offshore waters are known to consume more fish than cephalopods (see the preceding text in this section). An average of 760 fish otoliths were found in the 27 stomach samples (with a range of 4–2834). As these figures did not apparently distinguish between the left- and right-side otoliths, the number of fish represented by the otoliths could be smaller. The number of fish species in the stomachs was between 1 and 13 (7.3 on average). A total of 19,329 fish otoliths were identified to family: in decreasing order Myctophidae (80.6%), Gadidae (4.8%), Microstomatidae (3.3%), Notosudidae (0.8%), others (5%), and unidentified (5.8%). The top three families coincide with the results of Ohizumi et al. (2003) for Dall's porpoises in northern waters.

Wilke et al. (1953) were probably the first to describe the stomach contents of Dall's porpoises from off the Pacific coast of northern Japan. They examined 54 *truei*-types and 4 *dalli*-types caught in latitudes between 38°15′N and 42°N during March–May, 1949 and 1950. The southern part of this range could have been inhabited mostly by *truei*-types and the northern part by *dalli*-types. They identified the prey species using undigested food remains; they did not use squid beaks and fish otoliths for the analyses. The stomach contents of the 54 *truei*-types were represented mostly by Myctophidae (70% of total prey animals), small squid (18%), and one gadid. Myctophidae were not found in the stomachs of the 4 *dalli*-types, but they had squid (98% of total prey animals) and some gadids. Details on the location of capture of these animals are not available. The apparently distinct difference in prey items between *truei*- and *dalli*-types could be due to a latitudinal difference and small sample size for the latter. Wilke et al. (1953) also reported the stomach contents of striped dolphins that resembled those of the *truei*-types and were represented mostly by Myctophidae and some squid. Striped dolphins usually occur south of the Dall's porpoise range.

Walker (1996) reported the stomach contents of Dall's porpoises in the Okhotsk Sea based on samples obtained from the hand-harpoon fishery operated by a small-type whaler *Yasu-maru No.1* in July and August 1988. Out of 88 *dalli*-types examined, he obtained the stomach contents for 73. He identified fish species using left and right otoliths separately and used the larger figure for his analysis. In the same way, lower and upper beaks were separately identified and the greater figure was used. He found fish remains in all the 73 stomachs and squid remains in 54 (74%). A total of 2916 individual fish were represented by 13 species in 9 families, which were in decreasing order Clupeidae (90.1%, represented by the single species the Japanese sardine *Sardinops melanostictus*), Gadidae (7.4%, represented by the single species Alaska pollock *Theragra chalcogramma*), Zoarcidae (3 species identified), Engraulidae (the Japanese anchovy *Engraulis japonicus*), Moridae, Myctophidae, and the three families Microstomatidae, Ammodytidae, and Hexagrammidae, each represented by a single fish. Squid in the 54 porpoises comprised 733 individuals of 6 species in 3 families. The most abundant was Gonatidae (96.5% in number of individuals), of which the schoolmaster gonate squid *Berryteuthis magister* was the most abundant (86.9% of the total cephalopods). Thus, the major food items in the Okhotsk Sea in the summer of 1988, in terms of number of individuals, were the three species Japanese sardine, schoolmaster gonate squid, and Alaska pollock. Walker (1996) stated that the order remained the same on a nutritional basis. Time of capture and degree of digestion suggested that Japanese sardines were eaten during the day, while the schoolmaster gonate squid were consumed at night. Ohizumi (2008, in Japanese) reviewed recent information on the food habit of Dall's porpoises.

At an early stage of the study of food habits of Dall's porpoises, there was a general understanding that they were nocturnal feeders preying on Myctophidae and cephalopods that surface at night. Later, it became known that they were likely to change feeding behavior to meet the habits of prey species, for example, by feeding on Japanese sardines in the Okhotsk Sea. Amano et al. (1998d) attempted to confirm this feeding flexibility based on direct observation in the ocean. They assumed that Dall's porpoises were feeding if they frequently changed swimming direction while swimming at the surface and recorded this behavior on board the research vessel *Hoyo-maru No. 12* in the offshore North Pacific in 1986 and 1987 and on the hand-harpoon vessel *Man-ei-maru No. 5* in the Okhotsk Sea in 1988. In the offshore North Pacific, the feeding behavior was most frequent around sunrise (no observations available in night) and decreased with time during the day. On the other hand, in the Okhotsk Sea they encountered such feeding behavior at high frequency during the day.

Ohizumi et al. (2000) reported yearly change in food items of *dalli*-type Dall's porpoises taken by the hand-harpoon fishery off Hokkaido in the Okhotsk Sea and the Sea of Japan. The Japanese sardine was their major prey item in both the seas in the 1980s, but it almost disappeared from their stomachs in the 1990s and was replaced by Alaska pollock in the Sea of Japan and by Japanese anchovy and schoolmaster gonate squid in the Okhotsk Sea. Ohizumi et al. (2000) believed that this change was a response of Dall's porpoises to the decline of the sardine population around Japan and that Dall's porpoises were obliged to switch their prey items from epipelagic species to those in deeper waters.

We learned that Dall's porpoises probably change their prey with growth and with change in reproductive status. It is also evident that Dall's porpoises have the ability to switch prey items responding to change in abundance of prey species and to modify their feeding behavior to match the behavior of the prey species. However, it will require further study on the abundance of the earlier mentioned prey species and feeding energetics of Dall's porpoises before evaluating the effects on their physiology of the change in feeding environment after the decline of the Japanese sardine.

9.6 SOCIAL STRUCTURE

9.6.1 School Size

School size in cetaceans not only reflects social structure but is also influenced by environmental factors such as size and density of schools of prey species and chances of meeting predators. According to data obtained during the cruise of the *Hoyo-maru No. 12* in the area south of the western Aleutian Islands form August to September 1982 (Figure 9.21), Dall's porpoises sighted in waters between 44°N and 50°N and between 163°E and 174°E, which was close to the Aleutian Islands rarely rode the bow wave of the vessel; many of them suddenly disappeared after the first sighting, offering no opportunity for resighting. Within this ship-avoiding area, there was a slightly smaller area (46°N–50°N, 166°E–174°E) where we encountered numerous mother–calf pairs, which also avoided vessels. Thus, the northern part of the surveyed area near the Aleutian Islands was judged to be inhabited mostly by adult breeding individuals. To the southwest of this breeding ground, there was an area with approximate range of 41°N–47°N and 158°E–174°E, which was inhabited by young individuals that rode the bow wave and were harpooned (Kasuya and Jones 1984). The age composition of the sample obtained in 1982 and 1983 in the same area (Kasuya and Shiraga 1985) showed that there were almost no individuals of 1 year of age and that the peak frequency was at 2–4 years of age, which were of lower frequency in the drift-net sample from the Aleutian Islands area. Similar geographical differences in behavior have been reported in other parts of the range of Dall's porpoise (Sections 9.3.3 and 9.4.5; Kasuya and Ogi 1987; Yoshioka and Kasuya 1991; Amano and Kuramochi 1992).

The size of Dall's porpoise schools encountered during the *Hoyo-maru No. 12* cruise in 1982 ranged from 1 to 14 and had a modal frequency of 2 (29.8% to the north of 45°N and 27.4% to the south of that latitude). The mean school size, 3.51 (SD = 2.5) and 3.77 (SD = 2.1), was not different between the two latitudinal ranges where we observed great behavioral difference. These school sizes were obtained during the mating season (Section 9.4.4).

On the *Hoyo-maru No.12*, I encountered 28 schools of Dall's porpoises, mostly *truei*-types, off the Pacific coast of Hokkaido on September 16–18, 1982, before her arrival at the port of Kushiro (42°59′N, 144°23′E). School size ranged from 1 to 10 and had a mode at 3 and 4 (both represented by 6 schools). The mean schools size was 4.25 (SD = 2.3). Slightly greater mean school size in Japanese coastal waters could be a reflection of high productivity.

Miyashita (1991) provided information on school size of Dall's porpoises in summer off Japan, covering the whole Okhotsk Sea and the western North Pacific extending to 170°E (Figures 9.1 and 9.2). This area is inhabited in summer by one *truei*-type population and several *dalli*-type populations and is known to include several breeding grounds (Figure 9.10). His survey was conducted in August and September, 1989 and 1990, or in the same season as the cruise of the *Hoyo-maru No. 12* mentioned earlier, and reported mean school size of 4.29 for the *truei*-types, 5.13 for *dalli*-types in the Okhotsk Sea and Pacific area within 200 nm from the Kuril Islands, and 5.77 for the offshore *dalli*-types. The figure for *truei*-types was similar to the earlier figure, that is, 4.25, obtained off the Pacific coast of eastern Hokkaido, and slightly smaller than the mean school sizes for *dalli*-types.

Kasuya (1978) compared school size between 32 schools of *truei*-types off Sanriku (Pacific) and 54 schools of *dalli*-types in the Sea of Japan and southern Okhotsk Sea. The season was May–September for both. Both samples had a mode at 2 individuals, but the mean school size was 5.4 for the *truei*-types and 3.5 for the *dalli*-types. The proportion of schools of 4 or more was 59% for the *truei*-types and 37% for the *dalli*-types. The *dalli*-types had slightly smaller school sizes. There are no explanations available for the minor seasonal and geographical variation in group size.

9.6.2 Behavior in Relation to Age and Sex Structure

The social structure of some cetaceans has been studied based on the long-term observation of individually identified groups. Such study is suited to species that are individually identifiable, inhabit easily accessible coastal waters, and live in relatively small communities or populations. Such a situation does not always exist for the Dall's porpoise. The following is some information from the cruise of the *Hoyo-maru No.2* in 1982 that offers a glimpse of social structure in the species.

The *Hoyo-maru No.2* cruise was conducted in August and September, the season of peak sea surface temperature, when the southern limit of the Dall's porpoise was around 42°N in longitudes of 155°E–173°E and sea surface temperature was at 20°C. Warmwater species such as the common dolphin, striped dolphin, bottlenose dolphin, and short-finned pilot whale occurred in sea surface temperature over 17°C. The two intermediate species, Pacific white-sided dolphin and northern right whale dolphin, occurred in temperature of 12°C–19°C. Later, Myashita (1993) concluded, with additional data, that these two species in summer inhabit the same temperature range of 11°C–25°C and in latitudes of 40°N–50°N and estimated their abundance.

During the *Hoyo-maru No.2* cruise, the scientists noted that *dalli*-type Dall's porpoises responded to the vessel differently by sea surface temperature in the surveyed area (Figure 9.21). Out of 432 individuals sighted in waters above 11°C, 245 or 57% of the sighted individuals rode the bow wave, but only 4 or 2.3% of 171 sighted in waters below 11°C approached the bow. However, later surveys also identified the presence of a similar behavioral difference in other parts of the North Pacific and adjacent seas of different water temperature profiles, and it was found that the effect of sea surface temperature was not a primary factor (Section 9.3.3). Dall's porpoises in the southern survey area killed by harpooning contained a high proportion of 2–4-year-old individuals under the age of attainment of sexual maturity and a high male proportion of 83%. Currently, available information suggests that juveniles born in the summer are weaned at 4–6 months of age but stay at least until the second summer in the same waters inhabited by their mothers, and in the third summer or at the age of 2 years, they are likely to segregate from breeding individuals. The tendency of segregation from the breeding group is more pronounced in males, which has some similarity with the case in striped dolphins where weaned immature individuals, particularly males, tend to live in a segregated school of similar-age individuals (Section 10.5.14).

As mentioned earlier, we scientists on the research vessel *Hoyo-maru No.12* attempted to chase the *dalli*-type porpoises that avoided our vessel as long as possible, but we were unable to catch up with them. They repeatedly displayed spurts in speed followed by a pause at a distance of 300–500 m from our vessel and finally increased swimming speed and disappeared from sight. The vessel speed was about 10 knots (18.5 km/h), and duration of the chase was 12 min on average with a maximum of 48 min. This result was disappointing given the purpose of our cruise to collect unbiased samples.

One of the factors behind the geographical variation in behavior of the species was the presence of mother–calf pairs. Apart from one mother–calf pair sighted off the east coast of Hokkaido, we sighted 41 mother–calf pairs in offshore waters, and 36 (86%) of them occurred during 6 days from August 25–30 in surface of 9.8°C–12.8°C. The surface temperature in the remaining period was over 11°C. The area with mother–calf pairs was a part of the area where porpoises tended to avoid the research vessel. The whole area occupied by ship-avoiding individuals was tentatively designated as a "breeding area" (Section 9.3.3).

The *Hoyo-maru No.12* encountered 100 *dalli*-type Dall's porpoise schools during the 6-day period in the breeding area, of which only 29 schools were visually observed for their composition at a close distance (Kasuya and Jones 1984). Out of the 29 schools, 16 schools totaling 45 individuals did not contain mother–calf pairs, and 13 schools of 69 individuals contained a total of 36 mother–calf pairs. Since it has been over 20 years since the publication of Kasuya and Jones (1984), I cannot recall how we selected the 29 schools. Reexamination of original records for the 100 schools allows the conclusion that 22 schools of 70 individuals did not contain mother–calf pairs (mean school size was 5.57), that 14 schools of 78 individuals contained mother–calf pairs (mean school size was 3.18), and that the remaining 64 schools were not examined for the presence of mother–calf pairs.

The earlier observations suggest that about half of the individuals in the breeding area are found in schools that contain mother–calf pairs and that the mean size of schools with mother–calf pairs is slightly greater than those without them encountered in the same area. Only two Dall's porpoises were captured from the 16 schools containing mother–calf pairs; they were aged 3 and 10 years and were neither pregnant nor lactating but had a corpus luteum and corpora albicantia in the ovaries. These females could have been in estrus or in the very early stage of pregnancy. Adult males would have aggregated around such females near estrus for mating opportunities. A group of four individuals were visually identified as adults.

In the breeding area, we also encountered Dall's porpoise schools, which were composed only of smaller individuals, that is, schools apparently containing no individuals of adult size. They were two schools each containing three small individuals and one school of five small individuals. Such small individuals were also found in schools either with or without mother–calf pairs. They were probably 1-year-old juveniles born in the previous summer that had already become weaned before the summer but still remained in the breeding area living with individuals of the same age or with some adults. Many of them would segregate outside the breeding area (southern waters in this case) by the third summer or 2 years of age (Figure 9.19).

Thus, it appears that weaned calves tend to live together and increase distance from their mothers, and many of them (though not all) segregate outside the breeding area in the third summer. Examples of similar dispersal after weaning have been reported in striped dolphins (Section 10.5.14), bottlenose dolphins (Section 11.4.4), and male sperm whales (Best 1979).

Each of the previously mentioned 13 schools that contained at least one mother–calf pair and allowed some detailed observations contained 1–7 mother–calf pairs (Kasuya and Jones 1984). Seven of these 13 schools consisted of only mother–calf pairs. Their composition was 1 mother–calf pair (2 schools), 3 mother–calf pairs (2 schools), 4 mother–calf pairs (2 schools), and 7 mother–calf pairs (1 school). More surveys will find additional combinations. The last case with seven mother–calf pairs requires additional explanation. When it was first sighted, there were 2 schools close to each other, 1 school with 3 mother–calf pairs, and the other with 4 mother–calf pairs, but they joined to form a school of 7 mother–calf pairs when approached by our research vessel. These observations suggest that mother–calf pairs tend to merge to form larger aggregations. Mothers in these schools that contained no adult males are likely not in estrus.

The remaining 6 schools in the 13 schools contained individuals of adult size in addition to mother–calf pairs:

1 mother–calf pair + 1 adult: 3 schools
2 mother–calf pairs + 1 adult: 1 school
2 mother–calf pairs + 1 adult + 1 of unknown age: 1 school
3 mother–calf pairs + 2 adults: 1 school

The adult individuals seen with mother–calf pairs were probably adult males waiting for the opportunity to mate. Most females ovulate soon after parturition and enter into the next pregnancy while lactating (Sections 9.4.4 and 9.4.6).

Our current knowledge on social structure of Dall's porpoises can be summarized as follows. Calves are nursed for some period of more than 2 months and probably 4–6 months. Weaned calves leave their mothers by their second summer or before their mothers again give birth, which is deduced from the facts that most females are believed to breed annually, and there has been no sighting of a mother accompanied by two juveniles. These weaned juveniles spend the second summer in their mother's summering ground, but many of them seem to segregate from their mothers in the third summer when they are 2 years old to an area dominated by immature individuals. The tendency of segregation is more pronounced in males, judging from the high male proportion in the summering ground frequented by immature individuals and low male proportion in the summer breeding ground (Section 9.4.9). After these juveniles attain sexual maturity at ages around 4–5 years, they presumably return to their breeding ground, but some nonbreeding adults of both sexes are likely to summer with immature individuals.

The majority of females with their newborn calves live in a temporary aggregation of several mother–calf pairs in the summer breeding ground. The relatively small school size with a mode at two individuals and the earlier analysis of school structure suggest that it is unlikely that this species lives in cohesive schools that last for several years. Rather, the mother–calf bond, which is likely to last less than a year, seems to be the longest stable bond between individuals. The currently available limited observations do not support cooperation among adult males in gaining mating partners. Males are more likely to approach females on their own.

9.7 ABUNDANCE AND TRENDS

9.7.1 Populations around Japan (See Also Chapter 2)

Large-scale commercial hunting of Dall's porpoises started in Japan in the 1910s in Iwate Prefecture (in latitudes of 38°59′N–40°27′N) in the Sanriku Region on the Pacific coast of northern Japan, expanded geographically accompanied by increase in catches during the peri–World War II period, and since the 1950s withdrew to a local fishery along the Sanriku coast, particularly to the two prefectures of Iwate and Miyagi (37°44′N–38°59′N), as alternative winter fishing when other fishing objectives were limited. The fishery operation at the last stage also exhibited some fluctuation. The annual catch of around 10,000 in 1957–1965 declined to 5,000–6,000 in 1967–1975, accompanied by a decline in daily catch per vessel (Kasuya 1982). The cause of the decline in catch per vessel has not been clarified. Subsequently, the catch underwent two temporary declines (1979–1980 and 1984–1986) followed by rebounds. The first of the rebounds came with expansion of the winter fishery south to the coast off Ibaraki Prefecture (35°45′–36°50′N) and the spring to autumn operation off the Pacific coast and Sea of Japan coast of Hokkaido (41°30′–43°30′N) and the second with extensive summer operation in the southern Okhotsk Sea (43°50′–45°30′N). Such changes generated the speculation that the fishery retained catch level through the expansion of fishing ground and targeted populations (Kasuya and Miyashita 1989, in Japanese).

The Fisheries Agency of Japan reported increase in catch to over 10,000 in 1982 and a higher figure of over 13,000 in 1987, which was reported to the IWC in May 1988. Under these circumstances, I had an opportunity in August 1988 to travel the Hokkaido coasts to collect information on the Dall's porpoise hand-harpoon fishery and confirmed operations of 21 vessels in Abashiri (44°01′N, 144°16′E), a fishing port on the Okhotsk Sea coast, and 9 vessels in Kiritappu (43°05′N, 145°08′E), another fishing port on the Pacific coast of eastern Hokkaido. Nine of them were based in Hokkaido and the remaining 22 in Iwate Prefecture. Local fishermen on the Okhotsk Sea coast who were not operating the Dall's porpoise fishery stated that there were 40 vessels operating there in the summer. Dall's porpoise fishermen of the time told me that an ordinary vessel could catch 200 porpoises in a month and certainly over 1000 in a year, but that more catch was possible with a vessel of higher speed. Even if the hunters' estimate was halved, the estimated catch of these vessels exceeded the earlier mentioned official statistics for 1987, which threw doubt on the official statistics of 1987, unless it was assumed that there was a sudden expansion of the fishery in 1988. I reported this question to the Fisheries Agency and published it in Kasuya and Miyashita (1989, in Japanese) to be reviewed in the Scientific Committee Meeting of the IWC.

In 1989, the IWC Scientific Committee discussed the discrepancy in statistics and expressed concern about the reliability of the Japanese statistics and on the sustainability of the Dall's porpoise fishery (IWC 1990). Responding to the question on the government statistics, Kasuya (1992) examined landing records of fish markets in Iwate Prefecture, which were the basic source of the government statistics, and reached the conclusion (Kasuya 1992) that Iwate Prefecture underreported the catch of the 1987 season, presumably to avoid criticism of the rapid catch increase. Some attempt at overreporting was also seen for the 1988 catch, when introduction of a quota system was discussed for the fishery, with the possible purpose of getting a large share of the quota and with probable consent of Iwate Prefecture. The manipulation included use of an erroneous conversion factor that created a smaller-than-actual catch figure estimated from the weight of meat landed. By correcting the identified manipulations, Kasuya (1992) estimated Dall's porpoises taken in the Japanese hand-harpoon fishery at 37,200 in 1987, which was 145% of the official "revised" statistics for the year, and at 45,600 in 1988, which was 113% of the corresponding official figure (Table 2.9). The latter figure was believed to be the peak level for the fishery.

Change in the system of collecting catch statistics for Japanese small-cetacean fisheries, or improvement of coverage in later statistics, caused some additional confusion in evaluating the government statistics. The Offshore Division of the Fisheries Agency collected the statistics until 1987, when the task was transferred to the Coastal Division, which issued the 1988 statistics that were reported to the IWC in the spring 1989 and included levels much greater than reported for 1987. The Coastal Division, responding to the question raised by the Scientific Committee in 1989, in 1990 presented to the IWC the revised statistics for the years 1986 and 1987 (IWC 1991; Japan Progress Report 1991). Thus, there were two sets of catch figures for the 1986 and 1987 seasons, and the revised figures were all greater (Table 2.9). The disagreements between the two sets of statistics mainly came from incomplete geographical coverage by the earlier system, which did not include operation of the fishery in Hokkaido, Aomori, Ehime, and Oita Prefectures and catches of Iwate fishermen that were landed in Hokkaido (Kasuya 1992). The statistics of the Offshore Division could not keep up with the change in fisheries that occurred around 1985; the reason for the failure is unclear.

Another concern of the Scientific Committee was the population size and sustainability of the catch. Miyashita and Kasuya (1988) made the first attempt to estimate the abundance using sighting data from May to August in the area of the northeastern Sea of Japan north of 40°N and the coastal western North Pacific between 34°N and 45°N and obtained estimates of 46,400 for the Sea of Japan stock *dalli*-types (31,800 in the Sea of Japan and 14,600 in the Pacific) and 58,000 for *truei*-types. These figures certainly underestimated the abundance, because they did not include *truei*-types in the central Okhotsk Sea and *dalli*-types in the southern Okhotsk Sea, and were soon exceeded by total catches in several subsequent years in the late 1980s.

Miyashita (1991) resolved the problem with the earlier abundance estimates by surveying the Okhotsk Sea and adjacent area including the Russian EEZ outside of territorial waters using 3 vessels during July 26 to September 22, 1989 and 1990. The Sea of Japan was excluded because the Sea of Japan–Okhotsk Sea population remaining there was considered negligible during the season of the survey (Miyashita and Kasuya 1988). For the purpose of estimating abundance, the survey area was divided into four strata based on existing knowledge on the distribution of the stocks (Figures 9.1 and 9.2): (1) range of a putative *dalli*-type population in the western North Pacific (south of 51°N and in longitudes 150°E–165°E), (2) range of a *dalli*-type population summering in the northern Okhotsk Sea, (3) range of a Sea of Japan–southern Okhotsk Sea *dalli*-type population in the southern Okhotsk Sea, and (4) range of a *truei*-type population summering in the central part of the Okhotsk Sea, centered at the boundary between the earlier 2nd and 3rd populations. The boundary between the 2nd and 3rd population was the line connecting the north end of Sakhalin Island and the northern Kuril Islands (dotted line in Figure 9.2). The Pacific area west of 150°E contained both a *truei*-type population and Sea of Japan–Okhotsk Sea *dalli*-type population.

Abundances estimated from the previous analysis are in Table 9.9. The confidence intervals are tentative figures calculated by myself using the coefficient of variation presented in Miyashita (1991). Miyashita et al. (2007, in Japanese) reported another abundance estimate using sighting data obtained in 2003, which was not entirely an independent estimate because the density in unsurveyed areas was extrapolated using data in Miyashita (1991). The incomplete survey coverage occurred due to unavailability of a Russian permit for surveys in the Russian EEZ, which has always been the largest obstacle to the management of cetaceans off northern Japan.

One of the criticisms of the abundance estimates had to do with the effect of porpoise response to survey vessels. Dall's porpoises are often attracted to the vessel. If they approached the vessel before being sighted by observers, abundance would be overestimated, but if the animal fled from the vessel the abundance will be biased low. We know that response of Dall's porpoises toward vessels varies by region, reflecting behavioral difference and segregation by growth and reproductive status. Correction for such plausible biases was not made by Miyashita (1991) and Miyashita et al. (2007). These abundance estimates assumed g(0) = 1 as usual.

Among the 4 populations listed in Table 9.9, (1) and (2) have been subject to incidental kill in the salmon drift-net fishery since before World War II, (3) and (4) have been targeted by the Japanese hand-harpoon fishery, and (3) was once subject to incidental kill in the salmon drift-net fishery in the Sea of Japan. A goal is to evaluate the effect of these mortalities on the populations and to understand the current levels of the populations relative to those before exploitation. An attempt in this direction was made (Okamura et al. 2008), but further efforts are required for the construction of more realistic catch histories of the affected populations, including the large amount of hunting during the peri–World War II period and the 1980s.

The catch quotas given to the Japanese hand-harpoon fisheries for the 2008 season were 8396 *dalli*-types and 7916 *truei*-types, which were slight modifications of basic figures used in 1993 as 4% of the central estimate of abundance by Miyashita (1991) (Tables 6.2 and 6.4). The idea of the potential biological removal (PBR) is a method often used to judge whether the current or expected level of artificial removal of small cetaceans is within a safe level (Wade 1998), and I feel it places great weight on the safety of a population but not on the maximum utilization of the resource. This has been used in the United States to judge whether the known level of incidental mortality in fishing gear is acceptable. PBR is calculated by the following equation:

$$PBR = [\text{Minimum Abundance}, N_{min}] \times R_{max} \times 0.5 \times [\text{Safety Factor}, F_r]$$

TABLE 9.9
Abundance of Dall's Porpoises around Japan

	Miyashita (1991) in IWC (1992)			Miyashita et al. (2007)		
Geography/Population	**Abundance**	**CV**	**95% Confidence Interval (×10³)**	**Abundance**	**CV**	**95% Confidence Interval (×10³)**
(1) Individuals in the western North Pacific (*dalli*-type)	162,000					
(2) Northern Okhotsk Sea population (*dalli*-type)	111,000	0.29	49–173			
(3) Sea of Japan–South; Okhotsk Sea population (*dalli*-type)	226,000	0.15	158–294	173,638	0.21	115–261
(4) Sanriku–central Okhotsk Sea population (*truei*-type)	217,000	0.23	120–314	178,157	0.23	113–279

Instead of using the central estimate of abundance, the 20th percentile of the estimate is used for $N_{mim,}$ and F_r is arbitrarily selected from a range of 0.5–1.0 considering reliability of the catch statistics and the possibility of causes of additional mortality other than in fisheries (a smaller figure is used if such risk is considered great). R_{max} is the reproductive rate expected for the population at close to zero density, and 4% is assumed as a default for small cetaceans. The PBR, even with the most optimistic assumption of $F_r = 1$, would estimate as a safe level less than 3000 for either of the two Dall's porpoise populations targeted by the Japanese hand-harpoon fishery, which is less than half of the current catch quotas. If a smaller figure is assumed for F_r to compensate for possible underreporting, then the PBR value will be smaller. The Fisheries Agency of Japan stated in its publicity materials that even the PBR supported the current quota for the Dall's porpoise fishery. However, it used 0.08 or 0.09 for R_{max}, without presenting any supporting evidence for such an optimistic level. Thus, the current quota for the fishery is not supported by the PBR as used in the United States.

R_{max} is the maximum rate of increase expected for a population in the ideal environment with minimum density, which is difficult to measure for an actual population. I once attempted to determine the possible range of R_{max} for Dall's porpoises in collaboration with a population dynamist and obtained a range of 3%–9%. The Fisheries Agency felt that our calculation could not exclude the possibility of R_{max} being as low as 3%–4%, which was a figure at odds with the catch level of the time, and refused to allow our work to be presented to the Scientific Committee meeting of the IWC. I agree that our estimate of R_{max} is of limited scientific value but believe that such a biased attitude of the Fisheries Agency puts the rational management of fishery resources at risk. It should be noted that the current catch quotas for the Japanese Dall's porpoise fishery are supported by only very optimistic assumptions of R_{max}.

9.7.2 Offshore Populations

As a part of the activities of the INPFC to evaluate the effect of incidental mortality of Dall's porpoises in various drift-net fisheries on their population, the United States placed observers on board Japanese fishing vessels and their own research vessels to collect sighting data for an abundance estimate. Their data covered the months from June to August in 1987–1990. Using the data, Buckland et al. (1993) estimated the abundance for each 5° latitude by 5° longitude square and obtained a total abundance of 1,186,000 Dall's porpoises by summing the figures for each square (95% confidence interval was 991,000–1,420,000). This figure covered the range of the species in the Pacific Ocean from the Japanese coast to the U.S. coast and most of the Bering Sea area but did not cover the Okhotsk Sea, the Sea of Japan, and the part of the Bering Sea outside of the drift-net fishery operation. Combining the given figure with 550,000 that represents those areas 2, 3, and 4 in Table 9.9 results in a total of 1.74 million Dall's porpoises, which still does not include the porpoises in the Russian EEZ on the east coast of the Kamchatka Peninsula. This figure also ignores the response of Dall's porpoise toward the survey vessel and assumes g(0) = 1 as usual.

Abundance of the Pacific white-sided dolphin and northern right whale dolphin were estimated at 931,000 (with 95% confidence interval of 206,000–4,216,000) and 68,000 (20,000–239,000), respectively (Buckland et al. 1993). These figures cover the entire range of the species in the North Pacific proper.

Dall's porpoises were killed in the offshore North Pacific incidentally in the salmon drift-net fishery, squid drift-net fishery, and large-mesh drift-net fishery. Since a trial operation in 1914, Japan enjoyed free operation of the salmon drift-net fishery for a long period, but the International Convention for the High Seas Fisheries on the North Pacific Ocean signed in 1952 (Section 6.2) limited the Japanese operation to the international waters in the Bering Sea and western half of the North Pacific (Matsumoto 1980, in Japanese). Since 1956, a fishery agreement between Japan and USSR has required a USSR permit for the Japanese vessels operating in its EEZ. Regulation continued to be tightened thereafter, and only a small-scale operation was left after 1992 in the Russian EEZ, which was operated jointly with Russia (Morita 1994, in Japanese). The squid drift-net fishery was also started by Japanese fishermen in 1978 and operated off the coast of Sanriku and Hokkaido and expanded to 160°W in 1980 (Yatsu et al. 1993). This fishery was joined by vessels of Taiwan since the late 1970s and Korea since 1979 (Gong et al. 1993; Yeh and Tung 1993). The large-mesh drift-net fishery targets marlin and tunas. This fishery started in the 1840s and was operated in nearshore waters of Japan, expanded to the offshore waters of northern Japan in the early 1970s, and reached to 150°W in the 1980s (Nakano et al. 1993). Taiwanese fishermen also participated in this fishery, but details are not available. These large-scale high-seas drift-net fisheries ceased operation by the end of 1992 following the UN resolution of December 1989 (Section I.3.2).

It seems certain that the drift-net fisheries have influenced at least some of the numerous Dall's porpoise populations discussed previously and also the Pacific white-sided dolphin and northern right whale dolphin. For the latter two species, almost nothing is known about stock structure (Hobbs and Jones 1993).

At the occasion of a symposium in Tokyo convened by the INPFC in 1991, Hobbs and Jones (1993) estimated the incidental mortality of several dolphin species in the squid and large-mesh drift-net fisheries and their effects on the populations (Table 9.10). However, they did not make any significant mention of the mortality of Dall's porpoises in the salmon drift-net fishery and its effect on the population, which has been the major task of the small-cetacean-related activity of the INPFC for nearly 10 years. The salmon drift-net fishery operated since 1952 under international regulations, but the fishery deployed a several times greater number of nets than permitted by the regulations and underreported the resultant takes of salmon (Sano 1998, in Japanese). Under such circumstances, it must have been impossible to estimate with any certainty the number of Dall's porpoises killed in the fishery. It was also possible that the scientists had lost enthusiasm for

TABLE 9.10
Incidental Mortality of Dolphins and Porpoises in Drift-Net Fisheries in the Northern North Pacific Estimated for 1990

Fishing Type	Squid Drift Net			Large-Mesh Drift Net		
Country of Operation	**Japan**	**Korea**	**Taiwan**	**Japan**	**Taiwan**	**Total**
No. of operations observed	3,014	911	356	829	353	
Observed net length in 1,000 *tan*	2,364	669	170	513	194	
Total net deployed in 1,000 *tan*	22,769	24,589	3,452	2665	3,339	
Northern right whale dolphin	8,224	1,983	142	0	702	11,051
Pacific white-sided dolphin	4,459	1,065	101	31	103	5,759
Common dolphin	693	220	0	2492	805	4,210
Striped dolphin	58	37	0	2617	360	3,072
Dall's porpoise	2,937	843	41	0	17	3,838
Fur seal	4,960	147	0	0	206	5,313
Total marine mammals	22,169	4,370	385	6869	5,875	39,668

Source: Prepared from Tables 3 and 5 of Hobbs, R.C. and Jones, L.L., *Int. North Pacific Fish. Commn. Bull.*, 53(I), 409, 1993.
Note: *Tan* is the number of net pieces deployed; the length of a *tan* was standardized here into 50 m of float line.

pursuing the truth under the circumstance of the fishery going to be closed in accord with the UN resolution. The INPFC symposium in Tokyo concluded its scientific activity on Dall's porpoise mortality in the drift-net fishery.

REFERENCES

[*In Japanese Language]

Amano, M. and Hayano, A. 2007. Intermingling of *dalli*-type Dall's porpoises into a wintering *truei*-type population off Japan: Implication from color patterns. *Marine Mammal Sci.* 23(1): 1–14.

*Amano, M., Hayano, A., Ohizumi, H., and Tanaka, M. 1998a. Origin of *dalli*-type Dall's porpoises taken off Sanriku in winter. pp. 61–67. In: Miyazaki, N. (ed.). *Report of Contract Studies on Management of Dolphins and Porpoises.* Ocean Research Institute of University of Tokyo, Tokyo, Japan. 242pp.

*Amano, M., Hayano, A., Ohizumi, H., and Tanaka, M. 1998b. Origin of *dalli*-type Dall's porpoises taken off Sanriku in winter (II). pp. 69–75. In: Miyazaki, N. (ed.). *Report of Contract Studies on Management of Dolphins and Porpoises.* Ocean Research Institute of University of Tokyo, Tokyo, Japan. 242pp.

Amano, M. and Kuramochi, T. 1992. Segregated migration of Dall's porpoise (*Phocoenoides dalli*) in the Sea of Japan and Sea of Okhotsk. *Marine Mammal Sci.* 8: 143–151.

Amano, M., Marui, M, Guenther, T. Ohizumi, H., and Miyazaki, M. 2000. Re-evaluation of geographical variation in the white flank patch of *dalli*-type Dall's porpoise. *Marine Mammal Sci.* 16(3): 631–636.

Amano, M. and Miyazaki, N. 1992. Geographical variation and sexual dimorphism in the skull of Dall's porpoise, *Phocoenoides dalli. Marine Mammal Sci.* 8: 240–261.

Amano, M. and Miyazaki, N. 1993. External morphology of Dall's porpoise (*Phocoenoides dalli*): Growth and sexual dimorphism. *Can. J. Zool.* 71: 1124–1130.

Amano, M. and Miyazaki, N. 1996. Geographical variation in external morphology of Dall's porpoise, *Phocoenoides dalli. Aqu. Mammals* 22(3): 167–174.

*Amano, M., Ohizumi, H., Tanaka, M., Marui, M., and Amano, A. 1998c. Biological studies on Dall's porpoises landed at the Otsuchi Fish Market. pp. 33–49. In: Miyazaki, N. (ed.). *Report of Contract Studies on Management of Dolphins and Porpoises.* Ocean Research Institute of University of Tokyo, Tokyo, Japan. 242pp.

Amano, M., Yoshioka, M., Kuramochi, T., and Mori, K. 1998d. Diurnal feeding by Dall's porpoise, *Phocoenoides dalli. Marine Mammal Sci.* 14(1): 130–135.

Andrews, R.C. 1911. A new porpoise from Japan. *Bull. Am. Mus. Nat. Hist.* 30: 31–52.

Berta, A. and Sumich, J.L. 1999. *Marine Mammals: Evolutionary Biology.* Academic Press, San Diego, CA. 494pp.

Best, P.B. 1979. Social organization in sperm whales, *Physeter macrocephalus.* pp. 227–289. In: Winn, H.E. and Olla, B.L. (eds.). *Behavior of Marine Mammals: Current Perspectives in Research.* Plenum Press, New York. 438pp.

Brown, M.R., Cockeron, P.J., Hale, P.T., Shultz, K.W., and Bryden, M.M. 1995. Evidence for a sex segregated migration in the humpback whale (*Megaptera novaeangliae*). *Proc. R. Soc. Lond.*, Part B 259: 229–234.

Buckland, S.T., Cattanach, K.L., and Hobbs, R.C. 1993. Abundance estimates of Pacific white-sided dolphin, northern right whale dolphin, Dall's porpoise and northern fur seals in the North Pacific, 1987–1990. *Int. North Pacific Fish. Commn. Bull.* 53(III): 387–409.

Clapham, P.J. 2002. Humpback whale, *Megaptera novaeangliae.* pp. 589–592. In: Perrin, W.F., Würsig, B., and Thewissen, J.G.M. (eds.). *Encyclopedia of Marine Mammals.* Academic Press, San Diego, CA. 1414pp.

Dailey, M.D. and Brownell, R.L., Jr. 1972. A checklist of marine mammal parasites. pp. 528–589. In: S.H. Ridgway (ed.). *Mammals of the Sea, Biology and Medicine*. C.C. Thomas, Chicago, IL. 812pp.

DeMaster, D.P. 1978. Calculation of average age of sexual maturity in marine mammals. *J. Fish. Res. Bd, Canada* 35: 912–930.

Escorza-Treviño, S. and Dizon, S. 2000. Phylogeography, intraspecific structure, and sex-biased dispersal of Dall's porpoise, *Phocoenoides dalli*, revealed by mitochondrial and microsatellite DNA analysis. *Mol. Ecol.* 9: 1046–1060.

Escorza-Treviño, S., Pastene, L.A., and Dizon, A.E. 2004. Molecular analyses of the *truei* and *dalli* morphotypes of Dall's porpoise (*Phocoenoides dalli*). *J. Mammalogy* 85(2): 347–355.

Ferrero, R.C. and Walker, W.A. 1999. Age, growth, and reproductive patterns of Dall's porpoise (*Phocoenoides dalli*) in the central North Pacific Ocean. *Marine Mammal Sci.* 15(2): 273–313.

Fiscus, C.H. and Jones, L. 1999. A note on cephalopods from the stomach of Dall's porpoise (*Phocoenoides dalli*) from the Northwestern Pacific and Bering Sea, 1978–1982. *J. Cetacean Res. Manage.* 1(1): 101–107.

Gong, Y., Kim, Y.S., and Hwang, S.J. 1993. Outline of the Korean squid gillnet fishery in the North Pacific. *Int. North Pacific Fish. Commn. Bull.* 53(I): 45–69.

Harrison, R.J., Johnson, F.R., and Young, B.A. 1970. The oesophagus and stomach of dolphins (*Tursiops, Delphinus, Stenella*). *J. Zool. Lond.* 160: 377–390.

Hayano, A., Amano, M., and Miyazaki, N. 2003. Phylogeography and population structure of the Dall's porpoise, *Phocoenoides dalli*, in the Japanese waters revealed by mitochondrial DNA. *Genes Genet. Syst.* 78: 81–91.

Hershkovitz, H. 1966. *Catalogue of Living Whales*. Smithsonian Institute, Washington, DC. 259pp.

Hobbs, R.C. and Jones, L.L. 1993. Impact of high seas driftnet fisheries on marine mammal population in the North Pacific. *Int. North Pacific Fish. Commn. Bull.* 53(I): 409–432.

Houck, W.J. 1976. The taxonomic status of the species of the porpoise genus *Phocoenoides*. Paper ACMRR/MM/SC/114 presented to the FAO Scientific Consultation on Marine Mammals, Bergen, Norway. (unpublished).

INPFC. 1993. Symposium on biology, distribution and stock assessment of species caught in the high seas driftnet fisheries in the North Pacific Ocean. In: Ito, J., Shaw, W., and Burgner, R.L. (eds.). *Driftnet Fisheries of the North Pacific Ocean*. INPFC, Vancouver, British Columbia, Canada. 90pp.

IWC (International Whaling Commission). 1990. Report of the Scientific Committee. *Rep. Int. Whaling Commn.* 40: 39–93.

IWC. 1991. Report of the Sub-Committee on Small Cetaceans. *Rep. Int. Whal. Commn.* 41: 172–190.

IWC. 1992. Report of the Sub-Committee on Small Cetaceans. *Rep. Int. Whal. Commn.* 42: 178–234.

IWC. 2002. Report of the Standing Sub-Committee on Small Cetaceans. *J. Cetacean Res. Manage.* 4(Suppl.): 325–338.

Japan Progress Report. 1991. Progress report on cetacean research, May 1989 to May 1990. *Rep. Int. Whal. Commn.* 41: 239–243.

Jefferson, T.A. 1988. Phocoenoides dalli. *Mammal Species* 319: 1–7. American Society of Mammalogists.

Jefferson, T.A. 1989. Sexual dimorphism and development of external features in Dall's porpoise *Phocoenoides dalli*. *Fish. Bull.* 88: 119–132.

Jones, L., Kasuya, T., Gosho, M., and Miyazaki, N. 1985a. Variability by readers and method of preparation in Dall's porpoise age determination. Paper INPFC Doc. No. 2878 presented to the INPFC Meeting in March 1985, Tokyo, Japan (unpublished). 19pp.+5 text figures.

Jones, L.L., Rice, D.W., and Gosho, M.E. March 11–15, 1985b. Biological studies of Dall's porpoise: progress report. Document presented to the Scientific Subcommittee Meeting of the Ad Hoc Committee on Marine Mammals, International North Pacific Fisheries Commission, Tokyo, Japan. 18pp.

Joyce, G.G., Rosapepe, J.V., and Ogasawara, J. 1982. White Dall's porpoise sighted in the North Pacific. *Fish. Bull.* 80(2): 401–402.

Kasuya, T. 1978. The life history of Dall's porpoise with special reference to the stock off the Pacific coast of Japan. *Sci. Rep. Whales Res. Inst.* (Tokyo), 30: 1–63.

Kasuya, T. 1982. Preliminary report of the biology, catch and population of *Phocoenoides* in the western North Pacific. pp. 3–20. In: Clark, J.G., Goodman, J., and Soave, G.A. (eds.). *Mammals in the Sea*. Vol. IV (Small Cetaceans, Seals, Sirenians and Otters). FAO, Rome, Italy. 531pp. (This paper was presented in 1976 to the FAO Scientific Consultation on Marine Mammals, Bergen).

Kasuya, T. 1992. Examination of Japanese statistics for the Dall's porpoise hand harpoon fishery. *Rep. Int. Whal. Commn.* 42: 521–528.

Kasuya, T. 1995. Overview of cetacean life histories: an essay in their evolution. pp. 481–497. In: Blix, A.S., Walloe, L., and Ultang, O. (eds.). *Whales, Seals, Fish and Man*. Elsevier Science, Amsterdam, the Netherlands. 720pp.

Kasuya, T. and Jones, L.L. 1984. Behavior and segregation of the Dall's porpoise in the northwestern North Pacific Ocean. *Sci. Rep. Whales Res. Inst.* (Tokyo) 35: 107–128.

Kasuya, T. and Kureha, K. 1971. Sighting records of mammals and birds in the Sea of Japan. pp. 205–209. Preliminary Report of the Hakuho Maru Cruise KH-69-2, April 26-June 19, 1969. Ocean Research Institute, University of Tokyo, Tokyo, Japan. 209pp.

Kasuya, T. and Matsui, S. 1984. Age determination and growth of the short-finned pilot whale off the Pacific coast of Japan. *Sci. Rep. Whales Res. Inst.* (Tokyo) 35: 57–91.

*Kasuya, T. and Miyashita, T. 1989. Japanese small cetacean fisheries and problems in management. *Saishu-to Shiiku* 51(4): 154–160.

Kasuya, T. and Ogi, H. 1987. Distribution of mother-calf Dall's porpoise pairs and an indication of calving grounds and stock identity. *Sci. Rep. Whales Res. Inst.* (Tokyo) 38: 125–140.

Kasuya, T. and Shiraga, S. 1985. Growth of Dall's porpoise in the western North Pacific and suggested geographical growth differentiation. *Sci. Rep. Whales Res. Inst.* 36: 139–152.

Kasuya, T. and Tai, S. 1993. Life history of short-finned pilot whale stocks off Japan and a description of the fishery. *Rep. Int. Whal. Commn.* (Special Issue 14, Biology of Northern Hemisphere Pilot Whales): 439–473.

Kasuya, T., Tobayama, T., Saiga, T., and Kataoka, T. 1986. Perinatal growth of delphinoids: information from aquarium reared bottlenose dolphin and finless porpoises. *Sci. Rep. Whales Res. Inst.* (Tokyo) 37: 85–97.

*Kasuya, T. and Yamada, T. 1995. *List of Japanese Cetaceans*. Geiken Series 7. Nihon Geirui Kenkyusho (Institute of Cetacean Research), Tokyo, Japan. 90pp.

*Kawamura, A., Nakamura, H., Tanaka, H., Sato, N., Fujise, Y., and Nishida, K. 1983. Sightings of dolphins and porpoises from a railway ferry boat between Hakodate and Aomori. *Geiken Tsushin* 351/352: 29–51.

*Kishida, K. 1925. *Illustrated Guide of Mammals*. Bureau of Agriculture of Ministry of Agriculture and Commerce, Tokyo, Japan. 381+17+31pp.

*Koga, S. 1969. On Dall's porpoises killed in the drift-net salmon fishery in the northern North Pacific. *Bull. Tokyo Univ. Fisheries* 18(1): 53–63.

Kuramochi, T., Kubodera, T., and Miyazaki, N. 1991. Squids eaten by Dall's porpoise, *Phocoenoides dalli* in the northwestern North Pacific and in the Bering Sea. pp. 229–240. In: Okutani, T., O'dor, R.K., and Kubodera, T. (eds.). *Recent Advances in Fisheries Biology*. Tokai University Press, Tokyo, Japan. 752pp.

*Kuroda, N. 1938. *List of Japanese Mammals*. Herarudo-sha, Tokyo, Japan. 122pp.

*Kuroda, N. 1954. On the taxonomy of Dall's porpoise and True's porpoise. *Bull. Yamashina Inst. Ornithol.* 5: 44–46.

Lockyer, C. 1981a. Growth and energy budgets of large baleen whales for the southern hemisphere. pp. 379–481. In: Clark, J.G., Goodman, J., and Soave, G.A. (eds.). *Mammals in the Sea*. Vol. IV (General Papers and Large Cetaceans). FAO, Rome, Italy. 504pp.

Lockyer, C. 1981b. Estimation of the energy costs of growth, maintenance and reproduction in the female minke whale, (*Balaenoptera acutorostrata*), from the southern hemisphere. *Rep. Int. Whal. Commn.* 31: 337–343.

Lockyer, C. 1984. Review of baleen whale (Mysticeti) reproduction and implication for management. *Rep. Int. Whal. Commn* (Special Issue 6, Reproduction in Whales, Dolphins and Porpoises): 27–50.

Lockyer, C. 2003. Harbour porpoise (*Phocoena phocoena*) in the North Atlantic: Biological parameters. pp. 71–89. In: Haug, T., Desportes, G., Vikingsson, G.A., Witting, L., and Pike, D.G. (eds.). *Harbour Porpoise in the North Atlantic*. North Atlantic Marine mammal Commission, Tromso, Norway. 315pp.

*Marui, M., Amano, M., and Ohizumi, H. 1996. Between stocks difference of flank patch of *dalli*-type Dall's porpoises. pp. 51–60. In: Miyazaki, N. (ed.). *Report of Contract Studies on Management of Dolphins and Porpoises*. Ocean Research Institute of University of Tokyo, Tokyo, Japan. 242pp.

*Matsu-ura, Y. 1943. *Marine Mammals*. Ten-nen-sha, Tokyo, Japan. 298pp.

*Matsumoto, I. 1980. *Annotated Chronology of Modern Japanese Fisheries, after World War II*. Suisansha, Tokyo, Japan. 119pp.

Miller, G.S., Jr. 1929. The gums of porpoise *Phocoenoides dalli* (True). *Proc. U.S. National Mus.* 74: 1–4.

Miyashita, T. 1991. Stocks and abundance of Dall's porpoises in the Okhotsk Sea and adjacent waters. Paper IWC/SC/43/SM7 presented to the Scientific Committee Meeting of IWC (unpublished). 24pp. (Available from IWC Secretariat, Cambridge, U.K. The abundance estimate is cited in IWC 1992).

Miyashita, T. 1993. Distribution and abundance of some dolphins taken in the North Pacific driftnet fisheries. *Int. North Pacific. Fish. Commn. Bull.* 53(III): 435–449.

Miyashita, T. and Berzin, A.A. 1991. Report of the whale sighting survey in the Okhotsk Sea and adjacent waters in 1990. Paper SC/43/O5 presented to the Scientific Committee Meeting of IWC (unpublished). 14pp. (Available from IWC Secretariat, Cambridge, U.K.).

Miyashita, T. and Doroshenko, N. 1990. Report of the whale sighting survey in the Okhotsk Sea August, 1989. Paper SC/42/O18 presented to the Scientific Committee Meeting of IWC (unpublished). 16pp. (Available from IWC Secretariat, Cambridge, U.K.).

Miyashita, T. and Kasuya, T. 1988. Distribution and abundance of Dall's porpoise off Japan. *Sci. Rep. Whales Res. Inst.* (Tokyo) 39: 121–150.

Miyazaki, N., Jones, L.L., and Beach, R. 1984. Some observations on the school structure of *dalli*- and *truei*-type Dall's porpoises in the Northwestern Pacific. *Sci. Rep. Whales Res. Inst.* (Tokyo) 35: 93–105.

*Miyashita, T., Iwsaki, H., and Moronuki, H. 2007. Abundance of Dall's porpoise in the northwestern North Pacific. Abstract of autumn meeting of Nihon Suisan Gakkai [Japanese Society of Fisheries Science] in 2007, Tokyo, Japan. p. 164.

*Mizue, K. and Yoshida, K. 1965. On the porpoises caught by the Salmon fishing gill-nets in the Bering Sea and the North Pacific Ocean. *Bull. Faculty Fisheries Nagasaki Univ.* 19: 1–36.

*Mizue, K., Yoshida, K., and Takemura, A. 1966. On the ecology of the DALL's porpoise in the Bering Sea and the North Pacific Ocean. *Bull. Faculty Fisheries Nagasaki Univ.* 21: 1–21.

Morejohn, G.V. 1979. The natural history of Dall's porpoise in the North Pacific Ocean. pp. 45–83. In: Winn, H.E. and Olla, B.L. (eds.). *Behavior of Marine Animals, Current Perspectives in Research*. Vol. 3 (Cetaceans). Plenum Press, New York. 438pp.

*Morita, H. 1994. *History of Catcher Boats [of salmon drift net fisheries]*. Vol. 3. Nihon Sakemasu Gyogyo Kyodokumiai, Tokyo, Japan. 316pp.

*Nagasawa, R. 1916. Scientific names for 11 species of dolphins and porpoises. *Dobutsugaku Zasshi [Journal of Zoology*, published by Tokyo Zoological Society] 28(327): 35–39.

Nakano, H., Okada, K., Watanabe, Y., and Uosaki, K. 1993. Outline of the large-mesh driftnet fishery of Japan. *Int. North Pacific Fish. Commn. Bull.* 53(I): 25–37.

Newby, T.C. 1982. Life history of Dall's porpoise (Phocoenoides dalli, True 1885) incidentally taken by the Japanese high seas salmon mothership fishery in the northwestern North Pacific and western Bering Sea, 1978 and 1980. Doctoral Thesis, University of Washington, Washington, DC. 157pp.

*Noguchi, E. and Nakamura, T. 1946. *Utilization of Dolphins and Porpoises and Fisheries for Mackerel*. Kasumigaseki Shobo, Tokyo, Japan. 70pp.

*Ohizumi, H. 2008. Food of cetaceans around Japan. pp. 197–273. In: Murayama, T. (ed.). *Geirui-gaku [Cetology]*. Tokai University Press, Hadano, Japan. 402pp.

Ohizumi, H., Kuramochi, T., Amano, M., and Miyazaki, N. 2000. Prey switching of Dall's porpoise *Phocoenoides dalli* with population decline of Japanese pilchard *Sardinops melanostictus* around Hokkaido Japan. *Marine Ecol. Progress Ser.* 200: 265–275.

Ohizumi, H., Kuramochi, T., Kubodera, T., Yoshioka, M., and Miyazaki, N. 2003. Feeding habits of Dall's porpoise (*Phocoenoides dalli*) in the subarctic North Pacific and the Bering Sea basin and the impact of predation on mesopelagic micronekton. *Deep Sea Res., Part I* 50: 593–610.

Ohizumi, H. and Miyazaki, N. 1998. Feeding rate and daily energy intake of Dall's porpoise in the northeastern Sea of Japan. *Proc. National Inst. Polar Res. Symp. Polar Biol.* 11: 74–81.

Ohsumi, S. 1966. Allomorphosis between body length at sexual maturity and body length at birth in the cetacean. *J. Mamm. Soc. Jpn.* 3(1): 3–7.

Okamura, H., Iwasaki, T., and Miyashita, T. 2008. Toward sustainable management of small cetacean fisheries around Japan. *Fish. Sci.* 74(4): 718–729.

Olafsdottir, D., Vikingsson, G.A., Halldorsson, S.D., and Sigurjonsson, J. 2003. Growth and reproduction in harbour porpoise (*Phocoena phocoenoides*) in Icelandic waters. pp. 195–210. In: Haug, T., Desportes, G., Vikingsson, G.A., Witting, L., and Pike, D.G. (eds.). *Harbour Porpoise in the North Atlantic*. North Atlantic Marine Mammal Commission, Tromso, Norway. 315pp.

O'shea, T., Brownell, R.L., Jr., Clarke, D.R., Jr., Walker, W.A., Gay, M.L., and Lamont, T.G. 1980. Organochlorine pollutants in small cetaceans from the Pacific and south Atlantic oceans, November 1968-June 1976. *Pesticide. Monit. J.* 14: 35–46.

Perrin, W.F., Holts, D.B., and Miller, R.B. 1977. Growth and reproduction of the eastern spinner dolphin, a geographical form of *Stenella longirostris* in the eastern tropical Pacific. *Fish. Bull.* 75(4): 725–750, Cambridge, U.K.

Ridgway, S.H. and Johnston, D.G. 1966. Blood oxygen and ecology of porpoises of three genera. *Science* 151: 456–458.

Ridgway, S.H. and Kohn, S. 1995. The relationship between heart mass and body mass for three cetacean genera: Narrow allometry demonstrates interspecific differences. *Marine Mammal Sci.* 11(1): 72–80.

*Sano, O. 1998. *History of North Pacific Salmon Pelagic Fisheries.* Seizando Shoten, Tokyo, Japan. 188pp.

Sergeant, D.E. 1969. Feeding rates of cetacean. *Fiskeridirektoratets Skrifter,* Serie Havunderskoleser 15: 246–258.

Shimura, E. and Numachi, K. 1987. Genetic variability and differentiation in the toothed whales. *Sci. Rep. Whales Res. Inst.* (Tokyo) 38: 141–163.

Smith, T.D., Allen, J., Chapman, P.J., Hammond, P.S., Katona, S., Larsen, E., Lien, J. et al. 1999. An Ocean-basin-wide mark-recapture study of the North Atlantic humpback whale (*Megaptera novaeangliae*). *Marine Mammal Sci.* 15(1): 1–32.

Subramanian, A., Tanabe, S., Fujise, Y., and Tatsukawa, R. 1986. Organochlorine residues in Dall's and True's porpoises collected from Northwestern Pacific and adjacent waters. *Mem. Natl. Inst. Polar Res.* (Spec. Issue 44): 167–173.

Szczepaniak, I.D. and Webber, M.A. 1992. First record of a *Truei*-type Dall's porpoise from the eastern North Pacific. *Marine Mammal Sci.* 8(4): 425–428.

Tatsukawa, R., Tanabe, S., and Honda, K. 1979. Studies on some organochlorine compounds residues in the sea water surrounding Honshu Island. pp. 23–25. In: *Preliminary Report of the Hakuho Maru Cruise, KH-76-3.* Ocean Research Institute, University of Tokyo, Tokyo, Japan.

True, F.W. 1885. On a new species of porpoise, *Phocoena dalli*, from Alaska. *Proc. U.S. Natl. Mus.* 8: 95–98.

Wade, P.R. 1998. Calculating limits to the allowable human-caused mortality of cetaceans and pinnipeds. *Marine Mammal Sci.* 14(1): 1–37.

Walker, W. 1996. Summer feeding habits of Dall's porpoise, *Phocoenoides dalli*, in the southern Sea of Okhotsk. *Marine Mammal Sci.* 12(2): 167–181.

Walker, W.A. 2001. Geographical variation of the parasite, *Phyllobothrium delphini* (Cestoda), in Dall's porpoise, *Phocoenoides dalli*, in the northern North Pacific, Bering Sea, and Sea of Okhotsk. *Marine Mammal Sci.* 17(2): 264–275.

Wilke, F., Taniwaki, T., and Kuroda, N. 1953. *Phocoenoides* and *Lagenorhynchus* in Japan, with notes on hunting. *J. Mammalogy* 34(4): 488–497.

Winans, G.A. and Jones, L.L. 1988. Electric variability in Dall's porpoise (*Phocoenoides dalli*) in the North Pacific Ocean and Bering Sea. *J. Mammalogy* 69(1): 14–21.

Yatsu, A., Hiramatsu, K., and Hayase, S. 1993. Outline of the Japanese squid driftnet fishery of Japan. *Int. North Pacific Fish. Commn. Bull.* 53(1): 5–24.

Yeh, S.Y. and Tung, I.H. 1993. Review of Taiwanese pelagic squid fisheries in the North Pacific. *Int. North Pacific Fish. Commn. Bull.* 53(I): 71–76.

*Yoshioka, M. and Kasuya, T. 1991. Population structure of whales viewed from distribution and ecology. pp. 53–63. In: Sakuramoto, K., Tanaka, S., and Kato, H. (eds.). *Research and Management of Whale Stocks.* Koseisha Koseikaku, Tokyo, Japan. 273pp.

Yoshioka, M., Kasuya, T., and Aoki, M. 1990. Identity of Dalli-type Dall's porpoise stocks in the northern North pacific and adjacent seas. Paper IWC/SC/SM31 presented to the Scientific Committee of IWC (unpublished). 20pp. (Available from IWC Secretariat, Cambridge, U.K.)

10 Striped Dolphin

10.1 DESCRIPTION

The striped dolphin, *Stenella coeruleoalba* (Meyen, 1833), was described based on a specimen obtained in the estuary of the River La Plata under the name *Delphinus coeruleoalbus*. It was later moved by Oliver (1922) to the genus *Stenella*, which was created by Gray (1866), with accompanying inflection. The additional genera *Prodelphinus* and *Clymenia* were also used, in combination with the synonymous species *styx* and *euphrosyne* (Perrin et al. 1994; Rice 1998).

The striped dolphin has the typical dolphin body shape and attains a length of around 2.5 m. The projecting rostrum and fatty tissue of the forehead (melon) are bounded by a distinct groove. The rostrum projects beyond the melon for about 11 cm or 4.5%–5.8% of the body length and has a relatively small dorsal–ventral thickness. The rostrum, viewed from the top, is narrower than that of the bottlenose dolphin but broader than those of the common dolphins *Delphinus* spp. and the pantropical spotted dolphin *Stenella attenuata* (Kasuya 1994, in Japanese). The dorsal fin, located at about midlength of the body, a crescent shaped with a concave posterior edge and width of 8%–9% of body length (Okada 1936). It is blue-black dorsally and white below, with stripes between. One dark stripe extends anteriorly from a dark area surrounding the eye to the melon–rostrum boundary. A second dark stripe extends from the eye region to the base of the flipper and merges with a dark area on both sides of the flipper. A third dark stripe extends from the eye to the anal region; this stripe gives rise to the Japanese name *suji-iruka* (striped dolphin). At around the level of the flipper, a posterio-ventrally directed dark stripe branches from the third stripe and fades posteriorly to disappear before reaching the anal region. The area between the blue-black dorsum and the third stripe is gray and extends posteriorly to the lateral side of the tail peduncle. The dorsal edge of the tail peduncle and both sides of the tail flukes are blue-black. A broad swath of gray extends from the lateral thoracic area toward the base of the dorsal fin, fading into the dorsal blue-black pigmentation before reaching the base of the fin. This shoulder blaze, together with the thin broad rostrum, is a useful feature for the identification of this species at sea. Striped dolphins in Japanese waters have 43–50 teeth in each toothrow.

The earliest use of the Japanese vernacular name *suji-iruka* is found in *Koza-ura Hogei Emaki* (Pictorial Scroll of Whaling in Koza Inlet) from the eighteenth to nineteenth century depicting whaling activity at Koza (33°31′N, 135°49′E), which is about 13 km southwest of Taiji in Wakayama Prefecture on the Pacific coast of central Japan. Taiji (33°36′N, 135°57′E) is located about 300 km southwest of the Izu Peninsula (34°36′N–35°05′N, 138°45′E–139°10′E), which is about 60 km long and 35 km wide (Figure 3.1). It projects southward from the Pacific coast of central Japan and separates Sagami Bay to the east from Suruga Bay to the west.

Although Kasuya and Yamada (1995, in Japanese) list several older examples of this use of the name *suji-iruka*, some uncertainty remains because the attribution is not based on a figure but on a description of morphology. There apparently has been no confusion among biologists on the species name since the nineteenth century, but there has been confusion in some nonbiological publications. It was in the autumn of 1960 when I accompanied M. Nishiwaki and S. Ohsumi to Kawana (34°57′N, 139°07′E) on the east coast of the Izu Peninsula that I saw striped dolphins for the first time. The species was long called *ma-iruka* (right dolphin) by the Izu fishermen. In the 1960s, the dolphin fishermen still called the species *ma-iruka*, and such usage was common until the 1980s. However, in the 1990s, there was a report of catch-quota violations by the Izu fishermen, and I telephoned the Fishery Cooperative Union (FCU) in Futo (34°55′N, 139°08′E) to confirm the report. In my conversation with a member of the staff of the FCU, I asked if the *ma-iruka* he was talking about were striped dolphins. He angrily answered that he did not use that erroneous name. This incident suggested that the Izu fishermen in the 1990s used the two species names in the correct way, that is, *ma-iruka* for *Delphinus* dolphins and *suji-iruka* for *Stenella coeruleoalba*.

Shizuoka-ken Suisan Shi (Records of Fisheries of Shizuoka Prefecture) by Kawashima (1894, in Japanese) recorded large-scale drive-fishery operations for *ma-iruka* or *ma-yuruka* on the Izu coast in the nineteenth century (*iruka* and *yuruka* were used synonymously for dolphins and porpoises). Kasuya and Yamada (1995, in Japanese) concluded that the report referred to striped dolphins based on the vernacular names of the species used there in the 1960s and the description of the dolphins, which formed large schools and weighed 48–93 kg. However, they did not have firm evidence upon which to exclude similar species such as common or spotted dolphins from consideration.

Ma-iruka was used by Japanese fishermen in various locations, but it did not necessarily represent the same species (Ogawa 1950, in Japanese). The name was used for the species of a dolphin or porpoise that was most common or most preferred for hunting in each fishing region.

10.2 SEA SURFACE TEMPERATURE

The distribution of dolphins and porpoises around Japan is strongly influenced by sea surface temperature (SST) (Kasuya 1982). Table 10.1 lists the SST at the position of sightings of dolphins and porpoises, compiled by J. Yamamoto from year-round records obtained by research vessels in the western North Pacific west of 180° longitude. It is incorrect to assume that the

TABLE 10.1
Sea Surface Temperature (in Celsius) at the Position of Sighting of Small Cetaceans[a] in the Western North Pacific

SST (n)	*Truei*-Type Dall's Porpoise	*Dalli*-Type Dall's Porpoise	Harbor Porpoise	Finless Porpoise	N. Right Whale Dolphin	Pacific White-Sided Dolphin	Killer Whale	Risso's Dolphin	False Killer Whale	Northern-Form Pilot Whale	Southern-Form Pilot Whale	Bottlenose Dolphin	Striped Dolphin	Pantropical Spotted Dolphin	Total
5	1			2											3
6				5											5
7	1	1		13											15
8	3			10		1									14
9	2	19		24		1	2								48
10	5	6		5		3				1					20
11	2	3		10		3	1	1		1		1			22
12	1	19		1	1	12		1				1			36
13	1	42		32	3	10	1	1		5		1	1		97
14	3	14		15		2	1	3		4	1	3			46
15	9	14		17	1	2	1	3	2	2		2			53
16	14	22	6	12		6	1	2		6		1			70
17	2	12		7	1	5		4	3	7		2			43
18	2	9	3	21		2	6	5	4	5	2	4			63
19	1			26		3	3	23	3	4	2	2	1	2	70
20				40		3	2	31	1	5	2	5	6	1	96
21			2	42			7	17	4	1	5	5		1	84
22			1	44		1	2	11	2	2	4	9	1		77
23				35	1		1	9	2	7	3	11	8	1	78
24				57				10	1	4	2	3			77
25				73				6		8	2		1		90
26				56				11	3	1	3	5			79
27				59				6	3		3	5	1		77
28				27				3	2		2	1	1		36
29				3			1		3		1	7			15
30				6							1				7
Total	47	161	12	642	7	54	29	147	33	63	33	68	20	5	1321

Note: SST is grouped to $\geq n$ and $< n + 1$ (n being an integer).

[a] External pigmentation pattern distinguishes *dalli*-type and *truei*-type of Dall's porpoises and southern and northern forms of short-finned pilot whales. Two species of the genus *Tursiops* are not distinguished.

number of sightings recorded in the table correctly reflects the relative abundance of the species in the area because (1) the sighting effort was biased toward coastal waters while the distribution of cetaceans is also influenced by distance from the coast and (2) the sighting records covered the period from the 1970s to the 1990s, when the abundance of some species had been greatly reduced by hunting. Striped dolphins were common in warm coastal waters off the Pacific coast of Japan in the 1960s, but they are now less frequently seen in the area.

Small cetaceans around Japan can be divided into four groups by temperature preference. The first group contains the cold-water species Dall's porpoise and harbor porpoise. The second is a group of warmwater species that includes *Stenella* spp., *Tursiops* spp., and Risso's dolphin, among which pantropical spotted and spinner dolphins tend to occur near the upper range and bottlenose dolphins in the lower temperature range. The third group contains the Pacific white-sided dolphin and the northern right whale dolphin that inhabit the surface water temperature of 10°C–20°*C*. Some of the killer whale populations off Japan may belong to this group, but further information is needed for firm conclusions to be drawn because the species is likely to include numerous local populations that are each adapted to their own environment. The short-finned pilot whale off Japan includes two latitudinally segregating populations, so it exhibits a broad temperature preference as a species (Table 12.3). The fourth group includes only the finless porpoise, which inhabits coastal and inland waters less than 50 m in depth (Section 8.3). By enduring broad seasonal fluctuation of SST from 5°C to 30°C and accepting possible seasonal change in prey items, it probably enjoys a habitat of fewer predators and high productivity.

The SSTs in the ordinary habitat of the striped dolphin range from 19° to 29°*C*. In the western North Pacific, the lower temperature limit is found off Hokkaido, 42°N, in August, and off Shikoku, 32°N, in February. This range of temperature also occurs in summer in the Sea of Japan and the southern Okhotsk Sea extending from Soya or La Perouse Strait (45°40′N, 142°00′E) to off the coast of Abashiri (44°01′N, 144°16′E). This is due to the summer extension of the warm Tsushima current and the local solar heating off the northern coast of Hokkaido. It should be noted that in spite of the apparently suitable sea temperature in summer, striped dolphins are extremely rare in the Sea of Japan and in the Okhotsk Sea, which is an indication that SST alone does not determine the distribution of this species. More data will prove the lower temperature limit for the species to vary by season, that is, lower in winter and higher in summer, which explains the occurrence of the species off Taiji (33°36′N, 135°57′E) in winter. Seasonal change in habitat temperature is detailed for Dall's porpoises (Tables 9.1 and 9.2) and the two types of short-finned pilot whales off Japan (Table 12.3).

The seasonal change in the SST of their habitat suggests that there are additional environmental factors that influence the distribution of striped dolphins off Japan. The availability of prey could be such a factor, which is also influenced by the oceanography of deeper water. Deeper water exhibits smaller seasonal fluctuations in temperature than surface water. Thus, deeper water remains relatively cooler while the SST rises in summer, and the reverse will happen in the winter off the Pacific coast of Japan. Water depth or distance offshore could be another factor controlling the distribution of striped dolphins, which are rare in inland waters.

The upper limit of SST where striped dolphins are seen is around 29°C, while some other delphinids, such as false killer whales, short-finned pilot whales, and bottlenose dolphins, have been recorded in higher temperatures.

10.3 GEOGRAPHICAL DISTRIBUTION

Perrin et al. (1994), using the then available records of striped dolphins, showed that the species inhabits tropical and temperate waters of the world oceans. However, the range summarized by them should be considered the maximum geographical range of the species, including the occurrence of drifted carcasses or individuals straying outside the normal range. Records from the coasts of Greenland, Faroe Islands, and western Aleutian Islands represent such cases. Although SST allows the species to occur in the southern Kuril Islands area and in the Sea of Japan, occurrence in these regions seems to be rare. The species has not been recorded in the Sea of Japan during numerous sighting surveys. The ordinary range of the species off the Pacific coast of Japan is limited to the south of Hokkaido or to the south of 42°N.

The Far Seas Fisheries Research Laboratory of the Fisheries Agency beginning in the early 1980s operated whale-sighting cruises in a systematic way to investigate the distribution and abundance of cetaceans in the western North Pacific and adjacent seas. Although the targeted area was modified annually to meet research needs of the time and tended to place more weight on nearshore waters, the activity covered the whole area west of 180° longitude and provided valuable information on the distribution of cetaceans. Data on small cetaceans thus obtained have been analyzed by month (Miyashita 1993 and Figure 10.1).

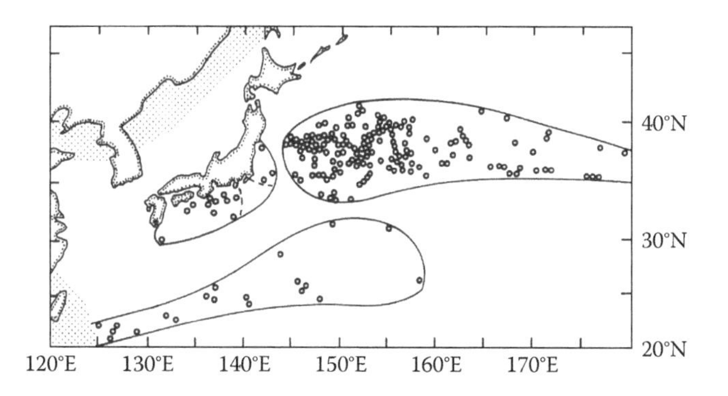

FIGURE 10.1 Positions of striped dolphin sightings (circles) in the western North Pacific during whale-sighting cruises of the Fisheries Agency in July–September, 1983–1991. Solid line delineates geographical group of sightings (or possible population) and dotted lines hypothetical boundaries of possible subgroups within the coastal area. The shaded area represents waters not surveyed by the cruises. The density of sighting effort is high in the western part of the range off the Pacific coast. See Figure 13.2 for a rough idea of the effort distribution. (Reproduced from Fig. 2 of Kasuya, T., *J. Cetacean Res. Manage.*, 1, 81, 1999.)

According to the results of these cruises, the northern limit of the striped dolphin in summer and in waters east of 145°E was around 42°N. There was limited survey effort in the Russian EEZ to the west of 145°E, but the presence of striped dolphins in the Russian EEZ off the southern Kuril Islands was unlikely because the cold Oyashio Current prevailed in the region (Kasuya 1999). It is worth noting that no striped dolphins have been recorded along the east coast of Hokkaido or south of the southern Kuril Islands, in spite of high sighting effort from July to September. Sighting surveys in the winter months, although the effort was limited, recorded striped dolphins in the waters of 10°N–35°N and 120°E–145°E.

The striped dolphin was not recorded in the eastern Sea of Japan during sighting cruises from July to September (e.g., Iwasaki et al. 1995; Miyashita et al. 1995) or in recent records of stranded cetaceans (Yamada 1993). The only significant record of the species there was of a school stranded near Shimonoseki (33°57′N, 130°56′E) near the opening to the East China Sea in May 1998. Thus, striped dolphins are uncommon in the Sea of Japan (Kasuya 1999).

Striped dolphins were not sighted in the East China Sea during a sighting cruise in summer (Iwasaki et al. 1995) or included in a collection of opportunistic sighting records of cetaceans (Uchida 1985, in Japanese). Another effort to investigate the cetacean fauna in the East China Sea in relation to the dolphin-fishery conflict in the Iki Island area (Section 3.4) also failed to confirm the presence of the species there. The only reliable record of striped dolphins in the East China Sea was a school sighted in winter to the north of the island Amami-oshima (28°15′N, 129°20′E) that bounds the East China Sea and the western North Pacific (Miyashita et al. 1995). Thus, even if striped dolphins may occur in the East China Sea, the species is believed to be rare there (Kasuya 1999). The scarcity of striped dolphins in the East China Sea and its possible absence in the Yellow Sea, further north of the East China Sea, might be due to the limited depth of those waters.

Striped dolphins are also considered rare off the coast of Taiwan. The east coast of Taiwan is edged by deep waters, but the only records of the species from there known to me are a skull in Taiwan Ocean University, one dolphin landed at Suao in northeastern Taiwan (Kasuya 1999), and a school sighted off Taiton in southeastern Taiwan over a water depth of 4000 m (Wang and Yang 2007, in English and Chinese).

In summary, the striped dolphin, a species of temperate and tropical waters, inhabits latitudes south of 42°N in the western North Pacific, but the species is uncommon in the peripheral seas. The distribution gaps within the western North Pacific are dealt with below (Section 10.4.1).

10.4 POPULATION STRUCTURE IN THE WESTERN NORTH PACIFIC

10.4.1 Population and Management

Because population dynamics proceed independently in each population or stock, identifying populations within a species range is important for management and conservation. While genetic mixing proceeds almost unrestricted within a population, genetic exchange is limited between populations. Even if there is a limited degree of genetic exchange between populations through permanent or temporally immigration of individuals, they should be managed as separate stocks if the immigration rate is negligibly low compared with the degree of human interactions. While a small amount of mixing between two nearby populations may mask genetic identification of the populations, they still need to be managed as independent populations because human-caused damage to such populations is likely to exceed the immigration rate and is likely biased toward one or the other population. Unless there is a very distinct difference in selection pressures on the two populations, which will not always be expected, such adjacent incompletely isolated populations will be difficult to identify by life history or genetic studies. Thus, the absence of firm life history or genetic evidence separating populations should not be considered as evidence of the absence of population structure for management purpose.

The range of two populations may overlap in a particular season or for individuals of particular growth or reproductive stages. Partial range overlap is known between Dall's porpoise populations (Section 9.3) and between offshore and coastal populations of bottlenose dolphins in the North Atlantic (Torres et al. 2003). Careful processing of available biological information is necessary to distinguish populations that occupy geographically overlapping habitats.

Striped dolphins live in schools of tens to hundreds of individuals. Although the schools are believed to be formed based on growth and reproductive stages and to exchange members, it is reasonable to expect that the probability of finding mother and calf or siblings in a single school will be higher than expected for random mixing. If between-school movement of individual striped dolphins is a function of geographical distance, then genetic similarity will be greater between nearby schools than between schools far distant from each other, and delineation of population boundaries will become difficult.

10.4.2 Geographical Distribution and Population Structure

Attempts to identify population structure often start with analysis of distribution. Using monthly records of sightings of small cetaceans in the western North Pacific published by Miyashita (1993), Kasuya (1999) constructed a map of the distribution of striped dolphins in summer (July–September), when northbound shift of the species was considered to have ceased (Figure 10.1). The sighting survey data used for these studies covered the period from 1983 to 1991 and the entire range of the species in the western North Pacific north of 20°N, but it reflected greater effort in waters inside the Japanese EEZ and less dense, albeit almost even, effort in farther offshore waters. Thus, the distribution of the species is overrepresented in nearshore waters compared with offshore waters.

Kasuya (1999) identified three concentrations in the summer distribution of striped dolphins in the western North Pacific. The first, the southern offshore concentration, is centered in the Kuroshio Counter Current area south of 30°N.

This concentration probably extends to the offshore waters of eastern Taiwan, but it should be noted that the species is absent or uncommon in Taiwanese coastal waters and in the East China Sea (Section 10.3). The second, the northern offshore concentration, is in offshore waters east of 145°E off northern Japan in latitudes 35°N–42°N. This concentration extends to at least longitude 180°E, but further eastern extension has not been confirmed. The third, the Japanese coastal concentration, is in nearshore waters off the Pacific coast of Japan, which covers the long coastline from the central part of the Sanriku Region (37°54′N–41°35′N) to Kyushu at around 30°N. There is no evidence to conclude that the third concentration is represented by a single population. The fact that survey effort was particularly dense in Japanese coastal waters in latitudes 34°N–43°N and west of 144°E suggests that the density of the species is particularly low in the coastal waters north of 35°N at least during the years surveyed (1980s and 1990s).

Abundance estimates for the three concentrations are 52,682 individuals for the southern offshore concentration, 497,725 for the northern offshore concentration, and 19,631 for the Japanese coastal concentration (for further details, see Table 3.13). The use of these figures for management requires caution because they are based on sightings during the 8 years from 1983 to 1991, when abundance of the species in Japanese coastal waters probably experienced a decline.

Based on the earlier observation on the distribution of striped dolphins in the western North Pacific, Kasuya (1999) concluded that each of the earlier-defined concentrations was likely to represent at least one population and that there were a minimum of three striped dolphin populations in the region. I thought that distribution of the species as shown in Figure 10.1 represents status at the climax of a northward shift of distribution and that the distributions would shift southward in autumn and winter. Thus, individuals in the southern offshore concentration would not migrate into Japanese coastal waters where the dolphin fishery operated. The possibility could not be excluded that some individuals in the northern offshore concentration visit Japanese coastal waters in autumn and winter and were hunted in some limited numbers by the fisheries off the Izu coast (drive) and Taiji (drive and hand harpoon) and perhaps also by a smaller-scale hand-harpoon fishery off Choshi (35°42′N, 140°51′E). However, they could not have been the major target of these fisheries. Annual catches of the dolphin drive fishery off Izu declined from 10,000–20,000 during the post–World War II period to less than 1,000 in the 1980s, while striped dolphins remained the main target of the fishery and the demands for the product remained high (see Sections 3.8 and 10.5.16). If individuals from the northern offshore concentration had constituted the main target of the Izu fishery, the collapse of the fishery would not have happened. The drive fisheries off Izu and hand-harpoon and drive fishery off Taiji most likely were supported by population(s) that constituted the Japanese coastal concentration, which was only about 20 thousand at the time of the surveys in 1983–1991 but could have been greater earlier on.

The question remains whether the Japanese coastal concentration contains more than one population. The warm Kuroshio Current turns eastward and leaves the Japanese coast around Choshi Point. Thus, the coastal environment north of 35°N exhibits a large seasonal change in oceanography, which is accompanied by seasonal alternation of the small cetacean fauna. Striped dolphins occur there only in summer; they are replaced by Dall's porpoises in winter. Most of these striped dolphins move south in autumn, passing off the southern tip of the Izu Peninsula (34°36′N, 138°51′E), and probably move further southwest. Thus, it is most likely that the Izu fishery in autumn to winter relied on striped dolphins that summered in the coastal waters north of the Izu Peninsula. It is unlikely that striped dolphins that summer in waters along the coast of Shikoku and Kyushu, further to the southwest of Taiji and Izu, were hunted by this fishery. This leads to speculation that the northern front area of the Kuroshio Current might harbor a dolphin population that is separated from individuals of the same species further south.

Two coastal populations of short-finned pilot whales off Japan, the northern and southern forms, segregate at around Choshi Point (Section 12.3).

10.4.3 Osteology

The Far Seas Fisheries Research Laboratory of the Fisheries Agency sampled striped dolphins by harpooning during its sighting cruises in the area inhabited by the northern offshore concentration of the species. Amano et al. (1997) compared these skulls with three samples collected from the coastal fishery:

1. 24 skulls from the Taiji and Izu fisheries in 1958–1979
2. 21 skulls from the Taiji fishery in 1992
3. 21 skulls from the northern offshore concentration taken in 1992

This grouping was probably a compromise, due to insufficient available materials, with the initial desire being to compare a total of five samples, that is, four samples representing the current and past operations at Taiji and Izu plus one recent sample from the northern offshore concentration. The results of multivariate analyses suggested that the three strata did not represent a single population.

Males of the three samples were separated from each other, but only the females of the second sample were separated from the specimens of the same sex of the first plus third samples. Although the authors did not reach a firm conclusion due to the insufficient sample sizes and uncertainty about females, they suggested that the three samples probably represented different populations. Such results would be expected (1) if the recent Taiji fishery targeted a population that was different from that exploited by the Taiji and Izu fisheries over 10 years ago, (2) if the proportions of populations in the catches of the Taiji and Izu fisheries changed over the period, or (3) if skull morphology has changed during the period of the collection. Their results also indicated that striped dolphins in the northern offshore concentration have not been hunted by the fisheries or that they have not been the major element targeted by the fisheries. It was unfortunate that they could not compare Taiji and Izu samples representing operations before the 1970s.

Amano et al. (1997) found sexual dimorphism in an early sample of specimens from Taiji and Izu obtained in 1958–1979 but not in the two more recent samples from Taiji and the northern offshore concentration obtained in 1992. This merits further investigation because the presence or absence of sexual dimorphism can be a population character (Archer 1996). In general, morphological studies such as osteology require caution relating to two possibilities. One is the inclusion of more than one population in a sample. Depending on the analytical method, it may not be detected, and different amount of mixing may lead to different conclusions. The second caution is that change in growth during the history of exploitation may have occurred, perhaps as a response of the population to density change.

10.4.4 Body Size

Iwasaki and Goto (1997) compared body size in three following samples:

1. Early sample from the Izu fishery reported by Kasuya (1972): 567 females and 391 males
2. Recent sample from the Taiji fishery collected in the 1991/1992–1994/1995 seasons: 412 females and 301 males
3. Recent sample from the northern offshore concentration taken in 1992: 42 individuals

The results are shown in Table 10.2. Although the sample size was small, it seems to be true that striped dolphins in the northern offshore concentration were greater in body size than those in the Taiji sample for similar years, while they were equal to or slightly greater in size than in the Izu sample from around 1970. Body size in the recent Taiji sample was about 10 cm smaller than in the older Izu sample, which is the reverse of the difference that might be expected for density-dependent growth change. These growth differences lead to the conclusion that the sample from the northern offshore concentration, the individuals sampled from the recent Taiji fishery, and those sampled from the past Izu fishery belong to different populations. This conclusion does not contradict the results from osteology.

TABLE 10.2
Temporal and Geographical Differences in the Body Size of Striped Dolphins in the Western North Pacific

Location, Time, and Methods of Take	Maximum Body Length		Modal Body Length	
	Female (cm)	Male (cm)	Female (cm)	Male (cm)
1. Izu[a] coast, 1960s–1970s, drive fishery	243–247	258–262	223–227	238–242
2. Taiji[b] coast, 1990s, drive fishery	238	249	210–219	220–229
3. Northern offshore[c], 1990s, hand harpoon	251	257		

Source: From Iwasaki, T. and Goto, M., Composition of driving samples of striped dolphins collected in Taiji during 1991/1992 to 1994/1995 fishing season, Paper IWC/SC/49/SM19 presented to the Scientific Committee Meeting of IWC in 1997, Available from IWC Secretariat, 1997, 20pp.

[a] Taken at Kawana (34°57′N, 139°07′E) and Futo (34°55′N, 139°08′N).
[b] Taiji, 33°36′N, 135°57′E.
[c] See Figure 10.1.

10.4.5 Genetics

Wada (1983) was the first to study population structure of striped dolphins off Japan using genetic information (isozymes in this case). He randomly collected 40 liver samples from a school of 431 striped dolphins driven at Kawana on the east coast of the Izu Peninsula in 1980 and analyzed 10 enzymes. He found that 5 of the 10 enzymes had two subtypes and were controlled by 15 loci in total. Only 7 of the 15 loci were polymorphic, and he found the average heterozygosity for all the 15 loci to be 0.021, which was extremely low compared with values for other mammals. He considered this to be due to inbreeding. Such results would be expected because the school was likely to contain genetically related individuals, such as siblings and mothers and offspring. Therefore, he recommended that multiple schools be examined. Later, Shimura and Numachi (1987) examined 12 enzymes for 370 striped dolphins taken from several schools and found higher heterozygosity similar to levels in Dall's porpoises and bottlenose dolphins. These two studies did not clarify population structure.

Yoshida and Iwasaki (1997) analyzed mitochondrial DNA (mtDNA) for the population structure of striped dolphins using 43 dolphins hand-harpooned from the northern offshore concentration in August 1992 and 34 collected from the Taiji fishery in September–November 1992. They identified a total of 61 haplotypes. Only 4 were shared by the two samples (represented by 4 offshore and 6 Taiji individuals), 36 haplotypes were unique to the offshore sample (in 36 individuals), and 21 were unique to the Taiji sample (in 27 individuals). Such differences might give the impression that the two samples were genetically different, but the difference was not statistically significant. This should not be interpreted to mean that the two samples came from the same population. Greater samples might confirm significant genetic difference.

10.4.6 Mercury Level

Molecules of persistent chemicals in marine water are absorbed by phytoplankton, which are moved through the food chain to zooplankton, fishes, and squids and finally to marine mammals. Striped dolphins, one of the top predators, are known to have high concentration of such chemicals (e.g., Miyazaki 1992, in Japanese). The concentration levels of such chemicals not only increase with age in general but also vary with food habits and pollutant levels in the habitat. Lipophilic pollutants such as DDT and PCB in females are discharged to the offspring through lactation, which causes difference in concentration levels between sexes at the same age.

Mercury, one of the persistent pollutants, is not lipophilic and its behavior is different from that of the organochlorines; it does not exhibit significant difference in level between the sexes (Honda et al. 1983). Itano et al. (1984) reported the relationship between levels of heavy metals and age in organs of striped dolphins using materials sampled from a school driven in Kawana (34°57′N, 139°07′E) on the east coast of the Izu Peninsula and from three schools driven in Taiji during the period 1977–1980. The levels of mercury and selenium in the liver were about 20 times higher than those in skeletal muscle, and the levels of total mercury and selenium showed a good correlation. Here, I will concentrate on the relationship between mercury levels in skeletal muscle and age because the sample size was the largest. I have ignored sex and reproductive status for this reason.

The total mercury level in the skeletal muscle of newborn animals was around 1 ppm/wet weight. It started to increase at around age 5 years, when body growth slowed and the dilution effect would have decreased (Figure 10.2). A group of striped dolphins had a total mercury of around 25–35 ppm at around 25 years of age and retained the level thereafter. Since excretion of mercury is a function of concentration level in the body, ingestion and excretion of mercury could have reached a balance at ages over 25 years, but the effect of historical change in mercury level in the environment could not be excluded. Mercury in another group of striped dolphins analyzed in Itano et al. (1984) leveled off at around 20 years of age and was at a lower level of 10–16 ppm. These two groups could already be separated at ages 10–15 years. On the other hand, the level of methyl mercury was reached at around 5 ppm at 10 years of age and retained thereafter, suggesting that its physiological activity was different from inorganic mercury in the body.

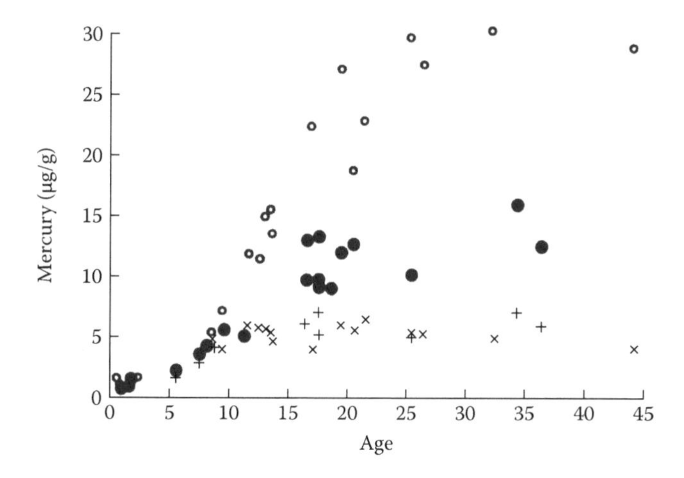

FIGURE 10.2 Scatterplot showing relationship between age and levels of total mercury (open and closed circles) and methyl mercury (+ and ×) in skeletal muscle (μg/g of wet weight) of striped dolphins in four schools driven on the Pacific coast of Japan. Open circles and ×'s represent samples collected from two schools, driven at Kawana (34°57′N, 139°07′E) in October 1977 and at Taiji (33°36′N, 135°57′E) in December 1980. Closed circles and +'s represent dolphins from two schools driven at Taiji in December 1978 and 1979. (Reproduced from Fig. 1 of Itano, K. et al., *Agric. Biol. Chem.*, 48(5), 1109, 1984. With permission; Also appeared in Kasuya, T., *J. Cetacean Res. Manage.*, 1, 81, 1999.)

As described earlier, full-grown striped dolphins in the two schools were divided into two groups by total mercury in the skeletal muscle, that is, 10–16 ppm versus 25–30 ppm. Miyazaki (1992) and Kasuya (1999) interpreted this to mean that multiple populations were being hunted by the Japanese drive fisheries. Itano et al. (1984) gave only location and month and year of the sample but did not state the number of dolphin schools from which their materials were collected. However, it seems to me that they were most likely collected from four schools in total as described in Figure 10.2. Thus, the high-mercury group was represented by a school driven at Kawana in October 1977 and another school driven at Taiji in December 1980, and the low-mercury group came from two schools driven at Taiji in December of 1978 and 1979. The striped dolphins in these two mercury-level groups must have used different feeding grounds or consumed different prey items for most of their life. It also suggested that no individuals switched their feeding habit during their lifetime.

Such striped dolphin groups with quite different feeding histories can be interpreted as representing different populations. Thus, during the late 1970s and early 1980s, the Taiji fishery targeted two populations of striped dolphins (high- and low-mercury groups) and one of them (high-mercury group) was also hunted in the Izu fishery (Kasuya 1999).

10.4.7 Fetal Size and Migration Timing

Figure 10.3 shows the fetal size composition of striped dolphins taken by the drive fishery on the east coast of the Izu Peninsula (October–December), the west coast (May), and at Taiji (January and June). In October there was one major fetal mode below 30 cm with a peak at 10 cm, in November there were two modes of similar size at around 0–40 cm and 60–100 cm, and in December there was one major mode above 60 cm with a peak at 85 cm. The gestation time of striped dolphins has not been precisely determined, but it can be safely assumed to be about 1 year based on information available for several other small toothed whales. For example, a gestation period of about 370 days is known for Indo-Pacific bottlenose dolphins in captivity, confirmed by monitoring of hormones in the blood serum (Brook et al. 2002). An assumption of 1-year gestation and average neonatal length of 100 cm suggest a very rough estimate of average fetal growth rate of about 8–9 cm/month. Thus, the apparent fetal growth of about 75 cm in the two-month period, from 10 cm in October to 85 cm in December, is too great to be explained by fetal growth. Rather, it can be explained by a difference in migration timing between females in early pregnancy and those that are near term. The two fetal modes are believed to represent two conception peaks, which are about 6 months apart. This species off Japan shows a weak seasonality in mating,

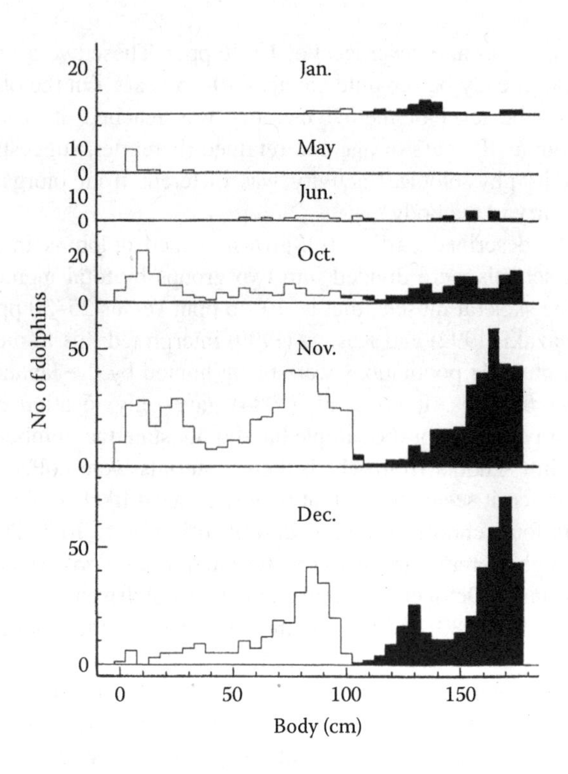

FIGURE 10.3 Monthly change in body-length composition of fetuses (white squares) and juveniles (black squares) of striped dolphins driven on the east coast (October–December) and west coast (May) of the Izu Peninsula (34°36′N–35°05′N, 138°45′E–139°10′E) and at Taiji (January and June). (Reproduced from Fig. 2 of Miyazaki, N., *Rep. Int. Whal. Commn.*, Special Issue 6, Reproduction in Whales, Dolphins and Porpoises, 343, 1984. With permission.)

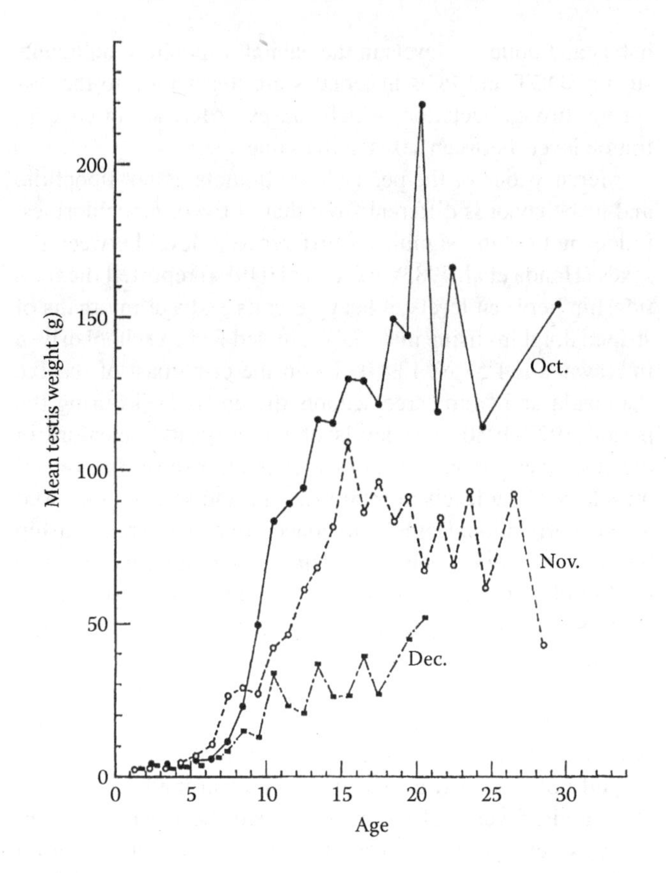

FIGURE 10.4 Monthly plots of the mean weight (g) of a single testis (epididymis removed) on the age of striped dolphins driven on the Izu coast. The testicular weights of the sexually mature individuals declined from October to December. Here, ages are estimated using only dentinal growth layers. This method is likely to underestimate the true ages of individuals over 15 years, which are certainly sexually mature (see Figure 10.7 and Section 10.5.5.3). (Reproduced from Fig. 8 of Miyazaki, N., *Sci. Rep. Whales Res. Inst.* (Tokyo), 29, 21, 1977. With permission.)

and the peaks have been estimated variously, but it has been agreed that one took place in winter (November/December or January) and the other in early summer (June or July) (Figure 10.12, Section 10.5.8).

The Izu fishery in October caught schools of striped dolphins that contained females in early pregnancy and perhaps those in estrus. Some of them had conceived in summer, and others were still involved in mating activity when they passed Sagami Bay off the east coast of the Izu Peninsula on their southbound migration. In November, another type of school arrived that contained females in near-term pregnancy conceived in the previous winter. In December, the first type of school almost disappeared and the second type predominated.

Miyazaki (1977) analyzed monthly change in the relationship between average weight of a single testis and age (Figure 10.4). The correlation between mean testis weight and age is lost at around 15 years of age, when males are likely to have completed testicular growth. Average testis weights of males over this age were 120–170 g in October, 60–100 g in November, and 25–50 g in December. Miyazaki (1977) interpreted this as a reflection of seasonal decline in male reproductive activity from October to December.

Adult females and adult males were always found in a school of striped dolphins, although the proportion varied among schools. It seems likely that the gonadal activity of males in these schools will have some correlation with female reproductive stage, that is, males with active testis will be found with females in estrus or in early pregnancy and inactive males with females in near-term pregnancy. We have already seen that testis weights of Dall's porpoises in the mating season almost double those in the nonmating season (Section 9.4.8). This allows for an interpretation of the data presented in Figure 10.4 different from that of Miyazaki (1977), that males caught in October had active heavier testes because they accompanied females in estrus or early pregnancy, and that males taken in December had lighter testes because they accompanied females in near-term pregnancy. I suspect that there were two types of males in the November catch, because there were two types of females in almost equal number caught off the east coast of the Izu Peninsula. Such a November sample could have an average testis weight intermediate between those of the October and December samples (Figure 10.4). To answer this question, it would be best to reanalyze the composition of the schools studied by Miyazaki (1977), but a quick hint could be obtained by confirming whether testis weights of adult males are bimodal in November as in the case of fetal body lengths. Such analysis has not been done to present (also see Section 10.5.9).

Females of striped dolphins in the same reproductive status (e.g., in estrus, late pregnancy, or lactation) are known to live in close association (Kasuya 1972; Miyazaki and Nishiwaki 1978). A similar pattern was observed in humpback whales that were killed along the coasts of Australia and New Zealand on the way from the feeding grounds in the Antarctic Ocean to the breeding grounds in lower latitudes (Dawbin 1966). The first to pass the coastal whaling ground in the middle latitudes were females in lactation or weaning, which were followed by immature females, and then by resting females. The last members that arrived from the Antarctic feeding ground were females with a near-term fetus, and their arrival almost agreed with the end of migration of the first element (weaning or lactating females). The time interval between the peaks of the first and the last elements was about 35 days. It is reasonable for pregnant females to remain longer in the feeding ground in order to prepare for parturition and subsequent lactation to be achieved in lower latitudes without feeding. The later arrival of striped dolphins in near-term pregnancy might be explained by some unknown physiological requirement of the females. This explanation assumes two breeding seasons, which are about six months apart, in a single coastal population of the species.

There is another plausible explanation for the migration patterns of pregnant striped dolphins, which assumes two populations for striped dolphins along the Pacific coast of Japan, breeding about 6 months apart. Different populations of minke whales (Best and Kato 1992; Kato 1992), short-finned pilot whales (Section 12.4.4), and finless porpoises (Section 8.5.1) around Japan are known with breeding seasons about 6 months apart. A similar case can be expected for the striped dolphins along the Pacific coast of Japan. Some supporting evidences for this hypothesis are found in the history of the drive fishery off the Izu coast.

The drive fishery for striped dolphins along the coast of the Izu Peninsula had two peaks in winter (December–January) and spring (March–June) (Table 10.3, Section 3.8.3). The winter season was conducted on both the east and west coasts of the peninsula and recorded a maximum catch in December in the 1953 and earlier seasons, although there was some small catch in the preceding months. This was reasonable from a fisheries point of view because price and demand for the dolphin meat increased toward winter. However, the season on the east coast tended to start earlier and beginning in the 1954 operation started in October, accompanied by decline of the catch in December and January. The west coast fishery did not show this seasonal shift but recorded a gradual catch decline and made its last regular operation in 1961. The east coast fishermen expanded the searching range to outside of Sagami Bay in 1962 (Kasuya 1985, 1999). The earlier start of the fishing season can be interpreted as evidence of shifting the weight of the fishing operation from one population to another, that is, until 1953 the fishery exploited mainly a population that arrived in the Izu area in December (with females in near-term pregnancy), and then the operation began moving to another population that arrived in the area in October (with females in early pregnancy and estrus). Miyazaki (1983) analyzed the operations of the drive fishery in 14 years from 1967/1968 to 1980/1981 by two driving groups, Kawana and Futo (34°55′N, 139°08′E), which were located on the east coast of the Izu Peninsula and operated in Sagami Bay (Figure 10.5). The amount of fishing effort expressed as number of searching days remained almost the same from early October to late December, but the catch in number of striped dolphins and number of schools, as well as average school size, had two peaks, one in late September to middle October and the other in middle to late November. These two peaks were separated by a trough in late October to early November.

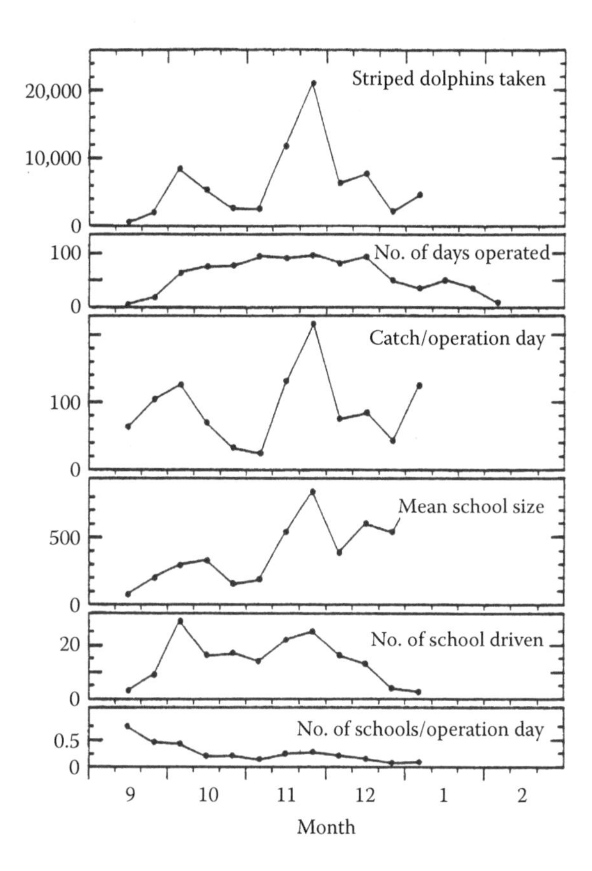

FIGURE 10.5 Analysis of dolphin drives in 14 seasons (1967/1968–1980/1981) at Kawana and Futo (34°55′N, 139°08′E) on the east coast of the Izu Peninsula that bounds western Sagami Bay (centered at 35°N, 139°20′E, see Figure 10.6), in an attempt to see the pattern of the movement of striped dolphins to the fishing ground. (Reproduced from Fig. 5 of Miyazaki, N., *Rep. Int. Whal. Commn.*, *Rep. Int. Whal. Commn.*, 33, 621, 1983. With permission.)

10.4.8 Suggested Population Structure

None of the previously discussed information on geographical distribution, skull morphology, growth, mtDNA, mercury concentration, and migration pattern allows definite conclusions to be made about the population structure of striped dolphins taken by fisheries off the Pacific coast of Japan. However, they suggest the possibility of involvement of multiple populations in the fishery rather than a single population. The evidence is summarized as follows, where "early operations" mean those in 1980 or before that date and "recent operations" mean those in 1990 or later.

1. The northern offshore concentration has not been a significant target of the Izu fishery, but the possibility of low-level involvement cannot be excluded (inferred from distribution, skull morphology, and trends in fishing operations).
2. Individuals from the northern offshore concentration have not been found among striped dolphins recently taken in the Taiji fishery (inferred from skull morphology, growth, and DNA) or among those taken by past operations (inferred from skull morphology).
3. Striped dolphins taken by the recent Taiji fishery contain individuals from a population that has not been recorded from the Izu fishery (inferred from skull morphology and summer distribution).
4. The past Taiji fishery targeted two populations of striped dolphins, and one of them was also taken by the Izu fishery (inferred from mercury concentration).
5. The Izu fishery could have targeted two populations that migrated through the fishing ground successively (inferred from fetal size and seasonality of the fishery).

The five conclusions do not contradict each other and allow the hypothesis of four populations in the western North Pacific north of 20°N, a slight modification of the population structure inferred from the distribution pattern (Section 10.4.2).

The first population is represented by the southern offshore concentration that inhabits the Kuroshio Counter Current area between 20°N and 30°N and has a range extending close to the east coast of Taiwan. This population was unlikely to have been exploited by the drive fisheries on the Taiji and Izu coasts.

The second population is the northern offshore population represented by the northern offshore concentration. This is a large population that summers over 150 nautical miles (280 km) offshore of northern Japan (35°N–42°N) with its eastern range extending close to 180° longitude. Even if the Japanese dolphin fisheries have occasionally killed migrants from this population, the number must have been limited.

The third population (temporarily named "northern coastal population") is hypothesized to summer in the coastal waters off northern Japan north of Choshi Point or north of the summering ground of the fourth population. Thus, it passes the Izu ground later in the season. The drive fishery off the east coast of the Izu Peninsula relied mainly on this population until the middle 1950s when it operated mainly in December and January. The west coast fishery probably hunted this population both in the winter operation (December–January) and in the spring operation (March–June) until the 1950s.

The fourth population ("southern coastal population") is hypothesized to summer in Japanese coastal waters south of Choshi Point. The southern limit of this population is unknown. The Izu Peninsula is located about 220 km southwest of Choshi Point. This population is likely to have been taken by the drive fishery off the east coast of the Izu Peninsula, in the early part of the autumn–winter operation (October and November), and probably by hand-harpoon and drive fisheries at Taiji in winter. Fishermen along the east coast of the Izu Peninsula probably increased pressure on this population beginning in the mid-1950s when the fishing season tended to start earlier. This population was not a major target of the west coast fishery that operated in December–January and March–June until 1961.

Fragmentary information on the oceanographic environment inhabited by striped dolphins in Japanese coastal waters is available from four cruises of the research vessel *Tansei-maru* of the Ocean Research Institute of the University of Tokyo, conducted from September to November 1970–1973 in waters around 35°N and west of 141°E. The results of the two cruises are shown in Figure 10.6. During the cruises, striped dolphins were sighted close to the northern boundary of the warm Kuroshio Current and in the meander of the current. As the season progressed, they moved southwest close to the Izu Peninsula and Oshima Island (34°45′N, 139°25′E, see bottom panel of Figure 10.6) at the entrance of Sagami Bay

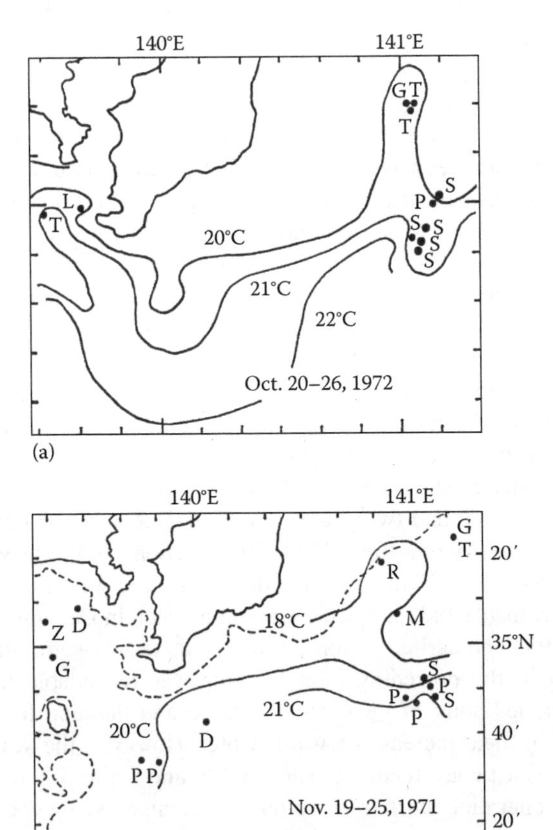

FIGURE 10.6 Distribution of striped dolphins and oceanographic features northeast of the dolphin hunting ground of Kawana and Futo in the early part of the season, observed during two sighting cruises of (a) October 20–26, 1972, and (b) November 19–25, 1971. The solid line indicates SST and the dotted line 200 m isobath. D, short-beaked common dolphin; G, short-finned pilot whale (southern form); L, Pacific white-sided dolphin; M, minke whale; P, sperm whale; R, Risso's dolphin; S, striped dolphin; T, common bottlenose dolphin; Z, Ziphiidae. (Reproduced from Fig. 8 of Miyazaki, N. et al., *Sci. Rep. Whales Res. Inst.* (Tokyo), 26, 227, 1974. With permission.)

and were located by dolphin-searching vessels from the Izu coast. The surface water temperature was 18°C–20°C when striped dolphins were sighted during the cruises. As this temperature moved between 42°N in summer and 32°N in winter (Section 10.2), these sightings might be allocated to the northern coastal population, one of the two hypothetical coastal populations, when further information becomes available on the distribution and population structure of the species along the Pacific coast.

The past drive fisheries along the Izu coast hunted striped dolphins in autumn and winter on their southbound migration and in the spring to early summer season on their northbound migration (Ohsumi 1972; Miyazaki and Nishiwaki 1978). This is shown in Table 10.3, which has been calculated from statistics in Miyazaki (1983) and indicates annual monthly mean landings of striped dolphins per drive team. Readers might consider that multiplying these figures by the number of drive teams gives an annual average of the catch on the Izu coast, but we must remember that the statistics are available only for the three drive teams of Kawana and Futo on the east coast and Arari (34°50′N, 138°46′E) on the west coast, that there are several teams that have not left catch statistics for the early post–World War II period (Table 3.11), and that we do not know if the magnitude of these missing operations was the same as those of the documented operations. The Kawana and Futo teams started cooperative driving beginning in 1967 or 1968 (Section 3.8.1) and drove dolphin schools alternatively into one of the two harbors. The effect of the changed operation pattern is ignored by counting them as one team. The catch at Arari after 1961 was excluded from Table 10.3, because Arari made its last systematic drives in 1961; later drives were opportunistic and irregular (Table 3.14). Table 10.3 contains a negligible number of other species (Kasuya 1976b, in Japanese). Tobayama (1969, in Japanese) reported that 96.7% of 64,797 dolphins taken at Kawana and Futo during the 1963–1968 seasons were striped dolphins (Section 3.8.3).

Until the 1961 season, fishermen on the west coast of the Izu Peninsula, that is, Suruga Bay coast, operated the dolphin drive fishery on a regular basis. Their catch showed two peaks from March to June and from December to January and a trough in February (Table 10.3). Before the mid-1950s, the operation from the east coast (including Futo and Kawana), that is, Sagami Bay coast, had a single peak in December, which overlapped with the winter peak of the west coast. This means that the drive fisheries along the coasts of the Izu Peninsula had two peaks of catch from March to June and from December to January before the mid-1950s (both aimed at the northern coastal population). From March to June, operation of the west coast fishery agrees with timing of the northbound migration of this population. However, beginning in the mid-1950s, the east coast fishery gradually shifted its operation to earlier months of the autumn, October and November, to target the southern coastal population and decreased operations from December to January targeting the northern coastal population.

In the early 1960s, dolphin fishermen at Kawana and Futo (east coast of the Izu Peninsula) told me that dolphin fishermen on the both coasts of the peninsula operated in the same season and in similar fishing grounds, that is, near the southern tip of the Izu Peninsula (Figure 3.1). This suggests that at a particular time of the year, perhaps from December to January, drive fishermen from the east coast occasionally encountered those from the other side of the Izu Peninsula in an area off the southern tip of the peninsula. Because the east coast fishermen did not operate the drive from March to June season, they either had no opportunity to encounter the

TABLE 10.3
Seasonal Operation Pattern of the Drive Fishery for Striped Dolphins on the East Coast (Kawana and Futo) and West Coast (Arari[a]) of the Izu Peninsula, Expressed as the Mean Number of Dolphins Taken/Month/Drive Group for the Given Period

Period	Area	Jan	Feb	Mar	Apr	May	Jun	Jul	Aug	Sep	Oct	Nov	Dec	Total[b]
1942–1953	East coast	81	126	3	195	0	0	0	0	59	0	148	1391	2002
1942–1953	West coast	797	177	705	1516	1542	659	270	13	22	126	9	1444	7280
1942–1953	Total	878	303	708	1711	1542	659	270	13	81	126	157	2835	9282
1954–1967	East coast	393	4	0	0	0	0	0	0	0	395	1048	1677	4731[b]
1954–1961	West coast	906	180	0	31	63	208	40	8	8	77	1	986	2508
1954–1967	Total	1299	184	0	31	63	208	40	8	8	472	1049	2663	6879
1968–1981	East coast	324	0	0	0	0	0	0	0	160	1157	2337	1060	5038
1968–1981	Total	324	0	0	0	0	0	0	0	160	1157	2337	1060	5038

Notes: Here are the periods of the following data: Kawana, 1942–1981; Futo, 1958–1981; and Arari, 1942–1961. Kawana was assumed as having no catch in the 1943 and 1947 seasons as interpreted by Kasuya (1976b), which disagreed with Miyazaki (1983), who thought only that data were missing for operations that occurred in those two seasons. The difference of interpretation does not significantly affect the seasonal pattern of the fishery operation.

[a] Arari, 34°50′N, 138°46′E.

[b] Some records give only annual total and no monthly figures, which caused disagreement between this figure and the total of monthly figures.

operation from the west coast or did not pay much attention to the operation of the west coast fishermen from March to June season.

If my hypothesis on the population structure is correct, it will be deduced that before the mid-1950s, the northern coastal population on the northbound migration was the main target of the spring operation off the west coast (for some unknown reason the east coast teams did not operate this spring hunt). The same population on the southbound migration was targeted by the autumn–winter (December–January) operations on both sides of the peninsula. This situation changed in the mid-1950s, when the east coast fishery started early autumn operation (September–November) on the southern coastal population on the southbound migration and maintained its previous catch level for a while, while the west coast fishery adhered to the northern coastal population and decreased its catch to finally stop the fishery in the early 1960s.

The currently available fragmentary information on striped dolphins off Japan may allow creation of more elaborate hypotheses about their population structure. It is also possible to ignore some of the unconvincing evidence and create simpler hypotheses. Simpler hypotheses allow easier fishery management but likely increase the risk of management failure. I believe that the hypotheses put forward here are probably the simplest that incorporate all the available biological information on the species. It is necessary to collect further evidences on the stock structure of the species and be ready to restructure the population hypotheses.

10.5 LIFE HISTORY

10.5.1 Age Determination

Fishery scientists have long used cetacean carcasses as a source of information necessary for management, and age estimation of sampled animals has always been one of the tasks to overcome. In Japan, attempts to estimate the age of toothed whales were started by Tasuke Amano, a whaling gunner, who examined various tissues of sperm whales in the early 1920s and reached the conclusion that layers in their teeth were promising and that females attained sexual maturity at around 7–8 dentinal layers. His manuscript was published posthumously in 1955 (Amano 1955, in Japanese).

Another attempt at age determination of toothed whales was started by Nishiwaki and Yagi (1953) with striped dolphins. In those days, seal scientists already used growth layers in dentine or growth ridges visible on the root of a canine tooth for aging, and medical physiologists studied the cycle of daily deposition of dentine layers by injecting lead acetate for time marking. Nishiwaki and Yagi (1953) injected lead acetate into striped dolphins driven by the Izu fishermen with the hope of determining the deposition rate of dentinal layers in the teeth. Their attempt was in the correct direction but failed to achieve its objective because the test animals lived for only a short period. Apart from the unknown effect of injected chemicals, the technology of maintaining dolphins in captivity was in its infancy at the time, particularly for striped dolphins, which in Japan are still considered difficult to maintain in captivity (now there are cases of over 1-year survival). Nishiwaki's group shifted its major effort in age determination to sperm and fin whales.

I received biological data for some striped dolphins examined earlier by M. Nishiwaki and S. Ohsumi and combined them with my own data on striped dolphins taken by the drive fishery on the Izu coast in 1968–1970 for a life history study (Kasuya 1972). In this study, thin polished tooth sections were observed under a light microscope to examine seasonal alternation of deposition of dentinal growth layers. This led to the conclusion that the alternation from translucent to opaque dentine occurred from late November to early December and that one pair of opaque and translucent layers represented a year. Later, Miyazaki (1977) reached the same conclusion using larger samples. Materials used in these studies covered only 5 months of the year from October to February, so they were unable to determine the time when the deposition alternated from opaque to translucent. This remains a weak point of the conclusion. However, the assumption of annual deposition of dentinal growth layers is currently accepted for striped dolphins, because the analysis of their life history has produced reasonable results and because the annual deposition of dentinal growth layers has been confirmed using a technique of time marking with tetracycline for other more tropical dolphins (Myrick et al. 1984). Japanese scientists have often expressed the estimated age of dolphins to the nearest $n + 0.5$ years (n being an integer) for convenience in analysis. This does not indicate the precision of age reading but only means that age is estimated at between n and $n + 1$ years.

The pulp cavity decreases in volume with increasing age as dentinal tissue is deposited on the pulp wall, and the growth layer deposition finally ceases at a certain age. Thus, counting dentinal growth layers only gives a minimum age for old individuals. Sergeant (1962) first pointed out this problem in his landmark study on long-finned pilot whales and suggested that cemental layers would continue to be deposited after cessation of dentinal deposition, but he still used dentinal layers in his analysis of the life history of the species. Later, I re-aged his materials by counting dentinal and cemental growth layers in decalcified and hematoxylin-stained thin tooth sections and confirmed that the deposition of dentinal growth layers ceased at ages between 15 and 25 layers (years) (Kasuya et al. 1988). With exception of the teeth of some odontocete species, for example, sperm whale, narwhal (tusk), and white whale, deposition of dentinal tissue ceases at a certain age, so ages of older individuals have to be estimated by counting cemental growth layers, which are thinner than dentinal layers but are believed to be deposited throughout life.

Striped dolphins are not an exception in this aging problem. Their dentine ceases to be deposited at ages between 13 and 20 layers (Figure 10.7). The oldest age determined by dentinal layers is 28 layers, while cemental layers gave an age of 49 layers for the same individual. Kasuya (1976a) observed for the pantropical spotted dolphin that deposition of dentine alters from a hematoxylin-stainable layer to an unstainable

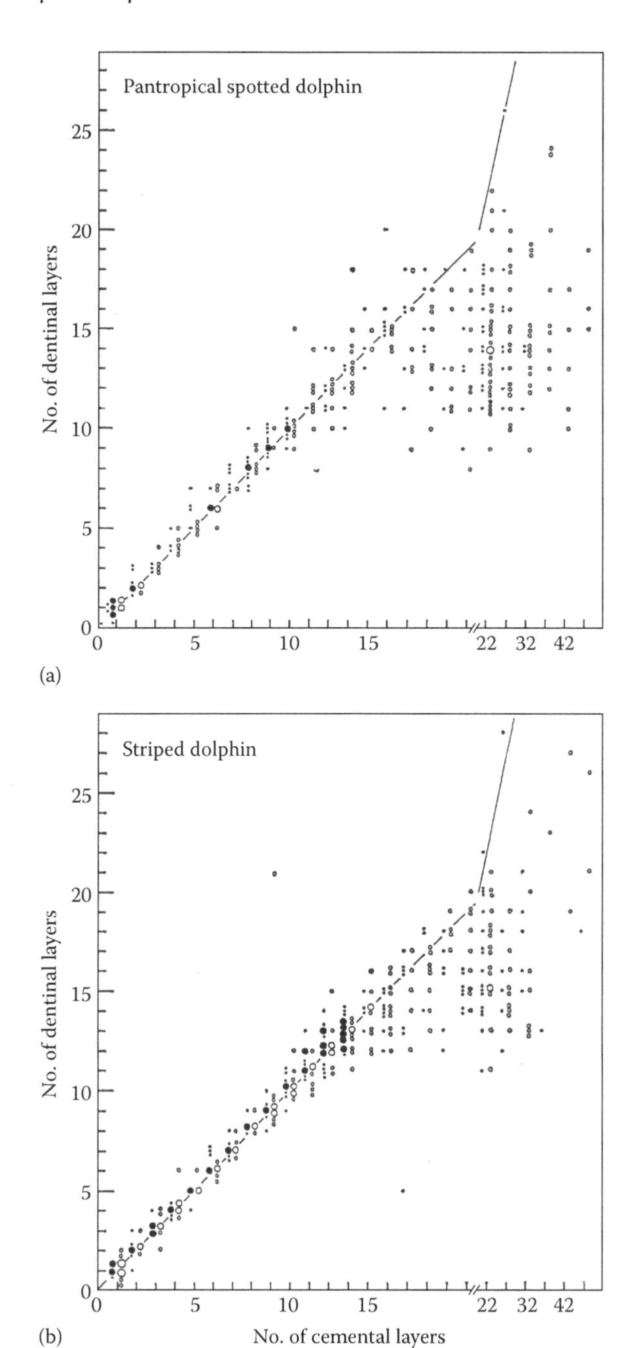

FIGURE 10.7 The comparison of the growth layer counts between dentine and cementum of the same tooth of (a) pantropical spotted and (b) striped dolphins, showing that dentine deposition in striped dolphins ceases at ages over 14–15 years. Open circles indicate females and closed circles males. For each sex, the larger symbol represents five individuals and the smaller symbol one individual. (Reproduced from Fig. 2 of Kasuya, T., *Sci. Rep. Whales Res. Inst.* (Tokyo), 28, 73, 1976a.)

layer from February to July and from an unstainable layer to a stainable layer in October. Such seasonal change has not been confirmed in the striped dolphin due to insufficient seasonal coverage of the material.

It was after publication of my study on the striped dolphin (Kasuya 1972) that I found a short statement in Sergeant (1962) on the deposition of cemental layers that are likely to continue after cessation of dentin deposition. I then reanalyzed the life history of striped dolphins using cemental layers (Kasuya 1976). Due to this error of mine, some life history analyses of dolphins in Japan used dentinal layers and contained problems in aging old individuals, that is, Kasuya (1972), Kasuya et al. (1974), Miyazaki (1977), and Miyazaki and Nishiwaki (1978).

Ages of dolphins and porpoises are currently estimated using layers in cementum or both dentine and cementum. Cementum is a tissue that connects a tooth and surrounding tissue and is accumulated throughout life. It is absent on the teeth of newborn dolphins but starts to be deposited at the age of several months when tooth eruption usually occurs and continues throughout life. Identifying and counting cemental growth layers are more difficult than dentine reading, because they are thinner and less regular in thickness. Thus, it helps to compare growth layers between dentine and cementum on the same tooth and establish a standard of age reading on younger animals before starting to age older individuals.

Age reading is better achieved on decalcified hematoxylin-stained thin sections of teeth rather than thin polished undecalcified sections. Two methods have been used for the preparation of thin decalcified sections before staining: (1) decalcifying a whole tooth or a part of a tooth trimmed to a suitable size and then sectioning it longitudinally with a cryostat and (2) gluing a longitudinal half tooth on a plastic plate, cutting off the outer portion and polishing down to a thin longitudinal section of a desirable thickness left on the plastic plate for decalcification. The thin decalcified tooth section is then stained with hematoxylin (double staining with hematoxylin and eosin is not recommended because it decreases readability). The first method can produce several slides from a single tooth. The second method can produce only one slide but is suitable for a laboratory without costly equipment. The preparations should be well rinsed to remove acid before mounting. Acid remaining on the slide often causes a quick fading of the stain. About 40 μm is suitable as a thickness for large teeth such as those of short-finned pilot whales and bottlenose dolphins and about 20 μm for species with small teeth such as the striped dolphin (Kasuya 1983, in Japanese). However, it is desirable to make the cementum portion slightly thinner than the dentine, which is technically easier with the second method mentioned earlier; a thick preparation obscures the growth layers.

10.5.2 Body Length at Birth

As measuring the body weight of cetaceans is difficult, their body size has been traditionally expressed by body length, which has the added benefit of being free from seasonal change due to physiological condition. Body length is measured in a straight line as the distance from the tip of the upper jaw to the bottom of the notch in the middle of the tail flukes, which corresponds to head-to-tail length of terrestrial mammals.

Cetacean neonates are born within a certain body-length range typical for each species or population, but a measurement at the instant of parturition is usually difficult to obtain. Alternatively, the average neonatal length of striped dolphins was estimated from the change in proportion of neonates to fetuses with increasing body length. If body lengths are grouped into intervals of 5 cm for a sample obtained from the drive fishery on the Izu coast, the largest fetus was in the length group of 102.5–107.5 cm and the smallest neonate in the 92.5–97.5 cm group, which suggests that neonates are born at body lengths between 92.5 and 107.5 cm. There were only 14 neonates within this range, while 58 fetuses were found in the range. Such disparity can happen if some newborns are lost during the drive or if the sample is obtained in the beginning of a parturition season, both of which are considered possible for the fishery.

Kasuya (1972) inflated the number of neonates to be equal to the number of fetuses within the length range, that is, 58, before calculating a point where fetuses and neonates were in equal frequency and obtained 99.8 cm as an estimate of mean neonatal body length. Miyazaki (1977) estimated the corresponding figure at 100.5 cm, using the same method and a smaller sample (a total of 52 fetuses and neonates in the range). The difference between the two estimates is statistically insignificant.

10.5.3 Body-Length Composition

The body-length frequency distributions of 567 female and 392 male striped dolphins taken by the drive fishery on the Izu coast from 1967 to 1970 had modes at 223–227 cm (females) and 238–242 cm (males). Maximum body lengths were in the length intervals of 243–247 cm (females) and 258–262 cm (males). Modal length should approximately agree with average body length at cessation of growth, that is, about 225 cm and 240 cm for females and males, respectively. The difference between sexes was about 15 cm. These figures are not different from corresponding figures for the small samples taken from the northern offshore concentration of the species (47 individuals of both sexes) (Table 10.2).

On the other hand, the body-length distributions of the same species for 412 females and 301 males taken by the drive fishery at Taiji during autumn of 1991 to spring of 1995 had modal lengths at 210–219 cm (females) and 220–229 cm (males) and maximum body lengths of 238 cm and 249 cm, respectively. These figures suggest that striped dolphins hunted by the Taiji fishery were about 10–15 cm smaller than the same species taken by the drive fishery off the east coast of the Izu Peninsula in autumn of 1967–1970 and those in the northern offshore concentration (Section 10.4.4).

10.5.4 Mean Growth Curve

Kasuya (1976a) reported the relationship between body length and age of striped dolphins taken by the drive fishery off the Izu coast during 1971–1975. Growth was rapid in the first year after birth with an average length of 164 cm reached at age 1 year or an increase of 64% in the first year of life. This was followed by a period of almost constant slower growth during ages 3–9 (females) or 3–10 (males). A much slower increase in average body length seemed to last until the age of around 17 (females) or 21 (males), when the correlation between body length and age was totally lost. Miyazaki (1984) confirmed a similar growth pattern using his own data collected mostly from the fishery on the Izu coast and some additional data from Taiji. Both studies used a similar age determination method, that is, reading growth layers in dentine and cementum in decalcified and stained thin tooth sections. If age is determined using only dentinal layers as in Kasuya (1972) and Miyazaki (1977), the method will underestimate the ages of animals older than 13 years and likely overestimate the growth of younger individuals. However, it will not result in significant error in the average body length of older individuals, because the age of 13–16 years when deposition of dentine layers ceases in most individuals agrees with the age when most striped dolphins cease growing.

Body lengths of female striped dolphins over age 17 were 200–250 cm, with a mean of 225.3 cm (n = 89, SE = 1.62) and those of males over age 21 were 220–248 cm, with a mean of 236.0 cm (n = 41, SE = 0.96) (Kasuya 1976a). These mean body lengths are estimates of mean body length at the attainment of physical maturity and are close to the figures estimated from the body-length frequency distributions. Miyazaki (1984) estimated the mean body length at the attainment of physical maturity as an average of the length of individuals over 11 years of age (females) or 16 years of age (males) and obtained the estimates of 225.7 cm (females) and 238.9 cm (males). The figure for males is about 3 cm greater than the corresponding value obtained by Kasuya (1976). One of the possible causes of the difference comes from different methods of measurement. While Kasuya (1976) measured body lengths by placing a person at each end of the tape, Miyazaki did it by himself. I have confirmed for some striped dolphins that error due to the difference in measurement method is greater for animals of greater body length.

The relationship between mean body length and age thus obtained is an apparent mean growth curve, but it differs from a real mean growth curve, which would be constructed by monitoring the growth of individuals born in a year for their whole growth period. The two type of mean growth curve will agree only when the growth pattern of the species has remained unchanged for the entire age range of the sample. Even when this condition is satisfied, the mean growth curve does not represent the growth of any particular individual. The age when the mean growth curve indicates cessation of growth must be considered, in a strict sense, as the age when all the individuals of the population cease growing (not the average age at cessation of growth). These problems have been detailed earlier (Section 8.5.2.4). Physical maturity is determined through the examination of fusion of vertebral epiphyses to the centra when a dolphin ceases growth in body length (Sections 9.4.10 and 13.4.5). Such a task required collecting of posterior thoracic or anterior lumbar vertebrae for microscopic examination, but it was not possible for the dolphin drive fishery because the carcasses were sent to market as whole bodies weighed after removing the viscera.

Cetaceans may change growth responding to decline in density or increase in per capita food availability. Male sperm whales in the western North Pacific measured below about 17 m until 1970, but some males started to exceed that size in the early 1970s, and in the early 1980s maximum male length reached about 19 m. This suggested about a 2 m increase in mean asymptotic length and was interpreted as a response of the population to a decline in density caused by whaling. Such a change in asymptotic length has not been identified among female sperm whales, presumably due to a different reproductive strategy of females, which are likely to maximize lifetime production through earlier attainment of sexual maturity rather than increasing asymptotic body length as in males (Kasuya 1991). Antarctic minke whales are thought to have increased body length in the prepubertal stage and decreased age at the attainment of sexual maturity, presumably responding to improved per capita food availability (Kato 1991, in Japanese). This nutritional improvement could be a result of decline of other baleen whales feeding in the Antarctic or decline of minke whales. There is no convincing evidence of the Antarctic minke whale changing asymptotic body length. It might be expected that there would be a response similar to that of female sperm whales to the density decline in the western North Pacific.

Kasuya (1985) reported for striped dolphins taken by the drive fishery on the Izu coast that presumably responded to a decline in density that mean female age at the attainment of sexual maturity declined from 9–10 years for individuals born in the late 1950s (which matured in the 1960s) to 7–8 years for individuals born around 1970 (which matured in the 1970s) (Section 10.5.7). I did not examine whether this change accompanied changes in the pattern of growth in body length. However, as far as can be seen in the apparent growth curve, there are no evident age-related differences of mean body length for ages above 17 years (Figure 10.8). This suggests that mean asymptotic body length has not changed, at least among individuals born in the period from 1930–1940 (about 40 years old at the time of sampling) to 1953–1963 (about 17 years old) or among individuals analyzed by Miyazaki (1984).

Selection by fisheries can cause bias in the apparent mean growth curve. It may be significant in the case of the hand-harpoon fishery, where fishermen may select larger individuals or when individuals of a particular growth or reproductive stage tend to approach fishing vessels and are vulnerable to being harpooned, but the effect seems to be minimal in the case of the drive fisheries. If natural mortality rate is a function of body size, the apparent mean growth curve can be biased, but this has not been examined for this species.

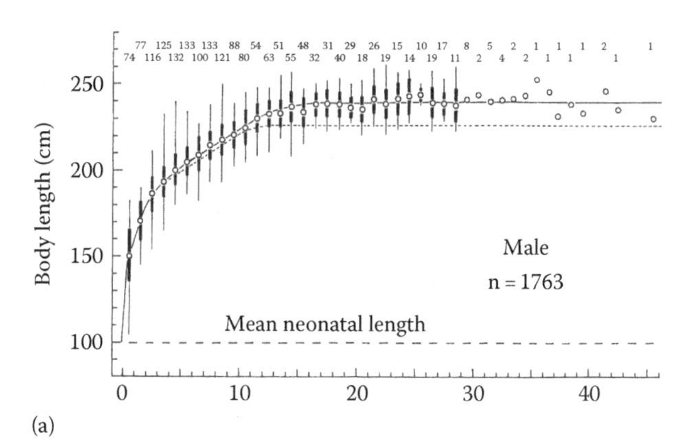

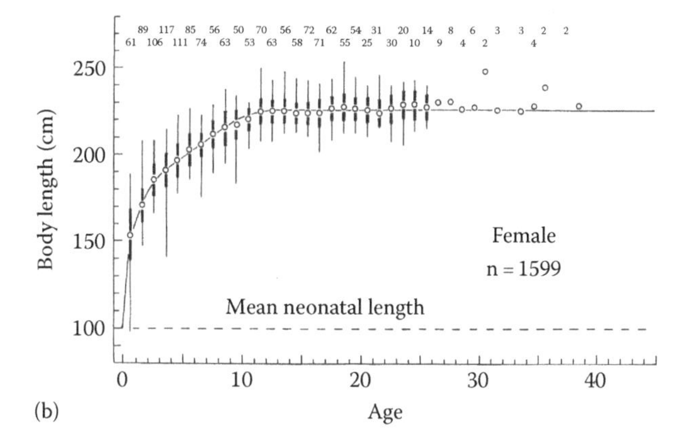

FIGURE 10.8 Mean growth curve of striped dolphins based on age determination using cemental and dentinal growth layers. (a) represents males (n = 1763), (b) females (n = 1599), and numerals on the top of each panel sample size. Circle, mean body length; black square, range of one standard deviation of the sample on each side of the mean; thin line, range of the sample. About 90% of the sample was obtained on the Izu coast in 1963–1973 and 10% at Taiji in 1973–1980. (Reproduced from Figs. 4 and 5 of Miyazaki, N., *Rep. Int. Whal. Commn.*, Special Issue 6, Reproduction in Whales, Dolphins and Porpoises, 343, 1984. With permission.)

10.5.5 Male Sexual Maturity

Sexual maturation in male cetaceans is accompanied by a process of anatomical development that takes several years to be completed and, at the same time, by changes in behavior or social activity. In addition to such growth-related changes, male reproductive tracts often undergo cyclic changes in activity reflecting the seasonality of reproduction. Although we know that striped dolphins apparently exhibit fourfold seasonal fluctuation in testis weight (Figure 10.4), there has been insufficient effort in the study of sexual maturity of striped dolphins off Japan to allow disentanglement of the seasonal cycle of the reproductive organs and unidirectional growth toward maturation. This also relates to the interpretation of reproductive seasonality of females hunted by the Izu fishery, that is, whether the two observed mating peaks, which are about 6 months apart, represent a single population or two different populations (Section 10.4.7). Seasonal change in a testicular activity is also known in the Dall's porpoise, which is a strict seasonal breeder and exhibits twofold seasonal fluctuation in the testicular weight and diameter of seminiferous tubules (Section 9.4.8), and in short-finned pilot whales off Japan, a weak seasonal breeder, where young males at ages 4–5 may undergo a low level of spermatogenesis only during the mating peak of the species but full-grown males at ages over 15 years undergo a high level of spermatogenesis throughout the year (Section 12.4.3.3).

Behavioral aspects of sexual maturation have not been well documented for the striped dolphin, but it is presumed that pubertal males will start approaching estrous females, be accepted by females at a certain maturity stage, and produce their first offspring. It is possible that males have to compete with other males for access to females. The stage where males gain access to estrous females is often called "social maturity." Identification of social maturity requires observation of behavior or paternity identification using genetic methodology, which has not been attempted for striped dolphins. Also, there have been no studies on female choice of mating partners.

My conclusions based on the following review of studies on sexual maturation in male striped dolphins off Japan can be summarized as follows: (1) at an age of about 2 years, some males may produce a limited quantity of spermatozoa, but they are unlikely to have the ability to reproduce, (2) at individually variable ages between 7 and 13, males attain the physiological ability to reproduce and participate in reproduction, and (3) the minimum weight of a single testis in such sexually mature males is about 20 g in the nonmating season and 30–70 g in the mating season.

10.5.5.1 Histology of Testicular Tissue

Hirose and Nishiwaki (1971) were the first to attempt to determine the criteria of sexual maturity in male striped dolphins off Japan. They examined the relationship between body length, testicular weight, and testicular histology and proposed the criteria of sexual maturity for each parameter. They first plotted the weight of a single testis on body length and found a sudden increase in weight at a certain body length. Although they did not mention it clearly, they seemed to have identified secondary spermatocytes only in testes weighing over 40 g. They noted that secondary spermatocytes would metamorphose to spermatozoa within several weeks and concluded that the presence of secondary spermatocytes or testicular weight of 40 g or more could safely be used as an indication of sexual maturity with negligible error. Later studies have shown that spermatogenesis occurs even at testicular weights of 10 g or less.

Kasuya (1972), like Hirose and Nishiwaki (1971), plotted testis weight on body length and noted that weight increased rapidly from less than 30 g for males smaller than 215 cm to 80–300 g for males over 225 cm, where correlation between testis weight and body length was lost. I thought that males were sexually mature if weight of a single testis was 35 g or more. This study relied only on the relationship between body length and testis weight and ignored age and testicular histology.

Miyazaki (1977) improved on the earlier studies by introducing information on testicular histology but used ages determined by reading dentinal growth layers. He confirmed that there is no bilateral asymmetry in testis weight (this is different from ovaries; see Section 10.5.6) and used for his subsequent analysis only one testis randomly selected from each individual. He examined the histology of a sample extracted from the midcenter of the testis and classified the growth into the following three stages:

1. Stage 1: Immature, neither spermatozoa nor spermatocytes present
2. Stage 2: Pubertal, with spermatocytes but no spermatozoa
3. Stage 3: Mature, with spermatozoa

I presume that he examined 20 seminiferous tubules for the earlier analysis as in the case mentioned as follows. Next, he examined 20 tubules of the same testicular section and classified "Stage 3" or the "mature" stage into the following three substages:

1. Stage 3.1: Maturity I, spermatozoa only in one tubule
2. Stage 3.1: Maturity II, spermatozoa in 2–19 tubules
3. Stage 3.3: Mature III, spermatozoa in all 20 tubules

In short, he classified maturity of male striped dolphins into five categories by observing 20 seminiferous tubules at the midcenter of a testis. Later, Komyo (1982, in Japanese) confirmed that the histological maturity stage of a testis did not vary between locations in the testis.

In classifying testicular growth into three stages of "immature," "pubertal," and "mature," Miyazaki (1977) did not take into account behavioral information on the species. Thus, his classification be considered an arbitrary classification of testicular growth stages, which is often the case in cetacean biology. There remains a question about the stability of the earlier classification that is based at times on the presence of a single cell of a particular type in 20 seminiferous tubules examined. Miyazaki (1977, 1984) defined the stage "pubertal" as males that had only spermatocytes and no spermatozoa in 20 observed tubules. I have seen such histology in the sample of Dall's porpoises obtained outside the mating season (Kasuya 1978) but not in bottlenose dolphins (Kasuya et al. 1997, in Japanese, cited in Section 11.4.7) or short-finned pilot whales (Kasuya and Marsh 1984; Kasuya and Tai 1993, cited in Section 12.4.3). In these two species, which are rather weak seasonal breeders, spermatocytes and spermatozoa have been always found together on a histological slide. The extremely broad range of age or testicular weight associated with each maturity stage (see the following text in this section) suggests that the earlier classification of maturity is confounded by cyclical seasonal change in testicular activity that can occur in both immature and mature individuals.

Next, Miyazaki (1977) tested his testicular stages against weight of testis or body length and noted that Stage 1 and Stage 2 were present only in testes weighing less than 40 g and Stage 3 (including Stages 3.1, 3.2, and 3.3) in testes weighing 7–225 g (Table 10.4). In the view of Komyo (1982), the occurrence of spermatogenesis in very small testes is possible. The testis weight when 50% of males were at Stage 3 was 15.5 g, and the body length when 50% of males were at Stage 3 was 219 cm. These are average testis weight and body length at the attainment of Stage 3, or sexual maturity as defined by Miyazaki (1977).

Miyazaki (1984) reanalyzed his male maturity classification with improved age estimation using growth layers in both dentine and cementum and reported that individuals of Stage 3 first appeared at age 6, reached 50% at 8.8 years,

TABLE 10.4
Two Definitions of Male Maturity of Striped Dolphins Off Japan

Miyazaki (1977, 1984)			Iwasaki and Goto (1997)	
Maturity Stages	Weight of a Testis (g)	Age	Maturity Stages	Weight of a Testis (g)
Immature	<10	<15	Immature	<10
Pubertal	4–40	2–14	Early maturing	5–37
Mature I	7–90	6–30	Late maturing	15–64
Mature II	15–200	6–30	Mature	12–229
Mature III	40–225	4–46		

Source: Weight of a single testis with epididymis removed.

and increased in proportion until age 14. Age 8.8 is the average age at formation of spermatozoa in at least one of the 20 seminiferous tubules at midcenter of the testis, or sexual maturity as defined by Miyazaki (1977). Age ranges were similar for Stage 3.1 (6–30 years), Stage 3.2 (6–30 years), and Stage 3.3 (4–46 years) (Table 10.4), but the proportion at Stage 3.1 decreased with increasing age, while that of Stage 3.3 increased. Stage 3.2 did not show such a trend. He analyzed age-related change in proportions of Stages 3.1, 3.2, and 3.3 within Stage 3 males (Stages 1 and 2 were excluded from this analysis) and found that the proportion of Stage 3.3 individuals reached 50% at age 16.5 years. The age would only be slightly higher if Stages 1 and 2 males were included in the calculation because Stages 1 and 2 were found at ages below 15 years (Table 10.4). The age of 16.5 is the average age of males at the attainment of "social maturity" of Miyazaki (1977), where he tacitly assumed that the difference among Stages 3.1 through 3.3 reflected only growth-related development and apparently ignored a possible effect of reproductive seasonality on the histological features.

Iwasaki and Goto (1997) classified the stage of testicular growth in a way different from that of Miyazaki (1977, 1984), using materials obtained from the drive fishery at Taiji in the 1993/1994 and 1994/1995 seasons. They classified maturity into four stages based on the proportion of seminiferous tubules that had undergone spermatogenesis:

1. Immature: 0 tubules were spermatogenic.
2. Early maturing: Over 0% and less than 50% of tubules were spermatogenic.
3. Late maturing: Over 50% but less than 100% were spermatogenic.
4. Mature: 100% of tubules were spermatogenic.

Spermatogenesis was identified by the presence of either spermatocytes or spermatozoa. Their method was similar to that of Kasuya and Marsh (1984) and is not free from problems accompanying arbitrary classification of maturity and the risk of relying on the observation of a limited number of tubules but has some improvement in robustness in classification over that of Miyazaki (1977, 1984). The comparison of the range of testicular weights with those of Miyazaki (1977, 1984) gives the impression that (1) the "immature" stage in the two studies appears to represent a similar growth stage; (2) "Stage 2, or pubertal," of Miyazaki (1977, 1984) corresponds with "early maturing" of Iwasaki and Goto (1997); (3) Stage 3.1 of Miyazaki (1977, 1984) almost agrees with "late maturing" of Iwasaki and Goto (1997) with the exception of a broader lower range of testicular weight in the former study; and (4) the "mature" stage of Iwasaki and Goto (1997) seems to include "Stages 3.2 and 3.3" of Miyazaki (1977, 1984) (Table 10.4).

Insufficient consideration has been given to the relationship between testis histology and male reproductive ability of striped dolphins. Miyazaki (1984) noted between-school differences in the reproductive status of adult female striped dolphins off the Izu coast. He called eight schools containing a high proportion of peri-estrus females as "mating schools," and another 10 schools dominated by different female status as "nonmating schools" (seven "mixed schools" were excluded from the analysis). Then he compared the age composition of males between the two school types and concluded that mating schools contained more males of age 16.5 years or older (I noted a peak of male frequency at ages around 13–20 years that tailed off to higher age) compared with the nonmating schools (I noted a peak in these at around 11–13 years). There was no apparent difference in the age range of males between the two school types. He interpreted this as evidence of movement of reproductive males into schools of estrous females. He then noted that the age of 16.5 years agreed with the average age of males at the attainment of Stage 3.3 or "full sexual maturity" and stated that male striped dolphins might attain social maturity at age 16.5 years or older. It is unclear why he limited this analysis only to a basis of age. If he had analyzed the histological stages of male maturity against the two school types of "mating" and "nonmating," his conclusion could have been more persuasive.

The study by Miyazaki (1984) had some value as an attempt to compare male behavior against age and testicular histology and to introduce an idea of social maturity in striped dolphins. However, I pay more attention to the vast overlap in ranges of testicular weight and age accompanying his maturity Stages 3.1 through 3.3 (Table 10.4) and suspect some factors behind them other than growth. An additional analysis is needed to be done examining seasonal change in testicular activity (including testicular histology as well as weight) of sexually mature males in relation to date of death and the reproductive status of females taken with them (Section 10.4.7).

My observations follow on Figure 10.9, which represents the relationships between the age and the testicular weight or diameter of seminiferous tubules obtained by Miyazaki (1984). (1) Below age 6, testis growth is slow with testes weighing only 10 g or less. (2) Some males start rapid growth in testicular weight at the age between 6 and 7. (3) Individual variation in testis weight apparently increases from age 6 to age 15. (4) Correlation between testis weight and age is lost

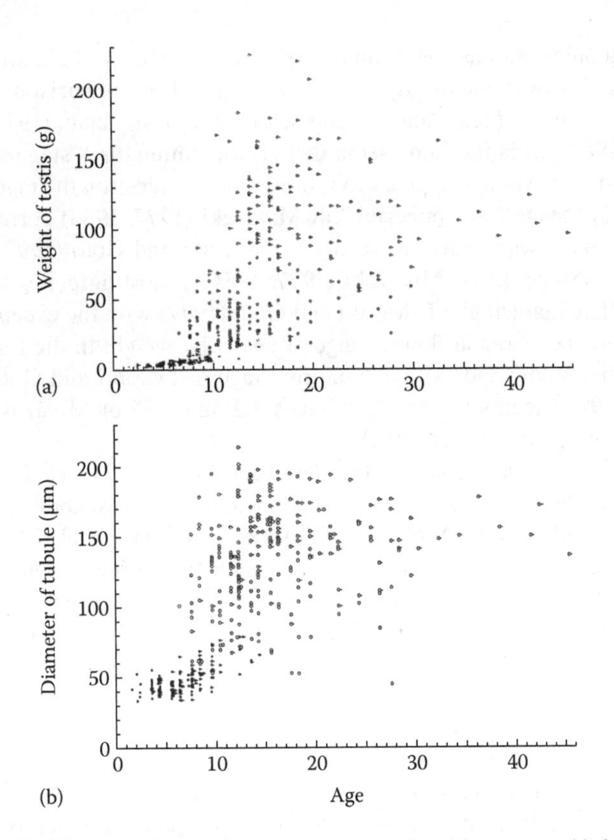

FIGURE 10.9 Scatterplot showing the relationship between (a) the age and the weight of a single testis (epididymis removed) and (b) the diameter of seminiferous tubules of striped dolphins from off Japan. Closed circle, immature; closed circle with bar, pubertal; open circle, mature I and mature II; open circle with bar, mature III. (Reproduced from Figs. 6 and 8 of Miyazaki, N., *Rep. Int. Whal. Commn.*, Special Issue 6, Reproduction in Whales, Dolphins and Porpoises, 343, 1984. With permission.)

after age 15, and it is at around 14–20 years when maximum testis weights (~220 g) are recorded. (5) All the males above age 15 must have reached the final stage of testicular growth, or full sexual maturity, and the broad range of testis weight (20–220 g) and the diameter of seminiferous tubules (60–230 μm) among them must reflect both individual variation and seasonal fluctuation. The minimum weight of the testes of individuals reproductive in the mating season is thought to be between the two extremes of the range. (6) The observations also apply in principle to the diameter of seminiferous tubules. (7) The minimum limits of the parameters given earlier (20 g and 60 μm) for ages over 15 years also apply to males aged 13–15 years. This suggests that all males at ages over 13 have attained the physiological capability of reproduction during the mating season. (8) Thus, assuming 20 g as the minimum testis weight of fully reproductive males, which likely represents weight in the nonmating season, the youngest males that have attained reproductive capacity must be among those aged over 6, when testis weight may exceed 20 g. However, because testis weight and level of physiological development have individual variation, it should not be concluded that reproductive ability is attained by all the young males with testis weights of more than 20 g in the nonmating season.

10.5.5.2 Spermatozoa in the Epididymis

The density of spermatozoa in the seminal fluid offers more direct information on the physiological ability of a male to reproduce than weight or histology of the testis. Cetacean spermatozoa produced in the seminiferous tubules are transported to the epididymis and stored there. To compensate for the difficulty of collecting seminal fluid from wild dolphins, Komyo (1982, in Japanese) analyzed the density of spermatozoa in epididymal fluid in relation to the weight of the testis and the diameter of seminiferous tubules (age was not examined) using striped dolphins taken by the drive fishery on the Izu coast. He identified three stages in the mean diameter of seminiferous tubules plotted on testicular weight: (1) a stage of rapid tubule growth (for testis weight below 40 g), (2) a stage of slow growth (testis weight of 40–110 g), and (3) an asymptotic stage with an average tubule diameter of 200 μm (testis weights over 110 g).

Testicular weight of 1.8 g was the minimum size where spermatozoa were detected in an epididymal smear, and 23.9 g was the maximum testicular weight where spermatozoa were not detected. This range was slightly lower than the corresponding figures, 7.2 and 57.2 g, obtained from spermatozoan counts in epididymal fluid. This suggests that the smear was more sensitive in detecting spermatozoa at low density, but its five density scales seem to be unsuitable for recording very high density (Table 10.5). It can also be concluded that some males transport spermatozoa to the epididymis when the testis weighs around 2–20 g, which does not disagree with Miyazaki's (1977) interpretation; he reported spermatogenesis in some animals at the age of 2 years. However, in view of their tendency to live in "juvenile schools" (Figure 10.16), such young spermatogenetic males are unlikely to approach estrous females, or it is unlikely that such low sperm density would accomplish fertilization.

We do not know how spermatozoan density fluctuates seasonally in the epididymal fluid of adult males. According to

TABLE 10.5
Spermatozoa Density in Epididymis

	Relative Density in a Smear[a]					
Absolute Density (×10^5/mL)	0	1	2	3	4	No. of Data
0	13	7	2			22
0.1–100	2	3	11	17	5	38
101–200				2	5	7
201–400					10	10
401–600					6	6
601–800					4	4
801–1000					1	1
≥1001					3	3

Source: Reproduced from Table 7 of Komyo, Y., On the process of sexual maturation of the male striped dolphin *Stenella coeruleoalba*, Master's thesis, Tokyo University of Fisheries, Tokyo, Japan, 1982, 46pp. With permission.

The comparison of two evaluation methods.

[a] The stage categories are almost the same as those used for the short-finned pilot whale (Table 12.8).

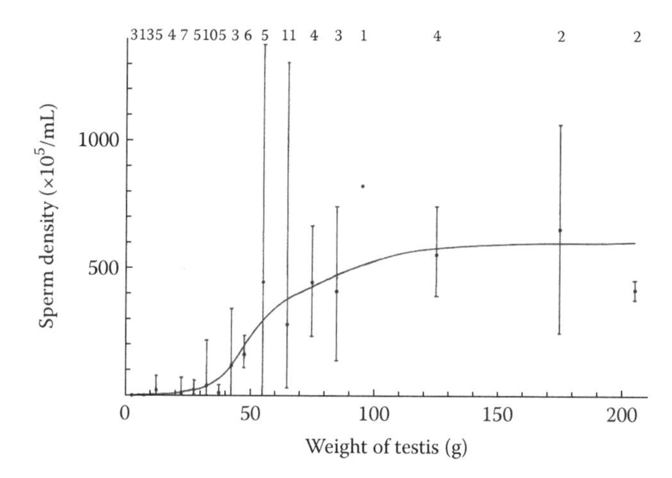

FIGURE 10.10 The relationship between the spermatozoan density (×10^5/ml) in the epididymal fluid and the weight of a testis (epididymis removed) off the Izu coast. Top numerals indicate sample size, closed circles mean density, and bars range of the sample. (Reproduced from Fig. 9 of Komyo, Y., On the process of sexual maturation of the male striped dolphin Stenella coeruleoalba, Master's thesis, Tokyo University of Fisheries, Tokyo, Japan, 1982, 46pp. With permission.)

Komyo (1982, in Japanese), density showed a rapid increase at testis weight of 30–60 g and attained a plateau of 25–1100 (×10^5/mL) at testis weights around 50–100 g. The approximate average density at the plateau was about 550 × 10^5/mL. A peak density occurred among males with testis weights of 50–70 g, where the sample size was large. Komyo concluded that the threshold of spermatozoan density for reproduction is at the lower range of the plateau, that is, 250 × 10^5/mL. The minimum testis weight with this sperm density was 35.5 g, and this density was found in all the males with testes weighing 82.5 g or more. In other words, male striped dolphins probably attain reproductive ability when their testis weight reaches between 35.5 g and 86.2 g. If one assumes a lower figure of 200 × 10^5/mL for the threshold sperm density, then the range of testis weight is narrowed to 30–70 g (Figure 10.10). These figures correspond to the state in the mating season.

10.5.5.3 Age at Attainment of Sexual Maturity

The attainment of sexual maturity in male striped dolphins must be determined by following the development of the reproductive tract, including testis tissue, which can be a function of both age and body size. Thus, males having fast-growing testes attain sexual maturity at a younger age, perhaps with heavier testes, while males having testes with slower growth attain sexual maturity at a greater age and perhaps with relatively smaller testes. This will cause a nonlinearity in the maturity–age relationship. Striped dolphins with a testis weight of 35 g are between 7 and 20 years old, and those with a testis weight of 60 g are between 8 and 27 years old, which does not change if a threshold weight of 86 g is used (Figure 10.9).

Some male striped dolphins off Japan are probably capable of reproduction at age 7–8 years when some testes may weigh over 20 g, and all of them attain reproductive ability by the age of 13 years. Although this physiological maturity is different from social maturity by definition, it is reasonable to assume that these males at physiological maturity make some effort at reproduction. The upper age range mentioned earlier is not significantly different from the age of 16.5 years proposed by Miyazaki (1984) for the age when males participate in reproduction. In the early 1970s, some Japanese scientists thought that striped dolphins would attain maturity at 3 years. Since then, scientists have made extensive effort at estimating age at sexual maturity of male striped dolphins and reached that understanding.

Now, it is necessary to give some consideration to the possibility of a difference between physiological capability and social maturity in male striped dolphins. As a parallel example, I will start with the breeding behavior of northern fur seals. Adult males of the species arrive at a breeding beach on an island and establish territories. Then females in near-term pregnancy land at the beach for parturition and subsequent estrus. Males must defend their territories against other males, who intrude into the territory from three directions, that is, the surf side and both the left and right boundaries. The inland side is protected by a cliff. Therefore, males who establish a larger territory on a better beach have greater reproductive success; young adult males of smaller body size have insufficient physical power to establish and defend their territories and thus are unable to participate in reproduction. In such a reproductive system, age at social maturity is significantly greater than the age at the attainment of the physiological capability to reproduce. The age difference between the two stages can be affected by the age structure of the male population, size of the breeding beach, and abundance of breeding females.

The reproductive system of sperm whales is slightly different from that of northern fur seals. Adult males in the breeding season successively visit nursing schools of adult females and their offspring to find estrous females. Aggressive behavior arises if two or more adult males meet at a nursing school (Kato 1984), but such cases are believed to be infrequent, at least in recent years (Whitehead 2003). The competition between males for mating opportunities may have been more frequent before the heavy exploitation of adult males by modern whaling. If an adult male wishes to monopolize a freely moving nursing school under such a mating system, he must defend his position in all directions below the sea surface, which is much broader than defense of the territory of a harem master fur seal. A third bull sperm whale might have the opportunity to sneak into the nursing school while two other bulls are fighting for an opportunity to mate. It seems to me that mating opportunities are more evenly shared among male sperm whales than in male northern fur seals.

Striped dolphins inhabit an oceanic environment like that of sperm whales, but they live in schools of tens or hundreds containing numerous individuals of both sexes that are physiologically reproductive (Figure 10.16). Under such circumstances, it would be almost impossible for a particular male to monopolize the opportunity for mating unless estrous females favored some particular males or if a group of males cooperated for their common benefit, both of which I believe unlikely, at least for the striped dolphin. Thus, reproductive opportunity must be more evenly distributed among males of the striped dolphin than in the sperm whale and northern fur seal. This is also supported by the less pronounced sexual dimorphism in the striped dolphin than in the two other marine mammal species. I believe that the physiological capability to reproduce and social maturity cannot be separated among male striped dolphins and that most males participate in reproduction when they attain the physiological ability to reproduce, that is, at around 10 years on average and by the age of 13 years.

10.5.6 Female Sexual Maturity

10.5.6.1 Ovarian Interpretation

Ovulation has been used as the marker of the attainment of sexual maturity of female striped dolphins, which has been believed identifiable by the presence of a corpus luteum or albicans in the ovaries. As most first ovulations are likely to be followed by conception, they are considered almost synonymous with the attainment of reproductive ability. Thus, sexual maturity of female cetaceans has been determined by examining their ovaries.

The corpus luteum is formed in the ovary after ovulation and is retained through pregnancy if the ovulation is followed by conception. After parturition or soon after ovulation not followed by conception, the corpus luteum degenerates into a corpus albicans. This means that both the corpus luteum of pregnancy and that of ovulation have been thought to remain in the ovary as corpora albicantia for life and that one can determine the number of ovulations experienced by a female cetacean by examining her ovaries (Perrin and Donovan 1984). However, this view has been questioned for certain species of toothed whales (Brook et al. 2002), based on the observation of an Indo-Pacific bottlenose dolphin kept in an aquarium that routinely was sampled for serum progesterone and examined by ultrasonography for the presence of a Graafian follicle and corpus luteum during her 12 years in the aquarium. This female was brought into the aquarium at age 3 years, had a first ovulation and formation of a corpus luteum in the 3rd year in the aquarium, and experienced 18 ovulations and 3 pregnancies during her life. Postmortem examination found only three corpora albicantia in her ovaries, which suggested that they represented only corpora lutea of pregnancy. A similar suggestion was based on the examination of stranded common dolphins (*Delphinus delphis*) (Dabin et al. 2008), and the problem was discussed at the Scientific Committee Meeting of the International Whaling Commission (IWC) in 2009 (IWC 2010). As counting degenerated old corpora albicantia is not an easy task, further investigation on ovarian histology and cross reading between biologists are needed before arriving at a firm conclusion. The relationship between pregnancy rate and accumulation rate of corpora in the ovaries of striped dolphins will be dealt with in a later section.

Some small toothed whales, including striped dolphins, are known to have precocious left ovaries (Ohsumi 1964). Hirose et al. (1970) examined this in detail using a large sample of striped dolphins and reported that the first ovulation occurred in the left ovary and ovulation in the right ovary first occurred after the left ovary had undergone 5–18 ovulations. After that point, ovulation occurred in both ovaries in equal frequency.

10.5.6.2 Age and Body Length at a First Ovulation

Kasuya (1972) reported that female striped dolphins experience the first ovulation at age 7–11 years (sample size in this age range was 26 females), and about 10 years later, Miyazaki (1984) reported a corresponding range of 4–13 years (sample size, 363 females). This broader range could have been due to the larger sample size as well as to a tendency of earlier maturation in more recent cohorts. These studies were based on age determination using dentinal layers, but the use of cemental layers reached a similar result of 5–13 years (Kasuya 1976a). Female striped dolphins attain sexual maturity at younger ages than males.

One way of estimating mean age at the attainment of sexual maturity is to follow a year class for years to record the age of first ovulation. Such a method is possible when the members of a population are individually identified but is currently inapplicable for the striped dolphin. Thus, Japanese scientists examined carcasses taken in drive fisheries, determined the relationship between age and proportion sexually mature, and calculated the age where 50% were sexually mature. It should be cautioned that this parameter is not free from bias due to the mathematical models applied to the calculation and is different in a strict sense from a mean of ages when females attained sexual maturity. Discussion of this problem in the Scientific Committee of IWC is detailed in Martin and Rothery (1993), DeMaster (1984), Cooke (1984), and Hohn (1989). Martin and Rothery (1993) offered a simple mathematical method to calculate unbiased mean age at the attainment of sexual maturity and the error accompanying the estimate. The last two papers cited have not been published but are available from the IWC Secretariat.

The smallest sexually mature female striped dolphin was in a body-length group of 188–192 cm and the largest immature female in the interval 218–222 cm, which means that individual variation in body length at the attainment of sexual maturity was about 30 cm (Kasuya 1972). The body length where half of the females were sexually mature was 212 cm (Kasuya 1972). This parameter may be used to estimate the proportion of sexually mature individuals in a catch of particular fisheries, but it is not free from bias due to selectivity of a fishery. Although it has little value in describing the growth of the species, for some time it has been erroneously called mean body length at sexual maturity.

Miyazaki (1984) estimated a similar, but theoretically distinct, figure for female striped dolphins. The figure, 216 cm, was the mean body length of females at age 8.8 years obtained from the mean growth curve. The age of 8.8 years was the mean age at the attainment of sexual maturity estimated separately. This body length has biological value as the mean body length when females reach sexual maturity and is about 4 cm greater than the body length where 50% of females are sexually mature in the fishery sample.

Striped dolphins, as well as other small toothed whales, show broad individual variation in growth. For example, we find an age range of 4–23 years for females at a body length of 216 cm (Figure 10.8). This relates to two factors (other than aging errors): broad variation in adult size, and growth rate that declines with increasing age soon after the attainment of sexual maturity. Below a certain body-length range, for example, less than around 216 cm, younger females are mostly immature and older individuals are mostly mature, and within a certain age range, for example, 8 years, large-sized individuals will be more likely mature than smaller individuals. This is the reason for the 4 cm disagreement between the two figures mentioned earlier.

More recently, Iwasaki and Goto (1997) analyzed sexual maturity in the species using 259 females obtained from the drive fishery at Taiji in the 1991/1992–1994/1995 seasons. They found the smallest mature female in a 190–199 cm group and the largest immature females in a 220–229 cm group and calculated 199.8 cm as the body length where 50% of females were sexually mature. Body-length range at maturation was about 10 cm broader and body length at 50% maturity about 10 cm smaller than the corresponding figures of Kasuya (1972) that were based on materials obtained from the Izu fishery in 1968–1971. These geographical or temporal differences were interpreted as an indication of involvement of more than one population in different proportions in the two drive fisheries at Izu and Taiji (Section 10.4.4).

10.5.7 Declining Age of Females at Sexual Maturity

Kasuya (1985) reported a decline in female age at the attainment of sexual maturity for striped dolphins taken by the drive fishery on the east coast of the Izu Peninsula from middle October to middle December over the period 1960–1980. Using sexual maturity and age data of females taken in the 19 drives made in the 20-year period, he grouped them into cohorts by year of birth and obtained the age range at sexual maturity and age of 50% maturity for each cohort. The age where half of the sample was sexually mature declined from 9.4 years for the 1956–1958 cohorts to 7.5 years for the 1968–1970 cohorts (Figure 10.11). The age range of the youngest mature female and the oldest immature female also changed during the period, that is, 7–9 years for the 1963–1967 cohorts and 5–9 years for the 1968–1971 cohorts, which means that the minimum age at the attainment of sexual maturity declined from 7 to 5 years, while the maximum age at maturity did not show significant change. In other words, early-maturing females occurred in the more recent cohorts followed by a decline in age at 50% mature.

Kasuya (1985) thought that the decline in female age at the attainment of sexual maturity was a response of the population to density decline caused by the dolphin fishery. The drive fishery on the Izu coast reported a gradual decline in the catch of striped dolphins from the 1960s to the 1980s, during which period the demand for dolphin meat remained unsatisfied by the supply (Section 3.8.1) and fishing effort did not decline but rather increased through the improvement

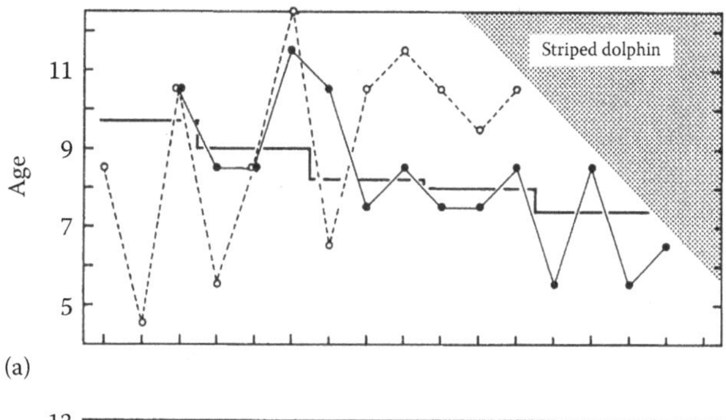

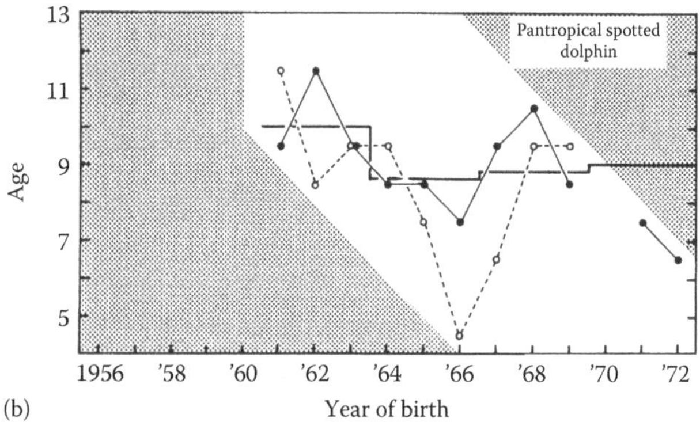

FIGURE 10.11 Age at the attainment of sexual maturity for cohorts (year of birth) of (a) striped dolphin and (b) pantropical spotted dolphin off Japan. The horizontal bar indicates the age where 50% of individuals are mature; closed circles and the thin solid line, the age of the youngest mature individuals; and open circles and the dotted line, the age of the oldest immature individual. The shaded area indicates no data. (Reproduced from Fig. 6 of Kasuya, T., *Sci. Rep. Whales Res. Inst.* (Tokyo), 36, 107, 1985.)

of equipment. This is strong evidence of decline in the availability of striped dolphins to the fishery, most likely a reflection of population decline. If this was the case and if we had been able to continue monitoring the trend into the future, the changes observed earlier could have been followed by a decline in the age of the oldest immature females. We have a similar example in North Pacific fin whales as reported by Ohsumi (1983), but it is also true that there remain several factors to be considered before arriving at a firm conclusion on the striped dolphin.

Ohsumi (1983) reported a decline in female age at the attainment of sexual maturity in North Pacific fin whales, which occurred accompanied by the progress of whaling. He examined the age of fin whales having only one corpus in the ovaries against the year of capture, which was slightly different from the analysis of Kasuya (1985) on striped dolphins. The range and mean ages of such females declined from 8–17 years (mean 12.4 years) in the 1957–1958 seasons to 4–11 years (mean 7.0 years) in the 1974–1975 seasons. This change was heralded by a decline in the youngest age of such mature females (i.e., females with one ovarian corpus) and then followed by a decline in the number of females maturing at higher ages. The decline in age of 50% mature occurred between the two changes. This was similar to the changes in age at sexual maturity of striped dolphins. Younger females are more sensitive to nutritional improvement or density decline than females that are already near puberty and will respond to environmental improvement by faster development.

The nutrition available for the striped dolphin population can be directly affected by the abundance of food resources. Among marine organisms consumed by striped dolphins off the Izu coast, that is, crustaceans, squids, and fishes, the Japanese flying squid, *Todarodes pacificus*, was the only species fished by Japanese fisheries of the time (Miyazaki et al. 1973). During the sampling period when we observed a decline in female age at the attainment of sexual maturity, the annual landing of this squid species in Japan decreased from 500–600 thousand tons in the 1960s to 100–200 thousand tons in the middle 1980s. If such a drastic decline in squid abundance influenced the nutrition of striped dolphins, the effect should have been opposite to that observed. Thus, the decline in female age at the attainment of sexual maturity cannot be explained by indirect interaction between dolphins and other fisheries.

Hunting of pantropical spotted dolphins is also worth some attention. It is believed that the Izu fishermen first drove this species in 1959 (Nishiwaki et al. 1965), but it was not preferred by the fishermen because of the low price due to small body size and inferior (tough) quality of the meat and blubber. The drive fishermen avoided driving the species until the late 1960s, but they started a low level of drive in the 1970s with the possible intent to supplement the decreased catch of striped dolphins (Table 3.16). Striped dolphins and pantropical spotted dolphins apparently occupy similar habitats and may have competed with each other for food resources. Thus, it is a possibility that a decline in pantropical spotted dolphins has improved the nutrition of striped dolphins, but exploitation of spotted dolphins cannot be the main factor of the observed change in age at the sexual maturity of striped dolphins because the change occurred before exploitation of pantropical spotted dolphins began.

Through the improvement of vessels, the operation range of the east coast fishery expanded to near the entrance of Sagami Bay and then further to outside the bay (Section 3.8.1 and Kasuya and Miyazaki 1982; Kasuya 1985, 1999). In the early twentieth century, the drive fishermen off the east coast of the Izu Peninsula searched for striped dolphins with rowed and sailing vessels in the nearshore waters of Sagami Bay. The introduction of motor-driven vessels around 1920 expanded the searching range to the entire Sagami Bay area (within about 30 km from the port), and the introduction of high-speed boats in 1962 by Kawana, followed by another team at Futo, allowed further expansion of the operation outside Sagami Bay, occasionally reaching a distance of 100 km from the port. The expansion of geographical range of the east coast drive fishery accompanied a gradual shift in the main fishing season from December–January (a peak in December) before the mid-1950s to October–December (a peak in November) from the late 1960s to 1981 (Table 10.3). These changes could have accompanied a shift of targeted population from a hypothetical "northern coastal population" to a "southern coastal population" (Section 10.4.8).

The changes in the operation pattern accompanied the decline in the catch of striped dolphins, which suggests decline in the availability of the species to the fishery during the 1960s through 1980s. However, a possibility that the fishery gradually shifted the target from one population to another (Section 10.4.8) causes difficulty in interpreting the observed decline in age at the attainment of sexual maturity of female striped dolphins off the Izu coast reported by Kasuya (1985). The apparent declining trend could be influenced by a possible difference of maturity/age relationship between the putative "northern" and "southern coastal" populations, on which we have no information. However, it seems unlikely that female age at the attainment of sexual maturity of the later-exploited "southern coastal population" was lower than that in a "northern coastal population" that had been exploited since an earlier time. Thus, the decline of female age at the attainment of sexual maturity observed in the catch of the drive fishery off the Izu coast was most likely a response of the exploited population to a density decline, although it could have a bias of unknown degree due to a possible shift of the fishery from one population to another.

10.5.8 Seasonality of Reproduction

Reproductive seasonality is usually detected from seasonal change in the occurrence of females in a particular reproductive status such as lactation or early pregnancy (Sections 9.4.4 and 9.4.5), but such observations were not useful for striped dolphins, which have extended or multiple breeding seasons and a long nursing period of over one year. Seasonal change in fetal body size in the catch of the drive fisheries offered another key for the analysis, although it was hampered by the short sampling season from October to December off the east coast of the Izu Peninsula from the 1960s to 1980s when biologists studied the fishery. The spring to early summer operation on the west coast had already ended, and only limited data on female reproductive tracts in the catch were collected by pioneer scientists and available for analyses.

Under these circumstances, Miyazaki (1984) analyzed the reproductive cycle of the species, using data from recent operations on the east coast of the Izu Peninsula, past operations on the west coast, and data from the more recent Taiji fishery (about 300 km southwest of the Izu Peninsula). He fitted a normal distribution to two fetal length modes identified in monthly catches off the east coast of the Izu Peninsula in October through December (Figure 10.3) and obtained six modal means. He also applied the same method to data from Taiji (January and June) and those from the west coast of the Izu Peninsula in May. By plotting the nine modal means on date (Figure 10.12), he identified two groups, each on a line, which were (1) a group of five points (one in May from the west coast of the Izu Peninsula; three in October, November, and December from the east coast of Izu; and one in January from Taiji), and (2) another group of four points (three in October, November, and December from the east coast of Izu and one in June from Taiji). He concluded that these two groups represented two mating seasons that were about 6 months apart. The linear regressions fitted to them

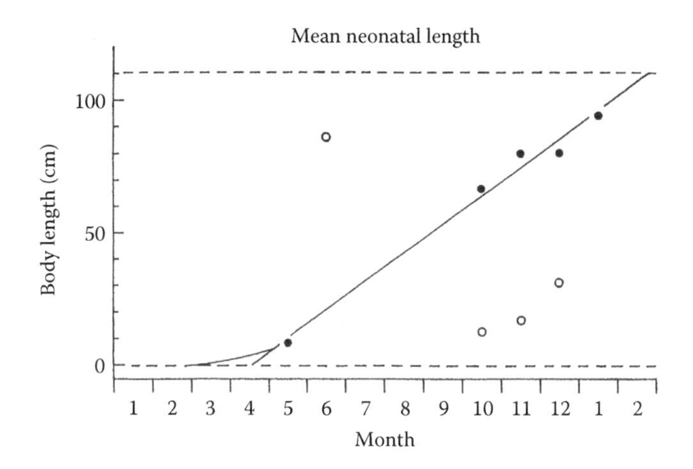

FIGURE 10.12 Fetal growth of striped dolphins estimated assuming two mating seasons for Taiji and Izu samples. Closed circles represent fetal modes from winter conceptions and open circles, those of summer conceptions. The calculation has correctly assumed neonatal length of 100 cm, although the figure apparently suggests greater neonatal length. (Reproduced from Fig. 3 of Miyazaki, N., *Rep. Int. Whal. Commn.*, Special Issue 6, Reproduction in Whales, Dolphins and Porpoises, 343, 1984. With permission.)

produced estimates of daily rate of fetal growth in the linear phase growth of 0.29 and 0.30 cm/day.

It is generally accepted that the growth of cetacean fetuses is linear except in the early stage of development (Laws 1959; Lockyer 1984; Perrin and Reilly 1984). The length of time from the date (t_0) when backward-extended linear growth cuts the axis of time to the date when forward-extended growth reaches the mean body length at birth (100 cm for striped dolphins) is a fraction of the total gestation time. The time from date of conception to t_0 is estimated to be about 14% of the total gestation time of mammals with gestation of approximately 1 year (Section 8.5.2). Using this rule, Miyazaki (1984) estimated the gestation time of the striped dolphin at 401 days or 13.2 months.

Although accuracy is a question, interspecies relationships have been used to supplement the paucity of data on gestation period of cetaceans, including the relationships among female body size and mean neonatal length, mean neonatal length and gestation length, and mean neonatal length and fetal growth rate (Ohsumi 1966; Kasuya et al. 1986; Kasuya 1995; Perrin et al. 1977). The relationship between mean neonatal length (X, cm) and gestation period (Y, months) is expressed by the following equation for toothed whales:

$$\log Y = 0.4586 \log X + 0.1659 \quad \text{(Perrin et al. 1977)}$$

Using X = 100, the gestation period of the striped dolphin is estimated at Y = 12.1 months. The relationship also gave an apparently reasonable estimate of gestation period for finless porpoises at 10.9 months (Section 8.5.2) but an estimate of 13.6 months (using $X = 128$) for bottlenose dolphins. The latter figure is about 1.4 months in excess of the observed gestation time of 370 days or 12.2 months (Section 11.4.2).

Many small toothed cetaceans have gestation periods of about 1 year. This can be explained by two factors: the time required for a fetus to grow to a certain minimum neonatal body size required for neonates to live a safe life in the aquatic environment and the annual cycle of the marine environment. If their reproduction were regulated only by the requirement of minimum body size, any gestation period through combination of a suitable fetal growth rate and neonatal body size above the minimum requirement would work. However, there is a practical difficulty in adopting a gestation period intermediate between 1 and 2 years under an annual cycle in the marine environment (Kasuya 1995). A gestation period slightly shorter than 12 months would benefit females for annual reproduction. The common porpoise, Dall's porpoise, and finless porpoise use this strategy. They have a relatively short nursing period, which suggests that they spend relatively less effort in rearing their offspring compared with other species. Only a few toothed whales have achieved a gestation time intermediate between 1 and 2 years, apparently accompanied by the augmentation of body size and stable social structure, for example, killer whales (observed gestation time of 17 months), sperm whales (estimated gestation time of 16 months), and short-finned pilot whales (estimated gestation time of 14.9 months). Cooperation within the group and longer parental care may compensate for the lower reproductive rate.

Kasuya (1972), assuming a gestation period of 12 months, indicated that parturitions of striped dolphins off Japan occur in every month of the year, with two peaks in June and November/December. Miyazaki (1984), using the fetal growth curve and mean body length at parturition described earlier, estimated conception peaks in July and January and parturition peaks around July and January/February. These two studies reached similar results on the parturition peaks. The unresolved question lies in the interpretation of the two peaks, which are about 6 months apart, as to whether a single population has two mating seasons or the two breeding seasons represent different populations that have been hunted by Taiji and Izu. The cases that are similar to the latter have been known for some other cetacean species off Japan (Section 15.4.1), that is, finless porpoises (Section 8.5.1), short-finned pilot whales (Section 12.4.4), and minke whales (Kato 1992; Kato et al. 1992).

10.5.9 Seasonality of Male Reproduction

Miyazaki (1977) presented information on seasonal change in the male reproductive tract of striped dolphins caught by the Izu fishery (Figure 10.4). The ages of most individuals above 13 years in Figure 10.4 may be underestimates of real age because they are based on reading growth layers in dentine (Figure 10.7). The average testis weight of males above 15 years, where the correlation between weight and age is lost, declined from 120–170 g in October to 30–50 g in December. The average weight in November was intermediate. Younger males between 9 and 15 years old also showed a similar seasonal trend in diminished testis weight, but the pattern was indistinct among males below 9 years of age. The proportion of males at maturity Stage 3.3, the most active or full-grown stage as defined by Miyazaki (1984), also declined from 75% in October to 28% in November. Miyazaki (1977) interpreted this as an indication of decline of reproductive activity from October to December that occurred in a striped dolphin population with two mating peaks in January and July. However, one might expect a rise in male reproductive activity in December if the population has one of the two mating peaks in January. The violation of this expectation suggests another possibility, that is, involvement of two populations in the fishery, with mating peaks 6 months apart.

10.5.10 Female Breeding Cycle and Annual Pregnancy Rate

Determination of pregnancy requires identification of a fetus in the uterus. It can also be determined with reasonable accuracy by histological examination of the endometrium (Kasuya and Tai 1993). However, in the early stage of Japanese cetacean biology, there were cases where the presence of a corpus luteum was used as an indication of pregnancy, which was incorrect. In Japanese coastal whaling a slit used to be made in the carcass for cooling during towing it to the land station. This often caused damage to the uterus and loss of the fetus and forced the scientists to determine pregnancy by the presence of a corpus luteum. This erroneous judgment of pregnancy was also used in the early stages of studies of striped dolphins taken by the drive fishery on the Izu coast. A correction for this was first attempted by Kasuya (1985). I examined 721 adult females in 10 schools driven during 1967–1979 and found that 242 of them were pregnant, 47 were pregnant and simultaneously lactating, 354 were in lactation, and 78 were resting. Among these, 14 resting females and one lactating female had a corpus luteum in their ovaries but no fetus. That is, 4.9% of females having corpora lutea in their ovaries were not found with fetuses in their uteri. Since a corpus luteum of ovulation lasts for a shorter period than that of pregnancy that persists for the entire period of pregnancy, it is safe to say that the average number of ovulations to be experienced before a conception is greater than 1.05. Such a difference will cause an important bias if one attempts to estimate the reproductive rate of the population, but I (Kasuya 1985) thought that it would not cause much of a problem for the task of analyzing the trend in pregnancy rate by year of capture or by age (age-dependent increase of ovulations not followed by conception is known for short-finned pilot whales, Table 12.16).

When a female gains the physiological capacity to maintain pregnancy, Graafian follicles start to grow in her ovaries. The period from shortly before to shortly after ovulation is called estrus, when the female accepts a male for copulation. If the ovulation is not followed by conception, the female will again ovulate after some time interval during the mating season, which is called the ovulation cycle or estrous cycle. We do not have an estimate of the estrous cycle for the striped dolphin, but average cycles of 42 days (killer whale) or 27 days (bottlenose dolphin) have been reported (Section 12.4.4.5).

A cycle of reproduction in a female is a sequence of incidents from ovulation, conception, parturition, and lactation to the next ovulation. There is usually a period of resting between lactation and ovulation. Some females undergo ovulation and subsequent conception during lactation, but such females are likely to cease lactation before the next parturition (deduced for short-finned pilot whales). The resting period is a period when the females are preparing for the next ovulation. Length of gestation is probably the least variable parameter among individuals within a species or a population; the duration of lactation and time after the end of lactation to the next estrus are more variable.

The average reproductive cycle of the striped dolphin was estimated using the proportion of individuals at each reproductive stage in the catch of the Izu fishery. This assumed that the proportion of females at each reproductive stage reflected average length of time at the stage. If an unbiased sample of adult females contains 50% pregnant, 30% lactating, and 20% resting individuals, then the mean length of lactation is estimated by $30p/50$ and mean resting period $20p/50$, where p is the mean gestation period and is about 1 year for striped dolphins. The total of the stages, $50p/50 + 30p/50 + 20p/50 = 2p$, is the mean reproductive cycle, and $1/(2p)$ is the annual pregnancy rate.

The annual pregnancy rate and apparent pregnancy rate are totally different. The latter is the proportion of pregnant females in a sample of adult females and does not take account of gestation period or any bias associated with the sample. The annual pregnancy rate is the probability of a female becoming pregnant in a given year, but the value calculated by the method mentioned earlier probably contain errors due to abortion and mortality of adult females during pregnancy. However, cetacean biologists often use the annual pregnancy rate as an annual parturition rate, because we do not currently have ways to identify annual conception rate and annual parturition rate through the studies of carcasses.

The apparent pregnancy rate will differ greatly between samples obtained before and after the mating season. Segregation of females by reproductive status and selection by fisheries for particular reproductive status affect the apparent pregnancy rate and cause bias in the annual pregnancy rate derived from it. Lactating females of large whales accompanied by suckling calves were protected from whaling operations in the twentieth century, and mother–calf pairs of Dall's porpoises tended to avoid hand-harpoon vessels, both of which caused problems in estimating annual pregnancy rates.

Because the drive fishery on the Izu coast operated for only 3 months in the early winter from the 1960s to 1980s, samples obtained from the fishery could not be free from these biases. Miyazaki (1984) reported the composition of 699 adult females killed by drive fisheries, mostly on the Izu coast with about 10% at Taiji. Reproductive composition of the sample was 33.8% in pregnancy, 1.9% in pregnancy and simultaneous lactation, 42.6% in lactation, and 21.7% in resting status. This composition and a gestation period of 12 months tentatively assumed for the species suggested that females started their pregnancy after a resting period of 7 months on average from the end of the preceding lactation that lasted about

TABLE 10.6
Reproductive Status of Female Striped Dolphins and the Average Reproductive Cycle in Months Calculated Assuming Gestation of 12 Months

Source	Sample Size	Pregnant	Pregnant and Lactating[a]	Lactating	Resting[b]	Total
Miyazaki (1984)	699	33.8%	1.9%	42.6%	21.7%	100%
		11.4 months	0.6 months	14.3 months	7.3 months	33.6 months
Kasuya (1985)	841	39.0%	5.6%	45.4%	10.0%	100%
		9.8–10.6 months	2.2–1.4 months	17.8–16.3 months	4.0–2.6 months	33.7–30.7 months

[a] Pregnancy with simultaneous lactation.
[b] Sexually mature females neither pregnant nor lactating.

15 months (Table 10.6). The average calving interval was 33.6 months. The use of gestation time of 12 months instead of the 13.2 months proposed by Miyazaki (1984) was just for convenience of comparison with other studies.

Kasuya (1985) used a sample of 841 adult females to compare the number of pregnant females and that of calves estimated as aged 1 year or younger in the same schools and found the latter was only 63% of the former. Such a disagreement could occur either if the sample were obtained in the beginning of the parturition season or if suckling calves were lost during the drive. The former is a possibility for the sample, but the latter would not explain a contrary trend observed among pantropical spotted dolphins taken in the same fishery and analyzed in the same study. I attempted to estimate a range of true annual reproductive rate by assuming two possibilities: (1) lactating females (as well as suckling calves) were underrepresented in the sample and (2) pregnant females were overrepresented; I corrected the sample for these possibilities. The result is compared with the figure calculated from the sample of Miyazaki (1984) (Table 10.6). There is no significant difference between the two estimates of the reproductive cycle. The minor differences between them are probably due to small sample size. The mean length of the reproductive cycle of striped dolphins off Japan was likely to have been 30–34 months (2.5–2.8 years) in the period from the late 1960s to the early 1980s. The length of lactation could have been a minimum of 15 months but more likely 18–20 months. The derived annual pregnancy rate was 0.356–0.398.

Kasuya (1976a) calculated a mean annual ovulation rate of 0.414 from regression of corpus count on age using specimens from the Izu fishery. The age interval of the sample used for this calculation was 5–25 years. The mean number of ovulations per conception was estimated at 1.035–1.173 by multiplying the annual ovulation rate by the reproductive cycle (30–34 months). This is close to an independent observation that a minimum of 1.05 ovulations were experienced before one conception in 1967–1979 (Section 10.5.10). The estimate of annual ovulation rate mentioned earlier can be underbiased in some unknown degree due to a decline of female age at the attainment of sexual maturity that occurred during the period covered by the sample (Section 10.5.7). Thus, the real number of ovulations experienced before a conception could be greater than the range given earlier, if estimation of the reproductive cycle is correct. This is in accord with an assumption of long persistence of both corpora lutea of ovulation and of pregnancy in the ovaries.

10.5.11 Female Reproductive Cycle and Age

Some cetacean species exhibit a decline in annual pregnancy rate with increasing age. A well-known example is the short-finned pilot whale off Japan, where females cease reproduction by the age of 37 years, leaving a maximum life of 25 additional years, and 25% of adult females are postreproductive (Tables 12.13 and 12.14). The oldest female striped dolphin was aged at 57 years and was resting. The next oldest 2 females were aged 49 years (one lactating and the other resting), and the oldest pregnant female was aged 48 years (Kasuya 1985). This suggests that female striped dolphins do not have significant length of life after cessation of reproductive activity. However, the oldest female that was pregnant and simultaneously lactating was aged 42 years, which was 15 years younger than the oldest female. It is believed that the total energy required for nursing during the first 6 months after parturition is about 20% greater than the total cost spent during the pregnancy of 1 year (Lockyer 1981). Conception during lactation can be difficult for old female striped dolphins.

Kasuya (1985) obtained the relationship, $Y = 0.0273X + 1.60$, between mean reproductive cycle (Y, years) and age (X, years) of striped dolphins. The correlation coefficient was significantly different from zero. Since the Y given earlier is not corrected for the sample bias mentioned in Section 10.5.10, the equation is good only for testing the age-related trend in the reproductive cycle (not for estimating the absolute length of the reproductive cycle). It shows that the mean length of the reproductive cycle of females increased by about 40% during 40 years of reproductive life (from 5 years of age to 45 years). It is likely that female striped dolphins experience decreased productivity with age. It is unclear if this change is due to lengthening of the lactation period or of the resting period.

10.5.12 Historical Change in Female Reproductive Cycle

A decline of age at the attainment of sexual maturity has been identified for female striped dolphins killed by the drive fishery on the east coast of the Izu Peninsula. The change is likely to be a response of the population to density decline due to exploitation, although the effect of a possible shift of targeted populations cannot be fully excluded (Section 10.5.7). In order to explore possible additional response of the population to the density decline, Kasuya (1985) analyzed historical change in the female reproductive cycle using striped dolphins killed by the Izu fishery from 1953 to 1980. Although the mean calving interval, mean resting period, and mean lactation period showed a decline with time, none of the changes were statistically significant. The only significant historical change was an increase in the proportion of females simultaneously lactating and pregnant. These changes, including the statistically insignificant ones, would be expected to occur when the proportion of young sexually mature (thus reproductively active) females increased in the population or when the annual pregnancy rate increased for all the age classes. Both such changes could be interpreted as a response of the population to exploitation.

Although the currently available evidence is insufficient, probably due to small sample size, it is likely that there has been response of striped dolphins exploited by the drive fishery along the Izu coast in a direction of increasing the population productivity. The length of the mean calving interval apparently decreased by about 1 year during 22 years of exploitation, that is, from 4.0 years in 1955 to 2.8 years in 1977, but the change was not statistically significant (Kasuya 1985).

10.5.13 Age Composition, Sex Ratio, and Apparent Mortality Rate

Table 10.7 shows the perinatal sex ratio of striped dolphins calculated using fetuses of 5 cm and greater and calves of 164 cm or shorter. Smaller fetuses were excluded to avoid uncertainty in sex identification. The calves at 164 cm were estimated to be 1-year old on average. These sex ratios do not differ from parity.

The proportion of females was 49.4% in 3100 postnatal striped dolphins randomly examined from 13 schools driven on the east coast of the Izu Peninsula. The proportion varied between 21.7% and 72.8% (mean 50.5%) in the 13 schools (Kasuya 1985). Thus, the postnatal sex ratio of the striped dolphin is near parity.

TABLE 10.7
Perinatal Sex Ratio of Striped Dolphins

Sex	Fetus	Neonate	Total
Male	152	265	417
Female	163	261	424
%	51.7%	49.6%	50.4%

Source: Prepared from Table 7 of Kasuya, T., *Sci. Rep. Whales Res. Inst.* (Tokyo), 36, 107, 1985.

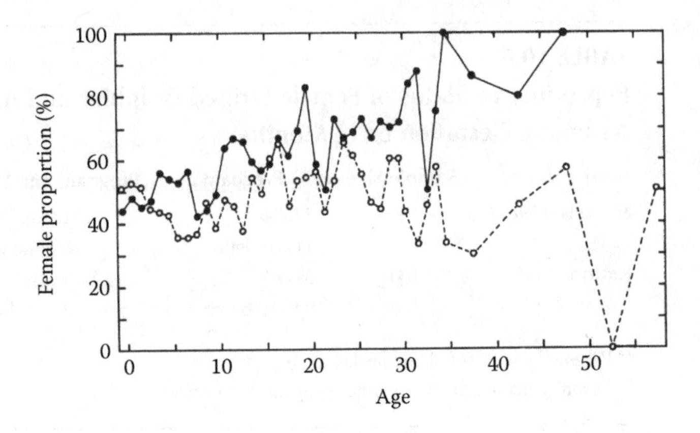

FIGURE 10.13 Age-dependent changes in sex ratio (females in %) of striped dolphins (open circles and dotted line) and pantropical spotted dolphins (closed circles and solid line). Striped dolphins (1110 males and 922 females) from 8 schools driven on the Izu coast in 1971–1977 and pantropical spotted dolphins (552 males and 767 females) from 11 schools driven at Izu and Taiji in 1970–1978. (Reproduced from Fig. 10 of Kasuya, T., *Sci. Rep. Whales Res. Inst.* (Tokyo), 36, 107, 1985.)

Out of the dolphins examined from 13 schools examined (Kasuya 1985), only 2032 individuals from 8 drives were aged; the female proportion in these was 21.7%–72.8% (mean 45.1%). The data from the 8 drives were used for the analysis of age-related change in sex ratio (Figure 10.13). The sex ratio was almost 1:1 up to the age of 2 years. From age 2 years to age 15 years, females tended to be fewer than males. This is explained by a difference in behavior between the sexes after weaning, that is, weaned calves tend to leave their mother's school to live with individuals of the same growth stage with the tendency more pronounced in males (Section 10.5.4), and the opportunity of scientists to examine particular school types. Females first ovulated at age 6–11 years (Figure 10.11) and males attained the physiological ability to reproduce at 7–13 years (Section 10.5.5 and Figure 10.9); subsequently, both sexes live in schools of adult individuals. The proportion of females started to decline at age 25 years, which presumably reflected higher female mortality. This age-related change in sex ratio differed from that of the pantropical spotted dolphin off Japan, where the female proportion increased from less than 50% at age below 3 years to over 70% at ages over 30 years. The biological background of the sex ratio in pantropical spotted dolphins is unknown.

The age composition of striped dolphins taken by the Izu fishery is shown in Figure 10.14. An age composition can be stable, and the right-hand slope can reflect the annual mortality rate only when abundance of the population is stable with constant and balanced recruitment and mortality. The age composition can be stable even when abundance changes at a constant rate (mortality and recruitment are not balanced), but the left-hand slope of such an age composition reflects the sum of mortality and rate of the population increase or decrease. The age composition will also be affected by the annual fluctuation in recruitment or selectivity of fishing gears, but the effect of gear selection can be ignored for the sample from

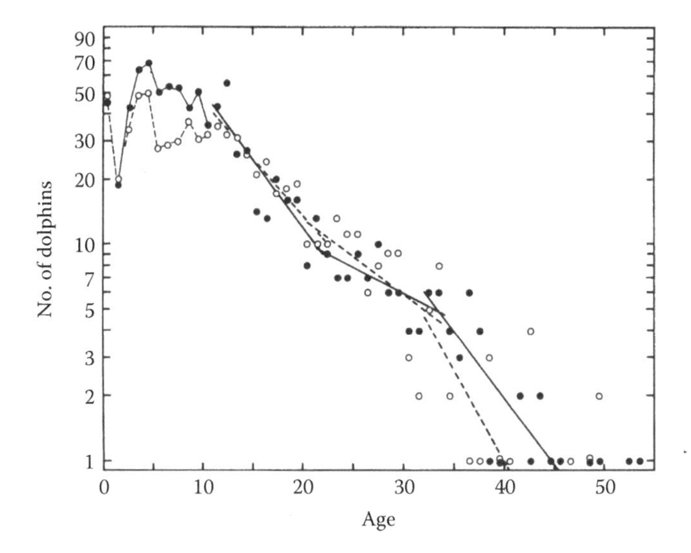

FIGURE 10.14 Age composition of striped dolphins using the same materials as in Figure 10.13. Closed circles represent males and open circles females. Solid lines (males) and dotted lines (females) represent linear regressions fitted to selected age ranges. (Reproduced from Fig. 10 of Kasuya, T., *Sci. Rep. Whales Res. Inst.* (Tokyo), 36, 107, 1985.)

the drive fishery, at least for individuals after weaning. The slope of the age composition is steeper for females of ages over 20 years than males of the same age (Figure 10.14). The number of males aged over 40 years was 11 versus 6 females of the same age range. This suggests that female mortality exceeds that of males at ages over 20 years. This is consistent with the apparent mortality rates calculated from the age composition (Table 10.8), although the difference is not statistically significant.

The age composition offers additional information on the biology of Japanese striped dolphins. There is an increase of mortality rate at higher ages, or over 35 years in this case, which is common among mammals. Also notable is the apparent mortality change at around 21 years of age. The boundary age, 21 years, roughly corresponds to individuals born around 1950–1956 and agrees with the time of great expansion of the catch by the Izu fishery after World War II that accompanied an earlier start of autumn operations and a broadened operation range (Section 10.4.8). This increase in apparent mortality around 1950–1956 reflects increased fishing mortality, the response of the dolphin population to the heavy exploitation, or both factors. Such a change in apparent mortality rate is not observed for the pantropical spotted dolphin, which was first exploited in a later period (since 1959) and only at a lower level (Kasuya 1985).

The regression coefficient of the slope in Figure 10.14 gives an estimate of the apparent instantaneous mortality rate. However, this method is not free from bias because it ignores points of zero frequency at higher ages. An alternative method is that proposed by Robson and Chapman (1961); results are presented in Table 10.8. These estimates, which are often called apparent mortality rates, require caution because they are a combination of natural mortality rate, fishing mortality rate, and change in the annual recruitment rate. The relationship between annual mortality rate (M) and instantaneous mortality rate (m) is expressed by $M = 1 - e^{-m}$. The value of m will be slightly greater than M, but the difference is negligible for M less than 0.1.

10.5.14 School Structure

Masaharu Nishiwaki of the Whales Research Institute, Tokyo, was probably the first fishery scientist who paid attention to the dolphin drive fishery on the Izu coast and started examining striped dolphins caught by the fishery with an intention to investigate their life history and school structure for fishery management. Shortly after this, his team obtained financial support from the Ministry of Education in 1961. An attempt at age determination was a part of their activity (Section 10.5.1). His work was succeeded by that of Seiji Ohsumi, Nobuyuki Miyazaki, and myself. Miyazaki stationed himself at Kawana on the east coast of the Izu Peninsula during several fishing seasons, joined the daily cruises in search of dolphin schools, and examined every school driven by the team. The last was particularly important because it offered him a chance to examine numerous small schools, which were processed before the arrival of scientists from Tokyo.

10.5.14.1 School Size

Drive fishermen on the Izu coast in the 1960s sent out searching vessels every early morning during the fishing season. If they found a suitable dolphin school, they immediately started driving it toward the port, and at the same time a loudspeaker placed at the FCU office in the port broadcasted a march called "Man of War March" to summon all the member fishermen to depart for assisting in the drive. Due to some unknown reason, they stopped playing the march sometime in the 1970s. The fishermen could combine several small dolphin groups to make one large drive or separate a portion of a large aggregation to make a smaller drive. There may have been cases where part of the original school escaped from the drive without being noticed by the fishermen. Several groups of striped dolphins may meet at a good feeding spot to form a temporary aggregation before being spotted by the fishermen. Thus, the school structure could have reflected technical or economic factors of the fishery. I used the term "school" for every unit of drive, ignoring biological or technical factors affecting the size or structure.

According to the records of the FCUs on the east coast of the Izu Peninsula or on the Sagami Bay coast, the size of striped dolphin schools ranged between 8 and 2136 dolphins, with a mean of 273 (n = 521) during the 1949–1974 seasons (Miyazaki and Nishiwaki 1978). The most frequent school size was in a range of 100–199, which made up 21.1% of the total number driven. Drives of less than 500 made up 85.8% of the drives. Only 4.3% of the drives were of 1000 or more dolphins.

Miyazaki and Nishiwaki (1978) compared the size of striped dolphin schools between operations on the east coast (Sagami Bay side) and west coast (Suruga Bay side) of the

TABLE 10.8
Instantaneous Mortality Rates of Striped Dolphins and Pantropical Spotted Dolphins Off Japan Calculated Using Data in Figure 10.12 and the Method of Robson and Chapman (1961)

	Striped Dolphin			Pantropical Spotted Dolphin		
Sex	Age	Mortality	95% Confidence Interval	Age	Mortality	95% Confidence Interval
Female	11–22	0.1074	±0.0376	11–26	0.0563	±0.0310
	20–34	0.0622	±0.0442	24–43	0.1026	±0.0418
	32–41	0.2132	±0.1453	>24	0.1348	±0.0282
	>32	0.1408	±0.0468			
Male	11–22	0.1486	±0.0366	7–22	0.0590	±0.0361
	20–34	0.0534	±0.0471	20–34	0.1494	±0.0734
	32–46	0.1489	±0.0771	>20	0.1618	±0.0439
	>32	0.1419	±0.0398			

Source: Extracted from Table 10 of Kasuya, T., *Sci. Rep. Whales Res. Inst.* (Tokyo), 36, 107, 1985.
Note: These are apparent mortality rates and can be skewed by changes in annual recruitment.

Izu Peninsula. The mean school size was 303 individuals for the east coast operation in September to February; this was not statistically different from 316 for the west coast operation in the same months. However, these figures were greater than the mean school size (156 individuals) in the spring/early summer operation on the west coast. We do not have an interpretation for the difference.

The school size changed with the time of the day in the autumn/winter operation on the east coast of the peninsula. Out of 78 schools found before 9:00 h, 61 (78%) contained over 100 dolphins, while only 58% (15 schools) of 26 schools found after that time were of that size. Miyazaki and Nishiwaki (1978) interpreted this to indicate that a large school was formed in the morning, but it could also be interpreted to mean that feeding aggregations formed during the night still remained in the morning and that they continued splitting into more stable groups of smaller size during the day. The school size is also known to change with the time of the day in pantropical spotted dolphins, spinner dolphins, and common dolphins in the eastern Pacific; school sizes were smaller before noon and larger in the afternoon (Scott and Cattanach 1998). The searching vessels of the Izu fishermen usually returned to port before noon, so no data were available on a school size in the afternoon.

10.5.14.2 Segregation of Weaned Immature Individuals

Some striped dolphin schools driven in the Izu fishery contained a broad age range of both sexes including sexually mature individuals and suckling calves but had a deficit in individuals of ages between weaning and sexual maturity. These were called "adult schools" (Figure 10.16). Although less common than the adult schools, there were also records of another type of school mostly composed of immature individuals aged 2–12 years; these tended to contain more males than females. These schools were called "juvenile schools."

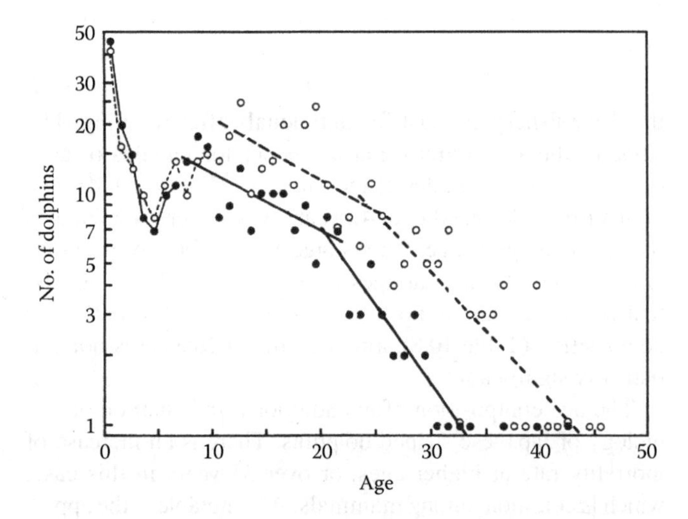

FIGURE 10.15 Age composition of the pantropical spotted dolphin using the same material as in Figure 10.13. Closed circles represent males and open circles females. Solid lines (males) and dotted lines (females) represent linear regressions fitted to selected age ranges. (Reproduced from Fig. 9 of Kasuya, T., *Sci. Rep. Whales Res. Inst.* (Tokyo), 36, 107, 1985.)

Miyazaki and Nishiwaki (1978) examined the proportion of sexually immature individuals in random samples from each school (the examination of the entire school was technically impossible), with individuals of suckling age or animals below 2.0 years of age excluded, and classified the schools into three types: 22 adult schools with a mature rate of 45%–100% (with a mode at 20%–40%), 10 mixed schools with a mature rate of 44%–81% (mode at 50%–60%), and 9 juvenile schools with a mature rate of 0%–25% (mode at 0%–10%). They concluded that "mixed schools" were temporary mixtures of "mature" and "juvenile schools." If this interpretation is rejected, then there was continuous variation in the mature rate from juvenile to adult schools.

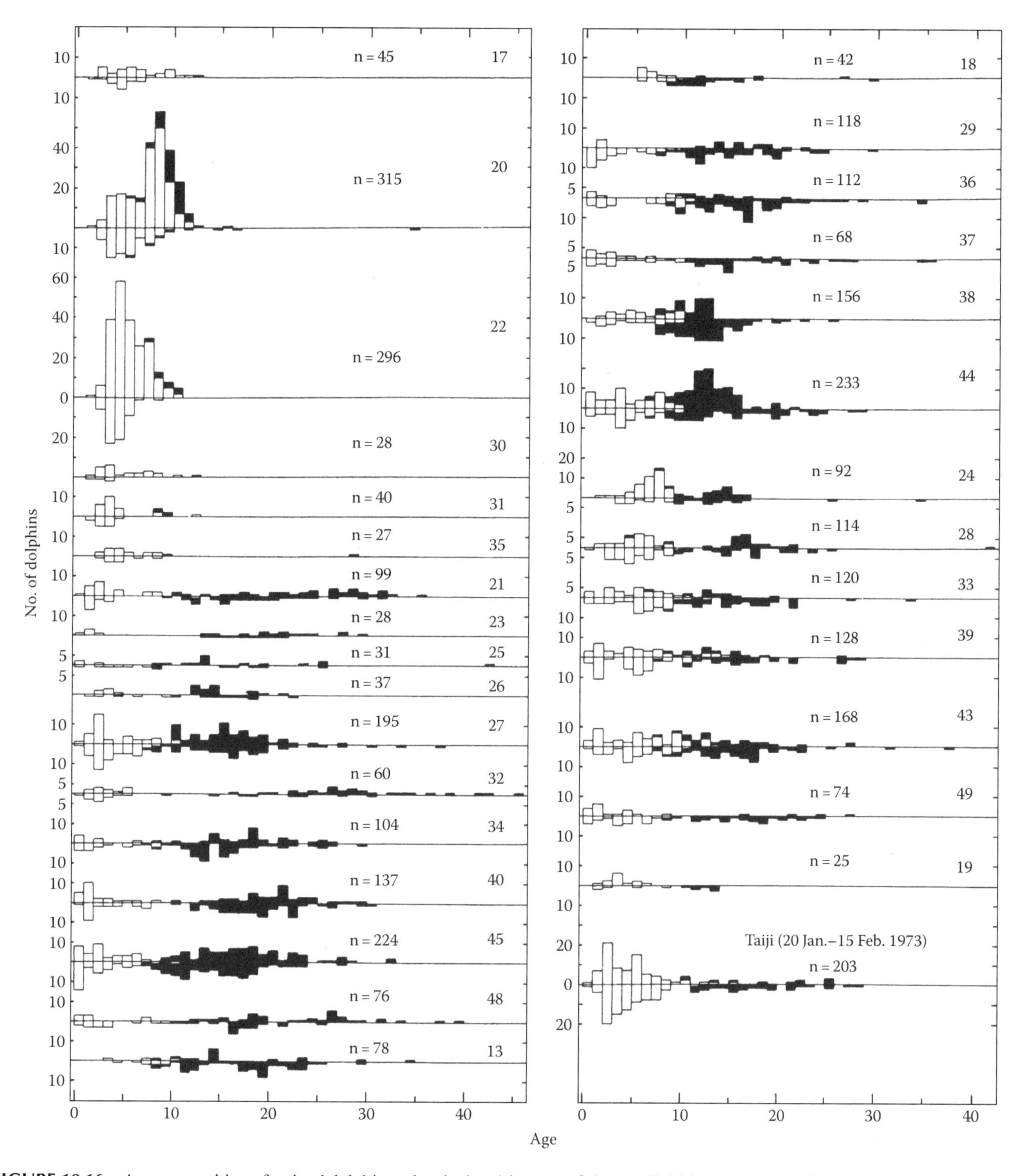

FIGURE 10.16 Age composition of striped dolphins taken by hand-harpoon fishery at Taiji from January to February 1973 (right bottom) and that of schools driven on the Izu coast (rest of panel). Males are shown above the horizontal line and females below. For both sexes, white squares represent sexually immature individuals and black squares mature individuals. (Reproduced from Fig. 1 of Miyazaki, N., *Rep. Int. Whal. Commn.*, Special Issue 6, Reproduction in Whales, Dolphins and Porpoises, 343, 1984. With permission.)

Even if there is a question on interpretation of the mixed schools, there is no doubt that weaned immature striped dolphins tended to leave their mother's school to live with other individuals of the same growth stage. The proportion of sexually mature females in the juvenile schools was extremely low. This suggested that females moved to adult schools at puberty or estrus, but not at the onset of pregnancy or lactation. Although Figure 10.16 may give an impression that males are more likely to remain in the immature school after sexual maturity, such a conclusion is premature in view of the ambiguity of the determination of male sexual maturity.

Age composition was variable among juvenile schools. In particular, the age of peak frequency varied among schools and between sexes. The size of juvenile schools was usually less than 100, although some exceptionally large juvenile schools of about 1000 were examined. This suggests several possibilities for the formation of juvenile schools, including cases where (1) weaned individuals in a school move out simultaneously from their mother's school, (2) weaned individuals in several adult schools become independent when adult schools temporarily aggregate, (3) weaned individuals in an adult school move to a juvenile school when two types of schools meet, and (4) two or more juvenile school merge to form a larger school.

The number of males tended to exceed that of females in a juvenile school, with an average male ratio of 75%. This is partly due to the fact that males attain sexual maturity at a greater age than females, while weaning age is not different between the sexes (see the following text in this section). Another reason for the imbalanced sex ratio could be the fact that weaned males have a stronger tendency to leave their mothers' school, which is shown by the male-biased sex ratio at any particular age in juvenile schools and by the contrary trend in adult schools (Figure 10.16). The behavior of weaned juveniles to live apart from their mothers is not limited to striped dolphins. Examination of age composition of wild dolphin schools finds schools that lack individuals at immature stages or contain mostly immature individuals, for example, in the Pacific white-sided dolphin (Takemura 1986, in Japanese) and Fraser's dolphin (Amano et al. 1996).

Weaning age may differ between the sexes in some toothed whales. Sons of bottlenose dolphins in Shark Bay in Western Australia wean by the age of 3 years, while daughters wean after the age of 4–5 years (Whitehead and Mann 2000). Some male long-finned pilot whales in the North Atlantic suckled at age 12 years, while females suckled only to age 7 years (Desportes and Mouritsen 1993). Sperm whales are known to suckle up to the age of 7 years (females) or 13 years (males) as determined by the detection of lactose in the stomach (Best et al. 1984). The longer suckling of male sperm whales is a reflection of their later attainment of puberty. The youngest individual in juvenile schools of striped dolphins was aged between 1.0 and 2.0 years of age (which is often expressed as 1.5 years), with no difference between the sexes. Thus, no evidence has been obtained on the sexual difference in the weaning age of striped dolphins.

Matrilineal social structure has been identified in killer whales and sperm whales through long-term observation, and the same was suggested by the DNA analysis of short-finned pilot whales off Japan and long-finned pilot whales in the North Atlantic, but the behavior of males after puberty is quite different among the species (Section 12.5.5). The stronger tendency of daughters in striped dolphins to remain with their mothers or the tendency of daughters of bottlenose dolphins to suckle longer than sons can be interpreted as a precursor of evolution toward a matrilineal social structure.

10.5.14.3 Adult Schools

The main components of adult schools described by Miyazaki and Nishiwaki (1978) were sexually mature individuals of both sexes and suckling calves accompanying their mothers, but these schools also contained some weaned juveniles, about 2–11 years of age. Of the 22 adult schools (number of drives), examined by these authors, 19 contained fewer than 600 dolphins and 3 contained over 1700.

Miyazaki and Nishiwaki (1978) examined random samples from 29 adult schools for age, sex, and reproductive status. Twenty-three schools contained females in all three reproductive stages of pregnancy, lactation, and resting. Although five of the remaining six schools had no resting females and one was found with only resting females, it seems hard to me to conclude that some of the elements were really absent in those schools, because the sample sizes were limited. The apparent pregnancy rate (proportion of pregnant females in sexually mature females) ranged from 0% to over 90%, lactation rate from less than 10% to 90%, and resting rate from 0% to 100%. The authors apparently did not obtain clear evidence supporting their hypothesis on the school structure of striped dolphins that estrous females and reproductive males assemble together to form a new mature school.

The idea that members of adult schools might come together at estrus come from observations of fetal body lengths, which varied among schools (Kasuya 1972). Fetal body length in each school appeared to have a narrower range and lesser variation than obtained by combining several schools taken in one fishing season. If school formation were independent of the incidence of estrus, that is, for example, schools were formed by females with suckling calves, there would be less school-specific fetal size structure. However, the narrow fetal length range can also be explained in another way, for example, females in a school synchronize their estrus in some way or mating by one individual causes estrus in other females. Best et al. (1984) proposed synchronized estrus among females in a sperm whale pod at the arrival of adult males to explain fetal body lengths unique to female pods, which may contain several matrilineal groups.

It is a difficult task to reconstruct the process of school formation of striped dolphins based on the observation of structure of schools taken by the drive fishery. The school, or unit of drive, varies in size from fewer than 50 individuals to over 1000. Some of the small- or medium-sized schools could have been part of larger schools, and some larger schools could have been forcibly created by the drive fishermen by combining closely located smaller schools. Some other large schools could have been formed through sequential merging of several smaller schools, or by temporary aggregation of smaller schools at a feeding ground. We do not have information on schools before the start of the drive operation.

Bottlenose dolphins (*Tursiops* spp.) have coastal and offshore populations, and some of the coastal populations, often called "communities," have been individually identified and studied to learn about maternal care and group formation. Connor et al. (2000) studied a community of about 200 individuals in a semienclosed bay. The dolphins live in many small groups that undergo fission–fusion group formation, where (1) lactating females gather to form a nursing school, (2) weaned calves join to live in a school of juveniles, (3) females return to their mother's school at sexual maturity, and (4) adult males live alone or together with other mature males. This community was almost isolated genetically from other nearby communities, but there were observed cases where some individuals visited another community to live there temporarily or cases where particular individuals moved permanently to another community. Thus, each community seems to be an element of a metapopulation containing several such coastal communities. Conserving each member community is important for the management of such a metapopulation.

The comparison of this information with what we know about striped dolphins allows several speculations: (1) The formation of nursing schools by of mother–calf pairs in the bottlenose dolphin has some similarity to the behavior of mother–calf pairs of Dall's porpoises (Section 9.6.2) but differs from the hypothesis for striped dolphins of Miyazaki and Nishiwaki (1978). The former possibility is worth consideration as a working hypothesis for striped dolphins. (2) The pattern of formation of juvenile schools has similarity between the bottlenose dolphin and the striped dolphin and is believed to be common among many other toothed whales. (3) The return of daughters to their mother's school at sexual maturity suggests a trend toward the formation of a matrilineal community; such a possibility should be investigated for striped dolphins. (4) The benefit of group living of adult males is an unanswered question. Male sperm whales are known to leave their mothers at puberty, spend the adolescent period with other males of similar age, and live a solitary life after attaining full sexual maturity (Best 1979), but males of some killer whale population live with their mothers for life (Bigg et al. 1990). The behavior of weaned male bottlenose dolphins has some similarity to that of the sperm whale, but male striped dolphins apparently behave differently.

The size of one ordinary school of striped dolphins off Japan is similar to the size of a community of coastal bottlenose dolphins studied by Connor et al. (2000). This could be interpreted in relation to the environment. Living in a large group is beneficial to avoid predators but unsuitable for living in an environment with limited food resources or that is inhabited by prey species in small schools. The offshore environment of striped dolphins may contain more predators and larger prey schools, which confers benefit of living in large schools. If striped dolphins off Japan have a structure similar to that identified in a community of bottlenose dolphins, then the size and geographical range of a community must be much greater, and we do not know how many such loosely isolated striped dolphin communities are contained in the range of the so-called "population" hunted by the Japanese drive fisheries at the Izu coast and Taiji.

10.5.15 Food Habits

The diet of striped dolphins off the Pacific coasts of Japan was studied by Miyazaki et al. (1973), using the contents of forestomachs collected from two schools driven at Kawana on the east coast of the Izu Peninsula. The sample sizes were 13 stomachs from school A and 14 from school B. Fish were identified to species using facial bones, mainly urohyals, and some otoliths. Squid and crustaceans were identified from the external morphology of relatively fresh food remains. The comparison of the numbers of squid beaks, fish otoliths, and urohyals in a stomach cannot be used as a key for the relative number of such food items consumed, because they may differ in the length of time retained in the stomach.

Miyazaki et al. (1973) identified 10 fish species in the stomachs. The commonest were the lantern fishes, Myctophidae, represented by 764 *Diaphus elicens* from both schools, 80 *Lampanyctus jordani* from both schools, 67 *Myctophum orientale* from school B, and 27 *Diaphus coeruleus* from both schools. These were followed by the members of other families: 190 *Nemichthys scolopaceus* from both schools, 117 *Erythrocles schlegelii* from both schools, and 89 *Chauliodus sloani* from both schools. The total lengths of these food items were estimated at 60–300 mm. The remains of squid were present in all the 27 stomachs and identified as two species with 42 *Todarodes pacificus* from both schools and 4 *Symplectoteuthis luminosa* from school B. Crustaceans, a total of 1971 individuals, were found in all the 27 stomachs but were not an important component in the weight of the total stomach contents because of a small body size of 38–130 mm. Of these prey items, the squid *Todarodes pacificus* is the only species fished in Japanese coastal fisheries.

Tobayama (1974; cited in Oizumi 2008, both in Japanese) reported fishes, squids, and shrimps in the stomachs of striped dolphins taken by the drive fishery on the east coast of the Izu Peninsula. He obtained results similar to those of Miyazaki et al. (1973) in the composition of fish species, that is, mainly lantern fishes with *Diaphus elicens* as the most dominant species of the group. He used squid beaks for species identification, which was an improvement over Miyazaki et al. (1973), and identified eight squid families. The most dominant squids were Enoploteuthidae followed by Loliginidae. These two studies reached similar results except for some differences in the squid fauna due to a difference in methodology.

10.5.16 Effect of Fisheries on the Population

The history of the dolphin fishery has been dealt with in Chapters 2 and 3. This section will summarize fishery activities that could have influenced striped dolphin populations off Japan. As striped dolphins often approach the bow waves of fishing vessels, they must have been hunted commercially and for personal use by the hand-harpoon method that is common in waters inhabited by the species. Choshi in Chiba Prefecture was once the most dominant locality in the hand-harpoon fishery for various marine organisms, and the occurrence of striped dolphin hunting was confirmed by information from fishermen. However, in spite of allocation of a catch quota of striped dolphins from the government, they reported almost no catch in recent years. Fishermen in the Sanriku Region have a long history of hand-harpoon fishing for various marine animals, so they could have hunted striped dolphins, at least during the peri–World War II period and during the summer, when the species migrated to the area.

As striped dolphins live in large schools, they are hunted with higher efficiency by driving. Since the fourteenth century, there are records of 52 groups conducting drives in Japan, including 6 on the Sea of Japan coast, 23 on the coast of the East China Sea and Tsushima Strait, 1 in Okinawa, and 22 on the Pacific coast (Kasuya 1999). Most of these locations did not obtain a license from the government in 1982, when the fisheries were placed under a license system. This indicates that these locations of historical drives had ceased operations before 1982. Striped dolphins are currently absent or extremely rare in the East China Sea and Sea of Japan; catch of the species could have been negligible in the past drive fisheries in these areas.

On the Pacific coast of northern Japan, Oura (39°27′N, 142°00′E) and Kamaishi (39°16′N, 141°54′E) on the Sanriku coast started dolphin driving in 1727 and made the last drive in 1920 (Kawabata 1986, in Japanese). Although it is a possibility that they took striped dolphins, available records are unable to confirm it.

In Okinawa, Nago (26°35′N, 127°59′E) has a long history of a drive fishery with particular preference for short-finned pilot whales and some additional species such as bottlenose dolphins and false killer whales. The striped dolphin is uncommon in waters around Okinawa, and there are no catch records of them (Miyazaki 1982; Uchida 1985, both in Japanese). There is some ambiguity about whether Nago obtained a prefecture license for the drive fishery in 1982 (Section 6.5). It did not obtain a catch quota under the system that began in 1993 for all the small cetacean fisheries in Japan. The crossbow fishery started at Nago in 1975 and supplied dolphin meat for the local market, but striped dolphins were not included in the target species.

The people of Hiratsuka (35°18′N, 139°21′E), which is on the northern coast of Sagami Bay, operated a dolphin fishery using a drag net in the late nineteenth century (Noshomu-sho Suisan-kyoku [Bureau of Fisheries of Ministry of Agriculture and Commerce] 1911, in Japanese). As the location is about 50 km NNW of Kawana and Futo on the east coast of the Izu Peninsula, they are likely to have taken some striped dolphins, but no records are currently available on the magnitude of operations or on the species taken.

Striped dolphins are currently hunted in three prefectures with government quotas: Wakayama (driving at Taiji and hand-harpoon fishery at Taiji and nearby ports), Shizuoka (driving at Futo on the east coast of the Izu Peninsula), and Chiba (hand-harpoon fishery at Choshi). The last two locations have almost ceased operations (Tables 2.11 and 3.16). Ohsumi (1972), Kasuya et al. (1984), Kishiro and Kasuya (1993), Kasuya (1994, 1997, both in Japanese), and Kasuya (1999) recorded the history of these operations from the view of population management.

Taiji and the nearby villages of Koza and Miwasaki (33°40′N, 135°59′E) have a history of net whaling before the mid-nineteenth century and all could have taken some striped dolphins as Taiji did (Section 10.1), but they left no catch records. The community of Taiji had a traditional system of opportunistic dolphin drives until the mid-1950s, but it did not leave catch statistics. The modern hand-harpoon operation at Taiji during the 10 years from 1963 to 1972 recorded an annual take of 331–1717 striped dolphins for a total of 7461 (Miyazaki et al. 1974). An active dolphin drive fishery using several searching vessels started at Taiji in 1969 for short-finned pilot whales and expanded its target to striped dolphins in 1973 (Kishiro and Kasuya 1993).

A hand-harpoon fishery operated for dolphins since an earlier date at Taiji to supply dolphin meat to the local market; it temporarily ceased operations when the higher-efficiency drive fishery started hunting striped dolphins in 1973 (Section 3.9.1), but it later resumed operations. In 1989, when the hand-harpoon fishery was placed under a license system, only 15 Taiji vessels obtained licenses in Wakayama Prefecture, but a total of 147 vessels from Taiji, Katsu-ura, and other villages in the prefecture joined the hand-harpoon fishery in 1991, attracted by the increased price of the products. These vessels subsequently obtained licenses and operate at present.

The oldest record of dolphin fisheries along the Izu coast is from 1619 (Shibusawa 1982, in Japanese). In those days, some driving teams at the northern part of the west coast of the Izu Peninsula, facing Suruga Bay, placed spotters at vantage points to search for schools of dolphins and fish such as tuna, but most of the teams operated opportunistically when a suitable school was found near the village. In the late nineteenth century, a total of 20 villages on both sides of the Izu Peninsula were known to operate drive fisheries, mainly for a dolphin species called *ma-iruka*, which means "true dolphin" or "correct dolphin" (Kawashima 1894, in Japanese). The species was recorded to weigh 48–93 kg and to form large schools of over 1000.

After World War II, Japanese scientists confirmed that the dolphin species taken in large numbers by the Izu fishery and called *ma-iruka* was the striped dolphin. Based on this information, Kasuya and Yamada (1995, in Japanese) thought that all the dolphins described by the Izu fishermen as *ma-iruka* since the nineteenth century were striped dolphins, but there is no firm evidence supporting it. The species called *ma-iruka* or *hase-iruka* along the coast of the Sea of Japan and the Goto Islands in Tsushima Strait is believed to be the common dolphin (*Delphinus* cf. *capensis*).

A stone monument of the dolphin drive in Kawana states that the village started the fishery in 1888. Kawashima (2008, in Japanese) reported the same data found on a painting donated in 1922 to the *Mishima Shrine* in Kawana. The village of Futo, which is located about 5 km south of Kawana, started driving dolphins around 1897–1898, or about 10 years after Kawana (Miyazaki 1983). None of these records mentioned the dolphin species hunted, and almost nothing is known on the dolphin drive fishery along the Izu coast from the early twentieth century to the 1930s.

The catch of dolphins along the Izu coast probably increased during World War II and the postwar period (Shizuoka-ken Kyoiku Iinkai [Board of Education of Shizuoka Prefecture] 1986, 1987, both in Japanese; Nakamura 1988, in Japanese). Shizuoka-ken Kyoiku Iinkai (1986, 1987, both in Japanese) stated that the Izu fishermen during the war shifted their operation to dolphin driving, which was possible in nearshore waters. Offshore operations of other fisheries became difficult due to a shortage of labor and lack of equipment during the war, and in addition, offshore operations became risky due to enemy attacks. The demand for dolphin meat increased during the postwar food shortage. These publications recorded the operation of drive fisheries at Inatori, Arari, Kawana, and Futo on the Izu coast (Figure 3.1). Some other villages on the Izu coast could have operated opportunistic dolphin drives during the postwar period but did not apparently leave any records (Kasuya 1976b, in Japanese).

The Ministry of Agriculture, Forestry and Fisheries since 1957 has published annual statistics of fishery products, including "dolphins and porpoises" and "blackfishes" without species identification. Using these publications, Ohsumi (1972) analyzed the catch of small cetaceans during the 14 years from 1957 to 1970. Of the 47 prefectures in Japan, only 5, Chiba, Shizuoka, Wakayama, Iwate, and Miyagi, recorded a mean annual catch of "dolphins and porpoises" of over 400. The first three prefectures were likely to have had striped dolphins in their major target. The mean annual catch and the ranges were Chiba (1,322; 453–3,110), Shizuoka (9,250; 3,601–21,342), and Wakayama (489; 34–1,879). Some other prefectures could have hunted smaller number of striped dolphins. Shizuoka Prefecture, where the Izu Peninsula is located, was operating the largest striped dolphin fishery in those days. Although the statistics analyzed by Ohsumi (1972) did not identify dolphin species taken, Tobayama (1969) reported that striped dolphins represented 97.7% of all dolphins landed on the east coast of the Izu Peninsula during the six seasons of 1963–1968 (Table 3.15). Thus, the take of species other than striped dolphins was negligible on the Izu coast. There were no official statistics for dolphins and porpoises before 1957.

Using this information and some ancillary information collected by themselves, Kasuya and Miyazaki (1982) made the following estimates of the mean annual catch of striped dolphins during the 10 years from 1950 to 1960: 1,500 at Choshi in Chiba Prefecture, 11,000 along the Izu coast in Shizuoka Prefecture, 630 at Taiji in Wakayama Prefecture, and 1,000 in other prefectures. The total annual take was estimated at 14,000 striped dolphins. Further, based on the available catch data for Kawana and Futo on the Izu coast from 1942 to 1953, they assumed that the total annual catch of the species from 1945 to 1950 could have been of a similar level. Further details for this estimate are given in Kasuya (1976b, in Japanese).

As requested by the IWC, the Japanese government in 1972 began its effort to collect statistics on small cetaceans by species and started reporting them to the IWC in 1979 (Section 1.4). These statistics were published in the annual report of the IWC. Before such a government system was established, individual scientists made efforts to collect catch statistics on striped dolphins at the landing ports; this included the work of Miyazaki (1980) at Taiji and the studies by Ohsumi (unpublished, cited in Kasuya 1976b, in Japanese), Tobayama (1969, in Japanese) and Miyazaki et al. (1974) on the Izu fishery. Miyazaki et al. (1974) erroneously interpreted *hasunaga-iruka* recorded at Arari on the west coast of the Izu Peninsula to be the pantropical spotted dolphin, for which I share the responsibility. However, the species is now considered to be the bottlenose dolphin (Section 3.8.2; Kasuya 1976b, in Japanese). Kishiro and Kasuya (1993) compiled previously published statistics and subsequent government statistics of catch of the species.

Inatori on the east coast of the Izu Peninsula had already ceased dolphin driving and the statistics were lost before 1960, when I first participated in studying dolphins taken by the Izu fishery. Catch statistics for Futo are missing for the years 1942–1957. It is urgent that catch statistics for many other undocumented operations on the Izu coast be recovered.

A brief description of the recent history of the drive fishery on the Izu coast follows. Shizuoka Prefecture placed all the "drive fisheries" under its license system in 1951. The "drive fishery" was defined as including driving of any marine organism, for example, driving tuna into a net set at a particular licensed location. Such fisheries were operated since at least the seventeenth century and were approved as fishing rights since early on (Sections 3.8.1 and 5.1). However, the actual targets of the drive fisheries have since been limited to dolphins, and the 1951 regulation was probably intended to regulate the establishment of new dolphin, drive teams in the prefecture. In 1959, Shizuoka Prefecture modified the 1951 regulation to be applied to dolphin drive fisheries and limited the season to 7 months from September 1 to March 31. Ten years after this, Arari village made the last regularly scheduled dolphin drive in 1961. In 1962, they returned to opportunistic driving only when suitable schools were found near the harbor or when they received requests from aquariums. The last dolphin drive at Arari occurred in 1973.

After 1962, Futo and Kawana remained as the only locations where regular dolphin drives occurred on the Izu coast. These two groups started cooperative driving in the 1967 or 1968 season (Section 3.8.1) and continued until the 1983 season. Kawana made the last drive in the 1983 season and left Futo as the only dolphin drive team on the Izu coast since 1984, but Futo appeared to have almost ceased driving in 2005 (Table 3.16).

While the two drive teams on the Izu coast operated the dolphin fishery using the same number of searching vessels during the 1960s and 1970s, the speed of the searching vessels increased, suggesting increased efficiency of the operation. Under this pressure, there was an apparent decline of the catch, although large annual fluctuation in the catch hinders a firm conclusion (Figure 10.17). I once attempted a primitive analysis of trend in the striped dolphin population, using natural and fishing mortality rates estimated from age composition and a decline of mean calving interval from 3.1 years in 1955 to 2.3 years in 1975 estimated from the catch composition (Kasuya 1976a; Kasuya 1976b, in Japanese). The calculation suggested that an initial population of 370–400 thousand declined to 260–310 thousand by around 1960 and further to 180–250 thousand by 1975. The replacement yield was estimated at 4000–6000 for the 1974 season, which was close to half of the annual catch of the time, that is, 7000–8000. I agree that the parameters used in the calculation contained uncertainties, that the model assumed sigmoid population dynamics that may be unsuitable for cetaceans, and that the

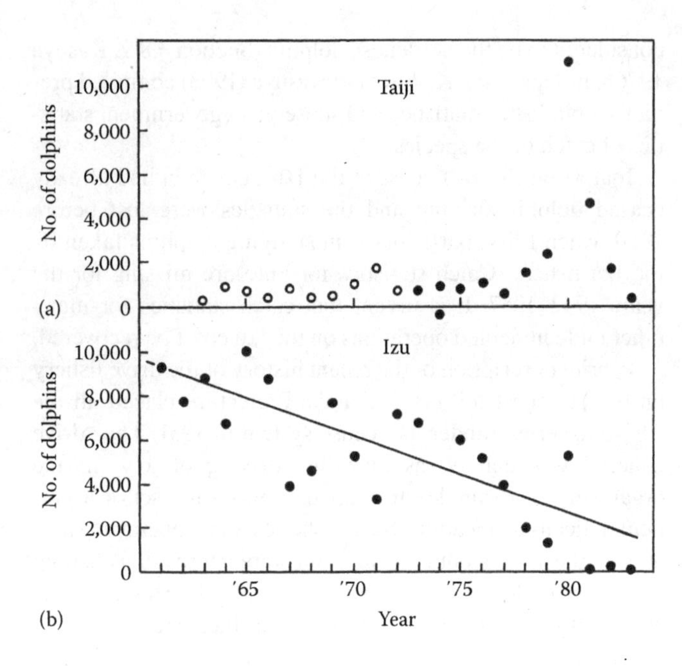

FIGURE 10.17 Annual catch of striped dolphins at Taiji (a) and on the Izu coast (b). Open circles indicate takes by hand-harpoon fishery and closed circles by drive fishery. During the period indicated in this figure, the Izu fishery was operated by two teams using a total of four vessels, and the efficiency of the vessels could have been improved through an increase of the vessel speed. (Reproduced from Fig. 11 of Kasuya, T., *Sci. Rep. Whales Res. Inst.* (Tokyo), 36, 107, 1985.)

results are of low reliability. However, I still believe, as a person who watched the fishery since 1960, that the exploited striped dolphin population was declining during the 1960s and 1970s (also see Section 15.3.1).

Although the catch of striped dolphins in the drive fishery on the Izu coast apparently continued to decline into the late 1970s, it was only in 1982 that the IWC accepted the probability of a statistically significant declining trend (IWC 1983). The committee in 1991 made several recommendations on the management of the stock. In 1992, it concluded that the population could not support exploitation at the current level and that there were no appropriate data on which to set a sustainable catch level. It recommended that Japan put in place an interim halt of all direct catches of the species until a proper assessment could be made of this population (Section 15.3.1).

The Fisheries Agency of Japan placed the dolphin drive fisheries under a license system in 1982 (the regulation of Shizuoka Prefecture preceded this) and the hand-harpoon fishery in 1989. It made a request to the prefecture governors to attempt to decrease the catch level and set catch limits (without species distinctions) for each fishery in 1990 (Table 6.1). These actions were finally followed by catch quotas set for each species and each fishery in the 1993 season (Tables 6.2 and 6.3), when a total quota of 1000 striped dolphins was shared between the three prefectures of Chiba, Shizuoka, and Wakayama. This quota was further reduced in subsequent years, reaching 670 for the 2008/2009 season, which was divided among the hand-harpoon fishery in Chiba (64), drive fishery in Shizuoka (56), and the hand-harpoon fishery (100) and drive fishery (450) in Wakayama. The total amounted to only about 5% of the catch at the height of the fisheries and is evidence of management failure.

The dolphin drive fisheries at Futo, on the Izu Peninsula in Shizuoka Prefecture, and at Taiji in Wakayama Prefecture now operate with a total quota of 2927 dolphins of seven species (striped dolphin, pantropical spotted dolphin, Pacific white-sided dolphin, bottlenose dolphin, Risso's dolphin, short-finned pilot whale, and false killer whale), 23% of which is striped dolphins (Table 6.3).

REFERENCES

[*In Japanese Language]

*Amano, T. 1955. On age determination of whales. *Geiken Tsushin* 45: 1–6.

Amano, M., Ito, H., and Miyazaki, N. 1997. Geographic and temporal comparison of skulls of striped dolphin off the Pacific coasts of Japan. Paper IWC/SC/49/SM39 presented to the Scientific Committee Meeting of IWC in 1997. 18pp. (Available from IWC Secretariat).

Amano, M., Miyazaki, N., and Yanagisawa, F. 1996. Life history of Fraser's dolphin, *Lagenodelphis hosei*, based on a school captured off the Pacific coasts of Japan. *Marine Mammal Sci.* 12(2): 199–214.

Archer, F.I. 1996. Morphological and genetic variation of striped dolphins (*Stenella coeruleoalba*, Meyen 1833). Doctoral thesis, University of California, San Diego, CA. 185pp.

Best, P.B. 1979. Social organization of sperm whales, *Physeter macrocephalus*. pp. 227–289. In: Winn, H.E. and Ola, B.L. (eds.). *Behavior of Marine Mammals: Current Perspective in Research*, Vol. 3 (Cetacea). Plenum, New York. 438pp.

Best, P.B., Canham, P.A., and Macleod, N. 1984. Patterns of reproduction in sperm whales, *Physeter macrocephalus*. *Rep. Int. Whal. Commn.* (Special Issue 6, Reproduction in Whales, Dolphins and Porpoises): 51–79.

Best, P.B. and Kato, H., 1992. Possible evidence from foetal length distributions of the mixing of different components of the Yellow Sea-East China Sea-Sea of Japan-Okhotsk Sea minke whale population(s). *Rep. Int. Whal. Commn.* 42: 166.

Bigg, M.A., Olesiuk, P.E., Ellis, G.M., Ford, J.K., and Balcomb, K. 1990. Social organization and genealogy of resident killer whales (*Orcinus orca*) in the coastal waters of British Columbia and Washington State. *Rep. Int. Whal. Commn.* (Special Issue 12, Individual Recognition of Cetacea): 383–405.

Brook, F.M., Kinoshita, R., and Benirschke, K. 2002. Histology of ovaries of a bottlenose dolphin, *Tursiops aduncus*, of known reproductive history. *Marine Mammal Sci.* 18(2): 540–544.

Connor, R.C., Wells, R.S., Mann, J., and Read, A.J. 2000. The bottlenose dolphin: Social relationship in a fission-fusion society. pp. 91–126. In: Mann, J., Connor, R.C., Tyack, P.L., and Whitehead, H. (eds.). *Cetacean Societies: Field Studies of Dolphins and Whales*. University of Chicago Press, Chicago, IL. 433pp.

Cooke, J.G. 1984. The estimation of ages at sexual maturity from age samples. Paper IWC/SC/36/O22 presented to the Scientific Committee Meeting of IWC in 1984. 8pp. (Available from IWC Secretariat).

Dabin, W., Cossais, F., Pierce, G.J., and Ridoux, V. 2008. Do ovarian scars persist with age in all cetaceans: New insight from the short-beaked common dolphin (*Delphinus delphis* Linnaeus, 1758). *Marine Biol.* 156: 127–139.

Dawbin, W.H. 1966. The seasonal migratory cycle of humpback whales. pp.145–169. In: Norris, K.S. (ed.). *Whales, Dolphins and Porpoises.* University of California Press, Berkeley, CA. 789pp.

DeMaster, D.P. 1984. Review of techniques used to estimate the average age at the attainment of sexual maturity in marine mammals. *Rep. Int. Whal. Commn.* (Special Issue 6, Reproduction in Whales, Dolphins and Porpoises): 175–179.

Desportes, G. and Mouritsen, R. 1993. Preliminary results on the diet of long-finned pilot whales off the Faroe Islands. *Rep. Int. Whal. Commn.* (Special Issue 14, Biology of Northern Hemisphere Pilot Whales): 305–324.

Gray, J.E. 1866. *Catalogue of Seals and Whales in the British Museum*, 2nd edn. British Museum, London, U.K. 402pp.

Hirose, K., Kasuya, T., Kazihara, T., and Nishiwaki, M. 1970. Biological study of the corpus luteum and the corpus albicans of blue white dolphin (*Stenella coeruleoalba*). *J. Mamm. Soc. Jpn.* 5(1): 33–40.

Hirose, K. and Nishiwaki, M. 1971. Biological study on the testis of the blue white dolphin, *Stenella coeruleoalba*. *J. Mamm. Soc. Jpn.* 5(3): 91–98.

Hohn, A.A. 1989. Comparison of methods to estimate the average age at sexual maturation in dolphins. Paper IWC/SC/O28 presented to the Scientific Committee Meeting of IWC in 1989. 20pp. (Available from IWC Secretariat).

Honda, K., Tatsukawa, R., Itano, K., Miyazaki, N., and Fujiyama, T. 1983. Heavy metal concentrations in muscle, liver and kidney tissue of striped dolphin, *Stenella coeruleoalba*, and their variations with body length, weight, age and sex. *Agric. Biol. Chem.* 47(6): 1219–1228.

IWC. 2010. Report of the Sub-Committee on Small Cetaceans. *J. Cetacean Res. Manage.* 12(suppl. 2): 306–330.

IWC (International Whaling Commission). 1983. Report of the Scientific Committee. Rep. Int.

Itano, K., Kawai, S., Miyazaki, N., Tatsukawa, R., and Fujiyama, T. 1984. Mercury and selenium levels in striped dolphins caught off the Pacific coast of Japan. *Agric. Biol. Chem.* 48(5): 1109–1116.

Iwasaki, T. and Goto, M. 1997. Composition of driving samples of striped dolphins collected in Taiji during 1991/92 to 1994/95 fishing season. Paper IWC/SC/49/SM19 presented to the Scientific Committee Meeting of IWC in 1997. 20pp. (Available from IWC Secretariat).

Iwasaki, T., Hwang, H.J., and Nishiwaki, S. 1995. Report of whale sighting surveys in waters off the Korean Peninsula and adjacent waters in 1994. Paper SC/47/NP18 presented to the Scientific Committee Meeting of IWC in 1995. 15pp. (Available from IWC Secretariat).

Kasuya, T. 1972. Growth and reproduction of *Stenella coeruleoalba* based on the age determination by means of dentinal growth layers. *Sci. Rep. Whales Res. Inst.* (Tokyo) 28: 73–106.

Kasuya, T. 1976a. Reconsideration of life history parameters of the spotted and striped dolphins based on cemental layers. *Sci. Rep. Whales Res. Inst.* (Tokyo) 28: 73–106.

*Kasuya, T. 1976b. Stock of striped dolphins, part I and II. *Geiken Tsushin* 295: 19–23 (part I); 296: 29–35 (part II).

Kasuya, T. 1978. The life history of Dall's porpoise with special reference to the stock off the Pacific coast of Japan. *Sci. Rep. Whales Res. Inst.* (Tokyo) 30: 1–63.

Kasuya, T. 1982. Preliminary report of the biology, catch and population of *Phocoenoides* in the western North Pacific. pp. 3–20. In: Clark, J.G., Goodman, J., and Soave, G.A. (eds.). *Mammals in the Sea.* Vol. IV (Small Cetaceans, Seals, Sirenians and Otters). FAO, Rome, Italy. 531pp. (This paper was presented in 1976 to the FAO Scientific Consultation on Marine Mammals, Bergen).

*Kasuya, T. 1983. Whale teeth and age determination, part I, II and III. *Kagaku-to Jikken* 34(4): 39–45 (part I); 34(5): 47–53(part II); 34(6): 55–62 (part III).

Kasuya, T. 1985. Effect of exploitation on reproductive parameters of the spotted and striped dolphins off the Pacific coasts of Japan. *Sci. Rep. Whales Res. Inst.* (Tokyo) 36: 107–138.

Kasuya, T. 1991. Density dependent growth in North Pacific sperm whales. *Marine Mammal Sci.* 7(3): 230–257.

*Kasuya, T. 1994. Striped dolphin. pp. 614–624. In: Odate,S. (ed.). *Basic Data of Rare Wild Aquatic Organisms of Japan* (I). Fisheries Agency, Tokyo, Japan. 696pp.

Kasuya, T. 1995. Overview of cetacean life histories: An essay in their evolution. pp. 481–497. In: Blix, A.S., Walloe, L., and Ultang, O. (eds.). *Whales, Seals, Fish and Man.* Elsevier, Amsterdam, the Netherlands. 720pp.

*Kasuya, T. 1997. Current status of Japanese whale fisheries and management of stocks. *Nihonkai Cetol.* 7: 37–49.

Kasuya, T. 1999. Review of the biology and exploitation of striped dolphins in Japan. *J. Cetacean Res. Manage.* 1(1): 81–100.

*Kasuya, T., Izumisawa, Y., Komyo, Y., Ishino, Y., and Maejima, Y. 1997. Life history parameters of common bottlenose dolphins off Japan. *IBI Report* (Kamogawa, Japan) 7: 71–107.

Kasuya, T. and Marsh, H. 1984. Life history and reproductive biology of the short-finned pilot whale, *Globicephala macrorhynchus*, off the Pacific coast of Japan. *Rep. int. Whal. Commn.* (Special Issue 6, Reproduction in Whales, Dolphins and Porpoises): 259–310.

Kasuya, T. and Miyazaki, N. 1982. The stock of *Stenella coeruleoalba* off the Pacific coast of Japan. pp. 21–37. In: Clark, J.G., Goodman, J., and Soave, G.A. (eds.). *Mammals in the Sea.* Vol. IV (Small Cetaceans, Seals, Sirenians and Otters). FAO, Rome, Italy. 531pp. (This paper was presented in 1976 to the FAO Scientific Consultation on Marine Mammals, Bergen).

Kasuya, T., Miyazaki, N., and Dawbin, W.F. 1974. Growth and reproduction of *Stenella attenuata* in the Pacific coast of Japan. *Sci. Rep. Whales Res. Inst.* (Tokyo) 26: 157–226.

Kasuya, T., Sergeant, D.E., and Tanaka, K. 1988. Re-examination of life history parameters of long-finned pilot whales in the Newfoundland waters. *Sci. Rep. Whales Res. Inst.* (Tokyo) 39: 103–119.

Kasuya, T. and Tai, S. 1993. Life history of short-finned pilot whale stocks off Japan and a description of the fishery. *Rep. Int. Whal. Commn.* (Special Issue 14, Biology of Northern Hemisphere Pilot Whales): 439–473.

Kasuya, T., Tobayama, T., and Matsui, S. 1984. Review of live-capture of small cetaceans in Japan. *Rep. Int. Whal. Commn.* 34: 597–600.

Kasuya, T., Tobayama, T., Saiga, T., and Kataoka, T. 1986. Perinatal growth of delphinids: Information from aquarium reared bottlenose dolphins and finless porpoises. *Sci. Rep. Whales Res. Inst.* (Tokyo) 37: 85–97.

*Kasuya, T. and Yamada, T. 1995. *List of Japanese Cetaceans.* Geiken Series 7. Nihon Geirui Kenkyusho (Institute of Cetacean Research), Tokyo, Japan. 90pp.

Kato, H. 1984. Observation of tooth scars on the head of male sperm whales, as an indication of intra-sexual fighting. *Sci. Rep. Whales Res. Inst.* (Tokyo) 35: 39–46.

*Kato, H. 1991. Density-dependent change in life history parameters of cetaceans. pp. 87–103. In: Sakuramoto, K., Tanaka, S., and Kato, H. (eds.). *Research and Management of Whale Stocks.* Koseisha Koseikaku, Tokyo, Japan. 273pp.

Kato, H. 1992. Body length, reproduction and stock separation of minke whales off northern Japan. *Rep. Int. Whal. Commn.* 42: 443–453.

Kato, H., Fujise, Y., and Wada, S. 1992. Morphology of minke whales in the Okhotsk Sea, Sea of Japan and off the east coast of Japan, with respect to stock identification. *Rep. Int. Whal. Commn.* 42: 437–453.

*Kawabata, H. 1986. Dolphin and porpoise fisheries. pp. 636–664. In: Yamada Choshi Hensan Iinkai (ed.). *History of Yamada Town*, Vol. 1. Yamada-cho Kyoiku Iinkai, Yamada-cho, Japan. 10+1095pp.

*Kawashima, T. 1894. *Records of Fisheries of Shizuoka Prefecture.* Shizuoka-ken Gyogyo-kumiai Torishimarisho, Shizuoka, Japan. Vol. 1: 288pp., Vol. 2: 182pp., Vol. 3: 406pp., Vol. 4: 362pp.

*Kawashima, S. 2008. *Drive Fisheries.* History of Human Culture and Materials, No. 142. Hosei Daigaku Shuppan-kyoku, Tokyo, Japan. 341pp.

Kishiro, T. and Kasuya, T. 1993. Review of Japanese dolphin drive fisheries and their status. *Rep. int. Whal. Commn.* 43: 439–452. 439–452.

*Komyo, Y. 1982. On the process of sexual maturation of the male striped dolphin *Stenella coeruleoalba.* Master's thesis, Tokyo University of Fisheries, Tokyo, Japan. 46pp.

Laws, R.M. 1959. The foetal growth of whales with special reference to the fin whale *Balaenoptera physalus* Linn. *Discovery Rep.* 29: 281–308.

Lockyer, C. 1981. Estimation of the energy costs of growth, maintenance and reproduction in the female minke whale, (*Balaenoptera acutorostrata*), from the southern hemisphere. *Rep. Int. Whal. Commn.* 31: 337–343.

Lockyer, C. 1984. Review of baleen whales (Mysticeti) reproduction and implications for management. *Rep. Int. Whal. Commn.* (Special Issue 6, Reproduction in Whales, Dolphins and Porpoises): 27–50.

Martin, A.R. and Rothery, P. 1993. Reproductive parameters of female long-finned pilot whales *(Globicephala melas)* around the Faroe Islands. *Rep. Int. Whal. Commn.* (Special Issue 14, Biology of Northern Hemisphere Pilot Whales): 263–304.

Miyashita, T. 1993. Abundance of dolphin stocks in the western North Pacific taken by Japanese drive fishery. *Rep. Int. Whal. Commn.* 43: 417–437.

Miyashita, T., Kishiro, T., Higashi, N., Sato, F., Mori, K., and Kato, H. 1996. Winter distribution of cetaceans in the western North Pacific inferred from sighting cruises 1993–1995. *Rep. Int. Whal. Commn.* 46: 437–441.

Miyashita, T., Peilie, W., Hua, C.J., and Guang, Y. 1995. Report of the Japan/China joint whale sighting cruise in the Yellow Sea and East China Sea in 1994 summer. Paper SC/47/NP17 presented to the Scientific Committee Meeting of IWC in 1995. 12pp. (Available from IWC secretariat).

Miyazaki, N. 1977. Growth and reproduction of *Stenella coeruleoalba* off the Pacific coast of Japan. *Sci. Rep. Whales Res. Inst.* (Tokyo) 29: 21–48.

Miyazaki, N. 1980. Catch records of cetaceans off the coast of the Kii Peninsula. *Mem. National Sci. Mus.* 13: 69–82.

*Miyazaki, N. 1982. Dugong and whales. pp. 157–166. In: Kisaki, K. (ed.). *Natural History of the Ryukyu Islands.* Tsukiji Shokan, Tokyo, Japan. 282pp. (First published in 1980).

Miyazaki, N. 1983. Catch statistics of small cetaceans taken in Japanese waters. *Rep. Int. Whal. Commn.* 33: 621–631.

Miyazaki, N. 1984. Further analyses of reproduction in the striped dolphin, *Stenella coeruleoalba*, off the Pacific coast of Japan. *Rep. Int. Whal. Commn.* (Special Issue 6, Reproduction in Whales, Dolphins and Porpoises): 343–353.

*Miyazaki, N. 1992. *Dreadful Marine Pollution: Marine Mammals Threatened by Pollutants.* Godo Shuppan, Tokyo, Japan. 190pp.

Miyazaki, N., Kasuya, T., and Nishiwaki, M. 1974. Distribution and migration of two species of *Stenella* in the Pacific coast of Japan. *Sci. Rep. Whales Res. Inst.* (Tokyo) 26: 227–243.

Miyazaki, N., Kusaka, T., and Nishiwaki, M. 1973. Food of *Stenella coeruleoalba. Sci. Rep. Whales Res. Inst.* (Tokyo) 25: 265–275.

Miyazaki, N. and Nishiwaki, M. 1978. School structure of the striped dolphin off the Pacific coast of Japan. *Sci. Rep. Whales Res. Inst.* (Tokyo) 30: 65–115.

Myrick, A.C., Shallenberger, E.W., Kang, I., and MacKay, D. 1984. Calibration of dental layers in seven captive Hawaiian spinner dolphins, *Stenella longirostris*, based on tetracycline labeling. *Fish. Bull.* 82: 207–225.

*Nakamura, Y. 1988. On dolphin and porpoise fisheries. pp. 92–136. In: Shizuoka-ken Minzoku Geino Kenkyukai (ed.). *Marine Folklore of Shizuoka: Culture in Kuroshio Current.* Shizuoka Shinbunsha, Shizuoka, Japan. 379pp.

Nishiwaki, M., Nakajima, M., and Kamiya, T. 1965. A rare species of dolphin (*Stenella attenuata*) from Arari, Japan. *Sci. Rep. Whales Res. Inst.* (Tokyo) 19: 53–64.

Nishiwaki, M. and Yagi, T. 1953. On the growth of teeth in a dolphin (*Prodelphinus coeruleoalba*) (I). *Sci. Rep. Whales Res. Inst.* (Tokyo) 8: 133–146.

*Noshomu-sho Suisan-kyoku 1911. *Fishery Activities of Japan,* Vol. 1. Suisan Shoin, Tokyo, Japan. 424pp.

*Ogawa, T. 1950. *Tale of Whales.* Chuo Koronsha, Tokyo, Japan. 260pp.

Ohsumi, S. 1964. Comparison of maturity and accumulation rate of corpora albicantia between the left and right ovaries in Cetacea. *Sci. Rep. Whales Res. Inst.* (Tokyo) 18: 123–153.

Ohsumi, S. 1966. Allomorphosis between body length at sexual maturity and body length at birth in Cetacea. *J. Mamma. Soc. Jpn.* 3(1): 3–7.

Ohsumi, S. 1972. Catch of marine mammals, mainly of small cetaceans, by local fisheries along the coast of Japan. *Bull. Far Seas Fish. Res. Lab.* 7: 137–166.

Ohsumi, S. 1983. Yearly change in age and body length at sexual maturity of the fin whale stock in the eastern North Pacific. Paper IWC/SC/A83/AW7 presented to the Scientific Committee Meeting of IWC in 1983. 19pp. (Available from IWC secretariat, Cambridge, U.K.).

*Ohizumi, H. 2008. Food of cetaceans around Japan. pp.197–273. In: Murayama, T. (ed.). *Geirui-gaku* [*Cetology*]. Tokai University Press, Hadano, Japan. 402pp.

Okada, Y. 1936. A study of Japanese Delphinidae (1). *Sci. Rep. Tokyo Bunrika Daigaku* 3(44): 1–16, pls. 1–5.

Oliver, W.R.B. 1922. A review of the cetacean of New Zealand Seas—I. *Proc. Zool. Soc. London* 1922: 558–585.

Perrin, W.F. and Donovan, G.P. 1984. Report of the Workshop. *Rep. Int. Whal. Commn* (Special Issue 6, Reproduction in Whales, Dolphins and Porpoises): 1–24.

Perrin, W.F., Holt, D.B., and Miller, R.B. 1977. Growth and reproduction of the eastern spinner dolphin, a geographical form of *Stenella longirostris* in the eastern tropical Pacific. *Fish. Bull.* 75(4): 725–750.

Perrin, W.F. and Reilly, S.B. 1984. Reproductive parameters of dolphins and small whales of the family Delphinidae. *Rep. Int. Whal. Commn* (Special Issue 6, Reproduction in Whales, Dolphins and Porpoises): 97–133.

Perrin, W.F., Wilson, C.E., and Archer, F.I. 1994. Striped dolphin *Stenella coeruleoalba* (Meyen, 1833). pp. 129–159. In: Ridgway, S.H. and Harrison, R. (eds.). *Handbook of Marine Mammals*, Vol. 5 (The First Book of Dolphins). Academic Press, London, U.K. 416pp.

Rice, D.W. 1998. *Marine Mammals of the World: Systematics and Distribution.* Special Publication No. 4. Society for Marine Mammalogy. Lawrence, KS. 231pp.

Robson, D.C. and Chapman, D.G. 1961. Catch curves and mortality rates. *Trans. Am. Fish. Soc.* 90: 181–189.

Scott, M.D. and Cattanach, K.L. 1998. Diel patterns in aggregations of pelagic dolphins and tunas in the eastern Pacific. *Marine Mammal Sci.* 14(3): 401–428.

Sergeant, D.E. 1962. The biology of the pilot whale or pothead whale *Globicephala melaena* (Traill) in Newfoundland waters. *Bull. Fish. Res. Bd. Can.* 132: 1–84+I-VII.

*Shibusawa, K. 1982. *Japanese Fishery Technology before the Meiji Revolution.* Noma Kagaku Igaku Kenkyu Shiryokan, Tokyo, Japan. 701pp.

Shimura, E. and Numachi, K. 1987. Genetic variability and differentiation in toothed whales. *Sci. Rep. Whales Res. Inst.* (Tokyo) 38: 161–163.

Shirakihara, M., Takemura, A., and Shirakihara, K. 1993. Age, growth and reproduction of the finless porpoise, *Neophocaena phocaenoides*, in the coastal waters of western Kyushu, Japan. *Marine Mammal Sci.* 9: 392–406.

*Shizuoka-ken Kyoiku Iinkai 1986. *Report of the Cultural Heritage of Shizuoka Prefecture 33: Manners and Customs of Fisheries in Izu.* Part 1. Shizuoka-ken Bunkazai Hozon Kyokai, Shizuoka, Japan. 211pp.

*Shizuoka-ken Kyoiku Iinkai 1987. *Report of the Cultural Heritage of Shizuoka Prefecture 39: Manners and Customs of Fisheries in Izu.* Part 2. Shizuoka-ken Bunkazai Hozon Kyokai, Shizuoka, Japan. 193pp.

*Takemura, A. 1986. Pacific white-sided dolphin and common bottlenose dolphin. pp. 161–177. In: Tamura, T., Ohsumi, S., and Arai R. (eds.). *Report of Contract Studies in Search of Countermeasures for Fishery Hazard (Culling of Harmful Organisms) (56th to 60th year of Showa).* Suisan-cho Gogyo Kogai (Yugaiseibutsu Kujo) Taisaku Chosa Kento Iinkai (Fishery Agency), Tokyo, Japan. 285pp.

*Tobayama, T. 1969. School size of striped dolphins and its variation viewed from fishing data in the Sagami Bay. *Geiken Tsushin* 217: 109–119.

*Tobayama, T. 1974. Study on feeding ecology of small cetaceans. Doctoral thesis, University of Tokyo, Tokyo, Japan. 231pp. (Not seen, cited in Ohizumi, H. 2008).

Torres, L.G., Rosel, P.E., D'Agrosa, C., and Read, A.J. 2003. Improving management of overlapping bottlenose dolphin ecotypes through spatial analysis and genetics. *Marine Mammal Sci.* 19(3): 502–514.

*Uchida, S. 1985. *Report of Surveys on Small Cetaceans for Aquarium Display.* Part 2. Kaiyohaku Kinen Koen Kanri Zaidan, Motobu, Japan. 36pp.

Wada, S. 1983. Genetic heterozygosity in the striped dolphin off Japan. *Rep. Int. Whal. Commn.* 33: 617–619.

Wang, J.Y. and Yang, S. 2007. *An Identification Guide to the Dolphins and Other Small Cetaceans of Taiwan.* Jen Jen Publishing and National Museum of Marine Biology, Taiwan, China. 207pp. (In Chinese and English).

Whitehead, H. 2003. *The Sperm Whales: Social Evolution in the Ocean.* University of Chicago Press, Chicago, IL. 431pp.

Whitehead, H. and Mann, J. 2000. Female reproductive strategies of cetaceans: Life histories and calf care. pp. 219–246. In: Mann, J., Connor, R.C., Tyack, P.L., and Whitehead, H. (eds.). *Cetacean Societies: Field Studies of Dolphins and Whales.* Univ. Chicago Press, Chicago, IL. 433pp.

Yamada, T. 1993. A brief introduction to cetacean stranding data base of our study group. *Nihonkai Cetology* 3: 43–65.

Yoshida, H. and Iwasaki, T. 1997. A preliminary analysis of mitochondrial DNA in striped dolphins off Japan. Paper IWC/SC/49/SM41 presented to the Scientific Committee Meeting of IWC in 1997. 7pp. (Available from IWC Secretariat).

11 Bottlenose Dolphins

11.1 TAXONOMY OF THE GENUS *TURSIOPS*

Species of the genus *Tursiops* inhabit temperate and tropical coastal and offshore waters of the world. They are among the most popular species displayed in aquariums and often met in their natural environments in coastal waters. Thus, they have offered us opportunities to study their morphology, behavior, and social structure. However, the taxonomy of this genus has not yet been fully settled. One of the reasons for this is the difficulty of obtaining specimen materials from their broad-ranging habitats, which is a common problem in cetacean studies. Another factor making their taxonomic study difficult is the existence of numerous local populations with various degrees of differentiation, as is usually the case in coastal small cetaceans. Under these circumstances, scientists have been able to identify differences between nearby populations but have had difficulty until recently in identifying and evaluating differences between geographically distant populations. This problem is being resolved by the use of genetics, which allows objective evaluation of differences. The genus contains two ecotypes—one inhabiting coastal waters and the other offshore waters—each of which contains numerous populations (Reeves et al. 2004), but the two ecotypes do not match up with the two species currently recognized in the genus.

One of the two currently recognized species, *T. aduncus* (Ehrenberg 1828, 1833), described from a specimen from the Red Sea, is known to inhabit coastal waters from the Indian Ocean to central Japan, including the coasts of Australia and Southeast Asia. It is smaller and has a relatively longer rostrum than the other species of the genus *T. truncatus* and the adults often have spots (Gao et al. 1995; Kasuya et al. 1997, in Japanese; Rice 1998; Wang et al. 1999, 2000a,b; Moller and Beheregaray 2001; Best 2007).

T. truncatus (Montagu 1821) was described based on a specimen from Great Britain. This species has greater body length, a relatively shorter and stouter rostrum than *T. aduncus,* and is not spotted. It inhabits offshore waters of the tropical and temperate regions of the Indian Ocean and the South and North Pacific (including Japanese waters), the North and South Atlantic, the Mediterranean Sea, and the Black Sea. It also inhabits coastal waters not inhabited by *T. aduncus*. Both species are known from Japanese waters.

The two species, *T. aduncus* and *T. truncatus*, interbreed easily in aquariums and produce fertile offspring (Hale et al. 2000). This suggests that they have not established a genetic mechanism to prevent interbreeding but that the isolation is achieved mainly by difference in habitat preference. The means of inhibiting interbreeding between such species will work differently between the sexes. Males are more likely to move outside of the ordinary range of a population and have a greater opportunity to interbreed with other species or populations. It will be valuable to further examine both mitochondrial and nuclear DNA for a better understanding of species and population structure within the genus *Tursiops*.

It was difficult to distinguish between the two species of *Tursiops* based on single morphological characters or measurements, but multivariate statistics applied in South Africa, China, and Australia successfully delineated the two species in comparisons between populations in neighboring habitats but not always between distant populations (Ross 1977; Wang et al. 2000b; Kemper 2004). This has suggested the possibility of some still undetermined taxonomic variations within each of the two species. Two forms, that is, coastal and offshore types, are known for *T. truncatus* in the western North Atlantic, where *T. aduncus* does not occur. The coastal type in Florida is smaller than conspecifics in offshore waters and also differs in skull morphology and physiology. The offshore type exhibits blood physiology better suited for deep diving (Hersh and Duffield 1990). Three populations, often called "communities," are known for the coastal *T. truncatus* on the west coast of Florida, each of which occupies tens of kilometers of the coastline (Wells and Scott 1990a; Weigle 1990). These coastal communities of *T. truncatus* apparently occupy a similar ecological niche as *T. aduncus* in other oceans, that is, in the coastal waters of the Indian Ocean and western Pacific. Future studies will reveal further details of geographical variation within the genus *Tursiops*.

Wang et al. (1999) analyzed mitochondrial DNA (mtDNA) to examine the geographical distribution of the two species of *Tursiops* in waters around eastern Asia and confirmed the presence of *T. aduncus* in the Taiwan Strait, Hainan Island, and Indonesia and *T. truncatus* in the Taiwan Strait, Hong Kong, the east coast of Taiwan, Mauritania, and Brazil. Thus, waters from the Taiwan Strait to Hong Kong are inhabited by both species. Before its prohibition in 1990, the dolphin-drive fishery in the Pescadores Islands off the west coast of Taiwan took both species of *Tursiops* (Wang and Yang 2007). Only *T. truncatus* has been reported from the Yellow Sea and the Bohai Sea, which are located to the north of Taiwan, and only *T. aduncus* from the waters south of Hong Kong (Zhou and Quian 1985; Wang et al. 2000a). However, in view of the segregation of the two species by the 30 m isobath off eastern Australia, there still remains the possibility of finding *T. truncatus* in the offshore waters of the South China Sea.

Perrin et al. (2007) compared partial mtDNA control-region sequences and morphology of the type specimen of *T. aduncus* with those of *Tursiops* spp. from other regions and found that mtDNA of the type specimen showed a perfect agreement with that of dolphins referred to the same species from the Atlantic coast of South Africa but that it did not match that of *T. truncatus*. Skull measurements of the type specimen of *T. aduncus* were also found to be closer to *T. aduncus* from South Africa (33 individuals) and Taiwan (19 individuals) than *T. truncatus* from South Africa (9 individuals) and Taiwan (50 individuals).

Before the study by Perrin et al. (2007), Natoli et al. (2004) analyzed mtDNA for a total of 11 samples of the two species of *Tursiops* from the Mediterranean, North Atlantic, Gulf of Mexico, Bahamas, West Africa, South Africa, and the Chinese coast. It was unfortunate that they did not analyze Japanese samples. Their conclusions were as follows: (1) *T. truncatus* and *T. aduncus* are distantly related, which was expected, (2) *T. aduncus* from South Africa and those from China represented two distantly related groups, and (3) *T. truncatus* in the Gulf of Mexico and U.S. east coast formed one group and those in the Mediterranean and offshore waters of eastern and western North Atlantic formed another group. Their results suggested that the genus *Tursiops* contains three groups and possibly more. It is possible that future studies will identify two species or subspecies within the current *T. aduncus* and the same in the current *T. truncatus*.

LeDuc et al. (1999) proposed that *T. aduncus* was genetically closer to *Stenella frontalis* in the Atlantic than to *T. truncatus*. The taxonomy of these genera has long relied upon easily identifiable external or skeletal morphology, which is likely subject to convergence, and restructuring of the classification using more objective genetic methodology is under way.

The common names currently in use are "common bottlenose dolphin" for *T. truncatus* and "Indo-Pacific bottlenose dolphin" for *T. aduncus*. The latter species was identified in Japan in 1965 by Nishiwaki (see Kasuya and Yamada 1995, in Japanese), although the presence of the species was fully confirmed later (Section 11.3). The Japanese common name *hando-iruka* or *bando-iruka* is used for the common bottlenose dolphin and *minami-hando-iruka* or *minami-bando-iruka* for the Indo-Pacific bottlenose dolphin.

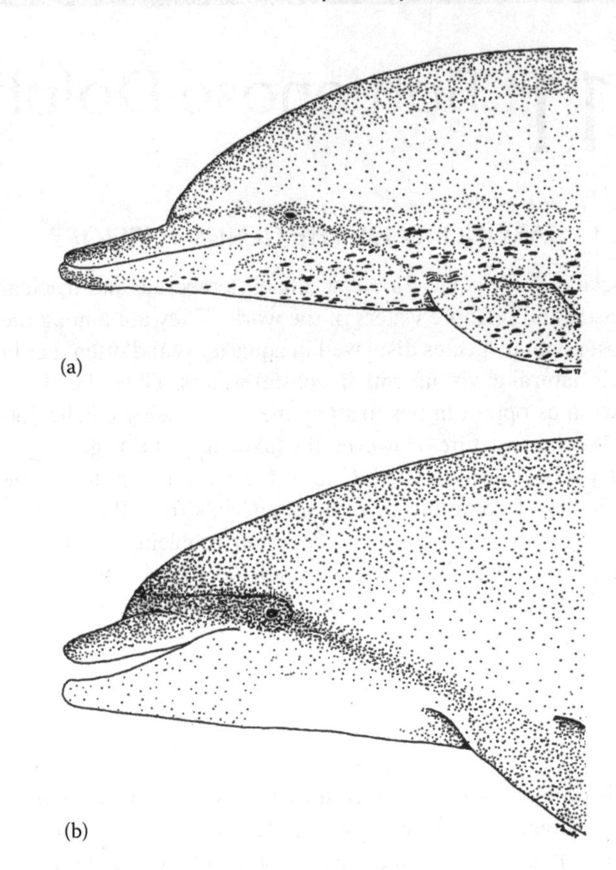

FIGURE 11.1 Two species of bottlenose dolphins in Chinese waters: (a) Indo-Pacific bottlenose dolphin and (b) common bottlenose dolphin. (Reproduced from Fig. 3 of Wang, Y.J. et al., *J. Mammal.*, 81(4), 1157, 2000a. With permission of the American Society of Mammalogists.)

11.2 IDENTIFICATION OF BOTTLENOSE DOLPHINS AROUND JAPAN

The Japanese common bottlenose dolphin, *T. truncatus*, can be distinguished from the Indo-Pacific bottlenose dolphin, *T. aduncus*, by greater body size, absence of spotting on adults, and shorter and more robust rostrum (Figure 11.1). Wang et al. (2000a) compared the external morphology of 40 common bottlenose dolphins and 17 Indo-Pacific bottlenose dolphins. Since his sample combined measurements of both sexes as well as juveniles, only maximum figures are cited in the following. The maximum body size for the common bottlenose dolphin was 295.5 cm, which exceeded by 28 cm the corresponding figure of 268.0 cm for the Indo-Pacific bottlenose dolphin. The maximum length of the rostrum of the common bottlenose dolphin was 13.0 cm, while that for the Indo-Pacific bottlenose dolphin was 13.7 cm. Although the absolute difference was negligible, the former species has a more robust rostrum. Tooth and vertebral counts also showed some differences, with range overlaps (Wang et al. 2000b; Kemper 2004; Table 11.1). The mean tooth diameter of the common bottlenose dolphin was 7.1 mm with a range of 6.0–8.4 mm (n = 12), while that of Indo-Pacific bottlenose dolphin was 6.2 mm with a range of 5.0–7.5 mm (n = 12), but the difference was not statistically significant (Kemper 2004). A small difference was probably masked by large individual variation. Gao et al. (1995) reported the maximum body size of the common bottlenose dolphin at 330 cm and that of the Indo-Pacific bottlenose dolphin at 254 cm. The former figure was about 30 cm greater than the maximum size of the same species reported by Wang et al. (2000a) as cited earlier. The difference is presumably due to inclusion of common bottlenose dolphins from the Bohai and Yellow Seas in Gao et al. (1995) as reported by Zhou and Qian (1985). The common bottlenose dolphin has broad geographical variation in body size reflecting environmental temperature or latitude. The largest individuals are found among populations in European waters, where males reach 380 cm and females 367 cm (Wells and Scott 1990a).

External morphology was observed and body lengths measured for bottlenose dolphins taken by the drive fisheries on the coast of the Izu Peninsula (34°36′N–35°05′N, 138°45′E–139°10′E) and at Taiji (33°36′N, 135°57′E) and culling operations at Katsumoto on Iki Island (33°45′N, 129°45′E), during the 11 years from 1973 to 1983, but none of them

TABLE 11.1
Comparison of Tooth and Vertebral Counts between Two Species of the Genus *Tursiops*

		Number of Teeth			Number of Vertebrae		
Location	**Species**	**Sample Size**	**Range**	**Mean**	**Sample Size**	**Range**	**Mean**
China and Taiwan	*T. truncatus*	54	80–106	93.9	20	64–67	65.5
South Africa	*T. truncatus*	9	88–96	93.1	4	64–65	64.5
China and Taiwan	*T. aduncus*	19	96–111	102.0	19	59–62	60.2
South Africa	*T. aduncus*	29	97–111	102.9	9	59–62	60.6

Source: Extracted from Table 3 of Wang, Y.J. et al., *J. Zool. Lond.*, 252, 147, 2000b, which gives detailed cranial measurements.
Note: No difference has been detected between tooth rows; total number of teeth in the four rami is given.

appeared to represent the Indo-Pacific bottlenose dolphin (Table 11.2). Iki Island is located in Tsushima Strait between the East China Sea and the southern Sea of Japan. Modal body lengths were 290–299 cm for both sexes, and the maximum body lengths were 336 cm (male) and 320 cm (female). These figures are close to those reported from the Yellow Sea and Bohai Sea but about 30 cm greater than the corresponding figures reported by Wang et al. (2000a) from Taiwan and to the south. Ogawa (1936, in Japanese) reported tooth counts of 21–24 in each jaw (total 88–96) with an average total count of 90 for four *Tursiops* sp. killed off the Sanriku Region of the Pacific coast of Japan in latitudes 37°44′N–41°30′N, which were closer to the range of the common bottlenose dolphin than that of the Indo-Pacific bottlenose dolphin shown in Table 11.1.

While examining common bottlenose dolphins listed in Table 11.2, Kasuya et al. (1997, in Japanese) collected several teeth at the center of the lower tooth row for age determination. Some smaller teeth were collected near the end of the tooth row from animals that had lost most of the teeth, presumably due to old age. The teeth were sectioned longitudinally and the diameter measured. Using only measurements of teeth from 203 females (over 10 years of age) from 10 schools with acceptable sample size (8–30 individuals/school), they got average diameters in a range of 7.2–8.3 mm for each

TABLE 11.2
Materials for Japanese Common Bottlenose Dolphins Analyzed by Kasuya et al. (1997, in Japanese)

					Number of Samples					
					Gonads		Age		Total Examined	
Location	**School Number**	**Driven At**	**Date of Sample**	**Dolphins Driven**	**Female**	**Male**	**Female**	**Male**	**Female**	**Male**
Pacific coast	P-1	Arari[a]	Aug. 23, 1973	—	9	8	14	8	28	9
	P-2	Taiji[b]	Apr. 27, 1981	66	26	29	27	33	27	34
	P-4	Taiji	Jan. 8, 1982	50	24	5	24	13	24	13
	P-5+6	Taiji	Jan. 24, 25, 1982	About 70	11	9	11	9	13	9
	P-7	Taiji	Mar. 7, 1982	30	10	0	0	0	10	17
	P-8	Taiji	Feb. 12, 1983	156	39	14	37	32	39	43
	P-8+9	Taiji	Feb. 12, 14, 1983	About 170	8	6	8	7	8	7
	Total	Pacific	1973–1983	542+	127	71	121	102	149	132
Iki Island, Tsushima Strait	I-1	Katsumoto[c]	Jan. 27, 1980	11	8	3	8	3	8	3
	I-2	Katsumoto	Feb. 22, 1980	191	43	19	43	35	43	36
	I-3	Katsumoto	Feb. 27, 1980	1,114	121	53	123	90	125	92
	I-4	Katsumoto	Mar. 6, 1980	359	19	9	19	21	19	21
	I-5	Katsumoto	Mar. 14, 1980	154	48	9	48	27	50	27
	I-10	Katsumoto	Mar. 15, 1979	394	1	1	1	1	1	1
	I-11	Katsumoto	Mar. 19, 1979	411	10	10	10	10	10	10
	Total	Katsumoto	1979–1980	2,634	240	104	252	187	256	190

Note: Dolphins were randomly selected from the catch of drive fishery.
[a] Arari, 34°50′, 138°46′E.
[b] Taiji, 33°36′N, 135°57′E.
[c] Katsumoto, 33°51′N, 129°42′E.

school and an average diameter of 7.9 mm for the sample of 203 (range 5.3–10.0 mm). Comparison between the samples from the Pacific coast (Izu and Taiji) and Iki did not find a significant difference in mean diameter. These values were about 1–2 mm greater than those reported by Kemper (2004) for both species (see the preceding text in this section), presumably reflecting different body size. Although there were some statistically significant differences observed between some schools from the Pacific coast, Kasuya et al. (1997, in Japanese) did not consider them important.

11.3 DISTRIBUTION OF BOTTLENOSE DOLPHINS AROUND JAPAN

The common bottlenose dolphin inhabits Japanese waters south of 43°N, or approximately south of southern Hokkaido (Figures 11.2 and 11.3). The Indo-Pacific bottlenose dolphin is resident in coastal waters around Amami Island (28°15′N, 129°20′E) (Miyazaki and Nakayama 1989; Rice 1998), Mikura Island (33°53′N, 139°35′E) (Kogi et al. 2004), the Bonin Islands (27°N, 142°E) (Ogasawara Whale Watching Association 2003, in Japanese), and Tsuji-shima Island (32°45′N, 130°20′E) in western Kyushu (Shirakihara et al. 2002) (Figure 11.2). Further investigation may find it to occur in other locations.

Shirakihara et al. (2002) identified 178 individuals in a population of Indo-Pacific bottlenose dolphins off Tsuji-shima Island from 1994 to 1998 and estimated the total population size at 218 after correcting for unidentifiable individuals and those that died during the study period. I had the impression during a shipboard observation off Tsuji-shima that bottlenose dolphins there are spotted to a lesser degree than those

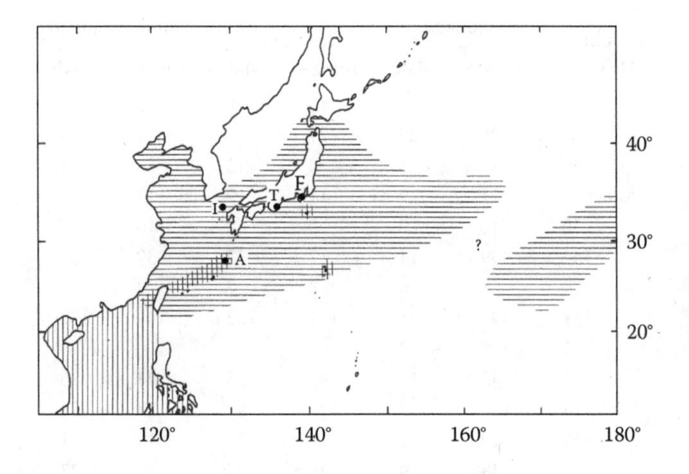

FIGURE 11.2 Distribution of Indo-Pacific bottlenose dolphin (vertical lines) and common bottlenose dolphin (horizontal lines) around Japan and in western North Pacific. A, Nago in Okinawa; F, Futo on the Izu coast; I, Iki in Tsushima Strait; T, Taiji. Records of small populations of or vagrant visitor Indo-Pacific common bottlenose dolphins are apparently increasing recently along the shores of southwestern Japan, but they are not included in this figure. (Reproduced from Fig. 1 of Kasuya, T. et al., *IBI Rep.* (Kamogawa, Japan), 7, 71, 1997, in Japanese.)

in Taiwanese waters. Judging from underwater photographs, the same species around Mikura Island has distinct spots.

The oceanographic environment and recent occasional records of the species suggest that resident populations of the Indo-Pacific bottlenose dolphin may have occurred along the Pacific coast of the Japanese islands south of Tokyo (ca. 35°50′N), but none have been recorded there. It is a possibility that such resident populations could have been exterminated by the local fisheries before being identified by scientists. In the 1960s, a small-type whaler and whale meat dealer in Taiji talked about a dolphin species called *hau-kasu* taken by the local fishery. The name meant a "hybrid with the spotted one." This species was said to be similar to the common bottlenose dolphin, which was locally called *kuro* (black), but had spots on the body. Thus, the fishermen thought them to be a hybrid between the common bottlenose dolphin and the *kasuri-iruka* (pantropical spotted dolphin). Kasuya and Yamada (1995, in Japanese) assumed considering additional information that the species had white lips and that it was probably the rough-toothed dolphin, *Steno bredanensis*. However, the possibility remains that the *hau-kasu* was the Indo-Pacific bottlenose dolphin (Section 2.7).

There have been several cases where a small number of Indo-Pacific bottlenose dolphins visited a Japanese coastal area and resided there for a period. Around 1976, one group lived for a time off Mi-irihama in Kamogawa City (35°07′N, 140°06′E) and then disappeared. A group of six with juveniles has inhabited the waters of Sunosaki in Kamogawa City since 1998, and four of them were identified as previously observed at Mikura Island (Fujita 2003, in Japanese). In July 2007, a male was caught in a trap net in Kamogawa City and transported to Kamogawa Sea World. This individual was reported to be from Mikura Island (Saeki 2007, in Japanese), but the relationship with those in Sunosaki was not reported.

According to Miki Shirakihara (personal communication in April 2008), most of the *T. aduncus* off Tsuji-shima traveled in the spring of 2000 southward for a distance of about 50–60 km to live in waters off Naga-shima Island (29°10′N, 130°10′E) for about one year. Then, leaving behind several tens of dolphins in this new habitat, most returned to their original habitat off Tsuji-shima to merge with old group members. Some members of the Tsuji-shima group were found in a group of several Indo-Pacific bottlenose dolphins that recently settled off Notojima (37°07′N, 137°00′E) on the Sea of Japan coast, which required travel of about 1100 km northeast along the coast of the sea. There must be more unreported similar cases.

These local movements of Indo-Pacific bottlenose dolphins along the coasts of Japan suggest a mechanism of maintaining population size. One of the direct results is to establish a new colony through emigration from a large, perhaps oversized, population. Another result is to supply new members or genetic variability to a small local population. The carrying capacity of a single coastal habitat may not be sufficient for a population of viable size, which is unknown but must certainly be over 200. Loss of a few individuals by emigration will not harm the survival of the local fully grown group, but it will contribute to founding a new colony or to supply members and new genetic variation to other smaller groups. If adult males temporarily visit nearby colonies to mate, that will also function to increase the genetic variability of the recipient group.

The Indo-Pacific bottlenose dolphin off Japan may have survived through repetition of extirpation and recolonization of sites and emigration, immigration, and genetic exchange between colonies. This structure containing a network of such numerous smaller colonies is called a metapopulation, which results in increasing population stability as a whole. Due to the limited carrying capacity of local habitats, many local colonies would have trouble maintaining population stability, but the physical and genetic interactions among the colonies enhance the stability of the whole complex of small populations. Further efforts to monitor the process of establishing new colonies along the coasts of Japan and the process of exchanging members between them will improve our understanding of their population structure.

Currently, we do not have records of resident populations of common bottlenose dolphins in Japanese coastal waters, although further effort will be required to confirm the absence. There are cases of resident populations of the species inhabiting tens or hundreds of kilometers of the coast of California in the eastern North Pacific (Wells et al. 1990). Similar cases are also known along the coast of Florida in the western North Atlantic, where each of the local population is called a "community." Some males of a community are known to temporarily visit other communities to mate, and some individuals of both sexes move permanently to nearby communities with an annual immigration rate of 2%–3% (Wells 1991; Connor et al. 2000). These common bottlenose dolphins occupy a habitat that is used by the Indo-Pacific bottlenose dolphin in other oceans. This is an indication of strong adaptability in the genus *Tursiops*.

Bottlenose dolphins, *Tursiops* spp., have been sighted around Japan in waters above 11°C in surface temperature (Table 10.1). The northern limit of their distribution in winter was thought to be in the Iki Island area near Tsushima Strait, which connects the East China Sea and the Sea of Japan, and at 36°N off the Pacific coast of Japan, expanding in summer to the coast of southern Hokkaido (Kasuya 1980; Miyashita 1986a, both in Japanese, Figure 11.3). However, the recent identification of a small resident group of Indo-Pacific bottlenose dolphins off Notojima on the Sea of Japan coast expanded the known winter range to 37°N. Knowing whether this small population at the northern limit of the range survives requires monitoring.

Fishery statistics give some additional information on distribution. Miyazaki (1980a) confirmed year-around landings of bottlenose dolphins at Taiji on the Pacific coast of central Japan, where all the *Tursiops* examined by scientists were common bottlenose dolphins. Catch statistics of a drive fishery at Arari (34°50′N, 138°46′E) on the west coast of the Izu Peninsula covering January 1950 to April 1957 record drives of *hasunaga*, which is believed to be the common bottlenose dolphin (Section 3.8.2), in March, May,

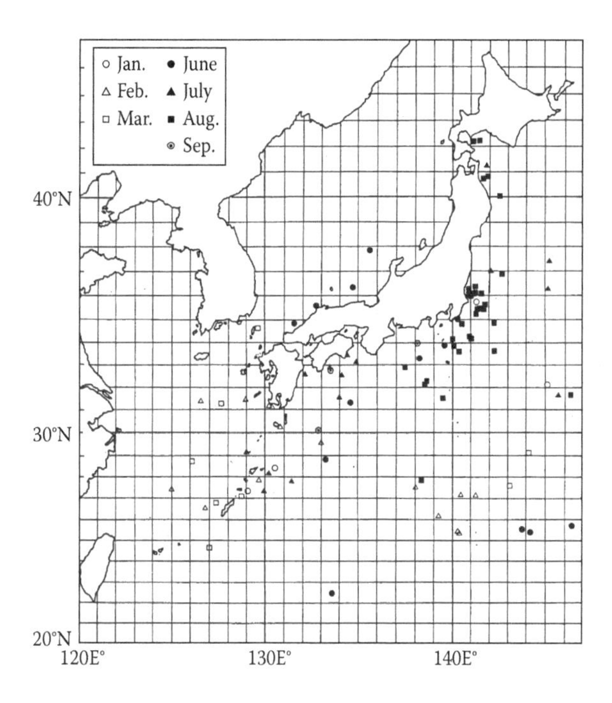

FIGURE 11.3 Monthly records of bottlenose dolphins recorded during cetacean-sighting cruises of Far Seas Fisheries Research Laboratory. (Reproduced from Fig. II-3-1a of Miyashita, T., Sightings from Research vessels, pp. 78–87, in Tamura, T., Ohsumi, S., and Arai, R., eds., *Report of Contract Studies in Search of Countermeasures for Fishery Hazard (Culling of Harmful Organisms) (56th to 60th year of Showa),* Suisan-cho Gogyo Kogai (Yugaiseibutsu Kujo) Taisaku Chosa Kento Iinkai (Fishery Agency), Tokyo, Japan, 1986a, 285pp, in Japanese. With permission.)

July, September, and December. These fishery data indicate year-around distribution of the common bottlenose dolphin at least in waters south of the Izu Peninsula.

Miyashita (1993) reported the occurrence of bottlenose dolphins during summer (from June to September) in a large area extending from the Japanese coast to the central North Pacific at 175°E based on data obtained through systematic sighting cruises conducted by the Far Seas Fisheries Research Laboratory of the Fisheries Agency in 1983 through 1991. These sighting surveys did not distinguish between the two species of the genus, but since the survey placed weight on the offshore area and the survey track lines almost avoided waters inside of the 100 m isobath, the possibility of including Indo-Pacific bottlenose dolphins must be negligible. These data, which are used in Figures 11.2 and 11.3, showed the distribution of the common bottlenose dolphin in June to extend down to latitudes 22°N–23°N off the east coast of Taiwan. Sighting effort south of these latitudes did not find the species, suggesting an absence of the species in offshore Pacific waters south of Taiwan. This apparent southern limit moved in July to latitudes of 25°N–26°N and remained there through August to September. Because the survey was limited to the south of 35°N in June, Miyashita (1993) did not provide information on the northern limit of the species in that month. In July, the survey covered latitudes from 12°N to 45°N and recorded common bottlenose dolphin from 25°N (off Okinawa) to 42°N–43°N (off southern Hokkaido). This northern limit remained the same through midsummer from August to September.

Miyashita (1993) confirmed an almost continuous distribution of the common bottlenose dolphin from the Pacific coast of Japan to 175°E, but distribution of the species to the east of this longitude remains to be studied. The density of the species was not uniform within this western North Pacific area. The density tended to be lower in more eastern waters, and there was an apparent distribution gap around 30°N, 165°E (Figure 11.2). This needs to be confirmed in future surveys.

The Fisheries Agency at one time collected sightings of cetaceans from fishery inspection vessels of Nagasaki Prefecture that operated in the East China Sea and the Iki Island area as part of research activities conducted in relation to the dolphin-fishery conflict in the Iki Island area. It recorded bottlenose dolphins, presumed to be common bottlenose dolphins, in the East China Sea in all months except August, with a peak frequency in April and May (Tamura et al. 1986, in Japanese). Similar records collected by Fisheries Agency vessels (Miyashita 1986a, in Japanese) confirmed the presence of the species in the East China Sea throughout the year, with the range extending from northern Kyushu to the Bohai Strait/western Yellow Sea in the northwest and to the southern Ryukyu Islands (ca. 24°30′N, 124°00′E) in the south. In the Sea of Japan, reported sighting of common bottlenose dolphins ranged from Tsushima Strait to off Aomori Prefecture (40°25′N–41°15′N), which did not disagree with the results of whale-sighting surveys presented in Figure 11.3. The northernmost record of the species in the Sea of Japan was off southern Hokkaido (42°N–43°N). The offshore distribution of the species in the Sea of Japan was limited to within about 120 nautical miles (about 220 km) off the coast. Only Dall's porpoises were sighted further offshore (Tamura et al. 1986, in Japanese). Warmer-water species such as the bottlenose dolphins, Pacific white-sided dolphin, and false killer whale occurred in the Sea of Japan only in the coastal waters of southern Korea and Japan (Tamura et al. 1986, in Japanese). Miyashita (1993) sighted bottlenose dolphins in June on the Sea of Japan coast between Tsushima Strait and Ishikawa Prefecture (*c.* 37°30′N, 137°15′E), but they were not recorded from October to December to the northeast of Ishikawa Prefecture. Seasonal north–south movement of the species is suggested between 37°N and 43°N. The residence of a small group of Indo-Pacific bottlenose dolphin in Notojima is at the northern limit of the genus in the winter months.

11.4 LIFE HISTORY

11.4.1 Age Determination

The age of bottlenose dolphins has been determined using the growth layers in dentine and cementum of the teeth. The common bottlenose dolphin is one of the best-studied species on the formation of growth layers in the teeth, and the annual deposition of the growth layers has been confirmed using wild or aquarium-reared animals of known age as well as time-marked animals in aquariums (Hohn et al. 1989; Myrick and Cornel 1990). Kasuya et al. (1986), using common bottlenose dolphins born in aquariums and those taken by fisheries of less than 4 years of estimated age, compared growth layers in dentine and cementum and established a standard for reading the annual growth layers in the teeth. Because deposition of dentinal growth layers ceases at a certain age, as in many other delphinids, the ages of old individuals have to be determined using cemental layers. Kasuya et al. (1997, in Japanese), on which the following description is based, counted layers in each tissue of common bottlenose dolphins three times and accepted the middle figure as the true value. If the cemental age exceeded the dentinal age, the former was used as the age of the individual. All the ages between n and n + 1 were expressed as n + 0.5 years (n being an integer). This was for the convenience of mathematical analysis and did not signify the precision of the readings. Because parturitions of Japanese common bottlenose dolphins occur from February to October with a peak in June (see Section 11.4.2.3) and most of the specimens used by Kasuya et al. (1997, in Japanese) were obtained during January through April, this grouping of ages would not cause a significant bias.

11.4.2 Gestation Period, Fetal Growth, and Breeding Season

11.4.2.1 Gestation Period

Many common bottlenose dolphins have been successfully maintained in aquariums, and their reproductive activities have been monitored with the help of modern technologies. A surge of estrous hormone or its metabolites in blood or urine indicates ovulation, and continuation of a high level of progesterone suggests pregnancy. Ultrasonography enables monitoring the growth of Graafian follicles and fetuses. Yoshioka et al. (1986) using such methodology confirmed that female common bottlenose dolphins were receptive only during a one- or two-day period before and after an ovulation in an aquarium (total receptive period was two to four days). Monitoring of 77 pregnancies in U.S. aquariums revealed that their gestation periods were within a range of seven days on both sides of a mean of 370 days (the 77 records were within a range of 14 days) (Asper et al. 1992). I expect a similar gestation period exists for common bottlenose dolphins around Japan.

Body lengths of neonates born in Japanese aquariums and measured within 11 days from birth were in a range of 116–140 cm (mean 128 cm, n = 20) (Kasuya et al. 1986), which was about 18 cm greater than the corresponding figure of 100–120 cm (mean 109.5 cm, n = 21) of neonates of presumed coastal forms stranded along the Texas coasts (Fernandez 1992, cited in Urian et al. 1996).

11.4.2.2 Fetal Growth

The fetal growth of cetaceans consists of an early phase of slow curvilinear growth and a later phase of linear growth. The period of the latter is a set proportion of the total gestation period. The length of time from conception to a point when the extended linear growth line cuts the axis of time is estimated at about 50 days, or 13.5% of the total gestation period for species with a gestation period of about 1 year (Section 8.5.2). Based on this and neonatal length of 128 cm, the fetal growth rate in the linear phase is estimated at about 0.4 cm/day or 12.3 cm/month for Japanese common bottlenose dolphins. Thus, the relationship between body length (*BL*, cm) and fetal age (*t*, days) is expressed by

$$BL = [(t - 0.135G)X]/[G(1 - 0.135)]$$

where

G is an average gestation period of 370 days
X is the mean neonatal length of 128 cm

This equation does not apply to the early phase of fetal growth, which lasts for about 2 months from conception.

11.4.2.3 Breeding Season

Kasuya et al. (1997) examined common bottlenose dolphins taken from January to April by drive fisheries at Taiji on the Pacific coast and at Katsumoto on Iki Island. They found fetal body length ranging from 10 to 120 cm and estimated the dates when these fetuses could have reached the mean neonatal body length if they had not been killed by the fishermen (Figure 11.4). One school taken at Arari on the Izu Peninsula in August (Table 11.2) was omitted from this analysis because it contained only two fetuses at 13 cm and 40 cm, but inclusion of that school would not change the following conclusion. The parturition dates estimated from fetuses over 30 cm in body length were similar between schools within the same geographical region, that is, the Pacific and Iki Island areas. The dates ranged from March to October for the Pacific sample and from February to October for the Iki Island sample, with a peak in June for both areas. The mean parturition date was July 6 (SD = 51 days, n = 46) for the Pacific sample and June 30 (SD = 65 days, n = 60) for the Iki sample (Kasuya et al. 1997, in Japanese). Thus, the

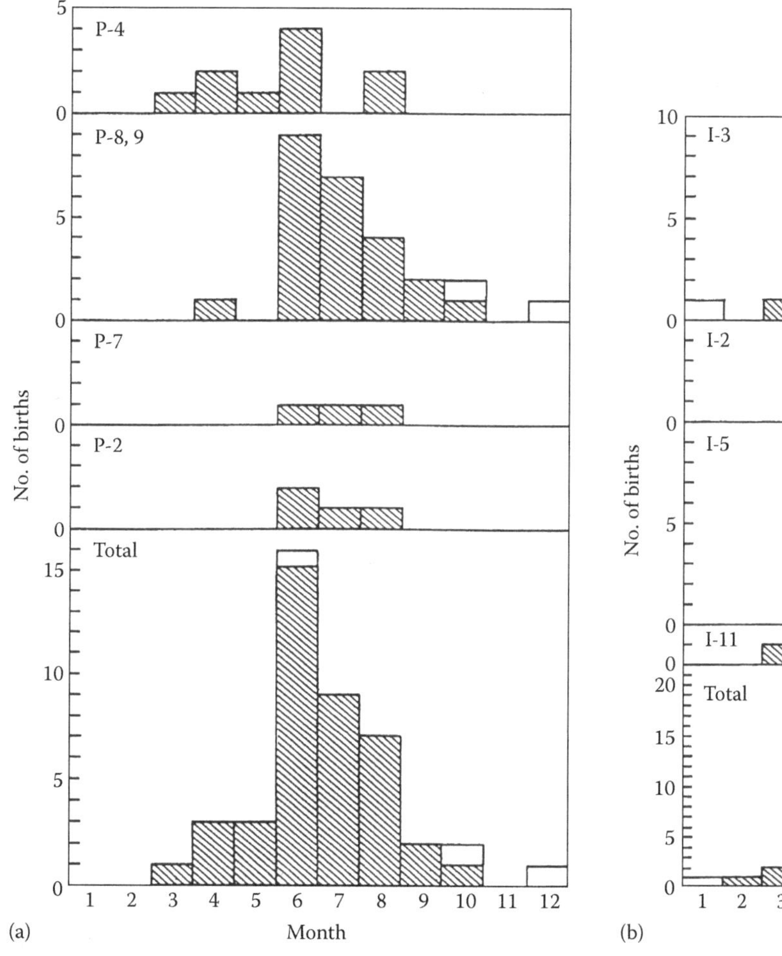

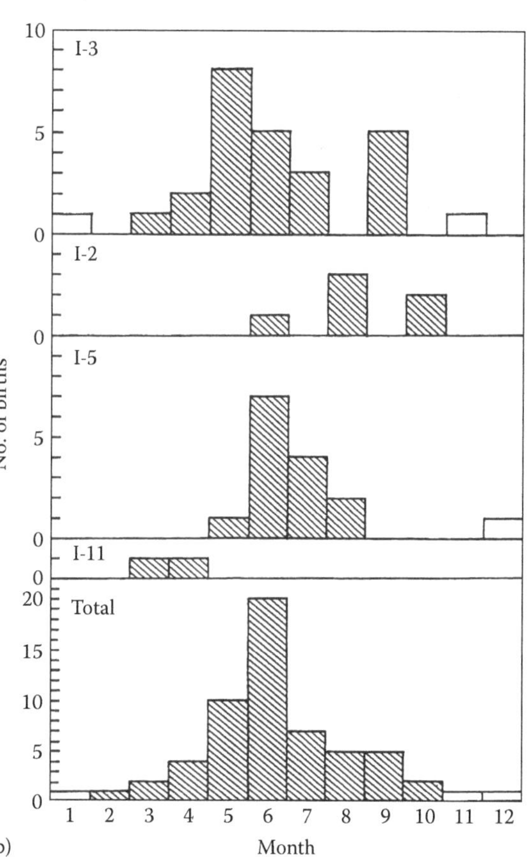

FIGURE 11.4 Monthly distribution of birth dates of the common bottlenose dolphins calculated from fetal body length and mean fetal growth of 4.0 mm/day in the linear part of fetal growth. Shaded squares represent birth dates calculated for fetuses greater than 30 cm and white squares smaller fetuses (date could be of lesser accuracy). (a) P represents Pacific sample from Taiji and (b) Iki Island sample in Tsushima Strait. Dates of samples are January (P-4), February (P-8, P-9, I-2, I-3), March (P-7, I-5), and April (P-2) (see Table 11.2). (Reproduced from Fig. 3a and 3b of Kasuya, T. et al., *IBI Rep.* (Kamogawa, Japan), 7, 71, 1997, in Japanese.)

breeding season was not significantly different between the Pacific and Iki areas. Fitting fetuses smaller than 30 cm to the linear growth curve would estimate the parturition date as earlier than the true one, with greater bias being caused by the smaller fetuses.

Steiner and Bossley (2008) recorded parturition dates for a resident population of Indo-Pacific bottlenose dolphins in an estuary area near Adelaide, Australia. They recorded 45 births during the period of 1989–2005, which occurred in 6 months from December to May (no births occurred from June to November). There was a peak in February (11 births) followed by January (7 births). The latitude of the study site was about 35°S, which was nearly opposite to the latitude of Japanese specimens (*c.* 33°35′N–33°45′N), and the mating peaks were about 4 months apart.

Urian et al. (1996) compared the parturition seasons of common bottlenose dolphins between wild individuals in Florida and Texas and between wild and aquarium animals from each of the regions. They found that the difference in parturition season was small between the east and west coasts of Florida. Births occurred in any month of the year, with a main peak from March through September. This pattern was very similar to that observed off Japan. However, parturitions on the Texas coast were limited to a short period from February to April. In order to investigate the regional difference in breeding seasonality, Urian et al. (1996) monitored females that experienced over five parturitions in an aquarium environment for a possible change in breeding season in the artificial environment. They obtained an unexpected result; the females retained the seasonality of the original habitat for at least up to the fifth parturition. Although the mechanism controlling the seasonality of breeding remained unknown, they found a tendency of parturition occurring slightly in the later part of the season and greater standard deviations for females that lived in an aquarium. These results indicate that the breeding seasonality of wild individuals not only is retained for a long period in an aquarium but also suggests a possibility that the seasonality of wild animals may be gradually lost during longer life in an artificial environment.

Breeding seasonality of Japanese common bottlenose dolphins in an artificial environment has been recorded for three females in Kamogawa Sea World (Yoshioka et al. 1986; Yoshioka 1990, in Japanese). These females were introduced from the wild in 1971–1978, presumably selected from drive hunts on the Izu coast. Their body lengths were 273–288 cm at the time of transfer, but their maturity was not determined from these measurements. Hormonal levels were analyzed 3–12 years after the transfer into the aquarium, or in 1982–1984, when two measured 282 and 285 cm and had already experienced parturition and another measured at 294 cm was considered sexually mature. One of the three females underwent several estrogen surges and subsequent rises of progesterone level in June–August of 1982 and 1983, but these did not result in pregnancies. The authors concluded that these estrous periods were not followed by conception. This female did not show such a cycle in 1984. The second female had estrous cycles in June–August of 1982 and 1984. The third female, which was the smallest with body length of 282 cm, started estrogen and progesterone cycles in April 1984 and repeated the cycle until October. The second female exhibited several months of pseudo-pregnancy after the estrous cycles, but the other two females did not show any features of pregnancy in spite of copulations during estrus. These records indicate that the breeding seasonality of females in the wild is retained for over 10 years in an aquarium.

Kamogawa Sea World also produced information on reproductive seasonality in a male common bottlenose dolphin (Katsumata et al. 1994, in Japanese). The specimen was caught at Futo (34°54′N, 139°06′E) on the west coast of the Izu Peninsula and brought to the aquarium in March 1985, when it was 292 cm in body length. It was monitored for serum testosterone level during 6 subsequent years, ending in 1991, when it measured 306 cm. The male could not be definitely determined to be mature based on body length at capture, 292 cm, but it was judged sexually mature in 1991, when it measured 306 cm. All 24 measurements of testosterone level during the first year in the aquarium were below 1 ng/mL, which was interpreted by Katsumata et al. (1994, in Japanese) as an indication of physiological pathology after introduction to the aquarium, but the possibility of the animal being sexually immature cannot be fully excluded in my view. In 1986, or in the second year at the aquarium, there were 13 measurements of testosterone level, with a peak of 10–20 ng/mL in July–August. Only 3–8 measurements were available for the subsequent years, which were insufficient for identifying annual peaks. Combining all the measurements of the male revealed low mean testosterone level of 5 ng/mL in January–February, a high value of over 15 ng/mL in April–August, and a gradual decline to November when it again reached a level of around 5 ng/mL. The seasonality of testosterone level in the captive male agreed with the mating season of females in the wild.

11.4.3 Sex Ratio

Mortality rate differs between the sexes throughout life; thus, the sex ratio can change during fetal and postnatal life. The sex ratio in our samples may also be influenced by gear selectivity in fisheries or by behavioral difference between the sexes.

The sex ratio of common bottlenose dolphins taken by drive fisheries is shown in Table 11.3. The fetal sex ratio of the Pacific sample is 22:22 and that of the Iki sample 20:37. Although the latter apparently indicates an excess of female fetuses, it is not statistically different from parity. The data are insufficient to see change in sex ratio with fetal growth.

Males slightly exceeded females in number among postnatal individuals below 10 years of age, but the difference is not statistically significant. The proportion of females increased with postnatal age in both the Pacific and Iki samples, that is, 45.5% (<10 years), 62.0% (10–20 years), and 58.6% (>30 years) for the Pacific sample and 48.3% (<10 years), 65.8% (10–20 years), and 70.3% (>30 years) for the Iki sample. This is likely to be due to a higher mortality rate in males. The age of the oldest female exceeded that of the oldest male in both

TABLE 11.3
Sex Ratio of Common Bottlenose Dolphins off Japan, Obtained from 7 Drives on the Pacific Coast and 4 Drives at Iki

Age (Year)	Fetus <60cm	Fetus >60cm	0–5	5–10	10–15	15–20	20–25	25–30	30–35	35–40	>40	Total Postnatal
Pacific												
Male	4	18	30	24	20	21	24	6	10	6	0	141
Female	8	14	26	19	32	35	35	18	10	1	1	177
Female (%)	66.7	43.8	46.4	44.2	61.5	62.5	59.3	75.0	50.0	14.3	100	55.7
Iki												
Male	3	17	91	31	17	21	14	5	7	0	1	187
Female	6	31	87	27	38	35	24	14	19	6	1	251
Female (%)	66.7	64.6	48.9	46.6	69.1	62.5	63.2	73.7	73.1	100	50.0	57.3

Source: Reproduced from Table 3 and text of Kasuya, T. et al., *IBI Rep.* (Kamogawa, Japan), 7, 71, 1997, in Japanese.

samples, 42.5 versus 39.5 years in the Pacific sample and 45.5 versus 43.5 years in the Iki sample, which also suggests a slightly higher survival rate in females.

11.4.4 Age Composition

Kasuya et al. (1997, in Japanese) obtained the age composition of a total of 15 schools of common bottlenose dolphins taken by drive fisheries on the Pacific coast and by a culling operation in the Iki Island area (Kasuya 1985). These schools are listed in Table 11.2, and the age composition is shown in Figure 11.5 and Table 11.3.

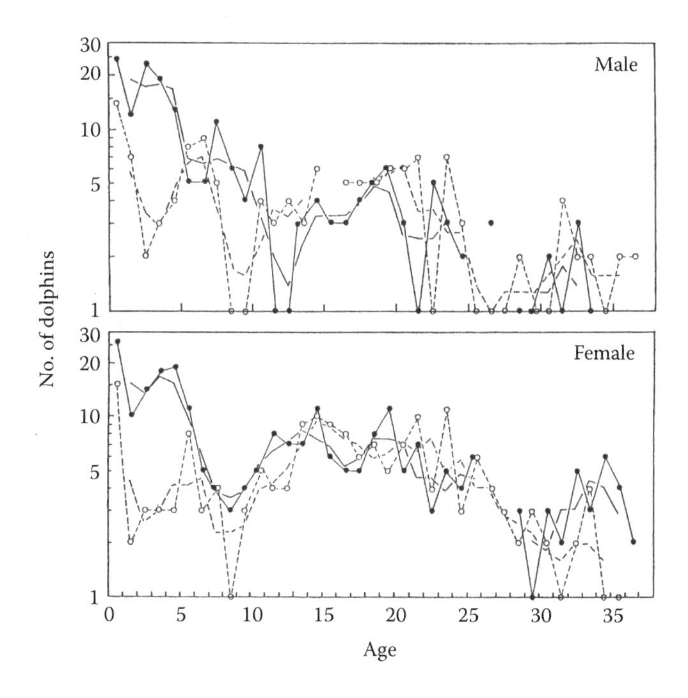

FIGURE 11.5 Age composition of common bottlenose dolphins driven on the Pacific coast (open circles and dotted line) and at Iki Island (closed circles and solid line). Three-year running geometric means are also shown. (Reproduced from Fig. 4 of Kasuya, T. et al., *IBI Rep.* (Kamogawa, Japan), 7, 71, 1997, in Japanese.)

Age frequencies were compared between the two geographical samples by grouping the dolphins into three age groups of 0–10, 10–20, and >20. The frequencies in the Pacific sample were 99, 108, and 111, or ratios of 1: 1.09: 1.12, while those in the Iki sample were 236, 111, and 91, or 1: 0.47: 0.39. The Iki sample contained an extremely large number of juveniles below 10 years of age. Such an age composition occurs if (1) the population has a high mortality rate, (2) the population is increasing with a high reproductive rate, or (3) a school of immature individuals has been taken by chance. I have the impression that the last is the most likely cause. This problem will be further discussed in Section 11.4.8.

As the age frequency is shown in Figure 11.5 on a logarithmic scale, the slope constitutes an "apparent mortality rate," which is a combination of mortality and increasing (or decreasing) growth rate of the population. The apparent mortality rate is equal to the real "mortality rate" of the population only if the population size remains constant with balanced mortality and recruitment. Because the common bottlenose dolphin may change in social behavior with age and the obtained age composition can be biased, caution is required in estimating the mortality rate from age composition. If we tentatively assume that the species exhibits a uniform mortality rate through life, which is certainly incorrect, and that 1% survive to 45 years, the maximum age observed for the species, then 9.7% is obtained as the average annual mortality rate for ages 0–45 years. An assumption of 5% surviving 45 years gives a mortality rate of 6.4%.

The observation of a resident community of common bottlenose dolphins in the coastal waters of Sarasota on the west coast of Florida during 1970 to 1987 revealed that the population was stable at a size of about 100 individuals and resulted in estimation of several parameters of population dynamics by Wells and Scott (1990b). They estimated the survival rate of calves in their first year at 0.803 (mortality rate of 0.197) and the annual mortality rate of individuals over age 1 year at 0.038. The average mortality rate for the whole age range must be between the two extremities, or between 0.197 and 0.038. They also calculated the annual recruitment rate to age 1 year at 0.048, which should be equal to the annual mortality rate of individuals over age 1 year given earlier, 0.038, if the population remained absolutely stable. The annual mortality rate estimated for individuals over 1 year of age included emigration to nearby communities. This emigration rate was not separated from mortality, but it could have been of similar level of immigration from other communities, which is known to be about 0.02. They calculated annual birth rate at 0.055, which was the number of newborns of both sexes produced by individuals of both sexes of all age classes. If the average mortality rate for all age classes were equal to this birth rate, the population would be stable. Dividing 0.144, which is the recruitment rate to age 1 year relative to the population of adult females calculated by Wells and Scott (1990b), by 0.803, a survival rate of newborns to age 1 year, gives 0.179, which is an approximate annual parturition rate of adult females. This annual parturition rate gives an average calving interval of 5.6 years. This annual parturition rate is smaller than the annual pregnancy rates of 0.33–0.41 estimated for Japanese populations (see Section 11.4.8.4). This suggests that the Sarasota population is almost stable with low rates of reproduction and mortality in a small habitat with limited predators. Although the annual pregnancy rates estimated for Japanese populations do not, in a strict sense, distinguish between annual conception rate and annual parturition rate, the effect would be negligible in the comparison between the two oceans.

The age composition of Japanese common bottlenose dolphins shows a deficiency of some age classes. Both sexes aged 1.5–12.5 years, which could actually include ages from 1 to 13 years, are apparently underrepresented in the Pacific sample. Since lactation in the population lasts for 1.7 years on average (Table 11.11) and both sexes attain reproductive ability by the age of 13 years (Tables 11.6 and 11.9), the underrepresented age classes roughly correspond with those after weaning and before sexual maturity. This can probably be interpreted as a result of growth-related behavioral change, where weaned immature individuals tend to live with individuals of similar age. Thus, the age composition of the sample depends on the chance of fishermen finding and driving particular types of schools. It is also a possibility that this may be due to geographical segregation or difference in school size, neither of which has been confirmed for this species.

Although the Iki sample does not exhibit a clear deficiency of ages just after weaning, it still exhibits a deficiency of females aged 5.5–10.5 years and males aged 5.5–13.5 years, which corresponds to the range of age around the attainment of sexual maturity. Thus, it is likely that dolphins of both sexes just before sexual maturity are underrepresented in the Iki sample.

It has been reported that striped and pantropical spotted dolphins between weaning and sexual maturity tend to leave the mother's school to live with individuals of the same growth stage (Kasuya et al. 1974; Miyazaki and Nishiwaki 1978; Kasuya 1999; see also Section 10.5.14). Common bottlenose dolphins in a resident population on the Florida coast are known to form groups of individuals of the same age, sex, and reproductive status (Shane et al. 1986; Wells 1991). In this Florida community, nursing females may live only with their suckling calves, but more often they live with other mother–calf pairs. Calves, particularly males, at age 5–6 years start to spend time apart from their mothers and finally leave their mothers to live with juveniles of the same sex. This behavior of common bottlenose dolphins has some similarity to what has been inferred for Japanese striped dolphins.

Female common bottlenose dolphins off Florida rejoin their mothers at sexual maturity, which suggests an evolutionary step toward the formation of a matrilineal social system. It is easy for the young females of this community to locate and join their mother's school at sexual maturity, because they live in a small community in a semienclosed habitat. Such long-term observations have not been achieved for a population of the same species in offshore open waters, and it remains to be confirmed if females in such offshore populations have the opportunity to return to their mother's group at sexual maturity.

Males in the Florida community at sexual maturity behave quite differently from females. They do not join their mothers at sexual maturity but may live alone or live in a coalition with several adult males and visit female schools. Such male coalitions may be formed by joining with other adult males or by maintaining a coalition after the juvenile period. Such male coalitions have also been observed in Indo-Pacific bottlenose dolphins in Shark Bay in Western Australia and some other waters and are attracting the attention of behavioral biologists attempting to determine the benefit of such a life (Connor et al. 2000). Some hypotheses are that the coalitions (1) protect against predators, (2) allow inferior males in the group the opportunity to mate, and (3) allow males to corral a female for mating. The answer is still unknown.

11.4.5 Body-Length Frequency

Figure 11.6 shows the body-length frequencies of common bottlenose dolphins off Japan. The Pacific sample is represented by 244 dolphins from 7 schools driven at Taiji during January through April and 37 from a group driven at Arari in August, and the Iki sample is represented by 446 dolphins in 7 schools driven from January through March at Katsumoto at Iki Island as a part of fishermen's activities in the "dolphin-fishery conflict" (Section 3.4) (Table 11.2). The main season of the sample, from January to April, is about 3–6 months after the end of the preceding parturition season, which extends from February to October with a peak in June (Figure 11.4).

The largest male was measured at 328 cm in the Pacific and 336 cm in the Iki Island area, and the largest female at 318 cm in the Pacific and 320 cm in the Iki Island Area. Males grow about 10–15 cm greater than females. Although the Iki sample contained a dolphin of larger size than the Pacific sample, the difference is statistically insignificant.

The body-length frequencies reveal three modes, that is, individuals of 130–140 cm, 180–210 cm, and over 220 cm. Since neonates of this species off Japan are born at 116–140 cm (Section 11.4.2), the smallest peak contains neonates born at the beginning of the parturition season. The second group,

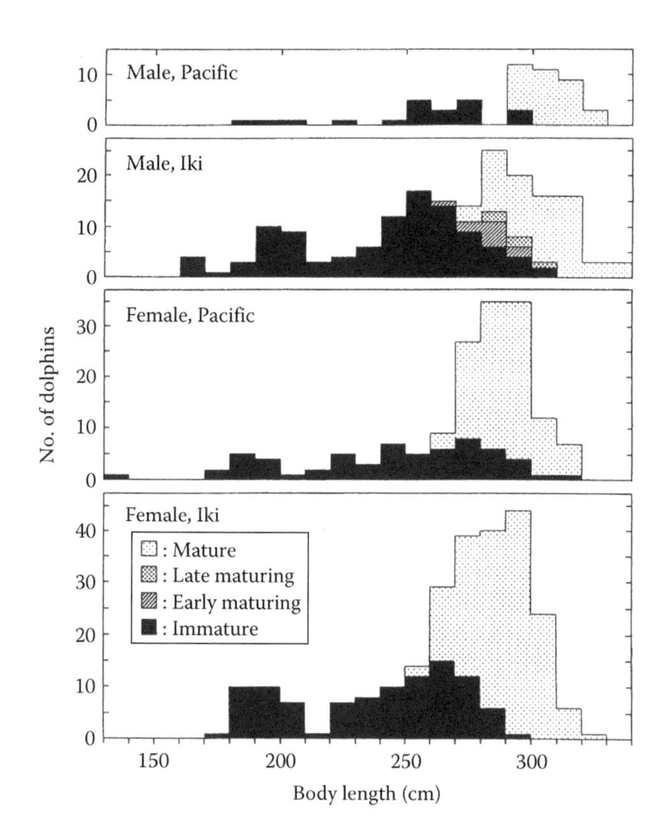

FIGURE 11.6 Body-length composition and maturity of common bottlenose dolphins taken on the Pacific coast and in the Iki Island area. (Reproduced from Fig. 5 of Kasuya, T. et al., *IBI Rep.* (Kamogawa, Japan), 7, 71, 1997, in Japanese.)

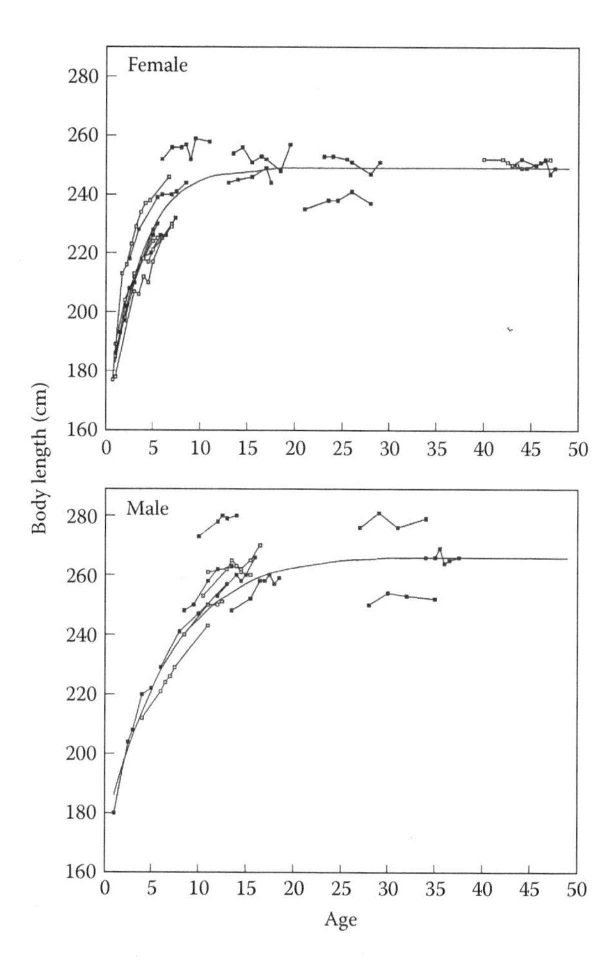

FIGURE 11.7 Growth patterns of individual common bottlenose dolphins (closed circles and solid line) in the inshore waters of Sarasota, Florida. Curvilinear line represents Gompertz curve fitted to all the available length-on-age data. (Reproduced from Fig. 4 of Read, A.J., Wells, R.S., Hohn, A.A., and Scott, M.D. Patterns of growth in wild bottlenose dolphins, *Tursiops truncatus*. *J. Zool. Lond.* 1993. 231. 107–123. Copyright Wiley-VCH Verlag GmbH & Co. KGaA. Reproduced with permission.)

180–210 cm, contains individuals born in the preceding parturition season that are 3–13 months of age. The largest group contains individuals of 2 or more years of age. It becomes difficult to separate age groups based on body length after the age of 2 years, because growth rate decreases and individual size variation increases with age. The age composition of the Pacific sample suggests underrepresentation of juveniles, which is also apparent in the body-length frequencies, particularly among males (Figure 11.6).

11.4.6 Growth Curve

The usual way to construct a mean growth curve is to base it on body lengths and estimated ages from catches or bycatches obtained during a certain length of time. As stated elsewhere, the mean growth curve thus created reflects the real mean growth curve only when the growth pattern in the population has remained stable during the whole lifetimes of the animals in the sample, that is, from the date of birth of the oldest individual to the date of the last sample. Information from aquarium-reared known-age individuals is free from this problem but is often accompanied by the problems of insufficient sample size and evaluating the effect of the artificial environment. It should also be noted that a mean growth curve does not represent the growth of any single individual, that is, an individual growth curve and the mean growth curve are likely to be quite different.

An attempt has been made to resolve the first problem by using repeated measurements of wild individuals in Sarasota, Florida (Read et al. 1993). Body lengths measured periodically were plotted on ages of animals known from birth or aged using teeth extracted from the individuals of unknown birth date. These materials were obtained mostly from young individuals below 10 years of age, covered less than 7 years, and lacked data on animals less than 1-year old (mother–calf pairs were not examined to avoid disturbance of their bond). The growth curves of individual dolphins thus obtained were used to construct a mean growth curve (Figure 11.7).

It appears that fast growth was uniform for 2–5 years, although the absolute growth rates were variable. Then, the growth rate declined at individually variable ages around 5 years in females and at a slightly higher age in males. This age variation contributes to the formation of a smooth upwardly convex mean growth curve at around age 5 years. Almost linear growth of a young individual was also reported for a common bottlenose dolphin of unknown age that

TABLE 11.4
Monthly and Annual Growth of Aquarium-Reared Common Bottlenose Dolphins, Calculated Assuming Neonatal Length of 128 cm, a Month of 30.4 Days, and a Year of 365 days

Dolphin	Sex	1st Year	2nd Year	3rd Year	4th Year
No. 24	Male	7.73 cm/month	1.69 cm/month	0.81 cm/month	—
		92.8 cm/year	20.3 cm/year	9.7 cm/year	
No. 30	Female	6.17 cm/month	2.30 cm/month	0.70 cm/month	—
		74.0 cm/year	27.6 cm/year	8.4 cm/year	
No. 31	Female	—	—	1.94 cm/month	0.22 cm/month
				23.3 cm/year	2.64 cm/year

Source: Kasuya, T. et al., *Sci. Rep. Whales Res. Inst.* (Tokyo), 37, 85, 1986.

approached humans on the English coast; it grew almost linearly from 228.6 to 270.9 cm during 400 days, for a mean growth rate of 3.2 cm/month (Lockyer and Morris 1987).

Kasuya et al. (1986) reported growth records of three aquarium-reared common bottlenose dolphins taken from Japanese waters (Table 11.4). By assuming a neonatal length of 128 cm, a body-length increment of 74–93 cm, or increase of 58–73%, in the first year after birth, is estimated.

Perrin et al. (1976) obtained the following interspecies relationship between mean fetal growth rate (X, cm/month), mean growth rate in a postnatal period that is equal to the gestation time (Y, cm/month), and mean neonatal body length (Z, cm):

$$\text{Log}\,(X - Y) = -1.33 + 0.997 \log Z$$

The mean monthly growth rate, Y = 4.77, is obtained for Japanese common bottlenose dolphins, with values of Z = 128 and X = Z/12 = 10.67. The value of Y suggests a body-length increment of 57.2 cm in the first year after birth, which is at the lower bound of the range calculated from aquarium specimens.

One of the growth models fitted by Read et al. (1993) to their mean age–length relationship is

$$\text{Female: } L = 249.2 \exp(-0.423 \exp(-0.314t))$$

$$\text{Male: } L = 266.4 \exp(-0.422 \exp(-0.164t))$$

where
L denotes body length in cm
t age in years

These equations suggest that female length will exceed that of males at ages 3–10 years, but that male asymptotic length is about 17 cm greater than that of females. It should be noted that a mean growth curve can be skewed by biased age composition or limited age range of input data. In the previous case, the input data are mostly from animals below age 15 years. The mean growth equation fitted to such data may correctly represent the mean growth of the young ages but may not represent real mean growth of older individuals. Extrapolation of such an equation to ages outside the range of the input data, for example, below the age of 1 year, should be avoided. Read et al. (1993) averaged body lengths of calves between ages 0.5 and 1.4 years and estimated the average length of calves at 1 year of age, which were 184 cm for females and 183 cm for males. Assuming mean neonatal length at 109.5 cm (Section 11.4.2) gives an estimate of about 68% as average body-length gain during the first year after birth for the coastal population of common bottlenose dolphins off Florida, which is similar to the 58%–73% estimated for Japanese waters.

The age–body length relationships for Japanese common bottlenose dolphins are shown in Figures 11.8 and 11.9, where ages have been estimated by counting growth layers

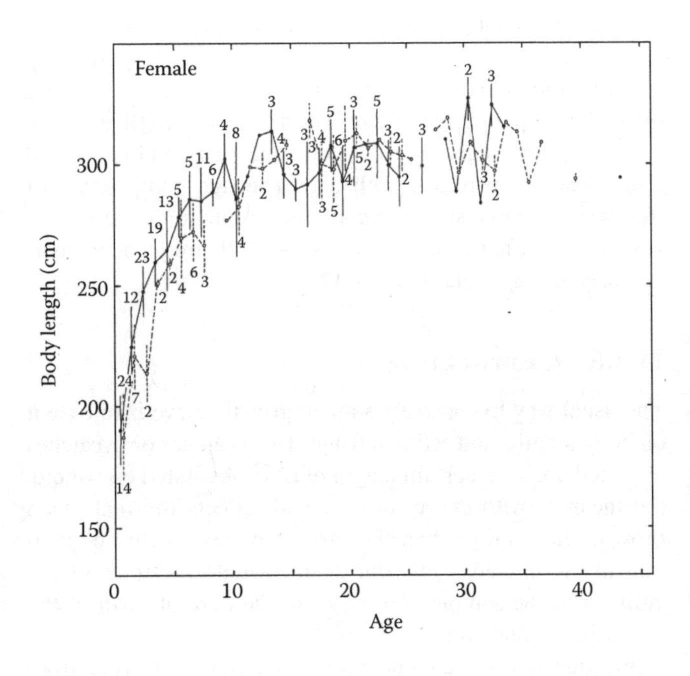

FIGURE 11.8 Relationship between body length and age of female common bottlenose dolphins. The sample size and one standard deviation of the sample are shown on each side of the mean body length. Open circles and the dotted line indicate the Pacific sample and closed circles and the solid line the Iki sample. (Reproduced from Fig. 7 of Kasuya, T. et al., *IBI Rep.* (Kamogawa, Japan), 7, 71, 1997, in Japanese.)

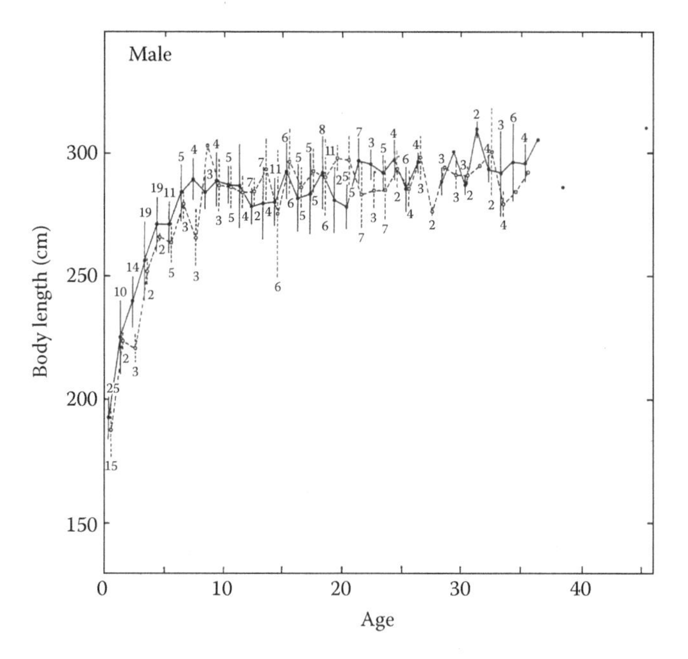

FIGURE 11.9 Relationship between body length and age of male common bottlenose dolphins. For further explanation, see Figure 11.8. (Reproduced from Fig. 6 of Kasuya, T. et al., *IBI Rep.* (Kamogawa, Japan), 7, 71, 1997, in Japanese.)

in decalcified and hematoxylin-stained dentine and cementum. The figures show only mean body length, sample size, and range of one standard deviation on each side of the mean. Those who are interested in growth equations are advised to attempt the task using the data in Table 11.5. The rapid increase in mean body length of juveniles ceases at around 5 years, and slower growth continues to around 10 years. Age-related increase in body length appears to cease at around 15 years in females and 20 years in males, which is expected to agree with ages where all the individuals cease growing.

The mean body length at cessation of growth, or asymptotic body length, was calculated as an average body length of individuals over 20 years of age, which gives the following figures (Kasuya et al. 1997, in Japanese):

Males, Pacific area:	305.3 cm (n = 32, SE = 1.5)
Males, Iki area:	305.2 cm (n = 27, SE = 2.4)
Females, Pacific area:	288.0 cm (n = 48, SE = 1.7)
Females, Iki area:	293.7 cm (n = 63, SE = 0.9)

Males did not show geographical differences in asymptotic lengths, but females showed statistically significant difference between the Pacific and Iki samples ($p < 0.01$). Females in the

TABLE 11.5
Age–Length Key for Japanese Common Bottlenose Dolphins, with Mean Body Length, Sample Size (N), and Standard Deviation (SD)

	Male						Female					
	Pacific			Iki			Pacific			Iki		
Age	Mean	N	SD	Mean	N	SD	Mean	N	SD	Mean	N	SD
0.5[a]	187.4	14	17.4	190.4	24	14.1	187.5	15	10.9	192.4	25	8.7
1.5	220.4	7	13.2	224.6	12	16.8	223.5	2	3.5	225.0	10	14.8
2.5	213.5	2	11.5	247.5	23	10.8	220.7	3	5.8	239.3	14	10.5
3.5	251.0	2	1.0	259.4	19	10.2	251.5	2	5.5	255.8	19	16.0
4.5	259.0	2	2.0	264.4	13	16.2	265.5	2	1.5	270.7	19	11.3
5.5	269.5	4	16.4	278.4	5	8.9	263.4	5	8.0	270.3	11	10.8
6.5	272.2	6	11.5	285.4	5	12.1	279.0	3	5.7	283.8	5	11.6
7.5	269.7	3	10.4	285.1	11	14.2	262.3	3	10.4	289.3	4	8.3
8.5	—	—	—	288.7	6	5.7	303.0	1	—	284.0	3	6.4
9.5	277.0	1	—	302.5	4	10.4	287.0	3	10.7	289.0	4	10.7
10.5	282.8	4	11.5	285.8	8	23.9	286.2	5	8.9	287.2	5	7.8
11.5	299.0	1	—	295.0	1	—	284.3	4	6.2	286.6	8	17.1
12.5	298.5	2	6.5	312.0	1	—	284.5	2	5.5	278.7	7	7.5
13.5	302.0	1	—	313.7	3	9.0	293.5	4	12.8	279.6	7	14.6
14.5	308.0	3	2.2	295.5	4	9.8	275.5	6	26.0	280.1	11	9.7
15.5	—	—	—	289.7	3	4.8	296.7	6	13.7	292.8	6	11.5
16.5	318.0	3	7.1	291.3	3	17.6	286.0	5	8.2	281.6	5	14.0
17.5	300.3	3	13.9	296.3	4	10.5	293.0	5	9.1	283.2	5	16.7
18.5	298.0	5	13.9	307.6	5	10.8	290.2	6	15.7	292.0	8	14.8
19.5	309.8	4	13.3	292.7	6	14.0	298.0	2	5.0	280.6	11	12.8
>20	305.3	32	8.2	305.2	27	12.6	288.0	48	11.8	293.7	63	7.3

Source: Constructed from Table 4 of Kasuya, T. et al., *IBI Rep.* (Kamogawa, Japan), 7, 71, 1997, in Japanese.
Note: Age is expressed to the nearest n + 0.5 years (n being an integer).
[a] Two newborns (one Pacific male and one Iki female below 130 cm) excluded and 160–230 cm individuals included.

Iki area apparently grew 5.7 cm larger than Pacific females. This can be interpreted as evidence that common bottlenose dolphins in these two geographical regions belong to different populations, although it is unclear whether the size difference is genetically determined or a reflection of difference in nutritional conditions.

Comparison between the Pacific and Iki samples showed that mean body lengths of Iki sample were always greater than those in the Pacific sample of the same age below 11 years (males) or 12 years (females). Except for 37 individuals obtained in August 1973 at Arari on the Pacific coast, materials in the comparison presented earlier were obtained in the same months and almost the same years, that is, January–March in 1979–1980 (Iki) and January–April in 1981–1983 (Pacific). Such a geographical difference in juvenile growth could be expected to occur if food availability in the Iki Island area were better than in the Pacific area in the 1970s and early 1980s.

11.4.7 Male Sexual Maturity

11.4.7.1 Identification of Maturity

In determining the sexual maturity of male reproductive organs, it is necessary to take into account the effect of reproductive seasonality and distinguish between a seasonal cycle and unidirectional development toward maturation. As Dall's porpoises mate in a few months of the summer and adult males almost cease spermatogenesis in winter, their testicular weight and diameter of the seminiferous tubules almost double in the mating season (Section 9.4.8). Full-grown males of the short-finned pilot whale off Japan are spermatogenic throughout the year, while some males in early stages of maturation become spermatogenic only seasonally. Kasuya et al. (1997, in Japanese) could not identify seasonal change in the male reproductive tracts of common bottlenose dolphins due to the limitation of the samples to a four-month period from January to April, which corresponded to the time of the lowest mating activity (January and February) and the early mating season (March and April) (Section 11.4.2). However, I expect that the seasonal change in male reproductive tracts of common bottlenose dolphins will not be as distinct as in the Dall's porpoise, because they have a much broader mating season extending for 8–9 months of the year (Figure 11.4) and the period of spermatogenic activity will extend longer than the mating season.

In classifying the maturity stages of testicular tissue of the common bottlenose dolphin, Kasuya et al. (1997, in Japanese) adopted the same categories used by Kasuya and Marsh (1984) for short-finned plot whales (Section 12.4.3.3), where maturity was classified into four histological stages by the proportion of spermatogenic seminiferous tubules at the mid-center of a testis. A seminiferous tubule was determined as spermatogenic if spermatocytes, spermatids, or spermatozoa were present, and a testis was determined as "immature" if no tubules were spermatogenic, as "early maturing" if more than 0% but less than 50% of the tubules were spermatogenic, as "late maturing" if less than 100% but 50% or more of the tubules were spermatogenic, and "mature" if 100% of the tubules were spermatogenic. The three types of cells occurred in all tubules identified as spermatogenic. The absence of spermatozoa would not cause significant bias in determining age at sexual maturity because it takes only a few weeks for a spermatocyte to develop into a spermatozoon.

11.4.7.2 Relationship between Age and Testis Weight

Figures 11.10 and 11.11 show the relationship between the age and the weight of a single testis. The testis weights are plotted on a logarithmic scale. This method covers a broad weight range and illustrates early testicular growth in detail but has the drawback of expressing a positive linear relationship between the weight and age for a curve with the slope that decreases with increasing age.

In the Iki sample (Figure 11.11), testis growth was slow at ages between 0.5 and 7.5 years, when weight increased from around 6 g to around 45 g. This corresponded to an increase of 5–6 g/year or about a seven fold increase in the 7 years. Then at around age 8 years, the testis started a rapid growth to reach about 200 g at 12 years of age, which was equivalent to an average increase of 39 g/year and a four- to fivefold increase during the 4 years. With exception of one male aged 20.5, all the testes weighing 150 g or more were classified as "mature." After the age of 12 years, the testis continued a similar annual weight increase until age 20, when it reached about 550 g, which was equivalent to an average increase of 44 g/year or a two- to threefold increase in the 8 years. Thus, testicular weight gain was almost constant at around

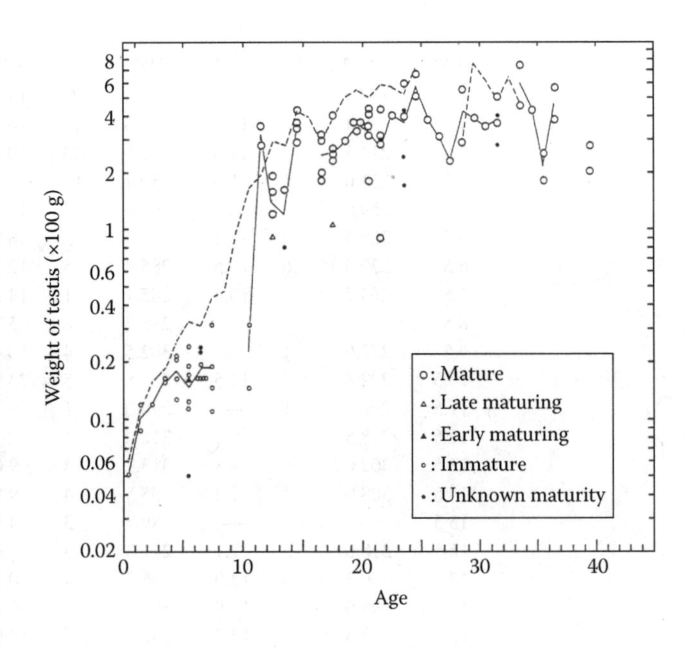

FIGURE 11.10 Scatterplot showing the relationship between the age, histological maturity, and weight of single testis (epididymis removed) of common bottlenose dolphins from the Pacific coast. The solid line represents arithmetic means of testis weights of the Pacific sample and the dotted line those of the Iki sample. (Reproduced from Fig. 8 of Kasuya, T. et al., *IBI Rep.* (Kamogawa, Japan), 7, 71, 1997, in Japanese.)

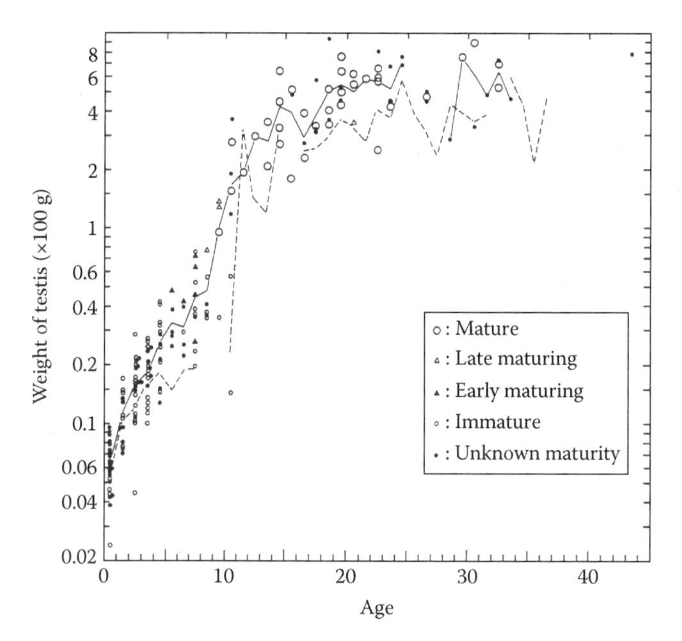

FIGURE 11.11 Scatterplot showing the relationship between the age, the histological maturity, and the weight of single testis (epididymis removed) of common bottlenose dolphins from the Iki Island area. The solid line represents arithmetic means of testis weights of the Iki sample and the dotted line those of the Pacific sample. (Reproduced from Fig. 9 of Kasuya, T. et al., *IBI Rep.* (Kamogawa, Japan), 7, 71, 1997, in Japanese.)

39–44 g/year for ages 8–20 years. Age-related testicular growth appeared to cease at the ages of 20–25 years.

The general pattern of testicular growth of the Pacific sample (Figure 11.10) appeared to be the same as that of the Iki sample, but the Iki individuals almost always had heavier testis compared to Pacific animals of the same age. The average testis weight of full-grown males from the Iki Island area was about 200 g greater than in the Pacific sample. This is further discussed in the following in relation to body length.

11.4.7.3 Relationship between Body Length and Testis Weight

Figures 11.12 and 11.13 show the relationship between body length and weight of a single testis plotted on a logarithmic scale. These figures appear slightly different from Figures 11.10 and 11.11, because increase in body length is small after the attainment of sexual maturity. Comparison between the Iki and Pacific samples showed that (1) both started rapid testicular growth at a similar testicular weight (about 200 g) but at a smaller body length in the Iki sample (260 vs. 285 cm) and (2) the Iki specimens in general had heavier testes than individuals of the same body length in the Pacific sample. Noting that the two samples were collected in similar months of the year and that the breeding season was the same between the two ocean areas, this can be interpreted to mean that males of the species in the Iki Island area start puberty at a smaller body length and perhaps at a younger age. There is as yet no explanation of why adult males in the Iki area have heavier testes, but it does indicate that the bottlenose dolphins in the two ocean areas belong to different populations.

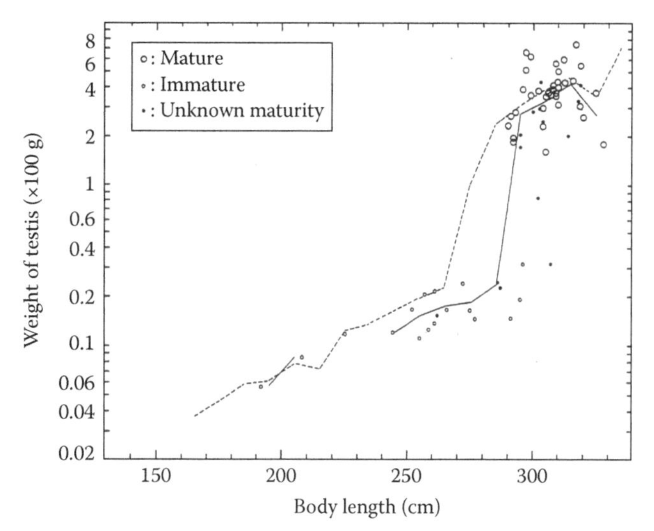

FIGURE 11.12 Scatterplot showing the relationship between the body length, the histological maturity, and the weight of single testis (epididymis removed) of common bottlenose dolphins from the Pacific area. The solid line represents arithmetic means of testis weights of the Pacific sample and the dotted line those of the Iki sample. (Reproduced from Fig. 10 Kasuya, T. et al., *IBI Rep.* (Kamogawa, Japan), 7, 71, 1997, in Japanese.)

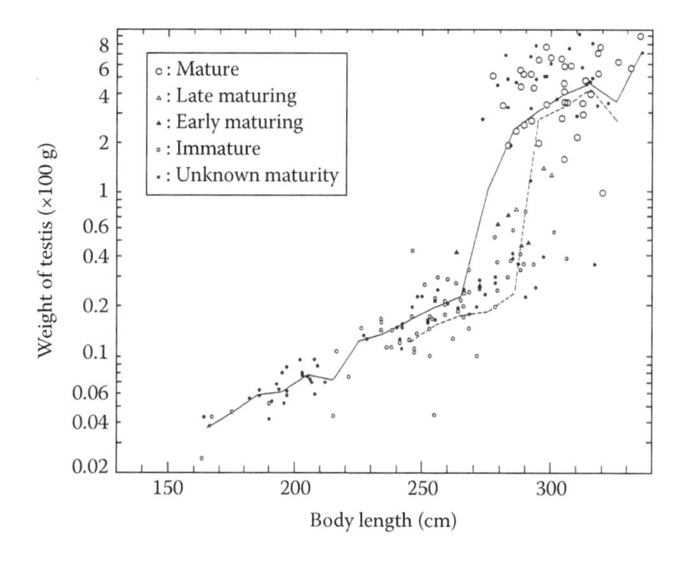

FIGURE 11.13 Scatterplot showing the relationship between the age, the histological maturity, and the weight of single testis (epididymis removed) of common bottlenose dolphins from the Iki Island area. The solid line represents arithmetic means of testis weights of the Iki sample and the dotted line those of the Pacific sample. (Reproduced from Fig. 11 of Kasuya, T. et al., *IBI Rep.* (Kamogawa, Japan), 7, 71, 1997, in Japanese.)

11.4.7.4 Age and Body Length at the Attainment of Sexual Maturity

Males start rapid testicular growth at around 8 years of age. The Iki dolphins had testis weight of about 45 g at age 8 years and attained about 200 g at age 12 years. This rapid testis growth is observed at body length of 275–310 cm. This body-length range corresponded also with the stage where the four maturity states of "immature," "early maturing," "late maturing," and "mature" coexisted and is interpreted as the period when males were rapidly growing toward sexual maturity. There was an apparent weight increase after body length of 300 cm, which could be a reflection of a positive correlation between testicular weight and body size. Such was also observed in short-finned pilot whales (Figure 12.12).

The period before age 8 years corresponded to an immature stage, when growth in testis weight was slow and spermatogenesis did not usually occur. However, some individuals underwent spermatogenesis in limited seminiferous tubules at the end of this stage (i.e., "early maturing" individuals). This was followed by the period of rapid testicular growth where some individuals underwent spermatogenesis in over 50% of the tubules (i.e., "late maturing" individual). By 11 years, almost all the males attained the "mature" stage where 100% of tubules were spermatogenic. However, some males remained in the "late maturing" stage until age 20 years in both the Iki and Pacific samples. Thus, the process of male maturation seemed to proceed slowly, lasting several years and with broad individual variation, and it was difficult to determine clear subdivisions within it, which is similar to the case in other delphinids.

Table 11.6 presents the relationship between maturity stage and age. The youngest males in "early maturing" stage was 5.5 years old in both the Iki and Pacific samples, those in "late maturing" stage 8.5 years old (Iki), and those in "mature stage" 9.5 years old (Iki) or 11.5 years old (Pacific). Thus, the minimum length of the period from the beginning of spermatogenesis to the attainment of the "mature" stage was estimated for the Iki sample at 4 years. Such estimation was not possible for the Pacific sample due to insufficient sample size. This was the case for precocious males. Slower maturation was represented by males at the "immature" stage at age 10.5 (both Iki and Pacific). These individuals could have reached the "early maturing" stage at age 11.5. Such individuals in the Pacific sample reached the "late maturing" stage at 12.5 years at the earliest or 17.5 years at the latest (Pacific) and would attain the "mature" stage at some age after 18.5 years (Pacific) or 20.5 years (Iki). These were cases, supported by age and maturity data, of males that entered puberty at a late age and proceeded to maturation with reasonable speed, suggesting that some males needed almost 10 years to achieve the maturation process. Although it is not supported by data, if a male that started the puberty at the earliest age (5.5 years) and proceeded to maturation process at the slowest speed, it would take almost 15 years to reach the mature stage at age 20.5. Such an extreme case is probably rare; male common bottlenose dolphins off Japan usually spend 4–10 years from early puberty to full sexual maturity.

The mean age at the attainment of sexual maturity is estimated with acceptable sample size only for the Iki sample. One method often used for this purpose is to fit a model to the relationship between age and the proportion of males sexually mature and to calculate an age when 50% are mature. Sigmoid or linear models have been used for this purpose. This method is not preferred because the estimate is not free from model-specific bias (DeMaster 1984). Kasuya et al. (1997, in Japanese) used the simple method of Hohn (1989), which is believed to be free from the bias. This method is to add to the age of the earliest "mature" individual the proportions of nonmature individuals in the older age classes. In the case of the Iki sample, this method calculates the mean age at the attainment of sexual maturity as

$$9.0 + 0.75 + 0.5 = 10.25 \text{ years (SE} = 0.382)$$

TABLE 11.6
Testicular Maturity and Age of Male Common Bottlenose Dolphins in Japan

	Pacific Sample				Iki Sample			
Age	Immature	Early Maturing	Late Maturing	Mature	Immature	Early Maturing	Late Maturing	Mature
4.5	4				8			
5.5	5	1				1		
6.5	5				1	1		
7.5	4				6	4		
8.5	1				3		1	
9.5					1		2	1
10.5	2				2			2
11.5				2				1
12.5			1	3				1
13.5–16.5				8				10
17.5			1	4				1
18.5–19.5				4				7
20.5				5			1	2

Source: Reproduced from Table 6 and text of Kasuya, T. et al., *IBI Rep.* (Kamogawa, Japan), 7, 71, 1997, in Japanese.
Note: Age is expressed to the nearest n + 0.5 years (*n* being an integer).

TABLE 11.7
Testicular Maturity and Body Length of Male Common Bottlenose Dolphins in Japan

	Pacific Sample[a]				Iki Sample[b]			
Body Length (cm)	Immature	Early Maturing	Late Maturing	Mature	Immature	Early Maturing	Late Maturing	Mature
250–259	5							
260–269	2				4	1		
270–279	4				4	2		
280–289					6	2	1	
290–299	3			4	2	1	1	1
300–309				4	2		1	2
310–319				1				
320–329				2				1

Source: Reproduced from Table 7 of Kasuya, T. et al., *IBI Rep.* (Kamogawa, Japan), 7, 71, 1997, in Japanese.
[a] Includes only individuals aged 4.5–18.5 years.
[b] Includes only individuals aged 4.5–11.5 years.

In the same way, the age at the attainment of the "late maturing" stage is calculated as 9.50 years (SE = 0.456), where the sum of the proportions of "late maturing" and "mature" stages is used in the calculation. Takemura (1986a, in Japanese) obtained 11.5 years as a mean age at the attainment of sexual maturity of males in the Iki Island area, but the definition of "sexual maturity" was not clarified.

Information was insufficient for the Pacific sample to analyze age-related change in male maturity (Table 11.6). The oldest "immature" male was aged 10.5 and the youngest "mature" male 11.5, but some males could have attained the "mature" stage at higher ages, as indicated by males of the "late maturing" stage at ages 12.5 years and 17.5 years. Similar exceptional late maturing cases occurred in the Iki sample, as in short-finned pilot whales off Japan (Figure 12.12). Kasuya et al. (1997, in Japanese) ignored these exceptional individuals and estimated that the Pacific individuals of the species usually attain the "mature" stage at ages 11–13 years. This was based on the estimation of the mean age at the attainment of the "mature" stage, 12.20 years (SE = 0.543), and that at the attainment of the "late maturing" stage, 11 years (SE = 0), using the method of Hohn (1989). It is possible that males in the Iki Island area attain sexual maturity at a younger age than those in the Pacific.

The mean body length at the attainment of sexual maturity was calculated as the point where the proportion of mature individuals reached 50%. The estimate varies with the age range of the sample used in the calculation, for example, either use of the whole sample with age range from 0 to the oldest individual or use of only individuals between the youngest mature and the oldest immature. The former method results in a smaller figure than the latter because (1) the dolphins exhibit broad individual variation in body length at sexual maturity, (2) increase in body length ceases soon after the attainment of sexual maturity, and (3) therefore numerous old and relatively small mature individuals will be included in the calculation. The first method also suffers from any historical change in growth.

Kasuya and Marsh (1984) and Kasuya et al. (1997, in Japanese) questioned the value of the first method as a way of obtaining a growth parameter for a particular dolphin population. Kasuya et al. (1997, in Japanese) extracted subsamples of individuals aged from 4.5 to 11.5 years (Iki) and from 4.5 to 18.5 years (Pacific), which nearly correspond to the ages where the process of maturation proceeds (Table 11.7). In the case of the Iki sample, the "early maturing" stage occurred at body length 260–309 cm and the "late maturing" stage at lengths over 290 cm. Half of the individuals are expected to be at "late maturing" or "mature" stage when they are between 290 and 309 cm; thus, the mean body length at the attainment of the "late maturing" stage is about 300 cm (for this task both the "late maturing" and "mature" stages should be combined). In the same way, the average body length where 50% of individuals are "mature" is expected to be between 309 and 320 cm. Takemura (1986a) obtained a body length of 299.1 cm as the body length where 50% of males were sexually mature in the Iki Island area. This figure is close to the mean body length at the attainment of the "late maturing" stage of Kasuya et al. (1997, in Japanese), although the method of calculation and the definition of maturity may not be the same.

Males of the Pacific area were estimated to attain the "mature" stage at body length 290–299 cm. Sample size was insufficient for further analyses of their growth parameters.

11.4.7.5 Male Reproductive Ability

The histology of 1–2 cm^2 tissue samples taken from midlength of testes of common bottlenose dolphins revealed various levels of spermatogenesis (Kasuya et al. 1997, in Japanese). Some precocious males had some seminiferous tubules that were spermatogenic (i.e., mature tubules) at age 5.5 years and body length 260–269 cm, but other individuals still had some immature tubules at age 20.5 years and body length 300–309 cm. The average age when all the seminiferous tubules in the examined tissue were spermatogenic (i.e., "mature" stage in the definition given earlier) was 10.25 years (Iki sample) or 12.20 years (Pacific sample). This definition of maturity stage was only a tentative one, not supported by evidence whether these "mature" males participated in mating and produced offspring. One can only expect correlation between the testicular histology and behavior of the male.

At some stage of the development of sexual maturation, a male cetacean will start to approach females for mating. However, before these efforts end with mating and production of offspring, some social or physiological conditions may need to be satisfied. If there is competition between males, a male must either win the competition or sneak in to mate with receptive females. Whether an estrous female accepts every approaching male or chooses a particular male for mating will depend on the social structure, but we do not know whether a female chooses a certain male when there are multiple courting males. Copulation must be concluded with ejaculation sufficient for fertilization, which would have some relevance to testicular histology. Such a reproductive environment for males will not be the same between wild and aquarium environments. The status of males who actually participate in reproduction is often called "social maturity." Some attempt has been made, without a clear conclusion, to match the state of social maturity with histological growth stages (Miyazaki and Nishiwaki 1978; Miyazaki 1984; Section 10.5.5).

Reproductive success and its individual variation among males remain as one of the important fields of cetacean biology to be investigated. Matching anatomical information on reproductive tracts against either reproductive behavior or genetic paternity evidence will help us reach understanding, but such information is currently unavailable for the wild common bottlenose dolphins around Japan. To substitute for such data, Kasuya et al. (1997, in Japanese) compared testicular histology against spermatozoan density in the complex lumen of the epididymis, which functions for storage of spermatozoa produced in the testis. They made a smear from the epididymis, air-dried it, and stained it with a water solution of toluidine blue for microscopy. The density of spermatozoa was classified into five grades of "absent" to "extremely dense," which was of a density found only in some epididymides but not in any testicular smear. This followed the classification used for short-finned pilot whales (Section 12.4.3.3 and Table 12.8). A difference of one density grade may sometimes be insignificant. A possible but unconfirmed effect of preceding ejaculations on the epididymal sperm density was ignored.

The epididymal smear was compared against testicular anatomy and the age of male common bottlenose dolphins (Table 11.8). The testis and epididymis were taken from the same side of each individual. About 23% of males classified as "immature" through testicular histology had a small amount of spermatozoa in the epididymal smear, which was an evidence of spermatogenesis in some portion of the testis not sampled for testicular histology. Similar cases, low levels of sperm production unidentified by testicular histology, were also reported in short-finned pilot whales aged 5–6 years (Kasuya and Marsh 1984) and in pantropical spotted dolphins aged 2 years (Kasuya et al. 1974). Testes of higher maturity

TABLE 11.8
Relative Spermatozoa Density in Epididymal Smear of Common Bottlenose Dolphins from Iki Island, Viewed against Histology of Testis, Weight of Single Testis Removed from Epididymis, and Age

Sperm Density Grade		0	1	2	3	4	Average Grade
Histology Testis	Immature	44	13				0.2
	Early maturing	3	1	2			0.8
	Late maturing	1			2	1	3.3
	Mature	1	7	11	10	6	2.4
Testicular weight (g)	<50	89	23	1			1.0
	50–100	6		1	1		0.3
	100–200		1	3	1	2	2.6
	200–300	1	2	1	2	2	2.3
	300–400		2	4	5	1	2.4
	400–600		2	6	7	7	2.9
	600–800			8	3	1	2.4
	800–1000				2	1	3.3
Age (year)	<5	73	18				0.2
	5–10	20	5	2	2	1	0.6
	10–15	3	5	4	1	4	1.9
	15–20		2	9	7	3	2.5
	20–30	1		5	7	5	2.8
	30–45		1	3	3	1	2.5

Source: Reproduced from Table 8 of Kasuya, T. et al., *IBI Rep.* (Kamogawa, Japan), 7, 71, 1997.
Note: Numerals indicate number of individuals examined.

stages in general had high epididymal sperm density, but the two testicular maturity stages of "late maturing" and "mature" showed no difference in epididymal sperm density. This suggests that males of the two maturity stages have a similar physiological capacity of reproduction. In other words, these males can participate in reproduction if the opportunity is available.

It should be noted that about 23% of males at the "late maturing" or "mature" stages had no spermatozoa or an extremely low level of spermatozoa in the epididymis. One of the possible explanations is that most of the sample was obtained during January to April, which covered the period of the lowest mating activity (December–January) to the beginning of the mating season that extended from February to October. Considering that one aquarium-reared adult male showed a seasonal testosterone cycle with the lowest level in January and February (Section 11.4.2), it remains possible that at least some males of the species cease or decrease spermatogenic activity in the nonmating season. Yoshioka et al. (1993) reported an interesting experiment in which a male common bottlenose dolphin was trained to ejaculate in response to a signal from the trainer. At the beginning of a series of such guided ejaculations, the semen did not contain spermatozoa, but the sperm density increased after more ejaculations. This suggests that external stimuli can accelerate spermatogenesis or transport of spermatozoa from the testis to the epididymis. This may happen in wild males of the species in response to visual or acoustic stimuli from other individuals.

The density of spermatozoa in the epididymis was compared with testicular weight (Table 11.8). The epididymal sperm density tended to be lower with the testes of less than 100 g compared to heavier testes, but there was no clear density difference, and the mean density indices remained the same among testes weighing over 100 g. It is noted that with the exception of one testis, all the testes weighing over 150 g were at the histological "mature" stage and that all the testes weighing less than 150 g were at the histological stage of "late maturing" or a lower stage (with the exception of one individual) (Figures 11.10 through 11.13). The lower bound for "mature" testis was 150 g and that for "late maturing" was about 100 g. From this, it can be concluded that common bottlenose dolphins in the Iki Island area have the physiological capacity of reproduction if a testis weighs over 100 g or if testicular histology identifies them as "late maturing." This is the minimum requirement for reproductive capacity; males not satisfying either of these criteria are unlikely to be reproductive.

This conclusion does not mean that male reproductive capacity is the same for testes weighing from 100 to 800 g or more. It remains to be investigated whether males with larger testis have higher levels of testosterone and are more active sexually. It seems to be a possibility that testis size has a positive correlation with body size and that males with larger testes are more likely to acquire more mating partners or that males with larger testes produce a larger quantity of spermatozoa and overwhelm ejaculates of other males. It is generally believed that larger testes will benefit reproductive success in random mating or in a polygamous community (Ralls and Mesnick 2002). The mating system of common bottlenose dolphins is close to random mating, as in the case of striped dolphins (Section 10.5.5.3).

The density of spermatozoa in the epididymis increased up to the age of 15. Then the correlation was lost (Table 11.8), which suggested that all the males over 15 years of age had a physiologically equal capacity to reproduce. Some males at ages 10–15 had a sperm density grade of 2 or more, suggesting that they had a similar reproductive capacity to that of individuals over 15 years of age.

The observations on common bottlenose dolphins in the Iki Island area can be summarized: (1) males having a testis of over 100 g or a testis at the histological maturity stages of "late maturing" or "mature" are physiologically capable of reproduction, (2) males aged 15 or older have the same capacity, and (3) males below age 15 can be considered to have the reproductive capacity if they satisfy the first condition. The same species in the Pacific area tended to have smaller testes and sexual maturity appeared to be reached about 3 years later than in the Iki individuals. However, the maturity criterion of 100 g will not cause a significant bias for the Pacific individuals because testicular development around the attainment of sexual maturity appears rapid.

11.4.8 Sexual Maturity and Reproductive Cycle of Females

11.4.8.1 Identification of Sexual Maturity

Biologists have classified female cetaceans as sexually mature if they have experienced ovulation, which is believed to be identifiable by the presence of a corpus luteum or corpus albicans in the ovaries. Most female cetaceans conceive at the first ovulation; a few become pregnant at the second or third ovulation. Although this is easily applicable by fishery scientists working on whale carcasses killed by a fishery, it is difficult to apply by scientists working on live wild cetaceans. It takes about one year or more for the first ovulation to be followed by the first parturition, which can be more significant for population dynamics. Further information on cetacean ovaries and morphology of the corpus luteum and albicans are available in Perrin and Donovan (1984). Recently, a question has been raised about the persistence of the corpus albicans of ovulation in some small toothed whales (see Sections 10.5.6.1 and 11.4.8.5).

In principle, pregnancy should be confirmed by the presence of a fetus in the uterus, but it can be determined with fair accuracy by the examination of the histology of the endometrium (Section 12.4.4.5). Stretch marks in nonpregnant uteri are possibly useful for the identification of past pregnancy but require further calibration. Lactation is usually identifiable through visual examination, but histological examination of the mammary gland increases accuracy. Light brown coloration of the mammary glands of nonlactating baleen and sperm whales allowed visual determination of past lactation, but the method is not applicable to dolphins due to the small thickness of their mammary glands.

Kasuya et al. (1997, in Japanese) confirmed field records of pregnancy and lactation using histology of the endometrium and mammary gland. There were some cases where a fetus was lost before examination by scientists and pregnancy was determined by histology of the endometrium. In principle, they identified female maturity by detecting a corpus luteum or albicans, but it was also determined by the presence of lactation or pregnancy for a small number of females where ovaries were removed by the fishermen. The following is based on the study by Kasuya et al. (1997, in Japanese) conducted using materials listed in Table 11.2.

11.4.8.2 Age at the Attainment of Sexual Maturity

Both immature and mature females occurred at ages 5.5–8.5 in the Iki sample and 6.5–12.5 in the Pacific sample (Table 11.9). The mean age at the attainment of sexual maturity was calculated in the same way as for males to obtain a value of 6.91 (SE = 0.53) for the Iki sample and 9.19 (SE = 0.716) for the Pacific sample. It seems probable that common bottlenose dolphins in the Iki Island area attained sexual maturity about 2 years earlier than those in the Pacific area around 1980.

Takemura (1986a, in Japanese) analyzed the reproduction of female common bottlenose dolphins by combining data of the 1979–1981 seasons contributed by my group and his own data for 1982–1985 and estimated the mean age at the attainment of

TABLE 11.9
Sexual Maturity and Age of Female Common Bottlenose Dolphins off Japan

	Pacific Sample			Iki Sample		
Age	Immature	Mature	Total	Immature	Mature	Total
4.5	3		3	19		19
5.5	8		8	8	3	11
6.5	1	2	3	3	2	5
7.5	3	1	4	1	3	4
8.5				1	2	3
9.5		3	3		4	4
10.5	2	3	5		5	5
11.5		4	4		8	8
12.5	2	1	3		7	7
13.5		9	9		7	7

Source: Reproduced from Table 9 of Kasuya, T. et al., *IBI Rep.* (Kamogawa, Japan), 7, 71, 1997, in Japanese.

Note: Age is expressed to the nearest n + 0.5 years (n being an integer).

TABLE 11.10
Sexual Maturity and Body Length of Female Common Bottlenose Dolphins off Japan[a]

Body Length	Pacific Sample			Iki Sample		
(cm)	Immature	Mature	Total	Immature	Mature	Total
230–239				1		1
240–249				7		7
250–259	4		4	5		5
260–269	2		2	11	1	12
270–279	5	3	8	9	1	10
280–289	1	5	6	6	4	10
290–299		6	6	1	6	7
300–309		1	1		2	2
310–319		1	1			

Source: Reproduced from Table 10 of Kasuya, T. et al., *IBI Rep.* (Kamogawa, Japan), 7, 71, 1997, in Japanese.

[a] Pacific sample includes only 5.5–3.5-year-old individuals and Iki sample 4.5–9.5-year-old individuals (see Table 11.9).

sexual maturity at "slightly below 7 years," which was in good agreement with the results of Kasuya et al. (1997, in Japanese).

11.4.8.3 Body Length at the Attainment of Sexual Maturity

The proportion of females sexually mature increased with increasing body length. However, the body length where the number of immature and mature individuals were equal in a sample, that is, the body length where 50% are mature, was not the same as the body length where half of the individuals of a cohort attained maturity, that is, the mean body length at the attainment of sexual maturity, because small cetaceans cease growing soon after attaining sexual maturity and the growth curve is upwardly convex. The former estimate is influenced by the age composition of the sample, that is, a sample biased toward older individuals will have smaller body length at 50% maturity. The latter method is a more suitable indicator of growth of a species.

In order to estimate the mean body length at the attainment of sexual maturity in the same way as used for males, Kasuya et al. (1997, in Japanese) selected individuals ranging from the age of 1 year below the youngest mature individual to the age of 1 year above the oldest immature individual, that is, 4.5–9.5 years for the Iki sample and 5.5–13.5 years for the Pacific sample (Tables 11.9 and 11.10). Such selected samples revealed that immature and mature individuals coexisted at body lengths 260–299 cm in the Iki sample and 270–289 cm in the Pacific sample. Around 1980, female common bottlenose dolphins attained sexual maturity at these body-length ranges. Females in the Iki Island area attained sexual maturity at a greater body length and with greater individual variation than the Pacific individuals.

The mean body length at the attainment of sexual maturity was estimated as a length where 50% of the females were sexually mature in the selected subsample mentioned earlier, which was about 290 cm for the Iki sample and about 280 cm for the Pacific sample (Table 11.10). Females in the Iki island area matured at a younger age (see above) and at greater body length.

Takemura et al. (1986a, in Japanese) obtained 272.5 cm for females in the Iki Island area as a body length where 50% of females are sexually mature. This calculation included all the age classes of the sample collected in 1979–1985, and so the estimate was understandably lower than the mean body length at the attainment of sexual maturity obtained by Kasuya et al. (1997, in Japanese), that is, ca. 290 cm, for the reason mentioned earlier.

11.4.8.4 Female Reproductive Cycle

Females of the short-finned pilot whales are known to exhibit distinct age-related decline in reproductive capacity (Tables 12.13 and 12.14), but striped dolphins show only a slight decline in pregnancy rate and a slight increase in resting females with age (Section 10.5.11). Kasuya et al. (1997, in Japanese) presented the oldest ages of the common bottlenose dolphins at each reproductive stage:

	Iki Sample	Pacific Sample
Pregnant and lactating	38.5	24.5
Pregnant	35.5	33.5
Resting	45.5	33.5
Lactating	36.5	42.5
Oldest female	45.5	42.5

Females ceased reproduction at the age 33.5–38.5 years, or 7–9 years before the age of the oldest females in each sample, but it seems to be premature to draw a firm conclusion on this topic from the limited analysis of this small sample. Table 11.11 shows age-related change in pregnancy rate for the Pacific and Iki samples. If high pregnancy rate for ages

TABLE 11.11
Reproductive Status of Female Common Bottlenose Dolphins off Japan and Mean Reproductive Cycle Calculated Assuming Gestation of 12 Months

									Total	
Age	5–10	10–15	15–20	20–25	25–30	30–35	>35	Unknown	No. of Dolphins	Average Period (Year)
Pacific sample										
Pregnant	1	8	8	10	3	5	1	5	41	0.85
Pregnant and lactating[a]			1	4	1			1	7	0.15
Lactating	2	13	20	18	10	3	2	8	76	1.58
Resting[b]	1	4	3	2	4	2		4	20	0.42
APR (%)[c]	25.0%	32.0%	28.1%	29.4%	26.3%	50.0%	33.3%	—	33.3%	3.00
Iki sample										
Pregnant	8	17	9	8	4	5	1	1	53	0.85
Pregnant and lactating[a]			4	1	1	1	2		9	0.15
Lactating	4	14	18	13	9	11	2	1	72	1.16
Resting[b]	3	7	4	1		2	1		18	0.29
APR (%)[c]	50.0%	44.7%	37.1%	39.1%	28.6%	33.3%	50.0%	—	40.8%	2.45

Source: Reproduced from Table 11 of Kasuya, T. et al., *IBI Rep.* (Kamogawa, Japan), 7, 71, 1997, in Japanese.

[a] Pregnant and simultaneously lactating.

[b] Mature females neither pregnant nor lactating.

[c] APR, apparent pregnancy rate, or the proportion of pregnant females in a sample of sexually mature females. Standard errors 3.9% (Pacific sample) and 4.0% (Iki sample) for the entire age range calculated as $SE = [APR(1-APR)/n]^{1/2}$.

below 15 years were omitted from the analysis because the age classes included females in their first pregnancy, then age-related change in pregnancy rate was not evident in the limited sample of ages above 15 years. Thus, age-related decline in reproductive capacity is not evident in this species.

Annual pregnancy rate is the probability of a sexually mature female becoming pregnant in a year, and the apparent pregnancy rate is the proportion of pregnant females in the total of sexually mature females. Therefore, if sample bias is ignored, the annual pregnancy rate is calculated by dividing the apparent pregnancy rate by the gestation period in years. As the gestation period of the common bottlenose dolphin was estimated at 1 year, annual pregnancy rate was 40.8% for the Iki sample and 33.3% for the Pacific sample. The mean calving interval is the reciprocal of the annual pregnancy rate, that is, 2.45 years for the Iki sample and 3 years for the Pacific sample. However, the standard errors of the annual pregnancy rates were about 4%, and the difference in pregnancy parameters between the two samples was not statistically significant ($0.1 < p < 0.2$).

The calculations of annual pregnancy rate and mean calving interval are valid only when the sample represents the composition of the population. It cannot be applied to samples from fisheries that may selectively take particular reproductive stages (modern whaling prohibited lactating females) or operate in particular seasons (many fisheries operate in a particular season, and cetaceans often breed seasonally). Although the drive fisheries for dolphins operated in particular seasons, the drive was almost nonselective except for a possible bias in the opportunity for encountering particular types of schools.

Both mating and parturition take place in an 8- to 9-month period from February/March to October among common bottlenose dolphins off Japan. Kasuya et al. (1997, in Japanese) used samples taken by the drive fisheries mostly from January to April (Table 11.2), which was about 6–10 months after the previous parturition/conception peak and agreed with the time from the lowest mating activity to the beginning of the parturition/conception season (Figure 11.4). Therefore, their sample could cover all females that conceived in one mating season and probably were not biased by a seasonal effect on pregnancy rate, although it is uncertain whether the samples correctly represented the proportion of lactating and resting females. We do not have information on the weaning season of bottlenose dolphins off Japan, which is expected to occur in a season of high food availability.

The mean lengths of the lactation and resting periods were calculated by ignoring the uncertainty about the weaning season (Table 11.11). The mean lactation period was 1.31 years for the Iki sample and 1.73 years for the Pacific sample. Kasuya and Miyazaki (1981, in Japanese) reported that a female calf of 182 cm killed at Katsumoto on Iki Island had only milk in the stomach and no trace of solid food. This calf was one of the individuals analyzed by Kasuya et al. (1997, in Japanese) and estimated at age below 1 year from body length. Cornell et al. (1987) stated that common bottlenose dolphins born in an aquarium started taking solid food at age 3–5 months and the solid food occupied a major portion of the diet at age 9–12 months. They also reported breeding intervals of 21–31 months with an average of 2.3 years for a female kept together with a male and their offspring. This probably represents a case close to the optimal environment. The estimated calving interval of common bottlenose dolphins off Japan, 2.45–3.00 years, is slightly longer than this but will reflect a difference in the nutritional environment.

Bottlenose dolphin calves do not necessarily separate from their mothers at completion of weaning, but we do not have good data to estimate the length of the mother–calf bond for the Japanese population in the wild. The age composition of the species taken by Japanese drive fisheries (Figure 11.5) has a trough at age 1.5 years (both sexes in the Iki sample and females in the Pacific sample) or 2.5 years (Pacific males). This trough is most probably created by the move of weaned calves to schools of immature individuals and suggests that departure of calves from the mother's school starts within less than 6 months from the end of suckling, which lasts 1.31–1.73 years on average, or before the next estrus of their mothers.

One of the resident communities of about 100 common bottlenose dolphins along the Florida coast was monitored for over 25 years. Females had calving intervals of 2–10 years with an average of 5 years (Scott et al. 1996). The mother–calf bond was gradually loosened after 4–5 years, but some bonds lasted for 7–8 years (Wells et al. 1987). Japanese populations of the same species seem to have a much shorter mother–calf association and a shorter female reproductive cycle. The difference is a reflection of habitat difference, that is, an open offshore habitat versus a closed inshore one. The life history of the Florida community has achieved a balance with a low reproductive rate and low mortality rate in inshore water with fewer predators, while the Japanese populations have taken a strategy of high reproductive rate to match high predation and probably hunting pressure.

The Indo-Pacific bottlenose dolphin inhabits coastal waters from Africa to central Japan. A population of the species in an estuarine area near Adelaide is one of the well-studied populations, containing 74 individuals (Steiner and Bossley 2008). Juvenile survival rate to weaning was only 54%, and 9 females were observed to have weaned their calves successfully. Their calving interval was 3 years (5 cases), 4 years (2 cases), 5 years (1 case), and 6 years (1 case). The average of the 9 intervals was 3.8 years. In addition to these 9 cases, one female underwent estrus while nursing her calf and gave birth to the next calf when the elder calf was 1.9 years old. The younger calf always accompanied the mother, but the older calf accompanied its mother with lower frequency after the birth. This was probably a case of exceptionally early weaning due to early estrus of the mother and subsequent conception. Weaned calves of this community usually accompany their mothers until the birth of the next calf, that is, for 2.8 years on average (pregnancy is assumed at 1 year). There were five cases where calves died before weaning, which caused early estrus, and the subsequent calving interval was only 1.7 years on average, which was strongly influenced by the timing of calf death. The total average of the 15 observed calving intervals was 2.9 years, which was close to the average calving interval in Japanese populations of common bottlenose dolphins. This suggests that the calving interval of Japanese common bottlenose dolphins has broad individual variation, although we do not have a way to confirm this.

Another study of the Indo-Pacific bottlenose dolphin was reported by Kogi et al. (2004), who worked with a group inhabiting waters near Mikura Island (33°5′N, 139°36′E), about 200 km south of Tokyo. Some of the members of this group were known to have visited or emigrated to the coast of Kamogawa and Tateyama (Section 11.3) near Tokyo or Toshima Island (35°44′N, 139°47′E), about 130 km south of Tokyo. There were 19 cases of successful weaning followed by conception, which resulted in calving intervals of 3–5 years with an average of 3.5 years. The total 26 known calving intervals, including 7 cases of death of calves before weaning, resulted in calving intervals of 1–6 years with an average of 3.4 years. Using the data in Kogi et al. (2004), I calculated a mean calving interval of 3.1 years for the seven cases where calves died before weaning. Thus, calving intervals were slightly shorter if the calves died before weaning. These seven cases included two cases of very short calving intervals of 1 and 2 years, due to the death of neonates at an early stage followed by conception within the same season or that of the next year. Kogi et al. (2004) recorded two adult females that did not breed in their study period of 8 years, inclusion of which could increase to some degree the mean calving interval calculated earlier. Neonates of Mikura Island always accompanied their mother when they were young but started to be separated at age 3–6 years with an average of 3.5 years.

The age when the mother–calf bond dissolved in the Mikura community was similar to the mean calving interval. This suggests that the end of the mother–calf bond and the next parturition are related and that, as suggested by Connor et al. (2000), the separation can be initiated by the near-term mother. Connor et al. (2000) stated that calves often tended to live apart from their pregnant mothers and that females in near-term pregnancy intentionally avoided their calves. The time between the end of suckling and the end of the mother–calf bond is still unknown. Weaning is a slow process of switching nutrition from milk to solid food, so it is difficult to confirm the end of suckling in the wild.

Although the difference was statistically insignificant, the data of Kasuya et al. (1997, in Japanese) suggested that common bottlenose dolphins in the Iki Island area had a shorter calving interval than those in the Pacific area, which could have been due to an about a 5-month shorter lactation period and a 1.5-month shorter resting period (see Table 11.11). Such differences as well as a growth difference in the juvenile stage between the two ocean areas (Section 11.4.6) were likely sensitive to the nutritional environment; it was possible that common bottlenose dolphins in the Iki Island area lived in a better nutritional environment than those in the Pacific area (Section 11.5.1).

11.4.8.5 Ovulation Rate

Belugas develop Graafian follicles during pregnancy. They do not disappear with ovulation but develop during pregnancy into numerous accessory corpora lutea, which are indistinguishable from a corpus luteum of ovulation or of pregnancy (Brodie 1972). Such a case is known among cetaceans only in the beluga, although it is not rare among mammals. The corpus luteum of ovulation degenerates into a corpus albicans and is believed to persist for life in a form indistinguishable from the degenerated corpus luteum of pregnancy (Takahashi et al. 2006). Cetacean biologists formerly believed that the total number of corpora could be used as an indicator of past ovulations. However, a recent study on an Indo-Pacific bottlenose dolphin that was monitored for the history of estrus and pregnancy in an aquarium suggested that only the corpus luteum of pregnancy would remain in the ovary as a corpus albicans (Brook et al. 2002). This important suggestion needs further confirmation (Section 10.5.6).

Keeping in mind this question, I will introduce the analysis of Kasuya et al. (1997, in Japanese). Figure 11.14 presents the size distribution of corpora lutea and albicantia with female age, where two or more age groups are combined into one. For example, females aged 5.5 and 6.5 years are represented at age 5 years, which actually included animals at ages between 5 and 7 (Section 11.4.1). The diameter was calculated as a geometric mean of three dimensions. The geometric mean is better suited for analysis than the arithmetic mean as an indication of volume of the compressed flat corpus albicans. Observing that the diameter of degenerated corpora albicantia peaked at around 5 mm with downward tailing to 2 mm and that the peak height increased with increasing age, Kasuya et al. (1997, in Japanese) concluded that the corpus albicans decreased in size with time to a diameter of around 5 mm on average; that is, the corpus albicans lasted in the ovaries for life. However, for their conclusion to be more convincing, they should have confirmed that the rate of age-related increase of corpora was explained by the annual ovulation rate. This may not be an easy task because the age composition of the female sample should also be taken into account.

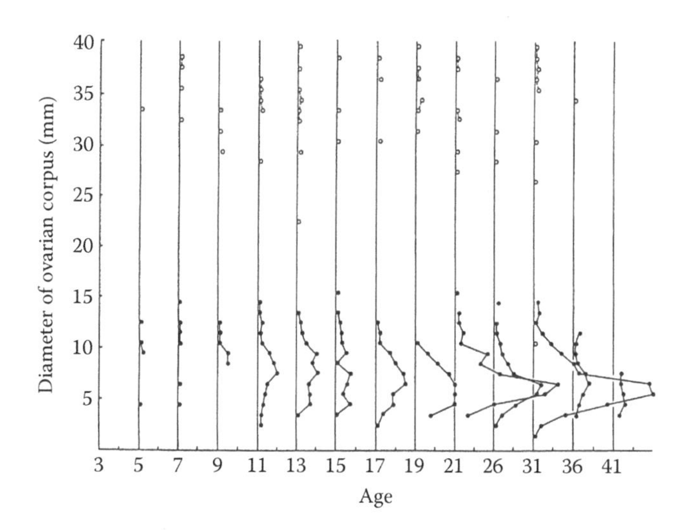

FIGURE 11.14 Distribution of diameters of corpora lutea (open circles) and albicantia (closed circles) on the age of common bottlenose dolphins in the Iki Island area. Individuals of the same age combined. (Reproduced from Fig. 13 of Kasuya, T. et al., *IBI Rep.* (Kamogawa, Japan), 7, 71, 1997, in Japanese.)

Their analyses could not reject the possibility that some small corpora albicantia were unidentifiable from surrounding tissues with the naked eye.

The average annual rate of accumulation of corpora in the ovaries was calculated by regression of the total number (y) of corpora on age (x, years), at 0.458 (SE = 0.052) for the Iki sample and 0.435/year (SE = 0.033) for the Pacific sample (Figure 11.15). The two figures were not significantly different statistically. The same method applied to each school separately gave a range of 0.44–0.55 (mean 0.48) for the Iki sample and 0.22–0.59 (mean 0.46) for the Pacific sample, and again there were no significant differences among schools (Kasuya et al. 1997, in Japanese). The regression equations were calculated using individuals aged over 10 years so that all were sexually mature. Inclusion of younger individuals causes a bias; immature individuals must be excluded from the calculation or dealt with as having zero corpora. The time difference from 1 corpus to 2 corpora is not the same with that from no corpus to 1 corpus.

The average annual ovulation rates thus calculated represent the true figures only if both age at sexual maturity and annual ovulation rate remained stable during the period covered by the sample, or from around 1940 to 1983. If during the 40-year period there was a decline in age at the attainment of sexual maturity or an increase in annual ovulation rate, the calculation will bias the recent ovulation rate downward.

Kasuya et al. (1997, in Japanese) compared the average annual ovulation rates estimated for the past 40 years with annual pregnancy rates during the sampling period, that is, the late 1970s to the early 1980s. The average number of ovulations per pregnancy is obtained by dividing the annual ovulation rate by the annual pregnancy rate:

Iki sample: 0.458/0.408 = 1.12
Pacific sample: 0.435/0.333 = 1.31

This means that the proportion of ovulations followed by pregnancy was 89.1% in the Iki Island area and 76.5% in the Pacific area. These figures are apparently reasonable and do not suggest rejection of the possibility that both the corpus luteum of pregnancy and that of ovulation remain as corpora albicantia in ovaries of the common bottlenose dolphin. This does not necessarily mean that Kasuya et al. (1997, in Japanese) identified all the ovulations as corpora.

11.5 COMPARISON BETWEEN THE PACIFIC AND THE IKI SAMPLES

11.5.1 Life History Parameters

Table 11.12 compares life history parameters of common bottlenose dolphins between the Iki and the Pacific samples, both collected from the catch of driving operations. The Iki sample of 446 individuals was from 2634 in 7 schools driven into a harbor near Katsumoto, Iki Island, in January–March from the Shichiriga-sone Bank (33°56′N, 129°30′E) between Iki Island and the Tsushima Islands. The sea surface temperature

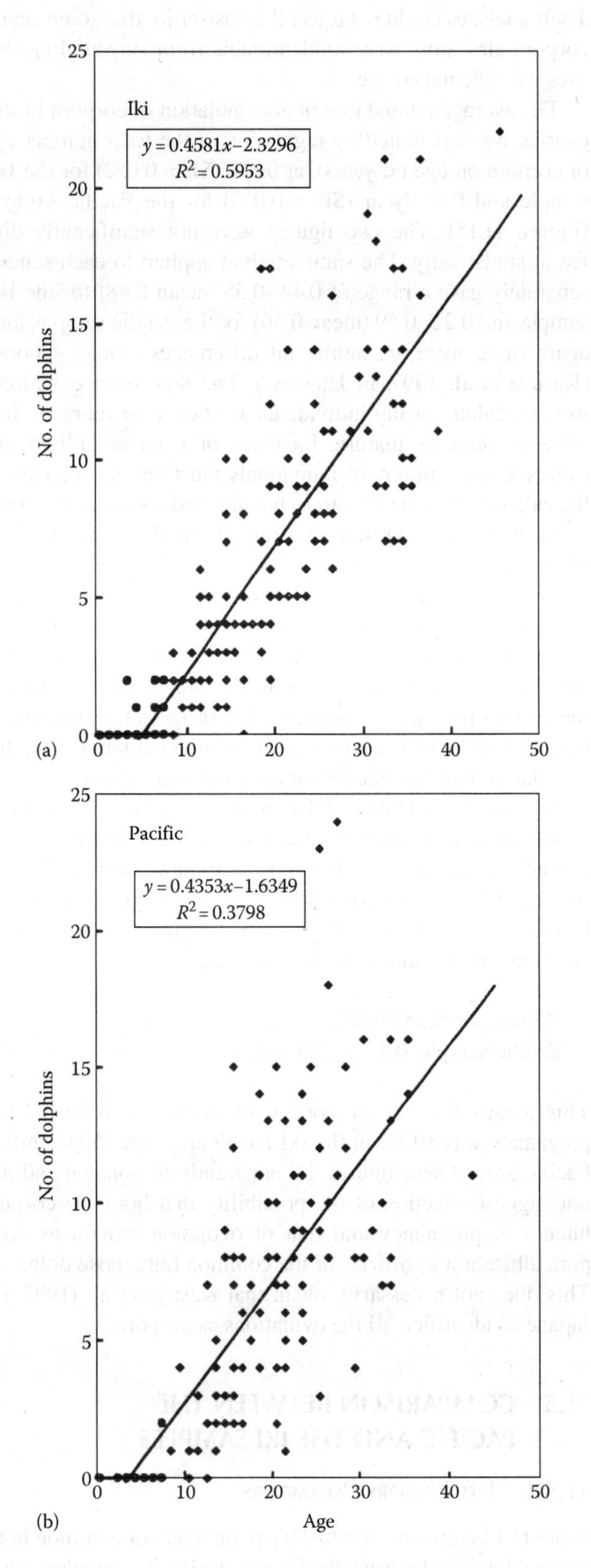

FIGURE 11.15 Scatterplots of ovarian corpora on the age, calculated for the Pacific (a) and Iki samples (b). The linear regressions were calculated using only females over 10 years of age. (Reproduced from Fig. 16 of Kasuya, T. et al., *IBI Rep.* (Kamogawa, Japan), 7, 71, 1997.), in Japanese.)

there in the season was 10°C–15°C, which was close to the lower bound of the temperature of waters inhabited by this species (Table 10.1); Dall's porpoises occurred about 150 km northeast of this area (Section 9.3.1 as illustrated in Fig. 1 of Miyashita and Kasuya 1988). Thus, the Iki sample was collected in waters close to the northern limit of the range of the species in the season in the East China Sea and the Sea of Japan (Figure 11.3).

The Pacific sample of 281 individuals was from 540 in 9 schools driven in Arari on the Izu coast and Taiji in Wakayama Prefecture. Taiji is located about 300 km southwest of Arari. The Arari sample (37) was taken in August, but the other 244 were obtained from 8 schools driven in Taiji in January–April. The northern limit of the range of this species off the Pacific coast of Japan extends in summer to latitudes 42°N–43°N off southern Hokkaido and retreats in winter to around latitude 35°N, which is also close to the southern limit of Dall's porpoises in winter. It is possible but has not been confirmed that the Arari and Taiji samples were from a single population.

The distribution pattern of the common bottlenose dolphin around Japan suggests that there is little opportunity for the schools inhabiting the Arari/Taiji area to meet with schools in the Iki Island area in winter, but it is possible that they meet in summer in the Tsugaru Strait area (41°30′N, 140°30′E), which is situated between Hokkaido and Honshu and connects the Sea of Japan and the Pacific. Thus, sample locations alone cannot be evidence that the Iki and Pacific samples belong to separate populations. However, some differences in life history parameters (Table 11.12) suggest that these samples represent different populations.

I classified life history parameters listed in Table 11.12 into the following three categories based on expected response to environmental variation, such as in nutrition:

1. Parameters that are likely to show a quick response to environmental change
2. Parameters in which response will occur quickly at the individual level but will take time before being identified at the population level
3. Parameters that are less affected by environmental change or those for which we do not know the direction of likely change

The first group of parameters includes annual pregnancy rate and growth in the juvenile stage.

The second group includes age at the attainment of sexual maturity, body length at the attainment of sexual maturity, and asymptotic body length. For example, nutritional improvement will quickly result in the improvement of body size of juveniles, but it will take nearly 10 years before its effect is identified as a change in body length at the attainment of sexual maturity.

The third group includes breeding season, age at physical maturity, age at cessation of testicular growth, testicular weight in adults, and annual ovulation rate. Improvement of nutrition of adult females will increase the proportion of estrus followed by conception, will shorten the resting period,

TABLE 11.12
Life History Parameters of Common Bottlenose Dolphins Compared between the Pacific and Iki Samples

Parameters	Description	Pacific	Difference	Iki
1. Parturition season	Range	Mar.–Oct.	Not detected	Feb.–Oct.
	Peak	Jun	Not detected	Jun
2. Age at physical maturity	Male	15–20 years	Not detected	15–20 years
	Female	10–15 years	Not detected	10–15 years
3. Asymptotic body length	Male	305.3 cm	Not detected	305.2 cm
	Female	288.0 cm	<	293.7 cm
4. Testicular growth	Male	About 20 years	Not detected	About 20 years
5. Testicular weight at maturity	Male[a]	40–100 g	Not detected	80–100 g
6. Testicular weight, range	>20 years	100–800 g	No changes made	300–900 g
Average	>20 years	200–500 g	No changes made	400–800 g
7. Body length at sex. maturity, range	Male[a]	270–309 cm	Not detected	280–319 cm
Mean	Male	290-299 cm	Not detected	300 cm
Range	Female	270–289 cm	<	260–299 cm
Mean	Female	280 cm	<	290 cm
8. Age at sexual maturity, range	Male[a]	Not available	Not detected	8–11 years
Mean	Male	11 years	>	9.5 years
Range	Female	6–13 years	No changes made	5–9 years
Mean	Female	9.2 years	No changes made	6.9 years
9. Body length of juveniles <7 years	Both sexes	Smaller	<	Larger
10. Annual pregnancy rate	Female	33.3%	<	40.8%
11. Annual ovulation rate	Female	0.435	Not detected	0.458

Source: Constructed from Table 13 and text in Kasuya, T. et al., *IBI Rep.* (Kamogawa, Japan), 7, 71, 1997, in Japanese.

[a] "Late maturing" and "mature" males are dealt with as mature for the purposes of this particular comparison.

and will reduce neonatal mortality, but it is unknown how the total of these responses would change annual ovulation rate.

Among the seven characters with difference between the two samples, five characters belonged to the first and the second groups, and all the differences observed were in the direction expected to occur if the Iki population had a more favorable nutritional environment. No geographical difference was identified in asymptotic body length. This suggests that some environmental change toward the improvement of nutrition of the common bottlenose dolphin in the Iki Island area occurred recently or during the 1970s to 1980s.

11.5.2 Background of the Change

11.5.2.1 Food of Common Bottlenose Dolphins off Japan

First, the nutritional requirement of common bottlenose dolphins is reviewed here. Tobayama and Shimizu (1973, in Japanese) kept 11 common bottlenose dolphins in an outdoor tank and examined per capita food consumption and change in body weight. The daily food requirement to maintain body weight varied from 6.08% of body weight for a young animal measuring 236 cm (n = 1) to 3.54% for larger animals measuring 290–310 cm (n = 5). For the latter five individuals, the daily food requirement decreased from about 4.5% to 3%, while the water temperature increased from 11.5°C to 24.3°C. The lower bound of the temperature range of the experiment was the lowest sea surface temperature recorded during sightings of this species in the wild in the western North Pacific (Table 10.1).

Takemura (1986b, in Japanese) reported the stomach contents of 56 common bottlenose dolphins, most of which were collected from those driven at Iki Island for culling purposes in February 1983–1985 or after the collecting activities of Kasuya et al. (1997) during 1979–1980 (Section 3.4). He removed food remains in the first and the second stomachs using a 32-mesh sieve. Many of the examined stomachs were almost empty, and most of the food remains were limited to fish otoliths and squid beaks, which will reflect time at the start of drive operations as well as the length of time between drive and slaughter. Thus, species composition could be determined but not the weight of the prey ingested.

Out of the 56 common bottlenose dolphins examined by Takemura (1986b, in Japanese), 15 had fish remains, 24 had squid remains, and 1 had shrimp remains (Table 11.13). The major food items consumed by common bottlenose dolphins in the Iki Island area were fish and squid. This was the same

TABLE 11.13
Food Habits of Common Bottlenose Dolphins in the Iki Island Area Indicated by the Number of Individuals Found with Particular Food Items

Dolphin Species	Fish	Cephalopods	Shrimp	No. of Dolphins Examined
Common bottlenose dolphin	15 (32)	24 (7)	1 (1)	56
False killer whale	23 (1)	26 (3)		32
Pacific white-sided dolphin	28 (3)	46 (3)		63
Risso's dolphin		2 (3)		3

Source: Extracted from Table VI-.2 of Takemura, A., Pacific white-sided dolphin and common bottlenose dolphin, pp. 161–177, in Tamura, T., Ohsumi, S., and Arai R. (eds.), *Report of Contract Studies in Search of Countermeasures for Fishery Hazard (Culling of Harmful Organisms) (56th to 60th year of Showa)*, Suisan-cho Gogyo Kogai (Yugaiseibutsu Kujo) Taisaku Chosa Kento Iinkai (Fishery Agency), Tokyo, Japan, 285pp, 1986a, in Japanese.

Note: In parentheses are the number of food species detected in the sample.

among three species of dolphins taken in the area: common bottlenose dolphins, false killer whales, and Pacific white-sided dolphins. The Risso's dolphins taken in the area were found with only squid in the stomachs. The cephalopod species listed in Table 11.13 were squids except for one octopus found in the stomach of a false killer whale. The common bottlenose dolphins had preyed on numerous species of fish and squid. The importance of fish and squid in the diet of common bottlenose dolphins is also known for the same species in coastal waters from Florida to Texas, but there was some geographical variation in the proportions of the two taxa (Barros and Odell 1990).

The common bottlenose dolphins consumed the highest variety of food items among the four species of dolphins culled in the Iki Island area (Table 11.13). This result may partly be exaggerated by the fact that materials examined by Takemura (1986a, in Japanese) included stomachs from a small number of specimens incidentally killed in a seine net fishery, but it is true that the species lives on a variety of food items. The jaws and teeth of bottlenose dolphins are intermediate between the extremely large and strong teeth of killer whales and false killer whales, and the delicate beak of *Stenella* spp. appears to be adapted for feeding on a broad spectrum of food species.

Takemura (1986b, in Japanese) reported total lengths of fish consumed by dolphins in the Iki Island area: 10–100 cm for bottlenose dolphins and 10–40 cm for Pacific white-sided dolphins. Although he did not find yellowtail in his sample, Kasuya (1985) reported that one Pacific white-sided dolphin and four false killer whales had remains of yellowtail in the stomach. The estimated total length of the yellowtail was 37 cm for the Pacific white-sided dolphin and 60–87 cm for the false killer whale.

TABLE 11.14
Geographical Difference in Food Habits of Common Bottlenose Dolphins Indicated by Proportion of the Number of Food Items Identified in the Sampled Stomachs (Individual Variation within the Sample Was Ignored)

Geography and Season of Sample	Pelagic Fish	Bottom Fish	Squid
Offshore, Dec.–Feb.	55.6%	27.7%	16.7%
Offshore, October	14.3%	57.1%	28.6%
Coastal, Jan.–May	0	71.4%	18.6%

Source: *Comprehensive Report of Basic Investigations in Search of Counteraction for Fishery Damages Caused by Small Cetaceans in Western Japan (42nd and 43rd Year of Showa)*, Research and Investigation Department of Fisheries Agency, Suisan-cho Chosa-kenkyu-bu, Tokyo, Japan, 108pp, 1969, in Japanese.

Geographical variation in food habits of the common bottlenose dolphin was reported by *Suisan-cho Chosa-kenkyu-bu* (Investigation and Research Department of Fisheries Agency) (1969, in Japanese), based on the results of the studies by scientists of the Saikai Regional Fisheries Research Laboratory (SRFRL) conducted in response to the first incident in the dolphin-fishery conflict in the Iki Island area that started around 1966. The second phase of the conflict started in 1978; the catches were investigated by my team in 1979–1980 (Kasuya 1985; Kasuya et al. 1997, in Japanese) and by a Fisheries Agency Team in 1981–1985 (Tamura et al. 1986, in Japanese; Takemura 1986a,b in Japanese) (Section 3.4). Scientists of the SRFRL collected the stomach contents of common bottlenose dolphins in the East China Sea and northern Kyushu area; grouped the contents into three major groups of squid, pelagic fish, and bottom fish; and examined geographical and seasonal variation in the composition (Table 11.14). Their findings were that the dolphins had consumed mainly bottom fish in coastal waters and pelagic and bottom fish in offshore waters. Squid made up 16%–28% of the total food items consumed.

All the available information on stomach contents of common bottlenose dolphins indicates a broad spectrum of prey and their great adaptability in feeding. Thus, a list of food items obtained in a particular season or in a particular area may not apply to a different season or area. It would also be difficult to evaluate the nutritional environment based on information on the availability of some limited food items in the ocean.

11.5.2.2 Effect of Dolphin Fisheries

If the size of a population of common bottlenose dolphins decreases due to hunting, per capita food availability will increase and the nutritional environment will be improved. I have suggested the possibility that overhunting of striped dolphins off the Izu coast resulted in the improvement of their growth and reproduction (Sections 10.5.7 and 10.5.12). The broad prey spectrum of the common bottlenose dolphin and possible overlap in food items with other dolphin species suggest that depletion of dolphin species other than common bottlenose dolphins would also help improve their nutritional environment.

The history of exploitation of bottlenose dolphins in Japan was reviewed by Kasuya (1996, in Japanese) based on then published statistics, including those in Kasuya (1985) and Kishiro and Kasuya (1993). This is revisited where using some additional statistical data. Common bottlenose dolphins have been hunted at least since the 1960s at Nago (26°35′N, 127°59′E) in Okinawa, Goto Islands (33°N, 129°E) in the East China Sea, Tsushima Island (34°30′N, 129°20′E) and Iki Island in Tsushima Strait, and Taiji and Izu on the Pacific coast of central Japan.

The community of Nago operated an opportunistic dolphin drive historically, but it was replaced in 1975 by a crossbow fishery (Section 2.8). The statistics for 20 years from 1960 to 1994 (except for 1971 and 1973) are available in Kasuya (1996, in Japanese). Additional statistics for 1960–1975 are available in Miyazaki (1980b, in Japanese) and those for 1960–1982 in Uchida (1985, in Japanese). Some disagreement exists among these sets of statistics, but it is negligible. These statistics are compiled in Table 2.13 for 1993 and later seasons when the fishery was operated with a catch quota and in Table 3.22 for years before that date. Before 1981, when the catch of bottlenose dolphins was sporadic, Nago recorded a total maximum estimated take of 209 common bottlenose dolphins in 21 years, or an annual mean of 10. This level of take was unlikely to have had a significant effect on the sample of Kasuya et al. (1997, in Japanese) collected in the far north, that is, the Pacific waters off Taiji and Izu or waters around Iki Island. Since 1981, when take of the species became a regularly activity, Nago recorded the annual takes of 0–77 common bottlenose dolphins with a total of 224 individuals in the 12 years 1981–1992. The annual average was only 19. The crossbow fishery was placed under a quota system in 1993: initially 10 bottlenose dolphins plus takes of several other delphinid species (Table 2.13)

Nagasaki Prefecture, including Goto, Tsushima, and Iki, had a long history of dolphin fisheries (Sections 3.3 through 3.6). Referring to the catch statistics of bottlenose dolphins in Nagasaki Prefecture since 1965, Kasuya et al. (1997, in Japanese) stated an impression that the catch could have been large enough to result in the local depletion of the species. The records of dolphin drives for 1944–1966 in Nagasaki Prefecture are available in Table 3.7. The table lists a total take of 9977 dolphins in 33 drives of 4 species during the 22-year period, which excludes one drive in 1963 for which number and species of dolphins were not reported. Twenty-seven drives (82%) of the 33 drives were in the 9 years 1958–1966, suggesting the possibility that statistics for the earlier 14 years 1944–1957 were incomplete. Out of the 33 drives, 10 were made on bottlenose dolphins, 2 on the mixtures of bottlenose dolphins and false killer whales (90 in total), and 1 on a mixture of bottlenose dolphins and Risso's dolphins (380 in total). By assuming that half of the animals in the mixed drives were common bottlenose dolphins, I estimate that 3498 dolphins, or 56.2% of those of known species, were bottlenose dolphins (13 drives). By assuming the same proportion for 3738 small cetaceans of unknown species (20 drives), a total take of 5599 bottlenose dolphins, or an annual mean of 243, is estimated for the 23 years from 1944 to 1966. This is not corrected for the possible large catches during 1944–1957, which includes the peri-World War II period (Sections 2.2, 2.3 and 3.8.3).

The statistics for subsequent years and prior to 1979 when I started sampling (Kasuya 1985; Kasuya et al. 1997, in Japanese) are available in Table 3.4, which included 1116 bottlenose dolphins as the total take of the species in Nagasaki Prefecture during the 12 years from 1967 to 1978. This figure plus an estimated additional 525 gives a figure of 1641 bottlenose dolphins, or an annual mean of 137 taken during the 12 years in Nagasaki Prefecture. The additional 525 was estimated by applying the proportion 56.2% obtained earlier to 934 dolphins recorded in Table 3.4 as *nezumi-iruka*, which is currently unidentifiable to species.

Thus, the mean annual catch of bottlenose dolphins in Nagasaki Prefecture in 1944–1966 could have been about 250, which is likely an underestimate due to incomplete statistics, and about 140 during 1967–1978. These figures are rough estimates of the fishing pressure on the species in the Tsushima Strait area before the drives for culling purposes that started in 1972 and expanded in 1977 with a large-scale drive operation under the leadership of the Iki fishermen and with financial support by the Fisheries Agency of Japan and the prefectural government (Section 3.4).

The abundance of bottlenose dolphins was estimated at 35,000 in winter for waters surrounded by 126°E–131°E and 25°N–35°N, which includes the eastern half of the East China Sea, the Tsushima Strait area, and the Pacific waters around the Okinawa Islands (Table 11.15; Miyashita 1986b, in Japanese). This estimate needs to be treated with caution before use for management purpose for two reasons: (1) a large coefficient of variation and (2) inclusion of the Pacific waters around the Okinawa Islands. It is questionable whether the dolphins that winter in the Okinawa Islands area belong to the same population that winters in the Iki Island area in Tsushima Strait further north. They might instead belong to another more southern population. Even if the possibility is accepted that the Okinawa dolphin might migrate in summer to the Tsushima Strait area or to the coast of Taiji or the Izu Peninsula, uncertainty remains whether they mix with dolphins wintering in these northern areas.

If the abundance of bottlenose dolphins is assumed to be proportional to size of the area of distribution, which is certainly incorrect, and if those in the Okinawa Islands area are excluded, the abundance of the species in the eastern East China and the Tsushima Strait area is nearly two-thirds of the central estimate obtained earlier, that is, 23,000. This figure and the broad confidence interval suggest the possibility that the annual take has exceeded 1% of the population and suppressed the abundance to some degree. Exploitation of other dolphin species could also have made some contribution to

TABLE 11.15
Abundance of Common Bottlenose Dolphins around Japan Estimated by Shipboard Sighting Surveys Assuming g(0) = 1

Area	Abundance (×10³)	95% Confidence Interval (×10³)	Season	Year
Okinawa–East China Sea–N. Kyushu	35	CV = 55%	Winter	1982–1985
Western North Pacific,				
North of 30°N, west of 145°E	37	22–60	Summer	1983–1991
North of 30°N, 145°E–180°E	100	52–192	Summer	1983–1991
23°N–30°N, 127°E–180°E	32	7–14	Summer	1983–1991

Sources: Table VI-2.3 of Miyashita, T., Abundance estimate (2): Using research vessels, pp. 202–213, in Tamura, T., Ohsumi, S., and Arai, R. (eds.), *Report of Contract Studies in Search of Countermeasures for Fishery Hazard (Culling of Harmful Organisms) (56th to 60th year of Showa)*, Suisan-cho Gogyo Kogai (Yugaiseibutsu Kujo) Taisaku Chosa Kento Iinkai (Fishery Agency), Tokyo, Japan, 1986b, 285pp and Table 15 of Miyashita, T., *Rep. Int. Whal. Commn.*, 43, 417, 1993.

improving the nutritional environment of bottlenose dolphins in the Iki Island area.

Bottlenose dolphins along the Pacific coast of Japan were hunted by fisheries along the Izu coasts and off Taiji. Statistics for Taiji in Wakayama Prefecture are given in Table 3.17 for 1963–1994 and those for the entire Wakayama Prefecture in Table 3.18. These two sets of statistics are from different sources and show some contradiction. Table 3.17, which I believe to be the more reliable, shows a total catch of 666 bottlenose dolphins (annual range of 3–103 and average 39) during the 17 years from 1963 to 1979, when the annual catch of short-finned pilot whales was 52–479 (average 166) and that of striped dolphins 331–2397 (average 916).

Because bottlenose dolphins were not preferred by the local people, the catch of the species was at a relatively low level in Taiji before 1980, which was the only location of significant dolphin hunting in Wakayama Prefecture. The current drive fishery in Taiji started in 1969 for short-finned pilot whales and expanded to striped dolphins in 1973 (Section 3.5). It was in 1980 that the fishermen found a new market for bottlenose dolphins for use as animal food in zoos and increased the catch of the species. With the increase of supply of bottlenose dolphins, sales for human consumption in the local market apparently increased (Kasuya et al. 1997, in Japanese). For these reasons, the catch of bottlenose dolphins increased beginning in 1980 and reached a peak of 1670 in 1987. Then the catch, reaching the quota that started in 1993 only in the year 2000, declined to less than 200 in recent years, which suggests decline in availability of the species to the fishery (Table 3.19 and 6.3).

The major target of the dolphin-drive fisheries on the Izu coast in Shizuoka Prefecture was striped dolphins, and the catch statistics were quite incomplete until 1971. Only Arari left catch statistics from January 1942 to April 1957, which were published in Kasuya (1976, in Japanese) and Kasuya (1996, in Japanese) and also cited in Table 3.14. During the period of about 14 years, catch figures are available by species for only about 8 years from 1950 to 1957, when a total of 44,672 dolphins of all species were caught, including 300 bottlenose dolphins (annual range 0–103, mean 43) (Table 3.14).

The statistics for the catch of dolphins in the entire Shizuoka Prefecture, which was almost the same as the catch on the Izu coast, are complete only for years since 1972, when the Fisheries Agency started collecting statistics (Table 3.16). During the 9 years from 1972 to 1980, Shizuoka Prefecture recorded a take of 274 bottlenose dolphins (annual range 0–120, mean 30).

Thus, the average annual catch of bottlenose dolphins along the Pacific coast was less than 50 in Shizuoka (Tables 3.14 and 3.16) or Wakayama Prefectures (Table 3.17) and the total average annual catch did not reach 100 during the years before 1981 when sampling by Kasuya et al. (1997, in Japanese) began. The catch of the species along the Pacific coast underwent a sudden increase to an annual mean of 569 during 1980–1984 with the contribution of the Taiji operation. This catch increase could have had some effect on pregnancy rate of the samples collected in 1981–1983, but the time could have been insufficient for other life history parameters estimated from the sample to be affected.

The abundance of common bottlenose dolphins was estimated at 37,000 individuals in summer (Table 11.15) for the Pacific area north of Yakushima Island (30°23′N, 130°35′E) and west of 145°E, which includes waters within 300–400 km off the Pacific coast of Japan. This range was much broader than the operation range of the drive fisheries, which extended at most to about 40 km from Taiji port (Kishiro and Kasuya 1993), and we do not know the population structure of the species within the broad geographical range used for the abundance estimation. However, it is possible to say that the abundance of common bottlenose dolphins in the Pacific area was of similar magnitude as or slightly greater than that of the same species in the eastern East China Sea and Tsushima Island area. This leaves the possibility that the common bottlenose dolphins in the Iki Island area received greater fishing pressure during 1960–1980 than the population of the same species off the Pacific coast of Japan. The striped dolphin population off the Pacific coast of Japan declined during the 1960s to the 1970s and could have influenced the feeding environment of the common bottlenose dolphins, which remains to be investigated.

In summary, it is fairly sure that some differences existed in the late 1970s and early 1980s in the life history parameters between populations of common bottlenose dolphin in the Iki Island area, the eastern East China Sea, and the Pacific area off Japan. The differences identified were in annual pregnancy rate, age at the attainment of sexual maturity, and body length of immature individuals but not on other life history parameters such as breeding season and body length and age at the attainment of physical maturity. The former group of parameters was more likely to respond to changes in the nutritional environment or in population density than the latter, and the difference was in the direction expected if nutritional environment was better in the Iki Island area. Although this apparently agreed with greater absolute catches of the species by local fisheries in the Iki Island area than those in the Pacific area before the time the samples were collected, full understanding of the background of the life history differences must await further information on the abundance of the species, the geographical range of the exploited populations, and the structure of the ecosystem, including other dolphin species as well as major fishery resources.

REFERENCES

[*In Japanese language]

Asper, E.D., Andrews, B.F., Antrim, J.E., and Young, W.G. 1992. Establishing and maintaining successful breeding programs for whales and dolphins in zoological environment. *IBI Rep.* (Kamogawa, Japan) 3: 71–84.

Barros, N.B. and Odell, D.K. 1990. Food habits of bottlenose dolphins in the southern United States. pp.309–328. In: Leatherwood, S. and Reeves, R.R. (eds.). *The Bottlenose Dolphin.* Academic Press, San Diego, CA. 653pp.

Best, P.B. 2007. *Whales and Dolphins of the Southern African Subregion.* Cambridge University Press, Cambridge, U.K. 338pp.

Brodie, P.F. 1972. Significance of accessory corpora lutea in odontocetes with reference to *Delphinapterus leucas. J. Mammal.* 53: 614–616.

Brook, F.M., Kinoshita, R., and Benirschke, K. 2002. Histology of the ovaries of a bottlenose dolphin, *Tursiops aduncus*, of known reproductive history. *Mar. Mammal Sci.* 18(2): 540–544.

Connor, R.C., Wells, R.S., Mann, J., and Read, A.J. 2000. The bottlenose dolphin: Social relationships in a fission-fusion society. pp. 91–126. In: Mann, J., Connor, R.C., Tyack, P.L., and Whitehead, H. *Cetacean Societies: Field Studies of Dolphins and Whales.* University of Chicago Press, Chicago, IL. 433pp.

Cornell, L.H., Asper, E.D., Antrim, J.E., Searles, S.S., Young, W.G., and Goff, T. 1987. Progress report: Results of a long-range captive breeding program for the bottlenose dolphin, *Tursiops truncatus* and *Tursiops truncatus gilli. Zoo. Biol.* 6:41–53.

DeMaster, D.P. 1984. Review of techniques used to estimate the average age at attainment of sexual maturity in marine mammals. *Rep. Int. Whales Commn.* (Special Issue 6, Reproduction in Whales, Dolphins and Porpoises): 175–179.

Ehrenberg, C.G. 1828, 1833. Mammalia, Decades I and II. Volume 1. In: Hemprich, F.W. and Ehrenberg, C.G. (eds.). *Symbolae Physicae, Pars Zoologica.* Officina Academica, Berlin.

Fernandez, S. 1992. *Composition de edad y sexo y paremetros del ciclo de viva de toninas (Tursiops truncatus) varadas en el noroeste del Golfo de Mexico.* M.S. thesis, Institute of Technology. Estudios Superiores Monterey, Guaymas, Mexico. 109pp. (cited in Urian et al. 1996).

*Fujita, K. 2003. Observation of [Indo-Pacific] bottlenose dolphins inhabiting southern Chiba. *Isana* 38: 85–89.

Gao, A., Zhou, K., and Wang, Y. 1995. Geographical variation in morphology of bottlenose dolphins (*Tursiops sp.*) in Chinese waters. *Aqu. Mammals* 21(2): 121–135.

Hale, P.T., Barreto, A.S., and Ross, G.J.B. 2000. Comparative morphology and distribution of the *aduncus* and *truncatus* forms of bottlenose dolphin *Tursiops* in the Indian and Western Pacific Oceans. *Aqu. Mammals* 26(2): 101–110.

Hersh, S.L. and Duffield, D.A. 1990. Distinction between Northwest Atlantic offshore and coastal bottlenose dolphins based on hemoglobin profile and morphology. pp. 129–139. In: Leatherwood, S. and Reeves, R.R. (eds.). *The Bottlenose Dolphin.* Academic Press, San Diego, CA. 653pp.

Hohn, A.A. 1989. Comparison of methods to estimate the average age at sexual maturation in dolphins. Paper IWC/SC/41/O28 presented to the Scientific Committee Meeting of IWC in 1989. 15pp. (Available from IWC secretariat, Cambridge, U.K.).

Hohn, A.A., Scott, M.D., Wells, R.S., Sweeney, J.C., and Irvine, A.B. 1989. Growth layers in teeth from known-age, free-ranging bottlenose dolphins. *Mar. Mammal Sci.* 5(4): 315–342.

*Kasuya, T. 1976. Stock of striped dolphins, part I and II. *Geiken Tsushin* 295: 19–23 (part I); 296: 29–35 (part II).

*Kasuya, T. 1980. Life history of dolphins and porpoises. *Anima* (Tokyo) 8(9): 13–23.

Kasuya, T. 1985. Fishery-dolphin conflict in the Iki Island area of Japan. pp. 253–272. In: Beddington, J.R., Beverton, R.J.H., and Lavigne, D.M. (eds.). *Marine Mammals and Fisheries.* George Allen and Unwin, London, U.K. 354pp.

*Kasuya, T. 1996. Bottlenose dolphins. pp. 334–339. In: Odate, S. (ed.). *Basic Data of Rare Wild Aquatic Organisms of Japan* (III). Nihon Suisan Shigen Hogo Kyokai, Tokyo, Japan. 582pp.

Kasuya, T. 1999. Review of the biology and reproduction of striped dolphins in Japan. *J. Cetacean Res. Manage.* 1(1): 81–100.

*Kasuya, T., Izumisawa, Y., Komyo, Y., Ishino, Y., and Maejima, Y. 1997. Life history parameters of common bottlenose dolphins off Japan. *IBI Rep.* (Kamogawa, Japan) 7: 71–107.

Kasuya, T. and Marsh, H. 1984. Life history and reproductive biology of the short-finned pilot whale, *Globicephala macrorhynchus*, off the Pacific coast of Japan. *Rep. Int. Whales Commn.* (Special Issue 6, Reproduction in Whales, Dolphins and Porpoises): 259–310.

*Kasuya, T. and Miyazaki, N. 1981. Dolphins and fishery dolphin conflict in Iki Island area: Preliminary report of three year study. *Geiken Tsushin* 340: 25–36.

Kasuya, T., Miyazaki, N., and Dawbin, W.H. 1974. Growth and reproduction of *Stenella attenuata* in the Pacific coast of Japan. *Sci. Rep. Whales Res. Inst.* (Tokyo) 26: 157–226.

Kasuya, T., Tobayama, T., Saiga, T., and Kataoka, T. 1986. Perinatal growth of delphinoids: Information from aquarium reared bottlenose dolphins and finless porpoises. *Sci. Rep. Whales Res. Inst.* (Tokyo) 37: 85–97.

*Kasuya, T. and Yamada, T. 1995. *List of Japanese Cetaceans.* Geiken Series 7. Nihon Geirui Kenkyusho (Institute of Cetacean Research), Tokyo, Japan. 90pp.

*Katsumata, E., Tobayama, T., Yoshioka, M., and Aida, K. 1994. Seasonal change in blood testosterone levels in a male common bottlenose dolphin in captivity. *Dobutsuen Suizokukan Zassi* 35(3): 73–78.

Kemper, C.M. 2004. Osteological variation and taxonomic affinities of bottlenose dolphins, *Tursiops spp.*, from south Australia. *Aust. J. Zool.* 52: 29–48.

Kishiro, T. and Kasuya, T. 1993. Review of Japanese dolphin drive fisheries and their status. *Rep. Int. Whales Commn.* 43: 439–542.

Kogi, K., Hishii, T., Imamura, A., Iwatani, T., and Dudzinski, K.M. 2004. Demographic parameters of Indo-Pacific bottlenose dolphins (*Tursiops aduncus*) around Mikura Island, Japan. *Marine Mammal Sci.* 20(3): 510–526.

LeDuc, R.G., Perrin, W.F., and Dizon, A.E. 1999. Phylogenetic relationship among the delphinid cetaceans based on full cytochrome *b* sequences. *Marine Mammal Sci.* 15: 619–648.

Lockyer, C. and Morris, R. 1987. Observed growth rate in a wild juvenile *Tursiops truncatus*. *Aqu. Mammals* 13(1): 27–30.

*Miyashita, T. 1986a. Sightings from research vessels. pp. 78–87. In: Tamura, T., Ohsumi, S., and Arai, R. (eds.). *Report of Contract Studies in Search of Countermeasures for Fishery Hazard (Culling of Harmful Organisms) (56th to 60th year of Showa)*. Suisan-cho Gogyo Kogai (Yugaiseibutsu Kujo) Taisaku Chosa Kento Iinkai (Fishery Agency), Tokyo, Japan. 285pp.

*Miyashita, T. 1986b. Abundance estimate (2): Using research vessels. pp. 202–213. In: Tamura, T., Ohsumi, S., and Arai, R. (eds.). *Report of Contract Studies in Search of Countermeasures for Fishery Hazard (Culling of Harmful Organisms) (56th to 60th year of Showa)*. Suisan-cho Gogyo Kogai (Yugaiseibutsu Kujo) Taisaku Chosa Kento Iinkai (Fishery Agency), Tokyo, Japan. 285pp.

Miyashita, T. 1993. Abundance of dolphin stocks in the western North Pacific taken by the Japanese dolphin fishery. *Rep. Int. Whal. Commn.* 43: 417–437.

Miyashita, T. and Kasuya, T. 1988. Distribution and abundance of Dall's porpoise off Japan. *Sci. Rep. Whales Res. Inst.* (Tokyo) 39:121–150.

Miyazaki, N. 1980a. Catch records of cetaceans off the coast of the Kii Peninsula. *Mem. Natl. Sci. Mus.* (Tokyo) 13: 69–82.

*Miyazaki, N. 1980b. Dugongs and whales. pp.157–166. In: Kisaki, K. (ed.). *Natural History of the Ryukyu Islands*. Tsukiji Shokan, Tokyo, Japan. 282p.

Miyazaki, N. 1984. Further analyses of reproduction in the striped dolphin, *Stenella coeruleoalba*, off the Pacific coast of Japan. *Rep. Int. Whales Commn.* (Special Issue 6, Reproduction in Whales, Dolphins and Porpoises): 343–353.

Miyazaki, N. and Nakayama, K. 1989. Records of cetaceans in the waters of the Amami Islands. *Mem. Natl. Sci. Mus.* (Tokyo) 22: 235–249.

Miyazaki, N. and Nishiwaki, M. 1978. School structure of the striped dolphin off the Pacific coast of Japan. *Sci. Rep. Whales Res. Inst.* (Tokyo) 30: 65–116.

Moller, L.M. and Beheregaray, L.B. 2001. Coastal bottlenose dolphins from southeastern Australia are *Tursiops aduncus* according to sequences of the mitochondrial DNA control region. *Mar. Mammal Sci.* 17(2): 249–263.

Myrick, A.C. Jr. and Cornell, L.H. 1990. Calibrating dental layers in captive bottlenose dolphins from serial tetracycline labels and tooth extractions. pp. 587–608. In: Leatherwood, S. and Reeves, R.R. (eds.). *The Bottlenose Dolphin*. Academic Press, London, U.K. 653pp.

Natoli, A., Peddemors, V.M., and Hoelzel, A.R. 2004. Population structure and speciation in the genus *Tursiops* based on microsatellites and mitochondrial DNA analysis. *J. Evolution. Biol.* 17: 363–375.

*Ogasawara Whale Watching Association: Dolphin Study Group. 2003. *Catalogue of Indo-Pacific Bottlenose Dolphin*. Ogasawara Whale Watching Association, Ogasawara, Japan. 43pp.

*Ogawa, T. 1936. Studies on Japanese toothed whales, part 2 and 3. *Shokubutsu-to Dobutsu [Plants and Animals]* 4(8): 15–22 (Part 1); 4(9): 1–10 (Part 2).

Perrin, W.F., Coe, J.M., and Zweifel, J.R. 1976. Growth and reproduction of spotted porpoise, *Stenella attenuate*, in the offshore eastern tropical Pacific. *Fish. Bull.* 74(2): 229–269.

Perrin, W.F. and Donovan, G.P. (eds.) 1984. Report of the workshop. *Rep. Int. Whales Commn.* (Special Issue 6, Reproduction in Whales, Dolphins and Porpoises): 1–24.

Perrin, W.F., Robertson, K.M., van Bree, P.J.H., and Mead, J.G. 2007. Cranial description and genetic identity of the holotype specimen of *Tursiops aduncus* (Ehrenberg, 1832). *Mar. Mammal Sci.* 23(2): 343–357.

Ralls, K. and Mesnick, S.L. 2002. Sexual dimorphism. pp. 1071–1078. In: Perrin, W.F., Würsig, B., and Thewissen, J.G.M. (eds.). *Encyclopedia of Marine Mammals*. Academic Press, San Diego, CA. 1414pp.

Read, A.J., Wells, R.S., Hohn, A.A., and Scott, M.D. 1993. Patterns of growth in wild bottlenose dolphins, *Tursiops truncatus*. *J. Zool., Lond.* 231: 107–123.

Reeves, R.R., Perrin, W.F., Taylor, B.L., Baker, C.S., and Mesnick, S.L. (eds.). 2004. Report of the Workshop on Short-coming of Cetacean Taxonomy in relation to Needs of Conservation and Management, April 30–May 2, 2004, La Jolla, CA. NOAA Technical Memorandum NMFS NOAA-TM-NMFS-SWFSC-363: 1–94.

Rice, D.W. 1998. *Marine Mammals of the World, Systematics and Distribution*. Society for Marine Mammalogy, Special Publication No.4. Lawrence, KS: 231pp.

Ross, G.J.B. 1977. The taxonomy of bottlenose dolphins *Tursiops* species in South African waters, with notes on their biology. *Ann. Cape Prov. Mus.* (Nat. Hist.) 11(9): 135–194.

*Saeki, H. 2007. Occurrence of an Indo-Pacific bottlenose dolphin on the Chiba coast. *Sakamata* (Kamogawa Sea World) 27: 4.

Scott, M.D., Wells, R.S., and Irvine, A.B. 1996. Long-term studies of bottlenose dolphins in Florida. *IBI Rep.* (Kamogawa, Japan) 6: 73–81.

Shane, S.H., Wells, R.S., and Würsig, B. 1986. Ecology, behavior and social organization of the bottlenose dolphin: A review. *Mar. Mammal Sci.* 2(1): 34–63.

Shirakihara, M., Shirakihara, K., Tomonaga, J., and Takatsuki, M. 2002. A resident population of Indo-Pacific bottlenose dolphins (*Tursiops aduncus*) in Amakusa, western Kyushu, Japan. *Mar. Mammal Sci.* 18(1): 31–41.

Steiner, A. and Bossley, M. 2008. Some reproductive parameters of an estuarine population of Indo-Pacific dolphins (*Tursiops aduncus*). *Aqu. Mammals* 34(1): 84–92.

*Suisan-cho Chosa-kenkyu-bu. 1969. *Comprehensive Report of Basic Investigations in Search of Counteraction for Fishery Damages Caused by Small Cetaceans in Western Japan (42nd and 43rd Year of Showa)*. Research and Investigation Department of Fisheries Agency, Suisan-cho Chosa-kenkyu-bu, Tokyo, Japan. 108pp.

Takahashi, Y., Ohwada, S., Watanabe, K., Ropert-Coudert, Y., Zenitani, R., Naito, Y., and Yamaguchi, T. 2006. Does elastin contribute to the persistence of corpora albicantia in the ovaries of the common dolphin (*Delphinus delphis*). *Mar. Mammal Sci.* 22(4): 819–830.

*Takemura, A. 1986a. Pacific white-sided dolphin and common bottlenose dolphin. pp. 161–177. In: Tamura, T., Ohsumi, S., and Arai R. (eds.). *Report of Contract Studies in Search of Countermeasures for Fishery Hazard (Culling of Harmful*

Organisms) (56th to 60th year of Showa). Suisan-cho Gogyo Kogai (Yugaiseibutsu Kujo) Taisaku Chosa Kento Iinkai (Fishery Agency), Tokyo, Japan. 285pp.

*Takemura, A. 1986b. Food habit and position in the ecosystem. pp.187–195. In: Tamura, T., Ohsumi, S., and Arai R. (eds.). *Report of Contract Studies in Search of Countermeasures for Fishery Hazard (Culling of Harmful Organisms) (56th to 60th year of Showa)*. Suisan-cho Gogyo Kogai (Yugaiseibutsu Kujo) Taisaku Chosa Kento Iinkai (Fishery Agency), Tokyo, Japan. 285pp.

*Tamura, T., Ohsumi, S., and Arai, R. (eds.). 1986. *Report of Contract Studies in Search of Countermeasures for Fishery Hazard (Culling of Harmful Organisms) (56th to 60th year of Showa)*. Suisan-cho Gogyo Kogai (Yugaiseibutsu Kujo) Taisaku Chosa Kento Iinkai (Fishery Agency), Tokyo, Japan. 285pp.

*Tobayama, T. and Shimizu, H. 1973. On relationship between food consumption and body weight of common bottlenose dolphins, *Tursiops gilli*, in captivity (food amount required for maintaining body weight). *Dobutsuen Suizokukan Kyokai Zasshi* 15(2): 37–39.

*Uchida, S. 1985. *Report of Surveys on Small Cetaceans for Aquarium Display,* Part 2. Kaiyohaku Kinen Koen Kanri Zaidan, Motobu, Japan. 36pp.

Urian, K.W., Duffield, D.A., Read, A.J., Wells, R.S., and Shell, E.D. 1996. Seasonality of reproduction in bottlenose dolphins, *Tursiops truncates. J. Mammal.* 77(2): 394–403.

Wang, Y.J., Chou, L.-S., and White, B.N. 1999. Mitochondrial DNA analysis of sympatric morphotypes of bottlenose dolphins (genus: *Tursiops*) in Chinese waters. *Mol. Ecol.* 8: 1603–1612.

Wang, Y.J., Chou, L.-S., and White, B.N. 2000a. Differences in the external morphology of two sympatric species of bottlenose dolphins (genus *Tursiops*) in the waters of China. *J. Mammal.* 81(4): 1157–1165.

Wang, Y.J., Chou, L.-S., and White, B.N. 2000b. Osteological difference between two sympatric forms of bottlenose dolphins (genus *Tursiops*) in Chinese waters. *J. Zool., Lond.* 252: 147–162.

Wang, J.Y. and Yang, S. 2007. *An Identification Guide to the Dolphins and Other Small Cetaceans of Taiwan*. Jen Jen Publishing and National Museum of Marine Biology, Taiwan, China. 207pp. (In Chinese and English).

Weigle, B. 1990. Abundance, distribution and movements of bottlenose dolphins (*Tursiops truncatus*) in lower Tampa Bay, Florida. *Rep. Int. Whales Commn.* (Special Issue 12, Individual Recognition of Cetaceans): 195–201.

Wells, R. 1991. The role of long-term study in understanding the social structure of bottlenose dolphin community. pp. 199–225. In: Prior, K. and Norris, K.S. (eds.). *Dolphin Societies*. University of California Press, Berkeley, CA. 397pp.

Wells, R.S., Hansen, L.J., Baldridge, A., Dohl, T.P., Kelly, D.L., and Defran, R.H. 1990. Northward extension of the range of bottlenose dolphins along the California coast. pp. 421–431. In: Leatherwood, S. and Reeves, R.R. (eds.). *The Bottlenose Dolphin*. Academic Press, London, U.K. 653pp.

Wells, R.S. and Scott, M.D. 1990a. Bottlenose Dolphin *Tursiops truncates* (Montagu, 1821). pp. 137–182. In: Ridgway, S.H. and Harrison, R. (eds.). *Handbook of Marine Mammals*, Vol. 6 (The Second Book of Dolphins and the Porpoises). Academic Press, San Diego, CA. 486pp.

Wells, R.S. and Scott, M.D. 1990b. Estimating bottlenose dolphin population parameters from individual identification and capture-release technique. *Rep. Int. Whales Commn.* (Special Issue 12, Individual Recognition of Cetaceans): 407–415.

Wells, R.S., Scott, M.D., and Irvine, A.B. 1987. The social structure of free-ranging bottlenose dolphins. pp. 247–305. In: Genoways, H. (ed.). *Current Mammalogy*, Vol. 1. Prenum Press, New York. 519pp.

*Yoshioka, M. 1990. Endocrine system. pp.14–22. In: Miyazaki, N. and Kasuya, T. (eds.). *Mammals of the Sea—Their Past, Present and Future*. Saientisutosha, Tokyo, Japan. 311pp.

Yoshioka, M., Mohri, E., Tobayama, T., Aida, K., and Hanyu, I. 1986. Annual change in serum reproductive hormone levels in the captive female bottlenose dolphins. *Bull. Jpn. Soc. Sci. Fish.* 52: 75–77.

Yoshioka, M., Tobayama, T., Ohara, S., and Aida, K. November 11–15, 1993. Ejaculation pattern of bottlenose dolphin. *Abst. Tenth Biennial Conf. Biol. Mar. Mammals.* 115.

Zhou, K. and Qian, W. 1985. Distribution of the dolphins of the genus *Tursiops* in the China seas. *Aqu. Mammals* 11(1): 16–19.

12 Short-Finned Pilot Whale

12.1 DESCRIPTION

12.1.1 Distinction between the *Globicephala* Species

The genus *Globicephala* contains two currently accepted species: the long-finned pilot whale *G. melas* (Trail 1809) and the short-finned pilot whale *G. macrorhynchus* Gray 1846. The former inhabits temperate and subpolar waters of the world, with exception of the North Pacific. This type of distribution pattern is called anti-tropical distribution (Davies 1963; Barnes 1985).

The long-finned pilot whale does not currently inhabit the North Pacific but is known to have once occurred there, based on a subfossil specimen and relics in archaeological sites in the North Pacific. Kasuya (1975) reported six skulls of the species from the western North Pacific and the northern Sea of Japan. One was found in a stratum in the bed of the Hekuri River in Tateyama City (35°00′N, 139°52′E). The ^{14}C dating of accompanying coral and oyster remains gave 6340 ± 140 years BP and 6430 ± 130 years BP, respectively, which corresponds in Japan to the era of the Jomon Culture (16,500–3000 years BP). The other five skulls were from two archaeological sites of the Okhotsk Sea Culture on Rebun Island (45°22′N, 141°02′E) in the northern Sea of Japan from the eleventh to thirteenth centuries. They accompanied numerous skeletal remains of various other marine mammals (Kasuya 1981, in Japanese). The species was also identified at an archaeological site on Unalaska Island in the eastern Aleutian Islands from 3500 to 2500 years BP (Frey et al. 2005). Thus, the long-finned pilot whale is believed to have been hunted by inhabitants of the northern North Pacific until relatively recently (Crockford 2008). Possible contribution of hunting activities to the disappearance from the North Pacific has not been evaluated. The short-finned pilot whale is the only current inhabitant of the genus *Globicephala* in the North Pacific.

Skull morphology, flipper length, and, to some extent, tooth counts help in morphologically distinguishing between the two species, but this cannot be accomplished by consideration of only body size or pigmentation, due to large geographical variation within the species (Kasuya and Matsui 1984; Kasuya et al. 1988). Measurements of 3400 long-finned pilot whales in the North Atlantic revealed that flipper length measured on a straight line from the anterior insertion to the tip increased almost linearly with body length but allometrically. Thus, the proportion flipper length to body length changed with body length. The proportion decreased from around 25% in newborns at body length 1.5 m to 20% at body length 2.5 m. It then increased proportionately to about 25% at body length of 6 m (Bloch et al. 1993a,b). The proportion was 16%–29% for all the postnatal animals and 18%–29% for adults. Yonekura et al. (1980) carried out a similar analysis on short-finned pilot whales off central Japan (Section 12.3) and showed that flipper length increased almost in proportion to body length of postnatal individuals; they presented the value of 15.8%–19.0% for 21 individuals, including newborns, and 15.8%–18.9% (mean 16.3%) for 18 adults. Thus, flipper length expressed as a proportion of body length showed some overlap between adults of the two species. Although Yonekura et al. (1980) also noted that short-finned pilot whales had a smaller proportional length from umbilicus to tail notch, the difference (56% vs. 60%–61%) seemed too small to be relied on for distinction between the two species.

Difference between the two species is most distinct in the shape of the premaxillae visible on the dorsal surface of the rostral portion of the skull. Kasuya (1975) compared 29 skulls of long-finned pilot whales from both hemispheres (skull length 580–712 cm) and 37 skulls of short-finned pilot whales from both hemispheres, including the North Pacific (skull length 540–748 cm). In the long-finned pilot whales, width of the rostrum at mid-length was 26.8%–33.0% (mean 29.2%) of total skull length, while the proportion was 28.3%–49.0% (mean 34.3%) in the short-finned pilot whales. The width of the premaxillae measured at the same point ranged from 21.1% to 28.5% (mean 24.9%) in the former species and from 26.7% to 47.2% (mean 32.8%) in the latter. The overlap between the two species was smaller in width across the premaxillae. These morphological differences in the rostral portion of the skull are easily recognized visually or by plotting the measurements on skull length (Figure 1 of Kasuya 1975) (Figure 12.1). The premaxillae are narrow with almost parallel lateral margins and the underlying maxillae are exposed outside the premaxillae on the rostrum of the long-finned pilot whale, while the premaxillae in the short-finned pilot whale expand laterally in the anterior portion of the rostrum and cover the maxillae. The short-finned pilot whale has slightly fewer teeth on average than the long-finned pilot whales. Kasuya (1975) compared tooth counts between the two species. Thirty-four upper tooth rows of long-finned pilot whales had 9–12 teeth (mean 10.1), while 74 tooth rows of short-finned pilot whales had 6–10 teeth (mean 7.9). However, it was an error to have treated the tooth rows of the same individual as independent data. The mandibular teeth were not examined because of uncertainty in species identification.

12.1.2 Description of Short-Finned Pilot Whale

The Japanese name *kobire-gondo* was a translation of "short-finned pilot whale" proposed for *G. macrorhynchus* by Nishiwaki (1965, in Japanese), when presence of this species

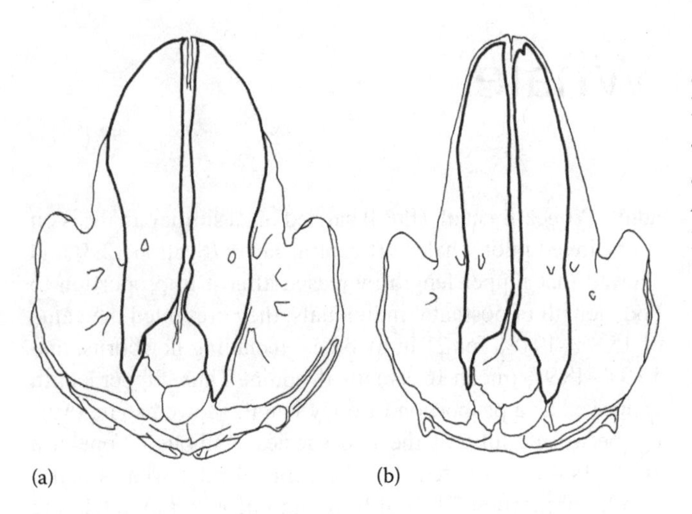

FIGURE 12.1 Dorsal view of skulls of a short-finned pilot whale (a) and a long-finned pilot whale (b). Premaxillae shown with enhanced contour. Premaxillae of the former species expand laterally over the dorsal surface of the anterior portion of the rostrum. (Reproduced from Kasuya, T., Short-finned pilot whale, in: Odate, S., ed., *Basic Data of Rare Wild Aquatic Organisms of Japan* (II), Nihon Suisan Shigen Hogo Kyokai, Tokyo, Japan, pp. 542–551, 751, 1995a, Figure 1, in Japanese.)

was yet not established in Japanese waters. Japanese whalers distinguished two forms of "pilot whales" segregating to the south and north with the boundary at around Choshi Point (35°42′N, 140°51′E) on the Pacific coast of Japan and called the northern one *tappa-naga* and the southern one *ma-gondo*, but we did not have a suitable name for the whole species *G. macrorhynchus*, which was later known to include these two forms off Japan. The two vernacular names are still used in Japan for the two geographical forms of the species off Japan (Section 12.3) and remain as candidates for Japanese common names for them in case in the future the two geographical forms may be recognized as separate species. They are distinguished in English as "northern form" and "southern form" respectively.

The Japanese words *gondo* and *goto* probably mean a large robust head (Otsuki 1808, in Japanese; Hattori 1887–1888, in Japanese), and names such as *gondo-kujira*, *goto-kujira*, *naisa-goto*, *ma-gondo*, and *shio-goto* have been used for the short-finned pilot whale or for one of the two geographical forms of the species. There are other examples of Japanese common names, including *gondo*: the Risso's dolphin (*hana-gondo*) and the false killer (*oki-gondo*, *oki-goto*, *dainan-goto*, or *o-onan-goto*). These species also have robust heads. The finless porpoise also has a round robust head and is called *ze-gon* or *ze-gondo* in the western Inland Sea, including Hiroshima Prefecture (Section 8.1). Other Japanese names, *nyudo-iruka* and *bozu-iruka*, were also used in various locations for the false killer whale or short-finned pilot whale (Sections 3.1, 3.7, 3.8.2, and Kasuya and Yamada 1995a, in Japanese). Both *nyudo* and *bozu* mean Japanese Buddhist monk, who usually has a shaved head, and *iruka* means dolphin or porpoise in Japanese.

The short-finned pilot whale belongs to the Delphinidae, attains a body length of 4–7 m, and has a drum-shaped round head and a broad fan-shaped dorsal fin. Okura (1826, in Japanese) described the fin as *amigasa-hire*, misprinted as *ashigasa-hire* in the reprint version of 1886, which has no meaning. *Amigasa* is a basketry helmet of varying type that was worn perhaps until the mid-nineteenth century by either sex to remain unidentified in the street or to avoid strong sunlight. *Hire* means fin. The large dorsal fin and rounded head are distinct when a school of short-finned pilot whales is met in the ocean. These two characters are indistinct on newborns, which have a dorsal fin only slightly broader than that of common dolphins and a head which is shaped like that of the Pacific white-sided dolphin or killer whale (Yonekura et al. 1980). The head of the newborn short-finned pilot whale has a slender melon and a short beak projecting beyond the melon. With growth the melon swells in three directions, that is, anteriorly, laterally and dorsally, and gradually covers the rostrum. On southern-form individuals off Taiji (33°36′N, 135°57′E) the melon covers the rostrum at age 2–7 years or at body length 2.7–3.0 m and the head becomes barrel-shaped. Development of the melon is particularly pronounced on adult males of the southern form, with the front end of the melon becoming flat as a drum (Figure 12.7). As short-finned pilot whales live in a school of 15–50 individuals of both sexes of various growth stages, including adult males, the form can be easily identified at a large distance, except in waters where both forms are expected.

The English language "black fishes" includes not only the two species of *Globicephala* but also other species such as false killer whales, pigmy killer whale, melon-headed whales, and probably Risso's dolphins. Most of the pilot whale's body is black, with various degree of pale portion behind the dorsal fin, posterior-dorsal area of eye, and ventral area from throat to anus. This pale pigmentation becomes darker with post-mortem time.

The three pale areas are present on both species of *Globicephala* but are indistinct in neonates and become more distinct with growth (Bloch et al. 1993b). The shape and degree of distinctness of the pale patches show geographical variation and may be useful for identification of populations within a species. However, caution is required in identifying the two species of *Globicephala* by pigmentation, because the geographical variation in pigmentation within a species can be of such degree that it is comparable to difference between the two species within the genus. Also, the pigmentation may change with growth.

The northern form of short-finned pilot whale off Japan has a broad saddle mark with distinct white pigmentation, and the southern form has only an indistinct slender saddle mark (Section 12.3; Figures 12.2 and 12.3; Kasuya et al. 1988), but similar kinds of geographical variation possibly exist among long-finned pilot whales. Davies (1960) reported that the saddle mark was common among long-finned pilot whales in the Southern Hemisphere but rare on the same species in the North Atlantic and proposed to deal with them as different subspecies. Bloch et al. (1993b) reported for the long-finned pilot whale in the North Atlantic that development of the saddle mark was a

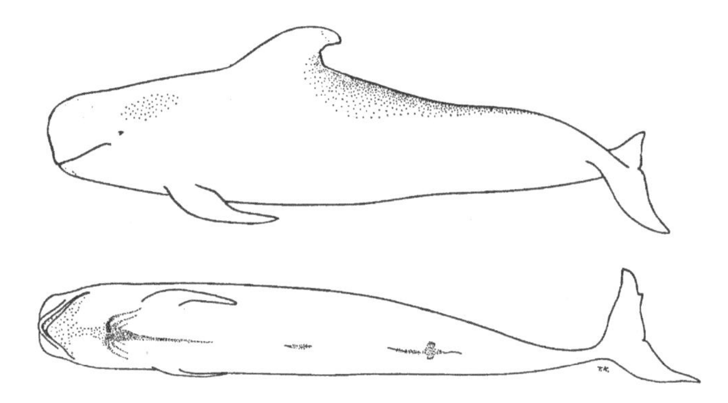

FIGURE 12.2 Body shape and pigmentation pattern of an adult female southern-form short-finned pilot whale taken off Taiji (33°36′N, 135°57′E) on the Pacific coast of central Japan. The shaded areas represent areas of pale pigmentation. (Reproduced from Yonekura, M. et al., *Sci. Rep. Whal. Res. Inst.*, 32, 67, 1980, Figure 2. With permission.)

(a)

(b)

FIGURE 12.3 Saddle marks of two geographical forms of short-finned pilot whale off Japan. (a) Left side of body of two northern-form individuals; (b) right side of body of a southern-form individual (small white spot behind dorsal fin is a reflection of sun). (Photo courtesy of T. Miyashita.)

function of age but not of sex. A saddle mark was present on 3% of individuals below age 5, 30% at ages 6–10, and 61% at ages over 11 years. In my view, the saddle mark identified by Bloch et al. (1993b) on the long-finned pilot whales in the North Atlantic was shaped like that of southern-form short-finned pilot whales off Japan. Thus, caution is required in analyzing geographical variation of pigmentation in the genus *Globicephala*.

12.2 DISTRIBUTION OF SHORT-FINNED PILOT WHALES

Short-finned pilot whales inhabit tropical and temperate waters of the world. The limit of their northern range extends in the North Pacific from around 43°N off eastern Hokkaido in northern Japan to around 50°N off Vancouver Island. The northern limit in the North Atlantic extends from the New Jersey coast (40°N) to the coast of France. The Mediterranean Sea is inhabited by long-finned pilot whales and not by short-finned pilot whales. Our information on the southern limit of the short-finned pilot whale is limited, but it is known from Sao Paulo (23°S), the coast of South Africa, west and east coasts of Australia, and the southern tip of the North Island of New Zealand (41°S) (Rice 1998). The South African coast is inhabited by both species (van Bree et al. 1978), with the boundary of the two species at around 30°S on the east coast, which is slightly north of Cape Agulhas, and at around 25°S on the west coast, where the cold Benguela Current runs northward. Short-finned pilot whales range to the north of these latitudes (Best 2007). Thus, the distribution of the species is currently discontinuous between the Indian Ocean and the South Atlantic.

Short-finned pilot whales are not evenly distributed within the range. In Japanese coastal waters they are common off the Pacific coast but rare in the East China Sea and the Sea of Japan. People around these seas use the word *gondo* for *oki-gondo* [false killer whale]. I have only one reliable record of sighting of a school of pilot whales, most likely short-finned pilot whales, in the central Sea of Japan, encountered by myself on the Yamato Bank (38°51′N, 135°00′E) in August 1978 during a cruise of the research vessel *Hakuho-maru*. In relation to the fishery/dolphin conflict in the Iki Island area, the Fisheries Agency requested records of cetacean sightings from vessels operated for fisheries research, fishery inspection, and fisheries education during September 1981 to August 1983, but there were no records identifiable as *Globicephala* (Nagasaki-ken Suisan Shikenjo [*Nagasaki Prefecture Fisheries Experimental Laboratory*] 1986, in Japanese). The Far Seas Fisheries Research Laboratory has annually operated cetacean sighting cruises around Japan since the early 1980s and reported the results of small-cetacean sightings to the Scientific Committee of the International Whaling Commission (IWC) until 2000; these cruises did not find pilot whales in the Sea of Japan.

Short-finned pilot whales are known from the Okinawa Islands area in southwestern Japan and have been hunted in opportunistic drives at Nago (26°35′N, 127°59′E) (Section 3.10). The short-finned pilot whale was last driven at Nago in 1982; subsequent take of the species was in the crossbow fishery (Section 2.8). The products of the crossbow fishery were sent to markets in Osaka (34°42′N, 135°30′E) and Fukuoka (33°35′N, 130°24′E) when the supply of whale meat declined around 1987 due to the moratorium on commercial whaling (Table 1.5 and Chapter 7).

Short-finned pilot whales seem to be scarce in the main part of the East China Sea and the Yellow Sea. Reliable records of the species are limited to two sightings in the Yellow Sea near the west coast of the Korean Peninsula (Wang 1999, in Chinese) and the holotype of *Delphinus globiceps* obtained off Nagasaki (32°45′N, 129°53′E) in northwestern Kyushu (Section 12.3.1). The species has not been recorded along the East China Sea coast of mainland China. Sightings of

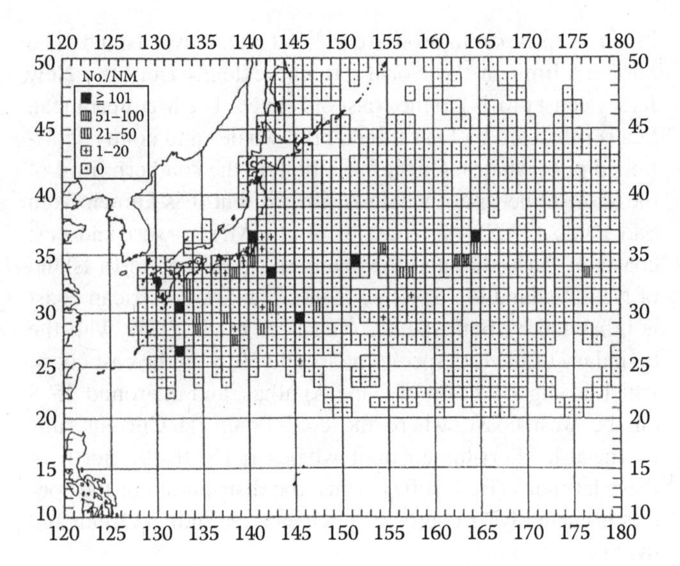

FIGURE 12.4 Density distribution of southern-form short-finned pilot whales in the western North Pacific observed in August and September, 1983–1991, during cetacean-sighting cruises operated by the Far Seas Fisheries Research Laboratory. Density is shown as number sighted per 100 nautical miles of survey track. (Reproduced from Miyashita, T., *Rep. Int. Whal. Commn.*, 43, 417, 1993, Figure 33. With permission of the author.)

the species off Taiwan are limited to the east coast in waters beyond the 1000 m isobath (Chou 1994, 2000, in Chinese; Wang and Yang 2007, in Chinese and English). A mounted skeleton of a short-finned pilot whale in Taiwan National University had a nameplate with the Japanese legend *Shimazu Seisakusho*, suggesting the possibility that it was imported for educational purposes during the Japanese occupation.

The Far Seas Fisheries Research Laboratory has recorded sightings of short-finned pilot whales around Japan during its routine cetacean-sighting cruises since the early 1980s (Miyashita 1993). Sightings of the species were concentrated in the western North Pacific in latitudes 22°N–43°N, east of the Southwestern Islands of Japan, which lie between Taiwan and Kyushu and contain the Okinawa Islands, and west of 155°E. Short-finned pilot whales are also known from the Hawaiian Islands area as well as off the west coast of North America, but these habitats are apparently discontinuous from the concentration in the western North Pacific off Japan (Figures 12.4 and 12.5).

12.3 NORTHERN AND SOUTHERN FORMS OFF JAPAN

In this section some details will be given on the two forms of short-finned pilot whales off Japan, where the "northern form" indicates a particular population of short-finned pilot whales inhabiting Pacific waters off northern Japan and the "southern form" a population in waters to the south. The eastern extent of these populations has not yet been fully determined. Although pilot whales similar to them may inhabit waters further east in the North Pacific (Section 12.3.7), their distribution is likely to be discontinuous from those of the Japanese populations.

12.3.1 Dawn of Japanese Cetology

First I will describe the long history of confusion in the taxonomy of the genus *Globicephala* in Japan. It was in 1758 when Linnaeus published the first volume of the 10th edition of *Systema Naturae*, which became the basis for modern animal taxonomy in Europe. Shortly after this date, in 1760, Harumasa Yamase published a book in Osaka titled *Geishi* [Natural History of Whales], which illustrated with descriptions 7 baleen and 6 toothed whale species found in his country in feudal times called *Kii* or *Ki-shu*, or the present Wakayama Prefecture (33°26′N–34°19′N, 135°00′E–135°59′E). He lived in the present Wakayama City (34°14′N, 135°10′E) and was also called Jiemon Kandoriya. Prior to this publication, Gensen Kanda in 1731 wrote a book *Nitto Gyofu* [Illustrated Fishes of Japan], but it was not published. Although he also listed the same number of species, the list of species was different and the drawings and descriptions were inferior to those of Yamase (1760, in Japanese). Species attributable to present-day blackfishes are one species—*goto-kujira*—of Kanda and three species—*naisa-goto*, *shiho-goto*, and *ohonan-goto*—of Yamase. Although the name *goto-kujira* of Kanda likely represented pilot whales, the drawing and description were totally confusing and thus it seems less reliable to me. Of the three species of Yamase, the first two were described as "upper jaw covers the lower jaw," while the third species *ohonan-goto* was described as having "both jaws meet," which is different from the first two species. Therefore, now *ohonan-goto* is thought to be the false killer whale (also see the following text in this section for my personal observation at Taiji in Wakayama Prefecture) and *naisa-goto* and *shiho-goto* to be some member of the genus *Globicephala*. Yamase also stated that the *shiho-goto* had a cloud-shaped, that is, irregularly shaped, white patch behind the dorsal fin, but that the *naisa-goto* did not (Figure 12.6).

A long time after the publication of Yamase (1760, in Japanese), around 1827 Jubei Kuroda wrote a book titled *Suizoku-shi* [Natural History of Aquatic Animals], which was published posthumously in 1884. His book listed *nagisa-gondo*, *shiho-gondo*, *dainan-gondo*, and *oho-uwokui* as species attributable in my view to the current blackfishes. The *ohonan-goto* of Yamase and *dainan-gondo* of Kuroda seem to represent the same species, that is, the false killer whale, because *oho* and *dai* have the same meaning of "large" or "great." The *Oho-uwokui* [great fish eater] of Kuroda is described as having a "slender body" of "length of 3–4.5 m," and being "black in color"; it also seems to represent the false killer whale, although Kuroda did not so state.

Almost in the same year when Kuroda wrote his book, around 1827 (published in 1884), Tsunenaga Okura in 1826 published a book *Joko-roku* [Leafhopper Control] with the intent of teaching Japanese farmers how to control leafhoppers in their rice paddies using whale oil. Whale oil was used widely in those days for pest control in rice paddies (Section 1.3). His book listed 11 species (5 baleen and

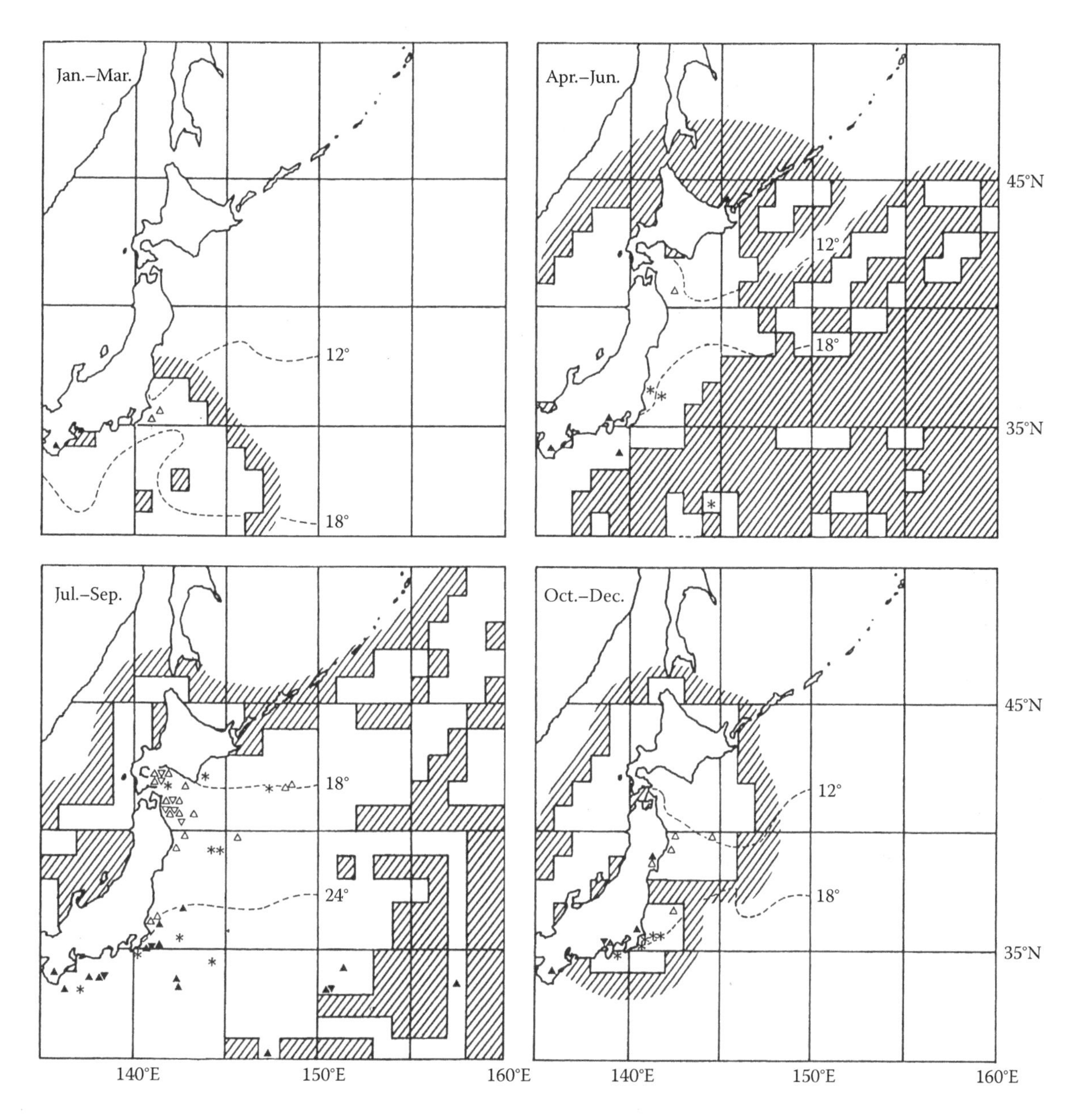

FIGURE 12.5 Geographical segregation of two forms of short-finned pilot whale off Japan observed during cetacean-sighting cruises operated by the Far Seas Fisheries Laboratory in 1982–1986. Sighting effort was absent in the shaded area. Open triangles indicate position of sighting of northern form, closed triangles southern forms, stars unidentified form, and dotted line sea surface temperature. (Reproduced from Kasuya, T. et al., *Sci. Rep. Whal. Res. Inst.*, 39, 77, 1988, Figure 2.)

6 toothed whales) of cetaceans with drawings and text, together with description of the quality of whale oil available from those species. His book contained a drawing attributable to *Globicephala*, particularly to the southern form short-finned pilot whale, with the name *koto-kujira*. He listed *shiho-goto*, *nai-goto*, *dainan-goto*, and *ohonan-goto* as synonyms of *koto-kujira*, but it seems to me that the last two names should be attributed to the false killer whale, which was shown by him with a drawing and the name *oho-iwokui*.

Nihon Hogei Iko [Miscellanea of Japanese Whaling] by To-oru Hattori (1887–1888, in Japanese) stated that *oho-iwokui* and *oki-goto* were used in Wakayama Prefecture for the same species and gave a drawing that was similar to a drawing labeled *oho-iwokui* by Okura (1826, in Japanese). He stated that the drawing was a copy from a whaling document of Atawa Village (33°48′N, 136°03′E) in Wakayama Prefecture.

These records suggest that there were three series of vernacular names for the false killer whale, that is, (1) *ohonan-goto* and *dainan-goto*, (2) *oki-goto* and *oki-gondo*, where *oki* means "offshore," and (3) *oho-iwokui* and *oho-uwokui*, which mean "large-" or "great-fish eater." I observed in Taiji in the 1970s that both *oho-iwokui* and *oki-goto* were used for *oki-gondo*, which meant false killer whale.

With cooperation of his students, Von Siebold brought whale specimens as well as their Japanese names that appeared in *Geishi* (Yamase 1760, in Japanese) to Europe (Ogawa 1950, in Japanese). Temminck and Schlegel (1844)

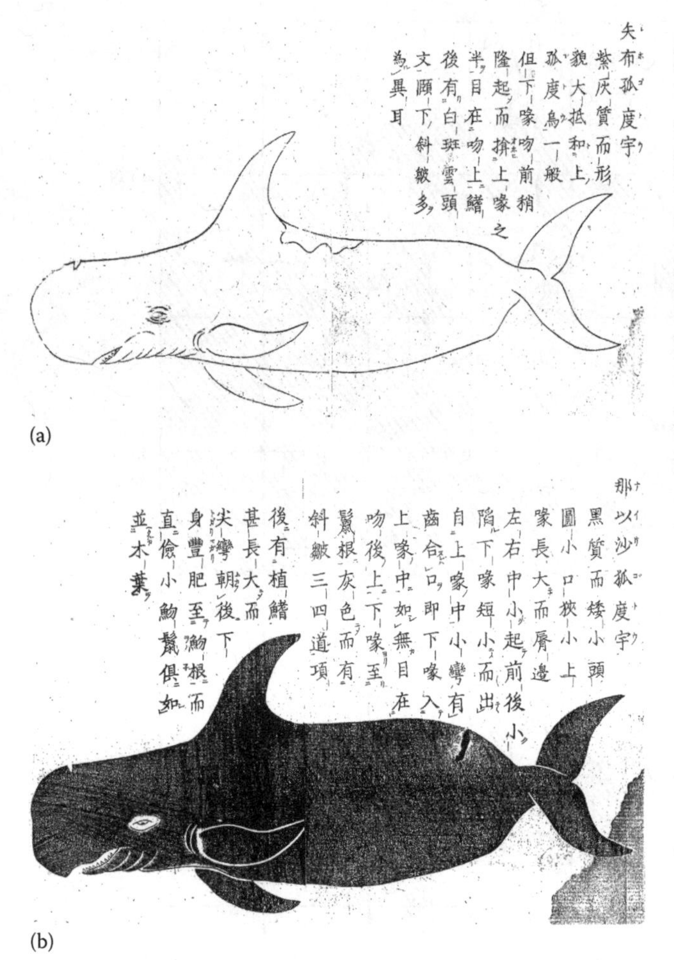

FIGURE 12.6 A *shiho-goto* (a, only with outline of the body) and a *naisa-goto* (b) described by Yamase (1760, in Japanese). The former has an irregular white saddle mark and is believed to correspond to the current northern form and the latter without the saddle mark to the current southern form. (a) it was reproduced from pp. 30–31, and (b) it was from pp. 28-29 of Yamase (1760, in Japanese). (Reproduced from two woodcut pages of Yamase 1760, in Japanese.)

identified a skull of a young pilot whale, 165 cm in body length, obtained by Siebold in Nagasaki with the Japanese name *naisa-goto*, as the same species as pilot whales in the northern North Atlantic (current *G. melas*) and used the name *Delphinus globiceps* for it. They also stated that there was another species called *siho-goto* in Japan. Later, Gray (1846) concluded that the Nagasaki specimen was different from the species in the northern North Atlantic and gave it a new name *Globicephalus siebodii*, and at the same time gave the name *Globicephalus chinensis* to a species said to inhabit the Chinese coast. Gray (1871) further proposed another species name *Globicephalus sibo* for a species called *sibo-golo* in Japan. At this stage he made two typographic errors, that is, changing *siho* to *sibo* and *goto* to *golo*. According to Dall (1874, in Scammon 1874), another species of pilot whale—*Globicephalus scammoni* (Cope, 1869)—was described based on a skull collected off California. Thus, there have been a total of four species described by modern taxonomists for the genus *Globicephala* in the North Pacific.

12.3.2 Confusion and Reorganization

After the Meiji Revolution (1864–1871), modern zoology flooded into Japan, and Japanese biologists started their efforts to match existing Japanese species names to the imported scientific names. Cetacean biology was not the exception but progressed at a slower pace than other zoological fields due to difficulties in accessing specimens and information. In the early twentieth century, R. C. Andrews was sent to Japan by the American Museum of Natural History in New York and collected specimens of large cetaceans at Japanese whaling stations. He found that Japanese whalers were still hunting gray whales, which had been thought to be extinct, on the Korean coast, and published a monograph on the species (Andrews 1914a). He also obtained a pilot whale at Ayukawa (38°18′N, 141°31′E) on the Pacific coast of northern Japan and used the name *Globicephala scammoni* for it (Andrews 1914b). Although he did not mention the basis for this conclusion, he must have considered the Ayukawa specimen to be the same species as pilot whales off California and apparently ignored *Globicephalus sieboldii* of Gray (1846). The Japanese zoologists Nagasawa (1916, in Japanese) and Kishida (1925, in Japanese) followed Andrews (1914b), but the latter cautiously stated that he would use the scientific name for the time being until the relationship between the scientific name and Japanese common names *siho*, *naisa*, and *ohonan* would be clarified (Kishida 1925, in Japanese). Taxonomy should be based on specimens, but that was not done by them.

While he was studying the brain anatomy of cetaceans, Teizo Ogawa of the Tokyo Imperial University (later renamed Tokyo University) noticed unresolved problems in the taxonomy of small cetaceans around Japan and collected taxonomic materials from northern Kyushu, Taiji, and the Sanriku Region (Pacific coast of northern Honshu in latitudes 37°54′N–41°35′N, where Ayukawa is located) (Ogawa 1950, in Japanese). Although he made a great contribution to the taxonomy of Japanese Delphinidae, he did not apparently complete his work on taxonomy of the genus *Globicephala* off Japan. He obtained a 236 cm juvenile pilot whale at Shiogama (38°19′N, 141°01E) on May 28, 1935. He identified it as *Globicephalus melas* (Traill) with the Japanese names *naisa-goto* and *ma-gondo*, and he further stated that he would tentatively deal with *G. melas* as a synonym of *G. sieboldii* Gray, which he believed to be very closely related (Ogawa 1937, in Japanese). In the same article Ogawa stated that he obtained two specimens of *Globicephalus scammoni* at Ayukawa and gave them the Japanese name *shiho-goto* and *tappa-naga*. It is probable that he had difficulty identifying the pigmentation pattern of those specimens due to postmortem change. It seems to me that Ogawa, as an anatomist, attached great importance to the morphology of individual specimens and ignored the possible range of individual variation and the zoogeography of animal species. The three Ogawa specimens collected in northern Japan were deposited in the National Science Museum in Tokyo and later identified as *Globicephala macrorhynchus* by Kasuya (1975) (see the following text in this section). Although it has not been confirmed, the collection locality of these three pilot whale specimens strongly suggests that they belonged to the northern form.

After Ogawa (1937, in Japanese), an erroneous concept became generally accepted in Japan, which was to assume the scientific name *G. melas* applied to pilot whales inhabiting southern waters from Chiba Prefecture (in latitudes of 34°55′N–35°44′N) to the Okinawa Islands (in latitudes 24°N–27°N) and locally called *naisa-goto* or *ma-gondo*, and to use *G. scammoni* for those inhabiting waters north of Chiba and locally called *tappa-naga* or *shio-goto*. This meant that *G. melas*, which inhabits the North Atlantic in latitudes from 35°N to 70°N, was thought to live in the western North Pacific south of 35°N, where the warm Kuroshio Current prevails. I remember one day in 1961–1964 when the late H. Omura of the Whales Research Institute expressed at lunchtime his doubt about the taxonomy of Japanese pilot whales by pointing out this zoogeographic contradiction.

In 1950, F. C. Fraser published the result of his osteological study on the genus *Globicephala*, concluding that the genus contains only the two species *G. melas* (Trail, 1809) and *G. macrorhynchus* (Gray, 1846), and that the latter name is a senior synonym of *G. scammoni*. Then the late M. Nishiwaki of the Ocean Research Institute started a U.S.-Japan cooperative study on cetacean fauna in the North Pacific. The project collected pilot whale specimens from the drive fishery at Arari (34°50′N, 138°46′E) and small-type whaling based at Taiji and Ayukawa, and concluded that all the pilot whales examined were *G. macrorhynchus* and that there was no trace of *G. melas* (Nishiwaki et al. 1967, in Japanese). This resolved one basic question on the taxonomy of pilot whales in Japanese waters.

One of the remaining problems was the identity of *G.sieboldii*. As this species was based on the skull of a young animal, characters of the species were indistinct, but Van Bree (1971) was able to conclude that this specimen belonged to *G. macrorhynchus*. During the process of identifying skulls of long-finned pilot whales, *G. melas*, excavated from archaeological sites in Japan, I examined all the pilot whale skulls in the National Science Museum, which included Ogawa's specimens, and confirmed that they were *G. macrorhynchus* (Kasuya 1975).

The pilot whale specimens used by Kasuya (1975) included a skull collected by M. Nishiwaki and R. L. Brownell in the early 1960s at Ayukawa. It seemed to me to belong to the short-finned pilot whale, but the size was extraordinary, larger than any pilot whale ever previously collected at Taiji or Arari. I failed to pursue this problem, but it was later investigated by Miyazaki and Amano (1994) (Section 12.3.7.5).

12.3.3 More about Japanese Pilot Whales

There still remained another question on the taxonomy of Japanese pilot whales. Yamase (1760, in Japanese) listed two species of pilot whales, and Japanese modern whalers also had a similar conception. In a study on long-finned pilot whales excavated in Japan, I analyzed the Monthly Report of Whaling Operation for the years 1949–1952, which was presented by small-type whalers to the Fisheries Agency, to see if there were records suggesting capture of long-finned pilot whale in northern Japan. I concluded from the zoogeographical point of view that the pilot whale called *tappa-naga* by small-type whalers must be the short-finned pilot whale and cannot be the long-finned pilot whale (Kasuya 1975).

This conclusion was correct, but it was a mistake to assume that the same individuals that wintered south of Chiba Prefecture would migrate in summer to the Sanriku coast. The main target of Japanese small-type whaling was not pilot whales but minke and Baird's beaked whales. Therefore many, but not all, of these whalers migrated seasonally following movements of the main target species. Those who recorded a peak pilot whale catch in early summer off Taiji were likely to have moved northward to reach the Sanriku coast in August and recorded another peak pilot whale catch in summer. This apparent shift in peak pilot whaling grounds did not in reality indicate movement of the pilot whales. Small-type whaling was allowed to take minke whales and toothed whales other than the sperm whale using a whaling cannon of less than 50 mm caliber and a vessel of less than 30 gross tons (later modified to 50 gross tons) (Chapter 4; Sections 5.4, 5.5, and 7.1; Ohsumi 1975). Ogawa also obtained his cetacean samples from such vessels.

Initially I overlooked or misunderstood some information from small-type whalers (Kasuya 1975). While I was analyzing catch statistics of small-type whalers, I interviewed ex-whalers in Taiji who hunted small cetaceans and asked what the *tappa-naga* and *ma-gondo* looked like. Their answers and my interpretation are summarized as follows. The word *tappa-naga* means "long flipper" and *ma-gondo* "the correct pilot whale."

1. "Generally speaking *tappa-naga* is larger than *ma-gondo*." The whalers did not pay attention to the sex. Males of the short-finned pilot whale grow to about 25% larger than the females. So I thought that if pilot whales segregated by sex it would explain the geographical size difference. This interpretation was not supported by data obtained later.
2. "*Tappa-naga* has a longer flipper than *ma-gondo*." Whalers did not clarify whether the difference was proportional or in absolute length. It was later found that the latter was the case.
3. "*Tappa-naga* is skinny and contains less oil than *ma-gondo*." This could be due to either seasonal physiological fluctuation or a population difference. Later it was found that the latter was the case.

It was in 1982 that the small-type whalers resumed hunting *tappa-naga* off Ayukawa, and the questions given earlier were resolved.

12.3.4 First Encounter with the Northern Form

Hereinafter I will refer to the two types of short-finned pilot whales off Japan, *tappa-naga* and *ma-gondo*, by our current English terminology of "northern form" and "southern form," respectively. It was in June 1975 that I saw a school of pilot whales from the research vessel *Tansei-maru* of the Ocean Research Institute, University of Tokyo, on a northbound cruise toward the Sanriku coast. Approaching the school for further observation, we noted the white saddle patch, and both my colleague N. Miyazaki and I were astonished by the beautiful patch shining in the sun. In 1982, I was occupied with studying short-finned pilot whales taken in Taiji. Every time I heard that Taiji had driven a school of pilot whales, I got on a night train and arrived at Taiji harbor early the next morning, examined carcasses the whole day, and got on the returning night train to Tokyo with samples in formalin. The fishermen usually slaughtered all the whales in a school in a day if it contained around 20 or fewer animals, but they spent 2 or 3 days for larger schools. The short-finned pilot whale I was examining at Taiji was called *ma-gondo* by fishermen, which is the "southern form" in our current terminology, and they did not have as distinct a saddle patch as I had seen in northern Japan in 1975. So I suspected that the pilot whales with a bright saddle patch were the *shiho-goto* of Yamase (1760, in Japanese) and *tappa-naga* of recent whalers and waited for the opportunity to examine them in detail.

Such an opportunity occurred in October 1982 when I was examining southern form short-finned pilot whales at Taiji. T. Isone, a gunner captain of a small-type whaling vessel in Taiji, told me that several small-type whalers at Ayukawa had started hunting pilot whales off the Sanriku coast. They used to operate on minke whales and made some additional catches of Baird's beaked whales, but they resumed taking pilot whales in order to compensate for a diminishing minke whale quota under international regulation. Since I was occupied with the task in Taiji, I called N. Miyazaki of the National Science Museum and advised him to examine the pilot whales being landed at Ayukawa. I told him that they could be the type of whales that we had encountered several years ago at sea. With the cooperation of whalers in Ayukawa, Miyazaki examined their catch and returned to Tokyo with several skulls and stomachs. These specimens formed the basis of studies on osteology (Miyazaki and Amano 1994) and food habits (Kubodera and Miyazaki 1993).

On returning to Tokyo after the work in Taiji, I visited Miyazaki for an update. He told me that the northern form which he examined at Ayukawa was easily distinguished from the southern form by a distinct saddle patch. I was very impressed by the size and shape of several heads that had just been delivered on a truck while I was in his office. The heads of both sexes were greater than those of adult males of the southern form, but the shape of the melon of the adult male was more like that of adult females or immature males of the southern form. The front contour of the melon is flat and square-shaped in adult males of the southern form if seen from the dorsal or ventral side, but the heads in front of us did not have such a shape (Figure 12.7). Another morphological difference between the two forms I noticed in subsequent years in Ayukawa is in the shape of the dorsal fin. In profile the front edge of the dorsal fin of the adult male southern form is expanded (Section 12.1.2), but this is less distinct in the northern form. In short, some of the secondary sexual characters other than body size are less pronounced in the northern form than the southern form. Some of these findings were reported by Miyazaki (1983, in Japanese).

Concerning the secondary sexual characters of the short-finned pilot whale, Y. Toba, owner of a whaling vessel in Ayukawa, told me in 1983 about the following interesting experience of his gunner in the 1982 season. This was the year when hunting of the northern form resumed. On one day of the season the whalers encountered a school of southern-form whales and took some of them. From the shape of the dorsal fin the gunner believed that the whales he took would be of great size but was greatly disappointed when he was informed at the port that his catches were less than 5 m long. A full-grown southern form male measures only 4.67 m on average (Table 12.1). This was the only take of southern-form

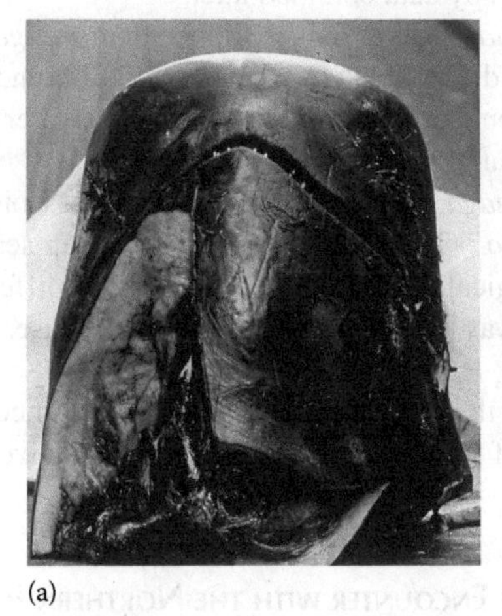
(a)

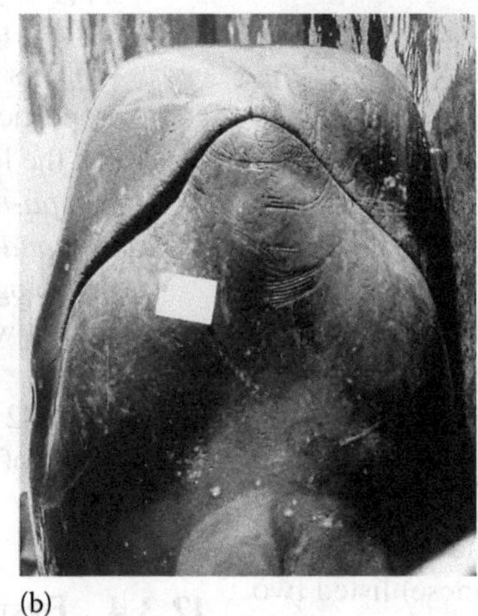
(b)

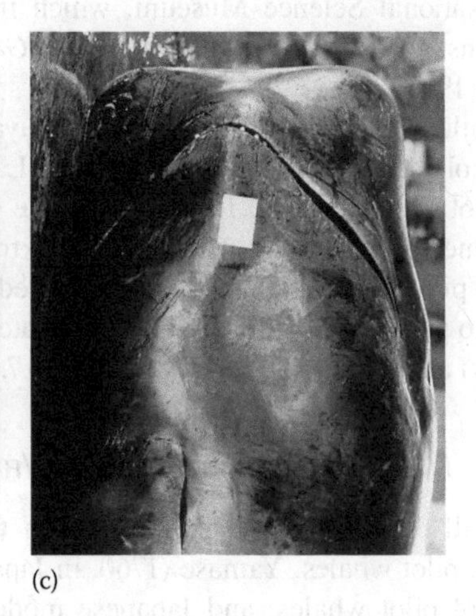
(c)

FIGURE 12.7 Ventral view of heads of short-finned pilot whales showing difference in shape of melon between two geographical forms of the species off Japan. (a) A full-grown male of the northern form (720 cm in body length); (b) an immature male of the southern form (409 cm in body length); (c) a full-grown male of the southern form (518 cm in body length). (Reproduced from Kasuya, T. and Tai, S., *Rep. Int. Whal. Commn.*, 339, 1993, Figure 2.)

TABLE 12.1
Mean Body Length of Two Forms of Short-Finned Pilot Whales off Japan

Growth Stage	Southern Form (m)	Northern Form (m)	Northern/Southern
Neonate	1.40	1.85	1.3
At sexual maturity			
Female	3.16	3.9–4.0	1.2–1.3
Male	4.22	5.6	1.3
At physical maturity			
Female	3.64	4.67	1.3
Male	4.74	6.5	1.4
Largest recorded			
Female	4.05	5.1	1.3
Male	5.26	7.2	1.4

Sources: Kasuya, T. and Marsh, H., *Rep. Int. Whal. Commn.*, 259, 1984; Kasuya, T. and Tai, S., *Rep. Int. Whal. Commn.*, 339, 1993.

whales recorded in the whaling operation off Sanriku coast in the 1982–1996 seasons when I was able to directly monitor the operation. I felt it interesting that Ayukawa whalers did not know about southern-form pilot whales, which indicated that the type was uncommon in northern waters.

The question remains whether the northern form has ever been taken by dolphin fisheries off the coasts of the Izu Peninsula (34°36′N–35°05′N, 138°45′E–139°10′E) and Wakayama Prefecture. Yamase (1760, in Japanese) described cetaceans taken in eighteenth-century whaling at Taiji and nearby Koza (33°31′N, 138°49′E), both in present-day Wakayama Prefecture, and listed *naisa-goto* and *siho-goto*, which are attributable to the southern and northern forms, respectively (Figure 12.6). This suggests past incidents of the northern form occurring off the Wakayama area, although the frequency is unknown. If this was the case, the northern form could have also occurred off the Izu Coast, which is located between Sanriku and Wakayama. However, I have not positively identified northern-form whales by external morphology among over 500 carcasses of both sexes of pilot whales examined at Taiji and Arari during 1962–1984 (Kasuya and Marsh 1984; Kasuya and Tai 1993), except for one ambiguous individual. This was a male in a group of southern-form whales driven at Taiji and had a body length of 5.8 m and weight of a single testis of 1.7 kg. Although these figures were reasonable for the northern form, the body size was too large and testis weight rather small for adult southern-form individuals, which have maximum body length of 5.2 m and testis weight in the range of 1.7–3.0 kg. No additional information (e.g., pigmentation and age) was available for this large male (Kasuya et al. 1988).

12.3.5 Distinction between the Two Forms

In April 1983, I moved from the Ocean Research Institute of the University of Tokyo to the Whale Resources Section of the Far Seas Fisheries Research Laboratory of the Fisheries Agency as a project leader. The task given to me included studying the biology and abundance of the northern-form short-finned pilot whale for fishery management. I was extremely interested in comparing its life history with that of the southern form. Because such a study was necessary for management, I requested a budget for the Fisheries Agency, but it was not forthcoming. So, as is usually the case for university scientists, I proposed to use my personal fund for the project. This seems to have caused a problem for the administration people in the laboratory. Finally the Whaling Section of the Fisheries Agency persuaded the Japanese Association of Small-Type Whaling and the Japanese Whaling Association to pay the cost of my travel and an assistant, which I appreciated greatly. The dependence on the industry continued for several seasons before the government became able to pay the full cost of the activity. Summarized in the following are differences between the two forms of short-finned pilot whales, including results obtained through the project and other activities.

12.3.5.1 Pigmentation

The saddle patch in the northern form shows broad individual variation, but is generally broader (dorsal-ventral width) and more distinct than that of the southern form. The posterior margin of the saddle patch is distinctly separated from the dark pigmentation of the dorsal area, and the margin extends anteroventrally with a more acute angle than in the southern form. Thus, the ventral-dorsal width of the patch is greater and its anterior-caudal length is shorter in the northern form. Compared with that of the northern form, the saddle patch of the southern form extends posteriorly beyond mid-length of the tail peduncle, the color is darker and indistinct, and the color fades into the dark pigmentation of the tail peduncle. Although the eye patch seems to be slightly more distinct in the northern form, it cannot be used to differentiate between the two forms. The saddle patch is not evident on neonates of the northern form but is distinct on 1.5-year-old juveniles (Kasuya et al. 1988).

12.3.5.2 Shape of the Head

Adult males of the southern form have a well-developed melon. In profile it appears as a round protuberance beyond the tip of the upper jaw, but if seen from the top or bottom the front contour it is almost flat or slightly concave and the lateral corners are square-shaped. This is different from females of the southern form and from adults of both sexes of the northern form (Figure 12.7). Thus secondary sexual characteristics of the melon are more distinct in the southern form.

12.3.5.3 Shape of the Dorsal Fin

The dorsal fin of the short-finned pilot whale becomes broad among adult males of both forms, but this secondary sexual character is more pronounced in the southern form.

12.3.5.4 External Proportions

Kishiro et al. (1990, in Japanese) compared mean measurements of external proportions between 119 northern-form and

23 southern-form pilot whales and found significant difference in two measurements: (1) mean distance from anterior end of the head to the blowhole measured along the body axis was slightly smaller among northern-form whales (males 8.2%, females 9.4%) than in the southern-form whales (males 9.5%, females 11.0%), and (2) mean distance from anterior insertion of the flipper to the center of the genital aperture (also measured parallel with body axis) was slightly greater in northern-form (males 41.8%, females 49.6%) than southern-form whales (males 38.9%, females 45.0%).

Other measurements of the head region showed a similar trend, although the differences were not statistically significant. This indicates that the northern form has a relatively shorter head and longer trunk. Relative flipper length did not differ between the two forms. Mean body length at several growth stages is compared between two forms in Table 12.1. The northern form has on average about a 30% greater body length than the southern form at the same growth stage. Fully grown northern-form whales are 1.7–1.8 m (males) and 1 m (females) greater than southern-form whales.

12.3.5.5 Skull

Miyazaki and Amano (1994) compared proportions of adult skulls between northern and southern forms of the same sex, as is detailed in Section 12.3.7.5. This analysis showed that the northern form had a proportionally broader rostrum and cranium compared with the southern form. Their multivariate analysis of 32 skull measurements with independent variables of form and sex clearly separated the forms and sexes but was mostly influenced by difference in skull size.

12.3.5.6 Body Weight

Body weight has not been measured for the northern form. The following relationships were obtained between body weight (W, kg) and body length (L, cm) of southern-form whales landed at Taiji.

Fetuses ($L \leq 125$cm): $\text{Log } W = 2.8772 \log L + \log 2.432 \times 10^{-5}$

Postnatals ($275\text{cm} \leq L \leq 400$cm): $\text{Log } W = 2.6642 \log L + \log 8.403 \times 10^{-5}$

Since these equations were very similar, the two sets of data were combined to obtain the following equation (Kasuya and Matsui 1984),

All data ($12.5\text{cm} \leq L \leq 400$cm): $\text{Log } W = 2.8873 \log L + \log 2.377 \times 10^{-5}$

This equation can be expressed also as $W = 2.377 \times 10^{-5} L^{2.8873}$. This is compared with the following weight–length relation obtained for the long-finned pilot whale in the North Atlantic.

G. melas: $W = 0.00023 L^{2.501}$, Bloch et al. (1993a).

Using this equation fitted to all data for short-finned pilot whales, average body weights of the southern and northern forms are estimated in Table 12.2. The northern-form neonates weigh about twice as much as those of the southern form and the adults about 2–2.5 times as much.

TABLE 12.2
Body Weight of Two Forms of Short-Finned Pilot Whales off Japan Calculated Using Single Length-Weight Equation (see text)

Growth Stage	Southern Form (kg)	Northern Form (kg)	Northern/Southern
Neonate	37.4	83.6	2.2
At sexual maturity			
Female	392	720–774	1.8–2.0
Male	904	2046	2.3
Testis weight	0.4	0.9	2.3
At physical maturity			
Female	590	1211	2.1
Male	1264	3146	2.5
Largest individual			
Female	802	1562	1.9
Male	1707	4227	2.5

Sources: Kasuya, T. and Marsh, H., *Rep. Int. Whal. Commn.*, 259, 1984; Kasuya, T. and Tai, S., *Rep. Int. Whal. Commn.*, 339, 1993.
Note: Testicular weights were obtained independently.

The northern and southern forms have minor differences in their body proportions and shape of the melon, and a question may arise whether the combined equation is appropriate for both forms. However, cetaceans exhibit large fluctuation in body weight reflecting seasonal changes in physiology or reproductive status. For example, females accumulate lipids in their body during pregnancy and those in near-term pregnancy may weigh more than lactating females. Because such fluctuation is ignored in the equation given earlier, body weight estimated from body length must be considered imprecise, and the effect of morphological difference on the estimated body weight of two forms can be ignored.

12.3.5.7 Meat

I did not pay much attention to the assertion by the whalers that the northern form has less oil, because oil content can vary seasonally and with physiology of animals. However, it was very evident on the flensing platform that the amount of adipose tissue in the muscle was in general much less in the northern form than the southern form. This difference was particularly distinct in adult males. Males of the southern form have lipid-rich "tail meat," but males of the northern form do not. The "tail meat" in Japanese whaling anatomy means a particular part of the skeletal muscle on the tail peduncle. Cetaceans have a total of four large muscle columns, with two columns above and two below the transverse processes of the dorsal and lumbar vertebrae. Japanese whalers call the former *se-niku* [dorsal muscle] and the latter *hara-niku* [ventral muscle]. The muscle above the transverse processes of the caudal vertebrae divides into two columns: upper and lower.

The upper column is the real *o-niku* or *o-no-mi* [tail meat] in whaling anatomy. The muscle of the tail meat contains thin layers of adipose tissue between layers of muscle tissue and is preferred by consumers. However, similar structure may be found also in the dorsal or ventral muscles of well-nourished whales such as fin whales in near-term pregnancy. In such cases the muscle can be classified commercially as "tail meat." Therefore whalers may say "I caught a whale full of tail meat" or "Minke whales usually have no tail meat."

Both forms of the short-finned pilot whale have the tail meat of whaling anatomy, but tail meat of commercial significance is limited to the tail part of adult males of the southern form, which is flavorful as *sashimi* [sliced raw meat], although it smells slightly strong and looks dark. The tail meat of the northern form, even of the adult male, does not have the structure of tail meat in commercial terms, with the exception of thin muscle a few millimeters thick near the vertebrae. This could be one of the reasons why the northern form has lower commercial value relative to body size.

A difference in fat content was evident between the two forms of short-finned pilot whale taken in the same season, that is, September to November. It is known that long-finned pilot whales in the North Atlantic accumulate lipid in autumn and winter, when they weigh 14%–23% more than in summer (Lockyer 1993). I was not able to identify such seasonal variation in physiology while I studied southern-form whales during 8 months of the year (Kasuya and Marsh 1984) or northern-form whales in 3 months of the year, September–November (Kasuya and Tai 1993). We have insufficient data to allow us to see seasonal change in nutrition in the northern form.

12.3.6 Adaptive Significance of Body Size

The warm Kuroshio Current has its origin in an area east of Luzon Island in the Philippines, runs along the east coast of Taiwan, and reaches the east coast of Honshu in Japan. It turns to the east at around Choshi Point to be called the Kuroshio Extension and then the North Pacific Current, and finally reaches the west coast of North America. Inside, or on the right side, of this clockwise semicircle of the Kuroshio and the Kuroshio Extension, there is a clockwise current called the Kuroshio Counter Current, which flows back toward the Kuroshio Current in the western North Pacific. The known range of the southern form short-finned pilot whale is within the area of the Kuroshio and the Kuroshio Counter Current system. The range of the northern form is limited to a smaller area bounded by northern Honshu and Hokkaido to the west, the nutrition-rich cold Oyashio Current to the north and north-east, and the front of the Kuroshio and Kuroshio Extension to the south. This area has an intrusion of an additional weak warm current called the Tsugaru Current in the northwestern corner (Figure 12.8). These currents create the habitat of the northern form as an area of high productivity, but they also make it an area of drastic seasonal change in the environment. In the summer the sea surface temperature rises to 18°C off Kushiro (42°59′N, 144°23′E) near the northern boundary of the habitat, but in winter the temperature there sinks to below 5°C, and sometimes ice floes can be seen. Thus, the sea surface temperatures ranged from 8°C to 23°C at the position of sightings of northern form whales, including some schools only suspected to be of the northern form (Table 12.3).

Sea surface temperature in the habitat of the northern form fluctuates with seasonal fluctuation of the two major currents (Kuroshio and Oyashio) as well as with solar radiation and wind, but the temperature of deeper water is more stable seasonally. Therefore, a water temperature of 15°C at a depth of 100 m has been used as the indicator of the position of the Kuroshio Front (Kawai 1972), which is stable seasonally at the latitude of Choshi Point. Similarly, a water temperature of 5°C (spring) or 8°C (autumn) at 100 m has been used as an indicator of the Oyashio Front, which is located in summer near Kushiro and in winter on a line extending from the Tsugaru Strait (41°30′N, 140°30′E) to the south-east. The northern limit of the northern form in summer almost agrees with the Oyashio Front and the southern limit with the Kuroshio Front in all seasons of the year (Figure 12.8).

The northern form makes maximum use of productivity of the area by expanding its range up to the front of the cold Oyashio Current in summer, and in winter it shifts its northern range down to the coast of Miyagi Prefecture (37°54′N–38°59′N) to avoid very low surface temperatures (Figure 12.8). However, it does not cross the Kuroshio Front to go further south in any season. The Kuroshio and Oyashio Currents often function to limit distribution of small cetaceans off Japan. Dall's porpoises and perhaps harbor porpoises usually remain in summer to the north of the front of the cold Oyashio Current and may extend their range in winter further south, where surface water becomes colder, but they do not usually cross the Kuroshio Front even in winter (Section 9.3.1). The southern-form pilot whale may extend its range to 37°N (Figure 12.8) and striped dolphins to 40°N (Section 10.4.2), between the Kuroshio Front (*c.* 34°N–35°N) and the summer front of the Oyashio Current (*c.* 43°N–44°N), but they retreat to south of the Kuroshio Front in winter (Miyashita 1993). The difference in the northern limit of the summer range between the two species probably reflects their differential dependence on deep water for feeding. Striped dolphins feed probably at lesser depths than southern-form pilot whales; thus their summer range responds more directly to expansion of warm surface waters.

Sea surface temperature at sightings differs between the northern and southern forms around Japan (Table 12.3). The temperature boundary between the two forms is at 24°C in summer. Insufficient data suggest that it is around 17°C–19°C in winter. Small-type whalers in recent years hunted northern-form whales in September through November and within a surface temperature range of 12°C–21°C, with a peak at 16°C–17°C. This suggests that northern-form whales do not usually enter surface water of 10°C–15°C, which is the southern boundary of the cold Oyashio Current in that season (Kasuya et al. 1988).

Both the northern and the southern forms undergo seasonal north-south shifts in their habitats. This functions to

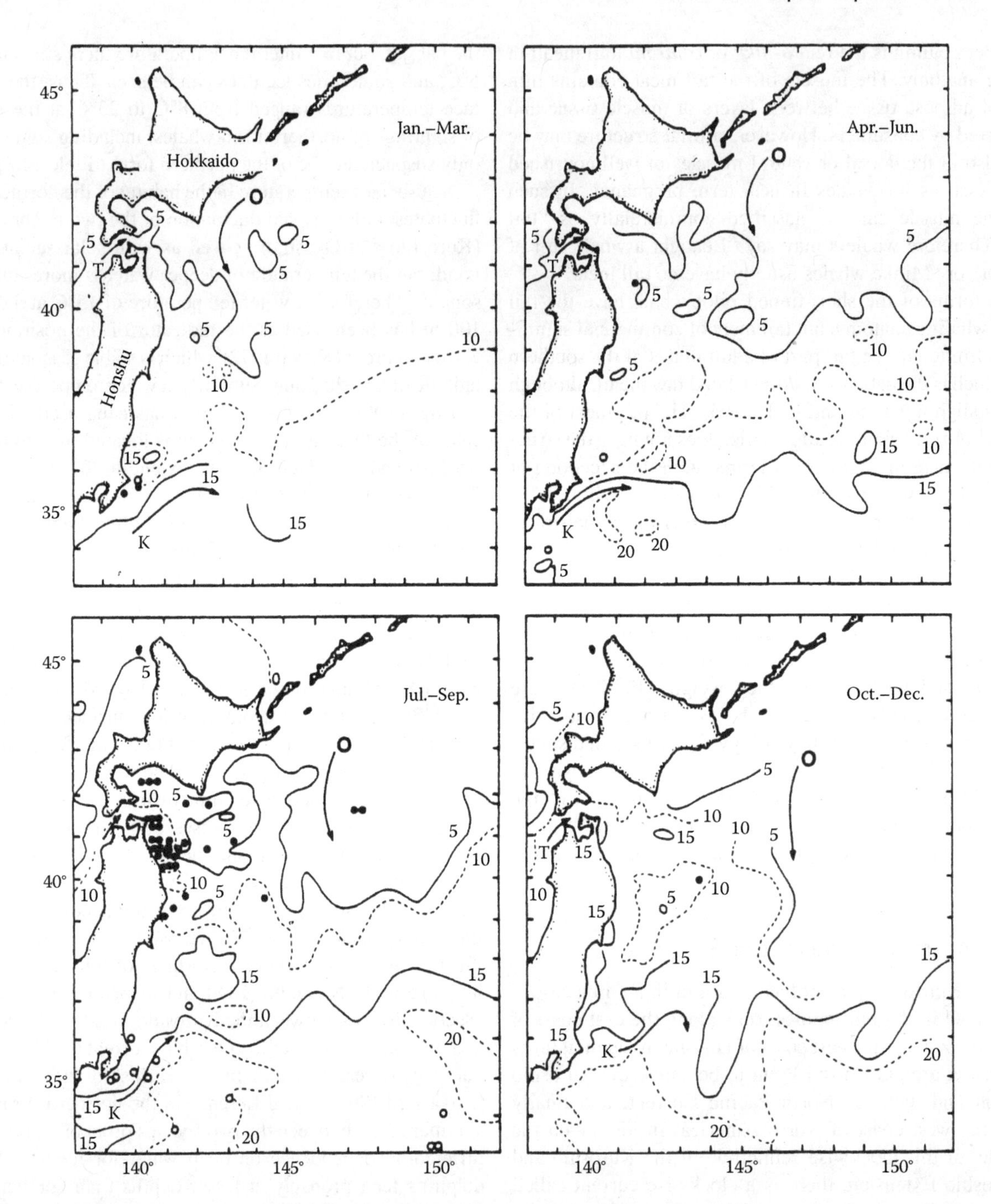

FIGURE 12.8 Segregation of southern-form (open circles) and northern-form (closed circles) short-finned pilot whales and subsurface water temperature (in Celsius) at 100 m in 1985. Front of the warm Kuroshio Current agrees with 15°C isotherm at 100 m and that of the cold Oyashio Current with 5°C (spring) and 8°C (autumn) isotherm at the same depth (Kawai 1972). Direction of cold Oyashio Current (O) and warm Kuroshio Current (K) indicated by arrows. Data used are those in Figure 12.5 plus additional records obtained by the end of 1987. (Reproduced from Kasuya, T. and Tai, S., *Rep. Int. Whal. Commn.*, 339, 1993, Figure 17.)

decrease seasonal change in their environmental temperature, but the degree of movement is insufficient to keep them in a particular temperature environment. It appears to me that their strategy is to minimize seasonal movement within their range of temperature tolerance. Since this strategy is more pronounced in the northern form that inhabits colder waters, it is likely to have a greater tolerance for a cold environment.

Hibernation is one of the ways adopted by mammals to cope with scarcity of food or low temperature; they lower body temperature and minimize activity in a den to decrease energy consumption until a season of high productivity arrives. They spend winter in a half-dead condition without significant growth and usually without reproduction. Cetaceans live in water and are unable to hibernate. Another way to cope with cold is to maintain body temperature constant with improved thermal insulation and perhaps increase heat production that relies in part on nutrition stored in the body. Cetaceans have lost body hair for insulation because

TABLE 12.3
Sea Surface Temperature (SST, in Celsius) at the Position of Sighting of Short-Finned Pilot Whales in the Western North Pacific

	Northern Form				Southern Form			Unknown Form		
	Month				Month			Month		
SST (n)	Jan–Mar	Apr–Jun	Jul–Sep	Oct–Dec	Jan–Mar	Jul–Sep	Oct–Dec	Apr–Jun	Jul–Sep	Oct–Dec
8								1		
14									1	
15		1								
16	1							1		
17	1			1						1
18										
19			6							
20			1		2					1
21			1		3			1		
22			4		2				1	2
23			4		1					
24						4				
25						1			1	
26						4	1		1	
27						12				
28						5		2	1	
29						2				
30						1				

Source: Reproduced from Kasuya, T. et al., *Sci. Rep. Whal. Res. Inst.*, 39, 77, 1988, Table 4.
Note: SST for ≥n and <n + 1 (n being an integer).

the maintenance of hair requires considerable cost and the insulation efficiency is extremely limited in deep water under high pressure. Thus, their thermal regulation must depend on control of heat loss through control of blood circulation, lipid deposited in the blubber, and heat production, which is a function of body size. Long-finned pilot whales are known to increase lipid deposition in the blubber during autumn and winter (see Section 12.3.5.7; Lockyer 1993). This functions as nutrition storage as well as for insulation. Larger body size also functions to increase heat production and to decrease relative heat loss by decreasing body surface relative to weight.

Body temperature of mammals is maintained by heat produced by muscles and internal organs. Thus heat production is proportional to body weight if body structure is constant. If body structure and external temperature are constant, heat loss is proportional to the body surface. Body weight is proportional to the cube of body length if the body shape is constant, while body surface is proportional to the square of body length. In other words, body surface per body weight is in inverse proportion to body length. The average body lengths of neonates of short-finned pilot whales, which are more sensitive to lower temperature than adults, are 185 cm in the northern form and 140 cm in the southern form (Section 12.4.3.1). This means that body surface per body weight of the northern-form neonate is only 76% of that of the southern-form neonate, which suggests greater tolerance of the northern form to low environmental temperature. This is probably an adaptation by the northern form to the colder northern environment, which possibly offers a benefit of high productivity.

Kasuya and Tai (1993) estimated the degree of resistance to a cold environment acquired by the northern form. Many mammal species maintain body temperature in a narrow range, except in the case of hibernation. If heat production increases through high activity, cetaceans increase heat dumping from the fins by increasing blood flow, and if heat production decreases or heat loss increases, they decrease blood flow to the body surface and increase retrieval of heat through a counter-current system of arteries and veins (Berta et al. 2006). However, if the environmental temperature goes below a critical temperature (Tl, °C) for the individual and maintaining body temperature at normal level (Tb, °C) becomes difficult, the individual will increase heat production to keep body temperature at a normal level. Shivering is one such response. Peters (1983) proposed the following equation among Tl, Tb, and body weight (W, kg):

$$Tl = Tb - 14.6W^{0.182}$$

The metabolic rate of cetaceans is probably the same as that of land mammals. Their normal body temperature is 36°C–37°C (Gaskin 1982). Applying this body temperature and the neonatal body weights of short-finned pilot whales in Table 12.2 in the equation presented earlier, the critical environmental temperature (Tl) for newborn northern-form short-finned pilot whales is estimated at 3.3°C–4.3°C and that of the southern form at 7.8°C–8.8°C. The critical environmental temperature of newborn northern-form whales is 4°C–5°C lower than that of southern-form neonates.

The equation of Peters (1983) is influenced by epidermis and subcutaneous fat, and the derived critical environmental temperature can vary by season in the same individuals. Even if validity of the absolute figures for the estimated critical environmental temperature is questioned, the 4°C–5°C difference between the two forms of short-finned pilot whale is real.

Parturition in the northern form is believed to be limited seasonally, mostly occurring in roughly a 4-month period from November to February. The average sea surface temperature at the northern limit of the population in this season is around 12°C (Figure 12.5; Kasuya et al. 1988). On the other hand, parturition can occur any time of the year in the southern form, with a peak period extending for 8 months from April to November. The mean sea surface temperature at the northern limit of the southern form is about 24°C in summer and 18°C in winter. Thus, the sea surface temperatures met by newborns of the two forms can differ by about 6°C, which is almost the same as the difference between the estimated critical environmental temperatures. It is likely that selection has worked toward augmentation of body size of neonates during evolution toward that of the current northern form.

Since body size of the mother is expected to have a positive correlation with that of the neonate, selection pressure could also have worked toward augmentation of female body size. This would also influence the body size of males because many genes controlling body size are on the autosomal chromosomes. The larger body size thus achieved contributes to increase in capacity for storage of nutrition and to reduction of heat loss, which both benefit life in colder waters and in an environment of seasonally fluctuating food supply. Such selection will be stronger in populations inhabiting colder area with greater seasonal fluctuation in the environment. Seasonal environment change is smaller in the Kuroshio and Kuroshio Counter Current area than in the area bound by the cold Oyashio Current and warm Kuroshio Current and its extension, which the northern form inhabits. Thus, there may have been stronger pressure for large body size in the northern form.

The next question to be answered is why the northern form has its parturition in a narrow period of the winter months. One way of avoiding the disadvantage of parturition in a cold temperature is to have parturition in the summer, but this does not occur in the northern form. The breeding season of mammals is controlled by several factors. One is the nutritional condition of females, which have to go through estrus, implantation, fetal growth, and lactation. A second is choice of a parturition season well suited for the growth and survival of neonates. Food availability for nursing females is a key element for the survival of neonates and success of breeding because the energy requirement of lactating females is believed to be greater than that of pregnant females (Lockyer 1981). This explains why many herbivorous animals accumulate food stores in summer, have estrus in the autumn, and produce neonates in early spring when plants start growing. Success in weaning is another key factor for breeding. Thus, the timing for switching from milk to solid food must be in a season of best food availability. The narrow breeding season of the northern form suggests that their reproductive success has been controlled seasonally more strongly than in the southern form.

As pregnancy of short-finned pilot whales lasts about 15 months for both forms, the mating peak is in August to November in the northern form and September to January in the southern form (Section 12.4.4.1; Kasuya and Tai 1993). Juveniles of the southern form start taking solid food at 6 months of age but continue suckling for a minimum of 2.5–3 years. Although we do not know how long suckling continues in the northern form, the time to start taking solid food will not be different between the two forms because it seems to be a common feature among several toothed whales at age 3–6 months. Juvenile northern-form whales probably start switching from milk to solid food in May–August. Switching from milk to solid food is important for later growth in juveniles. Once they succeed in this process, the importance of suckling decreases rapidly because their solid food is of high nutritional value.

Southern-form pilot whales off Taiji consumed only squid, while northern-form whales killed off the Sanriku coasts in early winter ate some octopus and fish in addition to the major food of squid. The squid species consumed by the northern form were, in decreasing order, Pacific flying squid (*Todarodes pacificus*), neon flying squid (*Ommastrephes bartramii*), and *Eucleoteus luminosa* (Kubodera and Miyazaki 1993). The Pacific flying squid, the most important food species for the northern form, is distributed in the most coastal waters, and the next important, the neon flying squid, is distributed more offshore. Both species feed in summer in Pacific waters off Sanriku and Hokkaido (41°30′N–44′N) and have a southward migration in autumn to the spawning grounds. The two squid species are fished off the Sanriku coast; the fishery has a peak catch in August–October. Monthly landing of these species in other months of the year has been less than 30% of that in a peak month. Thus, the peak of migration of the two most important food species for the northern form matches with the time when juveniles switch their major nutrition from milk to solid food. In short, the large neonatal size of the northern form is necessary for the survival of neonates produced in early winter, and placing their parturition season in early winter benefits calves with successful weaning in a squid-rich season. Having the mating season in August to November, when the food supply is abundant, also benefits females in successful maintenance of pregnancy and accumulation of food stores for subsequent nursing.

12.3.7 Taxonomy of the Two Forms of Short-Finned Pilot Whale

12.3.7.1 Background of the Problem

In my view, understanding the degree of difference between two animals is a part of biology, but classifying the animals is a product of human culture created for convenience in understanding the biology. Sometimes one may find certain characters only in one of two groups of animals in a common habitat. Such a case is often interpreted to mean that they have established some mechanism of reproductive isolation, and the two groups are usually dealt with as separate species. There might be a case where the habitats of the two animal groups do not overlap, as in the case of the northern and southern forms of short-finned pilot whales off Japan. Even in such a case the two forms might be classified as separate species based on discontinuity of characters. However, there may be another case where a character found to be discontinuous between two animal groups is of a type that is highly plastic or is only identifiable at a particular growth stage, or where the discontinuity becomes questionable if we assume a third group of intermediate type. Difficulty may arise in the classification of such animal groups, as is possibly the case we encounter for the two allopatric populations of the short-finned pilot whale off Japan.

We have generally accepted the current classification of the genus *Globicephala* to include the two recent species of long-finned (*G. melas*) and short-finned (*G. macrorhynchus*) pilot whales by using skull morphology and supplemental external characters. These morphological characters result in the classification of the two forms of pilot whale off Japan into the single species *G. macrorhynchus*. Two questions remain: (1) whether there is any basis for dealing with the two forms as different species and (2) what their taxonomic status is if they are to be dealt as of the same species.

12.3.7.2 Interpretation of Life History

Pigmentation is one of the difference in the external morphology of the northern and southern forms (Kasuya and Tai 1993), but it is too variable within the two species of the genus *Globicephala* and is unsuitable even for differentiating between the two species (Section 12.1.2).

Body size is the most distinct morphological difference between the two forms of short-finned pilot whales off Japan. The northern form grows to about 2 m (males) or 1 m (females) greater than the southern form, or 1.3–1.4 times the body length and 2.5–3 times the body weight of the southern form (sexes combined) (Tables 12.1 and 12.2). Short-finned pilot whales in the Indian Ocean are known with body size intermediate between those of the northern and southern forms off Japan (Kasuya and Matsui 1984). Another delphinid species, *Stenella longirostris*, is known with broad geographical variation in body length, where individuals in one population grow to 1.4 times greater than those in another population (Perrin et al. 1989). Mammals often exhibit large geographical variation in body size, and cetaceans and humans are no exception. The body size difference between the northern and southern forms is of a magnitude that is expected between populations within a species.

Life history is also a tool for describing the character of certain animal groups. Kasuya and Tai (1993) identified common features of life history between the two forms of short-finned pilot whales: (1) females cease reproduction by age 37, (2) females cease ovulation at ages 30–40, and (3) maximum longevity in female is 62 years and in male 45 years and the difference in maximum age between sexes is 17 years). Studies by Bloch et al. (1993a) and Martin and Rothery (1993) revealed that the long-finned pilot whale in the North Atlantic continues ovulation for life, which is different from the case in short-finned pilot whales off Japan, but there was some degree of similarity with short-finned pilot whales in longevity (female 59 years, male 46) and cessation of pregnancy, with some exceptions, after age 46. Thus, the life history characters of the two forms of short-finned pilot whales off Japan are quite similar and different from those of long-finned pilot whales in the North Atlantic. In addition to this, using the similarity in ecological niche occupied by two allopatric forms off Japan, Kasuya and Tai (1993) believed that the northern form of pilot whales off Japan constitutes a population within animals of similar morphological characters in a broader geographical range extending perhaps to the northwest coast of North America and that the two forms of the short-finned pilot whales in the North Pacific should be dealt with as geographically separate races or subspecies.

12.3.7.3 Isozymes

Compared with morphological characters, which often have problems in quantitative analysis and a risk of being subjective, genetic information can allow objective evaluation of evolutionary distance between animal groups. Isozymes have been used as a tool to evaluate genetic difference between the two pilot whale populations. Wada (1988) analyzed polymorphisms in 36 enzymes of 204 northern-form and 167 southern-forms pilot whales, and found that 31 enzymes were controlled by a single allele which was common between the two forms but that the remaining 5 enzymes were controlled by 2–3 alleles.

One of the parameters he derived from the analysis was genetic distance (D) between the two forms calculated using the frequency of alleles controlling the 36 enzymes. The value of D would be zero if genes and their frequency were the same between the two samples and infinite if they completely disagreed. The value of D was 0.0004 for the two samples of pilot whales, which is quite small compared with the genetic distance between the striped dolphin and the pantropical spotted dolphin (D = 0.026) or between the sei whale and Bryde's whale (D = 0.047). He noted that if sample size is over 50, the value of D is influenced by the number of genetic loci rather than by the sample size, calculated D assuming a 37th hypothetical locus that differed distinctly between the two samples, and confirmed that the D value was not significantly affected by such an extreme assumption.

Another point identified by Wada (1988) was the frequency of alleles of the five polymorphic loci. He found that the allele of the highest frequency for each of the five loci was common between the northern and southern forms. He noted the allele of the highest frequency for one species would not be the most frequent one for another similar species. He interpreted his finding as an indication that the two forms of pilot whales off Japan should be dealt with as a single species and that the distance between the two forms was of a level expected for different subspecies or even less.

In their study on skull morphology (see Section 12.3.7.5), Miyazaki and Amano (1994) expressed a concern about the samples analyzed by Wada (1988). They accepted the northern-form sample of 204 individuals, which were taken individually by whalers, but had a problem with the southern-form sample of 167, in that they would not represent the population because they were obtained from a limited number of matrilineal groups. If their concern was justified, the genetic difference calculated by Wada (1988) could be biased either way. In my view, the skull morphology analyzed by Miyazaki and Amano (1994) could also be biased. However, their concern could be unjustified because Wada (1988) stated that his southern-form sample was obtained from 5 drives and that he confirmed Hardy-Weinberg equilibrium in the sample, which suggested that reproduction could not be assumed nonrandom. Kage (1999, in Japanese) showed that each southern-form school driven at Taiji and examined by him contained multiple maternal lines and suggested that paternal genes in his sample were most likely to have derived from outside the maternal lines.

12.3.7.4 Mitochondrial DNA

Kage (1999) analyzed 375 bases in the control region of mitochondrial DNA (mtDNA) of short-finned pilot whales and long-finned pilot whales of the world. His material for the short-finned pilot whale was a total of 42 southern-form whales, including 37 from 7 schools driven at Taiji and 5 taken by the crossbow fishery at Nago in Okinawa, 4 northern-form whales taken by small-type whaling off Ayukawa, and data for 2 individuals from the eastern tropical Pacific obtained from GeneBank. He identified some school-specific variation, but it was smaller than the difference between the two forms. Difference in base sequence between individuals of the southern form was 0.3%–0.8% and that between individuals of two forms 1.4%, both much smaller than 3.7%, the corresponding figure between *G. melas* and *G. macrorhynchus*.

A dendrogram derived from his analysis placed the two forms of short-finned pilot whales off Japan in a group together with the same species in the eastern tropical Pacific. The North Atlantic long-finned pilot whales formed another group. These two groups formed a group of the genus *Globicephala*. Thus, Kage (1999) concluded that the northern and southern forms of pilot whales off Japan belong to the species *G. macrorhynchus*, which together with *G. melas* forms the genus *Globicephala*. Another interesting result of his study was the position of northern-form short-finned pilot whales off Japan, which formed a group with the same species in the eastern tropical Pacific as well as that in the central North Atlantic. This result allowed speculation on the evolution of short-finned pilot whales globally (see Section 12.3.7.7).

12.3.7.5 Skull Morphology

Miyazaki and Amano (1994) analyzed skull morphology to evaluate taxonomy of the two forms of short-finned pilot whale off Japan. First they eliminated variation due to age by examining specimens aged 10 and over (southern form) or 17 and over (northern form). They employed 32 measurements. The adult samples contained 25 southern-form skulls (10 males and 15 females) and 17 northern-form skulls (8 males and 9 females). Length of the southern-form skulls ranged from 61.2 to 70.6 cm (males) and from 55.2 to 59.8 cm (females) and those of northern-form skulls from 70.0 to 78.3 cm (males) and from 61.0 to 64.8 cm (females). There was only slight overlap for males. The maximum width of the southern-form skull was 45.9–51.9 cm (males) and 38.5–42.7 cm (females) and those for northern-form skulls 57.6–62.3 cm (males) and 44.9–48.9 cm (females); there was no overlap. Skull size strongly reflected difference in body size.

Miyazaki and Amano (1994) carried out an analysis of covariance using skull length as a covariate on log-transformed measurements. They first compared the mean values between sexes of the same form and found significant difference in some measurements, most of which related to the widths of the rostrum and the cranium. Males had larger values. This means that males of both forms had skulls that were broader than those of females of the same form. They compared the mean values between the two forms of the same sex and found significant difference in 20 (female) or 21 (male) measurements. Many of these measurements with differences were again of width of rostrum and cranium and showed that both sexes of the northern form had a broader skull than the same sex of the southern form.

They also carried out principal components analysis of the 32 measurements and made a two-dimensional scatter plot using two components. They found that the specimens were separated into a total of four groups correlated with sex and geographical form. However, as they stated, only the first component contributed to the separation of four groups (two sexes and two geographical forms), and the first component was most strongly influenced by the size of the skull, particularly skull length, rostral width, and cranial width. The second component only separated females of the northern form from males of the southern form but did not separate five other group pairs: sexes of the northern form, sexes of the southern form, males of the northern and southern forms, northern-form males and southern-form female, and females of the northern and southern forms. The significance of skull size, which is a reflection of body size, in taxonomy of the genus *Globicephala* is questionable.

Based on these analyses, Miyazaki and Amano (1994) proposed that the two forms of pilot whales off Japan should be dealt with as separate species. However, the information presented by them appears to be insufficient to support this conclusion. While their analysis indicating morphological difference between the two populations is convincing, I do not see a basis for classifying them into two species. For that particular purpose, they should have added to their analysis a minimum of two additional samples, which would include a third sample of *G. melas* and fourth sample of a different population of either *G. melas* or *G. macrorhynchus*. This addition could have clarified the magnitude of morphological difference between the currently accepted two species of *Globicephala* and that between populations within a species, which could have contributed to evaluating the difference Miyazaki and Amano (1994) found between the two forms off Japan.

12.3.7.6 Considerations on the Evolution of *G. macrorhynchus*

The northern form off Japan differs from the southern form in the presence of a distinct saddle patch, larger body size, and some other relatively minor morphological differences. Table 12.4 lists available information on the saddle patch of short-finned pilot whales in the world (see also photograph in frontispiece). The records include those made by individual authors through the process of individual identification and can be considered to be reliable. It is known that both pigmentation types occur and segregate geographically within the ocean basins of the North Pacific and North Atlantic.

The pigmentation of short-finned pilot whales off the west coast of North America is close to that of the northern form off Japan. Body length of a female stranded on the coast of British Columbia was 452 cm (Baird and Stacey 1993). Such large size occurs in the northern form off Japan but not in the southern form. Although the saddle patch was not described for this particular female, it was likely to have had a distinct saddle patch as observed in the species off Seattle and California. I interpret these observations to indicate that the northern form off Japan is morphologically closer to short-finned pilot whales off the northwest coast of North America.

Body lengths of 46 male and 108 female short-finned pilot whales were available in Ogden et al. (unpublished and cited in Kasuya and Matsui 1984). The geographical range seems to cover the entire Atlantic coast of the United States, including the coasts of the Gulf of Mexico and New Jersey where different pigmentation types are known to occur (Table 12.4). Therefore the measurements probably come from two pigmentation types, but the maximum body size of 535 cm (male) and 397 cm (female) are close to those for the Japanese southern form.

Polisini (1980) studied geographical skull variation in the genus *Globicephala* using two discrimination functions. One of the functions succeeded 100% in distinguishing between the two species in the genus. The second function was able to identify the origin of most of the short-finned pilot whales from the North Pacific and North Atlantic, but there was some overlap. Then he made a two-dimensional scatter plot using the two discriminate functions and found that the two species were fully separated from each other and that short-finned pilot whales from off California were fully separated from those from off the East Coast of the United States. As expected, two individuals from Bermuda and Florida were included in a group of U.S. East Coast individuals. I note that two short-finned pilot whales, one from Ayukawa in northern Japan and the other from Alaska, joined the California group. One animal from Arari on the coast of the Izu Peninsula was erroneously placed in the California group and another from California was misclassified with the Arari group, which I also felt was interesting.

The analyses of Polisini (1980) caught my attention particularly because short-finned pilot whales in the broad geographical range of the West Indies, the South Atlantic, the Indian Ocean, the South Pacific, and the western North Pacific off central Japan were grouped together, including one specimen from the Bahamas, four from West Africa, one from the Cape Region of South Africa, one from Malacca, one from Timor, one from the Philippines, one from the Marquesas, one from Hawaii, and three from central Japan (Arari on the Izu Peninsula and Wakayama Prefecture, possibly represented by Taiji). It is reasonable to assume from the location that the three from central Japan belonged to the southern

TABLE 12.4
Variation of Saddle Mark of Short-Finned Pilot Whales in the Northern Hemisphere

Location	Saddle Mark Similar to the Northern Form off Japan		References
North Pacific			
Off Northern Japan	Present		Kasuya et al. (1988)
Off Vancouver Island	Not mentioned		Baird and Stacey (1993)
Off Seattle	Present		Yoshioka et al. (1987)
Off California	Present		Mitchell (1975); Evans et al. (1984); Shane and McSweeney (1988)
West Coast of Baja California	Present		Kasuya et al. (1988)
East Coast of Baja California	Present		Kasuya et al. (1988)
Off Chiba-Okinawa		Absent	Kasuya et al. (1988)
Off Taiwan		Absent	Chou (2000, in Chinese); Wang and Yang (2007, in Chinese and English)
Philippines		Absent	Ma and Tan (1995)
New Guinea		Absent	Mitchell (1975)
Hawaii		Absent	Minasian et al. (1987); Shane and McSweeney (1988)
North Atlantic			
Gulf of Mexico	Present		Würsig et al. (2000)
Off New Jersey		Absent	See frontispiece
Azores Islands		Absent	Heimlich-Boran (1993)
Canary Islands		Absent	Reeves et al. (2002)

form. Polisini (1980) did not have material from the eastern tropical Atlantic. The result indicates that short-finned pilot whales in the broad area from the western North Atlantic, the Indian Ocean, the western tropical Pacific, and the coast of central Japan have similar morphological characters, while those in the North Pacific area from Baja California, the west coast of the United States, and northern Japan have common features of the skull.

During the summers of 1982 through 1987, sighting surveys for Dall's porpoises were conducted in the northern North Pacific north of 35°N, or north of the Kuroshio Extension and North Pacific Current (see Section 9.3.3 and Figure 9.10 for survey range), but the surveys did not encounter short-finned pilot whales in the central area in longitudes from 150°E to 135°W. This suggests that the northern form off Japan constitutes a population that is geographically separated from similar forms off the west coast of North America. Similarly, it is reasonable to expect multiple tropical North Pacific populations that share morphological characters with the southern-form short-finned pilot whales off Japan.

While Polisini (1980) addressed systematics of pilot whales using skull morphology, Oremus et al. (2009) used mtDNA for the same purpose. They analyzed a sequence of a maximum of 620 bases in the control region of mtDNA and calculated

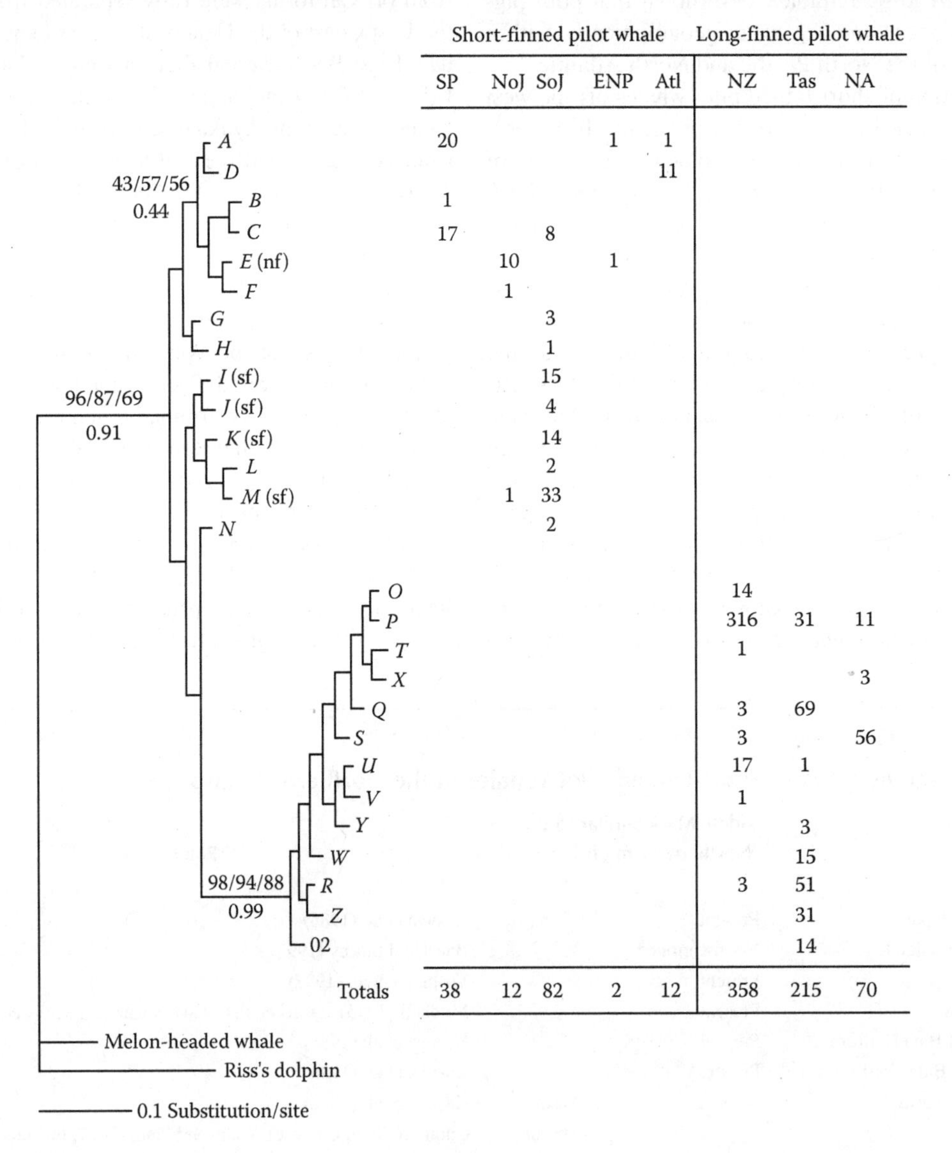

FIGURE 12.9 Phylogenetic relationship among mitochondrial DNA haplotypes of short-finned (14 haplotypes from A to N at top group) and long-finned (13 haplotypes from O to 02 at bottom group) pilot whales. Haplotypes are labeled with capitals. Small letters next to the haplotype, "nf" and "sf," indicate that the haplotype was obtained from Japanese individuals identified either as northern or southern form by external morphology. Numbers at right indicate sample size and the letters at top geographical location of the sample (Atl, Atlantic; ENP, eastern North Pacific; NA, North Atlantic; NoJ, northern Japan; NZ, New Zealand; SoJ, southern Japan; SP, South Pacific; Tas, Tasmania). Japanese material (labeled NoJ or SoJ) includes market samples of unknown or estimated origin that lack morphological identification to geographical form (see text). (From Oremus, M., Gales, R., Dalebout, M.L., Funahashi, N., Endo, T., Kage, T., Steel, D., and Baker, S.C.: Worldwide mitochondrial DNA diversity and phylogeography of pilot whales (*Globicephala* spp.). *Biol. J. Linnaean Soc.* 2009. 98. 729–744. Copyright Wiley-VCH Verlag GmbH & Co. KGaA. Reproduced with permission, Figure 3.)

similarities between geographical regions (Figure 12.9). This study was particularly interesting because it included Japanese specimens identified by Kage (1999) as of the northern and southern forms based on external morphology and location. However, the study included problematic Japanese market sample from unknown locations of capture and thus not identifiable to geographical form. They included all the specimens purchased in Chiba Prefecture or to the south in a southern Japan sample (SoJ) and those purchased in the north of the range into a northern Japan sample (NoJ). In view of the Japanese distribution system for whale products, such a classification seems to be too simplistic.

Oremus et al. (2009) created a dendrogram by a stepwise combination of closest haplotypes (Figure 12.9) and identified two distinct groups represented by the current *G. melas* and *G. macrorhynchus*. The authors did not find common haplotypes between the two species and accepted the current taxonomy that recognizes two species in the genus.

One of the two points of interest on the long-finned pilot whale in Oremus et al. (2009) was genetic difference between the neighboring waters of New Zealand and Tasmania, suggesting presence of local populations. A second point was the high genetic variability in the Southern Hemisphere (12 haplotypes), compared with only 3 haplotypes found in the North Atlantic, 2 of which were common with those in the Southern Hemisphere. Based on this observation the authors supported a classification dealing with long-finned pilot whales in the two hemispheres as separate subspecies and concluded that the species in the North Atlantic had its origin in immigrants from the Southern Hemisphere.

Long-finned pilot whales were present in the northern North Pacific up to at least about 6500 years BP to the thirteenth century (Section 12.1.1). Two possibilities have been proposed for their origin. One is that they crossed the equatorial region of the eastern tropical Pacific during the last glacial period that ended about 10,000 years ago. The other is that they crossed the Arctic Ocean during a warmer period after the last glacial period (Kasuya 1975). Genetic analysis may yield an answer.

Oremus et al. (2009) found the highest genetic variation in the short-finned pilot whale in the western North Pacific (11 haplotypes), followed by 3 in the South Pacific and 2 in both the eastern North Pacific and the North Atlantic. The authors stated that the southern form off Japan had 9 haplotypes and the northern forms 2 haplotypes, although their understanding is based, in part, on market samples that were of dubious geographical origin in my view.

The dendrogram of Oremus et al. (2009) separated the short-finned pilot whales into two indistinct subgroups. One of them included individuals of Japanese origin identified by pigmentation as of the southern form (marked “sf” in Figure 12.9) and had higher genetic variation than the second subgroup that included northern-form whales from off Japan (“nf” in Figure 12.9). Genetic variation of the northern form was low and agrees with our understanding that the sample is likely from a small local population. The second subgroup of short-finned pilot whales, including the northern form off Japan, covered a broad geographical range in the eastern North Pacific (ENP), South Pacific (SP), Atlantic (Atl), and Northern Japan (Noj), and common haplotypes were found among two or three geographical areas.

Oremus et al. (2009) paid attention to haplotype C, which was found in the second subgroup of short-finned pilot whales and was shared by 17 South Pacific samples and 8 market samples out of the 82 purchased in southern Japan (Soj). The authors noted that this haplotype had a large difference from other haplotypes known from southern-form whales from off Japan. They suggested two possibilities: (1) these 8 Japanese market samples were brought by fishermen from a distant fishing ground or (2) the southern form of short-finned pilot whales off Japan contained more than one population. I would propose some additional possibilities.

Although it can be true that most small cetaceans caught in southern Japan are consumed in southern Japan and that samples purchased in markets south of Chiba mostly represent animals taken in Chiba or to the south (Chiba, Izu, Taiji, and Nago in Okinawa), there is evidence that small cetaceans caught in northern Japan have been sold in southern Japan. Endo (2008, in Japanese) reported cases where a wholesale market in Fukuoka City (33°35′N, 130°24′E), northern Kyushu, received meat of northern-form pilot whales caught off the Sanriku Region northeastern Japan as well as that of southern forms taken off Nago in Okinawa, southwestern Japan. I have confirmed a case in the 1980s where a small-type whaler landed northern-form pilot whales at Ayukawa in the Sanriku Region and sent the meat to southwestern Japan, and another case in 2011 where Dall’s porpoises from the Sanriku Region as well as minke whales from Japanese scientific whaling were routinely processed by *Maruko Shoji* in Shimonoseki (33°57′N, 130°56′E) and distributed in Yamaguchi Prefecture (approximately 33°45′N–34°45N, 130°45′E–132°15′E) and the nearby area see Section 2.4 for Dall’s porpoise market).

Thus, there remains another possibility that the 8 of the 82 pilot whale meat samples purchased by Oremus et al. (2009) in southern Japan (labeled “SoJ” and not identified as “sf “in Figure 12.9) and found to have haplotype C actually came from northern-form whales landed in northern Japan. This disagrees with the interpretation of Oremus et al. (2009) and requires further consideration in interpreting the zoogeography of the species.

Oremus et al. (2009) did not present an opinion on the taxonomic status of the northern and southern forms off Japan. It is true that the genetic distance between the two subgroups in *G. macrorhynchus* was smaller than that between the two species of the genus *Globicephala*. Before we determine their taxonomic status we will have to reevaluate differences between two subspecies of *G. melas* and compare those with our knowledge of geographical variation within the species *G. macrorhynchus*.

12.3.7.7 Evolution and Zoogeography of Short-Finned Pilot Whales

The western tropical Pacific could have remained as an area of the highest sea surface temperatures in the entire Pacific area through past climate changes. It could have functioned as a refuge for warmwater species during cold climates and supplied outward emigrants when a warm climate returned. This applies to the short-finned pilot whale. The currently identified range of the southern-form pilot whale is located in the northern part of this refuge. This, together with the fact that pilot whales in this range exhibit the highest genetic variability, supports the hypothesis about their historical dispersal. The small body size of these whales should also be noted.

The current distribution of short-finned pilot whales in the Indian Ocean is discontinuous from that in the South Atlantic (Section 12.2), but contact is likely to have occurred during past periods of warmer climate, the last of which occurred after the last glacial period that ended about 10,000 years ago. South America extends south to almost 56°S and both coasts are bordered by cold currents, so the possibility of short-finned pilot whales passing around Cape Horn would have been smaller.

An early type of short-finned pilot whale could have remained in the Kuroshio and its Counter Current area and evolved into the present southern-form short-finned pilot whale, while others expanded their range into the South Pacific and the Indian Ocean and further to the South Atlantic, passing Cape Agulhas in a warm period. If this hypothesis is correct there should be a cline of genetic variability from the western tropical Pacific to the South Pacific, Indian Ocean, and South Atlantic, for which we do not have information.

There would have been another group of early short-finned pilot whales that moved to the east beyond the Hawaiian Islands and reached waters off Baja California. The gradient of sea surface temperature was small in the eastern tropical Pacific east of Hawaii and it was not a big obstacle for the dispersal; high ocean productivity could have invited the emigration. These animals, after reaching the California Current area, could have started evolving toward the northern form of short-finned pilot whale. Utilizing an abundant food supply and adjusting body size and reproductive cycle to the cold environment, they probably established a population off the Vancouver Island area. At the final stage some members of the population could have proceeded further west along the Arctic Convergence to settle in waters of high productivity off northern Japan between the cold Oyashio and warm Kuroshio Currents. These would be the current northern-form short-finned pilot whales off Japan. Timing of these incidents remains to be investigated.

This hypothesis assumes that the southern-form and northern-form pilot whales started from the same origin. One group of them remained in the original environment which was available in the Kuroshio Counter Current area and achieved relatively minor differentiation to become the current southern form, and the other expanded the range counterclockwise while acquiring greater modification in body and way of life to finally reach northern Japan. The two forms now face each other across the Kuroshio Front, where the environmental gradient is steep enough to be a barrier in their distribution. If this hypothesis is correct the genetic distance between the two forms off Japan should be greater than the distance between the Japanese northern-form and short-finned pilot whales off the west coast of North America, which is suggested by the small samples in Figure 12.9. The idea regarding the eastward expansion of short-finned pilot whales in the North Pacific emerged in 1967 in discussions with R.L. Brownell at Arari in Japan, where we were collecting skulls of short-finned pilot whales for taxonomic research.

12.4 LIFE HISTORY

12.4.1 Viewing Life through Carcasses

For study of any biological aspect of any mammal species, background information on the targeted individual such as age and maturity is important. The information would be the best if records of its whole life since birth were available. Individual identification has often been used for studies on behavior and social structure of cetaceans. The technique is particularly useful for species that have easily identifiable external characters, as in the case of northern-form short-finned pilot whales, but is applicable only in certain area of the range where agreement exists to not harm the animals being studied. I once attempted individual identification of the northern-form population off Sanriku on the Pacific coast of northern Japan but quickly realized the incompatibility of the methodology with the whaling operations, and I was obliged to switch my effort toward understanding life history through examination of whale carcasses brought back by whalers.

Relationships between biological events and accompanying behavior can be understood by following the daily life of particular individuals. Such efforts have increased our knowledge on individual variation in the breeding cycle of cetaceans and can contribute, with some caveats, to determining the variability of lifetime production among females and the causes of the individual variation. This methodology still has some limitation for species that inhabit broad pelagic waters or have large population sizes, whereas carcass studies can still make some contribution by yielding biological information mostly in the form of averages for the population being studied.

Their extended life span is another difficulty that accompanies studies on wild cetaceans. Depending on species, they reach sexual maturity at 3–17 years and live for 20–100 years or more, which is almost equivalent to the lifetime career of a scientist and is sufficiently long to expect large changes in the environment surrounding the targeted cetaceans. When we consider the magnitude of destruction of the marine environment that humans have perpetrated in the past 100 years, such as through increased underwater noise, accumulation of heavy metals and persistent organic compounds, and overexploitation of marine organisms, it is easy to imagine the possible additional destruction that may occur during the next 100 years. We will face the difficulty of separating age-dependent change in cetacean life history from response to environmental change.

Examination of cetacean carcasses available from whaling operations offers a quick view of a horizontal aspect of cetacean life history and can serve as a proxy for obtaining a vertical view. The current world atmosphere surrounding cetacean biology does not tolerate killing whales for science unless there is urgent need or a good contribution to conservation is expected. Japan is one of the few exceptions, where over 10,000 small cetaceans are killed annually for commercial purposes and a kill of over 1000 large whales is planned for the stated purpose of science.

All Japanese cetacean scientists enjoyed the generosity of dolphin and porpoise fishermen until 1997, during which time I worked for a university and then for the Fisheries Agency of Japan. However, the research environment changed in 1998, or 1 year after I left the Fisheries Agency. The fishery community, including both fishermen and related government sections, apparently started classifying scientists and controlling access to research opportunities. For example, a university student who wanted pieces of skin of short-finned pilot whales for genetic study was asked by the fishermen to obtain approval of the Fisheries Agency or scientists working for it before taking the samples. Some university scientists felt that their request for approval was ignored by the agency, or they received a counter proposal from agency scientists to have one of their names as a coauthor in exchange for the approval.

Although fisheries are currently allowed to pursue private profit by exploiting marine biological resources, the marine resources are, in my view, common property of the world human community. Therefore fisheries should offer opportunities for scientists to examine their catch for scientific purposes, and scientists in the community should make their results globally available.

12.4.2 Age Determination

12.4.2.1 Principle of Age Determination

Toothed whales are aged using growth layers in their teeth. The teeth of recent toothed whales are monophyodont and homodont with a single cusp and single root and usually of similar shape. They are not replaced during life (Section I.1.3). As in the case of other mammalian teeth, they are composed of the three elements of enamel, dentine, and cementum. Dentine constitutes a major part of the tooth. The enamel, the hardest tissue, covers the distal tip of the dentine. Cementum covers the root of the tooth and connects the dentine and surrounding soft tissue. Tooth germs, shaped like a crushed tennis ball, are formed in the jaw at an early stage of fetal life, and enamel and dentine layers are deposited within them. Deposition of these two tissues starts at a fetal size of 10–20 cm in the short-finned pilot whale. Deposition of enamel layers occurs at the distal end of the fetal tooth and ends shortly before birth. The dentine layers are deposited on the proximal side of the germ and continues for years after birth. The deposition of cementum, the third element of the tooth, starts in the short-finned pilot whale near the time of tooth eruption and continues for life on the root of the tooth. Each of the three tissues forms growth layers.

The enamel contains fine growth layers visible under a light microscope in a polished thin section of the tooth. These layers are likely to be a record of the daily physiological cycle and to have potential value for studying fetal growth, but such an attempt has not been published.

It is generally accepted that an annual cycle of growth is recorded in the dentine, but, depending on the species and on the method of preparation, additional finer layers have also been detected. Some of the fine structures have been interpreted as reflecting a monthly cycle (Myrick and Cornell 1990; Myrick 1991). As dentine is deposited on the wall of the pulp cavity, the cavity decreases in volume with age, finally leaving a narrow space for blood vessels when dentine deposition ceases. The age when dentine deposition ceases depends mainly on tooth size, which depends on species and position in the jaw. Aging of old individuals, where dentine deposition has ceased, has to rely on growth layers in the cementum.

Cementum is deposited on the surface of the tooth root for life, but the layers are usually thinner and harder to read than the dentinal layers. An exception is the cementum of the Baird's beaked whale. Cementum in the species is thick and contains numerous fine structures within an annual layer, which have been deduced to represent a cycle of about 30 days (Kasuya 1977; Section 13.4.1).

It is important for each reader to train his/her eyes to get the same age estimate from both dentine and cementum of the same tooth with an open pulp cavity. Short-finned pilot whales start to deposit cementum soon after birth, or at body length of 154–163 cm or estimated age of 1.5–3.5 months. The lag time is small enough to be ignored in aging (Kasuya and Matsui 1984).

12.4.2.2 Sampling and Preparation of the Tooth

I examined carcasses of short-finned pilot whales at the dockside of the fishing port while fishermen were processing their catch. The task included taking a measurement of body length, examination and collection of histological materials from reproductive tracts including fetuses and mammary-gland samples, and extraction of teeth for age determination. Such activity was possible only with the consent of the fishermen, and I appreciate the cooperation of the dolphin fishermen of Arari (34°57′N, 139°07′E), Ayukawa, Futo (34°55′N, 139°08′E), Kawana, and Taiji.

The lower teeth are more suited than the upper teeth for aging purposes. The upper teeth are curved in two directions and are unable to be cut in one plane through the center, which is required in preparations for aging. At the beginning of the work I used a hammer and chisel to extract 2 or 3 teeth from the jaw, which required daily training of my arms with a heavy hammer, but later I switched to a saw for the task. Some scientists have recently used an electric saw for the purpose. The teeth were placed in a perforated plastic bag together with a label and various histological samples and fixed in 10% formalin solution. Acid in the formalin solution may damage the tooth tissue to some degree, but it did not cause a problem because these teeth were scheduled to be decalcified in formic acid for age determination. The following processes, A to E, were carried out in the laboratory.

A. Using a knife, one tooth was cut out from the formalin-fixed block from the lower jaw. The tooth was cleaned of soft tissue, leaving a thin layer of gum tissue to allow later confirmation of the outermost, or most recent, cement layer on the tooth. Then, using an IsoMet low-speed saw, the tooth was cut longitudinally on a plane near the center of the tooth. The tooth had to be wet during the procedure; water was used as a lubricant. The sawed-off surface was polished flat to the center of the tooth using whetstones of 1000 and 2000 mesh. The surface of the whetstone had to be kept flat. Waterproof sandpaper on a flat plate can be used for the same purpose. If a tooth is polished through the center, the two rami of a growth layer meet with an acute point at the center of the tooth, but if the tooth is cut off center, the rami meet with a round curve at the center.

B. The tooth polished down to the center was glued on a clear plastic slide (e.g., 1 mm thick plate of polyvinyl chloride) using cyanoacrylate resin. During this operation the tooth had to be still wet, but the surface was wiped dry to keep the glue clear. If it was felt that the glue was hardening slowly, the slide was placed inside a glass bell together with vapor of a hardening catalyst. The half-cut tooth on the plastic slide was again placed on the diamond saw to cut off the other half, leaving a section about 1 mm thick on the slide. The thin section on the plastic slide was polished down to a thickness of 30–80 μm using the same set of whetstones as the previous one and water as a lubricant. The thickness was measured using the focusing dial of a microscope. It was desirable to polish the cementum portion thinner than the dentine in the same preparation because growth layers in the cementum are thinner than those in the dentine. This process must be completed in wet conditions before proceeding to the next step, C. The tooth tissue will shrink if it dries and will separate from the plastic slide. The reason for using a plastic slide plate was to compensate to some degree for this problem.

C. The thin section of the tooth glued on a plastic slide was placed in a 10% solution of formic acid for several hours. The tooth section was not harmed by being left in the acid, even overnight. The tooth tissue, when completely decalcified, appears transparent under the microscope, but the dentinal tubules will appear dark if decalcification is incomplete. If an insufficiently decalcified tooth section is put through the subsequent process, the completed preparation appears to have a whitish brown hue. The fully decalcified section was rinsed in running water overnight or for a full day to completely remove the formic acid. Because the tooth section was much thicker than an ordinary histological section, the rinsing procedure required a longer time. If acid remains in the tissue, the stain will fade quickly, perhaps in a few months.

D. The decalcified and rinsed tooth section can be stained with any type of hematoxylin. Double staining with eosin should be avoided because it decreases contrast of the growth layers. I stained the section with Mayer's hematoxylin for about 30 min, but slightly extending the time in this medium did not result in identifiable problems. The stained section was rinsed in running water for about 24 h for complete removal of acid. The process can be improved by placing the rinsed slide in water with a few drops of ammonia. If the slide was stained too dark, it can be bleached in formic acid and the procedure obtained earlier can be repeated. If darkly stained blotches are found on the slide, it is most probably due to partial separation of the tooth from the plastic plate. This is not attractive and causes some trouble in reading growth layers, but the slide is still usable.

E. At this stage the thin tooth section appeared blue on a plastic slide and was wet with water. It should go quickly through a series of alcohol rinses (i.e., 50% ethanol, pure ethanol I, and II) and then be mounted using a mounting medium dissolved in alcohol. Use of xylene should be avoided because it harms the plastic slide as well as the cyanoacrylate resin. When the mounting medium hardens, the preparation is complete. Caution was needed to avoid drying, particularly during step E. For longer storage of the preparations, they should be stored in a refrigerator.

This method of tooth preparation does not require costly machines but does require some experience, and it produces only one preparation from a single tooth. A powerful cryostat is needed to cut a decalcified cetacean tooth because it is still hard after decalcification, but the machine can produce several tooth sections from a single tooth. A strong decalcification agent is required for decalcification of a large tooth, and care should be taken to prevent overdecalcification of the outer portion of the tooth (e.g., cement layers).

I prepared dentine of short-finned pilot whales to a thickness of 40–80 μm and the cementum to the thickness of 30–40 μm, because growth layers in the former tissue were thicker than those in the latter (Kasuya 1983, in Japanese). Such a minor adjustment was easy if the teeth are polished by hand, but would be hard with a cryostat. Teeth as small as those of striped dolphins were able to be prepared with only a whetstone and without using a diamond saw. Reliable age determination is only possible with high-quality tooth preparations, which must be cut through the center of the tooth at a suitable thickness and stained to a suitable density. Unsatisfactory preparations should be discarded and replaced with good preparations from other teeth from the same animal.

12.4.2.3 Accumulation Rate of Growth Layers

Growth layers in the dentine were counted using a light-transmitting microscope at about 10× magnification and those in the cementum at about 100×. The microscope has to be well adjusted for lightning, and it is desirable to have a high-quality objective lens with a wide field and flat focal plane. The enamel layer is completely lost in a decalcified stained tooth section and remains only as a mold in the resin.

On the proximal side of the enamel mold there is dentine. The first dentine tissue next to the enamel is fetal dentine, which has an almost uniform structure. Except in teeth of neonates, the inner border of the fetal dentine is bordered by a lightly stained dentinal layer of 15–65 μm thickness. This is called neonatal dentine or the neonatal line, is dentin of weak calcification deposited just after birth, and is visually identifiable only after deposition of subsequent ordinary postnatal dentine. The neonatal line becomes visually identifiable within 1 month after birth (Kasuya and Matsui 1984). The task of age determination is to count the alternating differentially stainable structures in the postnatal dentine or in the cementum.

One growth layer is often described as a pair composed of a darkly stained and a less stained layer, but that definition is not precise enough and may be misleading. It is usual to find several alternations of stainability in shorter, and often irregular, cycles in a set of layers representing an "annual cycle." Some such fine layers have been interpreted as monthly layers (in the pantropical spotted dolphin) or marks of pregnancy or parturition (sperm whale), but most such fine structures remain without interpretation (Perrin and Myrick 1980). The regular "annual layer" has been called "growth layer group" and the fine structures within it as "accessory layers," although this terminology does not seem to me sufficiently satisfactory to avoid confusion or misunderstanding.

Identifying the neonatal line is the first step in reading annual layers in the dentine. The birth date of an individual dolphin probably affects the appearance of the dentine deposited next to the neonatal line, but this influence has not been well documented. In the southern-form short-finned pilot whale, the thickness of dentin deposited in the first year after birth is 0.5–1.0 mm and then it declines with increasing age. Some individuals cease deposition of dentine at age 25 and almost all individuals by age 40. Some individuals may still deposit dentine at age 30, but the thickness of the dentine of one cycle is only about 0.1 mm. Aging of animals that have ceased dentine deposition must rely on reading growth layers in the cementum. The total thickness of cementum at this stage will be 0.5–1 mm, wherein it is possible to count 30–40 growth layers.

The short-finned pilot whale is one of the easiest species for which to read annual growth layers in the dentine and cementum, although readability of the latter tissue is inferior to that of the former (Figure 12.10). Cautions in reading cemental layers are that they may contain some accessory layers and that annual layers deposited at age 1 or 2 years are identifiable only around the neck region of the tooth where layers of later deposition are not observed (see 5, 6, and 7 of Figure 12.10a). This is because the teeth of young individuals increase in length rapidly and the neck region soon becomes exposed above the gum. Deposition of cementum starts at age of a few months (see Section 12.4.2.1), and a neonatal line is not formed in it.

The accumulation rate of growth layers can be calibrated using teeth of known-age individuals or teeth that are stained *in vivo* using chemicals such as tetracycline, as was done with bottlenose dolphins (Myrick and Cornell 1990). In Japan, Nishiwaki and Yagi (1953) attempted this task by injecting lead acetate in striped dolphins, but they were unable to determine the accumulation rate because of short life of the dolphins in captivity. Until recently, striped dolphins have not been successfully kept in Japanese aquariums. Because such information was not available for short-finned pilot whales, the accumulation rate of growth layers in the dentine was estimated by examining seasonal alternation of the last dentinal layer in tooth samples that covered 8 months of the year.

Kasuya and Matsui (1984) observed the stainability of dentine being deposited on the pulp cavity wall at the time of death. In May–October 80%–90% of individuals were depositing dentine that was darkly stainable with hematoxylin, but the proportion decreased to 40% in December–January and to 20% in February. A 100% score could not be expected in the data because there can be some misclassification due to the presence of accessory layers. Taking this into account, the authors concluded that deposition of stainable dentine started in April–May and continued to October–November and unstainable dentine was deposited from November–December to March–April. They also observed that older individuals switched the deposition from stainable to unstainable layer in an earlier season than younger individuals, which suggested that the proportion of time represented by stainable and unstainable layers may change with age. Even accepting such minor uncertainties, it is safe to assume that a set of stainable and unstainable layers represents a period of one year.

12.4.2.4 Aging Error and Differences between Teeth

Every tooth in a jaw is under the influence of a common physiology, which can be recorded similarly in each tooth as growth layers. However, difference in tooth size between positions in a jaw may influence the readability of the layers and the age when dentinal deposition ceases.

Kasuya and Matsui (1984) examined readability difference of growth layers between positions in a jaw using teeth collected from three southern-form short-finned pilot whales aged between 20 and 45. The cusp in some teeth was lost by wear, leaving only a minor portion of the enamel layer, which meant that fetal dentine as well as the neonatal line was still present on the tooth for determination of the starting point for reading. One or two teeth near the end of the tooth row tended to yield lower dentinal counts than a larger tooth near the middle of the row. This was common in both a 45-year-old individual that had ceased dentinal deposition and in younger individuals with an open pulp cavity. The reason for lower counts in small teeth near the end of the tooth row was probably the difficulty of reading compressed growth layers in the dentine of such teeth. The readings for cemental layers showed reasonable agreement among teeth in a jaw. This result suggests that small teeth near the end of a tooth row should be avoided for age reading, and if there is no other choice, cemental layers should be carefully examined.

Kasuya and Matsui (1984) made three independent counts on the same tissue (dentine and cementum) of the same tooth and accepted the middle figure as the correct count for the tissue. The reading error was calculated as the difference between the middle figure and the one closest to it and was expressed as a percentage of the middle figure. The error was distributed on both sides of zero with a standard deviation of 2%. This suggested that the 95% confidence interval of their age reading was about 4% on each side of the estimated age. This was the major source of error in the age reading, but there remained another source of error contributed by rounding the age to nearest year by estimating by eye the thickness of the first and the last incomplete layers. If both errors are combined the true age of an individual was within ±0.9 year for animals estimated at 10 years old, ±1.8 years for 20-year-old

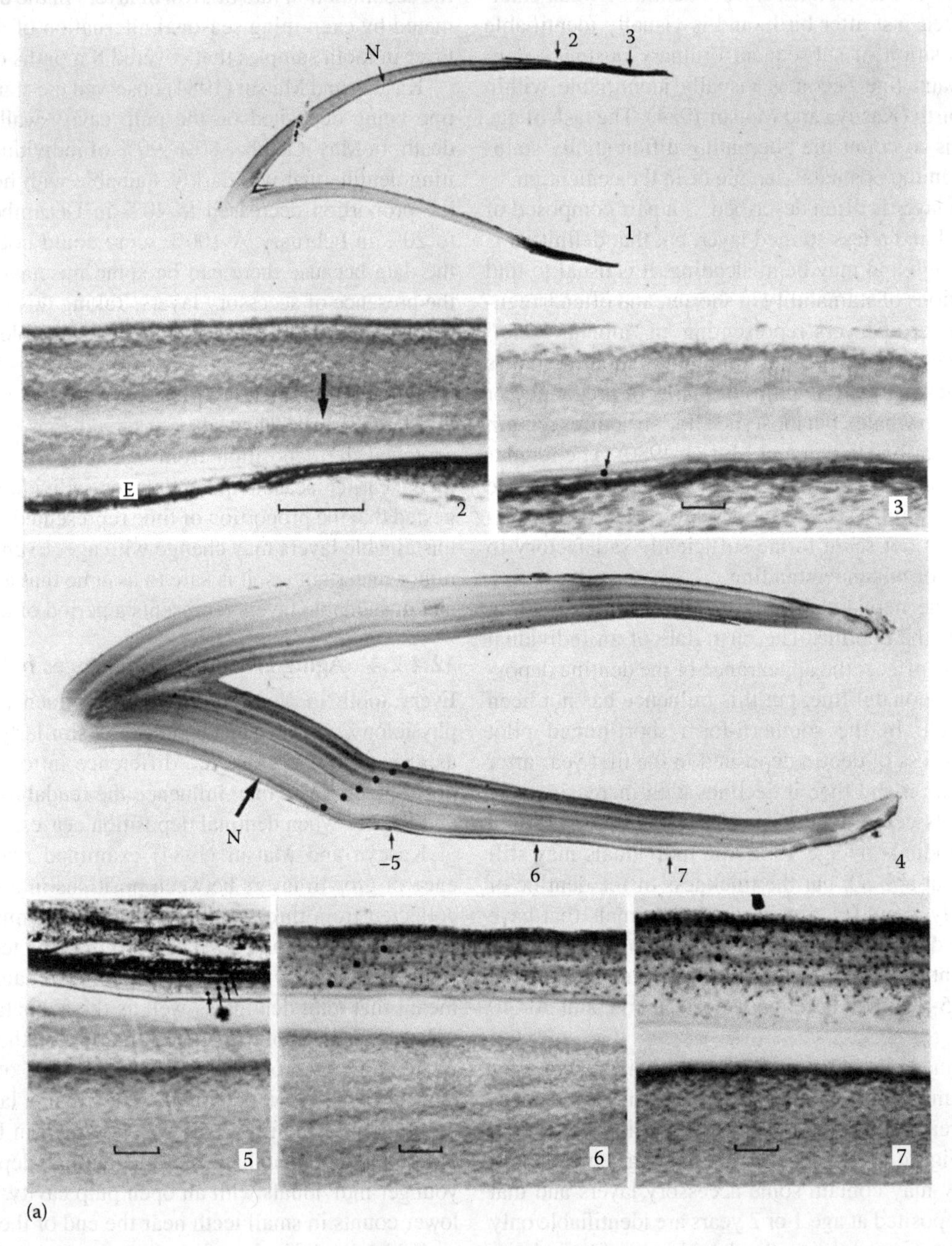

FIGURE 12.10 (a) Decalcified and hematoxylin-stained thin sections of teeth of southern-form short-finned pilot whales. E, cast of enamel that has been lost to acid; N, neonatal line; Black dot, hematoxylin-stainable layer of annual cycle. Scale represents 0.1 mm. (Reproduced from Kasuya, T. and Matsui, H., *Sci. Rep. Whal. Res. Inst.*, 35, 57, 1984, Plate I.) 1, Whole view of a tooth section of 163 cm female estimated at about 1/4 year of age. Partial enlargements of this tooth are shown in 2 and 3; 2, partial enlargement of position 2 in photo 1. Pulp cavity is on the top and arrow indicates neonatal line in dentine; 3, partial enlargement of position 3 in photo 1. Pulp cavity is at top. The first cemental layer is shown by a black dot and an arrow; 4, whole view of a tooth section of 326 cm male estimated at about 5 1/4 years of age. Black dots indicate first to sixth (incomplete) stainable layers of dentine. A constriction of the tooth formed soon after birth is indicated by a small arrow near number 5. Partial enlargements of this tooth are shown in 5–7; 5, partial enlargement of position 5 in photo 4 showing first to fourth cemental layers. The fifth and sixth cemental layers are not deposited here due to tooth eruption. Pulp cavity at bottom; 6, partial enlargement of position 6 in photo 4 showing second to fifth cemental layers. The sixth cemental layer is not deposited here due to tooth eruption. The pulp cavity at bottom; 7, partial enlargement of position 7 in photo 4 showing third to sixth cemental layers. This portion of the tooth was formed after formation of the first and second cemental layers. Pulp cavity at bottom. *(Continued)*

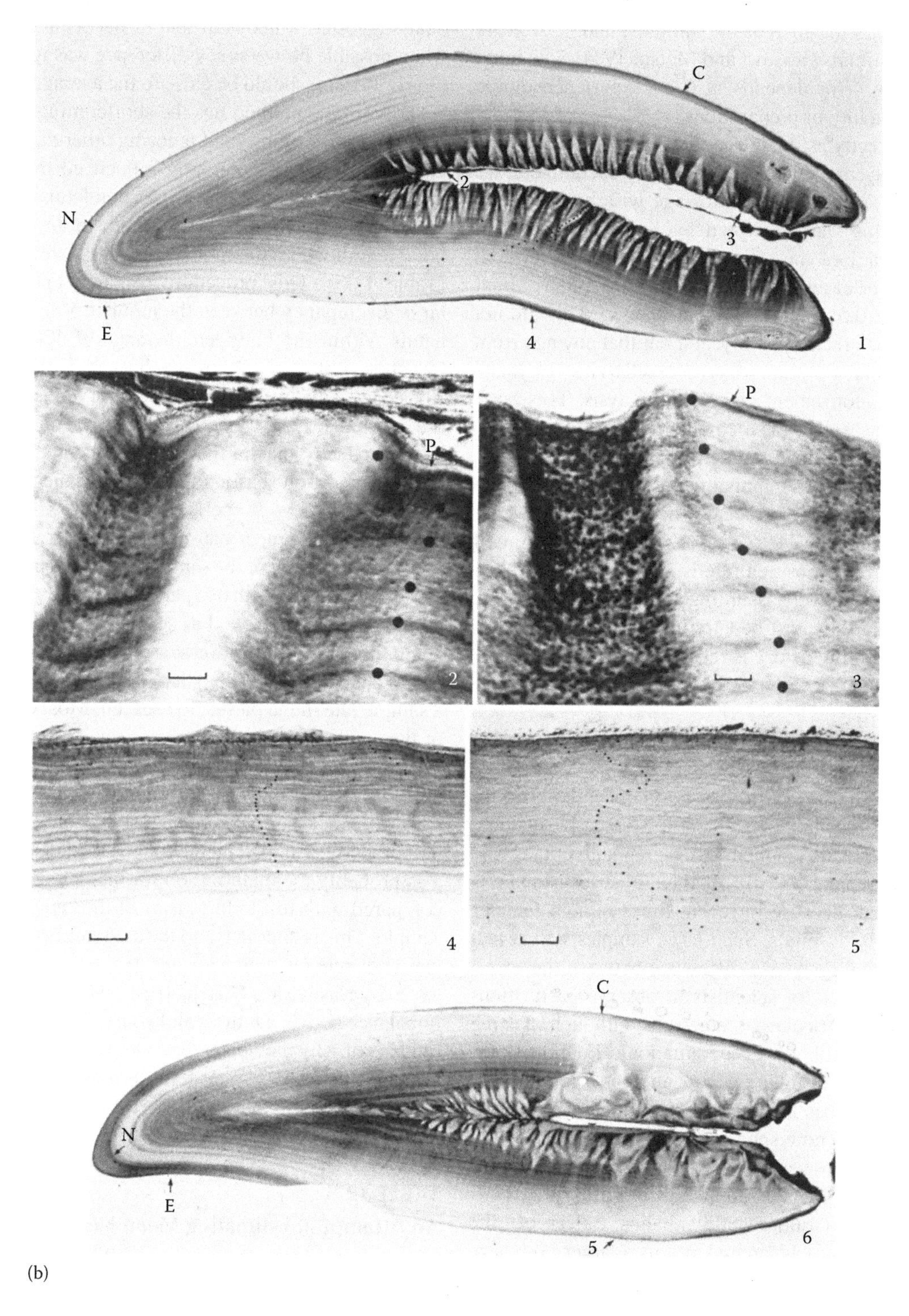

(b)

FIGURE 12.10 (*Continued*) (b) Decalcified and hematoxylin-stained thin sections of teeth of southern-form short-finned pilot whales. C, cemental layers; E, cast of enamel which has been lost to acid; N, neonatal line; P, predentin (young dentine waiting for subsequent calcification); black dot, hematoxylin-stainable layer of annual cycle. Scale represents 0.1 mm. (Reproduced from Kasuya, T. and Matsui, H., *Sci. Rep. Whal. Res. Inst.*, 35, 57, 1984, Plate II.) 1, Whole view of a tooth section of 366 cm female. Tooth length is 42.8 mm, and 28 hematoxylin-stainable annual layers have been deposited in the dentine. Dentine of weak calcification (less stainable with hematoxylin) and of higher calcification alternate irregularly near the pulp cavity, which is unrelated to annual cycle. Partial enlargements of this tooth are given in 2–4; 2, partial enlargement of position 2 in photo 1. Black spots indicate last 6 stainable dentine layers; 3, partial enlargement of position 3 in photo 1. Black spots indicate last 7 stainable dentine layers; 4, partial enlargement of position 4 in photo 1. Black spots indicate annual stainable layers in cementum. Pulp cavity at bottom; 5, partial enlargement of position 5 in photo 6. Black spots indicate 2nd to 36th annual stainable layers in cementum. Pulp cavity at bottom; 6, whole view of a tooth section of an adult female. Total tooth length is 39.5 mm, and 36 hematoxylin stainable annual layers have been deposited in the dentine. The pulp cavity is narrow, and deposition of dentinal layers has almost ceased in this tooth. A partial enlargement of cementum is given in 5.

animals, ±2.6 years for 40-year-old animals, and ±3.4 years for 60-year-old animals (Kasuya and Matsui 1984). The magnitude of reading error depends also on experience of the readers and the quality of preparations.

If we can correctly estimate the thickness of the first and the last growth layers, which are likely to be incomplete, we will be able to estimate the animal's age with higher precision. Although this was attempted by Kasuya and Matsui (1984), the accuracy of the estimates has not been verified. Thus, in most of the cases they expressed the age of an animal to the nearest n + 0.5 years, n being an integer. This did not mean that age was precise to 0.5 years, but that any age from n years to less than n + 1 years was expressed as n + 0.5 years. Such a method is convenient in growth analysis. This book uses this method as well as the method often used for humans, where age of n + 0.5 years is expressed by the integer n years.

It is reasonable to expect annual deposition of growth layers in mammals, including cetaceans, because their lives are governed by the annual cycle of seasons. However, there will be cases where layer deposition rate is unknown and life history analyzed by tentatively assuming annual deposition. The results of such analysis must be carefully evaluated to determine if the derived life history is reasonable.

12.4.3 Growth

12.4.3.1 Neonatal Body Length

The life of an animal is said to begin at conception, but for management purposes it is convenient to start at parturition. Thus, fetal life is included in the life of a pregnant female. The neonatal size of humans is easily obtained by measuring newborns at a hospital, because in recent times human females have given birth in hospitals. Such large samples will reveal even minor neonatal body size difference between the sexes. It is almost impossible for scientists to observe parturitions of wild animals, particularly cetaceans, although nesting animals may offer a slightly easier situation. The finless porpoise inhabits coastal or inland waters and has offered scientists better opportunities for obtaining stranded carcasses of neonates than the offshore species. As the short-finned pilot whale inhabits offshore waters, information on neonatal size has relied mostly on carcasses obtained from fisheries.

The northern-form pilot whale inhabits waters off the Pacific coast of northern Japan and attains a larger body size while the southern form inhabits Pacific waters off central and southern Japan and attains a smaller body size. Samples of the northern forms were obtained from the catch of small-type whaling, and those of the southern form from the catch of dolphin-drive fisheries at Arari, Futo, and Kawana along the coast of the Izu Peninsula and at Taiji in Wakayama Prefecture in the 1960s to the early 1980s. Arari and Kawana have now ceased operations and Futo appears to be very inactive in the fishery (Sections 3.8 and 3.9).

Among the southern-form individuals caught in these fisheries the smallest newborns were found at lengths of 142 cm (female) and 136 cm (male), and the largest fetuses at 144 cm (female) and 146 cm (male) (Kasuya and Marsh 1984). These data suggested a neonatal length between 136 and 146 cm, when possible between-sex difference was ignored. A middle figure, 141 cm, should be close to the average neonatal length.

This simple method has the shortcoming of depending on only the two extremes and ignoring other data between them, and the result will be greatly influenced by sample size. In order to overcome this, the proportion fetal and postnatal was examined in relation to body length in an attempt to find a body length where the ratios of the two categories were equal (Table 12.5). This procedure revealed another problem, a large discrepancy between the number of fetuses and of postnatals within the body-length range of 120–159 cm, where 29 fetuses, but only 7 postnatals (i.e., 4.1:1), were recorded. The exact same discrepancy was observed in striped dolphins (i.e., 58 fetuses and 14 neonates or 4.1:1) and was interpreted as either due to bias in the sample collected at the beginning of the parturition season or loss of neonates during the drive (Section 10.5.2).

Such disagreement can occur for several reasons, one of which is the timing of the sample relative to the parturition season. The timing of birth of a particular fetus is probably affected by fetal body size as well as season. A sample obtained at the beginning of the parturition season will have more fetuses than postnatals of the same body length. The reverse will be true for a sample late in the parturition season. Most of the parturitions of the southern-form pilot whale occur in June–October with a peak in August (see Section 12.4.4.2). Kasuya and Marsh (1984) obtained most of the samples in the early half of the parturition season in June–August and only a few in the last half of the season in September to October (Figure 12.23). This was probably the main reason for the deficiency of postnatals compared with full-term fetuses of the same body size in the sample. This is alternative to the assumption of offshore segregation of schools with neonates, that is, one of the hypotheses tested by Kasuya and Marsh (1984), which failed to explain the imbalance between fetuses and young calves. The question still remains of why the September–October sample of Kasuya and Marsh (1984) lacked near-term fetuses and neonates (Figures 12.23 and 12.24).

TABLE 12.5
An Attempt at Estimating Mean Neonatal Length of Southern-Form Short-Finned Pilot Whales Using Limited and Possibly Biased Samples

Body Length (cm)	120–129	130–139	140–149	150–159	Total
No. of fetuses observed	9	14	6	0	29
No. of neonates observed	0	2	3	2	7
No. of neonates corrected	0	8.3	12.4	8.3	29
%	0	37.2	67.9	100.0	50.0

Source: Abbreviated from Kasuya, T. and Marsh, H., *Rep. Int. Whal. Commn.*, 259, 1984, Table 12.

Other possible factors that may cause disagreement of fetal and postnatal counts are postnatal natural mortality and human-caused loss of postnatals during the drive, both of which can be higher in neonates than in near-term fetuses. However, Kasuya and Marsh (1984) failed to explain the imbalance with either of these single factors. I question whether loss of neonates during the drive can be of similar magnitude to that of striped dolphins that have often been driven in school of several hundred individuals, while short-finned pilot whales live in schools smaller in size and tighter in structure (Table 12.20). Fishermen can more easily keep their eyes on the whole pilot whale school and pay attention to not losing a single pilot whale, since they have higher per capita value.

Although the reason for the unbalanced fetus/postnatal ratio remains unresolved, Kasuya and Marsh (1984) raised the number of postnatals to the level of the fetuses (Table 12.5), fitted a linear regression to the proportion of neonates on body length, and obtained a body length where neonates were 50%. The average body length at birth thus estimated was 139.5 cm, which was almost the same as the 141 cm obtained earlier from the two extreme figures. Thus, 140 cm is currently used as the average neonatal length of southern-form short-finned pilot whales off Japan (Kasuya and Marsh 1984).

The northern-form pilot whales were taken by small-type whalers. They operated the fishery with a quota on the number of animals, so they were unlikely to catch neonates of low commercial value. The materials analyzed by Kasuya and Tai (1993) covered the season from late September to November, and the fetal length frequency distribution was bimodal, with one mode below 20 cm and another less distinct mode at 150–170 cm. The latter mode was considered to represent near-term fetuses. Ohsumi (1966) presented a relationship between average neonatal body length (Y, m) and average body length of females at attainment of sexual maturity (X, m):

$$Y = 0.532\ X^{0.916}$$

Applying average female length at sexual maturity of 3.9–4.0 m (see Section 12.4.3.2) to this equation, Kasuya and Tai (1993) obtained 185–189 cm as a range of average neonatal length of northern-form pilot whales. They further noted that Ohsumi's equation gave a slightly larger figure than the independent estimate of neonatal length of southern-form whales and concluded that the lower bound of the range, that is, 185 cm, would be suitable as the average neonatal length of northern-form whales. This was 45 cm greater than the mean neonatal length of southern-form whales estimated earlier.

12.4.3.2 Female Sexual Maturity

Female age at sexual maturity can be defined either as the age at the first ovulation or at first parturition. The latter is often used in population dynamics, because it directly indicates recruitment, albeit harder to be determined. However, the former has been used in the study of life history, because it is easily determined by the presence of a corpus luteum or albicans formed in the ovaries after ovulation. If an ovulation is followed by conception, the corpus luteum is maintained in the ovary during pregnancy and degenerates to a corpus albicans after parturition, but if the ovulation is not followed by conception the corpus luteum degenerates quickly to a corpus albicans (Marsh and Kasuya 1984). It has been believed that histology of the corpus albicans cannot distinguish between the two histories and that the corpora albicantia persist in the ovaries for life. Thus, counting these corpora in the ovaries will give the number of ovulations experienced by a female but not the number of pregnancies (see Section 10.5.6 for further discussion on this topic). As females will start behaving as adult individuals at around the first estrus, the first ovulation will be more suitable as a definition of attainment of sexual maturity for behavioral studies. Among cetaceans the first ovulation is usually followed by conception (see Section 12.4.4.8).

It is possible for some cetaceans to identify whether a female has had a past experience of pregnancy or lactation. One of the methods is to examine the hue of the mammary gland in balaenopterids, where mammary glands that have undergone lactation show a brownish color while those which have not are pink. This visual judgment is difficult for sperm whales, dolphins, and porpoises, but has potential to be tested histologically. The other method is to note the existence of stretch marks in the uterus that can be seen in nonpregnant females with past experience of conception (Benirschke et al. 1980). These stretch marks are apparently similar to the features seen in overstretched plastic film, and similar marks are seen on the breasts and abdomen of women with experience of childbirth.

Ovaries of young immature females of the short-finned pilot whale are flattened and oval in shape, 3–3.5 cm-long and 1–2 cm wide, and weigh 2–6 g (both sides combined). At varying age over 2 years females start to have heavier ovaries; thus individual variation in total weight increases to 2–12 g. This change is due to the development of Graafian follicles visible on the ovarian surface. No particular bilateral asymmetry of ovarian development exists in the short-finned pilot whale (Marsh and Kasuya 1984), which is different from several dolphins such as the striped dolphin where the left ovary usually matures before the right (Section 10.5.6.1; Ohsumi 1964).

The relationship between age and diameter of the largest Graafian follicle is shown in Figure 12.11. Follicle size is expressed as the cube root of the product of three diameters in order to maintain a unified rule for follicles, corpora lutea, and corpora albicantia. The geometric mean was considered better than the mathematic mean as an indicator of volume of these tissues which might be compressed by nearby structures through the process of degeneration (Marsh and Kasuya 1984). Some females aged over 2 years had the largest follicle measuring 5–8 mm in diameter, and this upper size limit remained unchanged among the immature individuals. Since the southern-form pilot whales first ovulate at ages between 7.0 and 12.0 (see the following text in this section), this observation suggests that some immature females develop Graafian follicles only to the size of 5–8 mm and cease further development for several years until a time when follicle growth resumes. However, it was unclear in the available data whether the follicle that remains dormant at 5–8 mm resumes growth for ovulation or another new follicle takes its place.

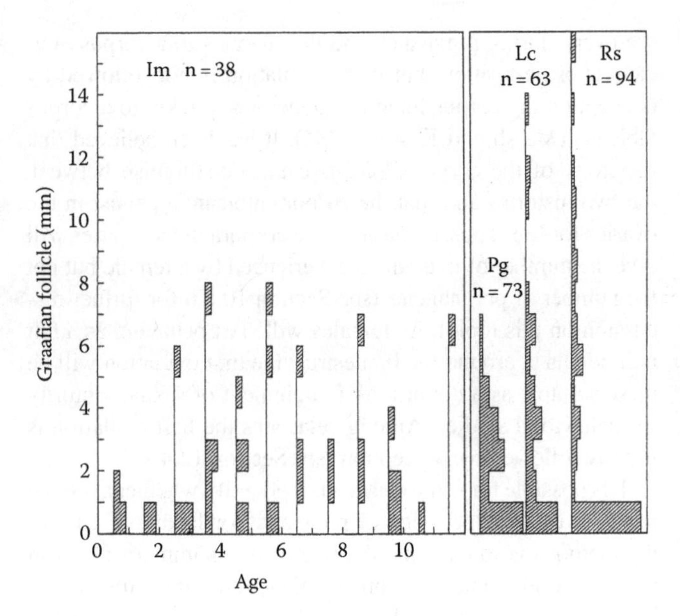

FIGURE 12.11 Age-related change in diameter of largest Graafian follicle of immature females (Im) and size frequencies of largest Graafian follicle of mature females in pregnancy (Pg), lactation (Lc), and resting status (Rs) in southern-form short-finned pilot whales taken by Japanese drive fisheries covering 8 months of the year, excluding March, April, September, and November, 1965–1981. Sample sizes also given. (Reproduced from Kasuya, T. and Marsh, H., *Rep. Int. Whal. Commn.*, 259, 1984, Figure 12.)

Maximum follicle size in resting females (mature females neither pregnant nor lactating) was 15–16 mm in diameter (Figure 12.11). Some lactating females also had large follicles. These females were probably preparing for the next ovulation, but follicle size at ovulation would probably be more than 20 mm (for further information see Section 12.4.4.7). The reason why we did not see such large follicles in immature females of maturing age would be explained by a rapid development of follicles before ovulation.

Puberty is a stage when secondary or tertiary sexual characters become identifiable accompanied by development of the reproductive organs (primary sexual characters) and can, but not always, be a process of gradual development. As cetacean females do not develop secondary sexual characters, identification of the pubertal stage must rely on tertiary sexual characters such as behavioral changes in the community. Future studies should be directed toward finding change in social behavior or in direct measurement of sexual hormones in the blood to understand the pubertal stage.

It is difficult, or impossible, to identify whether a particular female is in a pubertal stage by examining gonads extracted from a carcass of the short-finned pilot whale, other than inferring it from development of the Graafian follicles, which is a slow process of several years. Past attempts for some toothed whales to identify the pubertal stage from information in the gonads often lacked sufficient support from behavioral information.

Ovulation is known to occur only in females above a certain level of body mass, and it will cease if a female loses body mass below a certain nutritional level. Body mass is also a function of age. Thus, sexual maturity of mammals is a function of both age and body size, but there have been no attempts to disentangle the two factors in attainment of sexual maturity of cetaceans.

If sexual maturity is determined by experience of ovulation, the youngest mature female was aged at 8.5 and the oldest immature female at 11.5 years. For technical reasons, this age range included the ages between 8.0 and 12.0 years (Section 12.4.2.4). The proportion of females mature increased with increasing age, and the relation was expressed by an upwardly convex curve that transitioned to a horizontal line at higher ages. Observing that the age where 50% of individuals were sexually mature was between age 8.5 and 9.5 years, Kasuya and Marsh (1984) applied a linear regression to the relationship between maturity and age using individuals aged 7.5–10.5 years and obtained age 9.0 as the age when 50% of females were sexually mature. Individuals used in this calculation included four pregnant females that had only one corpus luteum and no corpora albicantia. Fetal size in these females suggested that they had first ovulated at ages 7.4–8.1 years. These four females became pregnant with their first ovulation and represent precocious cases. There were two additional pregnant females included in the calculation obtained earlier. One was aged 9.25, had three ovarian corpora, and was estimated to have conceived when she was 8.3 years old (fetal size was 100 cm). The other was aged 8.5 years, had four corpora, and was estimated to have conceived at age 8.4 (fetal size was 3 cm). Since these two females were not lactating, they were likely to have experienced their first conception at the third or fourth ovulation. These ages and the age of the oldest immature female of 11.5 years allow the conclusion that southern-form short-finned pilot whales first ovulate after age 7.0 years and before 12.0 years, which was about 5 years after the age when their ovaries started some degree of follicular development. The status of the six pregnant females mentioned earlier indicated that three of them became pregnant at the first ovulation and the remaining two females presumably at their third or fourth ovulations.

Body length on the mean growth curve (Figure 12.21) at age 9.0 years (mean age at attainment of sexual maturity) was approximately 320 cm. This could be affected by uncertainty in the mean growth curve. Body lengths of the smallest sexually mature female and the largest immature female were 300 and 344 cm, respectively. The proportion of sexually mature females plotted on body length showed, as in the case of the maturity–age relationship, an upwardly convex curve with a decreasing gradient at greater body length. The gradient was steep at the lower half of the range, or at 300–330 cm. Fitting a linear model to this range gave 315.6 cm as the body length where 50% of individuals were sexually mature (Kasuya and Marsh 1984). Although the influence of including all of the age range in this regression was considered small in this species, which is avoided in the above calculation, the above calculation is not totally free from this effect (this problem was discussed in detail in Sections 11.4.7.4 and 11.4.8.3). This partly explains why body length at 50% maturity was smaller than body length at age 9.0 years.

TABLE 12.6
Body Length and Age at the Attainment of Sexual Maturity of Female Short-Finned Pilot Whales off Japan

	Body Length (cm)		Age (Year)[a]	
Maturity	**Range[b]**	**Average[c]**	**Range[b]**	**Average[c]**
Southern form				
Immature	≤344		≤11.5	
Mature	≥300	315.6	≥7.5	9.0
Northern form				
Immature	≤420		≤11.5	
Mature	≥400	390–400	≥5.5	8.5

Sources: Kasuya, T. and Marsh, H., *Rep. Int. Whal. Commn.*, 259, 1984; Kasuya, T. and Tai, S., *Rep. Int. Whal. Commn.*, 339, 1993.

[a] Age to the nearest n + 0.5 years (n being an integer).

[b] Range of the sample examined.

[c] Average is the level where 50% of individuals have attained sexual maturity.

The corresponding growth parameters of the northern-form short-finned pilot whale are listed in Table 12.6 in comparison with those of the southern form.

12.4.3.3 Male Sexual Maturity

12.4.3.3.1 Definition

The male process of sexual maturation includes various stages, for example (1) start of spermatogenesis in the testis, (2) attainment of physiological ability to fertilize females, (3) development of desire to approach estrous females, (4) identification by estrous females as a mature male, or (5) obtaining the opportunity for reproduction through competition among males. The second stage has been used in cetaceans for estimating maturity, although the determination itself has problems. The third example is probably equivalent to the pubertal stage. The fourth and fifth stages will correspond to social maturity and require behavioral information for confirmation, which is hard to apply in carcass studies.

Female sexual maturity leaves various morphological marks in the body, such as corpora in the ovaries, pregnancy, or lactation, but male maturity usually does not leave such morphological clues. The following sections describe research to understand the process of sexual maturation of males using histological information. The results of such studies need to be compared against behavioral information.

12.4.3.3.2 Testicular Histology

If a whole testis is examined histologically from one end to the other, a firm conclusion can be reached on whether the individual has started spermatogenesis. However, such an attempt is like detecting a flea egg in a carpet and is certainly impractical. The proxy for this has to be examination of a small sample of the testicular tissue. Studies on sperm whales (Best 1969) and sei whales (Masaki 1976) revealed that spermatogenesis occurred first in a particular region of the testes of young males. The same was found for testes of Baird's beaked whales. Testes of the Baird's beaked whale were examined in detail on six males at early stages of sexual maturity and revealed that spermatogenesis started near the anterior end of the testis and extended subsequently toward the posterior end (Kasuya et al. 1997). However, similar examination of testes of the southern-form pilot whale did not find such polarity in young males, and Kasuya and Marsh (1984) examined a piece of tissue 5–10 mm^2 taken from the mid-center of the testis.

Histological examination showed various stages, from being spermatogenic throughout the sample, spermatogenic in only in part of the sample, or aspermatogenic. Kasuya and Marsh (1984) classified the maturity of the sampled tissue in southern-form whales based on proportion of seminiferous tubules that were spermatogenic into four categories: (1) immature, no spermatogenic tubules present, (2) early maturing, more than 0% but less than 50% of tubules spermatogenic, (3) late maturing, 50% or more but less than 100% spermatogenic, and (4) mature, 100% spermatogenic. This method had value in classifying maturity into four stages rather than two stages of "mature" and "immature," but it is not free from misclassification of maturity between adjacent categories by chance in encountering particular tubules. The risk can be avoided in part by analyzing other maturity-related parameters such as development determined from testicular and epididymal smears and by not placing great significance on difference between two adjacent categories, at least on an individual basis. The same procedures were applied for the northern-form short-finned pilot whale (Kasuya and Tai 1993).

The results of examination of testicular histology are summarized in Figures 12.12 and 12.13, where the mean weight of both testes is plotted on age or body length. Weight of a single testis was used if the other testis was not available for averaging. Testis weight differed slightly between left and right testes of the same individual, but there was no systematic bilateral weight difference detected in the southern-form pilot whale (Kasuya and Marsh 1984). In order to cover the weight range, two orders of magnitude from juveniles to adults, testicular weights are plotted on a logarithmic scale in Figures 12.12 and 12.13, where the equation $y = 10^{(ax + b)}$ is expressed as a straight line and the usual linear relationship as an upwardly convex curve. The following text relates to the southern-form pilot whale, but the northern form shows similar features, except for body size.

The testis of southern-form males weighed 15–20 g at age 2 years, slowly increased in weight to age 7–8 years or body length 3.3 m, when some males started a low level of spermatogenesis ("early maturing"), with testis weight of 50–60 g. Testis weight further continued a slow increase to reach an average weight of about 90 g at age 14 years. The average rate of weight increase was about 5.4 g/year during the 6–7 years from ages 7–8 to 14.

Rapid testicular growth occurred in most of the males between ages 14 and 18 and body length 3.8–4.2 m, which agrees with the time when southern-form pilot whales reached

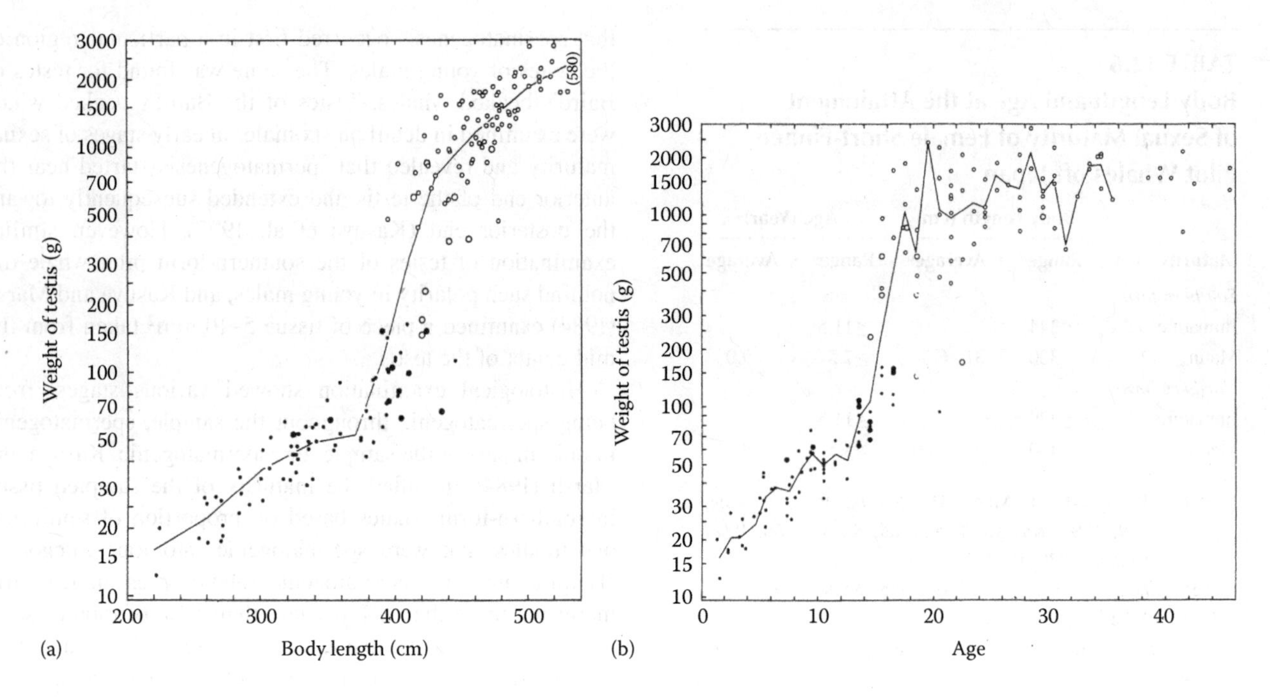

FIGURE 12.12 Scatter plots of weight of single testis (epididymis removed) (in g) and histological maturity on body length (a) and on age (b) of southern-form short-finned pilot whales taken by Japanese drive fisheries covering 8 months of the year, excluding March, April, September, and November (same material as in Figures 12.14, 12.15, 12.16, and 12.18). Small closed circles, immature; large closed circles, early maturing; large open circles, late maturing; small open circles, mature. (Reproduced from Kasuya, T. and Marsh, H., *Rep. Int. Whal. Commn.*, 259, 1984, Figure 6.)

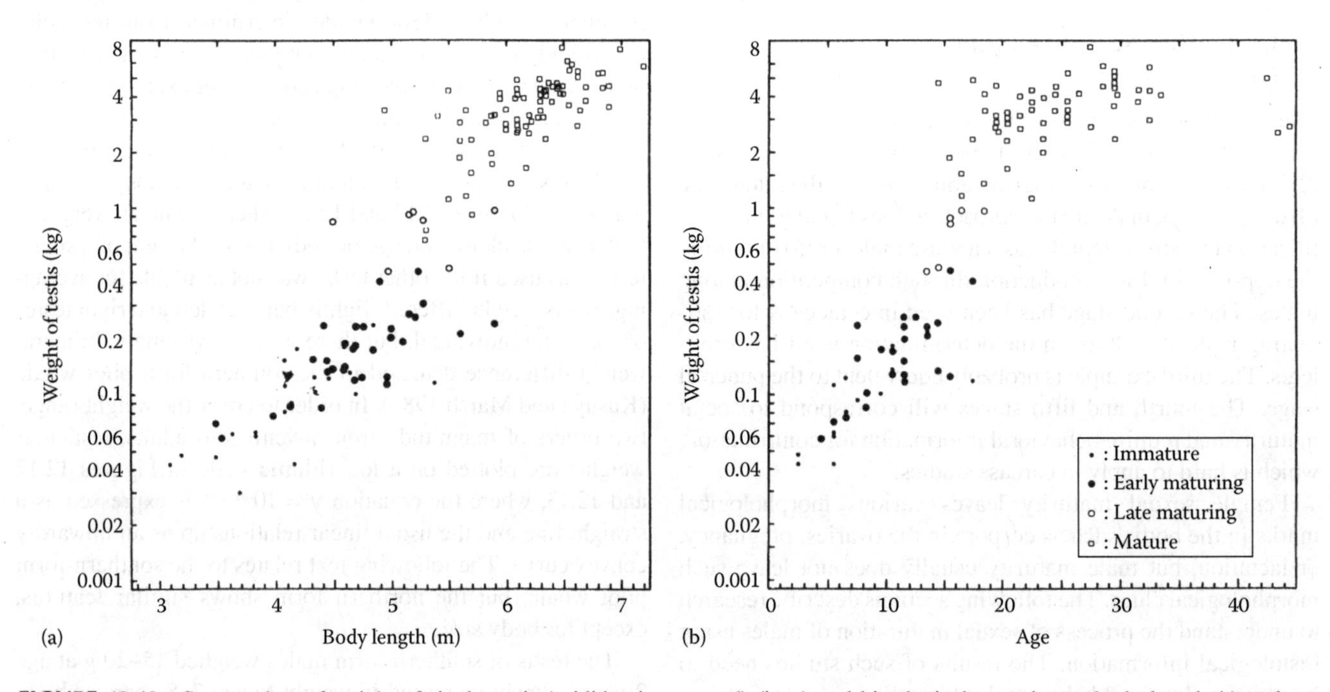

FIGURE 12.13 Scatter plots of weight of single testis (epididymis removed) (in g) and histological maturity on body length (a) and age (b) of northern-form short-finned pilot whales taken by Japanese small-type whaling in October and November (same material as in Figures 12.17 and 12.19). For further explanation see Figure 12.12. (Reproduced from Kasuya, T. and Marsh, H. *Rep. Int. Whal. Commn.*, 259, 1984, Figure 7.)

sexual maturity. Most males attained the "mature" stage at this age, but there seemed to be a lag of 6–7 years between precocious and slow-maturing males. This was expected from Figure 12.12 which suggests that rapid testicular growth started in some individuals at age 14 and was completed by age 16, while slower males started rapid testicular growth as late as age 23. Mean testis weight increased from about 90 g at age 14 to about 900 g at age 18, which was equivalent to increase of about 200 g/year. This rate of mean annual testicular growth probably biased testicular growth of individual animals downward because of broad individual variation in the timing of the start of the growth spurt. If the rapid process

TABLE 12.7
Body Length and Age at the Attainment of Sexual Maturity of Male Short-Finned Pilot Whales off Japan

	Body Length (cm)		Weight of Testis (g)[a]		Age (Year)[b]	
Maturity	**Range[c]**	**Average[d]**	**Range[c]**	**Average[d]**	**Range[c]**	**Average[d]**
Southern form						
Immature	≤409		<170		<20.5	
Early maturing	324–434	401.1	50–150	100	7.5–16.5	14.6
Late maturing	414–455	413.7	150–900	170	14.5–29.5	14.8
Mature	394–525	422.1	400–3000	450	15.5–45.5	17.0
Northern form						
Immature	≤540		<250		<17.5	
Early maturing	350–590		58–450		4.5–17.5	
Late maturing	450–590		450–960		13.5–18.5	
Mature	500–720	555	675–7200	900	14.5–44.5	16.5

Sources: Kasuya, T. and Marsh, H., *Rep. Int. Whal. Commn.*, 259, 1984; Kasuya, T. and Tai, S., *Rep. Int. Whal. Commn.*, 339, 1993.

[a] Weight of single testis (epididymis removed).
[b] Age to the nearest n + 0.5 years (n being an integer).
[c] Range of the sample examined.
[d] Average is the level where 50% have attained the indicated maturity stage.

of maturation is assumed to proceed within 2 years from 14 to 16 years of age, annual testicular growth could be 400 g/year. It is reasonable to expect some behavioral change in males at these ages.

This analysis suggests that a precocious male starts rapid testicular growth at age 14, but that slower males can start at age 23. These males are likely to achieve development from the "immature" stage to the "mature" stage within 2 years. Testis at the "immature" stage weighed <170 g, "early maturing" 50–150 g, "late maturing" 150–900 g, and "mature" 400–3000 g. Thus, males attained the "mature" stage at testicular weight of 400–900 g (single testis). Figure 12.12 also shows that 50% of males attained the "mature" stage at testicular weight of 450 g. Table 12.7 offers criteria to be used for determining male maturity using age, body length, or testicular weight.

Sexual maturity of short-finned pilot whales is a function of body size as well as age (Kasuya and Marsh 1984). Larger animals have a greater probability of being sexually mature than smaller individuals of the same age, and older individuals have a greater probability of being sexually mature than younger individuals of the same body size. Kasuya and Tai (1993) analyzed a sample of northern-form pilot whales obtained from small-type whaling, which selectively hunted larger individuals. Therefore it was likely that body length at the attainment of sexual maturity estimated by them was biased upward to an unknown degree and age at the attainment of sexual maturity biased downward. Because the differences in maturity-related age data in Tables 12.6 and 12.7 are in the direction to be expected due to the biases, the apparent difference between the two geographical forms in age at attainment of sexual maturity remains inconclusive.

The observations on testicular growth in southern-form whales can be summarized as follows. Weight of a single testis is about 15 g in neonates, reaches around 90 g at age 14 when rapid testicular growth starts, and attains the "mature" stage usually by age 18 at an average testicular weight of 450 g. The growth in testicular weight continues at the same annual rate to reach 1000 g at around age 20, and testicular growth finally ceases at around age 27 or at testis weighs of 1500–2000 g on average (Figure 12.12). The northern-form whales have heavier testes than the southern form, reflecting the difference in body size, but the testicular growth pattern is the same as in the southern form (Figure 12.13).

With the rapid growth of the testis associated with sexual maturity there is a change in the diameter of the seminiferous tubules. Kasuya and Marsh (1984) analyzed the relationship between testicular weight (X, g) and mean diameter of the seminiferous tubules (Y, μm) for the southern-form pilot whale (Figure 12.14). The mean tubule diameter was obtained by measuring 20 tubules at the mid-center of the testis. The relationship between the two was expressed by

$$\text{Log } Y = 0.1441 \log X + 36.5452, \quad X < 80 \text{ g}$$

$$\text{Log } Y = 0.3828 \log X + 12.7160, \quad X > 80 \text{ g}$$

The first equation indicates that testicular weight (X) is proportional to $Y^{0.07}$ when the weight is less than 80 g, indicating limited contribution of tubule diameter to weight increase. However, the latter equation indicates weight increase to be proportional to $Y^{2.6}$. The rapid weight increase after age 14 years is mainly due to increase in tubule diameter. The correlation was almost lost for testes weighing over 1500 g, which suggests that size of

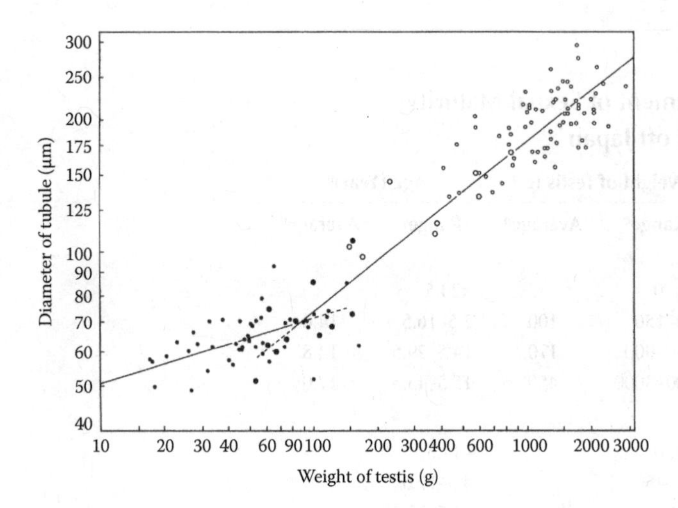

FIGURE 12.14 Scatter plot of diameter of seminiferous tubule and histological maturity of testis on weight of testis (epididymis removed) of southern-form short-finned pilot whales taken by Japanese drive fisheries (same material as in Figures 12.12, 12.15, 12.16, and 12.18). For testis maturity key see Figures 12.12 and 12.13. (Reproduced from Kasuya, T. and Marsh, H., *Rep. Int. Whal. Commn.*, 259, 1984, Figure 8.)

testes over 1500 g is likely a reflection of individual characteristics, not a reflection of growth stage or seasonal change. This agrees with the observation that adult males of large body size tend to have heavier testes (see the following text in this section).

This analysis suggested a strong correlation between sexual maturity and diameter of the seminiferous tubules. Figure 12.15 presents the relationship between age and seminiferous tubule

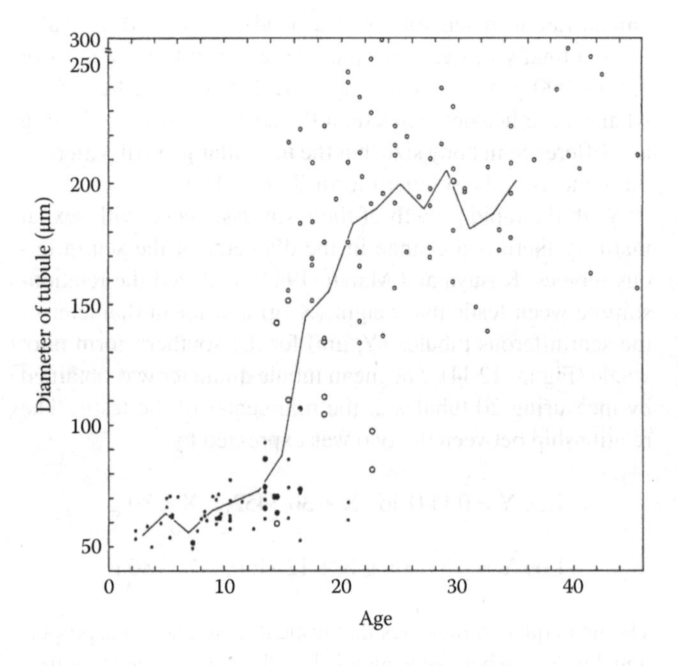

FIGURE 12.15 Scatter plot of diameter of seminiferous tubule and histological maturity of testis on age of southern-form short-finned pilot whales taken by Japanese drive fisheries (same material as in Figures 12.12, 12.14, 12.16, and 12.18). See Figures 12.12 and 12.13 for maturity key. (Reproduced from Kasuya, T. and Marsh, H., *Rep. Int. Whal. Commn.*, 259, 1984, Figure 7.)

diameter. Tubule diameter increased only slightly until the age of 15–16, when it started a rapid increase that continued until ages 18–20. Then a slower increase in the average diameter continued until around age 25. These features are similar to those observed in the growth in testis weight.

Testicular weight showed great individual variation, ranging from 700 to 3000 g among individuals over age 27 where testicular growth had been completed. One of the factors behind this was the tendency for males of larger body size to have heavier testes, which was supported by the facts that the correlation between age and body length was lost at age 27 and above (Figure 12.21) and that between age and testis weight was lost at age 24 and over (Figure 12.12), but there was a positive correlation between testis weight and body length among males of over 4.6 m (Figure 12.12), which were again likely to have ceased growing. Another factor behind the individual variation in testis weight among fully grown males was the possibility of seasonal fluctuation in reproductive activity, for which we did not have firm evidence but was suggested by wide individual variation of tubule diameter of 150–300 μm for males at age 27 or older (Figure 12.15). Future study of the male reproductive cycle should be directed toward both possibilities of a seasonally synchronous reproductive cycle as well as a nonsynchronous, less seasonal, reproductive cycle.

12.4.3.3.3 Ability to Reproduce

Spermatozoa produced in the testis are transported to the epididymis for maturation and are stored in the epididymal lumen, which opens to the seminal vesicle. Formation of spermatozoa alone is not evidence of the ability to reproduce. They have to be stored in sufficient quantity in the epididymis for reproduction.

Kasuya and Marsh (1984) examined development of the epididymal lumen with age in southern-form pilot whales. At "immature" the lumen was simple and the epithelium little folded, in "early maturing" it had slightly developed folds, and in "late maturing" and "mature" testis the lumen was fully folded and showed no difference between the two testicular stages. Thus maturation of the epididymis seemed to precede that of the testis.

Detection of spermatozoa in the epididymis and testis is considered more effective with a smear than through examination of histology of the tissues, presumably because sperm at low density could be lost during the preparation process or localized in distribution (Kasuya and Marsh 1984). A smear was taken from each formalin-fixed tissue sample, dried, and stained with a solution of toluidine blue for light microscopy. The testicular smear showed that some spermatozoa were present in 25% of testes that were classified as "immature" by histology. This contrasted with the fact that only 5% of individuals with a negative testicular smear were positive in testicular histology (i.e., classified at "early maturing" or later stages). This proved the higher efficiency of smears in detecting small amount of spermatozoa in the tissue.

Density of spermatozoa in the smear was classified into the following five stages. The microscopic field used in the classification was 1.82 mm in diameter.

1. *Absent*: No spermatozoa detected,
2. *Doubtfully present*: Very few, or only 1–2 spermatozoa in several fields,
3. *Scanty*: Fewer than 10 spermatozoa in each field,
4. *Intermediate*: Up to the maximum level found in an testicular smear of an adult,
5. *Copious*: Density usually found only in epididymal smear of fully grown male.

Histological testis maturity stages are compared with sperm density in testicular and epididymal smears in Table 12.8. Among 45 individuals that showed no spermatozoa in a testicular smear, only one (2%) had spermatozoa in an epididymal smear. This was reasonable, although the single case could indicate some deficiency of the method. This contrasted with the fact that out of 99 individuals that had spermatozoa in a testicular smear, 12 (12%) had no spermatozoa in an epididymal smear. These results indicated a lag between start of spermatogenesis in the testis and the arrival of spermatozoa to the epididymis.

Density of epididymal sperm is compared with testicular maturity in Table 12.8. The proportion of males having spermatozoa in the epididymis increased from "early maturing" (7 out of 11 males, or 70%) and "late maturing" stages (47 out of 9 males, or 78%) to the "mature" stage (100% of 69 males). This showed that spermatozoa were transported to the epididymis in most of the spermatogenic males. However, it was also true that sperm density in the epididymis increased with progress of histological maturity of the testis from "early maturing" to "mature." This was also shown by the fact that only 3 individuals (33%) out of 9 "late maturing" males showed epididymal sperm density of Stages 4 and 5, while 58 individuals (84%) out of 69 "mature" males showed that state. I note that many males with "late maturing" testes showed epididymal sperm density lower than the density found in males having "mature" testis (Table 12.8). This fact throws some doubt on the conclusion of Kasuya and Marsh (1984) that all the males at the "late maturing" stage would be as functionally mature as males at the "mature" stage.

Figure 12.16 shows the relationship between epididymal sperm density and age of southern-form pilot whales. A small number of spermatozoa first appeared in the epididymis at around 9 years, the highest Stage 5 appeared at age 15, and correlation between epididymal sperm density and age was almost lost at age 16. Individuals of the lower epididymal sperm densities, Stages 1 and 2, were not present at age 18 and higher. This suggested that males with epididymal sperm density Stages 4 and 5 were physiologically capable of reproduction (Kasuya and Marsh 1984). The age of 16–18 years when epididymal sperm density reached full densities (Stages 4 and 5) almost agreed with the age of 17, when testis histology reached the "mature" stage. Thus, it is correct to say that males of these stages, that is, histological "mature" stage or epididymal Stages 4 and 5, are physiologically capable of reproduction, and that about half of the males with testis at the "late maturing" stage also have reached the stage.

I have already questioned whether all the "late maturing" males are capable of reproduction to the same degree as "mature" males, but there remains another question, whether all the "mature" males and some "late maturing" with high epididymal sperm density really participate in reproduction.

TABLE 12.8
Sperm Density in Epididymal and Testicular Smears Compared with Histological Maturity of Testis of Southern-Form Short-Finned Pilot Whales off Japan

Testicular Smear and Histology		Epididymal Smear 1	2	3	4	5	Total
Smear	1. Absent	44	1				45
	2. Scanty	9	2		1		12
	3. Few	3	8	11	8	7	37
	4. Many			4	8	35	47
	5. Copious				2	1	3
	Total	56	11	15	19	43	144
Histology	Immature	47	3				50
	Early maturing	4	5	2			11
	Late maturing	2	2	2	1	2	9
	Mature		1	10	18	40	69
	Total	53	11	14	19	42	139

Source: Abbreviated from Kasuya, T. and Marsh, H., *Rep. Int. Whal. Commn.*, 259, 1984, Table 5.

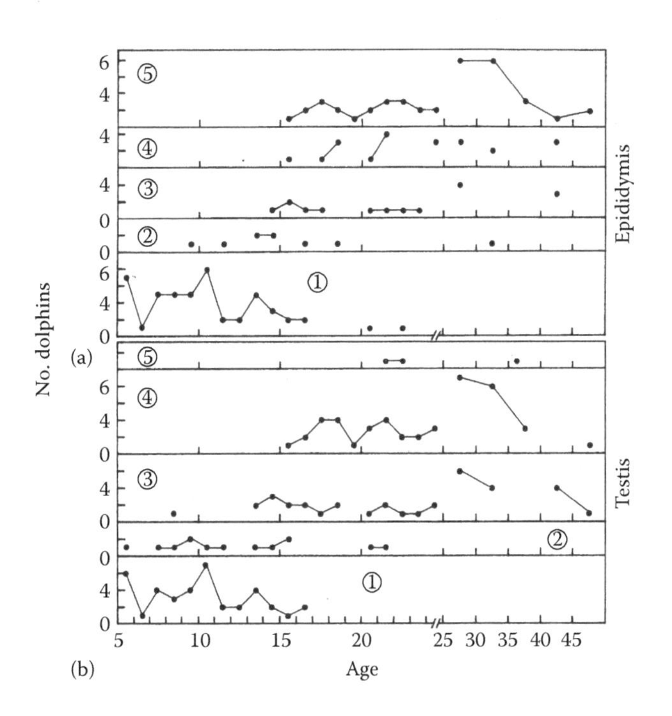

FIGURE 12.16 Age-related change in spermatozoa density in epididymal smear (a) and testicular smear (b) of southern-form short-finned pilot whales taken by Japanese drive fisheries (same material as in Figures 12.12, 12.14, 12.15, and 12.18). See Tables 12.8 and 12.9 for density codes for spermatozoa. (Reproduced from Kasuya, T. and Marsh, H., *Rep. Int. Whal. Commn.*, 259, 1984, Figure 9.)

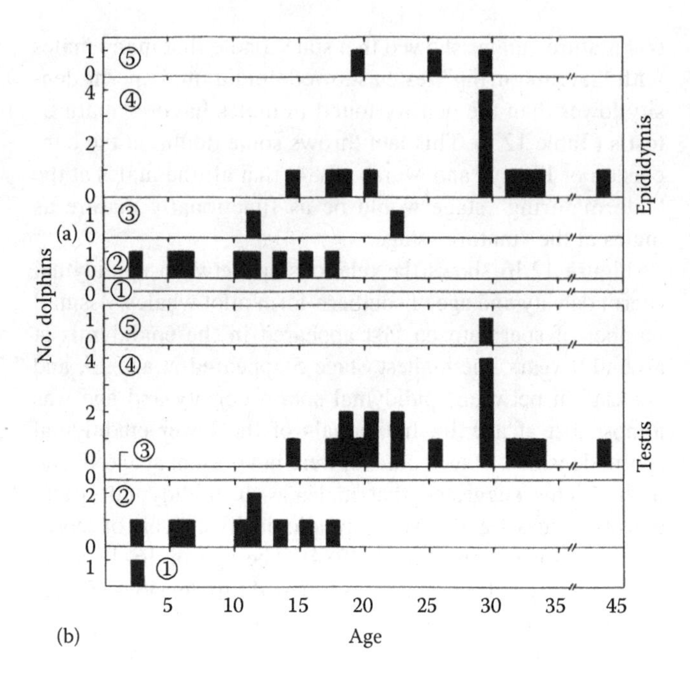

FIGURE 12.17 Age-related change in spermatozoa density in epididymal smear (a) and testicular smear (b) of northern-form short-finned pilot whales taken by Japanese small-type whaling (same material as in Figures 12.13 and 12.19). See Tables 12.8 and 12.9 for density codes for spermatozoa. (Reproduced from Kasuya, T. and Tai, S., *Rep. Int. Whal. Commn.*, 339, 1993, Figure 9.)

This question relates to so-called social maturity, which is a function of male interactions for access to estrous females as well as female choice of mating partners. Kasuya and Marsh (1984) concluded from analysis of school structure of southern-form plot whales that "early maturing" or "late maturing" males were not functioning equally with "mature" males in their community, which suggested that these maturing males were not participating in reproduction equally with "mature" males (Section 12.5.3.4). Northern-form pilot whales showed a similar feature as the southern form in age-related change of sperm density in testis and epididymal smears (Figure 12.17).

In summary, males of both northern-form and southern-form pilot whales by age 17–18 reached a stage where the entire testicular tissue underwent spermatogenesis and the sperm density in epididymis reached a plateau (i.e., correlation between sperm density and age was lost). They were considered to have attained the physiological capability of reproduction at that stage. We do not have evidence that spermatogenetic activity declines at higher age, but at the same time we do not know if male reproductive activity remains at the same level for their life from ages 17–18 to 45 years at maximum.

12.4.3.3.4 Seasonality in Reproduction

Analyses of fetal body size of southern-form whales revealed that conception occurred in all months of the year, with a peak in the 5 months from March to July and a trough in the 5 months from September to January (see Section 12.4.4.2). In order to determine whether males have this seasonal change in reproductive activity, Kasuya and Marsh (1984) examined seasonality in testis weight, seminiferous tubule diameter, and sperm density in testicular and epididymal smears. Average testis weight at each histological maturity stage did not show seasonal change, which could have been due to the limited sample size relative to the large individual variation in weight or to parallel seasonal change in maturity stages and weight. They next selected males measuring over 4 m, which excluded most "immature" and some "early maturing" males (Figure 12.12), and stratified them into two groups of December and May–July in an attempt to find seasonal change in testis weight. This analysis also failed to reveal seasonal change.

Mean diameter of the seminiferous tubules was compared between the high-conception months of May–July and low-conception month of December by testicular maturity stage and body-length groups. A significant but minor difference was identified in only one case. Mean tubule diameter of males over 460 cm, which were considered fully grown animals (Figures 12.12 and 12.21), was significantly greater ($p < 0.01$) in the high-conception months (217 μm, $n = 21$) than the low-conception month (193 μm, $n = 14$). This result suggests the possibility of some limited degree of seasonal changes in reproductive activity of adult males, but further confirmation is needed using larger samples.

The analysis presented earlier suggested the possibility of greater mean diameter of seminiferous tubules of adult males during the mating season, but it did not distinguish between two possible underlying mechanisms, that is, whether the proportion of active males increased or all males increased activity synchronously. We learned by examining males with testis weighing over 600 g (Table 12.9) that there were numerous active males even in a season of low conception frequency and that there were numerous males with high epididymal sperm density in every month of the year. This indicates that reproductively active males are available for estrous females throughout the year, which is reasonable for the population of southern-form whales where estrus occurs in every month of the year. Many males are ready, with some cost, for estrus of females that may occur only once in several years.

TABLE 12.9
Seasonal Fluctuation of Sperm Density in Epididymal Smear of Southern-Form Short-Finned Pilot Whales off Japan, for Males with Testis Weighing 600 g or More Which Can Be Judged Sexually Mature (see Figures 12.12 and 12.13)

Sperm Density	1. Absent	2. Scanty	3. Few	4. Many	5. Copious	Sample Size
Sep–Jan	0	6.2%	9.4%	28.1%	87.5%	32
Feb	0	0	33.3%	33.3%	33.3%	12
May–Jul	0	0	16.7%	8.3%	75.0%	24

Source: Abbreviated from Kasuya, T. and Marsh, H., *Rep. Int. Whal. Commn.*, 259, 1984, Table 11.

Kasuya and Marsh (1984) then examined seasonal change in the lower bound of the weight of a testis with spermatogenesis using sperm density in the testis and epididymal smears as indicators. They noted the range between minimum testis weight with spermatozoa and maximum testis weight without spermatozoa and tested whether the range differed between the high-conception months (May–July) and low-conception months (December–January). The range for the epididymal smear was 50–80 g in the high-conception months and 140–180 g in the low-conception months. The testicular smear also showed a similar seasonal trend: 40 g in the high-conception months and 90–160 g in the low-conception months. In a comparison between samples from the same season, the range for the epididymal smear was always greater than the range for the testicular smear. This reflects the lag between the start of spermatogenesis in the testis and transportation of the produced spermatozoa to the epididymis. It was unlikely that all males in the range had the physiological ability to reproduce, but their physiology was probably responding to seasonal change in the environment as in the case of adult males.

12.4.3.4 Growth Curve

A growth curve is a way of describing age-related change in body size. Body size can be expressed as body weight, head and body length, or head to tail length. As weight is difficult to obtain for cetaceans, the last is often used and customarily called "body length." This is a linear length measured on a body placed on a flat place from the anterior-most point of the upper jaw (or head) to the base of the notch at the center of the tail flukes. This principle has generally been accepted by biologists working on short-finned pilot whales but has caused trouble for some of them (Yonekura et al. 1980). The tip of the upper lip is located at the anterior-most point of the head in young pilot whales, but it becomes covered by a fatty tissue body called the melon that develops between the upper lip and the blowhole and is located aft of the foremost extent of the melon at a body length of 240–300 cm or ages of 2–5 years. Thus, the rule for measurement defined earlier causes anatomical inconsistency of the measurement occurring around the ages of 2–5 years.

If body length of a particular individual is recorded for life and plotted on age, a growth curve for the individual will be obtained. Growth curves thus obtained will be different between individuals. A fast-growing individual is larger than other individuals of the same age, will show a pubertal growth spurt at a younger age, and may cease growth at an earlier age. The opposite might be true for slow-growing individuals. Averaging these growth curves creates a mean growth curve, which will not match any individual's growth.

Even beyond the time and cost required to construct the growth curves of individual dolphins, it is technically difficult to maintain even one short-finned pilot whale in good health for tens of years in captivity. Some small cetaceans in captivity are known to have reached sexual maturity at ages younger than expected for wild animals, but those in captivity are also known to have a shorter life. In order to avoid such difficulties and utilizing available fishery data, there have been attempts to measure body size of numerous individuals killed in a fishery and to plot the average body sizes on ages estimated from tooth reading. Such samples may cover just a few seasons or several years. The growth curve thus created should be called a "pseudo"-mean growth curve, but cetacean biologists usually call it a "mean growth curve." This pseudo-mean growth curve can be accepted as a real mean growth curve only when the assumption is accepted that the growth pattern of the population did not change during the period covered by the sample, that is, the sampling period plus the lifetime of 50–60 years for short-finned pilot whales. If growth rate has increased through improvement of nutrition during the period, the pseudo-mean growth curve gives the impression that body size decreases at high age, which has been reported for sperm whales in the western North Pacific (Kasuya 1991).

Figure 12.18 shows body-length frequency of southern-form short-finned pilot whales randomly examined from the catch of drive fisheries in the 15 years from 1965 to 1980. It reveals features such as (1) sex ratio biased to females, (2) a female peak at around 360 cm and that of males at around 470 cm, and (3) male maximum size that is about 1 m greater than that of females. The first is explained by sexual segregation or longevity difference, and the latter two by sexual growth difference. The body-length composition of northern-form short-finned pilot whales has a similar difference between sexes in body size, where males are greater than females by about 1.7 m (Figures 12.19 and 12.20). The sex ratio is biased toward males due to fishery selection for larger individuals.

The "pseudo"-mean growth curves created using the data obtained earlier are shown in Figures 12.21 and 12.22. They are forced to pass through the mean neonatal lengths estimated earlier at age zero. Most of the individual body lengths come within a range of two standard deviations on each side of the mean, or within 6%–10% of the range on each side of the mean length of southern-form whales aged at or over 1 year.

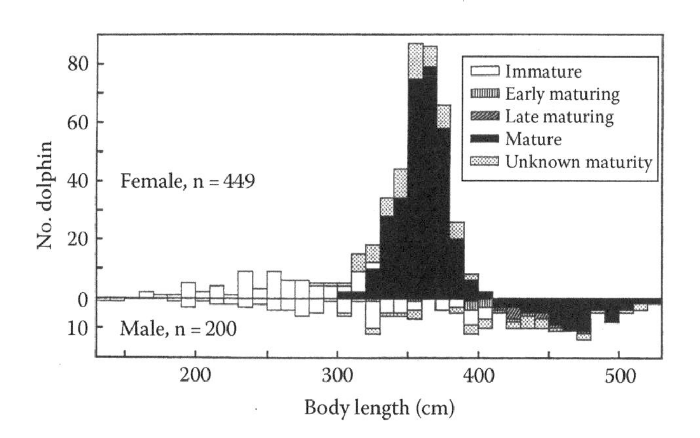

FIGURE 12.18 Body-length composition of females (top panel) and males (bottom panel) of southern-form short-finned pilot whales taken by Japanese drive fisheries covering 8 months of the year, excluding March, April, September, and November, 1965–1981. (Reproduced from Kasuya, T. and Matsui, S., *Sci. Rep. Whal. Res. Inst.*, 35, 57, 1984, Figure 11.)

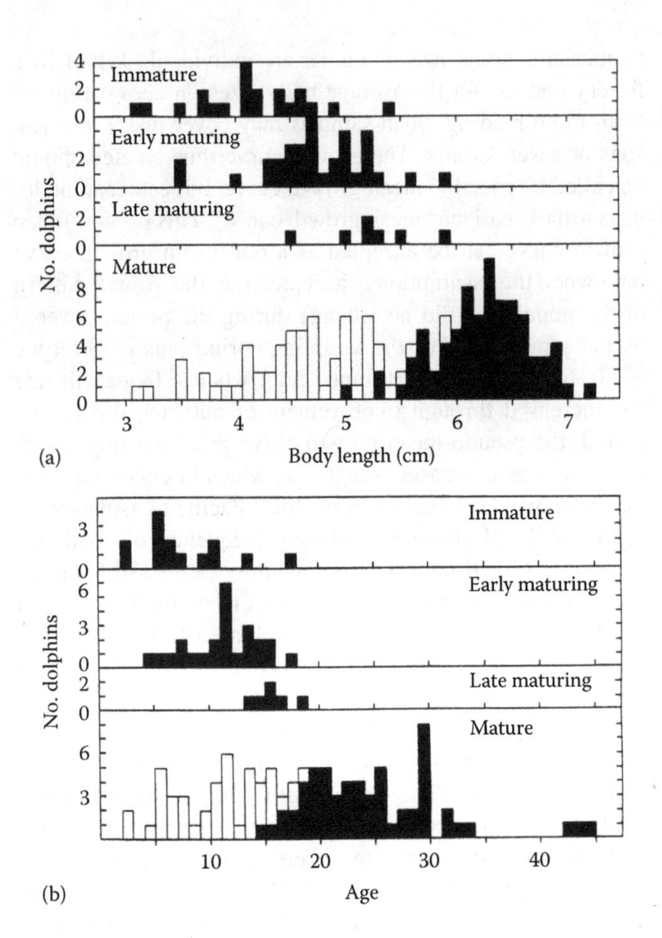

FIGURE 12.19 Body-length composition (a) and age composition (b) by maturity stage of male northern-form short-finned pilot whales taken by Japanese small-type whaling in October–November, 1983–1988. White boxes represent total of individuals other than mature individuals. (Reproduced from Kasuya, T. and Tai, S., *Rep. Int. Whal. Commn.*, 339, 1993, Figure 8.)

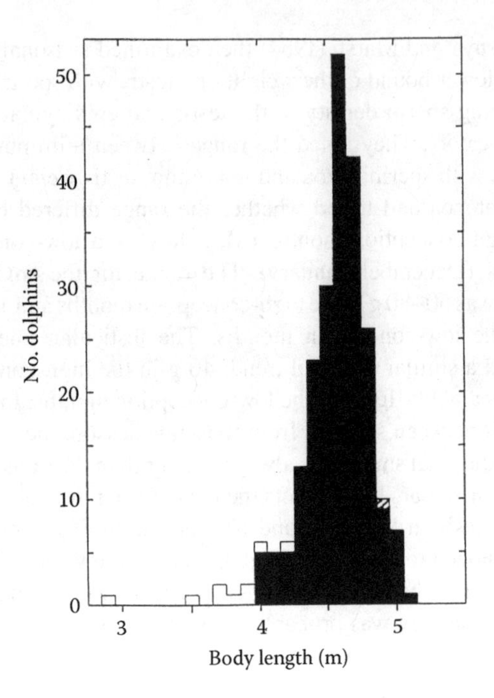

FIGURE 12.20 Body-length composition of female northern-form short-finned pilot whales taken by the same whaling operation as in Figure 12.19. White boxes represent sexually immature individuals, black boxes mature individuals, and shaded boxes individuals of unknown maturity. (Reproduced from Kasuya, T. and Tai, S., *Rep. Int. Whal. Commn.*, 339, 1993, Figure 11.)

However the standard deviation of southern-form whales aged between 0 and 1 was quite large, because their growth was rapid and any difference in birth time had a large influence on the body size of these individuals. Further details of these data on southern-form whales, including sample size, mean body length, and standard deviations, are available in Kasuya and Matsui (1984).

Kasuya and Matsui (1984) described the mean growth curve of southern-form pilot whales in three stages. The first stage was the period from birth to age of about 1.25 years, when growth was extremely fast and the whales reached an average length of 230 cm. The mean growth rate of this 1.25-year period was 72.4 cm/year, which was 62.1% of the growth rate in the linear phase of fetal growth. The available sample did not show evidence of sexual difference in growth in the period. This first phase gradually transitioned into the second phase, and the timing seemed to be later among males.

The second growth phase was a period of almost linear growth, begun at around age 2. The mean body length at age 2.5, shortly after the beginning of this stage, was about 254 cm for males, about 6 cm greater than that for females.

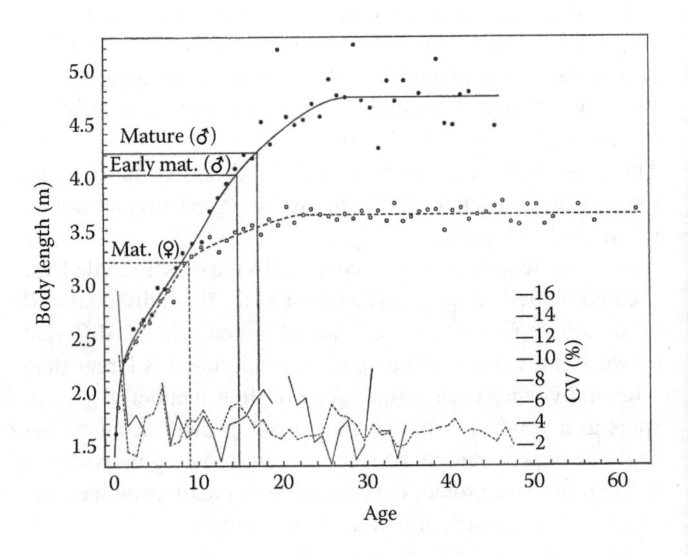

FIGURE 12.21 Mean growth curve of southern-form short-finned pilot whales. Closed circles represent mean male body length, solid line mean growth curve fitted to the means by eye, and open circles and dotted line those for females. Mean body length at the mean age at attainment of sexual maturity (Ma; for both sexes) and that at attainment of early maturing stage (Em; for males) are indicated. (Reproduced From Kasuya, T. and Matsui, S., *Sci. Rep. Whal. Res. Inst.*, 35, 57, 1984.)

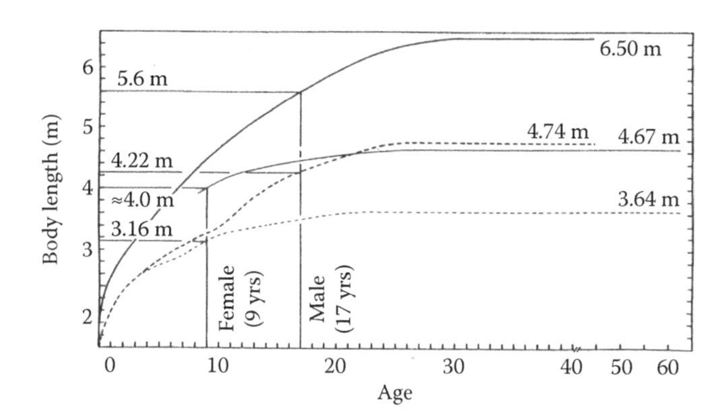

FIGURE 12.22 Comparison of mean growth curves of southern-form and northern-form short-finned pilot whales. Mean body length at attainment of sexual maturity (at 9 years for females and 17 years for males) indicated. The northern-form grows larger than the southern form of corresponding sex, and females live longer than males. (Reproduced from Kasuya, T. and Tai, S., *Rep. Int. Whal. Commn.*, 339, 1993, Figure 16.)

This stage lasted until the ages of 9–10, when females attained sexual maturity, and could be considered a juvenile period. The mean growth rate at this stage was 11–12 cm/year and did not show significant sexual difference. However, males of this stage were always larger than females of the same age, mainly because males remained in the previous stage for a longer period.

The third stage of the mean growth curve was characterized by a growth rate declining with age, which was interpreted as a reflection of decreasing growth rate of some individuals and increasing proportion of whales that had ceased growing. Females entered this stage at around age 9 and ended it at around age 22, while males entered it at 10 years and remained there until age 27. The end of this stage should agree with the ages when all whales of a sex have ceased growing.

The mean growth curve of males had one unique character. This was a slight increase in growth rate at around ages 11–14 years or at the beginning of the third stage, a reflection of the pubertal growth spurt. The available data were not sufficient to clarify individual variation of the timing and magnitude of the pubertal growth spurt, but the timing apparently agreed with the age when males enter the "early maturing" stage of testicular histology. The "mature" stage was attained at a later age, or at 15–22 years. The pubertal growth spurt is known in some other mammals, including pinnipeds and primates. Among humans the growth spurt is distinct in an individual growth curve, but it is indistinct in a mean growth curve because the spurt lasts for a short period with broad individual variation in timing. The same is expected for males of the short-finned pilot whale.

A pubertal growth spurt is often observed in mammals where males grow larger than females and is thought to have evolved accompanied by a polygynous mating system that increased inter-male competition for mating opportunities. However, the degree of sexual dimorphism does not necessarily reflect the degree of polygyny in a particular species or population. There is no reason to assume that the mating system and sexual dimorphism are in parallel. A pubertal growth spurt could have followed development of a polygynous mating system, and abandonment of polygyny for some reason could have preceded the change in growth pattern. In short, social structure and reproductive behavior in toothed whales are very flexible (Connor et al. 2000), probably more flexible than growth pattern. Killer whales are one of the species with distinct sexual dimorphism, but studies on the species off the Vancouver Island area has not produced evidence that they have strong inter-male competition for mating opportunities (Baird 2000).

TABLE 12.10
Mean and the 95% Confidence Interval of Body Length at Physical Maturity of Short-Finned Pilot Whales off Japan

Population	Southern Form		Northern Form	
Sex	Male	Female	Male	Female
Age range of sample	>27	>22	>30	>30
Sample size	35	181	11	58
Mean body length	473.5 cm	364.0 cm	650.4 cm	467.4 cm
95% CI	±9.1 cm	±1.9 cm	±10.1 cm	±7.5 cm

Sources: From Kasuya , T. and Marsh, H., *Rep. Int. Whal. Commn.*, 259, 1984; Kasuya, T. and Tai, S., *Rep. Int. Whal. Commn.*, 339, 1993.
Note: See text for a possible bias due to gear selection.

Figure 12.22 compares the mean growth curves of the two forms of short-finned pilot whale off Japan. The mean growth curves of the northern form are likely to be upwardly biased to an unknown degree, because the samples were obtained from small-type whaling that selectively took larger individuals (the fishery had no minimum size limit but a quota set by number). The bias could be greater for females than males, and greater for individuals close to the lower bound of length in the catch, which was about 3–3.5 m (Figures 12.19 and 12.20). Thus the mean growth curves for ages below 10 are of limited reliability. For both types of pilot whales the correlation between mean body length and age is lost at ages 25–30, when all the individuals are believed to have ceased growing (Kasuya and Tai 1993). Table 12.10 gives mean body lengths at this stage, which are indirect estimates of mean body length at physical maturity. Physical maturity is defined as evidenced by the complete fusion of the vertebral epiphyses to the centra, but examination of the vertebral epiphyses is often difficult for fishery-derived materials (Sections 9.4.10 and 13.4.5).

12.4.4 Female Reproductive Activity

12.4.4.1 Fetal Growth and Gestation Time

Huggett and Widdas (1951) found that the cube root of fetal weight plotted on time after conception falls on a line that is upwardly concave at the beginning and then linear until parturition. The time (t_0) when the left-ward extension of the linear growth phase crossed the axis of time was a function of total gestation time (t_g), where $t_0 = 0.1t_g$ for species with $t_g > 400$ days, but the proportion increased for species with shorter gestation (Section 8.5.2.2). Laws (1959) tested this rule on cetacean fetuses using body length as the indicator of body size and found that the value of t_0 was 0.9 of t_0 using the cube root of body weight. This means that $t_0 = 0.09t_g$ for gestation of $t_g > 400$ if fetal size is expressed by body length. This principle has been frequently used in the analysis of fetal growth of cetaceans.

Figure 12.23 presents the body-length frequencies of fetuses and juveniles of southern-form pilot whales taken by drive fisheries on the Izu coast and at Taiji. The June–August sample had two major modes, one of near-term fetuses at around 130 cm and another of small fetuses below 10 cm, but the October–February sample had only one major mode at 40–100 cm. With the assumption that these seasonal changes in fetal size reflected growth of the smaller fetuses and birth of the near-term fetuses, Kasuya and Marsh (1984) separated the fetuses into two arbitrary cohorts of conception divided by a dashed line in Figure 12.23 and calculated mean body length of the cohort for each month. Some neonates in the June–August sample were included in this calculation to obtain better resolution of seasonal growth.

Out of the 11 mean body lengths, 9 from October to the next October fell on a line, but the 2 means representing small fetuses in June–August appeared above the line. This was explained by the fact that the sample of small fetuses represented conceptions in the early part of the mating season. The possibility of overlooking some small fetuses cannot be completely excluded, but Kasuya and Marsh (1984) considered that the possibility was small because uteri accompanied by a corpus luteum were carefully examined. They fitted a linear regression to individual fetal lengths (Y, cm) representing the 9 means on number of days after the first of January (X), and obtained the following fetal growth equation:

$$Y = 0.3386\ (\pm 0.0425)\ X - 60.1, \quad r = 0.82$$

The 95% confidence interval is given in parentheses. (A regression fitted to the 9 mean values in Figure 12.23 was expressed by Y = 0.3398 X − 60.1, which was very close to the equation fitted to the individual fetuses.) The estimate of 0.3386 cm/day is the mean growth rate in the linear part of fetal growth. The equation cut the axis of time on June 26 and reached the mean neonatal length of 140 cm (Section 12.4.3.1) on August 11 (Figure 12.23). Thus, the value of ($t_g - t_0$) was calculated as 411 days and total gestation period as 452 days or 14.9 months (assuming a month of 30.4 days). Some possible problems that might arise in the process of identifying conception season and fetal growth rate are reviewed as follows.

The first problem was missing small fetuses during the field work. That could happen for two reasons, and it would cause loss or bias in information on the conception season. One potential reason was observation error by the scientists. This error could be avoided if the whole uterus were collected together with ovaries that contained a corpus luteum for later careful examination, including histology of the endometrium. The other potential reason was miscarriage at death in pregnant females. I have confirmed this in both forms of pilot whale by finding a fragment of the placenta remaining in the intact uterus or a small fetus in the vagina. Some fetuses, particularly those of early pregnancy, were delivered when the females were slaughtered by drive fishermen or shot by whaling cannon. Histological examination of the endometrium helps to minimize both types of misidentification. Kasuya and Marsh (1984) examined the endometrium of most of the southern-form females and Kasuya and Tai (1993) and Kasuya et al. (1993) that of all the sexually mature females (both forms). They identified early pregnancy with an embryo of 1.5 (Table 12.18) or 2.3 mm (Kasuya and Tai 1993) by a network of blood vessels developed beneath the epithelium, which was followed by development of epithelial villi.

The second problem was inclusion of some postnatal specimens in the calculation of the growth regression. It was theoretically correct to include some neonates that shared a conception peak with near-term fetuses, but inclusion of neonates of a broader age range would bias the fetal growth rate downward because the postnatal growth rate was slower than that of fetuses.

A third problem comes from the fact that the growth rate of fetuses in the early period of gestation is slower than that in the later linear growth phase. If data from small fetuses are included in the regression, it would underestimate the growth rate in the linear phase of growth. This could cause a problem in the same manner as in the fourth problem mentioned in the following. Kasuya and Marsh (1984) omitted fetal data that represented the ongoing conception peak to avoid this situation.

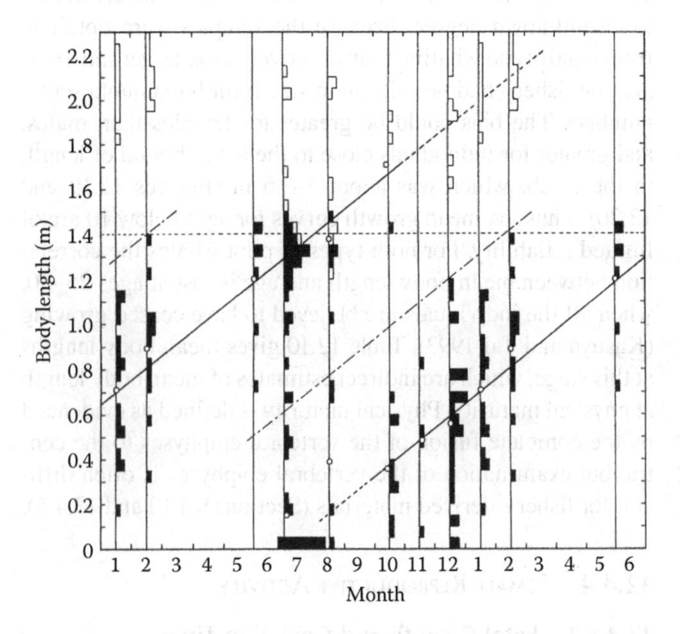

FIGURE 12.23 Body-length frequency of fetuses (black squares) and juveniles (white squares) of southern-form short-finned pilot whales taken by Japanese drive fisheries in 1965–1981. Modal means (circles) and linear regression fitted to them (solid line) indicated. (Reproduced from Kasuya, T. and Marsh, H., *Rep. Int. Whal. Commn.*, 259, 1984, Figure 16.)

A fourth problem was the effect of falling off of a conception season. As seen in the fetal records for southern-form whales in June–August (Figure 12.23), the sample obtained during a mating season underrepresented small fetuses to be conceived later in the season. The effect of this bias could be decreased by excluding particular samples. To the extent that we estimate fetal growth rate from the regression of body length on date, the obtained rate is not perfectly free from the possibility of such bias due to an extended breeding season (Martin and Rothery 1993). The problem of an extended mating season in the analysis of fetal growth also relates to grouping of fetuses by body length. If the boundary between two fetal groups, a dashed line in Figure 12.23, was incorrect, the resultant fetal growth rate and gestation time was incorrect.

Information on the gestation period was obtained for some cetaceans in captivity at Sea World in the United States, where the start of conception was determined by monitoring the progesterone level in the serum or urine (Asper et al. 1992). Mean gestation periods thus obtained were 345 days (n = 8) for Commerson's dolphin, *Cephalorhynchus commersonii*, that were born at 100 cm, 370 days (n = 77) for the larger bottlenose dolphin (born at 117–127 cm) and 515 days or about 16.9 months (n = 7) for the much larger killer whale (born at 219–235 cm).

Many baleen whale species have annual migrations between feeding grounds in higher latitudes and breeding grounds in lower latitudes, which places a constraint on the gestation period, maintaining it at around 12 months. Thus they are forced to increase their fetal growth rate while achieving augmentation of body size, both neonatal size and adult size. Augmentation of body size benefited their breeding that occurred during winter months of starvation. Some Bryde's whales apparently do not have large-scale annual migrations and perhaps no seasonality of breeding, but they still seem to maintain gestation of about 1 year. This is probably a reflection of their earlier life history. Some toothed whales, on the other hand, stay in a similar environment for the whole year, are free from a seasonal constraint on reproduction, and extend the gestation time while achieving augmentation of body size (Kasuya 1995b). Such toothed whale species include female sperm whales, Baird's beaked whales, short-finned pilot whales, and some killer whale populations.

The fetal growth curve of the southern-form pilot whale matched reasonably well with fetal length frequency, and the estimated gestation period was not unreasonable in view of the gestation period known from some aquarium-reared species. Perhaps the exclusion of small fetuses from the regression could have avoided the problem identified by Martin and Rothery (1993). They simulated the best fetal growth rate to meet both the observed fetal length frequency and assumed conception timing of long-finned pilot whales in the North Atlantic and concluded that the gestation period was 12 months. I cannot conceive a convincing explanation why two species of similar body size in a genus could exhibit such different gestation times. One possibility would be that a seasonal environmental cycle forced the long-finned pilot whale in higher latitudes to maintain parturition at a particular time of the year as in the case of migratory baleen whales (Kasuya 1995b).

12.4.4.2 Breeding Season

12.4.4.2.1 Southern Form

Linear fetal growth of the southern-form whales reached the mean neonatal length of 140 cm on August 11 (see above). This was an estimate of mean parturition date of the population. The monthly distribution of births was calculated by fitting the mean fetal growth rate to the body lengths of fetuses and date of the data, and a normal distribution was fitted to the distribution (Figure 12.24). Births had a peak in July and August and a trough in December–March. Another set of birth dates was obtained by back-calculating body lengths of juveniles below age 1.0 using their mean growth curve (bottom panel of Figure 12.24), but the seasonal pattern did not agree well with the pattern obtained from the fetuses. The disagreement was probably due to inferior accuracy of the postnatal growth curve.

Several factors could have caused bias in the estimate of parturition season. Error in the fetal growth rate biases birth date. The bias is greater for dates calculated from smaller fetuses than those calculated from larger fetuses. The fetal

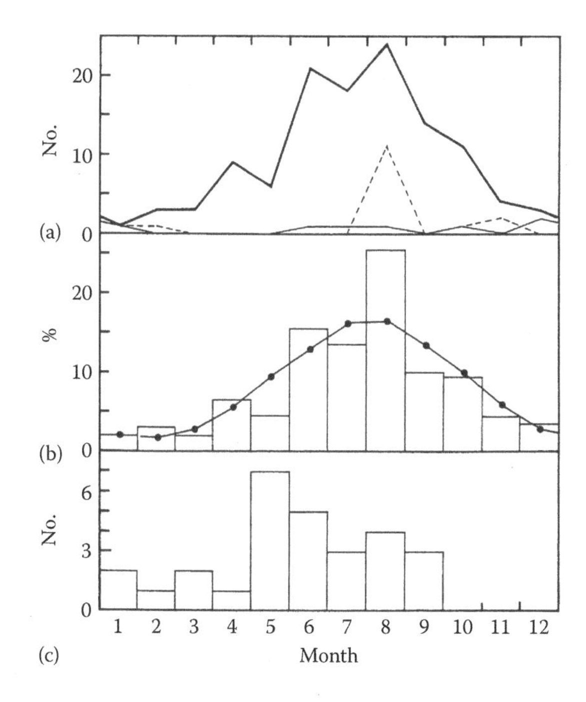

FIGURE 12.24 Monthly distribution of births calculated for southern-form short-finned pilot whales taken by Japanese drive fisheries. See Figure 12.23 for material used. (Reproduced from Kasuya, T. and Marsh, H., *Rep. Int. Whal. Commn.*, 259, 1984, Figure 17.) (a) Birth dates predicted or back-calculated by fitting a linear fetal growth rate to fetuses ≥20 cm and neonates ≤155 cm (thick solid line), to fetuses of 10–20 cm (thin solid line), and to fetuses ≤10 cm (dotted line); (b) sum of all the birth dates obtained in the upper panel (squares) and a normal distribution fitted to them (closed circles and solid line); (c) birth dates of dubious accuracy back-calculated by fitting mean age on body-length key (Kasuya and Matsui 1984) to juveniles judged at below age 1 year.

sample needs to represent the composition of the population for a correct estimate of the parturition season. If females of near-term pregnancy tended to live outside the fishing ground, the resultant parturition peak would come after the real peak. A seasonally biased sample may result in a biased parturition season. The sample used in Figure 12.24 had some seasonal bias and lacked data for March–May and September. The season March through May corresponds to the early mating season and the beginning of the parturition season, and the lack of data for near-term fetuses for these months had the effect of placing the calculated parturition peak artificially later. The absence of data for September had the opposite effect. The total effect of these deficits in the samples was not evaluated by Kasuya and Marsh (1984).

The monthly distribution of conceptions is estimated by sliding forward from the monthly parturition date by 14.9 months or the estimated gestation period, which suggests a conception peak in May and a trough in November. This is the best estimate currently available of the mating season of the southern-form pilot whale.

12.4.4.2.2 Northern Form

The biological data for the northern-form whale were obtained mostly in October and November, with a small number of samples in September due to the fishing season agreed between the whalers and the Fisheries Agency (Kasuya and Tai 1993). The small-type whalers used to make a slit in the abdomen of the whale carcass to cool it with seawater during on-deck transportation of several hours or towing for flensing at Ayukawa. This often caused loss of fetuses in the sea, large fetuses in particular. The whaling gunners retained some large fetuses found during the process for scientists when requested to do so (Kasuya and Tai 1993). This was one of the reasons why the histology of the endometrium had to be examined for confirmation of pregnancy. The short sampling season and the biased fetal size inhibited applying the method used for the southern form in the analysis of reproductive seasonality of northern-form pilot whales.

Kasuya and Tai (1993) analyzed fetuses measuring between 2.3 and 180 cm obtained in October and November, but their material should have contained two pregnant females later identified with embryos of 1.5 mm (Table 12.18; Kasuya et al. 1993). This suggested that conceptions occurred throughout the year as in the southern form. However, the fetal length distribution suggested that seasonality of breeding was likely to be more distinct in the northern form (Figure 12.25). Two peaks were identified in fetal size, one of small fetuses below body length of 15 cm and another with large fetuses of over 150 cm. The latter were considered underrepresented in the sample. Interpretation of the body-length difference between the two fetal modes, about 135 cm, was a basis for understanding reproductive seasonality.

Mean neonatal length was estimated at 140 cm for the southern form and 185 cm for the northern form (Section 12.4.3.1). The following equation between mean neonatal length (X, cm) and fetal growth rate in the linear phase of fetal

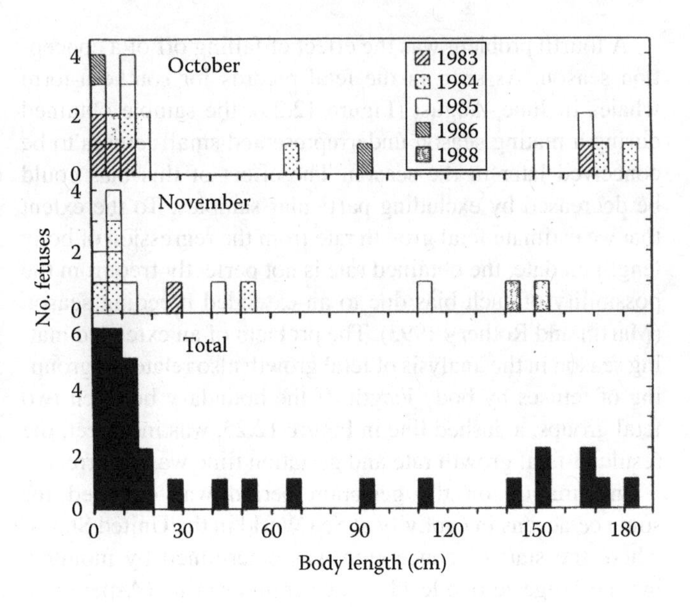

FIGURE 12.25 Fetal size frequency of northern-form short-finned pilot whales taken by Japanese small-type whaling in October and November, 1983–1988. (Reproduced from Kasuya, T. and Tai, S., *Rep. Int. Whal. Commn.*, 339, 1993, Figure 14.)

growth (Y, cm/day) applies to several species of Delphinoidea (Kasuya 1977):

$$Y = 0.001462\,X + 0.1622$$

Substituting the mean neonatal body lengths in this equation gave the following estimates of average fetal growth rate:

Southern form: 0.3668 cm/day
Northern form: 0.4327 cm/day.

The figure for the southern form was slightly greater than the corresponding figure of 0.3386 cm/day estimated directly from the fetal size distribution (Section 12.4.4.1) and corresponded to a gestation period of about 1 month shorter than the estimate based on the fetal size distribution. This suggests that the fetal growth rate predicted for the northern form likely underestimates the gestation period by about 1 month. Using this possibly upwardly biased fetal growth rate, a minimum length of time between the two fetal modes, that is, 15 and 150 cm, was estimated:

$$(150 - 15\text{ cm})/(0.4327\text{ cm/day}) = 312\text{ days or }10.3\text{ months.}$$

Even with the difficulty of determining the start and end of the mating season, it is safe to conclude from the given that the two fetal modes of the northern form observed in the October–November sample represented conceptions in 2 successive years (Kasuya and Tai 1993).

The calculation indicates that seasonality of mating in the northern form was unimodal as in the southern form, and that two successive mating peaks were 1 year apart. Comparison of fetal size between October and November revealed an additional feature of the mating season (Figure 12.25).

In October the number of fetuses below 5 cm exceeded that of fetuses at 5–10 cm, but this was reversed in November. This also suggests that the mating peak had already passed in November and that the mating peak for the population was short. However, if it is accepted that it took about 2 months for a fetus to grow to 15 cm (Kasuya et al. 1993), then the mating season could have already started in early September, had a peak in October, and ended in December. This allows about an 8-month interval between the two periods of high mating activity in successive years.

Although we currently have insufficient data to draw firm conclusions on details of the mating season of the northern form, it is possible to say that most conceptions occur from autumn to early winter with a peak in October/November and in a more seasonally limited period than in the southern form (Kasuya and Tai 1993).

The larger fetuses of the northern form had a mode at around 160 cm in October/November. It would take about 2 months before they grew to the mean neonatal length, that is, 58 days = (185 − 160)/0.433. This means that parturition peaked in December/January. This together with a possible conception peak in October/November suggests a gestation period of 14–15 months, which is reasonably close to the corresponding figure for the southern form.

Uncertainty remains about seasonality of reproduction of the northern form due to small sample size, the short season covered by the sample, difficulty in obtaining near-term fetuses, and unavailability of neonates. However, the currently available data suggest that the two forms of pilot whale off Japan mate about 5–6 months apart and that the northern form is a more seasonal breeder. No evidence exists for a difference in gestation period between the two forms (Kasuya and Tai 1993).

12.4.4.3 Weaning

Juvenile dolphins start taking solid food at the age of a few months, the start of the process of weaning. The amount of solid food they consume increases and the proportion of nutrition from milk decreases with time until suckling finally ceases, the completion of weaning. The period from the start of solid food ingestion to the end of suckling is called the weaning period. There is evidence that the weaning period of the southern form extends for several years. The weaning period functions as a period of learning foraging technique and adapting digestive physiology from milk to marine organisms. However, only 6–12 months is needed for this function, as seen in the finless porpoise (Section 8.5.2.3) and Dall's porpoise (Section 9.4.6). Even herbivorous mammals such as horses, which must undergo greater physiological adaptation during the weaning period, start taking solid food at 2 months and complete weaning by the age of 6 months.

Human infants start eruption of deciduous teeth at age 6 months, when taking of solid food may start, and complete the full set of teeth at around 2.5 years. Completion of weaning among humans is affected by culture and availability of other sources of nutrition. In the current Japanese community weaning is completed at around 1 year after birth, but it took longer in the past. In the agricultural community of Kawagoe in the 1930s, where I was raised, it was not uncommon to suckle until the age of 3–4 years and in rare cases until the age of 6–7 years, close to the age for beginning elementary school. The nutritional contribution of such extended suckling is questionable. As lactation may suppress estrus, extended nursing can adversely affect the reproductive success of mothers. The extended weaning period of some mammal species is likely explained by some function other than nutrition, as first proposed for cetaceans by Brodie (1969), who thought that the relatively long lactation period of toothed whales could function as a period of education or protection by the mother.

Humans have two sets of tooth germs at birth that grow and erupt successively, which is called diphyodonty. Sometime after completion of eruption of the deciduous teeth in humans, they start to be replaced by permanent teeth. The situation is different among the recent toothed whales, which have only one set of teeth, a state called monophyodonty (Section I.1.3). Tooth eruption starts in many toothed whales at the age of a few months, but it may occur much later in some species. In sperm whales, tooth eruption starts at age 5–18 years (mean 9 years), which agrees with female age at sexual maturity and male age at puberty (Ohsumi 1963, in Japanese). Beaked whales (Ziphiidae) have erupted teeth usually only in sexually mature males; females usually have no erupted teeth (McCann 1974). These species mainly feed on squid using suction feeding, which does not require functional teeth (Marshall 2009). The southern-form pilot whales off Japan feed almost exclusively on squid, and the structure of the broad oral cavity is apparently suited for suction feeding.

The drive fishery at Taiji in Wakayama Prefecture drove schools of southern-form pilot whales into a harbor for slaughter. Juveniles in the school often died in the enclosure before being killed by the fishermen and were found floating in the water sometime after death. The fishermen tied these juvenile carcasses together in the port before towing them to sea for dumping. The scientists used such occasions to examine tooth eruption (19 individuals of 136–271 cm) or stomach contents (8 individuals ranging from 180 to >258 cm) (Kasuya and Marsh 1984).

Five newborns (136–142 cm in body length) and three older individual (154–182 cm and aged 1/10–1/4 year) examined by Kasuya and Marsh (1984) had no erupted teeth. Three other individuals had erupted teeth only in the upper jaw at body lengths of 170, 190, and 207 cm and ages 1/4, 1/4, and 3/4 year, and eight individuals had erupted teeth in both jaws at body lengths of 163–272 cm and ages 1/4, 2/4, 3/4, and >2.0. Eruption started with teeth near the middle of the tooth row and advanced toward both ends of the row. The ages were estimates based on annual growth layers in the dentine, and precision was poor for ages below 1 year. The results suggest that upper teeth erupted in some individuals at age of 1/4 year and on all the individuals before the age of 3/4 year, soon followed by eruption of the lower teeth.

The eight individuals examined for stomach contents were aged >1/2 year and had body lengths of >180 cm. All of these individuals were found to have remains of solid food (squid beaks, shrimp, or both) in the stomach. The youngest of them, 180 cm long and aged 1/2 year, had squid beaks and shrimp remains in the stomach (no observation on tooth eruption). The squid beaks were seemingly smaller than those found in stomachs of adults driven in the same school. I should have examined details of prey size to confirm whether young calves in the weaning stage selectively consumed smaller squids than those taken by adults. Kasuya and Marsh (1984) did not examine individuals positively identified as having milk but no solid food in the stomach. Another four individuals had only squid beaks in the stomach and were aged at 2/4, 3/4, 2.5, and 3.5 years. Squid beaks and milk were found in a 2.5-year-old male, 2.5-year-old female, and two individuals of unknown sex whose ages were estimated at ≥2.5 and ≥3.0 years from teeth taken from sexually unidentifiable severed heads. Five of the eight individuals were examined for tooth eruption as well as stomach contents. One of them (aged 3/4 year) had erupted teeth only in the upper jaw and the other four had erupted teeth in both jaws.

From these data, Kasuya and Marsh (1984) concluded that southern-form pilot whales start taking solid food at around 6 months of age and continue suckling at least to age 3.0 years. They thought that the data were insufficient to determine the age at completion of weaning, because of the small sample size and difficulty in visually detecting small amounts of milk mixed with solid food in the stomach. As noted earlier, nutritional requirement for milk at such high ages is questionable.

Subsequent studies of long-finned pilot whales in the North Atlantic confirmed that they start taking solid food at age 6.5 months as in the case of short-finned pilot whales and that some individuals continue suckling until age 7 (males) and 12 (females) (Desportes and Mouritsen 1993). Such extended suckling has also been deduced for the southern-form short-finned pilot whale by an indirect method.

Examination of a limited number of juveniles led to the following conclusions on the early growth of southern-form pilot whales. Tooth eruption starts at around 3–6 months. Feeding on solid food is likely to accompany tooth eruption and starts by age 6 months, but some individuals continue suckling at least to the age of 3 years. The stomach contents analysis did not exclude the possibility of suckling at greater ages.

12.4.4.4 Lactation Period

12.4.4.4.1 Determining Lactation

Lactation in short-finned pilot whales is easily identifiable, in most cases, by pressing the lactation externally, but slitting the gland is recommended for better accuracy. If the situation permits, lactation status recorded in the field should be confirmed by histology of the mammary gland. Pressing the grand externally will not be sufficient for carcasses of whales near the end of lactation. Also, histology will be required for carcasses of whales that have been dead for a long time. Histological examination of the mammary glands was not carried out for some southern-form whales (Kasuya and Marsh 1984) but was performed for every female northern-form whale to verify the field records (Kasuya and Tai 1993). The mammary sinus in cetaceans may contain a brownish liquid of variable coloration and viscosity. Mammary glands of such females did not show secretory activity and were judged not lactating. The liquid could have been a remnant from a previous lactation.

Females with a full-term fetus often secrete colostrum. Colostrum is identified through examination of the uterus and ovaries or through histology of the mammary gland, where milk in the mammary sinus contains numerous eosin-stainable colostrum globules. If colostrum secreted before parturition is misidentified as remnants of lactation associated with a previous parturition, this causes a serious error in estimation of the reproductive cycle and pregnancy rate. A fetus may be aborted when the female is killed by harpooning. This can be identified histologically by presence of a developed endometrium and dense underlying blood vessels (Section 12.4.4.5). Although such an unusual abortion cannot be distinguished from an abortion due to a natural cause that has occurred just prior to the death in the fishery, such cases must be rare.

The color of the milk of short-finned plot whales varies from creamy white to a green color similar to that of vegetable juice. The viscosity is like that of cow milk and has no relationship with coloration. I have seen milk with the green tinge in both the southern-form and northern-form pilot whales but never in other cetaceans. The frequency of green milk did not change with postmortem time or time between drive and slaughter, but it apparently changed with season. Out of 72 southern-form females in lactation, 39 secreted milk with various degrees of green tinge, the proportion varying from 0% (n = 17) in December, 20% (n = 5) in January to 77% (n = 17) in February, 82% (n = 17) in June–July, and 69% (n = 16) in October. The proportion of green milk was highest in summer. Kasuya and Marsh (1984) thought from this seasonal change that the green pigmentation had some relationship with the food consumed by the female. Through my histological examination of mammary glands of most of the southern-form females, I identified only one case of pathology, suggesting that the green milk was not a pathological symptom. Ullrey et al. (1984) analyzed milk composition of a female *Mesoplodon stejnegeri*, which live-stranded 20–40 days after parturition. They noted that the milk had a blue-green color and identified the pigment as biliverdin, but they refrained from conclusion about the function of the pigment in the animal. Biliverdin is one of the bile pigments and is contained in the bile fluid of herbivorous mammals. The green pigment in milk of short-finned pilot whales could have been of the same type as observed in the beaked whale.

In conclusion, colostrum may be secreted when approaching parturition and it must be distinguished from lactation associated with previous parturition. Short-finned pilot whales and some other toothed whales may secrete milk with various degree of a green tinge. Histological examination of the mammary gland is recommended for final determination of lactation status.

12.4.4.4.2 Matching Juveniles and Lactating Females in a School

Kasuya and Marsh (1984) compared the age composition of juveniles with the number of lactating females driven together in a school of southern-form pilot whales and attempted to identify possible suckling calves. One example is shown in Table 12.11, which lists composition of a school of 20 individuals driven at Taiji, believed by the fishermen to have contained all the members of a school found at sea. We tentatively accepted the judgment of the fishermen, albeit without firm supporting evidence. Three whales in the school were slaughtered and processed before my arrival at the fishing village. I examined their viscera stored on the flensing platform and confirmed that they were sexually mature females either lactating or resting but certainly not pregnant. The ages and status of mammary glands of these three females were unknown because their heads and mammary glands were discarded before they could be examined.

Including six females positively identified as lactating and the three females not examined for mammary gland, this school could have contained 6–9 females in lactation. If it is assumed that sexually mature individuals were not suckling, there were seven candidates for suckling animals. The nursing system of the short-finned pilot whale was unknown. There were three possibilities: (1) female nurses only her own calf, (2) female nurses her own calf as well as calves of other females, (3) female nurses younger calves of other females after weaning her own calf. The first and second possibilities suggested that at least six calves among seven aged 0–16 years, or more, probably a minimum of six calves at 0–13 years of age, were suckling. The third alternative, which seemed less plausible, did not suggest anything about the length of suckling but still suggested that there was a female that continued lactation for 13 years.

Kasuya and Marsh (1984) carried out a similar analysis on 12 schools driven at Taiji as summarized in Figure 12.26, which shows the age composition of estimated suckling and weaned juveniles for driven groups that were believed by the fishermen to contain all the members of a school. Some individuals were judged as weaned at age 2–3 years. The proportion of suckling individuals decreased with age of the calf, reached 50% at age 4–5, and went to almost zero at ages over 10 years. However, some males were classified as suckling at age 13 (2 males) and 15 (1 male). This result alone does not indicate that males suckle longer than females, because of the assumption that sexually mature individuals are not suckling.

TABLE 12.11
An Example of School Composition of Southern-Form Short-Finned Pilot Whales off Japan (School No. 12 of Figures 12.26, 12.32, and 12.33)

Sex and Reproductive Status		No. of Individuals	Ages
Immature	Male[a]	4	0.5; 13.5; 13.5; 16.5
	Female	3	0.5; 7.5; 8.5
Mature female	Lactating	6	22.5; 33.5; 36.5; 42.5; 43.5; 48.5
	Lact. or resting	3	Not available
	Resting[b]	1	37.5
	Pregnant	2	11.5; 30.5
Mature male[a]		1	32.5
Total	Male	5	0–32.5
	Female	15	0–48.5

Source: Reproduced from Kasuya, T., Life history of toothed whales. pp. 80–127, in: Miyazaki, N. and Kasuya, T., eds., *Mammals of the Sea—Their Past, Present and Future*, Saientisutosha, Tokyo, Japan, 300pp, 1990, Table 3, in Japanese.

[a] This school did not contain males at "early" or "late maturing" stages.

[b] Mature female neither pregnant nor lactating.

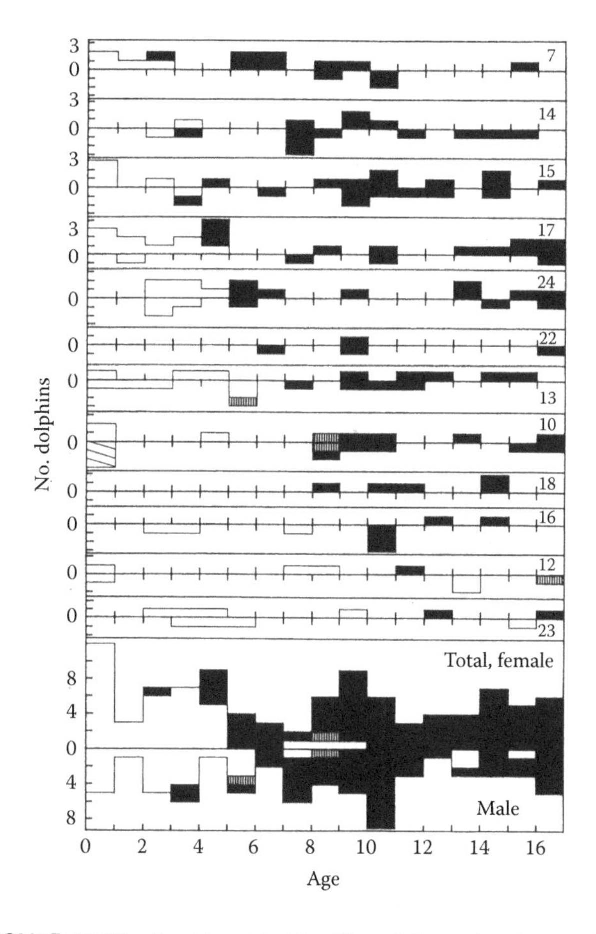

FIGURE 12.26 An attempt to identify suckling calves by matching number of lactating females and that of juveniles driven together of southern-form short-finned pilot whales taken by Japanese drive fisheries in 1965–1981. This method assumes that (1) lactating females and suckling calves have been driven together and (2) weaned individuals are always older than suckling calves in the same school. School number is at the right corner. White squares represent calves judged as suckling, black squares those judged as weaned, and shaded squares those undeterminable by the method due to presence of females of undetermined reproductive status. Male calves are placed above the horizontal line and females below. Calves of unknown sex are shown by white boxes with oblique lines. (Reproduced from Kasuya, T. and Marsh, H., *Rep. Int. Whal. Commn.*, 259, 1984, Figure 19.)

If this rider is removed, the oldest suckling whales would be age 11 (a female), 12 (a female), and 13 (a male).

Examination of stomach contents of long-finned pilot whales in the North Atlantic revealed incidents of high-age suckling, by a 7-year-old immature male and a 12-year-old pregnant female (Desportes and Mouritsen 1993). It would be extremely interesting if we could know the family structure where a sexually mature female suckled from some other female. It is not generally an easy task to identify small quantities of milk mixed with remains of solid food in the stomach. In order to overcome this difficulty, Best et al. (1984) used lactose in the stomach of sperm whales as evidence of suckling, because lactose occurs only in milk among marine organisms. The oldest cases of suckling thus confirmed were a 7.5-year-old female and a 13-year-old male. I note that these two species, particularly the sperm whale, are similar to the short-finned pilot whale in their life in matrilineal communities, presence of extended postreproductive life in females, and sexual maturity being attained in males much later than in females. Extended suckling seems to be a common feature among cetaceans within matrilineal school structure.

As the analysis of Kasuya and Marsh (1984) was based on several assumptions and lacked observation of the process of driving, it is hard to reach a firm conclusion on the extended suckling in the species. One critical problem was the possibility of missing juveniles during the drive without this being noticed by the fishermen. If the fishermen lost both mother and calf during the drive, it would not cause bias in the analysis presented earlier. However, loss of calves alone could cause overestimation of weaning age, and loss of lactating females could cause underestimation. The former possibility is greater than the latter. However, as pilot whales live in smaller schools and remain in tight aggregations at sea, the possibility of losing particular individuals is less than in the case of other species such as striped dolphins (see Sections 10.5.2 and 12.4.3.1).

A second problem was the assumption that weaned individuals were always older than suckling calves, which was not based on any data and was likely to be wrong as suggested by the fact that older females were likely to lactate longer (Section 12.4.4.4). This ungrounded assumption will not cause a bias in estimation of age when 50% of calves have weaned, but it will underestimate the upper range of length of suckling.

A third problem in the analysis was ignoring the possibility of communal nursing. Communal nursing is known in some matrilineal communities of mammals such as African elephants and hyenas, where females nurse their own calf as well as calves of other females. Because all the mothers are related, the behavior contributes to the total reproductive success of the matrilineal group. Even if this happened in the short-finned pilot whale, it would not cause bias in the estimation of the suckling period as far as each mother nursed her own calf as well. However, the estimate of suckling period would be unreliable if a female continued nursing calves of other females after weaning her own calf. If this happened, the estimate of Kasuya and Marsh (1984) would only have meaning as an indication of length of lactation.

A fourth problem would arise from a presumably rare but possible case, where a female nurses orphans by starting lactation without conception or parturition. Ridgway et al. (1995) reported cases where a captive female bottlenose dolphin placed together with an orphan calf of the same species started lactation. Such an incident could occur if short-finned pilot whales were placed in the same situation in an aquarium environment, and it could be possible even in wild individuals in a cohesive matrilineal group of the species.

In conclusion, matching the number of lactating females and the age composition of calves in the same school suggested that the suckling period in southern-form pilot whales lasts at least 2–3 years, half of the calves still suckle at age 4–5 years, and some few suckle to age 13–15. Even if this estimate is not accepted due to the technical questions summarized earlier, there is reason to believe that some females continue lactation for over 13 years after the last parturition.

12.4.4.4.3 Lactation of Old Females

Examination of carcasses of pilot whales killed in Japanese drive fisheries revealed that the oldest pregnant females were aged 35.5 for the southern form and 36.5 for the northern form (these ages may be expressed as 35 and 36 years, respectively). If they had not been killed, these pregnant females would have had their last parturition in the next year or at ages 36 and 37 (Kasuya and Marsh 1984; Marsh and Kasuya 1984; Kasuya and Tai 1993). School No. 12 of the southern form shown in Table 12.11 contained three females that were lactating at ages 36 or above. Their ages were 36, 42, 43, and 48, which meant that they had been lactating for a minimum of 0, 6, 7, and 12 years, respectively. This result matches quite well with the earlier conclusion that this school had females that lactated for 13 years and calves that perhaps suckled 13 years.

Kasuya and Marsh (1984) examined the age composition of lactating females in other southern-form schools for cases of extended lactation of old females. Using data common with those in Figure 12.26, they constructed Table 12.12 by the following procedure. Lactating females over the age of the oldest known pregnant females (35 years) were extracted, and their ages were listed in column (1) of Table 12.12. Their minimum lactation periods were estimated by subtracting 35 (not 36 as done earlier) from the ages as shown in column (3). This procedure ignores possible lactation before age 35. Then using the method of Table 12.11, the oldest possible suckling calves were allocated to these females and their age and sex were placed in column (4). Finally, ages of these lactating females at the time of their last parturition were calculated by subtracting the age of their assumed calves from age of the lactating females at capture (column (2) of Table 12.12).

School nos. 18 and 22 were excluded from the analysis because they had no lactating females, and school nos. 7, 14, and 15 were also excluded because they did not have lactating females aged over 35. The latter three schools did not have long-lactating females or long-suckling calves to be matched with them. Schools identified with old lactating females tended to have presumed long-suckling calves (school nos. 10, 12, 13, and 16). The oldest lactating female

TABLE 12.12
Females in Lactation at High Age (≥35.5 Years), Age at Presumable Last Parturition, and Age of Possible Last Calf in Southern-Form Short-Finned Pilot Whales off Japan

School No.	(1) Female Age at Lactation	(2) Age of Last Parturition, Which Is (1) − (4)	(3) Minimum Lactation Time, Which Is (1) − 35.5	(4) Age of the Calf[a]
17	47.5	37	12	10.5(♂)
24	50.5	36	15	14.5(♂)
13	42.5	37	7	5.5(♂)
	40.5	35	5	5.5(♂)
10	37.5	—	2	4–8
16	43.5	36	8	7.5(♂)
12	48.5	35	14	≥13.5(♂)
	43.5	30	8	≥13.5(♂)
	42.5	34	7	8.5(♀)
	36.5	29	1	7.5(♀)
9	45.5	—	10	
	41.5	—	6	

Source: Reproduced from Kasuya, T. and Marsh, H., *Rep. Int. Whal. Commn.*, 259, 1984, Table 19.

[a] A candidate for oldest suckling calf was selected through matching the number of lactating females and ages of immature individuals driven together. This procedure assumed all members of a school were driven and subsequently examined.

in Table 12.12 was aged 50 (which meant a minimum lactation period of 15 years), and her suspected calf was aged 14 by reading tooth layers. This analysis showed that the population had females in extended lactation and that such extended lactation was likely to occur in females who had experienced their last parturition.

Female southern-form pilot whales are likely to have extended lactation of 10–15 years after their last parturition.

12.4.4.5 Reproductive Cycle

Estrous cycles of cetaceans have been described in aquarium-reared specimens (Schroeder and Keller 1990; Asper et al. 1992; Robeck et al. 1993). Female cetaceans may ovulate in a mating season, mate if an adult male is available, and are likely to conceive. Such a type of ovulation is called spontaneous ovulation and is considered common among cetaceans. Ovulation usually occurs after weaning the calf of a previous pregnancy but may also occur soon after parturition during lactation as in the case of the Dall's porpoise (Section 9.4.6) or near the end of lactation as suspected for some northern-form short-finned pilot whales (see Section 12.4.4.5.2). If an ovulation is not followed by conception, the female may ovulate repeatedly during the same mating season, with an average interval of 42 days (killer whale: Robeck et al. 1993) or 27 days (common bottlenose dolphin: Schroeder and Keller 1990).

The interval between two series of ovulations varied from 4 to 12 months in a killer whale in an aquarium (Asper et al. 1992). Because calves were artificially separated from the female killer whale, the derived lactation period and calving intervals (35 months in average) cannot be considered to represent the process in the natural environment.

A female killer whale in captivity was reported to have mated in all periods of the estrous cycle and even during pregnancy (Asper et al. 1992), but most of the copulations occurred within a total of 72 h surrounding an ovulation (Robeck et al. 1993) or in a 5- to 10-day period around the ovulation (Asper et al. 1992). Mating in nonestrous females was confirmed in short-finned pilot whales and was considered to be an important key in understanding their social structure (see Sections 12.4.4.7 and 12. 4.4.8). However, interpreting mating activity among captive animals requires caution, because it can be different from that among wild individuals.

Seasonality of reproduction in the wild was maintained after introduction into an aquarium (Section 11.4.2.3). Bottlenose dolphins in aquariums had 70% of their estrous period in 42% of the year (July–November) (Schroeder and Keller 1990) and killer whales in February and August (Asper et al. 1992).

12.4.4.5.1 Reproductive Cycle of Southern-Form Females

Kasuya and Tai (1993) reported the composition by reproductive status of sexually mature females of southern-form and northern-form pilot whales taken by the drive fishery and small-type whaling, respectively (Tables 12.13 and 12.14). Biologists have analyzed these kinds of data for various

TABLE 12.13
Reproductive Status vs. Age of Southern-Form Short-Finned Pilot Whales off Japan, Showing a Rapid Decline in Reproductive Activity with Age

Age	Pregnant	Pregnant and Lactating[a]	Lactating	Resting[b]	Total
5.5–14.5	38 (60.3)	2 (3.2)	11 (17.5)	12 (19.0)	63 (100)
15.5–24.5	41 (43.2)	1 (1.1)	33 (34.7)	20 (21.1)	95 (100)
25.5–34.5	23 (25.6)	2 (2.2)	34 (37.8)	31 (34.4)	90 (100)
35.5–44.5	1 (1.6)	0	14 (15.6)	49 (54.4)	64 (100)
45.5–54.5	0	0	6 (19.4)	25 (80.6)	31 (100)
55.5–64.5	0	0	0	7 (100)	7 (100)
Unknown age	9 (25.7)	0	13 (37.1)	13 (37.1)	35 (100)
Total	112 (29.1)	5 (1.3)	111 (28.8)	157 (40.8)	385 (100)

Source: Reproduced from Kasuya, T. and Tai, S., *Rep. Int. Whal. Commn.*, 339, 1993, Table 9.

Note: The material was collected in 1974–1984 and covers 9 months of the year. Proportions in percent in parentheses.

[a] Pregnant and simultaneously lactating.

[b] Mature female neither pregnant nor lactating.

TABLE 12.14
Reproductive Status vs. Age of Northern-Form Short-Finned Pilot Whales off Japan, Showing a Rapid Decline in Reproductive Activity with Age

Age	Pregnant	Preg. and Lactating[a]	Lactating	Resting[b]	Total
5.5–14.5	5 (41.7)	2 (16.7)	3 (25.0)	2 (16.7)	12 (100)
15.5–24.5	14 (20.9)	4 (7.4)	26 (41.8)	10 (18.5)	54 (100)
25.5–34.5	12 (22.2)	6 (11.1)	21 (38.9)	15 (27.8)	54 (100)
35.5–44.5	2 (12.5)	0	1 (6.3)	13 (81.3)	16 (100)
45.5–54.5	0	0	0	8 (100)	8 (100)
55.5–64.5	0	0	0	2 (100)	2 (100)
Unknown age	7 (15.6)	2 (4.4)	24 (53.3)	11 (24.4)	45 (100)
Total	40 (20.9)	14 (7.3)	65(34.0)	61 (31.9)	191 (100)

Source: Reproduced from Kasuya, T. and Tai, S., *Rep. Int. Whal. Commn.*, 339, 1993, Table 8.

Note: Season of the material was October–November. Ages are available only for 1983–1985 sample; most individuals of unknown age are from 1986 and 1988 samples. Proportions in percent in parentheses

[a] Pregnant and simultaneously lactating.

[b] Mature female neither pregnant nor lactating.

cetacean species in an attempt to estimate their reproductive cycles. Such data are required to correctly represent the composition of the population. For example, let us assume a population with gestation of 14 months, lactation of 12 months, and 10-month resting period (neither pregnant nor lactating). We assume no conception during lactation. Then, the length of time from one conception to the next, or the mean reproductive cycle, will be 36 months, or 3 years. The annual pregnancy rate is the probability of a mature female conceiving during a year and is calculated as the reciprocal of the reproductive cycle, that is, $1/3 \approx 0.333$. If the reproductive cycles of females are not synchronized and proceed independent of season, then the proportion of pregnant, lactating, and resting females in the population will be 14:12:10 throughout the year. The apparent pregnancy rate is the proportion of pregnant females in the total number of mature females and is calculated as $14/36 \approx 0.389$. The annual pregnancy rate is calculated as [apparent pregnancy rate]/[gestation length], or $[14/36]/[14/12] \approx 0.333$ as obtained earlier. In a similar way we can estimate the mean length of lactating or resting period from the data. However, this method will be problematic if reproduction is seasonal and the sample is limited to a particular season or if females segregate geographically by reproductive status.

Cetacean reproduction is seasonal to various degree depending on species and environment inhabited. In order to understand the effect of this, we alter the condition of the hypothetical population given earlier to have all the conceptions in January and parturitions in March. Then the composition of female reproductive status changes with season. A sample in February contains pregnant females in two mating seasons, and the apparent pregnancy rate is double the annual pregnancy rate. On the other hand the apparent pregnancy rate is equal to the annual pregnancy rate in a sample obtained in an April–December period. The proportions of females in lactation and resting status also change seasonally. Therefore it will be almost impossible to estimate the mean reproductive cycle of such a population using a seasonally biased sample. The solution would be to obtain a seasonally unbiased sample, for example, a sample covering all months of the year equally.

The southern-form pilot whales in our samples had a parturition peak in July–August with a standard deviation of 73 days (Kasuya and Marsh 1984), which meant that 95% of the parturitions were within a 9.4-month period surrounding the peak (Figure 12.24). This was based on a sample containing some seasonal bias. Backward shifting of this parturition curve by 14.9 months, the length of gestation, gave the distribution of conception and offered a way to estimate seasonal change in apparent pregnancy rate by adding the pregnancies of two successive seasons and subtracting parturitions. The result showed that the apparent pregnancy rate would reach the minimum figure of 1.062 times the annual pregnancy rate in November and December and the maximum of 1.443 times the annual pregnancy rate in May and June (Kasuya and Marsh 1984). The apparent pregnancy rate of the southern form estimated by Kasuya and Marsh (1984), who lacked data for March, April, and September, was likely to contain a bias within this range. The possibility of geographical segregation was another problem, for which we do not have information to evaluate the effect.

Sexually mature females in 385 southern-form pilot whales consisted of 112 pregnant (29%), 5 pregnant and simultaneously lactating (1%), 111 lactating (29%), and 157 resting females (41%) (Table 12.13). The ratios of females in a particular reproductive status to pregnant females (117 in total) were 0.0427 for pregnant and simultaneous lactation, 0.9487 for simple pregnancy, and 1.3419 for resting females. These figures, together with a gestation period of 14.9 months, gave an apparent average reproductive cycle of 0.64 months for pregnancy accompanied by lactation, 14.14 months for lactation, and 19.99 months for the resting period. The total of these phases was 4.14 years, and the total lactation period was 14.8 months. This reproductive cycle was felt to be too short in view of the results obtained from the analysis of school structure.

Kasuya and Marsh (1984) compared the monthly estimates of apparent pregnancy rate of the sample against the theoretical monthly values of apparent pregnancy rate and found no correlation between them. The reasons for this result were unknown but could include (1) the monthly figures of apparent pregnancy rate were unreliable due to small sample size, (2) the sample was biased, and (3) estimation of breeding seasonality was wrong. The authors also found that the number of juveniles estimated to be younger than 14.9 months was only about 0.33 times the number of fetuses in the sample. This figure was only slightly improved to about 0.4 with later-obtained additional samples (Kasuya 1990, in Japanese; Kasuya and Tai 1993). These results suggested some important unknown elements in the reproductive sample of southern-form short-finned pilot whales, which could include (1) an extremely high juvenile mortality rate, (2) juveniles lost during the drive, (3) lactating females as well as their juveniles underrepresented in the sample for some unknown reason, (4) age of newborns overestimated, and (5) pregnancy rate increased during the sample period that covered 17 years. Kasuya and Marsh (1984) tested for these possible causes and found that no single effect could explain the observed deficiency in the sample. The authors did not evaluate a possible compound effect of several factors, and the cause of the bias was left uncertain. I have an impression that their sample was biased toward the beginning of the parturition season (Sections 12.4.3.1 and 12.4.4.1).

After failing in attempts to determine the cause of the unexplained bias, the authors simply multiplied the number of pregnant females by 0.315 (Kasuya and Marsh 1984) or 0.405 (Kasuya and Tai 1993) to match the number of pregnant females with the number of juveniles and arrived at their presumably reasonable apparent pregnancy rate. The apparent pregnancy rate thus created was divided by the gestation time of 14.9 months or 1.24 years to get an annual pregnancy rate. Although it was questionable as to how reliable the resultant figures obtained through such large corrections were, Kasuya and Tai (1993), using larger samples than those of Kasuya and Marsh (1984), estimated the annual pregnancy rate at 12.8% and the mean calving interval at 7.83 years for the southern form.

The estimations of mean pregnancy rate of 12.8% and mean calving interval of 7.83 years are figures for all females, including postreproductive females. The corresponding values for reproductive females alone have not been estimated; this would require knowledge on the age-dependent proportion of postreproductive females. Marsh and Kasuya (1984) concluded based on ovarian anatomy that the youngest postreproductive females occurred at ages 28–32 years, increased with age, and that all the females were postreproductive at least by age 40, but they presented evidence that females older than 36 years had already ceased calf production, that is, were postreproductive. This age-dependent decline in fecundity can easily be seen in the apparent pregnancy rates in Tables 12.13 and 12.14.

Most females aged 35.5 years or less were considered reproductive but must have included some proportion of postreproductive individuals. Applying the procedure presented earlier to females aged 35.5 years and less resulted in a mean calving interval of 5.21 years and annual pregnancy rate of 19.2% (Table 12.15). Integration of the annual pregnancy rate on age gave an estimate of 4–5 calves as the total production by a southern-form female that lived to postreproductive age (Kasuya and Marsh 1994).

Olesiuk et al. (1990) analyzed data obtained through a long individual-identification study of killer whales off Vancouver Island and reported reproductive parameters comparable to the estimates for the short-finned pilot whale. Their calving intervals ranged between 2 and 12 years with a mean interval of 5.3 years. Females were expected to produce 5–6 calves if they lived to the age of 40, when they ceased reproduction.

TABLE 12.15
Mean Reproductive Cycle of Two Forms of Short-Finned Pilot Whales of Japan, with Correction for Possible Sample Bias due to Seasonality of Reproduction

Population	Southern Form[a] (Years)	Northern Form[b] (Years)
Maximum reproductive age	36.5	37.5
Average calving interval, for all females[c]	7.83	5.1–7.1
Average calving interval, for reproductive females[d]	5.21	4.5–5.7
Pregnancy	1.10	0.92
Pregnancy and lactation	0.13	0.33
Lactation	2.23	2.00–2.78
Resting period	1.75	1.20–1.67

Source: Extract from Kasuya, T. and Tai, S., *Rep. Int. Whal. Commn.*, 339, 1993, Table 10.

[a] Based on samples obtained in 1974–1984.

[b] Using 1983–1985 sample for age-related estimates and 1983–1988 sample for other estimates.

[c] Calculated for all the sexually mature females including postreproductive.

[d] Calculated for mature females at or below the maximum reproductive age, which still includes some postreproductive females that entered into the status at a younger age. Lactation above maximum reproductive age ignored.

12.4.4.5.2 Reproductive Cycle of Northern-Form Females

Japanese small-type whaling resumed hunting these whales in 1982 after nearly 25 years of suspension. This provided an opportunity for studies of their taxonomy and life history. Kasuya and Tai (1994) reported the reproductive status of females collected in the five fishing seasons of 1983–1986 and 1988. Hunting was suspended in the 1987 season, because the whalers planned to use the 2 years' quotas in the 1988 season when hunting of minke whales was banned and small-type whalers started to depend on small cetaceans. The Government of Japan in July 1986 accepted the decision of the International Whaling Commission of 1982 and decided to close Antarctic whaling from the 1987–1988 season and North Pacific whaling from the season that would have started in 1988 (Chapter 7). It was the Japanese government position that hunting of small cetaceans was not bound by the IWC decision.

The process of estimating the female reproductive cycle from the catch composition required information on neonatal length, gestation length, and seasonality of reproduction, but insufficient data were available from the short fishing season for the northern-form population. A range of mean neonatal length in this population was obtained of 1.85–1.89 cm using an interspecies relationship between female body length and neonatal length, and the lower bound was accepted for the northern form based on experience that the method likely overestimated the true mean neonatal size for the species (Section 12.4.3.1). This neonatal length was about 45 cm greater than that of the southern form.

The 191 mature females examined in the 1983–1988 fishing seasons were composed of 40 pregnant (21%), 14 pregnant and simultaneously lactating (7%), 66 lactating (34%), and 61 resting females (32%). Out of the 54 gravid females only 32 had fetuses that were measured. Therefore it was not possible to know details of the proportion of fetuses representing two successive mating seasons. This was due to the fishermen slitting the carcass at sea for cooling. The body temperature of cetaceans is around 36°C–37°C (Section 12.3.6). Omura et al. (1942, in Japanese) reported for balaenopterids that the carcass temperature of around 37°C–38°C at 2–3 h after death rose to 40°C–41°C within 24 h after death, causing degradation of meat quality. Although we do not know how temperature changed in the pilot whale carcasses, it was evident that meat near the viscera changed in color to reddish brown within several hours after death. In order to decrease this damage the whalers made a slit on the body in which they poured seawater while the carcasses were on the deck or towed them in the water. Fetuses, particularly larger fetuses, were likely lost. Large fetuses were retained only through continuing requests to the gunners.

Kasuya and Tai (1993) reported body lengths of 32 fetuses in a range of 2.3–180 cm, but subsequent examination of small embryonic membranes identified two additional fetuses of 1.5 mm (Table 12.18). Due to fishery selection toward large individuals, we had no chance to study neonates and juveniles and could not compare the upper range of fetal length with neonatal length. This left some uncertainty in estimating mean neonatal length. Of the 34 fetuses, 21 were below 20 cm, 6 were above 150 cm, and 7 were in between. The mode of small fetuses represented a recent mating season and the mode at greater length mating in the previous year. The time required for a 20 cm fetus to reach 150 cm was estimated at about 10 months. Based on this information, gestation length of the northern form was assumed to be similar to that of the southern form, that is, about 15 months (Section 12.4.4.2).

With the given information in mind, let us look at the reproductive data for northern-form females. Of the 191 sexually mature females, 54 were pregnant, and 14 or 25.9% of the pregnant females were simultaneously lactating. This figure was greater than the figure of 4.4% observed for the southern form, where only 5 females out of 117 pregnant females were simultaneously lactating. All the 10 fetuses obtained from these 14 females in pregnancy and simultaneous lactation were below 60 cm in length, while 20 fetuses from the remaining 40 pregnant nonlactating females covered the entire fetal range from <10 to 180 cm and 12 were at or below 59 cm. This means that at least about half of the females were lactating when they conceived. The data also suggested that the females that became pregnant while in lactation would cease lactation by the next mating season. In other words, about half of the northern-form females that conceived in the early part (September–November) of the mating season were near the end of lactation. They would probably have ceased lactation shortly and certainly before the autumn of the next year. The proportion of females that entered estrus while they were lactating was apparently greater in the northern form than in the southern form.

A range of seasonal fluctuation in apparent pregnancy rate was estimated to be 1.062–1.443 times the annual pregnancy rate for the southern form using the seasonal distribution of conceptions and a gestation period of 14.9 months (see Section 12.4.4.5.1). The highest apparent pregnancy rate was expected for the months between the end of a mating season and the start of a parturition season. The upper limit of the range could be greater in a population with stronger seasonality of reproduction and could reach a value of 2 if seasons of mating and parturition were completely separated as hypothesized earlier. Also, as mentioned earlier, reproduction in the northern form was likely to be more seasonal than in the southern form, but the mating peak could still last 4–5 months surrounding the months of September–December. The reproductive data for the northern form presented earlier were obtained in the middle of the mating season and at the beginning of a parturition season. Thus the obtained apparent pregnancy rate must be near the highest peak, which is likely between 1.443 (a maximum possible figure in the southern form) and twice (hypothetical maximum) the annual pregnancy rate.

Applying the range of the apparent pregnancy rate to the data for the northern form in October and November, when 54 mature females out of 191 were pregnant, a possible range of annual pregnancy rate was calculated at between 54/191/2 ≈ 0.14 and 54/191/1.443 ≈ 0.20 (Kasuya and Tai 1993). The range of the mean calving interval was then estimated at 5–7 years as a reciprocal of the annual pregnancy rate. The corresponding figure for the southern form was 7.83 years. The currently available estimates of the reproductive cycles of the two short-finned pilot whale populations are compared in Table 12.15. In view of various uncertainties surrounding these estimates, the real magnitude of difference in reproductive cycle between the two populations remains undetermined.

12.4.4.6 Evidence of Postreproductive Females

Here I will discuss age-related change in female reproductive status of the two forms of short-finned pilot whale off Japan. The oldest females in the sample were aged at 62.5 years for the southern form and 61.5 years for the northern form (Kasuya and Marsh 1984; Kasuya and Tai 1993). These ages are expressed as 62 and 61 years, respectively, in the following parts of this chapter. The oldest pregnant females in these samples were 35 and 36. They would have given birth at the latest at ages 36 and 37 years, respectively. The 1-year difference between the two populations is insignificant. Thus female short-finned pilot whales in Japan attain sexual maturity at 7–12 years (Section 12.4.3.2), cease reproduction by age 36–37 at the latest, and live a subsequent maximum of 24–26 years. Females 35 or older have almost no possibility of reproduction. The proportion of such females in the total sexually mature female sample was 29% (southern form) and 18% (northern form) (Tables 12.13 and 12.14). This reproductive strategy has some similarity to that in human females, where menstruation starts at age 12–16 years (which does not necessarily mean the start of ovulation), production is repeated if not manipulated to age 40–55 years, and a postreproductive life of almost 30 years occurs before death at age 80–90 years. The transitional phase between the regular estrous cycle and cessation of the cycle is called menopause or climacterium. The term "menopause" is inappropriate for cetaceans because they do not have menstruation.

Some similarities have been identified between the human climacterium and the age-dependent decline in reproductive activities of female southern-form pilot whales (Marsh and Kasuya 1984). These are expected to hold also for the northern form. Table 12.13 shows the age-dependent change in apparent pregnancy rate of the southern form. The apparent pregnancy rate is extremely high, and the lactation rate appears low at ages 5–14 years, when females attain sexual maturity or experience their first ovulation. This is due to the terminology of attainment of sexual maturity defined as the first ovulation and the fact that in most cases the first ovulation is followed by conception. The apparent pregnancy rate declines with age, particularly after age 25, and reaches almost zero at ages over 35. This does not distinguish between the two components of (1) increase in calving interval of all females and (2) increase in proportion of females that cease ovulation. Distinction between the two requires examination of ovaries. A similar feature is observed in the northern form (Table 12.14), although its age-dependent decline appears slightly slower than in the southern form. The proportion of lactating females among total mature females showed a slight increase following a decrease in apparent pregnancy rate. The change is particularly distinct if compared to the number of pregnant females and indicates a shift of female effort from calf production to calf rearing as already indicated in the earlier section. The proportion of resting females increases with age. The function of postreproductive females will be discussed later.

Decline in the probability of ovulation followed by conception was identified in the southern-form population (Marsh and Kasuya 1984). Short-finned pilot whales ovulate spontaneously, and the ovulation is followed by formation of a corpus luteum of about 40 mm in diameter, called the corpus luteum of pregnancy or of ovulation. The corpus luteum of pregnancy retains its size for the entire period of pregnancy, but that of ovulation quickly degenerates into a corpus albicans. The corpora albicantia derived from either types of corpus luteum are believed to be indistinguishable and persist in the ovaries for life at a final average diameter of 5–6 mm (Marsh and Kasuya 1984), but some questions have been raised recently about this assumption (Section 10.5.6). Luteal tissue may be formed in some Graafian follicles without ovulation to form accessory corpora lutea, which are not uncommon in cetacean ovaries but are identifiable by their small size and absence of ovulation scars (Marsh and Kasuya 1984). Some ambiguity may remain in the distinction between a corpus luteum of ovulation and that of pregnancy at a very early stage.

Table 12.16 compares the ratio of corpora lutea of pregnancy to that of ovulation between two different age classes.

TABLE 12.16
Age-Related Increase of Ovulations Not Followed by Conception in Southern-Form Short-Finned Pilot Whales off Japan

Age	≤19.5 Years	≥20.5 Years	Total
Females with CL of ovulation[a]	2 (4%)	13 (31%)	15
Females with CL of pregnancy[b]	44 (96%)	29 (69%)	73
Total	46 (100%)	42 (100%)	88

Source: Reproduced from Marsh, H. and Kasuya, T., *Rep. Int. Whal. Commn.*, 311, 1984, Table 11.

[a] Females having corpus luteum in the ovaries but no sign of pregnancy in the uterus.

[b] Females having corpus luteum in the ovaries and fetus or signs of pregnancy in the endometrium.

Corpora lutea of pregnancy comprised 96% of total corpora lutea observed in females aged 19 years or less, but they totaled only 69% in females aged 20 years or older. The difference was statistically significant and indicated that ovulations of older females were less likely to be followed by conception. Male preference for younger females or lower male availability for older females can be excluded from possible causes of the apparent fertility decline, because males of the population are known to mate even with nonestrous females (see Section 12.4.4.7). Most of the corpora lutea identified as corpora lutea of ovulation in Table 12.16 must have been in the process of degeneration toward corpora albicantia after failure in conception or of implantation of the fertilized ovum. Such reproductive failures were considered to be a reflection of age-related change in female reproductive physiology (Marsh and Kasuya 1984).

This showed that ovulations of older females were less likely followed by conception than those of younger females. Next to be considered is the possibility of age-related change in frequency of ovulation. Figure 12.27 presents the total number of corpora lutea and albicantia plotted on age of individual animals. The scatter plot shows great individual variation, which is due to broad individual variation in age at the first ovulation and in the calving cycle. The plot suggests an average annual ovulation rate of around 0.5 for ages between 10 and 20 years, declining with increasing age to reach almost zero at age 40 (Marsh and Kasuya 1984). The average annual ovulation rate in the upper left corner of Figure 12.27 should be dealt with with caution, because it was calculated from the age-corpora regression and was model-dependent. The model does not fit for ages around 10 as well as for ages above 40.

In order to understand the background of accumulation of corpora in the ovaries, Marsh and Kasuya (1984) examined the process of regression of the corpus luteum in 245 southern-form females. First, they confirmed that diameter of the corpus luteum of pregnancy was in a range of 30.4–47.5 mm with a mean of 37.6 mm and that the diameter was retained through pregnancy. Then, based on histology, they classified

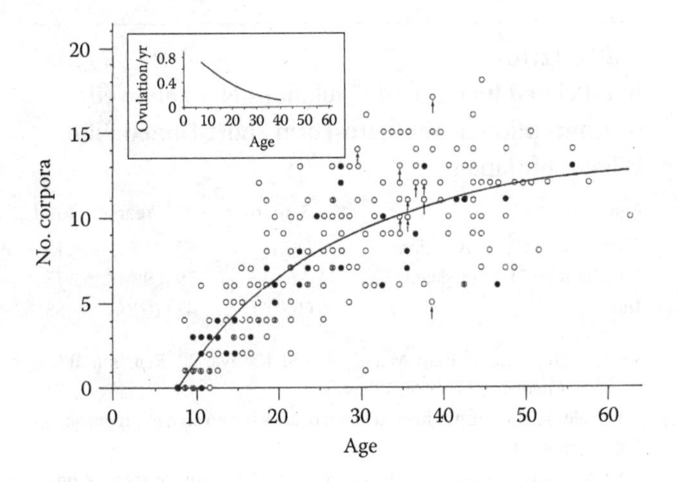

FIGURE 12.27 Scatter plot of number of corpora lutea and albicantia in ovaries on age of southern-form short-finned pilot whales taken by Japanese drive fisheries in 1965–1981. Open circles represent one individual, closed circles two, and number in circles 3 or more. Arrows indicate females <40 years old that have been identified as postreproductive by examination of ovaries (the cut-off for females of <40 years is conservative, and all the females of ≥40 years are safely judged as postreproductive). The curve $y = 13.39 - 19.65(0.95)^x$ is fitted to the plot, and the insert indicates mean annual ovulation rate calculated from the fit. (Reproduced from Marsh, H. and Kasuya, T., *Rep. Int. Whal. Commn.*, 311, 1984, Figure 25. With permission from the authors.)

the regression of corpora albicantia into three stages and gave their diameters, which were "young" with diameter of 8.5–28.5 mm (mean 15.2 mm), "medium" 5.5–16.5 mm (mean 10.4 mm), and "old" 2.5–12 mm (mean 6.4 mm). A conclusion was that degree of regression of a corpus albicans could not be determined by the diameter alone but required evaluation of the quantity of remaining luteal tissue and pigmentation. The stage of "old corpus luteum" was considered to be the final stage of regression. Marsh and Kasuya (1984) concluded that the corpus luteum in young females completed degeneration in 2 years, but that it took longer in older females. This was deduced from the fact that "medium" corpora albicantia were observed even in females aged 55 years, while corpora lutea of ovulation were observed only until age of 39 years. Regression was thought to be slower during pregnancy.

Marsh and Kasuya (1984) found females with only "old" corpora albicantia and no Graafian follicles greater than 1 mm in diameter among southern-form whales aged 28 or older. These females were likely to have not ovulated in the previous two years and were not preparing for ovulation in the near future, that is, were reproductively inactive. This suggests that some of the females aged 28 or older were likely to have ceased reproductive activity. Further they confirmed that "young" or "medium" corpora albicantia occurred only in females aged 40 or younger, which suggested that females ovulate the last time before age 40. We should note that the state of "young" corpus albicans will be retained some years after the last parturition or unsuccessful ovulation in old females. These observations confirmed that some females of the southern form experience the last ovulation before 28 years of age and all the females cease ovulation before age 40 and allowed the conclusion, with consideration of the rate of regression, that females in the population experience their last ovulation at ages between 26 and 39 years with broad individual variation extending for 13 years. This conclusion was further confirmed through observation of an age-related decline in density of oocytes in the ovarian cortex.

Mammalian oocytes are formed during the fetal stage and remain in a sac of single layers of cells called a primordial follicle until development for ovulation starts. At puberty, some of the primordial follicles start development to form an enlarged inner cavity with multiple cell layers surrounding it. This is visible on the ovarian surface and called a Graafian follicle. At the final stage of development, the Graafian follicle ruptures in ovulation. Ovulation is followed under certain circumstances by fertilization and implantation. Exact size of the Graafian follicle at ovulation has not been determined (see Section 12.4.4.7). Many oocytes degenerate before ovulation. The human female has about 7 million oocytes in the fetal stage, but only about 1%–2% of them at menopause, while only about 400–500 oocytes proceed to ovulation. We do not have such statistics for small cetaceans, except for the total number of ovulations in the life of a female as indicated by the number of corpora lutea and albicantia. For example, the maximum figure for the striped dolphin is 22 (Miyazaki 1984) and for the short-finned pilot whale 18 (Marsh and Kasuya 1984).

A deficiency of primordial follicles in the ovaries of old females was evidence of a climacterium among short-finned pilot whales. In some toothed whales, such as striped dolphins, ovulation starts in the left ovary and it is followed by an equal frequency of ovulation from both ovaries (Section 10.5.6.1; Ohsumi 1964). As short-finned pilot whales do not show a lateral difference at the age of first ovulation and ovulation frequency was not very different between the two sides (only 61% of total ovulations occurred on the left side, which was statistically significant), examination of either ovary is acceptable. Therefore, Marsh and Kasuya (1984) randomly selected one ovary from 30 females and counted the primordial follicles in 10 microscope fields with a field size of 2.7 mm^2 (Table 12.17). Two whales were immature and aged

TABLE 12.17
Age-Related Decline in Density of Primordial Follicle in Ovaries in Southern-Form Short-Finned Pilot Whales off Japan

Age	No. of Individuals	Primordial Follicles/10 Fields
4.5–14.5	4	>50
15.5–27.5	10	10–50
35.5, 40.5	2	10–50
29.5, 38.5	2	<2
40.5–62.5	12	<2

Source: Prepared from Marsh, H. and Kasuya, T., *Rep. Int. Whal. Commn.*, 311, 1984, Table 11, which lists individual whales.
Note: See text for the definition of follicle density.

4 years, and the other 28 were resting or lactating and aged 10–62 years. Four immature and mature females below the age of 14 had primordial follicle density of over 50 per 10 fields, and 12 females aged 15–40 had a density of 10–50. Two females aged 29 and 38 years and all the 12 females aged 40–62 had a low follicle density of less than 2 per 10 fields, which was less than 4% of the level found in immature and young mature females. This was similar to observations of human females in the climacterium, which was reached by short-finned pilot whales at various ages between the late 20s and around 40. This result agreed with information obtained from the earlier analyses.

In female southern-form pilot whales the probability of ovulation followed by conception decreases with increasing age and ovulation finally ceases at ages between 26 at the earliest and 39 at the latest. The oldest pregnant female was 35–36 years old in both northern- and southern-form whales, which was before the age of last ovulation. Age analysis showed that female short-finned pilot whales at age 36 still have an average life expectancy of 14 years. Marsh and Kasuya (1984) considered that this figure was comparable to 15 years life expectancy after the climacterium recorded for women of India in the nineteenth century, or before establishment of modern medical care. Marsh and Kasuya (1986) reviewed then available information on reproductive biology of cetaceans and felt that such long postreproductive lifetime was uncommon among mammals (see Section 12.5.7 for further details).

Marsh and Kasuya (1986) identified some degree of age-related decline in ovulation rate in fin and sei whales but did not consider that it necessarily indicated fecundity decline, because age-related decline in apparent pregnancy rate has not been identified among the baleen whales reviewed by them. Thus, they concluded that postreproductive females occurred rarely, if at all, in the population of baleen whales for which data are available. Presence of postreproductive females was not confirmed for any of the Phocoenidae species and most of the Delphinidae species reviewed by them, and a conclusion similar to that for the baleen whales was reached. The exception was sperm whales, false killer whales, and killer whales. An absence of pregnant females in 22 sperm whales aged over 42 years (maximum age 61; Best et al. 1984) and in 12 false killer whales aged over 44 years (maximum age 62; Kasuya 1986, in Japanese) allowed Marsh and Kasuya (1986) to conclude that these had a postreproductive lifetime. Photographic monitoring of killer whales off the Vancouver Island area has continued since 1973 and has revealed that 37 females, or 67% of adult females, did not experience parturition during a study of 10 years, and Marsh and Kasuya (1986) thought that these females included postreproductive individuals. Many of these nonreproductive killer whales were later estimated to be over 40 years old and considered postreproductive (Olesiuk et al. 1990).

12.4.4.7 Presence of Nonreproductive Mating

Copulation is one of the important social activities of wildlife, but observing it in wild cetaceans is difficult and rare. Confirmation of spermatozoa in the uteri of fishery-caught females has supplemented information on the social behavior of both types of short-finned pilot whales off Japan and confirmed that nonestrous females participate in copulation that concludes with ejaculation (Kasuya et al. 1993). The method was to make a small slit in an intact uterus for insertion of a pipette and collect a small amount of uterine fluid to be fixed in 10% formalin. If there was not sufficient fluid in the uterus, a small amount of formalin was injected and mucus mixed with the formalin was retrieved. The sample was allowed to settle for several days and the sediment was extracted with a micro-pipette, spread on a slide using a cover glass, air dried, and stained with a water solution of toluidine blue. In humans, spermatozoa ejaculated in the vagina move through the uterus and reach the fallopian tube in 15 min. Their maximum lifetime in female organs is believed to be 85 h. The position of ejaculation varies among livestock species, but the travel time of spermatozoa from the ejaculation point to the fallopian tube is believed to vary from several minutes to several hours. Dead spermatozoa are quickly ejected from the female body. The position of ejaculation is unknown for cetaceans, but an assumption that ejaculation occurs in the uterine cavity (Ms E. Katsumata, personal communication in 2009) seems to be reasonable in view of the lethal effect of seawater on ejaculated spermatozoa and the shape of the slender glans of the penis. Cetaceans have not been known with the ability to store spermatozoa for later ovulation as is known in some bat species. Therefore, presence of spermatozoa in the uteri of cetaceans is used as evidence of copulation that occurred within a few days and ended with ejaculation.

Kasuya et al. (1993) examined uterine mucus from 33 southern-form whales in three schools driven at Taiji and 53 northern-form whales taken individually from an unknown number of schools by small-type whalers off the Sanriku coast. The former sample was collected from individuals that were killed and processed after spending up to several days alive after the drive, and the latter sample was collected on the day of capture from carcasses towed to the land station at Ayukawa by the whalers. In the case of the southern-form whales, the proportion of females having spermatozoa in the uterus was higher among females killed within 3 days after being driven or within about 54 h from completion of the drive (11 females had sperm of 18 females examined), and lower among females killed on the fifth day after the drive (2 females of 15 females). No data were obtained on the fourth day after the drive. Two factors may have possibly affected the results: limited lifetime of spermatozoa in the female reproductive tract, or social or psychological effects that prevented copulations in the enclosure where the slaughter was progressing. Although each of the three schools of the southern-form whales contained 2–6 adult males at the time of completion of the drive, the timing of slaughter of these adult males relative to sampling of uterine mucus has not been analyzed. It is not known whether these adult males were mating partners of the females examined for uterine sperm. The information obtained from these females was insufficient for analysis of seasonality of copulation but offered some information on the relationship between female reproductive cycle and copulations as detailed as follows.

Kasuya et al. (1993) examined 3–5 slides under a light microscope for each female and calculated the mean number of spermatozoa per slide. This method had several sources of errors as an indicator of density or total abundance of spermatozoa in the uterus of a female: (1) position of ejaculation was ignored (most samples were taken from one uterine horn but ejaculation could have occurred in the other horn or in the vagina), (2) variation in the quantity of fluid contained in the uterus was not evaluated, (3) sampling and preparation was not quantitative. For example the authors collected uterine mucus samples from each horn of the uterus of a female in early pregnancy with a 3.5 cm fetus and compared sperm density between them using three slides from each of the horns. One horn showed an average sperm density of 11.0/slide (sd = 8.2), while the other showed 68.0 (sd = 27.8). The bilateral difference was statistically significant, but the cause of the difference was unknown. The probability of getting zero density depends on both the real sperm density as well as the number of slides examined. Although the study of Kasuya et al. (1993) contained such uncertainties, it still has some value as evidence of the timing of copulation relative to the female reproductive cycle if we concentrate on presence of uterine sperm rather than absence.

Female cetaceans ovulate when a Graafian follicle is at a certain size and releases the ovum together with fluid in the follicle. An estrogen surge occurs around this period and the female accepts males for copulation, which is called estrus. We do not know what the follicle size is at ovulation but do know that follicle size in adult females usually remains below 8 mm and that females with larger follicles are often seen in the mating peak of May–July when immature females also exhibit some follicle growth to over 4 mm (Marsh and Kasuya 1984). Bottlenose dolphins are known to ovulate at a follicle diameter of 20–24 mm through ultrasonography (Katsumata 2005, in Japanese). A southern-form female that had a maximum follicle size of 22.2 mm had a high count of uterine sperm at 1562 (Table 12.18). This female could have been in estrus. Another southern-form female with a corpus luteum of ovulation had the largest follicle of 27.5 mm in her ovary and showed a sperm density of 10.3/slide. The lack of histological examination of these large follicles precluded a firm conclusion about whether they were going to rupture or were or on the way to degeneration, but the high sperm counts suggest the females could have been near ovulation.

First we concentrated on females in lactation or resting stage. The latter category represents mature females neither pregnant nor lactating. Spermatozoa were confirmed in 8 (25.8%) of the 31 lactating females and 10 (37.5%) of the

TABLE 12.18
Female Reproductive Status and Density of Spermatozoa in Uterus (Mean No. Spermatozoa/Slide) as an Indication of Recent Copulation That Concluded with Ejaculation in Short-Finned Pilot Whales off Japan

Southern Form	Spermatozoa	No. of Individuals	Sperm Density	Follicle Diameter[a]	Ages
Resting females	Absent	11[b]	0	0–12.5 (1.8)	26.5–52.5
	Present	5[c]	0.25–1562	0–22.2 (7.2)	12.5–42.5
Females with CL of ovulation[d]	Present	3	4.0–10.3	0–27.5 (9.5)	27.5–32.5
Lactating	Absent	9	0	0–8.5 (2.3)	15.5–45.5
	Present	5	0.2–0.8	0–8.0 (4.8)	26.5–33.5
Immature	Absent	1	0	0.5	—
Northern Form	**Spermatozoa**	**No. of Individuals**	**Sperm Density**	**Follicle Diameter[a]**	**Ages**
Resting	Absent	7	0	0–14.5 (3.0)	28.5–61.5
	Present	5	0.3–2990	0–14.4 (5.4)	28.5–44.5
Females with CL of ovulation[d]	Absent	3	0	0–4.0 (1.7)	27.5–40.5
	Present	7	0.3–341	0–10.5 (3.4)	10.5–36.5
Pregnant[e]	Absent	1	0	5.0	32.5
	Present	11	1.3–57	0–6.9 (1.2)	6.5–36.5
Lactating	Absent	14	0	0–9.3 (2.4)	7.5–34.5
	Present	3	0.3–0.7	0–3.0 (1.5)	23.5
Immature	Absent	2	0	0–2.0 (0.8)	5.5–6.5

Source: Constructed from original data used in Kasuya, T. et al., *Rep. Int. Whal. Commn.*, 425, 1993.

[a] In parentheses are mean values calculated assuming 0.5 mm for any follicles less than 1 mm.

[b] Five judged postreproductive by ovarian examination.

[c] Three judged postreproductive by ovarian examination.

[d] Females having corpus luteum but no evidence of pregnancy, some of which were lactating.

[e] Fetal size ranged from about 1.5 mm (2 cases) to 13.3 cm (1 case), with average 6.3 cm. Some of the females were also lactating.

28 resting females (northern and southern forms combined). Their sperm counts covered a broad range from less than 1 to over 10/slide. It should be noted that there was no significant correlation between follicle size and sperm count, and spermatozoa were also found in females without measurable follicles (Table 12.18, Figure 12.28). This result indicates that copulations occur with nonestrous females in short-finned pilot whales.

Next we considered females with a corpus luteum of ovulation, which represents females having a corpus luteum but without fetus or features of the endometrium that accompany conception. These females had certainly recently ovulated, although it was uncertain if they were about to conceive or had experienced an unsuccessful ovulation. As reasonably expected for these females, 10 (76.9%) of the 13 with a corpus luteum of ovulation had varying density of spermatozoa in the uterus, from less than 1.0/slide to over 10/slide. The wide variety in sperm density reflects in part the qualitative nature of the methodology.

Most ovulations are followed by pregnancy. Pregnancy was identified by the presence of a fetus in the uterus or by histological features that are believed to accompany pregnancy, because some females could have aborted a fetus when killed by the fishermen. The pregnancy symptoms include development of blood vessels below the endometrium, which is followed by development of dendriform villi. Pregnant females examined for uterine sperm had fetuses in the range of 1.5 mm to 13.3 cm. Uteri with larger fetuses were not sampled due to technical difficulties of handling the large placental tissue filling the uterus. Eleven (91.7%) of the 12 pregnant females sampled, including a female with a 13.3 cm fetus, had spermatozoa in their uteri. The counts were variable from 1 to 10/slide (6 females) to over 10/slide (5 females with a maximum of 57/slide). The gestation period of short-finned pilot whales is estimated at 452 days, with slower growth in the early period (c. 40 days) and later growth of 10 cm/month (Section 12.4.4.1). Age of the fetus measured at 13.3 cm was probably between 1 and 2 months, when the parent female was still actively copulating.

The relationship between female age and uterine sperm density is shown in Figure 12.29. We have already determined that (1) the probability of an ovulation followed by conception declined with increasing age, (2) apparent pregnancy rate declined with increasing age and the oldest pregnant females was age 35 years (southern form) and 36 years (northern form), (3) the average annual ovulation rate declined with increasing age and the last ovulation occurred probably at age 39 years. Thus, estrus and conception are not expected for females over age 40. However, one southern-form resting female without measurable follicles (≥1 mm in diameter) and aged at 42 years had a uterine sperm count of 1.0/slide. There were two similar northern-form females in resting status; one with a follicle of 1.0 mm and age 44 had an uterine sperm count of 2990/slide and the other with follicle of 14.4 mm and aged 41 had an uterine sperm count of 172/slide. The follicles of the last female were not histologically examined. These three females, or at least the first two, were in nonestrous postreproductive status and had recently copulated.

12.4.4.8 Function of Nonreproductive Mating

Some mammalian females advertise their estrus. Females of the bonobo (pygmy chimpanzee) develop bright swollen genital skin as a visual sign of estrus, and dogs and horses use chemical signs for the purpose. Such signs are not developed in human females, and most males do not identify estrus in their partners. It is currently uncertain whether cetacean females have any clues to advertise their estrus, but the possibility of

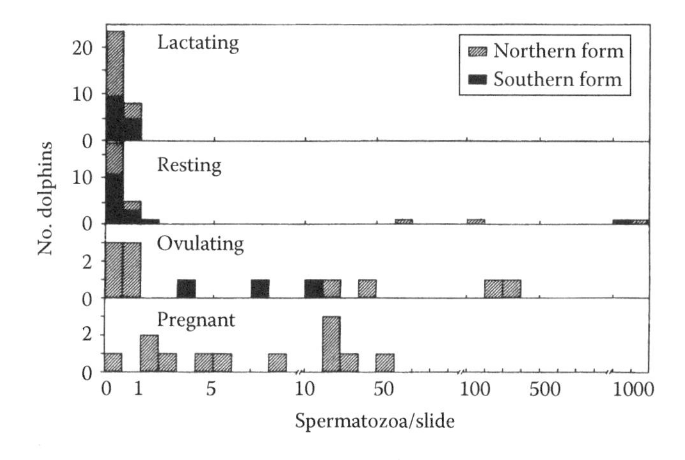

FIGURE 12.28 Spermatozoa density in uterine smear (spermatozoa/slide) by reproductive status of short-finned pilot whales. Shaded squares indicate northern forms taken by small-type whaling (October and November 1983–1985) and black squares southern forms taken by drive fishery (March and October 1982–1984) off Japan. (Reproduced from Kasuya, T. et al., *Rep. Int. Whal. Commn.*, 425, 1993, Figure 1. With correction of a labeling error.)

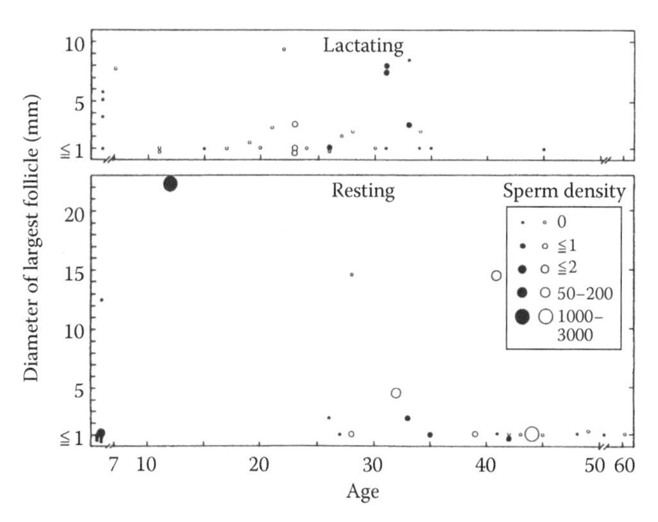

FIGURE 12.29 Scatter plots of density of uterine sperm (spermatozoa/slide) and diameter of the largest Graafian follicle on age of southern-form (closed circles) and northern-form (open circles) short-finned pilot whales off Japan. Only females in lactation or in resting status (mature female neither pregnant nor lactating) are included; females with corpus luteum of ovulation excluded. (Reproduced from Kasuya, T. et al., *Rep. Int. Whal. Commn.*, 425, 1993, Figure 2.)

acoustic or gestural signs for the purpose has been indicated (Pryor 1990). Kuznetzov (1990) reported pseudo-olfaction in the bottlenose dolphin. This has been described as a highly sensitive chemical receptor in the oral cavity and comparable to olfaction of land mammals, opening the possibility of a function in identifying chemical signs of female estrus.

Female receptivity unrelated to the estrous cycle is not limited to species where estrus is not advertised. Most mammalian females are receptive only around the period of estrus, but the restriction is lost in some species such as humans and bonobos, where copulations occur during any stage of the estrous cycle. Kuroda (1982, in Japanese) observed copulation bouts when a group of bonobos was placed in front of food and tension increased among the members for food sharing and interpreted this to mean that copulation functions to lessen tension. Waal (2005, Japanese translation) thought promiscuity of female bonobos unrelated to the estrous cycle functions to camouflage paternity and prevent infanticide. It is well known that most human copulations are not aimed at reproduction and have other functions. They usually accompany economic interactions and contribute to maintenance of partnership by lessening tensions between individuals or by increasing intimacy. Exchanging partners, which existed in Inuit communities, was believed to have contributed to constructing a cooperative network among individuals of the same sex who share partners (Houston 1999, Japanese translation). There are various examples known where brothers or sisters share a common partner of the opposite sex, or non-related multiple males share a wife, for example, Nynba in Nepal, Nayar in India, Sinhala in Sri Lanka, and some indigenous communities in South and North America (Kasuya 2008; Schultz and Lavenda 2003, Japanese translation). These social systems could function to strengthen bonds between partner sharers under particular social and economic situations. Extramarital copulations were allowed for community members following particular festivals of the communities in early Japan, which lasted until the 1920s in some locations (Ikeda 2003, in Japanese) and probably functioned to increase bonds among community members.

Bottlenose dolphins in captivity copulated within 1–2 days of the period of estrus (Yoshioka et al. 1986). Asper et al. (1992) reported copulation in captive killer whales to occur mostly in a 5- to 12-day period around estrus, but some also sometimes outside the estrus period, including during pregnancy. Interpretation of these observations on captive animals requires caution before extending them to wild individuals. Kenney (2002) reported frequent copulations between male and female right whales outside the breeding season and interpreted them to have a social function other than reproduction. These observations suggest that nonreproductive mating can exist in some cetacean species other than short-finned pilot whales. Variability of copulation behavior in cetaceans is probably comparable to that of primates and is worth further investigation.

Evidence of nonreproductive mating has been presented for short-finned pilot whales, and hypotheses formed to explain the function. One of the key pieces of information missing for evaluating the hypotheses is the origin of the spermatozoa found in female reproductive tracts. The questions include (1) the number of male partners for a female, (2) the genetic relationship between the female and her partner, (3) the social origin of the male partners, and (4) the age and growth stage of these males. Answers to some of these questions will be available by analyzing the DNA of individual spermatozoa in female reproductive tracts, and this will contribute to evaluation of the following hypotheses about the function of nonreproductive mating in the species. Some of the following hypotheses assume that short-finned pilot whales identify school members individually (Kasuya et al. 1993; Magnusson and Kasuya 1997; Kasuya 2008).

12.4.4.8.1 Education

Some primates such as chimpanzees and humans, particularly males, are believed to require learning or experiences for accomplishment of normal copulation (Nadler 1981). Tutin and McGinnis (1981) interpreted copulations between mature female and immature male chimpanzees as having an educational significance. Some precocious males of southern-form short-finned pilot whales produce spermatozoa at age 5 years and transport them to the epididymis at age 9 years, but they are believed to attain reproductive ability at age 17 on average. Among 139 southern-form males almost randomly selected, 69 were sexually mature and another 17 were not identified as sexually mature but had various quantities of spermatozoa in their epididymides. The remaining 53 males could have produced spermatozoa in the testis but did not have spermatozoa in their epididymides and were histologically classified as at the immature or maturing stages (Section 12.4.3.3). Some of these males that were in the process of sexual maturation could have left spermatozoa in the uteri of some females. If a nonestrous female gives mating education to young males in the same pod, members of which are likely to be genetically related with the female, it will contribute to total reproductive success of the female. For this hypothesis to stand, copulation within the pod has to be identified.

12.4.4.8.2 Enhancement of Bonds between Individuals

Sexual intercourse and gift exchange are inseparable activities in some mammalian species, including humans. Similar behavior is also known on some bird species as courtship feeding. Both sexual contact and gift exchange probably have a similar social effect in bond development between individuals of these animal species. Similar examples are known in bonobos as well as humans. Thus it remains as a possibility that copulation between members in a pod or members of different pods functions to establish and maintain a stable community.

12.4.4.8.3 Exchange of Greetings

Every human community has particular stereotypical behavior exchanged when two individuals of the community meet, which functions to show friendship or absence of hostility and is called greetings. Some human communities use handshakes for the purpose. External genitalia are organs of high tactile sense like hands, so it is reasonable to expect a similar function for them as in shaking hands and to assume that there is a copulation bout when two pods occasionally meet. This hypothesis agrees with our finding that a pregnant female with a 13.3 cm fetus had copious sperm in her uterus (Section 12.4.4.7), while another fetus of similar size (13.6 cm) did not have its father in the school (Section 12.5.4). Sexual intercourse in humans functions to diminish hostility or to increase friendship, as it does in bonobos. Sperm whales are known to increase accumulation of body scars from inter-male fighting for mating opportunities during the mating season (Kato 1984), but short-finned pilot whales do not show such scars and it is thought that incidents of inter-male fighting are uncommon among them. The reason for the low inter-male aggression is thought to be the generosity of females (Kasuya et al. 1993). The sexual generosity of females functions to decrease aggressions, which contributes to production and rearing of calves in the school. Gray whales (Jones and Swartz 2002) and right whales (Kenney 2002) are likely to have a similar mating system.

12.4.4.8.4 Camouflage of Paternity

Infanticide by males is known in various mammal species, which results in resumption of estrus of the mother and increases reproductive opportunities of the male but decreases that of the female. Although infanticide has not been confirmed in cetaceans, the possibility has been reported for bottlenose dolphins (Connor et al. 2000; Campagna 2002). The receptivity of nonestrous females observed in the short-finned pilot whale can camouflage paternity and decrease the incidence of infanticide, but it is unclear whether the social system of the species is likely to result in benefit to males of infanticide particularly in view of the study by Kage (1999, in Japanese), which did not find paternity of any fetuses within a school (Section 12.5.4).

12.4.4.8.5 Decoy Hypothesis

Adult male short-finned pilot whales are likely to be reproductive throughout the year. An average school of the southern form contains 2 adult males, 13 adult females, and several immature and maturing individuals of both sexes (Table 12.20). Twenty-five percent or more of adult females of the southern form are postreproductive. Although the proportion of postreproductive females is likely lower in the northern form, the annual pregnancy rate calculated including the postreproductive females is estimated at around 17% for either population.

Kasuya and Marsh (1984) estimated the number of ovulations experienced before the first conception of a female using corpus counts of 15 young females that were likely to be in their first pregnancy. Nine of the females were thought to have become pregnant at the first ovulation, three at the second ovulation, and one female each at the third, fourth, and sixth ovulations. These give an average of 1.9 ovulations before the first conception. Accepting this figure for all the reproductive females, the annual number of estrous periods expected for the average school per year is calculated at 13 (mature female) × 0.17 (annual pregnancy rate) × 1.9 = 4.1. If these reproductive females accepted males only during a total of 4 days including before and after the ovulation, the two mature males in the average school would have mating opportunities on only 16 days in a year. Males in such schools can have two strategies: a waiting strategy where males remain in a school for an average of 16 days of estrus in their school, and a searching strategy where males move between schools to seek estrous females. Magnusson and Kasuya (1997) mathematically tested which of the two strategies would more benefit male reproduction and found that a key element was the abundance of female schools to be encountered by the searching male. A reasonable range of the probability of encounter, which was not estimated from data, suggested either the possibility that waiting males have slightly better reproductive success or that the two strategies do not differ significantly.

The observed nonreproductive mating confirmed for the short-finned pilot whales creates 50–60 times greater mating opportunities; that is, males will find mating partners throughout the year in a school and have no reason to move out of the school to seek estrous females. From the female point of view, if postreproductive and young nonestrous females accept males and lure them to their school, it will help reproduction of genetically related females in real estrus. This is the decoy hypothesis.

Future scientists will determine whether any of the hypotheses mentioned earlier contain truth. The correct answer may be that more than one of them be valid, but they may all be invalid. The short-finned pilot whale is one of the few mammal species that has developed social functions in sexual contact between the two sexes. Such nonreproductive sexual contact functions in maintaining the community and was called "social sex" by Kasuya (2008). Such social sex likely exists in some cetacean species other than the short-finned pilot whale.

12.5 SCHOOL STRUCTURE

12.5.1 Mixed School

I call a school "mixed" if individuals of one cetacean species are found within or close to a group of another cetacean species. However, the spatial relationship between individual of the two cetacean species is not the same as that between conspecific individuals. The distances between conspecific individuals are usually smaller than those between individuals of different species in a mixed school.

Table 12.19 lists the incidence of mixed schools containing short-finned pilot whales sighted in the western North Pacific (Kasuya and Marsh 1984). Although the data do not distinguish between the two forms of short-finned pilot whales, the two forms have not been found together in a school except for an unconfirmed case (Section 12.3.4). Mixed schools of the two forms must be rare because they are geographically segregated.

The drive fishery offered information on other cetacean species found within or close to 17 schools of short-finned

TABLE 12.19
Cetacean Species Found Together with Schools of Short-Finned Pilot Whales

		Species			
Source	**No. Cases**	**None**	**Short-Finned Pilot Whale**	**Common Bottlenose Dolphin**	**Pacific White-Sided Dolphin**
Drive fishery	6	×			
	5		×		
	4		×	×	
	1			×	
	1			×	×
Research vessel	12	×			
	1		×		
	1		×	×	
	6			×	
	1				×

Source: Kasuya, T. and Marsh, H. *Rep. Int. Whal. Commn.*, 259, 1984.

pilot whales that were driven at Izu and Taiji. Six schools were found without any other cetacean species, and nine were found close to other school(s) of the same species. Four schools were found with schools of bottlenose dolphins (three or more schools of two species in one place). One school of southern-form pilot whales was found with a school of bottlenose dolphins, and another was found with bottlenose dolphins and Pacific white-sided dolphins.

Table 12.19 also lists 21 schools of short-finned pilot whales observed by scientists on research vessels. Twelve were found as solitary schools (without any nearby cetacean schools), two were found with an additional school of the same species, and eight with schools of other cetacean species, which included bottlenose dolphins and Pacific white-sided dolphins. The information on mixed schools obtained from the two separate sources was similar, except for the latter source giving a relatively lower incidence of multiple schools of short-finned pilot whales in the same place, about which further mention will be made later.

It is concluded that short-finned pilot whales off Japan often form a mixed school with common bottlenose dolphins and less frequently with Pacific white-sided dolphins.

12.5.2 School Size

Nonlethal studies on the social structure of short-finned pilot whales have recently started and are continuing along both coasts of North America and in the Azores Islands area using an individual-identification technique with photography (e.g., Heimlich-Boran 1993). These studies will produce valuable information on movement of individuals between schools as well as relationships between individuals within a school. Japanese studies on the school structure of short-finned pilot whales started in the 1970s by analyzing the composition of carcasses taken by drive fisheries (Kasuya and Marsh 1984). Some limited information was available from the drive fishermen on status of the school before the drive, but scientists were unable to observe the process of the drive and had difficulty in interpreting the data obtained.

The 6-year period from 1975 to 1980 spent by Kasuya and Marsh (1984) for data collection must be taken into consideration in interpreting results of the analyses. The species has a long history of exploitation at Izu, Taiji and Okinawa (Sections 3.8, 3.9, 3.10, 4.1, 4.2, and 12.6) and received high hunting pressure from the modern drive fishery that started at Taiji in 1969. Kasuya and Marsh (1984) were unable to identify historical change in the catch composition in their sample covering the 6 years, but it remains to be confirmed whether the fishery exploited a single population during the period. Kasuya and Marsh (1984) tentatively assumed that their material did not reflect historical change.

Kasuya and Marsh (1984) reported the school size in 21 drives, including 15 drives with details of the school at the time of sighting at sea. Eight of the latter were found alone, and all the members of the school were believed to have been driven. These schools were called single schools. The single schools were composed of 14–38 whales, with a mean of 24.6 and a mode of 21–30 (Figure 12.30).

Seven other drives of the 15 were made on a single group selected from an aggregation of several groups of the same species in an area. These schools ranged from 20 to 52 individuals with a mean of 35.1 and a mode of 21–40. These groups selected from an aggregation tended to contain more large males, which were preferred by the fishermen for greater

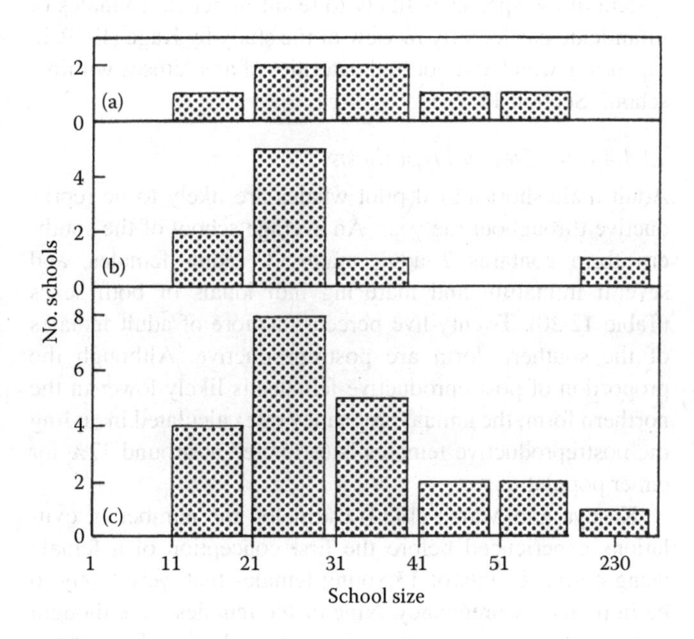

FIGURE 12.30 School size composition of southern-form short-finned pilot whales taken by Japanese drive fisheries in 1965–1981. (a), schools selected from aggregated schools of the same species; (b), schools found alone; (c), total of schools including those of unknown status. (Reproduced from Kasuya, T. and Marsh, H., *Rep. Int. Whal. Commn.*, 259, 1984, Figure 29.)

meat production and better meat quality with higher oil content (see Section 12.3.5).

Kasuya and Marsh (1984) examined only one drive where an entire aggregation of several groups was driven, on June 24, 1975, at Taiji (School No. 9 in Figures 12.32 and 12.33). The aggregation was found scattered loosely over a broad range and was composed of several groups of pilot whales as well as group(s) of about 100 bottlenose dolphins. Soon after the sighting the aggregation became tighter, which was believed by the fishermen to be a response to a killer whale group observed within a distance of 400–500 m, and made it possible for the fishermen to drive the entire aggregation. After the drive, the fishermen selectively killed 173 pilot whales of larger size and on July 4 released all the bottlenose dolphins and about 60 pilot whales that were mostly females and juveniles, because the animals started to lose weight and the market was saturated. Thus, the total number of pilot whales driven could have been about 230. This was about 10 times the size of a single school. Driving of such large aggregations was also known at Arari on the Izu coast in the post–World War II period (Section 3.8.2).

Miyashita (1993) reported sizes of 45 schools of southern-form pilot whales estimated from sighting survey vessels of the Fisheries Agency in the western North Pacific. The estimates ranged from 10–20 to 300–400 with an average of 49.8. Precision of these estimates was limited, but I would note that there were no sightings of schools smaller than 10 whales. My own shipboard observation on short-finned pilot whales off Japan recorded school sizes of 5–50 with a mean of 20.6 (n = 21, including both northern- and southern-form whales), which was close to the figure for a single group off Taiji. Two of the 21 sightings occurred about 100–200 m apart and could be considered sightings of aggregations.

The incidence of aggregation was higher in the records of the Taiji drive fishery than in my records. This is due either to a better than average feeding environment off Taiji or to the searching operation using numerous fishing vessels (7–14 vessels). Sergeant (1962) reported for the long-finned pilot whale that groups of about 20 scattered in a broad area tended to form an aggregation if frightened by an approaching aircraft. If such behavior were expected for the short-finned pilot whale, small groups would aggregate in response to drive vessels.

Now I will compare short-finned pilot whales off Japan and those off the Canary Islands area reported by Heimlich-Boran (1993). The whales in the Canary Islands area lived in "groups" of 3–33 that were slightly larger in summer than in winter. The "group" is a unit of encounter similar to "school" used for Japanese short-finned pilot whales. Data on movement of individuals between groups during the 2-year study revealed an existence of a stable core unit called a "pod." The pods were composed of 2–33 whales with a mean of 12.2. Most groups encountered were composed of a single pod (59%), but some contained 2–5 pods. The proportion of groups containing multiple pods slightly increased in the summer. In rare cases more than 10 pods formed a group. This was apparently similar to the aggregation mentioned earlier, but the group size was only 28, suggesting difficulty in identifying social structure from group size.

Forty-six pods were identified in the Canary Islands area, including 15 pods that were seen only once. Among the remaining 31 pods, some frequently formed a group with some other particular pods but rarely with others. However, there was no case where a particular pod completely avoided formation of a group with another particular pod. This feature has similarity with the social structure of the sperm whale, where a matrilineal unit (pod) of several to less than 20 individuals forms a basic stable unit of daily life, with units joining temporarily (Whitehead 2003).

Each of the resident killer whale communities off the Vancouver Island area is composed of several pods that are families of matrilineal descent. The pods in one community, which are believed to have derived from a single ancestral matriline, recognize each other and have temporary contact with each other, although it is rare for all the pods of one community to form one aggregation. In addition to the resident whales in the coastal waters, there is another form known as transient killer whale. Pods of the transient and resident killer whales have different vocal signals, do not merge with pods of the other form, and rather avoid or may show hostility toward each other (Baird 2000).

It is possible to assume a similar social structure for short-finned pilot whales off Japan, where matrilineal pods function as a basic social unit and occasionally merge with other pods of the common matrilineage. A single pod or an aggregation of several pods forms a "school" or the unit driven in the fishery. We know that the geographical range of southern-form pilot whales apparently extends to the longitude of 160°E (Figure 12.4), but we do not know the community structure possibly existing within the range. Information on community structure is urgently needed for safe management of the Japanese pilot whale fishery.

It is well known that a particular population of cetaceans inhabits a particular geographical area, but this does not necessarily mean that it attempts to monopolize a particular geographical range by rejecting other group of the same species, that is, formation of a territory. In the open marine environment where food species can move almost unpredictably, it would be technically difficult and economically unprofitable to monopolize a certain geographical range as a territory. The situation is different for killer whales that inhabit estuarine waters where an anadromous salmonid aggregates and offers a stable feeding ground. The whales benefit by monopolizing the feeding ground and rejecting members of other communities. This is the situation we see in resident communities of killer whales in the northeastern North Pacific.

12.5.3 Males in the School

12.5.3.1 Neonatal Sex Ratio

Southern-form pilot whales had a fetal sex ratio of 61 males (51.3%) and 58 females (48.7%) among 119 fetuses at or over 5 cm in body length; fetuses below 5 cm were excluded to avoid uncertainty of sex determination. The postnatal sex ratio for 36 juveniles at or below 220 cm was 19 males (52.8%) and 17 females (47.2%). Neither of these ratios was significantly different from parity. The body length of 220 cm corresponds approximately with ages of 1.1 years in males and 1.2 years in females. Combining these two sets of data gives a neonatal sex ratio (sex ratio can change through life) of 51.6% males (n = 155). This was again insignificantly different from parity (Kasuya and Marsh 1984). Although the possibility remains that males slightly exceed females at parturition, it has not been confirmed by the available data. Such an analysis was not carried out for northern-form whales due to an insufficient sample size (Kasuya and Tai 1993).

12.5.3.2 Age-Dependent Change in Sex Ratio

Males comprised 32.7% of 483 aged southern-form whales killed in the Japanese drive fishery. The sex ratio changed with age: females exceeded males at every age from 1 to 5 years, fluctuated at around 50% at ages between 6 and 16 years, again exceeded 50% at every age from 17 to 45, and finally came to 100% at ages above 45 (Kasuya and Marsh 1984). Further examination of this change shows a gradual decline of the female proportion with increasing age until 8 years of age, and the opposite trend for ages above 8 years. The age of 8 years is close to the average female age at sexual maturation and the male age at puberty. Kasuya and Marsh (1984) examined change in social behavior, segregation between sexes, and sexual difference in natural mortality rate as a plausible cause of the age-related change in sex ratio and concluded that (1) higher male mortality was the main factor in age-related decline in the male ratio after age 8 and (2) further behavioral information was needed to interpret the background of the age-related sex ratio below age 8.

The oldest individuals were age 45.5 (male) and 62.5 (female) in the sample from the drive fishery. The following equation was obtained for the relationship between the proportion of female (Y, %) and age (X, years) of southern-form pilot whales killed in the drive fisheries:

$Y = 0.991X + 41.74$, $10 < X < 47$ ($r = 0.59$) (Kasuya and Marsh 1984)

12.5.3.3 Mortality Rate

Figure 12.31 presents the age composition of southern-form whales taken by the Japanese drive fisheries during 1965–1981. It has been generally accepted that the mortality curve of mammals is U-shaped, high at young and old ages with a trough in the middle. A similar trend is observed in the age composition of whales over 10 years old, but it was not possible to explain the deficiency of young individuals below age 10 with such an age-related mortality rate pattern. A similar feature was met with in the age composition of striped dolphins (Figure 10.14), pantropical spotted dolphins (Figure 10.15), common bottlenose dolphin (Figure 11.5), and perhaps finless porpoises (Section 8.5.2.7), which suggests a behavioral change that could cause a bias in age composition of the sample.

The southern-form pilot whales off Japan have been under continued fishing pressure that varied in intensity over years. Thus the catch composition must reflect a combination of natural and fishing mortalities, which comprise total mortality. If the population responded to the fishing pressure by altering the recruitment rate, that would also affect the catch curve. Thus the right side slope of the catch curve is the sum of natural mortality, fishing mortality, and change in recruitment, disentanglement of which is not possible at present. By ignoring such complicating factors and using the method of Robson and Chapman (1961), which is free from model-dependent bias, Kasuya and Marsh (1984) estimated the apparent annual survival rates of the southern-form pilot whales as follows. The accompanying 95% confidence intervals take into account fluctuation of the age composition.

Female	18–47 years	0.9751 ± 0.0164
	45–63 years	0.8544 ± 0.0668
Male	9–30 years	0.9567 ± 0.0315
	27–46 years	0.9030 ± 0.0609

One can calculate the relationships $S = [1 - M]$ and $S = e^{-Z}$, where S is the annual survival rate, M the annual mortality rate, and Z the instantaneous total annual mortality that corresponds to the right side slope of the age frequency.

The apparent annual mortality rate of the population was about 2.5% (females) and 4.3% (males) at middle ages, which was lower than the figure for higher ages of 14.6% (females) and 9.7% (males). The mortality rate increased at certain ages, which differed between the sexes, that is, around 46 years in females and 29 years in males. When the sexes are compared at the same age, males always exceeded females in apparent mortality rate, which must reflect difference in the true natural mortality rate.

A similar mortality pattern is expected for the northern-form pilot whale, because of the similarity in maximum ages (44.5 years in males and 61.5 years in females) and between-sex difference in the age at sexual maturation (Kasuya and Tai 1993).

12.5.3.4 Social Behavior of Males

Male sperm whales visit breeding females in lower latitudes during the mating season and spend a solitary life in higher latitudes during the rest of the time (Best et al. 1984; Kasuya and Miyashita 1988). However, it is unknown whether these males participate in reproduction annually or skip breeding in some seasons. At the beginning of our study of short-finned pilot whales, we noticed a male deficiency in the age composition and thought about the possibility that adult males segregate from female schools. Information from the drive fishery did not answer the question because solitary males would not attract fishermen as a target for the drive. My own data confirmed the minimum school size in the species at five individuals, and study of the species in the Canary Island area revealed that they live in pods of 2–33, which may temporarily merge with other pods. Sighting cruises conducted by the Fisheries Agency of Japan did not always have experienced

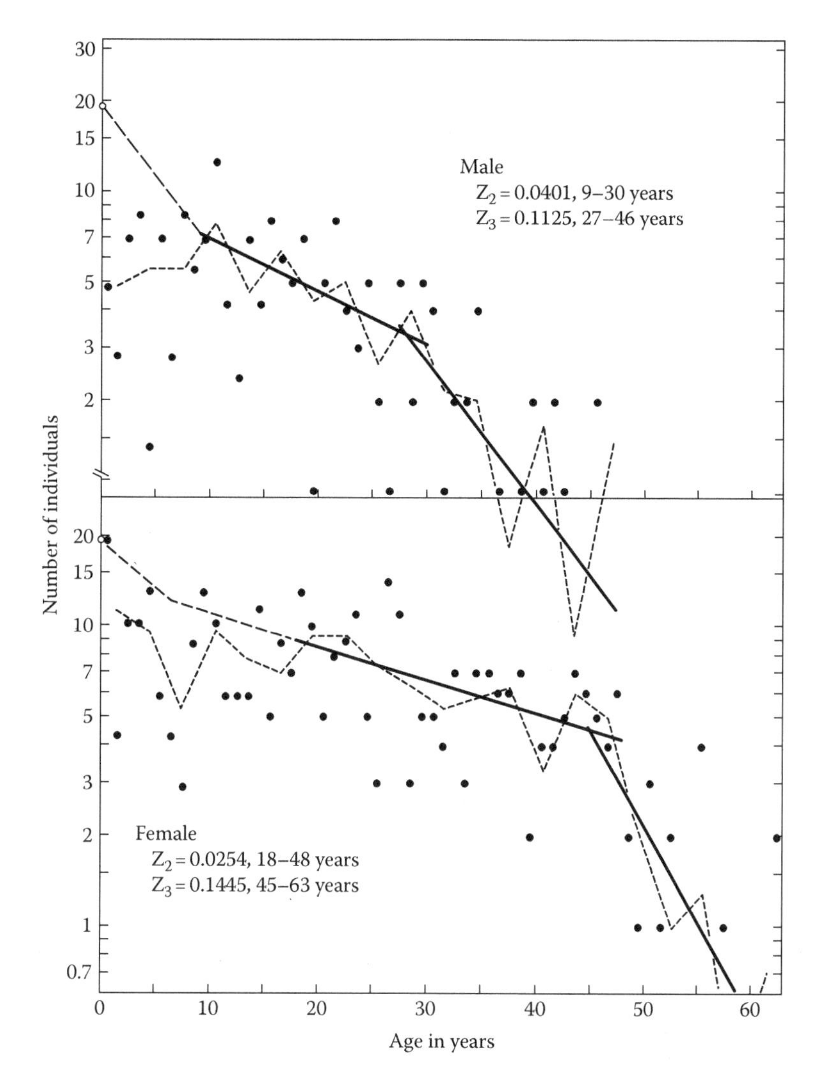

FIGURE 12.31 Age frequency (closed circles), 3-year average (fine dotted line), and total apparent mortality coefficient (Z) of southern-form short-finned pilot whales taken by Japanese drive fisheries in 1965–1981. Thick straight lines indicate least-squares regression fitted to the 3-year averages, open circle annual production at zero age calculated from a model, and dotted line connecting the open circle and thick strait line the hypothetical age frequencies of juveniles. (Reproduced from Kasuya, T. and Marsh, H., *Rep. Int. Whal. Commn*., 259, 1984, Figure 26.) *Note*: This figure is a new addition to the English version.

biologists on board, and their school-size estimates were often rounded to the nearest 10. However, the surveys covered a broad area of the North Pacific north of 25°N and west of 170°E, or most of the known range of the species in the western North Pacific. School size recorded during these cruises ranged from 10–19 to 300–399, and the mean school size of the 42 schools, excluding the three extremely large schools of over 200, was 49.8 (SE = 6) (Miyashita 1993). These observations suggested that male short-finned pilot whales do not usually spend a solitary life as observed in the sperm whale. Thus, the skewed sex ratio after puberty at around age 8 and the longevity gap observed between sexes are considered real and a reflection of difference in natural mortality rate.

The number of adult males ranged from 1 to 18 in the 13 drives with detailed biological information (Table 12.20). In addition, half of the schools contained 1–5 pubertal males (classified at the "early" or "late" maturing stages). The number of adult males differed slightly, but significantly, between drives of a single school and drives of selected portions of aggregated schools. In the former case the number of adult males per school ranged from 1 to 3 with a mean of 2.0 and median of 2.5, while in the latter case there were 1–18 males with a mean of 5.7 and median of 4. The difference was due to fishery selection for large males. Thus, it can be generalized that the usual number of adult males in a school of southern form short-finned pilot whales is 1–3 (observed in 9 cases), and the two cases with 8 or 18 adult males were unusual. In the last two cases, Schools Nos. 16 and 18 in Figures 12.32 and 12.33, also contained 13 and 5 adult females, respectively.

In addition to the 13 cases, another 14 schools offered some limited information on school structure. Thirteen of the 14 had a minimum of one sexually mature male. The highest number of adult males was found in 1 of the 14 schools (School No. 18 in Figures 12.32 and 12.33), 18 adult males (with 5 adult females). The remaining (School No. 20 in Figure 12.33) had

TABLE 12.20
School Composition of Southern-Form Short-Finned Pilot Whales Taken by Japanese Drive Fishery

School Types		Solitary School (6 Drives) No. of Whales	Mean	Selected from Aggregation (7 Drives) No. of Whales	Mean	Total (13 Drives) Mean
Males	Immature[a]	3–10	5.7	1–9	4.8	5.2
	Early maturing	0–2	0.3	0–1	0.3	0.3
	Late maturing	0–3	0.5	0–2	0.7	0.6
	Mature	1–3	2.0	1–18	6.0	4.0
	Total	5–12	8.5	5–21	12.5	10.1
Females	Immature[a]	1–10	3.8	0–12	5.2	4.5
	Resting[b]	3–8	6.0	0–15	6.4	6.2
	Lactating	0–8	3.8	0–10	5.6	4.8
	Pregnant	0–7	3.2	2–10	6.0	4.7
	Mature (no details)	0–4	0.8	0–3	0.4	0.6
	Total of mature	8–23	13.3	5–30	18.3	16.0
	Unknown maturity	0	0	0–3	0.5	0.2
	>35 years	0–7	3.5	0–8	5.6	4.6
	Total	9–23	17.7	7–42	23.5	20.8
Total of sexes		15–38	26.2	20–52	35.0	30.9

Notes: Comparison between schools found alone (solitary schools) and those selected from aggregation of several schools. Abbreviated from Kasuya and Marsh (1984), Table 38, which lists individual whales.

[a] Half of individuals of unknown sex were allocated to each sex.

[b] Mature females neither pregnant nor lactating.

no sexually mature males; this was a school of 14 driven at Taiji in January 1980. Thirteen whales were measured for body length and had sex confirmed, and the remaining one was judged an adult female based on estimated body length and external shape of the body. Only two individuals in the school were confirmed to be male; they were judged sexually immature (less than 3 m in body length). If we trust the observations of the drive fishermen that they had driven all the whales in the school, this School No. 20 was a rare school that contained no adult males but 8 adult females. However, I once experienced a drive at Futo, on the east coast of the Izu Peninsula, where one large male of adult size escaped at the entrance of the Futo harbor leaving the rest of the school to be driven into the harbor (School No. 14).

The ratio of adult males to adult females in the 13 schools mentioned earlier with better biological information ranged from 1:1.1 to 1:23, with 1:4 for the total. Exclusion of females at postreproductive ages (>35 years) still results in an excess of adult females in a range of 1:1.1 to 1:21. This is a reflection of later sexual maturity and shorter life span of males. As noted earlier, there has been no evidence for segregation of adult males from schools of adult females and juveniles of both sexes.

A question remains about the origin of adult males found with adult females and juveniles of both sexes, and it is still unclear whether all males attain maturity in their natal pod and live with their mother after sexual maturity or at least some of them move out to other pods, perhaps at around the time of attainment of sexual maturity. For the latter case, it is also unknown whether adult males remain with a school for more than one mating season or move between schools. When such information becomes available, we will be able to have better insight into the social structure of the species and obtain an answer to the broad variation in sex ratio among schools. Kasuya and Marsh (1984) did not find a correlation between male ratio and female reproductive status in a school, and some of the hypotheses presented by them remained speculative. The promising study of Kage (1999, in Japanese; see Sections 12.5.4 and 12.5.5) on genetics was based on sample sizes insufficient for full understanding of social structure. The history of related studies is outlined as follows.

Kasuya and Marsh (1984) noted that males at early or late maturing stages were always found in a school containing mature males, based on observation of seven schools that contained mature males but no maturing males and another six schools that contained both mature and maturing males (Figures 12.32 and 12.33). They deduced that maturing males were not functioning as adult males in a school and that they were not driven out by mature males, which was a feature different from that observed in sperm whales. They also noted the broad variation in number of maturing males in a school, that is, five maturing males in School No. 6, but none in School Nos. 13, 16, and 18, and speculated that pubertal males at ages between 10 and 20 perhaps tended to leave their natal schools and aggregate together. These authors did not estimate the probability at which such an imbalance in maturity composition of males could be created solely by chance.

The hypothesis of Kasuya and Marsh (1984) automatically assumed that adult males in a school were not always in the matrilineage of the school. The basis for this hypothesis was the clumped age structure of adult males in some schools, for example, School Nos. 7, 9, and 13. They thought that aggregation and movement of pubertal males between schools could explain such age structure as well as the extremely high (School No. 18) or low (School Nos. 13 and 20) number of adult males. There were some supporting evidences for this. In observation of short-finned pilot whales in the Canary Islands area Heimlich-Boran (1993) identified 46 pods in the vicinity and obtained detailed information on the structure of some of them. Twenty-eight pods contained adult individuals of both sexes and two other pods were known to contain only adult males: a pod of two males and another of six males. In a second instance of supporting evidence, one school of long-finned pilot whales driven in the Faroe Islands was composed of only eight males, including two immature males (3 and 10 years old) and six maturing or mature males (16–24 years old) (Desportes et al. 1992).

A weak point of the hypothesis presented earlier is that only data consistent with the hypothesis were used to construct it. Data that might indicate other possibilities are ignored or

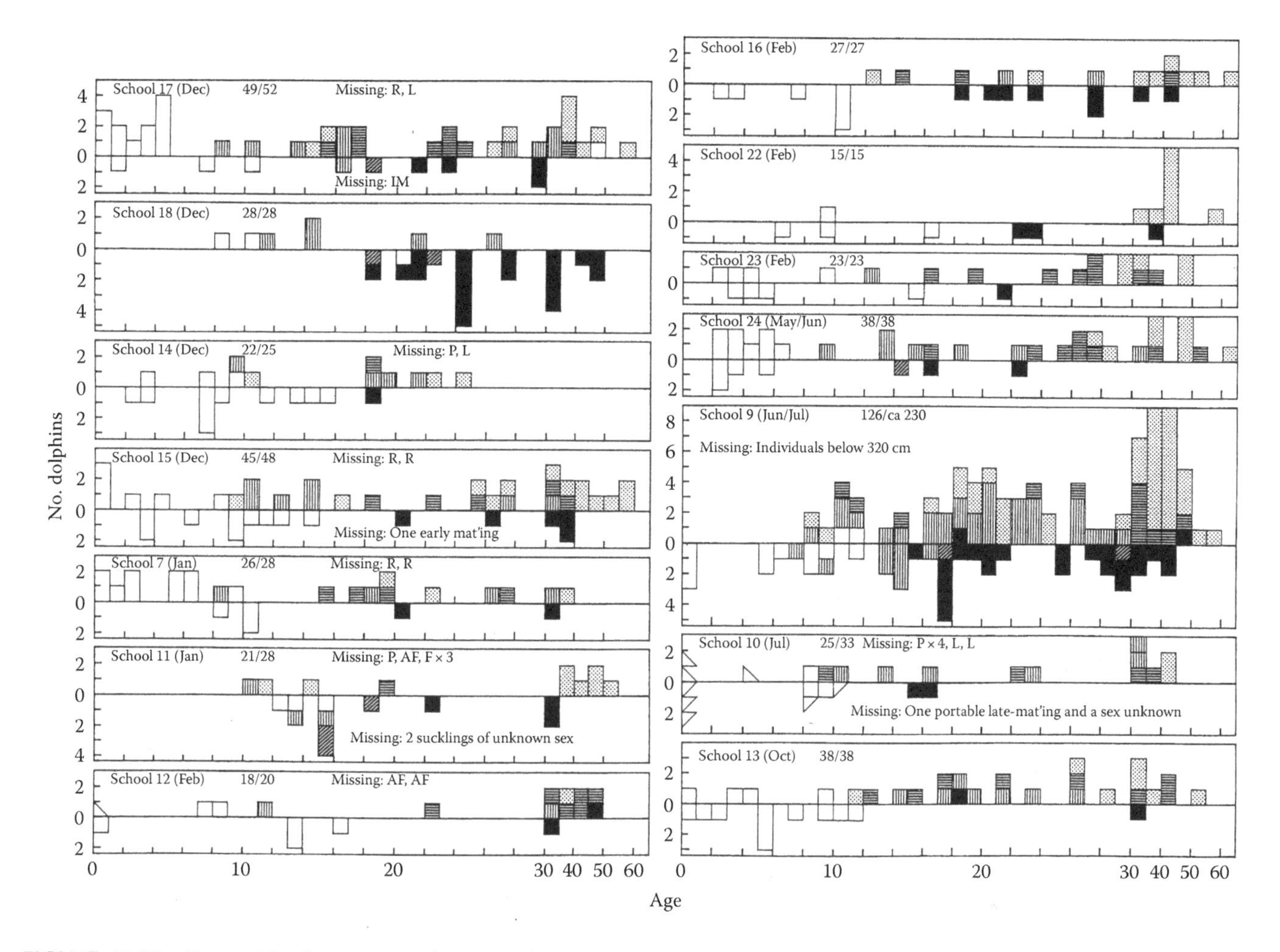

FIGURE 12.32 Composition by age, maturity, and reproductive status of schools of southern-form short-finned pilot whales taken by Japanese drive fisheries in 1965–1981. Females above the horizontal line and males below. School number, date, and sample size (no. of individuals examined/driven) given at the upper left. Triangle represents half of and individuals of unknown sex divided between the sexes. Missing individuals are those driven but not examined. Females: white, immature; horizontal lines, lactating; vertical lines, pregnant; dots, resting. Males: white, immature; vertical lines, early maturing; oblique lines, late maturing; black, mature. Missing individuals: M, immature; AF, mature female; AM, mature male; F, female; M, male; L, lactating; P, pregnant; R, resting. (Reproduced from Kasuya, T. and Marsh, H., *Rep. Int. Whal. Commn.*, 259, 1984, Figure 27.) *Note*: This was Figure 12.31 in the Japanese version.

forgotten. However, males not leaving their natal school have been recently suggested from genetic analyses of school structure of the two species of pilot whales (see Section 12.5.5).

12.5.4 Copulation Partner and Paternity

Sperm whale males leave their natal pods at puberty to live with other males of the same growth stage, which is followed by solitary life after attainment of full sexual maturity (Best et al. 1984). The departure of pubertal males from their natal pods is likely promoted by competition with adult males that visit the pod during the mating season. Young mature males of the species in a mating season tended to have scars caused by inter-male aggression for mating opportunities. This contrasts with resident killer whales off Vancouver Island, where males remain in their natal pods with their mothers for life and inter-male aggressions are apparently absent within the resident population. The situation is slightly different in the transient population of the same species that inhabits the same waters of the northeastern North Pacific. In the transients, the eldest son accompanies his mother, but younger males tend to leave their natal pods at puberty and live a solitary life. However, the possibility remains that some interaction or cooperation may persist between the solitary sons and their mothers living in the same vicinity. Baird (2000) suggested that the elder son obtains a benefit by living with the mother, who has greater experience and knowledge, and that younger sons are expelled by the elder son at puberty. It seems to me that males in both populations could benefit from living with their mother, but that it is not allowed for younger sons in the transient form for some reason. The reason is likely feeding economy and not a competition for mating opportunity as seen in the sperm whale. While killer whales of the resident form in coastal waters live on salmon, those in the transient population feed mainly on seals and Dall's porpoises. Seals and porpoises live in a small school and the energy available from a single feeding opportunity can satisfy only a small number of killer whales. So the transient killer whales are

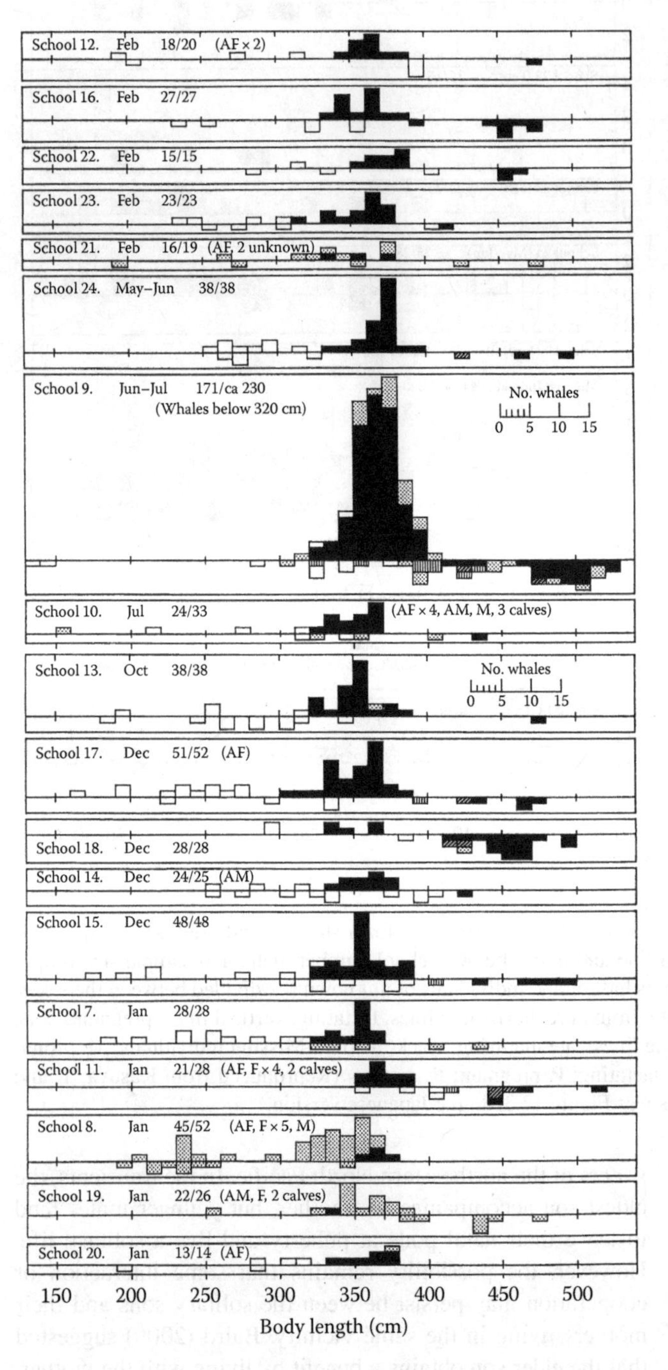

FIGURE 12.33 Composition of body length and maturity of southern-form short-finned pilot whales taken by Japanese drive fisheries in 1965–1981. School number, date, and sample size (number examined/driven) are in the upper left corner together with missing individuals in parentheses. Females: white, immature; black, mature; dots, unknown maturity. Males: white, immature; vertical lines, early maturing; oblique lines, late maturing; black, mature; dots, unknown maturity. For other symbols see Figure 12.32. (Reproduced from Kasuya, T. and Marsh, H., *Rep. Int. Whal. Commn.*, 259, 1984, Figure 28.) *Note*: This was Figure 12.32 in the Japanese version.

unable to live in large groups: the usual pod size is about three whales. Transient females leave their natal pods after maturity for the same reason, that is, because they cannot satisfy the appetites of their offspring if they remain in the natal pod. The distribution of squids, a main food of short-finned pilot whales, is less patchy than that of seals and porpoises and will enable short-finned pilot whales to live in a group larger than that of transient killer whales. Short-finned pilot whales, as in the resident killer whales, rarely have scars caused by teeth of the same species and are believed to have less inter-male competition.

Kage (1999, in Japanese) attempted to further our understanding of movement of males in the population of short-finned pilot whales by analyzing the DNA of four schools driven at Taiji, one of which was selected by fishermen from a greater aggregation and three were found as solitary schools at sea. The fishermen believed that they did not miss any whales from the solitary schools during the drive operation. Kage obtained a total of 12 fetuses (13.6–105.5 cm in body length) from the four schools and tested paternity against all the 18 candidates, including mature, late maturing, and early maturing males taken with the pregnant females. His method was to extract nuclear DNA from the fetus and process it with a restriction enzyme for electrophoresis. The fetal genotype was compared with that of the mother and then with that of male candidates. The fetal genotype must contain elements of both parents. The result was negative. He was unable to find possible fathers of the fetuses in the same school.

Kage made additional tests to determine if any combination of adult female and adult male in the same school could be the parents of any member of the same school, and found no matches, as expected from the previous results on the fetuses. As the smallest fetus of 13.6 cm could have been about 2 months of age or less (Sections 12.4.4.1 and 12.4.4.7), he concluded that the fathers of the fetuses had left the school of mating partners soon after estrus. A similar result has been reported for the long-finned plot whale in the North Atlantic (Amos et al. 1993). Two questions remain about the mating system of short-finned pilot whales. The first is the nature of the affiliation of adult males in a school, which will be dealt with in a later section (Section 12.5.5). The second question has to do with copulation: when and how it occurs. This relates to the finding of copious spermatozoa in the uterus of ovulating or pregnant females (fetuses $\leq$13.3 cm) (Section 12.4.4.7), while Kage's study on other schools failed to find paternity of fetuses (13.6–105.5 cm) there (see the preceding text in this section).

It is difficult to observe copulation in wild cetaceans; even the well-studied humpback and killer whales are not an exception. Toothed whales appear to have frequent social contact, and distinguishing between copulation for reproduction and for sexual play is sometimes difficult. I have not seen studies on relationships between male sexual behavior and resultant paternity in wild cetaceans. It seems to me that observation of the daily activity of a pod, which is likely a group of individuals of one matriline, will help us understand the mating behavior of short-finned pilot whales. Some pods join to form a group of tens of whales or even larger groups of over one hundred (Heimlich-Boran 1993). A unit of driving, which has been called a "school," corresponds to the pod or a group and "aggregated school" to a large group. In the case of resident killer whales, copulations are believed to occur between individuals from different pods of the same community.

It is likely that short-finned pilot whales can identify members of other pods of their community, and temporary merging of these pods may result in copulation bouts between members from different pods (Section 12.4.4.8). The copulation bout is not limited to estrous females and full-grown males but involves nonestrous females, as confirmed by presence of spermatozoa in the uterus. Although participation of maturing males has not been identified, it is also likely. Such copulations may function not only for reproduction but also for decreasing stress and maintaining peace between the pods involved. Identification of the origin of spermatozoa in female uteri will have a key role in evaluating this hypothesis.

12.5.5 Genetic Relatedness among School Members

Amos et al. (1993) analyzed this problem for two schools of long-finned pilot whales driven in the Faroe Islands. Because genetically testing parent–offspring relationship for all the possible pairs would be an impractically large task for a large school as in their case, they noted the genetic characters of a particular individual of either sex and calculated the probability of finding its mother or father in the school. The probability of having a father in the same school was almost negligible for individuals of any age, but the probability of having a mother was almost 100% for individuals below the age of 5 years and was still 30% for individuals aged over 20. The next analysis was to calculate the similarity of the genotype of a particular individual with that of other members of the school. The similarity was not affected by the sex of the particular individual, but it slightly increased significantly with age. It ranged from 0.08 to 0.42 for ages 5–10, and 0.17–0.7 for ages 30–45. Amos et al. (1993) concluded that the age-related change in similarity was due to the addition of new genetic traits of fathers from other school. These results indicate that fathers leave the school before the birth of their offspring, but the offspring accompany their mother even after attainment of sexual maturity. Such a community is called a matrilineal community. Their analyses did not indicate the number of matrilines in the schools examined. It also remains to be answered how frequently, if at all, offspring (sons and daughters) emigrate to other schools.

Andersen (1993) offered some information on school formation in long-finned pilot whales. She analyzed allele frequency of three polymorphic loci in 1948 whales in 31 schools driven in the Faroe Islands. She found that the allele frequencies of females, particularly those of adult females, contributed to the observed difference in allele frequency between schools. This result did not disagree with the result of Amos et al. (1993) and suggested that inter-school transfer of females is limited, if any. Then she tested whether the proportions of heterozygotes and homozygotes agreed with Hardy-Weinberg equilibrium. This was to see if mating was random within each school. The results did not reject randomness for many of the schools, but males in some schools were far from Hardy-Weinberg equilibrium, which she interpreted as an indication of inter-school movement of males or biased reproductive success of particular males.

The information on long-finned pilot whales suggested that matrilineal female groups formed a core element of schools of the species, with at least some males moving between schools. These results led us to revisit the study of Kage (1999, in Japanese) for the same purpose carried out on short-finned pilot whales driven at Taiji. He compared 13 alleles of nuclear DNA among four schools using a method called minisatellite fingerprinting. He found significant differences in allele frequency in four pair-wise comparison of the four schools, which would be expected if mother and calf lived together. He further examined the frequency of heterozygotes and found that it was higher than expected for a case of random mating, which he interpreted as a suggestion of (1) multiple genetic lineages in the school, (2) interbreeding between schools, or (3) inter-school movement of individuals. Because Kage (1999, in Japanese) combined both sexes in this analysis, it was unknown which sex had a higher tendency of movement between schools.

Kage (1999, in Japanese) also examined between-school variation in the control region of mitochondrial DNA (mtDNA), maternally inherited DNA, of southern-form pilot whale using samples from 248 individuals from 5 schools. He digested the control region of mtDNA with a restriction enzyme and analyzed the fragments with electrophoresis; he found two haplotypes, A and B. There could have been more variation in the portion of mtDNA, but the enzymes he used could only identify the two types. School No. 201 with 22 individuals contained both haplotypes, but the other four schools had only haplotype A. Using this result he separated the 22 individuals of School No. 201 into 2 matrilines, each containing a total of 11 individuals, with 1 mature male, 6 or 9 mature females, and 4 or 1 immature individuals. This result suggested that the two adult males could have remained with their mothers.

Kage applied similar electrophoresis to nuclear DNA to look for common elements between individuals (multi-locus DNA fingerprinting). In theory, half of the electrophoretic bands should be common between offspring and a parent, and values between 0.46 and 0.51 were identified between mother and fetus. This parameter will decrease with increasing genetic distance. Applying this analysis to all the pair-wise combinations in School No. 201 (22 individuals), he obtained a mean of common bands of 0.355, and the mean of all the combination of the 15 adult females (which probably contained two matrilines) was 0.35. The figure was higher between individuals of similar age (0.39–0.44) and lower between immature and mature individuals, 0.33–0.34. Comparisons between mature females and the two adult males gave similar figures (mean 0.39) with those between adult females (0.35). These results suggested that (1) males and females adhered similarly to their matriline, and (2) new genetic characters imported to the matriline increased with time and caused greater genetic distance between older whales (founders of the group) and their young descendants. Although this analysis alone did not indicate whether the new genetic characters were brought in by females or males, the mtDNA analysis suggested that the new genetic characters were due to sperm left by males in temporarily merging pods.

12.5.6 Summary on School Structure

Currently available information on the school composition of southern-form pilot whales driven at Taiji and on the Izu coast and some speculation allow the following hypothesis on the social structure of the species. The most basic unit of their life is a matrilineal group of around 10 individuals, which corresponds to a pod of resident killer whales off the Vancouver Island area and to haplotype groups A and B identified in the School No. 201 mentioned earlier. Each pod is composed of matrilineal members and usually contains adults of both sexes and immature individuals. The unit driven by the Japanese dolphin fisheries is a pod or an aggregation of pods. The aggregation is likely to be very fluid, as seen in the same species off the Canary Islands area (Heimlich-Boran 1993).

Males usually accompany their mother after maturation, but the possibility that some males leave their natal pod for another pod is not fully excluded. The timing of the possible male departure is not known, but could be at around attainment of sexual maturity, at death of their mother (a key individual in their pod identity), at splitting of a large pod into more than one pod, or at some still unknown time. Conception follows copulation with males of another pod on the occasion of temporary merging of pods. Such temporary merging of pods seems to be frequent, which is suggested by the fact that females in all the three schools examined for uterine fluid were found to contain sperm in the uterus (Section 12.4.4.7) and due to observation of the species in the Canary Islands area. A rather limited length of merging of multiple pods is suggested by the fact that a female pregnant with a 13.6 cm fetus (<2 months old) was not found with the father of the fetus in the same school (Section 12.5.4). However, it would be premature to conclude that all the nonreproductive "social sex" occurs between individuals of different pods. The possibility of copulation between matrilineal members remains to be investigated. There has been no evidence of emigration of females from their matrilineal groups.

12.5.7 Postreproductive Females

12.5.7.1 Identification

One of the two morphological types of short-finned pilot whale off Japan was studied in detail to examine age-related change in reproductive activity (Marsh and Kasuya 1984) and was found to have a high proportion of postreproductive females in the population, defined as females that had ceased reproduction due to great age. The study concluded that this stage was reached by some females at ages in the latter half of the 20s and before age 40 by all the females. The stage was characterized by near-complete depletion of primordial follicles in the ovaries, which occurred in some females by age 29, cessation of conception that occurred by age 36, and cessation of ovulation that occurred at the latest by age 39. Maximum life span was 62 years. Monitoring the physiology of a single female would reveal such age-related changes as a decline in conception following ovulation followed by complete cessation of ovulation. This is a process known as menopause in human females, and the stage after menopause is the postreproductive stage. Identification of the end of menopause and start of the postreproductive stage of an individual female is possible retrospectively only through continued observation of her physiology. It is almost impossible to determine with certainty for every female if she is in the postreproductive stage by examining the reproductive tracts of short-finned pilot whales, and such a method is likely to underestimate the proportion of such females in the population.

Noting that the oldest pregnant female was 35.5 years old in the southern form and 36.5 in the northern form (expressed in Table 12.21 as 35 and 36 years, respectively) and assuming that females at ages 35.5 years and over are postreproductive, we get approximate proportions of postreproductive females in the southern- and northern-form populations at 29% and 18% of sexually mature females, respectively (Tables 12.13 and 12.14). Marsh and Kasuya (1984) identified 24% of sexually mature southern-form females as postreproductive through examination of their ovaries. In view of the difficulty of ascertainment and difference in the sample sizes, the two figures for the southern form can be considered close. The difference of the two figures between the populations is not due to difference in reproductive strategy but more likely to a difference in age composition, which could reflect the history of exploitation and subsequent recovery.

The southern- and northern-form pilot whales had different histories of exploitation. The northern-form was vigorously hunted off Sanriku and Hokkaido by small-type whaling during the post–World War II period with an annual take of 200–400 (–1953), 11–127 (1954–1971), and 11 or fewer (1972–1979). The catch from 1954 to 1972 was low at an annual average of 47 (Table 12.22). This fishery selectively hunted large individuals, and a rapid decline of males was identified in the catch during the early operation (Section 12.6). This population was probably recovering during the nearly 27-year period from the middle 1950s to the 1982 season, when exploitation by small-type whaling resumed and the collection of reproductive data started (Kasuya and Tai 1993). Thus, there could have been an anomalously high proportion of young adult females in the population when their biology was first studied.

The history of hunting of southern-form whales was more complicated in fishing methodology and geography of the fishing grounds. Small-type whaling recorded an annual take of 200–400 in the post–World War II period (until 1953), and the drive fishery on the Izu coast recorded a take of 1891 in the seven seasons (1950–1956) for which statistics are available (Table 3.14). Although Nago in Okinawa operated an opportunistic drive, the catch statistics are available only since 1960 and recorded an annual take of over 100 until

TABLE 12.21
Life History Parameters of Selected Toothed Whales[a]

Species	Mean Age at Sexual Maturity	Age of Ovulation	Age of Conception	Age of Lactation	Maximum Age	Nursing Period (Years)	Average Calving Interval
*Sperm whale	9 (20)	≤41	≤41	≤48	61 (61)	Taking milk ≤7 (≤13)	5.2–6.5
*Southern form[b]	9 (17)	≤39	≤35	≤50	62 (45)	Lactating ≤9 (≤15)	7.8
*Northern form[b]	9 (16–18)	≤40	≤36	≤43	61 (44)		5.1–7.1
Long-finned pilot whale	8 (16.8)	≤55	≤55	≤55	59 (46)	Taking milk ≤7 (≤12)	5.1
*False killer whale	9 (18)		≤43	≤53	62 (56)		6.9
*Killer whale	13 (15)		≤41		80–90 (50–60)		6–10
Pantropical spotted d.	9 (11)		≤40	≤45	45 (42)	Mean lactation 1.7–2.1	3.0–3.6
Common bottlenose d.	6/9 (10/12)		≤38	≤42	45 (43)	Mean lactation 1.3/1.8	2.5/3.0
Baird's beaked whale	10–15 (6–11)[c]		≤49	≤54	54 (84)		3.4

Notes: Figures for males in parentheses. Females of species with asterisk are likely to have significant postreproductive life span. Mean calving interval calculated including postreproductive females. Also see Tables 12.12 through 12.14.

[a] References. Sperm whale: Nishiwaki et al. (1958) and Best et al. (1984); short-finned pilot whale (southern and northern forms): Kasuya and Marsh (1984) and Kasuya and Tai (1993); long-finned pilot whale: Martin and Rothery (1993), Desportes and Mouritsen (1993) and Desportes (1993); false killer whale: Kasuya (1986, in Japanese) and Kasuya (1990, in Japanese); killer whale: Olesiuk et al. (1990); pantropical spotted dolphin: Kasuya (1985b); common bottlenose dolphin: Kasuya et al. (1997, in Japanese), where two sets of figures represent those for the Pacific and Iki Island area; Baird's beaked whale: Kasuya et al. (1997).

[b] Two forms of short-finned pilot whales off Japan.

[c] Not average age, but range of individual variation.

1973 (Table 3.22). After these large takes, relatively low levels of catch were made by driving at Nago and on the Izu coast (Table 3.16) and by small-type whaling at Taiji (Table 12.22) until 1969, when a regular drive operation started for the southern form off Taiji (Section 3.9; Tables 3.17 and 3.18). Most of the reproductive information on the southern form was obtained from the drive fishery at Taiji. This history of exploitation as well as the uncertainty surrounding the mixing of individuals between the three major fishing grounds makes it difficult to speculate on the population trend before 1969. The effect of the catches on age structure of the short-finned pilot whales off Japan remains to be investigated.

Postreproductive status, *per se*, has no reproductive benefit for lifetime reproduction; rather it can have an adverse effect on the survival of a female's offspring through possible competition for nutrition. For such a genetic trait to be selected to affect such a significant proportion of a population as observed in the short-finned pilot whale, it will have to generate some beneficial trait or traits to compensate for the apparent adverse effect on total lifetime reproduction. Studying such beneficial traits, or the contribution of postreproductivity in females to the survival of their offspring, is an important research item in the study of behavior and social structure of short-finned pilot whales and can contribute to future management of the populations. Before we speculate on the topic, we will briefly review the presence of postreproductive females in cetacean communities using information presented by Marsh and Kasuya (1986) and subsequently obtained.

Table 12.21 presents information on age and reproductive activity of toothed whales. The ages of ovulation, pregnancy, and lactation represent values that have been confirmed. If the observed pregnancy ends with parturition, which is not observed, the age of oldest parturition will be about 1 year greater. The length of lactation is presented as the known maximum range, which may differ depending on the sex of the offspring, and there remains the possibility of communal nursing. The length of suckling is the maximum confirmed age of suckling, including the period when both milk and solid food are ingested. This does not distinguish between cases whether a calf suckles from its own mother or from other females. Maximum lengths of lactation are not available for the pantropical spotted dolphin and the common bottlenose dolphin, for which only values of the average duration of lactation are given.

Marsh and Kasuya (1986) reviewed age-related reproductive data for 12 toothed whale species and 6 baleen whale species and concluded that a significant postreproductive period was unlikely to exist in baleen whale populations but likely existed in short-finned pilot whales, resident killer whales off Vancouver Island, and false killer whales in the Iki Island area in western Japan. Some additional information then became available for sperm whales (Best et al. 1984) and long-finned pilot whales (Martin and Rothery 1993). This is reviewed in the following.

The information on false killer whales, 101 males and 116 females, was randomly collected from a larger number of carcasses driven and killed by the fishermen in the Iki Island area for culling purposes (Kasuya 1985a; Kasuya 1986, in Japanese). There were 64 mature females with age: 12 pregnant, 14 lactating, 31 resting and 7 of unknown status. The greatest age of females in pregnancy was 43 years and of those in lactation 53 years (the next oldest female in lactation was age 43), while the oldest resting female was 62. This suggested that females cease reproduction by age 44, leaving a maximum

TABLE 12.22
Number of Short-Finned Pilot Whales Taken by Japanese Small-Type Whaling Given by Sex and Geographical Regions for 1948–1957, 1965, and 1968–1979, Compiled from *Geiryo Geppo* [Monthly Report of Whaling Operation] Received by the Fisheries Agency

	Pacific Coast									Other Area[d]		
	Hokkaido[a]			Sanriku[b]			S/W Japan[c]					
Year	Male	Female	Total	Male	Female	Total	Male	Female	Total	Male	Female	Total
1948	0	0	0	200	121	321	69	35	104	0	0	0
1949	0	0	0	237	178	415	277	122	399	6	0	6
1950	0	0	0	126	163	289	231	130	361	3	1	4
1951	4	1	5	150	114	264	189	121	310	8	2	10
1952	3	4	7	64	56	120	118	76	194	6	3	9
1953	0	0	0	97	127	224	125	104	229	0	0	0
1954	1	0	1	25	18	43	24	12	36	0	0	0
1955	0	0	0	14	3	17	22	18	40	3	0	3
1956	53	29	82	11	8	19	99	70	169	2	1	3
1957	4	0	4	20	12	32	101	31	132	0	0	0
1965	7	7	14	76	37	113	101	10	121	0	0	0
1968	0	0	0	14	10	24	56	41	97	0	0	0
1969	0	0	0	7	4	11	53	22	75	0	0	0
1970	0	0	0	17	15	32	69	39	108	0	0	0
1971	0	0	0	15	14	29	51	60	101	5	5	10
1972	3	1	4	0	0	0	37	25	62	0	0	0
1973	5	4	9	0	0	0	37	29	66	0	0	0
1974	8	3	11	0	0	0	47	18	65	0	0	0
1975	0	0	0	1	0	1	25	28	53	0	0	0
1976	0	0	0	0	0	0	6	8	14	0	0	0
1977	0	0	0	0	0	0	3	3	6	0	0	0
1978	0	0	0	0	0	0	11	2	13	0	0	0
1979	0	0	0	0	0	0	1	2	3	0	0	0

Source: Reproduced from Kasuya, T. and Marsh, H. *Rep. Int. Whal. Commn.*, 259, 1984, Table 3. For geography, see Figure 12.34.
Note: This table is a new addition to the English version.
[a] Off the Pacific coast of Hokkaido in 42°N–43°N.
[b] Off Sanriku coast in 38°N–40°N.
[c] Off the Pacific coast of Izu, Wakayama, Shikoku, and Kyushu in 31°N–35°N.
[d] Off northern Kyushu (East China Sea) and the Sea of Japan.

prost-reproductive life of 18 years. The number of mature females ≥44 years old was 16 (1 lactating, 12 resting, and 3 of unknown reproductive status) and that of mature females aged ≤43 was 48 (12 pregnant, 13 lactating, 19 resting, and 4 of unknown reproductive status). This suggests that over 20% of sexually mature females in the Japanese false killer whale population were likely in a postreproductive status. As the ovaries were not examined in detail, it is unknown when ovulation ceases in this population (Kasuya 1986, in Japanese). The between-sex difference in observed maximum age was 6 years, which suggested a slightly greater longevity of females. Further details on this Japanese material has been published together with data from South Africa (Ferreira et al. 2013).

Killer whales off Vancouver Island have been studied since 1973 using individual identification with photography. Olesiuk et al. (1990) estimated growth and reproductive parameters using data thus obtained in the 15 years from 1973 to 1985. Numerous females did not breed for over 10 years. In view of the information on Japanese short-finned pilot whales, these females were believed postreproductive. Of particular interest was the age-dependent calving rate, which was the probability of an adult female calving in a year and identical with annual pregnancy rate if abortion can be ignored. The calving rate of this killer whale population rapidly declined from an extremely high value near 100% at age 12 years to a still high value of about 25% at age 18. These high calving rates can be explained by recruitment of young females to the reproductive population. I note that the calving rate declined linearly to below 5% at ages over 48. This age-related change was very similar to that observed in short-finned pilot whales. The reason why the calving rate did not reach zero even after age 48 could be explained possibly as a model-specific artifact of tailing.

Best et al. (1984) tabulated the age and reproductive status of sperm whales taken for research purposes off South Africa. Out of the 725 mature females there were 134 pregnant, 36 ovulating, 272 lactating, and 300 resting. Their data showed that both pregnant and ovulating females occurred at ages ≤41 and lactating females at ≤48 years, while females lived to a maximum age of 61 years. The number of females above age 41, the age of the oldest known pregnancy or ovulation, was 22 or about 3% of the total number of mature females. Although the proportion of likely postreproductive females in the population was not high, Kasuya (1995b) thought that the sperm whale has a postreproductive lifetime extending at least 20 years. Another interesting feature of the life history of this species was the similarity of longevity between the sexes (Table 12.21). Kasuya (1990, in Japanese) felt that such great male longevity has some relationship with the segregation of males to higher latitudes apart from nursing females in lower latitudes, thus avoiding competition for resources.

The reproductive data for long-finned pilot whales presented by Martin and Rothery (1993) are puzzling. Corpus count in the ovaries increased linearly until the age of 50. This means that ovulation occurred at almost a constant frequency to age 50. The relationship is unclear for ages over 50 due to scarcity of samples (1 individual at age 51, 2 at 55, and 1 at 59). The apparent pregnancy rate, that is, proportion of pregnant females in a sample of mature females, decreased from 40% at age 10 to 20% at age 35. It further decreased to almost zero at 41 years and over. However, among 41 females aged 40 or more (maximum age 59), there were two pregnant females aged 41 and 55 years. Martin and Rothery (1993) concluded from these data that "it is possible that some animals had ceased to ovulate altogether and/or that only a small percentage of ovulations in these older females lead to a successful fertilization. … The fact that one female…a 55 year old… was pregnant demonstrates that successful reproduction can potentially continue throughout life." It is possible to interpret the data given earlier to indicate that the whales live a rather short postreproductive stage at ages 50–59 years following a climactic stage at ages 41–50 when ovulations occur without conception. The probability of ovulations of these old females being followed by conception may change by improvement of nutrition. However, it seems to be also possible to interpret the data given earlier, by ignoring the pregnant female at age 55 as an outlier, to mean that long-finned pilot whales cease reproduction by age 41 and ovulation by age 50 while living to age 59. This interpretation means that females of long-finned and short-finned pilot whales are similar in having a considerable postreproductive life span, although the former species may have a slightly longer reproductive life and shorter postreproductive life. Detailed examination of ovaries of long-finned pilot whales will throw further light on their reproduction.

This allows the conclusion that the sperm whale, short-finned pilot whale, killer whale, false killer whale, and possibly the long-finned pilot whale have significant postreproductive life spans. The first three species are known to have a matrilineal social structure. Sperm whales live in a matrilineal pod of 10–50 individuals, which occasionally merge and form a network of pods (Whitehead 2003). Resident killer whales off Vancouver Island, known as an ecotype, live in several communities that contain multiple matrilineal pods. Pods may temporarily merge with other pods of the same community but not with those of other communities. This feature is apparently the same as in sperm whales and short-finned pilot whales in the Atlantic. Almost nothing is known on the social structure of false killer whales, but a matrilineal community is suspected.

12.5.7.2 Function of Postreproductive Females

Data in Table 12.21 make it clear that species with postreproductive females (I will call them type-2 species) have greater female longevity than those without them (type-1). The two groups have no significant difference in breeding lifetime of females. The female life history of type-2 species could have evolved from that of type-1 females by adding a life of non-reproduction to the life of type-1 females, while males were left unchanged (killer and short-finned pilot whales). For this view to be applied to females of the long-finned pilot whale the single old pregnant female aged 55 must be interpreted as an outlier. Additional explanation is required for sperm and false killer whales where males live as long as postreproductive females. The life history pattern of Baird's beaked whales is quite different from either of the two major types mentioned earlier in great extension of male longevity (Section 13.5.4). Their life history and social structure could have evolved in a direction quite different from those of the species mentioned earlier. The life history patterns of toothed whales are not limited to these. For example, the franciscana (Kasuya and Brownell 1979) and the porpoises (Chapter 9) apparently evolved early maturation, high annual reproductive rate, and short life expectancy (Kasuya 1995b).

Marsh and Kasuya (1986) suggested that postreproductive females would not exist if they did not have a positive function in the community and proposed that the function is a contribution to the welfare of juveniles. If the postreproductive females take care of the offspring of their daughters (their grandchildren), it contributes to their total lifetime production and allows for their genetic traits to survive in the population. Biological information of the time was already sufficient for Marsh and Kasuya (1984) to hypothesize matrilineal social structure for the killer and short-finned pilot whales. Now we have more information on social structure of toothed whales, and there are reasons to believe that the various types of social structure seen on the killer, sperm, and short-finned pilot whales are in a top group among many toothed whale species that have evolved toward formation of matrilineal social structure.

According to Lockyer (1981), who studied the energy requirements of cetaceans, the daily food consumption of a sperm whale is 3% of body weight in fully grown individual but 10% in young sexually mature females that are still growing. Sperm whales live mainly on squid, like short-finned pilot whales. The food requirement during pregnancy increases by 10% in young females and only 5% in fully grown females, which means that a pregnant female requires 20–40 kg of food daily for the fetus in addition to 420 kg necessary for maintenance of her own body weight. This is not a great increase in burden compared with the burden accompanying lactation. After parturition a mother must support the fast growth of her calf. The energy requirement increases by 63% in a young mother and 32% in a fully grown mother. This means that during lactation a young mother requires daily 685 kg of food and a fully grown mother 554 kg. Such a great increase in energy requirement cannot be an easy burden.

Young calves less than 1-year-old remain at the sea surface, while other members of the school dive for feeding, presumably because such young calves have limited diving ability. A young calf is often accompanied by an adult female other than its mother, called a babysitter (Whitehead 2003). The mother cannot stay with her young calf at the surface because she has to satisfy her energy requirement. The presence of a babysitter decreases the risk of the calf being attacked by predators such as killer whales or sharks, and offers a better feeding opportunity for the mother. However, there is no firm evidence that postreproductive females are participating in the task of babysitter.

As postreproductive females live with their daughters and grandchildren, they are able to increase their total lifetime production by taking care of survival and reproduction of their relatives in the pods, not by attempting production of their own offspring at a greater age. This was a hypothesis proposed to explain the evolution of postreproductive females (Marsh and Kasuya 1986).

Acting as a carrier of information is another function expected of older individuals. Accumulation of experience and information benefits animals in feeding, reproduction, and avoiding predators, and the importance is greater among the higher animals in general and probably in the aquatic environment. For example terrestrial primates have less difficulty in visiting particular trees in the right season for feeding on the fruit compared with dolphins that must locate prey species in the water and cooperate in attacking it. Fish schools move daily and their location is less predictable than the location of fruit trees. Experience and information on the environment accumulated by postreproductive females, which are usually old, benefits life and reproduction of the members of the matrilineal pod.

Information accumulated in some individuals and behavioral patterns derived from it will be transmitted to other members and maintained over generations to form a culture of the pod or community. Although identifying culture in a cetacean community is not easy, the presence has been accepted by cetacean biologists (Whitehead 2002; Krutzen 2006). If experience and knowledge are accumulated by long-lived females, the quantity and quality of culture will increase. Thus, postreproductive females function as carriers of culture of their community.

Both culture and genetic traits function as tools of adaptation of the carrier to the environment. Many animals and plants adapt themselves to changing or new environments by modifying their genetic traits, but the process requires a longer time to be accomplished. However, culture is characterized by the ability for quick modification or addition of new elements. In principle, new cultural traits will be acquired within less than one generation of a matrilineal pod of such species as short-finned pilot whales or killer whales and can be transmitted to other pods of the community with time. Culture is a powerful tool of a pod to adapt itself to a newly encountered environment and increases adaptability of the community. Cultural diversity within a species increases chances of the species to survive in a changing environment. A currently agreed goal of cetacean management is to maintain genetic diversity and certain levels of population size, but it does not consider maintaining cultural diversity and the ability to acquire new cultural elements. This simplistic management policy diminishes the ability of survival of some cetacean species, particularly species with long life spans.*

12.6 INTERACTION WITH HUMAN ACTIVITIES

12.6.1 Catch History

Human activities affecting survival of short-finned pilot whales off Japan are not limited to intentional take by fisheries, incidental mortality in fishing gear, and competition with fisheries for food resources. Other impacts include physiological and physical damage due to pollutants, underwater noise, ship strikes, and climate change (Introductory Chapter). Consideration of the latter four elements is beyond my expertise, and there are insufficient statistics on incidental mortality in fishing gear. Interaction between short-finned pilot whales and fishing activities for food resources has not been studied to a reliable degree. Currently available information suggests that the greatest factor affecting survival of the two forms of short-finned pilot whales is fishing activities targeting them, which are briefly summarized as follows. Further details on the related fisheries are available in Part I.

Short-finned pilot whales are rare in the Sea of Japan and the East China Sea (Section 12.2). Post–World War II operation of small-type whaling reported the take of a small number in the Sea of Japan off the west coast of the Oshima Peninsula (42°30′N–43°30′N) in southwestern Hokkaido (Figure 12.34, Table 12.22). Frequency of occurrence of the species in the region requires further confirmation to eliminate the possibility of confusion by whalers with the false killer whale, which is more common in the area and also called *gondo* (Sections 3.4 and 12.2). Here is a summary of the history of hunting of pilot whales off the Pacific coasts of Japan (from north to south).

Small-type whalers hunted pilot whales in the peri–World War II period from land stations at Kushiro, Akkeshi (43°03′N, 144°51′E), and Urakawa (42°10′N, 142°46′E) on the Pacific coast of Hokkaido (Figure 12.34). The main target of their operation was minke whales, and the catch of

* After publication of the Japanese version of this book, Foster et al. (2012) published evidence of contribution of postreproductive female killer whales to the survival of their sexually mature offspring of both sexes. The contribution was greater to sons than daughters, which suggested greater mother's investment in sons than daughters.

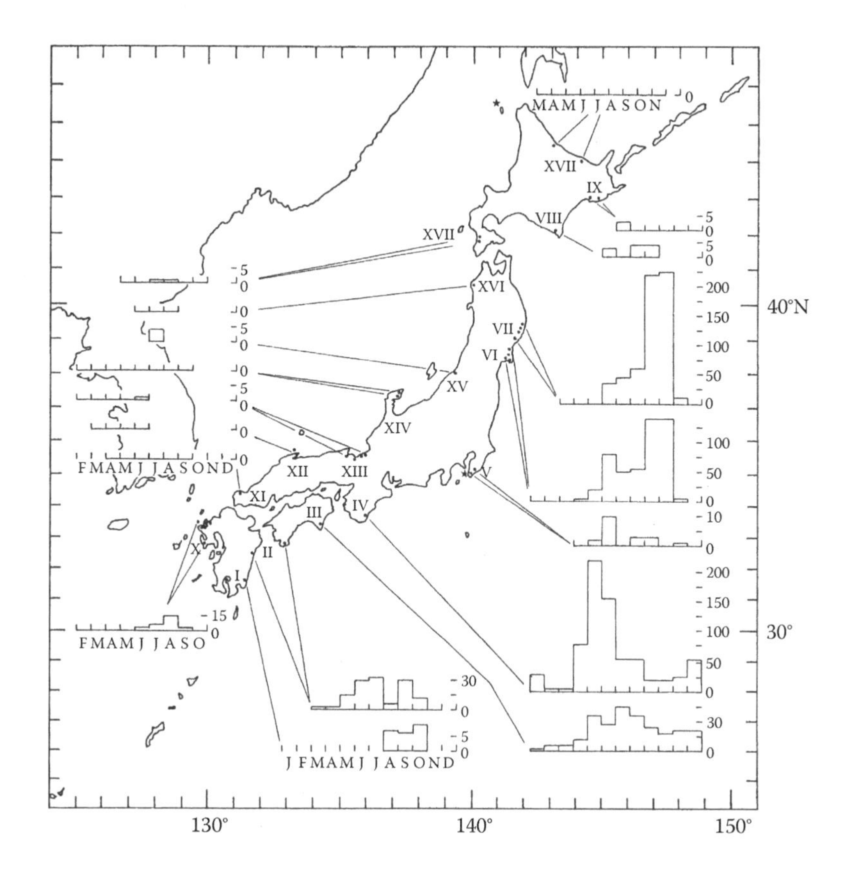

FIGURE 12.34 Monthly distribution of catch (in no. of individuals) of short-finned pilot whales based on Monthly Report of Operations by Japanese small-type whaling in 1949–1952. (Reproduced from Kasuya, T., *Sci. Rep. Whal. Res., Inst.* 27, 95, 1975, Figure 3.) *Note*: This was Figure 12.33 in the Japanese version.

short-finned pilot whales was small at an annual average of 4–5 in the 1949–1952 seasons. The low level of catch continued until the 1970s (Table 12.22). The location suggests that their catch was of the northern form.

Several villages on the Sanriku coast operated dolphin-drive fisheries since the eighteenth century (Section 3.7). They were apparently authorized to conduct the operation by the government in the mid-nineteenth century (Section 5.1). One village, Oura (39°27′N, 142°01′E) in Yamada Bay (39°28′N, 142°00′E), listed three species as their targets: *mai-ruka*, *nezumi-iruka,* and *goto-iruka* (Kawabata 1986, in Japanese). The last species is the short-finned pilot whale, presumed from the location to have been the northern form. These drive fisheries, which last operated in 1920, left no significant statistics (Section 3.7).

Small-type whaling and hand-harpoon fisheries increased during the peri–World War II period and caught northern-form pilot whales off the Sanriku Region (Table 12.22), while the hand-harpoon fisheries probably aimed mainly at smaller species. The number of small-type whaling vessels was 10 in Miyagi Prefecture and 4 in Iwate Prefecture (Pacific coast at 38°59′N–40°27′N) in November 1943 (Nihon Hogeigyo Suisan Kumiai [Whale Fisheries Association of Japan] 1943, in Japanese). The total number of Japanese small-type whaling vessels increased from 43 in 1942 to 73 in 1948 and reached a peak of 80 in 1950 (Ohsumi 1975) (Sections 4.1 and 7.1.1).

Kasuya (1975) analyzed statistics of the short-finned pilot whales taken by the small-type whaling for 1949–1952 (Figure 12.34, Table 12.22). The peak year of catch of northern-form pilot whales was in 1949, when a total of 415 were taken. The catch declined to about 200 in 1953 and to less than 50 during 1954–1971, when the fishery almost ceased, and continued until 1981. The change was most likely due to shift of operations to other species, for example, minke whales and perhaps poaching of sperm whales (Sections 13.7.2 and 15.3.2.3). However, Kasuya and Marsh (1984) noted that the change accompanied a decline in the proportion of large adult males in the catch, which was attributed to high hunting pressure (Section 12.6.2). The small-type whalers resumed exploitation of the northern form in 1982, and the operation has continued to the present at a low level (Table 4.3).

A record exists at Choshi in Chiba prefecture on the Pacific coast of a mass stranding of pilot whales in 1932 or 1933, which was utilized by the local people (Kanari 1983, in Japanese). This area is still known for occasional stranding of pilot whales. Whalers of the Chiba area before the early twentieth century took some short-finned pilot whales during their operation for Baird's beaked whales (Section 13.7.1). There were 11 (in 1941, Table 4.1) or 7 (in 1943, Nihon Hogeigyo Suisan Kumiai 1943, in Japanese) small-type whalers registered in Chiba Prefecture. They recorded an annual take of about five pilot whales in 1949–1952 (Figure 12.34).

In recent years some small-type whaling vessels operated from Wadaura (35°03′N, 140°01′E) for Baird's beaked whales (Section 13.7.2) and occasionally took small numbers of southern-form pilot whales. Pilot whales taken south of Choshi Pont (35°42′N, 140°51′E) could have been of the southern form (Kasuya et al. 1988).

Villages on the Izu coast operated a dolphin-drive fishery since the early seventeenth century. At least since the mid-nineteenth century the striped dolphin was the major target, but they also took cetaceans called *nyudo-iruka* or *o-iruka*, which are believed to have been southern-form pilot whales (Section 3.8.2 and 3.8.3). Currently, the drive fishery is operated by the Futo group at a very low level (Tables 3.16 and 6.3). One person in Inatori (34°46′N, 139°02′E) on the east coast of Izu peninsula in 1910 started small-type whaling for pilot whales using a five-barrel harpoon gun introduced from Taiji, which was subsequently replaced by a three-barrel gun and then a single-barrel harpoon gun; the operation continued until about 1955 (Nakamura 1988, in Japanese). Nihon Hogeigyo Suisan Kumiai (1943, in Japanese) recorded a similar but different story that two small-type whaling vessels registered in Shizuoka Prefecture operated from Inatori in November 1943. Most operations of small-type whalers were likely unrecorded before December 1947, when they were placed under control of the Japanese government (Section 7.1.1). Kasuya (1975) did not identify the operation at Inatori in Fisheries Agency records for the years 1947–1952 (Figure 12.34, Table 12.22).

Hand-harpoon whaling in Mikawa Bay (34°45′N, 137°00′E) around 1570 is the earliest available record of Japanese commercial whaling. This technology was transmitted to the present Chiba Prefecture in the east around 1558–1569 and survived as hand-harpoon whaling for Baird's beaked whales until the nineteenth century (Section 13.7). The technology was also transmitted via the Shima Peninsula (34°20′N, 136°45′E) to western regions: the present Wakayama Prefecture in 1606, Kochi Prefecture (33°30′N, 133°30′E) in 1624, coasts and islands off northwestern Kyushu (33°30′N, 129°45′E) in 1630, and western Yamaguchi Prefecture (34°15′N, 131°00′E) in 1672 (Hashiura 1969, in Japanese). The fishery changed into so-called net whaling and continued to the late nineteenth century (Section 1.3). Some pilot whales could have been taken during the operation (Section 3.9.1 and see below), but details of the catches are unknown.

Hand-harpoon whaling at Taiji in Wakayama Prefecture, where the southern form occurs, started in 1606, shifted to net whaling around 1677 (Sections I.2.1 and 1.3), and continued until ship wreck 1878. The operation slumped in the nineteenth century, and the whalers hunted pilot whales as a side business (Hamanaka 1979, in Japanese). Small vessels called *tento* in Taiji in the mid-nineteenth century were involved in pilot whaling (Section 3.9.1). The *tento* hunters first introduced a whaling cannon in 1903 and engines in 1913 (Section 3.9.1). This resulted in a rapid increase in the catch of pilot whales from 120 in 1912 to 381 the next year, when 11 vessels operated. In November 1943, there were 15 small-type whaling vessels registered in Wakayama Prefecture, and 11 of them belonged to Taiji (Nihon Hogeigyo Suisan Kumiai 1943, in Japanese). Hamanaka (1979, in Japanese) listed the names of 12 vessels engaged in small-type whaling (9 operating bodies) and 8 *tento* vessels (8 operating bodies) as operated from Taiji in the post–World War II period. The dates of statistics or distinction of the two types was not given, but the former type included vessels that were known to have seasonally migrated to other regions of Japan for minke whaling. Thus, the latter could have been smaller vessels operating only locally. All of these vessels were placed under a license system of the government in 1947. Responding to the policy of the Fisheries Agency to reduce the expanded fleet of small-type whaling vessels (Section 7.1.1), the licenses of all the small-type whaling vessels of Taiji, except one called *Katsu-maru*, were aggregated to be used for the establishment of a new large-type whaling company *Nihon Kinkai Hogei* [Japan Coastal Whaling] in 1950 (Hamanaka 1979, in Japanese). Supported by stable demand for the meat in Taiji, two small-type whalers currently operate from Taiji for pilot whales, *Katsu-maru No.7* and *Seiwa-maru* (Table 7.1) (see Section 12.6.3.2 for further details).

Taiji has a history of two types of dolphin-drive fisheries (Section 3.9.1). One was a traditional opportunistic drive for pilot whales, where community members joined in the activity when a school of the whales was sighted near the harbor. The earlier history is unknown, but the village was approved for the operation by the government after the Meiji Revolution in 1867–1872 (Section 5.1) and continued the operation until 1951. The license was not renewed in 1993, possibly because of cessation of the operation for a long period. In 1969, a group of fishermen succeeded in an attempt to drive a pilot whale school from offshore waters in order to supply live animals for the opening of the Taiji Whale Museum. This was the start of the current dolphin-drive fishery in Taiji. This operation is different from the earlier opportunistic drives in being based on daily searching activities in a radius of 15–20 nautical miles (28–37 km). Some Taiji fishermen once used seine nets for pilot whales, but the operation lasted only for a short period in 1933–1935.

A small number of short-finned pilot whales were also taken by small-type whaling in Kochi Prefecture (from the ports of Tosa-shimizu (32°47′N, 132°57′E), Mizu (33°17′E, 134°11′E) and nearby Shiina), Miyazaki Prefecture, and northwestern Kyushu during the post–World War II period (Figure 12.34 and Table 12.22). Nihon Hogeigyo Suisan Kumiai (1943, in Japanese) recorded two small-type whaling vessels registered in Yobuko (33°32′N, 129°54′E) in Saga Prefecture in northern Kyushu, four in Ise-yamada (34°31′N, 136°46′E) in Mie Prefecture, and nine in Tokyo. The last were owned by a whaling company in Tokyo and could have operated in various regions of Japan.

Nago in Okinawa Prefecture was known to have opportunistic dolphin drives for pilot whales and other toothed whales, in which the local people cooperated in driving dolphin schools found near the harbor and shared the catch (Section 3.10). Miyasato (1988, in Japanese) described Norwegian-type whaling at Nago under U.S. occupation that ended in 1972, as well as the dolphin-drive fishery in Nago.

Although Kishiro and Kasuya (1993) believed that Nago did not obtain a license for the drive fishery in 1982, when it came under the license system of the prefecture, a document of the Coastal Division of the Fisheries Agency (dated July 7, 1988) stated that it retained a prefectural license (Section 6.5). The last operations of the drive could have been on 9 short-finned pilot whales in 1982 and 20 bottlenose dolphins in 1989 (Kishiro and Kasuya 1993). As the supply of dolphin meat for Nago people almost ceased in the early 1970s, a new type of fishery called the crossbow fishery started in 1975 for pilot whales and other species. The crossbow fishery is operated by shooting a harpoon with cable and detachable harpoon head from a catapult powered by a rubber belt. This fishery supplied dolphin meat to the people in Nago and to markets in Shimonoseki and Osaka (Section 2.8; Endo 2008, in Japanese).

12.6.2 Abundance and Trends

Attempts to estimate the abundance of Japanese pilot whales go back to the 1980s. The IWC (1987) recorded a provisional estimate of northern-form pilot whales presented by T. Miyashita to the Scientific Committee of the IWC in 1986. Miyashita revised this estimate using additional sighting data and presented it to the Scientific Committee in 1991 (IWC 1992). The latter estimate was based on shipboard sightings in September and October, 1982–1988. The mean abundance was 4239 (Table 12.23). Although a confidence interval is not given, the large coefficient of variation of 0.61 suggests that the lower limit was close to zero.

The abundance of southern-form whales was reported to the Scientific Committee in 1991 and later published in Miyashita (1993); it was based on shipboard sightings in August and September in nine seasons, 1983–1991. During these months the northern limit of distribution of the southern form was at the latitude of 38°N. Miyashita arbitrarily divided the western North Pacific into three areas: northern coastal Pacific west of 145°E between 30°N and 38°N (which included all the coastal whaling grounds mentioned earlier except for Nago), northern offshore Pacific at 30°N–38°N and 145°E–180°E, and southern waters at 23°N–30°N and 127°E–180°E, which included Nago (Table 12.23). Although he did not specifically mention the basis for the division, the boundaries were close to the apparent distribution gap in the species in the western North Pacific (Figure 12.4), which separates the species into coastal waters west of the Kuroshio Current and offshore waters in the Kuroshio Counter Current area (southeast of the Kuroshio Current).

Minamikawa et al. (2007, in Japanese) estimated the abundance of southern-form pilot whales in the western North Pacific using fewer data obtained in 4 years (1998–2001). They obtained an estimate of 15,000 with a CV of 0.7 for the entire area, where Miyashita (1993) had estimated a total of about 50,000 (Table 12.23). The great CV of the more recent estimate makes it difficult to evaluate the difference. Minamikawa et al. (2007, in Japanese) did not give estimates by area, which also raises difficulty in using the results for management.

12.6.3 Problems in Management

12.6.3.1 Northern Form

Northern-form pilot whales inhabit coastal waters to the north of Choshi Point and are hunted only by small-type whalers. There was a considerable discussion in the Fisheries Agency of Japan about regulation of this fishery. Scientists of the Agency wished to limit the catch to within 1% of the

TABLE 12.23
Abundance of Short-Finned Pilot Whales off the Pacific Coast of Japan Estimated from Sighting Data Assuming g(0) = 1

Geographical Form	Area	Mean Estimate	95% Confidence Interval or CV	Survey	Reference
Northern form	Off northern Japan	5,344	819–9,669	Summer, 1984–1985	IWC (1987)[e]
	Off northern Japan	4,239	CV = 0.61	Sep–Oct 1982–1988	IWC (1992)[e]
Southern form	Coastal Japan[a]	14,012	8,996–21,824 (CV = 0.23)	Jun–Sep 1983–1991	Miyashita (1993)
	Offshore[b]	20,884	11,081–39,361 (CV = 0.33)	Jun–Sep 1983–1991	
	Southern area[c]	18,712	8,448–41,445 (CV = 0.42)	Jun–Sep 1983–1991	
	Broader area[d]	15,057	CV = 0.71	1998–2001	Minamikawa et al. (2007, in Japanese)

For additional analysis, see Appendix to this chapter.
Note: This was Table 12.22 in Japanese version.
[a] North of 30°N and west of 145°E. Drive fisheries on the Izu coast (34°36′N–35°05′N, 138°45′E–139°10′E) and at Taiji (33°36′N, 135°57′E) operate in the western part of this range.
[b] 30°N–38°N and 145°E–180°E.
[c] 23°N–30°N and 127°E–180°E. Drive fishery and crossbow fishery at Nago (26°35′N, 127°59′E) in Okinawa operate in the western part of this range.
[d] Approximately in a range of 10°N–40°N and 130°E–160°W.
[e] Based on Miyashita (unpublished).

estimated population size. This was based on a provisional understanding in the Scientific Committee of the IWC that a catch of 1% would not harm the stock and on our observations of the population dynamics of killer whale communities off Vancouver Island, which were once depleted to about 60% of the initial level and exhibited at that time an annual recovery rate of 1.3%–2.6%. If these killer whale populations have been experiencing decline in recent years, our earlier understanding should be reevaluated.

Post–World War II catches of the northern form off Japan reached a peak of over 400 in 1949 and then rapidly declined to 10–100 in 1955–1957 (Table 12.22). The decline arguably could be explained by a shift of the whaling operation to other cetacean species such as minke and sperm whales (see Sections 13.7.2 and 15.3.2.3). However, the decline was accompanied by a change in sex ratio; the male proportion declined from 68% in 1948 (n = 321) to 43% in 1953 (n = 224). Males are over twice as large as females and preferred by the whalers (Table 12.2). The same change was also observed in the catch of the more recent exploitation that started in 1982, that is, from 65% in 1983–1984 to 38% in 1985–1987 (Kasuya and Tai 1993). This change was noted by the Scientific Committee of IWC and interpreted as the reflection of a decline in male abundance or learning of males through experience with whaling vessels. Although it was unknown which interpretation was correct, the change was interpreted as a suggestion of high hunting pressure on the small population.

Under such circumstances the small-type whalers requested a large quota that was clearly unacceptable to us and unachievable by them (Section 7.1.2). They attempted to underreport the catch, which made us suspicious. They reported the catch of the 1982 season, the first year of the resumption of exploitation, as 85 northern-form whales. However, I felt that the catch was likely underreported and requested a correct figure. Then the Small-type Whaling Association corrected the figure to a still suspicious figure of 172 (Table 4.3). In the 1984 season, I was in Ayukawa to examine the northern-form pilot whales at the whaling stations of Gaibo Whaling, Nihon Whaling (a descendant of Nihon Kinkai Whaling), and Toba Whaling. As the season proceeded I had an impression that the daily catch became lower than expected for the good weather conditions that prevailed. Their flensing work and the biological examination of the carcasses usually ended in the early evening, and I used to revisit the stations the next morning to see subsequent activities of the whaling. One morning in a corner of the flensing platform of Gaibo Whaling I found a small fetus that I had not examined the previous evening. Re-examination of my field note and ovary samples failed to identify a candidate mother of the fetus. The chief flenser replied to my inquiry that no viscera had been brought from another whaling station. This incident further increased my suspicion. One cold evening the station manager of Gaibo Whaling announced the end of flensing for the day, and my assistant and I left the station for our hotel, which was about 2 km walk from the land station. Hearing the sound of the crusher of the village ice plant in the late evening, we jumped out of bed and rushed to a small hill to see through binoculars a pilot whale being towed up to the flensing platform of Gaibo Whaling. We rushed to the station to examine several additional carcasses of the northern form. Later the company made an amendment to the operation report of the day to the Fisheries Agency (Kasuya and Tai 1993). A villager later explained that a person in charge of watching us left his post due to the cold and failed to send an alarm to the station manager.

Another example of similar goings-on by the whaling people was in Taiji a few years before the end of the Japanese coastal sperm whale fishery. One day when I was in Taiji to examine sperm whales taken by Nihon Hogei, a station officer invited me for a drive to the Kumano Shrine saying that there was no sperm whale to be processed on that day. He made telephone contacts with the land station during the drive and took me to a restaurant in Katsuura Town for a long dinner. When I was allowed to leave and returned to my hotel in Taiji, I saw land station workers cleaning the flensing platform. Thus I missed a chance of collecting data in exchange for a luxurious dinner, and the company successfully processed some whales outside the view of the Fisheries Agency biologist. This kind of incident was more frequent for inspectors of the Fisheries Agency, who often left their duty for days at a hot spring (Section 15.3.2.3; Kasuya 1999a, b, in Japanese).

The recent regulation of small-type whaling for the northern-form pilot whale changed as follows. In the 1982 season there was no regulation, and the catch of 82 whales in the official report was corrected to 172 for the biologists (without official amendment). In the 1983 season 7 vessels operated with a quota of 175 whales and reported a take of 125. In the 1984 season 6 vessels operated with the same quota and reported a take of 160 whales. Based on the incident of hiding catches described earlier, for the 1985 season the biologists insisted that there should be no catch limit but only regulation of fleet size and the fishing season, which ended with a catch of 62 whales. In 1986 a quota of 50 was agreed and the take was 28. The 1988 season was expected to be the first year when Japanese small-type whalers would have to live without a minke whale quota. In order to smooth the way, the fishermen carried the quota of 50 for the 1987 season over to the 1988 season, when they operated with a 2-year quota of 100. Their catch in 1988 was 98.

The nearly stable annual catch of around 50 northern-form whales during 1988–2002 was followed by an apparent decline (Table 4.3). Examination of the operation pattern and catch composition has not been carried out (see Appendix at the end of this chapter). A possible cause of the decline is a shift of the whaling season for northern-form pilot whales from autumn/early winter to summer (Table 7.1), which was required for participation in the coastal element of Japanese scientific whaling for minke whales that started full-scale operations in 2003. This possibly also caused a conflict with the Baird's beaked whale hunt (Section 7.1.2).

The annual quota of 50 northern-form pilot whales remained unchanged until the 2005 season, when it was reduced to 36. The quota for the southern form allowed for small-type whaling also decreased from 50 to 36 in the 2006 season (Table 12.24).

TABLE 12.24
Quota of Short-Finned Pilot Whales Given to Japanese Fisheries[a]

Geographical Types	Fishery Types	1992	1993	1994–2004	2005	2006	2007	2008	2009
Northern form	Small-type whaling	50	50	50	36	36	36	36	36
Southern form	Small-type whaling	50	50	50	50	36	36	36	36
	Drive fishery (Taiji)	300	300	300	300	300	277	254	230
	Crossbow fishery (Nago)		100	100	100	100	92	85	77
	Government reserve		(50)	(50)					
	Total	350	450	450	450	436	405	375	343
Total for species		400	500	500	486	472	441	411	379

Note: This was Table 12.23 in Japanese version.

[a] Quota system changed in 1996 from calendar year to fishing season (Table 6.3). Here quota is indicated for the starting year of the season. Quota could be carried over to the next season. See Tables 4.3, 6.3, and 6.4 for subsequent years.

12.6.3.2 Southern Form

Along the Pacific coast of Japan southern-form pilot whales usually occur south of Choshi Point at 35°45′N; they are rare off Ayukawa at around 38°15′N (Section 12.3.4). In the offshore waters of the western North Pacific they occur south of 38°N in summer (Miyashita 1993). They have been hunted legally by small-type whaling, drive fisheries, and the crossbow fishery. Although the pilot whale (both forms) was not recorded in the catch of the hand-harpoon fishery since 1972 and was not allowed in that fishery since at least 1993, the head of a hand harpoon in the carcass of a northern-form whale (Kasuya and Tai 1993) suggested some unreported take of the species by the hand-harpoon fishery.

The current drive team in Taiji carried out its first operation in 1969 and increased to two teams in 1979 accompanied by doubling of the number of operating vessels. The two teams merged in 1982, but the number of operating vessels remained at almost the same level (Section 3.9.1). The Taiji drive fishery made its first autonomous catch limit of 500 pilot whales (southern form) and 5000 other unspecified small cetaceans in 1982. This was carried out with a prefectural license beginning in 1986. In 1991, the dolphin quota decreased to 2900, while the quota for pilot whales was unchanged. The quota was decreased in 1992 to a total of 2500 for all species of small cetaceans to be killed, which could include up to 300 pilot whales. In 1993, the Fisheries Agency set catch quotas for each of the 6 small-cetacean species, totaling 2380 including 300 pilot whales (Tables 3.19, 4.3, and 6.3). It was my understanding that the series of quotas had only the function of approving the catch level of the time and did not function to cap the catch.

There has been a change in the pilot whale fishery in Taiji since 1969. The annual number of southern-form pilot whales that landed at Taiji fluctuated between 50 and 150 before the establishment of a driving team in 1969. Since 1970 it increased to 91–479 during 1970–1979, and in 1980, when another team joined, it made the largest catch in the post–World War II history of the fishery in Taiji. The take in 1980 was 605 according to Fisheries Agency statistics (Table 3.18), but it could have been 841 (Table 3.17). The latter figure is based on statistics I have constructed from records of the Taiji Fishery Cooperative Union (FCU) with cooperation of a union staff member. After competing with the drive fishery, the only small-type whaling vessel (*Katsu-maru*) in Taiji shifted operations to the Chiba area in 1981, but it resumed the Taiji operation for pilot whales in 1988 together with a new small-type whaling vessel (*Seishin-maru*) (Table 4.3). (*Note*: the Taiji FCU started small-type whaling with the *Seishin-maru* for minke whales in the 1983 season, paused in the 1989–1991 seasons, and resumed operations with the *Seiwa-maru* in 1992; the *Katsu-maru* was replaced by the *Katsu-maru No. 7* in 1998.)

Thus, the large landings of pilot whales at Taiji were made by the drive fishery. After the peak catch in 1980 the landing of the species by the drive fishery declined annually, with considerable fluctuation, to 157 (or a total of 200, including 43 caught by small-type whaling) in 1993, when the quota was first set by species. The catch of the drive fishery reached the quota only once, in 1996, and exceeded 100 in only half of the 14 seasons (1993–2006) with a quota of 300 (Table 3.19). Decline in the availability of the species to the Taiji fishery seems to be evident in the catch trend.

The Fisheries Agency of Japan took the leadership in establishing a catch quota for 1993 for each fishery and species of small cetaceans (Section 6.5). The principle applied to the southern-form pilot whale was to multiply the abundance of 20,300 estimated for the coastal waters by 2%, and then add 50 to come to 450. The last portion, 50 individuals, was not shared among the fisheries but reserved by the government. The basis for the supposed replacement yield of 2% was given in a document dated January 25, 1993, which I received from the Coastal Division of the Fisheries Agency as a division director of the Far Seas Fisheries Research Laboratory. It was titled "Principle of the Fisheries Agency in [Managing] Small Cetacean Fisheries (Proposal)," and had "Strictly Confidential" handwritten in red ink.

The document first stated that the replacement yield of 1% would be retained for the Baird's beaked whale and northern-form pilot whale because it was accepted by the IWC, which was probably based on the judgment of the Scientific Committee that the current catch level of these species would not damage the population for the time being because the catch rate was about 1%. The document then noted that the United States assumed 4% as an average R_{max} for small cetaceans. R_{max} is the rate of increase expected for a population near zero density. The document assumed 2% (50% of the R_{max}) as a MSY rate for southern-form pilot whales and false killer whales and 3% (75% of the R_{max}) for other Delphinidae. The document assumed 6% as R_{max} for the Dall's porpoise after saying that Japanese biologists as well as those of other countries believe that the Dall's porpoise has a high R_{max} and used 4% (75% of the R_{max} assumed for the species) as the MSY rate of the species. The values of R_{max} and MSY rate were not based on scientific evidence but created by the government staff, supposedly to support the already existing figures of 2%, 3%, and 4% that were necessary to create quotas acceptable to the fisheries (Section 6.5). It should be noted that these arbitrarily created catch quotas were retained for a subsequent 14 years.

It was generally accepted among IWC scientists of the time for baleen whales that MSY is available at around 60% of the initial population level, and that setting a quota by multiplying current abundance by half of the R_{max} will, in theory, stabilize the population at a level close to MSY production. This assumes that R_{max} and the current abundance are correctly estimated, that correct catch statistics are available, and that the population does not oscillate. The population may start oscillating due to environmental as well as biological reasons. R_{max} has not been estimated reliably for any cetacean species, and U.S. scientists assumed 4% for all small cetaceans as a basis of management with a "safety factor" arbitrarily determined for each situation.

The Coastal Division probably needed the process presented earlier to create a quota that was close to the actual catch of the time and was acceptable by the fishermen and at the same time armed with quasi science. The Coastal Division did not mention much about the management of Baird's beaked whales and northern-form pilot whales because they were under control of the Offshore Division of the Fisheries Agency. Before the earlier described document was issued on January 25, 1993, we scientists received from the Coastal Division a document dated August 5, 1992, titled "On Setting Quotas for Dolphin Fisheries of the 1992 [*sic*] Season." The document mentioned only the striped dolphin and the Dall's porpoise and proposed zero quota for the striped dolphin and a total quota of 17,180 for two types of Dall's porpoises (Section 9.3.1), which was 3.9% of the abundance estimate of the time, with a condition to stop operation of the season if catch of one of the two types exceeded 9000. Scientists of the FSL responded in a letter drafted by myself dated August 6, 1992, which stated that (1) the catch quota for Dall's porpoises should be half of the proposed figure, which was close to the catch recorded before the explosion of the catch in late 1980s, and (2) management should be based on an abundance level smaller than the mean estimate to take into account uncertainty in the estimation. After the field season we discussed the problem and Miyashita sent up a revised version of our agreed opinion (dated December 8, 1992), which proposed a catch level of 1% and quota of 200 for southern-form pilot whales. Catch levels proposed in the letter for other species were 1% for the false killer whale, 2% for other delphinids, and 3% for Dall's porpoise.

The response of the Coastal Division to our suggestion was dated January 1, 1993, and titled "On Catch Quota of Small Cetacean for the 1993 Season." It stated that 1% was added to the catch level proposed by the FSL scientists, that is, 2% for the southern-form pilot whale, the false killer whale, and the striped dolphin; 4% for the Dall's porpoise; and 3% for other dolphins, which resulted in quotas of 406 southern-form pilot whales and 9000 for each of the two types of Dall's porpoise. FSL scientists then sent a letter signed by Kasuya and Miyashita and dated January 6, 1993, which stated that if a large quota is applied for economic or political reasons, the date and target level of catch to be achieved in the future should be clarified. This opinion was not accepted and resulted in the document of January 25, 1993, cited earlier.

It seems to be clear that logic was created to explain the already existing quota for southern-form pilot whales of 1993. The person who worked on the issue probably borrowed the idea of a potential biological removal (PBR) accepted in the United States. This was a tool developed to judge if an existing incidental mortality of cetaceans was at a safe level or not. Thus I am suspicious whether it is a suitable tool for management of dolphin fisheries, because it places weight on safety and at the same time allows arbitrary use of parameters (Section 1.2). It is calculated as

$$PBR = N_{min} \times R_{max} \times 0.5 \times F_r$$

where

N_{min} is minimum population size
F_r is a recovery factor

A lower limit of the 95% confidence interval was used for N_{min}, and 0.5 was used for F_r. Since the abundance of southern-form pilot whales targeted by Japanese fisheries was estimated at 20,300 with CV = 0.3, the N_{min} would be only 8,364 if a normal distribution of the error were assumed, and the PBR would be 83. The stock would be safe if the catch were less than this figure, which was probably correct but would not be accepted by the fishery. The PBR was further tuned by Wade (1998), who noted that the only remaining uncertainty in management was the population size if R_{max} were correctly estimated and correct catch statistics were available and showed by simulation that a population was safely managed by using $F_r = 1$ and using for N_{min} the lower 20th percentile of the estimated abundance. However, in many practical cases R_{max} is unknown and catch statistics are often biased, so $F_r = 0.5$ has to be accepted. The PBR is an extremely flexible method and has the risk of being used in an arbitrary way.

The abundance of 20,300 (CV = 0.3) was the basis used for the management of the short-finned pilot whale. The document dated December 8, 1992, stated that this figure was calculated by a staff member of the FSL, responding to a request by the Coastal Division of Fisheries Agency, by adding 6300 whales estimated for the Okinawa area in 20°N–30°N, 125°E–135°E to the abundance of 14,012 (Table 12.23) estimated for northern coastal waters off Japan. It should be noted that the most recent abundance estimate of southern-form pilot whales was only 15,000 for the entire western North Pacific (Minamikawa et al. 2007, in Japanese). We do not know the population structure of southern form short-finned pilot whales in this geographical range.

In addition to the abundance estimate and population structure, the geographically biased operation of the drive fishery has to be considered for management. The range of operation of the crossbow fishery (Okinawa) and the current drive fishery (Taiji) never exceeded 50 nautical miles (92.5 km) from the coast in the former and is likely to be within 15–20 nautical miles (28–37 km) from the port off Taiji, while the abundance estimate includes waters 200 nautical miles from shore and a broad latitudinal range. We know nothing about short-term movement of schools of pilot whales in the western North Pacific. Even if we were to assume long-term mixing based on genetic analyses or movement of marked individuals that would be available at some future time, it would be hard to assume free mixing of the species within the current management area inhabited by, for example, 20,300 individuals. If schools of short-finned pilot whales have some degree of site fidelity, which is very likely, a density gradient from offshore to coastal waters will be created after years of extensive harvest in nearshore waters.

12.6.3.3 Manageability of Short-Finned Pilot Whales

The previous section describes some difficulties to be met while managing short-finned pilot whales as a fishery resource; they relate to abundance, reproductive rate, and desire of the fishermen for larger takes. Here I will consider social structure and life history as elements that should be considered in the management of the species.

Current fishery biology assumes that a population decline due to a fishery improves quality of life of the remaining individuals and results in increase in reproductive rate through improvement of growth, survival, and fecundity, and that the population will move toward recovery. Thus, the fishery will arrive at a point where population size is stabilized under a certain catch (which is sustainable yield, SY). The ratio of the current population size to the initial population is called the population level or status, and the ratio of SY to the stabilized population size is called the sustainable yield rate (SYR). SYR is zero at initial population level but increases with decreasing population level and will reach a maximum value near zero population level, called R_{max}, which still remains to be estimated for any cetacean species. The product of the SYR and the population size is the SY and reaches a maximum value (MSY) at a certain population level called the maximum sustainable yield level (MSYL). The MSYL has not been determined for cetaceans, but the Scientific Committee of the IWC assumed it to be at 60% of the initial population for large baleen whales (not for toothed whales, which live with complex social systems). Various equations have been proposed for the relationship between SYR and population level.

It should be noted that the hypothesis presented earlier assumes a stable population under fishing pressure, where catch and reproduction are equal. Measuring density or abundance of cetaceans is difficult, and it is almost impossible to confirm whether a population is stabilized. We do not find in the history of whaling any case where catch and the composition (age and sex) have been stabilized. Even when the catch was apparently stable in number, the fishing effort and fishing ground could have changed. Thus, whale management under an assumption of a stable population is not based on evidence. Most fisheries record an explosion of catch at the beginning, and the stock could have declined before establishment of management. It is hard to expect an ideal situation in which catch starts from a low level and increases slowly allowing response of the population to the new density. The Taiji drive fishery for pilot whales saw an explosion of the catch in the 1970s, and nominal control started after the catch had started to decline by setting a high quota that was not often reached in subsequent operation of the fishery.

A delay in biological response of the population is one of the difficulties to be met in the management of short-finned pilot whales. The natural mortality rate may respond rather quickly, although measuring the change will be difficult. Full response in birth rate of mature female will take at least 4 years because the gestation time is about 15 months and the nursing period lasts a minimum of 3 years. Newborns thus produced might attain sexual maturity at a mean age of 7 years (currently 7–12 years). Therefore, it will take almost 11 years before the biology of the population completes response to a density decline caused by fisheries. The fishery is likely to further deplete the population without waiting for the full response of the population. Delayed response of a population to density change can cause oscillation of the population and further increase difficulty in management.

There is another element as a cause of delay in response of a population to density decline that is expected when the unit of management is not individuals but pods or matrilineal groups, as seen on drive fisheries. In a hypothetical situation where catches of pods and pod increase are balanced, the number of individuals (population size) and number of pods will be stable. If a drive fishery is established on a population of short-finned pilot whales, the number of pods as well as number of individuals in the population will decline by the amount removed by the drive fishery, but the size of the remaining pods is not directly affected by the fishery. This means that density of individuals in a pod remains the same, which is likely to cause further delay in populational response. Although evaluation of this effect awaits further understanding of feeding strategy, the following scenario is a possibility.

If a drive fishery reduces the number of pods in the population, the remaining pods will increase in size either at a faster speed by responding to the decline in pod density or at the same speed as before without responding to the change in pod density and finally split into more than one pod and achieve recovery of the number of pods in the population. The mechanism of pod splitting is unknown but could include (1) death of a core female, (2) limitation of social ability to maintain a large pod, or (3) feeding economy. There would be a large time lag between the loss of pods due to the fishery and the recovery by splitting the remaining pods. If the fishery management ignores the lag, then decline in the stock (in number of pods) will continue.

Fishery management should pay attention to conserving the social structure of short-finned pilot whales and the culture likely retained in the pod. Small-type whaling may selectively take large individuals for economic reasons, but most of the other dolphin fisheries are nonselective, and fishery management ignores a possible difference in effects of removing different component of the population. It is well known that contribution of members to reproduction differs by age and sex. I have pointed out that loss of a mother will adversely affect survival and production of her offspring either suckling or weaned (Section 12.5.7.2). There are other factors in the case of the short-finned pilot whale. Each pod must have accumulated memory and a behavioral pattern based on past experience, which is their culture. Females live longer, and older females are likely to have an important role as carriers of the culture. If such older females are killed in small-type whaling, the remaining members of the pod will be affected by loss of culture. A drive fishery, on the other hand, removes whole pods and wipes out all the culture contained in the pod. This process reduces the cultural diversity in a population and damages the adaptability of the population to a changing or fluctuating environment. I doubt if the current technology of managing fishery resources is successfully applicable to such highly social species as short-finned pilot whales and some primates.

APPENDIX

After publication of the Japanese version, Kanaji et al. (2011) analyzed the abundance of northern-form pilot whales off Japan using sighting data covering the period of 1984–2006. This study is new in its attempt to estimate annual abundance and its trend over the period and differs from past abundance estimates that combined data of several years to increase sample size and reach a single estimate. Their method included a multiplication of school density by mean school size to calculate density of individual whales. However, they identified two problems in their data due to observer difference between cruises: (1) a large annual fluctuation in school size that had an annual declining trend and (2) uncertainty in classifying primary and secondary sightings. In order to overcome these problems Kanaji et al. (2011) used two methods. One was to use a single common mean school size calculated from all the primary sightings as recorded by observers, and the other was to estimate annual mean school size based on regression of mean school size on year using the same data as given earlier. They also felt that some of the secondary sightings could have been erroneously classified as primary sighting. In order to correct this bias the authors applied the ratio of primary to secondary sightings obtained elsewhere. Thus they obtained four sets of estimates of abundance and annual trend. Because these correction procedures were made by combining all the data for 1984–2006, each estimate was not independent in a strict sense.

Each of the four abundance series was represented by eight estimates for the years 1985–1988, 1991, 1992, 1997, and 2006, where sample size was large. Thus, the number of abundance estimates was 32 in total. None of the members of the pairs were statistically different (CV of each estimate ranged between 0.49 and 0.80). The fluctuation must, at least in part, be due to limited opportunity of encountering schools of the species and limited sighting effort.

Kanaji et al. (2011) noted a trend common to the four series of abundance estimates: (1) high abundance figures for 1985 (6287–8646), (2) low figures for 1986–1988 (1086–1690), (3) a weak upward trend from 1991 to 2006 (2415–3971). In my view, the high figures for 1985 are likely reliable because they are close to the previous estimate of 5344 (IWC 1992) obtained by combining the data for 1982–1988 (Table 3.13). The smaller figures for the later years (1986–2006) may reflect a declining trend.

Kanaji et al. (2011) indicated that the annual catch since 1982, when exploitation resumed, exceeded the safe take level calculated by applying PBR to the annual abundance estimate and suggested that whaling contributed to the apparent decline. The parameters used for the PBR calculation were R_{max} = 4%, F_r = 0.5 and N_{min} as the 20th percentile as suggested by Wade (1998).

The total reported catch by the small-type whalers during 1982 through 1988 was 645 (Table 4.3), and it is still unconfirmed whether such a small take alone could explain the observed apparent declining trend.

REFERENCES

[* In Japanese Language]

Amos, B., Bloch, D., Desportes, G., Majerus, T.M.O., Bancroft, D.R., Barrett, J.A., and Dover, G.A. 1993. A review of molecular evidence relating to social organization and breeding system in the long-finned pilot whale. *Rep. Int. Whal. Commn.* (Special Issue 14, Biology of Northern Hemisphere Pilot Whales): 210–217.

Andersen, L.W. 1993. Further studies on the population structure of the long-finned pilot whale, *Globicephala melas*, off the Faroe Islands. *Rep. Int. Whal. Commn.* (Special Issue 14, Biology of Northern Hemisphere Pilot Whales): 219–231.

Andrews, R.C. 1914a. Monograph of the Pacific Cetacea, I—The Californian gray whale (*Rachianectes glaucus* Cope). *Mem. Am. Mus. Nat. Hist., New Ser.* 1(5): 227–287.

Andrews, R.C. 1914b. American Museum whale collection. *Am. Mus. J.* 14(8): 275–294.

Asper, E.D., Andrews, B.A., Antrim, J.E., and Young, W.G. 1992. Establishing and maintaining successful breeding programs for whales and dolphins in a zoological environment. *IBI Rep.* (Kamogawa, Japan) 3: 71–84.

Baird, R.W. 2000. The killer whale: Foraging specialization and group hunting. pp. 127–153. In: Mann, J., Tyack, P.L., and Whitehead, H. (eds.). *Cetacean Societies: Field Studies of Dolphins and Whales.* The University of Chicago Press, Chicago, IL. 433pp.

Baird, R.W. and Stacey, P.J. 1993. Sightings, strandings and incidental catches of short-finned pilot whales, *Globicephala macrorhynchus*, off the British Columbia coast. *Rep. Int. Whal. Commn.* (Special Issue 14, Biology of Northern Hemisphere Pilot Whales): 475–479.

Barnes, L.G. 1985. Evolution, taxonomy and antitropical distribution of the porpoise (Phocoenidae, Mammalia). *Marine Mammal Sci.* 1(2): 149–165.

Benirschke, K., Johnson, M.L., and Benirschke, R.J. 1980. Is ovulation in dolphins, *Stenella longirostris* and *Stenella attenuata*, always copulation-induced? *Fish. Bull. U.S.* 78(2): 507–528.

Berta, A., Sumich, J.L., and Kovacs, K.M. 2006. *Marine Mammals: Evolutionary Biology*. 2nd ed. Elsevier, Amsterdam, the Netherlands. 547pp.+16 plates.

Best, P.B. 1969. The sperm whale (*Physeter catodon*) off the west coast of South Africa 3, Reproduction in the male. *Invest. Rep. Div. Sea Fish. South Africa* 72: 1–20.

Best, P.B. 2007. *Whales and Dolphins of the Southern African Subregion*. Cambridge University Press, Cape Town, South Africa. 338pp.

Best, P.B., Canham, P.A.S., and MacLeod, N. 1984. Patterns of reproduction in sperm whales, *Physeter macrocephalus*. *Rep. Int. Whal. Commn.* (Special Issue 6, Reproduction in Whales, Dolphins and Porpoises): 51–79.

Bloch, D., Lockyer, C., and Zachariassen, M. 1993a. Age and growth of the long-finned pilot whale off the Faroe Islands. *Rep. Int. Whal. Commn.* (Special Issue 14, Biology of Northern Hemisphere Pilot Whales): 163–207.

Bloch, D., Zachariassen, M., and Zachariassen, P. 1993b. Some external characters of the long-finned pilot whale off the Faroe Islands and a comparison with the short-finned pilot whale. *Rep. Int. Whal. Commn.* (Special Issue 14, Biology of Northern Hemisphere Pilot Whales): 117–135.

Brodie, P.F. 1969. Duration of lactation in Cetacea: An indicator of required learning? *Am. Midland Nat.* 82(1): 312–314.

Campagna, C. 2002. Infanticide and abuse of young. pp. 625–629. In: Perrin, W.F., Würsig, B., and Thewissen, J.G.M. (eds.). *Encyclopedia of Marine Mammals*. Academic Press, San Diego, CA. 1414pp.

Chou, L.-S. 1994. *Guide to Cetaceans of Taiwan*. National Museum of Marine Organisms, Gaoxiong, Taiwan. 105pp. (In Chinese).

Chou, L.-S. 2000. *Ecological Photographs of Cetaceans off the East Coast of Taiwan*. Chinese Cetacean Association, Taipei, Taiwan. 107pp. (In Chinese).

Connor, R.C., Read, A.J., and Wrangham, R. 2000. Male reproductive strategies and social bonds. pp. 247–269. In: Mann, J., Tyack, P.L., and Whitehead, H. (eds.). *Cetacean Societies: Field Studies of Dolphins and Whales*. The University of Chicago Press, Chicago, IL. 433pp.

Crockford, S.J. 2008. Be careful what you ask for: Archaeological evidence of mid-holocene climate change in the Bering Sea and implications for the origins of Arctic Thule. pp. 113–131. In: Clark, G., Leach, F., and O'Connor, S. (eds.). *Islands of Inquiry: Colonization, Seafaring and the Archaeology of Maritime Landscapes, Terra Australis* 29. ANU E Press, Canberra, Australian Capital Territory, Australia. 510pp.

Dall, W.H. 1874. Catalogue of the Cetacea of the North Pacific Ocean. pp. 281–307. *In*: Scammon, C.M. *The Marine Mammals of the North-Western Coast of North America, Described and Illustrated, Together with an Account of the American Whale Fishery*. J.H. Carmany, San Francisco and G.P. Putnam's Sons, New York. 319+vpp.

Davies, J.L. 1960. The southern form of the pilot whale. *J. Mammal.* 41(1): 29–34.

Davies, J.L. 1963. The antitropical factor in cetacean speciation. *Evolution* 17(1): 107–116.

Desportes, G. 1993. Reproductive maturity and seasonality of male long-finned pilot whales off the Faroe Islands. *Rep. Int. Whal. Commn.* (Special Issue 14, Biology of Northern Hemisphere Pilot Whales): 233–262.

Desportes, G., Andersen, L.W., Aspholm, P.E., Bloch, D., and Mouritsen, R. 1992. A note about a male-only pilot whale school observed in Faroe Islands. *Frodskaparrit* 40: 31–37.

Desportes, G. and Mouritsen, R. 1993. Preliminary results on the diet of long-finned pilot whales off the Faroe Islands. *Rep. Int. Whal. Commn.* (Special Issue 14, Biology of Northern Hemisphere Pilot Whales): 303–324.

*Endo, A. 2008. Changes in small-type coastal whaling and utilization of cetaceans—Analyses of circulation systems of whale meat. PhD thesis, Hiroshima University, Higashihiroshima, Japan. 149pp.

Evans, W.E., Thomas, J.A., and Kent, D.B. 1984. A study of pilot whales (*Globicephala macrorhynchus*) in the southern California Bight. Hubbs-Sea World Research Institute, San Diego, CA. 47pp.

Ferreira, I.M., Kasuya, T., Marsh, H., and Best, P.B. 2013. False killer whales (*Pseudorca crassidens*) from Japan and South Africa: Difference in growth and reproduction. *Marine Mammal Sci.* 30(1): 64–81.

Foster, E.A, Franks, D.W., Mazzi, S., Darden, S.K., Balcomb, K.C., Ford, J.K.B., and Croft D.P. 2012. Adaptive prolonged postreproductive life span in killer whales. *Science* 337: 1313.

Fraser, F.C. 1950. Two skulls of *Globicephala macrorhyncha* (Gray) from Dakar. pp. 49–60, pls. 1–5. In: *Atlantide Report*, No. 1: Scientific Results of the Danish Expedition to the Coasts of Tropical West Africa, 1945–1946. Danish Science Press, Copenhagen, Denmark.

Frey, A., Crockford, S.J., Meyer, M., and O'Corry-Crowe, G. 2005. Genetic analysis of prehistoric marine mammal bones from an ancient Aleut village in the southeastern Bering Sea. p. 98. In: *Abstract of 16th Biennial Conference on the Biology of Marine Mammals in 2005*, San Diego, CA. 330pp.

Gaskin, D.E. 1982. *The Ecology of Whales and Dolphins*. Heinemann, London, U.K. 549pp.

Gray, J.E. 1846. On the Cetaceous Animals. pp. 13–53. In: Richardson, J. and Gray, J.E. (eds.). *The Zoology of the Voyage of H.M.S. Erebus and Terror*, Vol. 1, Mammalia. Janson, London, U.K. 53pp.+plates.

Gray, J.E. 1871. *Supplement to the Catalogue of Seals and Whales in British Museum*. Printed by order of the Trustees, London, U.K. 103pp.

*Hamanaka, E. 1979. *History of Taiji Town*. Taiji-cho, Taiji, Japan. 952pp.

*Hashiura, Y. 1969. *History of Whaling at Taiji in Kumano*. Heibonsha, Tokyo, Japan. 662pp.

*Hattori, T. 1887–1888. *Miscellanea of Whaling*. Dainihon Suisankai, Tokyo, Japan. Vol. 1, 109pp; Vol. 2, 210pp.

Heimlich-Boran, J.R. 1993. Social organization of the short-finned pilot whale, *Globicephala macrorhynchus*, with special reference to the comparative social ecology of delphinids. PhD thesis, University of Cambridge, Cambridge, U.K. 132pp.

*Houston, J. 1999. *Confessions of an Igloo Dweller*. Dobutsusha, Tokyo, Japan. 390pp. (Japanese translation, initially published in 1995).

Huggett, A. St. G., and Widdas, W.F. 1951. The relationship between mammalian foetal weight and conception age. *J. Physiol.* 114: 306–317.

*Ikeda, Y. 2003. *Folklore of Sex*. Kodansha, Tokyo, Japan. 264pp.

IWC (International Whaling Commission). 1987. Report of the Subcommittee of Small Cetaceans. *Rep. Int. Whal. Commn.* 37: 121–128.

IWC. 1992. Report of the Subcommittee of Small Cetaceans. *Rep. Int. Whal. Commn.* 42: 178–228.

Jones, M.L. and Swartz, S.L. 2002. Gray whale. pp. 524–536. In: Perrin, W.F., Würsig, B., and Thewissen, J.G.M. (eds.). *Encyclopedia of Marine Mammals*. Academic Press, San Diego, CA. 1414pp.

*Kage, T. 1999. Study on school structure of short-finned pilot whale (*Globicephala macrorhynchus*) using DNA polymorphism. PhD thesis, Mie University, Tsu, Japan. 141pp.

*Kanari, H. 1983. *Whaling of Boso (Chiba Prefecture)*. Ron Shobo, Nagareyama, Japan. 154pp.

*Kanda, G. 1731(Date of Preface). *Pictorial Catalogue of Japanese Fish*, Vol. 4 (fishes without scales). (unpublished manuscript).

Kanaji, Y., Okamura, H., and Miyashita, T. 2011. Long-term abundance trends of the northern form of the short-finned pilot whales (*Globicephala macrorhynchus*) along the Pacific coast of Japan. *Marine Mammal Sci.* 27(3): 477–492.

Kasuya, T. 1975. Past occurrence of *Globicephala melaena* in the western North Pacific. *Sci. Rep. Whal. Res. Inst.* 27: 95–110.

Kasuya, T. 1977. Age determination and growth of the Baird's beaked whales with a comment on the fetal growth rate. *Sci. Rep. Whal. Res. Inst.* 29: 1–20.

*Kasuya, T. 1981. Whale remains from the Kabukai A-site. pp. 675–683. In: Oba, T. and Oi, H. (eds.). *Study of Okhotsk Culture-3, Kabukai Site*, Vol. 2. Tokyo Daigaku Shuppan-kai (University of Tokyo Press), Tokyo, Japan. 727pp. +pp. 135–249.

*Kasuya, T. 1983. Whale teeth and age determination. *Kagaku-to Jikken* 34(4): 39–45 (part I); 34(5): 47–53(part II); 34(6): 55–62 (part III).

Kasuya, T. 1985a. Fishery-dolphin conflict in the Iki Island area of Japan. pp. 253–272. In: Beddington, J.R., Beverton, R.J.H., and Lavigne, D.M. (eds.). *Marine Mammals and Fisheries*. George Allen and Unwin, London, U.K. 354pp.

Kasuya, T. 1985b. Effect of exploitation on reproductive parameters of the spotted and striped dolphins off the Pacific coast of Japan. *Sci. Rep. Whal. Res. Inst.* 36: 107–138.

*Kasuya, T. 1986. False killer whale. pp. 178–187. In: Tamura, T., Ohsumi, S., and Arai, R. (eds.). *Report of Contract Studies in Search of Countermeasures for Fishery Hazard (Culling of Harmful Organisms) (56th to 60th year of Showa)*. Suisan-cho Gogyo Kogai (Yugaiseibutsu Kujo) Taisaku Chosa Kento Iinkai (Fishery Agency), Tokyo, Japan. 285pp.

*Kasuya, T. 1990. Life history of toothed whales. pp. 80–127. In: Miyazaki, N. and Kasuya, T. (eds.). *Mammals of the Sea—Their Past, Present and Future.* Saientisutosha, Tokyo, Japan. 300pp.

Kasuya, T. 1991. Density dependent growth in North Pacific sperm whales. *Marine Mammal Sci.* 7(3): 230–257.

*Kasuya, T. 1995a. Short-finned pilot whale. pp. 542–551. In: Odate, S. (ed.). *Basic Data of Rare Wild Aquatic Organisms of Japan* (II). Nihon Suisan Shigen Hogo Kyokai, Tokyo, Japan. 751pp.

Kasuya, T. 1995b. Overview of cetacean life histories: An essay in their evolution. pp. 481–497. In: Blix, A.S., Walloe, L., and Ultang, O. (eds.). *Whales, Seals, Fish and Man.* Elsvier, Amsterdam, the Netherlands. 720pp.

Kasuya, T. 1999a. Examination of reliability of catch statistics in the Japanese coastal sperm whale fishery. *J. Cetacean Res. Manage.* 1(1): 109–122.

*Kasuya, T. 1999b. Manipulation of statistics by Japanese coastal sperm whale fishery. *IBI Rep.* Kamogawa, Japan. 9: 75–92.

Kasuya, T. 2008. Cetacean biology and conservation: A Japanese scientist's perspective spanning 46 years. *Marine Mammal Sci.* 24(4): 749–773.

Kasuya, T. and Brownell, R.L. Jr. 1979. Age determination, reproduction and growth of the Franciscana dolphin, *Pontoporia blainvillei*. *Sci. Rep. Whal. Res. Inst.* 31: 45–67.

Kasuya, T., Brownell, R.L. Jr., and Balcomb, K.C. 1997. Life history of Baird's beaked whales off the Pacific coast of Japan. *Rep. Int. Whal. Commn.* 47: 969–979.

*Kasuya, T., Izumisawa, Y., Komyo, Y., Ishino, Y., and Maejima, Y. 1997. Life history parameters of common bottlenose dolphins off Japan. *IBI Rep.* 7: 71–107.

Kasuya, T. and Marsh, H. 1984. Life history and reproductive biology of short-finned-pilot whale, *Globicephala macrorhynchus*, off the Pacific coast of Japan. *Rep. Int. Whal. Commn.* (Special Issue 6, Reproduction in Whales, Dolphins and Porpoises): 259–310.

Kasuya, T., Marsh, H., and Amino, A. 1993. Non-reproductive matings in short-finned pilot whales. *Rep. Int. Whal. Commn.* (Special Issue 14, Biology of Northern Hemisphere Pilot Whales): 425–437.

Kasuya, T. and Matsui, S. 1984. Age determination and growth of the short-finned pilot whale off the Pacific coast of Japan. *Sci. Rep. Whal. Res. Inst.* 35: 57–59.

Kasuya, T. and Miyashita, T. 1988. Distribution of sperm whale stocks in the North Pacific. *Sci. Rep. Whal. Res. Inst.* 39: 31–75.

Kasuya, T., Miyashita, T., and Kasamatsu, F. 1988. Segregation of two forms of short-finned pilot whales off the Pacific coast of Japan. *Sci. Rep. Whal. Res. Inst.* 39: 77–90.

Kasuya, T. and Tai, S. 1993. Life history of short-finned pilot whale stocks off Japan and a description of the fishery. *Rep. Int. Whal. Commn.* (Special Issue 14, Biology of Northern Hemisphere Pilot Whales): 339–473.

*Kasuya, T. and Yamada, T. 1995. *List of Japanese Cetaceans,* Geiken Series 7. Nihon Geirui Kenkyusho (Institute of Cetacean Research), Tokyo, Japan. 90pp.

Kato, H. 1984. Observation of tooth scars on the head of male sperm whales, as an indication of intra-sexual fighting. *Sci. Rep. Whal. Res. Inst.* 35: 39–46.

*Katsumata, E. 2005. Study on reproduction of marine mammals in captivity. PhD thesis, Gifu University, Gifu, Japan. 145pp.

*Kawabata, H. 1986. Dolphin and porpoise fisheries. pp. 636–664. In: Yamada Choshi Hensan Iinkai (ed.). *History of Yamada Town*, Vol. 1. Yamada-cho Kyoiku Iinkai, Iwate, Japan. 10+1095pp.

Kawai, H. 1972. Oceanography of the Kuroshio and Oyashio currents. pp. 129–320. In: Iwashita, M., Komaki, T., Hoshino, S., Horibe, S., and Masuzawa, J. (eds.). *Physical Oceanography.* Tokai University Press, Tokyo, Japan. 328pp.

Kenney, R.D. 2002. North Atlantic, North Pacific, and southern right whales. pp. 806–813. In: Perrin, W.F., Würsig, B., and Thewissen, J.G.M. (eds.). *Encyclopedia of Marine Mammals.* Academic Press, San Diego, CA. 1414pp.

*Kishida, K. 1925. *Illustrated Guide of Mammals.* Bureau of Agriculture of Ministry of Agriculture and Commerce, Tokyo, Japan. 381+17+31pp.

*Kishiro, T., Kasuya, T., and Kato, H. 1990. Geographical variation of external morphology of short-finned pilot whales.*Abstract of the Spring Meeting of Japanese Society of Scientific Fisheries in 1990.* Tokyo, Japan. p. 31.

Kishiro, T. and Kasuya, T. 1993. Review of Japanese dolphin drive fisheries and their status. *Rep. Int. Whal. Commn.* 43: 439–452.

Krutzen, M. 2006. Dolphins join the culture club. *Aust. Sci.* 27(5): 26–28.

Kubodera, T. and Miyazaki, N. 1993. Cephalopods eaten by short-finned pilot whales, *Globicephala macrorhynchus*, caught off Ayukawa, Ojika (*sic*) Peninsula, in Japan, in 1982 and 1983. pp. 215–227. In: Okutani, T., O'Dor, R.K. and Kubodera, T. (eds.). *Recent Advances in Fisheries Biology*, Tokai University Press, Tokyo, Japan. 752pp.

*Kuroda, J. 1884. *Natural History of Aquatic Animals* [Suizoku-shi]. Bunkaisha, Tokyo, Japan. 316+33pp. (Manuscript dated 1827).

*Kuroda, S. 1982. *Pygmy Chimpanzee: an Unknown Ape*. Chikuma Shobo, Tokyo, Japan. 234pp.

Kuznetzov, V.B. 1990. Chemical sense of dolphins. pp. 481–503. In: Thomas, J.A. and Kastelein, R.A. (eds.). *Sensory Abilities of Cetaceans: Laboratory and Field Evidence*. Plenum Press, New York and London, U.K. 710pp.

Laws, R.M. 1959. The foetal growth rate of whales with special reference to the fin whale, *Balaenoptera physalus* Linn. *Discovery Rep.* 29: 281–308.

Lockyer, C. 1981. Estimates of growth and energy budget for the sperm whale, *Physeter catodon*. pp. 489–504. In: Clark, J.G., Goodman J., and Soave, G.A. (eds.). *Mammals in the Sea*, Vol. 3. FAO, Rome, Italy. 504pp.

Lockyer, C. 1993. Seasonal change in body fat condition of northeast Atlantic pilot whales, and their biological significance. *Rep. Int. Whal. Commn.* (Special Issue 14, Biology of Northern Hemisphere Pilot Whales): 325–350.

Ma, J. and Tan, L. 1995. *A Field Guide to Whales and Dolphins in the Philippines*. Bookmark, Manila, Philippines. 129pp.

Magnusson, K.G. and Kasuya, T. 1997. Mating strategies in whale populations: Searching strategy vs. harem strategy. *Ecol. Model.* 102: 225–242.

Marsh, H. and Kasuya, T. 1984. Change in the ovaries of the short-finned pilot whale, *Globicephala macrorhynchus*, with age and reproductive biology. *Rep. Int. Whal. Commn.* (Special Issue 6, Reproduction in Whales, Dolphins and Porpoises): 311–335.

Marsh, H. and Kasuya, T. 1986. Evidence of reproductive senescence in female cetaceans. *Rep. Int. Whal. Commn.* (Special Issue 8, Behavior of Whales in Relation to Management): 57–74.

Marshall, D.M. 2009. Feeding morphology. pp. 406–414. In: Perrin, W.F., Würsig, B., and Thewissen, J.G.M. (eds.). *Encyclopedia of Marine Mammals*. 2nd ed. Elsvier, Amsterdam, the Netherlands. 1316pp.

Martin, A.R. and Rothery, P. 1993. Reproductive parameters of female long-finned pilot whales (*Globicephala melas*) around the Faroe Islands. *Rep. Int. Whal. Commn.* (Special Issue 14, Biology of Northern Hemisphere Pilot Whales): 263–304.

Masaki, Y. 1976. Biological studies on the North Pacific sei whale. *Bull. Far. Seas Fish. Res. Lab.* 14: 1–104.

McCann, C. 1974. Body scarring on Cetacea-Odontocetes. *Sci. Rep. Whal. Res. Inst.* 26: 145–155.

*Minamikawa, S., Shimada, H., Miyashita, T., and Moronuki, H. 2007. Abundance of 6 cetacean species hunted by Japanese fisheries based on sighting data of 1998–2001. *Abstract of the Autumn Meeting of Japanese Society of Scientific Fisheries in 2007.* Hakodate, Japan. p. 151.

Minasian, S.M., Balcomb, K.C. III., and Foster, L. 1987. *The Whales of Hawaii*. Marine Mammal Fund, San Francisco, CA. 99pp.

Mitchell, E. 1975. Report of the meeting on smaller cetaceans, Montreal, April 1–11, 1974. *J. Fisheries Res. Bd. Can.* 32(7): 917–919.

*Miyasato, H. 1988. Whaling of Nago. *Ajima* (Nago Museum) 4: 15–54.

*Miyasato, H. 2008 (date of preface). *History of Fisheries in Nago*. Published by the Author, Nago, Japan. 94pp.

Miyashita, T. 1993. Abundance of dolphin stocks in the western North Pacific taken by the Japanese drive fishery. *Rep. Int. Whal. Commn.* 43: 417–437.

*Miyazaki, N. 1983. Pilot whales off Kinkazan [Sanriku Region]. *Abstract of the Autumn Meeting of Japanese Society of Scientific Fisheries in 1983*, Kyoto, Japan. p. 55.

Miyazaki, N. 1984. Further analyses of reproduction in the striped dolphin, *Stenella coeruleoalba*, off the Pacific coast of Japan. *Rep. Int. Whal. Commn.* (Special Issue 6, Reproduction in Whales, Dolphins and Porpoises): 343–353.

Miyazaki, N. and Amano, M. 1994. Skull morphology of two forms of short-finned pilot whales off the Pacific coast of Japan. *Rep. Int. Whal. Commn.* 44: 499–507.

Myrick, A.C. Jr. 1991. Some new and potential use of dental layers in studying delphinid populations. pp. 251–280. In: Pryor, K. and Norris, K.S. (eds.). *Dolphin Societies: Discovery and Puzzles*. University of California Press, Berkeley, CA. 397pp.

Myrick, A.C. Jr. and Cornell, L.H. 1990. Calibrating dental layers in captive bottlenose dolphins from serial tetracycline labels and tooth extraction. pp. 587–608. In: Leatherwood, S. and Reeves, R.R. (eds.). *The Bottlenose Dolphin*. Academic Press, San Diego, CA. 653pp.

Nadler, R.D. 1981. Laboratory research on sexual behaviour of the great apes. pp. 191–239. In: Graham, C. (ed.). *Reproductive Biology of the Great Apes, Comparative and Biomedical Perspectives*. Academic Press, New York. 435pp.

*Nagasaki-ken Suisan Shikenjo. 1986. Sightings from ferries and government vessels. pp. 43–54. In: Tamura, T., Ohsumi, S. and Arai, S. (eds.). *Report of Contract Studies in Search of Countermeasures for Fishery Hazard (Culling of Harmful Organisms) (56th to 60th year of Showa)*. Suisan-cho Gyogyo Kogai (Yugaiseibutsu Kujo) Taisaku Chosa Kento Iinkai (Fisheries Agency), Tokyo, Japan. 285pp.

*Nagasawa, R. 1916. Scientific names for 11 species of dolphins and porpoises. *Dobutsugaku Zasshi* (*J. Zool.*, published by Tokyo Zoological Society) 28(327): 35–39.

*Nakamura, Y. 1988. On dolphin and porpoise fisheries. pp. 92–136. In: Shizuoka-ken Minzoku Geino Kenkyukai (ed.). *Marine Folklore of Shizuoka: Culture in Kuroshio Current*. Shizuoka Shinbunsha, Shizuoka, Japan. 379pp.

*Nihon Hogei-gyo Suisan Kumiai. 1943. *Whaling Handbook*, Vol. 3. Nihon Hogeigyo Suisan Kumiai, Tokyo, Japan. 205pp.

*Nishiwaki, M. 1965. *Cetaceans and Pinnipeds*. Tokyo Daigaku Shuppankai (University of Tokyo Press), Tokyo, Japan. 439pp.

Nishiwaki, M., Hibiya, T., and Ohsumi (Kimura), S. 1958. Age study of sperm whales based on reading of tooth laminations. *Sci. Rep. Whal. Res. Inst.* 13: 135–170.

*Nishiwaki, M., Kasuya, T., Brownell, R.L. Jr., and Caldwell, D.K. 1967. On taxonomy of *Globicephala* in Japanese waters. *Abstract of the Autumn Meeting of Japanese Society of Scientific Fisheries in 1967.* p. 15.

Nishiwaki, M. and Yagi, T. 1953. On the growth of teeth in a dolphin (*Prodelphinus caeruleoalba*) (I). *Sci. Rep. Whal. Res. Inst.* 8: 133–146.

*Ogawa, T. 1937. Study on Japanese toothed whales, Part 9. *Shokubutsu-to Dobutsu* (*Plants Animals*) 5(3): 591–598.

*Ogawa, T. 1950. *Tale of Whales*. Chuo Koron Sha, Tokyo, Japan. 260pp.

Ogden, J.A. (no date). Flipper development in *Globicephala macrorhyncha*, I: Growth of flippers. 22pp. (unpublished manuscript cited in Kasuya and Matsui 1984).

*Ohsumi, S. 1963. Tale of sperm whale teeth. *Geiken Tsushin* 141: 84–100.

Ohsumi, S. 1964. Comparison of maturity and accumulation rate of corpora albicantia between the left and right ovaries in Cetacea. *Sci. Rep. Whal. Res. Inst.* 18: 123–158.

Ohsumi, S. 1966. Allomorphosis between body length at sexual maturity and body length at birth in Cetacea. *J. Mammal. Soc. Jpn.* 3(1): 3–7.

Ohsumi, S. 1975. Review of Japanese small-type whaling. *J. Fish. Res. Bd. Can.* 32(7): 1111–1121.

*Okura, N. 1826. *Leafhopper Control.* Kyoto Shorin, Kyoto, Japan. 56pp.

Olesiuk, P.F., Bigg, M.A., and Ellis, G.M. 1990. Life history and population dynamics of resident killer whales (*Orcinus orca*) in the coastal waters of British Columbia and Washington State. *Rep. Int. Whal. Commn.* (Special Issue 12, Individual Recognition of Cetaceans): 209–243.

*Omura, H., Matsuura, Y., and Miyazaki, I. 1942. *Whales: Their Biology and Practice of Whaling.* Suisan Sha, Tokyo, Japan. 319pp.

Oremus, M., Gales, R., Dalebout, M.L., Funahashi, N., Endo, T., Kage, T., Steel, D., and Baker, S.C. 2009. Worldwide mitochondrial DNA diversity and phylogeography of pilot whales (*Globicephala* spp.). *Biol. J. Linnaean Soc.* 98: 729–744.

*Otsuki, K. 1808 (manuscript). On natural history of whales. pp. 7–152. *In*: *Geishi Ko* (*On Natural History of Whales*). Kowa Shuppan, Tokyo, Japan. 538+31pp. (Published in 1983).

Perrin, W.F., Miyazaki, N., and Kasuya, T. 1989. A dwarf form of the spinner dolphin (*Stenella longirostris*) from Thailand. *Marine Mammal Sci.* 5(3): 213–227.

Perrin, W.F. and Myrick, A.C. 1980 (eds.). Age determination of toothed whales and sirenians. *Rep. Int. Whal. Commn.* (Special Issue 3): 229.

Peters, R.H. 1983. *The Ecological Implication of Body Size.* Cambridge University Press, Cambridge, U.K. 329pp.

Polisini, J.M. 1980. Comparison of *Globicephala macrorhyncha* (Gray, 1846) with the pilot whale of the North Pacific Ocean: An analysis of the skull of the broad-rostrum pilot whale of the genus *Globicephala.* PhD thesis, University of Southern California, Los Angeles, CA. 299pp.

Pryor, K. 1990. Concluding comments on vision, tactition, and chemoreception. pp. 561–569. In: Thomas, J.A. and Kastelein, R.A. (eds.). *Sensory Abilities of Cetaceans: Laboratory and Field Evidence.* Plenum Press, New York and London, U.K. 710pp.

Reeves, R.R., Stewart, B.S., Clapham, P.J., and Powell, J.A. 2002. *Guide to Marine Mammals of the World.* Alfred A. Knopf, New York. 525pp.

Rice, D.W. 1998. *Marine Mammals of the World, Systematics and Distribution.* Special Publication 4. Society for Marine Mammalogy. 231pp.

Ridgway, S., Kamolnick, T., Reddy, M., Curry, C., and Tarpley, R. 1995. Orphan-induced lactation in *Tursiops* and analysis of collected milk. *Marine Mammal Sci.* 11(2): 172–182.

Robeck, T.R., Schneyer, A.L., McBain, J.F., Dalton, M.L., Walsh, M.T., Czekala, N.M., and Kraemer, D.C. 1993. Analyses of urinary immunoreactive steroid metabolites and gonadotropines for characterization of the estrus cycle, breeding period, and seasonal estrous activity of captive killer whales (*Orcinus orca*). *Zoo Biol.* 12: 173–187.

Robson, D.C. and Chapman, D.G. 1961. Catch curve and mortality rates. *Trans. Am. Fish. Soc.* 90: 181–189.

Scammon, C.M. 1874. *The Marine Mammals of the North-western Coast of North America, Described and Illustrated, together with an Account of the American Whale-fishery.* J.H. Carmany, San Francisco, CA. 319+vpp.

Schroeder, J.P. and Keller, K.V. 1990. Artificial insemination of bottlenose dolphin. pp. 447–460. In: Leatherwood, S. and Reeves, R.R. (eds.). *The Bottlenose Dolphin.* Academic Press, San Diego, CA. 653pp.

*Schultz, E.A. and Lavenda, R.H. 2003. *Cultural Anthropology*, II. Kokon Shoin, Tokyo, Japan. 222pp. (Japanese translation, initially published in 1990 by West Publishing).

Sergeant, D.E. 1962. The biology of the pilot or pothead whale, *Globicephala melaena* (Traill), in Newfoundland waters. *Bull. Fish. Res. Bd. Can.* 132: 1–84.

Shane, S.H. and McSweeney, D. 1988. Pilot whale photoidentification: Potentials and comparison with other species. Document IWC/SC/A88/P20 presented to the *Scientific Committee Meeting of International Whaling Commission in 1988.* 10pp. and Figs. 1–2. (Available from IWC Secretariat, Cambridge, U.K.).

Temminck, C.J. and Schlegel, H. 1844. *Fauna Japonica.* Lugudini Batavorum, Arnz. pt. 3: 1–26 + pls. 21–29.

Tutin, C.E. and McGinnis, P.R. 1981. Chimpanzee reproduction in the wild. pp. 239–264. In: Graham, C. (ed.). *Reproductive Biology of the Great Apes, Comparative and Biomedical Perspectives.* Academic Press, New York. 435pp.

Ullrey, D.E., Schwartz, C.C., Whetter, P.A., Rajeshwar Rao, T., Euber, J.R., Cheng, S.G., and Brunner, J.R. 1984. Blue-green color and composition of Stejneger's beaked whale (*Mesoplodon stejnegeri*) milk. *Comp. Biochem. Physiol.* 79B(3): 349–352.

van Bree, P.J.H. 1971. On *Globicephala sieboldii* Gray, 1846, and other species of pilot whales (Notes on Cetacea, Delphinoidea III). *Beaufortia* 19(249): 79–87.

van Bree, P.J.H., Best, P.B., and Ross, G.J.B. 1978. Occurrence of the two species of pilot whales (genus *Globicephala*) on the coast of South Africa. *Mammalia* 42(3): 323–328.

*Waal, F. de 2005. *Our Inner Ape.* Hayakawa Shobo, Tokyo, Japan. 340pp. (Japanese translation, initially published in 2005 from Carlisle Co.).

Wada, S. 1988. Genetic differentiation between two forms of short-finned pilot whales off the Pacific coast of Japan. *Sci. Rep. Whal. Res. Inst.* 39: 91–101.

Wade, P.R. 1998. Calculating limits to the allowable human-caused mortality of cetaceans and pinnnipeds. *Marine Mammal Sci.* 14(1): 1–37.

Wang, J.Y. and Yang, S. 2007. *An Identification Guide to the Dolphins and Other Small Cetaceans of Taiwan.* Jen Jen Publishing and National Museum of Marine Biology, Taiwan. 207pp. (In Chinese and English).

Wang, P. 1999. *Chinese Cetaceans.* Ocean Enterprises, Hong Kong, People's Republic of China. 325pp.+Pl.1-VI. (In Chinese).

Whitehead, H. 2002. Culture in whales and dolphins. pp. 304–305. In: Perrin, W.F., Würsig, B. and Thewissen, J.G.M. (eds.). *Encyclopedia of Marine Mammals.* Academic Press, San Diego, CA. 1414pp.

Whitehead, H. 2003. *Sperm Whales: Social Evolution in the Ocean.* University of Chicago Press, Chicago, IL. 431pp.

Würsig, B., Jefferson, T.A., and Schmidly, D.J. 2000. *The Marine Mammals of the Gulf of Mexico.* Texas A&M University Press, College Station, TX. 232pp.

*Yamase, H. 1760. *Natural History of Whales.* Osaka Shorin, Osaka, Japan. 16+54pp.

Yonekura, M., Matsui, S., and Kasuya, T. 1980. On the external characters of *Globicephala macrorhynchus* off Taiji, Pacific coast of Japan. *Sci. Rep. Whal. Res. Inst.* 32: 67–95.

Yoshioka, M., Mohri, E., Tobayama, T., Aida, K., and Hanyu, I. 1986. Annual changes in serum reproductive hormone level in the captive female bottle-nosed dolphins. *Bull. Jpn. Soc. Sci. Fish.* 52: 1939–1946.

Yoshioka, M., Ogura, M., and Shikano, C. 1987. Results of the Transpacific Dall's Porpoise Research Cruise by Hoyomaru No. 12 in 1986. Document presented to the *Ad Hoc Committee on Marine Mammals of the Scientific Committee, International North Pacific Fishery Commission.* Tokyo, Japan. 22pp.

13 Baird's Beaked Whale

13.1 DESCRIPTION

Baird's beaked whale, *Berardius bairdii* Stejneger 1933, is known in Japan as *tsuchi, tsuchi-kujira or, tsuchin-bo.* In addition to these well-known names, whalers in Chiba Prefecture used *kokushira* for this species and *ko-kujira* (in two Chinese characters meaning small whale) probably for gray whales (Documents Nos. 182 and 183 cited in Kishinoue 1914, in Japanese; see Section 13.7.1), while the same *ko-kujira* was used for gray whale and Baird's beaked whales in the Wakayama area (Hattori 1887–1888, in Japanese). Thus, caution is required to avoid misinterpretation. The gray whale was called variously *koku, koku-kujira, ko-kujira, chigo-kujira, aosagi,* or *share.* Yamase (1760, in Japanese) used *asobi-kujira* with a drawing of a whale having a rostrum identifiable as a Baird's beaked whale but having teeth in both jaws in his book *Geishi* [*Natural History of Whales*]. Hattori (1887–1888, in Japanese) accepted the opinion in a book *Santei Geishi* [*Corrections to Geishi*] (not seen) by an anonymous author that the name *asobi-kujira* was due to a mix-up of the Chinese character for *tsuchi* with another character that looked slightly similar and considered the name incorrect (Kasuya and Yamada 1995, in Japanese). Although the species is dealt with by the International Whaling Commission (IWC) as a small-cetacean species for management purposes (Preface and Section I.2.2), it is the largest species of the family Ziphiidae and is comparable to female sperm whales in body size. It reaches 11–12 m; females grow slightly larger than males. The dorsal fin is 25–32 cm high, has a concave posterior contour, and is placed at a position about one-third of the body length from the posterior end. The head is much smaller than that of the sperm whale, and the proportion of length from the anterior end of the upper jaw to the anterior insertion of the flipper is only 15%–19% of the body length, which is much smaller than the 30% attained by sperm whales of the same size.

A slender beak projects 50–60 cm beyond the melon, and the severed head looks like a mallet (*tsuchi*), which gave rise to the Japanese name. No teeth are present in the upper jaw, but there are two or three pairs of flat triangular teeth near the tip of the lower jaw (Kirino 1956; Kuroe 1960, in Japanese). The anterior-most pair is the largest, attaining 10 cm in length and anterior-posterior width and 5 cm in lingual-buccal thickness (Kasuya 1995, in Japanese). The size of the posterior teeth is less than half that of the anterior-most teeth, and they are hidden in the mouth if the jaws are closed. The anterior-most teeth are placed beyond the anterior tip of the upper jaw; thus they do not function to grasp food but show abrasion, caused by contact with the sea floor, food, or the skin of other individuals of the same species.

The body is dark brown except for an irregular white area along the ventral midline. A pair of main throat grooves and accessory groves are seen in the throat region (Omura et al. 1955). These throat grooves are also present in other members of the Ziphiidae and in sperm whales and probably function in suction feeding to increase laryngeal capacity by expanding the throat region. The single blowhole is crescent-shaped and is posteriorly concave when closed, which is the same as in another member of the genus *B. arnuxii* Duvernoy, but differs from the state in other members of the Ziphiidae and all the Delphinoidea species.

The anterior end of the melon is often marked with scars, presumably caused by contact with the sea floor while feeding. Whales have more flexibility in bending the body ventrally than dorsally, which is similar to our condition. Swimming upside down would be a reasonable posture in which to approach the sea floor for feeding on benthic organisms and then return to the water column. This is the opposite of surfacing for respiration.

Baird's beaked whales have numerous tooth marks on the dorsal surface from the neck to the dorsal fin, presumably from contact with members of the same species. The dorsal surface of older individuals often appears almost whitish with heavy scarring (see frontispiece).

Body weights of a fetus of 169 cm and three postnatals at or below 10.8 m produced the following relationship between body length (L, cm) and body weight (W, kg) (Kasuya et al. 1997):

$$\text{Log } W = 3.081 \times \log L + \log(6.339 \times 10^{-6}), \quad r = 0.99$$

This can be rewritten as

$$W = (6.339 \times 10^{-6})L^{3.081}$$

Body weight in many cetacean species is proportional to the 2.6–2.8th power of body length (Bryden 1986), which means the body becomes more slender with growth. However, the equation above suggests that Baird's beaked whales do not change body shape with growth. It also suggests a body weight of 987 kg for neonates (4.56 m) and 12.7 tons for a full-grown female of average body length at 10.45 m.

Baird's beaked whales usually live in schools of 3–20 individuals and are found in a tight school at the sea surface. This together with body scars on the dorsal surface suggests that they live a social life.

Recent whaling for Baird's beaked whales in the Okhotsk Sea and whale-sighting surveys by Russian scientists in the region were unable to identify the presence of the northern bottlenose whale, *Hyperoodon ampullatus*, in the Okhotsk Sea, and the conclusion was reached that the species is absent there (Tomilin 1967). While supporting the conclusion of Tomilin (1967), Heptner et al. (1976) presented two photographs

(Figures 364 and 366) taken by Sleptsov in the [North] Pacific and thought to be Baird's beaked whale(s). However the shape of the melon as well as the direction of concavity of the closed blowhole disagreed with those of *Berardius*, and were more likely to be of a whale of the genus *Hyperoodon*. Either the species or the location of the photographs is in question.

Some Japanese whalers in Abashiri (44°01′N, 144°15′E) on the Okhotsk Sea coast of Hokkaido identified a particular type of whale similar to ordinary Baird's beaked whales but darker or smaller and more difficult to approach within shooting distance. They were called by the fishermen *kuro-tsuchi* [black Baird's beaked whale] or *karasu* [crow or raven]. I remember that biologists of the Whales Research Institute in the 1960s, when I was a young staff member there, often talked about the unidentified species in the Okhotsk Sea with speculation that it might represent a species of Ziphiidae yet to be reported from the sea. This idea was documented by Nishiwaki (1965, in Japanese). He noted that whales recorded by Okhotsk Sea whalers as Baird's beaked whales and were pregnant at a body length below 7.6 m could possibly represent the northern bottlenose whale *Hyperoodon ampullatus*. Two years later, now referring to the vernacular name *kuro-tsuchi*, Nishiwaki (1967) stated, "Considered from the record of the body length of a pregnant female, which is less than 25 ft, and body shape as observed by whalers, the whale cannot be presumed to be Baird's beaked whale. ... the author ventures to identify it with Foster's species [i.e., northern bottlenose whale]." The source of data on the small pregnant female was not given in either of the documents mentioned earlier. Kasuya (1986) questioned the presence of *H. ampullatus* in the Okhotsk Sea after examining skulls and teeth of Ziphiidae species available at the time from the sea.

T. Yamada of the National Science Museum in Tokyo presented new information on the unidentified whales. Three stranded unidentified toothed whales were reported on the Hokkaido coast of the southern Okhotsk Sea and Nemuro Strait since June 2008. They were similar to each other and had a body shape similar to that of ordinary Baird's beaked whale including shape of the blowhole, but one of them (a male) was physically mature at a body length of 6.6 m. Schools of the same small whales have been sighted annually in the Nemuro Strait (44°N, 145°20′E). In addition to the small body size, these whales lacked the tooth marks that are abundant on ordinary Baird's beaked whales, but they had white scars of cookie-cutter shark bites scattered on the dark body. None of the ziphiid species known from the North Pacific agree with these specimens. The information given earlier was in Asahi Shimbun Press (2010, in Japanese) and in a personal communication from Yamada in 2010.

The unidentified whales agree with *Berardius* in the direction of the closed blowhole but grow only to 60%–70% of the body length of the ordinary Baird's beaked whale. Less intense tooth marks on the body of these whales suggests some difference in their social behavior. I am not aware of any studies on the density of bites of cookie-cutter sharks on Baird's beaked whales, but such scars are also present on ordinary Baird's beaked whales taken off Wadaura (35°03′E, 140°01′E) in Chiba Prefecture (34°55′N–35°44′N, 139°45′–140°53′), on the Pacific coast of central Japan (Figure 13.13). The difference in density of bites, if confirmed, suggests some difference in habitat. It should be noted that both the recently identified *Berardius*-like whales (which are smaller) and ordinary Baird's beaked whales (which are larger) inhabit the southern Okhotsk Sea in summer.

An ex-whaler S. Fukuoka (personal communication in February 2014) of Abashiri, who has experienced whaling for minke and Baird's beaked whales in the Okhotsk Sea, identified three types of Baird's-beaked-whale-like whales: *aka-tsuchi* [red Baird's beaked whale], *kuro-tsuchi* [black Baird's beaked whale], and *karasu* [crow or raven]. He described the first type as having numerous scars on the back and swollen dorsal muscles and as the largest of the three types and stated that this type was the same as those taken off the Pacific coast of Japan. His description agreed well with our understanding of Baird's beaked whales dealt with in the main body of this chapter. The second type, *kuro-tsuchi*, was slender, slightly smaller than the first type, and had black pigmentation without such heavy tooth marks as seen on the first type. The third type of Fukuoka, *karasu*, was similar to the second type in coloration but distinct in small body size, growing only to half the size of the first type, which reached 11–12 m. Fukuoka also stated that this third type tended more to approach shallower waters if chased by whalers in the Okhotsk Sea than the ordinary Baird's beaked whales, which preferred deeper waters.

The description of the third type, or *karasu*, agrees with the whales that were stranded on the southern Okhotsk Sea and examined by Yamada and probably with the *kuro-tsuchi* as described by Nishiwaki (1967), but its taxonomic position is still left to be determined. The second type of Fukuoka, *kuro-tsuchi*, has not been examined by scientists. Validity of these two type of animals remains to be confirmed. Unless stated otherwise, the following parts of this chapter deal only with ordinary Baird's beaked whales.*

13.2 DISTRIBUTION

13.2.1 Outline of the Distribution

Baird's beaked whale is endemic to the northern North Pacific (Figure 13.1). It was described based on a specimen obtained on Bering Island (55°N, 166°15′E) in the western Bering Sea. Analysis of mitochondrial DNA has confirmed the taxonomic distinctness of the species from *Berardius arnuxii* in the southern hemisphere (Dalebout et al. 2004). The ordinary southern limit of this species in the eastern North Pacific is off the northern part of the Baja California Peninsula (ca. 30°N), but a mass stranding was reported from La Paz (24°N) in the

* Kitamura et al. (2013, in Marine. Mammal Science 29: 755–766) analyzed mtDNA of the dwarf or black-type individuals (n = 3) and that of ordinary Baird's beaked whales (n = 62). They found no common haplotypes between three sequences of the former and seven of the latter and found three independent clades containing the dwarf-type and the two existing species of genus *Berardius*.

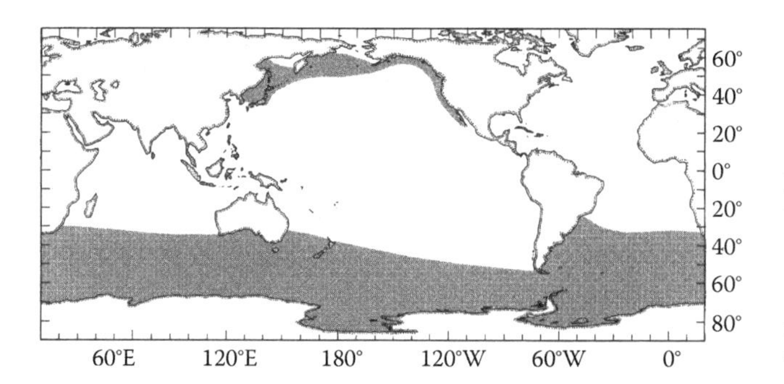

FIGURE 13.1 Ranges of Baird's beaked whale (North Pacific) and Anoux's beaked whale (Southern Hemisphere). (Reproduced from Kasuya, T.: Giant beaked whales. In Perrin, W.F., Würsig, B., and Thewissen, J.G.M. (eds.). *Encyclopedia of Marine Mammals.* Elsvier, Amsterdam, the Netherlands. pp. 498–500. 1316pp. 2009. Copyright Wiley-VCH Verlag GmbH & Co. KGaA. Reproduced with permission, Figure 2.

southern Gulf of California (Aurioles-Gamboa 1992). If the distribution of the species in the Gulf of California and in the Pacific is assumed to be continuous, which is without evidence, the southern limit of the species in the eastern North Pacific can be placed at around the tip of the Baja California Peninsula at about 23°N. The range of the species extends north along the west coast of North America to the Aleutian Islands and to Cape Navarin (62°N) in the Bering Sea (Reeves and Mitchell 1993).

The usual southern limit of this species on the Asian side of the Pacific is in Sagami Bay (34°45′–35°19′, 139°E–139°45′E), and the range extends north to the east coast of the Kamchatka Peninsula and the central Okhotsk Sea at latitude 57°N. The species is known from the entire Sea of Japan.

13.2.2 Regional Distribution and Population Structure

13.2.2.1 Chinese coast

Wang (1999, in Chinese) reported a Baird's beaked whale taken in the vicinity of the Zhoushan Islands (31°N, 122°E) in the East China Sea in the 1950s and collection of a specimen from the whale. The method of the take was not stated but it could have been captured by a whaling operation that existed in the area in those days. The species has not been reported from Taiwan. Cetacean-sighting cruises conducted by the Fisheries Agency of Japan in January–March and June–September did not record the species in the East China Sea (Kasuya and Miyashita 1997). The single record was likely a stray or from an unconfirmed population on the Chinese coast; this awaits further surveys in the Bohai and Yellow Seas. The shallow water depth, less than 50 m, in the western half of the Yellow Sea and the Bohai Sea and absence of sightings of the species in deeper waters off the Korean coast in the East China Sea suggest that the individual was a stray.

Wang (1999, in Chinese) recorded a Chinese whaling operation that lasted from 1953 to 1980. The operation in Daya Bay (23°16′N, 114°25′E) near Hong Kong started in 1953, expanded along the coast of Guangdong Province, and ended in 1970. The catch was mainly humpback whales but also included some gray and minke whales. Another whaling operation with hand harpoons started in 1953 on the Leizhou Peninsula (20°50′N, 110°05′E) and lasted for 2 years. Whaling in the northern Yellow Sea started in 1953 off Dalian (38°57′N, 121°38′E) and expanded to the southern Yellow Sea in 1963 using a newly introduced large catcher boat and five smaller vessels that had been in use since earlier years. The catch was of fin and minke whales, but it declined after 1965. Japanese whaling off the Goto Islands (33°N, 129°E) in the East China Sea started in 1955 for fin whales, and Korean whalers also took minke whales at the time. These operations could have influenced the decline of Chinese whaling at Dalian. China ceased all whaling operations in 1980 and joined the International Convention for Regulation of Whaling in the same year. Wang (1999, in Chinese) also noted that Taiwan started whaling in 1955 as a joint venture with Japanese whalers or by hiring one or more Japanese gunners, started shipboard flensing in 1979, and continued whaling until 1981. These whaling operations did not record take of Baird's beaked whales.

In the North Pacific, the take of bowhead, right, and gray whales has been prohibited by international convention since the pre–World War II period (Sections I.2.2 and 7.2.2), that of humpback and blue whales since 1966, and that of fin and sei whales since 1976 (Table 7.2). The only great whale species allowed to be taken in the North Pacific in September 1980 when China joined the convention were Bryde's and minke whales.

13.2.2.2 Sea of Japan

The Fisheries Agency of Japan carried out extensive cetacean surveys in the eastern part of the Sea of Japan in June to October and recorded sightings of Baird's beaked whales in June–August. They were sighted in almost the entire range of the surveyed area from west of Rebun Island (45°22′N, 141°02′E) to west of Oki Island (36°14′, 133°15′) (Kasuya and Miyashita 1997).

Japanese small-type whaling operated in the Sea of Japan during the post–World War II period and recorded takes of Baird's beaked whales. Omura et al. (1955) analyzed the catch statistics for the years 1948–1952. The main grounds for Baird's beaked whales were in Toyama Bay (36°45′N–37°15′N, 137°–137°30′) and off the west coast of the Oshima Peninsula (42°30′N–43°30′N) in south-western Hokkaido (Omura et al. 1955). These whaling grounds of past operations were part of the area where the species was recorded by recent surveys as reported by Kasuya and Miyashita (1997). Since minke whales were the main target of the small-type whaling of the time and Baird's beaked whales were a secondary target, it is uncertain if the observed season of Baird's beaked whales in the Sea of Japan, June to August with a peak in July, reflected a real seasonal change in their density. Nishiwaki and Oguro (1971) also analyzed similar records of small-type whaling in Japan for the 1965–1969 seasons, but hunting of Baird's beaked whales was not recorded in the Sea of Japan, presumably due to limited operations in the sea and economic reasons.

Presence of Baird's beaked whales in the eastern Sea of Japan in months other than June–August has been confirmed from strandings: on the coast of Kyoto Prefecture (35°30′N–35°46′N) in October, in Niigata Prefecture (36°59′N–38°33′N) in December and February, in Yamagata Prefecture (38°33′N–39°07′N) in October, and in Akita Prefecture (39°07′–40°20′) in February (Nishimura 1970; Ishikawa 1994, in Japanese). There is a record of the species in April in Peter the Great Bay on the west coast of the Sea of Japan (Tomilin 1967). Kasuya and Miyashita (1997) concluded from these records that Baird's beaked whales are present in the Sea of Japan year-round.

The whaling grounds for this species in the Sea of Japan, in Toyama Bay and off the coast of the Oshima Peninsula, had a common feature, the 1000 m isobath located close to the coast. On the Pacific coast of Japan this species at least in the summer season occurs on the continental slope between the 1000 and 3000 m isobaths (Kasuya 1986). It is likely that Baird's beaked whales in the Sea of Japan also prefer continental slope waters, and the fishing grounds were at places where deep submarine canyons approached the coast. If Baird's beaked whales in the Sea of Japan also prefer the continental slope, it is possible that there are two separate populations on the two sides of the sea. Although there is no evidence supporting such a population structure, the hypothesis could be a safe choice for management in the Sea of Japan.

Omura et al. (1955) compared the body length of Baird's beaked whales taken by small-type whaling in different grounds and reported that Sea of Japan whales had a smaller modal length than those off the coast of Chiba and Sanriku (37°54′N–41°35′N) on the Pacific coast: 30–32 ft (sexes combined) versus 34–35 ft (males) and 36 ft (females). A similar geographical difference was also observed in maximum body size, which would be more likely influenced by sample size. If we remember that the same whalers operated in those days in both the Sea of Japan and Pacific areas, the 3- to 4-ft difference in modal length can be assumed to be a real body-length difference.

Kishiro (2007) analyzed external measurements of Baird's beaked whales taken by small-type whaling in the Sea of Japan (21 taken during 1999–2004), Pacific area (47, 1992–2004), and Okhotsk Sea (34, 1988–2004). All the samples from the first two areas were measured by him. Although he did not mention it, his data showed a similar geographical body size difference as that noted by Omura et al. (1955). Both the mean body length (9.40 m) and maximum body length (10.15 m) in the Sea of Japan sample were smaller than the corresponding figures (mean 9.99 m, maximum 10.90 m) for the Pacific sample. It is very likely that Baird's beaked whales in the Sea of Japan are about 60–70 cm smaller than those off the Pacific coast of Japan.

Kishiro (2007) carried out a multivariate analysis of the external measurements of his sample, adjusting for body size difference, and concluded that the Sea of Japan sample taken off the Oshima Peninsula and the sample from the Pacific coasts of Chiba and Ibaraki Prefectures (35°45′N–36°50′N) represented different populations. The length and width of flippers contributed most to the observed morphological difference between the two samples. This analysis placed the Okhotsk Sea sample intermediate between the other two samples, but Kishiro refrained from conclusions about it because the measurements were collected by multiple scientists. The flipper lengths presented by him were 125.9 cm in the Pacific sample and 114.9 cm in the Sea of Japan sample (mean of both sexes with body length ≥8.4 m), which could have reflected the body size difference mentioned earlier. These measurements are 12.6% (Pacific) or 12.2% (Sea of Japan) of mean body length. Age-related change in the proportional size of flippers has not been compared between the two samples from the Pacific and the Sea of Japan.

This information supports the idea that the Sea of Japan is inhabited by a Baird's beaked whale population or populations different from those inhabiting the waters off Chiba and Ibaraki, near the southern range of the species in the western North Pacific. The identity of the Okhotsk Sea population is left for further investigation. Kasuya (1986) hypothesized that Baird's beaked whales in the Sea of Japan do not migrate through the four channels that connect the sea with the Okhotsk Sea, North Pacific, and East China Sea. The Sea of Japan has a maximum water depth of over 3000 m in the northern part and is connected with the outer seas by the four shallow channels. Two of them, the Tsugaru Strait (41°30′N, 140°30′E) opening to the Pacific and the Tsushima Strait (34°N, 129°E) opening to the East China Sea, are the deepest with maximum depths of 130 m. The remaining two—the Mamiya or Tatar Strait (50°N, 141°E) and the Soya or La Perouse Strait (45°46′N, 142°00′E), both opening to the Okhotsk Sea—are shallow, with maximum depths of about 20 and 30 m, respectively. We know that the species in the Sea of Japan does not usually, if at all, migrate to the East China Sea through the Tsushima Strait (see Section 13.2.2.1), but there is no evidence that denies movement of the species to the Okhotsk Sea or to the North Pacific. Further investigation is required, particularly on possible seasonal movement of the Sea of Japan individuals to the Pacific coast of Hokkaido through the Tsugaru Strait and to the southern Okhotsk Sea through the Soya Strait (Figure 13.2).

13.2.2.3 Western North Pacific and Bering Sea

Before examining the history of hunting Baird's beaked whales in this region, records of sighting surveys will be discussed. The Fisheries Agency of Japan in the early 1980s started a systematic sighting survey of commercially exploited cetaceans in the western North Pacific with the objective of estimating their abundances. Japanese waters known to be inhabited by Baird's beaked whales, a latitudinal range of 34°N–43°N or from the east coast of the Izu Peninsula (34°36′N–35°05′N, 138°45′E–139°10′E) to eastern Hokkaido, were surveyed in the months of July–November, while waters south of the range were surveyed through almost the whole year except for December and March (Figure 13.2). Using these data, Kasuya and Miyashita (1997) analyzed the distribution of Baird's beaked whales in the western North Pacific.

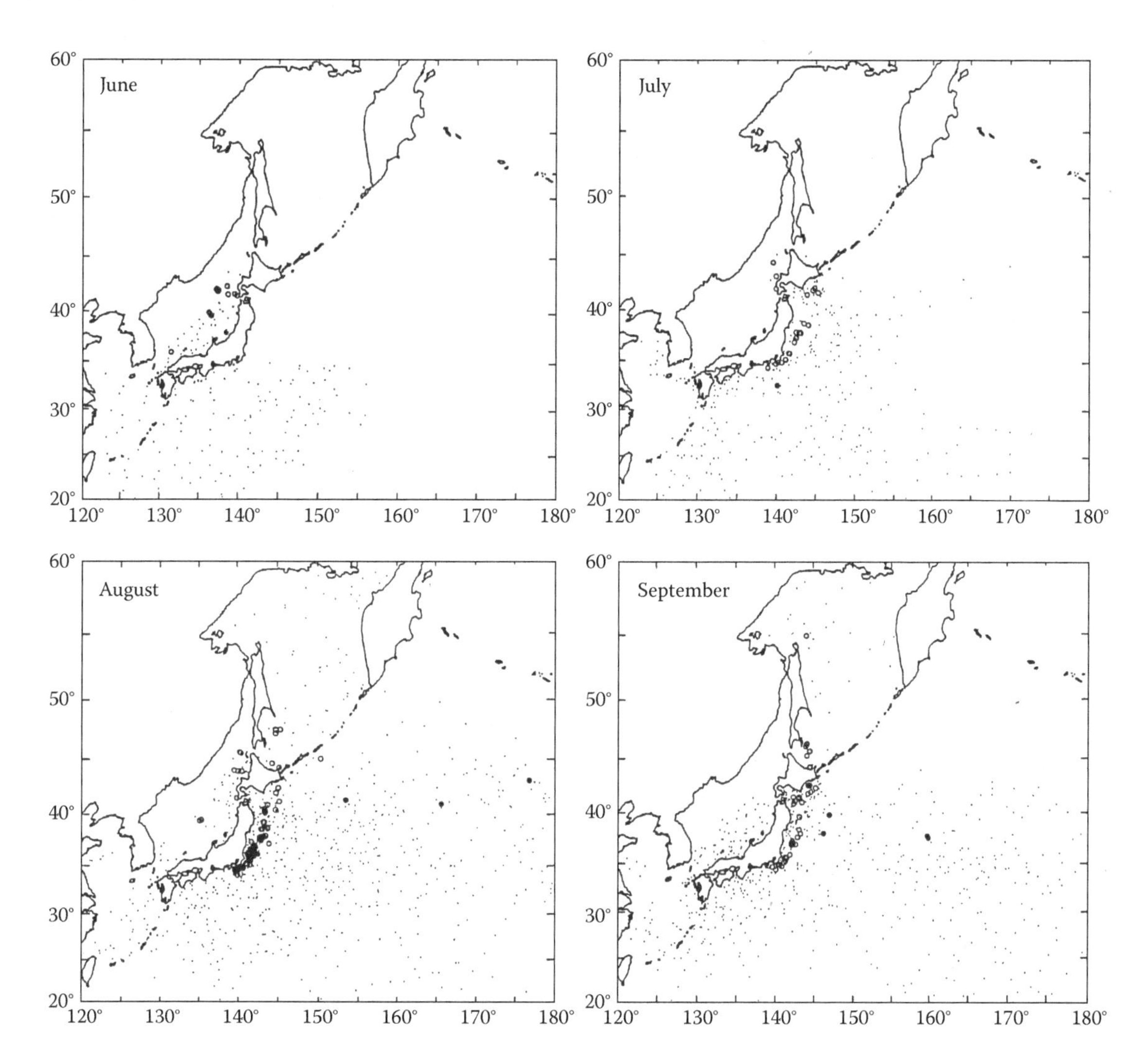

FIGURE 13.2 Positions of sightings of Baird's beaked whales (circles) in June–September, 1982–1994 during cetacean-sighting cruises of the Far Seas Fisheries Research Laboratory, Japan. Small dots are noon positions of survey vessels as an indication of distribution of sighting effort. Several offshore sightings (1 in July, 3 in August, and 3 in September) are likely to be species misidentifications (see text). Information for January–May and October–December is available in Kasuya and Miyashita (1997) but omitted here because of limited coverage of the survey area and few sightings. (Reproduced from Kasuya, T. and Miyashita, T., *Rep. Int. Whal. Commn.*, 47, 963, 1997, Figures 5 through 8)

With the exception of one record of this species in July in Suruga Bay off the west coast of the Izu Peninsula, which can be dealt as an outlier, all the other records occurred east of the peninsula and north of 34°N. This is the usual range of this species in the western North Pacific.

Japanese cetacean sighting cruises before the cruises analyzed by Kasuya and Miyashita (1997) often had only whalers on board and no scientists, and there remain uncertainties on the species identification of small cetaceans. Records of Baird's beaked whales are not the exception. There have been sightings reported in the North Pacific of an unidentified beaked whale apparently similar to the Baird's beaked whale, now considered to represent Longman's beaked whale, *Mesoplodon pacificus* (Pitman et al. 1999; Dalebout et al. 2003). Trouble with identification could relate to not only this species but other species as well. Careful observation of the pigmentation of the body, direction of the blowhole, and body size can distinguish Longman's beaked whales from Baird's beaked whales.

The *asobi-kujira* described by Yamase (1760) from the present Wakayama Prefecture (33°26′N–34°19′N, 135°00′E–135°59′E) could also be a ziphiid species other than Baird's beaked whale, because the area is outside the current usual range of that species. However, the presence of teeth in both jaws disagrees with any of the Ziphiidae except the Southern Hemisphere species *Tasmacetus townsendi*. Ishikawa (1994, in Japanese) reported a Baird's beaked whale from Tosa Bay (33°15′N, 133°30′E) or south of the usual range of the species. He stated that the species was identified based on skull, mandible, and mandibular teeth. Before this record can be evaluated, more information on the specific features used in the identification is needed, together with how the specimen was obtained, whether by stranding or hunting. Kasuya and Miyashita (1997) analyzed sightings of Baird's beaked whales during 1982–1994 and reserved confirmation of the species identification in eight sightings recorded by onboard scientists as Baird's beaked whales (Figure 13.2). These were sighted outside the usual range of the species or in offshore waters in 35°N–45°N and east of 145°E under unfavorable weather conditions or by inexperienced observers.

Analyses of earlier sighting records made by the crews of scouting boats attached to whaling fleets have concluded that Baird's beaked whales are distributed continuously from the Japanese coast to the coast of the United States (e.g., Ohsumi 1983; Kasuya and Ohsumi 1984), but such records are now believed to contain misidentifications of species and the conclusion has been rejected. These scouting boats could have correctly recorded species of large cetaceans hunted by them, but the observers probably did not pay sufficient attention to identification of other smaller species.

After rejecting the dubious records off the Pacific coast of Japan, there remain sightings of Baird's beaked whales that ranged from 34°N near the southern entrance of Sagami Bay on the east coast of the Izu Peninsula to the southern coast of Hokkaido at 43°N (Figure 13.2). In addition there is one record of a sighting off the southern Kuril Islands. Off the coast of Russia, the former USSR's whaling operated from land stations at the middle (47°N) and northern (52°N) Kuril Islands and recorded take of 109 Baird's beaked whales in May–November (Reeves and Mitchell 1993). Sleptzov (1955, in Japanese) reported records of Baird's beaked whales from both sides of the central Kuril Islands continuing to Pacific waters off the east coast of the middle of the Kamchatka Peninsula. The species has also been reported in the southeastern Okhotsk Sea around the Shiretoko Peninsula (43°40N′–44°19N′, 144°50′E–145°20′E) including the Nemuro Strait in August–October (Figures 13.2 and 13.4), but it is known to occur also in early spring in the Nemuro Strait (see Section 13.2.2.4). These records indicate distribution of the species at least in the summer off the east and west coasts of the entire Kuril Islands chain. The species probably passes through the Kuril Islands, which have several passes deeper than 1000 m.

Tomilin (1967) reported sightings of numerous Baird's beaked whales made during whaling operations in 1934 (Table 13.1). As the USSR's whaling of the time targeted fin, humpback, and sperm whales (Sleptzov 1955, in Japanese), the absence of sighting records of Baird's beaked whales does not necessarily mean absence of the species in the region. However, the recorded sightings will be used for present purposes as far as species identification is reliable. Tomilin (1967) indicated that the northernmost record was around Cape Navarin at 62°16′N, 179°06′E and that local marine mammal hunters in the Bering Strait area did not know about the species. He concluded that the northern limit of the species in the Bering Sea is on the continental shelf edge extending from Cape Navarin toward the east or southeast. Baird's beaked whales have been recorded from St. Matthew Island and the Pribilof Islands on the edge of the continental shelf (Rice 1998). This does not disagree with the opinion of Tomilin (1967). Thus, the northern limit of the distribution of this species in the Bering Sea is likely on a line connecting Cape Navarin, St. Matthew Island, the Pribilof Islands, and the tip of the Alaska Peninsula.

TABLE 13.1
Baird's Beaked Whales Recorded by Soviet Whalers Operating in the Western Bering Sea and off the East Coast of the Kamchatka Peninsula

Region and Approximate Latitude	May	Jun.	Jul.	Aug.	Sep.	Oct.	Nov.
Cape Navarin (62°N)				1			
Olyutorskii Bay (60°N)					65		
Commander Island (55°N)			85				
Kronotskii and Avacha Bays (52°N–55°N)	188	30	40			160	(+)

Source: Prepared from text of Tomilin, A.G., *Mammals of the USSR and Adjacent Countries, Cetacea*, Israel Program for Scientific Translations, Jerusalem, Israel. 717pp, 1967.

13.2.2.4 Okhotsk Sea

A triangular oceanic basin (Kuril Basin) is situated in the southeastern Okhotsk Sea. The eastern margin of the basin extends from around 50°N, 163°E off the northern Kuril Islands, along the Kuril Islands, to the southern Kuril Islands at around 44º30′N, 145º40′E. The southwestern border of the basin is near southern Sakhalin Island at around 47°N, 144ºE. Shallower water extends to the north of the Kuril Basin, with the depth decreasing from about 2000 m at the northern edge of the basin to 1000 m at 55°N and less further north. In winter most of the Okhotsk Sea is covered with ice, leaving a small open area in the southeastern part.

The Fisheries Agency of Japan operated systematic shipboard surveys for cetaceans in the Okhotsk Sea (Figure 13.2). The surveys covered most of the area outside the territorial waters of Russia (12 nautical miles) and months from May to October with major effort in August and September. Some sightings of Baird's beaked whales were made in each month (Kasuya and Miyashita 1997). They occurred mostly in the southern Okhotsk Sea, ranging from 48°N off southern Sakhalin Island to the Shiretoko Peninsula and the northern coast of Hokkaido (both at 44°N) (Figure 13.2). Another two sightings were in August near the northern coast of Sakhalin Island at 55°N and off the Pacific coast of the southern Kuril Islands, on the continental slope near the 1000 m isobath.

Fedoseev (1985) reported to the Scientific Committee of the IWC on the presence of Baird's beaked whales in the northern Okhotsk Sea, based on data obtained during aerial sighting surveys for seals. The surveys were made in April–May of 1979 and 1981 along track lines set at 20–30 km intervals and recorded a total of 70 Baird's beaked whales in waters between the northern tip of Sakhalin Island and south of Okhotsk City, at 54°N–59°N, where concentrations of bowhead whales and belugas were also observed. In addition to these sightings, Fedoseev recorded one sighting of Baird's beaked whales in the central Okhotsk Sea and two in waters west of the southern Kuril Islands. The whales were in dense schools of 7–11 traveling at high speed through open water in the sea-ice field and submerging synchronously.

This behavior pattern was apparently similar to that of Baird's beaked whales in the southern range of the species off the Pacific coast of Japan in the Chiba and Ibaraki areas, but their body length was not estimated and there remains a possibility that some of these observation, particularly those in northern waters, might represent the small "Baird's-beaked-whale-like" whale known from the southern Okhotsk Sea (Section 13.1).

Water depths at the sighting of Baird's beaked whales were estimated on a map to be in a range between 200 and 1000 m (Fedoseev 1985). Fedoseev (1985) noted three Baird's beaked whales sighted in the same area in December 1983, when the sea was not covered by ice, and concluded that the whales observed by him were not migrants proceeding toward the north through melting sea ice but year-around residents of the Okhotsk Sea.

The presence of Baird's beaked whales (or "Baird's-beaked-whale-like" whales) in the northern Okhotsk Sea throughout the year suggests that they represent a population to be distinguished from those in the Sea of Japan and in waters off the Pacific coasts of central Japan. We know that Baird's beaked whales are also present along the Hokkaido coast of the Okhotsk Sea and the Pacific in the summer through autumn and presumably in other seasons too (see Section 13.2.3) and that distribution of the species in the northern and southern Okhotsk Sea is possibly discontinuous. Based on these observations, Kasuya and Miyashita (1997) proposed a hypothesis of two Baird's beaked whale populations in the Okhotsk Sea, that is, a southern Okhotsk Sea population and a northern Okhotsk Sea population. Weighing of this hypothesis against the presence of "Baird's-beaked-whale-like whales" has not been carried out at the time of this writing. Baird's beaked whales taken from the southern concentration were analyzed by Kishiro (2007) and found to disagree, although inconclusively, with the same species inhabiting the Chiba-Ibaraki area in the Pacific and Sea of Japan (Section 13.2.2.2). The identity of the population off the Pacific coast of Hokkaido and hunted off Kushiro (Table 13.14) also needs to be analyzed in comparison with those in nearby waters, that is, the Sea of Japan, the southern Okhotsk Sea, and the Chiba-Ibaraki area, for population identity.

Baird's beaked whales are known to occur on the Okhotsk Sea side of the Nemuro Strait at around 44°N, 145°20′E, which connects the Okhotsk Sea and the Pacific in March–April in the presence of ice floes (Ishii 1987, in Japanese; personal communication of H. Sato in 1998) through to October (Photo in the frontispiece; Walker et al. 2002; Fukuoka 2014, in Japanese). Walker et al. (2002) examined 15 individuals taken in August and September off Rausu in the Nemuro Strait at 44°N–44°20′N for their stomach contents analyses, which indicated that these individuals were probably feeding at a water depth of about 200 m (Table 13.3). It should be noted that the species occurs there in early spring when sea ice can still remain. Although it still remains to be confirmed whether Baird's beaked whales pass through the Nemuro Strait (about 20 m deep) to travel between the Pacific and the Okhotsk Sea, there are sufficient sightings that suggest communication of the species between the two areas through the southern Kuril Islands near the Nemuro Strait.

We know that a population of Dall's porpoises winters in the Sea of Japan and summers in the southern Okhotsk Sea. One of their two northward migration routes is a short cut through the Soya Strait, and the other is a distant route to enter the Pacific through the Tsugaru Strait and then into the Okhotsk Sea through the southern Kuril Islands. However, another Dall's porpoise population that winters off the Sanriku Region does not use the Tsugaru/Soya route but migrates through the Kuril Islands to reach a summering ground in the central Okhotsk Sea. This type of asymmetry of migration route between populations might be a reflection of geological history and might also be expected for Baird's beaked whales off Japan.

13.2.3 Baird's Beaked Whales along the Pacific Coast of Japan

13.2.3.1 Distribution

Almost nothing is known about the winter range of Baird's beaked whales in the western North Pacific. The usual summer range on the Pacific coast of Japan is from east of the Izu Peninsula at 139°E to north of Miyake Island at 34°N. Figures 13.2 and 13.3 represent the summer distribution in this range. Sighting cruises presented in Figure 13.3 covered the months of June to September 1984 and were conducted with the purpose of investigating the distribution of sperm whales and Baird's beaked whales. I was on board the cruises from July 7 to August 6 to cover the area north of 34°N and west of

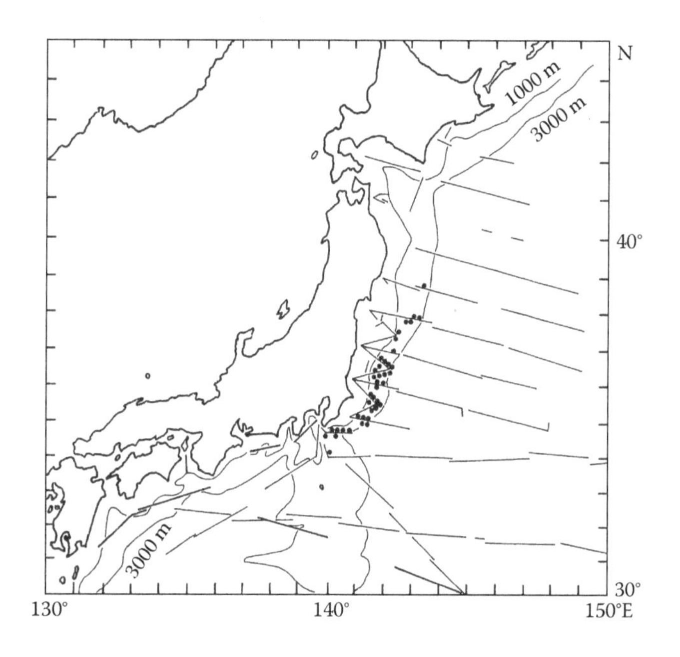

FIGURE 13.3 Survey track line (solid line) and sighting positions of Baird's beaked whales (closed circles) during the cruise of *Toshi-maru No. 25*, which surveyed waters north of 34°N and west of 150°E during July 7–August 6, 1984, and surrounding waters afterward. Isobaths of 1000 and 3000 m are indicated. (Reproduced from Kasuya, T., *Sci. Rep. Whales Res. Inst.*, 37, 61, 1986, Figure 2.)

150°E indicated in Figure 13.3. During this period there were numerous sighting of Baird's beaked whales but none of sperm whales. The track lines were set perpendicular to the coastline. There were also some additional saw-tooth tracks because of the statement by Nishiwaki and Oguro (1971), who analyzed whaling records, that Baird's beaked whales seemed to inhabit waters outside the 1000 m isobath. My cruise confirmed this and obtained the additional information that the distribution was limited to waters shallower than 3000 m. It was very impressive to find the species every time the vessel entered this depth range. The furthest eastern (not shown in Figure 13.3) and southern track lines were surveyed by other scientists and resulted in only one sighting of Baird's beaked whales (close to 34°N as shown in Figure 13.3) and a limited number of sperm whales. These cruises made it evident that Baird's beaked whales in summer inhabit very limited coastal waters off the Pacific coast of Japan but not the more offshore western North Pacific.

Baird's beaked whales taken off the Pacific coast of Chiba Prefecture and processed at a land station at Wadaura were known to feed on deep-water bottom fish (Section 13.3.1). Sometimes we found stones in their stomachs, which were often larger than a child's fist and could not have been acquired via a fish stomach. Additionally we found numerous scratches on the anterior portion of the melon caused by contact with some hard substance, perhaps the sea floor. These data suggest that Baird's beaked whales off the coast of Chiba and nearby Ibaraki Prefectures, at latitudes 34°N–37°N, feed on deep-sea fish on the sea floor. The continental slope area at depths of 1000–3000 m is probably inhabited by the fish species preferred by Baird's beaked whales (Section 13.3).

The continental slope with depth of 1000–3000 m is not limited to north of 34°N off the Pacific coast of Japan. It extends westward along the southern coast of Japan to reach the east coast of Taiwan through the Japanese Nansei Islands [Southwestern Islands]. It also extends southward along the submarine ridge to the Bonin Islands. However, we do not have confirmed records of Baird's beaked whales in these waters, and some factors other than water depth are expected to be working in the distribution of Baird's beaked whales in this case. The sea surface temperature, 23°C–29°C at the position of sighting the species in July–August, does not seem to be the controlling factor, but deep-sea temperature is a possibility. The southern limit of Baird's beaked whales in summer is correlated with water temperature of 15°C at a depth of 100 m. This water temperature is used by oceanographers as an indicator of the front of the warm Kuroshio Current (Section 12.3.6), and the position does not exhibit much seasonal fluctuation. The mid- or deep-water environment seems to regulate distribution of Baird's beaked whale or its prey species (Kasuya 1986).

13.2.3.2 Seasonal Movements and Population Structure

The southern coast of Hokkaido (42°N–43°30′N) is probably the place where Baird's beaked whales in May make their first appearance within the overall range of the species along the Japanese Pacific coast (Kasuya and Miyashita 1997). No survey data exist for June in the area. In July they were found along the entire extent of the coastline from Hokkaido (43°N) in the north to the Izu area (north of 34°N and east of 139°E) in the south, and this distribution remained the same through August–September (Figure 13.2). Limited sighting effort in October recorded them in a limited area off the coasts of Aomori (40°27′N–41°30′N) and Hokkaido (41°30′N–45°30′N) Prefectures and suggested possible disappearance from the area off the Chiba coast (south of 36°N). In November, our survey was limited to only south of 38°N and resulted in no sightings of the species. These observations allow the following deduction.

Baird's beaked whales first arrive off the Pacific coast of Hokkaido in May and extend their range south in summer to the Sagami Bay area east of the Izu Peninsula, which is the southwestern limit of the usual range of the species. After the summer, they disappear from the southwestern habitat, remaining only off the Hokkaido coast in October, and totally disappear from the Pacific coast of Japan in November. If this deduced pattern is accepted, questions remain to be answered: (1) what is their wintering ground and (2) what is the actual movement of individual whales (if they move north/south or vary in timing of arrival/departure from/to the unknown wintering ground)?

Matsuura (1942, in Japanese) analyzed the statistics of whaling operations in 1932–1942 in Chiba Prefecture and stated that Baird's beaked whales were actively hunted only by Chiba whalers and that whalers in the Ayukawa (38°18′N, 141°31′E) region did not hunt the species, although it was abundant there. A traditional culinary custom of the people in the southern part of Chiba Prefecture supported the Baird's beaked whale fishery. The local people since early times have consumed jerky made from beaked whale meat, red meat sliced, soaked in salt water and then dried. The whaling was operated without a catch quota or seasonal regulations in those days. The statistics analyzed by Matsuura (1942, in Japanese) covered the time when the government started an attempt to increase food production for the war with China that started expanding in 1937 (importation of whale meat from Antarctic was first permitted in the 1937/1938 season). The 8 years of statistics he analyzed showed that Baird's beaked whaling started in June and ended in September (1942) or October (1935–1941), which agreed with the results of the sighting surveys summarized earlier. The catch had a seasonal peak in July (1935, 1937–1939, 1941) or August (1936, 1940, 1942), and total catch over these years had a peak in July. Thus, the 10 years' statistics indicated that the peak of abundance of Baird's beaked whales off the Chiba area, near the southern limit of the range, was in July and perhaps also in August.

Omura et al. (1955) analyzed whaling statistics for the 1948–1952 seasons, when demand for whale meat was strong after World War II and Baird's beaked whales were hunted also off the Hokkaido and Sanriku coasts (Figures 13.4 and 13.5). The catch of Baird's beaked whales peaked off Chiba Prefecture in July as observed by Matsuura (1942, in Japanese), but the season started in May and ended in November. It must be questioned whether the lengthening fishing season can be explained

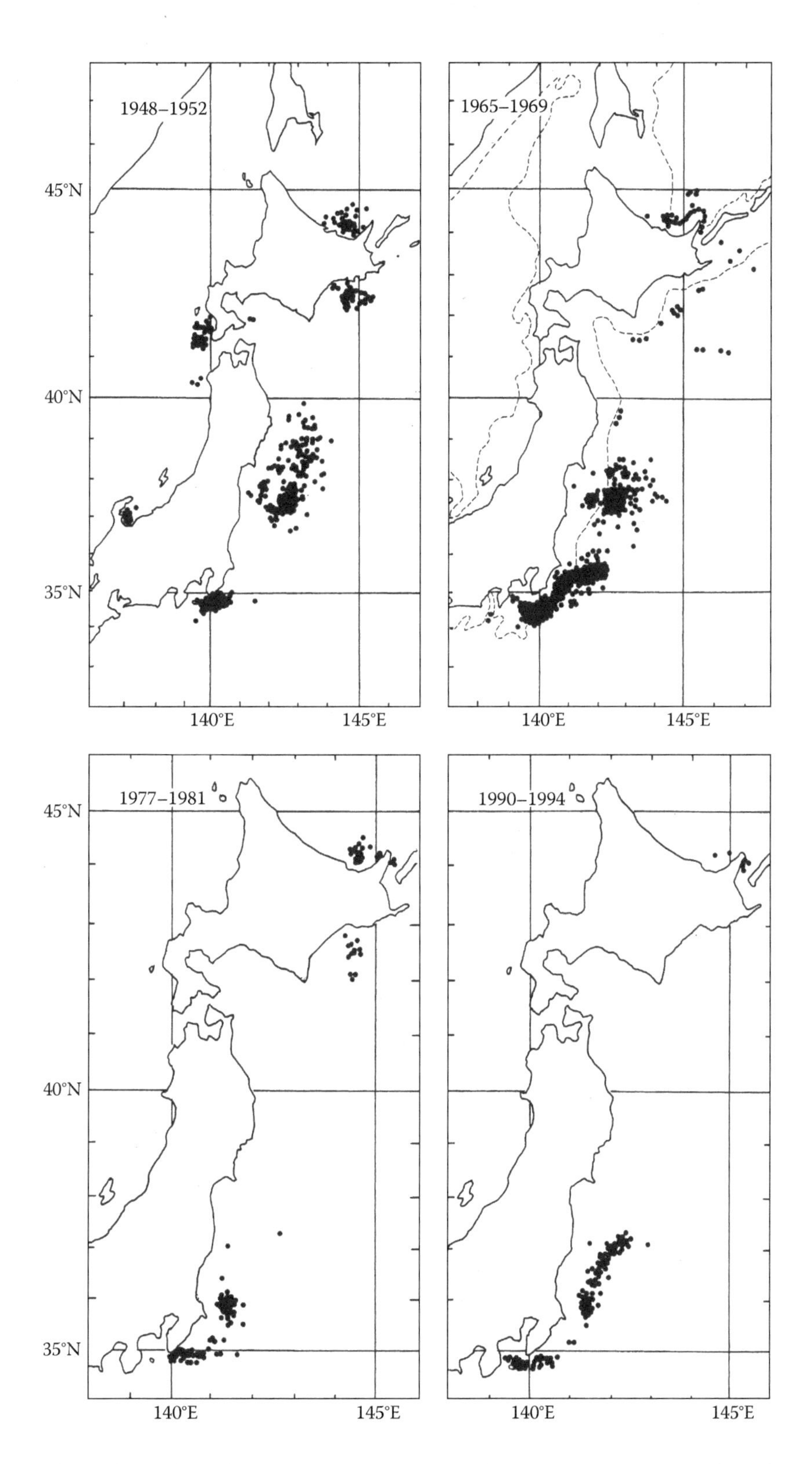

FIGURE 13.4 Catch positions of Baird's beaked whales reported by small-type whalers. (Redrawn from Omura, H. et al., *Sci. Rep. Whales Res. Inst.*, 10, 89, 1955, Figure 1 for 1948–1952; Nishiwaki, M. and Oguro, N., *Sci. Rep. Whales Res. Inst.*, 23, 111, 1971, Figure 2 for 1965–1969; Ohsumi, S., *Rep. Int. Whal. Commn.*, 33, 633, 1983, Figure 1 for 1977–1981; Kasuya, unpublished, from 1990 to 1994).

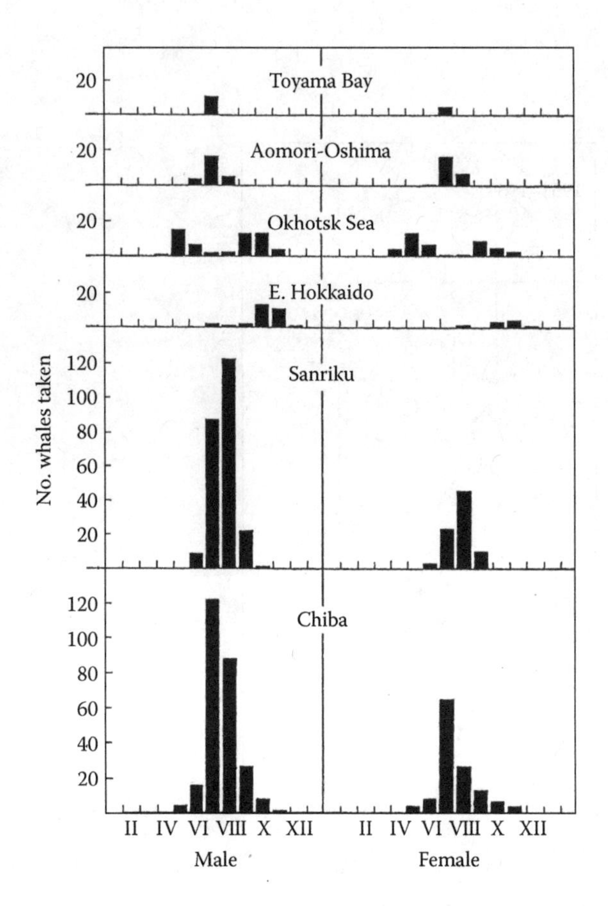

FIGURE 13.5 Seasons for hunting Baird's beaked whales during 1948–1952. From top to bottom: Toyama Bay, the Sea of Japan coast centered at 37°00′N, 137°15′E; Western Aomori-Oshima, Sea of Japan coast in latitudes 40°N–42°N; the Okhotsk Sea, in latitudes 44°N–45°N; eastern Hokkaido, Pacific coast in latitudes 42°N–43°N; Sanriku, Pacific coast in latitudes 36°30′N–40°N; Chiba, Pacific coast in latitudes 34°N–35°N. (Reproduced from Omura, H. et al., *Sci. Rep. Whales Res. Inst.*, 10, 89, 1955, Figure 2.)

solely by high demand for the products. The whaling operation for the species off the Sanriku Region, where Ayukawa is located, started in June, peaked in August, and closed in September. The tendency of narrowing of the fishing season was more distinct off the Pacific coast of Hokkaido, where the season started in July and ended in November (peak unclear). This seasonality will be discussed later.

Selectivity between whale species must be considered when interpreting fishing statistics. On the Chiba ground minke whales were uncommon and the whalers operated only with Baird's beaked whales as an authorized and attractive species to catch. On the Sanriku and Hokkaido grounds, the season for minke whales started in April. Because minke whales were easier to capture and sold at a higher price, whalers could have preferred that species over Baird's beaked whales. The minke whale ground moved northward with the season, and the main ground was in the Okhotsk Sea and off the Pacific coast of Hokkaido in August. Thus, the presence of minke whales could have biased Baird's beaked whale abundance in the catch statistics for the Sanriku and Hokkaido grounds.

The Japanese government set minimum size limits in 1938 for sperm whales and baleen whales other than minke whales, which were allowed to be taken in large-type (coastal) whaling (Omura et al. 1942, in Japanese). The whalers responded to this regulation by hiding some undersized whales by not reporting them or stretching their body length (Kasuya 1999). A catch quota was first set for coastal whaling in 1959 for sperm whales and in 1970 for baleen whales (Tato 1985, in Japanese). These regulations for large-type whaling and the resultant underreporting (Section 15.3.2) probably had a minimal effect on the statistics for the Baird's beaked whale fishery in the period 1948–1952, but it is generally believed that small-type whalers also poached sperm whales and reported some of them as Baird's beaked whales (Section 13.7.2). This must be considered in analyzing the catch statistics. Sperm whales were easier to hunt than Baird's beaked whales.

Anecdotes were documented of poaching of sperm whales and other large whales by small-type whalers so named in 1947 (Section 5.4) during World War II, and evidence was published on such activities by Japanese coastal whaling in the 1950s to 1970s (Sections 7.1.1, 13.5.3, 13.7.2 and 15.3.2.3). If the statistics for Baird's beaked whales contain poached and misallocated sperm whales, then the apparent extension of the season of Baird's beaked whales could be expected while the peak season remained unchanged. This feature is apparent in Table 13.2, where the catch of Baird's beaked whales (including misallocated sperm whales) occurred from May to December in the 1956–1972 seasons, while the season was shorter when poaching of sperm whales did not occur or was insignificant (1947–1955, 1973–1982).

Kasuya (1986) examined aerial sighting data on Baird's beaked whales recorded by the Fishery Aviation Co. Ltd. and

TABLE 13.2
Official Statistics for Baird's Beaked Whales Processed at Land Stations in Chiba Prefecture on the Pacific Coast in Latitudes 34°55′N–35°15′N

Seasons	Jan.	Feb.	Mar.	Apr.	May	Jun.	Jul.	Aug.	Sep.	Oct.	Nov.	Dec.
1947–1955	0	0	0	0	8	44	317	191	71	20	9	1
1956–1972	1	0	0	0	23	149	340	326	164	97	71	18
1973–1982	0	0	0	0	0	0	117	105	32	20	10	2

Source: From Kasuya, T. and Ohsumi, S., *Rep. Int. Whal. Commn.*, 34, 587, 1984.

the time budget of small-type whaling operations to investigate seasonal density change off Japan. The aerial survey was independent of the whaling operation, but the latter depended heavily on it as explained as follows. The time budget covering the period 1977–1982 recorded whales taken, whales sighted, and time spent searching, chasing, and towing separately. The main target of operations other than at Chiba was minke whales but in the Chiba area was Baird's beaked whales. These records were presented by whalers for discussion at the Scientific Committee Meeting of the IWC in the 1980s and used in Kasuya and Ohsumi (1984). Because the accuracy was not confirmed and there was a possibility of some manipulation of the records by the whalers to produce an optimistic annual abundance trend, its use for management was discontinued. The Committee of Three Scientists established in 1963 in the Scientific Committee of the IWC used the number of whales taken per number of operating days of whale catcher boats (CDW) as an index of abundance of whales, but the Scientific Committee later used searching hours instead of CDW as a measure of effort. In order to respond to this change, the whalers were requested to record the time budget of their daily operations. Such fishery-dependent data have been almost abandoned for management purpose in recent years.

The open circles and dotted line in Figure 13.6 represent the abundance of Baird's beaked whales expressed as the number of whales sighted per 100 h of sighting effort during the whaling operation (the time budget data for 1977–1982). The closed circles and solid line represent take of the species by small-type whaling during the post–World War II period (1948–1952) in Omura et al. (1955). The two types of abundance parameters show a similar seasonal trend, declining from July/August to October, in both the Chiba and Sanriku areas. However, the trend off the Hokkaido coast (Pacific and southern Okhotsk Sea) was different from that in the southern regions, that is, decreasing at a rate slower than in the south (catch series) or increasing toward the autumn (sighting rate). I note similarity of the trend between the two sets of density data, which represented periods that were about 25 years apart. The seasonal pattern of Baird's beaked whale migration to the Japanese coast seems to be different between northern and southern parts of the island chain.

Another set of data analyzed by Kasuya (1986) were collected by the Fishery Aviation Co. Ltd. in Tokyo during flights chartered for various fishery-related surveys along the Pacific coast of Japan which covered the period 1959–1983, most of which were unrelated to whaling (Figure 13.7). The offshore range of the flights was about 180–200 nautical miles (330–370 km), which fully covered the summer range of Baird's beaked whales and most of the months of the year with greatest effort, May through November. For the purpose of the analyses the sightings data from off the Pacific coast of Japan between 140°E and 150°E were stratified

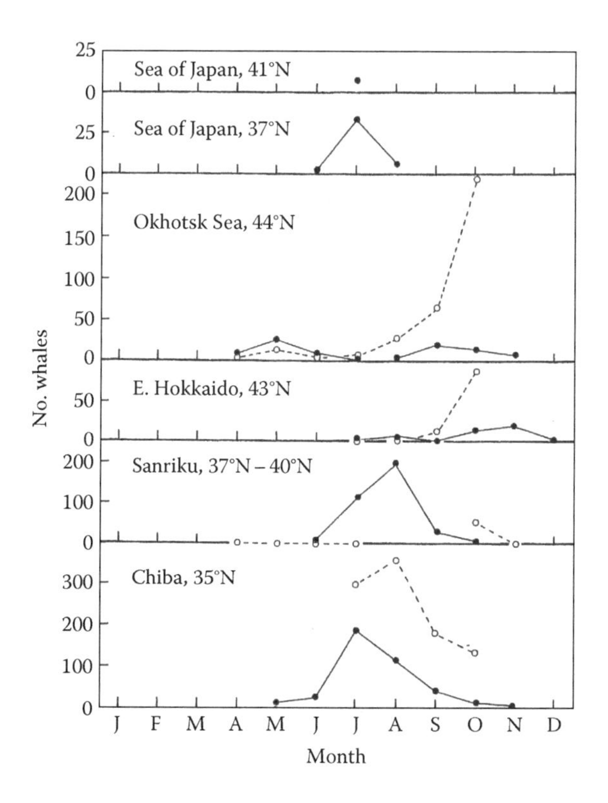

FIGURE 13.6 Number of Baird's beaked whales caught in 1948–1952 (closed circles and solid line, in Omura et al. (1955) and those sighted per 100 h of operation by Japanese small-type whaling in 1977–1982 (open circles and dotted line, in Kasuya and Ohsumi 1984). (Reproduced from Kasuya, T., *Sci. Rep. Whales Res. Inst.*, 37, 61, 1986, Figure 7.)

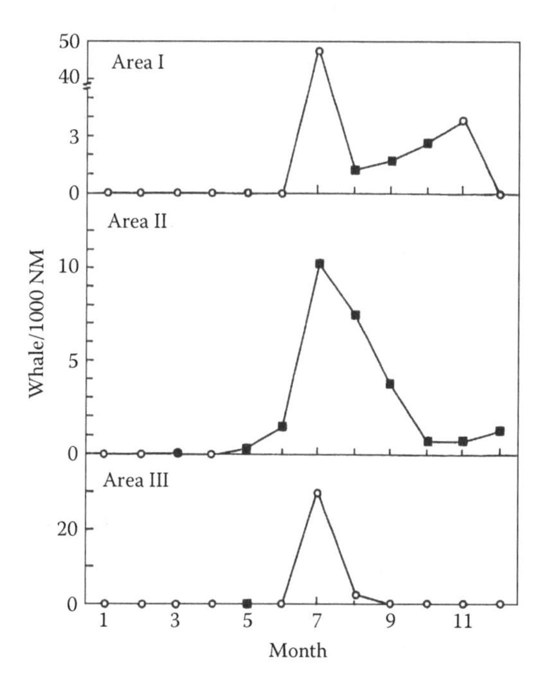

FIGURE 13.7 Sighting rate of Baird's beaked whales off the Pacific coast of Japan (no. of whales/100 nautical miles flown) by the Fisheries Aviation Co. Ltd., Tokyo, during April 1959–December 1983. Open circles represent results obtained from small sighting effort. I, 40°N–45°N west of 150°E; II, 35°N–40°N west of 150°E; III, 25°N–30°N and 140°E–150°E. (Reproduced from Kasuya, T., *Sci. Rep. Whales Res. Inst.*, 37, 61, 1986, Figure 5.)

into three latitudinal strata of 30°N–35°N, 35°N–40°N, and 40°N–45°N. The density of Baird's beaked whales, which was expressed as the number of individuals sighted per 100 nautical miles, had a peak in July in each of the three strata, although the peak off Hokkaido (Stratum I: 40°N–45°N) was based for 40 individuals in two schools sighted during a short flight (1550 km). More interesting was the trend after the peak month. In Stratum II (Chiba–Iwate: 35°N–40°N) the density steadily declined from a peak in July (10 whales/100 nautical miles) to 10% of that in October/November, which was based on sufficiently large monthly flight coverage ranging between 20,088 (August) and 32,973 (November) nautical miles. This was quite different, in high density and seasonal trend, from a feature seen in the northern Stratum I (Aomori–Hokkaido: 40°N–45°N) obtained from flight coverage of similar magnitude of 11,628 (August) to 37,876 (September) nautical miles, where the density trend slightly increased from only 1 whale/100 miles in August to a still low figure of 3 (October) or 4 (November).

Omura et al. (1955) concluded from monthly catches that Baird's beaked whales first arrived off Chiba in May/June, moved northward to arrive off the Sanriku Region in 1 month, and moved further off the coast of Hokkaido in October/November, as might be expected for Northern Hemisphere cetaceans. Kasuya (1986) also assumed from the continuity of distribution that all the Bird's beaked whales found off the Pacific coast of Japan from Chiba (34°N) to eastern Hokkaido (44°N) belong to a single population. The observed seasonality in distribution may be explained by these hypotheses but might fit other hypotheses as well.

Kasuya (1986) and Kasuya and Miyashita (1997) left two questions unresolved on the seasonal movement of Baird's beaked whales: (1) whether the apparent latitudinal shift of month of peak density from south to north along the Pacific coast of Japan is due to northward movement of individual whales, and (2) whether the whales off the Pacific coast of Hokkaido represent the same population as those off Chiba and the Sanriku area. Kishiro (2007) used external measurements to distinguish between the whales in the Sea of Japan and off the Chiba/Sanriku Region on the Pacific coast, but his sample did not include whales from further north, that is, off the Pacific coast of Hokkaido (Section 13.2.1). Baird's beaked whales off the Pacific coast of Hokkaido must be compared with the same species from the Sea of Japan and Okhotsk Sea as well as those from the Chiba/Sanriku Region in order to resolve the question of stock identity in the species off Japan. Future genetic studies and investigation of movements using radio transmitters hold promise. Walker et al. (2002) related a possible north-bound migration off Chiba/Sanriku Region to a spawning migration of *Laemonema longipes* (Moridae) (Section 13.3.1).

The wintering ground is another problem remaining unresolved for Baird's beaked whales off the Pacific coast of Japan, which are known to occur over the continental slope area during summer to autumn. Shipboard sighting surveys covered the area from the coast to longitude of 155°E in October and to 145°E in November but ended without sighting the species. Subramanian et al. (1988) used concentration levels of PCB and DDE in cetacean blubber to approach this problem. DDE is a degradation product of DDT. The same method was able to allocate Dall's porpoises found off the Pacific coast of Hokkaido in summer to a population that winters in the Sea of Japan (Section 9.3.4). The pollutants in the environment are accumulated through the food chain in cetaceans. The level in a cetacean reflects the pollution level in the feeding environment, food habits, age, and physiology of the whale. Transfer of these pollutants to the calf through the placenta and milk can cause large individual variation of pollutant levels in females. Therefore, Subramanian et al. (1988) selected only males for analysis of PCB/DDE ratio in Baird's beaked whales. They found that the PCB/DDE ratios of male whales taken off Chiba (i.e., southern part of the range of the species off the Pacific coast of Japan) were in a range of 0.18–0.41 (mean 0.24, n = 13), which was close to the corresponding value of male *dalli*-type Dall's porpoises, 0.37–0.41 (mean 0.39, n = 3), taken off the Pacific coast of Hokkaido (i.e., northern range of Baird's beaked whales off the Pacific coast of Japan), and concluded that Baird's beaked whales in these waters inhabit waters off the Pacific coast of Japan in all seasons.

The technique used by Subramanian et al. (1988) was new and probably promising but still retained some uncertainties in the analysis. It was known that their sample of *dalli*-type Dall's porpoises was from a population that winters in the Sea of Japan and summers off the Pacific coast of Hokkaido and in the southern Okhotsk Sea, which was distinguished from a *truei*-type population migrating between the Chiba/Sanriku coast and the central Okhotsk Sea (Section 9.3.4). The Dall's porpoises do feed through the year, which is different from some baleen whales that exhibit a cycle of summer feeding and winter fasting. Therefore, if the pollutant ratio was the same between Baird's beaked whales off Chiba and Dall's porpoise off the Pacific coast of Hokkaido, this does not support a simple conclusion that the former species spends the whole year in the coastal waters of the Pacific area. A second question is the significance of comparing the PCB/DDE level between two cetacean species that are very different in life history, food habits, and perhaps in physiology. A third question relates to interpretation of the results. Even accepting that a difference in the PCB/DDE ratio indicates difference in feeding ground or in food habits, similarity is not necessarily evidence of feeding in the same geographical area. Samples from two different locations may exhibit the same PCB/DDE ratio due to influence of a common current system or for other reasons. In order to use the PCB/DDE ratio for the purpose mentioned earlier, it will be necessary to have a better understanding of geographical distribution of the pollutants.

Careful inspection of the data presented by Subramanian et al. (1988) reveals some other interesting points not mentioned by the authors. The PCB/DDE ratios of the two cetacean species mentioned earlier are also close to the corresponding figures for northern-form short-finned pilot whales (range 0.39–0.41, mean 0.40, n = 2). The northern-form pilot whales inhabit Pacific waters from Chiba to the Pacific coast of Hokkaido and west of 150°E (Figure 12.8). The PCB/DDE ratios of these three species are different from those of *truei*-type Dall's porpoises taken off Sanriku (range = 0.71–1.15, mean = 0.91, n = 8) and *dalli*-types in offshore waters (range = 0.91–1.19, mean = 1.01, n = 13).

By ignoring between-species difference in food habits as done by Subramanian et al. (1988), the data given earlier allow the following interpretations:

1. Baird's beaked whales off the Chiba coast do not migrate to the central Okhotsk Sea; their PCB/DDE ratios disagree with those of *truei*-type Dall's porpoises known to summer in the central Okhotsk Sea.
2. Baird's beaked whales off the Chiba coast have their main feeding ground in coastal waters between Chiba and the Pacific coast of Hokkaido; their PCB/DDE ratio is close to that of northern-form short-finned pilot whales.
3. The habitat of Baird's beaked whales off Chiba is limited to west of 150°E; their PCB/DDE ratio is different from that of *dalli*-type Dall's porpoises in the offshore Pacific east of that longitude.

It is a possibility that future surveys on the wintering ground of Baird's beaked whales off the Pacific coast of Japan would find them in the waters suggested from the information given earlier, or in a range of 35°N–45°N and west of 150°E.

It is well known by whalers that Baird's beaked whales off the Chiba ground are extremely nervous about approaching vessels, including whale catcher boats and research vessels, but our scientists onboard sighting vessels in the Sea of Japan before resumption of a low level of exploitation in 1999 had the impression that the species were much more approachable in the northeastern Sea of Japan off the Oshima Peninsula in southern Hokkaido. This behavioral difference could be due to a different history of exploitation and is likely an additional indicator of difference in population identity. Such cultural differences can be maintained even under low levels of genetic mixing.

13.3 FOOD HABITS

13.3.1 Stomach Contents

Studies on the food habits of Baird's beaked whales have been few. I had no opportunity to see the analyses of eight stomachs reported by Pike (1953) and Rice (1963). Tomilin (1967) reported analysis of six stomachs from Russian waters together with the results of Pike (1953) and stated that the whales feed mainly on squid and to lesser degree on fish as observed on other Ziphiidae species. Heptner et al. (1976) reached the same conclusion using an analysis by Betesheva (1960, 1961), which listed four squid species (three *Gonatus* spp. and one *Ommatostrephes* sp.) and five fish species (two *Theragra* spp., *Podonema* sp., *Eleginus* sp., and *Coryphaenoides* sp.). Evaluation of the contribution of each species to the nutrition of Baird's beaked whales is difficult. Counting the food items consumed in a particular time period is difficult because the speed of digestion may vary among food species. Undigested fish otoliths and squid beaks in cetacean stomachs are used widely for species identification, but the length of time retained in a cetacean stomach is not the same for the two major food categories. Thus, it is not an easy task to estimate the total number or total weight of food items consumed during a particular time period through analysis of stomach contents.

Matsuura (1942, in Japanese) was the first to describe the stomach contents of Baird's beaked whales taken off Chiba Prefecture and noted eye lenses of squids in the stomachs. Nishiwaki and Oguro (1971) analyzed records of stomach contents of the species that appeared in *Geiryo Geppo* [Monthly Report of Whaling Operation] of small-type whalers submitted to the Fisheries Agency in 1965–1969. Out of the 701 individuals reported, 383 (55%) had food remains in the stomach: squid (found in 111 whales), mackerel (15), sardines (5), saury (1), and flatfish (1). In addition, 156 whales were identified with "deep-sea fish," and 93 with "fish of unknown species." The total was 383 whales reported with stomach contents, which meant that the whalers reported only one, presumably most dominant, item for each whale. It was also clear that they ignored fish otoliths and squid beaks. Using these data Nishiwaki and Oguro (1971) attempted to see geographical and seasonal differences in the stomach contents but succeeded only in the former objective because the available data were limited seasonally. They noted that most Baird's beaked whales taken in waters from south of latitude 37°20′N (close to Cape Shioya) to the Chiba coast were found with deep-sea fish (43 out of 44 whales), while 24 individuals taken in waters from that latitude north to Ayukawa had squid (13 whales), mackerel (7), and sardines (3). The two latitudinal areas were almost continuous as seen in the distribution of catch positions (Figure 13.4), and the presence of deep-sea fishes is also known in the northern area (see the following text in this section). The features of the northern sample were similar to those found for 26 whales from further north, off the Kushiro, Pacific coast of Hokkaido and off Abashiri and the Shiretoko Peninsula in the southern Okhotsk Sea, where squid (22 whales), deep-sea fish (2), and Alaskan pollack (1) were recorded.

Nishiwaki and Oguro (1971) concluded that the differences in stomach contents shown earlier reflected differences in marine fauna of the region (this is not supported by survey data) and that Baird's beaked whales exploited their prey without particular preferences. Although there is no firm evidence against this conclusion, the possibility remains that the observed geographical difference in stomach contents contains some artifacts. As the latitude of the boundary mentioned earlier is about 240 km from the whaling stations in Chiba and about 150 km from those in Ayukawa, the northern ground is more likely operated by Ayukawa whalers and the southern ground by Chiba whalers. Thus, personality or biological knowledge of the staff in charge of writing reports to the Fisheries Agency could have resulted in the geographical difference in stomach contents reported to the government.

In order to resolve this question, biologists must examine stomach contents by themselves, as done by Walker et al. (2002), who compared stomach contents of Baird's beaked whales between the southern Okhotsk Sea landed at Abashiri and those taken south of Cape Shioya and processed at Wadaura in Chiba. They confirmed the results obtained by Nishiwaki and Oguro (1971) that the main food was squid in the Okhotsk Sea and deep-sea fishes in waters south of Cape Shioya. It still remains to be confirmed whether Baird's beaked whales between Cape Shioya and Ayukawa area, or in intermediate latitudes, really depend on squid, mackerel, and sardines as indicated by Nishiwaki and Oguro (1971).

The study by Walker et al. (2002) was a landmark on food habits of Baird's beaked whales, where all the processes of collecting materials and sorting the stomach contents as well as identification of prey species were carried out by themselves (Table 13.3). Their materials came from two sources. One represented 107 individuals flensed at a land station of Gaibo Hogei [Gaibo Whaling] Co. Ltd. in Wadaura in Chiba Prefecture. The season was July and August in 1985–1987, 1989, and 1991, and the catch positions ranged from the entrance of Sagami Bay to Cape Shioya along the 1000 isobath. The latitudinal range was between 34°30′N and 37°N. The other source was 20 whales caught off Abashiri and in the Nemuro Strait, both in the southern Okhotsk Sea (Figure 13.4), and flensed at a land station in Abashiri in Hokkaido. The season was August and September of 1988 and 1989. Water depth at the position of catch was less than 1000 m with the exception of two whales, and the latitudinal range was 44°N–44°30′N.

Walker et al. (2002) identified the food species using the morphology of cephalopod beaks and fish otoliths. Upper and lower beaks, and left and right otoliths, were counted separately, and the larger figure was used as the number of individuals representing each species. Fish size was estimated from otolith size. Those fish with estimated body length of 10 cm or less were considered as coming from the stomachs of fish eaten by the whale. Cephalopod beaks smaller than 1 mm were excluded from the analysis for the same reason.

All the 127 Baird's beaked whales examined by Walker et al. (2002) had some food remains in their stomachs. Most of the Chiba samples (89.7% of the 107 whales) had remains of both fish and cephalopods, 7.5% only fish remains, and 2.8% only cephalopod remains. Most of the Okhotsk Sea sample (90% of the 20 whales) had both fish and cephalopod remains and 10% only cephalopod remains; there were no individuals found with only fish remains. The two samples differed in predominant food items. The 107 whales of the Chiba sample had a total of 8078 (81.8%) fish and 1782 (18.0%) cephalopods, while the reverse trend was the case for the 20 individuals from the Okhotsk Sea: 315 (12.9%) fish and 2117 (87.1%) cephalopods.

The whales taken off Chiba had eaten mostly fish, a total 8078 individual fish representing 13 species in 7 families. The most dominant species were *Laemonema longipes* (4297), *Coryphaenoides cinereus* (1469), and *C. longifilis* (1347). These three fish species represented 88% of the total fish identified in the stomachs and were benthic deep-water fish. I observed that most of the Baird's beaked whales processed at Wadaura had some stones or pebbles in the stomach, which could have been ingested during bottom feeding. None of the 13 fish species are targets of Japanese commercial fisheries. The 107 whales examined at Wadaura in Chiba had cephalopod beaks representing a total of 1782 individuals of 32 species in 14 families: *Taonius borealis* (388) in Cranchiidae, *Eogonatus tinro* (344) and *Gonatus berryi* (103) in Gonatidae, *Mastigoteuthis* sp. (cf. *M. dentate*) (218) in Mastigoteuthidae, and *Enoploteuthis chuni* in Enoploteuthidae (131). The total of these five species comprised only 66% of the total 1782 cephalopods, which means that Baird's beaked whales off

TABLE 13.3
Stomach Contents of Baird's Beaked Whales Processed at Wadaura and Abashiri

Taxonomic Group	Wadaura[a] No. of Species	Wadaura[a] No. of Individuals	Abashiri[b] No. of Species	Abashiri[b] No. of Individuals	No. of Common Species
Fish	14	8078	7	315	4
Synaphobranchidae	1	1			
Alepocephalidae	1	2			
Alepisauridae	1	2			
Moridae	2	4343	1	83	1
Macrouroididae	1	43			
Macrouridae	7[a]	3683	3	140	3
Gempylidae	1	4			
Gadidae			1	44	
Zoarcidae			2	48	
Cephalopods	32	1782	11	2,117	10
Enoploteuthidae	2	136			
Octopoteuthidae	1	17			
Onychoteuthidae	2	32			
Gonatidae	9	769	6	1,843	5
Histioteuthidae	2	29	1	1	1
Architeuthidae	1[c]	1			
Ommastrephidae	2[c]	61			
Chiroteuthidae	2	12			
Mastigoteuthidae	1[c]	218			
Cranchiidae	6[c]	487	3	265	3
Vampyroteuthidae	1	8			
Octopodidae	1	5	1	8	1
Alloposidae	1	5			
Ocythoidae	1	2			
Urochordata					
Pyrosoma atlanticum	1	20			

Source: Summarized from Walker, W.A. et al., *Mar. Mammal Sci.*, 18(4), 902, 2002, Tables 2 and 4, which gave species composition.

[a] Wadaura: On the Pacific coast at 35°03′N, 140°01′E.

[b] Abashiri: On the southern Okhotsk Sea coast at 44°01′N, 144°15′E.

[c] Samples that were identified to family or genus were counted as one species for the local summary but were included in the column of common species only when positively identified as same species.

Chiba feed on a broad range of cephalopod species. Japanese commercial fisheries target only two of the cephalopod species found in the stomachs of these whales: 58 Japanese flying squids *Tadarodes pacificus* of Ommastrephidae and 5 common octopuses *Octopus vulgaris* of Octopodidae.

Contrary to whales off Chiba, those in the southern Okhotsk Sea had more cephalopods than fishes in their stomachs in both number of species and number of individuals. The fish fauna was represented by seven species in four families, which were benthic as in the case of the Chiba sample. Five major species were *Laemonema longipes* (83) in Moridae, *Coryphaenoides cinereus* (74) and *Albatrossia pectoralis* (62) in Macrouridae, *Bathrocarina microcephala* (46) Zoarcidae, and *Theragra chalcogramma* (44) in Gadidae. The total number of fish identified in the stomach was 315, and these 5 species made up 98% of the total. Only the last species, *Theragra chalcogramma*, is commercially exploited by Japanese fisheries. Baird's beaked whales in the southern Okhotsk Sea relied mainly on cephalopods, including 11 species in 4 families, which were not limited to benthic species (Walker et al. 2007). The most important were four species of Gonatidae: *Berryteuthis magister* (771), *Gonatus madokai* (355), *Eogonatus tinro* (340), and *Gonatus beryi* (244). These four species comprised 80% of 2177 individual cephalopods identified in the whale stomachs. None of the four species are targets of Japanese fisheries. However, all the squid species identified in the stomachs of Baird's beaked whales either in the southern Okhotsk Sea or the Pacific area off Chiba are also consumed by sperm whales, and Walker et al. (2002) indicated a possibility of competition between the two cetacean species for nutrition.

One fish species of Moridae and three species of Macrouridae were common between the samples from the southern Okhotsk Sea and the Pacific off Chiba, of which two species, *Laemonema longipes* and *Coryphaenoides cinereus*, were important food items in both areas. Ten cephalopod species in four families were common between the two samples, but only two species, *Gonatus berryi* and *Eogonatus tinro*, appeared in the list of main food items. Mean body size (body length for fish and dorsal mantle length for cephalopods) and mean body weight were estimated from the size of fish otoliths and cephalopod beaks, respectively, for a limited number of species, which were in a range of 11–51 cm and 0.15–1.3 kg (fish) and 18–32 cm and 203–331 g (cephalopods).

The fish species *Laemonema longipes* (Moridae) was one of the two species important for Baird's beaked whales in both the Okhotsk Sea and Pacific, but size was different between the regions. Those eaten in the southern Okhotsk Sea had a mean body length of 36 cm and were mostly immature, but those consumed in the Pacific off Chiba had a mean body length of 51 cm and were mostly mature. Walker et al. (2002) suggested a relationship between the early summer to autumn shift of the range of Baird's beaked whales along the Pacific coast (Section 13.2.3) and northward migration of *Laemonema longipes* along the Pacific coast of Japan after spawning in the winter/spring season in waters south of Cape Choshi (35°42′N). However, Baird's beaked whales consume numerous food items. Even if we accept an apparent match between the postspawning migration of the food species and movement of Baird's beaked whales in early summer to autumn, I have some reservation about whether migration of the prey species is determining the seasonal movement of Baird's beaked whales off the Pacific coast of Japan. If movement of this fish species really affects the seasonal distribution of Baird's beaked whales, they would be better off feeding on the fish in the spawning season and on the spawning ground when the nutritional condition would be the best. It is unknown why Baird's beaked whales arrive at the spawning ground of the fish in May, that is, after the spawning season.

Walker et al. (2002) pointed out, based on records of trawl net surveys at latitudes around 38°N off the Pacific coast of Japan, that the most important fish species *Laemonema longipes* inhabits the sea floor at depths of 600–1000 m and the third most important fish species *Coryphaenoides longifilis* at depths of 1000–1500 m. This explains to some extent the presence of the whales in waters between the 1000 and 3000 m isobaths (Figure 13.3). The surveys did not cover deeper waters and were unable to confirm whether *Coryphaenoides cinereus* occurred there. It is interesting to note that the surveys indicated presence of Alaska pollack *Theragra chalcogramma* on the sea floor at depths of 100–300 m and Pacific cod *Gadus macrocephalus* at depths of 100–400 m. The abundance of the Alaska pollack was greater than the sum of abundances of the two prey species mentioned earlier (*L. longipes* and *C. longifilis*). Walker et al. (2002) concluded that this reflects the food preference of Baird's beaked whales inhabiting the Pacific coast of Japan at latitudes 34°30′N–37°N.

About 10 years after the sampling by Walker et al. (2002), Ohizumi et al. (2003) sampled stomach contents of Baird's beaked whales taken in July–August 1999: 24 off the Pacific coast of Japan and 2 in the southern Okhotsk Sea. The fishery types and latitudinal range of the samples were the same as those for the samples analyzed by Walker et al. (2002) but did not include whales taken in the Nemuro Strait. Ohizumi et al. (2003) found dominance in the number of fish remains over those of cephalopods in the Okhotsk Sea sample, which was the opposite of the findings by Walker et al. (2002), but the significance of the difference could not be ascertained because the more recent sample included only two whales from a small area off Abashiri. The top four fish species were among the top five species of Walker et al. (2002) from the same area, the southern Okhotsk Sea. Concerning the food of Baird's beaked whales off the Pacific coast, the two studies showed good agreement in predominance of fishes over cephalopods and in the composition of fish species, where the top three were common between the two studies. Ohizumi et al. (2003) accepted the analysis of benthic fish fauna and food preferences of Baird's beaked whales by Walker et al. (2002). The two studies did not show good agreement in species composition of cephalopods in either of the two sample areas. *Gonatus berryi* was the only cephalopod identified by both studies as important. Verification of species identification of the cephalopod beaks is needed before further consideration of the difference in cephalopod prey suggested in the two studies. Ohizumi et al. (2003) presented equations to estimate body size and body mass based on size of fish otoliths and cephalopod beaks.

13.3.2 Diving and Feeding

The skeletal muscles of marine mammals switch to anaerobic respiration when oxygen pressure in the blood decreases during a long dive. After a long dive they have to remain on the sea surface to process the lactic acid accumulated in the body through the anaerobic respiration. A way of shortening the sea surface time is to surface while the oxygen pressure is above a certain level or to delay the time of anaerobic respiration by decreasing heartbeat or lowering metabolic rate (Berta et al. 2006).

It was well known among whalers that Baird's beaked whales make long dives and perhaps dive very deep. Composition of their prey confirms that they dive deeply (Section 13.3.1). Kasuya (1986) observed surfacing intervals of Baird's beaked whales in July and August off the Pacific coast of Japan at latitudes shown in Figure 13.3 and identified three modes in duration of the subsurface time. These modes were 5–12 min (13 observations), 14–29 min (12), and 37–49 min (4). In addition to these he observed three dives of 1–2 min and a dive of 67 min. Duration on the surface, which was defined as the time when the blows of school members continued, ranged from 1 to 8 min (26 examples) to 13–14 min (2). The observations were made through binoculars from a sighting vessel that stayed at a distance that would not disturb the whales, and there were cases where schools were observed remaining near the surface while blows were made after short pauses (thus recorded as subsurface time). It is certainly incorrect to interpret the very short subsurface times as dives related to feeding. Rather, several surface times separated by short subsurface time should have been considered as a continuous period on the surface. Kasuya (1986) determined that short subsurface times of less than 15 min (likely representing shallow dives) were followed by surface times of less than 4 min (11 examples) and long subsurface times of 16–50 min followed by surface times of 3–14 min (9 examples).

Minamikawa et al. (2007) placed a recorder for depth, time, and temperature on a Baird's beaked whale of an estimated body length of 9–10 m. The position was about 90 km offshore from the Pacific coast of Chiba Prefecture at latitude of 34°55′N and between the 1000 m and 3000 m isobaths. The recorder later separated from the whale and was recovered 2 days after deployment. The position of recovery was near the 3000 m isobath and about 160 km northeast of the position of deployment. From these locations and direction of the Kuroshio Current, they concluded that the record of 29 h thus obtained represented dive profiles in waters between the 1000 and 3000 m isobaths.

The records showed extremely regular dive repetitions of one deep dive immediately followed by one series of intermediate dives and long stays on the surface which included shallow dives with a mean of 20 m (SD 12 m). The animal repeated six deep dives with intervals of 6 h in a continuous 24-h period. Duration of a deep dive was 45 min on average with depths of 1400–1700 m (mean 1566 m, SD 208 m). These deep dives were judged to have reached close to the bottom from existing knowledge of bottom topography and position of the release. The mean descent rate of 1.92 m/s was greater than the mean ascent rate of 1.45 m/s. A series of intermediate dives lasted for 2–3 h and contained 5–7 dives. Depth of an intermediate dive ranged from 200 to 700 m (mean 379 m, SD 158 m), and the animal usually remained at the depth for 20–30 min during one of these dives. After a series of intermediate dives, the animal remained near the surface for 80–200 min before the next deep dive. Near-surface time was always followed by a deep dive. This dive profile of a Baird's beaked whale has some similarity to that of a sperm whale, where several intermediate dives of about 500 m were followed by surface times of 4 h and subsequent deep dives of 1200 m (Watkins et al. 1993). Minamikawa et al. (2007) did not interpret their results.

Feeding of Baird's beaked whales at or near the sea floor is evident from food composition, scars on the head region, and stones often found in the stomach, but the results of the study by Minamikawa et al. (2007) strongly suggest that feeding is not limited to the bottom. The tagged animal could feed on the bottom only during a deep dives, when it spent 45 min on average on the round trip and was able to spend only 10–30 min at the bottom. While there may be some feeding at the bottom, it is likely that the deep dives have some other functions and that most of their prey is ingested during the intermediate dives that follow a deep dive. One possible function of the deep dive is to survey the vertical oceanography or distribution of prey organisms in order to determine the best depth for feeding during dives of intermediate depth. Thus, a feeding bout starts with a deep dive of about 45 min for surveying prey distribution followed by 5–7 dives of intermediate depth for the main foraging activities of 2–3 h, ending with a long rest period of 80–200 min.

13.4 LIFE HISTORY

13.4.1 Tooth Structure and Age Determination

Omura et al. (1955) were the first to report the presence of growth layers in the tooth cementum of Baird's beaked whales. Kuroe (1960, in Japanese) reported on the anatomy of the teeth. Following these pioneer studies Kasuya (1977) established a method of age determination for the species using growth layers in the cementum, and Kasuya et al. (1997) applied the method to analysis of the life history of the species using material obtained from small-type whaling off Chiba on the Pacific coast of Japan. Baird's beaked whales have 2–3 pairs of teeth in the anterior part of the mandible; the anterior-most pair is the largest and suited for age determination. Measurements of one very large and a moderate-sized example of the anterior-most teeth of adults (without sex data) ranged from 8.0 to 10.1 cm in anteroposterior width, 4.7–5.3 cm in lingua-buccal thickness, and 9.6–8.5 cm in total height, where 1–2 cm of cusp had been lost by abrasion. The posterior teeth have a structure similar to that of the anterior-most teeth, but are about half their size and the cemental growth layers are irregular and less suitable for age determination. The following description is of the anterior teeth.

The tooth of Baird's beaked whale is composed of three elements of enamel, dentine, and cementum, as is usually the case in mammalian teeth. Following formation of a tooth germ in the fetus, deposition of enamel starts on the distal side of the germ and continues until close to birth. Deposition of dentine starts at the same time as enamel on the proximal side of the tooth germ but continues for at least several years after birth. Formation of cementum starts sometime after birth and covers the dentine and connects the dentine and surrounding gum tissue. The enamel layer in this species is rudimentary and may be resorbed before eruption. It is lost by abrasion very quickly after eruption even if it may be covered by cementum before eruption. The cementum is extremely thick and apparently takes on part of the function of the dentine in other species in maintaining the shape and strength of the tooth. Deposition of cemental layers starts in many delphinids at around the beginning of ingestion of solid food or the time of tooth eruption, but the exact timing of the deposition in this species has not been determined. However, Kasuya (1977) found the same number of growth layers in dentine and cementum in the teeth of 2- to 3-year-old individuals, which was evidence of a relatively early start of cemental deposition, perhaps within 1 year from birth.

The dentine is covered by enamel on the distal end and cementum on the proximal portion and has an internal cavity. New dentine is accumulated on the wall of the pulp cavity. The dentine formed during the fetal period has the greatest thickness and is partially covered by enamel layers. The next layer is the one formed in the first year of birth and has the greatest length (measured from cusp to distal end of the tooth). The subsequent growth layers decrease in length as well as thickness, and deposition of ordinary dentine ends by 4–7 years after birth, leaving a large pulp cavity. After cessation of deposition of ordinary dentine, the pulp cavity is filled by osteodentine, which is then covered by thick cemental layers. Thus, the osteodentine and cementum occupy a large proportion of the tooth mass.

The osteodentine is a dentinal tissue of weak calcification and has blood vessels and cells scattered in it. A sperm whale tooth often has smaller globular inclusion of osteodentine in the ordinary dentine, but these are smaller and show circular growth layers. The structure observed in a cross section of osteodentine of Baird's beaked whale suggests that the formation starts at several deposition centers in the pulp tissue, replacing the pulp tissue with osteodentine, and finally fuses together to form a body of osteodentine that fills the entire pulp cavity. A mass of osteodentine almost filling the pulp cavity is seen in a tooth with open root of a whale with an estimated age of 21–30 months. Thus, the formation of this large osteodentine mass seems to start before the age of 2–3 years and completely fills the pulp cavity by age 4–7 years, when deposition of ordinary dentine ceases (Kasuya 1977). Soon after the pulp cavity is filled by osteodentine, cementum covers the proximal end of the tooth to complete the root. This is Baird's beaked whale's way to complete formation of the large tooth in a relatively short period. None of the whales off the Pacific coast of Chiba had completed the root before age 6, some had completed it age 6, and all had reached the stage by age 14. No difference between the sexes was identified in tooth growth (Table 13.4).

Although cemental layers are deposited throughout life on the teeth in the family Delphinidae, which includes striped dolphins and short-finned pilot whales, the contribution to tooth length is limited because of the small thickness of the cementum, and increase in tooth length ceases at completion of formation of the root. This is different from the teeth of Baird's beaked whales, where deposition of thick cemental layers contributes to increased tooth size (length and diameter) after formation of the root at ages 6–14. This is one of the specializations that could be derived from the general pattern seen in the Delphinidae. Additional examples of specialized cemental growth are seen in freshwater dolphins. Teeth of the South Asian river dolphin, *Platanista gangetica*, complete formation of the root at the age of four dentinal layers (presumably 4 years), but the increase in tooth length continues with accumulation of cementum. More than half of the tooth length, 2–3 cm, was occupied only by cemental layers at the age of 28 layers (Kasuya 1972). This species differs from Baird's beaked whale in the continuation of deposition

TABLE 13.4
Age at Occlusion of Root of the Front Mandibular Tooth of Baird's Beaked Whales Processed at Wadaura

Age (No. of Cement Layers)		5	6	7	8	9	10	11	12	13	14	≥15
Male	Open root		2		4		2		1	1		
	Occluded		1	2	1	3	2	1	3	5	2	33
Female	Open root	2		3		3		3				
	Occluded			1	1	1		3			1	7
Total	Open root	2	2	3	4	3	2	3	1	1	0	0
	Occluded		1	3	2	4	2	4	3	5	3	40

Source: Kasuya, unpublished.

of dentinal layers far beyond the age of root formation. The dolphin aged at 28 cemental layers had the same number of layers in the dentine, with the pulp cavity open for further dentine deposition. Another freshwater dolphin, the franciscana, *Pontoporia blainvillei*, has numerous small teeth about 10 mm in length. The tooth root is completed at age 4–5 years, followed by deposition of cemental layers around the cervical portion of the tooth that measures only about 2 mm at a young age. Age is estimated by counting layers in the cementum of this portion formed after cessation of dentine deposition (Kasuya and Brownell 1979). The brim-like structure of cementum doubles the diameter of the cervical portion of the tooth and likely increases the strength of the connection between tooth and alveolus. Dentinal deposition continues for life in the mandibular and maxillary teeth of sperm whales (Ohsumi et al. 1963) and white whales and in the task of narwhals (Perrin and Myrick 1980).

Baird's beaked whales are aged using growth layers in the cementum of the anterior-most mandibular teeth. The process starts with extraction of a front tooth from the mandible by inserting a slender knife between the tooth and the wall of the alveolus. The tooth can be stored in 10% neutral buffered formalin or frozen. Because the tissue is later decalcified in acid, damage of the tooth by acid in the formalin is not a problem. Next, the fixed tooth is cleaned with a knife by removing most of the soft tissue around the root but leaving a thin layer of gum tissue on the cementum. This helps in identifying the latest cemental layer on the completed histological slide and confirming that cemental deposition was continuing at the time of sampling. The cleaned tooth is cut on a lingual-buccal plane through the cusp and the middle of the root. The cut surface is polished flat using a whetstone or sand paper placed on a flat surface.

One way of reading growth layers is to place the polished tooth in acid, for example, 10% formic acid, for 30 min to several hours and dry it before examination with a light microscope under reflected light. The reader can determine the most suitable decalcification time. The cemental growth layers are observed as an alternation of grooves and ridges.

Another method of preparation is to make a thin decalcified and stained section for observation under transmitted light. This preparation is suitable for observation of the detailed structure of the growth layers. After polishing flat a longitudinal cut surface, the surface is dried and glued on a plastic slide of about 1 mm thickness with cyanoacrylate resin. It is unnecessary at this stage to dry the surface completely but it must be sufficiently free of water to avoid whitening the glue. The other half of the tooth is cut off, leaving a thin layer of tooth on the plastic slide. The tooth section on a slide is polished down to a thickness of 40–70 μm, and then placed in a 10% solution of formic acid overnight for decalcification. The decalcified section is rinsed in running water, stained in haematoxilin, and mounted for examination under transmitted light (see Section 12.4.2 for further information common to other species).

Baird's beaked whale exhibits at least two types of cycles in cemental deposition: those assumed to be annual and those

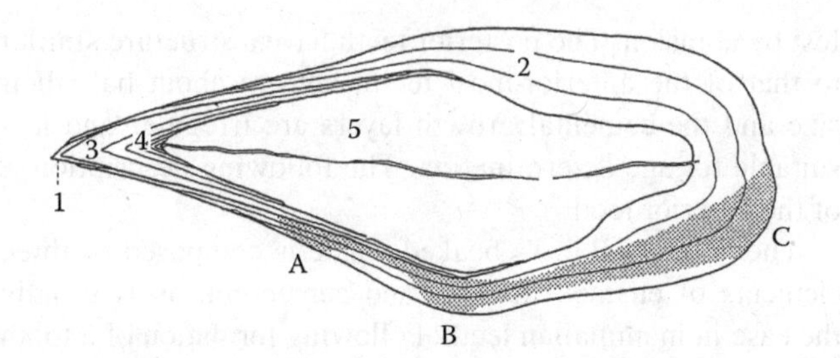

FIGURE 13.8 Schematic depiction of lingua-buccal longitudinal section of anterior mandibular tooth of a Baird's beaked whale. 1, enamel layer; 2, cemental layers (arrangement of 3 layers is shown); 3, dentine deposited during fetal stage; 4, postnatal layers of dentine; 5, osteodentine. Following tooth eruption, the neck of the tooth moves from position A to B; thus cemental layers have to be counted in the shaded area moving from A (for early layers) to B and then to C (later layers deposited on older individuals). (Reproduced from Kasuya, T., *Sci. Rep. Whales Res. Inst.*, 29, 1, 1977, Figure 1.)

of a shorter cycle. They must be distinguished from each other when reading growth layers in the tooth. The number of layers of shorter cycle within the annual layer varies with position in the tooth but is around 13 in fast-growing (i.e., thick) cementum and are assumed to represent a cycle of about 1 month. However, no evidence has been obtained to indicate synchrony with the lunar cycle. The longer cycle is assumed to represent an annual cycle, but this has not yet in a strict sense been confirmed for the species due to limitations in sample size and seasonal coverage of the sample.

Another caution needed in aging Baird's beaked whales is to select the best position in a tooth section. Cementum exposed above the gum line must be avoided because deposition of new cementum has already ceased at that position and the whole cemental age of the animal is not recorded. Growth layers may be compressed near the cervical portion even when deposition is continuing there, which is also unsuitable for aging. Thus, readers should start reading inner layers near the cusp and move toward outer layers near the base of the tooth. Reading the age of young animals should be done at position A in Figure 13.8, that of older individuals start at position A and move to B, and that of very old individuals start at position A and move to B and C. The oldest whales in the sample from individuals killed off Chiba were aged at 54 layers (a female) and 84 layers (male). In aging Baird's beaked whales, Kasuya et al. (1997) ignored thickness of the earliest and the last layers of possible incomplete thickness (i.e., assumed as having complete thickness) and expressed the number by an integer, which is used in this chapter as the age of the animal. The real age of an animal with n number of growth layers, which is counted in this way, is thought to be between n − 1.5 years and n + 0.5 years (n being an integer).

13.4.2 Sexual Maturity

The conservation biology of cetaceans has paid much attention to determination of sexual maturity and age at attainment of that state. Other related parameters for females are age at first ovulation and at first parturition, and the former has been used as an indicator of sexual maturity using the presence of a corpus luteum or albicans in the ovaries as the evidence of ovulation. Since conception usually occurs at the first ovulation, the time from the first ovulation to the first parturition is usually 1–2 years. Determining male maturity is more difficult and involves ambiguities. There are two different features involved in male maturation: the physiological capability of reproduction and actual participation in reproduction, which is called social maturity and is not easy to identify. Histology of the reproductive tract has been used for determination of physiological status (Section 12.4.3.3), but determination of social maturity requires understanding of the social structure or identification of the paternity of offspring.

After failure of conception or parturition, the corpus luteum in the ovary degenerates into a corpus albicans, which has been believed to persist for life in the ovaries and offer a key for easy identification of maturity. A corpus albicans derived from pregnancy and that from ovulation are believed to be indistinguishable (Perrin and Donovan 1984). Readers are advised to see Sections 10.5.6.1, 11.4.8.5, and 12.4.4.6 for discussion of recent finding concerning the persistence of the corpus albicans of ovulation in some delphinid species.

Japanese whalers slit the carcasses of Baird's beaked whales for the purpose of cooling them during towing to the land station, which often causes loss of fetuses and ovaries, particularly those of pregnant females. Thus, Kasuya et al. (1997) identified females as sexually mature if they were lactating or pregnant even when the ovaries were not available. This increased sample size, but had a small risk of overrepresenting mature females compared to immature ones and could have caused a slight underestimation of average age at attainment of sexual maturity. The magnitude of the bias has not been evaluated.

Kasuya et al. (1997) examined Baird's beaked whales taken off the Pacific coast of Japan and landed at Wadaura in Chiba Prefecture and found sexually mature females at the age of 10 years or older (25 individuals) and immature females at age 14 or younger (14 individuals). They concluded that females of the species attain sexual maturity at age 10–15 years. In the same sample the minimum body length of sexually mature females was 9.8 m and the maximum length of an immature female was 10.6 m. They concluded that females attain sexual maturity at 9.8–10.7 m.

As noted earlier, determination of male sexual maturity involves greater ambiguity. Kasuya et al. (1997) followed the criteria of testicular histology established by Kasuya and Marsh (1984) for the short-finned pilot whale, which was to classify male maturity into four stages of "immature," "early maturing," "late maturing," and "mature" based on the proportion of seminiferous tubules that were spermatogenic in a tissue sample from the mid-center of a testis (Section 12.4.3.3). They investigated how spermatogenesis proceeds within a testis using six males in the early and late maturing stages and confirmed that spermatogenesis starts at the anterior end of the testis and expands to the posterior portion. This feature was different from that observed in short-finned pilot whales (Section 12.4.3.3). Therefore, male Baird's beaked whale classified as "mature" by the previous criteria may contain some immature tubules in the posterior part of the testis, and some "immature" males may have had spermatogenesis in the anterior portion of the testis. Kasuya et al. (1997) examined the histology of testis tissue taken from the mid-center of the left testis or the right testis if the left was not available. They used the mean weight of both testes but substituted that of a single testis if only one was available.

Maturation of the epididymal tissue precedes that of the testis as in the short-finned pilot whale. Therefore the epididymal tissue of some maturing males exhibits a state similar to that of "mature" males. However, it was common to find only scanty spermatozoa in the epididymis of a "mature" testis of a Baird's beaked whale, which was quite different from the situation in the short-finned pilot whale (Table 12.8) but similar to that in the Dall's porpoise (Section 9.4.8). This suggests the possibility that Baird's beaked whales are seasonal breeders and that the season of the sample studied by Kasuya et al. (1997), July and August, was distant in time from the mating season.

Histologically "immature" testes occurred at age 8 and younger (4 individuals), "early maturing" at 6–9 (4 individuals), "late maturing" at 9–10 (2 individuals), and "mature" at 6 and older (40 individuals). The limited sample suggests that the "mature" stage is achieved at 6–11 years, or when a single testis weighs 1.4–1.5 kg and body length is 9.1–9.8 m (Table 13.5). We have almost no information on the social

TABLE 13.5
Sexual Maturity of Baird's Beaked Whales Processed at Wadaura on the Pacific Coast of Japan, by Age, Body Length, and Weight of Single Testis with Epididymis Removed

Sex, Maturity, and Date of Sample		Sample Size	Age in Cement Layers	Body Length (m)	Testis Weight (g)
Males (1985–1987)	Immature	4	3–8	7.8–8.9	278–370
	Early maturing	4	6–9	8.5–9.7	290–528
	Late maturing	2	9–10	9.5	1225–1465
	Mature	40	6–84	9.1–10.7	1360–8675
Females (1975, 1985–1987)	Immature	20	5–14	8.2–10.6	
	Mature	25	10–54	9.8–11.1	

Source: Prepared from Kasuya, T. et al., *Rep. Int. Whal. Commn.*, 47, 969, 1997, Tables 2 and 3.

behavior of this species, but analogy with other toothed whales suggests that males at the "mature" stage are physiologically capable of reproduction. Males attain sexual maturity at age 4 years younger and body length 0.7–0.9 m shorter than females. The dimorphism is small, but such reverse dimorphism is uncommon among the toothed whales (Ralls 1976).

Testicular weight, usually the mean of the two testes, was analyzed against maturity and age (Figure 13.9). All the males age 5 and younger had "immature" testes that weighed 250–400 g. We saw "early" and "late maturing" males and a small number of "mature" males at ages 6–10, when testicular weight showed a rapid increase. The increase from "early" to "late maturing" was particularly rapid, from 500 g at age 6 to 1.5–2 kg at age 10, which was equivalent to a mean annual increase of 0.25–0.4 kg/year or a three- to four-fold increase in the 4-year period. This stage of rapid testicular growth is the pubertal stage.

Body length showed almost no increase from age 10 to 30 (Section 13.4.4), but testicular weight continued increasing, probably at a decreasing rate: from 1.5–2 kg (age 10) to an average of 3 kg (age 20) or at a mean annual growth rate of 0.1–0.15 kg/year, then to about 4 kg in the next 10 years (equivalent to a mean annual growth rate of 0.1 kg/year). Testicular weight reached a plateau of 5.3 kg at age 40 (Figure 13.9, Table 13.6). Thus, males attain the histological "mature" stage at ages between 6 and 11 and continue to increase testicular mass for 30 more years until age 40. We do not know the biological meaning of such extended testicular growth, and we still do not know when the males participate in reproduction.

TABLE 13.6
Weight of Single Testis (Epididymis Removed) by Age of Baird's Beaked Whales Processed at Wadaura on the Pacific Coast of Japan

Age (No. of Cement Layers)	Sample Size	Weight Range (kg)	Mean Weight (kg)	95% Confidence Interval of the Mean (kg)
3–9	13	0.278–4.125	1.025	0.385–1.665
10–19	17	1.225–6.300	3.066	2.371–3.761
20–29	13	2.600–5.500	3.826	3.314–4.338
30–39	8	3.750–6.750	5.088	4.229–5.947
40–84	33	2.625–8.675	5.317	4.817–5.817

Source: Reproduced from Kasuya, T. et al., *Rep. Int. Whal. Commn.*, 47, 969, 1997, Table 4.
See Table 13.5 for histological maturity of testis.

Testicular weight showed great individual variation among "mature" males identified by testicular histology. It ranged from 1.4 to 8.67 kg (about a five- to six-fold variation) for individuals aged over 10 and from 2.63 to 8.67 kg (over 3-fold variation) for individuals over 40, which corresponded to the age where the extended testicular growth apparently ceased (Figure 13.9). Similar broad individual variation in testis weight, about four-fold, is known also among full-grown short-finned pilot whales of the two forms (Figures 12.12 and 12.13). Seasonal variation in reproductive activity alone did not explain this, because while one sample covered a broad seasonal range (southern-form pilot whales), other samples covered only 2 months of the year (Baird's beaked

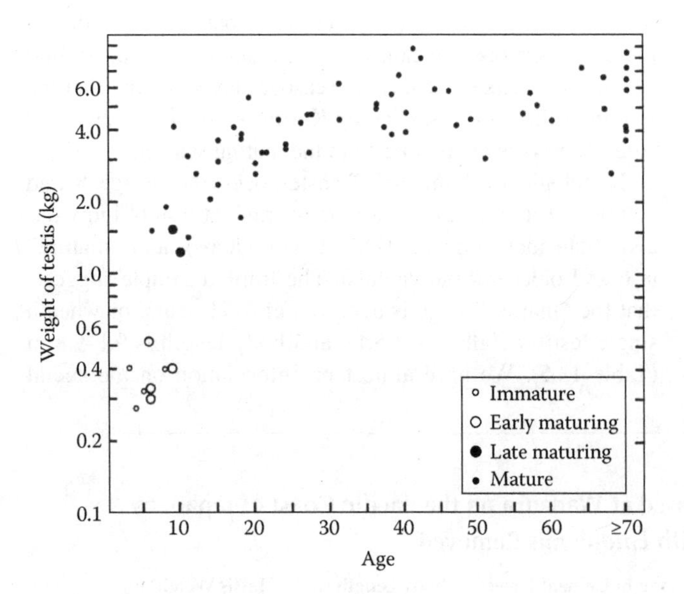

FIGURE 13.9 Scatter plot of weight of single testis (epididymis removed) and histological maturity on age of Baird's beaked whales processed at Wadaura (35°03′N, 140°01′E), Pacific coast of Chiba Prefecture, in July–August 1985–1987. Testis tissue develops in the order of immature (small open circles), early maturing (large open circles), late maturing (large closed circles), and mature stages (small closed circles), but the weight of mature testis shows further increase to age 30–40 years. (Reproduced from Kasuya, T. et al., *Rep. Int. Whal. Commn.*, 47, 969, 1997, Figure 2.)

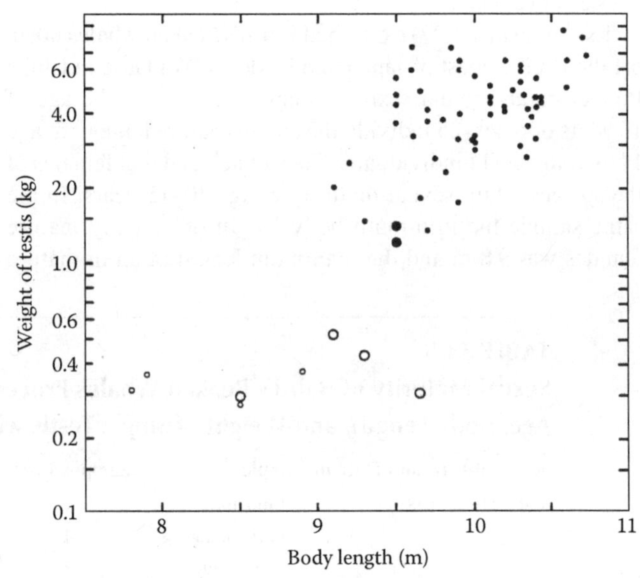

FIGURE 13.10 Scatter plot of weight of single testis (epididymis removed) and histological maturity on body length of Baird's beaked whales (same materials as in Figure 13.9). For maturity see Figure 13.9. (Reproduced from Kasuya, T. et al., *Rep. Int. Whal. Commn.*, 47, 969, 1997, Figure 3.)

whale and northern-form pilot whale). The individual variation in testis weight is explained, at least in part, by a positive correlation between body size and gonad weight. While animals over 30–40 years old were mostly over 10 m in body length, and the correlation between age and body length was lost in these whales (Figure 13.12), there was still some correlation between testis weight and body length among males with body length over 10 m. This suggests that males with greater body size have larger testes. Seasonal asynchrony in the male reproductive cycle remains as a possible cause of such testicular variation, but it has not been investigated.

13.4.3 Fetal Growth and Gestation Time

Our information on full-term fetuses or neonates of Baird's beaked whales is limited. Omura et al. (1955) reported maximum fetal size of 14 ft (about 4.2 m) in Japanese whaling statistics and another U.S. record of 15 ft 9 in. (4.81 m), which was measured along the dorsal surface and could be biased upward. They estimated neonatal length of the species at 15 ft (4.58 m). Recently I found a record of a 35-shaku (10.6 m) Baird's beaked whale pregnant with a 16-shaku (4.85 m) fetus killed on 8 October 1971 in the Nemuro Strait at around 44°20′N in the diary of S. Fukuoka, an ex-whaler who operated from Abashiri in the southern Okhotsk Sea (Fukuoka 2014, in Japanese). The *shaku* is 0.303 m and almost identical with a foot. These data suggest that neonatal body length in Baird's beaked whales is greater than that of the sperm whale of 4.0 m estimated by Best et al. (1984).

Japanese coastal whaling killed a large number of female sperm whales but left no reliable records of body length because of the practice of stretching the measurements to meet the minimum size limit of 10.6 m maintained until the 1965 season (Kasuya 1999; Kasuya 1999, in Japanese). Ohsumi and Satake (1977) reported body lengths of 68 female sperm whales taken under a special permit for scientific research. Their maximum size was in a length group of 12.0–12.1 m, and both sexually immature and mature females were found in a range of 9.4–9.9 m (n = 10). The latter body length is slightly smaller than the corresponding figure for female Baird's beaked whales of 9.8–10.6 m (Table 13.5), while the maximum female size of sperm whales exceeded that of the Baird's beaked whales (found in a length group of 11.0 and 11.1 m) with presumed effect of the disproportionately large head of full-grown sperm whales (Section 13.1). Thus, neonatal body length of Baird's beaked whales being greater than that of sperm whales is not unreasonable, although the current estimate of mean neonatal size requires improvement with further information.

Omura et al. (1955) analyzed seasonal change in fetal body length of Baird's beaked whales in the Japanese whaling statistics and estimated gestation time at 10 months. However, this figure should be questioned for several reasons. One of the problems in the estimate comes from the data. Fetal body length ranged from 4 to 9 ft during July and August, and identifying modal length was hard even for the months when the data were most abundant. This decreased accuracy of the estimate. The second problem was the extrapolations of the fetal growth curve beyond both sides of the data that covered July–October. This increased uncertainty about early and later fetal growth. A third problem was the assumption of curvilinear fetal growth, which is counter to the recent understanding of linear fetal growth of toothed whales after a period of early slow growth (Perrin and Reilly 1984; Section 8.5.2.2) and likely to lead to underestimation of the length of gestation.

In order to supplement the paucity of data on fetal growth of Baird's beaked whales, Kasuya (1977) used an interspecies relationship between neonatal size and fetal growth rate. The data used were from eight populations of seven species of Odontoceti, including the harbor porpoise and the sperm whale. The following relationship was obtained between daily fetal growth rate (Y, cm) in the linear part of fetal growth and mean neonatal length (X, cm):

$$Y = 0.001802\ X + 0.1234 \qquad (13.1)$$

The fetal growth rate Y = 0.949 cm/day was obtained by substituting for X 458 cm, the mean neonatal body length proposed by Omura et al. (1955). Thus, the number of days from the time (t_0), when the extended linear fetal growth curve cuts the axis of time, to the time (t_g), when the growth curve reaches the mean neonatal length, was calculated by

$$X/(0.001802\ X + 0.1234) \qquad (13.2)$$

The time was thus estimated at 483 days. Then the days from conception to t_0 had to be estimated. This was known to be a function of total gestation time (t_g) and to be about 7% of total gestation time for species with long gestation time such as Baird's beaked whale. In other words, time from t_0 to parturition, 483 days, comprises 93% of the total gestation time. This rule suggested t_0 = 36 days and total gestation time of 519 days or 17.0 months (Kasuya 1977). Killer whales in aquariums were observed with a gestation time of similar length, 16.9 months (Asper et al. 1992). Body length of these neonates ranged from 230 to 240 cm (Kasuya 1995). Adult size of killer whales varies among populations and is likely to affect neonatal size. Equation 13.2 converges at 555 days, or 18.2 months. This predicts that even with the additional slow growth stage from conception to t_0, the total gestation time of toothed whales does not exceed 19 months, or about 1.5 years. The killer whale is one of the species with the longest gestation time among toothed whales, as is possibly Baird's beaked whale.

The estimate of gestation time of Baird's beaked whale obtained earlier was based on an interspecies relationship among growth parameters in toothed whales. It is a hypothesis that awaits supporting evidence before being accepted as biological fact. The prediction is affected by the species used for obtaining the interspecies relationship. The relationship using particular taxa will disagree with the relationship for other group of cetaceans. For example, an interspecies relationship fitted to species of Delphinidae could be unsuitable for predicting parameters of Ziphiidae. Absence of ziphiid data is the weak point of Equation 13.1, which was based on data for the harbor porpoise, common bottlenose dolphin, pantropical spotted dolphin, striped dolphin, beluga (two populations), long-finned pilot whale, and sperm whale. Further discussion of interspecies relationships is available in Perrin and Reilly (1984), Kasuya et al. (1986), and Kasuya (1995).

13.4.4 Body Length and Age

Figure 13.11 presents body-length composition of Baird's beaked whales processed at Wadaura in Chiba, in 1975 and 1985–1987. There was no identifiable change over the 10-year period.

Individuals of body length 8–9 m were scarce in the catch and those below 8 m were entirely missing, which agrees with absence of individuals aged 2 years or younger (Figure 13.12). Segregation of such young whales and perhaps their mothers outside of the geographical range of the whaling operation or their ship-avoiding behavior is likely what results in such a catch composition, but currently there is no evidence supporting the possibility. Another more plausible cause of the underrepresentation of young whales is selection by whaling

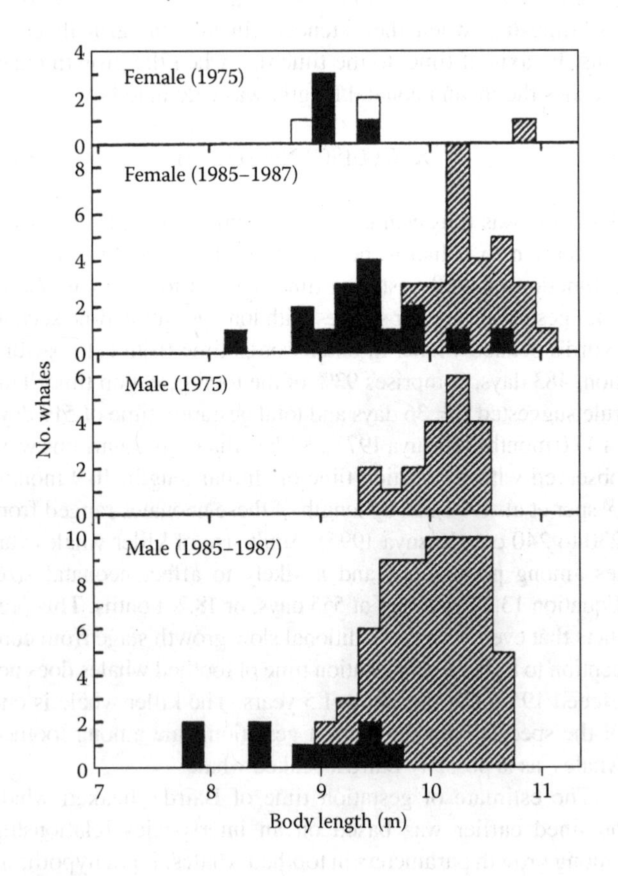

FIGURE 13.11 Body-length frequency of Baird's beaked whales processed at Wadaura, Pacific coast of Japan, in July and August, 1975 and 1985–1987. Shaded columns represent mature individuals, black columns other growth stages (i.e. immature, early, and late maturing), and white columns individuals of unknown maturity. Male maturity for 1975 samples was estimated from age and body length only for this figure. (Reproduced from Kasuya, T. et al., *Rep. Int. Whal. Commn.*, 47, 969, 1997, Figure 1.)

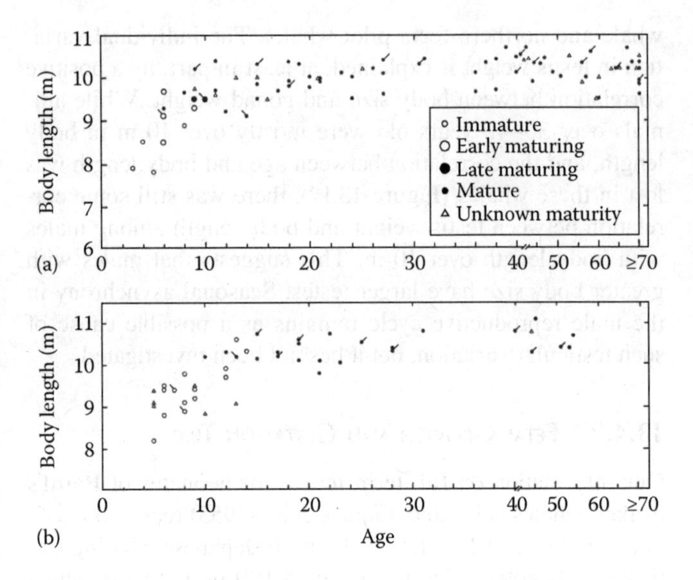

FIGURE 13.12 Scatter plot of body length and sexual and physical maturity on age of male (a) and female (b) Baird's beaked whales processed at Wadaura, Pacific coast of Japan in July and August, 1975 and 1985–1987. Key for sexual maturity is the same as in Figures 13.9 and 13.10. Physical maturity: bar on upper left, physically immature; bar on upper right, physically mature. (Reproduced from Kasuya, T. et al., *Rep. Int. Whal. Commn.*, 47, 969, 1997, Figures 4 and 5.)

gunners, who, if the situation permits, always prefer to shoot larger whales for greater production.

Peak body length, represented by 10.2 m and 10.3 m individuals, was not different between the sexes, but the range of the right-hand slope, which was mostly composed of full-grown individuals, was about 40 cm greater in females (≤11.1 m) than in males (≤10.7 m). This suggests that females on average attain slightly greater body length than males.

Figure 13.12 shows the relationship between body length and age for the same sample as in Figure 13.11. This is the apparent growth curve of the species and can represent the real growth curve if the assumption is accepted that growth in the species off the Pacific coast of Japan remained the same during the 80-year period lived by the sampled individuals. The age versus body length relationship suggests that mean body length ceases increasing at around age 15–16 in both sexes, although the trend for females is less clear due to smaller sample size. This age is likely to agree with the age when all individuals cease growing.

We know that the growth of male sperm whales in the North Pacific changed during the history of whaling (Kasuya 1991). Their maximum body size was around 16–17 m until the 1950s but increased to 18–19 m in the 1970s and thereafter. This change was accompanied by an increase in male body length at sexual maturation and is interpreted as the response of their physiology to per capita increase in food availability due to decline in the sperm whale population caused by whaling. Body-length-on-age data for the population do show an erroneous feature of growth in which body length apparently decreases at higher ages. This type of artifact has not been detected for Baird's beaked whales off Japan.

13.4.5 Physical Maturity

Cetacean biologists often also pay attention to physical maturity, which indicates a growth stage when growth in body length has ceased. It is identified by examining the vertebral centra and epiphyses and the cartilage layers between them. The length of the vertebral column increases by ossification of cartilage on the medial side and generation of it on the distal side. This vertebral growth ends when generation of the cartilage ceases and the ossified centrum meets the epiphyses. This process proceeds from both ends of the vertebral column and occurs last in the posterior thoracic and anterior lumbar vertebrae. Fishes and reptiles have no vertebral epiphyses, and their vertebral growth continues for life.

In order to examine the progress of physical maturity, a piece of epiphysis with centrum is extracted from the vertebrae around the last thoracic and the first lumbar vertebrae, fixed in 10% formalin solution, polished into a slice 1–2 mm thick, and stained in a water solution of toluidine blue before examination under a light microscope at low magnification. The cartilage appears in blue. Traditionally the completion of vertebral growth has been defined as the stage where the cartilage layer has completely disappeared between centrum and epiphysis. However, Kasuya et al. (1997) made a slight modification to this, where a vertebra is defined as to have ceased growing if the bone tissue of the centrum and epiphysis are fused together, even when some cartilage remains between them. This definition is suitable to determine the point where growth in vertebral length has ceased.

The result is shown in Figure 13.12. Among individuals examined for epiphyseal fusion, all the sexually immature individuals were physically immature and physically mature individuals were always sexually mature, which indicates that sexual maturity always precedes physical maturity. Physically mature males appeared at age 9 and older, and all the individuals examined for the character were physically mature at ages 15 or greater. The oldest physically immature male was 14. This limited sample indicated that males attain physical maturity at between 9 and 15 years. Physically immature females occurred at 14 or younger, and physically mature females at 15 years and greater. Thus, the limited sample of Kasuya et al. (1997) was unable to resolve between-sex difference in age at physical maturity.

Mean body length at physical maturity was estimated as the mean body lengths of whales at age 15 or older (Table 13.7). Females grow on average 35 cm longer than males.

TABLE 13.7
Average Body Length at Physical Maturity of Baird's Beaked Whales Processed at Wadaura on the Pacific Coast of Japan, Calculated Assuming Individuals over Age of 15 Cemental Layers (Years) Are Physically Mature

Sex	Sample Size	Mean (m)	95% Confidence Interval (m)
Male	66	10.10	10.02–10.18
Female	22	10.45	10.31–10.59

Source: Prepared from text of Kasuya, T. et al., *Rep. Int. Whal. Commn.*, 47, 969, 1997.

13.4.6 Female Reproductive Cycle

The whaling season for Baird's beaked whales off Wadaura in Chiba on the Pacific coast of central Japan was limited in recent years to July and August, and the body lengths of most of fetuses obtained from the operation were in a range of 4–9 ft long (Kasuya et al. 1997). This suggests that conceptions as well as parturitions are far outside the whaling season, perhaps in the winter including December–March.

One way of estimating the female reproductive cycle in cetaceans is to analyze the composition of reproductive status and its seasonal change, using large samples covering a broad range of seasons. The sample obtained by Kasuya et al. (1997) did not satisfy these conditions, covering only 4 years and limited to a total of 27 sexually mature females. There were 8 pregnant, 12 lactating, and 7 resting females, for an apparent pregnancy rate of 29.6%. If conception and parturition occurred before the whaling season, the pregnant females probably represented pregnancies in a single mating season and the apparent pregnancy rate was equal to the annual pregnancy rate. From this the mean length of a reproductive cycle was estimated at 3.4 years as a reciprocal of the annual pregnancy rate (the great uncertainty due to small sample size was ignored). This mean calving interval is slightly longer than those of the striped dolphin (Section 10.5.10) and common bottlenose dolphin (Section 11.4.8.4) but shorter than that of the short-finned pilot whale (Section 12.4.4.5).

Kasuya et al. (1997) obtained the following relationship between age (X, ≥15 years) and number of corpora lutea and albicantia (Y) for 19 females:

$$Y = 0.471\ X - 3.91,\ r = 0.80$$

The mean ovulation rate was 0.47/year or one ovulation/2.1 years. This suggests that the calving interval cannot be shorter than 2.1 years and that one of the 1.6 ovulations in 3.4 years, or 62% of total ovulations, end in conception.

Age-dependent decline in pregnancy rate and final cessation of conception at a certain age, which may or may not be accompanied by a decline in ovulation rate, is known in some toothed whales such as the sperm whale, short-finned pilot whale, long-finned pilot whale, and false killer whale (Section 12.5.7). Age of the oldest female Baird's beaked whale was 54 years, the oldest pregnant females were in their 40s, and the oldest females in lactation or resting stages in their 50s. Presence of a climacterium was not identified in the sample.

Length of the suckling period varies among Odontoceti species. Harbor porpoises and Dall's porpoises suckle for only several months, while short-finned pilot whales, long-finned pilot whales, and sperm whales suckle usually 3–4 years and even for over 10 years in a few cases (Sections 12.4.4.3 and 12.4.4.4). Material analyzed by Kasuya et al. (1997) was obtained in the summer or about half a year after the presumed parturition season. The following analysis of the reproductive cycle was made in consideration of the seasonality and ignoring the problem of the small sample size. Almost equal numbers of females in pregnancy (8) and resting stage (7) suggest that the total of these two stages is close to 2 × 17, or 34 months. Subtracting this total time from an average length of reproductive cycle of 3.4 years or 41 months, the average lactation period is estimated at 7 months. Even accepting the possibility that the females lactating in the summer continued to do so until the beginning of the next conception season, which was presumably in December–March, the lactation period was unlikely to extend over 1 year. This average lactation period is shorter than those of striped and common bottlenose dolphins and longer than those of harbor and Dall's porpoises. This estimate needs to be evaluated using further information.

Subramanian et al. (1988) noted that persistent organochlorines in female cetaceans are transmitted to their suckling calf through lactation. Thus, adult females should have lower pollutant level than males of the same species, and females of a species with longer lactation periods will have lower pollutant levels compared with the levels found in males of the same species. These authors obtained a female/male ratio of PCB at 0.55 and of DDE at 0.52 for Baird's beaked whales (n = 20, total of both sexes). They found that these figures were greater than the corresponding figures for Antarctic minke whales (PCB: 0.46; DDE: 0.43; n = 21), Dall's porpoises in the North Pacific (PCB: 0.23; DDE: 0.34; n = 27), and striped dolphins in the North Pacific (PCB: 0.22; DDE: 0.15; n = 28), and concluded that Baird's beaked whales have a shorter lactation period than those three species and suckle their calves for less than 6 months. It is generally accepted that Dall's porpoises (Section 9.4.6) and minke whales lactate for several months, while striped dolphins lactate a minimum of 1–2 years (Section 10.5.10).

The methodology used by Subramanian et al. (1988) is promising, but their assumption that between-sex difference in level of organochlorines was due solely to the length of lactation seems to be too simplistic, and some factors remain to be considered.

1. While the three species used by them as standards have nearly the same longevity in males and females, male Baird's beaked whales live 30 years longer than females. So their male sample could have been older on average than the female sample. This can create a small female–male ratio and overestimation of the lactation period, if correlation between pollutant level and age continues to great age against what we find for striped dolphins, where the correlation was lost at age around 20 years (Figure 10.2).
2. The total amount of pollutants transferred to an offspring during lactation must have a closer relationship with the total volume of secreted milk or more directly with the amount of secreted lipid rather than the total length of the lactation period. The daily quantity of secreted milk or lipid is not constant during the lactation period but decreases with increasing consumption of solid food by the calf or during the weaning period. Therefore their method underestimates length of the weaning period and thus the total lactation period. This explains why Dall's porpoises had a similar figure to that of striped dolphins that have a longer weaning period.
3. The pollutant level in a female relative to that in a male of the same age is a function of the proportion of her total life spent in lactation. Thus, it is influenced by the length of immature life and the proportion of lactation time in the reproductive cycle. Age at maturation and length of the calving interval are similar between Baird's beaked whales and striped dolphins off Japan, but the two species still show some difference in the pollutant parameter, suggesting a shorter lactation period in the former species.

The considerations presented earlier still support the general conclusion of Subramanian et al. (1988) of a relatively short lactation time in Baird's beaked whales, although estimation of the absolute length awaits further analysis. It would be a valuable exercise to carry out a similar analysis including short-finned pilot whales, which are known to have an extended weaning period and shorter male longevity. Inclusion of age information in the analysis is essential.

13.5 SOCIAL STRUCTURE

13.5.1 Sexual Maturity and Tooth Eruption

Tooth eruption starts by the age of 6 months among most Delphinidae and Phocoenidae, but it may occur at around the age at sexual maturation (sperm whales) or be limited to mature males (Ziphiidae). Omura et al. (1955) suggested that the teeth erupt in Baird's beaked whales at sexual maturation. I examine this here in some detail using data collected from carcasses processed at Wadaura in the 1985–1987 seasons, which were also used in Kasuya et al. (1997). A pair of the anterior-most teeth were visually examined and classified as "erupted" if any portion of a tooth on either side of the jaw was exposed in the gum. The erupted tooth would attain function slightly later. Some apparently very old individuals had traces of alveoli of lost teeth. These were also classified as "erupted."

The relationship between tooth eruption and age is shown in Table 13.8. Due to the small sample size it was not possible to detect minor sex difference in tooth eruption. With the exception of a female with unerupted teeth at age 14, whales with erupted tooth were found at ages 6 or older, and those with unerupted teeth at ages 10 or younger. Thus, tooth eruption usually occurs at ages 6–11 and all, with unusual exceptions, will have erupted teeth at age 11 or older.

Females attain sexual maturity at age 10–15. Some males start to produce spermatozoa at age 6. The "mature" stage

TABLE 13.8
Age at Eruption of Front Mandibular Tooth of Baird's Beaked Whales Processed at Wadaura on the Pacific Coast of Japan in 1985–1987

Sex	Age	4	5	6	7	8	9	10	11	12	13	14	15–19
Female	Unerupted		1	2	1	1						1	
	Erupted			2	2	2	2	1		3	2		6
Male	Unerupted	1		1		1		1					
	Erupted			2		1	3		2	1		1	7

Source: Kasuya, unpublished.
Note: Age is in number of cemental layers.

of testicular histology occurred at age 6 and older, and all the males were at that stage at 11 years or older (Section 13.4.2). Thus, it was confirmed that the age at tooth eruption almost agrees with age at the attainment of sexual maturity in both sexes.

Tooth eruption is directly compared with the status of the gonads in Table 13.9. Female maturity was determined by the indication of past ovulation. Most sexually mature females (82%) had erupted teeth, and most immature females (88%) did not have erupted teeth. Thus, tooth eruption and first ovulation are considered to occur almost at the same age in females.

Since some ambiguity is unavoidable in determining male sexual maturity, male growth stage was classified for this particular purpose into three stages of testis weight (<0.5, 0.5–1.5, ≥1.5 kg) in Table 13.5. Males with testis of less than 0.5 kg were composed mostly of individuals at the "immature" and some "early maturing" histological stage, and only 14% of them had erupted teeth. The testicular weight group of 0.5–1.5 kg was composed mostly of "late maturing" and some "mature" males, and about half of them had erupted teeth. Whales with testes weighing 1.5 kg or over were mostly composed of "mature" males, and only 5% of them had unerupted teeth. This allows the conclusion that some males have erupted teeth at the "early maturing" or early stage of puberty, about half of males have erupted teeth at the "late maturing" stage, and most males have erupted teeth at the "mature" stage.

Adult Baird's beaked whales have numerous scars on their dorsal surface caused by teeth of the same species

TABLE 13.9
Sexual Maturity and Eruption of Front Mandibular Tooth of Baird's Beaked Whales Processed at Wadaura on the Pacific Coast of Japan in 1985–1990

Sex	Male			Female	
Gonad	<0.5 kg	0.5–1.5 kg	≥1.5 kg	Immature	Mature
Unerupted	6	3	3	15	2
Erupted	1	3	57	4	18

Source: Kasuya, unpublished.

(a)

(b)

FIGURE 13.13 Tooth marks on the body surface of two Baird's beaked whales of adult size processed at Wadaura, Pacific coast of Japan. (a) Dorsal view (note erupted mandibular teeth); (b) ventral view. Also see frontispiece for tooth marks in the species. (Photo by T. Kasuya.)

(Figure 13.3). McCann (1975) interpreted tooth marks on the bodies of ziphiid species as derived from inter-male fighting for mating opportunities. Teeth erupt at a similar age in both sexes of Baird's beaked whales, and males live longer. Therefore, if accumulation of tooth marks is affected solely by age, males should have heavier scars in general. However, it was my impression that density of scars was not much different between sexes or that it was greater on females. Inter-male fighting as hypothesized by McCann (1975) does not explain tooth scars in Baird's beaked whales. Future analysis of age-related density of the scars, with distinction between scars from fighting and from other social contacts, will benefit understanding of social structure in the species.

Baird's beaked whales do not exhibit sexual dimorphism in the teeth, which erupt only 2–3 cm above the gum and are placed just at the end of the upper jaw. As such they would not be an effective tool in inter-male aggression. It would be interesting to understand how and when tooth marks are created on the backs of group members (Figure 13.13).

13.5.2 School Structure

Solitary Baird's beaked whales are uncommon. The whales usually live in a group and are found at the sea surface in a tight school (Balcomb 1989). I presented school size composition from two data sources (Kasuya 1986). One was my own observations assisted by observers in the crow's nest during a sighting cruise in the summer of 1984, and the other was data recorded by Fishery Aviation Co. Ltd. in Tokyo.

The shipboard observations recorded school size in a range of 1–25 individuals, with a mode at 4 (the next most frequent school size was 3) and mean of 7.2. The data from aerial observation were similar to the former in the range of school size, 1–30, but both the mode at 2 (the next most frequent school size was 1) and the mean of 5.9 were smaller than those from the shipboard observations. The possible causes of the difference include overlooking of small groups in the shipboard survey and overlooking of some group members remaining deeper in the water from the fast-moving airplane (Kasuya 1986).

The shipboard surveys, which were often used for abundance estimation, found that smaller schools were more likely to be overlooked than larger schools, particularly those at greater distances. While waiting for re-surfacing of Baird's beaked whales from a long dive during the cruise (Kasuya 1986), I experienced cases where smaller individuals surfaced first followed by a gradual increase of individuals at the surface. Timing of the dive is apparently synchronous among school members, but surfacing is not simultaneous. This means that correct estimation of school size requires some time, which is not possible for the usual aerial survey. The whales are in a very tight school on the surface, but they are probably more dispersed during feeding underwater, while maintaining some contact between school members.

Information on school composition of Baird's beaked whales is limited to two cases of mass stranding. One was on the coast of the Gulf of California (Section 13.2.1; Aurioles-Gamboa 1992) and the other was near the southern tip of the Miura Peninsula (35°08′N, 139°40′E) at the entrance of Tokyo Bay (Kasuya et al. 1997). The former stranding occurred in July 1986 and involved three females and four males. The smallest female was believed to be sexually immature because it measured 9.05 m in body length and had unerupted teeth. The remaining two females had erupted teeth with body lengths and ages of 9.85 m at 13 years and 10.55 m at 17 years, respectively. Thus they were believed to be sexually mature. All the four males were judged as sexually mature, because they had erupted teeth and were in a body-length range of 10.25–11.35 m and age range of 20–42 years.

The school stranded in July 1987 on the coast of the Miura Peninsula contained four whales when found by local people, but one escaped before arrival of whalers and scientists from Wadaura. Two were dead at our arrival and the third was killed by the whalers on the beach before the whales were examined by the biologists. They were a sexually mature female (10.8 m, 31 years old, lactating), immature female (8.2 m, 5 years old), and an "immature" male (8.5 m, 4 years old). Caution is needed in evaluating the younger ages, because they are expressed by the number of cement layers as an integer and may disagree to some degree with real age (Section 13.4.1). The female in lactation could have been the mother of one or both immature individuals. The whale that escaped was unlikely to have been a calf of the lactating female or mother of the juveniles.

The two mass strandings allow the following deductions about school structure in Baird's beaked whales. (1) Adult females may accompany multiple juveniles, which may or may not be her offspring. (2) Multiple mature females and multiple mature males may form a school, which may contain no juveniles. (3) Multiple mature males can live together in a school, which is similar to the case in killer whales and short-finned pilot whales. The same is true for narwhals, where some adult males with developed tusks swim together. Male sperm whales are believed to reject other adult males in a school. The genetic study of school members will be important if an opportunity becomes available for further understanding of the social structure of Baird's beaked whale.

13.5.3 Sex Ratio

Kasuya et al. (1997) analyzed sex ratio in 170 Baird's beaked whales taken off the Pacific coast of central Japan and processed at Wadaura in Chiba Prefecture during five whaling seasons. Sex was determined by scientists, who found only 32.4% to be females. The female deficiency was present in each of the five whaling seasons and in a small sample from the southern Okhotsk Sea taken in 1988 (sex was determined by a scientist). An earlier analysis by Omura et al. (1955) of whaling statistics for the 1948–1952 seasons found a similar female deficiency (Table 13.10).

Matsuura (1942, in Japanese) analyzed statistics from the same Chiba region in different seasons (1935–1939, 1942) and obtained a slightly different female proportion of 45.9%. The Japanese government made a 1-year emergency exception on January 15, 1944, to increase the food supply during World War II, which was to allocate small-type whalers to the companies of large-type whaling and allow them to hunt then-prohibited species, that is, sperm whales and large baleen whales other than minke whales (Maeda and Teraoka 1958; Kushiro-shi Somu-bu Chiiki Shiryo Shitsu 2006, both in Japanese). I do not have evidence to judge whether the government action was ratification after the act or creation of a new system, but there is circumstantial evidence supporting that the former was the case (Section 7.1.1). If the former was the case, it could have created an anomaly in the statistics analyzed by Matsuura (1942). Evaluation of the sex ratio reported by Matsuura (1942) should be suspended until further information on the actual situation becomes available.

A female deficiency such as that identified by Kasuya et al. (1997) and Omura et al. (1955) can occur for various reasons, including (1) a difference in migration pattern between the sexes (population sex ratio assumed even), (2) a difference in vulnerability to whaling (behavior difference on the whaling ground), or (3) the population containing fewer females than males.

Omura et al. (1955) thought that the female deficiency was due to a difference in migration pattern between the sexes. If this hypothesis were correct, one could expect the opposite sex ratio in some other location or season. In order to examine this possibility I examined available information on sex ratio but was unable to find a male deficiency in any geographical area (Table 13.10). Females were always fewer than males, suggesting that the female deficiency in the whaling statistics was not due to segregation between the sexes.

TABLE 13.10
Sex Ratio (Proportion of Females in %) of Baird's Beaked Whales Taken by Whaling or Stranded

Area	Kasuya et al. (1997)	Matsuura (1942)[a]	Omura et al. (1955)	Reeves and Mitchell (1993)	Auriolles-Gamboa (1992)
E. Sea of Japan			44.8 (67)		
Off Chiba[b]	32.4 (170)[f]	45.9 (244)	32.6 (393)		
Off Sanriku[c]			25.2 (322)		
Off Kushiro[d]			25.0 (40)		
Off Abashiri[e]	31.6 (19)		43.1 (102)		
Kuril Islands				22.6 (115)	
Aleutian Islands				9.5 (21)	
Off Canada				8.0 (25)	
Off USA				16.7 (18)[g]	
Off Mexico					42.9 (7)[h]

Note: Sample sizes in parentheses.

[a] Matsuura (1942) is in Japanese. Possibility of contamination by misreported sperm whales remains.

[b] Chiba: Pacific coast of Japan in latitudes of 34°55′N–36°52′N.

[c] Sanriku: Pacific coast of Japan in latitudes of 37°54′N–41°35′N.

[d] Kushiro: Pacific coast of Japan at 42°59′N, 144°23′E.

[e] Abashiri: Okhotsk Sea coast of Japan at 44°01′N, 144°16′E.

[f] Including three individuals (two females and one male) in a group of four stranded at Miura Cape near the entrance of Tokyo Bay.

[g] Including one individual taken by Soviet fleet.

[h] A group stranding.

Minimum size limits did not apply to Japanese small-type whaling. Female Baird's beaked whales grow to about 35 cm greater than males of the same species. Although it is doubtful that such a limited size difference affects the gunner's selection, the effect would have contributed to a higher, not lower, female proportion in the catch.

A prohibition on taking females accompanied by a calf was applied to small-type whaling but apparently was ignored by small-type whalers (my personal observation). A difference in catchability between the sexes, if it existed, could have caused the skewed sex ratio. When a school of Baird's beaked whales was sighted by a catcher boat operating from the Chiba coast, the vessel moved rapidly toward it to within a distance of about 400–500 m. Then it gradually lowered speed and finally approached by momentum to within 50 m (Kasuya and Ohsumi 1984). The vessel had to resume speed if it failed to move to within the shooting distance of less than 50 m, but this usually alerted the whales and caused them to dive. If the gunner had a good opportunity to fire within the 50 m distance, his priority was to attach a line to any position on the body using a cold harpoon head and then kill the animal with a second shot. In this type of whaling operation, there was limited opportunity for gunners to select particular individuals in a tight school, to select for larger individuals or to ignore mother–calf pairs (if so intended). Thus the harpooning process is not a likely cause to explain the female deficiency.

Baird's beaked whales off the Pacific coast of Japan were found in summer in waters between 1000 and 3000 m isobaths and were hunted by whalers traveling daily from port. Therefore, if males tended to inhabit the nearshore area, the whalers could have had a higher opportunity of killing males, resulting in the female deficiency in the catch statistics. In order to examine the possibility of sexual segregation by distance from the coast, I examined sex ratio in Baird's beaked whales taken by a factory ship of the former USSR that operated along the Aleutian Islands area. Out of the 28 individuals of the species taken by them and reported by Reeves and Mitchell (1993), there were only 6 females (21.5%). Thus, the hypothesis of offshore segregation of females is not supported.

If females are more nervous about approaching vessels and quicker to dive or tended to be more distant from an approaching catcher boat, we could expect fewer females in the catch statistics. This hypothesis has not been tested.

The remaining explanation for the female deficiency in the whaling statistics is the difference in longevity between the sexes. Kasuya et al. (1997) analyzed age composition of the whales processed at Wadaura (Table 13.11). The ages ranged from 3 to 84 years and had a mode at around 6–9 years for both sexes. The gunners, who operated with a catch quota, probably avoided shooting younger whales for economic reasons. The sex ratio was almost even for ages below 19. Then the proportion of females decreased with increasing age to reach zero at age 55 years and older. The oldest female was 54 and the oldest male 84 years old, and there were 22 males older than the oldest female (Table 13.11). Females experience their first ovulation at 10–15 years, and most of them are thought to conceive at the first ovulation, but males perhaps attain reproductive ability at 6–11 years (Section 13.4.2).

TABLE 13.11
Sex Ratio and Age of Baird's Beaked Whales Processed at Wadaura on the Pacific Coast of Japan, in 1975 and 1985–1988

Age	Males	Females	Total	Female %
3–9	13	16	29	55.2
10–19	17	16	33	48.5
20–29	13	7	20	35.0
30–39	8	1	9	11.1
40–54	12	7	19	36.8
55–84	22	0	22	0
No. age	30	8	38	21.1
Total	115	55	170	32.4

Source: Prepared from Kasuya, T. et al., *Rep. Int. Whal. Commn.*, 47, 969, 1997, Table 6.
Note: Age in number of cemental layers.
The 1988 samples not aged but included in the total.

The sex ratio was almost even before sexual maturity, but after sexual maturity females decreased in number at a higher rate than males. Thus the female deficiency of the species in the whaling statistics is explained by a higher natural mortality in sexually mature females and shorter female longevity.

To summarize: (1) a female deficiency exists in the entire range of the species (it cannot be explained by segregation), (2) the longevity difference of 30 years is hard to explain solely by an yet unconfirmed ship-avoidance behavior of old females, and (3) higher natural mortality of sexually mature females probably explains the female deficiency in the catch.

13.5.4 Why Do Males Live Longer?

Understanding the benefit the community of Baird's beaked whales receives from greater male longevity and the process by which the male trait has evolved requires knowledge on their social structure and behavior, but such information is almost completely unavailable to us. Compared with female members of the same species, males of most mammalian species grow larger, have a higher mortality rate, and die sooner. It is generally accepted that species with greater sexual dimorphism tend to have a polygynous reproductive system. In such polygynous species, females have adapted to increased lifetime production by extending reproductive lifetime, while males have achieved their objective by monopolizing opportunity of reproduction through fighting with other males (Ralls et al. 1980). The higher mortality and shorter life span of males in such species are thought to have arisen from higher stress due to competition and greater nutritional requirements for a larger body. Greater female size in some mammals has been interpreted as a result of adaptation for greater survival of larger neonates and for nursing during starvation. Some baleen whales are a good example, where mothers produce calves during winter starvation and bring them to higher latitudes for weaning.

Sperm whales are sexually dimorphic, and their social structure has been extensively studied. Unhealed tooth marks that increase on the male body during the mating season have been interpreted as evidence of inter-male aggression for mating opportunities (Kato 1984). Adult male sperm whales segregate in the nonmating season to higher latitudes, leaving breeding schools of females and juveniles in lower latitudes. Commercial whaling that ended in the 1980s selectively hunted adult males and caused a skewed sex ratio in the population. These factors have made it difficult to determine the natural sex ratio that existed in preexploitation times from either analyses of the past catch statistics or observation of current wild populations. However, some believe that longevity of sperm whales is not very different between the sexes (Ohsumi 1971; Best 1979). Age data for North Pacific sperm whales collected by Japanese scientists gave ages of oldest individuals in a range of 72–75 years for both sexes and suggested similar natural mortality rates for them. A view was presented that the extended (or similar to female) longevity by sexually dimorphic males became possible through segregation from breeding schools. Adult male sperm whales leave the breeding ground, where females and juveniles remain, to move in the nonbreeding season to higher latitudes of high productivity (Kasuya and Miyashita 1988). This decreases their nutritional stress, allowing faster recovery from reproductive activity, and decreases mortality through avoiding competition with nursing schools and utilizing the high productivity at higher latitudes (Kasuya 1995). Currently it is unknown whether these males participate in reproduction annually.

In summary, female Baird's beaked whales grow slightly larger than males and attain sexual maturity later. Their mean breeding cycle is estimated at 3–4 years, and there is no evidence to suggest age-related decline in female fecundity. Both sexes show reproductive capability for their entire life after maturation (although participation in reproduction has not been confirmed in a direct way for any male). The sex ratio is even before the age of sexual maturation, and then the female proportion decreases with increasing age and females live 30 years less than males. Therefore the population contains a greater number of sexually mature males than females. This together with an annual ovulation rate of 0.47 provides a large supply of mature males for the estrous females, but the available information does not support the possibility of inter-male competition for mating opportunities

If female maturity is determined by ovulation and that of males by testicular histology (Section 13.4.2), the sex ratio in mature individuals is 1 female to 3.3 males (Kasuya et al. 1997). This imbalance in sex ratio still remains even under the unrealistic assumption that males attain reproductive ability at age 30 or 40 years when age-related increase in testicular weight ceases, at a female to male ratio of 1:1.7 or 1:1.2. Thus, a hypothesis of female preference for old males does not fully explain the longevity difference.

The longevity pattern and social structure of short-finned pilot whales (Sections 12.4.3 and 12.5.3) do not fully match with those of either the sperm whale or Baird's beaked whale. In the short-finned pilot whale

1. Multiple adult males can coexist in a school, which is seen in Baird's beaked whale but not in the sperm whale
2. Sexual dimorphism is present, which is not seen in Baird's beaked whale but is seen in the sperm whale
3. Females live 20 years longer than males, which is not the case with Baird's beaked whale and is not the case with the sperm whale
4. Inter-male aggression seems to be uncommon, which is similar to behavior in Baird's beaked whale but not the sperm whale
5. Females have a significant postreproductive life span, which is similar to the sperm whale but not Baird's beaked whale.

Short-fined pilot whales could have developed a matrilineal social system as an extension of extended maternal care for the offspring, which offered the opportunity of cooperation among genetically related females. The contribution of these females to survival of their kin or to the rearing of offspring of their kin could have offered an opportunity for emergence of extended postreproductive life span in females (Kasuya and Marsh 1984; Marsh and Kasuya 1986). It is also noted that short-finned pilot whales have developed some way to reduce inter-male aggression for mating opportunities, which could include nonreproductive mating or generous receptivity by females (Section 12.4.4.8).

An extended female postreproductive life span is observed in the sperm whale, short-finned pilot whale, killer whale, and perhaps false killer whale. Using the communities of some primates as a model, this is interpreted to have probably evolved accompanying the formation of a matrilineal community (Section 12.5.7). The presence of old females in a school will reduce the reproductive burden on her daughters through cooperation in educating and protecting their grandchildren, which will increase the total lifetime production of old females. Experience and knowledge accumulated in the old females can benefit the lives of all the community members. If such a trait of long postreproductive lifetime emerges in a group, it will have the chance to increase in the community. The contribution of males to rearing their offspring is probably minimal because males live a shorter life (in killer whales and short-finned pilot whales), males do not father the calves in their school (killer whales and short-finned pilot whales), or because males segregate from breeding schools (sperm whales).

This reasoning led Kasuya and Brownell (1989) and Kasuya et al. (1997) to a hypothesis of patrilineal social structure for Baird's beaked whales, where males participate in rearing their offspring or grandchildren in the school. Baird's beaked whales feed in deep water (Section 13.3) and a lactating mother cannot abstain from deep diving, because she is in the peak period of nutritional requirement (Section 12.5.7.2). Her suckling calf will remain near the sea surface during feeding activity of the mother because it has limited diving ability and no need to dive to feed. If some other nonlactating individual accompanies the sucking calf while its mother is feeding deep, it will reduce the burden on the mother and enable her to breed in the short interval of 3–4 years. Such "babysitters" could also function, at some stage, in teaching the calf the technique for feeding in deep water. If genetically related males participate in these tasks of "babysitting," such traits will spread in the population and selection will work toward greater male longevity. For this mechanism to work, there must be a way for males to identify their offspring or a social structure where males live with their offspring with high probability. This was the basis of the hypothesis of existence of a patrilineal community in Baird's beaked whales proposed by Kasuya and Brownell (1989). Old males in the community could also function as carriers of culture.

In a patrilineal community, sons would remain in the paternal group after maturity but daughters could move to other groups. We know examples of such communities in humans and chimpanzees but not among cetacean species. Male cooperation in rearing calves is likely to occur in a community where parental pairing extends beyond the reproductive cycle of a female (which is unknown among cetaceans), where brothers cooperate in securing female(s) for reproduction (a similar example is known in bottlenose dolphins [Connor et al. 2000]), or where sexually mature daughters and sons live together with their mother (seen in the resident population of killer whales off Vancouver Island; Baird 2000).

Further effort is needed to understand the social structure and reproductive behavior of Baird's beaked whales and the process of evolution of such communities. It should also be noted that various factors may cause sexual dimorphism and sexual difference in longevity or mortality rate. Even if you can hypothesize the origin of these elements from the social structure of a species, you can never be sure about accurately estimating the social structure of another species based on observed sexual dimorphism or life history traits. Baird's beaked whale is no exception.

13.6 ABUNDANCE

Abundance of Baird's beaked whales has been estimated for only three regions around Japan: (1) Pacific waters off Japan, (2) eastern part of the Sea of Japan, and (3) southern Okhotsk Sea (Table 13.12).

Miyashita (1986) was the first to estimate abundance off the Pacific coast of Japan, and his is the only study published in a scientific manner. The data used by him were obtained from the sighting cruise of the *Toshimaru No. 25* in 1984, which was the same cruise used in the behavior study of Kasuya (1986). He used 29 primary sightings for the estimation. The remaining 13 sightings made during the cruise were secondary sightings obtained while confirming the primary sightings. In the first step he obtained a total abundance of 3547 as the sum of figures calculated for each 1° square of longitude and latitude of the surveyed area (Figure 13.3). This figure was corrected for the decline of sighting rate with increasing perpendicular distance from the track line but was based on an

TABLE 13.12
Abundance of Baird's Beaked Whales around Japan Estimated by Shipboard Sighting Surveys

Author	Year of Survey	Pacific Coast of Japan	Eastern Sea of Japan	Southern Okhotsk Sea
Miyashita (1986)[a]	1984	4,220 (0.295)		
Anon. (1990)[b]		2,500	1000[c]	
Miyashita (1990)[a], IWC (1991)	1983–1989	3,948 (0.276)	1468 (0.389)	663 (0.270)
Miyashita and Kato (1993), IWC (1994)	1991–1992	5,029 (1,801–14,085)		

Note: Coefficient of variation or 95% confidence interval in parentheses.
[a] The coefficient of variation predicts a 95% confidence interval extending over 50% on each side of the mean estimate.
[b] Not corrected for g(0). See text for correction of g(0) applied to other estimates.
[c] Abundance in Japanese EEZ extending from eastern Sea of Japan to southern Okhotsk Sea.

assumption of 100% probability of sighting of whales on the track line. Baird's beaked whales are usually found in a group of several animals, and even singletons. They repeat blowing several times at the surface. Thus the chance of observers missing whales at the surface is small. The observers scanned 180° in front of the survey vessel, which cruised at 11–12 knots (20–22 km/h). They were able to identify the presence of whales within a radius of 5–6 nautical miles (about 10 km).

Baird's beaked whales can remain underwater for a maximum of around 60 min (Section 13.3.2), while the research vessel can proceed about 10 km, that is, beyond the sighting range of the onboard observers. Thus, the assumption of sighting 100% of the whales on the track line seems to be violated. Miyashita (1986) used the surfacing and diving interval data of Kasuya (1986) to estimate the probability of the whales being passed by the vessel while they remained underwater. This suggested a 19% increase in the abundance estimate, for a corrected estimate of 4220 Baird's beaked whales in the surveyed area. His analysis assumed that all the whales that surfaced within 5.4 nautical miles on the track line (front side) were seen, but this could also be optimistic to a minor degree (and lead to an underestimate of abundance). The provisional estimate of Miyashita was reviewed and accepted at the meeting of the Scientific Committee of the IWC in 1985 and later finalized and published in Miyashita (1986).

The second estimate of the population appeared as a provisional figure in the Japanese progress report presented to the Scientific Committee Meeting of the IWC in 1989 (Anon 1990) and was cited in IWC (1990).

The third estimate was in a document presented to the Scientific Committee Meeting of the IWC in 1990 (Miyashita 1990), but this study has not been published as a scientific paper. The committee reviewed this document and cited the figure in IWC (1991) together with its comments. Miyashita (1990) used the same method as used in Miyashita (1986) and the sighting data covered the same geographical area but more years (1983 to 1989). He combined the data for several years and estimated the abundance for each month, which gave abundance estimates of 1935 (CV 0.349, n = 13) for July, 3947 (CV 0.276, n = 52) for August, and 1078 (CV 0.393, n = 26) for September, where n indicates the number of schools used for the calculation. The estimates for the months were not significantly different, but the figure for August appeared to be the greatest. Miyashita (1990) took the August figure as the abundance of the population.

The Scientific Committee made two statements about the estimate. If August was the migration peak of the species in the area, then the abundance estimate could be biased downward because some whales could have already left the area and some others could be on the way to the area. However, if the monthly figures represented only fluctuation of abundance due to limited sample size, then use of the largest figure, for August, would bias the abundance upward. Currently there are insufficient data on the migration of Baird's beaked whales off the Pacific coast of Japan to choose one of the two alternative interpretations. However, I have the impression that the first is the case, because of the similarity of the seasonal abundance trend with seasonal density change estimated using independent data (Section 13.2.3.2).

The fourth estimate of abundance of the species off the Pacific coast of Japan was in an unpublished document of Miyashita and Kato (1993) presented to the Scientific Committee Meeting of the IWC and cited in its report (IWC 1994). The mean average estimate was the same in these documents, but the accompanying confidence interval was greater in IWC (1994). The latter interval, which is shown in Table 13.12, was possibly a confidence interval recalculated reflecting discussion in the committee meeting.

Miyashita (1990) also made estimates of abundance of Baird's beaked whales in the eastern half of the Sea of Japan and in the southern Okhotsk Sea south of 45°N by combining sighting data of May–October (Table 13.12).

The abundance estimates are over 15 years old and were accompanied by broad confidence intervals. The broad range of the estimated abundances, from 2,000 to 14,000, made them difficult to use for management purposes. None of the differences among the four estimates for the Pacific population are statistically significant.

13.7 HISTORY OF EXPLOITATION

13.7.1 Types of Whaling Operation

Supported by the customs of people in the Awa Region, or the southern part of Chiba Prefecture who consumed dried meat of Baird's beaked whales, early hunting for Baird's beaked whale in Japan existed only around the Chiba area. Expansion of the fishery to other regions of Japan occurred after World War II (Ohsumi 1975, 1983).

I will list some publications useful for understanding the history of the Baird's beaked whale fishery in the Chiba area. Kishinoue (1914, in Japanese) published a book *Awagun Suisan Enkakushi* [History of Fisheries in Awa County], which collected original historical documents on early whaling in the Awa Region. An unpublished book by Takenaka (1887, in Japanese) *Bonan Hogei-shi* [History of Whaling in Southern Awa] covered the whaling system, whaling techniques, whaling ground, and whaling statistics in the period from 1703 to 1886. He also copied some historical documents on the whaling. Komaki (1969, 1994, and 1996, all in Japanese) added information on the recent operation of the fishery. Yoshihara (1976, in Japanese) gave the whaling history in the Chiba area, and Yoshihara (1982) compiled most of the whaling information in the references mentioned earlier. Recent publications on the fishery include Kanari (1983, in Japanese), Ohsumi (1975, 1983), and Kasuya (1995, in Japanese).

While people of the Chiba area pursued Baird's beaked whales since the seventeenth century, the fishing ground gradually shifted from the entrance of Tokyo Bay to the outer seas, and the whaling technique from hand-harpooning to Norwegian-type whaling. Yoshihara (1976, in Japanese) cited a text in *Keicho Kenbun Shu* [Incidents in the Keicho Era (1596–1615)] written by Joshin Miura, who wrote that Sukebei Maze of Owari (Ise and Mikawa Bays area on the Pacific coast at longitudes 136°40′E–137°15′E) came to the Miura Peninsula and started hand-harpoon whaling in 1592–1596 and that the fishery spread to the nearby area. Takenaka (1887, in Japanese) cited a letter of 1612 from Tadayosh Satomi, governor of Awa, to a Shinto priest of the Ise Shrine concerning donation of whale products. Miura and Awa were located on the western and eastern sides of the channel leading to Tokyo Bay, respectively. These records indicate that the fishery for Baird's beaked whales was established at the entrance of Tokyo Bay around 1600. However, all the old whaling records left in the region were lost to a tsunami in 1703.

According to Takenaka (1887, in Japanese) the oldest available record of whaling by the Daigo family in Kachiyama (present Katsuyama at 35°07′N, 139°50′E) on the eastern coast of the channel leading to Tokyo Bay went back to 1704. The Daigo family organized whale fishermen, who operated the whaling during June–August using their own harpoons and boats. The Daigo family offered only whale line and 0.9 L/day/person of rice for food to the member fishermen. They were allowed to pursue swordfish during the whaling season. Noting that their whaling tools were the same as those used in the hand-harpoon fishery for billfish and tuna, Komaki (1969, in Japanese) thought that the whaling originated from that fishery. The Daigo family purchased whales thus taken at a price it determined and subtracted the price of the rice from the price of the whale. The Daigo family was really in a lucrative situation in the whaling. This whaling group apparently carried out the last operation in 1869. Although there were several subsequent sporadic whaling records, for example, takes of 8 whales in 2 seasons of 1879 and 1880 (Kishinoue 1914, in Japanese) and several whales taken in 1884 as a response to an increased purchase price (Takenaka 1887, in Japanese), the reduced fishery could not be restored with the traditional whaling system.

Takenaka (1887, in Japanese) stated that there were two whaling groups in 1887, based at Kachiyama and nearby Iwaifukuro (35°07′N, 139°50′E), operating with 33 and 24 boats, respectively. He stated that Baird's beaked whales arrived from the outer sea along the submarine canyon to the entrance of a channel leading into Tokyo Bay in early summer to mid-summer and left the area in autumn for the outer eastern sea. The whaling was operated in an area of 20 × 30 km at the entrance of Tokyo Bay, or north of Cape Mera (34°55′N, 139°50′E), south of Takeoka (35°12′N, 139°50′E), and east of longitude of Jogashima Island (139°37′E). The eastern extent of the whaling ground was bounded by the coast of Chiba Prefecture.

A branch of a submarine canyon extends northward from Sagami Bay and reaches the whaling ground off Katsuyama; the maximum depth at the northern end of the whaling ground was about 200 m and at the southern limit about 1000 m. The current summer range of Baird's beaked whales along the Pacific coast of Japan is limited to between the 1000 and 3000 m isobaths. Distribution of the species extended to shallower waters in earlier times, when the abundance was probably greater than at present, and supported primitive whaling in the periphery of the range. Pioneers of modern whaling were surprised to find a high density of Baird's beaked whales south of the traditional whaling ground (Sekizawa 1892b, in Japanese). Yoshihara (1976, in Japanese) stated that Shinbei Daigo in 1903 requested permission to whale in a broader area between 139°18′E and 139°53′E north of 34°40′N, but this does not necessarily mean that the operation covered the whole area.

Document No. 182 reproduced in Kishinoue (1914, in Japanese) was written in 1881 by a villager of the Awa region for the Ministry of Finance and listed the whale species taken by the Awa whalers. The five species listed were *kokusira* or *tsuchin-bo* [Baird's beaked whale], *koto-kujira* [pilot whale], *uki-kujira* [*Kogia* spp.], *ohnan-kujira* [false killer whale], and *kaji-kujira* [Cuvier's beaked whale]. The document stated that "the local name *tsuchin-bo* was synonymous with the correct common name *kokushira*." It also stated there had been a decline of large species such as *makko* [sperm], *zato* [humpback], *sebi* [right], and *nagasu-kujira* [blue or fin whale (see Section 1.3)]. Another document No. 183 dated 1881 in Kishinoue (1914, in Japanese) added to the five species mentioned earlier a sixth target species *ko-kujira* (in two Chinese characters meaning small whale). Thus, these documents distinguished *kokushira* [Baird's beaked whale] and *ko-kujira*. The latter likely meant gray whale in the Chiba area (see Section 13.1). Takenaka (1887) stated that whale oil from the blubber was exported outside the area for lubricants, lighting oil, and a pesticide for use in rice paddies; it was exported to foreign countries in the late nineteenth century. He also stated that viscera, bones, and residues from the try works were used for fertilizer. The meat was given to the whalers as their salary or sold to local people, who made the meat into jerky.

The custom of making jerky from Baird's beaked whale still persists in the Awa region. People visit the whaling station at Wadaura for meat, or a retailer travels around the inland region on bicycle with the whale meat (Freeman 1989; Yamaguchi 1999, both in Japanese). Further details on the whaling equipment and methods of operation are given in Takenaka (1887, in Japanese) and Yoshihara (1976, in Japanese).

The 55 years (1815–1869) of catch statistics for Baird's beaked whales taken by the Daigo group were published by Takenaka (1887, in Japanese) and analyzed by Ohsumi (1983) and Kasuya (1995, in Japanese) (Figure 13.14). The statistics after this period are incomplete. Total catch of the species in the 55-year period was 504 whales, with an average of 9.2 per year. The broad annual fluctuation of 0–25 whales is a reflection of a fishery operating near the periphery of the distribution range.

Kasuya (1995, in Japanese) identified a declining trend in the catch history. There were 9 good years with catches of over 14 whales, and 19 poor years with fewer than 6 whales. The proportion of good years was high in the first 27 years (1815–1841) and low in the later 28 years (1842–1869), and the proportion of bad years increased with time (Table 13.13). This historical change could have reflected a decline in the

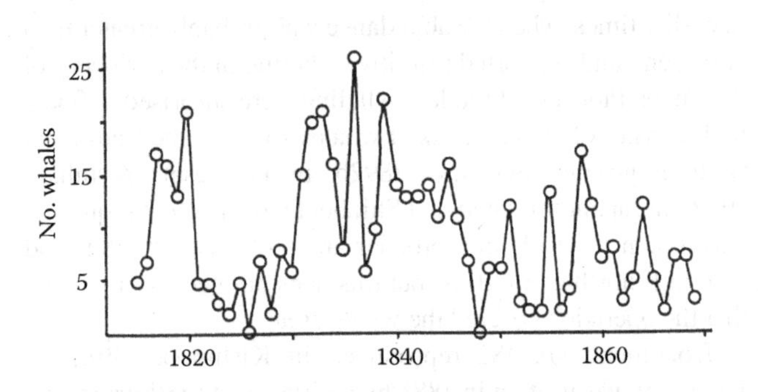

FIGURE 13.14 Annual number of Baird's beaked whales taken during 1815–1869 by hand-harpoon whaler Daigo Group, which was based at Katsuyama (35°07′N, 139°50′E) at the entrance of Tokyo Bay. (Reproduced from Ohsumi, S., *Rep. Int. Whal. Commn.*, 33, 633, 1983. With permission of the author, Figure 2.)

TABLE 13.13
An Early Symptom of Local Decline of Baird's Beaked Whales in the Coastal Whaling Operation of the Daigo Group at Katsuyama[a] in Chiba Prefecture, Where Baird's Beaked Whales Were the Major Target

Catch	1815–1841	1842–1869	Total
No. of whaling seasons	27	28	55
No. of good years (≥15 whales)	7	2	9
No. of bad years (≤5 whales)	8	11	19

Source: Compiled from statistics given in Takenaka, K., *History of Whaling in Southern Awa County*, Vols. 1 and 2, Unpublished, 1887, in Japanese.

[a] Katsuyama (35°07′N, 139°50′E) is at the entrance of Tokyo Bay.

number of Baird's beaked whales migrating to the whaling ground. The owner of the whaling team thought that the poor catch was due to the lack of enthusiasm by the fishermen and to the effect of increased operation of night drift nets (Document No. 183 dated 1881 in Kishinoue 1914, in Japanese). The latter idea assumed that the drift nets blocked migration of whales to the whaling ground. This was very similar to the claim made by some modern coastal whalers, who attributed a decline in the sperm whale fishery off the Pacific coast of Japan to the operation of large-mesh and squid gill-net fisheries. I do not totally deny the effect of net fisheries on migration of whales to nearshore waters, but I believe that overharvesting was more responsible for the change.

The Baird's beaked whale fishery at the entrance of Tokyo Bay showed an apparent catch decline during the nineteenth century. This was probably due to a decline in the number of whales visiting the whaling ground. The whaling ground was located at the periphery of the range of the population of Baird's beaked whales, and it is likely that reduction of the range of the population accompanied decline in population size, due perhaps to whaling. Thus the decline was more pronounced in the coastal whaling ground than in more offshore waters inhabited by the main body of the population.

Another possible explanation is the effect of behavior change in the species, which does not eliminate the possibility of population decline. The Baird's beaked whale is a species of long life extending to over 80 years (Table 13.11). The pattern of seasonal movement of individual whales in such a species will be determined by past experience and will vary among individuals. Humpback whale calves learn the way to their feeding ground by accompanying the mother in the first feeding summer and adhere to it during subsequent years. It is possible that Baird's beaked whales learn the way to their summering ground from other individuals of the species including their mothers and modify it based on their own later experience. They will revisit the summering ground if they had a positive experience there in the prior year. Due to whaling mortality, whales that summered in the coastal waters near the whaling ground had less opportunity for reproduction or for leading other individuals to the summering ground the next year, and whales that used the coastal summering ground gradually decreased in number while individuals that summered in the offshore waters were still abundant. This is a loss of the cultural trait of the population using the submarine canyon at the entrance of Tokyo Bay for summer feeding. Management of fishery resources tends to attach importance to conserving a population and ignores individuality. Such an approach is likely to fail in the management of cetaceans.

A third factor in the apparent population decline is the effect of unconfirmed operation of sailing-ship whalers from other countries, which first arrived in Japanese waters around 1820 and hunted mainly sperm whales but also took right whales, gray whales, and humpback whales. Although I do not have evidence of their operation on Baird's beaked whales, there are some suggestions of it. The *Kasshi Yawa,* written by Kiyoshi Matsuura (published posthumously in 1977–1982 in Japanese), a collection of his essays and descriptions of incidents during

1821 to 1841, recorded some incidents that suggested hunting of Baird's beaked whales by the foreign whalers. One English whaler in April–May 1822 entered the port of Uraga (35°15′N. 139°43′E) to get supplies. While the vessel stayed in the port the whalers sighted a whale and lowered for the chase, but the action was stopped by the government. Uraga faced the whaling ground for Baird's beaked whales of the time, and they could have attempted to take the species. Matsaura recorded a whale ship that anchored about 12 km off the Choshi Cape (35°42′N, 140°51′E) every year and hunted whales until August. They could have hunted short-finned pilot whales, Baird's beaked whales or sperm whales. As Baird's beaked whales are the same length as female sperm whales, they could have attracted the whalers. He also recorded several foreign whaling ships off Mito (36°20′N, 140°35′E) and off Onahama (36°57N, 140°54′E), both on the Pacific coast.

Although direct evidence is lacking, ship-avoiding behavior of Baird's beaked whales probably changes with experience. As whaling of the time used sailing or rowed boats, the whales were less nervous about approaching whalers than about the present-day catcher boats. Baird's -beaked whalers of Chiba first introduced motor driven catcher boats in 1907, and approaching the target was extremely difficult in the 1980s. In the Sea of Japan there was annual take of 2–26 Baird's beaked whales during 1948–1952 (Figure 13.4), and there has been almost no catch there since (Table 13.14). It was unfortunate that we have no record of whale behavior during the post–World War II hunting. After a pause of over 30 years our scientists on survey vessels in the Sea of Japan found that their vessel was approached by Baird's beaked whales (see Section 13.2.3.2). Their wariness of approaching vessels could have disappeared during the period. With an annual quota of eight Baird's beaked whales, Japanese small-type whalers resumed hunting the species off the Oshima Peninsula on the Sea of Japan coast of southwestern Hokkaido since 1999 (Table 13.15). The effect on behavior of the species will be worth watching.

Whaling for Baird's beaked whales by the Daigo group carried out its last operation in Chiba Prefecture in 1869 (Takenaka 1887, in Japanese). The cessation in whaling could not be attributed to a change in the human community such as a decrease in demand for whale meat or in enthusiasm for whaling, because the Daigo family sent out four exploratory whaling expeditions to Hokkaido and Sakhalin in 1802–1863 and attempted Baird's -beaked whaling in the offshore waters off Chiba using new equipment copied from imported whaling instruments (Takenaka 1887, in Japanese). One such instrument was similar to Pierce's harpoon bomb lance gun in Scammon (1874). This harpoon was shaped like an ordinary hand harpoon but had a barrel near the head. When the harpoon was placed in the whale's body, with the action of a trigger a bomb was discharged from the barrel to explode in the whale. The harpoon and boat were connected with a line; the maximum shooting distance was 25 yards (Scammon 1874).

In 1887 Akikiyo Sekizawa, with cooperation of the Daigo family and using a bomb lance, tested whaling off Oshima Island (37°44′N, 139°24′E), which is located at the entrance

TABLE 13.14
Official Statistics for Baird's Beaked Whales Taken by Japanese Small-type Whaling[a]

Year	Chiba	Sanriku	Kushiro	Abashiri	Sea of Japan	Total
1932	31	—	—	—	—	—
1933	31	—	—	—	—	—
1934	34	—	—	—	—	—
1935	35	—	—	—	—	—
1936	33	—	—	—	—	—
1937	50	—	—	—	—	—
1938	34	—	—	—	—	—
1939	50	—	—	—	—	—
1940	25	—	—	—	—	—
1941	23	—	—	—	—	—
1942	36	—	—	—	—	—
1943–1946	—	—	—	—	—	Missing
1947	60					
1948	43	0	2	24	4	76
1949	48	0	0	30	14	95
1950	122	18	1	19	26	197
1951	108	102	11	18	13	242
1952	72	202	26	11	10	322
1953	83	—	—	—	—	270
1954	76	—	—	—	—	230
1955	52	—	—	—	—	258
1956	53	—	—	—	—	297
1957	73	—	—	—	—	186
1958	82	—	—	—	—	229
1959	73	—	—	—	—	186
1960	79	—	—	—	—	147
1961	72	—	—	—	—	133
1962	64	—	—	—	—	145
1963	81	—	—	—	—	160
1964	68	—	—	—	—	189
1965	68	60	19	25	0	172
1966	85	54	17	15	0	171
1967	58	27	14	8	0	107
1968	80	27	9	1	0	117
1969	91	32	4	7	0	138
1970	54	—	—	—	—	113
1971	68	—	—	—	—	118
1972	79	—	—	—	—	86
1973	30	—	—	—	—	32
1974	32	—	—	—	—	32
1975	39	—	—	—	—	46
1976	11	—	—	—	—	13
1977	28	—	—	—	—	44
1978	33	—	—	—	—	36
1979	28	—	—	—	—	28
1980	31	—	—	—	—	31
1981	36	—	—	—	—	39
1982	57	0	0	3	0	60
1983	33	0	0	4	0	37
1984	35	0	0	3	0	38
1985	36	0	0	4	0	40

(Continued)

TABLE 13.14 (*Continued*)
Official Statistics for Baird's Beaked Whales Taken by Japanese Small-type Whaling

Year	Chiba	Sanriku	Kushiro	Abashiri	Sea of Japan	Total
1986	35	0	0	5	0	40
1987	35	0	0	5	0	40
1988	22	13	0	22	0	57
1989	27	22	0	5	0	54
1990	27	25	0	2	0	54
1991	27	25	0	2	0	54
1992	27	25	0	2	0	54
1993	27	25	0	2	0	54
1994	27	25	0	2	0	54
1995	27	25	0	2	0	54
1996	27	25	0	2	0	54
1997	27	26	0	1	0	54
1998	26	26	0	2	0	54
1999	26	26	0	2	8	62
2000	26	26	0	2	8	62
2001	26	26	0	2	8	62
2002	26	26	0	2	8	62
2003	26	26	0	2	8	62
2004	26	26	0	2	8	62
2005	26	26	0	4	10	66
2006	26	25	0	2	10	63
2007	26	27	0	4	10	67
2008	26	25	0	3	10	64
2009	26	27	0	4	10	67
2010	26	26	0	4	10	66
2011	26	5[b]	16[b]	4	10	61
2012	26	31[c]	0	4	10	71
2013	22	26	0	4	10	62

Note: Left-over quota could be carried over to the next year. This table has been revised with the addition of new data available since the original Japanese language edition appeared in 2011.

[a] Matsuura (1942, in Balcomb 1977) for 1932–1947; Omura et al. (1955) for regional figures of 1948–1952; Nishiwaki and Oguro (1971) for regional figures of 1965–1969; and Ohsumi (1983) and Kasuya and Ohsumi (1984) for national total of 1948–1981. Boundaries between the first three geographical areas may be disagreed between scientists.

[b] Temporary shift of operation due to damages of Ayukawa facilities caused by tsunami on March 11, 2011.

of Sagami Bay, and caught a mother Baird's beaked whale and calf (Sekizawa 1887a,b, in Japanese). He used a handheld gun of 33 mm caliber that discharged an explosive bomb with rubber stabilizing blades. He also used hand harpoons to connect the whale and the boat. This technique was called gun whaling (Torisu 1999, in Japanese).

In 1891 Sekizawa made a second attempt at whaling for Baird's beaked whales off Oshima Island using a "middle-sized whaling cannon," which was the same as a Greener's harpoon gun of about 30 mm caliber as described by Scammon (1874). The harpoon gun was fixed on a cradle at the bow of a rowed boat, and the harpoon was attached to a whaling line. The harpoon was not equipped with an explosive harpoon head. His attempt in 1891 was unsuccessful (Sekizawa 1892a, in Japanese), but another attempt the next year with 7 boats and 54 fishermen caught 7 Baird's beaked whales during the period from 7 July to 25 August (Sekizawa 1892b, in Japanese).

In those days the Japanese whaling industries made various attempts to develop whaling technology. Exhibited at the Second Fishery Exposition in 1898 were equipment of old-fashioned net whaling, a copy of an imported bomb lance, and a double-barreled modification of Greener's harpoon gun (Kaneda and Niwa 1899, in Japanese). The two barrels of the Greener's gun were side by side and had two triggers, which suggests that two harpoons were launched independently. Kenzo Maeda of Taiji was shown this type of gun by Sekizawa in Busan, Korea, and in 1903 developed a three-barreled whaling gun with the barrels arranged vertically and fired simultaneously, and in the next year he increased the barrels to five and obtained a patent for the "Multi-barrel pilot whale gun of Maeda" (Hamanaka 1979, in Japanese). Lines from the five harpoons were connected to a single main whaling line. This type of whaling cannon was used in the Taiji pilot whale fishery until around 1967.

With the success of Sekizawa in taking Baird's beaked whales off Chiba, the Nihon Suisan Kaisha was established in Tateyama (35°00′N, 139°52′E) in Chiba in 1888. It had a low level of take using bomb lances and hand harpoons and was dissolved in 1891 (Komaki 1969, in Japanese). The catch in 1890 was five Baird's beaked whales, considered good for the year. The operation must have been a wasteful one because the take of 5 whales was accompanied by 12 struck and lost whales (Daigo 1890, in Japanese).

The so-called American whaling, which used a mother sailing ship and rowed boats for harpooning, was already out of date in the late nineteenth century. However, Japanese whaling industries in the late nineteenth and early twentieth centuries made several attempts at this type of whaling and ended with failure.

The Boso Enyo Gyogyo Co. Ltd., established in 1899, was renamed the Tokai Gyogo Co. Ltd. in 1906. It retired temporarily from the whaling business by selling its whaling section to the Toyo Hogei Co. Ltd., which was established in 1909 by combining 12 whaling companies in Japan (Akashi 1910, in Japanese). The history of these 12 whaling companies lasted from 1 to 10 years (the Boso Enyo Gyogyo being the longest) before the merger, and some of them must have taken Baird's beaked whales in the Chiba area. Statistics of these operations are available in Akashi (1910, in Japanese) but are not given by species. Catch records of Baird's beaked whales in that period are not available at present.

After several attempts and failures of Norwegian-type whaling in Japan, the Nihon Enyo Ggyo-gyo Co. Ltd., founded in 1899, led the establishment of Japanese modern whaling for large cetaceans (Section I.2.2). When the Greener's harpoon gun was mounted on a steamship, this was Norwegian-type whaling in principle. Use of an explosive harpoon head

TABLE 13.15
National Quota of Baird's Beaked Whales for Small-type Whaling of Japan

Year	Total	Wadaura	Ayukawa[a]	Abashiri	Hakodate[b]	Notes
1983–1987	40	35	0	5	0	Start of national quota. Quota shared by 2 vessels.
1988	60[c]	22	13	22	0	Temporary increase[d]. Quota shared by 5 vessels.
1989	54	27	22	5	0	Decrease to meet the fleet decline. Shared by 4 vessels.
1990–1998	54	26	26	2	0	Quota shared by 4 vessels.
1999–2004	62	26	26	2	8	The Hakodate quota started as a temporary measure termed a special quota. Quota shared by 4 vessels.
2005–2013	66	26	26	4	10	The Hakodate quota became a regular one. Quota shared by 5 vessels.

Note: Cases existed where regional allocation was modified to meet operational needs. This table has been revised with the addition of new data available since the original Japanese language edition appeared in 2011.

[a] Ayukawa: Pacific coast of Japan in 38°18′N, 14°131′E, which is included in the Sanriku Region.

[b] Hakodate: Hunted in the Sea of Japan and landed at Hakodate (41°46′, 140°44′E). This quota was stated as a "special quota" (Small-type Whaling Association) or an "exceptional measure" (Kushiroshi Somubu Chiiki Shiryo Shitsu 2006, in Japanese), but such statements disappeared after 2005, presumably after no international criticism.

[c] Operated with a total quota of 60 and resulted in take of 57 as listed (regional allocation of the quota is unknown for this year).

[d] Stated as a temporary increase of quota to remedy difficulty by prohibition of minke whales after 1998, but became permanent. From this year forward, the quota of Baird's beaked whales was shared by all the small-type whalers.

must have been a simple technology, but it was not needed for Baird's beaked whales (Section 13.5.3). The Norwegian-type whaling was first introduced to the Baird's beaked whale fishery in 1907 by the Tokai Gyogyo Co. Ltd. using the *Amatomi-maru* (130 gross tons) and a land station in Tateyama, closer to the outer ocean than Katsuyama (Komaki 1969, in Japanese; Ohsumi 1983). Kanari (1983, in Japanese), however, stated that the whaling ship, a whaling cannon of 37 mm caliber, and the harpoons were ordered from Norway in 1907 for the first use in the 1908 season and that the cannon (pictured in his book) was used until World War II.

There apparently was pressure by the government behind the unification of Japanese whaling companies. The government promulgated Fisheries Law in 1901 and based on it put in force the Rule for Regulation of Whale Fisheries on October 21, 1909 (Order No. 41 of the Minister of Agriculture and Commerce) to place (large-type) whaling under a license system of the Minister (Section 5.2). Article 9 of the Rule gave to the Minister the right to decide the whale species to be taken and number of vessels to be given a license. First the government dictated a fleet size of only 30 (Notification No. 418 of the Ministry), but the Rule was revised in 1934 and 1936. The latter version stated that the vessels could take sperm whales and baleen whales other than minke whales (Omura et al. 1942, in Japanese; Nihon Hogei-gyo Suisan Kumiai 1943, in Japanese). I have not seen the version of 1934. This regulation applied to whaling defined in 1947 as "large-type whaling" but did not apply to other type of whaling for smaller cetacean species, which was defined in 1947 as "small-type whaling" (Sections 5.2, 5.4, and 5.5).

The number of so-called small-type whaling vessels was 22 in 1941 when Japan started the war with the United States but increased to over 30 or 33 in 1942 and 46 in 1943 during the war and further increased during the postwar period to 68 in 1951 (Ohsumi 1975) and 75 in April 1952 (Table 4.1) (for further details see Sections 4.1 and 7.1.1). The government attempted to stop this trend by a new Rule for Regulation of Steam Ship Fisheries and in 1947 placed the then unregulated type of whaling under a license system of the minister as "small-type whaling" that was allowed to hunt minke whales and toothed whales other than the sperm whale (Sections 5.4 and 5.5).

Before this action of the Japanese government, Chiba Prefecture called the fishery "steam ship and rowboat whaling," and placed it under its license system in 1920; 26 vessels received licenses. The number of licensed vessels decreased with time with some fluctuation to 12 in 1927 and 15 in 1940 (Komaki 1996, in Japanese). The number of licensed vessels did not necessarily indicate the real magnitude of the whaling effort or whaling efficiency. Komaki (1996) stated that only 5 of the 35 vessels operated in 1917 in the Chiba area were equipped with an engine and other 30 were rowed boats towed by motor-driven vessels to the offshore whaling ground. The rowed boats harpooned Baird's beaked whales, and the motor-driven vessels towed the boats and the catch. The proportion and function of motor driven vessels could have changed with time, but details are not available. All the licensees of the prefecture did not necessarily operate whaling every year. For example only 8 vessels operated in 1947 when there were 15 prefecture licensees (Komaki 1996, in Japanese). According to Komaki (1996, in Japanese), a whaling station at Otohama in Chiba last operated in 1972 and since 1973 there remained only the Wadaura station of the Gaibo Hogei Co. Ltd., a successor of Chiba Gyogyo Co. Ltd. established in 1948, as a small-type whaling station in the prefecture.

The shift of the whaling ground during the history of the fishery is worth noting. Hand-harpoon whaling was operated since the early seventeenth century from land stations at Katsuyama (35°07′N) at the entrance of Tokyo Bay called

Uraga Pass but ceased in 1869 due to a decline in catches. A new whaling company, Nihon Suisan Kaisha, established a station in 1888 in Tateyama (35°N) about 14 km south of Katsuyama and operated whaling further south around the entrance of Sagami Bay or near Oshima Island (latitude 34°45′N). This operation continued to 1891 (Komaki 1969, in Japanese). The Boso Enyo Gyogyo Co. Ltd. (established in 1899 and later renamed Tokai Gyogyo Co. Ltd.) started Norwegian-type whaling using a land station at Tateyama during 1907–1909. In 1913 it moved the station to Otohama (34°55′N, 139°56′E) on the southern coast of Chiba Prefecture facing the outer sea. All the remaining companies completed the shift by 1917. The only whaling station currently in operation in Chiba Prefecture is located at Wadaura on the coast facing the Pacific Ocean.

According to Kanari (1983, in Japanese), whalers from Taiji operated hand-harpoon whaling in the mid-seventeenth to early nineteenth centuries at Choshi on the Pacific coast of Chiba. Then, after a short pause, local people resumed whaling at Choshi and continued it for several years in the 1870s and 1880s. The Toyo Gyogyo Co. Ltd. of Yamaguchi Prefecture also operated whaling from Choshi for several years from 1906. This operation was the type defined as large-type whaling in 1909 (Section 5.2) and caught 68 whales of unknown species in 1907 and 42 in 1908. These operations could have caught Baird's beaked whales as well as larger species, but no further details are available.

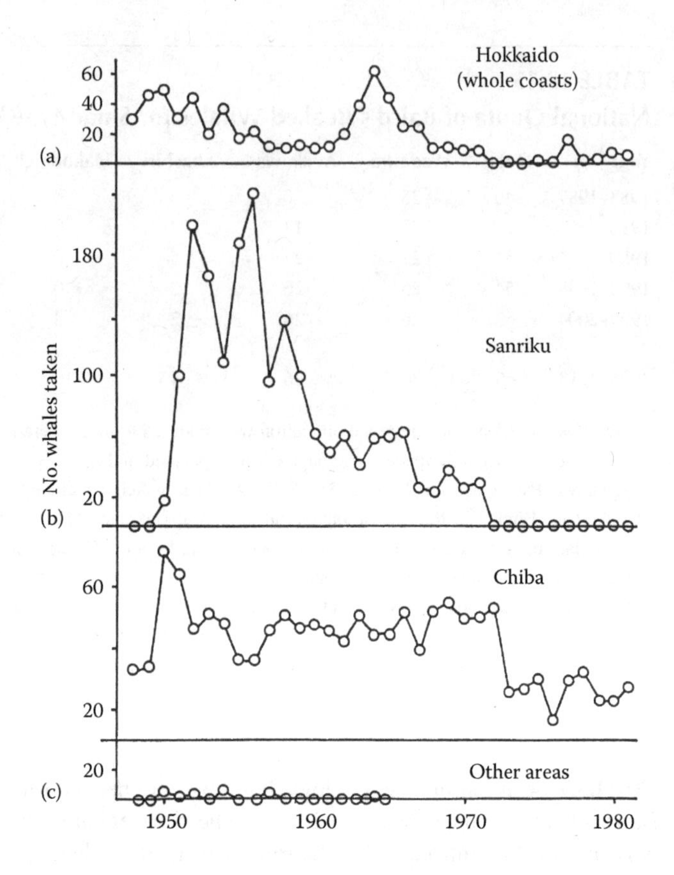

FIGURE 13.15 Annual trend of Baird's beaked whale fishery in Japan during 1948–1981 based on official statistics. (a) Hokkaido coasts (including the Sea of Japan, Okhotsk Sea, and Pacific). (b) Sanriku (Pacific coast in 37°54′N–41°35′N). (c) Chiba (Pacific coast in 34°55′N–36°06′N). (Reproduced from Ohsumi, S., *Rep. Int. Whal. Commn.*, 33, 633, 1983. With permission of the author, Figure 3.)

13.7.2 Catch Statistics

Ohsumi (1975, 1983) reported the statistics for 1948–1981; Kasuya and Ohsumi (1984) made minor correction to them. Omura et al. (1955) reported the geographical catch distribution of the species for 1948–1952, Ohsumi (1983) for 1947–1981, and Nishiwaki and Oguro (1971) for 1965–1969. These were in principle compiled from the Monthly Report of Whaling Operations presented by the whalers to the Fisheries Agency. In addition to these data, Balcomb and Goebel (1977) listed catches made by Gaibo Hogei in Chiba Prefecture in 1965–1975 together with statistics reported by earlier scientists for 1932–1942 and 1948–1972. Kasuya (1995, in Japanese) combined these statistics to prepare regional statistics for 1932–1993.

Table 13.14 is created from the publications and additional catch statistics of the Fisheries Agency for recent years but lacks figures for 1870–1931 and 1943–1946. Statistics for 1932–1942 were available only for Chiba, but the catch was probably small in other regions. Figure 13.15 shows the catches in 1948–1981 by region, some of which are not included in Table 13.14. The Japanese government stopped presenting catch statistics for small cetaceans to the IWC in 2001, but they are available on the home page of Research Division of the Fisheries Agency (http://kokushi.job.affrc.go.jp/index-2.html). The home page of the Nihon Small-type Whaling Association offered results of their operation compiled by each company but not by region.

World War II ended in 1945, and there was a rapid increase of take of Baird's beaked whales from 76 in 1948 to 197 in 1950 and 322 in 1952, the maximum in the history of Japanese small-type whaling. This was followed by a gradual decline to less than 100 in 1972 and thereafter. The 10-year average for 1972–1981 was only 38.7 whales. Although Chiba was not the exception, the rise and subsequent fall in catch applied to operations other than that at Chiba, in particular those in the Sanriku Region, which recorded an annual take of about 100–220 (Figure 13.15). The catch around Hokkaido remained lower, with two apparent peaks before 1955 and in 1962–1967.

The Japanese catch of Bard's beaked whales exhibited further decline until around 1980. The Scientific Committee of the IWC was concerned about this decline; it precipitated analysis of the population status. Japanese scientists, including myself, presented an analysis of the population status (Ohsumi 1975, 1983; Kasuya and Ohsumi 1984). However, now I believe that these analyses were based on questionable data and have almost no scientific value.

In response to the concerns of IWC member countries, at the 1982 meeting the Japanese government proposed an autonomous catch limit of 40 Baird's beaked whales beginning with the 1983 season. The figure of 40 whales was based on the average annual catch described earlier during 1972–1981. Japanese scientists made sighting surveys for the species in 1984 and presented the results to subsequent meetings of the Scientific Committee (Kasuya 1986; Miyashita 1986). The autonomous catch limit was by chance about 1% of the estimated population size, and the Scientific Committee felt that this level of catch will not harm the population status in the short term. In 1988, the Japanese government increased the quota to 60 as a temporary action to remedy the difficulty of whalers who lost their minke whale quota and in 1999 added a special quota of 8 whales for the Sea of Japan. These additions became a part of the ordinary quota in 2005; the total number increased to 66, with no criticism from the IWC (Table 13.15).

As of 2007, eight Japanese small-type whalers, who have a total of nine licensed vessels, jointly operate five whaling vessels (the whalers decreased to seven by a merger in 2008, but the number of operating vessels remained the same). Shares of the Baird's beaked whale quota are 14 whales each for *Koei-maru No. 75*, *Taisho-maru No. 28*, *Katsu-maru No. 7* and *Sumitomo-maru No.31*, and 10 whales for the *Seiwa-maru* of Taiji (Table 7.1). Currently only Japanese whalers exploit Baird's beaked whales.

Kasuya (1995, 1999, both in Japanese) questioned why the catch exploded in the postwar period and if such a catch was really possible. Even in Chiba, where there is a stable demand for Baird's beaked whales, the whalers poached sperm whales (Komaki 1994, in Japanese). I have learned from a whaler in Wadaura a term "Baird's beaked whale unit," which meant analogy with the Blue Whale Unit conversion of several sperm whales into one Baird's beaked whale for reporting purposes, although there was uncertainty how the custom prevailed in the industry (Kasuya 1995, in Japanese). Balcomb and Goebel (1977) reported information from H. Shoji, owner of Gaibo Hogei at Wadaura, that the annual catch of Baird's beaked whales off the Chiba coast remained at 25–30 whales for many years, and that the reported excess above that figure, when it happened, represented misreported other species. "Other species" cannot be other than sperm whales in my view. Nishiwaki and Oguro (1971) reported an annual take of 58–91 Bard's beaked whales (mean 76) off Chiba in the official statistics, with a total of 382 taken in the 5-year period 1965–1969 (Table 13.14). The true catch of the species during the same period reported by Balcomb and Goebel (1977) ranged from 21 to 39/year (mean 28) and totaled 140 whales. This figure was only 37% of the official figure. Earlier than this period, the official catch statistics for Chiba showed a 2- to 3-fold increase from 40 to 50 whales/year in 1948–1949 to 72–122 whales/year in the 1950–1952 seasons. A similar rapid increase was reported off Sanriku. Such a rapid increase appears unnatural to me.

Shoji (1996, in Japanese) stated that he had an extremely good catch of Baird's beaked whales off Chiba from 1946 to 1951, and then a poor catch concomitant with the start of hunting the species off Sanriku. The extremely good catch possibly indicates the take of 40–50 whales off Chiba in 1948–1949 mentioned earlier. The starting date of the Baird's beaked whale hunt off Sanriku in 1950 agrees with Table 13.14 and Figure 13.15, and it is reasonable to see interactions between the two whaling grounds off Chiba and Sanriku. So I do believe that a large catch of the species started off Sanriku around 1950, but a question remains about the magnitude. It is my impression that poaching of sperm whales started around 1950 off Sanriku and that some of them were misreported as Baird's beaked whales. The poaching could have ended in the 1970s.

Shoji (1996, in Japanese) gives annual statistic of whaling by his company Gaibo Hogei Co. Ltd. for 38 years from 1954 to 1991. These unique statistics are extremely valuable as real catch figures presented by the owner of a whaling company. Table 13.16 is a copy of his table in an abbreviated

TABLE 13.16
True Statistics for Whales Taken by the Gaibo Hogei Co. Ltd[a]

Season	Sperm w.	Baird's b. w.	Cuvier's b. w.	Short-finned p. w.	Killer w.	Minke w.	Bryde's w.	Humpback w.	Right w.
1954	0	29	1	0	0	0	0	0	0
1955–1959	6	125	12	40	3	2	8	0	0
1960–1964	108	125	41	55	10	2	12.5	0	1
1965–1969	272	163	7	5	17	84	7	1	0
1970–1974	302	133	4	0	0	234	3	0	0
1975–1979	0	101	1	0	0	199	1	0	0
1980–1984	0	193	1	195	1	578	0	0	0
1985–1989	0	113	0	62	15	573	0	0	0
1990–1991	0	31	1	16	0	0	0	0	0

Source: Compiled from Shoji, H., Small-type whaling in coastal Japan, in Komaki, K. (ed.). *Master of Yellowtail and Fisheries of Wada Town*, H. Shoji, Wada, Japan, pp. 100–127, 308pp, 1996, Table 13, which lists annual takes. Whales taken outside Chiba included. Cooperative take with another company is counted as half an individual, in Japanese.

[a] This table as well as Table 13 of Shoji (1996, in Japanese) does not include 3 fin whales taken by his company (2 in 1968 in the Tsushima Strait and 1 in 1969 in the Sea of Japan), which is stated on p. 111 of Shoji (1996, in Japanese). The latter catch agrees with my personal information about a fin whale landed in 1969 by a small-type whaler at a wholesale fish market on the east coast of Noto Peninsula (36°56′–37°30′, 136°40′E–137°20′E), which was obtained in the summer of 1969 through an interview with a market staff.

form with 5-year groups. It is unclear why take of three fin whales stated in the text is not listed in his table (see footnote to Table 13.16). His original table, which gives annual figures, clarified that his company started poaching sperm whales in 1959 and continued the practice until 1974. During this 16-year period he took 688 sperm whales and 441 Baird's beaked whales, with annual averages of 43.0 sperm whales and 27.6 Baird's beaked whales. The latter figure is the same as the average for the whole period of 1954–1991, which suggests that take of Baird's beaked whales in Chiba was about stable during the period.

Although it is very likely that the statistics of Baird's beaked whales reported to the Fisheries Agency contain some sperm whales poached by the small-type whalers, the situation does seem to be more complicated. Comparison of the official statistics (Kaiyo Dai-ikka [Ocean No.1 Division of Fisheries Agency] 1973 and Enyo-gyogyo-ka [Offshore Division of Fisheries Agency] 1977, both in Japanese) and the private statistics of Shoji (1996, in Japanese) for nine seasons (1968–1976), for which official statistics for an individual company are available to me, reveal that the official statistics recorded 268 Baird's beaked whales, which exceeded the true private figure of 244 by only 24. The difference was small probably because the statistics covered the period around the end of the poaching of sperm whales. The official figure exceeded the private figure in seven seasons and the reverse was true for the remaining two seasons. If the company attempted to maintain the balance between catch and products, it should have added sperm whales in the official report of Baird's beaked whales. It is unclear why the whaling operated without a catch quota had to underreport the catch.

The Fisheries Agency set catch quotas for Baird's beaked whales by region beginning in 1983 and allocated 35 whales to Chiba (Table 13.15). The true catches of Gaibo Hogei in Chiba recorded in Shoji (1996, in Japanese) was 39 in 1983 and 60 in 1984, which exceeded the quota for the two seasons by 29. Such violations, resultant of quotas and underreporting, were not identified in 1985 and later seasons, when scientists including myself were stationed at the whaling station in Wadaura.

In the earlier discussion I have described the inside workings of a whaling operation using a small-type whaler as an example, which was possible because the owner of the company released the true statistics. The illegal operations could not have been limited to this company. I had the impression while I worked close to Japanese coastal whaling that other small-type whalers as well as large-type whalers operated similar illegal whaling in Japanese coastal waters.

The next focus will be on Baird's beaked whaling off the Sanriku coast, which was also a good ground for sperm whales. The Japanese small-type whalers reported take of a total 322 Baird's beaked whales in the 1952 season, which comprised 72 from Chiba, 202 from Sanriku, and 26 from Kushiro (42°59′N, 144°23′E) on the Pacific coast of eastern Hokkaido (Figures 13.4 and 13.15, Table 13.14). A great amount of effort would have been needed by the small-type whalers in the Sanriku Region if they intended to take such a large number of Baird's beaked whales, which were not esteemed in the area and were more vigilant of approaching vessels than the sperm whales abundant in those days in the area. In the 1980s, when I was on board a small-type whaling vessel, *Koei-maru No. 7* stationed at Ayukawa in Sanriku, I was told by the gunner that he had poached 50–100 sperm whales annually until the 1970s (Kasuya and Miyazaki 1997, in Japanese). Kondo (2001, in Japanese), who worked as director of a land station of Nihon Kinkai Hogei Co. Ltd. (later Nihon Hogei Co. Ltd.) or one of three large-type whalers stationed at Ayukawa, wrote that small-type whalers of Ayukawa poached sperm whales and sold the catch to the large-type whalers, processing some of the remaining sperm whales at their own land stations. One of the large-type whaling operations (likely his land station) purchased a total of 1106 sperm whales during the 3 years of 1966–1968 from several small-type whalers. The annual purchases by the company declined during the 3 years to 297 in 1966, 149 in 1967, and 60 in 1968. He estimated that each of two other companies could have purchased similar numbers (total purchase three times the figures shown earlier).

I routinely witnessed skeletons of sperm whales mixed with those of small toothed whales in the bone dump of small-type whalers in Ayukawa in the 1960s. These whalers were certain to have taken both Baird's beaked whales and sperm whales and could have reported some of the latter as the former. Small-type whalers in the Sea of Japan and southern Okhotsk Sea could have been the exception, because sperm whales were uncommon there. The procedure, if it existed, of converting sperm whales into Baird's beaked whales requires further investigation.

Change in the fishery inspection system does not explain why the small-type whalers ceased poaching sperm whales, because it was my impression that an effective whaling inspection system was almost nonexistent from the 1960s to March 1988, when commercial hunting of baleen whales and sperm whales ended in Japan. Although international whaling observers were placed at some large-type whaling stations, their function was limited (Kasuya 1999, 1999, in Japanese).

A possible cause ending the poaching was the influence of the economy. Let us examine whaling statistics in Enyo-ka [Offshore Division of Fisheries Agency] (1984, 1987, both in Japanese). The main product of the sperm whale fishery was not the meat but the oil. The peak periods of sperm whale catch by Japanese whaling were the 1963–1964 Antarctic season when 4706 sperm whales were taken, the 1966–1969 North Pacific pelagic season when 3000 were taken, and the 1968 costal season when 3747 were taken. There was large-scale underreporting of the catch in the coastal fishery (Section 15.3.2.3). The 1964 season saw the maximum take of 8,800 sperm whales as well as oil production of 46,000 tons. Export of sperm-whale oil peaked in 1963 at 53,000 tons, which suggested that most oil was exported.

Next, we will compare the statistics between the 5-year periods of 1962–1966 and 1974–1978, which are 12 years apart. The totals were used to smooth annual fluctuations. During the 12-year period, the production of sperm-whale oil was halved from 190,000 to 89,000 tons, and exports decreased 78% from 152,000 to 33,000 tons. The proportion exported decreased from 80% to 37% and became zero after 1979. The price of sperm oil doubled during the 12-year period from 59,000 to 133,000 yen/ton. These changes together with shift of sperm whaling to offshore winter grounds following depletion of sperm whales in the nearshore summer ground (Section 15.3.2; Kasuya and Miyashita 1988) could have deprived small-type whalers of their opportunity to join the business of sperm whale hunting.

Demand for whale meat must be considered as another factor that could have attracted small-type whalers to illegal sperm whaling and later helped cause them to end it. The production of whale meat by Japanese whaling in the Antarctic, North Pacific, and coastal whaling peaked at 228,000 tons in 1962 and declined steadily from 214,000 in 1965 to 19,000 tons in 1979. Importation of whale meat from other countries remained in most years below 20,000 tons until 1972, then increased to 36,000 tons in 1977, and remained at a level between 20,000 and 30,000 tons until 1980. Thus, while national production of whale meat declined, importation increased in some degree and the total domestic supply of whale meat decreased during the period, from about 198,000 in 1965 to 102,000 tons in 1975 (adjusted for export). The reported catch of minke whales, the only baleen whales allowed for the small-type whalers, gradually increased from a low of 234 whales in 1968 to 400 in 1978, the quota of the time set by the IWC (Table 4.2). The composition of the baleen whales taken by Japanese whaling changed drastically between 1971 and 1976. While the catch of fin whales declined from 2215 to 116 and that of sei whales from 7114 to 1977, the catch of minke whales increased from 291 to 3377 in the same period (Antarctic catch included). It seems most probable that Japanese small-type whalers shifted their operations from sperm whales to minke whales, responding to the decline in demand for sperm-whale oil and increased demand for baleen whale meat due to the decrease in catch quotas for large baleen whales.

It should be remembered that the small-type whalers were placed in a difficult economic situation when they poached sperm whales. While large-type whaling in coastal waters and pelagic whaling in the North Pacific and Antarctic brought in great amount of whale products obtained using efficient equipment, the small-type whalers, who were politically less connected than the large-type whalers, were forced to take smaller less profitable whales using strictly limited equipment (Chapter 4). Take of a few minke whales and pilot whales, which was possible only on days of calm weather, was not enough to sustain their business. This situation is documented in Shoji (1988, in Japanese) and Komaki (1994, in Japanese), written by a whaler's wife and a whaler's daughter.

Facing the cessation of commercial whaling in March 1988, small-type whalers in Japan pursued their business by hunting Baird's beaked whales and pilot whales by utilizing a loophole in the International Convention of Regulation of Whaling. Although the Japanese government issued a permit in the autumn of 1987 to take Antarctic minke whales for scientific purposes, which is also called scientific whaling, its products were of a limited amount and the small-type whalers enjoyed a high price for their products during the subsequent 10 or 15 years. The situation has changed. Due to the expansion of the scientific whaling program and a decline in the demands for whale products, the recent prices of whale products are only 73% of the 1999 level, and the price of Baird's beaked whale products is less than 60% of the former level (Endo and Yamano 2007). The average value of one Baird's beaked whale decreased by 57% in the 7 years from 1998 to 2005 (Section 7.1.2). Small-type whalers are obliged to participate in the Japanese scientific whaling program to remedy this difficult situation caused by the program (Section 7.1).

13.7.3 Manipulation of Statistics and Management

In the autumn of 1984 I was in Ayukawa to inspect small-type whaling for northern-form short-finned pilot whales and to examine biology of the catch. Understanding of the catch level and the biology of the harvested whales was important for management. Whale carcasses were processed daily at three whaling stations from late afternoon to late evening, and I had an opportunity to observe underreporting of the catch (Section 12.6.3.1). Overreporting also existed in the cetacean fishery (Section 2.5).

Manipulation of catch statistics may occur for economic reasons, but underlying it is another difficult aspect in managing common property. Fishermen often excuses their illegal activity by saying that other fishermen will take the resource if they do not take it now, which is further elaborated by saying that there is no problem if it is not identified at sea, or that if everybody does it why not me, and finally they start to defend each other's illegal activities. At the last stage, people ignore regulations daily to take protected species or more than allocated quotas. Underreporting of catch can occur in order to escape criticism of increasing catches and delay introduction of catch regulations. Overreporting is a way of camouflaging population decline. I saw both of these while I was working on cetacean fisheries.

It is my impression that the statistics of the Japanese Baird's beaked whale fishery involved both underreporting and overreporting, but that the latter could have been greater than the former. The Scientific Committee of the IWC often back-calculated the initial size of a particular population based on current abundance estimated from sighting surveys and catch history of the population and calculated the depletion level (present abundance/initial abundance) as a basis for calculating a catch quota. This procedure was not applied to Baird's beaked whales. However, if it were carried out for Japanese Baird's beaked whales by assuming the official catch statistics as correct, it would have overestimated the past abundance and thus underestimated the current population level and likely have resulted in a small catch quota.

The opposite case applies to sperm whales off Japan, where the reported catch was greatly biased downward. This would underestimate the initial population level and overestimate the current level, thus leading to an optimistic conclusion. This could have happened in the IWC. At odds with the calculation of the scientists that suggested a recovering trend in the population, the whaling operation had increasing difficulty in finding sperm whales in the 1980s (Section 15.3.2). Japanese large-type whalers, which operated in coastal waters, hunted sperm whales with an annual quota that amounted to over 3500 whales in the peak season, but they actually caught two or three times that number for many years (Kondo 2001, in Japanese). An error in the assumption of population structure (Kasuya and Miyashita 1988) and the age–body length relationship (Kasuya 1991) could have enhanced the optimistic judgment in the stock assessment (Kasuya and Miyazaki 1997, in Japanese; Kasuya 2008). Poaching of sperm whales by small-type whalers also contributed to the optimistic error in the assessment.

REFERENCES

[*In Japanese Language]

*Akashi, K. 1910. *History of Norwegian-type Whaling in Japan.* Toyo Hogei Kabushiki Kaisha, Osaka, Japan. 280+40pp.

Anon. 1990. Progress Report-Japan. *Rep. Int. Whal. Commn.* 40: 198–205.

*Asahi Shinbun Press (ed.). 2010 *Exhibition of the Giant Mammals: Our Friends in the Sea.* Asahi Shinbunsha, Tokyo, Japan. 183pp. (Edited with advice of T. Yamada and Y. Tajima).

Asper, E.D., Andrews, B.F., Antrim, J.E., and Young, W.G., 1992. Establishing and maintaining successful breeding program for whales and dolphins in a zoological environment. *IBI Rep.* (Kamogawa, Japan) 3: 71–81.

Aurioles-Gamboa, D. 1992. Notes on a mass stranding of Baird's beaked whales in the Gulf of California, Mexico. *Calif. Fish. Game* 78(3): 116–123.

Baird, R. 2000. The killer whale: foraging specializations and group hunting. pp. 127–153. *In*: Mann, J., Connor, R.C., Tyack, P.L., and Whitehead, H. (eds.). *Cetacean Societies: Field Studies of Dolphins and Whales.* University of Chicago Press, Chicago, IL. 433pp.

Balcomb, K.C. III. 1989. Baird's beaked whale *Berardius bairdii* Stejneger, 1883: Arnoux's beaked whale *Berardius arnuxii* Duvernoy, 1851. pp. 261–288. *In*: Ridgway, S.H. and Harrison, R. (eds.). *Handbook of Marine Mammals, Vol. 4, River Dolphins and the Larger Toothed Whales.* Academic Press, London, U.K. 442pp.

Balcomb, K.C. III. and Goebel, C.A. 1977. Some information on a *Berardius bairdii* fishery in Japan. *Rep. Int. Whal. Commn.* 27: 485–486.

Berta, A., Sumich, J.L., and Kovacs, K.M. (eds.). 2006 *Marine Mammals: Evolutionary Biology.* Academic Press, London, U.K. 547pp.+16pls.

Best, P.B. 1979. Social organization of sperm whales, *Physeter macrocephalus.* pp. 227–2289. *In*: Winn, H.E. and Olla, B.L. (eds.). *Behavior of Marine Mammals.* Plenum Press, New York. 438pp.

Best, P.B., Canham, A.S., and Macleod, N. 1984. Patterns of reproduction in sperm whales, *Physeter macrocephalus. Rep. Int. Whal. Commn.* (Special Issue 6, Reproduction in Whales, Dolphins and Porpoises): 51–79.

Betesheva, E.I. 1960. Pitanie kashalota (*Physeter catodon* L.) i berardiusa (*Berardius bairdii* Stejneger) v raione Kuril'skoi gryady [Feeding of sperm whale (*Physeter catodon,* L.) and Baird's beaked whale (*Berardius bairdii* Stejneger) in the region of the Kuril range]. *Tr. Vses Gidrobiol. Ov-va,* 10. (not seen).

Betesheva, E.I. 1961. Pitanie promyslovikh kitov prikuril'skogo raiona [Feeding of game whales in the Kuril region]. *Tr. In-ta. Morf. Zhiv. Zhiv, AN SSSR.* No. 34. (not seen).

Bryden, M.M. 1986. Age and growth. pp. 212–224. *In*: Bryden, M.M. and Harrison, R. (eds.). *Research on Dolphins.* Clarendon Press, Oxford, U.K. 478pp.

Connor, R.C., Wells, R.S., Mann, J., and Read, A.J. 2000. The bottlenose dolphin: social relationship in a fission-fusion society. pp. 91–126. *In*: Mann, J., Connor, R.C., Tyack, P.L., and Whitehead, H. (eds.). *Cetacean Societies: Field Studies of Dolphins and Whales.* University of Chicago Press, Chicago, IL. 433pp.

*Daigo, S. 1890. Current state of whaling off Oshima. *Dainihon Suisankai Hokoku* 102: 506–509.

Dalebout, M.L., Baker, C.S., Mead, J.G., Cockroft, V.G., and Yamada, T.K. 2004. A comprehensive and validated molecular taxonomy of beaked whales, family Ziphiidae. *J. Heredity* 95(6): 459–473.

Dalebout, M.L., Ross, G.J.B., Baker, C.S., Anderson, R.C., Best, P.B., Cockroft, V.G., Hinsz, H.L., Peddemors, V., and Pitman, R.L. 2003. Appearance, distribution, and genetic distinctiveness of Longman's beaked whales, *Indopacetus pacificus. Mar. Mammal Sci.* 19(3): 421–461.

Endo, A. and Yamano, M. 2007. Policies governing the distribution of by-products from scientific and small-scale coastal whaling in Japan. *Marine Policy* 31: 169–181.

*Enyo-gyogyo-ka (Suisan-cho). 1977. *Information on Whaling.* Suisan-cho [Fisheries Agency], Tokyo, Japan. 359pp.

*Enyo-ka (Suisan-cho). 1984. *Outline of Whaling.* Suisan-cho [Fisheries Agency], Tokyo, Japan. 62pp.

*Enyo-ka (Suisan-cho). 1987. *Outline of Whaling.* Suisan-cho [Fisheries Agency], Tokyo, Japan. 62pp.

Fedoseev, G.A. 1985. Records of whales in ice conditions of the Okhotsk Sea. Paper IWC/SC/37/O4 presented to the *Scientific Committee Meeting of IWC in 1985.* 8pp. (Available from IWC secretariat, Cambridge, U.K.).

*Freeman, M.M.R. (ed.) 1989. *Cultural Anthropology of Whaling: Small-type Whaling of the Japanese Coast.* Kaimeisha, Tokyo, Japan. 216pp.

*Fukuoka, S. 2014. *Small-type Whaling of Abashiri: Journal of a Motorboat Driver (1971–1974).* Seibutsu Kenkyusha, Tokyo, Japan. 148pp.

*Hamanaka, E. 1979 (ed.). *History of Taiji Town.* Taiji Town, Taiji, Japan. 952pp.

*Hattori, T. 1887–1888. *Miscellanea of Whaling.* Dainihon Suisankai, Tokyo, Japan. Vol. 1: 110pp., Vol. 2: 210pp.

Heptner, V.G., Chapskii, K.K., and Sokolov, V.E. 1976. *Mammals of the Soviet Union,* Vol. II, Part 3. Science Publishers, Moscow, Russia. 995pp.

*Ishii, E. 1987. Fascinating Shiretoko, a home of wildlife. pp. 9–20. *In*: Niizuma, A., Ishii, E., and NHK Kushiro Station (eds.). *Animals that Accompany Drift Ice.* Nihon Hoso Kyokai, Tokyo, Japan. 173pp.

*Ishikawa, H. 1994. *Stranding Records of Cetaceans in Japan (1901–1993).* Geiken Series No. 6. Nihon Geirui Kenkyusho (Institute of Cetacean Research), Tokyo, Japan. 94pp.

IWC (International Whaling Commission). 1990. Report of the scientific committee. *Rep. Int. Whal. Commn.* 40: 39–79.

IWC. 1991. Report of the sub-committee on small cetaceans. *Rep. Int. Whal. Commn.* 41: 172–190.

IWC. 1994. Report of the sub-committee on small cetaceans. *Rep. Int. Whal. Commn.* 44: 108–119.

*Kaiyo Dai-ikka (Suisan-cho). 1973. *Information on Whaling.* Suisan-cho, Tokyo, Japan. 247pp.

*Kanari, H. 1983. *Whaling of Boso [Chiba Prefecture].* Ron Shobo, Nagareyama, Japan. 154pp.

*Kaneda, K. and Niwa, H. 1899. *Report of Evaluation of Items Exhibited at the Second Fishery Exposition*, Vol. 1, Part 2. Noshomu-sho Suisan-kyoku, Tokyo, Japan. 219pp.

Kasuya, T. 1972. Some information on the growth of the Ganges dolphin with a comment on the Indus dolphin. *Sci. Rep. Whales Res. Inst.* 24: 87–108.

Kasuya, T. 1977. Age determination and growth of the Baird's beaked whale with a comment on the fetal growth rate. *Sci. Rep. Whales Res. Inst.* 29: 1–20.

Kasuya, T. 1986. Distribution and behavior of Baird's beaked whales off the Pacific coast of Japan. *Sci. Rep. Whales Res. Inst.* 37: 61–83.

Kasuya, T. 1991. Density dependent growth in North Pacific sperm whales. *Mar. Mammal Sci.* 7(3): 230–257.

*Kasuya, T. 1995. Baird's beaked whale. pp. 521–529. *In*: Odate, S. (ed.). *Basic Data of Rare Wild Aquatic Organisms of Japan* (II). Nihon Suisan Shigen Hogo Kyokai, Tokyo, Japan. 751pp.

Kasuya, T. 1995. Overview of cetacean life histories: An essay in their evolution. pp. 481–497. *In*: Blix, A.S., Walloe, L. and Ultang, O. (eds.). *Whales, Seals, Fish and Man*. Elsvier, Amsterdam, the Netherlands. 720pp.

Kasuya, T. 1999. Examination of the reliability of catch statistics in the Japanese coastal sperm whale fishery. *J. Cetacean Res. Manage.* 1(1): 109–122.

*Kasuya, T. 1999. Manipulation of statistics by the Japanese Coastal Sperm Whale Fishery. *IBI Rep.* 9: 75–92.

Kasuya, T. 2008. Cetacean biology and conservation: A Japanese scientist's perspective spanning 46 years. *Mar. Mammal Sci.* 24(4): 749–773.

Kasuya, T. 2009. Giant beaked whales. pp. 498–500. *In*: Perrin, W.F., Würsig, B., and Thewissen, J.G.M. (eds.). *Encyclopedia of Marine Mammals*. Elsvier, Amsterdam, the Netherlands. 1316pp.

Kasuya, T. and Brownell, R.L.Jr. 1979. Age determination, reproduction and growth of Franciscana dolphin *Pontoporia blainvillei*. *Sci. Rep. Whales Res. Inst.* (Tokyo) 31: 45–65.

Kasuya, T. and Brownell, R.L. Jr. 1989. Male parental investment in Baird's beaked whales, an interpretation of the age data. pp. 5623–624. *In*: *Abstracts of the Fifth International Theriological Congress*. Rome, Italy. 1047+16pp.

Kasuya, T., Brownell, R.L. Jr., and Balcomb, K.C. III 1997. Life history of Baird's beaked whales off the Pacific coast of Japan. *Rep. Int. Whal. Commn.* 47: 969–979.

Kasuya, T. and Marsh, H. 1984. Life history and reproductive biology of the short-finned pilot whale, *Globicephala macrorhynchus*, off the Pacific coast of Japan. *Rep. Int. Whal. Commn.* (Special Issue 6, Reproduction in Whales, Dolphins and Porpoises): 259–310.

Kasuya, T. and Miyashita, T. 1988. Distribution of sperm whale stocks in the North Pacific. *Sci. Rep. Whales Res. Inst.* 39: 31–75.

Kasuya, T. and Miyashita, T. 1997. Distribution of Baird's beaked whales off Japan. *Rep. Int. Whal. Commn.* 47: 963–968.

*Kasuya, T. and Miyazaki, N. 1997. Cetacea. pp. 139–185. *In*: Kawamichi, T. (ed.). *Red Data of Japanese Mammals*. Bunichi Sogo Shuppan, Tokyo, Japan. 279pp.

Kasuya, T. and Ohsumi, S. 1984. Further analysis of the Baird's beaked whale stock in the western North Pacific. *Rep. Int. Whal. Commn.* 34: 587–595.

Kasuya, T., Tobayama, T., Saiga, T., and Kataoka, T. 1986. Perinatal growth of delphinoids: Information from aquarium reared bottlenose dolphins and finless porpoises. *Sci. Rep. Whales Res. Inst.* 37: 85–97.

*Kasuya, T. and Yamada, T. 1995. List of Japanese cetaceans. Geiken Series No.7. Nihon Geirui Kenkyusho (Institute of Cetacean Research), Tokyo, Japan. 90pp.

Kato, H. 1984. Observation of tooth scars on the head of male sperm whale, as an indication of intra-sexual fighting. *Sci. Rep. Whales Res. Inst.* 35: 39–46.

Kirino, T. 1956. On the number of teeth and its variability on *Berardius bairdii*, a genus of the beaked whale. *Okajimas Folia Anat. Jpn.* 28: 429–434.

*Kishinoue, K. 1914. History of fisheries in Awa County [Southern Chiba Prefecture]. Awagun Suisan Kumiai [Fishery Association of Awa County], Hojyo in Tateyama, Japan. 294+11pp. + 1 Plate.

Kishiro, T. 2007. Geographical variation in the external body proportions of the Baird's beaked whales (*Berardius bairdii*) off Japan. *J. Cetacean Res. Manage.* 9(2): 89–93.

*Komaki, K. 1969. Whaling in Boshu [southern Chiba Prefecture]. *Geiken Tsushin* 215: 83–86. Reproduction of an article of the same title published in 1958 in *Shiron* (Tokyo Women's Christian University) 6: 413–416.

*Komaki, K. 1994. Small-type whaling. pp. 280–291. *In*: Wadachoshi Hensan Shitsu (ed.). *Wadachoshi [History of Wada Town]*, Vol. 2. Wada Town, Wada, Japan. 1256pp.

*Komaki, K. (ed.). 1996 *Master of Yellowtail and Fisheries of Wada Town*. H. Shoji, Wada, Japan. 308pp.

*Kondo, I. 2001. *Rise and Fall of Japanese Coastal Whaling*. Sanyo-sha, Tokyo, Japan. 449pp.

*Kuroe, J. 1960. On teeth of Baird's beaked whales. *Hirosaki Igaku* 12(3): 460–477.

*Kushiro-shi Somu-bu Chiiki Shiryo Shitsu. 2006. *History of Whaling in Kushiro*. Kushiro City Office, Kushiro, Japan. 379pp.

*Maeda, K. and Teraoka, Y. 1958. *Whaling*. Isana Shobo, Tokyo, Japan. 346pp.

Marsh, H. and Kasuya, T. 1986. Evidence for reproductive senescence in female cetaceans. *Rep. Int. Whal. Commn* (Special Issue 8, Behavior of Whales in Relation to Management): 57–74.

*Matsuura, K. 1977–1982. *Kasshi Yawa [Miscellaneous Records for 1821–1841]*. *In:* Nakamura, Y. and Nakano, M. (eds.). *Kassi Yawa*. Heibonsha, Tokyo, Japan. Part I: Vols. 1–6, Part II: Vols. 1–8, Past III: Vols. 1–3.

*Matsuura, Y. 1942. On Baird's beaked whales off Boshu [southern Chiba Prefecture]. *Dobutsugaku Zasshi* [*J. Zool.*] (Tokyo Zoological Society) 54(12): 466–473.

McCann, C. 1975. Body scarring on cetacean-odontocetes. *Sci. Rep. Whales Res. Inst.* 26: 145–155.

Minamikawa, S., Iwasaki, T., and Kishiro, T. 2007. Diving behavior of a Baird's beaked whale, *Berardius bairdii*, in the slope water region of the western North Pacific: First dive records using a data logger. *Fish. Oceanography* 16(6): 573–577.

Miyashita, T. 1986. Abundance of Baird's beaked whales off the Pacific coast of Japan. *Rep. Int. Whal. Commn.* 36: 383–386.

Miyashita, T. 1990. *Population estimate of* Baird's *beaked whales off Japan*. Paper SC/42/SM28 presented to the *Scientific Committee Meeting of IWC in 1990*. (Available from IWC secretariat. The results are cited in IWC 1991).

Miyashita, T. and Kato, H. 1993. Population estimate of Baird's beaked whales off the pacific coasts of Japan using sighting data collected by R/V Shunyo Maru, 1991 and 1992. Paper IWC/SC/45/SM6 presented to the *Scientific Committee Meeting of IWC in 1993*. (Available from IWC secretariat. The results are cited in IWC 1994).

*Nihon Hogei-gyo Suisan Kumiai. 1943. *Whaling Handbook*, Vol. 4. Nihon Hogei-gyo Suisankumiai, Tokyo, Japan. 156pp.

Nishimura, S. 1970. Recent records of Baird's beaked whale in the Japan Sea. *Publ. Seto Mar. Biol. Lab.* 18(1): 61–68.

*Nishiwaki, M. 1965. *Cetaceans and Pinnipeds*. Tokyo Daigaku Shuppankai [University of Tokyo Press], Tokyo, Japan. 439pp.

Nishiwaki, M. 1967. Distribution and migration of marine mammals in the North Pacific area. *Bull. Ocean Res. Inst., Univ. Tokyo* 1: 1–64.

Nishiwaki, M. and Oguro, N. 1971. Baird's beaked whales caught on the coasts of Japan in recent 10 years. *Sci. Rep. Whales Res. Inst.* 23: 111–122.

Ohizumi, H., Isoda, T., Kishiro, T., and Kato, H. 2003. Feeding habits of Baird's beaked whale, *Berardius bairdii*, in the western North Pacific and Sea of Okhotsk off Japan. *Fish. Sci.* 69:11–20.

Ohsumi, S. 1971. Some investigations on the school structure of sperm whale. *Sci. Rep. Whales Res. Inst.* 23: 1–25.

Ohsumi, S. 1975. Review of Japanese small-type whaling. *J. Fish. Res. Bd. Canada* 32(7): 1111–1121.

Ohsumi, S. 1983. Population assessment of Baird's beaked whales in the waters adjacent to Japan. *Rep. Int. Whal. Commn.* 33: 633–641.

Ohsumi, S. and Satake, Y. 1977. Provisional report on investigation of sperm whales off the coast of Japan under a special permit. *Rep. Int. Whal. Commn.* 27:324–332.

Ohsumi, S., Kasuya, T., and Nishiwaki, M. 1963. The accumulation rate of dentinal growth layers in the maxillary tooth of the sperm whale. *Sci. Rep. Whales Res. Inst.* 17: 15–35.

Omura, H., Fujino, K., and Kimura, S. 1955. Beaked whale *Berardius bairdii* of Japan, with notes on *Ziphius cavirostris*. *Sci. Rep. Whales Res. Inst.* 10: 89–132.

*Omura, H., Matsuura, Y., and Miyazaki, I. 1942. *Whales: Their Biology and the Practice of Whaling*. Suisansha, Tokyo, Japan. 319pp.

Perrin, W.F. and Donovan, G.P. 1984. Report of the workshop. *Rep. Int. Whal. Commn.* (Special Issue 6, Reproduction in Whales, Dolphins and Porpoises): 1–24.

Perrin, W.F. and Myrick, A.C. Jr. 1980. Report of the workshop. *Rep. Int. Whal. Commn.* (Special Issue 3, Age Determination of Toothed Whales and Sirenians): 1–63.

Perrin, W.F. and Reilly, S.B. 1984. Reproductive parameters of dolphins and small whales of the family Delphinidae. *Rep. Int. Whal. Commn.* (Special Issue 6, Reproduction in Whales, Dolphins and Porpoises): 97–133.

Pike, G.C. 1953. Two records of *Berardius bairdii* from the coast of British Columbia. *J. Mammal.* 34: 102–107.

Pitman, R.L., Palacios, D.M., Brennan, P.L., Balcomb, K.C., and Miyashita, T. 1999. Sightings and possible identity of a bottlenose whale in the tropical Indo-Pacific: *Indopacetus pacificus? Mar. Mammal. Sci.* 15(2): 531–549.

Ralls, K. 1976. Mammals in which females are larger than males. *Q. Rev. Biol.* 51: 254–276.

Ralls, K., Brownell, R.L. Jr., and Ballou, J. 1980. Differential mortality by sex and age in mammals, with specific reference to the sperm whale. *Rep. Int. Whal. Commn.* (Special Issue 2, Sperm Whales): 233–243.

Reeves, R.R. and Mitchell, E. 1993. Status of beaked whale, *Berardius bairdii. Can. Field Nat.* 107: 509–523.

Rice, D.W. 1963. Progress report on biological studies of the larger cetacean in the waters of California. *Norsk Hvalfangst-Tidende* 52: 181–187.

Rice, D.W. 1998. *Marine Mammals of the World: Systematics and Distribution*, Special Publication Number 4. The Society for Marine Mammalogy, Lawrence, KS, 231pp.

Scammon, C.M. 1874. *The Marine Mammals of the North-Western Coast of North America together with an Account of the American Whale-Fishery*. J.H. Carmany, San Francisco, CA and G.P. Putnam's Sons, New York. 319+i-V pp.

*Sekizawa, A. 1887a. Results of testing whaling equipment. *Dainihon Suisankai Hokoku* 87: 11–22.

*Sekizawa, A. 1887b. Results of testing whaling equipment, Part 2. *Dainihon Suisankai Hokoku* 88: 4–16.

*Sekizawa, A. 1892a. Experiment of whaling gun. *Dainihon Suisankai Hokoku* 117: 4–25.

*Sekizawa, A. 1892b. Whaling in waters around Chiba and Oshima. *Dainihon Suisankai Hokoku* 117: 606–607.

*Shoji, H. 1996. Small-type whaling in coastal Japan. pp. 100–127. *In*: Komaki, K. (ed.). *Master of Yellowtail and Fisheries of Wada Town*. H. Shoji, Wada, Japan. 308pp.

*Shoji, M. 1988. For people waiting for summer whales. pp. 67–75. *In*: Takahashi, J. (ed.). *Women's Tales of Whaling: Life Stories of 11 Japanese Women who Live with Whaling*. Japan Whaling Association, Tokyo, Japan. 131pp. (Accompanying English text slightly disagrees with Japanese text).

*Sleptsov, M.M. 1955. *Whale Biology and Whaling in the Far East*. Geiken Series No.1. Geirui Kenkyusho (Whales Res. Inst.), Tokyo, Japan. 51pp.+ 1 plate. (Japanese translation, from Russian published in 1955).

Subramanian, A., Tanabe, S., and Tatsukawa, R. 1988. Estimating some biological parameters of Baird's beaked whales using PCBs and DDE as tracers. *Mar. Pollut. Bull.* 19(6): 284–287.

*Takenaka, K. 1887 (manuscript). *History of Whaling in Southern Awa County*, Vols. 1 and 2. (Unpublished).

*Tato, Y. 1985. *History of Whaling with Statistics*. Suisansha, Tokyo, Japan. 202pp.

Tomilin, A.G. 1967. *Mammals of the USSR and Adjacent Countries, Cetacea*. Israel Program for Scientific Translations, Jerusalem, Israel. 717pp. (Translation from Russian text published in 1957).

*Torisu, K. 1999. *Historical Study of Whaling in Western Japan*. Kyushu Daigaku Shuppankai, Fukuoka, Japan. 414+i-xxviii pp.

Walker, W.A., Mead, J.G., and Brownell, R.L. Jr. 2002. Diet of Baird's beaked whales, *Berardius bairdii*, in the southern Sea of Okhotsk and off the Pacific coasts of Honsyu, Japan. *Mar. Mammal Sci.* 18(4): 902–919.

Wang, P. 1999. *Chinese Cetaceans*. Ocean Enterprises, Hong Kong, People's Republic of China. 325pp.+Pl.1-VI. (In Chinese).

Watkins, W.A., Daher, M.A., Fristrup, K.M., Howard, T.J., and Sciara, G.N. 1993. Sperm whales tagged with transponders and tracked under water by sonar. *Mar. Mammal Sci.* 9: 565–67.

*Yamaguchi, E. 1999. *Whale Jerky: Traditional Food Custom and Chiba Fishermen*. Tamagawa Shinbunsha, Kawasaki, Japan. 358pp.

*Yamase, H. 1760. *Natural History of Whales*. Osaka Shorin, Osaka, Japan. 16+54pp.

*Yoshihara, T. 1976. Whaling in Southern Awa County. *Tokyo Suisan Daigaku Ronshu* 11: 15–114.

*Yoshihara, T. 1982. *Whaling in Southern Awa County, together with Tombstones of Whales* (*Historical Relicts of Chiba Area*). Kyodo Shiryo Toshokan Aizawa Bunko, Chiba, Japan. 227pp.

14 Pacific White-Sided Dolphin

14.1 DESCRIPTION

The Pacific white-sided dolphin *Lagenorhynchus obliquidens* Gill, 1865, is endemic to the North Pacific and one of the smallest marine delphinids occurring there. It attains adult size at less than 2.5 m. The Japanese common name *kama-iruka* [sickle dolphin] came from the falcate dorsal fin with gray trailing edge. The adult male has a dorsal fin with a round leading edge and drooping tip described in this chapter.

Maximum sea surface temperatures are reached in the northern North Pacific in August–September, which is about one month after the peak in terrestrial air temperature. During this time of the year, Pacific white-sided dolphins inhabit latitudes of 40°N–47°N with sea surface temperatures of 10°C–19°C in the central North Pacific (Iwasaki and Kasuya 1997). During the rest of the year, they shift their range to the south. The coldwater Dall's porpoise is found to the north of this latitudinal range and warmwater species such as the striped dolphin to the south. The habitat preference of the northern right whale dolphin is the same as that of the Pacific white-sided dolphin in the longitudinal range during that time of year. However, it tends to inhabit more offshore waters and is not usually found in coastal waters like the Pacific white-sided dolphin. Sea surface temperatures inhabited by these species are listed in Table 10.1.

14.2 POPULATIONS AND THEIR DISTRIBUTION

The Pacific white-sided dolphin inhabits Japanese coastal waters of the East China Sea coast of Kyushu, Sea of Japan coasts of Honshu and Hokkaido, Okhotsk Sea coast of Hokkaido, Pacific coast of Hokkaido, and south to central Honshu around the Kii Peninsula (33°26′N–34°15′N, 135°05′E–136°15′E). It is common within the range, but an apparently seasonal visitor to southern and northern part of the range.

The small cetaceans spend the spring to autumn seasons off the Sea of Japan coast of Japan, probably winter in waters around Iki Island (33°47′N, 129°43′E) and in the East China Sea. This was deduced from the predominance of Dall's porpoises in winter in the southern Sea of Japan off the coasts of Yamaguchi Prefecture (33°55′N–34°40′N, 130°50′E–131°40′E) (Section 9.3.1.1) and the shift of the former hand-harpoon fishery from the Dall's porpoise to the Pacific white-sided dolphin in middle May off Kinosaki (35°39′N, 134°50′E) (Noguchi 1946, in Japanese), although small numbers of Pacific white-sided dolphins could remain in this area in February and March as evidenced by a few individuals hunted during the Dall's porpoise season. The Pacific white-sided dolphins that winter in Tsushima Strait and the East China Sea start moving in the spring into the Sea of Japan along the southern coast of the sea.

Pacific white-sided dolphins are known to occur in the Tsugaru Strait (41°30′N, 140°40′E), through which the Sea of Japan opens to the Pacific, during April to July with a peak density in June (Kawamura et al. 1983, in Japanese), but their migration route is unknown.

Along the Pacific coast of Japan, the Pacific white-sided dolphin is known from the coast of the Kii Peninsula to the southern Kuril Islands, or in the approximate latitudinal range of 33°26′N–46°N (Hayano et al. 2004). This species is almost replaced by Dall's porpoise in winter in the northern habitat north of Choshi Point (35°42′N, 140°51′E). The pattern of north–south segregation of these two species is probably similar to that observed in the offshore central North Pacific, although the absolute latitudinal range might not be the same. Pacific white-sided dolphins are known to occur in the southern Okhotsk Sea in summer.

Hayano et al. (2004) identified a distribution gap around the longitude of 150°E or between Japanese coastal waters and the offshore western and central North Pacific, identified genetic differences between a Japanese coastal sample (Iki Island area, the Sea of Japan, the Tsugaru Strait, and the Pacific coasts of Hokkaido and Honshu) and an offshore sample from longitudes 160°E–160°W, and concluded that these two geographical samples came from different populations. However, it seems premature to me to conclude that the species around Japan constitutes a uniform single population. The question remains whether dolphins in the southern Sea of Japan and off the Pacific coast of central Japan really belong to a single population. A larger sample covering a broader geographical distribution and seasonal range is needed for further clarification of the genetic structure of the species around Japan. Hayano et al. (2004) stated that the low genetic difference could be due to a short time since reproductive isolation between the geographical populations off Japan.

Even if they are not genetically indistinguishable, it does not necessarily mean that individuals in the two wintering grounds—Iki–East China Sea and the Pacific coast of central Japan— have opportunities to freely intermingle and interbreed, which is key information needed for management. Under such a situation, genetics will be of limited use for the purpose of studying population structure. Studies on the geographical movement of individual dolphins or mixing of schools using natural marks or radio transmitters will offer more information.

14.3 GROWTH AND SEXUAL DIMORPHISM OF THE DORSAL FIN

14.3.1 Materials and Methods

14.3.1.1 Materials

Materials were obtained from dolphins killed at Katsumoto on Iki Island as a part of a culling operation on dolphins led by Iki fishermen (Section 3.4). A total of 441 Pacific

TABLE 14.1
Number of Pacific White-Sided Dolphins Analyzed for Growth and Dorsal Fin Morphology

Sex	Dorsal Fin Morphology	Sexual Maturity	Age	Body Length	Total
Male	10	97	73	117	117
Female	11	43	15	60	60
Total	21	140	88	177	177

white-sided dolphins were taken in four drives in February and March in 1979 and 1981; 177 were randomly sampled (Kasuya 1995) (Table 14.1). These schools were found on a bank called *Shichiriga Sone* (33°56′N, 129°30′E) located between Iki Island and the southern tip of Tsushima Island, which extends from 34°07′N, 129°10′E to 34°42′N, 129°24′E. Thus, the sample represents part of the putative southern Sea of Japan population spending the winter near the northern range of its winter distribution.

The sample was collected almost randomly with the cooperation of volunteers while the fishermen killed the dolphins and processed the carcasses for disposal. This explains the low sampling rate. No particular sex or growth stages were selected. A similar type of sampling was experienced for the common bottlenose dolphin in the Iki Island area (Chapter 12). Details of the culling operation and our research activities are in Kasuya and Miyazaki (1981, in Japanese) and Kasuya (1985). Our sampling activity at Iki Island ended with the 1981 winter season and was followed by operation of a Fisheries Agency team that was established in April 1981 and started sampling in winter 1982. Some biological information on the species from our study was given to the Fisheries Agency team at their request and analyzed by combining it with materials collected by the team (Takemura 1986, in Japanese). Takemura (1986, in Japanese) is occasionally cited in the following section to supplement my data.

14.3.1.2 Male Sexual Maturity

Male maturity was determined through histology of a testicular sample about 1 × 1 cm collected from the center of either the right or left testis. A seminiferous tubule containing any stage of reproductive cells from spermatocytes to spermatozoa was identified as spermatogenic, and four maturity stages were determined using the proportion of spermatogenic tubules in the histological slide: "immature" (0% spermatogenic), "early maturing" (>0% and <50% spermatogenic), "late maturing" (≥50% and <100% spermatogenic), and "mature" (100% spermatogenic). The maturity stages are the same as those used for the common bottlenose dolphin (Chapter 11), short-finned pilot whale (Chapter 12), and Baird's beaked whale (Chapter 13).

The testis was weighed after removing the epididymis.

14.3.1.3 Female Sexual Maturity

Female maturity was determined from the evidence of ovulation, that is, presence of a corpus luteum or albicans in the ovaries. If the ovaries were not available for examination, presence of lactation or pregnancy was used as evidence of sexual maturity. However, absence of these indicators was not used as evidence of immature status; the ovaries had to be examined for the identification of immature females. The consequences of such maturity determination are stated in Section 13.4.2.

14.3.1.4 Age Determination

The process of age determination was almost the same as for the striped dolphin (Chapter 10), common bottlenose dolphin (Chapter 11), and short-finned pilot whale (Chapter 12). The thickness of the tooth preparation, which was a function of tooth size or more directly of thickness of the growth layers, was close to that for the striped dolphin. A polished thin section of a mandibular tooth was decalcified and stained with hematoxylin for the examination of growth layers in the dentine and cementum, and the middle figure of three independent counts was used. If the pulp cavity was considered closed and the cemental tissue indicated a larger age, the number of cemental growth layers was used as the age of the individual. For other cases, the number of dentinal layers was used as the age. Accumulation of growth layers was assumed to be annual.

14.3.1.5 Measurement of the Dorsal Fin

Three measurements were taken: (1) height from the base of the fin to the tip measured perpendicular to the base, (2) total height of the fin measured from the base to the highest point, and (3) maximum depth of the posterior concavity of the fin measured in parallel with the base. The difference between (1) and (2) was used as an indicator of downward drooping of the fin tip.

14.3.2 Body-Length Composition and Neonatal Length

Maximum body length of 235–239 cm in males and 225–229 cm in females (Figure 14.1) was the same as given by Takemura (1986, in Japanese) for the same data plus data from two additional drives examined by him in 1983 and 1984 (one dolphin examined from another drive was ignored). These data indicated that males reach a body size about 10 cm greater than that of females.

The modal body length was 195–199 cm in males and 220–224 cm in females (Figure 14.1). This was considered not to represent the body-length composition of the population for three reasons. One reason was that the female mode in body length exceeded that of males, which was inconsistent with the observations on maximum body size. The second was that the location of the male mode agreed with the peak length of "nonmature" individuals, which could be explained by assuming that a significant proportion of "mature" individuals were missed in the sample. The third reason was that more reasonable modes, 200–209 cm for

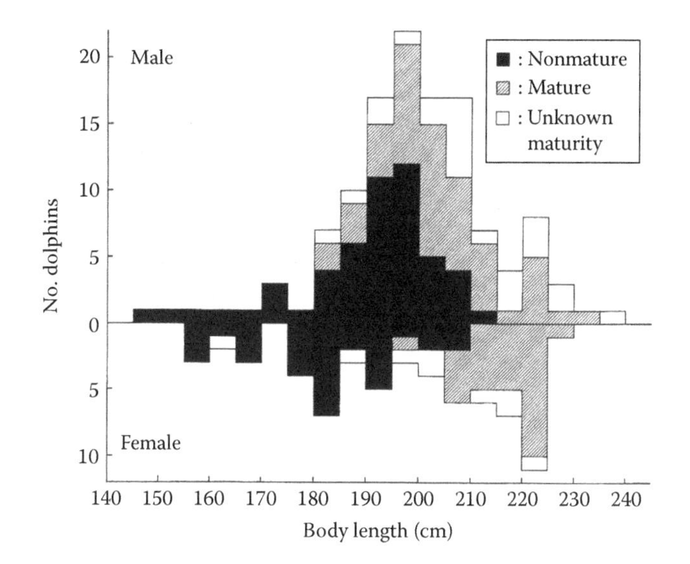

FIGURE 14.1 Body-length composition of male (top) and female (bottom) Pacific white-sided dolphins driven at Katsumoto (33°51′N, 129°42′E) at Iki Island, Japan. Shaded columns, sexually mature; black column, nonmature (immature, early maturing, and late maturing); white column, individuals of unknown maturity.

males and 190–199 cm for females, appeared in the body-length composition data reported by Takemura (1986, in Japanese) using a larger sample. His data showed great variation in the sex ratio among four drives, ranging from 1:5.6 to 1:0.4 (F:M). This suggests a social structure where groups are established by age, sex, and maturity, as suggested for striped dolphins (Section 10.5.14) and common bottlenose dolphins (Section 11.4.4).

Thus, the body-length composition of the sample used here (Figure 14.1) should be considered skewed with a possible deficiency of mature males as well as of immature and mature females, or a surplus of subadults.

Iwasaki and Kasuya (1997) presented the body-length composition of a population of the same species incidentally taken in the squid drift-net fishery in the offshore North Pacific in longitudes between 170°E and 145°W. Their maximum body lengths of males and females were both in a range of 230–239 cm and were not different from the figures from the Iki Island area.

Mean neonatal length for the offshore population was calculated at 91.8 cm (Ferrero and Walker 1996) or 93.7 cm (Iwasaki and Kasuya 1997), which were not statistically different. We do not have an estimate of mean neonatal body length of the Pacific white-sided dolphins off Japan, but it is expected to be similar to the figure estimated for the offshore population judging from the similarity in adult size.

In the offshore drift-net ground, full-term fetuses and neonates measuring 90–140 cm were obtained from May to September, within which period full-term fetuses occurred only in July and earlier months. This suggested that the parturition season was from May to August (Iwasaki and Kasuya 1997). Fetuses in the Iki Island area ranged from 30 to 50 cm in February–March or 30–80 cm in February–April (Takemura 1986, in Japanese), suggesting that these months were neither mating nor parturition seasons of the species in the Iki Island area. Fetal growth has not been analyzed.

14.3.3 Body Length and Age

Although there were sexually mature females in the sample of Kasuya (1995), females with estimated ages were limited to immature females of ages below 9 years. Males with estimated ages covered all the maturity stages of testicular histology. The oldest male was aged 44. The maximum age of the same species from the offshore North Pacific was 32–36 years for males and 27–40 years for females (Ferrero and Walker 1996; Iwasaki and Kasuya 1997). These figures obtained from limited samples suggest that the maximum longevity is around 45 years for both sexes.

Pacific white-sided dolphins have great individual variation in adult size, which is evidenced by a body-length range of 175–230 cm among males over 10 years old (Figure 14.2). Takemura (1986, in Japanese) reached the same conclusion and obtained the following equations between body length (L, cm) and age (t, year);

$$\text{Male: } L = 200.98(1 - \exp(-0.228(t + 5.679)))$$

$$\text{Female: } L = 196.97(1 - \exp(-0.292(t + 4.362)))$$

These equations have an asymptote at 201 cm (males) and 197 cm (females). Extrapolation of the equations to age zero has the risk of extrapolating outside of the data range.

Growth equations obtained by Ferrero and Walker (1996) and Iwasaki and Kasuya (1997) for the Pacific white-sided dolphin in offshore waters are similar. The following are

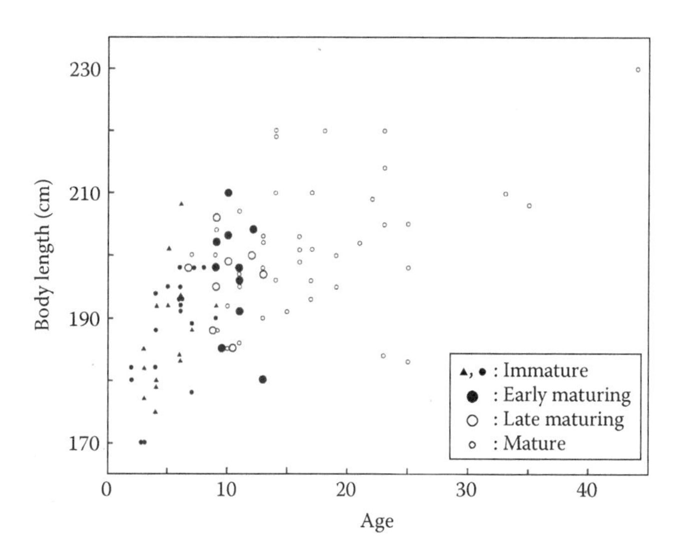

FIGURE 14.2 Scatterplot of body length and sexual maturity on the age of Pacific white-sided dolphins driven at Katsumoto, Iki Island, for females (which were all immature; triangles) and males (circles).

those of Iwasaki and Kasuya (1997), which are based on data with less of a skewed age composition.

$$\text{Male: } L = 93.7 \exp(0.6796(1 - \exp(-0.9451t)))$$

$$\text{Female: } L = 93.7 \exp(0.6709(1 - \exp(-1.2045t)))$$

The figure of 93.7 is their mean neonatal length in cm. Males have an asymptote at 184.9 cm and females at 183.3 cm. The equations are based on good samples of ages below 5 years but small samples of higher ages, that is, only 11 males and 10 females aged over 10 years. The validity of these equations, particularly for higher ages, should be reevaluated with larger samples.

14.3.4 Sexual Maturity

Takemura (1986, in Japanese) concluded that females attained sexual maturity at body length of 170–220 cm (average 183.0 cm) and at ages 6–9 years (average 8.2). The corresponding figures for males were 170–220 cm (average 184.5) and ages 7–9 years (average 8.3). How he determined male maturity is not explained.

For the offshore North Pacific, Ferrero and Walker (1996) and Iwasaki and Kasuya (1997) estimated male age at the attainment of sexual maturity to be 10–11 years and 9–12 years, respectively. They judged male maturity visually from the relationship between testis weight and age, and their sample was biased to younger individuals, presumably due to gear selection or segregation by age of animals. These factors, particularly the latter, result in overestimation of average age at the attainment of sexual maturity. They did not estimate female age at the attainment of sexual maturity.

Figure 14.3 shows relationships between testis weight, maturity, and age. Maturity correlates well with age; "immature" males occurred below age 10 years, "early maturing" in an age range of 7–13 years, "late maturing" 9–13 years, and "mature" males 7 years or over. Puberty seemed to be entered at around 7 years of age, and all males attained the "mature" stage by age 14. We are uncertain about the histological stage where males achieve the physiological ability to reproduce. However, in view of the extremely rapid testicular growth of males in the pubertal stage (Figure 14.3), it is safe to estimate that they attain reproductive ability at ages 7–13 years, with an average of about 10 years. This estimate is likely to be an overestimate to some degree due to underrepresentation of mature males in the sample and is closer to the estimates of Ferrero and Walker (1996) and Iwasaki and Kasuya (1997) than to that of Takemura (1986, in Japanese) based on different maturity criteria.

With the exception of one outlier, "immature" testes weighed 70 g or less, "early maturing" 60–185 g, "late maturing" 100–190 g, and "mature" 125 g and over. All the testes weighing over 190 g were safely identifiable as "mature." Testis weight increase between ages 7 and 13 is rapid.

Correlation between testicular maturity and body length is poor (Figure 14.4). "Immature" males occurred at body length of 209 cm or below, "early maturing" at 180–210 cm, "late maturing" at 185–207 cm, and "mature" at 183 cm and over. Males attain the "mature" stage at a broad body-length range of 183–210 cm.

All the testes weighing 190 g or more were found to be "mature." Weight of these testes did not show correlation with either age (Figure 14.3) or body length (Figure 14.4) and showed broad individual variation of 200–400 g. The sample covered the two months of February and March, which were

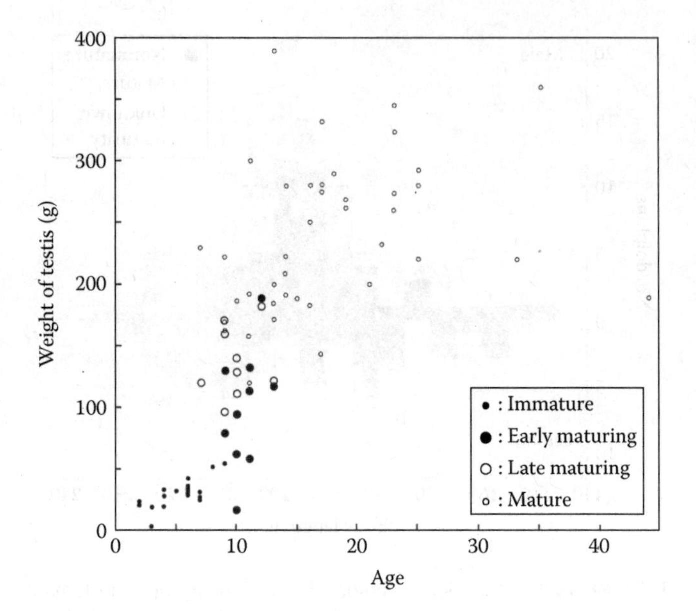

FIGURE 14.3 Scatterplot of weight of single testis (epididymis removed) and histological maturity on age of Pacific white-sided dolphins driven at Katsumoto, Iki Island.

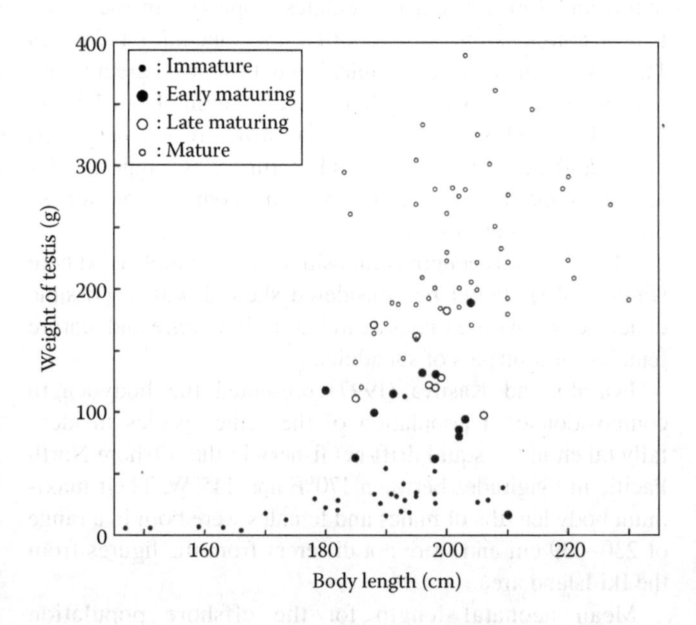

FIGURE 14.4 Scatterplot of weight of single testis (epididymis removed) and histological maturity on body length of Pacific white-sided dolphins driven at Katsumoto, Iki Island.

apparently outside the mating or parturition seasons. The reason for such broad individual variation in testicular weight has not been investigated.

14.3.5 Sexual Dimorphism of the Dorsal Fin

Males change the shape of their dorsal fin with growth (Figure 14.5). The dorsal fin of males with "immature" testes has a slightly convex leading edge and the tip pointing posterior-dorsally. These features are similar to those of immature and mature females. Males with "mature" testes have a strongly convex or rounded leading edge of the dorsal fin, with the tip drooping posterior-ventrally. This morphology results in the tip of the dorsal fin being situated below the highest point of the fin. This change is examined in the following in comparison with sex, maturity, and body length. Data are not available for this analysis on age and on males confirmed to be at the testicular histology stages of "early maturing" and "late maturing"

Figure 14.6b presents the total height of dorsal fin in relation to sex, maturity, and body length. Height increases almost linearly with body length until 190 cm and then increases at a decreasing rate. Females are likely to have dorsal fins of lesser height, but the sexual difference is small. This measurement is unsuitable for expressing sexual dimorphism of the dorsal fin as shown in Figure 14.5.

Figure 14.6a shows height from the base to the tip of the dorsal fin. The measurement increases linearly with increasing body length of immature individuals of both sexes but decreases in sexually "mature" males of over 215 cm. Some sexually mature females have a similar feature but of lesser magnitude.

Figure 14.6c further examines the change mentioned earlier using the value of (b–a), or the degree of droop of the dorsal fin tip. This parameter is small at less than

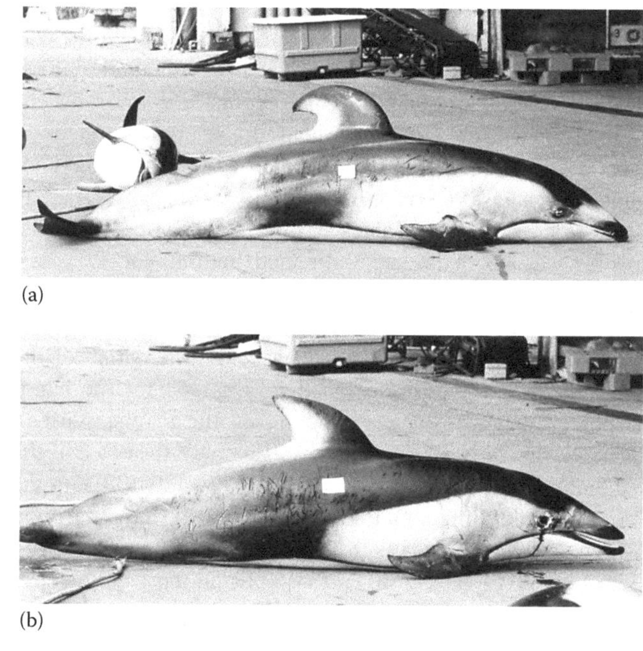

FIGURE 14.5 Secondary sexual character in the profile of dorsal fin of Pacific white-sided dolphins driven at Katsumoto, Iki Island. (a) 224 cm male with histologically mature testis weighing 261 g; (b) 209 cm male with histologically immature testis weighing 38 g; (c) 213 cm sexually mature female. See frontispiece for additional photographs.

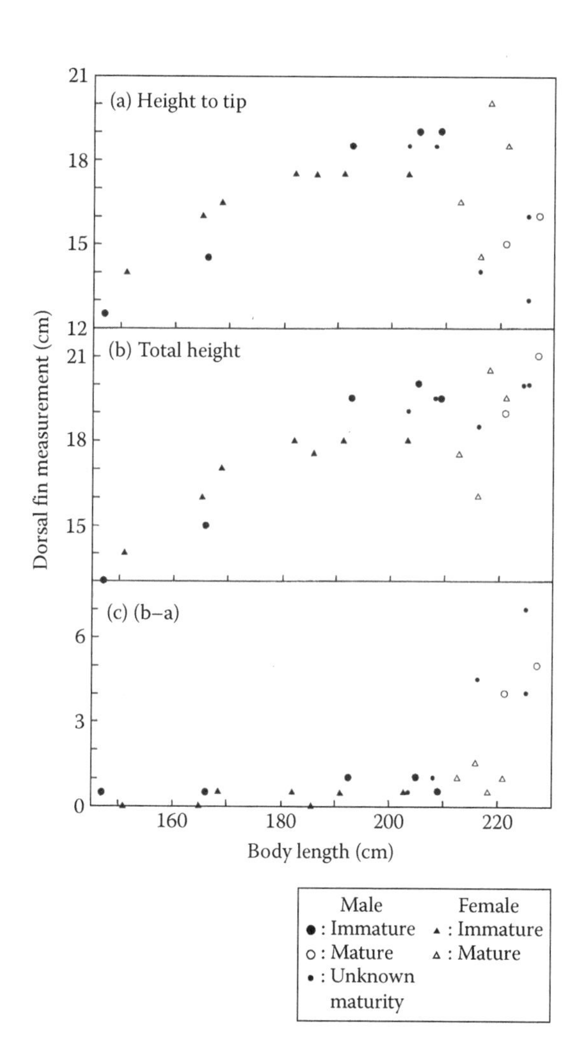

FIGURE 14.6 Scatterplot of morphological measurement of dorsal fin (cm) and histological maturity of testis on the body length of Pacific white-sided dolphins driven at Katsumoto, Iki Island. (a) Height of dorsal fin from the base to the tip; (b) total height of dorsal fin from the base to the highest point; (c) degree of downward droop of dorsal fin (b–c).

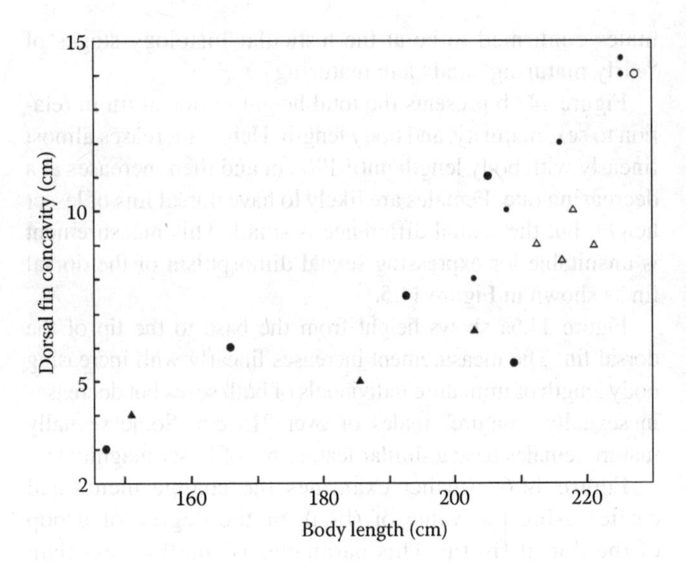

FIGURE 14.7 Scatterplot of degree of concavity of posterior margin of dorsal fin on body length of Pacific white-sided dolphins driven at Katsumoto, Iki Island. See Figure 14.6 for sex and maturity keys.

2 cm in "immature" males and females of all the growth stages but is great at 4–7 cm in "mature" males. "Early maturing" and "late maturing" males were not available for these analyses.

Concavity of the trailing edge of the dorsal fin increases in individuals over 210 cm, but the magnitude is greater in males (Figure 14.7).

14.3.6 Discussion

Significant morphological change of the dorsal fin is known in various cetaceans, such as the killer whale, short-finned pilot whale, spinner dolphin, and Dall's porpoise, and offers a key for the identification of sex and maturity. The analysis presented earlier proved that the Pacific white-sided dolphin is such a species. Among Pacific white-sided dolphins in the Iki Island area, only males measuring 215 cm or over had a dorsal fin with round leading edge and the tip situated more than 3 cm below the highest point of the fin. Such a dorsal fin was not seen on any males below 210 cm or females of any growth stage. The sample did not cover the body-length range of 210–215 cm.

This analysis of the morphology of the dorsal fin had two limitations. One was the lack of age data for direct comparison with the morphology of the dorsal fin, and the other was the lack of males determined with testicular histology at the "early maturing" and "late maturing" stages. Noting that "maturing" males are known in the body-length range of 180–210 cm and that all the males over 210 cm were classified as "mature," I would judge that three males of unknown testis maturity and measuring over 215 cm in Figure 14.6 were "mature" and two males at 200–210 cm were in the "immature" or "maturing" stages. Thus, it is concluded that the character of the dorsal fin described earlier is limited to sexually mature males and is not seen on pubertal and immature males, although a larger sample may provide some exceptions.

Van Waerebeek (1993) reported a round dorsal fin on male dusky dolphins, *Lagenorhynchus obscurus*. Walker et al. (1986) reported a similar character of the dorsal fin, as described earlier, on Pacific white-sided dolphins in the eastern North Pacific. Although they were unable to analyze the character against sex or growth stage, they suggested the possibility that it indicated a state of physical maturity. Age or body length at physical maturity has not been examined for the species from the Iki Island area, but limited increase in body length of this species after sexual maturity suggests that the body-length difference between attainment of sexual maturity and physical maturity is small and the time between the two stages short. It is my conclusion that the character of a round dorsal fin can be used as a sign of sexual maturity in males. This character has the potential to contribute to behavioral study of the species.

Pacific white-sided dolphins around Japan have been hunted by historical drive and hand-harpoon fisheries, by various types of dolphin fisheries during the peri–World War II period (Chapters 2 and 3), and by culling of a total of 4600 in the 1980s as a countermeasure against a dolphin-fishery conflict in the area of Iki Island in Nagasaki Prefecture (Sections 3.3 and 3.4). Subsequent low-level culling continued in Nagasaki Prefecture until 1995, but not of Pacific white-sided dolphins (Tables 3.5 and 3.6). After 1980, the species was not a target of directed hunting for 17 years. However, in 2007, the Fisheries Agency of Japan placed it on a list of species allowed for the dolphin fisheries and set an annual quota of 360 (Tables 6.3 and 6.4). Allocation of the quota to fishery types and locations was 154 for the hand-harpoon fishery in Iwate Prefecture on the Pacific coast in 38°59′N–40°27′N, 36 for the drive fishery on the coast of the Izu Peninsula (34°36′N–35°05′N, 138°45′E–139°10′E) on the Pacific coast of Shizuoka Prefecture, 36 for the hand-harpoon fishery, and 134 for drive fisheries in Wakayama Prefecture (33°26′N–34°19′N, 135°00′E–135°59′E) on the Pacific coast. Reasons for this action and the basis for the quota have not been given. The quota was left almost unused by the fisheries that received the allocations. The only possible reason for the policy could be demand from aquariums and the desire for live-capture fisheries. If drive fishermen have the opportunity, they are likely to drive schools of Pacific white-sided dolphins for selective transfer to aquariums.

REFERENCES

[*In Japanese Language]

Ferrero, R.C. and Walker, W.A. 1996. Age, growth, and reproductive patterns of the Pacific white-sided dolphin (*Lagenorhynchus obliquidens*) taken in high seas drift nets in the central North Pacific Ocean. *Can. J. Zool.* 74(9): 1673–1687.

Hayano, A., Yoshioka, M., Tanaka, M., and Amano, M. 2004. Population differentiation in the Pacific white-sided dolphin *Lagenorhynchus obliquidens* inferred from mitochondrial DNA and microsatellite analyses. *Zool. Sci.* 21: 989–999.

Iwasaki, T. and Kasuya, T. 1997. Life history and catch bias of Pacific white-sided (*Lagenorhynchus obliquidens*) and northern right whale dolphins (*Lissodelphis borealis*) incidentally taken by the Japanese high seas squid driftnet fishery. *Rep. Int. Whal. Commn.* 47: 683–692.

Kasuya, T. 1985. Fishery-dolphin conflict in the Iki Island area of Japan. pp. 253–272. *In*: Beddington, J.R., Beverton, R.J.H. and Lavigne, D.M. (eds.). *Marine Mammals and Fisheries.* George Allen and Unwin, London, U.K. 354pp.

Kasuya, T. 1995. Dorsal fin of *Lagenorhynchus obliquidens*: An indication of male sexual maturity. *Abstract of the 11th Biennial Conference on the Biology of Marine Mammals*, Orland, FL, December 14–18, 1995.

*Kasuya, T. and Miyazaki, N. 1981. Dolphins in the Iki Island area and conflict with fisheries—Preliminary report of three year study. *Geiken Tsushin* 340: 25–36.

*Kawamura, A., Nakano, H., Tanaka, H., Sato, O., Fujise, Y., and Nishida, K. 1983. Results of small cetacean surveys from a railway ferry boat between Aomori and Hakodate. *Geiken Tsushin* 351–352: 29–52.

*Noguchi, E. 1946. Dolphins and porpoises and their utilization. pp. 3–36. *In*: Noguchi, E. and Nakamura, T. (eds.). *Scomber Fishery and Utilization of Dolphins and Porpoises.* Kasumigaseki Shobo, Tokyo, Japan. 76pp.

*Takemura, A. 1986. Pacific white-sided dolphin and common bottlenose dolphin. pp. 161–177. *In*: Tamura, T., Ohsumi, S. and Arai R. (eds.). *Report of Contract Studies in Search of Countermeasures for Fishery Hazard (Culling of Harmful Organisms) (56th to 60th year of Showa).* Suisan-cho Gogyo Kogai (Yugaiseibutsu Kujo) Taisaku Chosa Kento Iinkai (Fishery Agency), Tokyo, Japan. 285pp.

Van Waerebeek, K. 1993. External features of the dusky dolphin *Lagenorhynchus obscurus* (Gray, 1828) from Peruvian waters. *Estud. Oceanol.* 12: 37–53.

Walker, W.A., Leatherwood, S., Goodrich, K.R., Perrin, W.F., and Stroud, R.K. 1986. Geographic variation and biology of the Pacific white-sided dolphin, *Lagenorhynchus obliquidens*, in the north-eastern Pacific. pp. 441–465. *In*: Bryden, M.M. and Harrison, R. (eds.). *Research on Dolphins.* Clarendon Press, Oxford, U.K. 478pp.

15 Cetacean Conservation and Biologists

This chapter is a summary of what I have learned through my activity as a scientist working on biology and conservation of cetaceans. It is based (with some additions) on my presentation upon receiving the Kenneth S. Norris Lifetime Achievement Award at the Biennial Conference of the Society for Marine Mammalogy in Cape Town in November 2007 (Kasuya 2008). This chapter covers a broader area than the earlier chapters of this book, although the readers will find some overlap with the earlier chapters.

Since the early twentieth century, Japanese cetacean scientists relied often on fisheries for their materials and opportunities for research. Therefore, their activities were limited and likely to be under the influence of whaling policy; I was not the exception. My life as a cetacean biologist started in the summer of 1960, when I chose age determination of sperm whales for my graduation thesis, worked mainly on cetaceans in the western North Pacific, and encountered the collapse of populations of striped dolphins and sperm whales due to overfishing. Along with the tragedies, I met with some interesting, but still unanswered, questions on the lives of the animals, such as (1) the reason for mating seasons half a year apart in two neighboring populations of the same species, (2) the function of postreproductive females that are abundant in some toothed whale populations, (3) the function of copulations by nonestrous female short-finned pilot whales, and (4) the function of excess males in the Baird's beaked whale community. New research methodology is needed to find answers to these questions. Management of cetacean populations needs to take into account the following points: (1) Conserving genetically identifiable populations is not sufficient; smaller units where individuals cooperate in daily life should be conserved. (2) The culture of cetaceans and its diversity should be conserved. (3) The function of individuals in a cetacean community needs further investigation. The current situation of marine mammals suggests that any marine mammal scientist is likely to encounter conservation-related problems in his/her activities. It is hoped that we will be ready for such situations.

15.1 THE COMMUNITY I GREW UP IN

Conservation is based on the values of community members, including scientists and policy makers. The values are influenced by personal experiences and memories. Thus, each person may respond differently to a single conservation problem, and a full understanding of the conservation-related opinions of a person is possible with knowledge on his history and his or her current position in the community. Therefore, I will describe briefly the community I grew up in.

I was born in 1937 in a farmhouse on the outskirts of Kawagoe City, about 40 km north of Tokyo. The thatched house was about 500 m from the Iruma River, and once I could hear the stream in my bed; this became impossible due to increased ambient noise. It was the year when Japan started a full-scale war against China. Although the war was characterized as the "China Incident" in order to avoid trade sanction from the United States, it ended up with the desperate declaration of war on the United States in 1941 (Osugi 2007, 2008, both in Japanese). Through this process, the country rapidly lost the democratic atmosphere thought to have been present in the early twentieth century. I started at a junior school in 1944, but the class was frequently interrupted by air raids by the United States. We were educated to die for Emperor Hirohito on the battlefield and trained to jump into enemy lines with a bomb. Only my grandmother said this was cruel and grieved over the situation. Other family members were unable to air their opinions for fear of being overheard.

This awful regime was maintained through information control and the activity of the secret police. Newspapers served as running dogs of the powerful state, radio broadcasts were controlled by the government, and long-range radio receivers were prohibited. The emperor and his government were controlled by the military, and political parties ceased to function. Now I think that the development of such a terrible situation might have been blocked if people had raised their voices at an early stage.

My father died in 1944 from illness and the war ended in 1945. The defeat was the saddest incident in my life as far as I remember. Soon after the cease-fire, there appeared new slogans such as "revive the country," "establish democracy," and "Japan without war." These slogans still have emotional power in my mind at the age of 77, but later I noticed that they did not contain any element of conservation. Although it was probably unavoidable at the time, Japan in the 1950s and 1960s succeeded in reviving the economy by sacrificing the environment. As a boy in middle school, I felt sad to see the construction of factories and houses in an area once covered by trees and a decrease in the bird population annually visiting my area. Fireflies were gone and fish disappeared from surrounding small streams due to the effect of poisonous fertilizers such as lime nitrogen and agricultural chemicals, and the fish remaining in larger stream became inedible, smelling bad due to pollutants from factories.

The Japanese sea lion *Zalophus japonicus* may have become extinct during this period, although some scientists are still reluctant to accept this (Ito and Shimazaki 1995, in Japanese; Ito 1997, in Japanese). I suspect that mortality incidental to the salmon gill-net fishery then operated in the Sea of Japan contributed to the extinction (Section 6.2). Pollution that progressed in coastal waters during the period could have contributed to the decline of a finless porpoise population in the Inland Sea (Section 8.4.5).

Japan's economic recovery is testified to by the fact that its gross domestic product is among the top 5 in the world, but a question still remains about the establishment of democracy. Reporters without Borders (2007) ranked Japan's press freedom at 51st in the world's 168 countries. I agree that Japan achieved considerable progress in over a half century, improving from a status similar to that in the current North Korea, which was ranked at the bottom of the 168 countries, but I have the impression that Japan is still far from the status I imagined from my postwar education. Such slow progress in Japanese democracy has some relationship with the fact that the Japanese prewar government system survived almost intact through postwar reform. The system aimed at benefiting itself and surviving by serving the power of the time. In the prewar period, it served the government of the emperor and the military behind it, and after the war, it switched to being the partner of industry. The language barrier helped the government to control information inflow by selecting only "good" information.

The achievement of peace is more difficult to evaluate. It is true that the Japanese postwar military that was founded in 1950 has not killed one enemy person in over 60 years. However, it is regrettable to see that the Japanese government values the military over conservation, as seen from its plan to construct a military base at the cost of the dugongs in Okinawa (Marsh et al. 2002). Ignoring paragraph 2 of Article 9 of the Japanese Constitution that reads "army, navy, air force and other war forces shall not be held, and right of warfare is not allowed," the current Japan maintains a military of a quarter million. If the government believes it necessary to have the military, it should modify the constitution. I am troubled by the government's attitude to pursue its desire by ignoring the constitution or making a convenient interpretation of it, and with the national character that allows such a government. Such a national character could have been created during the feudal period, when rules were given by the governor and considered by the people as something to be ignored or found a way around. The same seems to be behind the Japanese scientific whaling program, which utilizes a loophole in the International Convention of the Regulation of Whaling and plans to take more than 1000 whales a year for an indefinite time. A community that ignores such rules appears from the outside as an unpredictable enigmatic country.

15.2 INDUSTRY AND CONSERVATION

Maintaining scientists who are friendly to the whaling industry benefits the industry, at least for the time being. The Fisheries Agency of Japan has operated institutes and placed whale biologists there and also offered research funds to university scientists. The whaling industries and the government of Japan attempted to control the whale scientists as described next.

After completing a fisheries course at the University of Tokyo, I obtained a position in 1961 in the Whales Research Institute (WRI) in Tokyo. It was the largest institute in Japan dealing with whales, with eight scientists including the director Hideo Omura, and the staff studied large whales with financial support by the five major Japanese whaling companies of the time. This institute merged in 1987 with a part of the Nihon Kyodo Hogei (Japan Union Whaling) Co. Ltd. to form the Institute of Cetacean Research (ICR) as a structure dedicated to the Japanese scientific whaling program (Section 7.4; Kasuya 2002a, 2007). There were two other groups of scientists interested in cetaceans in Japan at the time. One was led by Teizo Ogawa, anatomist at the University of Tokyo, and the other by Kazuhiro Mizue at Nagasaki University. The former mainly targeted brain anatomy and also studied taxonomy, and the latter group studied the general biology of small cetaceans and their acoustics. Both groups relied on cetacean fisheries for materials and opportunities for research.

The training I received from WRI scientists during the 5 years I was there formed the base of my subsequent activities, and what I saw of the whaling activities decided my attitude toward whaling. Hideo Omura started in whale science in 1937 and worked mainly on the taxonomy and life history of large cetaceans. His taxonomic work was an extension of the efforts of Nagasawa (1915, in Japanese) and Matsuura (1935a,b, both in Japanese) to apply modern taxonomy to large cetaceans around Japan. Masaharu Nishiwaki joined the WRI in 1947 after 6 years of military service in the naval air force, which was recorded in his journal of an Antarctic whaling cruise published posthumously (Nishiwaki 1990, in Japanese). His taxonomic studies of small cetaceans around Japan were along the lines of Nagasawa (1916a,b, both in Japanese), Ogawa (1932, in Japanese), Kuroda (1940, in Japanese), and Okada and Hanaoka (1940). He was one of the pioneers in the field of age determination of cetaceans as well as the conservation biology of striped dolphins then hunted in great numbers along the coast of the Izu Peninsula (34°36′N–35°05′N, 138°45′E–139°10′E) (Section 10.5.1) and well known in the area of cetacean conservation. I also owe thanks to Seiji Ohsumi for the technical training in the WRI. He studied age determination and population dynamics of large cetaceans.

Omura was with the 1937/1938 Antarctic whaling fleet as an inspector. In his journal of the cruise published posthumously he expressed his concerns about the management of whale stocks (Omura and Kasuya 2000, in Japanese), but it was rare for him to do so in the WRI. As a director of the institute operated with financial support from the whaling companies, he must have had a limit on airing his personal opinion. One episode relates to this situation.

For many years, the WRI wanted to examine large whales caught by coastal whaling and succeeded in 1959 in obtaining funding from the Fisheries Agency of Japan for this activity, which continued until 1965. The coastal whaling companies were concerned about how to hide their misreporting of body lengths and underreporting of catch figures and set several conditions for accepting WRI scientists at their whaling land stations, including at least the following:

1. The name of the whaling station or of the whaling companies could not accompany biological data for individual whales. Thus, scientists recorded the data for a broad geographical region such as the Sanriku

coast (37°54′N–41°35′N) or the Pacific coast of Hokkaido, where several land stations operated.

2. Scientists could not measure whales. Thus, they copied body lengths from records made by the staff of the land station, or they estimated length by eye.
3. Scientists could not remain at a single whaling station but had to cover several nearby stations at the same time. This functioned to obscure responsibility of a particular whaling company for whales recorded by the biologists.

Whaling companies regularly submitted reports on their whaling operations to the Fisheries Agency, which included species, number, body length, and sex of each whale processed. We scientists noticed that these reports were totally unreliable but were unable to obtain firm evidence to confirm that. I have a clear memory of the following talk between Masaharu Nishiwaki and Director Hideo Omura. Nishiwaki stated: "We are currently editing an annual report of biological investigation of whales taken by coastal whaling. Careful examination of our report will identify disagreement with statistics reported by the whaling companies. Shall we be concerned about that?" Omura replied that we did not have to worry about it. Recently, I reanalyzed these reports of the WRI and was able to show manipulation of whaling statistics by the Japanese coastal whalers. I have made several additional attempts with the cooperation of some ex-whalers to reconstruct the true whaling statistics, but this is far from completion. The Japanese whaling community still attempts to hide past misconduct.

The heyday of Japanese whaling was long ago in the mid-1960s. The whaling companies reduced the activities of the WRI for reasons of economy, and some scientists moved to a whale section newly created in the Far Seas Fisheries Research Laboratory of the Fisheries Agency in Shimizu. Nishiwaki and Takahisa Nemoto, who studied whale diets, moved in 1965 to the Ocean Research Institute of the University of Tokyo, and I joined the laboratory of Nishiwaki in 1966 to study small cetaceans. University scientists had difficulty in working with large cetaceans that were almost monopolized by Fisheries Agency scientists, but the drive and hand-harpoon fishermen were still generous in allowing our investigation of their catches. This cooperation between the dolphin fishermen and scientists lasted for the next 30 years. After an attempt at cetacean systematics (Kasuya 1973), I moved to studies of the life history and social structure of toothed whales around Japan (Chapters 8–14) and some river dolphins (Kasuya 1972; Kasuya and Nishiwaki 1975; Kasuya and Brownell 1979). I had the impression that osteology was not a suitable tool for cetacean systematics due to morphological convergence. Now, DNA is widely in use for the purpose.

With the invitation of Seiji Ohsumi in April 1983, I joined the whale group of the Far Seas Fisheries Research Laboratory, where I worked for 14 years before moving to the Faculty of Bioresources at Mie University in April 1997. About one year after my shift to the university, there occurred a change in the attitude of dolphin fishermen toward university scientists. Drive fishermen in Taiji (33°36′N, 135°57′E) in Wakayama Prefecture and on the coast of the Izu Peninsula and hand-harpoon fishermen in Iwate Prefecture (38°59′N–40°27′N) on the Pacific coast rejected university scientists who wished to examine or purchase their catches, requiring prior approval by the Fisheries Agency or Fisheries Agency scientists. Requests by the university scientists for approval were ignored or were met with a counter-request from the Fisheries Agency scientists to have their names as coauthors of papers to be published in the future using the materials. University scientists or graduate students were unable to obtain even a piece of skin for DNA analyses and had to discontinue their studies that had lasted for several years. In the summer of 1998, a person in the Fishery Cooperative Union at Taiji replied to my inquiry by saying that it was their own idea to protect their fishery from unauthorized views of population structure. I felt this was untrue, and that the interference was on the advice of the government after our discussion on population structure of striped dolphins in the Scientific Committee Meeting of the International Whaling Commission (IWC) held in the autumn of 1997 (Section 10.4). This attitude of the dolphin fishermen was maintained throughout the period of my service to another university in 2001–2006. The fishery utilizes common resources of the ocean for personal benefit, and therefore, fishery activities as well as the opportunity for scientific studies should be open to every scientist.

The Scientific Committee of the IWC has had an important role in the management of cetaceans. The Committee is composed of scientists nominated by the member governments of the IWC, invited scientists attending by invitation of the chair, and NGO observers. The Scientific Committee presents advice to the Commission based on conclusions reached after reviewing papers submitted by the scientists. Papers by scientists nominated by Japan had to be reviewed by the Whaling Section of the Fisheries Agency and finally approved by the Japanese commissioner to the IWC before presenting them to the Committee. The effect of a paper on the industry was an important aspect of this procedure, and the scientists were often requested to modify their conclusions or to carry out further analysis, which meant delay in submission.

15.3 CETACEAN TRAGEDIES I HAVE WITNESSED

Many attempts to manage populations of large cetaceans have failed. With the exception of Bryde's and minke whales that have had a relatively short history of exploitation, large cetaceans in the North Pacific became protected by 1976 (Table 7.2). Some populations of these species may have been extinct before being identified by scientists (Clapham et al. 2008). The following are two examples of management failure I witnessed in Japan.

15.3.1 Striped Dolphins off the Pacific Coast of Japan

I first encountered striped dolphins in the autumn of 1960 when I accompanied Nishiwaki and Ohsumi on a sampling trip to Kawana (34°57′N, 139°07′E) on the Izu Peninsula.

Dolphin-drive fisheries were known on the Izu coast since the early seventeenth century. The operating bodies in the drive fisheries, which numbered 18 in the late nineteenth century, saw a decrease in the early twentieth century, had a short revival during the peri–World War II period, and again resumed further decline to only one group remaining at Futo in recent years (Sections 3.8 and 10.5.16). Only Kawana and Futo (34°55′N, 139°08′E) operated drive fisheries for 22 seasons (September to December) from 1962 to 1983. Each group used two searching and driving vessels (a total of four vessels), although vessel speed increased with time and survey range widened. The two groups operated jointly beginning in 1967 (Shizuoka-ken Kyoiku Iinkai [Board of Education of Shizuoka Prefecture] 1987, in Japanese). Kawana ceased operation in 1984, and only Futo continued to operate the fishery opportunistically with a small quota, but it has recorded no catch since 2005 (Section 3.8.3; Kishiro and Kasuya 1993). Most of the catch was consumed in the three prefectures of Kanagawa, Shizuoka, and Yamanashi.

Statistics of the early operations have been completely lost. Some fragmentary statistics are available for the late nineteenth and early twentieth centuries, and complete statistics are available only for the period since 1957 (Figure 15.1, Kasuya 1999b). Catches during the peri–World War II period could have been very large. Four of the 10 seasons from 1942 to 1951 recorded catches of 11,000–21,000, but the statistics did not cover all of the operating bodies and perhaps not completely cover the records of any single group. It is generally believed

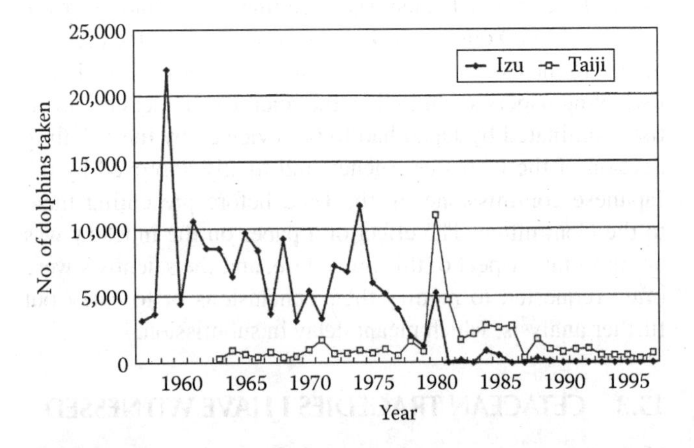

FIGURE 15.1 Annual catch of striped dolphins on the coast of the Izu Peninsula (34°36′N–35°05′N, 138°45′–139°10′E; closed circles and solid line) and Taiji (33°36′N, 135°57′E; open circles and dotted line), which have contributed most of the catch of the species in Japan. Although there is no doubt that larger catches occurred in the Izu area before the dates shown here, available statistics are fragmentary (see Section 3.8.3). The Izu fishery was operated during 1962–1983 by two drive groups with a total of 4 searching vessels (speed of the vessels improved during the period). The Taiji fishery used to hunt striped dolphins with hand harpoons, but it was replaced by the drive fishery that first drove the species in 1973 (see Section 3.9.1). (From Kasuya, T.: Cetacean biology and conservation: A Japanese scientist's perspective spanning 46 years, The Kenneth S. Norris Lifetime Achievement Award lecture presented on 29 November 2007, Cape Town, South Africa. *Marine Mammal Sci.* 2008. 24(4). 749–773. Copyright Wiley-VCH Verlag GmbH & Co. KGaA. Reproduced with permission, Figure 1.)

that striped dolphins constituted the major portion of the catch, at least in the twentieth century, but the catch composition was confirmed by scientists only for post–World War II operations.

While I studied the striped dolphin fishery at the Ocean Research Institute (1966–1982), I had telephone contact daily during the season with the Fishery Cooperative Union of Kawana to get information on their operation for opportunities to examine the catch and collect biological samples. Through this communication, I had an impression that the fishery was becoming increasingly difficult every year. The drive fishermen had a similar impression. So I presented a brief analysis of the population status to a meeting of the Working Party on Marine Mammals of the FAO Advisory Committee on Marine Resources Research, which was held in December 1975 in Bergen, Norway. The report concluded that the dolphin stock could not sustain the current level of hunting and that the population was declining (Kasuya and Miyazaki 1982). Subsequent reanalysis, which was published prior to the delayed publication of the preceding analysis, did not change the conclusion (Kasuya 1976). The methodology was primitive and the accuracy was questionable, but I believed that the conclusion was generally correct. Later analysis confirmed that the catch declined while the range of daily searching area broadened with improved vessel speed and that females shortened the prematurity period and mean calving interval through the history of exploitation (Kasuya 1985). Such changes are expected when the density of a population has declined.

Striped dolphins have also been hunted by the hand-harpoon fishery and probably by the opportunistic dolphin drive of the local community at Taiji, which is about 300 km southwest of the Izu Peninsula, but catch statistics are available only since 1963. In 1969, several hand-harpoon fishermen for dolphins learned about dolphin driving from the Izu fishery and succeeded in driving a school of short-finned pilot whales from offshore waters. This was the start of the current dolphin-drive fishery at Taiji, where fishermen go out to drive dolphin schools. They expanded their target to striped dolphins in 1973 and subsequently to other dolphins (Sections 2.7 and 3.9.1).

In the 1970s, the dolphin consumers in Shizuoka and nearby areas purchased striped dolphins from Taiji and Dall's porpoises from Iwate to supplement the reduced supply of striped dolphins. This means that the decline in striped dolphins landed on the Izu coast was not due to a decreased demand for dolphin meat.

In the mid-1970s, I met a team leader of the Whaling Section of the Fisheries Agency and explained to him how the Izu fishermen relied on striped dolphins and that the dolphin population was in an alarming situation. I requested some action to limit the catch to a sustainable level. After hearing my explanation, he responded by saying that "if the striped dolphin population is gone the Izu fishermen will find some other fishery items, so you do not have to worry about the fishery." I felt that the Whaling Section of the time had some concern about the whaling industry for large whales but that it did not care about the dolphin populations or the fisheries that utilized the species.

Although a catch decline was apparent in the late 1970s due to a large annual fluctuation in the catch, it was only in the 1982 meeting that the IWC Committee was convinced about the trend and recommended that Japan analyze natural and anthropological factors that might be behind the decline (IWC 1983). After spending 8 years in reviewing Japanese fisheries on Dall's porpoises, short-finned pilot whales, and Baird's beaked whales, the Scientific Committee returned in 1991 to the Japanese striped dolphin stocks, reiterated the 1982 recommendations, and further recommended that Japan establish a mandatory catch limit on a species and stock basis, undertake biological studies of striped dolphins killed in the fishery, and update description of the fishing operation (IWC 1992). The Committee of the next year noted that the catch declined to 10% of the level of the 1960s, accompanied by a catch per unit effort decline, and that reproductive parameters changed in a way expected to accompany stock decline. It noted that there was no scientific basis for the catch level currently recommended by the Japanese government for the fishery, that the population could not support continued exploitation at the current level, and that there were no appropriate data on which to set a sustainable catch level. Finally, against objection of some Japanese participants, it recommended an interim halt of all direct catches of the species until a proper assessment could be made of this population (IWC 1993). A Japanese Fishery Agency official who attended the Scientific Committee in 1991 as a scientist objected to the idea that the stock was depleted, stating (not for the record) that coastal dolphins may have moved to offshore waters.

The timing of decline of the striped dolphin fishery was different between Izu and Taiji. The decline occurred earlier in the Izu fishery, to record low annual take of less than 1000 in the early 1980s, but the Taiji fishery declined from the late 1980s to early 1990s (Figure 15.1). The difference in timing is not necessarily evidence against the possibility that the two fisheries operated on the same stock, but they could also have at least partially exploited different populations. Evidence suggesting the latter possibility has been collected (Section 10.4).

Miyashita (1993) estimated the abundance of striped dolphins off the Pacific coast of Japan: 19,600 (CV = 0.696) for the coastal stratum that contained both the Izu and Taiji grounds, 497,000 (CV = 0.179) for the northern offshore stratum, and 52,600 (CV = 0.952) for the southern offshore stratum (Figure 10.1 and Table 3.13). The estimate of abundance in the coastal stratum is equal to 1 or 2 years' catch in the heyday of the fishery. Dolphins in the two offshore strata are apparently separated from the coastal stratum by a density hiatus and probably have not been hunted by the Izu and Taiji fisheries (Kasuya 1999b). The question was raised whether the coastal stratum contains a single population (Section 10.4). The estimates presented earlier should be considered averages for the 9-year period 1982–1991 covered by the shipboard surveys. Therefore, if the coastal population declined during the period, the estimate was likely an overestimate of abundance at the end of the period.

In 1992, the Fisheries Agency of Japan set a quota of 1000 striped dolphins to be shared by two drive teams (Izu and Taiji) and the hand-harpoon fisheries at Choshi (35°42′N, 140°51′E) and in Wakayama Prefecture (which included Taiji). This total annual quota was almost equal to the average take (1053) in the preceding five seasons (1987–1991). The quota was likely created as a figure acceptable by the fisheries but was unlikely to be small enough to allow recovery of the population. The total quota was lowered to 725 in 1993 and maintained at that level until 2006, followed by minor annual decreases since 2007 (Tables 6.2 and 6.3).

There remain several unresolved problems in the management of the striped dolphin population off Japan. The first is the broad confidence interval of the abundance estimate, which makes it risky to use the estimate for management. The second is a still unresolved question on the population structure of the species in the coastal stratum. Last is the catch quota of 3.7% of the mean abundance estimate. Even ignoring the risk accompanying the broad confidence interval, this quota is too great to allow safe recovery of the population. The safe level of take calculated by a method called Potential Biological Removal (PBR) is only 1% of the mean average estimate (Section 12.6.3).

The Scientific Committee of the IWC took 18 years to reach the 1992 advice to temporarily halt the Japanese striped dolphin fishery after it reviewed the report of the first meeting of the small-cetacean subcommittee held in Montreal in 1974, where the Japanese dolphin fishery was one of the major topics. Reasons for the slow progress included uncertainty about the population structure, insufficient population analysis, and a statistically insignificant trend in the catch decline. I agree with these reservations, and I know the policy of the IWC is to manage whale population based on science. However, I learned through the process that the science required too many data to prove that the population was in danger. Some data could be hard to obtain for some populations, and fishery data could contain a large amount of noise obscuring the analysis. When the Scientific Committee in 1982 reached the conclusion that the catch of Japanese striped dolphins had declined over the past 19 years (IWC 1983), the population was in my view at the level where a fishery should not be permitted. The early IWC science attempted to prove that a stock was at risk, and scientists of the industry side raised points to be denied by other scientists. This process could have been a major reason for the slow response of the IWC to declining whale populations in general.

Recent IWC science appears to me to have changed the attitude toward seeking proof of safety of exploitation of a population, shifting the burden of proof from the conservation side to the fishery side.

15.3.2 Sperm Whales in the Western North Pacific

I first saw sperm whales at Ayukawa (38°18′N, 141°31′E) in the summer of 1961 when I visited the place with a team of several WRI scientists led by Hideo Omura. The object was to excavate and clean a sperm whale skeleton buried in the sand at nearby Kukunari Beach. This skeleton was cleaned, measured, and sent to a museum in Stuttgart in West Germany (Omura et al. 1962). Sperm whale carcasses were flensed day and night in those days at the three land stations of Taiyo Gyogyo (Taiyo Fisheries) Co. Ltd., Nihon Kinaki Hogei

[Japan Coastal Whaling] Co. Ltd., and Kyokuyo Hogei [Polar Whaling] Co. Ltd. at Ayukawa Town. Sperm whale carcasses waiting at the wharves for processing were inflated with the heat of the summer sun. Because the major product was sperm oil, the station did not apparently care about the freshness of the carcasses, and catcher boats brought in their catches one after another. On this occasion, I noticed the callosity on the tip of the female dorsal fin (Kasuya and Ohsumi 1966), which was used by some biologists as a marker of adult females.

My next encounter with sperm whales was in May 1962 as a whale biologist on board the whaling factory ship *Kinjyo-maru* of the Taiyo Gyogyo Company, which was on the way to the whaling ground in the Aleutian Islands, Bering Sea, and Gulf of Alaska (Kasuya 1963, in Japanese). Soon after the factory ship moved out of Sagami Bay and approached Katsu-ura (35°09′N, 140°19′E) on the Pacific coast of Chiba Prefecture, I sighted a group of sperm whales. There were numerous nursing schools of the species in these nearshore waters in the summer season, and the coastal whalers operated with an annual quota of 1800–2100 sperm whales (Kasuya 1999a).

The *Kinjo-maru* was a vessel renamed from *Nisshin-maru No. 1*, which was converted from a wartime standard ship and participated in the first Japanese postwar Antarctic expedition of the 1946/1947 season (Tokuyama 1992, in Japanese). Masaharu Nishiwaki was on board on her second expedition to the Antarctic in 1947/1948 season (Nishiwaki 1990, in Japanese).

In 1983, I moved to the whale section of the Far Seas Fisheries Research Laboratory of the Fisheries Agency and was again involved in the management of large cetaceans. I noticed that an unexpected change had happened in the coastal sperm whale fishery in the prior 17 years while I was working on small cetaceans. The sperm whale season had moved from summer in the 1960s to winter in the 1980s. The whale fishery in the 1960s hunted sperm whales off land stations at Sanriku (38°N–41°N) or eastern Hokkaido (43°N). The whaling moved from the summer operation off Sanriku and Hokkaido to a winter operation in the southern ground of the Bonin Islands and Yakushima Island in latitudes of 29°N–32°N and longitudes of 132°E–143°E.

In the summer of 1984, I organized a 2-month sighting cruise with two chartered whale catcher boats in the western North Pacific west of longitude 180° and in latitudes of 35°N–45°N (the eastern element of these cruises is in Figure 13.3). One of the objectives of the cruise was to confirm with my own eyes the current status of the previous sperm whale ground, and the other was to obtain information that would contribute to the discussion on Baird's beaked whales at the next Scientific Committee meeting of the IWC. I spent one month earlier in the summer whaling ground off Sanriku and Hokkaido and sighted no sperm whales. Another vessel that surveyed in offshore longitudes found only a few sperm whales. It was evident that the rich sperm whale ground off Sanriku and Hokkaido had disappeared, probably due to whaling. Illegal whaling by the USSR was known to have contributed to the depletion of North Pacific sperm whales (Ivashchenko et al. 2007), but the contribution of Japanese whalers was also clear.

After this cruise, I analyzed the movement of sperm whales marked with Discovery tags, their segregation by sex and growth stages, and shift of the whaling ground during the history of modern whaling. The IWC of the time assumed a single sperm whale population in the western North Pacific for management purposes, but the analysis did not support the hypothesis. It seemed to be more reasonable to assume existence of two populations in the western North Pacific, with two breeding aggregations, one summering off the Kuril Islands and the other off Sanriku and Hokkaido. The boundary between the two breeding aggregations roughly agreed with the boundary between the cold Oyashio and warm Kuroshio currents. The occupants of the Sanriku–Hokkaido ground seemed to alter with season (Kasuya and Miyashita 1988).

This hypothesis allows the assumption that Japanese coastal whaling in the 1960s exploited the breeding population of the southern stock in their summering ground off Sanriku/Hokkaido. In the 1970s, responding to the decline of the southern population, the fishery switched the season to winter (but without changing the location) and hunted the breeding population of the northern stock that was wintering off Sanriku/Hokkaido. In the 1980s, the northern breeding population was also depleted and the Japanese whalers had to fill their quota of a few hundred sperm whales by shifting operations to further south where the breeding stock of the southern population could have wintered. This last operation pattern was the worst for the whalers, because they were obliged to land their catch at an intermediate land station at Wadaura (35°03′N, 140°01′E) or at Taiji for crude butchering and transport by truck to Ayukawa or Yamada (39°28′N, 142°00′E) in the Sanriku Region for final processing.

Kasuya and Miyazaki (1997, in Japanese) made a simple analysis of the number of sperm whales sighted per unit of survey distance using survey data that appeared in the Japanese progress report to the IWC and found that the density parameter almost doubled during the 8 years after 1987 when Japan ceased hunting sperm whales. Such a rapid increase in density is easiest to explain as immigration from offshore less depleted waters and suggests a need to modify the stock hypothesis of Kasuya and Miyashita (1988). Further study is also needed to confirm the relationship between the breeding population that summered off Sanriku/Hokkaido and that which wintered in the Bonin Islands/Yakushima Island area.

The Scientific Committee of the IWC in the 1980s spent a large amount of effort to assess the status of the sperm whale stocks in the western North Pacific, but their mathematical model failed to reach a reliable conclusion (IWC 1981, 1988). The model predicted optimistic trajectories in the abundance, but the actual whaling operation worsened annually. I suspect several factors behind the contradiction:

15.3.2.1 Erroneous Population Hypothesis

The Scientific Committee of the IWC first assumed a single population of sperm whales in the western North Pacific, but an assumption of two populations better explains whale movements and the operation pattern of the fishery (see Section 15.3.1). If the fishery shifted from one population to a second following depletion of the first, the analysis of the whaling data under a single-population hypothesis was likely to mask the depletion of the first exploited stock. The Scientific Committee then placed a boundary between two hypothetical populations at the eastern edge of the western North Pacific stocks proposed by Kasuya and Miyashita (1988). The boundary should have been placed where the probability of encountering the eastern and western populations was equal. The erroneous population boundary functioned to allow greater than reasonable catch limits to whalers who operated in the western North Pacific.

15.3.2.2 Erroneous Growth Curve

The assessment of North Pacific Sperm whales by the Scientific Committee converted body lengths of male sperm whales captured by the fishery into ages, examined the annual trend of the age composition thus estimated, and then estimated the effect of hunting on the population (females were not used for the purpose). In converting the length composition into age composition it used a single age-length key, which was largely contributed by scientists of the WRI in the 1960s. However, later study found that the growth of male sperm whales changed following a decline in population density, possibly due to improvement of nutrition (Kasuya 1991). The maximum male size increased from 16.8 m in the early 1970s to 18.9 m in the late 1970s. This 2 m augmentation was preceded by increased growth rate of young males and was identified in the whole longitudinal range of the northern North Pacific except for the Bering Sea, where only full-grown males migrated and the hunting ceased around 1965 due to overexploitation. Sons grew larger than their fathers, as seen in some recent human populations. The use of a single age-length key could have overestimated the proportion of old males, thus underestimating the effect of the fisheries on the sperm whale population and leading to an optimistic population assessment.

It would have been ideal to obtain the age composition of the whales by reading growth layers in teeth collected from each sperm whale or to update the age-length key. Before I joined the group, the whale scientists of the Far Seas Fisheries Research Laboratory planned to do this. However, possibly due to an insufficient budget or shortage of technicians, they asked for tooth samples from the industry. The personnel working at the whaling stations possibly had trouble in allocating each tooth to a particular whale listed in their catch report (which was almost created from whole cloth; see Section 15.3.2.3), and therefore chose the easier way of collecting several teeth from a single whale and giving different numbers to them. This wreaked havoc in the attempts of scientists to carry out analyses. I identified this by finding several teeth of similar growth layer patterns in the sample, when I worked for the laboratory as an outside scientist. The Japanese whalers did not deny their malpractice, and the fact was reported to the IWC.

15.3.2.3 Erroneous Whaling Statistics

The Scientific Committee of the IWC used the official statistics of four whaling countries (Canada, Japan, the United States, USSR) for the assessment of sperm whale stocks. At least two of them (Japan and USSR) were identified by scientists to have committed underreporting (Brownell et al. 1999).

Scientists of the Whales Research Institute who had experienced field work at the land stations of Japanese coastal whaling felt that they were landing two to three times more sperm whales than reported in the official statistics. They were neither able to get firm evidence for this nor allowed to open the issue even if the evidence could be obtained (Section 15.2). Although the Fisheries Agency inspectors were in a position to detect and report the fraud, most of them did not have the courage for the task, and the Fisheries Agency did not have a budget sufficient to cover all the coastal operations. The inspectors visited the whaling stations only occasionally and used to disappear from the eyes of biologists by playing mahjong with the whaling station staff or going to the local hot spring by invitation of the station. Such incidents were confirmed by Kondo (2001, in Japanese), an ex-manager of the Ayukawa whaling station of Nihon Kikai Hogei. Some international observers from the United States noted underreporting of sperm whales or manipulation of their sex ratio and included it in their reports, but their coverage was extremely limited because they stayed in a distant comfortable hotel and depended for transportation on the whaling companies (Kasuya 1999a). Japanese whaling factory ships had three observers and seemed to me better inspected in the 1960s, except for determining body length and lactation status.

In the 1980s, at a meeting of Fisheries Agency officials and Japanese whale biologists, I proposed that the unreported sperm whales be incorporated in the official statistics for the assessment of the species, but this was rejected with the reason that there was no firm evidence of misreporting and no reliable figures on the magnitude. Subsequently, I examined details of the "Report of Investigation of North Pacific Whale Resources" prepared by scientists of the WRI and confirmed that the scientists often examined a greater number of sperm whales than reported in the official statistics of particular regions and months and that the proportion of females recorded by the scientists was greater than that in the official statistics. I was also able to collect true statistics of coastal whaling with the cooperation of some ex-whalers, although the coverage was still limited (Kasuya 1999a, Kasuya 1999, in Japanese; Kasuya and Brownell 1999, 2001; Kondo and Kasuya 2002). Results of these analyses are briefly presented as follows.

Japanese coastal whaling before World War II took limited numbers of sperm whales, several hundred annually, and the level of underreporting was considered low and limited to leaving undersized whales unreported (personal communication of H. Omura cited in Kasuya 1999a). The catch increased in the 1950s, and the magnitude of underreporting increased while there was no catch limit (Figure 15.2). I suspect that the objective of the underreporting could have been to camouflage the rapid increase with an intent to avoid criticism of overfishing and delay the introduction of a quota system. Underreporting for a similar purpose occurred in the Dall's porpoise fishery in Iwate Prefecture (Section 2.5). Once a quota system was introduced, the fishery had to continue underreporting in order to retain the previous catch level. The methods of underreporting catches could vary, from a simple process of deleting undersized whales, which were mostly females, to full falsification of the number of whales as well as amount of products.

The coastal sperm whale fishery of one company actually made a 1.3–1.5 times greater catch than reported in the

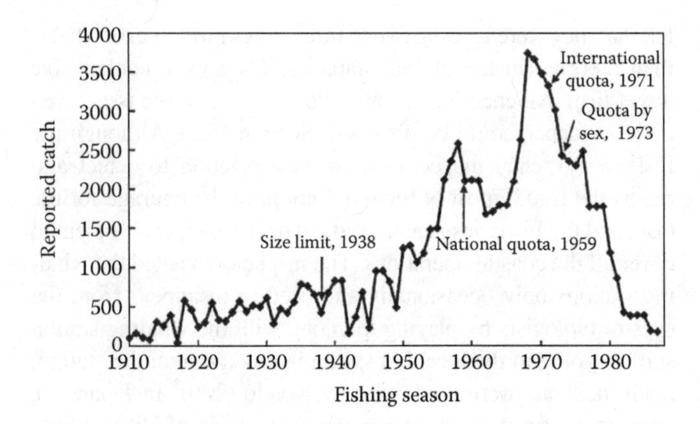

FIGURE 15.2 Annual catch of sperm whales by Japanese costal whaling as appeared in the official statistics. True catch could have exceeded the official statistics (see text). Minimum size limit started in 1938, autonomous national catch quota in 1959, quota by North Pacific whaling countries in 1971, and quota by IWC in 1972 and quota by sex in 1973. Catch ended in March 1988. (From Kasuya, T.: Cetacean biology and conservation: A Japanese scientist's perspective spanning 46 years, The Kenneth S. Norris Lifetime Achievement Award lecture presented on 29 November 2007, Cape Town, South Africa. *Marine Mammal Sci.* 2008. 24(4). 749–773. Copyright Wiley-VCH Verlag GmbH & Co. KGaA. Reproduced with permission, Figure 3.)

official records in the seven seasons from 1953 to 1958 and 1.6–3.0 times greater catch in the five seasons from 1959 to 1964 (the record is missing for 1963). Another company made a 1.8–3.3 times greater catch for the seasons 1965–1975, and the minimum figures of a third company was 1.1–1.3 times the reported catch for the seasons 1959–1961. Similar types of underreporting were not limited to the sperm whale fishery but occurred in the operations of a coastal whaling company on fin, Bryde's, and sei whales (Kondo and Kasuya 2002).

Russian and Western scientists published the true statistics of the sperm whale fishery by the USSR in the North Pacific (Brownell et al. 2000; Ivashchenko et al. 2007). Their true catch was 66,950 in total in six seasons (1966, 1967, 1970–1973), which was 1.8 times the catch reported in the official records. The sex ratio also disagreed with the official statistics. The true catch of males was 1.3 times the official figure (for 1966, 1967, 1970–1971), but that of females was 9.6 times greater than reported in the official statistics in the same period. The authors of these studies believed that such misreporting could have ceased at introduction of the international observer scheme by the IWC in 1972 (northern hemisphere) and 1972/1973 (Antarctic) seasons. However, a Russian scientist reported manipulation of body length and number of minke whales taken by a USSR fleet with an international observer on board (IWC 2010a). The manipulation was made to achieve the production plan within a given catch quota. The Japanese government sent whaling-related personnel to the USSR fleet as international observers, but they did not function as hoped by the IWC.

Sperm whales in the western North Pacific experienced overfishing twice. The first episode occurred in the nineteenth century by sailing-ship whaling of various countries, in which hand harpoons were thrown from rowed boats lowered from a sailing mother ship, "Yankee whaling." They tried out the oil on board, stowed the oil barrels in the hold, and continued a whaling cruise sometimes for 3–4 years. This fishery started in the late eighteenth century in the western North Atlantic, depleted sperm whales in nearby waters, and entered the south Pacific in 1789 rounding Cape Horn and reached the Hawaiian Islands in 1819 (Francis 1990). The whalers arrived in the western North Pacific region called the "Japan ground" around 1820 (Section I.2.1; Starbuck 1878). Damage to the sperm whale stock in the Japan ground is believed to have been considerable, but the details are unknown.

The second episode of overfishing of the sperm whale stocks was by Norwegian-type whaling, which involved killing whales using harpoons discharged from a cannon mounted on a steam ship. This method made it possible to also take blue and fin whales, which swim fast and sink after death. This type of whaling was first introduced into the western North Pacific by the Russian A.G. Didymov in 1889 (Tonnessen and Johnsen 1982). The fishery increased the take of sperm whales after World War II.

None of the past sperm whale fisheries left reliable catch statistics, and we lost valuable opportunities to learn about the mechanisms of population dynamics of sperm whales through the analysis of whaling operation data.

15.3.3 To Avoid Future Tragedies

The current Japanese fisheries receive quotas for nine small cetacean species, including striped dolphins (Tables 4.3, 6.3, 6.4, and 13.5). As long as exploitation continues, threats to these populations will continue, and monitoring of the fishery operations and their catches must be continued. Management policy must be reassessed based on the results of the monitoring and improved abundance estimates.

A large quota of Pacific white-sided dolphins was recently given to drive and hand-harpoon fisheries on the Pacific coast of Japan (Section 14.3.6), ignoring the absence of estimates of the abundance and information on stock structure. The catch currently remains at a low level, probably for economic reasons.

A relatively large catch, a total of 120, is allowed for the false killer whale, which has an estimated population of 2029 with confidence interval of 907–4541 (Table 3.13) in the coastal stratum off the Pacific coast. Wakayama fisheries recorded a high catch in the 1990s and lower takes afterward (Tables 3.18 and 3.19). The background of the change has not been analyzed. An effort is required to monitor the future trend of the catch as well as to improve the population estimate.

Catches of the Risso's dolphin have increased since the 1990s and still continue at a high level by the drive fishery in Taiji (Tables 3.17, 3.18 and 3.19), hand-harpoon fisheries in Wakayama Prefecture including Taiji (Table 2.12) and small-type whaling (Table 4.3). Almost nothing is known about the biology of this species except for an abundance estimated of 31,012 (Table 3.13) for the coastal stratum off the Pacific coast. The future trend should be carefully monitored.

The Dall's porpoise fishery should receive more attention because of the long-lasting large catches and questionable system of collecting statistics (Section 2.5).

Baird's beaked whales are currently hunted by small-type whaling in the Sea of Japan, the Okhotsk Sea, and off the Pacific coast of central and northern Japan at a relatively low and stable level (Tables 4.2, 4.3, and 13.14). Possible manipulation of past statistics and inclusion of poached sperm whales cause a problem in the assessment of the current level of the population (Section 13.7).

The catch statistics in Table 3.17 and Figure 15.3 were constructed with the cooperation of a staff member of the Taiji Fishery Cooperative Union. They disagree somewhat with the published official statistics, but I believe them to be closer to the truth (Section 3.9.2). The southern-form short-finned pilot whale has been hunted mostly by the drive fishery at Taiji and to a lesser degree by small-type whaling at Taiji and the crossbow fishery at Nago (26°35′N, 127°59′E), Okinawa (Table 6.3). The catch at Taiji reached the quota only once and was declining through the 1980s and 1990s (Tables 3.17 and 3.19) while the price increased (Table 3.20). Thus, quotas have not functioned to cap the catch and the stock of southern-form short-finned pilot whale is likely to follow the history of striped dolphins if the exploitation continues (also see Table 3.19 for recent trend).

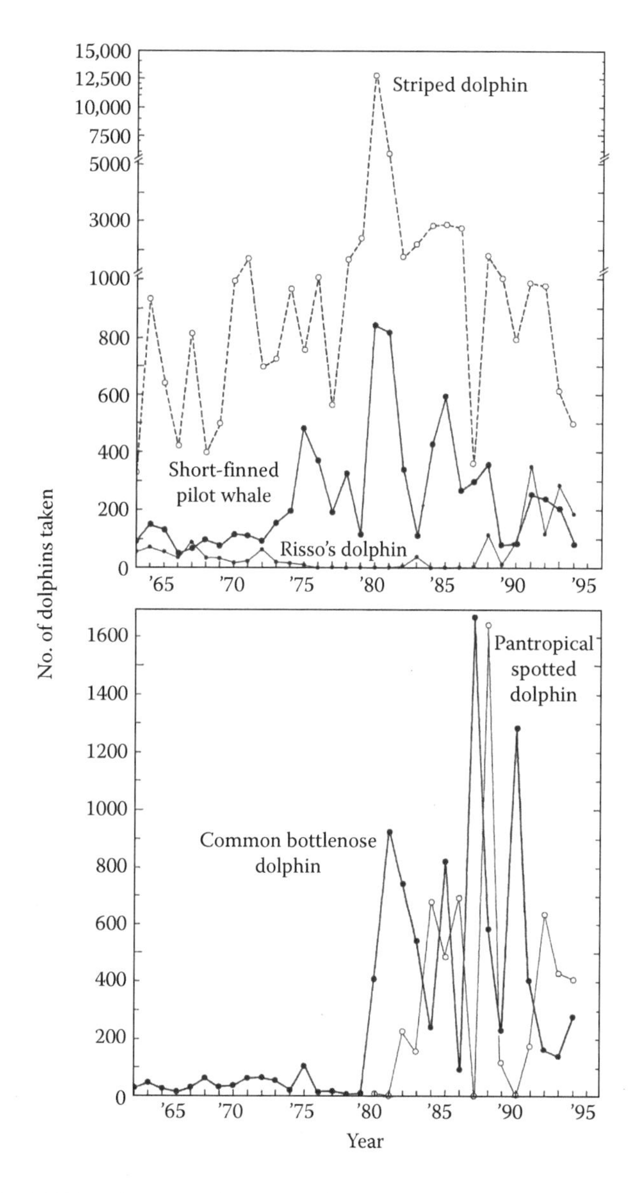

FIGURE 15.3 Expansion of magnitude and targeted dolphin species in the Taiji fishery as a result of introduction of drive fishery, based on statistics in Table 3.17. The statistics include dolphins taken by local small-type whaling. (From Kasuya, T.: Cetacean biology and conservation: A Japanese scientist's perspective spanning 46 years, The Kenneth S. Norris Lifetime Achievement Award lecture presented on 29 November 2007, Cape Town, South Africa. *Marine Mammal Sci.* 2008. 24(4). 749–773. Copyright Wiley-VCH Verlag GmbH & Co. KGaA. Reproduced with permission, Figure 3.)

Common bottlenose dolphins and pantropical spotted dolphins are currently hunted mostly in Wakayama Prefecture (Tables 2.12, 3.17, and 3.19) and to a lesser degree at Nago, Okinawa (only bottlenose dolphins; Table 2.13). Both species experienced a surge in catches in the 1980s, followed by a decline (Figure 15.3) with operations under almost nonfunctioning quotas (Table 6.3). Careful management is needed as in the case of the short-finned pilot whale.

Since 2001, the Japanese government has refused cooperation with the small-cetacean subcommittee of the Scientific Committee of the IWC and suspended reporting information on small-cetacean fisheries, including statistics and analysis of the populations. It also ignores the advice of the Committee on management of small cetaceans around Japan. Such practices are likely to result in neglect; thus, nine species of small cetaceans currently exploited by Japan face increased potential for mismanagement.

15.4 UNRESOLVED QUESTIONS IN CETACEAN BIOLOGY

15.4.1 Factors That Determine the Breeding Season

Several cetacean species around Japan are known to exist in a pair of populations that breed about 6 months apart (Table 15.1). One of the two minke whale populations inhabits mainly the western North Pacific and the Okhotsk Sea, and the other occupies the East China Sea, Yellow Sea, and Sea of Japan as its main habitat and is also known to occur in the southern Okhotsk Sea and off the Pacific coast of Japan. Although their breeding grounds and details of their summering grounds are still unknown (IWC 2010b), these two populations breed in seasons that are about 6 months apart.

Two morphologically distinct forms of short-finned pilot whales inhabit waters off the Pacific coast of Japan (Kasuya et al. 1988). One is called the northern form, inhabits the Pacific coast in approximate latitudes 35°N–44°N, and is hunted by Japanese small-type whaling (Chapter 4). The other is called the southern form and inhabits coastal waters south of the range of the northern form and offshore waters not occupied by the northern form. The range of the southern form extends to Taiwan, although it is uncertain and improbable that this broad area is occupied by a single population (Section 12.3.7). The southern-form whales have been hunted by various Japanese fisheries and are currently hunted in small-type whaling, a drive fishery (Section 3.9), and a crossbow fishery (Section 2.8).

TABLE 15.1
Paired Populations of Cetaceans off Japan with Mating Peaks about a Half Year Apart

Species	Population	Mating Seasons		Reference
		Peaks In	Peaks Separated By (months)	
Minke whale	Yellow Sea–Sea of Japan stock	July–Sep	5–6	Best and Kato (1992)
	West Pacific–Okhotsk Sea stock	Feb–Mar		
Finless porpoise	Ariake Sound–Tachibana Bay stock[a]	Nov–Dec	5–6	Chapter 8; Shirakihara et al. (2008)
	Stocks in the Omura Bay,[b] Inland Sea, Pacific	Apr–May		
Short-finned pilot whale	Northern-form stock off Sanriku[c]	Oct	5–6	Chapter 12; Kasuya and Tai (1993)
	A southern-form stock off Izu-Taiji area[d]	May		
Striped dolphin	Off the Pacific coast	June	5–6	Chapter 10; Kasuya (1999b)
		Nov–Dec		

[a] The Ariake Sound (32°30′N–33°20′N, 130°10′E–130°35′E) and Tachibana Bay (centered at 32°40′N, 130°05′E) are on the west coast of Kyusho facing the East China Sea. This population has another minor parturition peak in March (Section 8.5.1).
[b] Omura Bay (centered at 32°55′N, 129°53′E) is on the west coast of Kyushu and opens through a narrow channel to the East China Sea.
[c] Sanriku is on the Pacific coast of northern Honshu in latitudes 37°54′N–41°35′N.
[d] Izu (34°36′N–35°05′N, 138°45′E–139°10′E) and Taiji (33°36′N, 135°57′E) are on the Pacific coast of central Honshu.

There is about a 6-month difference in the breeding seasons of the two forms of short-finned pilot whales. The southern form has a broad parturition season with a peak in the early summer, and the northern form has a narrow parturition season with a peak in the winter (Section 12.4.4; Kasuya and Tai 1993). Calves of the northern form born in the winter have the benefit of starting to take solid food in the summer when the availability of squid is at its peak, and their greater neonatal body size increases resistance to a water temperature that is lower than the temperature met by neonates of the southern form (Section 12.3.6; Kasuya 1995).

Japanese finless porpoises inhabit coastal waters from western Kyushu (32°30′N, 129°30′E) to the coast of the Oshika Peninsula (38°30′N, 141°30′E) and are known to have a minimum of five populations (Section 8.3.4; Yoshida 2002): two in Omura Bay and Ariake Sound–Tachibana Bay on the East China Sea coast, and three along the Pacific coast in the Seto Inland Sea, Ise and Mikawa Bays, and from Tokyo Bay to the Oshika Peninsula including Sendai Bay (Figure 8.2). Future study will likely identify another population in Tokyo Bay and more than one population in the Chiba/Sendai Bay area (Kasuya et al. 2002). Only the second population, in the Ariake Sound and Tachibana Bay in latitudes of 32°30′N–33°10′N, has a main mating peak in November–December, while the other four populations in a greater latitudinal range of 32°50′N–38°30′N have a common mating peak in April–June. Thus, the difference cannot be explained by latitudinal difference.

The mating season and gestation period are thought to be designed to maximize the survival of the offspring (Kiltie 1984), where seasonal change in nutrition and physiology of the parent must also be reflected. Studies on environmental factors influencing female reproduction and offspring survival of the three cetacean species mentioned earlier will benefit understanding of factors controlling mating seasonality of cetaceans.

Striped dolphins hunted off the Pacific coast of Japan are known to have two mating peaks, in June and November–December. This was once interpreted as bimodal mating seasonality in one population, and it is also possible to interpret it to mean that the fishery exploited two populations with different mating seasons that migrated through the fishing ground successively (Section 10.4.7).

15.4.2 Function of Postreproductive Females

Marsh and Kasuya (1986) reviewed the literature and identified the presence of significant length of postreproductive lifetime in several toothed whales. The key information was obtained from over 10 years of observation of killer whale communities that identified broad individual variation in female reproduction together with the presence of some females that had not reproduced during the study period and from a carcass study of short-finned pilot whales that found age-related decline in apparent pregnancy rate to reach zero at middle age. Analogy to these life history factors suggested the presence of postreproductive life span also in false killer whales, sperm whales, and perhaps long-finned pilot whales (Section 12.4.4).

Tables 12.13 and 12.14 present age-related change in reproductive activities in the two stocks of short-finned pilot whales off Japan. The southern-form sample was obtained from drive fisheries and covered 9 months of the year in 1974–1984, and the northern-form sample came from the catch of Norwegian-type whaling called small-type whaling that operated in October and November of 1983–1985. The collection time of the latter sample corresponded with the beginning of the mating season and therefore could have contained a high proportion of females in lactation and simultaneous pregnancy. Many of the females in lactation and simultaneous pregnancy were in an early stage of pregnancy and suspected to have been in a late stage of lactation (Kasuya and Tai 1993).

The two populations of the short-finned pilot whale are similar in some important life history characteristics such as maximum longevity of females at 61 or 62 years and highest age of pregnant females at 35 or 36 years. The proportion of pregnant females decreased with increasing age to reach zero at age 37 years and over, while females in lactation were aged below 44 years (northern form) or 52 years (southern form). Thus, females at age 35 years and over are highly likely to have ceased reproduction; they accounted for 18% of adult females of the northern-form and 29% of the southern-form females. The observed difference in the proportion of postreproductive females could have reflected age composition of the population, which would again reflect the history of exploitation and recovery from earlier depletion (Section 12.5.7). Ovarian histology confirmed that a minimum of 25% of the sexually mature females in the southern form had already lost the ability to reproduce (Section 12.4.4.6; Marsh and Kasuya 1984).

Marsh and Kasuya (1986) interpreted the postreproductive life span of females as a trait that evolved accompanying matrilineal social structure. Postreproductive females may contribute to reproduction and survival of younger members in the group who are genetically related to them. For example a postreproductive female can take care of a juvenile in the group while its mother is feeding underwater. Older individuals have accumulated more experience and knowledge than the younger individuals and can use the information for the survival of kin in the group. Marsh and Kasuya (1986) thought that such activities of older females increased their total lifetime production and that selection worked to increase postreproductive lifetime. The trait of extended longevity will not have developed in males because they leave their natal school or do not live with their offspring, but the possibility of this is discussed in Section 15.4.4.

Culture in an animal community can be defined as information and behavior patterns maintained in the community by learning. Postreproductive females are old and are likely to function as a carrier of culture in their group. Some scientists have recently noted the importance of culture in a cetacean community (Whitehead 2002), and evidence exists on killer whales that postreproductive mothers are contributing to the survival of her sexually mature offsprings, particularly to that of sons (Foster et al. 2012). Such a perception will be accepted more widely in the future as our knowledge on behavior and social structure in cetaceans accumulates.

15.4.3 Social Sex in Cetaceans

The following is an outline of life history of the southern-form short-finned pilot whale obtained from the examination of carcasses from Japanese fisheries, but the features are almost same as those of the less-studied northern form (Chapter 12; Kasuya and Marsh 1984; Marsh and Kasuya 1984; Kasuya and Matsui 1984; Kasuya and Tai 1993).

Short-finned pilot whales live in a school of 15–40 individuals, which contains sexually mature males and females and immature individuals of both sexes. Females attain sexual maturity at age 7–11 years (9 years on average), cease reproduction by age 36, and live to 62. Some precocious males start spermatogenesis at age 5, but it is limited to the breeding season. Full sexual maturity is attained at ages 15–30 (17 on average) and such males are spermatogenic throughout the year. Short-finned pilot whales off Japan live in a matrilineal pod, and a school or unit driven by the fishery may contain more than one pod. The school contains a greater number of females (8–30) than males of full maturity (1–5). Limited information suggests that males can live with the mother for life as in the case of resident killer whales off the northwestern coast of North America, but the possibility that some males leave their natal pod at puberty remains to be investigated. The latter case is known in sperm whales. Short-finned pilot whales are spontaneous ovulators, and delayed implantation has not been identified in the species (Stewart and Stewart 2002).

Kasuya et al. (1993) examined the uterine mucus (or fluid in the uterus) of the two forms of short-finned pilot whales off Japan for the presence of spermatozoa to understand the reproductive activity of females. Spermatozoa reach the fallopian tube in humans and cattle within a few minutes or few hours after ejaculation and remain there for a maximum of 85 hours. The presence of spermatozoa in the pilot whale uterus confirms that the female had copulation that ended with ejaculation within a few days before slaughter. Kasuya et al. (1993) examined uterine mucus from 53 northern-form and 34 southern-form pilot whales and confirmed that females had copulated even when they were not in estrus:

1. Ten out of 13 nonpregnant females with a corpus luteum of ovulation were found with spermatozoa, as could be reasonably expected.
2. Eleven of 12 females in early pregnancy were found with spermatozoa. The largest fetus in a female with spermatozoa was 13.3 cm, estimated to be 30–60 days from conception. Females with larger fetuses were not examined due to technical reasons. The density of spermatozoa was variable from a few spermatozoa per slide to 50 per slide.
3. Eight of 31 lactating females had spermatozoa. Some of them had no Graafian follicles of measurable size (≥1 mm). The follicle size at ovulation is believed to be greater than 15 mm in diameter. The spermatozoa densities were low at less than one per slide. The presence of spermatozoa had no correlation with age or follicle size.
4. Ten of 28 resting females (mature females neither pregnant nor lactating) had spermatozoa in the uterus. The sperm densities were either very low at less than one per slide or very high at over 50 per slide and did not correlate with age or follicle size.
5. Uterine mucus was examined for eight resting female southern-form short-finned pilot whales that were identified as postreproductive by Marsh and Kasuya (1984) through ovarian histology. Three of the eight had spermatozoa in their uterus (0.25–1.5 spermatozoa per slide). One of the three females was aged 42 years and had no measurable follicles in the ovaries. No histology has been conducted on ovaries

of the northern form, but seven resting females aged 41–46 years were judged from age to be postreproductive, and two of them showed uterine sperm density of over 50 spermatozoa per slide. One of the two females had spermatozoa an extremely high density (2990 spermatozoa per slide) but had no measurable follicles in the ovaries.
6. Three sexually immature females had no spermatozoa in their uteri.

The incidence of uterine spermatozoa did not correlate with season but did with time after the drive. The proportion of females with uterine sperm became low after the fourth day after the drive. This could have been due to one or more of the following factors: (1) mating partners became unavailable through daily slaughter or were already absent at the time of drive, (2) stress from capture suppressed copulation, and (3) dead spermatozoa were discharged from the uterus (Kasuya et al. 1993).

High receptivity of females unrelated to estrus may decrease inter-male competition for mating opportunities and increase the stability of the school (Kasuya et al. 1993). A school of short-finned pilot whales contains multiple adult males and some pubertal males. The average number of adult females in a school was 13, and only 17% of them become pregnant after an average of 1.9 ovulations. Thus, an average school will experience about four bouts of estrus in a year. If females accepted males only during the 2–3 days of estrus, as known in some toothed whales, receptive females would be available only around 10 days in a year. However, if females showed the high receptivity as postulated by the presence of sperm in the uterus of nonestrous individuals, then the receptive days would increase by a minimum of 50–60 times (Section 12.4.4.8; Magnusson and Kasuya 1997). Kasuya et al. (1993) thought this could explain the absence of fighting scars on males of short-finned pilot whales, which is in contrast with the scarring observed on sperm whales (Section 12.4.4.8).

Copulation by nonestrous females has been called nonreproductive mating (Kasuya et al. 1993), and there have been several hypotheses presented for its function.

Paternity camouflage is one possibility. Generous receptivity of a female will obscure the paternity of her offspring and thus function to avoid infanticide by males who would expect reproductive benefit from the activity. Before accepting this hypothesis, we need to know whether the social structure is likely to invite infanticide if the generous female receptivity is absent.

A decoy hypothesis assumes that generous receptivity of females in a school functions to stop males from moving out of the school in search of estrous females, keeping males with the female school. Thus, young females in real estrus, which is infrequent in the community, will have a high probability of being fertilized.

Learning might be required for males for courting and successful copulation, as is known for some primates, and pubertal males might be learning the process from genetically related nonestrous females in the pod. The training would increase the future reproductive success of the young male and thus lifetime production of the educating female through kin selection.

In a community of highly social animals, copulation can function as a tool for relieving tension in the community or strengthening relationships between individuals. Sex aimed at such type of social purpose has been called social sex (Kasuya 2008). Examples of social sex are known in primates such as bonobos, *Pan paniscus*, and humans. Free exchange of copulation in a community of bonobos was thought to work for camouflaging paternity and avoiding infanticide by males and for relieving tension between the group members (Waal 2005, in Japanese). The latter function was supported by copulation bouts preceding food sharing, which was believed to increase tension between individuals (Kuroda 1982, in Japanese). Most of the copulations in the human community are not aimed at reproduction and function to strengthen economic cooperation and psychological solidarity in long-term or temporary couples. In some human communities, joint ownership of a partner or exchange of partners functions to strengthen solidarity between participants of the same sex (Sections 12.4.4.8, 12.5.4, and 12.5.5; Houston 1999; Schultz and Lavenda 1995, both in Japanese).

Before we evaluate the hypotheses given earlier, we need to identify males who are responsible for the spermatozoa in the uteri. The questions to be answered include (1) the number of males involved in the copulation with a female, (2) their age and genetic relationship with the female involved, and (3) the duration of their visit to the female school. It would be technically possible to obtain an answer to the first and perhaps the second question if there were an opportunity to obtain samples. The answers will greatly contribute to understanding the social structure of short-finned pilot whales.

15.4.4 Greater Male Longevity in Baird's Beaked Whales

Baird's beaked whales inhabit the continental slope of the North Pacific rim north of 34°N, although some vagrants may occur further south. No concentration of this species has been identified in offshore waters (Section 13.2; Kasuya 1986, 2002b; Kasuya and Miyashita 1997; Wang 1999 in Chinese). Their wintering ground has not been identified, but Subramanian et al. (1988) suggested, based on the analysis of pollutants in the blubber, that they remain in coastal waters.

Adult Baird's beaked whales reach a body length of 9.5–11 m and exhibit minimal sexual dimorphism, with female size exceeding that of males by about 35 cm on average (Kasuya et al. 1997). This trait has attracted the attention of some biologists (Ralls 1976). The small difference in body size leads one to expect minimum effect of gunner's selection for larger individuals on the sex ratio in whales caught in Japanese whaling.

Japanese scientists once attributed the deficiency of females in the catch in Japanese whaling to a difference in migration pattern (Omura et al. 1955), but later studies failed to find an opposite sex ratio in whaling operations in any other season or location (Table 13.10).

Age was determined for 84 males and 47 females taken off Wadaura in Chiba on the Pacific coast of central Japan in 1975 and 1985–1987 using the method of Kasuya (1977). Age ranged from 3 to 84 years. Absence of whales below the age of 3 years is due to gunner's selection for larger individuals. Sexual maturity was attained at 6–10 years in males and 10–15 years in females. The sex ratio was almost at parity at 24 males and 26 females for ages below 15, but the female proportion decreased with increasing age: 38 males to 20 females for ages 16–54 years and 22 males to 0 female for ages 55–84 years (Kasuya et al. 1997). The age of the oldest female was 54 and that of the oldest male 84 (Figure 15.4). Although there remained some ambiguities in defining male sexual maturity, males apparently attained maturity 4–5 years earlier and lived 30 years longer than females. The apparent pregnancy rate was 29.6% and did not decline with increasing age, which was different from the pattern seen in short-finned pilot whales (Section 13.4.6).

While testicular histology suggested that males attained reproductive ability when a single testis weighed 1.4–2.0 kg, which occurred at ages 6–10 years, testicular weight continued increasing until the age of around 30 years to reach 4–8 kg. We do not know the male age at first reproduction. Even assuming the most conservative age of 30 years, the sex ratio of mature males to females is 42 males (≥30 years of age) to 25 mature females. Such an inverse sex ratio is apparently rare among mammals (Ralls et al. 1980).

Greater male longevity is the only available explanation for the sex ratio of Baird's beaked whales off Japan. A similar longevity trend was reported for the northern bottlenose whale *Hyperoodon ampullatus* by Benjaminsen and Christensen (1979) and for Cuvier's beaked whale *Ziphius cavirostris* by Heyning (1989). These three species are the only toothed whales known with greater male longevity and

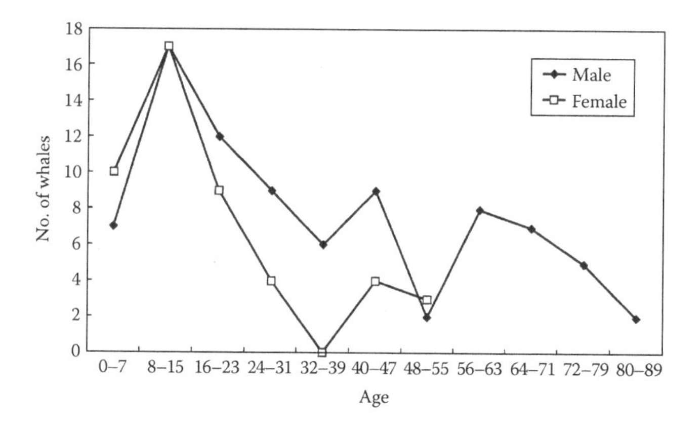

FIGURE 15.4 Age composition of Baird's beaked whales taken off the Pacific coast and processed at land station at Wadaura in July and August during four whaling seasons of 1975 and 1985–1987. (From Kasuya, T.: Cetacean biology and conservation: A Japanese scientist's perspective spanning 46 years, The Kenneth S. Norris Lifetime Achievement Award lecture presented on 29 November 2007, Cape Town, South Africa. *Marine Mammal Sci.* 2008. 24(4). 749–773. Copyright Wiley-VCH Verlag GmbH & Co. KGaA. Reproduced with permission, Figure 4.)

the only species of Ziphiidae for which the age structure of a population has been studied.

An explanation has not been reached for the greater male longevity in Baird's beaked whales and the two other Ziphiidae species. Baird's beaked whales usually live in a school that contains up to 25 individuals and make long dives extending for an hour, surfacing nearly synchronously within a viewing distance from the position of the preceding dive (Section 13.3.2). A hypothesis has been made by Kasuya et al. (1997) that the whales might live in a patrilineal society, with old experienced males contributing to the survival of their kin in the school, or that the old males participate in babysitting while the mother is feeding at depths of 800–1200 m or more (Walker et al. 2002; Minamikawa et al. 2007) not accessible to juveniles.

We currently know almost nothing about the social structure of Baird's beaked whales and are unable to evaluate the hypothesis given earlier. Further information on the social structure of this species will greatly contribute to broadening the spectrum of our knowledge of the diversity of cetacean societies.

15.5 MY VIEW ON CETACEAN CONSERVATION

15.5.1 Target for Conservation

People have various reasons to conserve cetaceans. My view is to consider them as an element of our environment. The conservation of cetaceans should aim at "guaranteeing their future survival and maintaining their within-species diversity." This objective will not be achieved if conservation action is targeted solely on cetaceans. Activity must be directed to conserve the ecosystem where cetaceans live in relation to other biological and nonbiological components of the environment. Conserving within-species diversity, such as subspecies, populations, communities and families, also needs to be kept in mind. In the following sections, I concentrate on the topic of within-species diversity.

15.5.2 Element to Conserve

The need to conserve genetic diversity is now widely accepted for cetaceans. Genetic diversity is important for a species to be able to utilize the broad spectrum of the natural environment and to maintain the ability to adjust to future environmental changes. However, I oppose the idea that a genetically identifiable population is sufficient as the unit to be conserved and propose an additional element to be considered, which is culture. Conservation should conserve the diversity of culture in populations, the system to maintain the culture in a community, and the ability to accumulate new elements of culture.

15.5.3 Why Culture?

Animal behavior includes two types of elements—those determined by heredity and those learned from experience or from other individuals. The importance of learning varies by species and with the environment the species lives in, but in general it increases in higher animals such as primates and cetaceans. Compared with some primates that live their daily lives by visiting particular fruit trees in particular seasons to satisfy their nutritional needs, cetaceans must overcome a seasonally more variable and less predictable environment in the ocean. The life of cetaceans in the ocean depends more on learning and cooperation among group members than for primates.

Some cetaceans are known to have high ability for learning, good memory, and long life in which to accumulate experiences (Würsig 2002). Many toothed whales live in groups, and such a life offers opportunities to have common experiences, exchange information, and form characteristic behavior patterns in the group. This is formation of the culture of the community, which I define approximately as a group of animals that possess a common culture. Such communities as units carrying culture will not necessarily be identifiable by currently available genetic methods. Genetic difference between communities depends on their reproductive system, the way of transmitting experience between individuals within a community, and the speed of cultural change.

Generally speaking, the speed of behavioral change due to genetic alteration will be extremely slow compared with change due to change in culture. New cultural traits will emerge in every generation and pass to the next generation as new elements, and old cultural elements can be lost if not functioning for a certain period. Thus, culture is a powerful tool of the community with which to adjust its life to a newly encountered environment. A broad spectrum of cultural traits in a species increases the chances of the species to survive through environmental changes.

Cultural diversity should be judged as an element in evaluating biological diversity.

15.5.4 Evidence of Cetacean Culture

Evidence of culture in cetacean communities is scanty. Examples include the vocalizations of killer whales, humpback whales, and sperm whales (Whitehead 2002, 2003); migration to summer feeding grounds learned by a calf from its mother humpback whale; and use of sponges for possible feeding purposes by a particular group of Indo-Pacific bottlenose dolphins (Holden 2005; Krutzen 2006). Identifying the presence of cetacean culture is a hard task because most cetaceans live far from our habitats and most of their activities occur underwater, invisible to us. The paucity of evidence on cetacean culture is due to technical difficulties in recognizing it.

A particular difficulty is in studying the behavior or culture of baleen whales. It is partly because of their great range of communication, which may extend several hundred kilometers or beyond the limit of our range of perception. If two individuals acoustically interact with each other at such a great distance, it will be hard for a scientist to even identify the fact. Cooperation among baleen whales can be different from that between toothed whales. We know that toothed whales often cooperate in hunting prey, and similar activity is also known in humpback whales. However, I would expect that inter-individual cooperation of baleen whales, if it exists, will contribute more to locating aggregations of food or a feeding ground than in capturing the food items.

The ability to maintain culture varies even among toothed whale species. Toothed whale schools vary among species in size, formation mechanism, membership, and stability. Even the ability to learn is likely to differ among species. Pirlot and Kamiya (1975) compared postnatal brain growth between the striped dolphin, a school-living species, and the franciscana (*Pontoporia blainvillei*), one of the river dolphin species believed to live a solitary life. Brain weight in the former species increased with growth from 400 g in neonates to 1200 g in adults. This postnatal brain growth of 200% is close to the 250% in humans. Brain weight of the latter species, the franciscana, increased from 150 g in neonates to 250 g in adults, postnatal increase of only 60%. The difference in the pattern of postnatal brain growth reflects the difference in learning abilities or the level of cognition that is required for social life. Primatologists call brains evolved for maintaining social life "social brains" (Dumbar and Shultz 2007). For further understanding of the relationship between social structure and brain growth in cetaceans, it is recommended that postnatal brain growth in the finless porpoise and Dall's porpoise, which is believed to live a less social life than either the striped dolphin or short-finned pilot whale, be investigated.

Given the information presented earlier, for management purposes, absence of evidence of culture should not be judged as evidence of absence of culture in cetacean species. Rather, it is safer to assume culture in a species if it has a life history or social structure suited for maintaining culture or shows behavior suggesting culture. Short-finned pilot whales and some other toothed whales mentioned in the earlier chapters meet these criteria. The contribution by individuals to their communities varies by sex, age, and personality, which will be evidenced in future studies based on the recognition of individuals. Those carrying out carcass studies, such as I am accustomed to, will find it hard to contribute to such a research field.

The management of dolphin fisheries must take into account both the social structure of the species to be hunted and the type of fishery. Simple calculation of numbers is inadequate. Some types of dolphin fisheries will find it difficult to coexist with some highly social cetaceans, if diversity of cultural traits is to be conserved. For example, driving schools of short-finned pilot whales into a harbor for subsequent slaughter is likely to eliminate one by one the cultural traits retained in the species and is acceptable only by ignoring the function of the school and that of individuals in it.

15.5.5 Responsibility of Cetacean Scientists

All citizens should be given opportunities to participate in making decisions on conservation policy. This process requires free access to unbiased scientific information and a free atmosphere in which to speak up. The Japanese government has often acted to favor industries. This has not been an exception in whale management and has led government to giving out pro-industry propaganda in an attempt to lead public opinion toward the benefit of the industries. This system, together with the language barrier, has placed Japanese citizens in a situation where it is difficult to make one's own judgment based on unbiased information.

The Japanese program of scientific whaling in the Antarctic that started in the 1987/1988 season first aimed at estimating age-specific natural mortality, in which I had some responsibility as a person who participated in the discussion at an early stage of the program. Japanese scientists later identified difficulty in achieving the initial goal and modified the objective to estimating average mortality rate and obtaining ecosystem-related information. Quantitative evaluation of the latter was hard. The draft report agreed upon at the review meeting on the project in December 2006 stated that the program failed to achieve most of the primary objectives, although this particular sentence was editorially removed in the published version of the report (IWC 2008). And the report made numerous critical statements on the results presented by the Japanese scientists, as is usually the case in such a meeting. Such critical statements were apparently kept hidden from the Japanese public, and the Fisheries Agency and the industry released their optimistic interpretations. The view of the government was reported by the media and influenced the formation of Japanese public views on the scientific whaling program.

I was educated with the view that the findings of scientists are common property of the community, with the scientists receiving honor for this. Such a perception of scientists appears to be changing, and the distinction between science and technology is becoming ambiguous, at least in Japan. Scientific findings are often evaluated for their monetary value even in universities, sold for money, monopolized through patents, or kept hidden for the benefit of a particular group of people. It has become increasingly easy for a particular group of scientists to monopolize opportunities for scientific studies. If we scientists in such a community agree to be a running dog of an employer or financial supporter, we may fail to accomplish our obligation to our community as scientists.

Conservation biology is not a science aimed at monetary benefit but is not free from financial constraints. I am afraid that if what I fear becomes a fact in the community of marine mammal science, it will destroy the credibility of our studies. We marine mammal scientists are responsible for the future of marine mammals. Some marine mammal scientists may be working in a field that is currently unrelated to conservation, or working for an employer currently unrelated to conservation. Even these scientists can have an opportunity in the future to face the conservation problems of the marine mammals they are working with. Thus, it is to be hoped that every marine mammal scientist would be motivated to have his/her own opinion on conservation and be ready to voice it.

REFERENCES

[*In Japanese Language]

Benjaminsen, T. and Christensen, I. 1979. The natural history of the bottlenose whale, *Hyperoodon ampullatus* Foster. pp. 143–164. *In*: Winn, H.E. and Olla, B. L. (eds.). *Behavior of Marine Animal, Vol. 3, Cetaceans*. Plenum Press, New York and London, U.K. 438pp.

Best, P.B. and Kato, H. 1992. Possible evidence from foetal length distributions of the mixing of different components of the Yellow Se-East China Sea-Sea of Japan-Okhotsk Sea minke whale population(s). *Rep. Int. Whal. Commn.* 42: 166.

Brownell, R.L. Jr., Kasuya, T., Kato, H., and Ohsumi, H. 1999. Report of the ad hoc intercessional sperm whale group meeting. *J. Cetacean Res. Manage.* 1(Suppl.): 147.

Brownell, R.L. Jr., Yablokov, A.V., and Zemsky, V.A. 2000. USSR pelagic catches of North Pacific sperm whale, 1949–1979: Conservation implications. pp. 123–131. *In*: Yablokov, A. V. and Zemsky, V. A. (eds.). *Soviet Whaling Data (1949–1979)*. Center for Russian Environmental Policy, Marine Mammal Council, Moscow, Russia. 408pp.

Clapham, P.J., Aguilar, A., and Hatch, L.T. 2008. Determining spatial and temporal scales for the management of cetaceans: Lessons from whaling. *Marine Mammal Sci.* 24: 183–201.

Dumbar, R.I.M. and Shultz, S. 2007. Evolution in social brain. *Science* 317: 1344–1347.

Foster, E.A., Franks, D.W., Mazzi, S., Darden, S.K., Balcomb, K.C., Ford, J.K.B., and Croft, D.P. 2012. Adaptive prolonged postreproductive life span in killer whales. *Science* 337: 1313.

Francis, D. 1990. *A History of World Whaling*. Viking of Penguin Group, Harmondsworth, Middlesex, U.K. 288pp.

Heyning, J.E. 1989. Cuvier's beaked whale, *Ziphius cavirostris* G. Cuvier, 1823. pp. 289–308. *In*: Ridgway, S. H. and Harrison, R. (eds.). *Handbook of Marine Mammals, Vol. 4, River Dolphins and the Larger Toothed Whales*. Academic Press, San Diego, CA. 442pp.

Holden, C. 2005. Cetacean culture? *Science* 308: 1545.

*Houston, J. 1999. *Confessions of an Igloo Dweller.* Dobutsu-sha, Tokyo, Japan. 380pp. Japanese translation, initially published in 1995.

*Ito, T. 1997. Japanese sea lion. pp. 118–119. *In*: Kawamichi, T. (ed.). *Red Data, Mammals of Japan*. Bun-ichi Sogo Shuppan, Tokyo, Japan. 279pp.

*Ito, T. and Shimazaki, K. 1995. Japanese sea lion. pp. 491–500. *In*: Odate, S. (ed.). *Basic Data of Rare Wild Aquatic Organisms of Japan* (II). Nihon Suisan Shigen Hogo Kyokai, Tokyo, Japan. 751pp.

Ivashchenko, Y.V., Clapham, P.J., and Brownell, R.L. Jr. 2007. Scientific reports of Soviet whaling expeditions in the North Pacific, 1955–1978. Alaska Fisheries Science Center, Seattle, WA. 36pp. + Appendix. Translation from Russian text of Y.V. Ivashchenko.

IWC (International Whaling Commission). 1981. Report of the sub-committee on sperm whales. *Rep. Int. Whal. Commn.* 31: 78–102.

IWC. 1983. Report of the Scientific Committee. *Rep. Int. Whal. Commn.* 33: 43–190.

IWC. 1988. Report of the sub-committee on sperm whales. *Rep. Int. Whal. Commn.* 38: 67–75.

*IWC. 1992. Report of the scientific committee. *Rep. Int. Whal. Commn.* 42: 51–79.

*IWC. 1993. Report of the scientific committee. *Rep. Int. Whal. Commn.* 43: 55–86.

IWC. 1993. Report of the sub-committee on small cetaceans. *Rep. Int. Whal. Commn.* 43: 130–145.

IWC. 2008. Report of the scientific committee. *J. Cetacean Res. Manage.* 10 (Suppl.): 1–74.

IWC. 2010a. Report of the sub-committee on in-depth assessments. *J. Cetacean Res. Manage.* 12(Suppl. 2): 180–197.

IWC. 2010b. Report of the working group on the in-depth assessment of Western North Pacific Common Minke Whales, with a Focus on J-stock. *J. Cetacean Res. Manage.* 12(Suppl. 2): 198–217.

*Kasuya, T. 1963. Report of biological investigations of whales taken during the 11th North Pacific (Pelagic) whaling expedition in 1962. Geirui Kankyusho (Whales Research Institute), Tokyo, Japan. 62pp.

Kasuya, T. 1972. Some information on the growth of the Ganges dolphin with a comment on the Indus dolphin. *Sci. Rep. Whales Res. Inst.* 24: 87–108.

Kasuya, T. 1973. Systematic consideration of recent toothed whales based on morphology of tympano-periotic bone. *Sci. Rep. Whales Res. Inst.* 25: 1–103 + Pls. 1–28.

Kasuya, T. 1976. Reconsideration of life history parameters of the spotted and striped dolphins based on cemental layers. *Sci. Rep. Whales Res. Inst.* (Tokyo) 28: 73–106.

Kasuya, T. 1977. Age determination and growth of the Baird's beaked whale with a comment on the fetal growth rate. *Sci. Rep. Whales Res. Inst.* 29: 1–20.

Kasuya, T. 1985. Effect of exploitation on reproductive parameters of the spotted and striped dolphins off the Pacific coast of Japan. *Sci. Rep. Whales Res. Inst.* 36:107–138.

Kasuya, T. 1986. Distribution and behavior of Baird's beaked whales off the Pacific coast of Japan. *Sci. Rep. Whales Res. Inst.* 37: 61–83.

Kasuya, T. 1991. Density dependent growth in North Pacific sperm whales. *Marine Mammal Sci.* 7: 230–257.

Kasuya, T. 1995. Overview of cetacean life histories: An essay in their evolution. pp. 481–497. *In*: Blix, A.S., Walloe, L. and Ultang, O. (eds). *Whales, Seals, Fish and Man.* Elsvier, Amsterdam, the Netherlands. 720pp.

Kasuya, T. 1999a. Examination of reliability of catch statistics in the Japanese coastal sperm whale fishery. *J. Cetacean Res. Manage.* 1: 109–122.

*Kasuya, T. 1999. Manipulation of statistics by Japanese coastal sperm whale fishery. *IBI Rep.* 9: 75–92.

Kasuya, T. 1999b. Review of the biology and exploitation of striped dolphins in Japan. *J. Cetacean Res. Manage.* 1: 81–100.

Kasuya, T. 2002a. Japanese whaling. pp. 655–662. *In*: Perrin, W.F., Würsig, B., and Thewissen, J.G.M. (eds.). *Encyclopedia of Marine Mammals.* Academic Press, San Diego, CA. 1414pp.

Kasuya, T. 2002b. Giant beaked whales. pp. 519–522. *In*: Perrin, W.F., Würsig, B., and Thewissen, J.G.M. (eds.). *Encyclopedia of Marine Mammals.* Academic Press, San Diego, CA. 1414pp.

Kasuya, T. 2007. Japanese whaling and other cetacean fisheries. *Environ. Sci. Pollut. Res.* 14: 39–48.

Kasuya, T. 2008. Cetacean biology and conservation: A Japanese scientist's perspective spanning 46 years, The Kenneth S. Norris Lifetime Achievement Award lecture presented on 29 November 2007, Cape Town, South Africa. *Marine Mammal Sci.* 24(4): 749–773.

Kasuya, T. and Brownell, R.L. Jr. 1979. Age determination, reproduction and growth of Franciscana dolphin, *Pontoporia blainvillei. Sci. Rep. Whales Res. Inst.* 31: 45–67.

Kasuya, T. and Brownell, R.L. Jr. 1999. Additional information on the reliability of Japanese coastal whaling statistics. Paper IWC/SC/51/O7 presented to the *Scientific Committee Meeting of IWC in 1999.* 15pp. Available from the IWC secretariat, Cambridge, U.K.

Kasuya, T. and Brownell, R.L. Jr. 2001. Illegal Japanese coastal whaling and other manipulation of catch records. Paper IWC/SC/53/RMP24 presented to the *Scientific Committee Meeting of IWC in 2001.* 4pp. Available from the IWC secretariat, Cambridge, U.K.

Kasuya, T. and Marsh, H. 1984. Life history and reproductive biology of the short-finned pilot whale, *Globicephala macrorhynchus*, off the Pacific coast of Japan. *Rep. Int. Whal. Commn.* (Special Issue 6, Biology of Northern Hemisphere Pilot Whales): 259–360.

Kasuya, T. and Matsui, S. 1984. Age determination and growth of the short-finned pilot whale off the Pacific coast of Japan. *Sci. Rep. Whales Res. Inst.* 35: 57–91.

Kasuya, T. and Miyashita, T. 1988. Distribution of sperm whale stocks in the North Pacific. *Sci. Rep. Whales Res. Inst.* 39: 31–75.

Kasuya, T. and Miyashita, T. 1997. Distribution of Baird's beaked whales off Japan. *Rep. Int. Whal. Commn.* 47: 963–968.

Kasuya, T. and Miyazaki, N. 1982. The stock of *Stenella coeruleoalba* off the Pacific coast of Japan. pp. 21–37. *In*: Clarke, J.G., Goodman, J. and Soave, G.A. (eds.). *Mammals of the Sea*, FAO Fisheries series No. 5, Vol. IV. FAO and UNEP, Rome, Italy. 531pp.

*Kasuya, T. and Miyazaki, N. 1997. Sperm whale. pp. 163–167. *In*: Kawamichi, T. (ed.). *Red Data, Mammals of Japan.* Bun-ichi Sogo Shuppan, Tokyo, Japan. 279pp.

Kasuya, T. and Nishiwaki, M. 1975. Recent status of the population of Indus dolphin. *Sci. Rep. Whales Res. Inst.* 27: 81–94.

Kasuya, T. and Ohsumi, S. 1966. A secondary sexual character of the sperm whale. *Sci. Rep. Whales Res. Inst.* 20: 89–94+Plate I.

Kasuya, T. and Tai, S. 1993. Life history of short-finned pilot whale stocks off Japan and description of the fishery. *Rep. Int. Whal. Commn.* (Special Issue 14, Biology of Northern Hemisphere Pilot Whales): 439–473.

Kasuya, T., Miyashita, T., and Kasamatsu, F. 1988. Segregation of two forms of short-finned pilot whales off the Pacific coast of Japan. *Sci. Rep. Whales Res. Inst.* 39: 77–90.

Kasuya, T., Marsh, H., and Amino, A. 1993. Non-reproductive mating in short-finned pilot whales. *Rep. Int. Whal. Commn.* (Special Issue 14, Biology of Northern Hemisphere Pilot Whales): 425–437.

Kasuya, T., Brownell, R.L. Jr., and Balcomb, K.C. III. 1997. Life history of Baird's beaked whales off the Pacific coast of Japan. *Rep. Int. Whal. Commn.* 47: 969–979.

Kasuya, T., Yamamoto, Y., and Iwasaki, T. 2002. Abundance decline in the finless porpoise population in the Inland Sea of Japan. *Raffles Bull. Zool.* Supplement 10: 57–65.

Kiltie, R. A. 1984. Seasonality, gestation time, and large mammal extinctions. pp. 299–314. *In*: Martin, P.S. and Klein, R.G. (eds.). *Quarternary Extinctions.* University of Arizona Press, Tucson, AZ. 892pp.

Kishiro, T. and Kasuya, T. 1993. Review of Japanese dolphin drive fisheries and their status. *Rep Int. Whal. Commn.* 43: 439–452.

*Kondo, I. 2001. *Rise and Fall of Japanese Coastal Whaling.* Sanyo-sha, Tokyo, Japan. 449pp.

Kondo, I. and Kasuya, T. 2002. True catch statistics for a Japanese Coastal Whaling Company in 1965–1978. Paper IWC/ SC/54/O13 presented to the *Scientific Committee Meeting of IWC in 2002.* 23pp. Available from the IWC secretariat, Cambridge, U.K.

Krutzen, M. 2006. Dolphins join the culture club. *Aust. Sci.* 27: 26–28.

*Kuroda, N. 1940. On cetaceans in Suruga Bay. *Shokubutsu-to Dobutsu* (*Plants Animals*) 8: 825–834.

*Kuroda, S. 1982. *Pygmy Chimpanzee: An Unknown Ape.* Chikuma Shobo, Tokyo, Japan. 234pp.

Magnusson, K.G. and Kasuya, T. 1997. Mating strategies in whale populations: Searching strategy vs. harem strategy. *Ecol. Model.* 102: 225–242.

Marsh, H. and Kasuya, T. 1984. Change in ovaries of the short-finned pilot whales, *Globicephala macrorhynchus*, with age and reproductive activity. *Rep. Int. Whal. Commn.* (Special Issue 6, Reproduction in Whales, Dolphins and Porpoises): 311–335.

Marsh, H. and Kasuya, T. 1986. Evidence of reproductive senescence in female cetaceans. *Rep. Int. Whal. Commn.* (Special Issue 8, Behavior of Whales in Relation to Management): 57–74.

Marsh, H., Penrose, H., Eros, C., and Hugues, J. 2002. *Dugong: Status Reports and Action Plans for Countries and Territories.* UNEP/DEWA/RS.02-1. UNEP, Cambridge, U.K. 162pp.

*Matsuura, Y. 1935a. Distribution and behavior of blue whales around Japan. *Dobutsugaku Zasshi* (*Zool. J.*) (Tokyo Zoological Society) 47: 742–759.

*Matsuura, Y. 1935b. Distribution and behavior of fin whales around Japan. *Dobutsugaku Zasshi* (*Zool. J.*) (Tokyo Zoological Society) 47: 355–371.

Minamikawa, S., Iwasaki, T., and Kishiro, T. 2007. Diving behavior of a Baird's beaked whale, *Berardius bairdii*, in the slope water region of the western North Pacific: First diving records using a data logger. *Fish. Oceanogr.* 16(6): 573–577.

Miyashita, T. 1993. Abundance of dolphin stocks in the western North Pacific taken by Japanese dolphin fishery. *Rep. Int. Whal. Commn.* 43: 417–437.

*Nagasawa, R. 1915. Scientific names of 14 whale species around Japan. *Dobutsugaku Zasshi* (*Zool. J.*) (Tokyo Zoological Society) 27: 404–410.

*Nagasawa, R. 1916a. Scientific names of dolphins and porpoises around Japan. *Dobutsugaku Zasshi* (*Zool. J.*) (Tokyo Zoological Society) 28: 35–39.

*Nagasawa, R. 1916b. Scientific names of cetaceans around Japan (revisit). *Dobutsugaku Zasshi* (*Zool. J.*) (Tokyo Zoological Society) 28: 445–447.

*Nishiwaki, M. 1990. *Cruise to Antarctic Ocean: Journal of an Antarctic Whaling Cruise in 1947/48.* Hokusensha, Tokyo, Japan. 133pp. Published posthumously.

*Ogawa, T. 1932. Taxonomy of Japanese dolphins and porpoises. *Saito Ho-onkai Jiho* 69+70: 1057.

Okada, Y. and Hanaoka, T. 1940. A study of Japanese Delphinidae (IV). *Sci. Rep. Tokyo Bunrika Daigaku*, Section B, 77: 285–306.

Omura, H., Fujino, K., and Kimura, S. 1955. Beaked whale *Berardius bairdi* off Japan, with notes on *Ziphius cavirostris*. *Sci. Rep. Whales Res. Inst.* 10: 89–132.

*Omura, H. and Kasuya, T. 2000. *Journal of an Antarctic Whaling Cruise: Record of Japanese Antarctic Whaling in Its Infancy in 1937/37 Season.* Toriumi Shobo, Tokyo, Japan. 204pp.

Omura, H., Nishiwaki, M., Ichihara, T., and Kasuya, T. 1962. Osteological note of a sperm whale. *Sci. Rep. Whales Res. Inst.* 16: 35–45 + pls. 108.

*Osugi, K. 2007. *The Way toward Chino-Japan War: Process from Conflict in the Chinese Northern Territory.* Kodansha, Tokyo, Japan. 452pp.

*Osugi, K. 2008. The way to Japan-US War: Nine choices to avoid the war. Kodansha, Tokyo, Japan. Vol. 1: 425pp. Vol. 2: 406pp.

Pirlot, P. and Kamiya, T. 1975. Comparison of ontogenetic brain growth in marine and coastal dolphin. *Growth* 39: 507–524.

Ralls, K. 1976. Mammals in which females are larger than males. *Q. Rev. Biol.* 51: 245–276.

Ralls, K., Brownell, R.L. Jr., and Ballou, J. 1980. Differential mortality by sex and age in mammals, with specific reference to the sperm whale. *Rep. Int. Whal. Commn.* (Special Issue 2, Sperm Whales): 233–243.

Reporters without Borders. 2007. Annual worldwide press freedom index 2006. http://www.rsf.org/article.php3?id_article=19388. Accessed January 20, 2008.

*Schultz, E.A. and Lavenda, R.H. 1995. *Cultural Anthropology*, II. Kokon Shoin, Tokyo, Japan. 222pp. Japanese translation, initially published in 1990.

Shirakihara, M., Seki, K., Takemura, A., Shirakihara, K., Yoshida, H., and Yamazaki, K. 2008. Food habits of finless porpoises *Neophocaena phocaenoides* in Western Kyushu, Japan. *J. Mammal.* 89(5): 1248–1256.

*Shizuoka-ken Kyoiku Iinkai (Board of Education of Shizuoka Prefecture) 1987. *Report of Cultural Heritage of Shizuoka Prefecture 39: Manners and Customs of Fisheries in Izu*, Part 2. Shizuoka-ken Bunkazai Hozon Kyokai, Shizuoka, Japan. 193pp.

Starbuck, A. 1878. *History of the American Whale Fishery*, Vol. 1. Based on reprint published in 1964 by Argosy-Antiquarian Ltd, New York. 407pp.

Stewart, R.E.A. and Stewart, B.E. 2002. Female reproductive system. pp. 422–428. *In*: Perrin, W.F., Würsig, B., and Thewissen, J.G.M. (eds.). *Encyclopedia of Marine Mammals.* Academic Press, San Diego, CA. 1414pp.

Subramanian, A., Tanabe, S., and Tatsukawa, R. 1988. Estimating some biological parameters of Baird's beaked whales using PCBs and DDE as tracers. *Marine Pollut. Bull.* 19: 284–287.

*Tokuyama, N. 1992. *History of Whaling Activities of the Taiyo Gyogyo* (*Taiyo Fishing*). Tokuyama, Yamaguchi, Fukuoka Prefecture, Japan. 825pp.

Tonnessen, J.N. and Johnsen, A.O. 1982. *The History of Modern Whaling.* University of California Press, Berkeley and Los Angeles, CA. 798pp.

*Waal, F.De 2005. *Our Inner Ape.* Hayakawa Shobo, Tokyo, Japan. 340pp. Japanese translation, initially published in 2005.

Walker, W.A., Mead, J.G., and Brownell, R.L. Jr. 2002. Diet of Baird's beaked whales, *Berardius bairdii*, in the southern Sea of Okhotsk and off the Pacific coast of Honshu, Japan. *Marine Mammal Sci.* 16: 902–919.

*Wang, P. 1999. *Chinese Cetaceans.* Ocean Enterprises, Hong Kong, People's Republic of China. 325pp. + Pl.1-VI. (In Chinese).

Whitehead, H. 2002. Culture in whales and dolphins. pp. 304–305. *In*: Perrin, W.F., Würsig, B., and Thewissen, J.G.M. (eds.). *Encyclopedia of Marine Mammals.* Academic Press, San Diego, CA. 1414pp.

Whitehead, H. 2003. *Sperm Whales.* University of Chicago Press, Chicago, IL. 431pp.

Würsig, B. 2002. Intelligence and cognition. pp. 628–637. *In*: Perrin, W.F., Würsig, B., and Thewissen, J.G.M. (eds.). *Encyclopedia of Marine Mammals.* Academic Press, San Diego, CA. 1414pp.

Yoshida, H. 2002. Population structure of finless porpoise (*Neophocaena phocaenoides*) in coastal waters of Japan. *Rafffles Bull. Zool.* Supplement No.10: 35–42.

Postscript

My first contact with whales occurred in the early summer of 1947, when I was nine years old, in an agricultural suburb of Kawagoe City. A piece of shriveled salted whale meat was rationed at my junior school. On the way home, while walking on a trail through yellow wheat fields, I ate a small piece of the meat from my pocket. It was tasty for a protein-deficient young boy. It was perhaps part of the product of the first postwar Antarctic whaling expedition in the 1946/1947 season. The edible oil rationed in those days to each family became solid in winter, and it was my work to thaw it in front of the hearth. I suspect it was whale oil. Later, I heard that my younger sister had fried whale meat in the school canteen. Memories of whale meat and the difficult life of the postwar years are inseparable.

In the last summer at university in 1960, I joined a group tour by my university class and was amazed by the tasty whale meat offered by a whaling company in western Kyushu. The taste was much superior to that of the dishes we students used to have at a popular restaurant in Tokyo. In order to join the one-week tour, I borrowed 3000 yen (US$8.30) from Prof. S. Konosu of the Faculty of Agriculture at the University of Tokyo and camped out in a baseball field in Hiroshima City to save one night's lodging cost of 400 yen. In those days, whale meat was used for sausage, baleen whale oil for margarine, and sperm oil for detergent; there was a high demand for whale products.

In the late 1980s, we frequently encountered unfamiliar dark-colored whale meat at shops in Tokyo. Whaling operations had shrunk due to the depletion of whale stocks, and the supply of baleen whale meat declined. This opened a new market for dolphin and porpoise meat to be sold under the label "whale meat," and the price rose. The good fortune of dolphin and porpoise hunters did not last long. Now the price of such products has declined, and products of the so-called scientific whaling from about 1000 baleen whales are having a hard time in the market. Demand for whale meat is declining in Japan as older whale consumers pass from the scene and younger consumers are not recruited in large numbers. The importance of the influence on demand of consumer education for wildlife conservation should be recognized.

My life as a whale biologist started in the summer of 1960, when I worked on age determination of sperm whales for my graduation thesis guided by Masaharu Nishiwaki of the Whales Research Institute (WRI). In the autumn of the same year, I accompanied Nishiwaki and Seiji Ohsumi to assist in their study of striped dolphins at Kawana on the Izu Peninsula and stayed at a small inn called Komeya (where I first tasted raw dolphin heart). These experiences led me to a position at the WRI in the spring of 1961. The WRI at the time had nine scientists, including the Director Hideo Omura and myself, and carried out research with financial support from the whaling companies. We worked on whale carcasses brought to coastal land stations and to whaling factory ships. My first visit to whaling stations was to those at Ayukawa, and I was astonished that every whaling company appeared to be misreporting its catch figures. This led me to have a strong suspicion about the whaling industry.

As the Japanese whaling industry declined, the WRI was forced to restructure itself, and in 1966, I moved to the Ocean Research Institute (ORI) of the University of Tokyo, as a Research Associate of Nishiwaki, who also moved there from the WRI. The 17 years at the ORI were memorable with unrestrained research activities on dolphins and porpoises around Japan, river dolphins in the world's great rivers, dugongs, and manatees. In 1983, I moved to the Far Seas Fisheries Research Laboratory (FSL) of the Japan Fisheries Agency and spent the following 14 years on the study of life history and management of sperm whales and small cetaceans taken in Japanese fisheries. I made an effort to explain the critical situation of those species to the Fisheries Agency and the Scientific Committee of the International Whaling Commission and tried to explain the need for better management. After leaving the FSL, I studied Japanese dugongs and finless porpoises and taught at the Faculty of Bio-Resources of Mie University (since 1997) and at the Division of Animal Sciences of Teikyo University of Science and Technology (since 2001). I retired from the last position in March 2006.

Studying the distribution and behavior of toothed whales from vessels was a part of my research, but most of my effort was directed toward carcasses of animals killed by fisheries, which was science with knife and notebook, as described by Ray Gambell. In brief, the WRI taught me the method, the ORI offered opportunities to use it, and the FSL gave me a chance to use the results to make recommendations for management. This is an easy way to study cetaceans, but possible only when many are killed by industry. The approach has disadvantages and cannot be the main method for future scientists. The carcass you observe does not tell you much about the life of the animal before death, and it does not tell you anything about the life it would have had if it had not been killed. Such information is available through nonlethal methods. An additional disadvantage of the old method arises from the recent change in the research environment in Japan, where only scientists who are favored by the Fisheries Agency and fishery groups have access to the carcasses.

I believe that scientists should try to benefit their community by transmitting their research results to it in addition to writing scientific papers. Conservation scientists must tell the community about their findings. I wrote articles on conservation biology of small cetaceans mostly in English. Japanese articles can disappear quickly, and English articles are difficult for Japanese readers. University students who are interested in cetacean biology or science-minded general citizens who are interested in conservation of cetaceans will find it difficult, under current situation, to obtain necessary information on small cetaceans around Japan. In a desire to improve this situation, in this book I have presented our current knowledge of small-cetacean biology, remaining questions, and existing conservation problems.

Another purpose of this book is to record the history of interactions between small cetaceans and the Japanese community. Commercial hunting of small cetaceans has continued uncontrolled since at least the fifteenth century. In modern Japan, small-cetacean fisheries have been overshadowed by the whaling industry and received little attention. Even with repeated expressions of concern by the Scientific Committee of the International Whaling Commission since 1975 on the large-scale hunting of Dall's porpoises and striped dolphins, the Japanese government kept the problem within the fishing community and made no attempt to control the fishery until the late 1980s. In 1993, when agreement was made among stakeholders to set catch quota by species, the striped dolphin fishery along the Izu coast had already collapsed, and there is still no sign of recovery. We know that the Dall's porpoise fishery started in the 1920s and experienced expansions in the 1940s and 1980s, but we do not have a good understanding on the current status of the exploited stocks. The government is primarily responsible for this, but we scientists are also to be blamed for not raising our voices enough. I believe it important to record such history for future generations.

I left most of the biological specimens used for my life history studies of small cetaceans at the Far Seas Fisheries Research Laboratory. I deposited my field notes, data sheets with age and reproductive status of each individual I had examined, and related photographs at the Animal Section of the National Science Museum of Japan (NSM). The museum also has skeletal specimens of small cetaceans received from the ORI. This collection was made by Masaharu Nishiwaki and his group, and I moved them to the NSM after his retirement. These osteological specimens carry specimen numbers starting with TK, which is an abbreviation of "Tokyo-daigaku Kaiyo-kenkyu-sho" (Ocean Research Institute, University of Tokyo) formulated before I joined the group. I expect that these specimens and biological information will be available for all interested scientists.

My activities as a scientist have been supported by numerous people. I owe much to eight scientists and three office staff of the WRI I worked with for 5 years. In particular, the education and training received from Director Hideo Omura, Masaharu Nishiwaki, and Seiji Ohsumi formed the foundation for my later research activities. I owe thanks to the crews of the two research vessels *Tansei-maru* and *Hakuho-maru* at ORI, as well as the non-ocean-going staff, and friends and students for help with field studies and preparation of materials. The head of my division, Masaharu Nishiwaki, continued to be my adviser after I left the WRI and offered me generous research opportunities. I owe thanks also to all the staff of the whale group at the FSL. In particular, I was fortunate to rejoin the group of Seiji Ohsumi there. He allowed me the maximum freedom for my research and conservation activities that he could offer within the system.

I received benefit for my research activities on cetaceans from the following whaling companies, factory ships, catcher boats, and land stations of each company: Taiyo Gyogyo (*Kinjyo-maru* in the North Pacific whaling ground, Ayukawa Station), Kyokuyo Hogei (*Kyokuyo-maru* in the North Pacific whaling ground, *Kyokuyo-maru No. 3* in the Antarctic whaling ground, Ayukawa Station), Nihon Kinkai Hogei (later Nihon Hogei) (Ayukawa and Taiji Stations), Nitto Hogei (Wadaura and Taiji Stations), Miyoshi Hogei (*Ginsei-maru No. 2*, Abashiri Station), Shimomichi Suisan (*Yasu-maru No. 1*, Abashiri Station), Toba Hogei (*Koei-maru No. 7*, Ayukawa Station), and Gaibo Hogei (Wadaura Station). My research activities on dolphins and porpoises received benefit from the following Fisheries Cooperative Unions (FCU) and local government facilities: Yamadawan FCU, Otsuchi FCU, Arari FCU, Futo FCU, Kawana FCU, Taiji FCU, Katsumoto FCU, Taiji Town Office, Shizuoka Fishery Experimental Laboratory (FEL), and Iwate FEL. Numerous volunteers helped my field work. Kenji Takeuchi, To-oru Hosoda, Nanami Kurasawa, and Naoko Funahashi helped me in obtaining whaling information and statistics. I received partial financial support from the International Whaling Commission, Japan Small-type Whaling Association, Japan Whaling Association, Nature Conservation Society of Japan, and U.S. Marine Mammal Commission.

The idea for this book was proposed in 1988 by Yoshifumi Komyo, editor of University of Tokyo Press, whom I have known since he was a student at the Tokyo University of Fisheries; his thesis is cited herein. He waited over 20 years for the manuscript. This publication owes much to his patience. The only benefit resulting from the delay would be in covering the progress of whale science during the period, which has improved the contents of the book. He made some proposals for further improvement of the structure of the book and also read the galley proof very carefully. Thus, the Japanese version is really a joint product of the editor and myself.

Publication of this English version was proposed by my old friend Bill Perrin, who I first met in January 1967 when he visited the *Sioux City* dockside in San Diego. The *Sioux City* was chartered by the U.S. government for marking and sighting whales in waters off Lower California, and I had an opportunity to join the cruise with Dale W. Rice of the Marine Mammals Laboratory in Seattle. William F. Perrin carried out editing the English text as well as all the tasks that accompanied this publication in the United States. This publication was not possible without his contribution. My sincere thanks are due to him.

Until my retirement in 2006, I often worked away from home during field studies and travel and left my family with two daughters in care of my wife Kazuko. I would like to thank all of my family members for their support of my activities; it allowed me to be a free-ranging scientist.

October 13, 2016
Toshio Kasuya

Appendix A: Fisheries Agency Notifications on Cetacean Management

1. Instructions for Handling Whales Incidentally Taken by Trap Net Fisheries. No. 2-1039, June 28, 1990. From Heads of Coastal Section, Division of Promotion, and of Offshore Section, Division of Marine Fisheries, to Head of Fisheries Division of the Prefecture Government. (Text omitted from this English version.)
2. Instructions for Handling Small Cetaceans (Dolphins and Porpoises) Incidentally Taken in Fisheries. No. 2-1050, September 20, 1990. From Head of Coastal Section, Division of Promotion, to Head of Fisheries Division of the Prefecture Government. (Text omitted from this English version.)
3. Instructions for Handling Small Cetaceans (Dolphins and Porpoises). No. 2-1066, November 30, 1990. From Head of Coastal Section, Division of Promotion, to Head of Fisheries Division of the Prefecture Government. (Text omitted from this English version.)
4. Instruction for Handling Small Cetaceans (Dolphins and Porpoises). No. 3-1022, March 28, 1991. From Head of Fisheries Division of the Prefecture Government (Text omitted from this English version.)
5. Instructions for Handling Whales Incidentally Taken in Trap Net Fisheries. No. 3-1022, March 28, 1991. From Heads of Coastal Section, Division of Promotion, and of Offshore Section, Division of Marine Fisheries, to Head of Fisheries Division of the Prefecture Government. (Text omitted from this English version. Two different notifications existed with the same document number and date.)
6. Instructions for Handling Cases of Capture and Incidental Take of Whales (including Small Cetaceans such as Dolphins and Porpoises) in Relation to Enforcement of Ministerial Order That Modifies a Ministerial Order on Permission and Regulation of Fisheries Operating with Ministerial License. No. 13 Sui-Kan 1004, July 1, 2001. From the Director General of Fisheries Agency to the prefecture governors. (The preface given in the Japanese version is omitted from this English version. The main text was not given in the Japanese version because it was revised on December 16, 2004 with additional regulations on cetaceans found stranded.)
7. Revision of No. 13 Sui-kan 1004, July 1, 2001 "Instructions for Handling Cases of Capture and Incidental Take of Whales (including Small Cetaceans such as Dolphins and Porpoises) in Relation to Enforcement of Ministerial Order That Modifies a Ministerial Order on Permission and Regulation of Fisheries Operating with Ministerial License," October 12, 2004. From Whaling Team, Offshore Section. (Text omitted from this English version.)

Appendix B: List of Cetacean Species Known from Japanese Waters

MYSTICETI

1. Bowhead whale, *Balaena mysticetus* Linnaeus, 1758. Family Balaenidae. Japanese name: *hokkyoku-kujira*. A vagrant was killed in Osaka Bay. This species inhabits Arctic waters including the Okhotsk Sea.
2. North Pacific right whale, *Eubalaena japonica* (Lacepede, 1818). Family Balaenidae. Japanese name: *semi-kujira*. This species was also called *semi* or *sebi* and hunted in traditional whaling in Japan. Right whales in the southern hemisphere and North Atlantic are currently considered separate species, *E. australis* and *E. glacialis*, respectively.
3. Gray whale, *Eschrichtius robustus* (Lilljeborg, 1861). Family Eschrichtiidae. Japanese name: *koku-kujira*. This species was called *koku*, *ko-kujira*, *aosagi*, *chigo-kujira*, and *sha-re* and was hunted in traditional and modern Japanese whaling. This species is recorded on both the Sea of Japan and the Pacific coasts of Japan and thought to be on migration between a feeding ground in the Okhotsk Sea and western Bering Sea and an unknown breeding ground suspected to be in Southeast Asia.
4. Minke whale, *Balaenoptera acutorostrata* Lacepede, 1804. Family Balaenopteridae. Japanese name: *minku-kujira*, but *koiwashi-kujira* was more common in the past. This species is common around Japan. A similar species in the Antarctic is considered a separate species, the Antarctic minke whale *B. bonaerensis*. As of the 2013/2014 season, both species were hunted in Japanese scientific whaling.
5. Sei whale, *Balaenoptera borealis* Lesson, 1828. Family Balaenopteridae. Japanese name: *iwashi-kujira*. Caution about vernacular names is needed because this Japanese name was used until the nineteenth century for "Bryde's whales" in southwestern Japan where *B. borealis* did not occur (Kasuya and Yamada 1995, in Japanese). Modern coastal whaling by Japan took this species off the Pacific coast of northern Japan. As of the 2013 season, this species was hunted in Japanese scientific whaling in the North Pacific.
6. Bryde's whale (small form), *Balaenoptera edeni* Anderson, 1879. Family Balaenopteridae. "Bryde's whales" or *nitari-kujira* in Japanese in the North Pacific is believed by some workers to include more than one species, and the relationship between *B. edeni* and *B. brydei* has not been settled. The Japanese name *katsuo-kujira* was recently proposed (without presenting supporting evidence) by Yamada and Ishikawa (in Ohdachi et al. 2009) for a smaller type that inhabits the coastal waters of southwestern Japan and believed by them to represent *B. edeni*. Caution is needed because *katsuo-kujira* as well as *iwashi-kujira* was likely used until the nineteenth century for "Bryde's whales" in southwestern Japan (Kasuya and Yamada 1995, in Japanese).
7. Bryde's whale (large form), *Balaenoptera brydei* Olsen, 1913. Family Balaenopteridae. Japanese name: *nitari-kujira*. Yamada and Ishikawa (in Ohdachi et al. 2009) thought this species to inhabit offshore temperate and tropical waters of the North Pacific. This vernacular name has been used since the 1950s for "Bryde's whale" in the offshore North Pacific, but caution is needed because *nitari* or *nitari-nagasu* was used for the blue whale by some whaling groups off northern Kyushu until the nineteenth century (Section 1.3; Kasuya and Yamada 1995, in Japanese). As of 2013, this species has been hunted in Japanese scientific whaling in the North Pacific. Some taxonomists consider this form to be a subspecies of *B. edeni*.
8. Omura's whale, *Balaenoptera omurai* Wada, Oishi, and Yamada, 2003. Family Balaenopteridae. Japanese name: *tsuno-shima-kujira*. This species occurs in coastal waters from southern Japan to Australia via Indonesia and New Guinea waters.
9. Blue whale, *Balaenoptera musculus* (Linnaeus, 1758). Family Balaenopteridae. Japanese name: *shironagasu-kujira*. This species was also called *nagasu* or *nitari-nagasu* until the nineteenth century by some whaling groups (Section 1.3) and was hunted in old and modern coastal whaling in Japan.
10. Fin whale, *Balaenoptera physalus* (Linnaeus, 1758). Family Balaenopteridae. Japanese name: *nagasu-kujira* or *nagasu*. This species was also called *noso* or *shironagasu* by some whaling groups in the nineteenth century (Section 1.3; Kasuya and Yamada 1995, in Japanese) and was hunted in old and modern coastal whaling in Japan. As of 2013/2014 season, this species was included as a target of Japanese scientific whaling in the Antarctic.
11. Humpback whale, *Megaptera novaeangliae* (Borowski, 1781). Family Balaenopteridae. Japanese name: *zato-kujira*. Old names include *Zato* and *biwahako*. This species has a long history of being hunted in Japan. It currently breeds in waters around the Bonin Islands and the Kerama Islands of Okinawa.

ODONTOCETI

12. Sperm whale, *Physeter macrocephalus* Linnaeus, 1758. Family Physeteridae. Japanese name: *makko-kujira*. This is a cosmopolitan species but uncommon in the Sea of Japan. As of 2013, this species has been hunted in Japanese scientific whaling in the North Pacific.
13. Pygmy sperm whale, *Kogia breviceps* (Blainville, 1838). Family Kogiidae. Japanese name: *ko-makko*. This species grows to 3.8 m in body length. It was also called *tsunabi* or *uki-kujira* (without distinction from the next species).
14. Dwarf sperm whale, *Kogia sima* (Owen, 1866). Family Kogiidae. Japanese name: *ogawa-komakko*. This species is smaller than 2.7 m in body length. It is also called *tsunabi* or *uki-kujira* (without distinction from the previously mentioned species).
15. White whale, *Delphinapterus leucas* (Pallas, 1776). Family Monodontidae. Japanese name: *shiro-iruka*. This species inhabits Arctic waters including the Okhotsk Sea. Vagrants are not rare in northern Japan.
16. Narwhal, *Monodon monoceros* Linnaeus, 1758. Family Monodontidae. Japanese name: *ikkaku*. This species inhabits the Arctic Ocean. One vagrant was recorded in the Sea of Japan in the eighteenth century.
17. Harbor porpoise, *Phocoena phocoena* (Linnaeus, 1758). Family Phocoenidae. Japanese name: *nezumi-iruka*. This species was also called *nuribo*. It inhabits the continental shelf area of northern Japan, north of around 37°N.
18. Finless porpoise, *Neophocaena phocaenoides* (G. Cuvier, 1829). Family Phocoenidae. Japanese name: *sunameri*. This species is called locally by various names including *suzame*, *name*, *nameno-uo*, *zegon*, or *ze-gondo*. Some taxonomists identify two species in this genus, *N. phocaenoides* (G. Cuvier, 1829) represented by individuals from the Persian Gulf to the Taiwan Strait and *N. asiaeorientalis* (Pilleri and Gihr, 1972) represented by those from the Taiwan Strait to Japan south of around 38°30′N. In Japan, this species inhabits shallow seas less than 50 m in depth.
19. Dall's porpoise, *Phocoenoides dalli* (True, 1885). Family Phocoenidae. Japanese name: *ishi-iruka*. It contains two distinct color types of *dalli*-type (*ishi-iruka*-type or *hankuro*) and *truei*-type (*rikuzen-iruka*-type), which were once dealt with as separate species and now considered single species. Some taxonomists consider them as subspecies with vernacular names of *ishi-iruka* and *rikuzen-iruka*, respectively. The species is also called in Japan *kamiyo* for both color types. This species is hunted by the Japanese hand-harpoon fishery.
20. Northern right whale dolphin, *Lissodelphis borealis* (Peale, 1848). Family Delphinidae. Japanese name: *semi-iruka*. This species is also called *semi*, *sao*, *sao-iruka*, or *tongrashi* and is an offshore species that summers in latitudes 40°N–46°N or between the range of the striped dolphin and Dall's porpoise.
21. Pacific white-sided dolphin, *Lagenorhynchus obliquidens* Gill, 1865. Family Delphinidae. Japanese name: *kama-iruka*. Also once called *shisumi-iruka* or *tengu-iruka*. This species is hunted by Japanese hand-harpoon and drive fisheries.
22. Risso's dolphin, *Grampus griseus* (G. Cuvier, 1812). Family Delphinidae. Japanese name: *hana-gondo*. This species was also called *kamabire-sakamata*, *matsuba-iruka*, and *shiro-shachi*. Northern range of this temperate and tropical species extends in summer to around 40°N in the western North Pacific. This species is hunted in Japanese small-type whaling, the hand-harpoon fishery, and drive fisheries.
23. False killer whale, *Pseudorca crassidens* (Owen, 1846). Family Delphinidae. Japanese name: *oki-gondo*. This species was called by various local names including *bo-iruka*, *bozu-iruka*, *dainan-gondo*, *kyuri-gondo*, *nyudo-iruka*, *ohnan-goto*, *ohiwo-kui*, *oki-goto*, *sakoma-iruka*, and *shachi-modoki*. This species is an inhabitant of tropical and temperate waters of the world and is hunted in Japanese small-type whaling, drive fisheries, and the crossbow fishery.
24. Killer whale, *Orcinus orca* (Linnaeus, 1758). Family Delphinidae. Japanese name: *shachi*. This species was also called *kurotonbo*, *sakamata*, *shachi-hoko*, *taka*, *taka-matsu*, and *kujira-toshi*. This species occurs all around Japan but at low density.
25. Common bottlenose dolphin, *Tursiops truncatus* (Montagu, 1821). Family Delphinidae. Japanese name: *hando-iruka* or *bando-iruka*. It was also called *hasunaga*, *kuro*, *nezumi-iruka*, and *uki-kuro*, but distinction from the next species is dubious. This species inhabits waters south of southern Hokkaido and is hunted in Japanese hand-harpoon, drive, and crossbow fisheries.
26. Indo-Pacific bottlenose dolphin, *Tursiops aduncus* (Ehrenberg, 1833). Family Delphinidae. Japanese name: *minami-hando-iruka* or *minami-bando-iruka*. This species occurs in coastal waters south of central Japan. A vernacular name *haukasu* was likely to have represented this species or rough toothed dolphin off Taiji (Kasuya and Yamada 1995, in Japanese).
27. Short-beaked common dolphin, *Delphinus delphis* (Linnaeus, 1758). Family Delphinidae. Japanese name: *ma-iruka*. This species locally called *kintama-iruka* (from their large testis), *suji-iruka*, *suzume-gata*, *suzume-iruka*, and *tsubame-iruka*, but the distinction from the next species is dubious. In Japan, this species is known from the Pacific coast of Shikoku to Hokkaido. It should also be noted that dolphin species locally called *ma-iruka* could have represented any commonest species in the region.

28. Long-beaked common dolphin, *Delphinus capensis* (Gray, 1828). Family Delphinidae. Japanese name: *hase-iruka*. This species inhabits the Indian Ocean, Southeast Asia, and southwestern Japan including northern Kyushu and the Sea of Japan where the other species in the genus does not occur.
29. Rough-toothed dolphin, *Steno bredanensis* G. Cuvier in Lesson, 1828. Family Delphinidae. Japanese name: *shiwaha-iruka*. This species inhabits tropical and subtropical waters of the world. The vernacular name *haukasu* was likely to have represented this species or the Indo-Pacific bottlenose dolphin off Taiji (Kasuya and Yamada 1995, in Japanese).
30. Striped dolphin, *Stenella coeruleoalba* (Meyen, 1833). Family Delphinidae. Japanese name: *suji-iruka*. This species was known by the local names *ma-iruka*, *suzume-iruka*, and *tsubame-iruka*. It inhabits tropical and temperate waters of the world and Japanese waters south of 40°N; it is unknown from the Sea of Japan. It was hunted in great numbers by Japanese drive fisheries, and low-level take still continues in hand-harpoon and drive fisheries.
31. Pantropical spotted dolphin, *Stenella attenuata* (Gray, 1846). Family Delphinidae. Japanese name: *madara-iruka*, but *arari-iruka* is also in use. This species was locally known by the names *hara-guro*, *kasuri-iruka*, and *ma-iruka*. It inhabits similar waters as the striped dolphin and is hunted in Japanese hand-harpoon and drive fisheries.
32. Spinner dolphin, *Stenella longirostris* (Gray, 1828). Family Delphinidae. Japanese name: *hashi-naga-iruka*. This is an offshore species occurring south of the range of the other two species of the genus off Japan.
33. Fraser's dolphin, *Lagenodelphis hosei* Fraser, 1956. Family Delphinidae. Japanese name: *sarawaku-iruka*. This species known from the Pacific coast of central and southern Japan has a worldwide distribution in tropical offshore waters.
34. Melon-headed whale, *Peponocephala electra* (Gray, 1846). Family Delphinidae. Japanese name: *kazuha-gondo*. This species is known from the Pacific coasts of central and southwestern Japan.
35. Pygmy killer whale, *Feresa attenuata* Gray, 1874. Family Delphinidae. Japanese name: *yume-gondo*. This species is known from the Pacific coast of central and southwestern Japan and the Tsushima Strait area.
36. Long-finned pilot whale, *Globicephala melas* (Trail, 1809). Family Delphinidae. Japanese name: *hire-naga-gondo*. This species became extinct from the North Pacific in around the twelfth century.
37. Short-finned pilot whale, *Globicephala macrorhynchus* Gray, 1846. Family Delphinidae. Japanese name: *kobire-gondo*. Off Japan, this species inhabits mainly Pacific waters and includes two geographical forms that are distinguishable by body size and saddle mark. One form, *tappa-naga*, is known from northern coastal waters between 35°N–43°N and the other, *ma-gondo*, inhabits southern waters south of 40°N not occupied by *tappa-naga*. Synonyms of *tappa-naga* include *shio* and *shio-goto* and those of *ma-gondo* include *hee-to*, *naisa*, *naisa-goto*, and *nagisa-gondo*. This species is hunted in Japanese small-type whaling, drive fisheries, and the cross-bow fishery.
38. Baird's beaked whale, *Berardius bairdii* Stejneger, 1883. Family Ziphiidae. Japanese name: *tsuchi-kujira*. This species is locally called *tsuchinbo*. It is endemic to the northern North Pacific and occurs in Japan off the coasts of the Sea of Japan, Okhotsk Sea, and the Pacific north of 34°N. This species is hunted in Japanese small-type whaling. See Section 13.1 for a still undetermined type known from the southern Okhotsk Sea.
39. Longman's beaked whale, *Indopacetus pacificus* (Longman, 1926). Family Ziphiidae. Japanese name: *taiheiyo-akabo-modoki*. This tropical species is recorded from Kyushu in Japan.
40. Hubbs' beaked whale, *Mesoplodon carlhubbsi* Moore, 1963. Family Ziphiidae. Japanese name: *habbusu-ohgiha-kujira*. This species, which is endemic to the North Pacific, is known from off the Pacific coast of central and northern Japan.
41. Blainville's beaked whale, *Mesoplodon densirostris* (de Blainville, 1817). Family Ziphiidae. Japanese name: *kobuha-kujira*. This cosmopolitan species is known from off central and southwestern Japan.
42. Ginkgo-toothed whale, *Mesoplodon ginkgodens* Nishiwaki and Kamiya, 1958. Family Ziphiidae. Japanese name: *icho-ha-kujira*. This species is limited to the Indian Ocean and the Pacific and is known from the south of central Japan.
43. Stejneger's beaked whale, *Mesoplodon stejnegeri* True, 1885. Family Ziphiidae. Japanese name: *ohgiha-kujira*. This species inhabits the northern North Pacific and is known in Japan from the Sea of Japan, the Okhotsk Sea, and the Pacific coast of northern Honshu.
44. Cuvier's beaked whale, *Ziphius cavirostris*. Family Ziphiidae. Japanese name: *akabo-kujira*. This species was locally called *kaji-kujira*, *kaji-ppo*, and *bo*. This cosmopolitan species is known from all the Japanese coasts except for the Sea of Japan.

REFERENCES

Kasuya, T. and Yamada, T. 1995. *List of Japanese Cetaceans*. Institute of Cetacean Research, Tokyo, Japan. 90pp. (in Japanese).

Ohdachi, S.D., Ishibashi, Y., Iwasa, M.A., and Saitoh, T. (eds.) 2009. *The Wild Mammals of Japan*. Kyoto, Japan. 544pp. (in Japanese).

Index

T

W

Y

Z